PROCEEDINGS OF THE
INTERNATIONAL CONGRESS OF
MATHEMATICIANS

AUGUST 3–11, 1986

INTERNATIONAL CONGRESS OF MATHEMATICIANS

Berkeley, California
August 3-11, 1986

EDITED BY
ANDREW M. GLEASON

Library of Congress-in-Publication Data

International Congress of Mathematicians (1986: Berkeley, Calif.)
   Proceedings of the International Congress of Mathematicians, 1986: August
3–11, 1986, Berkeley, California, USA.
   Includes index.
   1. Mathematics–Congresses. I. American Mathematical Society. II. Title.
QA1.I82 1986   510   87-24109
ISBN 0-8218-0110-4

The 1986 International Congress of Mathematicians was supported in part
by Grant INT-8417900 from the National Science Foundation including contributions from the Department of the Air Force, Air Force Office of Scientific
Research, and the Department of Energy, Office of Energy Research. Any opinions, findings, and conclusions or recommendations expressed in this publication
are those of the author(s) and do not necessarily reflect the views of the National Science Foundation, the Department of the Air Force, or the Department
of Energy.

This work relates to Department of Navy Grant N00014-85-G-0145 issued by
the Office of Naval Research. The United States Government has a royalty-free
license throughout the world in all copyrightable material contained herein.

The 1986 International Congress of Mathematicians was supported in part
by Grant DAAG29-85-G-0091 from the Department of the Army. The views,
opinions, and/or findings contained in this report are those of the author(s) and
should not be construed as an official Department of the Army position, policy,
or decision, unless so designated by other documentation.

Photo credits: William G. Chinn, Benedict H. Gross, Hikosaburo Komatsu,
Jill P. Mesirov, and Klaus Peters.

CONTENTS

INVITED FORTY-FIVE MINUTE ADDRESSES
AT THE SECTION MEETINGS

Proceedings of the International Congress of Mathematicians
Berkeley, California, USA, 1986

Geometry of Banach Spaces and Harmonic Analysis

J. BOURGAIN

1. Introduction. I intend to report on several results interrelating harmonic analysis, geometry of convex sets, and finite-dimensional Banach space theory. Part of this report, especially the work on high-dimensional convex sets, is related to the exposé of V. D. Milman in these proceedings. In §1 of my report, I will discuss recent research on the behavior of the maximal operator associated to a convex symmetric body in $\mathbf{R}^n$ for large n. It finds its origin in the work of E. Stein [St1] on the spherical maximal function, and the paper of E. Stein and J. O. Stromberg [St-Str]. In [B2] and [B3], we succeeded in developing part of this theory in its full generality. On the other hand, several interesting questions are still unsolved at present, even in the particular case of the euclidean ball.

It was proved in [B4] that the maximal operator corresponding to the circular means in the plane

$$\mathcal{M}_{(1)}f(x) = \sup_{t>0} \int |f(x+ty)|\sigma\,(dy) \tag{1.1}$$

(σ = normalized arclength measure of the circle) is a bounded operator on $L^p(\mathbf{R}^2)$ for $p > 2$. The two-dimensional case was left open in [St1] and [St-W]. The latter result and some of its consequences are discussed in §2 of this exposé.

In §3, progress on the theory of Sidon sets and the dichotomy problem for restriction algebras is described, starting from the remarkable discoveries of G. Pisier in the late seventies [P1]. The most recent result here is an affirmative solution to the so-called "cotype-dichotomy" for invariant subspaces of $\mathcal{C}(G)$, G a compact Abelian group, a problem proposed by A. Pelczynski [K-P].

2. Geometry of convex sets and maximal operators. We denote by B a convex symmetric body in $\mathbf{R}^n$ of n-dimensional volume $\mathrm{Vol}\,B = 1$. For a locally integrable function f on $\mathbf{R}^n$, define

$$M_B f(x) = \sup_{t>0} \int_B |f(x+ty)|\,dy \tag{2.1}$$

generalizing the classical Hardy-Littlewood maximal operator. Denoting

$$\|f\|_p = \left(\int_{\mathbf{R}^n} |f(x)|^p \, dx\right)^{1/p}, \qquad \|f\|_{1,\infty} = \sup_{\lambda>0} \lambda m[|f| > \lambda],$$

let $C_p(B)$ (resp. $C_{1,\infty}(B)$) be the smallest constant for which the following inequality holds:

$$\|M_B f\|_p \le C_p(B)\|f\|_p \qquad (p > 1), \tag{2.2}$$

resp.

$$\|M_B f\|_{1,\infty} \le C_{1,\infty}(B)\|f\|_1. \tag{2.3}$$

The finiteness of these constants is well known. Our attention goes here to their dependence on the set B and, in particular, the dimension n.

A first motivation for this study is the dimension-free character of the Littlewood-Paley theory [St2]. Let $(P_t)_{t>0}$ be the Poisson semigroup on $\mathbf{R}^n$, $\hat{P}_t(\xi) = e^{-t|\xi|}$. Define $u(x,t) = (f * P_t)(x)$. Then [St2]

$$\left\|\sup_{t>0}|f * P_t|\right\|_p \le c(p)\|f\|_p \qquad (1 < p \le \infty) \tag{2.4}$$

(maximal inequality)

and for $1 < p < \infty$

$$\left\|\left\{\int_0^\infty \left|\frac{\partial u}{\partial t}\right|^2 t\,dt\right\}^{1/2}\right\|_p \le \left\|\left\{\int_0^\infty |\nabla u(x,t)|^2 t\,dt\right\}^{1/2}\right\|_p \le c(p)\|f\|_p \tag{2.5}$$

(g-function inequality)

where the constant $c(p)$ is independent of dimension. These facts will play a role in the proof of the main result stated below. The second motivation is the dimension-free maximal inequality for the euclidean balls $\mathcal{E}_n = \{x \in \mathbf{R}^n : |x| = (\sum x_i^2)^{1/2} \le 1\}$ proved in [St-Str]. Thus we consider

$$M_{\mathcal{E}_n} f(x) = \sup_{t>0}\left\{\frac{1}{\text{Vol}\,\mathcal{E}_n}\int_{\mathcal{E}_n} |f(x+ty)|\,dy\right\}.$$

It is shown in [St-Str] that

$$C_p(\mathcal{E}_n) \le C(p) \quad \text{independent of } n \text{ for } p > 1. \tag{2.6}$$

Their argument uses the geometry of the euclidean ball and does not generalize to arbitrary convex symmetric bodies. We will come back on it in §2.

THEOREM 1 [B2, B3]. $C_2(B)$ is bounded by an absolute constant. Also $C_p(B) \le C(p)$ for $p > \frac{3}{2}$.
Here B refers to a general convex symmetric body in $\mathbf{R}^n$.

The restriction $p > \frac{3}{2}$ seems to be of a technical nature. However, even for the cartesian cube $[\frac{1}{2},\frac{1}{2}]^n$, the maximal inequality with constant independent of n is not proved for $p > \frac{3}{2}$.

Notice that if $v \in \text{GL}(\mathbf{R}^n)$ is a linear permutation of $\mathbf{R}^n$, then $M_{v(B)}f = M_B(f \circ v) \circ v^{-1}$. Hence $C_p(B) = C_p(v(B))$ is an affine invariant.

We now invoke the following geometrical fact.

LEMMA 1. *There is an affine position B_1 of B, $\operatorname{Vol} B_1 = 1$, such that for some constant L (depending a priori on B) the Fourier transform*

$$\hat{\mathcal{X}}(\xi) = \int_{B_1} e^{-2\pi i\langle x,\xi\rangle}\, dx$$

of the indicator function $\mathcal{X}$ of B_1 satisfies the estimates

$$|1 - \hat{\mathcal{X}}(\xi)| \leq CL|\xi|, \qquad |\hat{\mathcal{X}}(\xi)| \leq C(L|\xi|)^{-1},$$
$$|\langle \nabla\hat{\mathcal{X}}(\xi),\xi\rangle| \leq C, \tag{2.8}$$

where C is an absolute constant.

This special position B_1 of B is obtained by diagonalization of the covariance matrix

$$\beta = L^2 I_{\mathbf{R}^n} \quad \text{where } \beta_{ij} = \int_{B_1} x_i x_j\, dx.$$

The proof of the inequalities (2.8) then involves the Brunn-Minkowski theorem [**B-Z**].

Write now $\mathcal{X} = P_L + (\mathcal{X} - P_L)$ and apply to $K = \mathcal{X} - P_L$ the following general L^2-estimate.

LEMMA 2. *Let $K \in L^1(\mathbf{R}^n)$ and define for $j \in \mathbf{Z}$*

$$\alpha_j = \sup_{2^j \leq |\xi| \leq 2^{j+1}} |\hat{K}(\xi)|, \qquad \beta_j = \sup_{2^j \leq |\xi| \leq 2^{j+1}} |\langle \nabla\hat{K}(\xi),\xi\rangle|.$$

Then

$$\left\|\sup_{t>0}|f * K_t|\right\|_2 \leq C\Gamma(K)\|f\|_2 \quad \text{where } \Gamma(K) = \sum \alpha_j^{1/2}(\alpha_j + \beta_j)^{1/2}. \tag{2.9}$$

This fact is a simple consequence of Parseval's formula.

The boundedness of $C_2(B)$ in Theorem 1 then follows from (2.4), (2.8), and (2.9). To obtain the bound on $C_p(B)$, we use an L^p-variant of Lemma 2 which is derived from (2.5), see [**B3**].

LEMMA 3. *Assume $K \in L^1(\mathbf{R}^n)$, $K \geq 0$, satisfies in addition to*

$$|\hat{K}(\xi)| \leq C|\xi|^{-1}, \qquad |1 - \hat{K}(\xi)| \leq C|\xi|,$$

the condition

$$\left|\partial_t^{(j)}\hat{K}(t\xi)|_{t=1}\right| \leq C \quad \text{for } j \leq \eta. \tag{2.10}$$

Then

$$\left\|\sup_{t>0}|f * K_t|\right\|_p \leq A(C,p)\|f\|_p \quad \text{for } 1 + \frac{1}{2\eta} < p \leq 2. \tag{2.11}$$

Again let us point out that the estimate (2.11) is dimension-free. In the context of Theorem 1, it is applied with $\eta = 1$.

REMARKS. 1. The estimate on $C_p(B)$ for $\frac{3}{2} < p < 2$ is due independently to T. Carbery.

2. The number L appearing in Lemma 1 is inverse-proportional to the volume of the central sections of B_1 in codimension 1 (up to an absolute constant).

One may choose $L \geq 1$ while the problem of the upper estimate is unclarified presently. This amounts to the following geometrical question: Is there a fixed constant $c > 0$ such that every centrally symmetric convex set of volume 1 has a codimension 1 central section of volume at least c.

3. The use of Lemma 3 enables us, in certain cases, to remove the restriction $p > \frac{3}{2}$ in Theorem 1. This is the case for the euclidean ball (providing an alternative proof of (2.6)) and more generally the l_n^r-ball $[x \in \mathbf{R}^n : (\sum x_i^r)^{1/r} \leq 1]$ $(n \to \infty)$ where r is an even number.

Based on a refinement of Vitali's covering lemma, E. Stein and J. O. Stromberg showed in [St-Str] that

THEOREM 2. $C_{1,\infty}(B) \leq Cn \log n$ for any convex symmetric body B in $\mathbf{R}^n$.

This result is purely geometrical and derived from

LEMMA 4. Let $\{B_\alpha\}_\alpha$ be a finite collection of "balls" (= translates of dilates of the given B). Denote by B_α^* the ball with the same center as B_α and n-fold radius. Then there is a subcollection $B_1, \ldots, B_N$ satisfying

$$m\left(\bigcup_\alpha B_\alpha\right) \leq Cm\left(\bigcup_{j=1}^N B_j\right), \tag{2.12}$$

$$\sum_{j=1}^N \frac{m(I_j)}{m(B_j^*)} \chi_{B_j^*} \leq Cn \log n \tag{2.13}$$

$(C = constant)$ where $I_j = B_j \backslash (B_1 \cup \cdots \cup B_{j-1})$.

One might hope to improve the estimate (2.13).

3. Spherical averages and applications. Let $k \geq 2$ and denote by σ_{k-1} the normalized invariant measure of the unit sphere $S^{(k-1)} = [x \in \mathbf{R}^k : |x| = 1]$ in $\mathbf{R}^k$. Consider the spherical averages

$$A_t f(x) = \int f(x + ty)\sigma_{k-1}(dy), \qquad t > 0,$$

where f is a bounded measurable function on $\mathbf{R}^k$. Let also

$$\mathcal{M}_{(k-1)}f = \sup_{t>0} |A_t f|$$

be the corresponding maximal operator.

In [St1], E. Stein proved

THEOREM 3. If $k \geq 3$ and $p > k/(k-1)$, then there is an inequality

$$\|\mathcal{M}_{(k-1)}f\|_p \leq c(k,p)\|f\|_p. \tag{3.1}$$

This inequality fails for $p = k/(k-1)$.

Stein applied estimates related to (3.1) in studying the behavior of solutions $u = u(t,x)$ of the wave equation

$$Lu = 0, \qquad L = \frac{\partial^2}{(\partial t)^2} - \sum_{j=1}^k \frac{\partial^2}{(\partial x_j)^2} \qquad \text{when } t \to 0.$$

Recent developments in this direction appear in the joint work of E. Stein and C. Sogge.

Notice that if $k \geq 3$ is a fixed dimension and $n > k$, then integration on the Grassmannian of k-dimensional subspaces of $\mathbf{R}^n$ leads to the inequality

$$\|\mathcal{M}_{(n-1)}f\|_p \leq \int_{O(n)} \|\mathcal{M}_{(k-1)}(f \circ U)\|_p \, dU \qquad (3.2)$$

where $O(n)$ is the orthogonal group and $\mathcal{M}_{k-1}$ is defined with respect to the first k variables. Thus by (3.2), $c(k,p)$ in (3.1) may be replaced by $c(p)$. Integration in polar coordinates then implies (2.6) for the euclidean ball.

The proof (3.1) relies mainly on the L^2-estimate given by Fourier analysis. In the circular case, $\mathcal{M}_{(1)}$ is unbounded on $L^2(\mathbf{R}^2)$ explaining the difficulty in obtaining the two-dimensional statement. The next fact was only proved recently, combining Fourier analysis with geometrical techniques.

THEOREM 4.
$$\|\mathcal{M}_{(1)}f\|_p \leq C(p)\|f\|_p \quad for \ p > 2. \qquad (3.3)$$

This enables us to answer some rather old questions such as

COROLLARY 5. *No planar measure zero set contains circles with arbitrary center.*

A subset A of $\mathbf{R}^k$ has positive upper density provided

$$\delta(A) = \lim_R \frac{|B(0,R) \cap A|}{|B(0,R)|} > 0$$

where $B(0,R) = \{x \in \mathbf{R}^k, |x| < R\}$.

Y. Katznelson and B. Weiss obtained the following fact combining methods from probability and ergodic theory.

THEOREM 6. *Whenever A is a subset of $\mathbf{R}^2$ with positive upper density, there is a number $l = l(A)$ such that $|x - y| = l'$ for some $x, y \in A$, whenever $l' > l$.*

In [B5], a simple proof is given using circular means. The latter argument permits several variants, for instance, the following higher-dimensional version of Theorem 6.

THEOREM 7. *Assume $A \subset \mathbf{R}^k$, $\delta(A) > 0$, and V is a set of k points spanning a $(k-1)$-dimensional hyperplane. There exists some number l such that A contains an isometric copy of $l'V$ whenever $l' > l$.*

Some problems arising in this context are open. For instance, does the statement in Theorem 7 hold when $k = 2$ and V is a nondegenerate triangle?

4. Sidon sets and the dichotomy problem for restriction algebras. A still open problem in Harmonic Analysis is the dichotomy conjecture for restriction algebras (see [G-M] for a discussion). Here we will consider the following version of it. Let G be a compact Abelian group, $\Gamma = \hat{G}$ the dual group of G. For

$\Lambda \subset \Gamma$, define $A(\Lambda) = \{\hat{f}|_\Lambda : f \in L^1(G)\}$, i.e., the restriction algebra of Fourier transforms of L^1-functions on G.

CONJECTURE ($*$). *Either Λ is a Sidon set or any function operating on $A(\Lambda)$ is analytic.*

Recall that $F : [-1, 1] \to \mathbf{C}$ operates on $A(\Lambda)$ provided $\varphi \in A(\Lambda)$, $\varphi(\gamma) \in \,] -1, 1[$ for $\gamma \in \Lambda \Rightarrow F \circ \varphi \in A(\Lambda)$. For $A(\Lambda)$ to be analytic, it is necessary and sufficient that for some $c > 0$ and for t large enough

$$N(\Lambda, t) = \sup_{\substack{\varphi \in A(\Lambda), \varphi \text{ real} \\ \|\varphi\| \leq 1}} \|e^{it\varphi}\| \geq e^{ct}.$$

Here $\|\varphi\|$ stands for the quotient norm

$$\|\varphi\|_{A(\Lambda)} = \inf\{\|f\|_1 : f \in L^1(G), \hat{f}(\gamma) = \varphi(\gamma) \text{ for } \gamma \in \Lambda\}.$$

Despite the counterexamples of M. Zafran [**Z**] and G. Pisier [**P2**] to the dichotomy problem in the context of homogeneous Banach algebras, I think that the results mentioned in this section give significant support to the conjecture stated above.

A subset Λ of Γ is a Sidon set provided there is a constant $C > 0$ such that for all scalar sequences $(a_\gamma)_{\gamma \in \Lambda}$,

$$C \left\| \sum_\Lambda a_\gamma \gamma \right\|_{\mathcal{C}(G)} \geq \sum |a_\gamma| \tag{4.2}$$

holds. Here $\mathcal{C}(G)$ is the space of continuous functions on G equipped with the supremum norm, and the characters Γ are seen as functions on G. The smallest C satisfying (4.2) will be denoted $S(\Lambda)$, the Sidon constant of Λ. Concrete examples of Sidon sets are dissociated sets. A subset Λ of Γ is dissociated provided that any $(\pm 1, 0)$-relation of the elements of Λ is trivial. Thus

$$\sum_\Lambda \varepsilon_\gamma \gamma = 0 \quad (\text{in } \Gamma), \qquad \varepsilon_\gamma = 0, 1, -1 \Rightarrow \varepsilon_\gamma = 0 \quad \text{for } \gamma \neq 0.$$

In the case of dissociated sets, the interpolating measures are given by the standard Riesz-products. The following theorem, due to G. Pisier [**P1**] makes the connection with the more abstract notion of Sidon sets (see also [**B6**] for a different approach).

THEOREM 8. *For a subset Λ of Γ, the following conditions are equivalent:*
(i) Λ *is a Sidon set.*
(ii) $\|f\|_{L^{\psi_2}(G)} \leq C\|f\|_{L^2(G)}$ *for* supp $\hat{f} \subset \Lambda$.
(iii) *There is $\delta > 0$ such that any finite subset Λ_1 of Λ contains a dissociated subset $\Lambda_2 \subset \Lambda_1$ with $|\Lambda_2| > \delta|\Lambda_1|$.*

Here $L^{\psi_2}(G)$ refers to the Orlicz function $e^{x^2} - 1$.

In [**G**], C. Graham pointed out that an affirmative solution to ($*$) implies the following fact, proved in [**B7**].

THEOREM 9. *Let Λ be a subset of Γ for which the restriction algebras of Radon measures and discrete measures coincide, i.e., $B(\Lambda) = B_d(\Lambda)$. Then Λ is a Sidon set and hence a Helson set in the Bohr compactification.*

Lower estimates on $N(\Lambda, t)$ can be obtained from

LEMMA 5. *Let $l > 0$ be an integer. Suppose there exist f with $\hat{f}$ finitely supported by Λ and points $x_1, \ldots, x_l \in G$ such that*

$$f(0) = \|f\|_\infty = 1, \tag{4.3}$$

$$\sum_{S \subset \{1,\ldots,l\}} \left| f\left(y - \sum_{k \in S} x_k \right) \right| \leq B \quad \textit{for all } y \in G. \tag{4.4}$$

Then, for $0 < t < cl$, $N(\Lambda, t) \geq e^{ct}/B$ where $c > 0$ is numerical.

This criterion applies to solve the dichotomy problem for the tensor-algebras (see [G-M] for background). Thus [B8]

THEOREM 10. *Let I, J be discrete spaces and E a subset of $I \times J$. Then either E is a V-Sidon set ($=$ a finite union of sections) or the restriction algebra $V(E) = c_0(I) \hat{\otimes} c_0(J)/E^\perp$ is analytic.*

For a subset Λ of Γ, define $C_\Lambda = \{f \in \mathcal{C}(G): \operatorname{supp} \hat{f} \in \Lambda\}$, the subspace of $\mathcal{C}(G)$ of those functions with Fourier transform supported by Λ. Results of N. Varopoulos [V] and G. Pisier [P3] led to the so-called "cotype dichotomy" problem; see [K-P]. In a sense, this dichotomy is a Banach space version of $(*)$. It was solved recently in joint work with V. D. Milman [B-M].

THEOREM 11. *Either Λ is a Sidon set ($\Leftrightarrow C_\Lambda$ has cotype 2) or C_Λ contains l_n^∞-subspaces (uniformly) of arbitrary large dimension ($\Leftrightarrow C_\Lambda$ has no finite cotype).*

We recall the definition of cotype. A normed space X has cotype $2 \leq q < \infty$ with constant $C_q(X)$ provided

$$C_q(X) \operatorname*{Aver}_{\pm 1} \left\| \sum_{1 \leq i \leq r} \pm x_i \right\|_X \geq \left(\sum \|x_i\|^q \right)^{1/q}$$

whenever $\{x_i\}_{i=1}^r$ is a finite sequence of vectors in X. This notion plays an important role in the finite-dimensional theory of normed spaces (cf. the exposé of V. D. Milman). The proof of Theorem 11 is nonconstructive and combines results from harmonic analysis and geometry of Banach spaces. One of the facts involved is

LEMMA 6. *For a finite subset E of Γ, define*

$$d(E) = \min\{d \mid C_E \text{ is 2-isomorphic to a subspace of } l_d^\infty\}.$$

Then $\Lambda \subset \Gamma$ is a Sidon set if and only if

$$\tau \equiv \inf_{\substack{E \subset \Lambda \\ E \text{ finite}}} \frac{\log d(E)}{|E|} > 0.$$

The notion $\alpha(E)$ is closely related to the concept of arithmetical diameter considered by Katznelson and Malliavin in [K-M].

REFERENCES

[B2] _____, *On high dimensional maximal functions associated to convex bodies*, Amer. J. Math. (to appear).

[B3] _____, *On the L^p-bounds for maximal functions associated to convex bodies in $\mathbf{R}^n$*, Israel J. Math. **54** (1986), no. 3, 257–265.

[B4] _____, *Averages in the plane over convex curves and maximal operators*, J. Analyse Math. (to appear).

[B5] _____, *A Semeredi type theorem for sets positive density in $\mathbf{R}^k$*, Israel J. Math. **54** (1986), no. 3, 307–316.

[B6] _____, *Sidon sets and Riesz products*, Ann. Inst. Fourier (Grenoble) **35** (1985), no. 1, 1937–148.

[B7] _____, *Sur les ensembles d'interpolation pour les mesures discrètes*, C. R. Acad. Sci. Paris Sér. I. Math. **296** (1982), 149–151.

[B8] _____, *On the dichotomy problem for tensor algebras*, Trans. Amer. Math. Soc. **293** (1986), no. 2, 793–798.

[B-M] J. Bourgain and V. D. Milman, *Dichotomie du cotype pour les espaces invariants*, C. R. Acad. Sci. Paris Sér. I Math. **300** (1985), no. 9, 263–266.

[B-Z] Y. D. Burago and V. A. Zalgaler, *Geometric inequalities*, "Nauka", Leningrad, 1980. (Russian)

[G] C. Graham, *Sur un théorème de Katznelson et McGehee*, C. R. Acad Sci. Paris Sér. A-B **276** (1973), A37–A40.

[G-M] C. Graham and O. C. McGehee, *Essays in commutative harmonic analysis*, Grundlehren Math. Wiss. **238** (1979).

[K-M] Y. Katznelson and P. Malliavin, *Un critère d'analyticité pour les algèbres de restriction*, C. R. Acad Sci. Paris Sér A-B **261** (1965), A4964–A4967.

[K-P] S. Kwapien and A. Pelczynski, *Absolutely summing operators and translation invariant spaces of functions on compact abelian groups*, Math. Nachr. **94** (1980), 303–340.

[P1] G. Pisier, *De nouvelles caractérisations des ensembles de Sidon*, Math. Anal. and Appl., Part B, Adv. in Math., Suppl. Studies, 7b, Academic Press, New York, 1981, pp. 685–726.

[P2] _____, *A remarkable homogeneous Banach algebra*, Israel J. Math. **34** (1979), no. 1-2, 38–44.

[P3] _____, *Ensembles de Sidon et espaces de cotype 2*, Sém. sur la Géométrie des Espaces de Banach 1977–78, Exp. 14, Centre de Mathématiques, Palaiseau, 1978.

[St1] E. M. Stein, *Maximal functions, spherical means*, Proc. Nat. Acad. Sci. U.S.A. **73** (1976), 2174–2175.

[St2] _____, *Topics in harmonic analysis*, Ann. of Math. Studies, No. 63, Princeton Univ. Press, Princeton. N.J., 1970.

[St-Str] E. M. Stein and J. O. Stromberg, *Behavior of maximal functions in $\mathbf{R}^n$ for large n*, Ark. Mat. **21** (1983), no. 2, 259–269.

[St-W] E. M. Stein and S. Waigner, *Problems in harmonic analysis related to curvature*, Bull. Amer. Math. Soc. **84** (1978), 1239–1295.

[V] N. T. Varopoulos, *Sous-espaces de $C(G)$ invariant par translation et de type $\mathcal{L}^1$*, Sém. Maurey-Schwartz 1975/76, Exp. 12, Ecole Polytechnique.

[Z] M. Zafran, *The dichotomy problem for homogeneous Banach algebras*, Ann. of Math. **108** (1978), 97–105.

INSTITUT DES HAUTES ÉTUDES SCIENTIFIQUE, BURES-SUR-YVETTE, FRANCE

UNIVERSITY OF ILLINOIS, URBANA, ILLINOIS 61801, USA

Cyclic Cohomology and
Noncommutative Differential Geometry

A. CONNES

Cyclic cohomology appeared independently from two different streams of ideas, algebraic K-theory and noncommutative differential geometry. I shall try to explain in this paper the meaning of noncommutative differential geometry. Its main object is a new notion of space. The need for considering such spaces and developing for them the analogues of the usual tools of differential geometry is best understood in the following two examples. In both, one tries to prove a result of classical differential geometry, and a heuristic proof is possible provided one accepts the new notion of space.

First example.

THEOREM (LICHNEROWICZ, 1961). *If M is a compact spin manifold whose $\hat{A}$ genus is nonzero, then it is impossible to endow M with a Riemannian metric of strictly positive scalar curvature.*

The proof of the result uses a simple global idea. By the Lichnerowicz identity, the square of the Dirac operator is $\nabla^*\nabla + \frac{1}{4}\chi$ where $\nabla^*\nabla$ is a positive operator and χ is the scalar curvature. Thus for $\chi > 0$, the Dirac operator has index equal to zero. But by the index theorem index (Dirac) $= \hat{A}(M) \neq 0$. Q.E.D.

A stronger result about the nonexistence of metrics with positive scalar curvature is the following

THEOREM [14]. *Let M be a compact oriented manifold with $\hat{A}(M) \neq 0$. Then there is no integrable spin subbundle F of TM with strictly positive scalar curvature.*

Let me give a heuristic proof of this result which will work when we get the right tools. The idea is the following: Given an integrable subbundle F of the tangent bundle of M, one can a priori integrate it and get a foliation of M which creates a new space B of leaves of this foliated manifold. (See Figure 1.)

Now $\hat{A}(M)$ is the index of the Dirac operator, at least if M is spin, or, equivalently, it is the pushforward $\pi!(L)$ of the trivial line bundle L on M by the map $\pi \colon M \to \mathrm{pt}$. As $\pi = \pi_1 \circ \pi_2$, where π_2 is the projection of M on the space

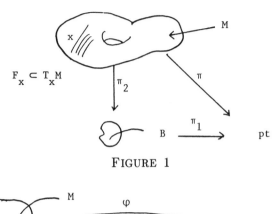

$$F_x \subset T_x M$$

FIGURE 1

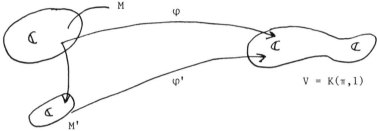

FIGURE 2

of leaves B, one has $\pi!(L) = \pi_1!(\pi_2!(L))$, but $\pi_2!(L) \in K(B)$ is the index of the family of Dirac operators along the leaves and hence is zero since the scalar curvature of leaves is strictly positive. This reasoning does work if one has just a fibration; one then applies the index theorem for families. However, in general, given an integrable subbundle F it is impossible to decide whether it creates a fibration or a foliation. For instance, on the two torus $\mathbf{T}^2 = \mathbf{R}^2/\mathbf{Z}^2$ the equation $dy = \theta\,dx$ defines a fibration iff θ is rational. Thus it is impossible to restrict to the case of fibrations, and one needs to handle spaces such as the space B of leaves of an arbitrary foliation. One needs new tools to understand and use such spaces because when just viewed as ordinary topological spaces they are of no use; in general they would carry the coarse topology and $K(B)$ would be trivial.

Second example. We now pass from the space of leaves of a foliation to another example related to discrete groups. It comes from a problem stated by Novikov— the homotopy invariance of the higher signatures. Let M be a compact oriented manifold and φ a map from M to a $K(\pi, 1)$ space V. For instance, one could take for φ the map which classifies the universal cover of M. For each cohomology class $\omega \in H^*(V, \mathbf{C}) = H^*(\pi, \mathbf{C})$, the higher signature of the pair (M, φ) is given by the scalar $\langle \mathcal{L}_M \cdot \varphi^*(\omega), [M] \rangle$ where $\mathcal{L}_M$ is the L genus of M and $\varphi^*(\omega)$ the pullback of ω by φ. The problem is the following: Is the well-defined number above a homotopy invariant of the pair (M, φ)? (See Figure 2.)

When $V = \mathrm{pt}$, one gets the ordinary signature of M, which is a homotopy invariant. By the work of Wall and Miscenko, on equivariant surgery theory, one can assign a π-equivariant signature to the covering $\tilde{M}$ of M pullback by φ of the universal cover $\tilde{V}$ of V. Moreover, this equivariant signature belongs (neglecting torsion) to the Witt group of the group ring $\mathbf{C}\pi$ and is a homotopy invariant, $\text{Signature}_\pi(M) \in \text{Witt}(\mathbf{C}\pi)$. When π is *commutative*, one can prove the homotopy invariance of higher signatures as follows. There is indeed a *space* assigned to the group π, the space of characters, i.e., the dual $\hat{\pi}$, which is Hausdorff and compact, finite-dimensional if π is finitely generated. Then the group ring $\mathbf{C}\pi$ embeds as a subring of the ring $C(\hat{\pi})$ of continuous functions on $\hat{\pi}$:

$$\mathbf{C}\pi \subset C(\hat{\pi}).$$

The diagonalization of quadratic forms on $C(\hat{\pi})$ yields a map from the Witt group of $\mathbf{C}\pi$ to the K^0 group of $\hat{\pi}$:

$$\text{Witt}\,\mathbf{C}\pi \to K^0(\hat{\pi}).$$

Now any

$$\omega \in H^n(V, \mathbf{C}) = H^n(\pi, \mathbf{C})$$

is represented by a group cocycle $\omega(g^1, \ldots, g^n)$ totally antisymmetric in the g^i's. One then defines uniquely a current C on $\hat{\pi}$ by the equality:

$$\langle c, f^0\, df^1 \wedge \cdots \wedge df^n \rangle = \sum_{\prod_0^n g^i = 1} \hat{f}^0(g^0)\hat{f}^1(g^1) \cdots \hat{f}^n(g^n)\omega(g^1, \ldots, g^n)$$

where the f^i are functions on $\hat{\pi}$ so that their Fourier transform $\hat{f}^i$ are functions on the group π itself. The current C is closed because ω is a group cocycle.

The main lemma, then, which is a corollary of the index theorem for families, says that if you pair C with the Chern character of the equivariant signature you get the higher signature:

$$\langle C, \text{Ch}(\text{Signature}_\pi(\tilde{M})) \rangle = \langle \mathcal{L}_M \cdot \varphi^*(\omega), [M] \rangle.$$

Thus the right-hand side is a homotopy invariant.   Q.E.D.

In general, *when π is not commutative*, there is no interesting space of characters and one cannot really talk about the dual of π as a space. However, and this will be the key to this discussion, one can assign a noncommutative C^*-algebra to π; it is the completion of the group ring $\mathbf{C}\pi$ acting in the Hilbert space $l^2(\pi)$.

A careful scrutiny of the two previous examples reveals that one needs, in order to proceed, a suitable generalization of the notion of space, which would allow one to handle both leaf spaces and duals of noncommutative groups as if they were ordinary spaces.

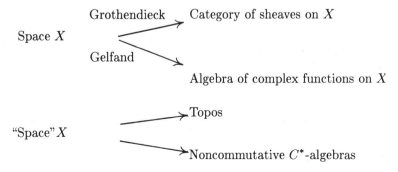

The basic idea underlying the new notion of space discovered by Grothendieck —and which he named "topos"—is that in an ordinary topological space the main part is not so much played by the points and their proximity relations, but by the category of sheaves on the space. Indeed the original topological space can be recovered from this category, and, moreover, if one keeps only the truly relevant conditions satisfied by such categories one obtains the notion of topos which plays a fundamental implicit role in the new algebraic geometry. The new notion of space that we shall deal with is based on a similar idea, but assigns a specific role to the complex numbers $\mathbf{C}$ or, equivalently, to functional analysis. It goes back to Gelfand's theory of C^*-algebras. It asserts that a compact topological space X is characterized by the $*$-algebra $C(X)$ of complex-valued continuous functions on X and that such algebras are the most general *commutative* C^*-algebras. That there is no good reason to restrict oneself to commutative C^*-algebras versus noncommutative ones, goes back to the early development of quantum mechanics with the discovery by Heisenberg of *matrix mechanics*. In understanding, from a very positivistic point of view highly enforced by experimental evidence in spectroscopy, the interaction of matter with the radiation field, Heisenberg showed that the usual observables of classical mechanics have to be replaced by matrices which violate the commutativity of mutliplication. Thus the phase space of quantum particles is an early example of the new type of spaces that we shall deal with. To take this second idea of space further, we need many examples, each being used as a small laboratory in which to test ideas and to see what works. We summarize a few examples in the following table:

Space	*Algebra*
X	$C(X)$
$X = \hat{\pi}$ dual of	$C^*(\pi) \supset \mathbf{C}\pi$
a discrete group	(completion in $l^2(\pi)$)
$X = M/F$ leaf space	$C^*(M, F)$
Example: Kronecker foliation	$VU = (\exp 2\pi_i \theta)UV$
$X = \Omega/G$ orbit space	$C_0(\Omega) \rtimes G$ crossed product

We have already discussed the first example. The second comes from foliations. There is a very natural C^*-algebra coming from operators which differentiate only in the leaf direction, and are elliptic in that direction. These turn out to have natural parametrices; they are invertible modulo operators which are smoothing in the leaf direction. These operators constitute a C^*-algebra, $C^*(M, F)$. An example would be to take the Kronecker foliation of the two torus, which is induced by the equation $dy = \theta\, dx$ where θ is irrational. In that case you get a C^*-algebra generated by two unitary elements which do not commute, but do commute up to a phase $\lambda = \exp 2\pi i\theta$. This is an algebra with which one may do many computations, exactly as if one were computing with the ordinary functions on the two torus using Fourier analysis.

Another very important example was discovered by Bellissard [6] from solid state physics and the quantum Hall effect. In the study of disordered systems, the Hamiltonian H_ω is labelled by a parameter $\omega \in \Omega$. Moreover, H_ω fails to commute with $H_{T_x(\omega)}$ where T is the action of the translation group on the parameter space Ω. Thus the translates of the Hamiltonian generate a noncommutative C^*-algebra, which corresponds to the "energy spectrum" of the system.

Given these examples one needs the right tools. The first comes from my original field of study: "von Neumann algebras." These algebras together constitute exactly the noncommutative analogue of measure theory. Their classification and understanding have now reached a fairly complete and satisfactory stage.

But what we need then is a little more than just measure theory; we need topology. I will now describe the basic tool in topology, first introduced by Grothendieck in algebraic geometry, and then by Atiyah for the purposes of topology. That tool is K-theory. There is a quite simple relation between complex vector bundles over the space X and projective modules over the algebra $A = C(X)$; this is the Serre-Swann theorem:

$$K^i(X) = K_i(A = C(X)).$$

It allows us to do K-theory of spaces by doing linear algebra where the field $\mathbf{C}$ is replaced by the ring A. Then the group of dimensions of finite projective modules is the K-group $K_0(A)$. The Bott periodicity theorem tells us that the K-groups of a C^*-algebra A are the homotopy groups of the gauge group, i.e., of the unitary group $\mathcal{U}$ of infinite matrices over A:

$$K_i(A) = \pi_{i+1}(\mathcal{U}).$$

Whenever a space is constructed by patching together two spaces, such that one has a short exact sequence of algebras, there is a corresponding long exact sequence of K-groups, which is shortened thanks to periodicity:

$$0 \to J \to A \to B \to 0 \Rightarrow \quad
\begin{array}{ccc}
 & K_0(J) \longrightarrow K_0(A) & \\
K_1(B) \nearrow & & \searrow K_0(B) \\
 & K_1(A) \longleftarrow K_1(J) &
\end{array}$$

Moreover, there is a general principle which is absolutely crucial. Above, we used twice the index theorem for families. Now the principle is that a "space" X will be described by a noncommutative algebra A, and that when one has a family (D_x), $x \in X$ indexed by X, such as the family of leafwise Dirac operators indexed by the space of leaves, then the index of this family belongs to $K^0(X) = K_0(A)$. This principle is very important because it allows us to translate into K-theoretic terms the basic analytical properties such as:

• The vanishing of the index of the family of leafwise Dirac operators:

$$\text{Index}(\text{Dirac}_L)_{L \in M/F} = 0$$

when the scalar curvature of leaves is strictly positive.

• The homotopy invariance of the π-equivariant signature: $\text{Signature}_\pi(\tilde{M}) \in K_0(C^*(\pi))$.

The first vanishing above takes place in the K-group $K_0(C^*(M, F))$. Both K-groups are countable abelian groups but are at first extremely mysterious objects, being defined through the above C^*-algebras. When dealing with ordinary spaces one gets some intuition about their K-groups, but this is less clear when dealing with C^*-algebras. The first real breakthrough which got everything started was done by Pimsner and Voiculescu [26] who, in particular, computed the K-groups for the Kronecker flow foliation discussed above. It allowed P. Baum and the author to guess what the answer should be in both general and geometric terms. The situation is described as follows: We construct both a geometric group, the K-homology of the classifying space, and a map μ to the K-group of the C^*-algebra. The classifying space makes sense in all the above situations since topologists have a way of making sense, up to homotopy, of spaces like the leaf space of a foliation or the orbit space of a group action. What they do is to amplify the space, say M, on which the group Γ acts, by crossing M with a contractible space $E\Gamma$ on which Γ acts freely; then the quotient $M \times_\Gamma E\Gamma$ makes sense and is "homotopic to M/Γ."

$$K_*(\text{Classifying space}) \xrightarrow{\mu} K(C^*\text{-algebra})$$

The map μ is difficult to construct [5, 4, 12, 24] and even when one deals with a one point space, its mere existence is the Atiyah-Singer index theorem [5]. It is essentially a Poincaré duality map to the extent that it reverses functorialities. The main problem of the theory is to handle this map μ; all computations so far indicate that it is a bijection [4, 23, 24, 27]. An important tool developed by the Russian school, by Miscenko and Kasparov in particular, and also by Atiyah, Brown, Douglas, and Fillmore (cf. [23, 24, 1, 7]), is K-homology for C^*-algebras. Since this theory played a crucial role in the understanding of the analogue of de Rham's theory of currents for the above spaces, I shall sketch it briefly. For ordinary spaces, K-homology is defined, using duality, by a general theorem which states that given any cohomology theory (such as K-theory) there is a corresponding homology theory, called here K-homology. One wants to realize this homology theory concretely. It is quite striking that if one was very conservative and wanted to stick to ordinary spaces, not accepting "spaces," one would not be able to describe the theory K-homology (X) (there is a K_{even} and K_{odd}) as homotopy classes of maps from spaces $Z_{\text{even}}, Z_{\text{odd}}$ to the space X. However, with "spaces" this is possible; Z_{ev} is obtained by glueing together two contractible "spaces," and the C^*-algebra $C(Z_{\text{ev}})$ is the noncommutative algebra A_{ev} of pairs of operators (x, y) in Hilbert space h whose difference $x - y$ is a compact operator. Similarly $C(Z_{\text{odd}}) = A_{\text{odd}}$, which also appears in *Beyond Affine Lie Algebras*, by I. Frenkel, is the algebra of 2×2 matrices (x_{ij}) of operators, such that x_{12} and x_{21} are compact. Of course a "continuous map" from Z_{ev} to X is given by a homomorphism from $C(X)$ to $C(Z_{\text{ev}})$, i.e., a homomorphism from the C^*-algebra $A = C(X)$ to A_{ev}. This is called a *Fredholm module* over

A because it amounts to giving a $\mathbf{Z}/2$ graded Hilbert space, $h \oplus h$ with grading $\varepsilon = \begin{bmatrix} 1 & 0 \\ 0 & -1 \end{bmatrix}$, with a left A-module structure such that

(1) $\varepsilon a = a\varepsilon \ \forall a \in A$,

(2) $[F, a]$ is compact $\forall a \in A$ where $F = \begin{bmatrix} 0 & 1 \\ 1 & 0 \end{bmatrix}$.

There is a similar notion of odd Fredholm module. On any even-dimensional compact spin^c manifold, the module of L^2 spinors, with $\mathbf{Z}/2$ grading given by the γ_5 matrix [5] and operator F given by the phase $F = D|D|^{-1}$ of the Dirac operator, is a Fredholm module which represents the *fundamental class* of the manifold in K-homology [5]. If one puts together this notion of a Fredholm module with the ideas of Helton and Howe, Carey and Pincus [18, 9] on operators commuting modulo trace ideals, one is led to the noncommutative analogue of de Rham's theory: cyclic cohomology. Helton and Howe associated to any operator T, normal modulo trace class operators, a de Rham current on $\mathbf{R}^2$ with boundary carried by the essential spectrum of T. Their work was very inspiring because it showed that the calculus of differential forms could be born from purely operator theoretic considerations in Hilbert space. This is what is done in [11]; given a Fredholm module over A, one can define *differential forms* on the corresponding "space," not by using local charts and patching these together but directly as operators in h. It is exactly the same step as the replacement, in quantum mechanics, of Poisson brackets by commutators. Thus

$$da = i[F, a] \quad \forall a \in A$$

defines the differential of a function. The forms of degree q are obtained as sums of products of 1-forms: $\Omega^q = \{\sum x^0 \, dx^1 \cdots dx^q, \ x^j \in A\}$. In this way, one gets a graded differential algebra; the product is the product of operators and the differential is given by

$$d\omega = i(F\omega - (-1)^q \omega F) \quad \text{for } \omega \in \Omega^q.$$

One has $d^2 = 0$, and the main point is to obtain an integration of forms $\omega \to \int \omega \in \mathbf{C}$ satisfying $\int d\omega = 0$ and $\int \omega_2 \omega_1 = (-1)^{q_1 q_2} \int \omega_1 \omega_2$.

The formula which works is quite simple: $\int \omega = \text{Trace}(\varepsilon\omega)$. This is where the *dimension* appears, the trace only makes sense if ω is a *trace class* operator. By the Holder inequality this holds, for any $\omega \in \Omega^n$, provided $[F, a] \in \mathcal{L}^n \ \forall a \in A$. Here, for every *real* number $p \in [1, \infty]$, $\mathcal{L}^p$ is the ideal of compact operators T with $\sum \lambda_q(|T|)^p < \infty$, where $\lambda_q(|T|)$ is the qth eigenvalue of the absolute value of T. The *dimension* of a Fredholm module over an algebra is the infimum of the p's for which $[F, a] \in \mathcal{L}^p \ \forall a \in A$. For the fundamental class of a manifold M described above, it yields the dimension of M. In general it need not be an integer. Given an even Fredholm module of dimension p on A one can integrate only the forms $\omega \in \Omega^n$ of degree $\geq p$. Moreover, odd forms have integral 0. Thus the above construction yields for each even integer $n \geq p$, the functional τ_n called the n-dimensional character of the Fredholm module:

$$\tau_n(a^0, \ldots, a^n) = \int a^0 \, da^1 \cdots da^n \quad \forall a^i \in A.$$

Carefully analyzing these functionals led me to discover cyclic cohomology in 1981. It was discovered independently from algebraic K-theory by Feigin and Tsigan [**17**, **30**], replacing group homology by Lie alagebra homology in the basic construction of Quillen's algebraic K-theory. It also appeared, at least in implicit form, in the work of Hsiang and Staffeld on the algebraic K-theory of spaces [**20**]. It is of course quite striking that from different streams of ideas one gets to the same theory: *cyclic cohomology.*

A crucial and simple lemma is the following.

LEMMA.   *Let A be an algebra and τ an $(n+1)$-linear map $A \times A \times \cdots \times A \to \mathbf{C}$ such that*
   (1) $\tau(a^1, \ldots, a^n, a^0) = (-1)^n \tau(a^0, \ldots, a^n)$ $\forall a^i \in A$;
   (2) $\sum_0^n (-1)^j \tau(a^0, \ldots, a^j a^{j+1}, \ldots, a^{n+1}) + (-1)^{n+1} \tau(a^{n+1} a^0, \ldots, a^n) = 0$ $\forall a^i$
$\in A$.
Then the map $e \in A$, $e^2 = e \to \tau(e, \ldots, e)$, gives a morphism of $K_0(A)$ to $\mathbf{C}$.

In fact $K_0(A)$ is generated by idempotents $e^2 = e$ in matrices over A, $M_q(A) = M_q(\mathbf{C}) \otimes A$, and one has to extend τ to $M_q(A)$ by the equality:

$$\tau_q(m^0 \otimes a^0, \ldots, m^n \otimes a^n) = \mathrm{Trace}(m^0 \cdots m^n) \tau(a^0, \ldots, a^n) \qquad (*)$$

$\forall m^j \in M_q(\mathbf{C})$, $a^j \in A$.

Here are a few examples of functionals τ satisfying (1) and (2):

EXAMPLE α. Let $A = C^\infty(M)$, the algebra of smooth functions on a compact manifold, and C a closed current on M of dimension k. Then $\tau(f^0, \ldots, f^k) = \langle C, f^0 \, df^1 \wedge \cdots \wedge df^k \rangle$ $\forall f^j \in A$ has exactly the properties (1), (2) of a cyclic cocycle. In fact τ satisfies $\tau^\sigma = \mathrm{sign}(\sigma)\tau$ for *any* permutation of $\{0, 1, \ldots, k\}$, but since $\mathrm{Trace}(m^0 \cdots m^k)$ is invariant only under *cyclic* permutations it is only (1) which is satisfied by all τ_q. One has $K_0(A) = K^0(M)$ and the lemma gives back the ordinary Chern character, viewed as a pairing with the homology of M.

EXAMPLE β. Let π be a discrete group, $A = \mathbf{C}\pi$ the group ring, and $\omega \in Z^n(\pi, \mathbf{C})$ a group cocycle suitably normalized so that $\omega(g^1, \ldots, g^n) = 0$ if $g^1 \cdots g^n = 1$. Then the equality

$$\tau(g^0, \ldots, g^n) = 0 \text{ if } g^0 \cdots g^n \neq 1 \quad \forall g^i \in \pi,$$
$$\tau(g^0, \ldots, g^n) = \omega(g^1, \ldots, g^n) \text{ if } g^0 g^1 \cdots g^n = 1 \quad \forall g^i \in \pi,$$

defines an n-cyclic cocycle τ on A. Moreover, extending τ to infinite matrices over A one can show that

$$\langle \tau, \mathrm{Signature}_\pi(\tilde{M}) \rangle = \langle \mathcal{L}_M \cdot \varphi^*(\omega), [M] \rangle$$

with the notations of the higher signature problem. The cyclic cohomology of group rings is computed by Burghelea in [**8**].

EXAMPLE γ. For each even $n \geq p$, the n-dimensional character τ_n of a Fredholm module over A is a *cyclic cocycle.* Moreover, the pairing with $K_0(A)$, $\langle \tau_n, e \rangle$, is given for any idempotent e by the *index of a Fredholm operator,* and, in particular, lands in $\mathbf{Z} \subset \mathbf{C}$. It corresponds to the $\mathbf{Z}$-valued pairing between K-theory and K-homology, which ensures that it is highly nontrivial.

Given any algebra $\mathcal{A}$, there is a trivial way to construct cyclic cocycles on $\mathcal{A}$, namely $\tau = b\varphi$ where $\varphi \in C_\lambda^{n-1}$ is an n-linear functional on $\mathcal{A}$ satisfying (1), and $b\varphi$ its Hochschild coboundary given by formula (2). The relevant group is the quotient $H_\lambda^n(\mathcal{A}) = Z_\lambda^n(\mathcal{A})/bC_\lambda^{n-1}$, where $Z_\lambda^n = \operatorname{Ker} b$, and is called *cyclic cohomology* of $\mathcal{A}$. It turns out that just by working with Example γ, of Fredholm modules, all the properties of cyclic cohomology fall into one's lap. First a Fredholm module has many characters τ_q, one for each even integer $q \geq p$, and it would be unreasonable to expect that $\tau_{q+2}, \tau_{q+4}, \ldots$ bring new information not contained in τ_q. Explicit computations show that there is a natural periodicity operator

$$S \colon H_\lambda^n(\mathcal{A}) \to H_\lambda^{n+2}(\mathcal{A})$$

given in fact by cup product by the generator of $H_\lambda^2(\mathbf{C})$ and such that $\tau_{q+2k} = S^k\tau_q$ in $H_\lambda^{q+2k}(\mathcal{A})$. Then, in order to find the smallest n for which τ_n is defined, one needs to determine the image of S. But by construction, the complex (C_λ^n, b) is a subcomplex of the Hochschild complex (C^n, b) where C^n is the space of all $(n+1)$-linear functionals on $\mathcal{A}$. It turns out that $\tau \in \operatorname{Im} S$ iff τ is trivial in the latter complex, whose cohomology $H^n(\mathcal{A}, \mathcal{A}^*)$, the Hochschild cohomology of $\mathcal{A}$ with coefficients in the bimodule of linear forms on $\mathcal{A}$, is computable by the general methods of homological algebra. The final point is the construction of a natural operator B from Hochschild cohomology $H^n(\mathcal{A}, \mathcal{A}^*)$ to $H_\lambda^{n-1}(\mathcal{A})$ and the proof of the exactness of the following sequence:

$$H_\lambda^n(\mathcal{A}) \xrightarrow{S} H_\lambda^{n+2}(\mathcal{A}) \xrightarrow{I} H^{n+2}(\mathcal{A}, A^*)$$

$$B \overset{\frown}{H_\lambda^{n+1}}(\mathcal{A}) \xrightarrow{S} H_\lambda^{n+3}(\mathcal{A}) \xrightarrow{I} H^{n+3}(\mathcal{A}, \mathcal{A}^*)$$

Thus Hochschild cohomology and cyclic cohomology from an exact couple which together with the associated spectral sequence becomes a basic tool to compute cyclic cohomology of algebras. The power of this tool is illustrated by two examples:

EXAMPLE a. Let M be a compact manifold, $\mathcal{A} = C^\infty(M)$. Imposing obvious continuity conditions to cochains one finds that the Hochschild cohomology groups $H^q(\mathcal{A}, \mathcal{A}^*)$ are identified with the space Ω_q of de Rham currents of dimension q on M. The map $I \circ B$ of the exact couple is the de Rham boundary d^t, and one gets

$$H_\lambda^q(\mathcal{A}) = \{\operatorname{Ker} d^t \subset \Omega_q\} + H_{q-2}(M, \mathbf{C}) + H_{q-4}(M, \mathbf{C}) + \cdots.$$

The de Rham homology of M identifies with the periodic cyclic cohomology of $\mathcal{A}$: $H_{\mathrm{Per}}^*(\mathcal{A}) = \varinjlim(H_\lambda^n(\mathcal{A}), S)$.

EXAMPLE b. Let (M, F) be a foliated manifold, $A = C^*(M, F)$ the corresponding C^*-algebra. In A there is a natural dense subalgebra $\mathcal{A}$ of smooth elements and one has to compute its cyclic cohomology. One has

$$H_{\mathrm{Per}}^*(\mathcal{A}) \cong H_\tau^*(\text{Classifying space})$$

where the right-hand side is the cohomology with complex coefficients of the classifying space of the holonomy groupoid or graph of the foliation. The index τ means that this cohomology is twisted by the orientation of the transverse bundle τ of the foliation. Using sheaves on M and the naturality of the construction of $\mathcal{A}$, one constructs a localization morphism λ_M, which is a far-reaching generalization of the Ruelle-Sullivan current:

$$\lambda_V : H^*_{\mathrm{Per}}(\mathcal{A}) \xrightarrow{\lambda_M} H^*_\tau(M, \mathbf{C})$$

and one reaches the following cohomological formulation of the longitudinal index theorem for foliations [12].

THEOREM. *Let (M, F) be a compact foliated manifold, D a longitudinal elliptic operator, and τ a cyclic cocycle on $\mathcal{A}$. Then*

$$\langle \tau, \mathrm{Index}(D) \rangle = \langle \lambda_M(\tau) \, \mathrm{Td}(F_{\mathbf{C}}) \, \mathrm{Ch}\, \sigma_D, [M] \rangle$$

where σ_D is the longitudinal symbol of D.

There is, however, still a really hard step in order to use cyclic cohomology as ordinary de Rham theory for our "spaces"—such as the space of leaves of a foliation—and to prove Theorem 2 of this paper, for instance [14]. The point is that $\mathcal{A} \subset A$, in Example b, is not in general an isomorphism in K-theory, and the analytic information lies in $K(A)$ not $K(\mathcal{A})$. This problem is fully resolved in [14] for the transverse fundamental class of M/F and all classes coming by pullback of the Gelfand-Fuchs cohomology by the map $B(\text{Classifying space}) \to B\Gamma_q$.

The difficulty is that for a general foliation it is impossible to reduce the transverse structure group to a compact group. Equivalently, for a group of diffeomorphisms acting on a manifold, one cannot find an invariant Riemannian metric. The result implies, in particular, the Novikov conjecture for Gelfand-Fuchs cohomology classes on $B(\mathrm{Diff}\, N)$ for any N.

BIBLIOGRAPHY

Bibliographical comments. For a survey of cyclic cohomology and an extensive bibliography, see [10, 17]. The computation of the homology of the traces on the universal differential algebra $\Omega(\mathcal{A})$ of an algebra $\mathcal{A}$ was explicitly done in [11, Theorem 33, p. 118]. Once dualized, it gives the computation [22] in terms of cyclic homology of the cohomology of the universal de Rham complex introduced by Karoubi in [21]. The computation of the homology of the Lie algebra of matrices in terms of cyclic homology is done in [25, 30]. The idea of using the Lichnerowicz identity in a C^*-algebra context is due to J. Rosenberg [27].

1. M. F. Atiyah, *Global theory of elliptic operators*, Proc. Internat. Conf. on Functional Analysis and Related Topics, Univ. of Tokyo Press, Tokyo, 1970, pp. 21–29.

2. _____, *K-theory*, Benjamin, New York, 1967.

3. M. F. Atiyah and I. Singer, *The index of elliptic operators.* IV, Ann. of Math. **93** (1971), 119–138.

4. P. Baum and A. Connes, *Geometric K-theory for Lie groups and foliations*, Preprint I.H.E.S., 1982.

5. P. Baum and R. Douglas, *K-homology and index theory*, Operator Algebras and Applications, Proc. Sympos. Pure Math., vol. 38, part I, Amer. Math. Soc., Providence, R.I., 1982, pp. 117–173.

6. J. Bellissard, *K-theory of C-algebras in solid state physics* (Conf. on Statistical Mechanics and Field Theory, Mathematical Aspects, Groningen, August 26–30, 1985).

7. L. G. Brown, R. Douglas, and P. A. Fillmore, *Extensions of C^*-algebras and K-homology*, Ann. of Math. (2) **105** (1977), 265–324.

8. D. Burghelea, *The cyclic cohomology of the group rings*, Comment. Math. Helv. **60** (1985), 354–365.

9. R. Carey and J. D. Pincus, *Almost commuting algebras, K-theory and operator algebras*, Lecture Notes in Math., vol. 575, Springer, Berlin-New York, 1977.

10. P. Cartier, *Homologie cyclique*, Exposè 621, Seminaire Bourbaki Fevrier 1984.

11. A. Connes, *Noncommutative differential geometry*. Part I, *The Chern character in K-homology*, Part II, *de Rham homology and noncommutative algebra*, (Preprint), Inst. Hautes Études Sci. Publ. Math. **62** (1986), 44–144.

12. A. Connes and G. Skandalis, *The longitudinal index theorem for foliations*, Publ. Res. Inst. Math. Sci. **20** (1984), 1139–1183.

13. A. Connes, *C^* algèbres et géométrie différentielle*, C. R. Acad. Sci. Paris Sér. I Math. **290** (1980), 599–604.

14. ____, *Cyclic cohomology and the transverse fundamental class of a foliation*, Preprint I.H.E.S. M/84/7, 1984.

15. ____, *Spectral sequence and homology of currents for operator algebras*, Math. Forschungsinstitute Oberwolfach Tagungsbericht 42/81, Funktionalanalysis and C^*-Algebren, 27–9/3–10–1981.

16. R. Douglas and D. Voiculescu, *On the smoothness of sphere extensions*, J. Operator Theory **6** (1981), no. 1, 103–111.

17. B. L. Feigin and B. L. Tsygan, *Additive K-theory*, Preprint, 1985.

18. J. Helton and R. Howe, *Integral operators, commutators, traces, index and homology*, Proc. Conf. on Operator Theory, Lecture Notes in Math., vol. 345, Springer, Berlin-New York, 1973.

19. ____, *Traces of commutators of integral operators*, Acta Math. **135** (1975), 271–305.

20. W. C. Hsiang and R. E. Staffeldt, *A model for computing rational algebraic K-theory of simply connected spaces*, Invent. Math. **68** (1982), 227–239.

21. M. Karoubi, *Connexions, courbures et classes caractéristiques en K-théorie algébrique*, CMS Conf. Proc., 2, Amer. Math. Soc., Providence, R. I., 1982.

22. ____, *Homologie cyclique et K-théorie algébrique*. I, II, C.R. Acad. Sci. Sér. I Math. **297** (1983), 447–450, 513–516.

23. G. Kasparov, *K-functor and extensions of C^*-algebras*, Izv. Akad. Nauk SSSR Ser. Math. **44** (1980), 571–636.

24. ____, *K-theory, group C^*-algebras and higher signataures*, Conspectus, Chernogolovka, 1983.

25. J. L. Loday and D. Quillen, *Cyclic homology and the Lie algebra of matrices*, C.R. Acad. Sci. Paris Sér. I Math. **296** (1983), 295–297.

26. M. Pimsner and D. Voiculescu, *Exact sequences for K-groups and Ext-groups of certain cross-product C^*-algebras*, J. Operator Theory **4** (1980), 93–118.

27. J. Rosenberg, *C^*-algebras, positive scalar curvature and the Novikov conjecture*, Inst. Hautes Études Sci. Publ. Math. **58** (1984), 409–424.

28. I. Segal, *Quantized differential forms*, Topology **7** (1968), 147–172.

29. ____, *Quantizatain of the de Rham complex*, Global Analysis, Proc. Sympos. Pure Math., vol. 16, Amer. Math. Soc., Providence, R.I., 1970, pp. 205–210.

20. B. L. Tsigan, *Homology of matrix Lie algebras over rings and Hochschild homology*, Uspekhi Mat. Nauk **38** (1983), 217–218.

COLLEGE DE FRANCE, 75005 PARIS, FRANCE

Opérateurs de Calderón-Zygmund

GUY DAVID

I. Introduction. Des opérateurs d'intégrale singulière apparaissent naturellement dans de nombreux problèmes, notamment d'équations aux dérivées
partielles ou de théorie du potentiel. Nous nous bornerons dans cet exposé à
étudier une classe d'opérateurs d'intégrale singulière introduite par Calderón et
Zygmund [**CZ**]. Les méthodes utilisées pour prouver la continuité sur L^2 de tels
opérateurs sont souvent très proches de celles avec lesquelles on traite d'autres
intégrales singulières (comme la transformée de Hilbert le long d'une courbe, ou
certaines intégrales oscillantes, ou encore la fonction maximale sphérique). Nous
renvoyons à ce sujet à l'exposé de E. M. Stein.

Précisons un peu le type d'opérateurs dont nous allons parler. Par "noyau
standard" sur $\mathbf{R}^n$, nous entendrons une fonction K, définie sur $\mathbf{R}^n \times \mathbf{R}^n$ privé
de la diagonale, et telle que, pour un $C \geq 0$ et un $\delta \in]0,1]$,

$$|K(x,y)| \leq C|x-y|^{-n}, \tag{1}$$

et

$$|K(x',y) - K(x,y)| + |K(y,x') - K(y,x)| \leq C|x'-x|^{\delta}/|x-y|^{n+\delta} \tag{2}$$

pour $|x'-x| < \frac{1}{2}|x-y|$.

DEFINITION 1. Nous appellerons dans la suite opérateur d'intégrale singulière
(en abrégé SIO) tout opérateur linéaire continu de $\mathcal{C}_c^{\infty}(\mathbf{R}^n)$ dans $(\mathcal{C}_0^{\infty}(\mathbf{R}^n))'$ qui
a la propriété suivante: il existe un noyau standard K tel que, si f et g sont deux
fonctions-test à supports disjoints, la distribution Tf appliquée à g soit donnée
par $\langle Tf, g \rangle = \iint K(x,y)f(y)g(x)\,dx\,dy$.

Cette définition, introduite par Coifman et Meyer, a l'avantage sur la définition
habituelle d'être bien plus flexible. Notons que l'identité est un SIO, associé au
noyau $K \equiv 0$.

Il est maintenant classique que, si T est un SIO et si de plus T est continu sur
L^2 (c.à.d. peut être prolongé en un opérateur borné sur L^2), alors T est aussi
continu sur L^p pour $1 < p < +\infty$, peut être étendu en un opérateur continu de
l'espace atomique H^1 dans L^1 et, dualement, de L^∞ dans BMO (les fonctions à
oscillation moyenne bornée de John et Nirenberg). Il faut encore pouvoir décider
si un opérateur donné est borné sur L^2.

II. Le théorème de Coifman, McIntosh et Meyer. L'exemple fondamental est donné par le noyau de Cauchy sur un graphe lipschitzien. Si $A \colon \mathbf{R} \to \mathbf{R}$ est une fonction lipschitzienne, l'opérateur de Cauchy sur le graphe de A peut être défini par

$$\mathcal{C}_A f(x) = \text{v.p.} \int [x - y + i(A(x) - A(y))]^{-1} f(y) \, dy \quad \text{pour } f \text{ assez régulière.}$$

Après des résultats partiels de Calderón [**C1**], Coifman-Meyer [**CM1**] et Calderón [**C2**], Coifman, McIntosh et Meyer ont montré en 1981 le résultat suivant.

THÉORÈME 1 [**CMM**]. *L'opérateur $\mathcal{C}_A$ est borné sur $L^2(\mathbf{R})$ pour toute fonction lipschitzienne A.*

Ce résultat a une importance considérable car, en plus de ses conséquences directes (citons par exemple l'existence, pour toute fonction f de carré intégrable sur le graphe de A, d'une décomposition $f = f_+ + f_-$, où f_+ (resp. f_-) s'étend en une fonction analytique au-dessus (resp. au-dessous) du graphe) il permet de démontrer la continuité de nombreux autres opérateurs, comme le potentiel de double-couche associé à un graphe lipschitzien $\subset \mathbf{R}^{n+1}$ (voir [**CDM**]). Le théorème 1 a eu très rapidement de nombreuses applications, notamment à la solution d'équations aux dérivées partielles sur des domaines peu réguliers. Citons seulement [**DK1, DKV1, DKV2, V1**].

En plus des démonstrations originales de [**C2**] et [**CMM**], on dispose maintenant de plusieurs manières d'aborder la continuité de $\mathcal{C}_A$. Signalons une démonstration de S. Semmes, qui ne donne pour le moment que le cas où $\|A'\|_\infty$ est assez petite, mais qui peut être utilisée pour d'autres problèmes. Donnons une idée de la stratégie.

Soient Γ le graphe de A et $H^2_+(\Gamma)$ (resp. $H^2_-(\Gamma)$) l'espace de Hardy des traces sur Γ de fonctions analytiques au-dessus (resp. au-dessous) de Γ, et qui sont uniformément dans $L^2(\Gamma + i\varepsilon)$ pour ε positif (resp. négatif). La continuité de $\mathcal{C}_A$ est équivalente à l'existence, pour toute fonction $f \in L^2(\Gamma)$, d'une décomposition $f = f_+ + f_-$, où $f_\pm \in H^2_\pm(\Gamma)$. La fonction F, égale au-dessus de Γ à l'extension de f_+ et au-dessous de Γ à l'extension de $-f_-$, est donc analytique hors de Γ et a un saut égal à f sur Γ. On décide de chercher F sous la forme $F = G \circ \rho^{-1}$, où ρ est un homéomorphisme bi-lipschitzien de $\mathbf{C}$ qui envoie $\mathbf{R}$ sur Γ. Soit $\mu = \overline{\partial}\rho(\partial\rho)^{-1}$ la dilatation complexe de ρ; la fonction G vérifie alors $(\overline{\partial} - \mu\partial)G = 0$ hors de $\mathbf{R}$, et a le saut $g = f \circ \rho^{-1}$ sur $\mathbf{R}$. On appelle $C(g)$ l'intégrale de Cauchy de g; $C(g)$ est donc analytique hors de $\mathbf{R}$; on note $C(g)'$ sa dérivée et $h(z) = \mu(z)C(g)'(z)$ pour $z \notin \mathbf{R}$. Alors $H = G - C(g)$ n'a plus de saut sur $\mathbf{R}$, et vérifie $(\overline{\partial} - \mu\partial)H = h$.

Il s'avère aussi qu'on peut choisir ρ assez régulière pour que μ vérifie certaines estimations quadratiques du type "mesures de Carleson," qui à leur tour permettent de résoudre l'équation $(\overline{\partial} - \mu\partial)H = h$ quand $g \in L^2$. Voir [**Se1**] pour plus de détails.

On sait aussi démontrer le théorème 1, à partir du cas où $\|A'\|_\infty$ est assez petit, par une méthode de perturbations. L'idée est de trouver, pour tout A

et tout intervalle compact, une fonction $\tilde{A}$ telle que $\|\tilde{A}'\|_\infty \leq \frac{9}{10}\|A'\|_\infty$, et qui coincide avec A sur une part significative de l'intervalle. Si l'on sait déjà que $\mathcal{C}_{\tilde{A}}$ est borné dès que $\|\tilde{A}'\|_\infty \leq \frac{9}{10}\|A'\|_\infty$, on peut en déduire un contrôle local de $\mathcal{C}_A$, que l'on transforme en contrôle global en utilisant des inégalités aux bons λ, ou tout autre outil équivalent. Cette idée, déjà utilisée dans [**CM4**] pour l'étude des courbes corde-arc, puis dans [**Dv**] pour les graphes lipschitziens, a été bien perfectionnée par T. Murai et P. Tchamitchian. C'est ainsi que T. Murai démontre le

THÉORÈME 2 [**Mu2**]. *La norme de* $\mathcal{C}_A$ *sur* $L^2(\mathbf{R})$ *est inférieure à* $C(1 + \|A'\|_\infty)^{1/2}$.

Notons que cette estimation est la meilleure possible. Il est assez surprenant que, de toutes les méthodes connues pour prouver la continuité de $\mathcal{C}_A$, ce soit une méthode de perturbations successives qui donne le meilleur résultat. Comme l'a remarqué S. Semmes, le théorème 2 permet de montrer que le noyau $|(A(x) - A(y))(x - y)^{-1}|(x - y)^{-1}$ définit un opérateur borné sur $L^2(\mathbf{R})$ lorsque $A\colon \mathbf{R} \to \mathbf{R}$ est lipschitzienne.

III. Critères de continuité sur L^2. Une autre voie de recherches est de trouver des conditions générales qui entraînent la continuité sur L^2 d'un opérateur d'intégrale singulière T. Lorsque T n'est pas un opérateur de convolution, on ne peut pas appliquer Plancherel directement. On peut par contre utiliser avec une certaine efficacité le lemme de Cotlar, Knapp et Stein sur les sommes "presque-othogonales" d'opérateurs sur un espace de Hilbert (il était déjà question d'appliquer ce lemme à des opérateurs d'intégrale singulière dans [**F**]). On obtient ainsi un premier critère de continuité sur L^2.

THÉORÈME 3 [**DJ**]. *Soit* T *un SIO. Alors* T *s'étend en un opérateur borné sur* $L^2(\mathbf{R}^n)$ *si et seulement si* $T1 \in BMO$, ${}^tT1 \in BMO$ *et* T *est faiblement borné.*

Rappelons que, bien que 1 ne soit pas à support compact, $T1$ peut être défini à une constante additive près; le transposé ${}^tT$ est donné par $\langle {}^tTf, g\rangle = \langle Tg, f\rangle$. Enfin, pour $x \in \mathbf{R}^n$ et $t > 0$, notons $\mathcal{A}(x, t)$ l'opérateur de translation-dilatation défini par $\mathcal{A}(x, t)f(y) = t^{-n/2}f((y - x)/t)$. Nous dirons que T est faiblement borné si les opérateurs $\mathcal{A}^{-1}(x, t)T\mathcal{A}(x, t)$, $(x, t) \in \mathbf{R}_+^{n+1}$, sont uniformément bornés de $\mathcal{C}_c^\infty(\mathbf{R}^n)$ dans $(\mathcal{C}_c^\infty(\mathbf{R}^n))'$. Ainsi, "T est faiblement borné" traduit seulement une certaine stabilité, vis à vis à des translations et des dilatations, des inégalités qui permettent de définir T.

Disons deux mots de la démonstration du théorème 3. On commence par soustraire à T deux opérateurs que l'on sait traiter (des paraproduits, par exemple), pour se réduire au cas où $T1 = {}^tT1 = 0$ dans BMO. Ensuite, fidèle à la tradition, on découpe T en petits morceaux de la manière suivante. On se donne une fonction $\varphi \in \mathcal{C}_c^\infty(\mathbf{R}^n)$ positive, radiale, d'intégrale 1 et l'on note P_t l'opérateur de convolution par $\varphi_t(x) = t^{-n}\varphi(x/t)$ et $Q_t = -t(\partial P_t/\partial t)$. Ainsi,

Q_t est l'opérateur de convolution par la fonction $\psi_t(x) = t^{-n}\psi(x/t)$, où ψ est une fonction semblable à φ, mais d'intégrale nulle.

On écrit

$$T = \lim_{t \to 0} P_t T P_t = -\int_0^\infty \frac{\partial}{\partial t}(P_t T P_t)\, dt = \int_0^\infty Q_t T P_t \frac{dt}{t} + \int_0^\infty P_t T Q_t \frac{dt}{t}.$$

Les propriétés $T1 = 0$ et ${}^t T1 = 0$ permettent de montrer que le découpage de l'opérateur

$$\int_0^\infty Q_t T P_t \frac{dt}{t} \quad \text{en} \quad \sum_{k \in \mathbf{Z}} \int_{2^k}^{2^{k+1}} Q_t T P_t \frac{dt}{t}$$

satisfait aux hypothèses du lemme de Cotlar-Knapp-Stein; on en déduit le théorème.

Notons que, bien que le théorème 3 permette de réduire à quelques intégrations par parties la démonstration de la continuité de $\mathcal{C}_A$ lorsque $\|A'\|_\infty$ est assez petit, il ne permet pas de traiter le cas général où $\|A'\|_\infty < +\infty$.

On peut aller plus loin dans la direction du théorème 3. A. McIntosh et Y. Meyer ont remarqué que, si $b \in L^\infty$ est une fonction telle que, pour un $\delta > 0$, $\operatorname{Re} b(x) \geq \delta$ pour tout x, alors on a le résultat suivant [**McM**]. Si T est un SIO, si $Tb = 0$ et ${}^t T b = 0$, et si $\{b\}T\{b\}$ est faiblement borné (où $\{b\}$ désigne l'opérateur de multiplication par $b(x)$), alors T est borné sur L^2. Ce résultat entraîne le théorème 1: on prend $b = 1 + iA'$ et on remarque que $Tb = {}^t T b = 0$ par la formule de Cauchy.

En fait, on peut démontrer un résultat un peu plus général encore. On dit que la fonction bornée $b: \mathbf{R}^n \to \mathbf{C}$ est "para-accrétive" si, pour tout $x \in \mathbf{R}^n$ et tout $d > 0$, il existe un cube Q, dont le centre x' vérifie $|x' - x| \leq Cd$, dont le côté r vérifie $(1/C)d \leq r \leq Cd$, et tel que $|(1/|Q|) \int_Q b(u)\, du| \geq \delta$. (Bien entendu, les constantes C et $\delta > 0$ sont indépendantes de x et de d.)

THÉORÈME 4 [**DJS**]. *Soient b_1 et b_2 deux fonctions para-accrétives et T un SIO. Alors T est borné sur $L^2(\mathbf{R}^n)$ si et seulement si $Tb_1 \in BMO$, ${}^t T b_2 \in BMO$ et $\{b_2\}T\{b_1\}$ est faiblement borné.*

Ce théorème se démontre un peu comme le théorème 3. Il faut cependant remplacer les opérateurs P_t et Q_t définis plus haut par des décompositions de l'identité adaptées aux fonctions b_i, ce qui soulève quelques difficultés techniques. On est ainsi amené à utiliser les techniques de décomposition de Littlewood-Paley sur un espace de nature homogène, et à modifier un peu le lemme de Cotlar-Knapp-Stein.

IV. Opérateurs multilinéaires et estimations polynomiales. La manière la plus classique d'attaquer l'opérateur $\mathcal{C}_A$ est de le décomposer en série de puissances (les commutateurs de Calderón). Ainsi,

$$\mathcal{C}_A f = \sum_{k=0}^\infty i^k T_k(A', \ldots, A', f),$$

où

$$T_k(a_1, a_2, \ldots, a_k, f) = \text{v.p.} \int \frac{A_1(x) - A_1(y)}{x - y} \cdots \frac{A_k(x) - A_k(y)}{x - y} \frac{f(y)}{x - y} \, dy,$$

où l'on a noté $A_i(x) = \int_0^x a_i(t) \, dt$.

Lorsqu'on applique le théorème 3 dans cette situation, on utilise le fait que remplacer f par 1 permet de ramener l'étude d'un opérateur $(k+1)$-linéaire à celle d'un opérateur k-linéaire, ce qui donne une chaîne d'estimations comme

$$\begin{aligned}
\|T_k(A', \ldots, A', .)\|_{L^2, L^2} &\leq C \|T_k(A', \ldots, A', 1)\|_{\text{BMO}} \\
&= C \|T_{k-1}(A', \ldots, A')\|_{\text{BMO}} \\
&\leq C \|T_{k-1}(A', \ldots, A', .)\|_{L^\infty, \text{BMO}} \\
&\leq C C' \|T_{k-1}(A', \ldots, A', .)\|_{L^2, L^2} \leq \cdots.
\end{aligned}$$

On obtient une majoration de T_k en k étapes et, à chaque étape, on perd une constante multiplicative. On pourrait donc craindre que ce type de démonstration ne puisse fournir que des estimations de normes en C^k. Il n'en est rien, et M. Christ et J.-L. Journé ont pu, avec des méthodes similaires, obtenir des estimations polynomiales en k de la norme certains opérateurs k-linéaires (dont le k ième commutateur de Calderón).

Assez curieusement, certaines améliorations du formalisme jouent un rôle important. Ainsi, on a vu que pour prouver que T est continu, on avait envie d'écrire $T = \int_0^\infty T_t \, dt/t$ avec, par exemple, $T_t = P_t T Q_t + Q_t T P_t$. On a en fait intérêt à étudier directement ce que les auteurs de [**CJ**] appellent une ε-famille.

DÉFINITION 2. Soit $0 < \varepsilon \leq 1$. On appelle ε-famille une famille $\mathcal{T} = (T_t)_{t>0}$ d'opérateurs donnés par leur noyau $T_t(x, y)$, où

$$|T_t(x, y)| \leq C \frac{t^\varepsilon}{(t + |x - y|)^{n+\varepsilon}}, \tag{3}$$

et

$$|T_t(x, y) - T_t(x, y')| \leq C \left(\frac{|y' - y|}{t + |x - y|} \right)^\varepsilon \frac{t^\varepsilon}{(t + |x - y|)^{n+\varepsilon}} \tag{4}$$

pour tous x, y, et $y' \in \mathbf{R}^n$ tels que $|y' - y| < \frac{1}{2}(t + |x - y|)$.

On dira que $\mathcal{T}$ est bornée si, pour tout $f \in L^2$, $\int_0^\infty \|T_t f\|_2^2 \, dt/t \leq C \|f\|_2^2$.

Il y a une correspondance entre SIO faiblement bornés d'une part, et ε-familles d'autre part. Ainsi si T est un SIO faiblement borné, alors $(Q_t T P_t)_{t>0}$ et $(P_t T Q_t)_{t>0}$ sont des ε-familles. Réciproquement, si $(T_t)_{t>0}$ est une famille d'opérateurs donnés par des noyaux $T_t(x, y)$ vérifiant (3) et

$$|\nabla_x T_t(x, y)| + |\nabla_y T_t(x, y)| \leq \frac{C}{t} \frac{t^\varepsilon}{(t + |x - y|)^{n+\varepsilon}}, \tag{5}$$

alors $T = \int T_t \, dt/t$ est un SIO faiblement borné.

Le Théorème 3 a un analogue dans le langage des ε-familles.

THÉORÈME 5 [**CJ**]. *Soit* $\mathcal{T} = (T_t)_{t>0}$ *une* ε-*famille. Alors* $\mathcal{T}$ *est bornée si et seulement si* $\mu = |T_t 1(x)|^2 \, dx \, dt/t$ *est une mesure de Carleson sur* $\mathbf{R}_+^{n+1}$

(c.à.d. que $\|\mu\|_C = \sup_B \mu(B\times]0,r[)/|B|$, où la borne supérieure est prise sur toutes les boules de B de $\mathbf{R}^n$ et où r est le rayon de B, est fini). De plus,

$$\left\|\,|T_t a(x)|^2 \frac{dx\,dt}{t}\,\right\|_C \leq \|a\|_\infty \left\|\,|T_t 1(x)|^2 \frac{dx\,dt}{t}\,\right\|_C + C\|a\|_\infty \qquad (6)$$

où la constante C dépend des estimations (3) et (4).

Nous attirons l'attention du lecteur sur le fait que, dans (6), le premier terme du second membre n'est précédé d'aucune constante. C'est un peu comme si, dans le théorème 3, on avait montré que $\|T\|_{L^\infty,\mathrm{BMO}} \leq \|T1\|_{\mathrm{BMO}} + C$ pour, disons, tout SIO antisymétrique faiblement borné!

La démonstration du théorème 5 est encore une application relativement directe du lemme de Cotlar-Knapp-Stein.

On peut maintenant donner une idée de la manière dont Christ et Journé obtiennent des estimations polynômiales en k sur la norme du $k$ième commutateur de Calderòn. On considère la forme $(k+2)$-linéaire F définie par

$$F(f_1,\ldots,f_k,f_{k+1},f_{k+2}) = \langle T_k(f_1,\ldots,f_k,f_{k+1}), f_{k+2}\rangle$$

$$= \lim_{\varepsilon\searrow 0}\iint_{|x-y|>\varepsilon}\left[\prod_{i=1}^k \int_y^x f_i(t)\,dt\right]f_{k+1}(y)f_{k+2}(x)\frac{dy\,dx}{(x-y)^{k+1}}.$$

On remarque aisément que, dans la forme F, les $(k+2)$ fonctions jouent des rôles semblables. Plus précisément, si l'on fixe k d'entre elles dans L^∞, alors F définit une forme bilinéaire entre les deux dernières fonctions, et cette forme bilinéaire provient d'un SIO. (Dans le langage des auteurs de [**CJ**], F est une forme $(k+2)$-linéaire d'intégrale singulière.)

On transforme maintenant $T_k(a,a,\ldots,a,.)$ en somme d'ε-familles en écrivant

$$\langle T_k(a,a,\ldots,a,f),g\rangle$$

$$= F(a,\ldots,a,f,g)$$

$$= -\int_0^\infty t\frac{\partial}{\partial t}[F(P_t a, P_t a,\ldots,P_t a, P_t f, P_t g)]\frac{dt}{t}$$

$$= \int_0^\infty F(Q_t a, P_t a,\ldots,P_t g)\frac{dt}{t} + \cdots + \int_0^\infty F(P_t a,\ldots,P_t f, Q_t g)\frac{dt}{t}.$$

On peut maintenant utiliser le fait que F est une forme $(k+2)$-linéaire d'intégrale singulière pour appliquer le théorème 5 en cascade et remplacer les fonctions a par 1 les unes après les autres. Pour chacun des $(k+2)$ termes de la somme, il faut moins de $(k+1)$ applications du théorème 5 pour se réduire au cas où $a = 1$. De plus, les constantes qui interviennent dans (3) et (4) sont de l'ordre de $(1+k)^{2\varepsilon}$; on obtient donc une estimation $(1+k)^{2+2\varepsilon}\|a\|_\infty^k$ pour la norme du $k$ième commutateur de Calderón. En utilisant un peu mieux l'antisymétrie des commutateurs, les auteurs de [**CJ**] obtiennent en fait

$$\|T_k(a,a,\ldots,a,.)\|_{L^2,L^2} \leq C_\delta(1+k)^{1+\delta}\|a\|_\infty^k \qquad (7)$$

pour tout $\delta > 0$. Cette estimation est d'ailleurs la meilleure connue actuellement.

Le lecteur a deviné que le théorème 5 peut être appliqué à d'autres opérateurs multilinéaires. On peut même, avec plus de travail, l'utiliser pour contrôler des opérateurs dont le noyau ne vérifie pas (2).

On se place en dimension $n \geq 2$, et on se donne un opérateur de convolution de Calderón-Zygmund T. On note $K(x - y)$ son noyau-distribution (hors de la diagonale, $K(x - y)$ est une fonction qui vérifie (1) et (2), et de plus T est borné sur L^2). Pour toute fonction bornée a et tous $x, y \in \mathbf{R}^n$, on note $m_{[x,y]}a = \int_{t=0}^{1} a(x + t(y - x)) \, dt$ la moyenne de a sur le segment $[x, y]$. Bien que $m_{[x,y]}a$ soit en général une fonction très peu régulière de x et de y, le théorème 5 permet de démontrer le résultat suivant.

THÉORÈME 6 [CJ]. *Si $a_1, \ldots, a_k$ sont k fonctions bornées définies sur $\mathbf{R}^n$, alors la formule*

$$\langle T_k f, g \rangle = \iint K(x, y) \left[\prod_{i=1}^{k} m_{[x,y]} a_i \right] f(y) g(x) \, dx \, dy$$

définit un opérateur borné sur L^p pour $1 < p < +\infty$, et

$$\|T_k\|_{L^p, L^p} \leq C(p, \delta)(1 + k)^{2+\delta} \prod_{i=1}^{k} \|a_i\|_\infty$$

pour tout $\delta > 0$.

V. Ondelettes et algèbre de Lemarié. Rappelons que la famille des opérateurs d'intégrale singulière T bornés sur $L^2(\mathbf{R}^n)$, et tels que $T1 = {}^t T1 = 0$ dans BMO, est une algèbre, que nous noterons $\mathcal{A}$ (voir [L1] ou [M1]). La découverte par Y. Meyer d'une nouvelle base de L^2 a permis de construire quelques opérateurs non banaux de $\mathcal{A}$.

Soient $\psi \colon \mathbf{R} \to \mathbf{C}$ une fonction, et $I = [k2^{-j}, (k+1)2^{-j}]$ un intervalle dyadique. Nous noterons $\psi_I(x) = 2^{j/2} \psi(2^j x - k)$ (cette notation peut être justifiée par le fait que si ψ était à support dans $[0, 1]$, alors ψ_I serait à support dans I).

THÉORÈME 7 [LM]. *Il existe une fonction $\psi \in \mathcal{S}(\mathbf{R})$ telle que les ψ_I, I intervalle dyadique, forment une base orthonormée de $L^2(\mathbf{R})$.*

Si on oubliait la contrainte "ψ est dans classe de Schwartz," le système de Haar conviendrait. Signalons aussi qu'on ne peut pas choisir $\psi \in \mathcal{C}_c^\infty(\mathbf{R})$.

Pour faciliter l'exposé, nous avons pris le parti de rester en dimension 1. Toutefois, les résultats que nous mentionnons sont encore valables en dimension $n \geq 2$. On construit une base de $L^2(\mathbf{R}^n)$ à partir de ψ comme on le fait avec le système de Haar. La base en question est obtenue à partir de $(2^n - 1)$ fonctions de base, qui sont elles-mêmes des produits tensoriels de k fonctions ψ (pour $1 \leq k \leq n$) et de $(n - k)$ fonctions φ, où φ est une autre fonction de $\mathcal{S}(\mathbf{R})$, mais qui vérifie $\int \varphi = 1$ alors que $\int \psi = 0$. (La fonction φ joue le rôle de $1_{[0,1]}$ dans le système de Haar.)

L'avantage principal des ψ_I (qu'il est convenu d'appeler une base d'ondelettes) sur le système de Haar, par exemple, est que les ψ_I sont dans tous les espaces fonctionnels raisonnables. On en déduit aisément que $(\psi_I)_I$ est aussi une base inconditionnelle de L^p, $1 < p < +\infty$, et même de l'espace atomique H^1. De même, l'appartenance d'une fonction f à un espace fonctionnel donné est, pour la plupart des espaces fonctionnels usuels, caractérisée de manière simple par la taille (ou la décroissance) des coefficients $\int f(x)\overline{\psi_I}(x)\,dx$.

Revenons à l'algèbre $\mathcal{A}$, et donnons un exemple d'opérateur d'intégrale singulière qui peut être construit à partir d'ondelettes. Il s'agit d'un opérateur $T \in \mathcal{A}$, inversible sur $L^2(\mathbf{R})$, mais dont l'inverse n'est pas un opérateur d'intégrale singulière, et n'est d'ailleurs pas continu sur L^p pour un $p \neq 2$ fixé à l'avance. (Le premier exemple d'un tel opérateur a été construit, toujours en utilisant des ondelettes, par P. Tchamitchian [**T2**]; nous présentons ici un exemple de P. G. Lemarié.)

Soit ψ_I, I intervalle dyadique, la base du théorème 7. Pour tout intervalle dyadique I, on note $\overline{I}$ le fils de gauche de I (c.à.d. que si $I = [k2^{-j}, (k+1)2^{-j}]$, alors $\overline{I} = [k2^{-j}, (k+\frac{1}{2})2^{-j}]$). Soit S l'opérateur de $L^2(\mathbf{R})$ défini par $S(\psi_I) = \psi_{\overline{I}}$ pour tout I dyadique; S est clairement une isométrie de $L^2(\mathbf{R})$ sur son image. On vérifie aisément que le noyau de S est

$$K(x,y) = \sum_I \psi_{\overline{I}}(x)\psi_I(y) = \sum_j \sum_k 2^j \sqrt{2}\psi(2^{j+1}x - 2k)\psi(2^j y - k)$$

et est un noyau standard. Comme ψ est d'intégrale nulle, on a aussi $S1 = {}^tS1 = 0$, donc $S \in \mathcal{A}$.

Choisissons $T = 1 - rS$. Clairement, $T \in \mathcal{A}$, et T est inversible sur $L^2(\mathbf{R})$ pour $|r| < 1$. De plus,

$$T^{-1}(\psi) = \psi + rS(\psi) + r^2 S^2(\psi) + \cdots$$
$$= \psi + r\psi_{[0,\frac{1}{2}]} + r^2\psi_{[0,\frac{1}{4}]} + \cdots.$$

Si $p > 2$ est fixé à l'avance, on peut choisir r assez proche de 1 pour que $T^{-1}(\psi)$ ne soit pas dans $L^p(\mathbf{R})$, et T est le contre-exemple cherché.

BIBLIOGRAPHIE

[**Bd**] G. Bourdaud, *Sur les opérateurs pseudo-différentiels à coefficients peu réguliers*, Thèse d'Etat, Univ. Paris VII, 1983.

[**B1**] J. Bourgain, *High dimensional maximal functions associated to convex sets* (à paraître).

[**B2**] ——, *On the spherical maximal function in the plane* (à paraître).

[**C1**] A. P. Calderón, *Commutators of singular integral operators*, Proc. Nat. Acad. Sci. U.S.A. **53** (1965), 1092–1099.

[**C2**] ——, *Cauchy integrals on Lipschitz curves and related operators*, Proc. Nat. Acad. Sci. U.S.A. **74** (1977), 1324–1327.

[**C3**] ——, *Commutators, singular integrals on Lipschitz curves and applications*, Proc. Internat. Congr. Math. (Helsinki, 1978), Acad. Sci. Fennica, Helsinki, 1980, pp. 85–96.

[**CZ**] A. P. Calderón et A. Zygmund, *On the existence of certain singular integrals*, Acta Math. **88** (1952), 85–139.

[**C...W**] H. Carlsson, M. Christ, A. Cordoba, J. Duoandikoetxea, J.-L. Rubio de Francia, J. Vance, S. Wainger et D. Weinberg, *L^p-estimates for maximal functions and Hilbert transforms along flat convex curves in $\mathbf{R}^2$*, Bull. Amer. Math. Soc. **14** (1986), 263–267.

[Ch] M. Christ, *Hilbert transforms along curves. II: A flat case*, Amer. J. Math. (à paraître).

[CJ] M. Christ et J.-L. Journé, *Polynomial growth estimates for multilinear singular integral operators*, Acta Math. (à paraître).

[CG1] J. Cohen et J. Cosselin, *The Dirichlet problem for the biharmonic equation in a bounded C^1 domain in the plane*, Indiana Univ. Math. J. **32** (1985), 635–685.

[CG2] ___, *Adjoint boundary value problems for the biharmonic equation on C^1 domains in the plane*, Ark. Mat. **23** (1985), 217–240.

[CDM] R. R. Coifman, G. David et Y. Meyer, *La solution des conjectures de Calderón*, Adv. in Math. **48** (1983), 144–148.

[CDeM] R. R. Coifman, P. G. Deng et Y. Meyer, *Domaine de la racine carrée de certains opérateurs différentiels accrétifs*, Ann. Inst. Fourier (Grenoble) **33** (1983), no. 2, 123–134.

[CMM] R. R. Coifman, A. McIntosh et Y. Meyer, *L'intégrale de Cauchy définit un opérateur borné sur L^2 pour les courbes lipschitziennes*, Ann. of Math. **116** (1982), 361–388.

[CM1] R. R. Coifman et Y. Meyer, *Commutateurs d'intégrales singulières et opérateurs multilinéaires*, Ann. Inst. Fourier (Grenoble) **28** (1978), no. 3, 177–202.

[CM2] ___, *Au-delà des opérateurs pseudo-différentiels*, Astérisque 57, Société Mathématique de France, Paris, 1978.

[CM3] ___, *Le théorème de Calderón par les méthodes de variable réelle*, C. R. Acad. Sci. Paris Sér. A-B **289** (1979), no. 7, 425–428.

[CM4] ___, *Une généralisation du théorème de Calderón sur l'intégrale de Cauchy*, Actes de la Conférence d'Analyse de Fourier, El Escorial, 1979.

[Dh] B. Dahlberg, *Weighted norm inequalities for the Lusin integral and the non-tangential maximal function for functions harmonic in a Lipschitz domain*, Studia Math. **67** (1980), 297–314.

[DFK] B. Dahlberg, E. Fabes et C. Kenig, *A Fatou theorem for solutions of the porous medium equation*, Proc. Amer. Math. Soc. **91** (1984), 205–212.

[DJK] B. Dahlberg, D. Jerison et C. Kenig, *Area integral estimates for elliptic differential operators with non-smooth coefficients*, Ark. Mat. **22** (1984), 97–108.

[DK1] B. Dahlbert et C. Kenig, *Hardy spaces and the L^p-Neumann problem for Laplace's equation in a Lipschitz domain*, Preprint.

[DK2] ___, *Area integral estimates for higher order boundary value problems on Lipschitz domains* (à paraître).

[DKV1] B. Dahlberg, C. Kenig et C. Verchota, *Boundary value problems for the systems of linear elastostatics on Lipschitz domains*, Preprint.

[DKV2] ___, *The Dirichlet problem for the bilaplacian in a Lipschitz domain*, Ann. Inst. Fourier (Grenoble) (à paraître).

[Dv] G. David, *Opérateurs intégraux singuliers sur certaines courbes du plan complexe*, Ann. Sci. École Norm. Sup. **17** (1984), 157–189.

[DJ] G. David et J-L. Journé, *A boundedness criterion for generalized Calderón-Zygmund operators*, Ann. of Math. **120** (1984), 371–397.

[DJS] G. David, J-L. Journé et S. Semmes, *Opérateurs de Calderón-Zygmund, fonctions para-accrétives et interpolation*, Rev. Math. Ibero-Americana (à paraître).

[DxR] J. Duoandikoetxea et J-L. Rubio de Francia, *Maximal and singular integral operators via Fourier transform estimates*, Invent. Math. **84** (1986), fasc. 3, 541–561.

[FJK] E. Fabes, D. Jerison et C. Kenig, *Multilinear Littlewood-Paley estimates with applications to partial differential equations*, Proc. Nat. Acad. Sci. U.S.A. **79** (1982), 5746–5750.

[FJR] E. Fabes, M. Jordeit et N. Rivière, *Potential techniques for boundary value problems on C^1 domains*, Acta Math. **141** (1978), 165–186.

[F] C. Fefferman, *Recent progress in classical Fourier Analysis*, Proc. Internat. Congr. Math. (Vancouver, B.C., 1974), Vol. 1, Canad. Math. Congress, Montreal, Que., 1975, pp. 95–118.

[JN] F. John et L. Nirenberg, *On functions of bounded mean oscillation*, Comm. Pure Appl. Math. **14** (1961), 415–426.

[J1] J-L. Journé, *Calderón-Zygmund operators, pseudo-differential operators, and the Cauchy integral of Calderón*, Lecture Notes in Math., Vol. 994, Springer-Verlag, 1983.

[KS] A. W. Knapp et E. M. Stein, *Intertwining operators on semisimple Lie groups*, Ann. Math. **93** (1971), 489–578.

[L1] P. G. Lemarié, *Algèbre d'opérateurs et semi-groupes de Poisson sur un espace de nature homogène*, Thèse de 3ème Cycle, Orsay, 1984.

[L2] ____, *Continuité sur les espaces de Besov des opérateurs définis par des intégrales singulières*, Ann. Inst. Fourier (Grenoble) **35** (1985), fasc. 4, 175–187.

[LM] P. G. Lemarié et Y. Meyer, *Ondelettes et bases hilbertiennes*, Rev. Math. Ibero-Americana (à paraître).

[McM] A. McIntosh et Y. Meyer, *Algèbre d'opérateurs définis par des intégrales singulières*, C. R. Acad. Sci. Paris Sér. I Math., **301** (1985), no. 8, 395–397.

[M1] Y. Meyer, *Les nouveaux opérateurs de Calderón-Zygmund*, Colloque en l'honneur de L. Schwartz, Vol. 1, Astérisque **131** (1985), 237–254.

[M2] ____, *Intégrales singulières, opérateurs multilinéaires, analyse complexe et équations aux dérivées partielles*, Proc. Internat. Congr. Math. (Varsovie, 1982), Polish Scientific Publishers.

[M3] ____, *Real analysis and operator theory*, Proceedings Notre-Dame 1984, Sympos. (à paraître).

[Mu1] T. Murai, *Boundedness of singular integral operators of Calderón type V*, Adv. in Math. **59** (1986), 71–81.

[Mu2] T. Murai, *Boundedness of singular integral operators of Calderón type VI*, Preprint Series, College of General Education, Nagoya, no. 12, 1984.

[NVWW] A. Nagel, J. Vance, S. Wainger et D. Weinberg, *Hilbert transforms for convex curves*, Duke Math. J. **50** (1983), 735–744.

[R] J-L. Rubio de Francia, *A. Littlewood-Paley inequality for arbitrary intervals*, Rev. Math. Ibero-Americana, **1** (1985), no. 2, 1–13.

[Se1] S. Semmes, *Estimates for $(\bar{\partial} - \mu\partial)^{-1}$ and Calderón's theorem on the Cauchy integral* and *Quasiconformal mappings and chord-arc curves*, Trans. Amer. Math. Soc. (à paraître).

[Se2] ____, *The Cauchy integral, chord-arc curves and quasiconformal mappings*, Proceedings of the Symposium on the Occasion of the Proof of the Bieberbach Conjecture, Math. Surveys Monogr., Vol. 21, 1986, pp. 167–184.

[St] E. M. Stein, *Singular integrals and differentiability properties of functions*, Princeton University Press, Princeton, N.J., 1970.

[SW] E. M. Stein et S. Wainger, *Problems in harmonic analysis related to curvature*, Bull. Amer. Math. Soc. **84** (1978), 1239–1295.

[T1] P. Tchamitchian, *Estimations précises des normes de certains opérateurs d'intégrales singulières*, preprint, Ecole Polytechnique 91128 Palaiseau Cedex, France.

[T2] ____, *Calcul symbolique sur les opérateurs de Calderón-Zygmund et bases inconditionnelles de $L^2(\mathbf{R})$*, C. R. Acad. Sci. Paris (à paraître).

[V1] G. Verchota, *Layer potentials and boundary value problems for Laplace's equation in Lipschitz domains*, Thesis, Univ. of Minnesota, 1982; J. of Funct. Anal. **59** (1984), 572–611.

[V2] ____, *The dirichlet problem for polyharmonic functions on Lipschitz domains*, Preprint.

CENTRE DE MATHÉMATIQUES, ECOLE POLYTECHNIQUE, 91128 PALAISEAU CEDEX, FRANCE

Proceedings of the International Congress of Mathematicians
Berkeley, California, USA, 1986

Singular Minimizers in the Calculus of Variations

A. M. DAVIE

1. Introduction. Consider the following variational problem: find an absolutely continuous function u on $[a, b]$ which minimizes the functional

$$I(u) = \int_a^b f(x, u(x), u'(x)) \, dx$$

subject to the boundary conditions $u(a) = \alpha$, $u(b) = \beta$. Here $f(x, u, p)$ is always assumed to be a nonnegative C^∞ function on $[a, b] \times \mathbf{R} \times \mathbf{R}$.

The subject of this article is the occurrence of singularities in minimizers of such functionals. The classical theory developed by Tonelli and others in the 1920s and 1930s gave conditions for existence and regularity of solutions. Assuming f is strictly convex in p, Tonelli proved a partial regularity theorem which asserted that any minimizer is smooth outside a closed set of measure zero. Examples where this singular set is nonempty were first described recently by Ball and Mizel [1, 2], who gave detailed analyses of a number of examples. Closely related are some older examples of Lavrentiev [6] and Mania [7], showing that the infimum over the absolutely continuous functions may differ from that over the smooth functions. We shall describe some of these results and some recent further developments.

We denote by Ω the set of real-valued absolutely continuous functions u on $[a, b]$ satisfying $u(a) = \alpha$, $u(b) = \beta$. The partial derivatives of f will be denoted by f_x, f_u, f_p, etc. We say f has *superlinear growth* if there is a function $\varphi(p)$ such that $f(x, u, p) \geq \varphi(p)$ for all x, u, p and

$$\varphi(p)/|p| \to \infty \quad \text{as } |p| \to \infty.$$

2. Existence and regularity theorems. The classical existence theorem of Tonelli [8] is as follows:

PROPOSITION 2.1. *Suppose $f_{pp} \geq 0$ (i.e., f is convex in p), and f has superlinear growth. Then there exists $u \in \Omega$ minimizing the functional I.*

Briefly, the proof runs as follows: pick a sequence $u_n \in \Omega$ with $I(u_n)$ converging to the infimum. The superlinear growth condition implies that the sequence

u'_n is weakly compact in $L^1[a, b]$, so passing to a subsequence we have $u'_n \to u'$ weakly in L^1. The convexity in p gives a lower semicontinuity property of I so that $I(u) \le \lim \inf I(u_n)$. Thus u is a minimizer.

Tonelli and others have proved stronger versions of Proposition 2.1, in which f is only required to have superlinear growth in p outside a suitably small set of (x, u) values. For details, see Cesari [**3**, Chapter 12].

Turning to regularity theorems, we first mention a standard result [**3**, §2.6]. LIP denotes the set of Lipschitz continuous functions on $[a, b]$.

PROPOSITION 2.2. *Suppose* $f_{pp}(x, u, p) > 0$ *for all* x, u, p *and suppose the infimum of* I *over* $\Omega \cap \mathrm{LIP}$ *is attained by* $u_0 \in \Omega \cap \mathrm{LIP}$. *Then* $u_0 \in C^\infty[a, b]$ *and satisfies the Euler-Lagrange equation*

$$f_u(x, u_0(x), u'_0(x)) = (d/dx) f_p(x, u_0(x), u'_0(x)).$$

In other words, any Lipschitz continuous minimizer is smooth. However, we have no analogue of Proposition 2.1 asserting the existence of Lipschitz minimizers. For general minimizers we have the following partial regularity theorem of Tonelli [**8**].

PROPOSITION 2.3. *Suppose again that* $f_{pp} > 0$ *and now suppose* u_0 *attains the infimum of* I *over* Ω. *Then* u_0 *has a (possibly infinite) derivative at each point of* $[a, b]$, *and* u'_0: $[a, b] \to \mathbf{R} \cup \{\infty, -\infty\}$ *is continuous. Moreover, if* $E = \{x: u'_0(x) = \infty$ *or* $-\infty\}$, *then* E *is closed and has measure zero, and* u_0 *is* C^∞ *outside* E.

For a proof see [**2**, Theorem 2.7].

In the following sections various examples are described which show that any closed set of measure zero can occur as the singular set E of a minimizer for a suitable problem satisfying the hypotheses of Proposition 2.1. The Euler-Lagrange equation always holds outside E, but the examples show that it may not be possible to regard f_p in any meaningful way as an integral of f_u.

We note some further results in special cases, without making any attempt to give the strongest statements known:

PROPOSITION 2.4. (a) *If* f *depends only on* u *and* p, *then for any minimizer* u_0 *of* I *over* Ω *we have*

$$f(u_0(x), u'_0(x)) - u'_0(x) f_p(u_0(x), u'_0(x)) = \text{const.}$$

(b) *If* f *depends only on* x *and* p, *then for any minimizer* u_0 *of* I *over* Ω *we have*

$$f_p(x, u'_0(x)) = \text{const.}$$

For this and stronger results due to Tonelli, see Ball and Mizel [**2**, Corollaries 2.4 and 2.5].

COROLLARY 2.5. *Suppose* f *depends only on* u *and* p, *and* $f_{pp} > 0$. *Then for any minimizer* u_0 *we have that* $f_p(u_0(x), u'_0(x))$ *is continuous on* $[a, b]$.

This follows by an elementary argument from Proposition 2.4(a) using the convexity of f and the fact, which follows from Proposition 2.3, that $|u_0'(x)| \to \infty$ as x approaches the singular set E.

If f has superlinear growth, then $f_p \to \infty$ as $p \to \infty$. Hence from Corollary 2.5 and Proposition 2.4(a) we obtain

PROPOSITION 2.6. *Suppose $f_{pp} > 0$, f has superlinear growth, and either f depends only on u and p or f depends only on x and p. Then any minimizer of I over Ω is in $C^\infty[a,b]$.*

See [2, Theorem 2.10] for stronger results. See also [9] for related results.

Finally we mention a result of Clarke and Vinter [4]: if f is a polynomial in x, u, and p and $f_{pp} > 0$, then for any minimizer the exceptional set E is at most countable with finitely many limit points.

3. Examples related to the Lavrentiev phenomena.
Lavrentiev [6] gave an example of a problem for which the infimum over smooth u is strictly greater than that over all absolutely continuous u. Mania [7] gave a simpler example, as follows:

Let $f_0(x, u, p) = (x - u^3)^2 p^6$ on the interval $[0, 1]$ with boundary conditions $u(0) = 0$, $u(1) = 1$.

Clearly the corresponding functional I_0 has infimum zero over Ω, obtained by $u_0(x) = x^{1/3}$. Mania showed however that there exists $c > 0$ such that $I_0(u) \geq c$ for all $u \in \Omega$ such that u has a finite derivative at $x = 0$. Proof of this and further results on the Lavrentiev phenomenon can be found in Cesari [3, Chapter 18].

Modifying arguments of Ball and Mizel [2], we show how Mania's example yields examples of nonempty singular sets E in the situation of Proposition 2.3. To do this we add a small term to f_0 to obtain $f_{pp} > 0$:

Let $f(x, u, p) = f_0(x, u, p) + \varepsilon f_1(p)$, where $f_1(p) = (1 + p^2)^{2/3}$.

Note that $I_1(u_0) < \infty$, where I is the functional corresponding to f.

Now $f_{pp} > 0$, f has superlinear growth, and Proposition 2.1 implies the existence of $v \in \Omega$ minimizing $I = I_0 + \varepsilon I_1$. If $v'(0)$ is finite, then by Mania's result $I(v) \geq I_0(v) \geq c$; since $I(u_0) = \varepsilon I_1(u_0)$, this contradicts the minimality of v, provided ε is small enough. Thus v has a singularity at 0, i.e., the set E contains 0 (in fact $E = \{0\}$).

The singularity obtained thus is at an end-point of the interval; we can get an internal singularity, using the same f on the interval $[-1, 1]$, with boundary conditions $u(1) = 1$, $u(-1) = -1$. Again we find a minimizer with $E = \{0\}$.

Ball and Mizel [2] give a detailed analysis of a similar problem:

$f(x, u, p) = (x^2 - u^3)^2 p^{14} + \varepsilon p^2$ on the interval $[0, 1]$ with boundary conditions $u(0) = 0$, $u(1) = k$.

Amongst other things it is proved [2] that for ε small enough any minimizer has infinite derivative at 0. While the proof is similar to Mania's argument, this problem does not exhibit the Lavrentiev phenomenon (to get which the p^{14} term should be replaced by $|p|^s$ for $s \geq 15$).

Ball and Mizel also discuss the problem:

$f(x, u, p) = (x^4 - u^6)^2 p^s + \varepsilon p^2$ on $[-1, 1]$ with suitable boundary conditions. They show that for ε small enough, this example gives singular minimizers for $s \geq 26$ and the Lavrentiev phenomenon for $s \geq 27$. Moreover for $s > 27$ the infimum of I over $\Omega \cap \mathrm{LIP}$ is obtained by a C^∞ function u (a "pseudominimizer").

The arguments described above can be modified to construct, given a closed subset E of $[a, b]$ having measure zero, an example of a minimizer whose singular set is precisely E. To do this we first find a function v on $[a, b]$ with $v' \in L^2$, $v' > 0$, $v'(x) \to \infty$ as $x \to E$, and $v \in C^\infty$ outside E. Then we construct f of the form

$$f(x, u, p) = [\varphi(u) - \varphi(v(x))]^2 \psi(p) + \varepsilon p^2,$$

where φ, ψ are suitable smooth functions, such that $\varphi \circ v$ is smooth, and ψ is convex and rapidly increasing as $p \to \infty$. For details see [5].

In all the examples described in this section f has superlinear growth, and consequently for any minimizer $f_p(x, u, u') \to \infty$ as $x \to E$. Thus $f_u(x, u, u')$ cannot be in L^1 if E is nonempty, so the Euler-Lagrange equations fail to the extent that f_p cannot be expressed as the Lebesgue integral of f_u. Ball and Mizel [2] give detailed analyses of the Euler-Lagrange equations in their examples.

4. Examples where f depends only on u and p. Such examples were given by Ball and Mizel [1, 2] who showed that given any closed set $E \subseteq [a, b]$ of measure zero, one could construct $f(u, p)$ with $f_{pp} > 0$ and

$$|p| \leq f(u, p) \leq \mathrm{const}(1 + p)^2 \tag{1}$$

such that I has a unique minimizer u_0 in Ω with singular set E; moreover $f_u \notin L^1[a, b]$, implying failure of the Euler-Lagrange equations.

Note that by Proposition 2.6 no such example can have superlinear growth, so the lower bound in (1) is essentially optimal.

The construction allows one to specify u_0 in advance, provided it satisfies the following condition:

$u_0 \in \Omega$, u_0': $[a, b] \to \mathbf{R} \cup \{\infty\}$ is continuous and strictly positive, and u_0 is C^∞ outside $E = \{u_0' = \infty\}$.   (*)

The idea of the construction is as follows:

Let g be the inverse function to u_0', defined on $[\alpha, \beta]$. Then u_0 minimizes $\int_a^b [g'(u(x)) u'(x)]^2 \, dx$. $f(u, p)$ is constructed so that

$$f(u, p) \geq [g'(u) p]^2 + p \tag{2}$$

with equality when $u = u_0(x)$, $p = u_0'(x)$, i.e., when $p = 1/g'(u)$. f also satisfies $f_{pp} > 0$ and (1).

It then follows that u_0 is the unique minimizer of the corresponding functional I.

The construction is generalized in [5] by replacing the last term p in (2) by $\rho(u)p$, where ρ is a given positive continuous function, smooth outside $u_0(E)$. In this way the following is proved:

Suppose u_0 satisfies $(*)$ and σ is a positive continuous function on $[a, b]$, which is C^∞ outside E. Then we can find $f(u, p)$ with $f_{pp} > 0$ such that u_0 is the unique minimizer of the corresponding functional I and moreover

$$f_p(u_0(x), u_0'(x)) = \sigma(x), \qquad x \in [a, b] \backslash E.$$

The hypothesis that σ be continuous is necessary in view of Corollary 2.5. This result illustrates the failure of the Euler-Lagrange equations; for example, one could choose E to be a perfect set and σ to be constant on each complementary interval of E—then $f_u(u_0, u_0') = 0$ a.e. but $f_p = \sigma$ need not be constant.

5. Examples where f depends only on x and p. In this case Proposition 2.4(b) states that all minimizers satisfy the Euler-Lagrange equation, which reduces to

$$f_p(x, u'(x)) = \text{constant}. \tag{3}$$

Conversely, a straightforward argument using the convexity of f shows that any $u \in \Omega$ satisfying (3) is a minimizer; moreover if $f_{pp} > 0$ there can be at most one minimizer.

We describe some examples from [5] where $f_{pp} > 0$ but there is a minimizer with nonempty singular set. Note that in view of Proposition 2.6 such examples cannot have superlinear growth.

EXAMPLE 5.1. $f(x, p) = (1+p^2)^{1/2} + x^2 p^2$ on $[-1, 1]$ with boundary conditions $u(-1) = -\alpha$, $u(1) = \alpha$ $(\alpha > 0)$.

In this example, analysis of the equation $f_p = C$ yields the following conclusions: there is $\alpha_0 > 0$ (corresponding to $C = 1$) such that if $\alpha > \alpha_0$ there is no minimizer in Ω, whilst if $\alpha \leq \alpha_0$ there is a unique minimizer, which is smooth if $\alpha < \alpha_0$ but if $\alpha = \alpha_0$ it has singular set $E = \{0\}$.

EXAMPLE 5.2. $f(x, p) = (1+p^2)^{1/2} + x^2 p^4$ on $[-1, 1]$ with the same boundary conditions as 5.1.

In this case a unique minimizer exists for all $\alpha > 0$. There exists $\alpha_1 > 0$ such that if $\alpha < \alpha_1$, then the minimizer is smooth, whilst if $\alpha \geq \alpha_1$, then $E = \{0\}$.

These examples can be extended to yield an arbitrary closed set of measure zero as singular set. More precisely, given u_0 satisfying condition $(*)$ on $[a, b]$ one can find $\varphi \in C^\infty[a, b]$ and $\psi \in C^\infty(\mathbf{R})$ such that

$$f(x, p) = (1 + p^2)^{1/2} + \varphi(x)\psi(p)$$

satisfies $f_{pp} > 0$ and u_0 is the unique minimizer of the corresponding functional I over Ω.

6. A jump discontinuity in the Euler-Lagrange equations. Finally we show how the methods of §§3 and 5 can be combined to obtain an example of a minimizer satisfying $f_u = 0$ and whose singular set consists of a single point at which f_p has a jump discontinuity. Note that in view of the results of §2, in any such example f must depend on all three variables $x, u,$ and p and cannot have superlinear growth.

We start with Example 5.2,

$$f_0(x, p) = (1 + p^2)^{1/2} + x^2 p^4,$$

and denote the corresponding functional by I_0. We try to produce a discontinuity in f_p by combining different minimizers on $[-1, 0]$ and $[0, 1]$. Choose α_2 and α_3 with $\alpha_3 > \alpha_2 > \alpha_1$, where α_1 comes from Example 5.2, and let u_2, u_3 be the corresponding minima. Now define v on $[-1, 1]$ by

$$v(x) = \begin{cases} u_2(x), & x \geq 0, \\ u_3(x), & x < 0. \end{cases}$$

Then v minimizes $I_0(u)$ subject to the boundary conditions $u(-1) = -\alpha_3$, $u(1) = \alpha_2$ together with the internal condition $u(0) = 0$. To eliminate the internal condition we consider

$$f(x, u, p) = f_0(x, p) + K f_1(x, u, p),$$

where $f_1(x, u, p) = (h_2(u) - x)^2 (h_3(u) - x)^2 p^8$, where h_i, defined on $[-\alpha_i, \alpha_i]$, is inverse to u_i; one can check that h_2 and h_3 are smooth, and extend them to **R**.

Then $I_1(v) = 0$ and by methods similar to those of §3 it can be shown that $I_1(u) \geq x > 0$ if $u(0) \neq 0$ and u satisfies the boundary conditions $u(-1) = -\alpha_3$, $u(1) = \alpha_2$. It follows that if K is large enough, then v minimizes $I = I_0 + K I_1$, subject to these boundary conditions.

Moreover $f_p(x, v, v') = (f_0)_p(x, v, v')$ is constant on both $[-1, 0)$ and $(0, 1]$ but the two constants differ.

REFERENCES

1. J. M. Ball and V. J. Mizel, *Singular minimisers for regular one-dimensional problems in the calculus of variations*, Bull. Amer. Math. Soc. (N.S.) **11** (1984), 143–146.

2. ____, *One-dimensional variational problems whose minimisers do not satisfy the Euler-Lagrange equation*, Arch. Rational Mech. Anal. **90** (1985), 325–388.

3. L. Cesari, *Optimisation—theory and applications*, Springer-Verlag, New York, 1983.

4. F. Clarke and R. Vinter, *Regularity of solutions to variational problems with polynomial Lagrangians* (to appear).

5. A. M. Davie (in preparation).

6. M. Lavrentiev, *Sur quelques problemes due calcul des variations*, Ann. Mat. Appl. **4** (1926), 7–28.

7. B. Mania, *Sopra un esempio di Lavrentieff*, Bull. Un. Mat. Ital. **13** (1934), 147–153.

8. L. Tonelli, *Fondamenti di Calcolo delle Variazioni*. Vol. II, Zanichelli, Bologna, 1923, p. 359.

9. F. Clarke and R. Vinter, *Regularity properties of solutions to the basic problem in the calculus of variations*, Trans. Amer. Math. Soc. **289** (1985), 73–98.

UNIVERSITY OF EDINBURGH, EDINBURGH, SCOTLAND

Proceedings of the International Congress of Mathematicians
Berkeley, California, USA, 1986

Advances in Quantized Functional Analysis

EDWARD G. EFFROS

Perhaps the most profound distinction between classical and quantum physics is Heisenberg's principle that one must represent the basic variables of physics, such as energy or momentum, by operators rather than by functions. It was von Neumann's conviction that a corresponding "quantized mathematics" should be formulated. Over the last five decades, it has been the goal of operator algebraists to "quantize" functional analysis and, more recently, topological and differential geometry. The procedures for accomplishing this are only partially understood, but they invariably entail replacing functions by operators.

The most obvious novelty of operators is that they need not commute, this being a reflection of the uncertainty principle (see [23, pp. 78–79]). There is, however, a more subtle phenomenon associated with operators: *a matrix of operators may be interpreted as another operator.* If one keeps track of this additional matricial structure, with its attendant matrix orders and norms, one discovers that much of classical functional analysis has quantized analogues. Although the usual order and norm structures are potentially of interest (see [19, 1], and the beautiful work of Friedman and Russo [32, 33]), it is the matricial orderings and norms that have played a decisive rôle in the algebraic classification of the operator algebras (see [7] and the recent work of Cowling and Haagerup introducing new invariants for factors).

In this report we shall outline some of the recent developments in the matricial norm theory, concentrating especially on multilinear functions. As pointed out by Paulsen and Smith [25], the quantized multilinear theory is in some ways much simpler than the classical theory. We shall consider this phenomenon in §2. The multilinear theory has proved to be especially useful for proving new results on the Johnson-Kadison-Ringrose versions of Hochschild cohomology [6]. This application is discussed in §3. Finally we indicate in §4 how completely bounded bilinear functionals may be regarded as the key ingredient in Haagerup's proof that nuclear C^*-algebras are amenable.

We will assume all vector spaces to be complex. Given a vector space V, we let $\mathbf{M}_{m,n}(V)$ be the vector space of $m \times n$ matrices $[v_{ij}]$, $v_{ij} \in V$,

---

Supported in part by the National Science Foundation.

$\mathbf{M}_n(V) = \mathbf{M}_{n,n}(V)$, and $\mathbf{M}_n = \mathbf{M}_n(\mathbf{C})$. Given a linear map of vector spaces $\varphi\colon V \to W$, we define $\varphi_{mn}\colon \mathbf{M}_{mn}(V) \to \mathbf{M}_{mn}(W)$ by $\varphi_{mn}([v_{ij}]) = [\varphi(v_{ij})]$, and we let $\varphi_n = \varphi_{nn}$. Given complex Hilbert spaces H, K we let $\mathcal{B}(H, K)$ denote the Banach space of bounded operators $T\colon H \to K$, and $\mathcal{B}(H) = \mathcal{B}(H, H)$. As stressed above, we may interpret an $m \times n$ matrix $T = [T_{ij}]$, $T_{ij} \in \mathcal{B}(H, K)$, as an element of $\mathcal{B}(H^n, K^m)$: given $x = (x^j) \in H^n$, we have that

$$(Tx)^i = \sum_j T_{ij} x^j.$$

1. Quantized linear functionals. We begin by contrasting function spaces with operator spaces. A *function space* V is a linear space of continuous complex functions on a compact Hausdorff space X, provided with the usual uniform norm

$$\|f\| = \sup\{|f(x)|\colon x \in X\}. \tag{1}$$

As is well known, any normed vector space is linearly isometric to a function space; hence

• *normed vector spaces are just abstract function spaces.*

As suggested by its name,

• *the first purpose of functional analysis is to determine the functionals, i.e., the bounded linear functions $F\colon V \to \mathbf{C}$ on a given normed vector space V.*

For the Banach space $C(X)$ of all complex continuous functions on X, the functionals are just the complex Radon measures μ. Thus

• *the strategy for determining the functionals $F\colon V \to \mathbf{C}$ is to fix an embedding $V \hookrightarrow C(X)$, and to then use the Hahn-Banach Theorem to extend F to a measure $\mu\colon C(X) \to \mathbf{C}$.*

An *operator space* is defined to be a linear space $\mathcal{V}$ of operators on a Hilbert space H. We may impose the standard operator norms

$$\|T\| = \sup\{\|T(\xi)\|\colon \|\xi\| = 1,\ \xi \in H^n\} \tag{2}$$

on each of the spaces $\mathbf{M}_n(\mathcal{V}) \subseteq \mathcal{B}(H^n)$ $(n \in \mathbf{N})$. We say that a vector space is *matricially normed* if, as in this case, norms are provided on the matrix spaces $\mathbf{M}_n(\mathcal{V})$ $(n \in \mathbf{N})$. Using matrix multiplication to define αv and $v\alpha$ for $\alpha \in \mathbf{M}_n$ and $v \in \mathbf{M}_n(\mathcal{V})$, we may regard $\mathbf{M}_n(\mathcal{V})$ as an $\mathbf{M}_n$ bimodule. In the following we let $\mathbf{M}_n = \mathcal{B}(\mathbf{C}^n)$ have the usual operator norm.

THEOREM 1 (RUAN [28]). *Suppose that $\mathcal{V}$ is a matricially normed vector space. Then $\mathcal{V}$ is completely isometric (see below) to an operator space if and only if these norms satisfy*

$$\|\alpha v\| \leq \|\alpha\|\,\|v\|, \qquad \|v\alpha\| \leq \|v\|\,\|\alpha\|, \tag{R_1}$$

$$\|v \oplus w\| = \max\{\|v\|, \|w\|\} \tag{R_2}$$

for all $v \in \mathbf{M}_n(\mathcal{V})$, $w \in \mathbf{M}_m(\mathcal{V})$, and $\alpha \in \mathbf{M}_n = \mathbf{M}_n(\mathbf{C})$.

Thus

• *matricially normed spaces satisfying* (R₁) *and* (R₂) *are just the abstract operator spaces.*

In contrast to normed vector spaces, the dual of an operator space with the dual matricial norms will not be an operator space, since (R₂) is not self-dual. By inserting zero entries, we may enlarge $m \times n$ matrices to square matrices. This provides an unambiguous definition for norms on each of the spaces $\mathbf{M}_{m,n}(\mathcal{V})$ when $\mathcal{V}$ is an operator space.

The appropriate morphisms for operator spaces take into account the matricial norms. Given a linear map $\varphi \colon \mathcal{V} \to \mathcal{W}$, we define *the completely bounded norm of* φ by

$$\|\varphi\|_{\mathrm{cb}} = \sup\{\|\varphi_n\| \colon n \in \mathbf{N}\},$$

and we say that φ is *completely bounded* (resp., a *complete contraction*, resp., a *complete isometry*) if $\|\varphi\|_{\mathrm{cb}} < \infty$ (resp., $\|\varphi\|_{\mathrm{cb}} \leq 1$, resp., each of the maps φ_n is an isometry).

In quantized functional analysis, we replace functionals by operator-valued functions. As formulated by Arveson (who, however, is not to be blamed for the terminology):

• *the first purpose of quantized functional analysis is to determine the "operatorals," i.e., the completely bounded linear operator-valued functions* $\varphi \colon \mathcal{V} \to \mathcal{B}(H)$ *for a given operator space* $\mathcal{V}$ *and general Hilbert spaces* H.

A (concrete) unital C^*-*algebra* $\mathcal{A}$ is an operator space that is closed under multiplication, the *-operation, and the norm-topology and contains the identity operator. If $\mathcal{A}$ is also closed in the weak operator topology it is said to be a von Neumann algebra. C^*-algebras play the part of $C(X)$ in the quantized realm. In this analogy, the operatorals $\varphi \colon \mathcal{A} \to \mathcal{B}(H)$ on a C^*-algebra $\mathcal{A}$ correspond to "quantized measures." Once again, these are in a sense "known." In fact, C^*-algebra theory began with the determination of a very special class of quantized measures. A bounded linear map $\varphi \colon \mathcal{A} \to \mathbf{C} = \mathcal{B}(\mathbf{C})$ is a *state* if it preserves order and satisfies $\varphi(1) = 1$. (For commutative C^*-algebras these are just the probability measures.) The states on a C^*-algebra are completely bounded, and they are described by the *Gelfand-Naimark-Segal Theorem*: they must have the form $\varphi(a) = \pi(a)\xi \cdot \xi$ where $\pi \colon \mathcal{A} \to \mathcal{B}(K)$ is a *-homomorphism for some Hilbert space K (i.e., a "representation" of $\mathcal{A}$ on K) and $\xi \in K$. It has only recently been realized that this result has a generalization to all "operatorals":

THEOREM 2 (SEE [**26**, **31**]). *Suppose that* $\mathcal{A}$ *is a* C^*-*algebra and that* $\varphi \colon \mathcal{A} \to \mathcal{B}(H)$ *is a completely contractive linear map. Then*

$$\varphi(x) = S_0 \pi(x) S_1 \qquad (x \in \mathcal{A}),$$

where $\pi \colon \mathcal{A} \to \mathcal{B}(K)$ *is a representation, and*

$$H \xrightarrow{S_1} K \xrightarrow{S_0} H$$

is a diagram of contractions.

The Hahn-Banach Theorem has an operator space analogue. This important result was first proved by Arveson [**2**] for operator spaces that contain the identity operator and unital maps, and was subsequently extended to the general case by Wittstock [**30, 31**] (see also [**24**]).

THEOREM 3. *Given operator spaces* $\mathcal{V} \subseteq \mathcal{W}$ *we may extend any complete contraction* $\varphi\colon \mathcal{V} \to \mathcal{B}(H)$, *to a complete contraction* $\psi\colon \mathcal{W} \to \mathcal{B}(H)$.

Thus as was emphasized by Arveson [**2**] we have:

• *the strategy for determining the operatorals* $\varphi\colon \mathcal{V} \to \mathcal{B}(H)$ *is to fix an embedding* $\mathcal{V} \hookrightarrow \mathcal{B}(G)$ *and to then use the matricial form of the Hahn-Banach Theorem* (*Theorem 3*) *and Theorem 2 to extend* φ *to a complete contraction* $\psi\colon \mathcal{B}(G) \to \mathcal{B}(H)$ *of the form* $\psi(x) = S_0\pi(x)S_1$.

2. Quantized multilinear functionals. The norm of a multilinear map of normed vector spaces $\varphi\colon V_1 \times \cdots \times V_r \to W$ is defined by

$$\|\varphi\| = \sup\{\|\varphi(v_1,\ldots,v_r)\|\colon \|v_1\|,\ldots,\|v_r\| = 1\}.$$

We may regard φ as a linear function on the tensor product $V_1 \otimes \cdots \otimes V_r$. We then find that $\|\varphi\|$ is just the usual operator norm of the linear map

$$\varphi\colon V_1\hat{\otimes}\cdots\hat{\otimes}V_r \to W,$$

where $V_1\hat{\otimes}\cdots\hat{\otimes}V_r$ denotes $V_1 \otimes \cdots \otimes V_r$ with the *projective norm*

$$\|u\|^{\widehat{}} = \inf\left\{\sum_i \|v_1^i\|\cdots\|v_r^i\|\colon u = \sum_i v_1^i \otimes \cdots \otimes v_r^i\right\}$$

(we do not bother to take the completion). This tensor product is associative, and given contractions, $\varphi_k\colon V_k \to W_k$, the corresponding linear map

$$\varphi_1 \otimes \cdots \otimes \varphi_r\colon V_1\hat{\otimes}\cdots\hat{\otimes}V_r \to W_1\hat{\otimes}\cdots\hat{\otimes}W_r$$

is again a contraction. The strategy that we used to determine the functionals on a normed vector space does not extend to multilinear functions for two reasons:

• *the bounded bilinear maps* (*or "bimeasures"*) $F\colon C(X) \times C(Y) \to \mathbf{C}$ *are generally not described by measures on* $X \times Y$ (*see* [**12**]),

and one cannot use the Hahn-Banach Theorem to extend multilinear functions since

• *given inclusions* $V_k \hookrightarrow W_k$, *the corresponding map* $V_1\hat{\otimes}\cdots\hat{\otimes}V_r \to W_1\hat{\otimes}\cdots\hat{\otimes}W_r$ *need not be isometric.*

Given a multilinear map of operator spaces

$$\varphi\colon \mathcal{V}_1 \times \cdots \times \mathcal{V}_r \to \mathcal{W},$$

we again extend it linearly to a map $\varphi\colon \mathcal{V}_1 \otimes \cdots \otimes \mathcal{V}_r \to \mathcal{W}$. Given a sequence of integers $n = n_0,\ldots,n_r = n$ and matrices $v_j \in M_{n_{j-1},n_j}(\mathcal{V}_j)$, we define

$$v_1 \odot \cdots \odot v_r \in \mathbf{M}_n(\mathcal{V}_1 \otimes \cdots \otimes \mathcal{V}_r)$$

by "matrix multiplication":

$$v_1 \odot \cdots \odot v_r(i,j) = \sum_{k_1,\ldots,k_{r-1}} v_1(i,k_1) \otimes \cdots \otimes v_r(k_{r-1},j).$$

The *completely bounded norm* of a multilinear map φ is defined by

$$\|\varphi\|_{\mathrm{cb}} = \sup\{\|\varphi_n(v_1 \odot \cdots \odot v_r)\|: \|v_1\|, \cdots \|v_r\| \le 1,\ n \in \mathbf{N}\}.$$

This definition is due to Christensen and Sinclair [5]. The *Haagerup norm* $\| \ \|_h$ on $\mathbf{M}_n(\mathcal{V}_1 \otimes \cdots \otimes \mathcal{V}_r)$ is given by

$$\|u\|_h = \inf \left\{ \sum_j \|v_1^j\| \cdots \|v_r^j\|: u = \sum_j v_1^j \odot \cdots \odot v_r^j \right\} \tag{3}$$

$$= \inf\{\|v_1\| \cdots \|v_r\|: u = v_1 \odot \cdots \odot v_r\}$$

([**10, 11**]—this definition was motivated by results in [**15**]). Letting $\mathcal{V}_1 \otimes_h \cdots \otimes_h \mathcal{V}_r$ denote $\mathcal{V}_1 \otimes \cdots \otimes \mathcal{V}_r$ with these matricial norms, we have that $\|\varphi\|_{\mathrm{cb}}$ is precisely the completely bounded norm of the linear map

$$\varphi: \mathcal{V}_1 \otimes_h \cdots \otimes_h \mathcal{V}_r \to \mathcal{W}.$$

The Haagerup tensor product is associative, and using Ruan's criterion (Theorem 1), it is a simple matter to prove that the Haagerup tensor product of operator spaces is again an operator space. (This was first proved in [**25**].) It should be noted that the equality in (3) requires some care. It suffices to show that the second definition indeed gives a norm. The case $r = 2$ is discussed in [**10**], and the general result follows from associativity.

If one is given complete contractions, $\varphi_k: \mathcal{V}_k \to \mathcal{W}_k$, the corresponding linear map

$$\varphi_1 \otimes \cdots \otimes \varphi_r: \mathcal{V}_1 \otimes_h \cdots \otimes_h \mathcal{V}_r \to \mathcal{W}_1 \otimes_h \cdots \otimes_h \mathcal{W}_r$$

is a complete contraction. If the φ_k are only contractive, then the tensor product map need not be bounded. For a C^*-algebra $\mathcal{A}$, the multiplication map $\mathcal{A} \times \mathcal{A} \to \mathcal{A}$ is completely contractive; hence given completely contractive maps $\varphi_1, \varphi_2: \mathcal{V}_k \to \mathcal{A}$, the product map $\varphi_1\varphi_2: \mathcal{V}_1 \otimes_h \mathcal{V}_2 \to \mathcal{A}$ is again a complete contraction. This is generally not the case for any of the usual C^*-algebraic tensor product norms, and this is why C^*-algebraic tensor products are inappropriate for the study of cohomology.

In striking contrast to the function case, one can describe the multilinear operatorals on C^*-algebras, and the Haagerup tensor product is "injective":

THEOREM 4 [5]. *Suppose that $\mathcal{A}_k$ $(k = 1,\ldots,r)$ are C^*-algebras and that $\varphi: \mathcal{A}_1 \times \cdots \times \mathcal{A}_r \to \mathcal{B}(H)$ is a complete contraction. Then*

$$< \varphi(x_1,\ldots,x_r) = S_0\pi_1(x_1)S_1 \cdots S_{r-1}\pi_r(x_r)S_r, \tag{4}$$

where $\pi_k : \mathcal{A}_k \to \mathcal{B}(K_k)$ are $*$-representations, and

$$H \overset{S_r}{\to} K_r \overset{S_{r-1}}{\to} K_{r-1} \to \cdots \to K_1 \overset{S_0}{\to} H$$

is a diagram of contractions. (The S_k are referred to as "bridging maps" between the representation spaces.)

THEOREM 5. *Suppose that one is given operator spaces* $V_k \hookrightarrow W_k$. *Then the resulting injection* $V_1 \otimes_h \cdots \otimes_h V_r \to W_1 \otimes_h \cdots \otimes_h W_r$ *is completely isometric.*

Thus it is clear that

• *one may use the one variable strategy to determine the completely bounded multilinear operatorals on operator spaces.*

A recent discovery has shed considerable light on Theorem 4. As mentioned above, the tensor product $A_1 \otimes_h \cdots \otimes_h A_r$ is an operator space. In fact there is a canonical realization of this tensor product. The (nonamalgamated) *free product* $A_1 * \cdots * A_r$ of the C^*-algebras $A_1, \ldots, A_r$ is the C^*-algebraic envelope of the algebraic free product

$$\left[\sum_i^{\oplus} A_i \right] \oplus \left[\sum_{i \neq j}^{\oplus} A_i \otimes A_j \right] \oplus \cdots \oplus \left[\sum_{(i_1, \cdots, i_p) \in D_p}^{\oplus} A_{i_1} \otimes \cdots \otimes A_{i_p} \right] \oplus \cdots,$$

where $D_p = \{(i_1, \ldots, i_p) : i_k \neq i_{k-1}\}$, and the *-algebraic operations are given by

$$(a_{i_1} \otimes \cdots \otimes a_{i_p})(b_{j_1} \otimes \cdots \otimes b_{j_q}) = \begin{cases} a_{i_1} \otimes \cdots \otimes a_{i_p} b_{j_1} \otimes \cdots \otimes b_{j_q}, & i_p = j_1, \\ a_{i_1} \otimes \cdots \otimes a_{i_p} \otimes b_{j_1} \otimes \cdots \otimes b_{j_q}, & i_p \neq j_1, \end{cases}$$

where $a_{i_h} \in A_h$ and $b_{j_k} \in B_k$, and

$$(a_{i_1} \otimes \cdots \otimes a_{i_p})^* = a_{i_p}^* \otimes \cdots \otimes a_{i_1}^*.$$

THEOREM 6 [6]. *The natural inclusion*

$$A_1 \otimes_h \cdots \otimes_h A_r \to A_1 * \cdots * A_r$$

is completely isometric.

Using Theorem 3, a complete contraction $\varphi : A_1 \times \cdots \times A_r \to B(H)$ may therefore be extended to a complete contraction $\psi : A_1 * \cdots * A_r \to B(H)$. From Theorem 2, we may then write $\psi(u) = S_0 \pi(u) S_r$ for some representation $\pi : A_1 * \cdots * A_r \to B(K)$. Using the universal property of free products, we have corresponding noncommuting representations $\pi_k : A_k \to B(K)$ with

$$\pi(a_1 \otimes \cdots \otimes a_r) = \pi_1(a_1) \cdots \pi_r(a_r).$$

Thus we obtain Theorem 4 with $S_1 = \cdots = S_{r-1} = I$.

3. Bounded and completely bounded cohomology. The time development of a quantum system may be thought of as a one-parameter automorphism group on the C^*-algebra of observables. Since such groups are generated by derivations of the C^*-algebra, considerable attention has been devoted to the theory of derivations. In the 1960s, this culminated in the theorem of Kadison and Sakai that a bounded derivation of a von Neumann algebra must be inner [20, 29]. In the 1970s, Johnson, Kadison, and Ringrose developed norm and weak continuous Hochschild cohomology theories for operator algebras and their bimodules (see [18, 21, 22, 27]). These have been used to study liftings of derivations, as well as perturbations of representations and algebraic structure.

The incorporation of the norm and weak topologies into Hochschild cohomology by Johnson, Kadison, and Ringrose required considerable ingenuity, and a number of basic problems remain open. Among the questions considered by these authors were:

• *Given a von Neumann algebra $\mathcal{R}$, does one have that $H_c^r(\mathcal{R}, \mathcal{R}) = 0$ for $r \geq 2$? (For $r = 1$ this is just the Kadison-Sakai result that all bounded derivations of $\mathcal{R}$ into itself are inner.)*

• *Given a von Neumann algebra $\mathcal{R} \subseteq \mathcal{B}(H)$, are all bounded derivations of $\mathcal{R}$ into $\mathcal{B}(H)$ inner?*

They showed that the first statement is true if $\mathcal{R}$ is injective. Erik Christensen [3, 4] showed that the second is true if $\mathcal{R}$ is either properly infinite, or has a cyclic vector.

There are many cases in which one can restrict to completely bounded cocyles [6]. Thus one is reduced to studying *completely bounded cohomology*, which in some cases is much more computable than the bounded theory. It is this phenomenon which we will briefly illustrate.

Let us suppose that $\mathcal{A} \subseteq \mathcal{B}(H)$ is a concrete C^*-algebra on a Hilbert space H. We define the *completely bounded cochains* $C_{cb}^r = C_{cb}^r(\mathcal{A}, \mathcal{B}(H))$ to be the completely bounded r-linear maps

$$\varphi: \mathcal{A} \times \cdots \times \mathcal{A} \to \mathcal{B}(H).$$

We define the *coboundary map* $\delta = \delta_r: C_{cb}^r \to C_{cb}^{r+1}$ in the usual manner:

$$\begin{aligned}
\delta\varphi(a_1, \ldots, a_{r+1}) &= a_1\varphi(a_2, \ldots, a_{r+1}) - \varphi(a_1 a_2, \ldots, a_{r+1}) \\
&\quad + \varphi(a_1, a_2 a_3, \ldots, a_{r+1}) - \cdots + (-1)^{r+1}\varphi(a_1, a_2, \ldots, a_r)a_{r+1}.
\end{aligned}$$

The *completely bounded r-cocyles* $Z_{cb}^r = Z_{cb}^r(\mathcal{A}, \mathcal{B}(H)) \subseteq C_{cb}^r(\mathcal{A}, \mathcal{B}(H))$, and r-*coboundaries* $B_{cb}^r = B_{cb}^r(\mathcal{A}, \mathcal{B}(H)) \subseteq C_{cb}^r(\mathcal{A}, \mathcal{B}(H))$ are defined by $Z_{cb}^r = \ker \delta_r$, and $B_{cb}^r = \operatorname{range} \delta_{r-1}$. The *completely bounded Hochschild cohomology groups* are defined by $H_{cb}^r = H_{cb}^r(\mathcal{A}, \mathcal{B}(H)) = Z_{cb}^r / B_{cb}^r$. If we instead consider bounded cochains, we obtain the bounded Hochschild cohomology $H_c^r(\mathcal{A}, \mathcal{B}(H))$ of Johnson, Kadison, and Ringrose.

It was shown in [4] that completely bounded derivations of $\mathcal{A}$ into $\mathcal{B}(H)$ are necessarily inner. Since derivations are just 1-cocycles, this is equivalent to the statement that $H_{cb}^1(\mathcal{A}, \mathcal{B}(H)) = 0$. More generally we have:

THEOREM 7 [6]. *Suppose that $\mathcal{A} \subseteq \mathcal{B}(H)$ is a unital C^*-algebra. Then for all $r \geq 1$, $H_{cb}^r(\mathcal{A}, \mathcal{B}(H)) = 0$.*

COROLLARY 8 [6]. *Suppose $\mathcal{A} \subseteq \mathcal{B}(H)$ is a unital C^*-algebra such that the weak closure $\mathcal{R} = \overline{\mathcal{A}}$ is properly infinite or $\mathcal{R} = \mathcal{R}_0 \otimes_{vn} \mathcal{N}$, where $\mathcal{N}$ is the injective II_1 factor. Then $H_c^r(\mathcal{A}, \mathcal{B}(H)) = 0$.*

Corollary 8 follows from the fact that for the specified algebras, the bounded and completely bounded cohomology in dual modules often coincide. This requires some careful averaging techniques, which we shall not consider here. Since it is a particularly beautiful application of Theorem 4, we shall outline the proof of Theorem 7.

Given a completely bounded cochain $\varphi \colon \mathcal{A} \times \cdots \times \mathcal{A} \to \mathcal{B}(H)$, a standard cohomological argument shows that we may assume that φ is *reduced*, i.e., $\varphi(a_1, \ldots, a_r) = 0$ if $a_k = 1$ for any k. From Theorem 4 we may let

$$\varphi(a_1, \ldots, a_r) = S_0 \pi_1(a_1) S_1 \cdots S_{r-1} \pi_r(a_r) S_r,$$

for suitable representations $\pi_k \colon \mathcal{A} \to \mathcal{B}(K_k)$ and operators S_k. Regarding K_k as an $\mathcal{A}$ module, we may notationally suppress the π_k, giving us

$$\varphi(a_1, \ldots, a_r) = S_0 a_1 S_1 \cdots S_{r-1} a_r S_r.$$

Since φ is reduced, we may, after modifying the S_k, assume that $S_{k-1}S_k = 0$, $k = 1, \ldots, r$. Due to the cancellation of adjacent S_k, a simple computation shows

$$\delta\varphi(a_1, \ldots, a_{r+1}) = [S_0, a_1][S_1, a_2] \cdots [S_{r-1}, a_r][S_r, a_{r+1}], \qquad (5)$$

where we use the usual commutator notation. Since φ is a cocycle,

$$[S_0, a_1][S_1, a_2] \cdots [S_{r-1}, a_r][S_r, a_{r+1}] = 0$$

for all $a_k \in \mathcal{A}$. Letting Q be the projection on the join of the ranges of the operators $[S_r, a_{r+1}]$, $a_{r+1} \in \mathcal{A}$, we conclude that

$$[S_0, a_1][S_1, a_2] \cdots [S_{r-1}, a_r]Q = 0. \qquad (6)$$

Recalling that the commutator is a derivation in each variable, we have for any $a_r \in \mathcal{A}$,

$$a_r[S_r, a_{r+1}] = [S_r, a_r a_{r+1}] - [S_r, a_r]a_{r+1};$$

hence it is evident that a_r leaves QK_r invariant, i.e., $[a_r, Q] = 0$. Thus

$$[(1-Q)S_r, a_r] = [1 - Q, a_r]S_r + (1 - Q)[S_r, a_r] = 0.$$

Letting

$$\psi(a_1, \ldots, a_{r-1}) = S_0 a_1 S_1 \cdots a_{r-1} S_{r-1}(1 - Q)S_r,$$

the calculation for (5) gives us that

$$\begin{aligned}
(\delta\psi)(a_1, \ldots, a_r) &= [S_0, a_1] \cdots [S_{r-2}, a_{r-1}][S_{r-1}(1-Q)S_r, a_r] \\
&= [S_0, a_1] \cdots [S_{r-2}, a_{r-1}] \\
&\quad \times \{[S_{r-1}, a_r](1-Q)S_r + S_{r-1}[(1-Q)S_r, a_r]\} \\
&= [S_0, a_1] \cdots [S_{r-2}, a_{r-1}]\{[S_{r-1}, a_r](1-Q)S_r\} \\
&= [S_0, a_1] \cdots [S_{r-2}, a_{r-1}][S_{r-1}, a_r]S_r \\
&= \varphi(a_1, \ldots, a_r),
\end{aligned}$$

where in the second to last step we used (6).

4. Amenability. The notion of amenability appears in various guises in algebraic analysis. Suppose that G acts isometrically on a normed vector space V, i.e., V is a left Banach G module. Then G has a dual action on the dual Banach space V^*. G is said to be *amenable* if it has either of the following equivalent properties:

(a) Any bounded derivation of G into a dual Banach G-module V^* is inner.

(b) G has an invariant finitely additive probability measure $\mu \in 1^1(G)^{**}$.

Amenability provides an important dichotomy for discrete groups. In particular, a group G is amenable if and only if the regular representation weakly contains all of the irreducible unitary representations (see [13]).

The notion of an amenable Banach algebra was formulated by Johnson [16, 17]. Let us suppose that V is a normed vector bimodule for a unital Banach algebra $\mathcal{A}$. (We assume that the operations $\mathcal{A} \times V \to V$, $V \times \mathcal{A} \to V$ are contractions.) Then under the dual operations, V^* is a dual Banach bimodule. An element $v^* \in V^*$ is *central* if $av^* = v^*a$ for all $a \in \mathcal{A}$. We say that $\mathcal{A}$ is *amenable* if it has either of the following equivalent properties:

(a) Any bounded derivation of $\mathcal{A}$ into a dual Banach bimodule is inner.

(b) The A bimodule $(\mathcal{A}\hat{\otimes}\mathcal{A})^{**}$ has a central element M such that $\pi(M) = 1$, where $\pi: (\mathcal{A}\hat{\otimes}\mathcal{A})^{**} \to \mathcal{A}^{**}$ is the second adjoint of the multiplication map $(a, b) \to ab$. M is called a "virtual diagonal."

Connes and Haagerup proved that if $\mathcal{A}$ is a C^*-algebra, then $\mathcal{A}$ is amenable if and only if it is nuclear, i.e., the von Neumann algebra $\mathcal{A}^{**}$ is injective [8, 14]. This result was surprisingly difficult to prove, and in particular, the implication $\mathcal{A}$ nuclear $\Rightarrow$ $\mathcal{A}$ amenable remained open for several years. The proofs for both implications require that one consider the *von Neumann algebraic analogue of amenability.*

A dual bimodule V^* for a von Neumann algebra is said to be *normal* if the maps $r \mapsto rv^*$ and $r \mapsto v^*r$ are σ-weak, weak* continuous $(v^* \in V^*)$, and a derivation $\delta: \mathcal{R} \to V^*$ is *normal* if it is also continuous in those topologies. In order to define the analogue of a virtual diagonal, we must introduce a normal version of the projective tensor product [9]. To do this, we define $\mathrm{Bil}^\sigma(\mathcal{R}, \mathcal{R}) \subseteq (\mathcal{R}\hat{\otimes}\mathcal{R})^*$ to be the bounded bilinear functionals $F(r, s)$ which are normal in each variable, and we define the *normal projective product* by

$$\mathcal{R}\hat{\otimes}^\sigma \mathcal{R} = [\mathrm{Bil}^\sigma(\mathcal{R}, \mathcal{R})]^*.$$

A von Neumann algebra $\mathcal{R}$ is said to be *normally amenable* if it satisfies either of the following equivalent properties:

(a) Any bounded normal derivation of $\mathcal{R}$ into a normal dual Banach bimodule is inner.

(b) The $\mathcal{R}$ bimodule $\mathcal{R}\hat{\otimes}^\sigma \mathcal{R}$ has a central element M such that $\pi(M) = 1$, where $\pi: \mathcal{R}\hat{\otimes}^\sigma \mathcal{R} \to \mathcal{R}$ is the unique weak* continuous extension of the multiplication map $\mathcal{R} \otimes \mathcal{R} \to \mathcal{R}$.

As opposed to the Banach algebra case, it is hard to prove that these conditions are equivalent. The difficulty is that $\mathcal{R}\hat{\otimes}^\sigma \mathcal{R}$ is not a normal dual module; hence one cannot directly apply (a). This problem is circumvented by considering instead the *normal Haagerup tensor product*

$$\mathcal{R} \otimes_h^\sigma \mathcal{R} = [\mathrm{Bil}_h^\sigma(\mathcal{R}, \mathcal{R})]^*,$$

where $\mathrm{Bil}_h^\sigma(\mathcal{R}, \mathcal{R}) \subseteq (\mathcal{R} \otimes_h \mathcal{R})^*$ consists of the completely bounded bilinear functionals $F(r, s)$ which are normal in each variable. This is a normal dual bimodule, and thus (a) may be applied to prove the existence of a central element $M \in \mathcal{R} \otimes_h^\sigma \mathcal{R}$ for which $\pi(M) = 1$. A tricky application of the Grothendieck-Pisier-Haagerup inequality enables one to "lift" this to a normal virtual diagonal $M \in \mathcal{R} \hat{\otimes}^\sigma \mathcal{R}$. We refer the reader to [**9, 10**] for the details.

REFERENCES

1. E. Alfsen, H. Hanche-Olsen, and F. Schultz, *State spaces of C^*-algebras*, Acta Math. **144** (1980), 267–365.

2. W. Arveson, *Subalgebras of C^*-algebras*, Acta Math. **123** (1969), 141–224.

3. E. Christensen, *Extensions of derivations* (to appear).

4. ____, *Extensions of derivations*. II, Math. Scand. **50** (1982), 111–122.

5. E. Christensen and A. Sinclair, *Representations of completely bounded maps*, J. Funct. Anal. (to appear).

6. E. Christensen, E. Effros, and A. Sinclair, *The Haagerup tensor product and completely bounded cohomology*, Invent. Math. (to appear).

7. A. Connes, *Classification of injective factors*, Ann. of Math. (2) **104** (1976), 73–116.

8. ____, *On the cohomology of operator algebras*, J. Funct. Anal. **28** (1978), 248–253.

9. E. Effros, *Amenability and virtual diagonals for von Neumann algebras*, J. Funct. Anal. (to appear).

10. ____, *On multilinear completely bounded module maps*, Contemp. Math., vol. 62, Amer. Math. Soc., Providence, R. I., 1987, pp. 479–501.

11. E. Effros and A. Kishimoto, *Module maps and bounded Hochschild cohomology*, Indian J. Math. (to appear).

12. C. Graham and B. Schreiber, *Bimeasure algebras on locally compact groups*, Pacific J. Math. **115** (1984), 91–127.

13. F. Greenleaf, *Invariant means on topological groups*, Van Nostrand, Princeton, N.J., 1969.

14. U. Haagerup, *All nuclear C^*-algebras are amenable*, Invent. Math. **74** (1983), 305–319.

15. ____, *Decomposition of completely bounded maps on operator algebras*, unpublished, 1980.

16. B. E. Johnson, *Cohomology in Banach algebras*, Mem. Amer. Math. Soc. No. 127 (1972).

17. ____, *Approximate diagonals and cohomology of certain annihilator Banach algebras*, Amer. J. Math. **94** (1972), 685–698.

18. B. E. Johnson, R. Kadison, and J. Ringrose, *Cohomology of operator algebras*. III: *Reduction to normal cohomology*, Bull. Soc. Math. France **100** (1972), 73–96.

19. R. Kadison, *Isometries of operator algebras*, Ann. of Math. (2) **54** (1951), 325–338.

20. ____, *Derivations of operator algebras*, Ann. of Math. (2) **83** (1966), 280–293.

21. R. Kadison and J. Ringrose, *Cohomology of operator algebras*. I: *Type I von Neumann algebras*, Acta Math. **126** (1971), 227–243.

22. ____, *Cohomology of operator algebras*. II: *Extending cobounding and the hyperfinite case*, Ark. Mat. **9** (1971), 53–63.

23. G. Mackey, *Mathematical foundations of quantum mechanics*, W. A. Benjamin, New York, 1963.

24. V. Paulsen, *Completely bounded maps on C^*-algebras and invariant operator ranges*, Proc. Amer. Math. Soc. **86** (1982), 91–96.

25. V. Paulsen and R. Smith, *Multilinear maps and tensor norms on operator systems* (to appear).

26. V. Paulsen and C. Y. Suen, *Commutant representations of completely bounded maps*, J. Operator Theory **13** (1985), 87–101.

27. J. Ringrose, *Cohomology theory for operator algebras*, Proc. Sympos. Pure Math., vol. 38, part 2, Amer. Math. Soc., Providence, R.I., 1982, pp. 229–252.

28. Z.-J. Ruan, *Subspaces of C^*-algebras* (to appear).

29. S. Sakai, *Derivations of W^*-algebras*, Ann. of Math. (2) **83** (1966), 273–279.

30. G. Wittstock, *Ein operatorwertiger Hahn-Banach Satz*, J. Funct. Anal. **40** (1981), 127–150.

31. _____, *On matrix order and convexity*, Functional Analysis Surveys and Recent Results, Math. Studies, vol. 90, North Holland, Amsterdam, 1984.

32. Y. Friedman and B. Russo, *Algebraic structure in normed spaces without order* (to appear).

33. _____, *The Gelfand-Neumark theorem for JB^* triples* (to appear).

UNIVERSITY OF CALIFORNIA, LOS ANGELES, CALIFORNIA 90024, USA

Corona Problems, Interpolation Problems, and Inhomogeneous Cauchy-Riemann Equations

JOHN B. GARNETT

Let Ω be a complex analytic manifold and let $f_1, f_2, \ldots, f_N$ be bounded analytic functions on Ω, written $f_j \in H^\infty(\Omega)$. The *corona theorem* for Ω is the statement: there exist $g_1, g_2, \ldots, g_N \in H^\infty(\Omega)$ such that

$$f_1 g_1 + \cdots + f_N g_N = 1, \tag{1}$$

if the necessary condition

$$\sum |f_j(z)| \geq \eta > 0, \qquad z \in \Omega, \tag{2}$$

holds. This lecture surveys what is known to date on corona problems in general and outlines the proof of a recent corona theorem in the plane. We emphasize the connections with interpolating sequences and inhomogeneous Cauchy-Riemann equations.

With (2) and a partition of unity it is easy to find C^∞ bounded functions $\phi_1, \phi_2, \ldots, \phi_N$ such that

$$f_1 \phi_1 + \cdots + f_N \phi_N = 1, \tag{3}$$

but the functions ϕ_j may not be analytic. Following Hörmander [8], we write $g_j = \phi_j - \alpha_j$ and seek bounded functions $\alpha_1, \alpha_2, \ldots, \alpha_N$ on Ω satisfying

$$\sum_{j=1}^{N} f_j \alpha_j = 0 \tag{4}$$

and solving the *inhomogeneous Cauchy-Riemann equations*

$$\bar{\partial} \alpha_j = \bar{\partial} \phi_j, \qquad 1 \leq j \leq N, \tag{5}$$

where, in $\mathbf{C}^d$,

$$\bar{\partial} h = \sum_{k=1}^{d} \frac{\partial h}{\partial \bar{z}_k} \, d\bar{z}_k$$

Research supported by National Science Foundation Grant MCS-8002955.

and

$$\frac{\partial}{\partial \bar{z}_k} = \frac{1}{2} \left(\frac{\partial}{\partial x_k} + i \frac{\partial}{\partial y_k} \right),$$

so that $\bar{\partial} h = 0$ exactly when h is analytic.

To reduce the algebra we fix $N = 2$. In that case (4) is equivalent to

$$\left. \begin{array}{l} \alpha_1 = f_2 R, \\ \alpha_2 = -f_1 R, \end{array} \right\} \tag{4'}$$

and by (2) R is bounded if and only if both α_1 and α_2 are bounded. Then (5) holds if and only if

$$\bar{\partial} R = \frac{\bar{\partial} \phi_1}{f_2} = -\frac{\bar{\partial} \phi_2}{f_1}, \tag{5'}$$

in which the two right sides are equal by (3). Thus for $N = 2$ the corona theorem holds for Ω if and only if every equation (5′), arising from (2) and (3), has a solution bounded in Ω.

1. For the unit disc, or upper half plane $\mathcal{U}$, Carleson proved the corona theorem in 1962 [2]. His proof amounts to constructing ϕ_1 and ϕ_2 so that

$$G(x, y) = \frac{1}{f_2} \frac{\partial \phi_1}{\partial \bar{z}} = \frac{-1}{f_1} \frac{\partial \phi_2}{\partial \bar{z}}$$

satisfies the hypothesis of his:

THEOREM 1. *Assume* $y |G(x, y)| \in L^{\infty}(\mathcal{U})$ *and assume* $|G(x, y)| \, dx \, dy$ *is a Carleson measure:* $\iint_Q |G| \, dx \, dy \leq A h$ *whenever* $Q = \{x_0 < x < x_0 + h, \ 0 < y < h\}$. *Then* $\partial R / \partial \bar{z} = G$ *has a solution in* $\mathcal{U}$ *such that* $\|R\|_{L^{\infty}(\mathcal{U})} \leq C_1 A$.

In 1980 T. Wolff found another solution to the $\bar{\partial}$-problem (5) with

THEOREM 2. *Assume* $G(x, y) \in C^1(\mathcal{U})$ *and assume*

$$y |G|^2 \, dx \, dy \quad and \quad y |\partial G / \partial z| \, dx \, dy$$

are Carleson measures:

$$\iint_Q y |G|^2 \, dx \, dy \leq A_1 h, \qquad \iint_Q y |\partial G / \partial z| \, dx \, dy \leq A_2 h$$

for all Q *as in Theorem 1. Then* $\partial R / \partial \bar{z} = G$ *has a solution in* $\mathcal{U}$ *satisfying* $\|R\|_{L^{\infty}(\mathcal{U})} \leq C_2 \sqrt{A_1} + C_3 A_2$.

Theorem 2 yields Wolff's surprising proof of the corona theorem for $\mathcal{U}$ and a generalization.

THEOREM 3. *Assume* $f_1, f_2, \ldots, f_N$ *and* g *are* H^{∞} *functions* $\mathcal{U}$ *such that* $|g| \leq |f_1| + |f_2| + \cdots + |f_N|$. *Then* g^3 *lies in the ideal generated by* $\{f_1, f_2, \ldots, f_N\}$, *that is, there are* $g_1, g_2, \ldots, g_N \in H^{\infty}(\mathcal{U})$ *such that* $g^3 = g_1 f_1 + g_2 f_2 + \cdots + g_N f_N$.

Problem 1. Under the hypothesis of Theorem 3, is g^2 in the ideal generated by $\{f_1, \ldots, f_N\}$? Rao [16] has shown that g need not be.

A difference between Theorem 1 and Theorem 2 is that the hypothesis of Theorem 1 depends only on $|G|$, whereas Theorem 2 can hold for G and not for $\overline{G}$. See [6] and [11] for more on these results. We use a form of Theorem 1 below.

The corona theorem for $\mathcal{U}$ has two other proofs. Varopoulos [21] has a probabilistic argument somewhat akin to Wolff's. Berndtsson and Ransford [1], and also Slodkowski [18], have found a new point of view in the theory of analytic multifunctions.

2. In 1970 B. Cole exhibited an open Riemann surface for which the corona theorem is false. This surface can be enlarged to make a counterexample in $\mathbf{C}^3$ that is a domain of holomorphy and smooth and strictly pseudoconvex at all but one boundary point. See [5]. Sibony [17] has found domains of holomorphy $\Omega_1 \subsetneqq \Omega_2 \subset \mathbf{C}^2$ such that all function in $H^\infty(\Omega_1)$ have extensions in $H^\infty(\Omega_2)$. Recently M. Hayashi found such Riemann surfaces $\Omega_1 \subsetneqq \Omega_2$. We have no nontrivial example of a domain in $\mathbf{C}^2$ for which the corona theorem is true.

In the ball of $\mathbf{C}^d$ Varopoulos [19] has shown (1) has analytic solutions in the class $\bigcap_{p<\infty} L^p$, and a similar result is true for polydiscs, see [20], [4] and [14]. For both the ball and the polydisc generalizations of Theorem 1 or Theorem 2 are known but they have the weaker conclusion $R(z) \in \bigcap_{p<\infty} L^p$. Examples with $R(z) \notin L^\infty$ exist, but those for bad $R, G = \bar{\partial}R$ has not come from a corona problem.

Problem 2. In the ball or polydisc find a counterexample to the corona theorem.

3. For open subsets of compact Riemann surfaces the corona question has not been resolved (the examples lie on no compact surface), and since the $\bar{\partial}$-problem is easy when the partition of unity has compact support in Ω, it is enough to consider plane domains. Here much is known (see [12] and its bibliography), but the general case is still beyond reach. However we can solve the problem assuming a symmetry, and with Peter Jones we have proved:

THEOREM 4. *Assume $\Omega = \mathbf{C} \backslash E$ where E is a compact subset of the real axis. Then the corona theorem is true for Ω.*

Carleson [3] proved Theorem 4 under the extra hypothesis

$$m(E \cap (x - t, \ x + t)) \geq ct \qquad (6)$$

for all $x \in E$ and all $t > 0$. Write $\Omega = \mathcal{U}/\Gamma$ with Γ a Fuchsian group. Then $H^\infty(\Omega) = \{f \in H^\infty(\mathcal{U}): f \circ S = f \text{ for all } S \in \Gamma\}$. Using (6) and harmonic measures, Carleson constructed a "partition function" $P \in H^\infty(\mathcal{U})$ such that $\sum_{S \in \Gamma} P(Sz) = 1$, $\sum_{S \in \Gamma} |P(Sz)| \leq$ Const. For a corona problem on Ω solutions $G_j(z)$ on $\mathcal{U}$ then become solutions $g_j(z) \in H^\infty(\Omega)$ via $g_j(z) = \sum_{S \in \Gamma} P(Sz)G_j(Sz)$. See [12] for a different proof and a refinement. However, partition functions sometimes do not exist for the domains in Theorem 4.

4. Theorem 4 motivates the constructive attitude taken on results like Theorem 1 in, for instance, [**6**] and [**10**]. We have two half planes, but (5′) must also hold on $\mathbf{R}\backslash E$. The paper [**7**] with P. Jones has the details of the sketch below.

Suppose $f_1, f_2 \in H^\infty(\Omega)$ satisfy (2) and $\|f_j\|_{L^\infty(\Omega)} \le 1$. Set $\varepsilon = \eta/4$ and

$$\Omega_1(\varepsilon) = \left\{ z \in \Omega \colon \frac{1}{\pi} \int_E \frac{|y|}{(x-t)^2 + y^2}\, dt > \varepsilon \right\}, \qquad \Omega_2(\varepsilon) = \Omega\backslash\overline{\Omega}_1(\varepsilon).$$

LEMMA 1. *On $\Omega_k(\varepsilon)$ there are solutions $G_{1,\varepsilon}^{(k)}$ and $G_{2,\varepsilon}^{(k)}$ of (1) such that* $\|G_{j,\varepsilon}^{(k)}\|_{L^\infty(\Omega_k)} \le C(\varepsilon)$.

PROOF. For $\Omega_1(\varepsilon)$ this follows from Carleson's theorem for $\mathcal{U}$ because all components of $\Omega_1(\varepsilon)$ are simply connected. For $\Omega_2(\varepsilon)$ we symmetrize:

$$f_j^+(z) = \frac{f_j(z) + \overline{f_j(\bar z)}}{2}, \qquad f_j^-(z) = \frac{f_j(z) - \overline{f_j(\bar z)}}{2i}.$$

Then $f_j^\pm \in H^\infty(\Omega)$, $\|f_j^\pm\| \le 1$ and $\mathrm{Im}(f_j^\pm) = 0$ on $\mathbf{R}\backslash E$. Thus by the Poisson formula

$$|\mathrm{Im}\, f_j^\pm(z)| < \varepsilon, \qquad z \in \Omega_2(\varepsilon),\ j = 1, 2.$$

Consequently

$$H(z) = (f_1^+(z))^2 + (f_1^-(z))^2 + (f_2^+(z))^2 + (f_2^-(z))^2$$

satisfies $\mathrm{Re}\, H(z) \ge \eta^2/4$ on $\Omega_2(\varepsilon)$ and $G_{j,\varepsilon}^{(2)} = (f_j^+ - if_j^-)/H$ are solutions on $\Omega_2(\varepsilon)$.

By Carleson's original construction there is a symmetric contour $\Gamma \subset \Omega_1(\varepsilon/2)$ $\cap\, \Omega_2(\varepsilon)$ such that arc length on $\Gamma^+ = \Gamma \cap \mathcal{U}$ is a Carleson measure with absolute constant A and such that Γ separates $\Omega_1(\varepsilon)$ from $\Omega_2(\varepsilon/2)$. For $\alpha = \alpha(\varepsilon)$ let $\mathcal{D}^+$ be the hyperbolic neighborhood of Γ^+,

$$\mathcal{D}^+ = \{ z \in \mathcal{U} \colon \inf_{\varsigma \in \Gamma^+} |z - \varsigma|/y < \alpha \}$$

and set $\mathcal{D} = \mathcal{D}^+ \cup \{\bar z \colon z \in \mathcal{D}^+\}$. Take symmetric $\psi \in C^\infty(\Omega)$; $0 \le \psi \le 1$, $\psi = 1$ on $\Omega_2(\varepsilon/2)$, $\psi = 0$ on $\Omega_1(\varepsilon)$, with $|y\nabla\psi| \le c(\alpha)\chi_{\mathcal{D}}$. Then $\chi_{\mathcal{U}}|\nabla\psi|\, dx\, dy$ is a Carleson measure, with constant $C(\varepsilon)A$, having support in $\mathcal{D}^+$. By Lemma 1 there are smooth solutions of the form $\phi_j = G_{j,\varepsilon/2}^{(1)}(1-\psi) + G_{j,\varepsilon}^{(2)}\psi$, and for these the right side $G(x,y)$ of (5′) satisfies

$$|G(x,y)| \le \frac{C(\varepsilon)}{\eta}|\nabla\psi(x,y)| \le \frac{C'(\varepsilon)}{|y|}\chi_{\mathcal{D}}(x,y).$$

Therefore Theorem 4 is a consequence of this extension of Theorem 1:

LEMMA 2. *Assume* $|yG(x,y)| \le C'(\varepsilon)\chi_{\mathcal{D}}(x,y)$. *Then* $\partial R/\partial \bar z = G$ *has a solution in the domain* Ω *satisfying*

$$\|R\|_{L^\infty(\Omega)} \le C(\varepsilon)A. \tag{7}$$

To prove Lemma 2 we may replace $G\, dx\, dy$ by $\chi_{\mathcal{U}}G(x,y)\, dx\, dy$. Cover $\mathcal{D}^+$ by squares

$$S_n = \{ z_n = z_n(\varsigma) = z_n(0) + \varsigma y_n(0) \colon 0 \le \mathrm{Re}\, \varsigma,\ \mathrm{Im}\, \varsigma \le \alpha/2 \}$$

with disjoint interiors. Then $\chi_\mathcal{U} G(x, y)\, dx\, dy$ is an average of discrete measures

$$\mu_\varsigma = \sum a_n(\varsigma) y_n \delta_{z_n}, \qquad z_n = z_n(\varsigma),$$

with $|a_n(\varsigma)| \leq C(\varepsilon)$.

For each ς,

$$|z_n - z_m|/y_n \geq \alpha/4, \qquad m \neq n, \tag{8}$$

so that by the Carleson measure condition $\{z_n\}$ is an *interpolating sequence* for $H^\infty(\mathcal{U})$: given $\{b_n\} \in l^\infty$ there is $F \in H^\infty(\mathcal{U})$ such that

$$F(z_n) = b_n, \qquad n = 1, 2, \ldots. \tag{9}$$

LEMMA 3. *If* (8) *holds and if* $z_n \in \mathcal{D}^+$ *for all* n, *then* $\{z_n\}$ *is an interpolating sequence for* $H^\infty(\Omega)$, *that is,* (9) *has solution* $F \in H^\infty(\Omega)$, *and* $\|F\|_{H^\infty(\Omega)} \leq C(\varepsilon) \sup_n |b_n|$.

Accept Lemma 3 for now. Let $B = B_\varsigma$ be the half plane Blaschke product with zeros $z_n(\varsigma)$ and let $F = F_\varsigma \in H^\infty(\Omega)$ solve (9) with

$$b_n = -a_n(\varsigma) y_n B'(z_n)/2\pi i.$$

Then F/B has an analytic continuation to $\Omega \setminus \{z_n\}$ and, as distributions on Ω,

$$\partial(F/B)/\partial \bar{z} = \mu_\varsigma.$$

Moreover

$$|F/B| \leq c(\varepsilon, \alpha) y_n/|z - z_n|, \qquad z \text{ near } z_n,$$

$$|F/B| \leq c(\varepsilon, \alpha) \quad \text{on } \bigcap \{|z - z_n| > 2\alpha y_n\},$$

and, by its construction, F/B is measurable in ς. Consequently the average

$$R(z) = \iint (F_\varsigma/B_\varsigma)\, d\xi\, d\eta, \qquad \varsigma = \xi + i\eta,$$

satisfies (7) and $\partial R/\partial \bar{z} = G$.

Thus the corona theorem for Ω depends on the $H^\infty(\Omega)$ interpolation result of Lemma 3. We outline its proof in the special case

$$\inf_n \prod_{k; k \neq n} |(z_k - z_n)/(\bar{z}_k - z_n)| \geq 1 - \gamma(\varepsilon),$$

with $\gamma(\varepsilon)$ small; that is sufficient for the Theorem because $\{z_n\}$ can be partitioned into a finite union of such sequences.

If $\gamma(\varepsilon)$ is small then by the definition of $\Omega_1(\varepsilon)$ there are disjoint sets $E_n \subset E \cap \{|t - x_n| < 3y_n/\varepsilon\}$ such that

$$\omega(z_n, E_n) = \frac{1}{\pi} \int_{E_n} \frac{y_n}{(x_n - t)^2 + y_n^2}\, dt \geq \frac{\varepsilon}{3}$$

and $\omega(z_n, \bigcup_{k \neq n} E_k) \leq \beta(\varepsilon)$. Then there exist real functions u_n such that

$$\|u_n\|_\infty \leq C(\varepsilon), \tag{10}$$

$$u_n \text{ has support } E_n, \tag{11}$$

and

$$\int u_n \, dt = 0. \tag{12}$$

Writing

$$u_n(z) = \frac{1}{\pi} \int \frac{y}{(x-t)^2 + y^2} u_n(t) \, dt$$

and

$$\tilde{u}_n(z) = \frac{1}{\pi} \int \frac{(x-t)u_n(t)}{(x-t)^2 + y^2} \, dt$$

for the Poisson integral and its conjugate, we can also get

$$u_n(z_n) = 0, \tag{13}$$

$$\tilde{u}_n(z_n) = \pi, \tag{14}$$

and

$$\sum_{y_k ; y_k < y_n} |\tilde{u}_k(z_n)| \le C_1(\varepsilon)\beta. \tag{15}$$

Let $-1 \le c_n \le 1$ and set $F(z) = \exp \sum c_n(u_n(z) + i(\tilde{u}_n(z)))$. Since the E_n are disjoint $e^{-C(\varepsilon)} \le |F(z)| \le e^{C(\varepsilon)}$, by (10), and since $\sum c_n u_n(t)$ is supported on E, $F(z)$ reflects to be analytic on Ω and $|F(z)| \le e^{C(\varepsilon)}$, $z \in \Omega$. Moreover using (12)–(15) we can choose the c_n inductively so that

$$|F(z_n) - (b_n / \|b_n\|_\infty)| < \tfrac{1}{2}$$

if $\beta(\varepsilon)$ is small. An iteration then yields (9). An argument without Blaschke products is given in [**7**].

To eliminate the use of symmetry we pose

Problem 3. If E is a compact subset of a Lipschitz graph in the plane, prove the corona theorem for $\Omega = \mathbf{C} \backslash E$.

This is true if E has the density condition (6) for arc length on the graph. C. Moore has treated the case of a $\mathbf{C}^{1+\varepsilon}$ graph [**15**].

Problem 4. Let $E = K \times K$ where K is the usual Cantor set. Is the theorem true for $\Omega = \mathbf{C} \backslash K$?

REFERENCES

1. B. Berndtsson and T. J. Ransford, *Analytic multifunctions, the $\bar{\partial}$-equation and a proof of the corona theorem*, Pacific J. Math. **124** (1986), 57–72.

2. L. Carleson, *Interpolation by bounded analytic functions and the corona theorem*, Ann. of Math. **76** (1962), 547–559.

3. _____, *On H^∞ in multiply connected domains*, Conference on Harmonic Analysis in Honor of Antoni Zygmund (Chicago, Ill., 1981), Vol. II, Wadsworth Math. Ser., Wadsworth, Belmont, Calif., 1983, pp. 349–372.

4. S. Y. A. Chang, *Two remarks on H^1 and BMO on the bidisc*, Conference on Harmonic Analysis in Honor of Antoni Zygmund (Chicago, Ill., 1981), Vol. II, Wadsworth Math. Ser., Wadsworth, Belmont, Calif., 1983, pp. 373–393.

5. T. W. Gamelin, *Uniform algebras and Jensen measures*, London Math. Soc. Lecture Notes Ser., no. 32, Cambridge University Press, Cambridge-New York, 1978.

6. J. B. Garnett, *Bounded analytic functions*, Academic Press, New York, 1981.

7. J. B. Garnett and P. W. Jones, *The corona theorem for Denjoy domains*, Acta Math. **155** (1985), 27–40.

8. L. Hörmander, *Generators for some rings of analytic functions*, Bull. Amer. Math. Soc. **73** (1967), 943–949.

9. P. W. Jones, *Carleson measures and the Fefferman-Stein decomposition of* BMO($\mathbf{R}^n$), Ann. of Math. **111** (1980), 197–208.

10. ____, *L^∞ estimates for the $\bar{\partial}$-problem in a half-plane*. Acta Math. **150** (1983), 137–152.

11. ____, *Recent advances in the theory of Hardy spaces*, Proc. Internat. Congr. Math. (Warsaw, 1983), Vol. I, North-Holland, New York, 1984, pp. 829–838.

12. P. W. Jones and D. E. Marshall, *Critical points of Green's function, harmonic measure, and the corona problem*, Ark. Mat. **23** (1985), 281–314.

13. P. Koosis, *Lectures on H_p spaces*, London Math. Soc. Lecture Notes Ser., no. 40, Cambridge University Press, Cambridge-New York, 1980.

14. K. C. Lin, *H^p solutions for the corona problem on the polydisc in $\mathbf{C}^n$* (to appear).

15. C. N. Moore, *The corona theorem for domains whose boundaries lie on smooth curves* (to appear).

16. K. V. R. Rao, *On a generalized corona problem*, J. Analyse Math. **18** (1967), 277–278.

17. N. Sibony, *Prolongement des fonctions holomorphes bornées et métrique de Carathéodory*, Invent. Math. **29** (1975), 205–230.

18. Z. Slodkowski, *An analytic set-valued selection and its applications to the corona theorem, to polynomial hulls and joint spectra* (to appear).

19. N. Th. Varopoulos, *BMO functions and the $\bar{\partial}$-equation*, Pacific J. Math. **71** (1977), 221–273.

20. ____, *Probabilistic approach to some problems in complex analysis*, Preprint, Univeristé de Paris-Sud, 1980.

21. ____, *The Helson-Szegö theorem and A_p-functions for Brownian motion and several variables*, J. Funct. Anal. **39** (1980), 85–121.

UNIVERSITY OF CALIFORNIA, LOS ANGELES, CALIFORNIA 90024, USA

Вероятность в геометрии банаховых пространств

Е. Д. ГЛУСКИН

Эта статья посвящена некоторым результатам геометрии банаховых пространств, полученных благодаря привлечению теоретико-вероятностных соображений. Классический способ доказательства существования объектов, обладающих специальными свойствами, выявлением того, что они заполняют множество большой меры, стал интенсивно использоваться в геометрии банаховых пространств с середины 70-х годов. К этому времени были найдены достаточно прозрачные вероятностные доказательства теоремы Дворецкого о почти евклидовых сечениях выпуклых тел (В. Д. Мильман [38], Фигель [14], Шанковский [52]) и появился знаменитый пример Энфло [13] пространства без свойства аппроксимации, конструкция которого почти сразу приобрела вероятностный характер (А. Дэви [11], см. также Митягин [39]). Примерно в то же время появился значительный интерес к изучению конечномерных нормированных пространств (иначе пространств Минковского). Многие вопросы теории банаховых пространств получили содержательную интерпретацию на конечномерном уровне, а решение ряда бесконечномерных проблем было получено «склейкой» конечномерных результатов. При этом на первый план выдвинулась проблема существования конечномерных пространств и операторов с заданными свойствами. Изучая нормы случайных матриц, элементы которых независимо принимают значения ± 1 с вероятностью $1/2$, Беннет, Гудмэн и Ньюмен [3] завершили описание классов (p,q)-абсолютно суммирующих операторов в гильбертовом пространстве. Дальнейшее изучение случайного подпространства — образа такой матрицы позволило им совместно с Дором и Джонсоном [2] построить пример недополняемого гильбертова подпространства в пространстве L_p при $1 < p < 2$. Воспользовавшись для равномерной аппроксимации евклидова шара случайно выбранными подпространствами, Б. С. Кашин [32] сумел получить точные в степенной шкале оценки (а при $n \asymp N$ и точные порядки) поперечников $d_n(B_p^N, l_q^N)$ (см. определения ниже). Тем самым был исчерпан старый вопрос о порядках поперечников соболевских классов. Исследование вложений, возникших в работе Б. С. Кашина, позволило Фигелю, Квапеню и Пелчинскому [16] построить для каждого n n-мерное пространство, у которого любой базис имеет константу безусловности порядка не ниже $\sqrt{n}$; т.е. n-мерное пространство с наихудшей по порядку безусловной базисной константой. (См. также развитие этого результата в работе Фигеля и Джонсона [15].) Список результатов банаховой геометрии, в которых решающую роль играют вероятностные соображения можно продолжать еще очень долго. Не имея возможности это сделать, остановимся подробнее на следующих двух вопросах: (1) стохастический выбор пространств Минковского со специальными свойствами; (2) использование случайных подпространств для оценок поперечников. К сожалению, за рамками этого обзора остаются исследования по факторизации случайных операторов (см., например, [4,10]); исследования, примыкающие к теореме Дворецкого (см. [17], работы В. Д. Мильмана последних лет и библиографию к ним), в частности, изящное доказательство гипотезы Малера данное Бургейном и Мильманом [8]; восходящие к работе М. И. Кадеца [29] замечательные результаты Джонсона, Шехтмана [28] и Пизье [45] о вложениях пространства и др.

0. Стандартные обозначения. Ниже $\langle \cdot, \cdot \rangle$ — скалярное произведение, vol или vol_n — лебегов объем в $\mathbf{R}^n$; $S^{n-1} \subset \mathbf{R}^n$ — единичная сфера, μ_n-лебегова мера на S^{n-1}, нормированная условием $\mu_n(S^{n-1}) = 1$.

Стандартная гауссова мера в пространстве $\mathbf{R}^n$ определяется плотностью $(2\pi)^{-n/2} \exp(-\sum_{i=1}^n x_i^2/2)$ и обозначается γ_n. $G_{n,N}$ — многообразие всех n-мерных подпространств пространства $\mathbf{R}^N$; $\mu_{n,N}$ — унитарно-инвариантная вероятностная мера на $G_{n,N}$.

l_p^n — n-мерное (в зависимости от контекста вещественное или комплексное) нормированное пространство с нормой

$$\|x\|_p = \left(\sum_{i=1}^n |x_i|^p \right)^{1/p}, \quad 1 \le p < \infty; \qquad \|x\|_\infty = \max_{1 \le i \le n} |x_i|.$$

B_p^n — единичный шар этого пространства. Буквой c обозначаются различные в разных местах положительные константы. Стандартным образом используются обозначения $O(\cdot)$ и $o(\cdot)$. Иногда вместо $\varphi = O(\psi)$ пишем $\varphi \prec \psi$. Запись $\varphi \asymp \psi$ означает, что $\varphi \prec \psi$ и $\psi \prec \varphi$.

1. Конечномерные пространства со специальными свойствами. Пусть X, Y — изоморфные банаховы пространства. Величина

$$d(X, Y) = \inf\{\|T\| \, \|T^{-1}\| : T \colon X \to Y \text{ — линейный изоморфизм}\}$$

называется дистанцией Банаха-Мазура между пространствами X и Y. Совокупность всех нормированных пространств фиксированной размерности n с мерой близости $\log d(\cdot, \cdot)$ является метрическим компактом. Он называется компактом Минковского и будет обозначаться $\mathfrak{M}_n$.

ВОПРОС. Как ведет себя с ростом n величина

$$\operatorname{diam} \mathfrak{M}_n = \sup\{d(X, Y) \colon X, Y \in \mathfrak{M}_n\}? \tag{1}$$

Точное значение $\operatorname{diam} \mathfrak{M}_n$ известно только для случая $n = 2 : \operatorname{diam} \mathfrak{M}_2 = 3/2$ (Стромквист [**49**]). Классический результат Джона [**27**] показывает, что $\sup\{d(X, l_2^n) \colon X \in \mathfrak{M}_n\} = \sqrt{n}$. Отсюда немедленно получается, что для любых $X, Y \in \mathfrak{M}_n$

$$d(X, Y) \leq d(X, l_2^n) d(l_2^n, Y) \leq n.$$

Тем самым $\sqrt{n} \leq \operatorname{diam} \mathfrak{M}_n \leq n$. Ограничение супремума в (1) на те или иные подмножества $\mathfrak{M}_n$ приводит ко многим интересным задачам. При этом обычно удается получить оценку порядка $\sqrt{n}$ (см., например, [**58,59, 12, 9**]). Так, например, обстоит дело в случае, когда пространства X, Y обладают 1-симметричным базисом (Н. Томчак-Ягерманн [**59**]). Для получения таких оценок очень плодотворным оказалось использование вероятностных соображений (см., например, [**58, 12**]): на множестве операторов T, осуществляющих изоморфизм между пространствами X и Y, специальным образом вводится вероятностная мера. При удачном ее выборе задача оценки среднего относительно этой меры значения $\|T\| \, \|T^{-1}\|$ оказывается сравнительно простой и приводит к нужному результату о величине $d(X, Y)$.

Попытаемся наивно применить этот подход к случаю $X = Y = l_1^n$, задав на множестве линейных изоморфизмов гауссову меру (последнее означает, что оператор T задается матрицей (t_{ij}), элементы t_{ij} которой — независимые стандартные гауссовы величины). Нетрудно заметить, что в этом случае с большой вероятностью выполняется неравенство $\|T\| \, \|T^{-1}\| \succ n$. Это наталкивает на мысль, что немного «подпортив» пространство l_1^n, можно построить пространства X и Y так, чтобы для любого изоморфизма $T \colon X \to Y$ выполнялось неравенство $\|T\| \, \|T^{-1}\| > cn$ ($c > 0$) — абсолютная константа). Это действительно так.

ТЕОРЕМА 1 [**19**]. *Справедлива оценка* $\operatorname{diam} \mathfrak{M}_n \asymp n$.

Как уже отмечалось, оценка $\operatorname{diam} \mathfrak{M}_n \prec n$ вытекает из теоремы Джона. Явно описать пространства $X_n, Y_n \in \mathfrak{M}_n$, для которых с некоторой абсолютной константой $c > 0$ справедливо неравенство $d(X_n, Y_n) > cn$, однако до сих пор не удается. Для доказательства теоремы 1 используется стохастический выбор пространств X_n и Y_n. При реализации такого подхода на $\mathfrak{M}_n \times \mathfrak{M}_n$ нужно ввести вероятностную меру и доказать, что неравенство $d(X, Y) > cn$ выполняется с положительной вероятностью. В настоящее время структура компакта $\mathfrak{M}_n$ еще мало изучена, в частности, не понятно, каковы наиболее естественные меры на $\mathfrak{M}_n$ (и на $\mathfrak{M}_n \times \mathfrak{M}_n$). Для наших целей удобной оказывается следующая вероятность $\mathcal{P}$ на $\mathfrak{M}_n \times \mathfrak{M}_n$, определение которой инспирировано рассуждениями предпосланными формулировке теоремы 1.

Сопоставим последовательности $F = (f_1, \dots, f_{2n})$ элементов $\mathbf{R}^n$ пространство $X_F \in \mathfrak{M}_n$, единичный шар которого совпадает с множеством $A_F = \mathrm{conv}\{B_1^n, \pm f_1, \dots, \pm f_{2n}\}$. Мера ν на $\mathfrak{M}_n$ индуцируется при отображении $F \to X_F$, когда элементы $f_1, \dots, f_{2n}$ случайно и независимо выбираются из (S^{n-1}, μ_n). Мера $\mathcal{P}$ равна $\nu \times \nu$.

ПРЕДЛОЖЕНИЕ 1. *Существует абсолютная константа $c > 0$ такая, что, если пространство Y принадлежит носителю меры ν, то ν — мера тех $X \in \mathfrak{M}_n$, для которых найдется оператор T из X в Y с нормой $\|T\| \le c\sqrt{n}\,|\det T|^{1/n}$ меньше 2^{-n^2}.*

Как X так и Y в предложении 1 являются пространствами вида X_F. Такое пространство — это специальным образом нормированное $\mathbf{R}^n$. Тем самым и в X и в Y имеется фиксированный базис, что позволяет отождествить любой оператор из X в Y с соответствующей матрицей. Поэтому, можно говорить об определителе оператора и о его действии в другой паре пространств того же вида.

Теорема 1 немедленно выводится из предложения 1. Его доказательство делится на две относительно независимые части: во-первых, доказывается, что для любого оператора S, $|\det S| = 1$, и любого $\varepsilon > 0$ выполнено (A — абсолютная константа)

$$\nu\{X \in \mathfrak{M}_n \colon \|S\|_{X \to Y} \le 2\varepsilon\sqrt{n}\} \le (A\varepsilon^2)^{n^2}.$$

Во-вторых подбирается конечное множество операторов $\mathcal{M}$, $\mathrm{Card}\,\mathcal{M} \le (A/\varepsilon)^{n^2}$, зависящее только от Y и ε, такое, что как только найдутся пространство X из носителя ν и оператор T из X в Y с нормой $\|T\|_{X \to Y} \le \varepsilon\sqrt{n}\,|\det T|^{1/n}$, то найдется оператор $S \in \mathcal{M}$ со следующими свойствами: $|\det S| = 1$ и $\|S - |\det T|^{-1/n}T\|_{X \to Y} \le \varepsilon\sqrt{n}$ (при этом автоматически $\|S\|_{X \to Y} \le 2\varepsilon\sqrt{n}$).

Сравнение (2) с мощностью $\mathcal{M}$ при $\varepsilon = c = \frac{1}{2}A^2$ доказывает предложение 1.

Доказательство (2) легко выводится из независимости векторов $f_1, \dots, f_{2n}$, определяющих единичный шар случайного пространства $X = X_F$, и из следующей оценки распределения на (S^{n-1}, μ_n) случайной величины $\|Sf\|_Y$ через лебегов объем $\mathrm{vol}\,BY$ единичного шара пространства Y:

$$\mu_n\{f \in S^{n-1} \colon \|Sf\|_Y \le \lambda\} \le \lambda^n |\det S|^{-1}\,\mathrm{vol}\,BY/\mathrm{vol}\,B_2^n. \qquad (3)$$

При этом неравенство (3) применяется при $\lambda = 2\varepsilon\sqrt{n}$, а $\mathrm{vol}\,BY$ оценивается за счет того, что единичный шар пространства Y имеет небольшое число крайних точек (см. [33]).

Неравенство (3) было впервые обнаружено и использовано в банаховой геометрии Шареком [56]. Оно легло в основу введенного Шареком и Томчак-Яегерманн [**57**] важного понятия «объемное отношение пространств Минковского».

В качестве множества $\mathcal{M}$ берется минимальная ε-сеть (в метрике, порожденной l_2^n-операторной нормой) множества матриц T таких, что $\|T\|_{l_1^n \to Y} \leq 1$ и $|\det T| = 1$.

Замечания. 1. Носитель меры ν — очень специальное подмножество $\mathfrak{M}_n$ — это фактор-пространство пространства l_1^{3n}. Иначе говоря, меру ν можно рассматривать, как вероятность на многообразии $G_{2n,3n}$ всех $2n$-мерных подпространств l_1^{3n}. При таком взгляде привычнее рассматривать меру $\tilde{\nu}$ на $G_{2n,3n}$ унитарно-инвариантную в l_2^{3n}-смысле. Она отличается от меры ν, но довольно близка к ней и также может использоваться для доказательства теоремы 1. Для определения меры $\tilde{\nu}$ в терминах, аналогичных определению ν, надо вместо множеств A_F рассмотреть множества $\mathrm{conv}\{\pm f_1, \ldots, \pm f_{3n}\}$, а элементы $f_1, \ldots, f_{3n}$ независимо выбирать из $(\mathbf{R}^n, \gamma_n)$. Близость мер ν и $\tilde{\nu}$ обусловлена тем, что во-первых, при больших n гауссов вектор f распределен в основном в окрестности сферы $\sqrt{n} S^{n-1}$, а во-вторых, когда $f_1, \ldots, f_n$ независимо пробегают $(\mathbf{R}^n, \gamma_n)$, объем «октаэдрального» множества $\mathrm{conv}\{\pm f_1, \ldots, \pm f_n\}$ оказывается довольно большим с вероятностью близкой к 1.

2. По аналогии с определением мер ν, $\tilde{\nu}$ естественно ввести меры $\nu_{k,B}$, $\tilde{\nu}_{k,B}$ на $\mathfrak{M}_n$. Их определение отличается тем, что вместо $2n$ точек $f_1, \ldots, f_{2n}$ берется k точек, а «октаэдр» B_1^n заменяется другим выпуклым центрально-симметричным телом B. Интересны различные вопросы относительно появляющихся таким образом случайных пространств Минковского. Один из простейших среди них — вопрос об оценке средних значений интересных по тем или иным причинам функционалов на $\mathfrak{M}_n$.

В [**24**] Гордон ввел следующее понятие, описывающее степень симметричности пространств Минковского. Константой асимметрии $s(X)$ $X \in \mathfrak{M}_n$ называется нижняя грань чисел λ, обладающих следующим свойством: существует группа G обратимых линейных операторов в X такая, что каждый оператор, коммутирующий со всеми $g \in G$, имеет вид $a\,\mathrm{Id}$ (Id — тождественный оператор) и $\sup\{\|g\| \colon g \in G\} \leq \lambda$.

Так как $d(X, l_2^n) \leq \sqrt{n}$ для любого $X \in \mathfrak{M}_n$, то и $s(X) \leq \sqrt{n}$. Точный порядок роста величины $s_n = \sup\{s(X) \colon X \in \mathfrak{M}_n\}$ вычислен Манкевичем [**36**].

ТЕОРЕМА 2 [**36**]. *Справедлива оценка* $s_n \asymp n^{1/2}$.

В основе доказательства этого результата лежит следующая конструкция. Пусть X, $Y \in (\mathfrak{M}_n, \nu_{20n}, B_1^n)$ — независимые случайные пространства. Положим $Z = X \oplus_2 Y$ (т.е. Z — пространство пар $z = (x, y)$, $x \in X$, $y \in Y$, снабженное нормой $\|z\| = (\|x\|^2 + \|y\|^2)^{1/2}$).

Вычисление математического ожидания величины $s(Z)$ и приводит к требуемому результату.

Пример Энфло пространства без свойства аппроксимации породил следующую конечномерную проблему: существует ли последовательность пространств $X_n \in \mathfrak{M}_n$ с неограниченно растущей базисной константой? Напомним соответствующие определения. С любым базисом $x_1, \ldots, x_n$ пространства X ассоциируется последовательность проекторов $P_1, \ldots, P_n$:

$$P_k \left(\sum \alpha_i x_i \right) = \sum_{i=1}^{k} \alpha_i x_i.$$

Базисная константа системы $x_1, \ldots, x_n$ определяется как $b(\{x_i\}_{i=1}^n) = \sup_{1 \leq k \leq n} \|P_k\|$. Нижняя грань величин $b(\{x_i\}_{i=1}^n)$ по всем возможным базисам $\{x_i\}_{i=1}^n$ называется базисной константой пространства X и обозначается $b(X)$. Из теоремы Джона [27] следует, что $b(X) \leq \sqrt{n}$.

В [20] доказано, что для каждого n найдется пространство $X \in \mathfrak{M}_n$, обладающее следующим свойством ($c > 0$ — абсолютная константа)

$$\|P\| > c \frac{\min\{\operatorname{rank} P, n\text{-}\operatorname{rank} P\}}{\sqrt{n \log n}}, \quad \text{для любого проектора } P \text{ в } X. \quad (4)$$

В частности, $b(X) > 2^{-1}c\sqrt{n/\log n}$. Более того, справедлив следующий результат.

ТЕОРЕМА 3 [20]. *Пусть $X \in \mathfrak{M}_n$ случайное пространство с распределением $\nu_{n^2, n^{-1/2}B_2^n}$. В этом случае вероятность события (4) стремится к 1 с ростом n.*

Используя немного другую вероятностную конструкцию, Шарек [53] независимо доказал следующий результат, устанавливающий правильный порядок величины

ТЕОРЕМА 4 [53]. *Для каждого n существует $X_n \in \mathfrak{M}_n$ такое, что для любого проектора P в X_n ранга $[n/2]$ выполнено $\|P\| \geq \delta n^{1/2}$ ($\delta > 0$ — абсолютная константа). В частности, $b(X_n) \geq \delta n^{1/2}$.*

Развивая свою конструкцию, Шарек [54] сумел «склеить» из конечномерных пространств бесконечномерное банахово пространство без базиса со свойством ограниченной аппроксимации и решить тем самым одну из последних проблем, порожденных примером Энфло. (В силу теоремы Пелчинского [42] наличие свойства ограниченной аппроксимации у банахова пространства означает, что оно изоморфно дополняемому подпространству пространства с базисом.)

Классический результат Кадеца и Снобара утверждает, что каково бы ни было n-мерное подпространство L банахова пространства X, найдется проектор $P : X \to L$ такой, что $\|P\| \leq \sqrt{n}$. Теорема 4 доказывает существование n-мерного пространства, в котором эта оценка не может быть существенно улучшена ни для какого подпространства размерности $[n/2]$. Пример теоремы 3 приводит к нетривиальным оценкам снизу норм всех проекторов, ранг и коранг которых больше $\sqrt{n \log n}$. До сих пор, однако, не ясно, существует ли последовательность пространств $X_n \in \mathfrak{M}_n$ и неограниченно возрастающая функция φ такие, что для любого проектора P в X_n ранга не выше $n/2 + 1$ выполнено $\|P\| > \varphi(\operatorname{rank} P)$.

В замечательной работе Пизье [44] построен пример бесконечномерного банахова пространства, которое, в частности, обладает следующим свойством: для любого конечномерного проектора P в нем с некоторой абсолютной константой $c > 0$ выполнено $\|P\| > c\sqrt{\operatorname{rank} P}$.

До сих пор все рассмотрения были вещественными. Их перенесение на комплексный случай не сложно. Надо просто смотреть на n-мерное комплексное пространство, как на вещественное пространство размерности $2n$. Например, определение мер $\nu_{k,B}$ модифицируется при этом следующим образом. Пусть для $m = 1, \ldots, k$ ξ_m, $\eta_m \in \mathbf{R}^n$, а $f_m = \xi_m + i\eta_m \in \mathbf{C}^n$. Далее $B \subset \mathbf{C}^n$ — фиксированное абсолютно-выпуклое подмножество $\mathbf{C}^n$. Определим множество $A_F^{\mathbf{C}}$ $(F = (f_1, \ldots, f_k))$ равенством

$$A_F^{\mathbf{C}} = \left\{ z \in \mathbf{C}^n \colon \exists \lambda_0, \ldots, \lambda_k \in \mathbf{C}, \ \sum_{m=0}^{k} |\lambda_m| \leq 1, \ z \in \lambda_0 B + \sum_{m=1}^{k} \lambda_m f_m \right\}$$

и пусть X_F — n-мерное над полем $\mathbf{C}$ нормированное пространство с единичным шаром $A_F^{\mathbf{C}}$. Мера $\nu_{k,B}^{\mathbf{C}}$ индуцируется при отображении $F \to X_F$, когда ξ_m, η_m независимо пробегают (S^{n-1}, μ_n).

При «вещественном» подходе к комплексным пространствам естественно возникает вопрос, не определяется ли комплексная структура вещественной геометрией пространства. Иначе говоря, существуют ли комплексно неизоморфные банаховы пространства изоморфные в вещественном смысле. Для формулировки аналогичного вопроса в конечномерном случае введем следующие обозначения. Пусть X, Y — n-мерные над полем комплексных чисел нормированные пространства. Положим

$$d_{\mathbf{C}}(X,Y) = \inf \left\{ \|T\| \, \|T^{-1}\| \colon T \colon X \to Y \text{ — (комплексно-)линейный} \right.$$
$$\left. \text{обратимый оператор} \right\},$$

$$d_{\mathbf{R}}(X,Y) = \inf \left\{ \|S\| \, \|S^{-1}\| \colon S \colon X \to Y \text{ — вещественно-линейный} \right.$$
$$\left. \text{обратимый оператор} \right\}.$$

ВОПРОС. Верно ли, что равенство $d_{\mathbf{R}}(X,Y) = 1$ влечет $d_{\mathbf{C}}(X,Y) < K$ (K — абсолютная константа)?

Отрицательный ответ на этот вопрос дал Бургейн [7]. Для точной формулировки его результата понадобятся следующие определения. Пусть

X — комплексное банахово пространство. Через $\overline{X}$ обозначим комплексное банахово пространство, совпадающее с X по запасу элементов, в котором произведение $\lambda \circ x$ скаляра λ на элемент $x \in \overline{X}$ определяется равенством $\lambda \circ x = \bar{\lambda}x$ ($\bar{\lambda}$ — комплексно сопряженное с λ). Операция сложения и норма в $\overline{X}$ наследуются из X (тем самым в вещественном смысле X и $\overline{X}$ изометричны).

Пусть банахово пространство X отождествлено с линейным подмножеством пространства l_2 (или l_2^n). Через X_θ, $0 < \theta < 1$, будем обозначать пространство $[X, l_2]_{\theta,2}$, получающееся путем вещественной интерполяции (определение см., например, в [5]). Положим $X_0 = X$, $X_1 = l_2$ (или l_2^n).

ТЕОРЕМА 5 [7]. *Пусть X^n — комплексное n-мерное случайное пространство с распределением $\nu_{n^4,\varnothing}^{\mathrm{C}}$. Тогда с некоторой абсолютной константой $\delta > 0$ вероятность того, что $d_{\mathrm{C}}(X_\theta^n, \overline{X}_\theta^n) > (\delta n / \log n)^{1-\theta}$ положительна и стремится к 1 с ростом n каково бы ни было $\theta \in [0, 1]$.*

Кроме того, Бургейну удалось также показать, что с большой вероятностью пространство X_θ^n нельзя «хорошо» вложить в пространство $\overline{X}_\theta^n + l_2$, что позволило ему «склеить» из них бесконечномерное банахово пространство Z не изорморфное $\overline{Z}$.

Судя по ссылкам, близкий к теореме 5 результат доказан Шареком [**55**]. Манкевичу [**37**] удалось обнаружить общий факт, из которого, как частные случаи получаются утверждения типа теорем 2–5, правда, с несколько худшими оценками (отличающимися от наилучших логарифмическим множителем).

Формулируя теоремы 1–5, мы не уточняли, с какой скоростью мера соответствующих множеств стремится к 1. На самом деле при доказательстве этих утверждений получаются экспоненциальные оценки для меры дополнительного множества. Мы не останавливаемся на этом подробно, поскольку для геометрических приложений важна лишь положительность меры, дающая теорему существования.

2. Аппроксимация случайными подпространствами. Пусть L — линейное подпространство банахова пространства X, а B — некоторое подмножество X. Отклонением множества B от L называется величина

$$\rho(B, L) = \rho_X(B, L) = \sup_{x \in B} \inf_{y \in L} \|x - y\|.$$

Колмогоровские поперечники множества B определяются равенством

$$d_n(B, X) = \inf_{L_n : \dim L_n \leq n} \rho_X(B, L);$$

инфимум берется по множеству всех подпространств размерности не выше n. Это понятие введено в 1936 г. А. Н. Колмогоровым [**34**]. Вопрос о скорости убывания поперечников естественным образом обобщает классическую задачу теории приближений о степени аппроксимации класса функций фиксированными конечномерными подпространствами (алгебраическими или тригонометрическими полиномами и т.п.). Исторически первой задачей этой области явился вопрос об оценке поперечников соболевских классов W_p^l в L_q. Уже в первых работах на эту тему (С. Б. Стечкин [48], М. З. Соломяк и В. М. Тихомиров [46]) выявилась тесная связь между задачей о поперечниках $d_n(W_p^l, L_q)$ и конечномерным вопросом о равномерной оценке величины $d_n(B_p^N, l_q^N)$. Сравнительно несложно вычислить точные значения поперечников $d_n(B_p^N, l_q^N)$ при $1 \leq q \leq p \leq \infty$ (см., например, [43, 47]). Однако даже сколько-нибудь содержательная оценка величин $d_n(B_p^N, l_q^N)$ при $\max\{2, p\} < q \leq \infty$ оказывается трудной. Лишь в 1974 г., воспользовавшись для аппроксимации «октаэдра» B_1^N специальными инвариантными относительно циклического сдвига подпространствами, Р. С. Исмагилов [26] сумел получить содержательную в широкой области изменения n и N оценку величины $d_n(B_1^N, l_\infty^N)$. При этом им было введено важное понятие — тригонометрический поперечник — до сих пор еще мало изученное (автору известны лишь две публикации на эту тему: Э. С. Белинский [1], И. Маковоз [35]).

Очень плодотворным в этом круге вопросов оказался вероятностный способ рассуждения, впервые примененный для оценки поперечников Б. С. Кашиным [30, 32]. Схема его рассуждений такова: введем на многообразии всех n-мерных подпространств пространства X вероятностную меру. Для оценки поперечника $d_n(B, X)$ будем использовать среднее относительно этой меры значение отклонения $\rho_X(B, L)$. При удачном выборе вероятностной меры, среднее значение, о котором идет речь, будет близко к $d_n(B, X)$, а его вычисление может оказаться значительно проще непосредственной оценки поперечника. При реализации такого подхода для оценки $d_n(B_2^N, l_\infty^N)$ Б. С. Кашин следующим образом ввел меру на $G_{n,N}$. Пусть $A = (a_{ij})$ — случайная $n \times N$-матрица, элементы которой независимо принимают значения ± 1 с вероятностью $1/2$. Нужная мера на многообразии $G_{n,N}$ индуцируется при переходе от матрицы A к n-мерному подпространству $\mathbf{R}^N$ — ее образу. Б. С. Кашину удалось доказать, что при таком выборе меры на $G_{n,N}$ с большой вероятностью выполнено (c — абсолютная константа)

$$\rho_{l_\infty^N}(B_2^N, L) \leq c(\log(N/n) + 1)^{3/2}/\sqrt{n}.$$

Как уже отмечалось во введении, это дает точную в степенной шкале оценку (а при $n \asymp N$ и точный порядок) поперечников $d_n(B_2^N, l_\infty^N)$.

Рассуждения Б. С. Кашина упрощаются, если перейти к унитарно-инвариантной мере $\mu_{n,N}$ на многообразии $G_{n,N}$ всех n-мерных подпространств пространства $\mathbf{R}^N$. При этом удается получить и более точные оценки. Полезно иметь в виду, что мера $\mu_{n,N}$ индуцируется (при переходе от матрицы к ее образу) случайной $n \times N$-матрицей $\Gamma = (g_{ij})$, элементы которой независимые гауссовы случайные величины.

ТЕОРЕМА 6 ([**21, 18**], см. также [**60**]). *Для $L \in G_{n,N}$ с положительной $\mu_{n,N}$-вероятностью (стремящейся к 1 с ростом n) выполняются неравенства ($c < \infty$ — абсолютная константа)*

$$\rho_{l_q^N}(B_2^N, L) \leq c \min\{1, N^{1/q}/\sqrt{n}\} \quad \text{при } q < \infty; \tag{5}$$

$$\rho_{l_\infty^N}(B_2^N, L) \leq c \min\{1, \sqrt{(\log(N/n)+1)/n}\}. \tag{6}$$

Эти оценки дают правильный порядок соответствующих поперечников [**21, 18**]. Несложные рассуждения интерполяционного характера [**22**] позволяют с их помощью завершить исследование вопроса о порядках поперечников $d_n(B_p^N, l_q^N)$ при $1 \leq p < q < \infty$ и $2 \leq p < q = \infty$. Следующий результат показывает, что с точки зрения аппроксимации в l_∞^N случайными подпространствами множества B_p^N при $1 \leq p \leq 2$ оказываются неразличимыми. В то же время, как показывает известный результат Б. С. Кашина [**31**] (см. также Хеллиг [**25**]), при $n = o(N)$ поперечники $d_n(B_1^N, l_\infty^N)$ и $d_n(B_2^N, l_\infty^N)$ имеют разный порядок. Иначе говоря, при $1 \leq p < q < \infty$ или при $2 \leq p < q = \infty$ подпространство из $(G_{n,N}, \mu_{n,N})$ общего положения почти экстремально для поперечника $d_n(B_p^N, l_q^N)$, а при $1 \leq p < 2 < q = \infty$ нет.

ТЕОРЕМА 7 [**23**]. *Для $L \in G_{n,N}$ с положительной $\mu_{n,N}$-вероятностью (стремящейся к 1 с ростом n) выполняются неравенства ($0 < c_1 < c_2 < \infty$ — абсолютные константы)*

$$c_1 \min\{1, \sqrt{(\log(N/n)+1)/n}\} \leq \rho_{l_\infty^N}(B_p^N, L) \leq c_2 \min\{1, \sqrt{(\log(N/n)+1)/n}\}$$

каково бы ни было p, $1 \leq p \leq 2$.

Недавно Б. Карл (личное сообщение) передоказал основной результат [**18**] $d_n(B_2^N, l_\infty^N) \geq c \min\{1, \sqrt{(\log(N/n)+1)/n}\}$ и доказал неравенство

$$d_n(TB_2^N, l_\infty^N) \leq c\|T\| \min\{1, \sqrt{(\log(N/n)+1)/n}\},$$

где T — произвольный оператор, действующий из l_2^N в l_∞^N.

Доказательство неравенств (5), (6) удобно проводить в двойственных терминах. Такой подход использовался Б. С. Митягиным [**40**] при переизложении результата Б. С. Кашина [**32**]. Пусть X, Y — N-мерные банаховы пространства, причем X фиксированным образом вложено в Y, B — единичный шар пространства X. Используя теорему Хана-Банаха, несложно показать (см., например, [**26**]) что для любого подпространства $L \subset Y$

$$\rho_Y(B, L) = \sup\{\|z\|_{X^*}/\|z\|_{Y^*} : z \neq 0, z \in L^\perp\}.$$

Здесь и далее X^*, Y^* — сопряженные к X, Y пространства, а $L^\perp \subset Y^*$ — подпространство функционалов, аннулирующих L. Положим

$$K(\lambda) = \{z \in Y^* : \|z\|_{X^*} \leq \lambda \|z\|_{Y^*}\}. \tag{7}$$

В этих обозначениях утверждение $d_n(B,Y) \leq \lambda$ эквивалентно тому, что найдется подпространство $L^n \subset Y^*$ коразмерности не выше n такое, что $L^n \cap K(\lambda) = 0$. Таким образом, задача об оценке поперечников сводится к вопросу о том, насколько малой может быть коразмерность подпространства, не пересекающего заданный конус. Ответить на этот вопрос помогает следующее простое утверждение.

ПРЕДЛОЖЕНИЕ 2. *Пусть* $f_1, \ldots, f_m \in B_2^N$, $m \leq 2^n$, *а вектор* $z \in \mathbf{R}^N$ *таков, что* $\|z\|_2 > 12$. *Если* $V = \operatorname{conv}\{f_1, \ldots, f_m\}$ — *выпуклый многогранник с вершинами* $f_1, \ldots, f_m$, *а*

$$K = \{x \in \mathbf{R}^N : x = \alpha(z+v),\ \alpha > 0,\ v \in V\}$$

— *коническая оболочка множества* $z + V$, *то выполняется оценка*

$$\mu_{n,N}\{L \in G_{n,N} : L^\perp \cap K \neq \varnothing\} \leq 2^{-n}.$$

ДОКАЗАТЕЛЬСТВО. Напомним сначала, что вероятностная мера $\mu_{n,N}$ индуцируется при отображении $\Gamma \to \operatorname{Im}\Gamma$ гауссовой мерой $\gamma_{n,N}$ на пространстве $n \times N$-матриц, отождествленном стандартным образом с $\mathbf{R}^{nN}$. Если подпространство L определено, как $\operatorname{Im}\Gamma$, то условие $L^\perp \cap K \neq \varnothing$ эквивалентно тому, что найдется $v \in V$ такое, что (Γ^* — транспонированная матрица) $v + z \in \ker\Gamma^*$. Иначе говоря, $\Gamma^* v = -\Gamma^* z$. Тем более $\|\Gamma^* v\|_2 = \|\Gamma^* z\|_2$ и, по определению множества V,

$$\max_{1 \leq i \leq m} \|\Gamma^* f_i\|_2 \geq \|\Gamma^* z\|_2.$$

Так как сумма независимых гауссовых случайных величин снова гауссова случайная величина, векторы $\Gamma^* f_i / \|f_i\|_2$ и $\Gamma^* z / \|z\|_2$ имеют распределение γ_n. Для наших целей достаточно следующих элементарных оценок распределения нормы гауссова вектора $g \in (\mathbf{R}^n, \gamma_n)$.

$$\gamma_n\{g : \|g\|_2^2 \geq 4n\} \leq 16^{-n};$$

$$\gamma_n\{g : \|g\|_2^2 \leq n/36\} \leq 2(2\sqrt{2})^{-n}.$$

Остается заметить, что, если $\max_{1 \leq i \leq m} \|\Gamma^* f_i\|_2 \geq \|\Gamma^* z\|_2$, то либо

$$\max_{1 \leq i \leq m} \|\Gamma^* f_i\|_2 / \|f_i\|_2 \geq 2\sqrt{n},$$

либо

$$\|\Gamma^* z\|_2 / \|z\|_2 < \sqrt{n}/6.$$

СЛЕДСТВИЕ. *Пусть для* $i = 1, \ldots, 2^n - 1$ *заданы многогранники* $V_i \subset B_2^N$ *и элементы* $z_i \in \mathbf{R}^N$ *такие, как в предложении 2.* $K_i = \operatorname{con}\{z_i + V_i\}$ — *коническая оболочка множества* $z_i + V_i$. *Тогда существует подпространство* $L^n \subset \mathbf{R}^n$ *коразмерности* n *такое, что* $L^n \cap (\bigcup_{i=1}^{2^n-1} K_i) = \varnothing$.

Тем самым для оценки поперечников достаточно покрыть конус $K(\lambda)$ из (7) «небольшим» числом конусов специального вида. Именно на этом пути была доказана теорема 6.

В формулировке предложения 2 предполагалось, что множество V — многогранник с «небольшим» числом вершин. Геометрически это очень наглядно, однако, при доказательстве использовалось лишь следующее свойство таких множеств (как и раньше Γ — гауссова $n \times N$-матрица):

$$\gamma_{nN}\{\Gamma\colon \sup\{\|\Gamma^* x\|_2 \colon x \in V\} > 4\sqrt{n}\} \le 8^{-n}.$$

Изопериметрическое свойство гауссовой меры [51] или [6] показывает, что такая оценка имеет место, если справедливо неравенство

$$h_1(V) \overset{\text{def}}{=} \int_{\mathbf{R}^n} \sup\{|\langle \xi, x \rangle| \colon x \in V\} d\gamma_n(\xi) \le c\sqrt{n}$$

($c > 0$ — некоторая абсолютная константа).

Возникает вопрос, какие внутренние характеристики конуса K позволяют построить его покрытие $K = \bigcup_{i=1}^{2^n-1} K_i$ такое, что $K_i = \mathrm{con}\{V_i + z_i\}$, причем $\|z_i\|_2 > 12$, $V_i \subset B_2^N$ и $h_1(V_i) < c\sqrt{n}$? Ответ на него удалось получить Пажору и Томчак-Ягерманн, что привело их к следующему изящному результату.

ТЕОРЕМА 8 [41]. *Пусть K — конус в $\mathbf{R}^N$. Существует абсолютная константа $c < \infty$ такая, что, как только при некотором n $h_1(K \cap B_2^N) < c\sqrt{n}$ справедлива оценка*

$$\mu_{n,N}\{L \in G_{n,N} \colon L^\perp \cap K \ne \varnothing\} \le 2^{-n}.$$

Доказательство теоремы существенно опирается на энтропийные оценки для гауссовых случайных процессов, полученные В. Н. Судаковым [50].

Следующее простое утверждение, вытекающее из изопериметрического свойства гауссовой меры показывает, что, с точностью до константы, условие $h_1(K \cap B_2^N) \le c\sqrt{n}$ является необходимым для реализации изложенной схемы.

ПРЕДЛОЖЕНИЕ 3. *Пусть $A \subset \bigcup_{i \in I} A_i \subset B_2^N$. Тогда с некоторой абсолютной константой c справедлива оценка*

$$h_1(A) \le \max_{i \in I} h_1(A_i) + c\sqrt{\log \mathrm{Card}\, I}.$$

Представляется интересным следующий вопрос.

Верно ли, что если конус $K \subset \mathbf{R}^N$ таков, что $\mu_{n,N}$-мера подпространств $L \in G_{n,N}$ таких, что $K \cap L^\perp \ne \varnothing$, достаточно мала (например, меньше $1/2$ или $1/2^n$), то $h_1(K \cap B_2^N) \le c\sqrt{n}$ (c — абсолютная константа).

В некотором очень слабом смысле это действительно так, как показывает следующее элементарное наблюдение автора и Б. С. Цирельсона. Пусть конус K и подпространство $L \in G_{n,N}$ таковы, что $K \cap L^\perp = \varnothing$. Тогда для любого $\varepsilon > 0$ найдется линейное преобразование T такое, что $TK \cap B_2^N$ лежит в ε-окрестности некоторого n-мерного подпространства. Следовательно, $h_1(TK \cap B_2^N) \le c(\sqrt{n} + \varepsilon\sqrt{N})$ с некоторой абсолютной константой c. Другими словами, требуемый результат получается, если вместо обычного скалярного произведения в $\mathbf{R}^N$ рассматривать форму $\langle Tx, Ty \rangle$ с надлежащим T.

ЛИТЕРАТУРА

1. Э. С. Белинский, *Приближение периодических функций многих переменных «плавающей» системой экспонент и тригонометрические поперечники*, Докл. АН СССР **284**, № 6 (1985), 1294–1297.

2. G. Bennett, L. E. Dor, V. Goodman, W. B. Johnson, and C. M. Newman, *On uncomplemented subspaces of L_p*, $1 < p < 2$, Israel J. Math. **26** (1977), 178–187.

3. G. Bennett, V. Goodman, and C. M. Newman, *Norms of random matrices*, Pacific J. Math. **59** (1975), 359–365.

4. Y. Benyaminy and Y. Gordon, *Random factorisation of operators between Banach spaces*, J. Analyse Math. **39** (1981), 45–74.

5. И. Берг и И. Лефстрем, *Интерполяционные пространства*, «Мир», Москва, 1976.

6. C. Borell, *The Brunn-Minkowski inequality in Gauss space*, Invent. Math. **30** (1975), 207–216.

7. J. Bourgain, *A complex Banach space such that X and $\overline{X}$ are not isomorphic*, Preprint.

8. J. Bourgain and V. D. Milman, *On Mahler's conjecture on the volume of a convex symmetric body and its polar*, Preprint, IHES (85).

9. ____, *Distances between normed spaces, their subspaces and quotient spaces*, Preprint, IHES (84).

10. S. Chevet, *Séries de variables aléatoires Gaussiennes à valuers dans $E \hat{\otimes}_\varepsilon F$*. Application aux produits d'espaces de Wiener abstraits, Seminaire Maurey-Schwartz (1977/78), exposé XIX, 1978.

11. A. M. Davie, *The approximation problem for Banach spaces*, École Polytech., Palaiseau, Bull. London Math. Soc. **5** (1973), 261–266.

12. W. J. Davis, V. D. Milman, and N. Tomczak-Jaegermann, *The distance between certain n-dimensional spaces*, Israel J. Math. **39** (1981), 1–15.

13. P. Enflo, *A counterexample to the approximation problem in Banach spaces*, Acta Math. **130** (1973), 309–317.

14. T. Figiel, *A short proof of Dvoretzky's theorem on almost spherical sections*, Compositio Math. **33** (1976), 297–301.

15. T. Figiel and W. B. Johnson, *Large subspaces of l_∞^n and estimates of the Gordon-Lewis constant*, Israel J. Math **37** (1980), 92–112.

16. T. Figiel, S. Kwapien, and A. Pelczynski, *Sharp estimates for the constants of local unconditional structure of Minkowski spaces*, Bull. Acad. Polon. Sci. **25** (1977), 1221–1226.

17. T. Figiel, J. Lindenstrauss, and V. D. Milman, *The dimension of almost spherical sections of convex bodies*, Acta Math. **139** (1977), 53–94.

18. А. Ю. Гарнаев и Е. Д. Глускин, *О поперечниках евклидова шара*, Докл. АН СССР **277**, № 5 (1984), 1048–1052.

19. Е. Д. Глускин, *Диаметр компакта Минковского примерно равен n*, Функц. анализ и его прилож. **15**, № 1 (1981), 72–73.

20. ____, *Конечномерные аналоги пространств без базиса*, Докл. АН СССР **261**, № 5 (1981), 1046–1050.

21. ____, *Нормы случайных матриц и поперечники конечномерных множеств*, Мат. сборник **120 (162)**, № 2 (1983), 180–189.

22. ____, *О некоторых конечномерных задачах теории поперечников*, Вестник ЛГУ, № 13 (1981), 5–10.

23. ____, *Октаэдр плохо приближается случайными подпространствами*, Функц. анализ и его прилож. **20**, № 1 (1986), 14–20.

24. Y. Gordon, *Asymmetry and projection constants in Banach spaces*, Israel J. Math. **14** (1973), 50–62.

25. K. Hölig, *Approximationzahlen von Sobolev-Einbettungen*, Math. Ann. **242** (1979), 273–281.

26. Р. С. Исмагилов, *Поперечники множеств в линейных нормированных пространствах и приближение функций тригонометрическими многочленами*, Успехи матем. наук **29**, № 3 (1974), 161–178.

27. F. John, *Extremum problems with inequalities as subsidiary conditions*, Courant Anniversary Volume, Interscience, New York, 1948, 187–204.

28. W. B. Johnson and G. Schechtman, *Embedding l_p^m into l_1^n*, Acta Math. **149** (1983), 71–85.

29. М. И. Кадец, *О линейной размерности пространств L_p и l_q*, Успехи матем. наук **13** № 6, (1958), 95–98.

30. Б. С. Кашин, *О колмогоровских поперечниках октаэдров*, Докл. АН СССР **214**, № 5 (1974), 1024–1026.

31. ____, *О поперечниках октаэдров*, Успехи матем. наук **30**, № 4 (1975), 251–252.

32. ____, *Поперечники некоторых конечномерных множеств и классов гладких функций*, Изв. АН СССР, Серия матем. **41** (1977), 334–351.

33. S. V. Kisliakov, *What is needed for a 0-absolutely summing operator to be nuclear?*, Lecture Notes in Math., vol. 864, Springer-Verlag, Berlin and New York, 1981, pp. 336–364.

34. A. Kolmogoroff, *Über die beste Annäherung von Funktionen einer gegebener Funktionen-Klassen*, Ann. of Math. **37** (1936), 107–111.

35. Y. Makovoz, *On trigonometric n-widths and their generalization*, J. Approx. Theory **41** (1984), 361–366.

36. P. Mankiewicz, *Finite dimensional Banach spaces with symmetry constant of order $\sqrt{n}$*, Studia Math. **79** (1984), 193–200.

37. ____, *Subspace mixing properties of operators in   with applications to pathological properties of Gluskin spaces*, Inst. Math. PAN, Preprint N 363, Warszawa 1986.

38. В. Д. Мильман, *Новое доказательство теоремы А. Дворецкого о сечениях выпуклых тел*, Функц. анализ и его прилож. **5**, № 4 (1971), 28–37.

39. Б. С. Митягин, *Функциональные банаховы пространства типа Энфло*, Труды шестой зимней школы по математическому программированию и смежным вопросам (Дрогобыч, январь 1973 г.), Москва, 1975, 7–27.

40. ____, *Случайные матрицы и подпространства*, в кн.: Геометрия линейных пространств и теория операторов, Ярославль, 1977, 175–202.

41. A. Pajor and N. Tomczak-Jaegermann, *Subspaces of small codimension of finite-dimensional Banach spaces*, Preprint.

42. A. Pelczynski, *Any separable Banach space with the bounded approximation property is a complemented subspace of a Banach space with a basis*, Studia Math. **40** (1971), 239–242.

43. A. Pietsch, *s-numbers of operators in Banach spaces*, Studia Math. **51** (1974), 201–223.

44. G. Pisier, *Counterexamples to a conjecture of Grothendieck*, Acta Math. **151** (1983), 181–208.

45. ____, *On the dimension of the l_p^n-subspaces of Banach spaces for $1 \le p < 2$*, Trans. Amer. Math. Soc. **276** (1983), 201–211.

46. М. З. Соломяк и В. М. Тихомиров, *О геометрических характеристиках вложения классов W_p^α в C*, Изв. вузов, Математика, № 10 (65) (1967), 76–81.

47. М. И. Стесин, *Александровские поперечники конечномерных множеств и классов гладких функций*, Докл. АН СССР **220**, № 6 (1975), 1278–1281.

48. С. Б. Стечкин, *О наилучшем приближении заданных классов функций любыми полиномами*, Успехи матем. наук **9**, № 1 (1954), 133–134.

49. W. Stromquist, *The maximum distance between two-dimensional spaces*, Math. Scand. **48** (1981), 205–225.

50. В. Н. Судаков, *Гауссовские случайные процессы и меры телесных углов в гильбертовом пространстве*, Докл. АН СССР **197**, № 1 (1971), 43–45.

51. В. Н. Судаков и Б. С. Цурельсон, *Экстремальные свойства полупространств для сферически инвариантных мер*, Записки научных семинаров ЛОМИ, **41**, «Наука», Ленинград, 1974, 14–24.

52. A. Szankowski, *On Dvoretzky's theorem on almost spherical sections of convex bodies*, Israel J. Math. **17** (1974), 325–338.

53. S. J. Szarek, *The finite dimensional basis problem with appendix on nets of Grassmann manifolds*, Acta Math. **151** (1983), 153–179.

54. _____, *A Banach space without a basis which has the bounded approximation property*, Preprint.

55. _____, *On the existence and uniqueness of complex structure and spaces with few operators*, Trans. Amer. Math. Soc. (to appear).

56. _____, *On Kashin's almost Euclidean orthogonal decomposition of l_1^n*, Bull. Acad. Polon. Sci **26** (1978), 691–694.

57. S. J. Szarek and N. Tomczak-Jaegermann, *On nearly Euclidean decomposition for some classes of Banach spaces*, Compositio Math. **40** (1980), 367–387.

58. N. Tomczak-Jaegermann, *The Banach-Mazur distance between trace classes c_p^n*, Proc. Amer. Math. Soc. **72** (1978), 305–308.

59. _____, *The Banach-Mazur distance between symmetric spaces*, Israel J. Math. **46** (1983), 40–66.

60. _____, *On n-widths of finite-dimensional spaces*, Preprint.

Ленинградский финансово-экономический институт им. Н. А. Вознесенского, Ленинград 191023, СССР

Subfactors of Type II$_1$ Factors
and Related Topics

V. F. R. JONES

1. Introduction. In this article we use the definitions of von Neumann algebra theory appearing in Haagerup's paper [**H**] in these proceedings.

So let M be a type II$_1$ factor with (unique) normalized trace tr: $M \to \mathbf{C}$ $(\mathrm{tr}(1) = 1, \mathrm{tr}(ab) = \mathrm{tr}(ba))$. Whenever M acts as a von Neumann algebra on a Hilbert space $\mathcal{H}$, there is a uniquely defined number $\dim_M(\mathcal{H}) \in [0, \infty]$ satisfying $\dim_M(\bigoplus_{i=1}^{\infty} \mathcal{H}_i) = \sum_{i=1}^{\infty} \dim_M(\mathcal{H}_i)$, $\dim_M(\mathcal{H}) = \dim_M(\mathcal{H}')$ iff $\mathcal{H}$ and $\mathcal{H}'$ are isomorphic M-modules, and $\dim_M(L^2(M, \mathrm{tr})) = 1$ where $L^2(M, \mathrm{tr})$ is the Hilbert space obtained from M by completion with respect to the inner product $\langle a, b \rangle = \mathrm{tr}(b^*a)$. This number $\dim_M(\mathcal{H})$ is the coupling constant of Murray and von Neumann [**MvN**] and was originally defined as $\mathrm{tr}_M(P_{[M'\xi]})/\mathrm{tr}_{M'}(P_{[M\xi]})$ where $\xi \neq 0$ is an arbitrary vector in $\mathcal{H}$ and $P_{[M\xi]}$ denotes the orthogonal projection onto the closure of the subspace $M\xi \subseteq \mathcal{H}$. (The other symbols have their obvious meanings, M' being the commutant of M.) This definition presupposes that M' is also a II$_1$ factor. If this is not so, one puts $\dim_M(\mathcal{H}) = \infty$.

It is important to note that, for any II$_1$ factor M, $\{\dim_M(\mathcal{H}) \mid \mathcal{H}$ a Hilbert space over $M\} = [0, \infty]$. This is a formulation of the "continuous dimensionality" that so fascinated Murray and von Neumann. It is also important in Connes's noncommutative integration theory [**C1**] where real-valued Betti numbers are associated to foliated compact manifolds with invariant transverse measure.

The elementary example of this paragraph will serve as motivation for the definition that follows. If Γ is a discrete group, all of whose (nonidentity) conjugacy classes are infinite (an i.c.c. group), and Γ_0 is an i.c.c. subgroup of Γ, then the von Neumann algebras $U\Gamma$ and $U\Gamma_0$ on $l^2(\Gamma)$, generated by left translations by the appropriate group elements, are both II$_1$ factors, and the coset decomposition of Γ over Γ_0 shows immediately that $\dim_{U\Gamma_0}(l^2(\Gamma)) = [\Gamma : \Gamma_0]$. Note also that $l^2(\Gamma)$ is the same as $L^2(U\Gamma)$ so we have

$$\dim_{U\Gamma_0}(L^2(U\Gamma)) = [\Gamma : \Gamma_0]. \tag{1}$$

Note that the left-hand side of equation (1) only involves II_1 factors. So we are led to make the following definition.

DEFINITION 2. *If $N \subseteq M$ are II_1 factors (with the same identity), define $[M:N]$, the index of N in M, by*

$$[M:N] = \dim_N(L^2(M,\mathrm{tr})).$$

With this definition one may interpret the following result of M. Goldman [G] as being an analogue of the fact that a subgroup of index 2 of a group is normal.

THEOREM 3 [G]. *If $N \subseteq M$ are as in Definition 2 and $[M:N] = 2$, then there is a $u \in M$ with $uNu^* = N$, $u^2 = 1$, and $M = N \oplus Nu$.*

Both the examples from $\Gamma_0 < \Gamma$ and Goldman's theorem give the impression that the index $[M:N]$ is a discrete object, but its definition suggests an arbitrary real number between 1 and ∞. The next result shows that neither impression is correct.

THEOREM 4 [J1]. (a) *If $[M:N] < 4$, there is an $n \in \mathbf{Z}$, $n \geq 3$, with $[M:N] = 4\cos^2 \pi/n$.*

(b) *If $r = 4\cos^2 \pi/n$, n as above, or $r \in \mathbf{R}$, $r \geq 4$, there is a pair $N \subseteq M$ of II_1 factors with $[M:N] = r$. One may suppose N and M hyperfinite.*

The appearance of these numbers $4\cos^2 \pi/n$ was not at all expected a priori. Note that $1 = 4\cos^2 \pi/3$, $2 = 4\cos^2 \pi/4$, and $3 = 4\cos^2 \pi/6$. The first "new" index value is $4\cos^2 \pi/5 \simeq 2.6180339$, the square of the golden ratio.

We would like to add that, as first pointed out by Connes, the continuous variation of $[M:N]$ may be illusory since the examples that realize the numbers between 4 and ∞ are "reducible" in the sense that $N' \cap M$ contains elements other than scalars. If one imposes the irreducibility condition $N' \cap M = \mathbf{C}$, then the smallest known value of $[M:N]$ (greater than 4) is $3 + \sqrt{3}$. The current feeling is that there should be a gap between 4 and the next irreducible index value.

2. Proof of Theorem 4(a). We shall outline a proof of Theorem 4 which makes a connection with Coxeter-Dynkin diagrams. The proof relies on an analysis of inclusions of finite-dimensional von Neumann algebras.

A finite-dimensional von Neumann algebra is semisimple, so is a direct sum of full matrix algebras over $\mathbf{C}$. We shall represent such an algebra by a finite set of vertices corresponding to the simple direct summands, together with integers giving the size of the matrix algebras. For instance, $\mathbf{C} \oplus M_3(\mathbf{C}) \oplus M_2(\mathbf{C})$ would be represented by $\overset{1}{\cdot} \ \overset{3}{\cdot} \ \overset{2}{\cdot}$. With this convention a pair $A \subseteq B$ of such algebras can be represented by a graph (Bratteli diagram) where the number of edges connecting a vertex of the smaller algebra to a vertex of the larger one has the obvious "multiplicity" meaning. For instance, the diagonal inclusion of $M_2(\mathbf{C}) \oplus M_3(\mathbf{C})$

in $M_5(\mathbf{C})$ would be represented by the diagram

and the inclusion

$$a \longmapsto \begin{pmatrix} a & 0 & 0 \\ 0 & a & 0 \\ 0 & 0 & a \end{pmatrix}$$

of $M_2(\mathbf{C})$ in $M_6(\mathbf{C})$ would be represented by the diagram:

In general the matrix Λ_A^B is defined as the matrix whose rows are indexed by the vertices of A, whose columns are indexed by those of B, and whose entries are the multiplicities.

The other ingredient of the proof of Theorem 4 is the iteration of a certain basic construction which is made as follows: given $N \subseteq M$ finite von Neumann algebras with the same identity, and a faithful normal trace tr on M, one lets N and M act on $L^2(M, \text{tr})$ as before and one considers the von Neumann algebra $JN'J$ where J is the extension to $L^2(M, \text{tr})$ of the $*$ operation. Since $M = JM'J$ [D], one has $N \subseteq M \subseteq JN'J$. If there are several algebras present we will use J_M to denote the J on $L^2(M, \text{tr})$.

In the case that N and M are finite factors, one has

PROPOSITION 5. (i) $JN'J$ is a II$_1$ factor $\Leftrightarrow [M:N] < \infty$.

(ii) If (i) is satisfied, then (a) $[JN'J:M] = [M:N]$; (b) $N' \cap M$ is finite-dimensional; (c) $[M:N] \geq \dim(N' \cap M)$.

In the case that M is finite-dimensional, the following result holds.

PROPOSITION 6. $\Lambda_M^{JN'J} = (\Lambda_N^M)^t$ (independent of the trace). (Here we have identified the center of N with that of N' and so $JN'J$, which allows us to make the correspondence between rows of $\Lambda_M^{JN'J}$ and columns of Λ_N^M and vice versa.)

To illustrate Proposition 6, suppose $N \subseteq M$ were given by

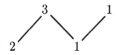

Then $N \subseteq M \subseteq JN'J$ would have the diagram

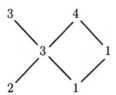

Given a subfactor $N \subseteq M$ of finite index, the next step in the proof is to iterate the above construction to obtain a tower M_i of II_1 factors with $M_0 = N$, $M_1 = M$, and $M_{i+1} = J_{M_i} M'_{i-1} J_{M_i}$, where J_{M_i} is the involution on $L^2(M_i, \mathrm{tr})$. By finite dimensionality one obtains a tower $\partial M_i = N' \cap M_i$ of finite-dimensional von Neumann algebras with corresponding matrix $\Lambda_i = \Lambda_{\partial M_i}^{\partial M_{i+1}}$. The proof of Theorem 4(a) will follow easily from the following result.

PROPOSITION 7. (i) *There is an isomorphism of $J_{\partial M_i}(\partial M_{i-1})' J_{\partial M_i}$ onto a two-sided ideal of ∂M_{i+1} which gives a containment of $(\Lambda_i)^t$ as a submatrix of Λ_{i+1}.*

(ii) $\dim \partial M_i = \mathrm{trace}((\prod_{k=1}^{i} \Lambda_k)(\prod_{k=1}^{i} \Lambda_k)^t)$.

(iii) *If $\Lambda_k = (\Lambda_{k-1})^t$, then $\Lambda_p = (\Lambda_{p-1})^t$ for all $p \geq k$ and $[M:N] = \|\Lambda_k\|^2$.*

It follows from Proposition 7(i) that $\|\Lambda_i\|$ is a nondecreasing function of i, and then from 7(ii) that $\dim M_k$ grows asymptotically at least as fast as $\|\Lambda_i\|^{2k}$ for any i. But $[M_k: N] = [M:N]^k$, so by 5(c) $\|\Lambda_i\|^2 \leq [M:N]$ for all i. If $[M:N] < 4$, the Λ_i are then a nondecreasing sequence of 0–1 matrices of norm < 2. By [**Bo, GHJ**] the possible values of these norms are precisely the set $\{2\cos\pi/n \mid n = 3, 4, \ldots\}$. So by strict monotonicity of the norm there must be a k for which $\Lambda_k = (\Lambda_{k-1})^t$. By 7(iii) we are through.

In fact, one obtains more information from this proof than just the values $4\cos^2\pi/n$. It follows from [**GHJ**] and the connectedness of the Bratteli diagram that if k is such that $\Lambda_k = (\Lambda_{k-1})^t$, then Λ_k must be the adjacency matrix for a bipartite structure on one of the Coxeter-Dynkin diagrams A_n, $n \geq 3$, D_n, $n \geq 4$, E_6, E_7, or E_8, and then $[M:N] = \|\Lambda_k\|^2 = 4\cos^2\pi/r$ where r is the Coxeter number of the diagram. For instance, one might have the inclusion $(\partial M)_2 \subseteq (\partial M_3)$ given by the Bratteli diagram

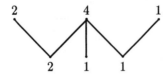

which corresponds to the Coxeter-Dynkin diagram E_6, $[M:N] = 4\cos^2\pi/12$.

The question of which diagrams arise from subfactors is interesting. Ocneanu has added to ∂M a "comultiplication" coming from the J_i's which completely axiomatizes their structure even in index ≥ 4. He claims that D_5 is impossible, whereas A_n and D_4 are relatively easy to construct. Bion-Nadal has shown that the construction of the next section realizes E_6.

Proof of Theorem 4(b). Let us first dispose of the case $r \geq 4$. The hyperfinite II₁ factor R has fundamental group $= \mathbf{R}$ [**D**], so choose a projection $p \in R$ with $\text{tr}(p)^{-1} + \text{tr}(1-p)^{-1} = r$ and an isomorphism $\theta: pRp \to (1-p)R(1-p)$. Let $M = R$ and $N = \{x + \theta(x) \mid x \in pMp\}$. One checks $[M: N] = r$. Notice though that this proof relies on the fundamental group. Pimsner and Popa [**PP**] have shown that for II₁ factors with Connes' property T (and hence countable fundamental group—see [**C2**]), the set of index values for subfactors is countable!

We now suppose $r = 4\cos^2 \pi/n$, $n = 4, 5, 6, \ldots$. Let $A \subseteq B$ be an inclusion of finite-dimensional von Neumann algebras whose Bratteli diagram is a Coxeter-Dynkin diagram with Coxeter number n (to obtain $4\cos^2 \pi/n$). Let $q = e^{2\pi i/n}$.

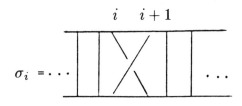

FIGURE 13

There is a unique trace tr on B which admits an extension to $J_B A' J_B$ satisfying

$$\text{tr}(xg_B) = z\,\text{tr}(x) \quad \text{for } x \in B, \tag{8}$$

where $g_B: B \to B$ is q on A and -1 on the orthogonal complement of A, and $z = (q+1)^{-1}$.

Iterating this process as before, one obtains a tower B_i of C^*-algebras together with elements $g_i = g_{B_i}$ which define an endomorphism $\Phi: \bigcup_i B_i \to \bigcup_i B_i$ by $\Phi(x) = \lim_{k\to\infty}(g_1 g_2 \cdots g_k)x(g_1 g_2 \cdots g_k)^*$. Then Φ preserves the trace on $\bigcup_i B_i$, so applying the GNS construction one obtains a II₁ factor M from $\bigcup_i B_i$ and $\Phi: M \to M$. One may show that $[M: \Phi(M)] = 4\cos^2 \pi/n$.

Hecke algebras and braids. In the proof of Theorem 4(b) we used a sequence of elements g_i in the tower construction. It is easy to see that they satisfy the relations

$$g_i^2 = (q-1)g_i + q, \tag{9}$$

$$g_i g_{i+1} g_i = g_{i+1} g_i g_{i+1}, \tag{10}$$

$$g_i g_j = g_j g_i \quad \text{for } |i-j| \geq 2. \tag{11}$$

If q is a prime power, relations (9), (10), and (11) are known [**Bo**] to present the commutant of $G = \mathrm{GL}_n(\mathbf{F}_q)$ acting on the complex-valued functions on G/B, B being the subgroup of upper triangular matrices. This is called the Hecke algebra $H(q, n+1)$ of type A_n; the name also applies to the algebra presented by (9), (10), (11) for any value of q. Thus the Hecke algebra is represented (not faithfully) in the tower. When $q = 1$, relations (9), (10), and (11) present the group algebra of the symmetric group, and one may deduce much of the structure of the Hecke algebra from that of the symmetric group.

Relation (8) suggests that there might be traces on $H(q, n+1)$ defined by $\mathrm{tr}(1) = 1$ and

$$\mathrm{tr}(xg_n) = z\,\mathrm{tr}(x) \quad \text{for } x \in H(q, n) \tag{12}$$

for arbitrary values of z. This was proved by Ocneanu (see [**HKW**, **J2**, **W**]) who also determined the values of (q, z) for which the Hecke algebra admits a von Neumann algebra structure for which $\mathrm{tr}(a^*a) \geq 0$. Wenzl calculated the indices of the corresponding subfactors defined using $\lim_{k \to \infty} g_1 g_2 \cdots g_k$ as in the proof of Theorem 4(b). See [**W**].

Artin showed that relations (10) and (11) present the braid group B_n on n strings where the $n - 1$ g_i's correspond to the $n - 1$ σ_i's as in Figure 13.

$$\alpha \in B_3$$

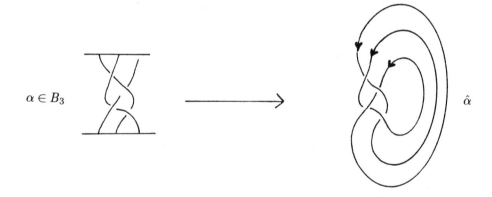

$$\hat{\alpha}$$

FIGURE 14

Thus for $q \neq 0$ there is a representation π of B_n in $H(q, n)$ defined by $\pi(\sigma_i) = g_i$. At present there is no geometric interpretation of this representation though it does contain (as a direct summand) the Burau representation which can be deduced from the action of the braid group on the homology of an infinite cyclic cover of the disc minus n points. It is not known whether or not π is faithful for $n > 3$. The special values $q = e^{\pm 2\pi i/n}$ correspond to the values for which π may be unitarized (though one must take a quotient of the Hecke algebra—see [**W**]).

The definition of Φ in the proof of Theorem 4(b) was suggested by the well-known braid group relation $(\sigma_1 \cdots \sigma_n)\sigma_i(\sigma_1 \cdots \sigma_n)^{-1} = \sigma_{i+1}$ for $i < n$.

Braids and links. A braid $\alpha \in B_n$ may be closed to give the oriented link $\hat{\alpha}$ as in Figure 14.

Any tame oriented link in S^3 may be obtained in this way (Alexander) and the equivalence relation on braids defined by isotopy of their closures was described algebraically by Markov. (For a general reference see [**Bi**].) It is generated by types I and II Markov moves which are the following:

$$\text{type I: } \alpha \in B_n \Leftrightarrow \beta\alpha\beta^{-1} \in B_n, \qquad \text{type II: } \alpha \in B_n \Leftrightarrow \alpha\sigma_n^{\pm 1} \in B_{n+1}.$$

One may consider the function of q and z on the disjoint union of the braid groups defined by $\alpha \to \text{tr}(\pi(\alpha))$ where tr is defined by relation (12). This function is invariant under type I Markov moves, and because of the similarity of (12) and type II Markov moves, it may be renormalized to give a link invariant. It is convenient to change variables by putting $\lambda = (1 - q + z)/qz$. One then has the result that

$$X_\alpha(q, \lambda) = \left(\frac{\lambda q - 1}{\sqrt{\lambda}(1 - q)} \right)^{n-1} (\sqrt{\lambda})^e \, \text{tr}(\pi(\alpha))$$

depends only on $\hat{\alpha}$ where the image of $\alpha \in B_n$ in $\mathbf{Z}$ under abelianization is e.

Relation (9) translates into the fact that if L_+, L_-, and L_0 are three links with projections differing in only one crossing where they are as in Figure 15, then

$$\frac{1}{\sqrt{\lambda}\sqrt{q}}X_{L_+} - \sqrt{\lambda}\sqrt{q}X_{L_-} = \left(\sqrt{q} - \frac{1}{\sqrt{q}} \right) X_{L_0}. \tag{16}$$

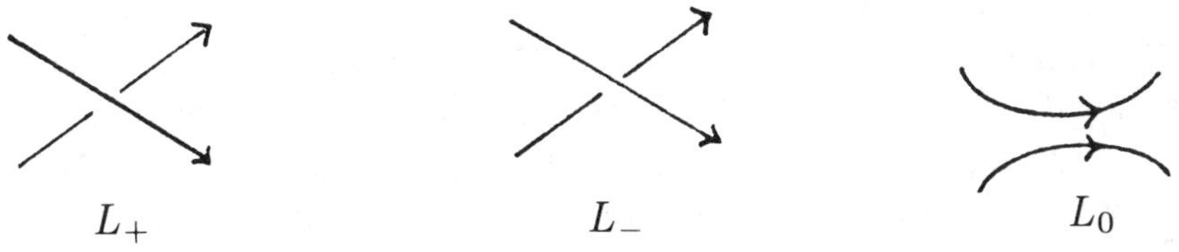

$$L_+ \qquad\qquad L_- \qquad\qquad L_0$$

FIGURE 15

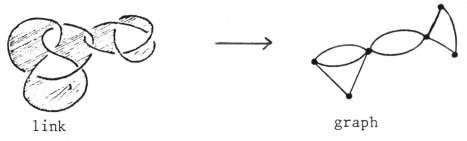

link   graph

FIGURE 17

The existence of such a link invariant was proved in [**F**+], Ocneanu using the approach outlined above. Note that if L is the unlink with c components,

$$X_L = \left(\frac{\lambda q - 1}{\sqrt{\lambda}(1 - q)} \right)^{c-1}.$$

The invariant X_L has been found to be quite powerful. It contains the Alexander polynomial via the specialization $\Delta_L(t) = X_L(t, 1/t)$, but is also very sensitive to mirror image asymmetry. There is no known nontrivial knot with the same X as the unknot. It is easy to see that X can be made into a two variable Laurent polynomial P_L via the popular substitution $m = i(\sqrt{q} - 1/\sqrt{q})$, $l = i/\sqrt{\lambda}\sqrt{q}$. The invariant is multiplicative under connected sums and is unaltered if all orientations of a link are reversed. But it is very sensitive to reversal of the orientation of a single component.

There is another specialization of X_L which is proving to be of particular interest. It is $V_L(t) = X_L(t, t)$ and has the remarkable property that it only changes by a power of t if the orientation of any component of a link is changed. $V_L(t)$ comes from precisely the Hecke algebra quotient given by the proof of Theorem 4(b), and was actually noticed before the discovery of X_L (see [**J3**]).

Kauffman has given an explicit formula for $V_L(t)$ (a "states model") which can be calculated from an arbitrary link projection. In the case of an alternating link, this exhibits V_L as a specialization of the Tutte polynomial of the graph associated to the checkerboard shading of the link projection (Figure 17).

These ideas enabled Kauffman, Murasugi, and Thistlethwaite to develop powerful methods for handling alternating projections of a link—in particular, solving some century-old problems of Tait.

In another line of development, Figure 15 was extended to include the "L_∞" case:$)($. This is not orientable and Brandt, Lickorish, Millett, and Ho defined a polynomial $Q_L(x)$ of unoriented links by $Q_{L_+} + Q_{L_-} = x(Q_{L_0} + Q_{L_\infty})$. Kauffman improved on this by first defining an invariant R of regular isotropy (the move $\mathcal{Q} \Leftrightarrow$ ___ is not allowed) by the same formula as for Q and $R(\mathcal{Q}) = aR(-)$, $R(\mathcal{Q}) = a^{-1}R(-)$. If one then gives the link diagram an orientation, one defines $F_L(a, x) = a^{-w(L)}R(a, x)$ which is a link invariant (where $w(L)$ is the sum of the signs of the crossings).

REFERENCES

[**Bi**] J. Birman, *Braids, links and mapping class groups*, Ann. of Math. Studies, No. 82, Princeton Univ. Press, Princeton, N. J., 1974.

[**Bo**] N. Bourbaki, *Groupes et algèbres de Lie*. IV, V, VI, Hermann, Paris, 1968.

[**C1**] A. Connes, *Sur la théorie non-commutative de l'intégration*, Lecture Notes in Math., vol. 725, Springer-Verlag, Berlin-New York, 1979.

[**C2**] ____, *A II$_1$ factor with countable fundamental group*, J. Operator Theory **4** (1980), 151–153.

[**D**] J. Dixmier, *Les algèbres d'opérateurs dans l'espace hilbertien*, Gauthier-Villars, Paris, 1957.

[F+] P. Freyd, D. Yetter, J. Hoste, W. Lickorish, K. Millett, and A. Ocneanu, *A new polynomial invariant of knots and links*, Bull. Amer. Math. Soc. (N.S.) **12** (1985), 183–312.

[G] M. Goldman, *On subfactors of factors of type* II₁, Michigan Math. J. **7** (1960), 167–172.

[GHJ] F. Goodman, P. de la Harpe, and V. Jones, *Coxeter graphs and towers of algebras* (to appear).

[H] U. Haagerup, *Classification of hyperfinite von Neumann algebras*, Proc. Internat. Congr. Math. (Berkeley, Calif., 1986).

[HKW] P. de la Harpe, M. Kervaire, and C. Weber, *On the Jones polynomial*, Enseign. Math. (to appear).

[J1] V. Jones, *Index for subfactors*, Invent. Math. **72** (1983), 1–25.

[J2] ____, *Hecke Algebra Representations of braid groups and link polynomials*, Univ. Calif., Berkeley, Preprint.

[J3] ____, *A polynomial invariant for knots via von Neumann algebras*, Bull. Amer. Math. Soc. (N.S.) **12** (1985), 103–111.

[MuN] F. Murray and J. von Neumann, *On rings of operators*, Ann. of Math. **37** (1936), 116–229.

[PP] M. Pimsner and S. Popa, *Sur les sous-facteurs d'indice fini d'un facteur de type* II₁ *ayant la propriété T*, C. R. Acad. Sci. Paris **303** (1986), 359–361.

[W] H. Wenzl, *Representations of Hecke algebras and subfactors*, Thesis, Univ. of Pennsylvania, 1985.

INSTITUT DES HAUTES ÉTUDES SCIENTIFIQUES, 35, ROUTE DE CHARTRES, 91440 BURES-SUR-YVETTE, FRANCE

Proceedings of the International Congress of Mathematicians
Berkeley, California, USA, 1986

Carleman Estimates, Uniform Sobolev Inequalities for Second-Order Differential Operators, and Unique Continuation Theorems

CARLOS E. KENIG

1. Introduction, background, and history. It is well known that if $P(x, D)$ is an elliptic differential operator, with real analytic coefficients, and $P(x, D)u = 0$ in an open, connected set $\Omega \subset \mathbf{R}^n$, then u is real analytic in Ω. Hence, if there exists $x_0 \in \Omega$ such that u vanishes of ∞ order at x_0, u must be identically 0. If a differential operator $P(x, D)$ has the above property, we say that $P(x, D)$ has the strong unique continuation property (s.u.c.p.). If, on the other hand, $P(x, D)u = 0$ in Ω, and $u \equiv 0$ in Ω', an open subset of Ω, implies that $u \equiv 0$ in Ω, we say that $P(x, D)$ has the unique continuation property (u.c.p.). Finally, if $P(x, D)u = 0$ in Ω, and supp $u \subset K \Subset \Omega$ implies that $u \equiv 0$ in Ω, we say that $P(x, D)$ has the weak unique property (w.u.c.p.).

Through the work of Hadamard [19] on the uniqueness for the Cauchy problem, and Holmgren's uniqueness theorem (which are strongly related to the unique continuation property), it became clear that it would be desirable to establish the unique continuation property for operators whose coefficients are not necessarily real analytic, or even of class C^∞. The first results in this direction are to be found in the pioneering work of T. Carleman [12] in 1939. He was able to show that $P(x, D) = \Delta + V(x)$ in $\mathbf{R}^2$ has the s.u.c.p. whenever the function $V(x)$ is in $L^\infty_{\text{loc}}(\mathbf{R}^2)$. In order to prove this result he introduced a method, the so-called Carleman estimates, which has permeated almost all the subsequent work in the subject. In this context, a Carleman estimate is, roughly speaking, an inequality of the form

$$\|e^{\lambda\phi} f\|_{L^2(U)} \le C \|e^{\lambda\phi} \Delta f\|_{L^2(U)},$$

for all $f \in C_0^\infty(U)$, U an open subset of $\mathbf{R}^2$, and a suitable function ϕ, where the constant C is independent of λ, for a sequence of real values of λ tending to ∞. As we shall see later on, such estimates readily give the s.u.c.p. for $\Delta + V(x)$,

Supported by the National Science Foundation.

$V \in L^\infty_{\text{loc}}(\mathbf{R}^2)$. By now, there is a large literature of applications of Carleman-type estimates to uniqueness questions. (See Hörmander [25].)

Carleman's result was extended to $\mathbf{R}^n$ by C. Müller [34] in 1954. Subsequently, there was a great flurry of activity in the subject, in the late fifties and early sixties. Most notable among these works are the contributions of H. O. Cordes [15], N. Aronszajn [5], L. Nirenberg [36], the fundamental work of A. P. Calderón [11] on the uniqueness for the Cauchy problem, the work of L. Hörmander [20, 21], of N. Aronszajn, A. Krzywicki, and J. Szarski [6], and of S. Agmon [1]. In the context of the strong unique continuation property, the best result was the one of Aronszajn, Krzywicki, and Szarski [6], who showed that if $\{a_{jk}(x)\}$ is a real, symmetric, positive definite, $n \times n$ matrix that is Lipschitz continuous for $x \in \mathbf{R}^n$, and if u verifies the differential inequality

$$\left| \sum a_{jk}(x) \frac{\partial^2}{\partial x_j \partial x_k} u(x) \right| \leq C \sum_{|\alpha| \leq 1} C_\alpha(x) \left| \frac{\partial^\alpha}{\partial x^\alpha} u(x) \right| \qquad (*)$$

in a connected neighborhood of 0 in $\mathbf{R}^n$, and u vanishes of infinite order at 0, then u must be identically 0 if $C_\alpha(x) \in L^\infty_{\text{loc}}(\mathbf{R}^n)$. They established their result by means of an appropriate Carleman estimate. Moreover, an example of Pliš shows that the regularity assumed on the coefficients $\{a_{jk}(x)\}$ is optimal. In fact A. Pliš [37] showed the existence of a nonzero solution u, vanishing in an open set, whenever $\{a_{jk}(x)\}$ are Hölder continuous of any order < 1. At this point we would like to mention the recent work of N. Garofalo and F. H. Lin [17], who established the result of Aronszajn, Krzywicki, and Szarski without using Carleman estimates, but using instead real variable methods and the theory of A_∞ weights developed by B. Muckenhoupt and others [33, 14].

Recently, there has been growing interest in establishing unique continuation results for solutions of differential inequalities such as $(*)$, with L^p conditions, $p < \infty$, on the lower-order coefficients (see Simon [44] and Kenig [29] for surveys of these problems). The reason for interest in these questions comes from mathematical physics. Suppose, for example, that we consider the Schrödinger operator $H = -\Delta + V(x)$ as a selfadjoint operator on the Hilbert space $L^2(\mathbf{R}^n)$. Here $H_0 = -\Delta$ is the kinetic energy, and $V(x)$ the potential energy. $V(x)$ is not supposed to be smooth, continuous, or even locally bounded. In fact, the Coulomb potentials $(V(x) = 1/|x|)$ are unbounded, and appear in models for hydrogen atoms.

A useful division of the spectrum of H (see, for example, Reed and Simon [38] for the precise definitions) distinguishes among σ_{p}, the point spectrum, consisting of eigenvalues, isolated or not, and σ_{cont}, the continuous spectrum, which is associated to the restriction of H to the part of $L^2(\mathbf{R}^n)$ orthogonal to eigenvectors. The definitions allow for the possibility that σ_{cont} and σ_{p} intersect. In one-body physics, the potential V tends to 0 at ∞, and typically

$$\sigma_{\text{cont}} = [0, \infty), \qquad \sigma_{\text{p}} \cap \sigma_{\text{cont}} = \{0\} \text{ or } \varnothing.$$

This is expected on physical grounds. The relationship of this decomposition to quantum physics is that σ_p comprises the energy of bound states, and the spectral subspace associated with σ_{cont} consists of dynamical states that may participate in scattering. The reason why one expects on physical grounds that $\sigma_p \cap \sigma_{cont} = \{0\}$ or $\varnothing$ is that if the potential V tends to 0 at ∞ and the energy of a particle is positive, one expects that quantum fluctuations would eventually propel the particle to a place where its motion would not be confined, and this would of course make the bound state large, and hence, not in $L^2(\mathbf{R}^n)$. In 1929, von Neumann and Wigner [54] constructed an example of a one-dimensional potential $V(x)$ that goes to 0 at ∞, but with positive eigenvalues. (See also Reed and Simon [38] for a study of this example, as well as Reed-Simon [38, Vol. 4, Ex. 2, XIII] for a different kind of 3-dimensional example.) Thus we see that the problem of determining for which potentials we have no positive eigenvalues is a difficult one. The most successful philosophy for eliminating positive eigenvalues in dimension greater than one has been developed by T. Kato [28], S. Agmon [2, 3], B. Simon [43], and others. To illustrate this philosophy, let us assume that V has compact support, say supp $V \subset \{|x| < R\}$. Suppose that $u \in L^2(\mathbf{R}^n)$, $E > 0$, and $[-\Delta + V(x)]u = Eu$ in $\mathbf{R}^n$. By our support assumption on V, we have

$$-\Delta u - Eu = 0 \quad \text{in } |x| > R, E > 0, u \in L^2(|x| > R).$$

A classical theorem of Rellich [39] now shows that $u \equiv 0$ for $|x| > R$. We then have $\{-\Delta + [V(x) - E]\}u = 0$ in $|x| < 2R$, supp $u \subset \{|x| \leq R\}$. If we knew that $\{-\Delta + [V(x) - E]\}$ has the w.u.c.p., we would then conclude that $u \equiv 0$, and we would be done. As was mentioned before, V need not be locally bounded, and this leads us to study the unique continuation property when the lower-order coefficients are in L^p classes. Similar considerations, also connected with mathematical physics, lead us to also study operators where we replace the Laplacian Δ by the Dirac operator D.

   The first results on the unique continuation property for operators of the form $\Delta + V(x)$, $V \in L^p_{loc}(\mathbf{R}^n), p < \infty$, seem to be due to A. M. Bérthier [8, 9] in 1979 and to V. Georgescu [18] also in 1979. Bérthier proved the w.u.c.p. if $p > \max(n - 2, n/2)$, while Georgescu proved the u.c.p. if $p \geq \max(2, (2n-1)/3)$. M. Schechter and B. Simon [42] proved the u.c.p. if $p > 1$ for $n = 1, 2$, $p > (2n-1)/3$ for $n = 3, 4, 5$, and $p \geq n - 2$ if $n \geq 6$, while Saut and Scheurer [40] proved the u.c.p. for $p > 2n/3$. Also, Amrein, Bérthier, and Georgescu [4] proved the s.u.c.p. for $p > n/2$ if $n = 2, 3, 4$ and $p \geq n - 2$ for $n \geq 5$. More recently (1983), Hörmander [24] showed the u.c.p. for solutions of $(*)$, with $(a_{jk}(x))$ Lipschitz continuous, $C_\alpha(x) \in L^{p(|\alpha|)}_{loc}(\mathbf{R}^n)$, where $p(0) = 2$, when $n < 4$; $p(0) > 2$ when $n = 4$, and $p(0) = (4n - 2)/7$ when $n > 4$; and $p(1) > 2$ when $n > 2$ and $p(1) > (3n - 2)/2$ when $n > 2$.

   (See also the work of E. Sawyer [41] in $\mathbf{R}^3$, under a different kind of assumption on V.)

To clarify this myriad of results, let us restrict our attention to $\Delta + V(x)$, $V \in L^p_{\text{loc}}(\mathbf{R}^n)$. For $n = 2, 3, 4$, the best results are those of Amrein, Bérthier, and Georgescu [4], who proved the s.u.c.p. for $p > n/2$. For n large, the best results are those of Hörmander [24], who proved the u.c.p. when $p \geq (4n-2)/7$.

EXAMPLE [27]. Let $u(x) = \exp(-\log 1/|x|)^{1+\varepsilon}$, $\varepsilon > 0$. Then u vanishes at 0 of ∞ order, while $V(x) = -\Delta u(x)/u(x) \sim (\log 1/|x|)^{2\varepsilon} \cdot 1/|x|^2 \in L^p_{\text{loc}}(\mathbf{R}^n)$ for all $p < n/2$. Hence the s.u.c.p. cannot hold for $p < n/2$.

In 1984, D. Jerison and C. Kenig [27] were able to show that $V \in L^{n/2}_{\text{loc}}(\mathbf{R}^n)$ implies that $\Delta + V(x)$ has the s.u.c.p. This shows that for the s.u.c.p., $L^{n/2}_{\text{loc}}(\mathbf{R}^n)$ is the sharp class in the scale of L^p spaces.

In 1985, D. Jerison [26] gave an alternative proof of the result of Jerison and Kenig [27], and was also able to show that the operator $D + V(x)$, where D is the Dirac operator, and $V(x) \in L^\gamma_{\text{loc}}(\mathbf{R}^n)$, $\gamma = (3n-2)/2$, has the u.c.p. In the case, $n = 3$, $\gamma = 7/2$. This improved previous results of Bérthier and Georgescu (1980) [10], who had obtained $\gamma = 5$. The natural conjecture for the u.c.p. for $D + V(x)$ is that $V \in L^n_{\text{loc}}(\mathbf{R}^n)$. Jerison's result falls short of this conjecture, but he proved that his result is the best one can obtain by the method of Carleman estimates.

Finally, in 1986, C. Kenig, A. Ruiz, and C. Sogge [30] observed that to prove the w.u.c.p. for $\Delta + V(x)$, $V(x) \in L^{n/2}(\mathbf{R}^n)$, it sufficed to prove the Carleman estimate

$$\|e^{\lambda x_n} u\|_{L^{p'}(\mathbf{R}^n)} \leq C\|e^{\lambda x_n} \Delta u\|_{L^p(\mathbf{R}^n)}, \qquad \frac{1}{p} - \frac{1}{p'} = \frac{2}{n}, \qquad (**)$$

whose proof is simpler than the Carleman estimate proved by Jerison and Kenig in [27], to obtain the s.u.c.p. (See §2 for this result.)

If one makes the substitution $e^{\lambda x_n} u = v$, one sees that $(**)$ is equivalent to the estimate

$$\|v\|_{L^{p'}(\mathbf{R}^n)} \leq C \left\| \left[\Delta - 2\lambda \frac{\partial}{\partial x_n} + \lambda^2 \right] v \right\|_{L^p(\mathbf{R}^n)}, \qquad \frac{1}{p} - \frac{1}{p'} = \frac{2}{n}. \qquad (**')$$

Thus one is led to the idea of proving Sobolev-type estimates for second-order constant coefficient operators, which are uniform in the lower-order terms. This was accomplished for operators $P(D)$ with constant complex coefficients, and principal part $Q(D)$, where $Q(\xi)$ is a nonsingular real quadratic form on $\mathbf{R}^n$, i.e., $Q(\xi) = -\xi_1^2 - \cdots - \xi_j^2 + \xi_{j+1}^2 + \cdots + \xi_n^2$. In this setting, Kenig, Ruiz, and Sogge [31] proved the uniform Sobolev inequality

$$\|v\|_{L^{p'}(\mathbf{R}^n)} \leq C\|P(D)v\|_{L^p(\mathbf{R}^n)}, \qquad \frac{1}{p} - \frac{1}{p'} = \frac{2}{n}, \qquad (S)$$

where C depends only on n.

(S) in turn implies Carleman estimates and unique continuation theorems for operators whose top-order terms are not necessarily elliptic, and include, for example, the wave operator

$$\square = \frac{\partial^2}{\partial x_1^2} - \frac{\partial^2}{\partial x_2^2} - \cdots - \frac{\partial^2}{\partial x_n^2}.$$

(S) yields global and local unique continuation theorems. For the global ones, let p be as in (S), and suppose that $(\partial^\alpha u/\partial x^\alpha) \in L^p(\mathbf{R}^n)$, $|\alpha| = 2$. Assume also that u vanishes on one side of a hyperplane, that $P(D)$ is as in (S), and that $|P(D)u| \le |Vu|$, where $V \in L^{n/2}(\mathbf{R}^n)$. Then $u \equiv 0$. If the hyperplane is characteristic for $P(D)$ (and hence we are in the nonelliptic case), there are well known examples (see, e.g., [25, vol. I, pp. 310–311]) of C^∞ functions u, with $P(D)u = 0$, and which vanish on one side of the hyperplane. These examples however do not have the growth property $(\partial^\alpha u/\partial x^\alpha)$, $|\alpha| = 2 \in L^p(\mathbf{R}^n)$. As far as local theorems, one can use a reflection across convex "spheres" as in Nirenberg [36], when the principal part of $P(D)$ is Δ or $\square$. In the case when it is Δ, this shows that the u.c.p. for $\Delta + V(x)$, $V \in L^{n/2}_{loc}(\mathbf{R}^n)$, follows from (**). When it is $\square$, if Γ is the light cone $\Gamma = \{x: x_1 > |x'|, x' = (x_2, \ldots, x_n)\}$, and $(\partial^\alpha u/\partial x^\alpha)$, $|\alpha| = 2$, belong to $L^p_{loc}(\Gamma)$ (p as in (S)), $|P(D)u| \le |Vu|$ in Γ, with $V \in L^{n/2}_{loc}(\Gamma)$, then, if u vanishes for large x_1, it must vanish identically. Similar results, under the stronger assumption $V \in L^\infty_{loc}(\Gamma)$, were obtained by Hörmander [23].

Let us now illustrate the mechanism for passing from Carleman estimates to unique continuation theorems. For example, let us show that if $\Delta u \in L^p(\mathbf{R}^n)$, $1/p - 1/p' = 2/n$, $V \in L^{n/2}(\mathbf{R}^n)$, $|\Delta u| \le |Vu|$, and supp $u \subset \{x_n > 0\}$, then $u \equiv 0$. We will use (**) for $\lambda < 0$. It is enough to show that there exists $\rho > 0$ so that $u \equiv 0$ in S_ρ, where $S_\rho = \{x \in \mathbf{R}^n: 0 \le x_n \le \rho\}$. Choose ρ so small that if C is as in (**), $C\|V\|_{L^{n/2}(S_\rho)} \le 1/2$. Then

$$
\begin{aligned}
\|e^{\lambda x_n} u\|_{L^{p'}(S_\rho)} &\le C\|e^{\lambda x_n} \Delta u\|_{L^p(\mathbf{R}^n)} \\
&\le C\|e^{\lambda x_n} Vu\|_{L^p(S_\rho)} + C\|e^{\lambda x_n} \Delta u\|_{L^p(\mathbf{R}^n \setminus S_\rho)} \\
&\le \tfrac{1}{2}\|e^{\lambda x_n} u\|_{L^{p'}(S_\rho)} + C\|e^{\lambda x_n} \Delta u\|_{L^p(\mathbf{R}^n \setminus S_\rho)}
\end{aligned}
$$

The third inequality follows by Hölder's inequality, our choice of ρ, and $1/p - 1/p' = 2/n$. Hence, $\|e^{\lambda(x_n-\rho)}u\|_{L^{p'}(S_\rho)} \le 2C\|\Delta u\|_{L^p(\mathbf{R}^n)}$, uniformly for all $\lambda < 0$, which shows that $u \equiv 0$ in S_ρ.

In the rest of the paper, we will attempt to describe some of the main points in the proofs of the Carleman estimates in the works of Jerison and Kenig [27], Jerison [26], and Kenig, Ruiz, and Sogge [31].

The main underlying theme is the application of the ideas and methods of classical Fourier analysis, such as oscillatory integrals [13, 22, 49], restriction theorems for the Fourier transform [52, 51, 49], complex interpolation [46], and the uncertainty principle [16], to problems arising in mathematical physics and partial differential equations.

The connection between restriction lemmas for the Fourier transform and Carleman estimates seems to have been first observed by Hörmander [24]. He used the L^2-restriction theorem for $\mathbf{R}^n$. His Carleman estimates, unlike (**), the estimate in [27], and (S), involve L^2 norms in the right-hand side (and the "convex" weights $e^{\lambda(x_n+x_n^2)}$), and for this reason his unique continuation theorems involve potentials in "worse" L^p spaces. On the other hand, the use of the L^2 norm expediates the passage to variable coefficients. This is accomplished exploiting the "convexity" of the weights, via the so-called "Trèves identity" [53]. This is an example of the uncertainty principle, as was pointed out by Jerison [26].

The Carleman estimate in [27] was proved by complex interpolation in a manner resembling the proof of the L^2 restriction theorem. Later, D. Jerison [26] used the discrete restriction lemma of [45] for S^{n-1}, to give a new proof of [27]. Jerison also combined this with ideas related to the uncertainty principle and "convex" weights to prove Carleman estimates for the Dirac operator.

Finally, C. Kenig, A. Ruiz, and C. Sogge [30] used the mapping properties of the Stein-Tomas operator (which is the main tool in proving the L^2 restriction theorem) in $\mathbf{R}^n$ (as opposed to $\mathbf{R}^{n-1}$) to prove (∗∗). Similar ideas, involving Strichartz's [51] generalization of the Stein-Tomas operator for quadratic forms of arbitrary signature, led to the proof of (S) in [31].

2. Schrödinger operators of the form $\Delta + V(x)$, $V \in L_{\mathrm{loc}}^{n/2}(\mathbf{R}^n)$. The main point in the work of Jerison and Kenig [27] was the following Carleman estimate:

THEOREM 2.1. *Let* $n \geq 3$, $1/p - 1/p' = 2/n$. *Suppose that* $\lambda > 0$ *is not an integer, and let* δ *be the distance from* λ *to the nearest integer. There is a constant* C, *depending only on* δ *and* n, *such that for every* $u \in C_0^\infty(\mathbf{R}^n \backslash 0)$,

$$\||x|^{-\lambda} u\|_{L^{p'}(\mathbf{R}^n, dx/|x|^n)} \leq C \||x|^{-\lambda+2} \Delta u\|_{L^p(\mathbf{R}^n, dx/|x|^n)}.$$

This inequality was proved by complex interpolation. Fix λ, and consider the analytic family of operators T_z, depending on λ, given by

$$T_z g = |x|^{-\lambda} \Delta^{-z/2}(|x|^{\lambda-z} g),$$

modified by a Taylor series of order the integer part of λ. Theorem 2.1 follows from estimates for T_z. By E. Stein's interpolation theorem [46], we need an $L^2 \rightarrow L^2$ estimate when Re $z = 0$. Because of rotation and dilation invariance, using polar coordinates (r, ω), we study $T_z(r^{i\eta} P_k(\omega))$, where P_k is a spherical harmonic of degree k. We have:

$$T_z(r^{i\eta} P_k(\omega))$$
$$= \left[\frac{2^{-z}}{\Gamma(\frac{n-z}{2})} \frac{\Gamma(\frac{1}{2}(k-\lambda-i\eta)) \cdot \Gamma(\frac{1}{2}(n+k+\lambda-z+i\eta))}{\Gamma(\frac{1}{2}(k-\lambda+z-i\eta)) \cdot \Gamma(\frac{1}{2}(n+k+\lambda+i\eta))} \right] r^{i\eta} P_k(\omega).$$

When Re $z = 0$, Stirling's formula shows that the expression in brackets is bounded independently of k, λ, and η, with bound depending only on δ. This gives the desired $L^2 \rightarrow L^2$ estimate.

At the other end point of the interpolation, Re $z = n$, and T_z is essentially a logarithmic potential. It was proved in [27] that in this case $T_z: L^1 \rightarrow X$, where X is an enlargement of L^∞, which has the same complex interpolation properties as L^∞, but allows logarithmic singularities. This estimate follows by a uniform asymptotic estimate for the hypergeometric function. (See [27].)

E. M. Stein [48] observed that one could bypass the space X and the hypergeometric function, by noting that when $n - 1 < \text{Re } z < n$, the kernel of T_z is pointwise dominated by the one for fractional integration of order $\text{Re } z$. This also gives Lorentz space estimates.

3. The Schrödinger operator revisited, and the Dirac operator. We will now discuss Jerison's work [26]. Introduce polar coordinates in $\mathbf{R}^n$, $x = e^y \omega$, $\omega \in S^{n-1}$, $y \in \mathbf{R}$. In those coordinates,

$$\Delta = e^{-2y} \left[\frac{\partial^2}{\partial y^2} + (n - 2)\frac{\partial}{\partial y} + \Delta_S \right],$$

where Δ_S denotes the spherical Laplacian. Let us start out by outlining Jerison's proof of Theorem 2.1. Because of the above formula, if P_k is a spherical harmonic of degree k, then

$$|x|^{-\lambda+2}\Delta(|x|^\lambda e^{iy\eta} P_k(\omega)) = -(k - (\lambda + i\eta)) \cdot (k + n - 2 + \lambda + i\eta)e^{iy\eta} P_k(\omega).$$

(Notice that this is the same formula as the case $\text{Re } z = 2$ of T_z in §2.)

Let $\sigma_\lambda(\eta, k) = -1/(k - \lambda + i\eta)(k + n - 2 + \lambda + i\eta)$, and let ξ_k denote the projection operator from $L^2(S^{n-1})$ to the space of spherical harmonics of degree k. Also, let $\tilde{f}(\eta, \omega) = \int_{-\infty}^{+\infty} e^{iy\eta} f(y, \omega)dy$ denote the partial Fourier transform in y. For $f \in C_0^\infty(\mathbf{R} \times S^{n-1})$, let

$$R_\lambda f(y, \omega) = \sum_{k=0}^{\infty} \frac{1}{2\pi} \int_{-\infty}^{\infty} \sigma_\lambda(\eta, k) e^{-iy\eta} \xi_k \tilde{f}(\eta, -)(\omega)d\eta.$$

It is then easy to see that Theorem 2.1 is equivalent to

$$\|R_\lambda f\|_{L^{p'}(\mathbf{R} \times S^{n-1})} \leq C\|f\|_{L^p(\mathbf{R} \times S^{n-1})},$$

for all $f \in C_0^\infty(\mathbf{R} \times S^{n-1})$. Now let $\{\theta_\beta\}_{\beta=0}^n$ be a partition of unity of the positive real axis with $\text{supp } \theta_0 \subset \{r: r \leq 1\}$, $\text{supp } \theta_N \subset \{r: r > \lambda/400\}$, $\text{supp } \theta_\beta \subset \{r: 2^{\beta-2} \leq r \leq 2^\beta\}$, $2^N \leq \lambda/10 < 2^{N+1}$. Consider the operator R_λ^β, analogous to R_λ, but with symbol $\sigma_\lambda^\beta(\eta, k) = \theta_\beta(|k - \lambda + i\eta|)\sigma_\lambda(\eta, k)$. Note that $\sigma_\lambda^\beta(\eta, k)$ for $\beta \leq N - 1$ is supported where $|k - \lambda + i\eta| \leq 2^\beta$, and hence there are at most $2^{\beta+1}$ nonzero terms in the sum over k which defines R_λ^β, and the value of k is in each case comparable to λ. We can now apply the following "discrete" version of the restriction theorem, due to C. Sogge [45].

LEMMA 3.1. *There is a constant C such that*

$$\|\xi_k g\|_{L^{p'}(S^{n-1})} \leq Ck^{1-2/n}\|g\|_{L^p(S^{n-1})},$$

where p is as in Theorem 2.1.

In fact, using the formula

$$R_\lambda^\beta f(y, \omega) = \int_{-\infty}^{+\beta} K_\lambda^\beta(y - y')f(y', \cdot)(\omega)(dy'),$$

where

$$K_\lambda^\beta(s) = \sum_{k=0}^{\infty} \frac{1}{2\pi} \int_{-\infty}^{+\infty} \sigma_\lambda^\beta(\eta, k) e^{is\eta} d\eta \xi_k,$$

we see that for $\beta < N-1$, the integration in η is over an interval of length $\leq 2^{\beta+1}$. Hence, by Lemma 3.1 and integration by parts in η, it follows that $K_\lambda^\beta(s)$ is a bounded operator from $L^p(S^{n-1})$ to $L^{p'}(S^{n-1})$ whose norm is bounded by $C2^\beta \lambda^{-2/n}(1+|2^\beta s|)^{-10}$. If we now let $1/r + 1/p = 1/p' + 1$, i.e., $1/r = 1 - 2/n$, Minkowski's integral inequality and Young's inequality show that

$$\|R_\lambda^\beta f\|_{L^{p'}(\mathbf{R} \times S^{n-1})} \leq C\lambda^{-2/n} 2^{2\beta/n} \|f\|_{L^p(\mathbf{R} \times S^{n-1})}.$$

But $\sum_{\beta=0}^{N-1} 2^{2\beta/N} \leq \lambda^{2/n}$, while $R_\lambda^N f$ can be controlled by ordinary fractional integration, and hence Theorem 2.1 follows.

The Dirac operator is $D = \sum_{j=1}^n \alpha_j \partial/\partial x_j$, where α_j are skew hermitian matrices with $\alpha_j^* = -\alpha_j$, $\alpha_j \alpha_k + \alpha_k \alpha_j = -2\delta_{jk}$, $\alpha_j \in \mathrm{GL}(m, \mathbf{C})$, where $m = 2^{n/2}$ if n is even, and $m = 2^{(n+1)/2}$ if n is odd. It is easy to see that $D^* = D$, and $D^2 = -\Delta$. Jerison observed that the analogue of Theorem 2.1 fails for D. In fact if

$$\||x|^{-\lambda} u\|_{L^q(U, \mathbf{C}^m)} \leq C\||x|^{-\lambda+1} Du\|_{L^p(U, \mathbf{C}^m)}$$

for all $u \in C_0^\infty(U, \mathbf{C}^m)$, $U = \{x \in \mathbf{R}^n : a < |x| < b\}$, $0 < a < b$, uniformly for a sequence of $\lambda \to \infty$, then $q \leq p$. Serious difficulties remain even if one replaces $|x|^{-\lambda}$ by $e^{\lambda\phi(x)}$, where ϕ is any smooth real-valued function, not identically 0. The corresponding inequality can then only hold if $1/p - 1/q \leq 1/\gamma$, $\gamma = (3n-2)/2$. This is unfortunate, because the conjectured gap was $1/\gamma = 1/n$. On the positive side, Jerison proved

THEOREM 3.2. Let $0 < a < b < 1$, $n \geq 3$, $U = \{x \in \mathbf{R}^n : a < |x| < b\}$. Let $\phi(x) = (\log|x|)^2/2$, $q = (6n-4)/(3n-6)$, i.e., $1/2 - 1/q = 1/\gamma$, $\gamma = (3n-2)/2$. Then there exist $C = C(n, a, b)$ such that, for all $\lambda \in \mathbf{R}$,

$$\|e^{\lambda\phi} u\|_{L^q(U, \mathbf{C}^m)} \leq C\|e^{\lambda\phi} Du\|_{L^2(U, \mathbf{C}^m)},$$

for all $u \in C_0^\infty(U, \mathbf{C}^m)$.

The above theorem easily implies the u.c.p. for solutions of $Du = Vu$, $V \in L_{\mathrm{loc}}^\gamma(\mathbf{R}^n, \mathrm{GL}(m, \mathbf{C}))$, $\gamma = (3n-2)/2$, with $Du \in L^2(\Omega, \mathbf{C}^m)$.

In order to prove Theorem 3.2, Jerison considered $e^{\lambda y^2} e^y D e^{-\lambda y^2}$. This equals $\hat{\alpha} A_\lambda$, where $\hat{\alpha}$ is unitary, $A_\lambda = \partial/\partial y - (\lambda y + L)$, and L is a first-order operator on the ω variables. A_λ is now variable coefficient, and this gives the improved estimate $A_\lambda \geq C\lambda^{1/2}$. (This is an instance of the uncertainty principle.) However, to find a left inverse for A_λ, one then has to use "pseudodifferential" operators. Jerison found an exact left inverse, using the formula for the left inverse of $(d/dy - y)$ on $\mathbf{R}$, given in Nagel and Stein [35]. He then bound the left inverse using Lemma 3.1 and a device of P. Tomas [52] to obtain $L^q \to L^2$ estimates from $L^p \to L^{p'}$ estimates. See [26] for the details.

**4. Uniform Sobolev inequalities for second order constant coeffi-
cients operators.** In this section we will outline some of the ideas in the work
of C. Kenig, A. Ruiz, and C. Sogge [**31**]. Let $Q(\xi) = -\xi_1^2 - \cdots - \xi_j^2 + \xi_{j+1}^2 + \cdots + \xi_n^2$
be a nonsingular real quadratic form on $\mathbf{R}^n$. Let $P(D)$ be a constant coefficient
operator, with complex coefficient lower-order terms, and whose principal part
is $Q(D)$.

THEOREM 4.1. *Let $n \geq 3$, and let $1/p - 1/p' = 2/n$. Then there exists a
constant C depending only on n such that, for all $u \in C_0^\infty(\mathbf{R}^n)$, we have*

$$\|u\|_{L^p(\mathbf{R}^n)} \leq C\|P(D)u\|_{L^{p'}(\mathbf{R}^n)}.$$

As was remarked in the introduction, Theorem 4.1 yields unique continuation
theorems for operators whose principal part is not necessarily elliptic.

Let H_+^n and H_-^n be the open subsets of $\mathbf{R}^n$ on which Q is strictly positive
and negative respectively. Also, let S_+^{n-1} and S_-^{n-1} be the level sets $S_\pm^{n-1} =
\{\xi : Q(\xi) = \pm 1\}$. There are canonical measures $d\omega_\pm$ on $S_\pm^{n-1}$ so that on $H_\pm^n$,
$d\xi = \rho^{n-1} d\rho\, d\omega_\pm$. The key ingredient for Theorem 4.1 is

LEMMA 4.2. *Let $n \geq 3$, Q, p as above. Then, for $f \in C_0^\infty(\mathbf{R}^n)$*
(a)

$$\left\| \int_{S_\pm^{n-1}} \hat{f}(\omega) e^{ix\omega} d\omega_\pm \right\|_{L^{p'}(\mathbf{R}^n)} \leq C\|f\|_{L^p(\mathbf{R}^n)};$$

(b) *there exists an absolute constant C, such that, for all $z \in \mathbf{C}$,*

$$\|u\|_{L^{p'}(\mathbf{R}^n)} \leq C\|[Q(D) + z]u\|_{L^p(\mathbf{R}^n)}, \qquad u \in C_0^\infty(\mathbf{R}^n).$$

In the Euclidean case (when Q is elliptic), (a) is due to Stein-Tomas [**52**],
while the other cases are due to Strichartz [**51**]. (b) does not seem to be in the
literature; however, its proof involves only simple modifications of the proof of
(a).

The difficulty in establishing Theorem 4.1 comes from the fact that the symbol
of $P(D)$ may vanish away from the origin. However, if this is the case, the zero
set of $P(\xi)$ always lies on a "sphere," which explains the relevance of Lemma
4.2.

It is not hard to see that it is enough to prove Theorem 4.1 in the case when
$P(D) = Q(D) + \sigma + \varepsilon\{\partial/\partial x_j + i\beta\}$, where $\sigma = \pm 1$, $\varepsilon, \beta \in \mathbf{R}\setminus\{0\}$, $j = 1$ or n.
We will deal with the case $j = n$, $\sigma = 1$, the other ones being similar. We are
thus reduced to proving the multiplier theorem

$$\left\| \left\{ \frac{\hat{f}(\xi)}{Q(\xi) + 1 + i\varepsilon(\xi_n + \beta)} \right\}^\vee \right\|_{p'} \leq C\|f\|_p.$$

Let $m(\xi) = (Q(\xi) + 1 + i\varepsilon(\xi_n + \beta))^{-1}$, and let $\chi(t) = 1$, for $|t| \in [1, 2]$, and
0 otherwise. Set $\chi_k(\xi_n) = \chi(2^k(\xi_n + \beta))$, and define $m_k(\xi) = \chi_k(\xi_n)m(\xi)$.
Because of Littlewood-Paley theory (see [**47**]), the fact that $p < 2 < p'$, and

Minkowski's integral inequality, it suffices to show that there is a constant C, independent of k, ε, and β, for which

$$\|\{m_k(\xi)\hat{f}(\xi)\}^{\vee}\|_{p'} \le C\|f\|_p.$$

To prove this last estimate, we use (b) in Lemma 4.2, with $z = 1 + i\varepsilon 2^{-k}$.

We are then reduced to showing that

$$\left\|\left\{\frac{\chi_k(\xi_n)[i\varepsilon(\xi_n + \beta - 2^{-k})]\hat{f}(\xi)}{(Q(\xi) + 1 + i\varepsilon(\xi_n + \beta))(Q(\xi) + 1 + i\varepsilon 2^{-k})}\right\}^{\vee}\right\|_{p'} \le C\|f\|_p.$$

Let T_k be the above multiplier, and use polar coordinates $\xi = \rho\omega$ associated to Q. It is easy to see that Minkowski's integral inequality and Lemma 4.2(a) give

$$\|T_k f\|_{p'}$$

$$\le \sum_{\pm} \int_0^{\infty} \left\| \int_{S_{\pm}^{n-1}} \frac{\varepsilon\hat{f}(\rho\omega)(\xi_n + \beta - 2^{-k})\chi_k(\xi_n)e^{i\rho\omega x}}{(\pm\rho^2 + 1 + i\varepsilon(\xi_n + \beta))(\pm\rho^2 + 1 + i\varepsilon 2^{-k})} d\omega_{\pm} \right\|_{p'} \rho^{n-1}\, d\rho$$

$$\le \sum_{\pm} \int_0^{\infty} \rho^{n-1}\rho^{-2n/p'} \left\| \left\{ \frac{\varepsilon\hat{f}(\xi)\chi_k(\xi_n)(\xi_n + \beta - 2^{-k})}{(\pm\rho^2 + 1 + i\varepsilon(\xi_n + \beta))(\pm\rho^2 + 1 + i\varepsilon 2^{-k})} \right\}^{\vee} \right\|_p d\rho.$$

Since $1 = n - 1 - 2n/p'$, the definition of χ_k shows that this last term is bounded by

$$C\|f\|_p \int_0^{\infty} \frac{\varepsilon 2^{-k}\rho}{(\rho^2 - 1)^2 + (\varepsilon 2^{-k})^2}\, d\rho,$$

which gives the desired inequality.

In the case when $Q(D) = \Delta$, one can prove that

$$\|u\|_{L^s(\mathbf{R}^n)} \le C\|P(D)u\|_{L^r(\mathbf{R}^n)},$$

for the optimal range of s and r (see [31]). Also, if $P(D) = Q(D) + \sum_{j=1}^n a_j \partial/\partial x_j$ $+ b$, a_j's real, $|\mathrm{Re}\, b| \ge 1$, then $\|u\|_{L^{r'}(\mathbf{R}^n)} \le C\|P(D)u\|_{L^r(\mathbf{R}^n)}$ for $2/(n+1) \le 1/r - 1/r' \le 2/n$ (see [31]). This generalizes some results in [50] and [32].

5. Some open problems. To conclude, we would like to point out a few open problems.

(a) Does the unique continuation property hold for

$$D + V, \qquad V \in L_{\mathrm{loc}}^n(\Omega, \mathrm{GL}(m, \mathbf{C})),$$

where D is the Dirac operator of §3? As was pointed out in §3, the corresponding Carleman estimates are false.

(b) Does the unique continuation property hold for $\Delta + \sum v_i(x)\partial/\partial x_i$, where $v_i \in L_{\mathrm{loc}}^n(\mathbf{R}^n)$? This question is closely tied to the previous one. It is known that the u.c.p. holds if the $v_i \in L_{\mathrm{loc}}^{\gamma}(\mathbf{R}^n)$, $\gamma = (3n - 2)/2$ (see [24, 7]). However, it is also known, just as for the Dirac operator, that this is the best exponent that Carleman estimates can yield.

(c) Variable coefficient problems: For example, does the s.u.c.p. or even the w.u.c.p. hold for operators of the form $P(x, D) = \sum a_{jk}(x)(\partial^2/\partial x_j\, \partial x_k) + V(x)$,

where $a_{jk}(x)$ is an elliptic, Lipschitz continuous real symmetric matrix, and
$V(x) \in L_{\text{loc}}^{n/2}(\mathbf{R}^n)$? We could also ask whether the uniform Sobolev inequalities
of §4 hold for operators with $\sum a_{jk}(x)(\partial^2/\partial x_j \, \partial x_k)$ as principal part. Part of
the difficulty comes from the need of a very precise knowledge of the left inverses
of such operators.

(d) The last question bears on the distinction between the s.u.c.p., w.u.c.p.,
and u.c.p. As was pointed out in the introduction, there are examples of poten-
tials $V \in L_{\text{loc}}^p(\mathbf{R}^n)$, $p < n/2$, for which the s.u.c.p. for $\Delta + V$ does not hold. As
far as we know, there are no examples known of potentials $V \in L_{\text{loc}}^1(\mathbf{R}^n)$ such
that the u.c.p. for $\Delta + V$ does not hold. Indeed, it is possible that the u.c.p.
holds for $\Delta + V$, whenever $V \in L_{\text{loc}}^1(\mathbf{R}^n)$. This would be of interest for the
application to the absence of positive eigenvalues.

REFERENCES

1. S. Agmon, *Unicité et convexité dans les problèmes différentials*, Sém. Math. Sup.,
No. 13, Presses Univ. Montréal, Montréal, 1966.

2. _____, *Lower bounds for solutions of Schrödinger type equations in unbounded
domains*, Proc. Internat. Conf. on Functional Analysis and Related Topics (Tokyo, 1969),
Univ. of Tokyo Press, Tokyo, 1970, pp. 216–224.

3. _____, *Lower bounds for solutions of Schrödinger equations*, J. Analyse Math. **23**
(1970), 1–25.

4. W. Amrein, A. Bérthier, and V. Georgescu, *L^p inequalities for the Laplacian and
unique continuation*, Ann. Inst. Fourier (Grenoble) **31** (1981), 153–168.

5. N. Aronszajn, *A unique continuation theorem for solutions of elliptic partial differential equations or inequalities of second order*, J. Math. Pures Appl. **36** (1957), 235–249.

6. N. Aronszajn, A. Krzywicki, and J. Szarski, *A unique continuation theorem for
exterior differential forms on Riemannian manifolds*, Ark. Mat. **4** (1962), 417–453.

7. B. Barceló, C. Kenig, A. Ruiz, and C. Sogge, *Unique continuation properties for
solutions of inequalities between the Laplacian and the gradient*, Preprint.

8. A. M. Bérthier, *Sur le spectre ponctuel de l'operateur de Schrödinger*, C. R. Acad.
Sci. Paris Ser. A **290** (1980), 393–395.

9. _____, *On the point spectrum of Schrödinger operators*, Ann. Sci. Ecole Norm. Sup.
(4) **15** (1982), 1–15.

10. A. M. Bérthier and V. Georgescu, *Sur le proprieté de prolongement unique pour
l'operateur de Dirac*, C. R. Acad. Sci. Paris Ser. A **291** (1980), 603–606.

11. A. P. Calderón, *Uniqueness in the Cauchy problem for partial differential equations*, Amer. J. Math. **80** (1958), 16–36.

12. T. Carleman, *Sur un problème d'unicité pour les systèmes d'equations aux derivées
partielles à deux variables indépendantes*, Ark. Mat. **26B** (1939), 1–9.

13. L. Carleson and P. Sjölin, *Oscillatory integrals and a multiplier problem for the
disc*, Studia Math. **44** (1972), 287–299.

14. R. R. Coifman and C. Fefferman, *Weighted norm inequalities for maximal functions
and singular integrals*, Studia Math. **51** (1974), 241–250.

15. H. O. Cordes, *Über die Bestimmtheit der Lösunger elliptischer Differentiel-
gleichungen durch Anfangsvorgaben*, Nachr Akad. Wiss. Göttingen Math. Phys. Kl. IIa **11**
(1956), 239–258.

16. C. Fefferman, *The uncertainty principle*, Bull. Amer. Math. Soc. (N.S.) **9** (1983),
129–206.

17. N. Garofalo and F. H. Lin, *Monotonicity properties of variational integrals, A_p
weights and unique continuation*, Indiana Univ. Math. J. (to appear).

18. V. Georgescu, *On the unique continuation property for Schrödinger Hamiltonians*,
Helv. Phys. Acta **52** (1979), 655–670.

19. J. Hadamard, *Le problème de Cauchy et les equations aux derivées partielles linéaires hyperboliques*, Hermann, Paris, 1932.

20. L. Hörmander, *On the uniqueness of the Cauchy problem*, Math. Scand. **6** (1958), 213–225.

21. ____, *On the uniqueness of the Cauchy problem. II*, Math. Scand. **7** (1959), 177–190.

22. ____, *Oscillatory integrals and multipliers on FL^p*, Ark. Mat. **11** (1971), 1–11.

23. ____, *Théorie de la diffusion à courte portée pour des opérateurs à caractéristiques simples*, Sém. Goulaovic-Meyer-Schwartz, École Polytechnique, 1980–81, Exp. XIV.

24. ____, *Uniqueness theorems for second order elliptic differential operators*, Comm. Partial Differential Equations **8** (1983), 21–64.

25. ____, *The analysis of linear partial differential operators*. Vols. I, III, Springer-Verlag, New York, Berlin, 1983, 1985.

26. D. Jerison, *Carleman inequalities for the Dirac and Laplace operators, and unique continuation*, Adv. in Math. (to appear).

27. D. Jerison and C. Kenig, *Unique continuation and absence of positive eigenvalues for Schrödinger operators*, Ann. of Math. (2) **121** (1985), 463–494.

28. T. Kato, *Growth properties of solutions of the reduced wave equation with variable coefficients*, Comm. Pure Appl. Math. **12** (1959), 403–425.

29. C. Kenig, *Continuation theorems for Schrödinger operators*, Proc. Workshop on Metastability and Partial Differential Equations (I.M.A., Univ. of Minn., 1985) (to appear).

30. C. Kenig, A. Ruiz, and C. Sogge, *Remarks on unique continuation theorems*, Seminarios, U.A.M. Madrid, 1986.

31. ____, *Sobolev inequalities and unique continuation for second order constant coefficient differential operators*, Preprint.

32. B. Marshall, W. Strauss, and S. Wainger, *$L^p - L^q$ estimates for the Klein-Gordon equation*, J. Math. Pures Appl. **59** (1980), 417–440.

33. B. Muckenhoupt, *The equivalence of two conditions for weight functions*, Studia Math. **49** (1974), 101–106.

34. C. Müller, *On the behavior of the solution of the differential equation $\Delta u = f(x, u)$ in the neighborhood of a point*, Comm. Pure Appl. Math. **1** (1954), 505–515.

35. A. Nagel and E. Stein, *Lectures on pseudo-differential operators: Regularity theorems and applications to non-elliptic problems*, Math. Notes, no. 24, Princeton Univ. Press, Princeton, N. J., and Univ. of Tokyo Press, Tokyo, 1979.

36. L. Nirenberg, *Uniqueness in Caucy problems for differential equations with constant leading coefficient*, Comm. Pure Appl. Math. **10** (1957), 89–106.

37. A. Pliš, *On non-uniqueness in Cauchy problem for an elliptic second order differential operator*, Bull. Acad. Polon. Sci. **11** (1963), 95–100.

38. M. Reed and B. Simon, *Methods of modern mathematical physics*. Vols. I–IV, Academic Press, New York, 1978.

39. F. Rellich, *Über des asymptotische Verhalten der Lösumgen von $\Delta u + \lambda u = 0$ in unendlichen Gebieten*, Über Deutsch. Math. Verein **53** (1943), 57–65.

40. J. Saut and B. Scheurer, *Un théorème de prolongement unique pour des opérateurs elliptiques dont les coefficients ne sont pas localement bornés*, C. R. Acad. Sci. Paris Ser. A **290** (1980), 598–599.

41. E. Sawyer, *Unique continuation for Schrödinger operators in dimension three or less*, Ann. Inst. Fourier (Grenoble) **33** (1984), 189–200.

42. M. Schechter and B. Simon, *Unique continuation for Schrödinger operators with unbounded potential*, J. Math. Anal. Appl. **77** (1980), 482–492.

43. B. Simon, *On positive eigenvalues of one body Schrödinger operators*, Comm. Pure Appl. Math. **22** (1969), 531–538.

44. ____, *Schrödinger semigroups*, Bull. Amer. Math. Soc. (N.S.) **7** (1982), 447–526.

45. C. Sogge, *Oscillatory integrals and spherical harmonics*, Duke Math. J. **53** (1986), 43–65.

46. E. Stein, *Interpolation of linear operators*, Trans. Amer. Math. Soc. **83** (1956), 482–492.

47. _____, *Singular integrals and differentiability properties of functions*, Princeton Univ. Press, Princeton, N. J., 1972.

48. _____, *Appendix to "Unique continuation,"* Ann. of Math. (2) **121** (1985), 489–494.

49. _____, *Oscillatory integrals in Fourier analysis*, Beijing Lectures in Harmonic Analysis, Ann. of Math. Studies, Princeton Univ Press (to appear).

50. R. Strichartz, *A priori estimates for the wave equation and some applications*, J. Funct. Anal. **5** (1970), 218–235.

51. _____, *Restriction of Fourier transforms to quadratic surfaces*, Duke Math. J. **44** (1977), 705–714.

52. P. Tomas, *A restriction theorem for the Fourier transform*, Bull. Amer. Math. Soc. **81** (1975), 477–478.

53. F. Trèves, *Relations de domination entre opérateurs differentiels*, Acta Math. **101** (1959), 1–139.

54. J. von Neumann and E. Wigner, *Über merkürdige diskrete Eigenwerte*, Phys. Z. **30** (1929), 465–467.

UNIVERSITY OF CHICAGO, CHICAGO, ILLINOIS 60637, USA

The Concentration Phenomenon and Linear Structure of Finite-Dimensional Normed Spaces

VITALI D. MILMAN

1. Notations, basic notions, and introductory results. The first part of this presentation will deal with what we call today the Local Theory of Normed Spaces, i.e., the theory of the linear structure of finite-dimensional normed spaces with main emphasis on asymptotic properties when the dimension of the spaces increases to infinity. A family of such spaces often represents, in our view, a useful substitution for the notion of an infinite-dimensional normed space and possesses a remarkable and rich structure.

Let $X = (\mathbb{R}^n, \| \cdot \|)$ be an n-dimensional normed space $(n \geq 2)$ equipped also with a euclidean norm $(\mathbb{R}^n, | \cdot |)$ and, as a consequence, with the inner product (x, y) such that $(x, x) = |x|^2$. Then the dual norm $\| \cdot \|^*$ is naturally defined by $\|x\|^* = \sup\{|(x, y)| : \|y\| \leq 1\}$ and the dual space $X^* = (\mathbb{R}^n, \| \cdot \|^*)$.

The following family of n-dimensional spaces (ℓ_p-spaces for $1 \leq p \leq \infty$) plays a special role in the Local Theory: $\ell_p^n = (\mathbb{R}^n, \| \cdot \|_p)$ where, for $p < \infty$, $\|(a_i)_{i=1}^n \in \mathbb{R}^n\|_p = (\sum_1^n |a_i|^p)^{1/p}$ and, for $p = \infty$, $\|(a_i)_1^n\|_\infty = \max_{1 \leq i \leq n} |a_i|$.

Let the multiplicative distance (the Banach-Mazur distance) between two normed spaces X and Y be (assuming $\dim X = \dim Y < \infty$)

$$d(X, Y) := \inf\{\|T\| \cdot \|T^{-1}\| \text{ over all linear isomorphism } T : X \to Y\};$$

and $d_X := d(X, \ell_2^{\dim X})$. Obviously $d(X, Y) \geq 1$ while $d(X, Y) \leq 1 + \varepsilon$ means that X and Y are close (we say $(1 + \varepsilon)$-isomorphic). In geometrical language this means that the two unit balls $K(X) = \{x \in X : \|x\| \leq 1\}$ and $K(Y)$ may be put, by an affine transform (say $\varphi : Y \to X$), in the same linear space in a position so that $K(X) \subset \varphi(K(Y)) \subset d(X, Y)K(X) \subset (1+\varepsilon)K(X)$. Hence, after such an affine transform the two convex bodies $K(X)$ and $K(Y)$ become close in the geometric sense. We also denote by D the unit ball of the euclidean space $(\mathbb{R}^n, | \cdot |)$. Throughout this note we denote by c and C some numerical positive constants.

First, we give a picture of a geometrical structure of a convex symmetric body in $\mathbb{R}^n$ in terms of its central linear sections.

We write $\ell_p^k \overset{1+\varepsilon}{\hookrightarrow} X$ if X contains a k-dimensional subspace E such that $d(E, \ell_p^k) \le 1 + \varepsilon$ (i.e., X contains a $(1 + \varepsilon)$-isomorphic copy of ℓ_p^k). To describe the linear structure of X, we introduce the following integer functions:

$$k_p(X; \varepsilon) := \sup\{k | \ell_p^k \overset{1+\varepsilon}{\hookrightarrow} X\} .$$

The first result of the Local Theory is Dvoretzky's Theorem [D] of 1960 which was conjectured also by A. Grothendieck [G]:

1.1. DVORETZKY THEOREM. (*Real case* [D]; *another proof which covers also the complex case* [Mi1].)
(a) *For any $\varepsilon > 0$ the integer function*

$$k_2(n; \varepsilon) = \inf\{k_2(X; \varepsilon) | X = (\mathbb{R}^n, \| \cdot \|)\} \to \infty \qquad (n \to \infty) .$$

(b) [Mi1] *Moreover, $k_2(n; \varepsilon) \ge c(\varepsilon) \log n$ where $c(\varepsilon) > 0$ depends only on $\varepsilon > 0$ and this estimate is exact: $k_2(\ell_\infty^n; 1) \le c \log n$.*

The logarithmic estimate in Theorem 1.1, although exact in the general case, can be improved significantly in most cases.

1.2. THEOREM [Mi1]. $k_2(X; \varepsilon) \ge c(\varepsilon) n / d_X^2$ *where* $c(\varepsilon) \ge c\varepsilon^2 / \log 1/\varepsilon$.

Recently Gordon [Go] took out the logarithmic factor and showed that, indeed, $c(\varepsilon) \ge c\varepsilon^2$.

By Theorem 1.2, if a family of spaces $\{X_n\}$ has a uniformly bounded distance from ℓ_2^n (i.e., $d_{X_n} \le \text{Const.}$), then $k_2(X_n; \varepsilon)$ is proportional to n; by this reason we shall estimate only $k_2(X_n; 1)$ which is simply denoted as $k(X_n)$ or, also, $k_2(X_n)$.

1.3. THEOREM [FLM]. $k(X)k(X^*) \ge cn^2 / d_X^2$ (*recall, $n = \dim X$*).

In comparison with 1.2 it means that either $k(X)$ or $k(X^*)$ is at least cn/d_X.

1.4. It is well known [J] that $d_X \le \sqrt{\dim X}$ and therefore in Theorem 1.3 we also have $k(X) \cdot k(X^*) \ge cn$. Note also that if d_X is close to the extremal case, that is $d_X \ge c\sqrt{n}$ (for some fixed constant $c > 0$ and n large enough) then, by [MiW], X contains a $(1 + \varepsilon)$-isomorphic copy of ℓ_1^k; to be precise, $k_1(X; \varepsilon) \ge c(\log n)^{\alpha(d_X/\sqrt{n}, \varepsilon)}$ where a function $\alpha(\lambda, \varepsilon) > 0$ for $\lambda > 0$, $\varepsilon > 0$, and $\alpha(1, \varepsilon) = 1$ for any $\varepsilon > 0$.

(The above result was extended along the same lines by W.B. Johnson, Pisier [P1], Kashin [KS], J. Bourgain, Tomczak-Jaegermann [T1].)

It is clear today that this theorem represents an example of a very general phenomenon: an "extremely bad" property of a space usually involves ℓ_1^k-subspaces.

1.5. One of the most important new ingredients in the modern study of the linear structure of normed spaces (finite- and infinite-dimensional) are the (probabilistic) notions of type and cotype. These concepts were introduced by Hoffmann-Jørgensen [HJ] but they became of key importance through results of Maurey and Pisier (see, for example, [MP]).

DEFINITION. X has cotype q (type p) with cotype q constant $C_q(X)$ (type p constant $T_p(X)$) if, for any k and $x_1, ..., x_k \subset X$,

$$\left(\sum_1^k \|x_i\|^q \right)^{1/q} \leq C_q(X) \left(\underset{\varepsilon_i = \pm 1}{\text{Avg}} \left\| \sum_1^k \varepsilon_i x_i \right\|^2 \right)^{1/2}.$$

$$\left(\text{Similarly,} \quad T_p(X) \left(\sum_1^k \|x_i\|^p \right)^{1/p} \geq \left(\underset{\varepsilon_i = \pm 1}{\text{Avg}} \left\| \sum_1^k \varepsilon_i x_i \right\|^2 \right)^{1/2}. \right)$$

We also use these definitions for a family $\mathbf{X}$ of spaces and, e.g., $C_q(\mathbf{X})$ means the supremum of $C_q(X)$ for $X \in \mathbf{X}$. (Examples for intuition: $C_2(\ell_2) = T_2(\ell_2) = 1$ and, conversely, [**Kw**] $d_X \leq C_2(X)T_2(X)$; for $1 \leq p \leq 2$, L_p-spaces have cotype 2 and type p; for $\infty > p \geq 2$, L_p-spaces have type 2 and cotype p.) Note that by [**T2**] it is enough, in the above definition, to consider $k \leq n = \dim X$ to define $C_2(X)$ and $T_2(X)$ up to a factor, say 4. (See also the extension of this result to C_q and T_p by [**Kö**].) For information on basic results involving notions of type and cotype we refer to the monographs [**MiSch, T1, Pi2**] and the reports [**F, Pi3**, and **Pel**]. The notions of type and cotype becomes an important key which shows what kind of tools have to be applied to the study of a given normed space (or a family of spaces). These notions often enter into proofs in a significant way even when the result does not involve these notions. The following theorems give such examples.

1.6. *Dichotomy principle* [**FLM**]. Let $\mathbf{X} = \{X_t\}$ be a family of finite-dimensional normed spaces (with $\sup \dim X_t = \infty$). Then there either exists an $\alpha > 0$ such that $k_2(X_t) \geq c(\dim X_t)^\alpha$, or a sequence of spaces X_{t_n} such that $k_\infty(X_{t_n}; \varepsilon) \to \infty$ $(n \to \infty)$ for any $\varepsilon > 0$ (i.e., in the second case, X_{t_n} contains a $(1 + \varepsilon)$-isomorphic copy of ℓ_∞^k of an increasing (with $n \to \infty$) dimension).

1.7. [**AlM1**] For any $X = (\mathbb{R}^n, \|\cdot\|)$, $k_2(X) \cdot k_\infty(X; 1) \geq e^{c\sqrt{\log n}}$.

(We note that a geometrical interpretation of the results 1.1–1.7 gives the existence of special central symmetric sections of symmetric compact bodies.)

The following few formulas demonstrate, in a more straightforward way, the connections of type-cotype notions with the linear structure of a space:

1.8. (i) [**FLM**] $k_2(X) \geq c(\dim X)/C_2(X)^2$ and also $\geq c(\dim X)^{2/q}/[C_q(X)]^2$ (Examples: for $1 \leq p < 2$, $k_2(\ell_p^n) \geq cn$; for $p > 2$, $k_2(\ell_p^n) \simeq cn^{2/p}$.)

The dual version of the case $p = 1$ has an interesting geometrical interpretation: Fix $\varepsilon > 0$ and let $n > Ck/\varepsilon^2$ (C is a numerical constant); then there exist intervals $\{I_i\}_{i=1}^n \subset \mathbb{R}^k$ of length 1 centered at $O \in \mathbb{R}^k$ such that the convex body $K = \sum_1^n I_i$ has Hausdorff distance at most ε from some euclidean ball.

(ii) [**Pi4**] $k_p(X; \varepsilon) \geq c(p; \varepsilon)T_p(X)^q$, where $\frac{1}{p} + \frac{1}{q} = 1$ $(1 < p < 2)$; an even stronger result is true: $T_p(X)$ can be substituted by a larger number $ST_p(X)$-stable type p constant of X which is not introduced here. (The case $p = 2$ is also true; see [**MiSch**].)

The last result generalizes the [**JSch**]-formula $k_p(\ell_1^n, \varepsilon) \geq c(p, \varepsilon) \cdot n$ for $1 \leq p \leq 2$. Another generalization of this formula is the following recent theorem of

Schechtman:

1.9. [**Sch1**] (i) *Let $1 \leq p < 2$ and X_k be any k-dimensional subspace of $L_p[0,1]$. Then for any $0 < r \leq p$, ℓ_r^n contains a $(1 + \varepsilon)$-isomorphic copy of X_k for all $k^{1+r/p} \leq c(r; \varepsilon) \cdot n$.*

(ii) *Let $2 < p < \infty$ and X_k be any k-dimensional subspace of $L_p[0,1]$. Then $X_k \overset{1+\varepsilon}{\hookrightarrow} \ell_p^n$ for all $k^{1+p/2} \leq c(p, \varepsilon)n$.*

1.10. An important tool of the type-cotype theory is *Krivine's theorem* [**Kr**] which we state in the quantitative variant, following [**AmM2**]:

$$k_p(X, \varepsilon) \geq c_1(\varepsilon; T)[k_p(X, T)]^{c_2 \cdot (\varepsilon/T)^p} .$$

1.11. *Rademacher projection.* Let

$$L_2(X) = \{f : [0,1] \to X = (\mathbb{R}^n, \|\cdot\|) \mid \|f\|_{L_2} = \left(\int_0^1 \|f(t)\|^2 dt\right)^{1/2} < \infty\}.$$

Consider the subspace $\mathrm{Rad}(X) = \{f \in L_2(X) : f(t) = \sum_1^\infty r_i(t)x_i$ where $x_i \in X$ and $r_i(t)$ is the ith Rademacher function on $[0,1]\}$. The following natural projection is called the Rademacher projection $\mathrm{Rad}_X : L_2(X) \to \mathrm{Rad}(X)$, $\mathrm{Rad}_X f = \sum r_i(t)x_i$ where $x_i = \langle f, r_i \rangle = \int_0^1 f(t)r_i(t)dt \in X$. Note that this projection is, in a sense, a type of linearization (see, e.g., [**BMW** and **Pi2**]). Because of this the logarithmic bound 1.12(a) below is extremely useful.

It follows straightforwardly from the definitions of type, cotype, and Rad_X that $C_q(X) \leq T_p(X^*) \leq \|\mathrm{Rad}_X\| C_q(X)$ for $1/q + 1/p = 1$ (i.e., a duality relation up to $\|\mathrm{Rad}_X\|$).

1.12. The following bounds of Pisier [**Pi7** and **Pi5**] have a crucial importance in the theory:

(a) $\|\mathrm{Rad}_X\| \leq c \log(d_X + 1)$, (and this is, in general, exact [**B1**]);

(b) $\|\mathrm{Rad}_X\| \leq f(k_1(X, 1))$, where $f(t)$ is defined and finite for any $t < \infty$ and $f(t) \to \infty$ if $t \to \infty$.

All previous examples could lead to the conclusion that the power-type estimates on dimensions of subspaces which we are looking for are typical, while a proportional type (which we had in some examples) is an exceptional one. However, we will show next that there exists a nontrivial theory of subspaces of a proportional dimension.

2. A proportional theory; applications to classical convexity theory.
In this section we study linear subspaces and quotient spaces of $X = (\mathbb{R}^n, \|\cdot\|)$ of a *proportional dimension* λn. We also equip $\mathbb{R}^n$ with a euclidean norm $|\cdot|$ and $D = \{x \in \mathbb{R}^n \mid |x| \leq 1\}$ is the euclidean ball. Denote sX a subspace of X and qY a quotient space of a space Y.

2.1. THEOREM [**Mi2, Mi3**]. *For any space $X = (\mathbb{R}^n, \|\cdot\|)$ and any $1/2 < \lambda < 1$, there exists qsX—a quotient of a subspace of X—such that $\dim qsX \geq \lambda n$ and $d_{qsX} \leq C(1 - \lambda)^{-1} \log(1 - \lambda)^{-1}$.*

One of the direct consequences of this theorem is the inverse form of Santalo's inequality [**S**]. Let $K \subset \mathbb{R}^n$ be a convex compact body with the origin 0 belonging to the interior $\overset{\circ}{K}$ of K. Then $K^\circ = \{x \in \mathbb{R}^n \mid (x, y) \leq 1$ for any $y \in K\}$ is the dual body. Consider

$$\text{v.r. } K/D \overset{\text{def}}{\equiv} (\text{Vol } K/\text{Vol } D)^{1/n}$$

(the volume ratio of K by D).

2.2. THEOREM [**BMi1**]. *There exists a numerical constant $c > 0$ such that for any integer n and any convex body $K \subset \mathbb{R}^n$, $0 \in \overset{\circ}{K}$,*

$$c \leq (\text{v.r. } K/D) \cdot (\text{v.r. } K^\circ/D) .$$

2.3. Theorem 2.1 opens a new opportunity to study different kinds of volume inequalities. An important observation in this approach is a *"weak dependence" of volume* on dimension in the following sense. Let K be a symmetric convex compact body in the euclidean space $(\mathbb{R}^n, |\cdot|)$. Denote $sK = K \cap (s\mathbb{R}^n)$ (as before $s\mathbb{R}^n$ is a subspace of $\mathbb{R}^n$), $qK = \text{Proj } K$ (the orthogonal projection of K on a space $q\mathbb{R}^n$), and

$$d_K \equiv d(K, D) \overset{\text{def}}{=} \inf\{a \cdot b \mid D \subset bK \subset abD\}.$$

Then [**BMi1**] v.r.$(sK/sD) = $ v.r.$(K/D)(1 + o(1))$ (where $o(1) \to 0$ if $n \to \infty$) if $(n - \dim(s\mathbb{R}^n))/n \simeq o(1/\log d_K)$. (Similarly for qK.)

2.4. *Isomorphic symmetrization.* By §2.3, Theorem 2.1 allows us, with no essential change of volume and with a small decrease of dimension, to replace the symmetric convex body K (viewed as the unit ball of a space X) by another convex body K_1 (the unit ball of $X_1 = qsX$) which is much closer to a euclidean ball than the original K (say $d_{X_1} \leq C(\log d_X)^2$), and so on. Finally, after a finite number of steps, we pass to a space Y in some fixed neighborhood of the euclidean space (again with no essential change of v.r. of the unit ball of Y with respect to K). Therefore we have a symmetrization type procedure, but in an *isomorphic* sense (and *not the isometric one*), because the numerical constant C in Theorem 2.1 does not allow us to approach the euclidean space isometrically. We come to some (fixed, independent of n) neighborhood of an ellipsoid but cannot come (as some examples indeed show) too close to it. The following result gives an example of an application of such a symmetrization.

2.5. THEOREM. ([**Mi4**] *The inverse form of the Brunn-Minkowski inequality.*) *For every convex compact body $K \subset \mathbb{R}^n$ there corresponds an affine transform $u_K : \mathbb{R}^n \to \mathbb{R}^n (\det u_K = 1)$ such that for any two such bodies K_1 and K_2, and any $\varepsilon > 0$,*

$$\{\text{Vol}(u_{K_1} K_1 + \varepsilon u_{K_2} K_2)\}^{1/n} \leq C\{(\text{Vol } K_1)^{1/n} + \varepsilon (\text{Vol } K_2)^{1/n}\}$$

*for some numerical constant C independent of n and the bodies K_i. At the same time this inequality is satisfied also for the polar bodies $(u_{K_1} K_1)^\circ$ and $(u_{K_2} K_2)^\circ$. (For a new simplified proof of this theorem see [**Pi8**].)*

2.6. One important special case of the well-known Pietsch problem (about duality of entropy numbers) can be solved using Theorem 2.5. Let $N(K_1, K_2)$ denote the covering number of K_1 by K_2, i.e.,

$$N(K_1, K_2) = \min \left\{ N \in \mathbb{N} | \exists x_1, \ldots, x_N \in \mathbb{R}^n, \text{ and } K_1 \subset \bigcup_1^N (x_i + K_2) \right\}.$$

(K_1 and K_2 are compact convex symmetric bodies in $\mathbb{R}^n$.) Then, by [KMi], there exist constants $0 < c$ and C such that for every n and every two bodies K_1 and K_2 in $\mathbb{R}^n$,

$$c \leq [N(K_1, K_2)/N(K_2^\circ, K_1^\circ)]^{1/n} \leq C .$$

Note that a proof of Theorem 2.5 uses Theorem 2.1 and all techniques involved in 2.1–2.4. However, being proved, Theorem 2.5 implies 2.1 in a stronger, "probabilistic" version. We state a consequence of this improvement.

2.7. [Mi4, Mi5] *There exists a function $C(\lambda, \mu)$ depending on $0 < \lambda < 1$ and $0 < \mu < 1$ only such that any $X = (\mathbb{R}^n, \|\cdot\|)$ has a basis $e = \{e_i\}_1^n$ with the following property: for every $A \subset [1, \ldots, n]$, $|A| = [\lambda n]$, and every subset $B \subset A$, $k = |B| = [\mu |A|]$,*

$$d(\mathrm{span}\{e_i\}_{i \in A}/\mathrm{span}\{e_i\}_{i \in A \setminus B}, \ell_2^k) \leq C(\lambda, \mu) .$$

2.8. *Problems.* We would like to raise a few questions in the direction of a "proportional" theory, which we consider to be a current important direction of investigation.

PROBLEM 1. Is there an absolute constant C such that every $X = (\mathbb{R}^n, \|\cdot\|)$ contains a subspace E, $\dim E \geq n/2$, such that $C_2(E^*) \leq C$?

(In other words, does any X have a qX such that $\dim qX \geq n/2$ and $C_2(qX) \leq C$?)

If this problem has a positive answer then a number of open problems in Local Theory would be solved. We list some of them below.

PROBLEM 2. Does any $X = (\mathbb{R}^n, \|\cdot\|)$ contain a subspace E, $\dim E \geq n/2$, such that $T_2(E) \leq f(q; C_q(X))$?

(I.e., $T_2(E)$ depends only on the cotype q constant of X for $q < \infty$.)

Problem 2 follows from Problem 1 using [Pi6]. An even weaker version of this problem is open:

PROBLEM 3. Does any $X = (\mathbb{R}^n, \|\cdot\|)$ contain a subspace E, $\dim E \geq n/2$, such that $\|\mathrm{Rad}_E\| \leq f(q; C_q(X))$?

Pisier's following question would follow from Problem 3 (see [Pi7]).

PROBLEM 4. Let $X = (\mathbb{R}^n, \|\cdot\|) \overset{2}{\hookrightarrow} \ell_\infty^N$. Is it true that $N \geq \exp(cn)$ for $c = c(q; C_q(X)) > 0$ if $q < \infty$, $C_q(X) < \infty$?

We recall also the well-known problem (closely related to the above discussion).

PROBLEM 5. Do a number C and an integer function $k = k(n) \to \infty$ $(n \to \infty)$ exist such that every $X = (\mathbb{R}^n, \|\cdot\|)$ contains a k-dimensional subspace

$E \hookrightarrow X$ such that there exists a projection $P : X \to E$, $\|P\| \leq C$, and E has a 2-symmetric basis?

The positive solution for Problem 5 (in a slightly weaker form) would also follow from the positive solution of the following question. Let $u : \ell_2^n \to X$ and $e = \{e_i\}_1^n$ be an orthonormal basis in ℓ_2^n. Define $r_e(u) = \text{Avg}_{\varepsilon_i = \pm 1} \|\sum_1^n \varepsilon_i u e_i\|$.

PROBLEM 6. Is it true that there exists a numerical constant C such that for every $X = (\mathbf{R}^n, \| \cdot \|)$ we may find a linear map $u : \ell_2^n \to X$ and orthonormal bases $e = \{e_i\}_1^n$ and $\eta = \{\eta_i\}_1^n$ such that

$$r_e(u) \cdot r_\eta((u^{-1})^*) \leq Cn \ ?$$

Another extremely interesting direction (connected with §1.8) is reflected in the following question:

PROBLEM 7. Construct explicitly large ℓ_2-subspaces in ℓ_1^n (or, similarly, in ℓ_p^n).

Note that Problem 4 was solved in one special case by [BMi2] and it implies a solution of the so-called dichotomy of cotype problem for invariant spaces:

2.9. THEOREM [BMi2]. *Let Λ be a subset of the dual group Γ of a compact Abelian group G. Then either Λ is a Sidon set (and then the space C_Λ of functions f in $C(G)$ whose Fourier transform $\hat{f}$ is supported by Λ has the cotype 2 property), or this space C_Λ has no finite cotype.*

(In the "local" form, using notations from §1, Theorem 2.9 states that

$$k_\infty(C_\Lambda; 1) \geq f(d(C_\Lambda, \ell_1^{|\Lambda|}))$$

where $f(t) \to \infty$ if $t \to \infty$.)

2.10. The first steps of the proportional theory have already led to a new development. Starting from some knowledge about a linear structure of a space X we are able in some cases to recover its global geometrical properties. For example:

THEOREM [MiP]. *Let a normed space X, $\dim X = \infty$, satisfy the following condition (so-called a weak cotype 2 property): there exists a constant $0 < \varepsilon < 1$ such that $k_2(E) \geq \varepsilon \dim E$ for every finite-dimensional subspace $E \hookrightarrow X$. Then there exists a constant $C(= C(\varepsilon))$ such that the volume ratio*

$$\text{v.r. } E \leq C \tag{$*$}$$

for every finite-dimensional subspace $E \hookrightarrow X$.

Recall a definition of v.r. $E \overset{\text{def}}{=} (\text{Vol } K(E)/\text{Vol } D(E))^{1/\dim E}$ where $K(E)$ is the unit ball of E and $D(E)$ is the maximal volume ellipsoid inscribed in $K(E)$. This definition was introduced by [SzT] to generalize the important property of ℓ_1^n noted by [K](see also [Sz]). In particular, $(*)$ leads to a conclusion that for every $\lambda < 1$ every subspace $E \hookrightarrow X$ contains a $C(\lambda)$-isomorphic copy of ℓ_2^k for $k \geq \lambda \dim E$.

3. Applications to finite metric spaces. The spirit of the theory discussed in the first two sections, and the techniques developed, found unexpected applications in the study of finite metric spaces and (nonlinear) Lipschitz maps. There has been a growing interest in understanding the similarity between asymptotic properties of finite-dimensional normed spaces and finite metric spaces. Some known examples ([**Gr2**, **MarP**, **JL**, **BMW**, **JLSc**, **B2**, **BFM**]) indicate that there is a close analogy between the results obtained for metric spaces and the previously-known results from the Local Theory of Normed Spaces, where the role of the dimension of a normed space is played by the logarithm of the cardinality of the finite metric space. We present some results.

Let (X, ρ_X) and (Y, ρ_Y) be metric spaces. Define

$$\|f : X \to Y\|_{Lip} = \sup\{\rho_Y(f(x_1), f(x_2))/\rho_X(x_1 x_2) \mid x_1 \neq x_2\} \,.$$

3.1. THEOREM [**MarP**]. *Let* $1 \leq p \leq 2$ *and a finite set* $F \subset L_p[0,1]$, $|F| = k$. *Let* $f : F \to \ell_2$. *Then there exists an extension* $\tilde{f} : L_p \to \ell_2$ (*i.e.,* $\tilde{f}|_F = f$) *such that*

$$\|\tilde{f}\|_{Lip} \leq C(\log k)^{1/p-1/2} \cdot \|f\|_{Lip} \,.$$

3.2. THEOREM [**JL**]. *Let* X *be any normed space and a set* $F \subset X$, $|F| = k$. *For any map* $f : F \to \ell_2$ *there exists an extension* $\tilde{f} : X \to \ell_2$ *such that* $\|\tilde{f}\|_{Lip} \leq C(\log k)^{1/2} \cdot \|f\|_{Lip}$.

3.3. THEOREM [**JLSc**]. *Let* S *be a metric space and* X *be a normed space. Let* $T \subset S$, $|T| = k$. *Then any map* $f : T \to X$ *has an extension* $\tilde{f} : S \to X$ *such that* $\|\tilde{f}\|_{Lip} \leq C(\log k)\|f\|_{Lip}$.

3.4. The last result involves $\log k$ instead of $(\log k)^{1/2}$ which is expected from the Local Theory. It could correspond to another difference between the linear and nonlinear theories discovered by Bourgain [**B3**]: The Lipschitz distance of an n-point metric space to the (best chosen) n point subset of the Hilbert space is at most $\sim \log n$ and this is *the right order* (up to a $\log \log n$) and not $(\log n)^{1/2}$ as would be expected from the Local Theory (compare with §1.4).

3.5. The notion of type (see §1.5) can be put in a nonlinear context and be considered for any metric space (a "metric type" by [**BMW**]). Such a view brings new connections also for linear spaces. For example:

THEOREM [**BMW**]. *Let* $C_2^n = \{-1,1\}^n$ *be the* n-*cube with the Hamming metric (i.e.,* $\rho(\bar{\varepsilon}, \bar{\eta}) = \frac{1}{2}\sum_1^n |\varepsilon_i - \eta_i|$ *for* $\bar{\varepsilon}, \bar{\eta} \subset C_2^n$) *and let a normed space* X *contain a* C-*Lipschitz copy of* C_2^n, *i.e., there exists a map* $f : C_2^n \to X$ *such that* $\|f\|_{Lip} \cdot \|f^{-1}|_{\text{Im } f}\|_{Lip} \leq C$. *Then (using the terminology of §1)* $k_1(X;1) \geq \psi(n;C)$ *where* $\psi(n;C) \to \infty$ *if* $n \to \infty$ *for any fixed* $C < \infty$.

3.6. Strangely, type and cotype behave differently when we try to extend these notions to the nonlinear theory. The metric type (corresponding to Rademacher type) can be extended, as we noted in §3.5. However, this is not the case with the notion of cotype. *There is no* notion of a nonlinear cotype (corresponding to

Rademacher cotype). However, it is reasonable to assume that a right notion of nonlinear cotype would be connected with (nonlinear) maps of euclidean spheres S_n (or ε-nets on such spheres) into metric spaces (such antipodal maps were considered in [**AmM1** and **2**]). Therefore it would be an extension of Gaussian cotype. (Note that Gaussian type cannot be extended to a nonlinear case.)

Most questions in this direction are still far from being clarified at this moment.

4. The concentration phenomenon, Levy families.

In this section we describe one of the main concepts used to develop the asymptotic theory of finite-dimensional normed spaces—the so-called concentration phenomenon of invariant measures on high-dimensional structures. Let (X, ρ, μ) be a metric compact set with a metric ρ, diam. $X \geq 1$ and a probability measure μ. Define the concentration function $\alpha(X; \varepsilon)$ of X by

$$\alpha(X; \varepsilon) = 1 - \inf\{\mu(A_\varepsilon) | A \text{ be a Borel subset of } X, \ \mu(A) \geq \tfrac{1}{2}\} \qquad (4.1)$$

(here $A_\varepsilon = \{x \in X | \rho(x, A) \leq \varepsilon\}$). The isoperimetric inequality for the euclidean sphere S^n, equipped with the geodesic distance ρ and a rotation invariant probability measure μ_n, implies that

$$\alpha(S^{n+1}; \varepsilon) \leq \sqrt{\pi/8} \exp(-\varepsilon^2 n/2) \to 0 \quad \text{for } n \to \infty \qquad (4.2)$$

for any fixed $\varepsilon > 0$. Following this example (observed in the twenties by P. Levy [**L**]), we call a family (X_n, ρ_n, μ_n) of metric probability spaces a *Levy family* [**GrM1**] if for any $\varepsilon > 0$, $\alpha(X_n, \varepsilon \cdot \text{diam } X_n) \to 0$ for $n \to \infty$, and a *normal Levy family* [**AmM2**] with constant $(c_1; c_2)$ if

$$\alpha(X_n; \varepsilon) \leq c_1 \exp(-c_2 \varepsilon^2 n) . \qquad (4.3)$$

(The factor diam X_n in (4.3) is omitted because this way most of the examples below become normal Levy families with their natural metric and natural enumeration.)

Let $f \in C(X)$ be a continuous function on a space X with the modulus of continuity $w_f(\varepsilon)$. Define a median L_f (called in Local Theory also a *Levy mean*) as a number such that $\mu\{x \in X : f(x) \geq L_f\} \geq \tfrac{1}{2}$ and $\mu\{x \in X : f(x) \leq L_f\} \geq \tfrac{1}{2}$. Then $\mu\{x : |f(x) - L_f| \leq w_f(\varepsilon)\} \geq 1 - 2\alpha(X, \varepsilon)$. This means that if $\alpha(X, \varepsilon)$ is small, then "most" of the measure of X is concentrated "around" one value of $f(x)$.

Note that back in 1911, E. Borel pointed out the following geometric interpretation of the law of large numbers: let $C^n = [-1, 1]^n$ be the cube in $\mathbf{R}^n$ with the standard euclidean distance "dist." Then diam $C^n = 2\sqrt{n}$. Consider a linear functional f, $f(x) = \sum_1^n x_i$ (i.e., Ker $f = (1, \ldots, 1)^\perp$). Then $(1/2^n)\text{Vol}\{x \in C^n : \text{dist}(x, \text{Ker } f) \geq \varepsilon\sqrt{n}\} = P\{|\tfrac{1}{n}\sum_1^n \xi_i| > \varepsilon | (\xi_i)_{i=1}^n$ are uniformly distributed in $[-1, 1]$ independent random variables$\} \leq c_1 \exp(-c_2 \varepsilon^2 n)$. Therefore, "most" of the volume of C^n is concentrated near a "small slice" (relative to the diameter).

Comparing this example with the definition of a Levy family we see that the concept of a Levy family (and especially a normal Levy family) generalizes the concept behind the law of large numbers in two directions: (a) The measures are not necessarily the product of measures (i.e., no condition of "independence") and (b) Any Lipschitz function on the space is considered instead of linear functionals only.

It still surprises me that normal Levy families are a typical phenomenon and not a very rare one. We shall list below many examples of such families. Most of them have found an intensive use in Local Theory and, moreover, were discovered for Local Theory purposes. It is usually a nontrivial task to estimate the concentration function $\alpha(X; \varepsilon)$ and, consequently, to prove that a family $\{X_n\}$ is a Levy family. At this point we have to use different techniques.

4.A. *Isoperimetric inequalities technique.*

4.A.1. *Riemannian case.* Let μ_X be a normalized riemannian volume element on a connected riemannian manifold without boundary X and let $R(X)$ be a Ricci curvature of X.

THEOREM. (*Gromov* [**Gr1**]; *see also* [**GrM1**] *and App. 1 in* [**MiSch**]). *Let $A \subseteq X$ be measurable and let $\varepsilon > 0$; then $\mu_X(A_\varepsilon) \geq \mu(B_\varepsilon)$ where B is a ball on the sphere $r \cdot S^n$ with $n = \dim X$, and r such that $R(X) = R(r \cdot S^n)(= (n-1)/r^2)$ and $\mu_X(A) = \mu(B)$, μ being the normalized Haar measure on $r \cdot S^n$.*

4.A.2. *Examples* [**GrM1**]. (a) The above theorem shows that the family of orthogonal groups $\{SO(n)\}_{n \in \mathbb{N}}$ equipped with the Hilbert-Schmidt operator metrics and the Haar normalized measures is a normal Levy family with constants $c_1 = \sqrt{\pi/8}$, $c_2 = 1/8$. (This example of a Levy family was applied in topology by [**GrM1**].)

(b) Similarly the family $X_n = \Pi S^n$ (m_n-times), $n = 1, 2, \ldots$, with the product measure and the natural Riemannian metric is a normal Levy family with constants $c_1 = \sqrt{\pi/8}$, $c_2 = 1/2$.

(c) Of course, homogeneous spaces of $SO(n)$ inherit the property of being a Levy family. It follows that any family of Stiefel manifolds $\{W_{n,k_n}\}_{n=1}^\infty$ with $1 \leq k_n \leq n$, or any family of Grassman manifolds $\{G_{n,k_n}\}_{n=1}^\infty$ with $1 \leq k_n \leq n$, is a normal Levy family with constants $c_1 = \sqrt{\pi/8}$ and $c_2 = 1/8$. The last families were also considered earlier (without exact estimate on the concentration functions $\alpha(X; \varepsilon)$) and applied to infinite-dimensional integration by [**Mi6**] and [**Mi7**].

(d) It could be shown also that the family of cubes C^n (and torus T^n) is a normal Levy family (after normalizing their natural ℓ_2-metrics in such a way that diam $C^n = 1$).

4.A.3. The family $\{\mathbb{R}^n\}$ equipped with the Gaussian measure and ℓ_2-metric is a normal Levy family. This useful translation of the Levy result (4.2) was done by C. Borell [**Bo**].

4.A.4. *Combinatorial case.* If $E_2^n = \{-1; 1\}^n$ has the normalized Hamming metric $d(s, t) = |\{i : s_i \neq t_i\}|/n$ and the normalized counting measure then

$\alpha(E_2^n, \varepsilon) \leq \frac{1}{2}\exp(-2\varepsilon^2 n)$. (See [**AmM1**], which uses an isoperimetric fact from [**Har**].)

4.A.5. *Convex bodies in* $\mathbb{R}^n$ *with the Lebesgue probability measure*. The use of Brunn's theorem and the Brunn-Minkowski inequality gives us partial results in this direction (see [**Bo, GrM2**]). However, in the case of uniformly convex bodies, an isoperimetric inequality from [**GrM3**] estimates their concentration functions and gives new examples of Levy families.

4.B. *Martingale approach*. This approach was initiated by Maurey [**Ma**], who showed that

4.B.1. The group Π_n of permutations of $\{1, \ldots, n\}$ with the normalized Hamming metric $d(\pi_1, \pi_2) = |\{i : \pi_1(i) \neq \pi_2(i)\}|/n$ and the normalized counting measure has $\alpha(\Pi_n, \varepsilon) \leq \exp(-\varepsilon^2 n/64)$ (i.e., $\{\Pi_n\}$ is a normal Levy family).

(A number of interesting averaging formulas for vector-valued functions on Π_n were considered and used in Local Theory; see [**Sc, KwSc**].)

4.B.2. The approach of §4.B.1. was developed and extended by G. Schechtman [**Sch2, Sch3**] with many new examples of Levy families.

EXAMPLE (see [**MiSch**], Chapter 7.12). Let G be a group, compact with respect to a translation invariant metric. Let $G = G_0 \supseteq G_1 \supseteq \cdots \supseteq G_n = \{1\}$ be a sequence of closed subgroups of G. If $a_k = \operatorname{diam}(G_{k-1}/G_k)$, $k = 1, \ldots, n$, then $\alpha(G; \varepsilon) \leq 2\exp(-\varepsilon^2/16 \sum_1^n a_k^2)$.

4.C. *Laplacian operator approach*. Let M be a compact connected Riemannian manifold with μ being the normalized riemannian volume element of M. Then the *Laplacian* $-\Delta$ on M has its spectrum consisting of eigenvalues $0 = \lambda_0 < \lambda_1(M) \leq \cdots$.

4.C.1. THEOREM [**GrM1**]. *The concentration function*

$$\alpha(M; \varepsilon) \leq \frac{3}{4}\exp(-\varepsilon\sqrt{\lambda_1(M)}\log 3/2) .$$

4.C.2. The approach can be interpreted in the discrete case (especially for the Cayley graphs); see [**AlM2**]. The use of representation theory and the T-property of Kazhdan [**Kaz**] gives us new Levy families different in spirit from any discussed above. These families found interesting applications in computer science for explicit constructions of superconcentrators [**AlM2**].

5. The concept of a spectrum. The concept of concentration described in §4, is often used in Local Theory through another concept of "spectrum" of uniformly continuous functions on high-dimensional structures. It will be easier to emphasize the main idea in an infinite-dimensional language.

5.1. Let X be an infinite-dimensional Banach space, $S = S(X) = \{x \in X \mid \|x\| = 1\}$ and let $UC(S)$ be the space of all uniformly continuous functions on S. If $f \in UC(S)$ then $a \in \mathbb{R}$ belongs to *the spectrum* $S(f)$ of the function f iff for every $\varepsilon > 0$ and every $n \in \mathbb{N}$ there exists a subspace $E \hookrightarrow X$, $\dim E = n$ and $|f(x) - a| < \varepsilon$ for any $x \in S \cap E$.

THEOREM [**Mi8**]. *For every* $f \in UC(S)$, *the spectrum* $S(f) \neq \emptyset$.

(Using this theorem, we choose a function f in such a way that $f = \text{Const.}$ means a given geometrical property; then, by the theorem, we find subspaces of any large dimension, where this property is "almost" satisfied; see [**Mi8**] for such examples.)

A few more examples in the same spirit; let $S^\infty = S(\ell_2)$:

5.2. If $W_{\infty,2} = \{(x;y) | x \text{ and } y \in S^\infty \text{ and } x \perp y\}$ is a 2-Stiefel manifold then we may similarly define the spectrum $S(f)$ for $f \in UC(W_{\infty,2})$ and, again, $S(f) \neq \emptyset$ for every $f \in UC(W_{\infty,2})$.

5.3. THEOREM [**Mi7**]. *Let* $f \in UC(S^\infty \times S^\infty)$. *There exists a continuous function* $\varphi(t)$, $-1 \leq t \leq 1$, *such that for every* $\varepsilon > 0$ *and any* $n \in \mathbb{N}$ *there exists a subspace* $E \hookrightarrow \ell_2$, $\dim E = n$, *and* $|f(x;y) - \varphi((x,y))| < \varepsilon$ *for every* x *and* $y \in S^\infty \cap E$ (*here* (x,y) *is the inner product of* x *and* y).

5.4. The main observation beyond the notion of spectrum is that the uniformly continuous function on infinite-dimensional G-spaces depends "essentially" only on orbits (as is the case in Theorem 5.3). Of course we may interpret the word "essentially" in different ways. An interpretation by measure will bring us back to the Levy family-notion. However, considering substructures (say, linear subspaces in §5.1–5.3) where a function is "almost" constant, we come to the concept of spectrum. This notion is discussed precisely in [**Mi7, Mi8**] and, in a more general context of G-spaces, in [**GrM1**]. (For the case of riemannian manifolds, see Gromov [**Gr2**].) Note, that the well-known Ramsey type theorems in Combinatorics are very close in spirit to the discussed notion of spectrum.

References

[**AlM1**] N. Alon and V. D. Milman, *Embedding of* ℓ_∞^k *in finite dimensional Banach spaces*, Israel J. Math. **45** (1983), 265–280.

[**AlM2**] _____, λ_1, *isoperimetric inequalities for graphs, and superconcentrators*, J. Combin. Theory Ser. B **38** (1985), 73–88.

[**AmM1**] D. Amir and V. D. Milman, *Unconditional and symmetric sets in n-dimensional normed spaces*, Israel J. Math. **37** (1980), 3–20.

[**AmM2**] _____, *A quantitative finite-dimensional Krivine theorem*, Israel J. Math. **50** (1985), 1–12.

[**B1**] J. Bourgain, *On martingales transforms in finite dimensional lattices with an appendix on the K-convexity constant*, Math. Nachr. **119** (1984), 41–53.

[**B2**] _____, *The metrical interpretation of superreflexivity in Banach spaces*, Israel J. Math. **56** (1986), 222–230.

[**B3**] _____, *On Lipschitz embedding of finite metric spaces*, Israel J. Math. **52** (1985), 46–52.

[**BFM**] J. Bourgain, T. Figiel, and V. D. Milman, *On Hilbertian subsets of finite metric spaces*, Israel J. Math. **55** (1986), 147–152.

[**BMi1**] J. Bourgain and V. D. Milman, *Sections euclidiennes et volume des corps symetriques convexes dans* $\mathbb{R}^n$, C. R. Acad. Sci. Paris, Ser. I **300** (1985), 435–438. (See also: *On Mahler's conjecture on the volume of a convex symmetric body and its polar*, Preprint, I.H.E.S., 1985.)

[**BMi2**] _____, *Dichotomie du cotype pour les espace, invariants*, C.R. Acad. Sci. Paris Ser. A. **300** (1985), 263–266.

[**BMW**] J. Bourgain, V. D. Milman, and H. Wolfson, *On type of metric spaces*, Trans. Amer. Math. Soc. **294** (1986), 295–317.

[**Bo**] C. Borell, *The Brunn-Minkowski inequality in Gauss spaces*, Invent. Math. **30** (1975), 207–216.

[**D**] A. Dvoretzky, *Some results on convex bodies and Banach spaces*, Proc. Internat. Sympos. Linear Spaces (Jerusalem 1960), Jerusalem Academic Press, Jerusalem; Pergamon, Oxford, 1961, pp. 123–160.

[**F**] T. Figiel, *Local theory of Banach spaces and some operator ideal*, Proc. Internat. Congr. Math. (Warsaw, 1982), pp. 961–976.

[**FLM**] T. Figiel, J. Lindenstrauss, and V. D. Milman, *The dimensions of almost spherical sections of convex bodies*, Acta Math. **139** (1977), 53–94.

[**G**] A. Grothendieck, *Sur certaines classes de suites dans les espaces de Banach et le théorèm de Dvoretsky-Rogers*, Bol. Soc. Mat. Sao Paulo **8** (1953), 83–110.

[**Go**] Y. Gordon, *Some inequalities for gaussian processes and applications*, Israel J. Math. **50** (1985), 265–289.

[**Gr1**] M. Gromov, *Paul Levy's isoperimetric inequality*, Preprint, I.H.E.S., 1980.

[**Gr2**] ——, *Filling Riemannian manifolds*, J. Differential Geom. **18** (1983), 1–147.

[**GrM1**] M. Gromov and V. D. Milman, *A topological application of the isoperimetric inequality*, Amer. J. Math. **105** (1983), 843–854.

[**GrM2**] ——, *Brunn theorem and a concentration of volume of convex bodies*, GAFA Seminar Notes, Israel, 1983–1984.

[**GrM3**] ——, *Generalization of the spherical isoperimetric inequality to the uniformly convex Banach spaces*, Compositio (1987) (to appear).

[**Har**] L. H. Harper, *Optimal numberings and isoperimetric problems on graphs*, J. Combin. Theory **1** (1966), 385–393.

[**H-J**] J. Hoffmann-Jorgensen, *Sums of independent Banach space valued random variables*, Aarhus Universitat 1972/73.

[**J**] F. John, *Extremum problems with inequalities as subsidiary conditions*, Courant Anniversary Volume, Interscience, New York, 1948, pp. 187–204.

[**JL**] W. B. Johnson and J. Lindenstrauss, *Extensions of Lipschitz mappings into a Hilbert space*, Contemp. Math. **26** (1984), 189–206.

[**JLS**] W. B. Johnson, J. Lindenstrauss and G. Schechtman, *Extensions of Lipschitz maps into Banach spaces*, Israel J. Math. **54** (1986), 129–138.

[**JS**] W. B. Johnson and G. Schechtman, *Embedding ℓ_p^m into ℓ_1^n*, Acta Math. **149** (1982), 71–85.

[**K**] B. S. Kashin, *Sections of some finite dimensional sets and classes of smooth functions*, Izv. Akad. Nauk. SSSR Ser. Mat **41** (1977), 334–351. (Russian)

[**KMi**] H. König and V. D. Milman, *On the covering numbers of convex bodies*, GAFA-Israel Seminar 85–86, Lecture Notes in Math., vol. 1267, Springer-Verlag, Berlin-New York, pp. 82–95.

[**KS**] B. S. Kashin and S. S. Saakyan, *Orthogonal series*, "Nauka", Moscow, 1984. (Russian)

[**Kaz**] D. Kazhdan, *Connection of the dual space of a group with the structure of its closed subgroups*, Functional Anal. Appl. **1** (1969), 63–65.

[**Kö**] H. König, *Type constants and $(q,2)$-summing norms defined by n vectors*, Israel J. Math. **37** (1980), 130–138.

[**Kr**] J.-L. Krivine, *Sous-espaces de dimension finie des espaces de Banach reticules*, Ann. of Math. **104** (1976), 1–29.

[**Kw**] S. Kwapień, *Isomorphic characterizations of inner product spaces by orthogonal series with vector valued coefficients*, Studia Math. **44** (1972), 583–595.

[**KwSc**] S. Kwapień and C. Schütt, *Some combinatorial and probabilistic inequalities and their application to Banach space theory*, Studia Math. **82** (1985), 91–106.

[**L**] P. Levy, *Problèmes concrets d'analyse fonctionnelle*, Gauthier-Villars, Paris, 1951.

[**Ma**] B. Maurey, *Constructions de suites symétriques*, C.R. Acad. Sci. Paris Sér A-B **288** (1979), 679–681.

[**MarP**] M. B. Marcus and G. Pisier, *Characterizations of almost surely continuous p-stable random Fourier series and strongly stationary processes*, Acta Math. **152** (1984), 245–301.

[**MP**] B. Maurey and G. Pisier, *Series de variables aléatoires vectoriélles indépendantes et propriétés géométriques des espaces de Banach*, Studia Math. **58** (1976), 45–90.

[**Mi1**] V. D. Milman, *A new proof of the theorem of A. Dvoretzky on sections of convex bodies*, Funktsional. Anal. i Prilozhen. **5** (1971), 28–37. (Russian)

[**Mi2**] ____, *Almost Euclidean quotient spaces of subspaces of finite dimensional normed spaces*, Proc. Amer. Math. Soc. **94** (1985), 445–449.

[**Mi3**] ____, *Random subspaces of proportional dimension of finite dimensional normed spaces: Approach through the isoperimetric inequality*, Banach Spaces (Proc. Missouri Conference, 1984), Lecture Notes in Math., vol. 1166, Springer-Verlag, Berlin-New York, pp. 106–115.

[**Mi4**] ____, *An inverse form of the Brunn-Minkowski inequality with applications to Local Theory of Normed Spaces*, C.R. Acad. Sci. Paris, Sér. I **302** (1986), 25–28.

[**Mi5**] ____, *Geometrical inequalities and mixed volumes in Local Theory of Banach Spaces*, Astérisque **131** (1985), 373–400.

[**Mi6**] ____, *Asymptotic properties of functions of several variables defined on homogeneous spaces*, Soviet Math. Dokl. **12** (1971), 1277–1281.

[**Mi7**] ____, *On a property of functions defined on infinite-dimensional manifolds*, Soviet Math. Dokl. **12** (1971), 1487–1491.

[**Mi8**] ____, *Geometric theory of Banach spaces. II, Geometry of the unit sphere*, Russian Math. Survey **26**, no. 6 (1971), 80–159. (Translated from Russian)

[**MiP**] V. D. Milman and Pisier G., *Banach spaces with a weak cotype 2 property*, Israel J. Math. **54** (1986), 139–158.

[**MiSch**] V. D. Milman and G. Schechtman, *Asymptotic theory of finite dimensional normed spaces*, Lecture Notes in Math., vol. 1200, Springer-Verlag, Berlin-New York, 1986.

[**MiW**] V. D. Milman and H. Wolfson, *Minkowski spaces with extremal distance from the euclidian space*, Israel J. Math. **29** (1978), 113–131.

[**Pel**] A. Pelczynski, *Structural theory of Banach spaces and its interplay with analysis and probability*, Proc. Internat. Congr. Math. (Warsaw, 1982), pp. 237–270.

[**Pi1**] G. Pisier, *Sur les espaces de Banach de dimension finie à distance extrémale d'un espace euclidien, d'après V.D. Milman et H. Wolfson*, Exposé no. 16, Séminaire d'Analyse Fonctionnelle 78/79, Ecole Polytechnique-Palaiseau.

[**Pi2**] ____, *Probabilistic methods in the geometry of Banach spaces* (CIME, Summer School, Varenna 85), Lecture Notes in Math., vol. 1206, Springer-Verlag, Berlin-New York, pp. 167–241.

[**Pi3**] ____, *Finite rank projections on Banach spaces and a conjecture of Grothendieck*, Proc. Internat. Congr. Math. (Warsaw, 1982), pp. 1027–1039.

[**Pi4**] ____, *On the dimension of the ℓ_p^n-subspaces of Banach spaces, for $1 \leq p < 2$*, Trans. Amer. Math. Soc. **276** (1983), 201–211.

[**Pi5**] ____, *Holomorphic semi-groups and the geometry of Banach spaces*, Ann. of Math. **15** (1982), 375–392.

[**Pi6**] ____, *Martingale theory in harmonic analysis and Banach spaces* (Cleveland, 1981), Lecture Notes in Math., vol. 939, Springer-Verlag, Berlin-New York.

[**Pi7**] ____, *Remarques sur un résultat non publié de B. Maurey*, Sem. d'Anal. Fontionnelle 1980/81, Ecole Polytechnique, Paris.

[**Pi8**] ____, *A simpler proof of several results of V. Milman*, Preprint.

[**S**] L. A. Santalo, *Un invariante afin pasa los cuerpos convexos del espacio de n dimensiones*, Portugal. Math. **8** (1949), 155–161.

[**Sc**] C. Schütt, *On the positive projection constant*, Studia Math. **78** (1984), 185–198.

[**Sch1**] G. Schechtman, *More on embedding subspaces of L_p in ℓ_r^n* (to appear)

[**Sch2**] ____, *Levy type inequality for a class of metric spaces*, Martingale theory in harmonic analysis and Banach spaces, Springer-Verlag, Berlin-New York, 1981, pp. 211–215.

[**Sch3**] ____, *Random embedding of euclidean spaces in sequence spaces*, Israel J. Math. **40** (1981), 187–192.

[**Sz**] S. J. Szarek, *On Kashin almost Euclidean orthogonal decomposition of ℓ_1^n*, Bull. Acad. Polon. Sci. Sér. Sci. Tech. **26** (1978), 691–694.

[SzT] S. J. Szarek and N. Tomczak-Jaegermann, *On nearly Euclidean decompositions for some classes of Banach spaces*, Compositio Math. **40** (1980), 367–385.

[T1] N. Tomczak-Jaegermann, *Finite dimensional operator ideals and the Banach-Mazur distance* (to appear).

[T2] ——, *Computing 2-summing norm with few vectors*, Ark. Mat. **17** (1979), 273–277.

TEL AVIV UNIVERSITY, RAMAT-AVIV, TEL AVIV, ISRAEL

Гомеоморфизмы окружности, модификации функций и ряды Фурье

А. М. ОЛЕВСКИЙ

Введение. Мы рассматриваем ряды Фурье непрерывных функций, определенных на окружности $T = \mathbf{R}/2\pi\mathbf{Z}$:

$$f \sim \sum_{n \in \mathbf{Z}} \hat{f}(n)e^{int}. \tag{1}$$

Тема нашего доклада состоит в следующем: как соотносятся основные свойства разложений (1), такие как равномерная сходимость ряда или суммируемость преобразования Фурье, с топологической и метрической структурой функции. В какой мере упомянутые свойства могут быть улучшены посредством преобразований, сохраняющих или мало меняющих эту структуру. В качестве таких преобразований рассматриваются:

гомеоморфизмы окружности на себя; важно различать среди них абсолютно непрерывные и сингулярные гомеоморфизмы;

так называемые «исправления», т.е. изменения значений функции на произвольных множествах малой лебеговой меры.

Согласно теореме Ж. Пала (1914), улучшенной Г. Бором (1935) для каждой действительной функции $f \in C(T)$ можно указать гомеоморфизм $\varphi\colon T \to T$ такой, что суперпозиция $F = f \circ \varphi$ имеет равномерно сходящееся разложение (1). Метод доказательства позволяет одновременно достичь достаточно быстрого убывания коэффициентов Фурье, в частности, условия

$$\hat{F} \in l_p(\mathbf{Z}) \qquad \forall p > 1. \tag{2}$$

Долгое время оставалось неясным как обстоит дело при $p = 1$. Можно ли произвольную вещественную функцию $f \in C(T)$ привести надлежащей заменой переменной в алгебру $A(T)$ абсолютно сходящихся рядов Фурье? Эта задача, поставленная Н. Н. Лузиным, в значительной мере стимулировала исследования последних лет по данной теме.

В задаче исправления важную роль играет теорема Д. Е. Меньшова (1940), усиливающая классическое C-свойство Лузина. Согласно этой теореме для каждой функции $f \in C(T)$ можно указать функцию F с равномерно сходящимся рядом Фурье, отличающуюся от f на множестве, мера которого меньше наперед заданного положительного ε.

Можно ли подобным образом достичь условия (2) при каком-нибудь $p < 2$, или, как говорят, устранить особенность Карлемана — этот вопрос также долго оставался открытым.

Ниже дается обзор результатов докладчика, содержащих, в частности, решения обеих поставленных задач, а также работ других авторов по смежным вопросам. Мы отметим некоторые нерешенные задачи (различной степени трудности). В заключительной части мы затронем некоторые аспекты общих ортогональных разложений.

1. Пространства U и A_p.

1. Через $U(T)$ обозначается банахово пространство функций, представимых равномерно сходящимся рядом (1), с нормой

$$\|f\| = \sup_{\nu} \| \sum_{|n|<\nu} \hat{f}(n)e^{int} \|_{C(T)}.$$

Через $A_p(T)$ — преобразование Фурье пространства $l_p(\mathbf{Z})$, с нормой

$$\|f\| = \|\hat{f}\|_{l_p}.$$

Возникнув в рамках гармонического анализа, эти пространства представляют интерес с различных точек зрения. В частности $A_1 (= A)$ является модельным примером в теории банаховых алгебр, на котором впервые были осмыслены некоторые идеи этой теории. В исследовании данных пространств с успехом используются стохастические методы; некоторые свойства обнаруживают арифметическую природу.

Классический аспект рядов Фурье состоит в исследовании метрических и дифференциальных свойств функции, обеспечивающих ее принадлежность рассматриваемым пространствам. Например, теорема Хаусдорфа-Юнга: $L^q \subset A_{q/q-1}$, $1 \le q \le 2$ показывает, как с увеличением q повышается степень суммируемости преобразования Фурье. При $q > 2$ подобный эффект не имеет места: даже непрерывная функция может не принадлежать никакому классу A_p с $p < 2$. В таких случаях говорят, что f обладает *карлемановской особенностью*. Движение по гёльдеровской шкале $H^\alpha(T)$ приводит к дальнейшему улучшению скорости убывания преобразования Фурье:

$$H^\alpha \subset A_p, \qquad \alpha > \alpha(p) = 1/p - 1/2, \tag{3}$$

причём результат окончателен: для каждого p, $1 \le p < 2$, класс A_p уже не содержит целиком класса Гёльдера с *критическим показателем* $\alpha(p)$ (С. Н. Бернштейн, О. Сас). Принадлежность $U(T)$ требует меньшей гладкости; соответствующее неулучшаемое условие выражается в терминах молуля непрерывности: $\omega_f(\delta) = o(\ln 1/\delta)^{-1}$; оно тесно связано с логарифмическим ростом констант Лебега.

2. Условия гладкости, впрочем, схватывают лишь «поверхностный слой» пространств U и A_p. Наиболее тонкие их свойства определяются функциями с медленно убывающим модулем непрерывности. Характерна с этой точки зрения неустойчивость относительно гладкой замены переменной. Согласно теореме А. Бёрлинга–Н. Хелсона, З. Л. Лейбензона гомеоморфизмы окружности, инвариантно действующие в алгебре $A(T)$, сводятся к повороту и симметрии. Представляется вероятным, что подобный результат справедлив и для U. Первые примеры в этом направлении указаны П. Тураном. Дальнейшие результаты принадлежат Ж. Клуни и Л. Алпару. Последний, в частности, доказал [1], что для любого нелинейного аналитического диффеоморфизма φ окружности найдется функция $f \in U$, $\omega_f(\delta) = O(\ln 1/\delta)^{-\alpha}$, такая, что $f \circ \varphi \notin U$. В отношении A, кажется, неизвестно, можно ли сопроводить упомянутую выше теорему трёх авторов соответствующей оценкой гладкости; например, верно ли, что для любого нелинейного гомеоморфизма φ можно указать $f \in A \cap H^{1/2}$ такую, что $f \circ \varphi \notin A$? Нам неизвестно также, рассматривался ли вопрос о гомеоморфизмах, действующих в A_p, $p > 1$.

Отметим еще следующую теорему Р. Кауфмана (1974), см. [11], в аналитическом случае доказанную ранее Алпаром: пусть φ — гомеоморфизм класса C^ν, $\nu \geq 3$, причём производные $\varphi^{(k)}$, $2 \leq k \leq \nu$, не обращаются в нуль одновременно. Тогда φ переводит $A(T)$ в $U(T)$.

3. Пространство U, в отличие от A не образует алгебры относительно поточечного умножения (Р. Салем). Возникает вопрос: какой запас функций можно получить, исходя из элементов U, применяя конечное число операций умножения и сложения? Иными словами, какова минимальная алгебра, содержащая U? Ответ даёт следующая

ТЕОРЕМА 1 [25]. *Алгебра функций, порожденная* $U(T)$, *совпадает с алгеброй всех непрерывных функций на окружности. Точнее, каждая* $f \in C(T)$ *представима в виде*

$$f = \varphi_1 \cdot \varphi_2 + \varphi_3, \qquad \varphi_i \in U(T), \ 1 \leq i \leq 3. \tag{4}$$

При этом спектр гармоник, входящих в разложение (1) *сомножителей* $\varphi_{1,2}$ *можно сосредоточить, соответственно, на положительной и отрицательной полуосях. Если* f *пробегает заданный компакт в* $C(T)$, *то* φ_1 *в* (4) *можно выбрать универсальной.*

Наводящим соображением (не дающим, впрочем, никаких указаний к доказательству) может служить сопоставление двух классических фактов: произведение двух сходящихся рядов — суммируется методом Чезаро; ряд Фурье любой непрерывной функции суммируется этим методом. Не исключено, что каждая $f \in C(T)$ факторизуется в виде $f = \varphi_1 \cdot \varphi_2$, $\varphi_i \in U$.

2. Неустранимые особенности. Приводимые ниже результаты показывают, что особенности карлемановского типа в общем положении обнаруживают устойчивость по отношению к произвольному варьированию функции, сохраняющему ее значения на каком-либо множестве положительной лебеговой меры. В этом отношении классы A_p, $p < 2$, принципиально отличаются от U.

1. Теорема Меньшова привела к постановке следующего вопроса (см. [**27**]): можно ли для любой непрерывной функции f указать функцию F, отличающуюся от нее на множестве малой меры и принадлежащую классу $A_p(T)$ при каком-либо $p < 2$. Оставался неясным даже случай $p = 1$, хотя в отношении этого случая П. Л. Ульяновым была высказана гипотеза, что ответ — отрицательный. Доказательство было получено спустя 10 лет различными методами в [**13**] и [**22**]. Первая публикация принадлежит И. Кацнелсону [**13**], построившему пример функции $f \in C(T)$, не исправимой в $A(T)$. Метод этой работы, основанный на свойствах так называемой меры Рудина-Шапиро, специфичен и не применим при $p > 1$; построенная там функция f, как отмечено в [**13**], имеет лишь логарифмический модуль непрерывности.

Наш подход позволил получить более сильные результаты.

ТЕОРЕМА 2 ([**22**]; подробнее см. [**24**]). *Для каждого p, $1 \le p < 2$, существует функция f с критическим показателем гладкости: $f \in H^{\alpha(p)}$, которая никаким исправлением на множестве неполной меры не может быть приведена в класс $A_p(T)$.*

При этом оказывается, что эффект неисправимости является типичным. Грубо говоря, объем класса A_p слишком мал, чтобы в него можно было «исправить» любую непрерывную или даже гёльдерову функцию.

Поясним метод доказательства в относительно более простом случае $p = 1$. Мы рассматриваем дискретный вариант задачи исправления. Пусть в $\mathbf{C}^n$ наряду с каноническим базисом $\{e_k\}$ произвольно фиксирован еще один ортонормированный базис $\{\tau_\nu\}$. Рассмотрим куб $Q = \{x; \ |(x, e_k)| \le 1/n \ \forall k\}$ (играющий роль единичного шара в пространстве $H^{1/2}(T)$). Очевидно, $\sum_\nu |(x, \tau_\nu)| \le 1 \ (x \in Q)$. Оказывается, для большинства векторов куба эта оценка не может быть существенно улучшена посредством произвольного изменения фиксированной доли координат (x, e_k). Точнее, при заданных положительных θ, h отнесём вектор $x \in Q$ к множеству $X(n, \theta, h)$, если найдется вектор $\tilde{x} \in \mathbf{C}^n$, для которого

$$\sum_\nu |(\tilde{x}, \tau_\nu)| < h,$$

причём выполняется условие

$$\operatorname{card}\{k; (x - \tilde{x}, e_k) = 0\} > \theta n. \tag{5}$$

Можно показать, что при каждом θ и достаточно малом $h(\theta)$ объём множества $X(n, \theta, h(\theta))$ в овеществленном пространстве $\mathbf{R}^{2n}$ экспоненциально мал по параметру n.

Переход от дискретной задачи к исходной осуществляется на основе связи между преобразованием Фурье на группе окружности и на её дискретной подгруппе. В роли τ_ν выступают характеры последней.

Варьирование p позволяет получить следующий результат.

ТЕОРЕМА 3 [**22**, **24**]. *Существует функция* $f \in C(T)$, *обладающая тем свойством, что каждая функция* $F \in C(T)$, *совпадающая с ней на каком-нибудь множестве положительной меры, обладает особенностью Карлемана.*

Функция f может быть выбрана из любого компакта H^ω с заданной мажорантой $\omega(\delta)$ модуля непрерывности, имеющей порядок убывания к нулю медленнее любой степени.

Из метода доказательства стандартными приемами выводится, что множество функций f в теореме 3 — топологически массивно: дополнение к нему есть счётная сумма нигде не плотных множеств.

2. В дальнейшем эта тема развивалась С. В. Хрущевым [**30**], результат которого относится к случаю $p = 1$. Пользуясь методом работы [**22**], этот автор показал, что траектории некоторых естественных с вероятностной точки зрения случайных процессов почти наверное неисправимы в класс $A(T)$. Впрочем гладкость этих траекторий несколько уступает условию $f \in H^{1/2}$ теоремы 2.

Отметим еще, что теорема 2, будучи окончательной в смысле гладкости по Гёльдеру, оставляет возможность для некоторого уточнения в более подробной шкале классов H^ω. В каждом ли классе H^ω, не вложенном в $A(T)$, можно указать неисправимую функцию — этот вопрос остается открытым.

3. Конечномерный вариант задачи исправления, указанный выше, привёл к следующей постановке вопроса. Пусть попрежнему $\{e_k\}$ — канонический, а $\{\tau_\nu\}$ — любой другой ортонормированный базис в $\mathbf{C}^n$. Введем аналоги норм в C и U:

$$\|x\|_C = \max_k |(x, e_k)|; \qquad \|x\|_U = \max_q \|\sum_{\nu < q} (x, \tau_\nu)\tau_\nu\|_C.$$

Можно ли для любого $x \in \mathbf{C}^n$ и любого θ, $0 < \theta < 1$, указать «исправленный» вектор $\tilde{x}$ с условием (5) так, чтобы выполнялось неравенство: $\|\tilde{x}\|_U \le K(\theta)\|x\|_C$? В случае дискретной тригонометрической системы $\{\tau_\nu\}$ положительный ответ на наш вопрос дал Б. С. Кашин [**14**], использовавший метод случайной расстановки знаков. Общий случай остается открытым.

4. Отметим еще одну нерешенную задачу, относящуюся к данному кругу вопросов. Нами была доказана (1961) следующая теорема: су-

ществует функция $f \in C(T)$, ряд Фурье которой (1) после некоторой перестановки членов расходится почти всюду. Эта теорема имеет давнюю историю: с условием $f \in L^2$ она была сформулирована А. Н. Колмогоровым еще в 1926 году (подробности см. в [21, гл. 3]). Из общей теории ортогональных рядов следует, что $\hat{f}$ не может лежать в l_p, $p < 2$. Мы имеем дело, таким образом, с особенностью более сильной, чем карлемановская. Является ли она устранимой, т.е. каждую ли функцию можно исправить на множестве малой меры так, чтобы ряд Фурье полученной функции сходился почти всюду при любой перестановке? Положительный ответ кажется более вероятным.

3. Продолжение с компактов меры нуль. В отличие от предыдущего в этом параграфе мы рассматриваем ситуацию, когда значения функции не разрешается менять на заданном компакте $E \subset T$. Речь идёт о возможности интерполяции произвольной функции $f \in C(E)$ в класс $U(T)$ или $A_p(T)$, $p < 2$. Хорошо известно, что в обоих случаях для положительного решения необходимо, чтобы E имел меру нуль. Через $B(E)$ обозначаем сужение заданного функционального пространства $B(T)$ на компакт E.

1. В случае $A(T)$ уже для счётных компактов могут возникать препятствия арифметического характера. Компакт E называется хелсоновским, если $A(E) = C(E)$. Соображения двойственности дают эквивалентную формулировку в терминах преобразования Фурье мер, сосредоточенных на E. Множества Хелсона подробно изучались (см., например, [10]). Сколько-нибудь эффективной их характеризации не существует. Известно, что хелсоновость тесно связана с рациональной независимостью.

Аналогичное понятие вводится и на k-мерном торе T^k. Здесь возникают интересные связи с геометрическими характеристиками множеств. Например, Макджи и Вудворд доказали [19], что каждая выпуклая кривая в T^2 — не хелсонова; в то же время график липшицевой функции может быть множеством Хелсона.

2. При интерполяции в A_p, $p > 1$, следует дополнительно требовать непрерывность. Соответствующее банахово пространство $A_p^c(T)$ возникает как пересечение $A_p(T)$ с $C(T)$ с нормой, равной максимуму из двух норм. Соображения двойственности здесь еще менее эффективны, поскольку сопряженное к A_p^c явно не описано. Скажем, что компакт $E \subset T^k$ p-хелсоновский, если $A_p^c(E) = C(E)$. Легко показать, что любой счётный компакт p-хелсоновский при любом $p > 1$. Для компактов меры нуль это уже не так. Более того, справедлива

ТЕОРЕМА 4 [21]. *Для любой непрерывной функции $f \notin A_p(T)$ ($p < 2$ — фиксировано) найдется компакт E меры нуль такой, что сужение $f \mid_E$ не принадлежит $A_p^c(E)$.*

Это означает, что карлемановские особенности *всегда* предопределяются значениями функции на некоторых компактах нулевой меры.

В этих вопросах естественно привлекать более тонкие, чем лебегова мера, метрические характеристики, например, хаусдорфову размерность. Приведем результат, недавно полученный в этом направлении В. Н. Деменко.

ТЕОРЕМА 5. *Если компакт $E \subset T^k$ имеет хаусдорфову размерность λ, то E — p-хелсоновский при $p > p_0 = 2k/(2k - \lambda)$. При $p < p_0$ существует контрпример.*

3. Аналогичная задача для $U(T)$ (см. [21, стр. 79]) была полностью решена Д. Оберлином.

ТЕОРЕМА 6 [20]. *Для любого компакта $E \subset T$ меры нуль имеет место равенство $U(E) = C(E)$.*

Интересно, и, быть может, неожиданно, что этот результат получен как следствие теоремы Л. Карлесона о сходимости почти всюду L^2-рядов Фурье. Возможность использование этой теоремы при исследовании пространства U была обнаружена С. А. Виноградовым [5], показавшим, что неравенство Л. Карлесона–Р. Ханта (ограниченность оператора мажоранты частичных сумм) позволяет дать нетривиальную оценку снизу для нормы функционала $\Phi\colon f \to \int_T f \, d\mu$ над U:

$$\|\Phi\|_{U^*} \geq K \sup_{s>0} \mathrm{mes}\{y; (G\mu)(y) > s\}; \tag{6}$$

здесь $G\mu$ — преобразование Коши-Стилтьеса меры μ, $K > 0$ — абсолютная постоянная. Это неравенство вместе с асимптотикой преобразования Коши сингулярных мер (Г. Буль, О. Д. Церетели и др.) приводит к теореме 6. Подробный обзор по этой теме, содержащий ряд дополнительных результатов, дан С. А. Виноградовым и С. В. Хрущевым [6].

Вопрос о возможности конструктивного доказательство теоремы Оберлина остается открытым. Было бы интересно, по крайней мере, выделить достаточно широкие классы компактов, для которых можно указать явную конструкцию интерполирующего оператора. Отметим в связи с этим следующее предложение (Ж.-П. Кахан, И. Кацнелсон, см. [11]): если E — канторовское множество на окружности, то каждая функция $f \in C(E)$, после линейной интерполяции на смежных интервалах попадает в $U(T)$.

Из других приложений неравенства (6) отметим результат С. В. Кислякова, дающий точную оценку нормы оператора «исправления» в смысле Меньшова.

ТЕОРЕМА 7 [15]. *Для любой функции $f \in C(T)$ и любого $\varepsilon > 0$ существует функция F, отличная от f на множестве меры $< \varepsilon$, причём выполняется неравенство: $\|F\|_U \leq K \ln(1/\varepsilon)\|f\|_{C(T)}$.*

Метод Меньшова даёт не оптимальную оценку: $\|F\|_U \le K(1/\varepsilon)\|f\|_C$, но зато он предъявляет прозрачную конструкцию исправляющего оператора (см. [**2, 21, 30**]).

4. Гомеоморфизмы окружности и алгебра $A(T)$.

1. Теорема Пала-Бора, приведенная во введении, показывает, что принадлежность вещественной непрерывной функции f пространствам U или A_p, $p > 1$, не накладывает на нее с топологической точки зрения никаких ограничений. Аналогичная задача для $p = 1$ была поставлена Н. Н. Лузиным (см. [**2**, стр. 306]). Решение было получено в нашей работе [**23**]. Оказалось, что существует топологический инвариант, способный различать классы $C(T)$ и $A(T)$. Он учитывает характер осцилляции функции в левой и правой полуокрестностях произвольной точки. Основную роль играет следующий факт: *если F представляет собой высокочастотное биение единичной амплитуды при $t > 0$, и аннулируется при $t < 0$, то $\|F\|_{A(T)}$ неизбежно велика, причём оценка снизу этой нормы зависит только от частоты.* Точная формулировка такова (см. [**24**, лемма 3.3]): пусть в некоторой γ-окрестности нуля функция F имеет вид:

$$F(t) = \begin{cases} \sin N g(t) & 0 \le t < \gamma, \\ 0 & -\gamma < t < 0, \end{cases}$$

где g — произвольная строго возрастаяющая функция на $[0, \gamma]$, $g(0) = 0$, $g(\gamma) = \pi$. Тогда выполняется неравенство:

$$\|F\|_{A(T)} > K \ln^\alpha N. \tag{7}$$

где K и α — абсолютные положительные постоянные.

Стоит заметить, что если выход на нуль происходит в режиме затухающих колебаний, то эффект теряется. Ясно, например, что функция $F_N(t) = a(t) \sin Nt$ где $a(t)$ — индикатор отрезка $[\delta, \pi - \delta]$, линейно интерполированный на участках $[0, \delta]$ и $[\pi - \delta, \pi]$, удовлетворяет оценке $\|F_N\|_{A(T)} < K(\delta)$.

Отметим, что оценки снизу для величины $\|e^{iNg}\|_A$ возникают при исследовании автоморфизмов алгебры $A(T)$; в частности, для гладких нелинейных g они основаны на лемме ван дер Корпута. Однако, в отличие от (7) эти оценки носят асимптотический характер по N при фиксированной g; нам же необходима именно равномерность по g. Важную роль в доказательстве неравенства (7) играет аналогия с методом П. Коэна оценки L_1-норм экспоненциальных сумм [**16**].

Неравенство (7) приводит к теореме, дающей ответ на задачу Лузина:

ТЕОРЕМА 8 ([**23**], подробнее см. [**24**]). *Существует вещественная функция $f \in C(T)$, обладающая тем свойством, что суперпозиция её $F = f \circ \varphi$ с произвольным гомеоморфизмом φ окружности T на себя не принадлежит алгебре $A(T)$.*

Функция f, которую мы строим, имеет модуль непрерывности порядка $O(\ln 1/\delta)^{-\alpha}$. Неизвестно, в какой мере эта гладкость близка к максимально возможной; в частности, не исключено, что гёльдеровы функции посредством надлежащей замены приводятся в $A(T)$.

2. Параллельно с нашей работой [**23**] Ж-П. Кахан и И. Кацнелсон [**12**] развили другой подход к проблеме Лузина, позволивший, впрочем, получить лишь следующий результат: существует пара вещественных функций f_1 и f_2, не приводимых в $A(T)$ одновременно одной и той же заменой переменной. Этот подход основан на следующих соображениях: «плохие» функции могут приводиться в $A(T)$ только с помощью весьма негладких замен; однако, такие замены выводят «хорошие» функции из этого пространства (подробнее см. [**11, 24**]).

Интересно, что этот результат (и приводящие к нему соображения) сохраняются при малых возмущениях класса $A(T)$. Опеределим для заданной числовой последовательности $\varepsilon(n)$, $n \in \mathbf{Z}$, класс функций

$$A_\varepsilon = \left\{ F \colon \sum |\hat{F}(n)\varepsilon(n)| < \infty \right\}.$$

ТЕОРЕМА 9 (Кахан, Кацнелсон, см. [**11, 24**]. *Существует последовательность* $\varepsilon(n) \to 0$ $(|n| \to \infty)$ *и пара вещественных функций* $f_1, f_2 \in C(T)$, *не приводимых одним и тем же гомеоморфизмом* φ *в класс* A_ε.

В то же время теорема 8 с этой точки зрения окончательна и не допускает подобного усиления. Это следует из результата, принадлежащего А. А. Саакяну:

ТЕОРЕМА 10 [**26**]. *Для любой последовательности* $\{\alpha(n)\} \notin l_1(\mathbf{N})$ *(при минимальных условиях регулярности) и для любой вещественной функции* $f \in C(T)$ *существует гомеоморфизм* φ, *такой, что* $F = f \circ \varphi$ *удовлетворяет условию* $\hat{F}(n) = O(\alpha(|n|))$.

Что касается класса U, то в него можно одновременно привести не только пару функций, но даже любой компакт пространства $C(T)$.

ТЕОРЕМА 11 (Кахан, Кацнелсон, 1978, см. [**11**]). *Для любого модуля непрерывности* ω *можно указать гомеоморфизм* φ *окружности, такой что* $f \circ \varphi \in U$ *при всех* $f \in H^\omega$.

Последние две теоремы основаны на идейно близких соображениях, напоминающих метод исправления Меньшова.

Было бы интересно выяснить, можно ли в теореме 11 достичь коэффициентного условия: $\widehat{f \circ \varphi}(n) = o(1/|n|)$, обеспечивающего принадлежность пространству $U(T)$.

5. Роль сингулярных гомеоморфизмов. Замена переменной стирает грань, отличающую функцию класса $A_p^c(T)$, $p > 1$, от произвольной непрерывной. Оказывается, однако, что этот эффект носит чисто топологический характер. Метрическая структура функции под действием «улучшающего» гомеоморфизма в общем положении неизбежно искажается.

Гомеоморфизм φ называется *сингулярным*, если он переводит некоторое множество полной меры в множество меры нуль.

ТЕОРЕМА 12 [24]. 1. *Для каждого $p \in [1, 2[$ можно указать функцию f, принадлежащую классу Гёльдера с критическим показателем $\alpha(p)$ (3) такую, что если $f \circ \varphi \in A_p(T)$, то гомеоморфизм φ — сингулярный.*

2. *Существует функция $f \in C(T)$, суперпозиция которой с любым несингулярным гомеоморфизмом обладает особенностью Карлемана.*

Этот результат идейно близок теоремам §2; доказательство основано на соответствующей модификации использованного там метода. Можно осуществить синтез упомянутых результатов. Скажем, что функции f и g, определенные на окружности, *сопряжены*, если можно указать компакты $F, G \subset T$ положительной меры и несингулярный гомеоморфизм $\varphi \colon F \to G$ так, что выполняется равенство: $f \mid_F = g \circ \varphi$.

ТЕОРЕМА 13. 1. *Для каждого $p \in [1, 2[$ существует функция $f \in H^{\alpha(p)}(T)$, не сопряженная ни с какой функцией класса $A_p(T)$.*

2. *Существует $f \in C(T)$ такая, что каждая сопряженная с ней функция обладает особенностью Карлемана.*

Остается открытым вопрос, также указанный Лузиным (см. [2]): можно ли произвольную $f \in C(T)$ привести в $U(T)$ посредством абсолютно непрерывного гомеоморфизма. Положительный ответ кажется более вероятным.

6. Инвариантные подмножества. Представляет интерес описание инвариантных частей рассматриваемых функциональных классов, т.е. совокупности элементов, остающихся в заданном классе при действии любого гомеоморфизма окружности. Иными словами, речь идет об неулучшаемых условиях принадлежности функции данному классу, выраженных в топологически инвариантных терминах. В отношении U эта задача решена А. Бернштейном и Д. Ватерманом [3] в терминах гармонической вариации. Легко показать, что инвариантная часть алгебры $A(T)$ — тривиальна: состоит из констант. Для классов A_p, или A_p^c, $p > 1$, вопрос не решен.

Подобные вопросы могут быть содержательны и для других классов функций и для других групп преобразований. В частности, О. Д. Церетели [31] изучал «перестановки» функций, т.е. унитарные операторы, отвечающие обратимым сохраняющим меру отображениям окружности на себя. Один из его результатов состоит в том, что инвариантная часть класса $\operatorname{Re} H^1(T)$ (H^1 — класс Харди) по отношению к указанной группе операторов совпадает с классом $L \ln^+ L$. Интересное метрическое условие возникает при описании инвариантной части дополнения предыдущего класса до $L(T)$.

Отметим еще слелующий результат А. Б. Гулисашвили [8]: каждая функция $f \in L(T)$ может быть приведена в класс $\bigcap_{p>2} A_p(T)$ «перестановкой», тождественной вне множества малой меры.

К этому же кругу относится вопрос о соотношении топологических и дифференциальных свойств функции. А. Брукнер и К. Гофман впервые рассмотрели вопрос об условиях приводимости функции посредством гомеорфизма окружности в класс $C^1(T)$. Оказалось [4], что для этого необходимо и достаточно выполнение следующих двух условий:

1° f имеет ограниченную вариацию;

2° f-образ множества критических точек E_f имеет меру нуль. (По определению $t_0 \in E_f$, если f не является постоянной или строго монотонной ни в какой окрестности t_0.)   Обобщение на другие классы гладкости получено В. В. Лебедевым [18].   Приведем его результат, впервые опубликованный в [24].

ТЕОРЕМА 14. *Функция $C(T)$ приводится посредством гомеоморфизма в класс $C^k(T)$ (натуральное k — фиксировано) если и только если выполнено предыдущее условие 2° и $\sum_\nu |\omega_\nu|^{1/k} < \infty$, где ω_ν — колебания f на смежных с E_f интервалах. В случае, если это имеет место при всех k, функция f приводится в C^∞.*

Аналогичный результат независимо получили М. Лашкович и Д. Прейс [17].

Интересно было бы найти многомерные аналоги этих результатов. Возможно, что это задача связана с методами А. С. Кронрода–А. Г. Витушкина, см. [7].

7. Об ортогональных разложениях.

1. Один из итогов развития теории ортогональных рядов за последние 25 лет состоит в более отчётливом осознании грани, отделяющей общие закономерности, присущие всем полным или всем ограниченным ортонормированным системам функций (ОНС), от других свойств, где возможны, подчас, неожиданные контрпримеры. Так было установлено, что *ОНС Ψ_n не может быть равномерно ограничена одновременно со своими функциями Лебега $\mathcal{L}_n$* (Олевский, 1965). Точнее, если

$$|\varphi_n(x)| \le K, \tag{8}$$

то на некотором множестве положительной меры выполняется соотношение

$$\varlimsup_{n \to \infty} \mathcal{L}_n(x) = \infty. \tag{9}$$

Это означает невозможность ограниченной ОНС, доставляющей каждой непрерывной функции всюду сходящееся разложение Фурье, или каждой функции класса L — разложение, сходящееся в среднем.   Чуть позже (1966) нами была дана точная оценка скорости роста функций Лебега равномерно ограниченных ОНС: для бесконечно многих n выполняется неравенство

$$\mathcal{L}_n(x) > \alpha \ln n \qquad (\forall x \in E, \ \text{mes}\, E > 0). \tag{10}$$

Она дает оценку снизу скорости расходимости рядов Фурье по системам (8). Однако, в отличие от тригонометрического случая, соотношение (10) не влечёт за собой соответствуюшей гладкости функции с расходящимся рядом Фурье: можно построить полную равномерно ограниченную ОНС, доставляющую каждой функции, принадлежащей наперед заданному классу H^ω, равномерно сходящееся разложение. Подробно об этих результатах см. [**21**, гл. 1].

Эта тема в дальнейшем развивалась другими авторами (С. В. Бочкарёв, С. Шарек, Ж. Бургейн и др.). В частности, первый из них показал (1975), что неравенство (9) вместе с методом А. Н. Колмогорова построения расходящихся рядов Фурье позволяет строить по системам (8) ряды класса L, расходящиеся на множествах положительной меры.

Для полных ОНС был развит подход, основанный на некоторых специальных свойствах классической системы Хаара χ_n. Впервые он был апробирован при доказательстве невозможности полной ОНС безусловной сходимости (Олевский, Ульянов, 1961). В дальнейшем обнаружилась экстремальная роль системы Хаара среди всех полных систем. Грубо говоря, нами было показано (1966), что если в некотором классе функций, инвариантном относительно сохраняющих меру преобразований отрезка, имеет место некоторое явление расходимости ряда Фурье, то такое явление неизбежно для любой полной ОНС, см. [**21**, гл. 3]. Этот результат имеет ряд приложений (одно из них — приведенная выше теорема 4). Подход, основанный на системе Хаара, был впоследствии (1973) применен Ф. Г. Арутюняном в задаче представления измеримых функций почти всюду сходящимися рядами; им был выделен широкий класс ОНС, включающий классические системы и их перестановки, для которых эта задача имеет положительное решение. В то же время существует пример полной ОНС, для которой такое представление имеют лишь функции класса L^2 (Б. С. Кашин, 1977). Отметим еще, что свойство «если ряд Фурье функции f по полной ОНС сходится почти всюду, то сумма его равна f» может нарушаться в пространствах, чуть расширяющих L^2 [**21**, гл. 4].

2. Некоторые результаты, приведенные в докладе, также могут представить интерес с точки зрения возможности их распространения на более или менее общие классы ОНС. Пусть

$$A_p^\Psi = \{f; (f, \Psi_n) \in l_p\}.$$

Дискретная задача исправления (§2), не предъявляя никаких требований к взаимному расположению базисов, наводит на мысль, что аналоги теорем 2, 3 справедливы в более общей ситуации. В частности, нетрудно распространить эти теоремы на систему Уолша и близкие к ней. Для системы Хаара соответствующие аналоги на основе более простого подхода были еще раньше получены Ю. С. Фридляндом [**29**]. С другой стороны, существует пример А. А. Талаляна (1964) полной ОНС Ψ, по которой каждая функция исправима в класс A_1^Ψ. Такой пример можно организовать из равномерно ограниченных тригонометрических полиномов [**29**]. Было бы интересно выделить достаточно общий класс ОНС, для которых имеют место теоремы о «неисправимости». Следует отметить, что непрерывные функции с карлемановскими особенностями существуют по любым полным ОНС (Олевский, 1961). Также для таких систем классы $H^{\alpha(p)}$ не вложены в A_p^Ψ (Б. С. Митягин, С. В. Бочкарёв, 1964).

В отношении задачи Лузина о гомеоморфизме отметим, что для системы характеров группы p-адических целых чисел соотвествующий аналог этой задачи имеет положительное решение (М. Гатесуп [**32**]).

По поводу теоремы 12 заметим, что не исключена возможность ее распространения на любые полные ОНС.

Вопрос о гомеоморфизмах, действующих в классах A_p, повидимому, не исследовался для систем, отличных от тригонометрической (в частности, для системы Хаара). П. Л. Ульянов, осуществивший, начиная с 60-х годов, широкое исследование свойств системы Хаара, изучил вопрос о внешних суперпозициях, действующих в классах A_1. Он доказал [**28**], что функция $\Phi: \mathbf{R} \to \mathbf{R}$ обладает тем свойством, что $f \in A_1^\chi \Rightarrow \Phi \circ f \in A_1^\chi$ если и только если она липшицева. Это условие существенно слабее (локально) условия аналитичности, фигурирующего в соответствующей теореме П. Леви–И. Кацнелсона, относящейся к $A(T)$. Ульянов предположил, что дальнейшее ослабление условия на Φ невозможно в классе полных ОНС; это согласуется с упомянутой выше экстремальной ролью системы Хаара.

ЛИТЕРАТУРА

1. L. Alpar, *Convergence uniform et changement de variable*, Studio Sci. Math. Hungarica **9** (1974), 267–275.

2. Н. К. Бари, *Тригонометрические ряды*, Физматгиз, 1961.

3. A. Baernstein and D. Waterman, *Functions, whose Fourier series converge uniformly for every change of variable*, Indiana Univ. Math. J. **22** (1972), 569–576.

4. A. Bruckner, *Differentiation of real functions*, Lect. Notes in Math., v. 659, Springer-Verlag, Berlin and New York, 1978.

5. С. А. Виноградов, *Сходимость почти всюду рядов Фурье функций и поведение коэффициентов равномерно сходящихся рядов Фурье*, ДАН СССР, 1976.

6. С. А. Виноградов и С. В. Хрущев, *Свободная интерполяция в пространстве равномерно сходящихся рядов Фурье*, ЛОМИ, 1980.

7. А. Г. Витушкин, *О многомерных вариациях*, Гостехиздат, 1955.

8. А. Б. Гулисашвили, *Об особенностях суммируемых функций*, Записки семинаров ЛОМИ **113** (1981), 76–96.

9. Б. Н. Деменко, *О p-хелсоновых кривых на плоскости*, Матем. заметки **39** (1986), 349–359.

10. Ж.-П. Кахан, *Абсолютно сходящиеся ряды Фурье*, Мир, 1975.

11. J.-P. Kahane, *Quatre leçons sur les homéomorphismes du cercle et les séries de Fourier*, Topics in Modern Harmonic Analysis, v. II, Ist. Naz. Alta Mat. Francesco Severi, Rome, 1983, pp. 955–990.

12. J.-P. Kahane and Y. Katznelson, *Homéomorphismes du cercle et séries de Fourier absolument convergent*, C. R. Acad. Sci. Paris sér. 1 **292** (1981), 271–273.

13. Y. Katznelson, *On a theorem of Menchoff*, Proc. Amer. Math. Soc. **53** (1975), 396–398.

14. Б. С. Кашин, *О некоторых свойствах пространства тригонометрических полиномов, связанных с равномерной сходимостью*, Сообщ. АН Груз. ССР **93** (1979), 281–284.

15. С. В. Кисляков, *Количественный аспект теорем об исправлении*, Записки семинаров ЛОМИ **92** (1979), 182–191.

16. P. Cohen, *On a conjecture of Littlewood and idempotent measures*, Amer. J. Math. **82** (1960), 191–212.

17. M. Laczkovich and D. Preiss, *α-variation and transformation into C^n functions*, Indiana Univ. Math. J. **34** (1985), 405–424.

18. В. В. Лебедев, *Гомеоморфизмы отрезка и гладкость функции*, Матем. заметки **40** (1986).

19. O. McGehee and G. Woodward, *Continuous manifolds in $\mathbf{R}^n$ that are sets of interpolation for the Fourier algebra*, Ark. Mat. **20** (1982), 169–199.

20. D. Oberlin, *A Rudin-Carleson theorem for uniformly convergent Taylor series*, Mich. Math. J. **27** (1980), 307–313 (preprint 1978).

21. A. M. Olevskiĭ, *Fourier series with respect to general orthogonal systems*, Springer-Verlag, Berlin and New York, 1975.

22. А. М. Олевский, *Существование функций с неустранимыми особенностями Карлемана*, ДАН СССР **238** (1978), 796–799.

23. ____, *Замена переменной и абсолютная сходимость ряда Фурье*, ДАН СССР **256** (1981), 284–288.

24. ____, *Модификация функций и ряды Фурье*, Успехи матем. наук **40** (1985), 157–193.

25. ____, *Об алгебрах, порожденных равномерно сходящимися рядами Фурье*, ДАН СССР (1987).

26. А. А. Саакян, *Интегральные модули гладкости и коэффициенты Фурье суперпозиций*, Матем. сборн. **110** (1979), 597–608.

27. П. Л. Ульянов, *Решенные и нерешенные проблемы теории тригонометрических и ортогональных рядов*, Успехи матем. наук 1топ **9** (1964), 3–69.

28. ____, *Абсолютная сходимость рядов Фурье-Хаара от суперпозиций функций*, Analysis Math. **4** (1978), 225–236.

29. Ю. С. Фридлянд, *О неустранимой особенности Карлемана для системы Хаара*, Матем. заметки **14** (1973), 799–807.

30. С. В. Хрущев, *Теорема Меньшова об исправлении и гауссовские процессы*, Труды МИАН **155** (1981), 151-183.

31. О. Д. Церетели, *О сопряженных функциях*, Матем. заметки **22** (1977), 771–783.

32. M. Gatesoupe, *Topics in harmonic analysis on abelian compact groups homeomorphic with the Cantor set*, Topics in Modern Harm. Anal., Vol. II, Ist. Naz. Alta Mat. Francesco Severi, Rome, 1983, pp. 991–1009.

Московский институт электронного машиностроения, Москва 109028, СССР

Proceedings of the International Congress of Mathematicians
Berkeley, California, USA, 1986

Generalizations of Fatou's Theorem

THOMAS H. WOLFF

By Fatou's theorem we mean that a positive harmonic function on the unit
ball $B \subset \mathbf{R}^n$ has radial limits almost everywhere. It is tempting to try to
generalize this result to other elliptic equations, and we will describe what is
known along these lines for equations with bad coefficients and certain nonlinear
equations.

1. Linear equations. Consider, e.g., a divergence-form elliptic equation
$Lu = \text{div}(A\nabla u) = 0$ where $A\colon B \to$ positive symmetric $n \times n$ matrices is
measurable. Suppose L is uniformly elliptic on compact subsets of B (by uniformly elliptic we mean here that the eigenvalues of $A(x)$ are bounded from 0
and ∞ independently of x). Suppose also that the Dirichlet problem is solvable with continuous boundary data. For fixed $z \in B$ one can then define
harmonic measure ω_z by the rule $\int_{\partial B} u \, d\omega_Z = u(z)$ when u is continuous on $\overline{B}$
and $Lu = 0$ on B. For two different z's the corresponding ω_z's are boundedly
absolutely continuous to each other by (Moser's) Harnack inequality, and the
density $K_z = d\omega_z/d\omega_0\colon \partial B \to \mathbf{R}^+$ is called the kernel function.

In many cases it is possible to prove a result of the following type: if $u\colon B \to$
$\mathbf{R}^+$ and $Lu = 0$ then $\lim_{r\to 1} u(rz)$ exists a.e. $(d\omega)$. This result has been
proved, e.g., for equations uniformly elliptic on B by Ancona [1], Caffarelli-Fabes-
Mortola-Salsa [5], and for the Laplacian of a metric with curvature bounded
between two negative constants by Anderson-Schoen [3]. (In the second case,
coordinates should be chosen so that the geodesic spheres centered at 0 are Euclidean spheres.) Recently Ancona [2] has proved a very general version which
includes all or most of the previous results. There are two parts to the arguments:

(i) a representation formula, generally $u(z) = \int K_z(Q) \, d\mu(Q)$ for some positive measure μ on ∂B;

(ii) an analysis of K_z leading to an $L^1(d\omega_0) \to$ weak $L^1(d\omega_0)$ estimate for
the radial maximal function $N\mu(Q) = \sup_r u(rQ)$, from which Fatou's theorem
follows as in the classical case.

The author is an Alfred P. Sloan Foundation Fellow. Research supported by National
Science Foundation Grant DMS-84-07099.

All this is highly nontrivial; I'll just try to give the formal outline. For $|z| < r < 1$ define also $K_z^r = d\omega_z^r/d\omega_0^r$ where ω_z^r is harmonic measure on $|z| = r$ for the Dirichlet problem for L on rB. Since u is continuous on $|z| < 1$ it is immediate that $u(z) = \int_{|z|=r} K_z^r \, d\mu_r$ where $d\mu_r = u \, d\omega_0^r$. The μ_r are positive measures with mass $u(0)$ so they have a w^* cluster point as $r \to 1$. That means that the representation formula follows provided $K_z^r(rQ) \to K_z(Q)$ for fixed $z \in B$, uniformly over $Q \in \partial B$ as $r \to 1$. This convergence is proved in the above papers; the tool seems to be the *boundary Harnack* type theorem according to which two positive solutions vanishing continuously on an open set of ∂B must vanish at the same rate on any smaller open set. To prove the weak type 1 estimate, one defines also the Hardy-Littlewood type maximal function $M\mu(Q) = \sup_D \mu(D)/\omega_0(D)$, where D denotes a surface ball (intersection of a ball with ∂B) centered at Q. Two estimates are then required: the *doubling property* $\omega_0(2D) \le C\omega_0(D)$, which implies by a standard covering argument that $M\mu \in$ weak $L^1(d\omega_0)$, and a suitable decay property of $K_z(Q)$ as z moves away from Q in order to bound $N\mu$ by $M\mu$ pointwise. Again we refer to the papers.

If there is a *natural* measure $d\theta$ on ∂B (e.g., Lebesgue measure, in the case of a uniformly elliptic equation) then one can try to prove the results with respect to this measure or to show, in other words, that $d\omega$ and $d\theta$ are mutually absolutely continuous. For the Laplacian on a Lipschitz domain, this is a famous result of Dahlberg. For general uniformly elliptic equations it is not true (see [4]; they pull back the Laplace equation in $\mathbf{R}^2$ by a quasiconformal mapping whose restriction to the unit circle is a singular homeomorphism of the circle to itself), but a theorem of Fabes-Jerison-Kenig [8] (also [6]) gives optimal regularity on the coefficients for it to hold. Formally speaking again, if the operators $u(z) \to u(rz)$ are bounded in $L^2(d\theta)$ then $d\omega < d\theta$ with $d\omega/d\theta \in L^2$, and the L^2 boundedness is proved in [8] using, among other things, extensions of the techniques used for the Cauchy integral on Lipschitz curves. This even gives that $d\omega$ is an A_∞ weight with respect to $d\theta$, so $d\theta < d\omega$ also. In the setting of the Anderson-Schoen results, one can make sense of the absolute continuity question by taking $d\theta =$ geodesic angle at a fixed point p, at least if one knows that the different $d\theta_p$'s are mutually absolutely continuous. For negatively curved two-dimensional surfaces with compact quotients the $d\theta_p$'s are absolutely continuous to each other, but usually they are all singular to $d\omega$ (communicated by A. Katok).

Fatou-type theorems have also been proved for divergence-form parabolic equations taking limits as the time $t \to 0$. See references in [7]. In this case the *parabolic measure* is always absolutely continuous since the necessary L^2 bound is just the usual L^2-decay for selfadjoint parabolic problems.

2. Some nonlinear situations. We want to discuss counterexamples for the $p - Laplacian$ $\mathcal{L}_p u = \text{div}(\nabla u|\nabla u|^{p-2}) = 0$, $1 < p < \infty$, $p \ne 2$. First we note that there is (at least) one nonlinear equation for which a Fatou theorem has been proved, the parabolic *porous medium* equation $u_t = \text{div}(u^{k-1}\nabla u)$ [7]. Comparing this equation with the p-Laplacian one sees why $\mathcal{L}_p$ should behave worse. The porous medium equation remains uniformly parabolic except as $u \to 0$ or ∞, so (especially since the parabolic meaure is absolutely continuous) one can hope to apply the linear theory and this is actually done in [**7**]. On the other hand, the p-Laplacian is far from uniformly elliptic if u behaves at all badly at the boundary.

The following result is shown for $p > 2$ in [**12**] and for $p < 2$ by John Lewis [**9**]: there are bounded functions $u\colon \mathbf{R}_+^2 (= \{(x,y) \in \mathbf{R}^2\colon y > 0\}) \to R$ with $\mathcal{L}_p u = 0$ such that $\lim_{y\to 0} u(x,y)$ exists almost nowhere. To prove this, one uses the principle that nontrivial a.e. convergence theorems are possible only when there is some sort of cancellation. This is seen clearly in the classical Khinchin theorem that a series $\sum a_k r_k$, r_k bounded iid's, $\{a_k\} \in l^2$, converges almost surely when the $\{r_k\}$ have mean zero, but certainly not otherwise. For harmonic functions the cancellation is given by the mean value property. For the p-Laplacian the necessary cancellation fails by the following lemma [**12, 9**]: there are smooth bounded functions $f\colon \overline{\mathbf{R}_+^2} \to \mathbf{R}$, $\mathcal{L}_p f = 0$, 1-periodic in the first variable, and such that $\int_0^1 f(x,0)\,dx \neq 0$, $\lim_{y\to\infty} f(x,y) = 0$. One can do a type of rudimentary harmonic analysis with dilates of the functions f playing the role of characters: let $\phi(x) = f(x,0)$. If $T_n \to \infty$ rapidly, then (because $\int \phi \neq 0$) there are series of the form $\sigma(x) = \sum a_j L_j(x)\phi(T_j x)$, $\{a_j\} \in l^2$, $\|L_j\|_\infty \leq 1$, $\|L_j\|_{\text{Lip }1} \leq T_{j-1}$, which have bounded partial sums but diverge almost everywhere. For $n < \infty$, let $\hat{\sigma}_n$ be the solution of the Dirichlet problem for $\mathcal{L}_p$ with boundary values the nth partial sum of the series σ. The fact that $f(x,y) \to 0$ as $y \to \infty$ can be used to show that the $\hat{\sigma}_n$ converge uniformly on compacts of $\mathbf{R}_+^2$. The limit function will be a solution of $\mathcal{L}_p = 0$ whose limiting behavior as $y \to 0$ is asymptotic to the series σ, hence boundedly divergent a.e., thus finishing the proof.

Attempts have been made to define harmonic measure for the p-Laplacian, although due to the nonlinearity of the equation one does not expect this set function to be additive. The good approach seems to be that of Martio and coworkers (e.g., [**10**]) based on the Perron process: if Ω is a domain and $E \subset \partial\Omega$ then $\omega_p(E, z) = \inf(u(z)\colon \mathcal{L}_p u \geq 0 \text{ on } \Omega, \liminf_{z\to E} u(z) \geq 1, \liminf_{z\to\partial\Omega} u(z) \geq 0)$. In this context the construction of [**12, 9**] can be used to show that sets with full Lebesgue measure on the boundary of a half-space may have zero p-harmonic measure. On the other hand, it is not known whether radial limits exist almost everywhere with respect to p-harmonic measure.

3. The Laplacian. Recently there has been work on absolute continuity and singularity properties of harmonic measure for the Laplacian as domains worse than Lipschitz—in fact, completely arbitrary. The results are mostly due to Makarov ([**11**] and his article in these proceedings). When $n = 2$, harmonic measure always puts full mass on some set with σ-finite one-dimensional Hausdorff measure, and in the simply-connected case it puts no mass on any set with dimension less than one. Partial results have been proved in higher dimensions by Bourgain. One could try to make sense of nontangential behavior on these wild domains using the lines of steepest descent of the Green's function (I learned of this possibility from P. Jones) but apparently no results are known.

REFERENCES

1. A. Ancona, *Principe de Harnack a la frontier et theoreme de Fatou pour un opérateur elliptique dans un domaine Lipschitzien* Ann. Inst. Fourier (Grenoble) **28** (1978), 169–213.

2. _____, *Variétes a courbure negative, operateurs elliptiques et frontiere de Martin,* Orsay Preprint, 1985.

3. M. Anderson and R. Schoen, *Positive harmonic functions on complete manifolds of negative curvature,* Ann. of Math. **121** (1985), no. 3, 429–462.

4. L. Caffarelli, E. Fabes, and C. Kenig, *Completely singular elliptic-harmonic measures,* Indiana Univ. Math. J. **30** (1981), 917–924.

5. L. Caffarelli, E. Fabes, S. Mortola, and S. Salsa, *Boundary behavior of nonnegative solutions of elliptic equations in divergence form,* Indiana Univ. Math. J. **30** (1981), 621–640.

6. B. Dahlberg, *Absolute continuity of elliptic measures,* Preprint, 1984.

7. B. Dahlberg, E. Fabes, and C. Kenig, *A Fatou theorem for the porous medium equation,* Proc. Amer. Math. Soc. **91** (1984), 205–212.

8. E. Fabes, D. Jerison, and C. Kenig, *Necessary and sufficient conditions for absolute continuity of elliptic-harmonic measures,* Ann. of Math. **119** (1984), no. 1, 121–143.

9. J. L. Lewis, *Note on a theorem of Wolff* (to appear).

10. P. Lindquist and O. Martio, *Two theorms of N. Wiener for quasilinear elliptic equations,* Acta Math. **155** (1985), 153–171.

11. N. G. Makarov, *Distortion of boundary sets under conformal mapping,* Proc. London Math. Soc. **51** (1985), 369–384.

12. T. Wolff, *Gap series constructions for the p-Laplacian,* Ann. Scuola Norm. Sup. Pisa Cl. Sci. (4) (to appear).

NEW YORK UNIVERSITY, COURANT INSTITUTE OF MATHEMATICAL SCIENCES, NEW YORK, NEW YORK 10012, USA

A Class of Markov Fields with Finite Range

TAIVO ARAK

1. Introduction. The theory of Markov random fields with continuous parameter has been developed by Wong [10], Pitt [8], Molchan [4], and Rozanov [9] (for other references see, e.g., [9]). Interest in the theory especially increased after Nelson [5–7] had shown its connection with quantum field theory. From this new point of view the Gaussian case as studied before seems at present to be of less importance, as it corresponds to physical systems without interaction.

One can get non-Gaussian fields from a Gaussian one by multiplying the corresponding probability measure by a multiplicative density functional (see [5, 9]). And until quite recently this was the only way known, except for a small number of specific examples of other kinds.

In this paper, construction is given which leads to Markov fields with finite number of values. The fields to be considered are defined on a bounded convex region $T \subset \mathbf{R}^2$, and they belong to a class of fields called *polygonal*, for their realizations having constancy domains of polygonal form.

Some polygonal Markov fields can be extended to the whole plane $\mathbf{R}^2$ in such a way that the resulting field will be invariant under all Euclidean transformations of $\mathbf{R}^2$. But none of them satisfy another condition necessary for applications in quantum field theory, namely, the condition called by Nelson [6] the *reflection property*. Nevertheless this new class of fields may prove to be useful in other applications, if only because it contains isotropic fields—the property which is so often present in real physical situations, and which cannot be achieved in lattice models.

Some particular cases of constructions to be dealt with (including those given in §4) have been published in [1] and [2]. In the present extent of generality the results have been obtained together by D. Surgailis and the author. The complete proofs will be published in [3].

2. Polygonal fields.

2.1. *The space of realizations.* We define the main probability space Ω_T (the space of realizations) as a set of functions of special type given on a bounded convex open domain $T \subset \mathbf{R}^2$ and taking values from a finite ordered set J. First let $\tilde{\Omega}_T$ be the set of all measurable functions $\omega\colon T \to J$ such that

$$\omega(z) = \inf_X \lim_{\varepsilon \to 0} \sup\{\omega(z')\colon \|z' - z\| < \varepsilon,\ z' \in T \setminus X\}, \tag{1}$$

where the infimum is taken over all sets $X \subset T$ with Lebesgue measure 0. For any functions $\omega \in \tilde{\Omega}_T$ consider the set $\partial\omega \subset T$ of discontinuity points:

$$\partial\omega = \{z \in T\colon \limsup_{z' \to z} \omega(z') \neq \liminf_{z' \to z} \omega(z')\}.$$

Let $\mathfrak{L}_T$ be the set of straight lines $l \subset \mathbb{R}^2$ which have nonempty intersection with T, and let $\mathfrak{L}_T^n$ be the n-fold direct product of $\mathfrak{L}_T$ with itself. For $(l_1, \ldots, l_n) \in \mathfrak{L}_T^n$ consider the subset $\Omega_T(l_1, \ldots, l_n) \subset \tilde{\Omega}_T$ consisting of all functions ω such that there exist closed intervals $[l_i] \subset l_i$ $(i = 1, \ldots, n)$ with positive length and satisfying the following conditions:

(ω1) $[l_i] \subset \overline{T}$, where $\overline{T}$ is the topological closure of T.
(ω2) $\partial\omega = \bigcup_{i=1}^n [l_i] \cap T$.
(ω3) If $i \neq j$ but $l_i = l_j$, then $[l_i] \cap [l_j] = \varnothing$.

The set $\Omega_T(l_1, \ldots, l_n)$ is obviously finite. For $n = 0$ it consists of constants: $\omega(z) \equiv \mathrm{const} \in J$. We set

$$\Omega_T^{(n)} = \bigcup_{(l_1, \ldots, l_n) \in \mathfrak{L}_T^n} \Omega_T(l_1, \ldots, l_n),$$

$$\Omega_T = \bigcup_{n=0}^{\infty} \Omega_T^{(n)}.$$

A typical realization $\omega \in \Omega_T(l_1, l_2, l_3, l_4)$ is represented in Figure 1.

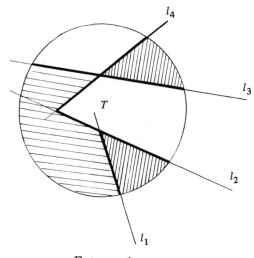

FIGURE 1

A topology in the set Ω_T may be introduced in a natural way, for example defining a distance $d(\omega, \omega')$ between two elements $\omega \in \Omega_T^{(n)}$ and $\omega' \in \Omega_T^{(m)}$ by setting

$$d(\omega, \omega') = \mathrm{meas}\,\{z \in T\colon \omega'(z) \neq \omega(z)\}$$

if $n = m$ and

$$d(\omega, \omega') = +\infty$$

if $n \neq m$, where meas $(\cdot)$ is the Lebesgue measure. Let $\mathfrak{B}_T$ be the Borel σ-algebra corresponding to this topology.

REMARKS. 1. The condition (1) is needed for the uniqueness of definition of ω on discontinuity lines.

2. It is easy to see that the set $A_{z,j} = \{\omega \colon \omega(z) \leq j\}$ is open and therefore measurable.

2.2. *The probability measure.* Let $\mu = \mu(dl)$ be a measure in the space of straight lines $l \subset \mathbb{R}^2$ finite on compacts. First, for the sake of simplicity we shall assume that μ has no atoms. For fixed T and μ we introduce the class $\Phi_{T,\mu}$ of all measurable functions $F \colon \Omega_T \to \mathbb{R} \cup \{+\infty\}$ such that

$$Z_T = Z_{T,F,\mu} = \sum_{n=0}^{\infty} \int_{\mathfrak{L}_T^n} \frac{\mu(dl_1) \cdots \mu(dl_n)}{n!} \sum_{\omega \in \Omega_T(l_1,\ldots,l_n)} e^{-F(\omega)} < \infty. \quad (2)$$

Further on the functions $F \in \Phi_{T,\mu}$ will be called potentials.

DEFINITION 1. The *polygonal field* corresponding to μ and F is the probability measure $\mathbf{P}_T = \mathbf{P}_{T,F,\mu}$ on $(\Omega_T, \mathfrak{B}_T)$ with

$$\mathbf{P}_T(A) = Z_T^{-1} \sum_{n=0}^{\infty} \int_{\mathfrak{L}_T^n} \frac{\mu(dl_1) \cdots \mu(dl_n)}{n!} \sum_{\omega \in A \cap \Omega_T(l_1,\ldots,l_n)} e^{-F(\omega)} \quad (3)$$

for any $A \in \mathfrak{B}_T$.

REMARKS. 3. If μ is allowed to have atoms, the denominator $n!$ in (2) and (3) must be replaced by the quantity $\kappa(l_1, \ldots, l_n)$ defined as the number of different sequences which can be obtained from the sequence $(l_1, \ldots, l_n)$ by permutation of its elements. For example,

$$\kappa(l,l,l,l,l',l',l',l'',l'') = \frac{9!}{4!\, 3!\, 2!}$$

if l, l', l'' are different.

3. Additive potentials and Markov property.

For $S \subset \mathbb{R}^2$ and $\varepsilon > 0$, we denote by $(S)_\varepsilon$ the ε-neighborhood of S, by ∂S the topological boundary of S, and by $\mathfrak{B}_T(S)$ the σ-algebra generated by the random variables $\{\omega(z)\}_{z \in S \cap T}$. A function $F \colon \Omega_T \to \mathbb{R} \cup \{+\infty\}$ is said to be *additive* if for any $S \subset T$ and for any $\varepsilon > 0$ there exist two functions $F_1(\omega)$ and $F_2(\omega)$ measurable with respect to $\mathfrak{B}_T((S)_\varepsilon)$ and $\mathfrak{B}_T((T \setminus S)_\varepsilon)$ correspondingly, with $F = F_1 + F_2$.

THEOREM 1. *A polygonal field corresponding to an additive potential satisfies the Markov property*

$$\mathbf{P}_{T,F,\mu}(A|\mathfrak{B}_T((T \setminus S)_\varepsilon)) = \mathbf{P}_{T,F,\mu}(A|\mathfrak{B}_T((\partial S)_\varepsilon))$$

for any $S \subset T, \varepsilon > 0, A \in \mathfrak{B}_T(S)$.

REMARKS. 4. The assertion of Theorem 1 may be made more accurate, as follows. Consider the σ-algebra $\mathfrak{B}_T^0(\partial S) \subset \mathfrak{B}_T((\partial S)_\varepsilon)$ generated by the random variables $\{\omega(z)\}_{z \in \partial S \cap T}, \nu, \alpha_1, \ldots, \alpha_\nu$, where ν is the number of intervals $[l_i]$ forming the discontinuity set $\partial \omega$ and intersecting $\partial S \cap T$; $\alpha_1, \ldots, \alpha_\nu$ are the angle between these intervals and the y-axis. Then under the conditions of Theorem 1

$$\mathbf{P}_{T,F,\mu}(A | \mathfrak{B}_T(T \setminus S)) = \mathbf{P}_{T,F,\mu}(A | \mathfrak{B}_T^0(\partial S))$$

for any $A \in \mathfrak{B}_T(S)$.

4. Consistent fields. Let $\mathfrak{G}$ be a family of bounded convex regions $T \subset \mathbb{R}^2$ including a subsequence $T_n \nearrow \mathbb{R}^2$. A family of probability measures $(\mathbf{P}_T)_{T \in \mathfrak{G}}$ defined on $(\Omega_T, \mathfrak{B}_T)$ is said to be *consistent* if for any $S, T \in \mathfrak{G}$ such that $S \subset T$ and for any $A \in \mathfrak{B}_S$ the equality $\mathbf{P}_S(A) = \mathbf{P}_T(\pi_S^{-1} A)$ is valid, where π_S is the restriction operator from Ω_T onto Ω_S. By Kolmogorov's theorem any consistent family defines a random field on the whole plane $\mathbb{R}^2$.

In this section we give a simple example of a consistent family of Markov fields with two values, which in addition admits a simple formula for Z_T.

We introduce the following parametrization of the straight lines l on the plane $\mathbb{R}^2 = \{(t, y) : t, y \in \mathbb{R}\}$:

$$l = l(p, \alpha), \qquad (p, \alpha) \in \mathbb{R} \times [0, \pi),$$

where p is the alternating length of the perpendicular from the origin to l and α is the angle between this perpendicular and the abscissa-axis (t-axis).

Let

$$\mu_0(dl) = dp \, d\alpha, \tag{4}$$

$J = \{0, 1\}$, and let $\mathfrak{G}_0$ be the family of bounded convex polygons. Consider the additive functional

$$F_0(\omega) = 2L(\partial \omega), \tag{5}$$

where $L(\cdot)$ is the length. According to Theorem 1 the field $\mathbf{P}_{T,F_0,\mu_0}$ (if it exists) is Markovian.

THEOREM 2. (a) *The polygonal field* $\mathbf{P}_{T,F_0,\mu_0}$ *corresponding to* μ_0 *and* F_0 *defined by* (4) *and* (5) *exists for any* $T \in \mathfrak{G}_0$.

(b) $Z_{T,F_0,\mu_0} = 2 \exp\{L(\partial T) + \pi \operatorname{meas}(T)\}$.

(c) *The family* $\{\mathbf{P}_T\}_{T \in \mathfrak{G}_0}$ *is consistent.*

(d) *The corresponding field on* $\mathbb{R}^2$ *is homogeneous and isotropic, its restriction to any straight line* l *is a Markov process, and the constancy intervals of this process have exponential distribution with parameter* 2.

The field $\mathbf{P}_{T,F_0,\mu_0}$ admits a simple description in terms of evolution of a system of one-dimensional particles in space-time region $\overline{T} \subset \{(t, y) : t, y \in \mathbb{R}\}$. In subsequent description the t-coordinate will be interpreted as time and the y-coordinate as spatial.

We shall call a particle a point moving on $\mathbb{R}$ with piecewise constant speed. We shall assume that the evolution of the speed is a Markov process with probability of jumping from the state v during the time dt to the set du equal to

$$q(v, du)\, dt = \frac{|u - v|}{(1 + u^2)^{3/2}}\, du\, dt.$$

Let

$$\xi_0 = ((t_j^0, y_j^0, v_j^0): j = 1, \ldots, \nu_0),$$
$$\xi_1 = ((t_j, y_j, v_j', v_j''): j = 1, \ldots, \nu_1)$$

be two independent Poisson point processes corespondingly on $\partial T \times \mathbb{R}$ and $T \times \mathbb{R}^2$ with intensities $|dy - v\, dt|\gamma(dv)$ and $|v' - v''|\, dt\, dy\gamma(dv')\gamma(dv'')$, where $\gamma(dv) = (1 + v^2)^{-3/2}\, dv$. Suppose that at any moment t_j^0 $(j = 1, \ldots, \nu_0)$ at the point y_j^0 a particle is born with initial speed v_j^0, and at any moment t_j $(j = 1, \ldots, \nu_1)$ a couple of particles are born with initial speeds v_j' and v_j''. These $\nu_0 + 2\nu_1$ particles begin to evolve independently from each other according to the law described above and in compliance with the following rule: A particle vanishes after exit from the region $\overline{T}$ as well as after collision with another particle. (Particles born on ∂T with initial speed directed out of $\overline{T}$ vanish immediately.)

Such an evolution obviously determines a random partition Ξ of T into sets of polygonal form (see Figure 2(a)). Obviously, with probability 1 this partition is such that there exist two and only two functions from Ω_T satisfying the condition $\partial\omega = \Xi$. Denote these functions by ω^+ and ω^- (see Figure 2(b), (c)). Assigning to both of them equal probability, we obtain a random element of Ω_T. Let $\mathbf{Q}_T$ denote the probability induced by this random element.

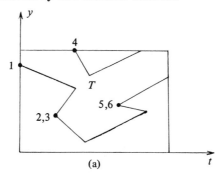

(a)

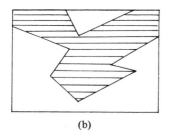

(b)

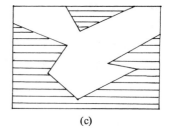

(c)

FIGURE 2. (a) Trajectories of six particles in a rectangle T. The birth points are marked by numbers 1–6. (b), (c) The functions ω^+ and ω^-.

THEOREM 3. $\mathbf{Q}_T = \mathbf{P}_T$.

In Figure 3 a realization of the field $\mathbf{P}_T$ on a rectangle is represented.

FIGURE 3

A similar construction will be given in [3] for the case of more than two values.

REFERENCES

1. T. Arak, *Some examples of Markovian random fields with piecewise constant realizations*, 12th European Metting of Statisticians, Varna, September 3–7, 1979, Abstracts, p. 39.

2. ____, *On Markovian random fields with finite number of values*, IV USSR-Japan Sympos. on Prob. Theory and Math. Stat., Tbilisi 1982, Abstracts of Communications, vol. I, Tbilisi, 1982, pp. 110–111.

3. T. Arak and D. Surgailis, *Markov random fields with polygonal trajectories*, Probability Theory and Related Fields (to appear).

4. G. M. Molchan, *On characterization of Gaussian fields with Markovian property*, Dokl. Akad. Nauk SSSR **197** (1971) no. 4, 784–787.

5. E. Nelson, *Quantum fields and Markoff fields*, Partial Differential Equations, Proc. Sympos. Pure Math., vol. 23, Amer. Math. Soc., Providence, R.I., 1973, pp. 413–420.

6. ____, *Construction of quantum fields from Markoff fields*, J. Funct. Anal. **12** (1973), 97–112.

7. ____, *The free Markoff field*, J. Funct. Anal. **12** (1973), 211–227.

8. L. D. Pitt, *A Markov property for Gaussian processes with a multidimensional parameter*, Arch. Rational Mech. Anal. **43** (1971), 367–391.

9. Yu. A. Rozanov, *Markov random fields*, Nauka, Moscow, 1981.

10. E. Wong, *Homogeneous Gauss-Markov fields*, Ann. Math. Statist. **40** (1969), 1625–1634.

TARTU STATE UNIVERSITY, TARTU 202400, ESTONIA, USSR

Nonlinear Filtering and Stochastic Flows

M. H. A. DAVIS

1. Introduction. The aim of this article is to describe some recent issues in filtering theory, mainly related to robustness and continuity properties. In general terms, *nonlinear filtering* refers to the problem of calculating the conditional distribution of a "signal" x_t given "observations" $\{Y_s, 0 \le s \le t\}$, where $\{x_s\}$, $\{Y_s\}$, $s \in [0, T]$, are stochastic processes defined on the same probability space, denoted throughout $(\Omega, \mathcal{F}, P)$. In this generality, very little can be said, and the vast majority of work in this area has concerned the case where $\{x_t\}$ is a Markov process and $\{Y_t\}$ is given by

$$Y_t = h(x_t) + n_t, \tag{1.1}$$

where $\{n_t\}$ is some form of "wide band noise." The most familiar model is the "additive white noise" model where we define

$$y_t = \int_0^t Y_s \, ds, \qquad w_t^0 = \int_0^t n_s \, ds$$

and take $\{w_t^0\}$ to be Brownian motion (BM), giving an observation model usually written in differential form as

$$dy_t = h(x_t) \, dt + dw_t^0. \tag{1.2}$$

The best known result in this area is of course the Kalman filter, where $\{x_t, y_t\}$ satisfy linear stochastic equations

$$dx_t = Ax_t \, dt + C dw_t, \qquad dy_t = Hx_t dt + G dw_t \tag{1.3}$$

where $\{w_t\}$ is a vector BM and it is assumed that $GG^T > 0$ and that the initial state x_0 is independent of $\{w_t\}$ with normal distribution $N(m_0, P_0)$. Then the conditional distribution of x_t given $\{y_s, 0 \le s \le t\}$ is $N(\hat{x}_t, P(t))$ where $\{\hat{x}_t\}$ and $P(t)$ satisfy a linear stochastic differential equation and a deterministic Riccati equation respectively; see [6]. The immense success of the Kalman filter in applications is largely due to its modest computational complexity: $P(t)$ is nonrandom, so the conditional distribution $N(\hat{x}_t, P(t))$ is parametrized by the low-dimensional sufficient statistic $\hat{x}_t$.

The *extended generator* of a homogeneous Markov process on a state space E is an operator $(A, \mathcal{D}(A))$ such that for each $f \in \mathcal{D}(A)$, the process

$$C_t^f := f(x_t) - f(x_0) - \int_0^t Af(x_s)\, ds \tag{1.4}$$

is a martingale. The standard results in nonlinear filtering concern a situation where the signal is a Markov process with given extended generator and initial distribution π^0, and the linear observation equation (1.3) of the Kalman filter is replaced by the nonlinear equation (1.2). We assume that E is a Polish space and that the sample paths of $\{x_t\}$ are in $D_E[0, T]$ (the space of right-continuous E-valued functions on $[0, T]$ with left-hand limits). y_t and w_t^0 take values in R^m and $\{w_t^0\}$ is m-dimensional BM; $y_0 = w_0^0 = 0$. Let us denote by $\mathcal{Y}_t = \sigma\{y_s, 0 \le s \le t\}$ the natural filtration of the observation process, and by π_t the conditional distribution of x_t given $\mathcal{Y}_t$; we also write $\pi_t(f) = \int_E f(x)\pi_t(dx)$ for $f \in B(E)$. In this case it is generally too much to expect that there will be any low-dimensional sufficient statistic for π_t, which should be thought of as a $\mathcal{Y}_t$-adapted process taking values in $\mathcal{P}(E)$, the set of probability measures on E. Its evolution is described in two equivalent ways. The first of these involves the innovations process $d\nu_t = dy_t - \pi_t(h)\, dt$ which, as in the Kalman filter, is a standard BM. The direct nonlinear equivalent of the Kalman filter is the *Kushner-Stratonovich* or *Fujisaki-Kallianpur-Kunita* equation, a nonlinear stochastic differential equation (SDE) satisfied by π_t. For $f \in \mathcal{D}(A)$,

$$d\pi_t(f) = \pi_t(Af)\, dt + (\pi_t(hf) - \pi_t(h)\pi_t(f) + E[\alpha_t^f \mid \mathcal{Y}_t])\, d\nu_t, \tag{FKK}$$
$$\pi_0(f) = \pi_0(f).$$

Here $\alpha_t^f = d\langle C^f, w^0 \rangle_t / dt$ where $\langle M, N \rangle_t$ denotes the "joint variation" process for square integrable martingales. In many cases of interest, the state space E is a manifold and α_t^f is given by

$$\alpha_t^f = Zf(x_t) \tag{1.5}$$

where Z is a vector field on E. Then $E[\alpha_t^f \mid \mathcal{Y}_t] = \pi_t(Zf)$. Of course, $\alpha^f = 0$ if $\{x_t\}$ and $\{w_t^0\}$ are independent, as is commonly the case. Derivations of this equation can be found in full detail in the textbooks [17, 23]. A quick account, also covering the Zakai equation below, is given in [10].

It turns out that the awkwardly nonlinear coefficient of dy_t in (FKK) is occasioned by the requirement that $\pi_t \in \mathcal{P}(E)$, i.e., $\pi_t(\mathbf{1}) = 1$ (here $\mathbf{1}$ denotes the function $\mathbf{1}(x) = 1$). For $f \in \mathcal{D}(A)$ define

$$\sigma_t(f) = \pi_t(f) \exp\left(\int_0^t \pi_s(h)\, d\nu_s + \frac{1}{2} \int_0^t [\pi_s(h)]^2\, ds \right).$$

Since $\pi_t(\mathbf{1}) = 1$, the exponential term is equal to $\sigma_t(\mathbf{1})$ and $\pi_t(f) = \sigma_t(f)/\sigma_t(\mathbf{1})$. σ_t is an *unnormalized conditioned distribution*: it is a $\mathcal{Y}_t$-adapted $\mathcal{M}_+(E)$-valued process, where $\mathcal{M}_+(E)$ denotes the set of positive measures on E. It satisfies the *Zakai equation* [10, 28]

$$d\sigma_t(f) = \sigma_t(Af)\, dt + \sigma_t(Df)\, dy_t, \qquad \sigma_0(f) = \pi^0(f) \tag{Z}$$

where $Df = Zf + hf$, assuming α_t^f is given by (1.5). When $\alpha_t^f = 0$ this reduces to

$$d\sigma_t(f) = \sigma_t(Af)\,dt + \sigma_t(hf)\,dy_t. \qquad (\text{Z}')$$

This is substantially simpler than (FKK) since each coefficient is linear in σ_t, and furthermore, the equation is driven by the observation process $\{y_t\}$ directly rather than by the indirectly-defined innovations process $\{\nu_t\}$. For these reasons recent work has concentrated almost entirely on analysis of (Z) rather than (FKK).

An integral expression for σ_t can be obtained using the so-called reference probability method. Let us suppose that $\{x_t\}$ and $\{w_t^0\}$ are independent and that $h \in \mathcal{B}(E)$, i.e., h is bounded. Then according to Girsanov's theorem, the formula

$$\frac{dP_0}{dP} = \exp\left(-\int_0^T h(x_s)\,dw_s^0 - \frac{1}{2}\int_0^T h^2(x_s)\,ds\right)$$

defines a probability measure on $(\Omega, \mathcal{F})$ under which $\{y_t\}$ is a BM and $\{x_t\}$, $\{y_t\}$ are independent. The inverse Radon-Nikodým derivative is given by

$$\frac{dP}{dP_0} := \Lambda_T = \exp\left(\int_0^T h(x_s)\,dy_s - \frac{1}{2}\int_0^T h^2(x_s)\,ds\right).$$

Let E_0 denote expectation with respect to P_0. By a standard formula of conditional expectations,

$$\pi_t(f) = E[f(x_t) \mid \mathcal{Y}_t] = \frac{E_0[f(x_t)\Lambda_t \mid \mathcal{Y}_t]}{E_0[\Lambda_t \mid \mathcal{Y}_t]} \quad P\text{-a.s.} \qquad (1.6)$$

It is not hard to show that the numerator of this expression is equal to $\sigma_t(f)$ (and hence the denominator is $\sigma_t(1)$). Since $\{x_t\}$ is independent of $\mathcal{Y}_t$ under P_0, the conditional expectation can be evaluated by integrating with respect to the sample space measure μ_t of $\{x_s, 0 \le s \le t\}$ on $D_E[0,t]$:

$$E_0[f(x_t)\Lambda_t \mid \mathcal{Y}_t] = \int_{D_E[0,t]} f(x_t)\exp\left(\int_0^t h(x_s)\,dy_s - \frac{1}{2}\int_0^t h^2(x_s)\,ds\right)\mu_t\,(dx).$$

$$(1.7)$$

(1.6), (1.7) are known as the *Kallianpur-Striebel* (KS) formula. When $\alpha_t^f \ne 0$ the KS formula is substantially more complicated; this will be referred to in §2.

2. Uniqueness. In applications, one is going to compute a solution π_t or σ_t to (FKK) or (Z) and then claim that this solution is the conditional distribution (normalized or not). To substantiate this claim it is of course necessary to show that the solutions to these equations are unique.

One route to such results is through the theory of stochastic partial differential equations (PDEs) [25]. Suppose that $E = R^n$ and that $\{x_t\}$ is a diffusion process with generator[1]

$$Af = \frac{1}{2}a^{ij}(x)\frac{\partial^2 f}{\partial x^i \partial x^j} - b^i(x)\frac{\partial f}{\partial x^i}.$$

---

[1]The Einstein summation convention is used here and throughout the rest of this paper.

If the coefficients a^{ij}, b^i are smooth and σ_t has a smooth density, i.e., there is a random function $\rho(t,x)$ such that $\sigma_t(f) = \int_E f(x)\rho(t,x)\,dx$, then (Z) can be written in "strong" form as

$$d\rho(t,x) = A^*\rho(t,x)\,dt + D^*\rho(t,x)\,dy_t, \qquad \rho(0,x) = \rho^0(x), \qquad (1.8)$$

where A^*, D^* are the formal adjoints of A, D and ρ^0 is the density of the initial measure π^0. Under the assumption that h and b are bounded measurable, a^{ij} is continuous with first derivatives in L^∞, and $[a^{ij}] \geq \delta I$ for some $\delta > 0$, $\rho^0 \in L_2(R^n)$, it is shown in [25] that (1.8) has a unique solution

$$\rho \in L_2(\Omega \times [0,T]; H^1) \cap L_2(\Omega; C([0,T]; L_2(R^n))),$$

and that ρ is the density of σ_t. Another approach, which treats the problem in much greater generality, is the so-called "filtered martingale problem" idea of Kurtz and Ocone [22]. Recall that in the martingale problem (MP) approach to Markov process theory, initiated by Stroock and Varadhan, the martingale property of C^f in (1.4) is regarded as encapsulating the connection between operator A and process $\{x_t\}$. The MP $(A, \mathcal{D})$ is said to be *well posed* if for each initial measure π^0 there is a unique probability measure P_{π_0} on $D_E(R_+)$ such that x_0 has distribution π^0 and C^f is a P_{π_0}-martingale for each $f \in \mathcal{D}$. If uniqueness holds it is generally not hard to show that x_t is a homogeneous Markov process with transition measure $P_x = P_{\delta_x}$ (δ_x = Dirac measure at x).

Kurtz' and Ocone's approach [21] is to show that if an MP is well posed, then so is an associated family of "filtered" MPs. Note from (FKK) that

$$\pi_t(f) - \int_0^t \pi_s(Af)\,ds$$

is a stochastic integral with respect to the innovations process, and hence, it is a local martingale; under suitable conditions it is a martingale. More generally,

$$\pi_t f(\cdot, y_t) - \int_0^t \pi_s Af(\cdot, y_s)\,ds$$

is a martingale for functions $f : E \times R^m \to R$, where

$$\pi_t f(\cdot, y) = \int_E f(x,y)\pi_t\,(dx).$$

Think of $\{x_t, y_t\}$ as a joint Markov process with extended generator $(\tilde{A}, \mathcal{D}(\tilde{A}))$ where $\mathcal{D}(\tilde{A}) \subset B(E \times R^m)$. A process (μ, U) with sample paths in $D_{P(E) \times R^m}(R_+)$ is a *solution of the filtered MP for* $(A, \mathcal{D})$ if μ is $\mathcal{F}_{t+}^U$-adapted and

$$\mu_t f(\cdot, U_t) - \int_0^t \mu_s Af(\cdot, U_s)\,ds$$

is an $\mathcal{F}_{t+}^U$-martingale for each $f \in D(\tilde{A})$. Here $\mathcal{F}_t^U$ denotes the natural filtration of $\{U_t\}$. Uniqueness holds if any two solutions have the same finite-dimensional distributions. If uniqueness holds and

$$E[\mu_0 f(\cdot, U_0)] = E[f(x_0, y_0)] = \int_E f(x,0)\pi^0(dx),$$

then (π, Y) has the same distribution as (μ, U). Since μ is $\mathcal{F}^U_{t+}$-adapted, there is a function $H: R_+ \times C_{R^m}(R_+) \to P(E)$ such that $\mu_t = H(t, U)$ a.s. and hence uniqueness implies $\pi_t = H(t, y)$ a.s. Thus any algorithm that solves (FKK) does indeed generate the required conditional distribution. The following result gives uniqueness for the Zakai equation (Z′) (the case $\alpha^f_t \neq 0$ is also covered but we do not give details here).

THEOREM 2.1 [21]. *Suppose*
(i) $E \int_0^T |h(x_s)|^2 \, ds < \infty$;
(ii) A *maps* $C_b(E)$ *into* $C_b(E)$, $D(A)$ *is a dense subalgebra of* $C_b(E)$, *and the MP for A is well posed;*
(iii) $f(x)h_i(x) \in C_b(E)$ *for all* $f \in D(A)$, $i = 1, \ldots, m$.
If $\{\rho_t\}$ *is a* Y_t-adapted cadlag M_+-valued process satisfying

$$\rho_t(f) = \pi^0(f) + \int_0^t \rho_s(Af) \, ds + \int_0^t \rho_s(hf) \, dy_s,$$

$$\rho_t(1) = 1 + \int_0^t \rho_s(h) \, dy_s,$$

for all $f \in D(A)$, $t \in T$, *then* $\rho_t = \sigma_t$ *for all* $t \in T$, a.s.

Uniqueness holds for (FKK) under the same conditions.

3. Pathwise filtering. For the remainder of the paper we shall consider only signal processes $\{x_t\}$ with continuous sample paths, i.e., sample paths in the space $C_E[0, T]$, although many results can be generalized to paths in $D_E[0, T]$, possibly at the expense of some complication. We then have the following simple result.

PROPOSITION 3.1 [16]. *Suppose* $D(A)$ *is an algebra; i.e.,* $f, g \in D(A)$ *implies* $fg \in D(A)$ *where* $fg(x) := f(x)g(x)$. *Then* $\langle C^f, C^g \rangle_t = \int_0^t \Delta^{fg}(x_s) \, ds$ *where*

$$\Delta^{fg}(x) = A(fg) - fAg - gAf. \tag{3.1}$$

The idea behind "pathwise filtering" is to recast the equations of nonlinear filtering in a form in which *no stochastic integration is involved*. Apart from its intrinsic interest, this is important from the point of view of mathematical modelling; see Clark [5] or Davis [7] for discussions of this point.

Let us consider first the independent signal and noise case: $\{x_t\}$ and $\{w^0_t\}$ are independent. Then $\pi_t(f) = \sigma_t(f)/\sigma_t(1)$, where $\sigma_t(f)$ is given by the KS formula (1.8), and we have the following result. For notational simplicity we assume for the moment that $m = 1$, i.e., y_t and w^0_t are scalar.

THEOREM 3.2 [5]. *Suppose that the process* $t \to h(x_t)$ *is a semimartingale and define a function* $\phi: [0, T] \times C_{R^m}[0, T] \times B(E)$ *by*

$$\phi(t, \xi, f) := E^{(x)} \Bigg[f(x_t) \exp(\xi(t) h(x_t))$$

$$\times \exp \left(-\int_0^t \xi(s) \, dh(x_s) - \frac{1}{2} \int_0^t h^2(x_s) \, ds \right) \Bigg]$$

where $E^{(x)}$ denotes integration with respect to the distribution of $\{x_t\}$. Then

(i) *For each t, f the function $\xi \to \phi(t, \xi, f)$ is locally Lipschitz continuous with respect to the uniform norm on $C_{R^m}[0, T]$.*

(ii) *$\phi(t, y, f) = E_0[f(x_t) \mid \mathcal{Y}_t]$ a.s., i.e., $\phi(t, y, f)$ is a version of $\sigma_t(f)$.*

REMARKS. 1. The functional $\phi(t, \xi, f)$ is obtained simply by integrating by parts the stochastic integral in the KS formula (1.8).

2. Property (i) is a "robustness" property of the filter; see [5, 7].

From now on, $y(\cdot)$ will denote an arbitrary but fixed continuous function, and any integration is always over the distribution of $\{x_t\}$. We will write $\sigma_t(f)$ for $\sigma(t, y, f)$; i.e., we always choose the robust version.

Theorem 3.2 provides a pathwise formula in integral form, but we would like to get it in a differential equation form similar to the Zakai equation. The key to this is to notice that the functional

$$\alpha_s^t(y) := \exp\left(\int_0^t y(s)\, dh(x_t) - \frac{1}{2}\int^t h^2(x_s)\, ds\right)$$

is a multiplicative functional (m.f.) of $\{x_t\}$ and hence defines a two-parameter semigroup of operators $T_{s,t}^y$ on $\mathcal{B}(E)$ by

$$T_{s,t}^y f(x) = E_{s,x}[f(x_t)\alpha_s^t(y)].$$

Thus $\sigma_t(f)$ can be expressed in the form

$$\sigma_t(f) = (T_{s,t}^y(e^{y(t)h}f), \pi^0). \tag{3.2}$$

The following is the main result of pathwise filtering for the independent signal and noise case.

THEOREM 3.3 [8]. *Suppose $\mathcal{D} \subset \mathcal{D}(A)$ is a set such that $h \in \mathcal{D}$ and $hf \in \mathcal{D}(A)$ for all $f \in \mathcal{D}$. Then the extended generator A_t^y of the semigroup $T_{s,t}^y$ is given by*

$$A_t^y f(\nu) = e^{y(t)h(x)} A(e^{-y(t)h}f) - \frac{1}{2}h^2(x)f(x) \tag{3.3}$$

$$= Af(x) - y(t)\Delta^{hf}(x) + \left[\frac{1}{2}y^2(t)\Delta^{hf}(x) - y(t)Ah(x) - \frac{1}{2}h^2(x)\right]f(x)$$

where Δ^{hf} is given by (3.1).

This is proved by factoring $\alpha_s^t(y)$ into the product of "Girsanov" m.f. and a "potential" m.f.

The significance of the result is that $\sigma_t(f)$ can be calculated, in principle (by considering the adjoint semigroup $(T_s^y)^*$ in (3.2)), by the following procedure: let ρ_t be the solution of the "Fokker-Planck" equation

$$\frac{d}{dt}\rho_t = (A_t^y)^*\rho_t, \qquad \rho_t = \rho. \tag{3.4}$$

Then

$$\sigma_t(f) = \int_E f(x)e^{y(t)h(x)}\rho_t\,(dx).$$

The exact interpretation of (3.4) depends on the context, but generally it will have the same interpretation as the Fokker-Planck equation for the $\{x_t\}$ process itself since, from (3.3), A_t^y is a "first order" perturbation of A.

EXAMPLE 3.4. Suppose that E is a C^∞ manifold, that $X_0, X_1, \ldots, X_n$ are vector fields, $w_t^1, \ldots, w_t^n$ are independent BMs, and $\{x_t\}$ is the solution of the SDE

$$df(x_t) = X_0 f(x_t)dt + X_i f(x_t) \circ dw_t^i, \qquad f \in C^\infty(E), \qquad (3.5)$$

where "$\circ$" denotes the Stratonovich stochastic integral. Then $\{x_t\}$ is a diffusion process with generator $A = X_0 + \frac{1}{2}\sum_{i=1}^n X_i^2$ and $\Delta^{hf}(x) = \sum_{i=1}^n X_i h X_i f$. Thus

$$A_t^y f(x) = \frac{1}{2}\sum_1^n X_i^2 f(x) + X_0 - y(t)\sum X_i h X_i f + \psi(y(t), x)f(x)$$

where

$$\psi(y, x) = \frac{1}{2}y^2\sum_1^n (X_i h)^2 - y\left(X_0 h + \frac{1}{2}\sum X_i^2 h\right) - \frac{1}{2}h^2(x).$$

Thus A_t^y has the same second-order part as A but differs from A in the first and zero-order terms. In particular, ρ_t of (3.4) has a smooth density if A is nondegenerate, i.e., $X_1, \ldots, X_n$ span $T_x(E)$ at each $x \in E$.

All of the above results extend immediately to the case of multidimensional observation $n > 1$.

4. Pathwise filtering with noise correlation [11]. This cannot be handled at the same level of generality as above, and we restrict our attention to the situation considered in Example 3.4 where $\{x_t\}$ is a diffusion on a manifold specified by equation (3.5). It will not be necessary, however, to suppose that $\{x_t\}$ is nondegenerate. We take $\{y_t\}$ to be scalar; in contrast to the situation in §3, this assumption is needed for validity of most results described below. Noise correlation arises when the BMs w^i in (3.5) are not independent of w^0. Specifically, we assume that w^i and w^j are independent for $i \neq j \neq 0$ but that

$$\langle w^i, w^0 \rangle_t = \int_0^t a^i(x_s)\, ds.$$

Referring back to §1, (1.5) then holds with $Z = a^i(x)X_i$. What makes the present case more complicated is that when we introduce the measure P_0 via the Radon-Nikodým derivative (1.7), $\{y_t\}$ becomes a BM but the distributions of $\{x_t\}$ are not preserved. More precisely we have the following results.

THEOREM 4.1 [11]. *There are vector fields* $Y_0, Y_1, \ldots, Y_n$ *and independent BMs* $\{b_t^1, \ldots, b_t^n\}$ *on* $(\Omega, \mathcal{F}, P_0)$, *independent of* $\{y_t\}$ *such that* $\{x_t\}$ *satisfies the following SDE*

$$df(x_t) = Y_0 f(x_t)\, dt + Z f(x_t) \circ dy_t + Y_i f(x_t) \circ db_t^i. \qquad (4.1)$$

In [11] the vector fields Y_i and processes b^i are expressed explicitly in terms of the original coefficients and processes. We now decompose (4.1) in a manner pioneered by Kunita [19]. Let $\varsigma(t, x)$ be the flow of the vector field Z, i.e.,

$$\frac{d}{dt} f(\varsigma(t, x)) = Z f(\varsigma(t, x)), \qquad f \in C^\infty(E),$$

$$\varsigma(0, x) = x.$$

We suppose that Z is complete; i.e., $\varsigma(t, x)$ is defined for all $t \in R$. Now define

$$\xi_t(x) = \varsigma(y_t, x). \tag{4.2}$$

Then ξ_t is a diffeomorphism for all $t > 0$ and satisfies

$$df(\xi_t(x)) = Z f(\xi_t(x)) \circ dy_t. \tag{4.3}$$

Now consider the equation

$$df(\eta_t) = \xi_{t*}^{-1} Y_0 f(\eta_t) \, dt + \xi_{t*}^{-1} Y_i f(\eta_t) \circ db_t^i \tag{4.4}$$

where ξ_{t*}^{-1} is the differential map:

$$\xi_{t*}^{-1} Y_0 f(x) = Y_0(f \circ \xi_t^{-1})(\xi_t(x)).$$

This SDE uniquely defines a process η_t, and applying the Kunita-Bismut extended Ito formula [1, 19] we find that

$$x_t(x) = \xi_t \circ \eta_t(x) = \varsigma(y_t, \eta_t(x)). \tag{4.5}$$

Now in (4.2), (4.4), $\{y_t\}$ appears simply as a parameter and the b^i are independent of $\{y_t\}$. Thus conditioned on $\mathcal{Y}_t$, η_t is a diffusion process with generator $A_t^* = \xi_{t*}^{-1} Y_0 + \frac{1}{2} \sum (\xi_{t*}^{-1} Y_i)^2$ and x_t is diffeomorphically related to η_t via (4.5). With this information in hand we can derive a pathwise filtering formula by decomposition of multiplicative functionals, much as before. The result is

THEOREM 4.2 [11]. $\sigma_t(f) = (T_{0,t}^y(B_{y(t)}f), \pi^0)$ where $T_{s,t}^y$ is a two-parameter semigroup with extended generator

$$A_t^y = \exp(H_{y(t)}) A_t^* \exp(-H_{y(t)}) - \frac{1}{2} \xi_t^*(Zh + h^2).$$

Here $H_t(x) := \int_0^t \varsigma_s^* h(s) \, ds$ and B_t is the group of operators

$$B_t f(x) = \varsigma_t^* f(x) \exp\left(\int_0^t \varsigma_t^* h(x) \, du\right)$$

(notation: $\varsigma_t^* f(x) = f \circ \varsigma(t, x)$).

As before, the significance of this result is that σ_t can be computed by solving the Fokker-Planck equation (3.4) and then performing an integration:

$$\sigma_t(f) = \int_E B_{y(t)} f(x) \pi_t \, (dx).$$

All of this is done pathwise, i.e., separately for each sample path $y(\cdot, w)$.

5. Observations on a manifold. Several authors [12, 13, 26] have considered filtering problems where the observations take values in a manifold N (example: measurement of an angle). The observation equation (1.2) is then replaced by an equation of the form

$$df(y_t) = L_0 f(x_t, y_t)\, dt + L_i f(y_t) \circ dw_t^{0i} \tag{5.1}$$

where L_i are vector fields on N, or by the requirement that y_t be a nondegenerate diffusion whose generator G has an x_t-dependent first-order term. It is shown in [12] that pathwise filtering is possible if L_0 in (5.1) takes the form

$$L_0 f(x, y) = \sum_i L_i g(x, y) L_i f(y)$$

for some scalar function g, or, equivalently, if $L_0 = \operatorname{grad}_y g(x, y)$ (the gradient with respect to the Riemannian metric on N determined by G).

6. Continuity. We have shown in previous sections that when there is no noise correlation it is possible to choose a version of the conditional expectation $\sigma_t(f)$ such that the map $y \to \sigma_t(f)(y)$ is continuous with respect to the uniform norm on $C_{R^m}[0, T]$ and that this is still possible with noise correlation if the observations are scalar, $m = 1$. When $m > 1$, (4.3) is replaced by an equation of the form

$$df(\xi_t(x)) = Z_i f(\xi_t(x)) \circ dy_t^i \tag{6.1}$$

and the mapping $y \to \xi_t$ is no longer continuous unless the vector fields Z_i commute [20].

Nonetheless, with smooth coefficients the map $x \to \xi_t(x)$ is almost surely a diffeomorphism, so the decomposition (4.4) and other formulas of §4 are still valid, but only almost surely.

Weaker notions of continuity have been studied by Chaleyat-Maurel and Michel [3, 4]. In [3] it is shown that $\sigma_t(f)(y)$ is infinitely differentiable in the sense of Malliavin calculus whereas in [4] a notion of continuity related to the Stroock-Varadhan support theorem [27, 15, §6.8] is introduced. Let $\sigma_t(f)$ be the solution of the Zakai equation when the signal is a diffusion as in Example 3.4, and let $\sigma_t^u(f)$ be the (deterministic) solution when $y(t)$ is replaced by an arbitrary H^1 function $u(t)$. Then under smoothness and growth conditions, for each $\eta > 0$,

$$\lim_{\delta \to 0} P_0 \left[\sup_t |\sigma_t(f) - \sigma_t^u(f)| > \eta \, \Big| \, \sup_t |y_t - u_t| \le \delta \right] = 0.$$

7. Existence of conditional densities. Space limitations unfortunately preclude any discussion of this important topic, which is concerned with determining conditions under which the measure σ_t has a (smooth) density when the signal process $\{x_t\}$ is a diffusion. Most of the cases discussed in this paper are best handled by using decompositions similar to (4.5), which reduce the question to one of *unconditional* diffusions; see [18, 20]. Then "classical" results obtained using Hörmander's theorem or Malliavin calculus can be applied. More general cases have been studied by a number of authors using extensions of Malliavin calculus [2, 22, 24].

REFERENCES

[1] J. M. Bismut, *A generalized formula of Ito and some other properties of stochastic flows*, Z. Wahrsch. Verw. Gebiete **55** (1981), 331-350.

[2] J. M. Bismut and D. Michel, *Diffusions conditionelles*. I, II, J. Funct. Anal. **44** (1981), 174-211; **45** (1982), 274-292.

[3] M. Chaleyat-Maurel, *Robustesse du filtre et calcul des variations stochastique*, J. Funct. Anal. (to appear).

[4] M. Chaleyat-Maurel and D. Michel, *Une propriété de continuité en filtrage non-linéaire*, Stochastics **19** (1986), 11-40.

[5] J. M. C. Clark, *The design of robust approximations to the stochastic differential equations of nonlinear filtering*, Communication Systems and Random Process Theory (J. K. Skwirzynski, ed.), NATO Advanced Study Institute Series, Sijthoff and Noordhoff, Alphen aan den Rijn, 1978.

[6] M. H. A. Davis, *Linear estimation and stochastic control*, Chapman and Hall, London; Halsted Press, New York, 1977.

[7] ____, *Pathwise nonlinear filtering*, Stochastic Systems: The Mathematics of Filtering and Identification and Applications, Reidel, Dordrecht, 1981, pp. 505-528.

[8] ____, *On a multiplicative functional transformation arising in nonlinear filtering theory*, Z. Wahrsch. Verw. Gebiete **54** (1980), 125-139.

[9] ____, *Lectures on nonlinear filtering and stochastic control*, Tata Institute for Fundamental Research Series, Narosa, New Delhi; Springer-Verlag, Berlin, 1984.

[10] M. H. A. Davis and S. I. Marcus, *An introduction to nonlinear filtering*, Stochastic Systems: The Mathematics of Filtering and Identification and Applications, Reidel, Dordrecht, 1981, pp. 53-75.

[11] M. H. A. Davis and M. P. Spathopoulos, *Pathwise nonlinear filtering for nondegenerate diffusions with noise correlation*, SIAM J. Control Optim. (1986) (to appear).

[12] ____, *Pathwise nonlinear filtering with observations on a manifold*, Proc. 25th IEEE Conference on Decision and Control, Athens, 1986, pp. 1337-1338.

[13] T. E. Duncan, *Some filtering results in Riemannian manifolds*, Inform. and Control **35** (1977), 182-195.

[14] M. Hazewinkel and J. C. Willems (Editors), *Stochastic systems: the mathematics of filtering and identification and applications*, Reidel, Dordrecht, 1981.

[15] N. Ikeda and S. Watanabe, *Stochastic differential equations and diffusion processes*, North-Holland, Amsterdam, 1981.

[16] J. Jacod, *Calcul stochastique et problèmes de martingales*, Lecture Notes in Math., vol. 714, Springer-Verlag, Berlin-Heidelberg-New York, 1979.

[17] G. Kallianpur, *Stochastic filtering theory*, Springer-Verlag, Berlin-Heidelberg-New York, 1980.

[18] H. Kunita, *Densities of a measure-valued process governed by a stochastic partial differential equation*, Systems Control Lett. **1** (1981), 100-104.

[19] ____, *On the decomposition of solutions of stochastic differential equations*, Stochastic Integrals (D. Williams, ed.), Lecture Notes in Math., vol. 851, Springer-Verlag, Berlin-Heidelberg-New York, 1981.

[20] ____, *Stochastic partial differential equations connected with nonlinear filtering*, Nonlinear Filtering and Stochastic Control (S. K. Mitter and R. Conti, eds.), Lecture Notes in Math., vol. 972, Springer-Verlag, Berlin-Heidelberg-New York, 1982.

[21] T. G. Kurtz and D. Ocone, *Unique characterization of conditional distributions in nonlinear filtering*, Ann. Probab. (to appear).

[22] S. Kusuoka and D. Stroock, *The partial Malliavin calculus with applications to filtering*, Stochastics **12** (1984), 83-142.

[23] R. S. Liptser and A. N. Shiryaev, *Statistics of random processes*. I, II, Springer-Verlag, Berlin-Heidelberg-New York, 1977, 1978.

[24] D. Ocone, *Stochastic calculus of variations for stochastic partial differential equations*, Preprint, 1986.

[25] E. Pardoux, *Equations du filtrage nonlinéaire de la prediction et du lissage*, Stochastics **6** (1982), 193–231.

[26] M. Pontier and J. Szpirglas, *Filtrage non-linéaire avec observation sur une varieté*, Stochastics **15** (1985), 121–148.

[27] D. W. Stroock and S. R. S. Varadhan, *Multidimensional diffusion processes*, Springer-Verlag, Berlin-Heidelberg-New York, 1979.

[28] M. Zakai, *On the optimal filtering of diffusion processes*, Z. Wahrsch. Verw. Gebiete **11** (1969), 230–243.

IMPERIAL COLLEGE, LONDON SW7 2BT, ENGLAND

Proceedings of the International Congress of Mathematicians
Berkeley, California, USA, 1986

Conditionally Positive Definite Functions in Quantum Probability

A. S. HOLEVO

Introduction. There is a well-known result, going back to Schoenberg [1], that if $l(g)$ is a complex-valued function on a group G, then $\varphi_t(g) = \exp tl(g)$, $g \in G$, is positive definite for all $t \geq 0$ if and only if $l(g^{-1}) = \overline{l(g)}$ and $l(g)$ is conditionally positive definite [2, 3]. In probability theory (where G is abelian) the family $\{\varphi_t(g), t \geq 0\}$ arises as the Fourier transform of a convolution semigroup, closely related to the limit theorems for triangular arrays of random variables with values in the dual group $\hat{G}$ [4, 5]. The function $\varphi(g) = \exp l(g)$ is the characteristic function of an infinitely divisible distribution on $\hat{G}$, and $\varphi[g(\cdot)] = \exp \int l(g(t)) \, dt$ is the characteristic functional of a generalized stochastic process with independent values [2]. A complete description of conditionally positive definite functions and related probabilistic objects is given by the Levy-Khinchin formula [6, 4].

On the other hand, recent study of irreversible Markov evolutions in the theory of open quantum systems [7–10] has led to the concept of dynamical semigroup, which is defined as a semigroup

$$\phi_t = \exp t\mathcal{L}, \qquad t \geq 0,$$

of completely positive maps of a corresponding C^*-algebra of observables. A necessary and sufficient condition for the generator $\mathcal{L}$ of a dynamical semigroup, found by Lindblad [9], is that $\mathcal{L}$ should be completely dissipative. The general form of such maps $\mathcal{L}$ is given by the formula found in [8, 9].

The purpose of this work is to develop a theory which embraces both cases and makes explicit the strong connection between the underlying mathematical structures. In §1 we give the main definitions and state the corresponding Schoenberg's type theorem. In §2 the principal theorem is formulated, which gives a representation for an arbitrary conditionally positive definite function with values in the space of bounded linear maps of a C^*-algebra. The Levy-Khinchin type formulas can be obtained from this representation by using the established results of cohomology of groups.

In §3 we consider applications to the problem of continuous measurement in quantum mechanics [**11–13**]. It turns out that this problem is surprisingly closely related to such classical topics of probability theory as infinite divisibility and limit theorems for stochastic processes.

1. Positive definite and conditionally positive definite functions.

1.1. Let G be a group with the neutral element e, $\mathcal{H}$ a Hilbert space, $\mathcal{B}(\mathcal{H})$ the algebra of all bounded operators in $\mathcal{H}$, $\mathcal{A}$ a C^*-subalgebra of $\mathcal{B}(\mathcal{H})$ containing the identity operator I, $\mathcal{F}$ the Banach algebra of bounded linear maps of $\mathcal{A}$ into itself [**14**]. For $\phi \in \mathcal{F}$ we put $\phi^*[X] \equiv (\phi[X^*])^*$, $X \in \mathcal{A}$. Let $\phi(g)$, $g \in G$, be a function with values in $\mathcal{F}$. We call it *hermitean* if $\phi(g^{-1}) = \phi(g)^*$, $g \in G$, *positive definite* if

$$\sum_{j,k}(\psi_j \mid \phi(g_j^{-1}g_k)[X_j^*X_k]\psi_k) \geq 0 \tag{1}$$

for all finite sets $\{\psi_j\} \subset \mathcal{H}$, $\{g_j\} \subset G$, $\{X_j\} \subset \mathcal{A}$, and *conditionally positive definite* if (1) holds for sets satisfying $\sum_j X_j\psi_j = 0$.

PROPOSITION. *Let $\mathcal{L}(g)$, $g \in G$, be a function with values in $\mathcal{F}$. The following conditions are equivalent:*

(1) *the functions $\exp t\mathcal{L}(g)$, $g \in G$, are positive definite (p.d.) for all $t \geq 0$;*

(2) *the function $\mathcal{L}(g)$, $g \in G$, is hermitean conditionally positive definite (h.c.p.d.);*

(3) *the function $\mathcal{L}(g)$, $g \in G$, is hermitean and satisfies*

$$\sum_{j,k}(\psi_j \mid D\mathcal{L}(g_j, g_k; X_j, X_k)\psi_k) \geq 0 \tag{2}$$

for all finite sets $\{\psi_j\} \subset \mathcal{H}$, $\{g_j\} \subset G$, $\{X_j\} \subset \mathcal{A}$, where

$$D\mathcal{L}(g,h;X,Y) = \mathcal{L}(g^{-1}h)[X^*Y] - \mathcal{L}(g^{-1})[X^*]Y$$
$$- X^*\mathcal{L}(h)[Y] + X^*\mathcal{L}(e)[I]Y.$$

1.2. In the case $\mathcal{H} = \mathcal{A} = \mathcal{F} = \mathbf{C}$, the proposition reduces to the classical Schoenberg's type result for scalar functions. The other extreme case is $G = \{e\}$, where one deals with a single map $\phi \in \mathcal{F}$. The condition (1) then means that ϕ is *completely positive* (see, e.g., [**10**]):

$$\sum_{j,k}(\psi_j \mid \phi[X_j^*X_k]\psi_k) \geq 0$$

for all $\{\psi_j\} \subset \mathcal{H}$, $\{X_j\} \subset \mathcal{A}$, and Proposition 1.1 reduces to the result obtained by Lindblad [**9**]: the maps $\phi_t = \exp t\mathcal{L}$ are completely positive for all $t \geq 0$ if and only if $\mathcal{L} = \mathcal{L}^*$ and

$$\sum_{j,k}(\psi_j \mid D\mathcal{L}(X_j, X_k)\psi_k) \geq 0$$

for all $\{\psi_j\} \subset \mathcal{H}$, $\{X_j\} \subset \mathcal{A}$, where $D\mathcal{L}(X,Y) = \mathcal{L}[X^*Y] - \mathcal{L}[X^*]Y - X^*\mathcal{L}[Y] + X^*\mathcal{L}[I]Y$. Such maps $\mathcal{L}$ are called *completely dissipative*.

If $\mathcal{A}$ is finite-dimensional, then Proposition 1.1 can be deduced from [15], where the general algebraic background of the notion of positive definiteness is investigated. The proof in the general case, given in [17], uses the results of [9].

1.3. The family $\phi_t(g) = \exp t\mathcal{L}(g)$, $t \geq 0$, is a semigroup of p.d. functions, satisfying $\phi_0(g) = \mathrm{Id}$, where Id is the identity map of $\mathcal{A}$. Each function $\phi_t(g)$ is infinitely divisible in the sense that $\phi_t(g) = [\phi_{t/n}(g)]^n$, where $\phi_{t/n}(g)$ is again a p.d. function.

An example of an h.c.p.d. function is $\mathcal{L}(g) = \phi(g) - c\,\mathrm{Id}$, where $\phi(g)$ is a p.d. function, $c \in \mathbf{R}$. The pointwise limit (in the norm of $\mathcal{F}$) of such functions is again an h.c.p.d. function. Proposition 1.1 implies the converse: any h.c.p.d. function can be obtained in this way, since $\mathcal{L}(g) = \lim_{n\to\infty} n(\phi_{1/n}(g) - \mathrm{Id})$. These facts underlie the probabilistic applications of h.c.p.d. functions to be given in §3.

2. The representation theorems.

2.1. For p.d. functions with values in $\mathcal{F}$ we have the canonical representations, which includes both the Gelfand-Raikov representation for the scalar positive-definite functions and the Stinespring representation for completely positive maps. We denote by $\mathfrak{B}(\mathcal{H}, \mathcal{K})$ the space of all bounded linear operators from a Hilbert space $\mathcal{H}$ to a Hilbert space $\mathcal{K}$.

PROPOSITION. *Let* $\phi(g)$, $g \in G$, *be a p.d. function with values in* $\mathcal{F}$. *There exist a Hilbert space* $\mathcal{K}$, *a unitary representation* $g \to V_g$ *of the group* G *in* $\mathcal{K}$, *a* *-representation* $X \to \rho[X]$ *of the algebra* $\mathcal{A}$ *in* $\mathcal{K}$, *such that*

$$V_g \rho[X] = \rho[X] V_g, \qquad g \in G, \ X \in \mathcal{A},$$

and an operator $F \in \mathfrak{B}(\mathcal{H}, \mathcal{K})$, *such that*

$$\phi(g)[X] = F^* V_g \rho[X] F, \qquad g \in G, \ X \in \mathcal{A}. \tag{3}$$

The representation (3) implies some useful inequalities for p.d. functions and, in particular, the Kadison type inequality

$$\phi(g)[X]^* \phi(g)[X] \leq \|\phi(e)[I]\| \cdot \phi(e)[X^* X]. \tag{4}$$

It follows that if $\phi(g)$ satisfies the normalization condition $\phi(e)[I] = I$, then $\|\phi(g)\| \leq 1$, $g \in G$.

2.2. As usual, $\mathcal{A}'$ denotes the commutant of $\mathcal{A}$, and $\mathcal{A}''$ the bicommutant, or the von Neumann algebra, generated by $\mathcal{A}$.

THEOREM. *Let* $\mathcal{L}(g)$, $g \in G$, *be an h.c.p.d. function with values in* $\mathcal{F}$. *There exist*

(a) *a Hilbert space* $\mathcal{K}$, *a unitary representation* $g \to W_g$ *of* G, *and a* *-representation* $X \to \pi[X]$ *of* $\mathcal{A}$ *in* $\mathcal{K}$, *satisfying* $W_g \pi[X] = \pi[X] W_g$, $g \in G$, $X \in \mathcal{A}$;

(b) *an operator* $A \in \mathfrak{B}(\mathcal{H}, \mathcal{K})$ *and a function* $B(g)$, $g \in G$, *with values in* $\mathfrak{B}(\mathcal{H}, \mathcal{K})$, *such that* $B(g)X = \pi[X]B(g)$, $g \in G$, $X \in \mathcal{A}$, *and satisfying the cocycle equation*

$$B(hg) = W_h B(g) + B(h), \qquad g, h \in G; \tag{4}$$

(c) *a function* $Z(g)$, $g \in G$, *with values in the center* $Z = A' \cap A''$ *of the algebra* A'', *satisfying* $Z(g)^* = Z(g)$, $g \in G$, *and*

$$Z(gh) - Z(g) - Z(h) = \operatorname{Im} B(g^{-1})^* B(h), \qquad g, h \in G; \tag{5}$$

and an operator $C \in A''$, *such that*

$$\begin{aligned}
\mathcal{L}(g)[X] &= A^* W_g \pi[X] A + A^* B(g) X + X B(g^{-1})^* A \\
&\quad + X[-\tfrac{1}{2} B(g)^* B(g) + Z(g)] + C^* X + X C.
\end{aligned} \tag{6}$$

2.3. For a scalar h.c.p.d. function the representation reduces to the term in squared brackets; the corresponding result can be found in Guichardet [19] (see also Parthasarathy and Schmidt [4]). In the case $G = \{e\}$, Theorem 2.2 and Proposition 2.1 give the general form of a completely dissipative map [9, 16]: $\mathcal{L}[X] = \phi[X] + C^* X + X C$, where ϕ is an arbitrary completely positive map. The main ingredients of the proof of Theorem 2.2 and Proposition 2.1 are given in [18].

2.4. Let A be a von Neumann algebra, A_* its predual, $\mathcal{F}_\sigma$ the closed subalgebra of $\mathcal{F}$, consisting of ultraweakly continuous maps [14]. For any $\phi \in \mathcal{F}_\sigma$ there is a uniquely defined ϕ_*, which is a bounded linear map of A_* into A_*, such that $\langle \phi_*[T], X \rangle = \langle T, \phi[X] \rangle$, $T \in A_*$, $X \in A$. We say that a net $\{\phi_\alpha\} \subset \mathcal{F}_\sigma$ converges *-strongly if the net $\{(\phi_\alpha)_*\}$ converges strongly on A_*.

Let G be a topological group and $\phi(g)$ in Proposition 2.1 be *-strongly continuous. Then the representations V and ρ may be chosen continuous. The corresponding remark applies also to Theorem 2.2. By using the cocycle properties (4), (5) we can prove that a continuous h.c.p.d. function grows "not faster than a quadratic function." We shall give a precise statement for the case $G = \mathbf{R}^s$.

LEMMA. *Let* $\mathcal{L}(g)$, $g \in \mathbf{R}^s$, *be a* *-strongly continuous h.c.p.d. function, then*

$$\|\mathcal{L}(g)\| \leq c(1 + |g|^2), \tag{7}$$

where $|\cdot|$ *is a norm on* $\mathbf{R}^s$, *and the constant* c *depends only on the value* $\sup_{|g| \leq 1} \|\mathcal{L}(g)\|$.

2.5. By using the known structure of representations and cocycles for concrete classes of groups [4, 19], we can derive Levy-Khinchin type formulas from Theorem 2.2. Let us state the result for the case $A = \mathcal{B}(\mathcal{H})$, where $\mathcal{H}$ is a separable Hilbert space, G-abelian separable locally compact group. In this case, an arbitrary *-strongly continuous h.c.p.d. function has the form

$$\mathcal{L}(g)[X] = \mathcal{L}_1(g)[X] + \mathcal{L}_2(g)[X] + C^* X + X C + i\lambda(g) X,$$

where $C \in \mathcal{B}(\mathcal{H})$, $\lambda(g)$ is a continuous morphism of G into $\mathbf{R}$; the last term represents a general solution for the equation $D\mathcal{L}(g, h; X, Y) \equiv 0$. The first term is

$$\mathcal{L}_1(g)[X] = \sum_{j,k} \gamma_{jk} R_j^* X R_k + \sum_j [R_j^* X r_j(g) - \overline{r_j(g)} X R_j]$$

$$- \frac{1}{2} \sum_{j,k} \gamma^{jk} \overline{r_j(g)} r_k(g), \tag{8}$$

where $R_j \in \mathcal{B}(\mathcal{H})$ and $r_j(g)$ are continuous morphisms of G into $\mathbf{C}$; $[\gamma_{jk}]$ is the Gram matrix of an unconditional basis in a Hilbert space; $[\gamma^{jk}]$ is the Gram matrix of the conjugate basis. Moreover, the series $\sum \gamma_{jk} R_j^* R_k$ converges ultraweakly, and the series $\sum \gamma^{jk} \overline{r_j(g)} r_k(g)$ absolutely for all $g \in G$. The second term is

$$\mathcal{L}_2(g)[X] = \sum_{j,k} V_j^* X V_k \int_{\mathcal{X}} g(x) \nu_{jk}(dx)$$

$$+ \sum_j \left\{ V_j^* X \int_{\mathcal{X}} [g(x) - 1] \nu_j(dx) + \int_{\mathcal{X}} [g(x) - 1] \overline{\nu_j(dx)} X V_j \right\}$$

$$+ \left\{ \int_{\mathcal{X}} [g(x) - 1 - i\Im(g,x)] \nu(dx) \right\} X,$$

where $V_j \in \mathcal{B}(\mathcal{H})$; $\mathcal{X} = \hat{G}$ is the dual group, $g(x)$ is the value of the character $g \in G$ of the group $\mathcal{X}$ on the element x; $\Im(g,x)$ is the standard function described in [4]; $\nu_{jk}(dx)$, $\nu_j(dx)$, $\nu(dx)$ are complex-valued measures on $\mathcal{X}$ having no mass at the neutral element of $\mathcal{X}$ and such that the matrix

$$\begin{bmatrix} \nu_{jk}(dx) & \overline{\nu_k(dx)} \\ \nu_j(dx) & \nu(dx) \end{bmatrix}$$

is positive definite. Moreover, the series $\sum V_j^* V_k \nu_{jk}(\mathcal{X})$ converges ultraweakly and the integral $\int |g(x) - 1|^2 \nu(dx)$ converges for all g and tends to zero as $g \to e$.

The derivation of these formulas from Theorem 2.2 is given in [**18**]. The "Gaussian" functions of the form (8) were introduced in the case $G = \mathbf{R}^s$ in the works of Barchielli, Lanz, Prosperi [**11, 12**] in connection to the problem of continuous quantum measurement. More general expressions including "Poisson" terms were considered in [**13, 20**]; in view of the present work the result of these papers may be interpreted as a construction of a special representation (3) (of the type of Araki-Woods imbedding [**4, 19**]) for factorizable p.d. functions given by the time-ordered exponentials $\mathcal{T} \exp \int \mathcal{L}(g(t)) \, dt$.

3. Limit theorems for repeated measurements in quantum probability.

3.1. Let $\mathcal{A}$ be a von Neumann algebra, $(\mathcal{X}, \mathcal{B})$ a standard Borel space. *Instrument* with values in $(\mathcal{X}, \mathcal{B})$ is a set function $\underline{\mathcal{U}} = \{\mathcal{U}(B), B \in \mathcal{B}\}$ such that: (1) for any $B \in \mathcal{B}$, $\mathcal{U}(B)$ is a completely positive element of $\mathcal{F}_\sigma$; (2) $\mathcal{U}(\mathcal{X})[I] = I$; (3) $\mathcal{U}$ is σ-additive in the $*$-strong topology. The notion of instrument, being a far-reaching extension of the von Neumann's "projection postulate," is introduced for the description of repeated measurements in quantum mechanics [**7**].

If S is a state (a positive element of $\mathcal{A}_*$, satisfying $\langle S, I \rangle = 1$), then the probability distribution of a sequence of measurements described by the instruments $\underline{\mathcal{U}}_1, \ldots, \underline{\mathcal{U}}_n$, is determined by the relation

$$\mathcal{U}_S(dx_1 \cdots dx_n) = \langle S, \mathcal{U}_1(dx_1)[\cdots \mathcal{U}_n(dx_n)[I] \cdots] \rangle. \tag{9}$$

3.2. Let $\mathcal{X}$ be an abelian separable locally compact group (we shall use the additive notation from now on). Then *the convolution of the instruments* can be defined as

$$(\underline{\mathcal{U}}_1 * \cdots * \underline{\mathcal{U}}_n)(B)[X] = \int \cdots \int_{x_1 + \cdots + x_n \in B} \mathcal{U}_1(dx_1)[\cdots \mathcal{U}_n(dx_n)[X] \cdots]. \tag{10}$$

We shall investigate the problem of describing measurements, which durate continuously for some interval of time and correspond to a continual analog of expression (9). The question is closely related to the limit behavior of the convolution (10), and its solution relies upon the results of previous sections.

3.3 Let $G = \hat{\mathcal{X}}$ be the dual group. *The characteristic function of the instrument* $\mathcal{U}$ is defined as

$$\phi(g)[X] = \int_{\mathcal{X}} g(x)\mathcal{U}(dx)[X], \qquad g \in G.$$

A function $\phi(g)$, $g \in G$, with values in $\mathcal{F}_\sigma$ is a characteristic function of an instrument if and only if: (1) $\phi(e)[I] = I$; (2) $\phi(g)$ is *-strongly continuous; (3) $\phi(g)$ is positive definite in the sense of 1.1 (see [**17**]). Operation of convolution corresponds to the pointwise composition of the characteristic functions.

In what follows we restrict to the case $G = \mathcal{X} = \mathbf{R}^s$ with $|x| = \max_{1 \le i \le s} |x^i|$, where $x = [x^1, \ldots, x^s]$. The following inequality is useful: for any $\lambda > 0$,

$$\mathcal{U}(x\colon |x| \ge 2\lambda) \le 2(\lambda/2)^s \int_{-\lambda^{-1}}^{\lambda^{-1}} \cdots \int_{-\lambda^{-1}}^{\lambda^{-1}} [\phi(0) - \phi(g)] \, dg_1 \cdots dg_s \tag{11}$$

in the sense that the difference of the right- and left-hand sides is completely positive.

3.4. By a *quasi-characteristic function* we mean a *-strongly continuous h.c.p.d. function with values in $\mathcal{F}_\sigma$, such that $\mathcal{L}(e)[I] = 0$. If $\mathcal{L}(g)$ is such a function, then $\exp \mathcal{L}(g)$ is the characteristic function of an instrument $\underline{\mathcal{U}}$ which is infinitely divisible in the sense that for any $n = 1, 2, \ldots$ there is a representation $\underline{\mathcal{U}} = \underline{\mathcal{U}}_n^{*n} \equiv \underline{\mathcal{U}}_n * \cdots * \underline{\mathcal{U}}_n$ (where $\underline{\mathcal{U}}_n$ is the instrument with the characteristic function $\exp \frac{1}{n} \mathcal{L}(g)$, cf. 1.3). Now let there be given, for any n, an instrument $\underline{\mathcal{U}}_n$ with the characteristic function $\phi_n(g)$, and consider the instrument $\underline{\mathcal{U}}_n^{*n}$ with the characteristic function $\phi_n(g)^n$. Assume that *the functions* $n(\phi_n(g) - \mathrm{Id})$ *converge pointwise *-strongly to a *-strongly continuous function* $\mathcal{L}(g)$ (which is thus a quasicharacteristic function). Then we can show that $\phi_n(g)^n \to \exp \mathcal{L}(g)$, which implies that *for any state S the probability distribution of the instrument* $\underline{\mathcal{U}}_n^{*n}$ *converges weakly to the probability distribution of the instrument with the characteristic function* $\exp \mathcal{L}(g)$.

3.5. Let $\underline{\mathcal{U}}_t$ be the instrument with the characteristic function

$$\phi_t(g) = \exp t\mathcal{L}(g), \qquad (t \geq 0).$$

The family $\{\underline{\mathcal{U}}_t, t \geq 0\}$ forms a convolution semigroup of the instruments, satisfying the continuity condition

$$\|\mathcal{U}_t(x\colon |x| \geq \lambda)\| \leq ct(1 + \lambda^{-2}), \tag{12}$$

which is derived from (7), (11), and the fact that $\|\phi_t(g)\| \leq 1$ (see 2.1). Conversely, if $\{\underline{\mathcal{U}}_t, t \geq 0\}$ is a convolution semigroup of instruments such that $\lim_{t \to 0} \|\mathcal{U}_t(x\colon |x| \geq \lambda)\| = 0$ for any $\lambda > 0$, then $\underline{\mathcal{U}}_t$ has the characteristic function $\exp t\mathcal{L}(g)$, where $\mathcal{L}(g)$ is a quasicharacteristic function.

3.6. The following definition is motivated by the description of continuous process of quantum measurement in [12], but uses a much smaller trajectory space; in fact, we show that the trajectories of continuous quantum measurement lie in the space of generalized derivatives of the functions without discontinuities of the second kind. Let $\mathbf{R}$ be the time axis, $\mathcal{Y}$ the space of all functions on $\mathbf{R}$ with values in $\mathbf{R}^s$, $\mathcal{B}_{a,b}$ the σ-algebra of subsets of $\mathcal{Y}$, generated by the increments $y(s) - y(t)$, $a \leq t \leq s \leq b$. By *instrumental process with independent increments* (*i-process*) we call the family $\{\underline{\mathcal{U}}_{a,b}; a \leq b; a, b \in \mathbf{R}\}$, where $\underline{\mathcal{U}}_{a,b}$ is an instrument with values in $(\mathcal{Y}, \mathcal{B}_{a,b})$, satisfying

$$\mathcal{U}_{a,b}(E) \cdot \mathcal{U}_{b,c}(F) = \mathcal{U}_{a,c}(E \cap F), \tag{13}$$

if $a \leq b \leq c$ and $E \in \mathcal{B}_{a,b}$, $F \in \mathcal{B}_{b,c}$. The i-process $\{\underline{\mathcal{U}}_{a,b}\}$ is *homogeneous* if $\mathcal{U}_{a+\tau,b+\tau}(E) = \mathcal{U}_{a,b}(T_\tau^{-1}(E))$, where $(T_\tau y)(t) = y(t + \tau)$, for all $\tau, a, b \in \mathbf{R}$. If $\{\underline{\mathcal{U}}_{a,b}\}$ is a homogeneous i-process, then the instruments with values in $\mathbf{R}^s$, defined by the relation

$$\mathcal{U}_t(B) = \mathcal{U}_{a,a+t}(y(\cdot)\colon y(a + t) - y(a) \in B), \qquad t \geq 0,$$

form a convolution semigroup, which determines uniquely the "finite-dimensional distributions" of the process. Namely, if $E = \{y(\cdot)\colon y(\tau_1) - y(\tau_0) \in B_1, \ldots, y(\tau_p) - y(\tau_{p-1}) \in B_p\}$, where $a = \tau_0 < \tau_1 < \cdots < \tau_p = b$, is a cylindric set from $\mathcal{B}_{a,b}$, then

$$\mathcal{U}_{a,b}(E) = \mathcal{U}_{\tau_1 - \tau_0}(B_1) \cdots \mathcal{U}_{\tau_p - \tau_{p-1}}(B_p). \tag{14}$$

Conversely, if a convolution semigroup of instruments $\{\underline{\mathcal{U}}_t, t \geq 0\}$ is given, then (12) determines finite-dimensional distributions which, by a generalization of the Kolmogorov theorem, extend to a homogeneous i-process with values in $\mathcal{Y}$.

3.7. Let $\mathcal{D} \subset \mathcal{Y}$ be the space of functions on $\mathbf{R}$ with values in $\mathbf{R}^s$, which are right continuous and have limits from the left. I-process with values in $\mathcal{D}$ is defined as in §3.6 with the σ-algebras $\mathcal{B}_{a,b}$ replaced by $\mathcal{B}_{a,b} \cap \mathcal{D}$. Let $\mathcal{L}(g)$ be a quasicharacteristic function, $\{\underline{\mathcal{U}}_t\}$ the corresponding convolution semigroup. The characteristic function of a finite-dimensional distribution defined by (14) is equal to

$$\phi_{\tau_0,\ldots,\tau_p}(g_1,\ldots,g_p) = \exp(\tau_1 - \tau_0)\mathcal{L}(g_1) \cdots \exp(\tau_p - \tau_{p-1})\mathcal{L}(g_p). \tag{15}$$

Using the estimate (12), the "independence of the increments" property (13), and the known criterion for a probability measure to be concentrated on D (see Billingsley [**21**, §15]), we can prove that *there exists a homogeneous i-process with values in D, the finite-dimensional distributions of which are defined by the relation* (15). We call $\mathcal{L}(g)$ the *generator* of this i-process.

3.8. Consider now the question of convergence of a sequence of repeated measurements to the process of continuous measurements. For $n = 1, 2, \ldots$, let there be given a division of the time axis $\mathbf{R}$ into intervals $[t_i^{(n)}, t_{i+1}^{(n)})$ of length $1/n$, and to each moment $t_i^{(n)}$ let a measurement correspond described by an instrument $\mathcal{U}_n$ with the characteristic function $\phi_n(g)$. Fix n and the interval $[a, b)$ and let i_a be the first and i_b the last of the numbers i such that $a < t_i^{(n)} \le b$. Let $x_i = y(t_i^{(n)}) - y(t_i^{(n)} - 0)$ and consider the $(\mathcal{B}_{a,b} \cap D)$-measurable mapping $y(\cdot) \to [x_i; i = i_a, \ldots, i_b]$. Denote by $\underline{\mathcal{U}}_{a,b}^{(n)}$ the image of the set function $\mathcal{U}_n(dx_{i_a}) \cdots \mathcal{U}_n(dx_{i_b})$ under the inverse mapping. The family $\{\underline{\mathcal{U}}_{a,b}^{(n)}\}$ is an i-process with values in D. The finite-dimensional distributions of this i-process have the characteristic functions

$$\phi_{\tau_0, \ldots, \tau_p}^{(n)}(g_1, \ldots, g_p) = \phi_n(g_1)^{m_1} \cdots \phi_n(g_p)^{m_p}, \tag{16}$$

where m_r is the number of points $t_i^{(n)}$ in the interval $(\tau_{r-1}, \tau_r]$.

3.9. THEOREM. *Let the functions $n(\phi_n(g) - \mathrm{Id})$ ∗-strongly converge, as $n \to \infty$, to a ∗-strongly continuous function $\mathcal{L}(g)$, and moreover*

$$\sup_n \sup_{|g| \le 1} n\|\phi_n(g) - \mathrm{Id}\| < \infty.$$

Then for any state S the probability distributions of the i-process $\{\underline{\mathcal{U}}_{a,b}^{(n)}\}$ weakly converge in the sense of the Skorohod topology in D to the probability distributions of the homogeneous i-process $\{\underline{\mathcal{U}}_{a,b}\}$ with the generator $\mathcal{L}(g)$.

The convergence of the finite-dimensional distributions reduces to the pointwise convergence of the characteristic functions (16) to (15). The latter is established as in 3.4, taking into account that $|m_r/n - (\tau_r - \tau_{r-1})| \le 1/n$. The proof of tightness is based on the inequality

$$\|\mathcal{U}_{a,b}^{(n)}(y(\cdot): |y(b) - y(a)| \ge \lambda)\| \le c(m/n)(1 + \lambda^{-2})$$

where m is the number of the points $t_i^{(n)}$ in the interval $(a, b]$, which can be deduced from (11) and Lemma 2.4. The proof is accomplished by using property (13) and the criterion of tightness of probability measures in D [**21**].

3.10. Let $R^1, \ldots, R^s$ be commuting selfadjoint operators from $\mathcal{A}$, $p(x^1, \ldots, x^s)$ a probability density with zero mean in $\mathbf{R}^s$. The instrument

$$\mathcal{U}_n(dx^1 \cdots dx^s)[X] = p\left(\sqrt{n}x^1 - \frac{1}{\sqrt{n}}R^1, \ldots, \sqrt{n}x^s - \frac{1}{\sqrt{n}}R^s\right)^{1/2} \cdot X \tag{17}$$

$$\cdot p\left(\sqrt{n}x^1 - \frac{1}{\sqrt{n}}R^1, \ldots, \sqrt{n}x^s - \frac{1}{\sqrt{n}}R^s\right)^{1/2} \cdot n^{s/2}\, dx^1 \cdots dx^s,$$

describes "joint approximate measurement of observables $R^1, \ldots, R^s$" (cf. [11]). If $p(x^1, \ldots, x^n)$ is smooth and decreases fast enough at infinity, then

$$n(\phi_n(g) - \mathrm{Id}) = \mathcal{L}(g) + \varepsilon_n(g),$$

where

$$\mathcal{L}(g)[X] = \frac{1}{4} \sum_{j,k} \mathfrak{I}_{jk} \left[R^j X R^k - \frac{1}{2}(R^j R^k X + X R^j R^k) \right]$$

$$+ \frac{i}{2} \sum_j g_j (R^j X + X R^j) - \frac{1}{2} \left(\sum_{j,k} \gamma^{jk} g_j g_k \right) X \tag{18}$$

is a "Gaussian" generator with

$$\mathfrak{I}_{jk} = \int \frac{\partial \ln p}{\partial x^j} \cdot \frac{\partial \ln p}{\partial x^k} \cdot p \cdot dx^1 \cdots dx^s,$$

$$\gamma^{jk} = \int x^j x^k p \, dx^1 \cdots dx^s,$$

and $\lim_{n\to\infty} \sup_{|g|\leq\kappa} \|\varepsilon_n(g)\| = 0$ for any κ. Theorem 3.9 then implies that the series of measurements described by (17) converge in the specified sense to the homogeneous i-process on $\mathcal{D}$ with the generator (18). By using the Kolmogorov criterion of continuity of the trajectories and the estimates of momenta of the instrument (17) one can establish that the limit process is defined in fact on the space of continuous functions, under suitable conditions on $p(x^1, \ldots, x^s)$ (which hold, e.g., for the normal probability density).

REFERENCES

1. I. J. Schoenberg, *Metric spaces and positive definite functions*, Trans. Amer. Math. Soc. 4 (1938), 522–530.

2. I. M. Gel'fand and N. J. Vilenkin, *Generalized functions*, Vol. 4: Applications of harmonic analysis, Translated by Amiel Feinstein, Academic Press, New York-London, 1964.

3. S. Johansen, *An application of extreme point method to the representation of infinitely divisible distributions*, Z. Wahrsch. Verw. Gebiete 5 (1966), 304–316.

4. K. R. Parthasarathy and K. Schmidt, *Positive definite kernels, continuous tensor products, and central limit theorems of probability theory*, Lecture Notes in Math., Vol. 272, Springer-Verlag, Berlin-New York, 1972.

5. H. Heyer, *Probability measures on locally compact groups*, Springer-Verlag, 1977.

6. B. V. Gnedenko and A. N. Kolmogorov, Translated and annotated by K. L. Chung, *Limit distributions for sums of independent random variables*, Addison-Wesley, Cambridge, Mass., 1954.

7. E. B. Davies, *Quantum theory of open systems*, Academic Press, London, 1976.

8. V. Gorini, A. Kossakowski, and E. C. G. Sudarshan, *Completely positive dynamical semigroups of N-level systems*, J. Math. Phys. 17 (1976), 821–825.

9. G. Lindblad, *On the generator of quantum dynamical semigroups*, Comm. Math. Phys. 48 (1976), 119–130.

10. D. E. Evans and J. T. Lewis, *Dilations of irreversible evolutions in algebraic quantum theory*, Comm. Dublin Inst. Adv. Stud. Ser. A. 24 (1977).

11. A. Barchielli, L. Lanz, and G. M. Prosperi, *A model for macroscopic description and continuous observations in quantum mechanics*, Nuovo Cimento B 72 (1982), 79–121.

12. ____, *Statistics of continuous trajectories in quantum mechanics: operation-valued stochastic processes*, Found. Phys. 13 (1983), 779–812.

13. A. Barchielli and G. Lupieri, *Dilations of operation-valued stochastic processes*, Lecture Notes in Math., vol. 1136, Springer-Verlag, Berlin-New York, 1985, pp. 57–66.

14. S. Sakai, *C*-algebras and W*-algebras*, Springer-Verlag, 1971.

15. M. Schürmann, *Positive and conditionally positive linear functionals on coalgebras*, Lecture Notes in Math., vol. 1136, Springer-Verlag, Berlin-New York, 1985, pp. 475–492.

16. E. Christensen and D. E. Evans, *Cohomology of operator algebras and quantum dynamical semigroups*, J. London Math. Soc. **20** (1979), 358–368.

17. A. S. Holevo, *Infinitely divisible measurements in quantum probability*, Theory Probab. Appl. **31** (1986), no. 3, 560–569. (Russian)

18. ____, *An analog of the Levy-Khinchin formula in quantum probability*, Theory Probab. Appl. **31** (1987), no. 1. (Russian)

19. A. Guichardet, *Symmetric Hilbert spaces and related topics*, Lecture Notes in Math., vol. 261, Springer-Verlag, Berlin-New York, 1972.

20. K. R. Parthasarathy, *One parameter semigroups of completely positive maps on groups arising from quantum stochastic differential equations*, Preprint ISI, 1985.

21. P. Billingsley, *Convergence of probability measures*, Wiley, New York, 1968.

STEKLOV MATHEMATICAL INSTITUTE, ACADEMY OF SCIENCES OF THE USSR, MOSCOW 117966, USSR

Proceedings of the International Congress of Mathematicians
Berkeley, California, USA, 1986

Stochastic Flows and
Stochastic Partial Differential Equations

HIROSHI KUNITA

Introduction. In recent years the relationship between stochastic differential equations and stochastic flows of diffeomorphisms has been studied thoroughly. See Elworthy [3], Bismut [2], Malliavin [12], Ikeda-Watanabe [5], Baxendale [1], Le Jan [10], Le Jan-Watanabe [11], Kunita [9], etc. Some of the basic facts will be surveyed in §1.

We shall apply the theory to stochastic parabolic partial differential equations. In §2 we briefly discuss a first order equation following partly from Kunita [8]. The solution will be represented by a stochastic characteristic curve or a certain stochastic flow.

It is a known fact that solutions to a certain second order parabolic partial differential equation are represented by means of a diffusion process or a stochastic flow. The equations are so-called Kolmogorov's backward equations; these are solved backward with given terminal conditions, while the associated diffusion processes proceed forward. In order to dissolve this forward-backward dichotomy, we want to make use of the inverse flow. This will provide a better probabilistic interpolation of parabolic partial differential equations.

In §3 we shall realize the above idea in a certain second order stochastic partial differential equation. It includes the parabolic partial differential equation mentioned above. One should note that the stochastic partial differential equation originated from nonlinear filtering problems. See, e.g., Pardoux [13], Krylov-Rozovsky [6], and Kunita [7].

1. Stochastic differential equations and stochastic flows. We shall survey the relationship between stochastic differential equations and stochastic flows of diffeomorphisms following [9, 11]. We begin by introducing function spaces. Let k and l be nonnegative integers. We denote by $C^{k,l} = C^{k,l}(R^d \times [0,T]; R^d)$ the space of maps $f: R^d \times [0,T] \to R^d$ which are k-times differentiable in x, l-times differentiable in t, and such that the derivatives are continuous in (x,t). $C_b^{k,l}$ is the subset of f in $C^{k,l}$ such that f and its derivatives are bounded functions.

Let $X(x, t, \omega)$, $x \in R^d$, $t \in [0, T]$ be a continuous random field with values in R^d such that for almost all ω, it is of $C^{k,0}$. Then $X(t) \equiv X(\cdot, t)$ may be considered a continuous C^k-valued process, where $C^k = C^k(R^d; R^d)$. In particular, if $X(t_{i+1}) - X(t_i)$, $i = 0, \dots, n-1$, are independent for any $0 \le t_0 < \cdots < t_n \le T$, it is called a C^k-valued Brownian motion. In the sequel we assume that the following limits exist uniformly in t.

$$m(x, t) \equiv \lim_{h \to 0} \frac{1}{h} E[X(x, t+h) - X(x, t)], \tag{1.1}$$

$$v_{ij}(x, y, t) \equiv \lim_{h \to 0} \frac{1}{h} E[(X_i(x, t+h) - X_i(x, t))(X_j(y, t+h) - X_j(y, t))]. \tag{1.2}$$

Obviously $v = (v_{ij})$ is symmetric, i.e., $v_{ij}(x, y, t) = v_{ji}(y, x, t)$ and nonnegative definite. The pair (m, v) is called the *local characteristic* of $X(x, t)$. A typical example of a C^k-valued Brownian motion is given by

$$X(x, t) = \sum_{m=1}^{n} \int_0^t f^m(x, r) \, dB^m(r) + \int_0^t f^0(x, r) \, dr, \tag{1.3}$$

where $f^m(x, r)$ are of $C^{k+1,0}$ and $(B^1(t), \dots, B^n(t))$ is a standard Brownian motion. In this case, the local characteristic is given by $m(x, t) = f^0(x, t)$, $v_{ij}(x, y, t) = \sum_{m=1}^{n} f_i^m(x, t) f_j^m(y, t)$. We introduce an assumption for the local characteristic.

A_k. The local characteristic (v, m) is of $C_b^{k,0}$.

Let $X(x, t)$ be a C^k-valued Brownian motion satisfying A_{k+1}, $k > 1$. We consider a stochastic differential equation

$$d\phi_t = X(\phi_t, \circ dt), \tag{1.4}$$

where the right-hand side is the Stratonovich differential. If $X(t)$ is of the form (1.3), the equation coincides with the classical stochastic differential equation

$$d\phi_t = \sum_{m=1}^{n} f^m(\phi_t, t) \circ dB^m(t) + f^0(\phi_t, t) \, dt. \tag{1.5}$$

REMARK. Denote the Itô differential by $X(\phi_t, dt)$. Then the two integrals are related by

$$\int_0^t X(\phi_r, \circ dr) = \int_0^t X(\phi_r, dr) + \frac{1}{2} \sum_{i=1}^{d} \left\langle \int_0^t \frac{\partial}{\partial x_i} X(\phi_r, dr), \phi_t^i \right\rangle. \tag{1.6}$$

In particular if ϕ_t is a solution to equation (1.4), then

$$\int_0^t X(\phi_r, \circ dr) = \int_0^t X(\phi_r, dr) + \int_0^t k(\phi_r, r) \, dr, \tag{1.7}$$

where

$$k(x, t) = \frac{1}{2} \sum_{j=1}^{d} \frac{\partial}{\partial x_j} v_{\cdot j}(x, y, t)|_{y=x}. \tag{1.8}$$

Now for any given $(x, s) \in R^d \times [0, T]$ the equation (1.4) has a unique solution starting at x at time s. We denote it by $\phi_{s,t}(x)$. Then it has a modification satisfying (a)–(e).

(a) *For almost all ω, $\phi_{s,t}(x, \omega)$ is $(k-1)$-times differentiable in x and the derivatives $D_x^\alpha \phi_{s,t}(x, \omega)$ are continuous in (s, t, x) for any $|\alpha| \leq k - 1$.*

(b) *For almost all ω, $\phi_{t,u}(\phi_{s,t}(x, \omega), \omega) = \phi_{s,u}(x, \omega)$ is satisfied for all $s < t < u$ and x.*

(c) *For almost all ω, the map $\phi_{s,t}(\cdot, \omega); R^d \to R^d$ is a C^{k-1}-diffeomorphism for any $s < t$.*

(d) *For any $0 \leq t_0 < t_1 < \cdots < t_n \leq T$, $\phi_{t_i,t_{i+1}}$, $i = 0, \ldots, n - 1$, are independent of each other.*

(e) *There is a positive function $\varepsilon(h)$ with $\varepsilon(h) \downarrow 0$ as $h \downarrow 0$ such that*

$$|\tfrac{1}{h} E[\phi_{t,t+h}(x) - x] - m(x, t) - k(x, t)| \leq \varepsilon(h)(1 + |x|),$$
$$|\tfrac{1}{h} E[(\phi_{t,t+h}(x) - x)(\phi_{t,t+h}(y) - y)] - v(x, y, t)| \leq \varepsilon(h)(1 + |x|)(1 + |y|).$$

A random field $\phi_{s,t}(x, \omega)$, $0 \leq s \leq t \leq T$, $x \in R^d$, with the properties (a)–(c), is called a *stochastic flow of C^{k-1}-diffeomorphisms*. Further, if (d) is satisfied, it is called a *Brownian flow (of C^{k-1}-diffeomorphisms)*.

(f) *Let $\phi_{s,t}$ be a Brownian flow of C^{k-1}-diffeomorphisms satisfying (e) where $v_{ij}(x, y, t)$ and $m(x, t)$ are of $C_b^{k+1,0}$. Then there is a unique C^k-valued Brownian motion $X(x, t)$ with the local characteristic (v, m) such that $\phi_{s,t}(x)$ is the solution to the stochastic differential equation (1.4).*

DEFINITION. The C^k-valued Brownian motion $X(x, t)$ is said to be the *infinitesimal generator* of the flow $\phi_{s,t}$, and $\phi_{s,t}$ is said to be *generated by $X(x, t)$*.

Let $\phi_{s,t}$ be a Brownian flow of C^{k-1}-diffeomorphisms generated by a C^k-Brownian motion $X(x, t)$. Then for each $s < t$, the map $\phi_{s,t}(\cdot, \omega)$ has the inverse map $\psi_{s,t} = (\phi_{s,t})^{-1}$. It has the following properties.

(g) *For almost all ω, $\psi_{s,t}(x)$ are $(k-1)$-times differentiable and the derivatives are continuous in (s, t, x).*

(h) *For almost all ω, $\psi_{s,t}(\psi_{t,u}(x, \omega), \omega) = \psi_{s,u}(x, \omega)$ is satisfied for any $s < t < u$ and x.*

The properties (g), (h) together with (c) show that $\psi_{s,t}$ is a stochastic flow of C^{k-1}-diffeomorphisms to the backward direction. Indeed we have the following.

(i) *For each fixed t and x, $\psi_{s,t}(x)$, $s \in [0, t]$, satisfies the backward stochastic differential equation $\psi_{s,t}(x) = x - \int_s^t X(\psi_{r,t}(x), \hat{o}dr)$, where the last member is a backward Stratonovich integral.*

Here the *backward Stratonovich integral* is defined by

$$\int_s^t X(f_r, \hat{o}dr) = \frac{1}{2} \lim_{|\Delta| \to 0} \sum_{i=1}^{n-1} \{ X(f_{t_{i+1}}, t_{i+1}) - X(f_{t_{i+1}}, t_i)$$
$$+ X(f_{t_i}, t_{i+1}) - X(f_{t_i}, t_i) \}$$

where $\Delta = \{s = t_0 < t_1 < \cdots < t_n = t\}$ and $|\Delta| = \max|t_{i+1} - t_i|$. The *backward Itô integral* is defined by

$$\int_s^t X(f_r, \hat{d}r) = \lim_{|\Delta|\to 0} \sum_{i=0}^{n-1} X(f_{t_{i+1}}, t_{i+1}) - X(f_{t_{i+1}}, t_i).$$

Finally we study a growth property of the stochastic flow.

LEMMA 1. *Let $X(x,t)$ be a C^k-valued Brownian motion satisfying A_{k+1} where $k > 1$ and let $\phi_{s,t}$ be the stochastic flow generated by $X(x,t)$. Then for any α with $|\alpha| \leq k - 2$, we have*

$$\lim_{|x|\to\infty} \frac{|D^\alpha \phi_{s,t}(x)|}{1 + |x|^\varepsilon} = 0, \qquad \lim_{|x|\to\infty} \frac{|D^\alpha \psi_{s,t}(x)|}{1 + |x|^\varepsilon} = 0$$

for any $\varepsilon > 0$.

PROOF. For any $p \geq 2$ there is a positive constant $K = K(p)$ such that $E[|D^\alpha \phi_{s,t}(x)|^p] \leq K$ holds for any $x \in R^d$ and $|\alpha| \leq k - 2$. See [9]. Set $x = y/|y|^2$. Then for any $\varepsilon > 0$ and p with $p\varepsilon > d$,

$$E\left[\left|D^\alpha \phi_{s,t}\left(\frac{y}{|y|^2}\right)\right|^p \Big/ \left(1 + \frac{1}{|y|}\right)^{p\varepsilon}\right] \leq K|y|^{p\varepsilon}.$$

Then Kolmogorov's criterion for the continuity of the random field shows that $D^\alpha \phi_{s,t}(y/|y|^2)/(1 + \frac{1}{|y|})^\varepsilon$ converges to 0 as $y \to 0$ a.s. for any $s < t$. This is equivalent to the first assertion. The second can be shown similarly.

2. Inverse flows and first order stochastic partial differential equations.
Let $(X_1(x,t),\ldots,X_d(x,t),Y(x,t),Z(x,t))$ be a $C^k(R^d; R^{d+2})$-valued Brownian motion satisfying A_{k+1}, $k \geq 5$. Given a function f of C^3, we consider a first order stochastic partial differential equation of the form

$$v(x,t) = f(x) + \sum_{i=1}^d \int_0^t X_i(x, \circ dr)\frac{\partial v}{\partial x_i}(x,r) + \int_0^t Y(x, \circ dr)v(x,r) + Z(x,t). \quad (2.1)$$

Here $X_i(x, \circ dr)$ is the Stratonovich differential. The equation can be solved similarly as a first order (deterministic) partial differential equation, making use of stochastic characteristic curves. Indeed, let $(\phi_{s,t}(x), \eta_{s,t}(x,y))$ $(x \in R^d, y \in R^1)$ be the stochastic flows in R^{d+1} generated by

$$(-X_1(x,t),\ldots,-X_d(x,t), yY(x,t) + Z(x,t)).$$

When the initial point (s,x,y) is fixed, the trajectories $(\phi_{s,t}(x), \eta_{s,t}(x,y))$, $t \in [s,T]$, are called the characteristic curves of the equation (2.1). The first component is the stochastic flow generated by $-(X_1,\ldots,X_d)$ and the second component is represented by

$$\eta_{s,t}(x,y) = \left(\exp\int_s^t Y(\phi_{s,r}(x), \circ dr)\right)$$

$$\cdot \left\{y + \int_s^t \left(\exp - \int_s^r Y(\phi_{s,u}(x), \circ du)\right) Z(\phi_{s,r}(x), \circ dr)\right\}. \quad (2.2)$$

THEOREM 1 (CF. [8]). *The equation (2.1) has a unique solution. It is represented by*

$$v(x,t) = \eta_{0,t}(\psi_{0,t}(x), f(\psi_{0,t}(x)))$$

$$= f(\psi_{0,t}(x)) \exp \int_0^t Y(\psi_{r,t}(x), o\hat{d}r) \tag{2.3}$$

$$+ \int_0^t \left(\exp \int_s^t Y(\psi_{r,t}(x), o\hat{d}r) \right) Z(\psi_{s,t}(x), o\hat{d}s),$$

where $\psi_{s,t}$ is the inverse flow of $\phi_{s,t}$.

REMARK. If $Y = Z = 0$, equation (2.1) can be considered as a formula for the change of the variables or "Itô's formula" for the inverse flow $\psi_{0,t}$. In fact (2.1) is written as

$$f(\psi_{0,t}(x)) = f(x) + \sum_{i=1}^d \int_0^t X_i(x, o dr) \frac{\partial}{\partial x_i} (f \circ \psi_{0,r})(x). \tag{2.4}$$

The following will be used in the next section.

COROLLARY. *Suppose that f is of C_b^3. Then the solution $v(x,t)$ and its derivatives $D_x^\alpha v(x,t)$, $|\alpha| \le 2$, have bounded moments of any order, i.e.,*

$$\sup_{x,t} E[|D^\alpha v(x,t)|^p] < \infty \qquad \forall p > 1. \tag{2.5}$$

Furthermore $v(x,t)$ has a following growth property:

$$\lim_{x \to \infty} v(x,t)/(1 + |x|)^\varepsilon = 0 \tag{2.6}$$

holds a.s. for any $\varepsilon > 0$.

PROOF. The second property (2.6) follows from (2.5) immediately, making use of a device similar to that in Lemma 1. For the proof of the first property (2.5), it is sufficient to show the same property for each term of the right-hand side of (2.3). We shall check it for

$$\hat{\xi}_{s,t}(x) = \exp \int_s^t Y(\phi_{r,t}(x), o\hat{d}r) \tag{2.7}$$

in case $|\alpha| = 0$ only. Others can be shown similarly.

Rewrite the exponent of (2.7) using the Itô integral. Then $\hat{\xi}_{s,t}(x)$ is written as

$$\hat{\xi}_{s,t}(x) = \exp \left\{ M_{s,t}(x) + \int_s^t h(\psi_{r,t}(x), r)\, dr \right\},$$

where $M_{s,t}(x)$ is a backward martingale satisfying

$$\langle M_{s,t}(x) \rangle = \int_s^t a_{d+1,d+1}(\psi_{r,t}(x), \psi_{r,t}(x), r)\, dr.$$

Let $c = c(p)$ be a positive constant bounding $\frac{1}{2}p^2 a_{d+1,d+1} + p|h|$. Then we have

$$E[|\hat{\xi}_{s,t}(x)|^p] = E\left[\exp \left\{ pM_{s,t}(x) + p\int_s^t h(\psi_{r,t}(x), r)\, dr \right\} \right]$$

$$\le E[\exp pM_{s,t}(x) - \tfrac{1}{2}p^2 \langle M_{s,t}(x) \rangle] e^{ct}.$$

The expectation at the last member is equal to 1 since the exponential of $pM_{s,t} - \frac{1}{2}p^2\langle M_{s,t}\rangle$ is a backward martingale. Therefore $E[|\hat{\xi}_{s,t}(x)|^p]$ is bounded by e^{ct}. The proof is complete.

3. Second order stochastic partial differential equation. Next we shall study a second order parabolic partial differential equation with random coefficients:

$$u(\cdot,t) = f + \int_0^t L_s f \, ds + \sum_{i=1}^d \int_0^t F_i(\cdot, \circ ds)\frac{\partial u}{\partial x_i} \tag{3.1}$$

$$+ \int_0^t F_{d+1}(\cdot, \circ ds)u + F_{d+2}(\cdot, t).$$

Here L_t is an elliptic operator of the form

$$L_t u = \frac{1}{2}\sum_{i,j} a_{ij}(x,t)\frac{\partial^2}{\partial x_i \partial x_j}u + \sum_i b_i(x,t)\frac{\partial u}{\partial x_i} + c(x,t)u + d(x,t), \tag{3.2}$$

where $a_{ij}(x,t)$ is symmetric and nonnegative definite. In the sequel we assume that coefficients are of $C_b^{k,0}$ for $k \geq 6$ and there is a nonnegative function $v_{ij}(x,y,t)$ of $C_b^{k,0}$ such that $v_{ij}(x,x,t) = a_{ij}(x,t)$. The random field $(F_1(x,t),\ldots,F_{d+2}(x,t))$ is a $C^k(R^d; R^{d+2})$-valued Brownian motion satisfying A_{k+1}, $k \geq 5$.

We shall construct a solution of (3.1) by a purely probabilistic method, attaching a certain first order stochastic partial differential equation. Let $(W, \mathbf{B}, Q)$ be another probability space where a C^k-valued Brownian motion $X(x,t)$ with the local characteristic $v_{ij}(x,y,t)$ and $m(x,t) = b(x,t) - k(x,t)$ is given. ($k(x,y)$ is defined by (1.8).) On the product probability space $(\Omega \times W, \mathbf{F} \times \mathbf{B}, P \times Q)$ we consider a first order stochastic partial differential equation

$$v(\cdot,t) = f + \sum_{i=1}^d \int_0^t \{X_i(\cdot, \circ ds) + F_i(\cdot, \circ ds)\}\frac{\partial v}{\partial x_i} \tag{3.3}$$

$$+ \int_0^t (c(\cdot,s)\,ds + F_{d+1}(\cdot, \circ ds))v + \int_0^t d(\cdot,s)\,ds + F_{d+2}(\cdot,t).$$

If f is of C_b^3 it has a unique solution by Theorem 1. It is represented by

$$v(x,t) = f(\psi_{0,t}(x))\hat{\xi}_{0,t}(x) + \hat{s}_{0,t}(x), \tag{3.4}$$

where $\psi_{s,t}$ is the inverse of the stochastic flow generated by $-(X_1 + F_1,\ldots, X_d + F_d)$ and

$$\hat{\xi}_{s,t}(x) = \exp\left\{\int_s^t c(\psi_{r,t}(x),r)\,dr + \int_s^t F_{d+1}(\psi_{r,t}(x), \circ d\hat{r})\right\}, \tag{3.5}$$

$$\hat{s}_{s,t}(x) = \int_s^t \hat{\xi}_{r,t}(x)\{d(\psi_{r,t}(x),r)\,dr + F_{d+2}(\psi_{r,t}(x), \circ d\hat{r})\}. \tag{3.6}$$

THEOREM 2. *The conditional expectation*

$$u(x,t,\omega) = E_Q[f(\psi_{0,t}(x))\hat{\xi}_{0,t}(x) + \hat{s}_{0,t}(x)](\omega)$$

is well defined and is a solution to equation (3.1). Furthermore it has the follow-ing property. For any $\varepsilon > 0$,

$$\lim_{x \to \infty} \frac{u(x,t)}{1 + |x|^\varepsilon} = 0 \quad a.s. \tag{3.7}$$

PROOF. Rewrite the term $\int_0^t X_i(x, \circ ds)(\partial v/\partial x_i)$ in (3.3) using the Itô integral. Then

$$\int_0^t X_i(x, \circ ds)\frac{\partial v}{\partial x_i} = \int_0^t X_i(x, ds)\frac{\partial v}{\partial x_i} + \frac{1}{2}\langle X_i(x,t), \frac{\partial v}{\partial x_i}(x,t)\rangle.$$

Since v satisfies (3.3), its derivative satisfies

$$\frac{\partial v}{\partial x_i}(x,t) = \sum_j \int_0^t \frac{\partial X_j}{\partial x_i}(x, \circ ds)\frac{\partial v}{\partial x_j} + \sum_j \int_0^t X_j(x, \circ ds)\frac{\partial^2 v}{\partial x_i \partial x_j} + R(x,t),$$

where $R(x,t)$ is the remainder involving stochastic integrals by $F_i(x, \circ ds)$ etc. Then we have

$$\frac{1}{2}\sum_i \langle X_i(x,t), \frac{\partial v}{\partial x_i}(x,t)\rangle = \sum_j \int_0^t k_j(x,s)\frac{\partial v}{\partial x_j}\, ds$$

$$+ \frac{1}{2}\sum_{i,j}\int_0^t v_{ij}(x,x,s)\frac{\partial^2 v}{\partial x_i \partial x_j}\, ds.$$

Therefore, setting $M_i(x,t) = F_i(x,t) - \int_0^t m_i(x,r)\, dr$, we have

$$v(x,t) = f(x) + \int_0^t L_s v(x,s)\, ds + \sum_{i=1}^d \int_0^t M_i(x, ds)\frac{\partial v}{\partial x_i}$$

$$+ \sum_{i=1}^d \int_0^t F_i(x, \circ ds)\frac{\partial v}{\partial x_i} + \int_0^t F_{d+1}(x, \circ ds)v + F_{d+2}(x,t).$$

Each term of the above has a finite expectation with respect to Q by the Corollary to Theorem 1. Further we can change the order of the integration by Q and the stochastic integrals. Note that $\int_0^t M_i(x, ds)\frac{\partial v}{\partial x_i}$ is of mean 0 with respect to Q a.s. ω. Then we obtain

$$E_Q[v(x,t)] = f(x) + \int_0^t E_Q[L_s v(x,s)]\, ds$$

$$+ \sum_{i=1}^d \int_0^t F_i(x, \circ ds)E_Q\left[\frac{\partial v}{\partial x_i}(x,s)\right] \tag{3.8}$$

$$+ \int_0^t F_{d+1}(x, \circ ds)E_Q[v(x,s)] + F_{d+2}(x,t).$$

Set $u(x,t) = E_Q[v(x,t)]$ and change the order of integral E_Q and derivation L_s. Then we find that $u(x,t)$ is a solution to equation (3.1).

Now the solution $u(x,t)$ has a bounded moment of any order by the Corollary to Theorem 1. Then u has the growth property (3.7) as before. The proof is complete.

REMARK. If $F_1, \ldots, F_{d+2}$ are identically 0 in (3.1), equation (3.1) is a second order (deterministic) partial differential equation

$$\frac{\partial u}{\partial t} = L_t u, \qquad u|_{t=0} = f. \tag{3.9}$$

The solution is then represented by

$$u(x, t) = E_Q \left[f(\psi_{0,t}(x)) \left(\int_0^t c(\psi_{s,t}(x), s) \, ds \right) \right. \tag{3.10}$$
$$\left. + \int_0^t \left(\exp \int_s^t c(\psi_{r,t}(x), r) \, dr \right) d(\psi_{s,t}(x), s) \, ds \right]$$

where $\psi_{s,t}$ is the inverse of the stochastic flow generated by $-(X_1(x, t), \ldots, X_d(x, t))$.

The use of the inverse flow will provide a better probabilistic interpretation for diffusion equation or heat equation. To see this we consider the case $c = d = 0$. Then $u(x, t) = E[f(\psi_{0,t}(x))]$ is a solution to a heat equation with the initial condition $u(x, 0) = f(x)$. Now a particle starting from $\psi_{0,t}(x)$ and moving along the trajectory $\phi_{0,s}(\psi_{0,t}(x))$, $0 \leq s \leq t$, will arrive at x at time t since $\phi_{0,t}(\psi_{0,t}(x)) = x$. Hence $f(\psi_{0,t}(x))$ can be interpreted as the temperature at the state x at time t, which is carried through the above trajectory from the point $\psi_{0,t}(x)$. Then by the law of large numbers, its expectation $u(x, t)$ indicates the temperature at the state x at time t.

We next study the uniqueness of the solution to equation (3.1).

THEOREM 3. *Any solution to equation* (3.1) *satisfying the growth property* (3.7) *is at most unique.*

PROOF. We have shown the uniqueness in some special case in [7]. However, the method cannot be applied to the present cases. So we shall take another approach.

It is enough to consider the case $F_{d+2} \equiv 0$. Let $w_t(f)$ be the solution to the first order equation (3.1) in case $L_t \equiv 0$ and $F_{d+2} \equiv 0$. Then by Theorem 1 it is represented by $w_t(f) = f(\psi_{0,t}(x))\xi_{0,t}(\psi_{0,t}(x))$ where $\psi_{0,t}$ is the inverse of the stochastic flow $\phi_{0,t}$ generated by $-(F_1, \ldots, F_d)$ and $\xi_{0,t}(x) = \exp \int_0^t F_{d+1}(\phi_{0,r}(x), \circ dr)$. We may consider that for almost all ω, w_t is a linear map on $C(R^d; R^1)$. It is one-to-one and onto. The inverse map is given by $w_t^{-1}(f)(x) = f(\phi_{0,t}(x))\xi_{0,t}(x)^{-1}$.

Now suppose that $u(x, t)$ is a solution to (3.1) with $F_{d+2} \equiv 0$. Set $\tilde{u}(x, t) \equiv w_t^{-1}(u(x, t)) = u(\phi_{0,t}(x), t)\xi_{0,t}(x)^{-1}$. Then, using generalized Itô's formula [9], we have

$$d_t \tilde{u}(x, t) = \left\{ d_t u(\phi_{0,t}(x), t) + \sum_i \frac{\partial u}{\partial x_i}(\phi_{0,t}(x), t) \circ d\phi_{0,t}^i(x) \right\} \xi_{0,t}(x)^{-1}$$
$$- \tilde{u}(x, t) F_{d+1}(\phi_{0,t}(x), \circ dt)$$
$$= L_t u(\phi_{0,t}(x), t)\xi_{0,t}(x)^{-1} \, dt = w_t^{-1}(L_t u) \, dt.$$

Therefore setting $L_t^w \equiv w_t^{-1} L_t w_t$, $\tilde{u}(x,t)$ satifies

$$\frac{d}{dt}\tilde{u} = L_t^w \tilde{u}, \qquad \tilde{u}|_{t=0} = f. \qquad (3.11)$$

A direct computation yields

$$L_t^w = \frac{1}{2}\sum a_{ij}^w(x,t)\frac{\partial^2}{\partial x_i \partial x_j} + \sum b_i^w(x,t)\frac{\partial}{\partial x_i} + c^w(x,t), \qquad (3.12)$$

where a_{ij}^w, b_i^w, c^w are smooth functions with random parameters ω defined by

$$a_{ij}^w(x,t) = \sum_{k,l} a_{kl}(\phi_t(x),t)\partial_k(\psi_t^i)(\phi_t(x))\partial_l(\psi_t^j)(\phi_t(x)),$$

$$b_i^w(x,t)$$
$$= \frac{1}{2}\sum_{k,l} a_{kl}(\phi_t(x),t)\partial_k\partial_l(\psi_t^i)(\phi_t(x))$$
$$\quad + \sum_k \left\{ b_k(\phi_t(x),t) + \sum_l a_{kl}(y,t)\int_0^t \partial_l(F_{d+1}\circ\psi_{r,t})(y,\circ\hat{d}r)\Big|_{y=\phi_t(x)} \right\}$$
$$\quad \times \partial_k(\psi_t^i)(\phi_t(x)),$$

$$c^w(x,t)$$
$$= \left\{ \sum_{k,l} a_{kl}(y,t)\int_0^t \partial_k(F_{d+1}\circ\psi_{r,t})\,(y,\circ\hat{d}r)\int_0^t \partial_l(F_{d+1}\circ\psi_{r,t})(y,\circ\hat{d}r) \right.$$
$$\quad + \frac{1}{2}\sum_{k,l} a_{kl}(y,t)\int_0^t \partial_k\partial_l(F_{d+1}\circ\psi_{r,t})(y,\circ\hat{d}r)$$
$$\quad \left. + \sum_k b_k(y,t)\int_0^t \partial_k(F_{d+1}\circ\psi_{r,t})(y,\circ\hat{d}r) + c(y,t) \right\}\Big|_{y=\phi_t(x)}.$$

Here we set $\phi_t = \phi_{0,t}$ and $\psi_t = \psi_{0,t}$. These are not bounded but satisfy the following growth condition by Lemma 1:

$$|a_{ij}(x,t)| \le G(1+|x|)^{\gamma_1},$$
$$|b_i(x,t)| \le G(1+|x|)^{\gamma_2}, \qquad (3.13)$$
$$|c(x,t)| \le G(1+|x|)^{\gamma_3},$$

where γ_1, γ_2, γ_3 are positive constants such that γ_1, $\gamma_2 < 1$ and $\gamma_1 + \gamma_3 < 2$. Then the solution to equation (3.11) with the growth condition (3.7) is at most unique by the following lemma. This proves the uniqueness of the solution to equation (3.1).

LEMMA 2. *Let L_t be the elliptic operator of (3.2) such that coefficients a, b, c, d are of $C^{3,0}$ and satisfy the growth condition (3.13). Let $f(x,t)$ be of $C^{2,1}$ with the polynomial growth. If it satisfies*

$$-\frac{\partial f}{\partial s} + L_s f \le 0, \qquad s \ge 0, \qquad (3.14)$$

$$\lim_{s\downarrow 0} f(\cdot, s) \geq 0, \tag{3.15}$$

then f is nonnegative.

PROOF. In the case where $c(x,t)$ of L_t is nonpositive and f is bounded from below, the assertion can be proven similar to Stroock-Varadhan [14], Theorem 3.1, keeping in mind the growth condition (3.13).

In the general case we shall follow Friedman [4]. Introduce the function

$$H(x,t) = \exp\left[-\frac{(1+k|x|^2)^\gamma}{1-\mu t}\right]$$

where γ satisfies $\gamma_3 < 2\gamma$ and $\gamma_1 + 2\gamma < 2$. Define

$$\tilde{L}_t u = H(x,t)\left(-\frac{\partial}{\partial t} + L_t\right) H(x,t)^{-1} u.$$

Then we have

$$\tilde{L}_t = -\frac{\partial}{\partial t} + \frac{1}{2}\sum a_{ij}\frac{\partial^2}{\partial x_i \partial x_j} + \sum \tilde{b}_i \frac{\partial}{\partial x_i} + \tilde{c}$$

where

$$\tilde{c} = c + 2k^2\gamma^2\left\{\frac{(1+k|x|^2)^{2(\gamma-1)}}{(1-\mu t)^2} + \frac{\gamma-1}{\gamma}\frac{(1+k|x|^2)^{\gamma-2}}{1-\mu t}\right\}\sum_{i,j} a_{ij}x_i x_j$$

$$+ \frac{k\gamma(1+k|x|^2)^{\gamma-1}}{1-\mu t}\sum_i a_{ii} + 2k\frac{\gamma(1+k|x|^2)^{\gamma-1}}{1-\mu t}\sum b_i x_i$$

$$- \frac{\mu(1+k|x|^2)^\gamma}{(1-\mu t)^2}.$$

We can choose positive constants k, μ and t_0 such that $\tilde{c}(x,t) \leq 0$ for $t \leq t_0$.

Now if f is of $C^{2,1}$ with the polynomial growth, and if it satisfies (3.14) and (3.15), then $g \equiv Hf$ satisfies $\tilde{L}_t g \leq 0$ and $\lim_{s\downarrow 0} g(\cdot,s) \geq 0$. Since g is a bounded function, $g(x,t) \geq 0$ holds for $t \leq t_0$, which proves $f(x,t) \geq 0$ for $t \leq t_0$. Repeating this argument, we see that f is nonnegative in $R^d \times [0,T]$. The proof is complete.

REFERENCES

1. P. Baxendale, *Brownian motions in the diffeomorphisms group* I, Compositio Math. **53** (1984), 19–50.

2. J. M. Bismut, *A generalized formula of Itô and some other properties of stochastic flows*, Z. Wahrsch. Verw. Gebiete **55** (1981), 331–350.

3. K. D. Elworthy, *Stochastic dynamical systems and their flows*, Stochastic Analysis, A. Friedman and M. Pinsky, eds., Academic Press, New York, 1978, pp. 79–95.

4. A. Friedman, *Partial differential equations of parabolic type*, Prentice-Hall, Englewood Cliffs, N.J., 1964.

5. N. Ikeda and S. Watanabe, *Stochastic differential equations and diffusion processes*, North Holland-Kodansha, 1981.

6. N. V. Krylov and B. L. Rozovsky, *On the Cauchy problem for linear stochastic partial differential equations*, Izv. Akad. Nauk SSSR **41** (1977), 1329–1343.

7. H. Kunita, *Cauchy problem for stochastic partial differential equation arising in nonlinear filtering theory*, Systems Control Lett. **1** (1981), 37–41.

8. ____, *First order stochastic partial differential equations*, Taniguchi Symp. SA Katata 1982, 249–269.

9. ____, *On stochastic flows and applications*, Tata Institute of Fundamental Research, Bombay, (to appear).

10. Y. Le Jan, *Flots de diffusions dans R^d*, C. R. Acad. Sci. Paris Sér I Math. **294** (1982), 697–699.

11. Y. Le Jan and S. Watanabe, *Stochastic flows of diffeomorphisms*, Taniguchi Symp. SA Katata 1982, 307–332.

12. P. Malliavin, *Stochastic calculus of variation and hypoelliptic operators*, Proceedings of the International Symposium on Stochastic Differential Equations (Res. Inst. Math. Sci., Kyoto Univ., Kyoto, 1976), Wiley, New York, 1978, pp. 195–263.

13. E. Pardoux, *Stochastic partial differential equations and filtering of diffusion processes*, Stochastics **3** (1979), 127–167.

14. D. W. Stroock and S.R.S. Varadhan, *Multi-dimensional diffusion processes*, Springer-Verlag, New York, 1979.

KYUSHU UNIVERSITY, HAKOZAKI, FUKUOKA 812, JAPAN

Spatial Stochastic Growth Models— Survival and Critical Behavior

THOMAS M. LIGGETT

1. Introduction. The classical analogue of the class of models we will discuss is the branching process. The simplest example is a continuous time Markov chain X_t on the set of nonnegative integers in which the transitions $n \to n+1$ and $n \to n-1$ occur at rates $n\lambda$ and n respectively. Here X_t represents the size of a population in which individuals die at rate one and give birth to a single offspring at rate $\lambda > 0$. The first natural question involves the survival or extinction of the population. The simple answer is that extinction occurs with probability one if and only if $\lambda \leq 1$, and the survival probability

$$\rho(\lambda) = P(X_t \neq 0 \text{ for all } t | X_0 = 1)$$

is given by $\rho(\lambda) = 1 - \lambda^{-1}$ for $\lambda > 1$. This process and its generalizations have been studied extensively for more than three decades. Progress in each of these decades has been reported in [**8**], [**2**], and [**1**].

Motivated in part by problems in Physics and Biology, the past decade has seen the development of a new field in which a spatial element is introduced into these growth models. Individuals are no longer simply counted—they have spatial locations as well. These locations play a role in determining the birth rates at unoccupied sites. The spatial dependence has several implications for the study of these models. From a technical point of view, the principal one is that the size of the population does not have the Markov property. Therefore one is forced to consider Markov chains on larger and richer state spaces. The behavior of the process is significantly affected as well. For example, while supercritical branching processes generally grow at an exponential rate, the spatial constraints in the new setup mean that systems can grow at most at rate t^d, where d is the dimension of the set of sites.

While there are many ways in which a spatial dependence can be introduced, we will discuss a particular class of models A_t which fit into the framework of interacting particle systems [**12**]. Specifically, A_t is a continuous time Markov

Research supported by National Science Foundation Grants MCS 83-00836 and DMS 86-01800.

chain on the collection of finite subsets of the d-dimensional integer lattice $\mathbf{Z}^d$. To describe it, let $\beta(x, A)$ be a nonnegative function defined for nonempty finite subsets A of $\mathbf{Z}^d$ and for $x \in \mathbf{Z}^d \backslash A$. Assume that $\beta(x, A) = \beta(0, A - x)$ and that $\sum_{x \notin A} \beta(x, A) < \infty$. Then A_t has the following transition rates: (a) The empty set $\varnothing$ is an absorbing state, (b) $A \to A \backslash \{x\}$ occurs at rate 1 for each $x \in A$, and (c) $A \to A \cup \{x\}$ occurs at rate $\beta(x, A)$ if $A \neq \varnothing$ and $x \notin A$.

The chain A_t is said to survive if the survival probability

$$\rho = P^{\{0\}}(A_t \neq \varnothing \text{ for all } t \geq 0) \tag{1.1}$$

is positive, and it is said to die out if $\rho = 0$. The chain may only be defined until a finite explosion time. In this case, the event that explosion occurs is taken to be contained in the survival event in (1.1). We will be concerned primarily with the following two questions: (a) For what choices of the birth rates $\beta(x, A)$ does A_t survive, and for what choices does it die out? (b) When A_t survives, how large is ρ?

Often we will consider one-parameter families of models, and then $\rho(\lambda)$ is the survival probability for the model with parameter $\lambda > 0$. In most cases, $\rho(\lambda)$ is increasing in λ, and we define the critical value by $\lambda_c = \inf\{\lambda \colon \rho(\lambda) > 0\}$. Then question (a) asks for bounds on λ_c, while question (b) involves the behavior of $\rho(\lambda)$ as $\lambda \downarrow \lambda_c$. If, for example, $\rho(\lambda) \sim L(\lambda - \lambda_c)^\alpha$, then α would be called the critical exponent for $\rho(\lambda)$.

EXAMPLE 1.2. THE CONTACT PROCESS. Let

$$\beta(x, A) = \lambda |\{y \in A \colon |y - x| = 1\}|,$$

where $|\cdot|$ denotes the cardinality. This example was introduced by Harris [9] and can be viewed as a model for the spread of infection. It has also come up in high energy physics. A general treatment of the contact process can be found in Chapter VI of [12].

EXAMPLE 1.3. FINITE NEAREST PARTICLE SYSTEMS. Here $d = 1$ and $\beta(x, A) = \beta(l_x(A), r_x(A))$, where $l_x(A) = x - \max\{y \in \mathbf{Z} \colon y \leq x \text{ and } y \in A\}$, $r_x(A) = \min\{y \in \mathbf{Z} \colon y \geq x \text{ and } y \in A\} - x$, $\beta(l, r) = \beta(r, l) \geq 0$ for $1 \leq l, r \leq \infty$, $\sum_n \beta(n, \infty) < \infty$, and $\beta(\infty, \infty) = 0$. This is the finite version of an infinite system which was introduced by Spitzer [16]. A treatment of nearest particle systems can be found in Chapter VII of [12].

Our understanding of these growth models is much more complete when A_t is reversible than when it is not. The contact process is not reversible. The finite nearest particle system is reversible if and only if $\beta(l, r)$ has the form

$$\begin{aligned}
\beta(l, r) &= \lambda \beta(l) \beta(r) / \beta(l + r) && \text{if } 1 \leq l, r < \infty, \text{ and} \\
\beta(\infty, n) &= \beta(n, \infty) = \lambda \beta(n) && \text{if } 1 \leq n < \infty,
\end{aligned} \tag{1.4}$$

where $\lambda > 0$ and $\beta(n)$ is a strictly positive probability density on the positive integers. Results for general systems are surveyed in the next section. The more precise results which are available for reversible systems are discussed in §3.

2. General systems. In this section, we will give some easily verifiable sufficient conditions for extinction and for survival of general growth models. For many parametric families of models, these results give upper and lower bounds on the critical value of the system. First we consider the case of extinction.

THEOREM 2.1. *Suppose that* $\sum_{x \notin A} \beta(x, A) \leq \beta(|A|)$ *for all* A, *where* $\beta(\cdot)$ *is a positive function satisfying*

$$\sum_{n=1}^{\infty} \frac{n!}{\prod_{k=1}^{n} \beta(k)} = \infty. \tag{2.2}$$

Then A_t *dies out.*

PROOF. The hypothesis guarantees that A_t can be coupled with a Markov chain X_t on the nonnegative integers with transitions $n \to n - 1$ and $n \to n + 1$ at rates n and $\beta(n)$ respectively, in such a way that $|A_t| \leq X_t$ a.s. for all t. Assumption (2.2) implies that $P(X_t \neq 0 \text{ for all } t) = 0$. Therefore

$$P(A_t \neq \varnothing \text{ for all } t) = 0.$$

Clearly, the same proof would show that if $\sum_{x \notin A} \beta(x, A) \geq \beta(|A|)$ for all $A \neq \varnothing$, where the series in (2.2) converges, then A_t survives. This result is not very useful, however. In most cases of interest, including Examples 1.2 and 1.3,

$$\lim_{n \to \infty} \frac{1}{n} \inf_{|A|=n} \sum_{x \notin A} \beta(x, A) = 0.$$

This is a consequence of the crowding which occurs in $\mathbf{Z}^d$. To obtain a useful sufficient condition for survival, one must work harder. Define $T: \mathbf{Z}^d \to \mathbf{Z}$ by $Tx =$ sum of the coordinates of x. The following result and its proof are based on techniques developed in [**10**].

THEOREM 2.3. *Suppose that for each* A, *the birth rates satisfy*

$$\sum_{x_i < Tx < x_{i+1}} \beta(x, A) \frac{Tx - x_i - 1}{x_{i+1} - x_i - 2} \geq 2 \quad \text{for } 0 \leq i \leq n - 1,$$

and

$$\sum_{x_i < Tx < x_{i+1}} \beta(x, A) \frac{x_{i+1} - Tx - 1}{x_{i+1} - x_i - 2} \geq 2 \quad \text{for } 1 \leq i \leq n,$$

where $\{x_i, 0 \leq i \leq n+1\}$ *are defined by* $x_0 = -\infty$, $x_{n+1} = +\infty$, $x_1 < x_2 < \cdots < x_n$, *and* $\{Tx: x \in A\} = \{x_1, \ldots, x_n\}$. *(If* $x_{i+1} = x_i + 1$, *there is no assumption; the ratios* $\frac{0}{0}$ *and* $\frac{\infty}{\infty}$ *are interpreted as 1.) Then* $\rho \geq \frac{1}{2}$, *so* A_t *survives.*

PROOF. Let ν be the probability measure on $\{0, 1\}^{\mathbf{Z}}$ which is the distribution of the stationary renewal process on $\mathbf{Z}$ whose inter-arrival times τ have the distribution given by $P(\tau \geq n) = (2n - 2)! 4^{-n+1}/(n - 1)! n!$. Define a function h on the finite subsets of $\mathbf{Z}$ by $h(A) = \nu\{\eta: \eta(x) = 1 \text{ for some } x \in A\}$, and a function g on the finite subsets of $\mathbf{Z}^d$ by $g(A) = h(\{Tx: x \in A\})$. We will show that $E^A g(A_t)$ is nondecreasing in t for each A. The desired conclusion follows from this, since then

$$\rho = \lim_{t\to\infty} P^{\{0\}}(A_t \neq \varnothing) \geq \lim_{t\to\infty} E^{\{0\}} g(A_t) \geq g(\{0\}) = \tfrac{1}{2}.$$

To prove the monotonicity of $E^A g(A_t)$, we need the following two facts:

(a) $\sum_{x\in A}[h(A) - h(A\backslash\{x\})] \leq 2\sum_{x\notin A}[1_A(x+1) + 1_A(x-1)][h(A\cup\{x\}) - h(A)]$.

(b) $h(A \cup \{x\})$ is a concave function of x in each connected component of the complement of A.

These appear in [12] as Lemma 1.25 of Chapter VI and Lemma 2.8 of Chapter VII. The proof of (b) is quite easy; that of (a) is substantially more difficult.

Now fix a finite subset A of $\mathbf{Z}^d$, and let $B = \{Tx : x \in A\}$. Then

$$\sum_{x\in A}[g(A) - g(A\backslash\{x\})] \leq \sum_{y\in B}[h(B) - h(B\backslash\{y\})], \qquad (2.4)$$

since all the summands are nonnegative, and each positive summand on the left agrees with one of the summands on the right. Let $\{x_i, 0 \leq i \leq n+1\}$ be as in the statement of the theorem, and put $f(y) = h(B \cup \{y\}) - h(B)$ for $y \in \mathbf{Z}\backslash B$. Then

$$\sum_{x\notin A} \beta(x, A)[g(A \cup \{x\}) - g(A)]$$

$$\geq \sum_{i=0}^{n} \sum_{x_i < Tx < x_{i+1}} \beta(x, A) f(Tx)$$

$$\geq \sum_{Tx < x_1} \beta(x, A) f(x_1 - 1) + \sum_{Tx > x_n} \beta(x, A) f(x_n + 1)$$

$$+ \sum_{i=1}^{n-1} \sum_{x_i < Tx < x_{i+1}} \beta(x, A)$$

$$\cdot \frac{(x_{i+1} - Tx - 1)f(x_i + 1) + (Tx - x_i - 1)f(x_{i+1} - 1)}{x_{i+1} - x_i - 2}$$

by property (b). Note that this property implies that f is a decreasing function to the left of B and an increasing function to the right of B. Using the hypothesis of the theorem, it follows that

$$\sum_{x\notin A} \beta(x, A)[g(A \cup \{x\}) - g(A)] \geq 2 \sum_{y\notin B}[1_B(y+1) + 1_B(y-1)]f(y).$$

Combining this with (a) and (2.4) yields

$$\sum_{x\notin A} \beta(x, A)[g(A \cup \{x\}) - g(A)] \geq \sum_{x\in A}[g(A) - g(A\backslash\{x\})],$$

or equivalently, that $d(E^A g(A_t))/dt \geq 0$ as claimed.

Applying Theorems 2.1 and 2.3 to the contact process gives the bounds $1/2d \leq \lambda_c \leq 2/d$. More is known in this case: $\lambda_c \geq 3/2$ if $d = 1$, and λ_c is asymptotic to $(2d)^{-1}$ as $d \to \infty$. (See pp. 289 and 308 of [12].) When applied to finite nearest particle systems, Theorems 2.1 and 2.3 guarantee survival whenever $\sum_{l+r=n} \beta(l,r) \geq 4$ for $2 \leq n < \infty$, and $\sum_{r=1}^{\infty} \beta(\infty, r) \geq 2$, and extinction whenever $\sum_{l+r=n} \beta(l,r) \leq 1$ for $2 \leq n < \infty$, and $\sum_{r=1}^{\infty} \beta(\infty, r) \leq 1$.

In this section we have concentrated on the problem of survival and extinction of general systems, saying nothing about the behavior of the process when it does survive. If the birth rates $\beta(x, A)$ have additional properties, much more can be proved. For example, if the rates have certain additivity properties, then percolation techniques can be applied to prove limit theorems in the supercritical case. For examples of this, see [5] and [6]. Instead of discussing additive systems further, we will turn in the next section to the case in which A_t is reversible.

3. Reversible systems. This section is devoted to a discussion of the more complete results which can be proved for growth models which are reversible. In the reversible case, the birth rates are given by

$$\beta(x, A) = \pi(A \cup \{x\})/\pi(A) \tag{3.1}$$

for $x \notin A$, $A \neq \varnothing$, $\pi(A) > 0$, and $\beta(x, A) = 0$ otherwise. Here π is a nonnegative function on the nonempty finite subsets of $\mathbf{Z}^d$ which is assumed to satisfy
   (a) $\pi(A) = \pi(B)$ whenever A and B are translates of each other,
   (b) $B \supset A$ and $\pi(A) = 0$ imply $\pi(B) = 0$,
   (c) $\sum_x \pi(A \cup \{x\}) < \infty$ for each $A \neq \varnothing$,
   (d) $\pi(A) = 1$ if $|A| = 1$, and
   (e) there are sets of arbitrarily large cardinality with $\pi(A) > 0$.
The state space of the chain is $\{\varnothing\} \cup \{A: \pi(A) > 0\}$.

The most important new tool which is available in the reversible case is the Dirichlet principle, which provides a variational characterization of the survival probability. In order to state it, let H be the collection of all functions h on the finite subsets of $\mathbf{Z}^d$ which satisfy
   (a) $0 \leq h(A) \leq 1$ for all A, $h(\varnothing) = 0$,
   (b) $h(A) = h(B)$ if A and B are translates of each other, and
   (c) $\lim_{n \to \infty} \inf_{|A|=n} h(A) = 1$.
For $h \in H$, define

$$\Phi(h) = \sum_{A \ni 0} \frac{\pi(A)}{|A|} \sum_{x \in A} [h(A) - h(A \setminus \{x\})]^2.$$

The proofs of the following theorem and its corollaries can be found in [13]. See also [7].

THEOREM 3.2. *With the above definitions, $\rho = \inf_{h \in H} \Phi(h)$.*

This result has the following consequences. The first is almost immediate; the second is obtained by using a particular $h \in H$ which depends on A only through its cardinality; the proof of the third involves two applications of the Schwarz inequality.

COROLLARY 3.3. *Let ρ and ρ' be the survival probabilities for the systems corresponding to the weights $\pi(\cdot)$ and $\pi'(\cdot)$ respectively. If $\pi(A) \leq \pi'(A)$ for all A, then $\rho \leq \rho'$.*

COROLLARY 3.4. *For $n \geq 1$, let $\pi_n = \sum_{|A|=n, A \ni 0} \pi(A)$. Then*

$$\rho \leq \left[\sum_{n=1}^{\infty} \pi_n^{-1} \right]^{-1},$$

where the bound is interpreted as zero if the series diverges.

COROLLARY 3.5. *Suppose that*

$$\pi(A) = c_n \sum_{x \in A} \pi(A \backslash \{x\}) \gamma_{A \backslash \{x\}}(x) \quad \text{if } |A| = n \geq 2,$$

where $c_n > 0$, $\gamma_A(x) \geq 0$, $\gamma_A(x) = 0$ if $x \in A$, $\sum_x \gamma_A(x) = 1$, and $\gamma_{A+y}(x+y) = \gamma_A(x)$ for all $y \in \mathbf{Z}^d$. Then $\pi_n = n c_n c_{n-1} \cdots c_2$ and $\rho \geq [\sum_{n=1}^{\infty} n \pi_n^{-1}]^{-1}$.

For one-parameter systems in which the weights $\pi(\cdot)$ are obtained via the special construction in Corollary 3.5, Corollaries 3.4 and 3.5 can often be used to determine the critical value exactly. Unfortunately, many interesting systems do not arise in this way, as was shown in §3 of [14]. Even so, these special systems can often be used as comparison models to obtain bounds on the survival probability for a more interesting system via Corollary 3.3. This type of application can be found in §6 of [13] and §§3 and 5 of [14].

Next we consider reversible nearest particle systems in one dimension, since very explicit results can be obtained for them using Theorem 3.2 and its corollaries. These are the growth models described in Example 1.3 with birth rates chosen as in (1.4). Proofs of the following results can be found in [15]. They improve results in [7].

The birth rates for reversible nearest particle systems have the form (3.1) with $\pi(\cdot)$ given by

$$\pi(\{x_1, \ldots, x_n\}) = \lambda^{n-1} \prod_{k=1}^{n-1} \beta(x_{k+1} - x_k) \tag{3.6}$$

for $x_1 < x_2 < \cdots < x_n$. These weights result from the construction in Corollary 3.5, using $\gamma_A(x) = \beta(l_x(A))$ for x to the right of A, and $c_n = \lambda$. Furthermore, $\pi_n = n\lambda^{n-1}$. Therefore, Corollaries 3.4 and 3.5 provide the following bounds on the survival probability:

$$\frac{\lambda - 1}{\lambda} \leq \rho(\lambda) \leq \left| \lambda \log \frac{\lambda - 1}{\lambda} \right|^{-1} \quad \text{for } \lambda > 1. \tag{3.7}$$

Corollary 3.3 implies that $\rho(\lambda)$ is an increasing function of λ, so the critical value is well defined. It then follows from (3.7) that $\lambda_c = 1$.

The next question involves the behavior of $\rho(\lambda)$ for λ just to the right of the critical value. The dependence of $\rho(\lambda)$ on the probability density $\beta(\cdot)$ becomes more evident in the answers to this question. A more careful analysis of the expression for the survival probability given in Theorem 3.2 than was used in its corollaries gives

THEOREM 3.8. *For reversible nearest particle systems,*

$$\liminf_{\lambda \downarrow 1} \frac{\rho(\lambda)}{\lambda - 1} \geq 2, \qquad (3.9)$$

with the inequality being strict if $\sup_n \sum_{k=1}^{n-1} \beta(k)\beta(n-k)/\beta(n) < \infty$. *If* $\sum_n e^{-\theta n}\beta(n) = 1 - c\theta + \gamma\theta^\alpha + o(\theta^\alpha)$ *as* $\theta \downarrow 0$ *for some* $c > 0$, $\gamma > 0$, *and* $1 < \alpha \leq 2$, *then*

$$\limsup_{\lambda \downarrow 1} \frac{\rho(\lambda)}{(\lambda - 1)^{\alpha - 1}} \leq 4\gamma c^{-\alpha}.$$

If $\beta(\cdot)$ has a finite variance, the second part of Theorem 3.8 with $\alpha = 2$ implies that

$$\limsup_{\lambda \downarrow 1} \frac{\rho(\lambda)}{\lambda - 1} \leq 2 \sum_{n=1}^{\infty} n^2 \beta(n) \left[\sum_{n=1}^{\infty} n\beta(n) \right]^{-2}. \qquad (3.10)$$

From (3.9) and (3.10), we can conclude that the critical exponent for $\rho(\lambda)$ is 1 in this case.

The above results suggest that (a) $L = \lim_{\lambda \downarrow 1} \rho(\lambda)(\lambda - 1)^{-1}$ exists always, (b) L is finite if and only if $\sum_n n^2 \beta(n) < \infty$, and (c) $\lim_{\lambda \downarrow 1} \rho(\lambda)(\lambda - 1)^{1-\alpha}$ exists and is positive and finite whenever $\beta(\cdot)$ is in the domain of normal attraction of a one-sided stable law of index $\alpha \in (1, 2)$. These remain open problems. Note that (3.10), and the strict inequality which is asserted in (3.9) under the extra hypothesis, imply that L, if it exists, cannot be the same for all densities $\beta(\cdot)$.

We conclude with a discussion of a new class of models in higher dimensions which generalize reversible nearest particle systems. One particularly interesting feature of these models is that there is a close connection between them and another important area of probability theory: self-avoiding random walks. Among the many recent papers in this area are [3] and [4].

Let $\beta(x)$ be a probability density on $\mathbf{Z}^d$ for which $\beta(0) = 0$. Suppose that $\pi(A)$ is given by

$$\pi(A) = \lambda^{n-1} \sum \prod_{k=1}^{n-1} \beta(x_{k+1} - x_k), \qquad (3.11)$$

where the sum is taken over all orderings $x_1, \ldots, x_n$ of the n elements of A. To see that these models generalize reversible nearest particle systems, take $d = 1$ and $\beta(\cdot)$ concentrating on the positive integers. Then the right side of (3.11) agrees with (3.6). In general, one can think of the birth rates of these systems as being certain averages of nearest particle birth rates.

Now let S_n be the random walk on $\mathbf{Z}^d$ whose increments have density β. One of the quantities of interest in the theory of self-avoiding random walks is $\sigma_n = P(B_n)$, where B_n is the event that $S_0, S_1, \ldots, S_n$ are all distinct. It is well known and easy to check that $\log \sigma_n$ is subadditive, so that

$$\gamma = \lim_{n \to \infty} \sigma_n^{1/n} \qquad (3.12)$$

exists and is positive.

THEOREM 3.13. *Suppose that the growth model A_t has the birth rates given in (3.1) where $\pi(\cdot)$ is given by (3.11). Then*

$$\frac{\lambda\gamma - 1}{\lambda\gamma} \leq \rho(\lambda) \leq \left[\sum_{n=0}^{\infty} \frac{1}{(n+1)\lambda^n \sigma_n}\right]^{-1}, \qquad (3.14)$$

so that $\lambda_c = \gamma^{-1}$.

PROOF. To prove the right inequality in (3.14), we will compute π_n and then will apply Corollary 3.4:

$$\pi_n = \sum_{|A|=n, A \ni 0} \pi(A) = \lambda^{n-1} \sum_{|A|=n, A \ni 0} \sum_{\text{orderings of } A} \prod_{k=1}^{n-1} \beta(x_{k+1} - x_k)$$

$$= n\lambda^{n-1} \sum_{F} \prod_{k=1}^{n-1} \beta(u_k) = n\lambda^{n-1} \sigma_{n-1},$$

where $F = \{(u_1, \ldots, u_{n-1}): 0, u_1, u_1 + u_2, \ldots, u_1 + \cdots + u_{n-1} \text{ are distinct}\}$. Unfortunately, Corollary 3.5 does not apply to these examples to give a lower bound for $\rho(\lambda)$. Therefore, we need to use Theorem 3.2 directly:

$$\Phi(h) = \sum_{n=1}^{\infty} \frac{\lambda^{n-1}}{n} \sum_{|A|=n, A \ni 0} \sum_{\text{orderingsof } A} \prod_{k=1}^{n-1} \beta(x_{k+1} - x_k)$$

$$\cdot \sum_{j=1}^{n} [h(A) - h(A \setminus \{x_j\})]^2$$

$$\geq \sum_{n=1}^{\infty} \lambda^{n-1} \sum_{F} \prod_{k=1}^{n-1} \beta(u_k)[h(\{0, u_1, u_1 + u_2, \ldots, u_1 + \cdots + u_{n-1}\})$$
$$\qquad\qquad\qquad - h(\{0, u_1, u_1 + u_2, \ldots, u_1 + \cdots + u_{n-2}\})]^2 \qquad (3.15)$$

$$= \sum_{n=0}^{\infty} \lambda^n \sigma_n E\{[h(\{S_0, \ldots, S_n\}) - h(\{S_0, \ldots, S_{n-1}\})]^2 | B_n\}.$$

For $n \leq N$ and distinct $0, x_1, \ldots, x_n$, $P(B_N | S_1 = x_1, \ldots, S_n = x_n) \leq \sigma_{N-n}$, so that

$$P(S_1 = x_1, \ldots, S_n = x_n | B_N) \leq \frac{\sigma_n \sigma_{N-n}}{\sigma_N} P(S_1 = x_1, \ldots, S_n = x_n | B_n).$$

Using this in (3.15), and then applying the Schwarz inequality gives

$$\Phi(h) \geq \sum_{n=0}^{N} \lambda^n \frac{\sigma_N}{\sigma_{N-n}} E\{[h(\{S_0, \ldots, S_n\}) - h(\{S_0, \ldots, S_{n-1}\})]^2 | B_N\}$$

$$\geq \sum_{n=0}^{N} \lambda^n \frac{\sigma_N}{\sigma_{N-n}} [h_n - h_{n-1}]^2,$$

where $h_n = E[h(\{S_0, \ldots, S_n\})|B_N]$. Another application of the Schwarz inequality yields

$$h_N^2 = \left[\sum_{n=0}^N [h_n - h_{n-1}]\right]^2 \leq \sum_{n=0}^N \lambda^n \frac{\sigma_N}{\sigma_{N-n}} [h_n - h_{n-1}]^2 \sum_{n=0}^N \lambda^{-n} \frac{\sigma_{N-n}}{\sigma_N}.$$

Therefore

$$\Phi(h) \geq h_N^2 \left[\sum_{n=0}^N \lambda^{-n} \frac{\sigma_{N-n}}{\sigma_N}\right]^{-1} \tag{3.16}$$

for each N. Now let

$$\delta_N = \max_{1 \leq n \leq N} [\sigma_{N-n}/\sigma_N]^{1/n} \geq \sigma_N^{-1/N} \quad \text{and} \quad \delta = \liminf_{N \to \infty} \delta_N.$$

By (3.12), $\delta \geq \gamma^{-1}$. On the other hand, a recursive argument using the definition of the δ_N's implies that for each N there are positive integers $n_1, \ldots, n_k$ so that $n_1 + \cdots + n_k = N$ and $\sigma_N^{-1} = \prod_{i=1}^k \delta_{n_k + \cdots + n_i}^{n_i}$, from which it follows that $\delta \leq \gamma^{-1}$. Therefore $\delta = \gamma^{-1}$, so that we can pass to the limit in (3.16) along a subsequence of N's, using the fact that $\lim_{N \to \infty} h_N = 1$, to conclude that $\Phi(h) \geq [\sum_{n=0}^\infty (\lambda\gamma)^{-n}]^{-1}$ for every $h \in H$. Now use Theorem 3.2 to obtain the lower bound in (3.14).

REMARKS. (a) In the nearest particle case, $\sigma_n = 1$ and $\gamma = 1$, so (3.14) and (3.7) agree in this case.

(b) In order to identify the critical value in Theorem 3.13, it is enough to know that the sequence σ_n satisfies (3.12). More information about this sequence would be required in order to reach other conclusions. For example, it would be nice to know whether $\rho(\lambda_c) = 0$, as is the case for reversible nearest particle systems. In [11], Kesten proved for the simple random walk that σ_{n+2}/σ_n converges to γ^2. He tells me that he has recently proved similar results for more general random walks.

There are a number of other classes of reversible growth models in higher dimensions which can be partially analyzed with the help of the Dirichlet principle. Some of these are discussed in [13] and [14]. Much remains to be done, however.

REFERENCES

1. S. Asmussen and H. Hering, *Branching processes*, Prog. Prob. Statist., 3, Birkhäuser, Boston, Mass., 1983.

2. K. Athreya and P. Ney, *Branching processes*, Springer-Verlag, New York-Heidelberg, 1972.

3. A. Berretti and A. Sokal, *New Monte Carlo method for the self-avoiding walk*, J. Statist. Phys. **40** (1985), 483–531.

4. D. Brydges and T. Spencer, *Self-avoiding walk in 5 or more dimensions*, Comm. Math. Phys. **97** (1985), 125–148.

5. R. Durrett and D. Griffeath, *Supercritical contact processes on* **Z**, Ann. Probab. **11** (1983), 1–15.

6. R. Durrett and R. Schonmann, *Stochastic growth models*, Preprint.

7. D. Griffeath and T. Liggett, *Critical phenomena for Spitzer's reversible nearest particle systems*, Ann. Probab. **10** (1982), 881–895.

8. T. Harris, *The theory of branching processes*, Springer-Verlag, New York-Heidelberg, 1963.

9. ——, *Contact interactions on a lattice*, Ann. Probab. **2** (1974), 969–988.

10. R. Holley and T. Liggett, *The survival of contact processes*, Ann. Probab. **6** (1978), 198–206.

11. H. Kesten, *On the number of self-avoiding walks*, J. Math. Phys. **4** (1963), 960–969.

12. T. Liggett, *Interacting particle systems*, Springer-Verlag, New York-Heidelberg, 1985.

13. ——, *Reversible growth models on symmetric sets*, Proceedings of the 1985 Taniguchi Symposium (to appear).

14. ——, *Reversible growth models on $\mathbf{Z}^d$: some examples*, Proceedings of a 1986 IMA Workshop (to appear).

15. ——, *Applications of the Dirichlet principle to finite reversible nearest particle systems*, Prob. Th. and Rel. Fields (to appear).

16. F. Spitzer, *Stochastic time evolution of one dimensional infinite particle systems*, Bull. Amer. Math. Soc. **83** (1977), 880–890.

UNIVERSITY OF CALIFORNIA, LOS ANGELES, CALIFORNIA 90024, USA

Wave Propagation and Heat Conduction
in Random Media

GEORGE C. PAPANICOLAOU

1. Introduction. The mathematical analysis of propagation phenomena in random media has advanced substantially in the last ten or fifteen years. One reason for this is that techniques in the modern theory of stochastic processes such as asymptotics for stochastic equations, the theory of interacting particle systems, etc., have improved to where really difficult problems can be solved. Problems in random media are difficult because randomness has a spatial character: a process such as diffusion takes place in a medium whose properties vary randomly from point to point in space. Does this diffusion process behave after a long time like a diffusion process in a deterministic medium? Can it get trapped so that its long time behavior is different from that of a usual diffusion process? What about wave processes in a random medium? When do we have effective propagation and when do the waves get trapped so that we have localization? Even in one space dimension these questions are not easy to answer.

We shall consider here only one aspect of propagation in random media that makes use of asymptotic methods. It is one-dimensional wave propagation: reflection and transmission of monochromatic, single frequency, waves and pulses. The asymptotic methods that we use have been developed in the last several years primarily for the study of such problems.

2. Monochromatic waves. Let $u(x)$ be the complex-valued wave field at x in R^1, let k be the free space wave number, and let $n(x)$ be the index of refraction of the medium. The wave field satisfies the reduced wave equation

$$u_{xx} + k^2 n^2 u = 0 \tag{2.1}$$

in an interval $0 < x < L$. We assume that a wave of unit amplitude is incident from the left so that

$$u = e^{ikx} + Re^{-ikx} \quad \text{for } x < 0 \tag{2.2}$$

and

$$u = Te^{ikx} \quad \text{for } x > L. \tag{2.3}$$

Here R and T are the complex-valued reflection and transmission coefficients, respectively. At $x = 0$ and $x = L$ the wave functions u and u_x are continuous. Since k and n are real it is easily seen that the total power scattered by the slab in $0 < x < L$ is equal to one

$$|R|^2 + |T|^2 = 1. \tag{2.4}$$

It is also easily seen that the continuity condition and (2.2), (2.3) imply that the solution of equation (2.1) satisfies the two-point boundary conditions

$$u + \frac{1}{ik}u_x = 2 \quad \text{at } x = 0, \qquad u - \frac{1}{ik}u_x = 0 \quad \text{at } x = L. \tag{2.5}$$

Reflection and transmission by a random slab is modelled by an index of refraction $n(x)$ which is a random process defined on some probability space. If $n(x)$ is positive, bounded, and piecewise continuous (measurable) the solution of (2.1) is a well-defined process. The reflection and transmission coefficients are then random variables depending on the wave number k and the slab width L. The problem is now to find qualitative or quantitative properties of R and T given some properties of the random medium through the index of refraction— in particular, to find properties that require randomness and do not hold for periodic or almost periodic media, for example.

A basic qualitative property that has been known for some time [1] is the exponential decay of the transmission coefficient as the slab width L goes to infinity. Furstenberg treated discrete versions of the problem in which the analog of $n(x)$ is independent identically distributed random variables. Kotani [2] recently gave a proof of this result for the present problem with minimal hypotheses. If $n(x)$ is a stationary, ergodic process that is bounded with probability one and is nondeterministic (that is, it has trivial tail sigma field) then

$$\lim_{L \to \infty} \frac{1}{L} \log |T|^2 = -2\gamma(k) \tag{2.6}$$

with probability one, and $\gamma(k)$ is a positive constant. The content of the theorem is the positivity which gives the exponential decay of the transmission coefficient with the size of the slab. The decay constant $\gamma(k)$ depends on the wave number and is the (maximal) Lyapounov exponent of (2.1). Its reciprocal is called the localization length because it characterizes the depth of penetration (skin depth) of the wave into a random medium occupying a half-space. A random half-space is a perfectly reflecting medium by (2.4) and (2.6).

The positivity of the Lyapounov exponent $\gamma(k)$ is at the root of nearly every theorem dealing with qualitative properties (2.1). An example is the discreteness of the spectrum in an infinite random medium with bounded, stationary ergodic index of refraction (with trivial tail field) first proved by Goldsheid, Molcanov, and Pastur [3] (with many more hypotheses) and in great generality by Kotani [4].

If we want to get quantitative information about scattering by a random medium, we must look into interesting asymptotic limits: large or small wave number, large or small variance of $n(x)$ about its mean, etc. Let us assume that

$$n^2(x) = 1 + \mu(x) \tag{2.7}$$

where $\mu(x)$ is a zero mean stationary random process with rapidly decaying correlation functions. Let

$$\alpha = \int_0^\infty E\{\mu(x)\mu(0)\}\, dx.$$

Here E denotes expectation for the stationary process μ. The parameter α has dimensions of length and can be thought of as a correlation length of the refractive index. We can now investigate the behavior of the Lyapounov exponent $\gamma(k)$ for small $k\alpha$, or large $k\alpha$ [5]. Small $k\alpha$ will be discussed here since we are aiming at the pulse problem of the next section where this case is important. Small $k\alpha$ means of course that the wave length of the incident wave is large compared to the typical "size" of inhomogeneities in the medium. We expect the random medium to have an effective behavior independent of the detailed characterization of $n(x)$. In fact

$$\gamma(k) \approx \gamma_0 k^2 \tag{2.8}$$

as $k \to 0$ where $\gamma_0 = \alpha/4$. Whereas $\gamma(k)$ is a complicated functional of the process μ, γ_0 depends only on α. We have a limit theorem.

The simple result (2.8) gives us a way to assess quantitatively the transmission coefficient $T = T(k, L)$ which is a complicated functional of μ. We see in fact from (2.6) and (2.8) that if as $L \to \infty$, $k \to 0$ like $1/\sqrt{L}$, then roughly

$$|T| \approx e^{\gamma(k)L} \approx e^{\gamma_0}. \tag{2.9}$$

This rough estimate is meant to indicate only that T should have a nontrivial distribution in this limit. When the slab of random medium is large (compared to the correlation length α) then only long wavelengths of order $\sqrt{L}$ will be transmitted. Short ones are blocked. And T has a limit distribution that can be computed explicitly.

To describe this result we must introduce some notation. First it is convenient to consider the reflection coefficient R instead of the transmission coefficient. They are related by (2.4). Let us denote R by $R(L, k, \alpha)$ to indicate dependence on the slab width, the wave number, and the correlation length. Let $\varepsilon > 0$ be a small dimensionless parameter. Define

$$R^\varepsilon(L) = R(L/\varepsilon^2, \varepsilon k, \alpha) \tag{2.10}$$

and consider $R^\varepsilon(L), L \geq 0$ for each $\varepsilon > 0$, as a process on the hyperbolic disc $H = \{R \text{ in } C | |R| = 1\}$. The scaling (2.10) is a convenient way to express the limit described before (2.9). We have

THEOREM 1. *Assume that $\mu(x)$ is a stationary process that has finite moments of all orders and is rapidly mixing (see [6] for definitions and sharp* *conditions). Then the process R^ε converges weakly as $\varepsilon \to 0$ to Brownian motion on the hyperbolic disc H with infinitesimal variance αk^2.*

If we parametrize H by polar coordinates

$$R = e^{-i\psi}\tanh(\theta/2), \tag{2.11}$$

then the theorem says that the limit process $R(L), L > 0$, is a Markov process with generator $\frac{1}{2}\alpha k^2 \Delta$ where Δ is the Laplace-Beltrami operator on H,

$$\Delta = \frac{\partial^2}{\partial\theta^2} + \coth\theta\frac{\partial}{\partial\theta} + \operatorname{csch}^2\theta\frac{\partial^2}{\partial\psi^2}. \tag{2.12}$$

This result was first obtained by Gertenstein and Vasiliev [7] who realized that in a discrete medium the reflection coefficient (defined in a slightly different way) transforms by linear fractional transformations as L changes in discrete units. Thus $R(L)$ does some kind of random walk in H as L varies. The simplest diffusion approximation to a random walk is, naturally, Brownian motion. Hence the result. A more complete derivation was given in [8] (for a slightly different but very similar problem). It was noticed then, and subsequently in much greater detail in [9, 10] and also [11–13], that although this theorem is indeed a diffusion approximation to a complicated random motion in H, there is a reason why the limit is Brownian motion and not a more complicated process. For example, another reasonable scaling limit for the reflection coefficient is the white noise limit where L and k are fixed as the process $\mu(x)$ tends to white noise. That could be done by replacing $\mu(x)$ by $\mu^\varepsilon(x) = (1/\varepsilon)\mu(x/\varepsilon^2)$. We again have a diffusion approximation, but the limit is not Brownian motion now. The scaling (2.10) is special for it leads to a rapid deterministic rotation in H that isotropizes the limiting process.

This observation is simple and is contained in the general limit theorems for stochastic equations, as we describe in the following paragraphs. It enhanced profoundly our ability to calculate statistics of interesting scattering quantities (as in part II of [9], for example, and in [14]) but it seemed to be just good luck (if one likes computing). It was when we looked recently into pulse propagation, the subject of the next section, that we realized that this simplification due to rapid rotation leads to striking results in the time domain that are almost entirely due to this phenomenon.

Let us now describe how Theorem 1 is obtained. Let A and B be defined by

$$u(x) = e^{ikx}A(x) + e^{-ikx}B(x), \tag{2.13}$$

$$\frac{du(x)}{dx} = ik[e^{ikx}A(x) - e^{-ikx}B(x)].$$

Then from (2.1) and (2.7) we see that A and B solve the system

$$\frac{d}{dx}\begin{bmatrix} A(x) \\ B(x) \end{bmatrix} = \frac{ik\mu(x)}{2}\begin{bmatrix} 1 & e^{-2ikx} \\ -e^{2ikx} & 1 \end{bmatrix}\begin{bmatrix} A(x) \\ B(x) \end{bmatrix} \tag{2.14}$$

in $0 < x < L$ with the two-point boundary condition

$$A(0) = 1, \qquad B(L) = 0, \tag{2.15}$$

The reflection and transmission coefficients are given by

$$R(L, k) = B(0), \qquad T(L, k) = A(L). \tag{2.16}$$

Let $Y(x, y)$ be the complex two-by-two fundamental matrix of system (2.14) with $0 \le y \le x \le L$ and $Y(x, x) = I$. It is easily seen that this matrix has the form

$$Y = \begin{bmatrix} a & b \\ \bar{b} & \bar{a} \end{bmatrix}, \qquad |a|^2 - |b|^2 = 1, \tag{2.17}$$

which means that Y belongs to $SU(1,1)$, the Lie group of such matrices. The fundamental matrix is thus a random process with values on the group manifold. The reflection and transmission coefficients are obtained from the equation

$$\begin{bmatrix} a(L, 0) & b(L, 0) \\ \bar{b}(L, 0) & \bar{a}(L, 0) \end{bmatrix} \begin{bmatrix} 1 \\ R \end{bmatrix} = \begin{bmatrix} T \\ 0 \end{bmatrix} \tag{2.18}$$

which gives

$$R = -\bar{b}/\bar{a}, \qquad T = 1/\bar{a}. \tag{2.19}$$

It is clear from the discussion above that all relevant information about monochromatic (single frequency) scattering is contained in the fundamental matrix process Y. The scaling introduced in (2.9), along with the parameter ε in (2.10), can be implemented simply by letting $Y^\varepsilon(x) = Y(x/\varepsilon^2, 0; \varepsilon k)$; that is, replace x by x/ε^2 and k by εk. The matrix $Y^\varepsilon(x)$ solves the system

$$\frac{d}{dx} Y^\varepsilon = \frac{ik\mu(x/\varepsilon^2)}{2\varepsilon} \begin{bmatrix} 1 & e^{-2ikx/\varepsilon} \\ -e^{2ikx/\varepsilon} & 1 \end{bmatrix} Y^\varepsilon. \tag{2.20}$$

We can now apply to the system (2.20) one of several theorems that characterize the limit process $Y(x)$ as $\varepsilon \to 0$. The theorems differ mostly in their generality and technical level of proof, and they tell us that $Y(x)$ is a diffusion process on $SU(1,1)$. In our case here we find that if, for example, the process $\mu(x)$ is bounded and sufficiently mixing (has rapidly decaying correlations), then $Y^\varepsilon(x), x \ge 0$, converges weakly to a diffusion process on $SU(1,1)$ as $\varepsilon \to 0$.

If we introduce polar coordinates in $SU(1,1)$,

$$a = e^{i(\phi+\psi)/2} \cosh(\theta/2), \qquad b = e^{i(\phi-\psi)/2} \sinh(\theta/2), \tag{2.21}$$

then the generator of the limit diffusion process $Y(x)$ has the form

$$\frac{k^2 \alpha}{2} \left[\Delta + (2 + \coth^2 \theta) \frac{\partial^2}{\partial \phi^2} - \coth \theta \operatorname{csch} \theta \frac{\partial^2}{\partial \phi \partial \psi} \right] \tag{2.22}$$

where Δ is defined by (2.12). Note that this is not the Laplace-Beltrami operator on $SU(1,1)$, which is not of course elliptic and cannot arise from the limit theorem. Note also that Theorem 1 follows immediately now from (2.22), (2.21), and (2.19). The remarks we made regarding the rapid rotation are associated with the trigonometric terms on the right-hand side of (2.20).

Perturbation expansions to analyze limits of stochastic equations like (2.20) were developed formally by Stratonovich [15], and the first theorem was proved by Khasminskii [16]. By the time [6] appeared, results of this kind had been obtained by several methods and with considerable generality. These and related developments are sketched in the survey [17] and in a recent book by Kushner [18]. Wave propagation in a one-dimensional random medium is an area to which these methods can be used very effectively. But they are limited to one-dimensional problems where the approach via a fundamental matrix (or a transfer matrix) is possible. Even for one-dimensional vector waves, elastic waves for example, the dimensionality of the spaces goes up and the effectiveness of the limit theorems for computations is diminished. A lot more can no doubt be obtained from the multidimensional diffusions that result and that could be very interesting for the scattering problem. Multidimensional reflection-transmission problems are essentially untouched without some kind of forward scattering approximation that makes the problem tractable (but not easy). Reflection is, of course, gone in this approximation. Some remarks on this, as well as references to other literature, are given in [19].

3. Propagation of pulses. In the previous section we described some problems and results that cover many aspects of one-dimensional wave propagation, which are well understood, and have been around for some time. Recently we came across an interesting paper by Richards and Menke [20] where extensive numerical simulations of pulse reflection from a one-dimensional random half-space are carried out. The questions they asked are motivated by geophysical exploration problems. They wanted to understand, for example, how to distinguish multiple scattering effects by small-scale inhomogeneities from dissipation in the medium, when one has access to reflected signals, seismograms in their case. Of course the more general basic question here is: what can one say about the medium from the reflected signal if there are small-scale inhomogeneities present that one would like to ignore in an intelligent way?

We have not answered this question yet, but in the course of analyzing related ones we have found several results that will be described in the following paragraphs. It seems, at present, that we are not so far from a reasonable answer to this basic question. What follows is joint work with Burridge, Sheng, and White [21–24].

Let $u(x,t)$ and $p(x,t)$ be the velocity and pressure of an acoustic wave propagating in a medium with density $\rho(x)$ and bulk modulus $K(x)$. The equations of acoustics are

$$\rho u_t + p_x = 0, \qquad \frac{1}{K} p_t + u_x = 0 \tag{3.1}$$

for $t > 0, x > 0$. Let us suppose that the medium in $x < 0$ is homogeneous with $\rho(x) = \rho_0$ and $K(x) = K_0$. Let $f(s)$ be a smooth function of compact support in $0 < s < \infty$. We then let

$$u = \tfrac{1}{2} f(t - x/c_0), \qquad p = \tfrac{1}{2} f(t - x/c_0) \qquad (3.2)$$

for $t < 0$ large and for all x, and we assume that u and p are continuous. Here $c_0 = \sqrt{K_0/\rho_0}$ is the sound speed in the homogeneous half-space. This completes the formulation of the pulse problem if the density ρ, bulk modulus K, and pulse form f are known. Note that the pulse strikes the interface $x = 0$ between the homogeneous and inhomogeneous half-spaces at time $t = 0$. The observed quantities are the pressure and velocity at the interface $x = 0$ at $t > 0$.

When $\rho(x)$ and $K(x)$ are random functions, the reflected signals are stochastic processes, and we want to know what their properties are. In this complete generality, without any reference scales to distinguish phenomena, nothing very interesting can be obtained. We would like to model a situation where the pulse width is large compared to small-scale inhomogeneities in the medium, but small compared to the large scale variations, so as to resolve them while averaging out the noise. Thus, we expect that the reflected signal will contain all large scale information and a minimal amount of the small scale stuff. Such a framework can be formulated with a small parameter ε as follows.

We pass to dimensionless variables and assume that

$$\rho = \rho_0(x)\left(1 + \eta\left(\frac{x}{\varepsilon^2}\right)\right), \qquad \frac{1}{K} = \frac{1}{K_0(x)}\left(1 + \nu\left(\frac{x}{\varepsilon^2}\right)\right) \qquad (3.3)$$

where η and ν are bounded, stationary stochastic processes with mean zero. The functions $\rho_0(x)$ and $K_0(x)$ are the mean density and bulk modulus of the half-space. The acoustic equations are now

$$\rho_0(x)\left(1 + \eta\left(\frac{x}{\varepsilon^2}\right)\right) u_t + p_x = 0,$$
$$\frac{1}{K_0(x)}\left(1 + \nu\left(\frac{x}{\varepsilon^2}\right)\right) p_t + u_x = 0 \qquad (3.4)$$

for $x > 0$ and $t > 0$ and

$$u = \frac{1}{2\sqrt{\varepsilon}} f\left(\frac{t - x}{\varepsilon}\right), \qquad p = \frac{1}{2\sqrt{\varepsilon}} f\left(\frac{t - x}{\varepsilon}\right) \qquad (3.5)$$

for $t < 0$, large.

Note that there are three length scales in (3.4), (3.5). The order one scale measures macroscopic variations, the order ε scale is the pulse width (mean speed is equal to one in the dimensionless formulation) and the order ε^2 scale measures the size of the random inhomogeneities. We have also scaled up the amplitude of the incident pulse by $1/\sqrt{\varepsilon}$. This makes the total energy incident on the random half space of order one.

We shall now describe our results for problem (3.4), (3.5) when $\rho_0 = 1$ and $K_0 = 1$. The results in the general case, including discontinuities of the mean density and bulk modulus, are similar but more complicated. In the homogeneous-in-the-mean case we let

$$A(x,t) = u(x,t) + p(x,t), \qquad B(x,t) = u(x,t) - p(x,t),$$
$$m(x) = (\eta(x) + \nu(x))/2, \qquad n(x) = (\eta(x) - \nu(x))/2. \qquad (3.6)$$

Then A and B solve the problem

$$A_t + A_x + m(x/\varepsilon^2)A_t + n(x/\varepsilon^2)B_t = 0,$$
$$B_t - B_x + n(x/\varepsilon^2)A_t + m(x/\varepsilon^2)B_t = 0 \qquad (3.7)$$

for $x > 0$ and $t > 0$ and

$$B = 0, \qquad A = \frac{1}{\sqrt{\varepsilon}}f\left(\frac{t-x}{\varepsilon}\right) \qquad (3.8)$$

for $t < 0$, large. By analogy with the homogeneous case we call A and B the right and left travelling wave amplitudes, respectively. The reflected signal is $B(0,t)$, $t > 0$. We introduce the notation

$$B_{f,t}^{\varepsilon}(\sigma) = B(0, t + \varepsilon\sigma) \qquad (3.9)$$

to emphasize dependence of the reflected signal on the pulse form f and to introduce what we call the windowed process. That is, we look at the reflected signal in a time window centered at time t with width of order ε. The variable σ is the window time variable.

THEOREM 2. *Assume that $m(x)$ and $n(x)$ are bounded, zero mean, stationary, rapidly strongly mixing (see [6]) processes. Then for each smooth, compactly supported f and $t > 0$ the process $B_{f,t}^{\varepsilon}(\cdot)$ converges weakly as $\varepsilon \to 0$ to a stationary, zero mean, Gaussian process with power spectral density*

$$S_{f,t}(\omega) = |\hat{f}(\omega)|^2 \frac{1}{t}\mu(\sqrt{\alpha t}\omega) \qquad (3.10)$$

where

$$\alpha = \int_0^{\infty} E\{n(s)n(0)\}\,ds, \qquad (3.11)$$

$\hat{f}$ is the Fourier transform of f, and $\mu(\omega)$ is a universal function that is equal to $\omega^2(1 + \omega^2)^{-2}$.

Let us make a few remarks about this theorem. It provides a complete characterization for reflected signal processes from a half-space that is homogeneously random; there are no macroscopic variations in the medium properties. The statistical properties of the medium affect the reflected signal only through the parameter α given by (3.11). This is, of course, just like the results we discussed in §2. It is surprising though that the limit process has this windowed structure and, in particular, that it is Gaussian.

Theorem 2 is proved by taking Fourier transforms in time and working with the resulting stochastic ordinary differential equations. The new twist here is that we must analyze the transformed amplitudes for all frequencies simultaneously. Otherwise we cannot reconstruct the signal and show that it is Gaussian. This requires infinite-dimensional versions of the limit theorems that we discussed in §2. The remarkable thing is that in this infinite-dimensional setting, in the time domain actually, the results seem to be simpler than the ones for the monochromatic situation.

The reflected process has the following Fourier representation

$$B^{\varepsilon}_{f,t}(\sigma) = \frac{1}{\sqrt{\varepsilon}} \int_{-x}^{x} e^{-i\omega(\sigma + t/\varepsilon)} \hat{f}(\omega) R^{\varepsilon}(L, \omega) \, d\omega. \qquad (3.12)$$

Here R^{ε} is the reflection coefficient as defined in §2 by (2.10). Since (3.7) is a hyperbolic system with finite propagation speeds, we may convert the half-space to a slab of width L large enough. In the notation for the reflection coefficient we also indicate dependence on the frequency ω (it is k in §2). From (3.12) we see that moments of the reflected process can be analyzed by the limit theorems of §2 applied to finite sets of reflection coefficients at different frequencies. This works formally for the second moment, the covariance, but fails completely for higher moments and can never give the Gaussian law. The reason for this is that the trigonometric factor on the right-hand side of (2.20) introduces nonuniformities as $\varepsilon \to 0$. Reflection coefficients are asymptotically independent at different frequencies, but when the frequencies are order ε apart, dependence persists. This point is essential for the analysis of the integral in (3.12) and so is the oscillatory factor in that integral. The usual methods for limit theorems for sums of weakly dependent random variables, coupled with the older limit theorems for stochastic equations, do not work. The process (3.12) must be studied directly as it is. It is here that the rapidly rotating trigonometric terms play an essential role.

4. Concluding remarks. The study so far of pulse propagation in a one-dimensional random medium has shown that there is an enormous amount of detailed structure that can be found and analyzed very well. Our understanding of problems in the frequency domain, although extensive, is not sufficient for pulses and can at times be even confusing. It is best to study pulses in the time domain as much as possible. There are also statistical and inverse aspects to the pulse problem that are quite interesting and will keep us happily going for a while.

REFERENCES

1. H. Furstenberg, *Noncommuting random products*, Trans. Amer. Math. Soc. **108** (1963), 377–428.

2. S. Kotani, *Lyapounov indices determine absolutely continuous spectra of stationary random one-dimensional Schroedinger operators*, Stochastic Analysis, K. Ito, editor, North-Holland, Amsterdam, 1984, pp. 225–247.

3. I. J. Goldsheid, S. A. Molchanov, and L. A. Pastur, *A random one-dimensional Schroedinger operator has pure point spectrum*, Functional Anal. Appl. **11** (1977), 1–10.

4. S. Kotani, Preprint, 1985.

5. L. Arnold, G. Papanicolaou, and V. Wihstutz, *Asymptotic analysis of the Lyapounov exponent and rotation number of the random oscillator and applications*, SIAM J. Appl. Math. **46** (1986), 427–450.

6. H. Kesten and G. Papanicolaou, *A limit theorem for turbulent diffusion*, Comm. Math. Phys. **65** (1979), 97–128.

7. M. E. Gertsenstein and V. B. Vasiliev, *Wave guides with random inhomogeneities and Brownian motion in the Lobachevski plane*, Theory Probab. Appl. **11** (1966), 390–406.

8. G. Papanicolaou, *Wave propagation in a one-dimensional random medium*, SIAM J. Appl. Math. **21** (1971), 13–18.

9. W. Kohler and G. Papanicolaou, *Power statistics for wave propagation in one dimension and comparison with transport theory*, J. Math. Phys. **14** (1973), 1733–1745; **15** (1974), 2186–2197.

10. J. B. Keller, G. Papanicolaou, and J. Weilenmann, *Heat conduction in a one-dimensional random medium*, Comm. Pure Appl. Math. **32** (1978), 583–592.

11. Yu. L. Gazarian, Sov. Phys. JETP **29** (1969), 996.

12. R. H. Lang, J. Math. Phys. **14** (1973), 1921.

13. J. A. Morrison, J. Math. Anal. Appl. **39** (1972), 13.

14. V. I. Klyatskin, *Stochastic equations and waves in random media*, Nauka, Moscow, 1980.

15. R. L. Stratonovich, *Topics in the theory of noise*, Elsevier, New York, 1963.

16. R. Z. Khasminskii, *A limit theorem for solutions of differential equations with random right hand side*, Theory Probab. Appl. **11** (1966), 390–406.

17. G. Papanicolaou, *Asymptotic analysis of stochastic equations*, MAA Studies in Mathematics, vol. 18, M. Rosenblatt, editor, Math. Assoc. America, Washington, D.C., 1978.

18. H. J. Kushner, *Approximation and weak convergence methods for random processes with applications to stochastic systems theory*, MIT Press, Cambridge, Mass., 1984.

19. D. A. Dawson and G. Papanicolaou, *A random wave process*, Appl. Math. Optim. **12** (1984), 97–114.

20. P. G. Richards and W. Menke, *The apparent attenuation of a scattering medium*, Bull. Seismol. Soc. Amer. **73** (1983), 1005–1021.

21. R. Burridge, G. Papanicolaou, and B. White, *Statistics for pulse reflection from a randomly layered medium*, SIAM J. Appl. Math. **47** (1987), 146–168.

22. P. Sheng, Z.-Q. Zhang, B. White, and G. Papanicolaou, *Multiple scattering noise in one dimension: universality through localization length scaling*, Phys. Rev. Lett. **57** (1986), 1000–1003.

23. ____, *Minimum wave localization length in a one dimensional random medium*, Phys. Rev. B. **34** (1986), 4757–4761.

24. R. Burridge, G. Papanicolaou, and B. White, *One dimensional wave propagation in a highly discontinuous medium* (to appear).

COURANT INSTITUTE, NEW YORK UNIVERSITY, NEW YORK, NEW YORK 10012, USA

A Nonparametric Framework for Statistical Modelling

CHARLES J. STONE

1. Introduction. Much of mathematical statistics deals with inference concerning the unknowns in a stochastic model for a random phenomenon. In the parametric approach the unknowns are a specific finite number of real parameters. In the nonparametric approach they are functions, perhaps subject to smoothness or other regularity conditions. These functions can be approximated by means of a flexible finite-dimensional function space. To some extent, this reduces the nonparametric approach to the parametric approach. But the asymptotic theory is different when the error of approximation is taken into account and the dimension of the approximating function space is allowed to tend to infinity with the sample size. We will illustrate this theory by means of three examples: density estimation, logistic regression, and additive logistic regression.

2. Density estimation. Let Y be a d-dimensional random variable taking on values from a known compact subset C of $\mathbf{R}^d$. It is assumed that the distribution of Y has a density function f which is continuous and positive on C. By definition $\int f = 1$. Set $g = \log f$. Let S_n denote a p_n-dimensional vector space of functions on C having basis B_{nj}, $1 \leq j \leq p_n$. It is assumed that $\sum_j B_{nj} = 1$ on C and that no nontrivial linear combination of B_{nj}, $1 \leq j \leq p_n$, is almost everywhere equal to zero on C. Given $\theta \in \Theta_n$, the space of p_n-dimensional vectors, set

$$C_n(\theta) = \log \left(\int \exp \left(\sum_j \theta_j B_{nj} \right) \right)$$

and

$$f_n(\,\cdot\,;\theta) = \exp \left(\sum_j \theta_j B_{nj} - C_n(\theta) \right).$$

Then $\int f_n(\,\cdot\,;\theta) = 1$ for $\theta \in \Theta_n$. Observe that $f_n(\,\cdot\,;\theta)$, $\theta \in \Theta_n$, is an exponential family in canonical form. Let Θ_{n0} denote the $(p_n - 1)$-dimensional space consisting of those $\theta \in \Theta_n$, the sum of whose elements is zero. Let θ_n denote

This research was supported in part by National Science Foundation Grant DMS-8600409.

the unique value of $\theta \in \Theta_{n0}$ that maximizes the expected log-likelihood function $\Lambda_n(\,\cdot\,)$, defined by

$$\Lambda_n(\theta) = E\left[\sum_j \theta_j B_{nj}(Y) - C_n(\theta)\right] = \sum_j \theta_j \int B_{nj} f - C_n(\theta), \qquad \theta \in \Theta_{n0}.$$

Consider the *loglinear density approximation* $f_n = f_n(\,\cdot\,; \theta_n)$ to f.

Let $Y_1, \ldots, Y_n$ be independent random variables each having density f. The log-likelihood function for the parametric model is given by '

$$l_n(\theta) = \sum_j \theta_j \sum_i B_{nj}(Y_i) - nC_n(\theta).$$

The maximum likelihood estimator (MLE) $\hat\theta_n$ is the value of $\theta \in \Theta_{n0}$ that maximizes $l_n(\,\cdot\,)$. Since $l_n(\,\cdot\,)$ is a strictly concave function on Θ_{n0}, the MLE is unique if it exists. The corresponding estimate $\hat f_n = f_n(\,\cdot\,; \hat\theta_n)$ of f is called a *loglinear density estimate*, since $\log \hat f_n \in S_n$.

Let $\| \ \|_2$ and $\| \ \|_\infty$ denote the usual L_2 and L_∞ norms of functions on C. Let $\| \ \|_{\infty,A}$ denote the L_∞ norm of functions on A. It is assumed that $p_n \to \infty$ as $n \to \infty$, that $p_n = o(n^{.5-\varepsilon})$ for some $\varepsilon > 0$, and that $\lim_{n\to\infty} \inf_{s \in S_n} \|s - h\|_\infty = 0$ if h is continuous on C. Let $\Pi_n h$ denote the orthogonal projection of h onto S_n with respect to $L_2(C)$. It is assumed that there is a positive constant M such that $\|\Pi_n h\|_\infty \leq M\|h\|_\infty$ for $n \geq 1$ and $h \in L_\infty(C)$; $|B_{nj}| \leq M$ on C for $n \geq 1$ and $1 \leq j \leq p_n$; for $1 \leq j \leq p_n$, $B_{nj} = 0$ outside a set C_{nj} having diameter at most $Mp_n^{-1/d}$ and $C_{nj} \cap C_{nk}$ is nonempty for at most M values of k; and

$$M^{-1}|\theta_j|^2 \leq \left\| \sum_k \theta_k B_{nk} \right\|_{\infty,C_{nj}}^2 \leq Mp_n \int_{C_{nj}} \left(\sum_k \theta_k B_{nk} \right)^2$$

for $n \geq 1$, $\theta \in \Theta_n$, and $1 \leq j \leq p_n$.

These properties can be satisfied with C_{nj}, $1 \leq j \leq p_n$, a partition of C and B_{nj} the indicator function for C_{nj}. Here $\hat f_n$ is the corresponding histogram density estimate. They can also be satisfied with $d = 1$, S_n a space of splines and B_{nj}, $1 \leq j \leq p_n$, a basis consisting of B-splines; see de Boor [1, 2] or Stone [7, 8, 9] for details. Presumably, they can also be satisfied with tensor product spaces of splines and with spaces of the type that arise in the use of the finite element method (see arguments in Descloux [3] and de Boor [1]).

Some asymptotic properties of loglinear density estimation will now be summarized. The proofs follow from arguments in Stone [9]. Set

$$\delta_n = \inf_{s \in S_n} \|s - g\|_\infty.$$

THEOREM 1. (i) $\|f_n - f\|_\infty = O(\delta_n)$;
(ii) $\hat f_n$ *exists except on an event whose probability tends to zero with* n;
(iii) $\|\hat f_n - f_n\|_2 = O_{\mathrm{pr}}((n^{-1}p_n)^{1/2})$; *and*
(iv) $\|\hat f_n - f_n\|_\infty = O_{\mathrm{pr}}((n^{-1}p_n \log(p_n))^{1/2})$.

Write $\hat{f}_n - f = f_n - f + \hat{f}_n - f_n$. The quantity $f_n - f$ is a "bias" term, while $\hat{f}_n - f_n$ is a "noise" term whose magnitude is indicated by its asymptotic variance. Under typical smoothness assumptions on g, $\delta_n = O(p_n^{-q/d})$ for some positive number q (this holds with $q = m$ if g has a bounded mth derivative). Set $\gamma = 1/(2q + d)$ and $r = q\gamma$. Suppose that $\gamma d < 1/2$ or, equivalently, that $q > d/2$. To get the optimal rate of convergence of $\|\hat{f}_n - f\|_2$ to zero, choose $p_n \sim n^{\gamma d}$. Then $\delta_n^2 \sim n^{-1} p_n \sim n^{-2r}$ and hence

$$\|\hat{f}_n - f\|_2 = O_{\mathrm{pr}}(n^{-r}).$$

To get the optimal rate of convergence of $\|\hat{f}_n - f\|_\infty$ to zero choose $p_n \sim (n/\log(n))^{\gamma d}$. Then $\delta_n^2 \sim n^{-1} p_n \log(p_n) \sim (n^{-1} \log(n))^{2r}$ and hence

$$\|\hat{f}_n - f\|_\infty = O_{\mathrm{pr}}((n^{-1} \log(n))^r).$$

(See Stone [5] for a precise definition of *optimal rate of convergence*.)

Let $I_n(\cdot)$ denote the information function based on the random sample of size n. Then $I_n(\theta)$ is the Hessian matrix of $nC_n(\cdot)$ at θ; that is, the $p_n \times p_n$ matrix whose (j, k)th element is $n(\partial^2 C_n(\theta)/\partial\theta_j\partial\theta_k)$. Let $I_n^{-1}(\theta)$ denote the inverse to $I_n(\theta)$ viewed as a linear transformation of Θ_{n0}. Set $I_n^{-1} = I_n^{-1}(\theta_n)$ and $\hat{I}_n^{-1} = I_n^{-1}(\hat{\theta}_n)$. Let $G_n(y)$, $\hat{G}_n(y) \in \Theta_{n0}$, denote the p_n-dimensional vectors having elements

$$G_{nj}(y) = B_{nj}(y) - \frac{\partial C_n}{\partial\theta_j}(\theta_n) \quad \text{and} \quad \hat{G}_{nj}(y) = B_{nj}(y) - \frac{\partial C_n}{\partial\theta_j}\left(\hat{\theta}_n\right),$$

respectively. Set

$$SE(\hat{f}_n(y)) = f_n(y)(G_n(y)' I_n^{-1} G_n(y))^{1/2}$$

and

$$\widehat{SE}(\hat{f}_n(y)) = \hat{f}_n(y)(\hat{G}_n(y)' \hat{I}_n^{-1} \hat{G}_n(y))^{1/2}.$$

THEOREM 2. *Uniformly in $y \in I$,*

$$SE(\hat{f}_n(y)) \sim (n^{-1} p_n)^{1/2}, \qquad \widehat{SE}(\hat{f}_n(y))/SE(\hat{f}_n(y)) = 1 + o_{\mathrm{pr}}(1),$$

and

$$\mathcal{L}\left(\frac{\hat{f}_n(y) - f_n(y)}{SE(\hat{f}_n(y))}\right) \to \mathcal{N}(0, 1).$$

It follows from Theorem 2 that $\hat{f}_n(y) \pm z_{1-.5\alpha}\widehat{SE}(\hat{f}_n(y))$ is an asymptotic $(1 - \alpha)$-level confidence interval for $f_n(y)$; if $\delta_n = o((n^{-1} p_n)^{1/2})$, it is also an asymptotic $(1 - \alpha)$-level confidence interval for $f(y)$. Here $\Phi(z_q) = q$, Φ being the standard normal distribution function.

Let P denote the distribution corresponding to f, defined by $P(A) = \int_A f$, and let P_n and $\hat{P}_n$ be defined similarly in terms of f_n and $\hat{f}_n$. Let $\mathcal{A}$ denote a class of subsets of C. Given distributions Q_1 and Q_2 on C set

$\|Q_1 - Q_2\|_\infty = \sup_{A \in \mathcal{A}} |Q_1(A) - Q_2(A)|$. Under reasonable conditions on $\mathcal{A}, S_n$, and f, $\|P_n - P\|_\infty = O(p_n^{-1/d}\delta_n)$ and

$$\mathcal{L}\left(\frac{\hat{P}_n(A) - P_n(A)}{SE(\hat{P}_n(A))}\right) \to \mathcal{N}(0,1) \quad \text{with } SE(\hat{P}_n(A)) = \left(\frac{P(A)(1 - P(A))}{n}\right)^{1/2}.$$

It was shown in Stone [9] for the special case of $d = 1$, bases consisting of B-splines, and $\mathcal{A}$ the collection of subintervals of a compact interval C, that $\|\hat{P}_n - P_n\|_\infty = O_{\text{pr}}(n^{-1/2})$. What is a corresponding result in the more general context of the present paper?

3. Logistic regression. Let X, Y be a pair of random variables such that X ranges over a known compact subset C of $\mathbf{R}^d$ and Y takes on only two values, 0 and 1. It is assumed that the distribution of X is absolutely continuous and that its density is bounded away from zero and infinity on C. Let f be the regression function, defined on C by

$$f(x) = \Pr(Y = 1 | X = x).$$

It is assumed that f is continuous and that $0 < f < 1$. Let g denote the corresponding logistic regression function, defined by

$$g = \text{logit}(f) = \log(f/(1 - f));$$

so that $f = \exp(g)/(1 + \exp(g))$.

We can approximate g by a member of a p_n-dimensional vector space S_n of functions on C. Let B_{nj}, $1 \le j \le p_n$, denote a basis of S_n. Then the expected log-likelihood function $\Lambda_n(\cdot)$ is defined by

$$\Lambda_n(\theta) = E\left[\sum_j \theta_j B_{nj}(X)Y - \log\left(1 + \exp\left(\sum_j \theta_j B_{nj}(X)\right)\right)\right], \qquad \theta \in \Theta_n.$$

Let θ_n be the unique $\theta \in \Theta_n$ that maximizes $\Lambda_n(\theta)$ and set

$$g_n = \sum_j \theta_{nj} B_{nj} \quad \text{and} \quad f_n = \exp(g_n)/(1 + \exp(g_n)).$$

Let $(X_1, Y_1), \ldots, (X_n, Y_n)$ be independent random pairs, each having the same distribution as (X, Y). The log-likelihood function for the parametric model is given by

$$l_n(\theta) = \sum_j \theta_j \sum_i Y_i B_{nj}(X_i) - \sum_i \log\left(1 + \exp\left(\sum_j \theta_j B_{nj}(X_i)\right)\right),$$

which corresponds to an exponential family in canonical form. Let $\hat{\theta}_n$ denote the MLE of θ and set $\hat{g}_n = \sum_j \hat{\theta}_{nj} B_{nj}$ and $\hat{f}_n = \exp(\hat{g}_n)/(1 + \exp(\hat{g}_n))$. Under appropriate regularity conditions, analogs of Theorems 1 and 2 of §2 should hold.

4. Additive logistic regression. Let C be a rectangle, say, $C = [0,1]^d$. It is then useful in practice to assume that g is additive or, more generally, to replace g by its best additive approximation g^*; this is defined to be the unique additive function h on C that maximizes the expected log-likelihood

$$E[h(X)Y - \log(1 + e^{h(X)})].$$

Set $f^* = \exp(g^*)/(1 + \exp(g^*))$. If g itself is additive, then $g^* = g$ and $f^* = f$.

To obtain a p_n-dimensional space of additive approximations to g^*, we consider p_{nk}-dimensional vector spaces S_{nk} of functions on $[0,1]$ for $1 \leq k \leq d$, each containing the constant functions, and let S_n be the collection of all functions of the form $s(x_1, \ldots, x_d) = \sum_k s_k(x_k)$, where $s_k \in S_{nk}$ for $1 \leq k \leq d$. Then $p_n = 1 + \sum_k (p_{nk} - 1)$. Analogs of Theorem 1 of §2 and its consequences for optimal rates of convergence should hold with f replaced by f^*, g replaced by g^*, and d replaced by 1; see Stone [**7, 8**] for what has been rigorously verified to date. An analog to Theorem 2 should also hold if g itself is additive. Otherwise, a more complicated standard error formula would be required since $\Pr(Y = 1|X = x)$ would not be exactly equal to $f^*(x)$.

REFERENCES

1. C. de Boor, *A bound on the L_∞-norm of the L_2-approximation by splines in terms of a global mesh ratio*, Math. Comp. **30** (1976), 765–771.

2. ____, *A practical guide to splines*, Springer-Verlag, New York, 1978.

3. J. Descloux, *On finite element matrices*, SIAM J. Numer. Anal. **9** (1972), 260–265.

4. C. J. Stone, *Optimal rates of convergence for nonparametric estimators*, Ann. Statist. **8** (1980), 1348–1360.

5. ____, *Optimal global rates of convergence for nonparametric regression*, Ann. Statist. **10** (1982), 1040–1053.

6. ____, *Optimal uniform rate of convergence for nonparametric estimators of a density function or its derivatives*, Recent Advances in Statistics: Papers in Honor of Herman Chernoff on his Sixtieth Birthday (M. H. Rezvi, J. S. Rustagi, and D. Siegmund, editors), Academic Press, New York, 1983, pp. 393–406.

7. ____, *Additive regression and other nonparametric models*, Ann. Statist. **13** (1985), 689–705.

8. ____, *The dimensionality reduction principle for generalized additive models*, Ann. Statist. **14** (1966), 590–606.

9. ____, *Asymptotic properties of logspline density estimation*, Technical Report No. 69, Department of Statistics, Univ. of California, Berkeley, 1986.

UNIVERSITY OF CALIFORNIA, BERKELEY, CALIFORNIA 94720, USA

Proceedings of the International Congress of Mathematicians
Berkeley, California, USA, 1986

Compactness of Solutions to Nonlinear PDE

RONALD J. DIPERNA

We shall describe some recent work dealing with oscillations and concentrations in solutions to nonlinear partial differential equations arising in mechanics.

1. Compensated compactness. In the context of hyperbolic and elliptic systems of conservation laws,

$$\sum_{j=1}^{m} \partial_j g_j(u) = 0, \qquad \partial_j = \partial/\partial y_j, \tag{1}$$

a general problem is to assess the compactness properties of sequences of solutions u_ε. For the prototypical models of mechanics, maximum principles and/or energy arguments frequently ensure a uniform pointwise or integral bound on the amplitude of the field u_ε, i.e.,

$$|u_\varepsilon(y)| \le c \quad \text{or} \quad \int |u_\varepsilon(y)|^p \, dy \le c \tag{2}$$

where the constant c depends only on the data. Uniform bounds of the form (2) guarantee weak compactness, specifically the existence of a subsequence converging weakly to a field u: w-lim $u_\varepsilon = u$ in the sense that the local average of u_ε converges to the local average of u,

$$\lim \int_\Omega u_\varepsilon \, dy = \int_\Omega u \, dy, \tag{3}$$

for all bounded domains Ω. One problem is to determine whether or not u_ε converges strongly to u by virtue of the geometric and algebraic structure of (1). Do we have s-lim $u_\varepsilon = u$ in the sense that

$$\lim \int_\Omega |u_\varepsilon - u| \, dy = 0 \tag{4}$$

for all bounded Ω?

Strong convergence (4) implies nearly uniform convergence while weak convergence (3) allows persistent fluctuations. Do the geometric and algebraic constraints encoded in (1) admit or exclude oscillations in solution sequences? If the convergence is strong, then the limiting field u is a solution of (1): nonlinear maps are continuous in the strong topology s-lim $g(u_\varepsilon) = g(\text{s-lim } u_\varepsilon)$ under mild restrictions on g. On the other hand, if the convergence is weak the limiting field u may not be a solution: in the weak topology there appears a nontrivial commutator w-lim $g(u_\varepsilon) - g(\text{w-lim } u_\varepsilon) \neq 0$. In the latter situation the goal is to describe the static structure and dynamic behavior of oscillations.

One of the recently developed approaches involves the theory of compensated compactness initiated by L. Tartar and F. Murat [24–14, 20, 21]. In this framework the structure of oscillations is recorded in the structure of composite weak limits as follows. For simplicity let us assume that v_ε is an arbitrary sequence of vector fields from R^m to R^n which is uniformly bounded in L^∞. Let g denote a state variable, a continuous real-valued map on the state space R^n. As a consequence of the L^∞ bound, there exists a subsequence of v_ε and a family ν of probability measures ν_y over the state space R^n such that

$$\text{w-lim } g\{v_\varepsilon(y)\} = \int_{R^n} g(\lambda)\, d\nu_y(\lambda), \qquad (5)$$

for all g. Roughly speaking, (5) asserts that the physical space average given by the left side coincides with the state space average given by the right side, namely, the expected value of $g = g(\lambda)$ with respect to the representing measure $\nu_y = \nu_y(\lambda)$. The Young measure ν was first introduced into partial differential equations by L. Tartar for the purpose of representing oscillations [20].

It is not difficult to show that v_ε converges strongly if and only if the Young measure reduces to a Dirac mass at almost all y: $\nu_y = \delta_{v(y)}$. Weak convergence manifests itself in state space R^n through the non-Dirac structure of the Young measure, reflecting the presence of oscillations in the physical space R^m. In the context of both steady and time dependent solutions of compressible and incompressible media, one of the problems is to analyze the structure of the Young measure associated with general sequences.

For the purpose of determining whether or not the Young measure ν of a solution sequence u_ε reduces to a Dirac mass, one is lead to classify mappings g which are insensitive to oscillations in the sense that w-lim $g(u_\varepsilon) = g(\text{w-lim } u_\varepsilon)$. For which state variables g does the macroscopic value of u determine the macroscopic value of $g(u)$? Knowledge of such robust state variables may be used to restrict the permissible oscillations and in certain circumstances to eliminate them entirely. We shall briefly describe some recent work dealing with the structure of the Young measure for systems of conservation laws of hyperbolic, elliptic, and mixed type. The results make use of the div-curl lemma of Tartar and Murat [12, 20] which describes special weakly continuous quadratic state variables.

In the context of 2×2 systems of hyperbolic conservation laws in one space dimension,

$$\partial_t u + \partial_x f(u) = 0, \qquad u \in R^2, \qquad (6)$$

the Young measure associated with uniformly bounded sequences of entropy solutions reduces to a Dirac mass if the eigenvalues are nondegenerate [3, 4, 6].

From each sequence one may extract a subsequence that converges strongly. The physical source of the compactification lies in the absorption of acoustic waves by shock waves. As a corollary it can be shown that the viscosity method for the associated parabolic system $\partial_t u_\varepsilon + \partial_x f(u_\varepsilon) = \varepsilon \partial_x^2 u_\varepsilon$ converges strongly as the diffusion coefficient ε tends to zero [3, 4].

In the setting of general 2×2 hyperbolic systems (6), a definitive characterization of the Young measure has recently been give by M. Rascle [15] verifying a conjecture of D. Serre [16, 19]. If the eigenvalues are degenerate, the system may support oscillations and a non-Dirac Young measure.

In the setting of the transonic equations of gas dynamics, a system of two conservation laws of mixed (hyperbolic-elliptic) type, C. Morawetz [11] has recently established reduction of the Young measure and strong convergence for regularized sequences. Transonic flows which avoid the vacuum and stagnation states cannot sustain persistent oscillations in the presence of the entropy condition.

For mixed systems of hyperbolic-parabolic type arising in the continuum mechanical description of phase transitions, M. Slemrod and V. Rotyburd have described the structure of the Young measure under certain mild hypotheses [17, 18]. If the eigenvalues of a 2×2 system coalesce on a strip in state space, then the Young measure is supported on the strip.

The analysis in each of the situations above makes use of an infinite sequence of weakly continuous state variables formed from Lax entropy pairs [8]. Thus, a problem arises of determining whether or not a finite number of physical entropy pairs suffices to exclude all oscillations. The problem is of particular interest in connection with multidimensional systems which typically admit just one entropy pair. A positive result for the equations of gas dynamics in one space dimension has been established in [6]: a sequence of compressible fluids with uniformly small amplitude does not admit persistent oscillations if just the two Noether entropy pairs are considered.

We refer the reader to [3, 4] for applications of the compensated compactness method to large data existence problems for hyperbolic conservation laws and to convergence problems for conservative finite difference schemes. A discussion of the underlying geometry can be found in [6] together with reduction theorems for the Young measure associated with general elliptic systems. The seminal article of L. Tartar [20] discusses the Young measure in the context of a scalar conservation law.

2. Measure-valued solutions. The Young measure has motivated a notion of measure-valued solution to conservative equations of general type [5]. The goal is to represent and analyze weak limits of classical (vector-valued) solutions. We shall briefly recall the definition of measure-valued solution in the setting of systems of conservation laws in one space dimension (6). A measure-valued solution of (6) is a mapping ν from the physical domain R^2 to the space of probability measures over the state space R^n, $\nu: y \to \nu_y \in \mathrm{Prob}(R^n)$, such that the divergence of the expected value of the field (u, f) vanishes in the sense of distributions:

$$\partial_t \langle \nu_y, \lambda \rangle + \partial_x \langle \nu_y, f(\lambda) \rangle = 0, \qquad y = (x, t). \tag{7}$$

The expected value of a mapping g is denoted by a bracket: $\langle \nu_y, g(\lambda) \rangle = \int_{R^n} g(\lambda) \, d\nu_y(\lambda)$. In this framework, classical distribution solutions of (6) are represented as Dirac masses: $\nu: y \to \delta_{u(y)}$ is a measure-valued solution of (7) if and only if the point of concentration $u(y)$ is a classical solution. More generally, each sequence of classical solution u_ε of (6) generates a mv-solution of (7). Taking a weak limit in the equation $\partial_t u_\varepsilon + \partial_x f(u_\varepsilon) = 0$ and using the defining property (5) of the Young measure ν reveals that ν is a measure-valued solution of (6). Furthermore, any stable approximation

$$\partial_t u_\varepsilon + \partial_x f(u_\varepsilon) = \varepsilon L u^\varepsilon \tag{8}$$

based on a conservative (higher order) differential operator L generates a measure-valued solution ν of the unperturbed system (6).

THEOREM. *If $\{u_\varepsilon\}$ is a uniformly bounded sequence of classical solutions of (8), then there exists a measure-valued solution of (6) which represents the limiting oscillations in the sense of (5): all composite weak limits are expressed as expected values as ε tends to zero.*

Two examples of special interest are given by the zero diffusion limit $L = \partial_x^2 u$ and the zero dispersion limit $L = \partial_x^3 u$. In the former case, one anticipates that the associated measure-valued solution reduces to a Dirac mass. In the latter case, strong convergence is maintained for a short period of time after which the measure-valued solution expands. Using the spectral representation the zero dispersion limit of the KdV equation has been analyzed in substantial detail by P. Lax and D. Levermore [7] and by S. Venakides [24, 25]. In the context of general modulation theory we refer the reader to the work of Flaschka, Forest, and McLaughlin [26, 27] and to the references cited therein for an analysis of slowly modulated wavetrains and for related examples of measure-valued solutions.

A general problem is to classify measure-valued solutions according to the structure of the regularizing operator L. Which measure-valued solutions of (6) arise as a Young measure of a diffusion limit? Which arise from a dispersion limit? It may turn out that some measure-valued solutions are not associated with any L.

In the case of a scalar conservation law, measure-valued solutions associated with the zero diffusion limit have been characterized through an averaged Lax entropy inequality [8].

THEOREM. *If ν is a measure-valued solution of a scalar conservation law which is a Dirac mass at time $t = 0$ and if $\partial_t \langle \nu_y, \eta(\lambda) \rangle + \partial_x \langle \nu_y, q(\lambda) \rangle \le 0$ for all convex entropy pairs (η, q), then ν reduces to a Dirac mass in $t > 0$.*

In short, no oscillations can develop in a sequence of scalar entropy solutions if none are present in the initial data. As a corollary, one obtains convergence of the viscosity method for a scalar conservation law in one and several space dimensions using only the natural L^∞ bound supplied by the maximum principle [5].

We refer the reader to the work of M. Slemrod and V. Rotyburd [17] for a treatment of measure-valued solutions in the context of dynamic phase transitions.

3. Concentration compactness. In the setting of the incompressible Euler equations

$$\partial_t u + \operatorname{div} u \otimes u + \nabla p = 0, \qquad \operatorname{div} v = 0, \tag{9}$$

solution sequences u_ε with uniformly bounded kinetic energy may exhibit a variety of defects in the space L^2. An ongoing program of the author and Andy Majda deals with the geometry of L^2 defects associated with incompressible flow in two space dimensions. Sequences of solutions with uniformly bounded kinetic energy need not be precompact in L^2. Explicit examples exhibiting L^2 defects have been constructed from Rankine and Kirchoff vortices.

The physical motivation comes from the study of vortex structures, in particular vortex sheets. In this situation, the initial configuration for the Cauchy problem has vorticity represented by a Borel measure with finite total mass. One of the problems is to determine whether or not there exists a globally defined two-dimensional Euler solution with initial vorticity given by a bounded measure. A second problem is to determine whether or not the associated Navier-Stokes solutions u^ε satisfying

$$\partial_t u_\varepsilon + \operatorname{div} u_\varepsilon \otimes u_\varepsilon + \nabla p_\varepsilon = \varepsilon \Delta u_\varepsilon, \qquad \operatorname{div} u_\varepsilon = 0$$

converge strongly in L^2 as the diffusion coefficient ε tends to zero.

It can be shown that the Navier-Stokes limit generates a measure-valued solution of the Euler equations (9) in the form of a measure-valued distribution. Concentration effects associated with vortex collapse are recorded in the singular part of ν while oscillation effects are recorded in the absolutely continuous part.

Some of the functional analytic motivation comes from concentration compactness work of P. L. Lions which provides a systematic study of losses of compactness in minimizing sequences of elliptic variational problems due to the action of noncompact groups [9, 10]. This work includes several results that quantify the loss of compactness in certain classical function inequalities, such as the Sobolev inequality and the Hardy-Littlewood-Sobolev inequality, at the critical exponent. Critical structure also arises in the context of the Euler equations.

4. Background. One of the central problems in the theory of homogenization concerns the study of oscillations in sequences of solutions to linear elliptic equations associated with highly heterogeneous media. How do oscillations in the coefficients of a linear elliptic equation induce oscillations in the solution? Major progress in this area has been made by Tartar and Murat using aspects of compensated compactness [14, 22, 23] to estimate effective parameters.

In the context of nonlinear elasticity we refer the reader to the basic work of J. Ball [1, 2] dealing with energy minimizing sequences and the weak topology. The results involve nonlinear elliptic systems of Euler-Lagrange type.

References

1. J. M. Ball, *On the calculus of variations and sequentially weakly continuous maps*, Lecture Notes in Math., vol. 564, Springer-Verlag, 1976.

2. ____, *Convexity conditions and existence theorems in nonlinear elasticity*, Arch. Rational Mech. Anal. **63** (1977), 337–403.

3. R. J. DiPerna, *Convergence of approximate solutions to conservation laws*, Arch. Rational Mech. Anal. **82** (1983), 27–70.

4. ____, *Convergence of the viscosity method for isentropic gas dynamics*, Comm. Math. Phys. **91** (1983), 1–30.

5. ____, *Measure-valued solutions to conservation laws*, Arch. Rational Mech. Anal. **88** (1985), 223–270.

6. ____, *Compensated compactness and general systems of conservation laws*, Trans. Amer. Math. Soc. **292** (1985), 383–420.

7. P. D. Lax and C. D. Levermore, *The small dispersion limit for the KdV equation.* I, II, Comm. Pure. Appl. Math. **36** (1983), 253–290, 571–594.

8. P. D. Lax, *Shock waves and entropy*, Contributions to Nonlinear Functional Analysis (E. A. Zarantonello, ed.), Academic Press, 1971, pp. 603–634.

9. P. L. Lions, *The concentration-compactness principle in the calculus of variations, the locally compact case.* I, II, Ann. Inst. Henri Poincaré **1** (1984), 109–145.

10. ____, *The concentration-compactness principle in the calculus of variations, the limit case.* I, II, Riv. Mat. Iberoamericana **1** (1984), 145–201; **1** (1985), 45–121.

11. C. Morawetz, *On a weak solution for a transonic flow problem*, Comm. Pure Appl. Math. (to appear, 1986 or 1987).

12. F. Murat, *Compacité par compensation*, Ann. Scuola Norm. Sup. Pisa **5** (1978), 489–507.

13. ____, *Compacité par compensation: condition necessaire et suffisante de continuité faible sous une hypotheses de rang constant*, Ann. Scuola Norm. Sup. Pisa **8** (1981), 69–102.

14. ____, *H-convergence*, Séminarie d'Analyse Fonctionelle et Numerique, 1977/1978, multigraphed.

15. M. Rascle, preprint.

16. M. Rascle and D. Serre, *Compacité par compensation et systems hyperboliques de lois de conservation*, C. R. Acad. Sci. **299** (1984), 673–676.

17. M. Slemrod and V. Roytburd, *Measure-valued solutions to a problem in dynamic phase transitions*, preprint.

18. ____, private communication.

19. D. Serre, *La compacité par compensation pour les systemes hyperbolique nonlineares de deux équations a une dimension d'espace*, Equipe d'Analyse Numerique, Université de St. Etienne, France, 1985, preprint.

20. L. Tartar, *Compensated compactness and applications to partial differential equations*, Nonlinear Analysis and Mechanics: Heriot-Watt Symposium (R. J. Knops, ed.), Research Notes in Math., vol. 39, Pitman Press, 1979.

21. ____, *The compensated compactness method applied to systems of conservation laws*, Systems of Nonlinear Partial Differential Equations (J. M. Ball, ed.), NATO ASI Series, Reidel, 1983.

22. ____, *Estimations de coefficients homogeneises*, Computer Methods in Applied Sciences and Engineering (Proc. Third Internat. Sympos., Versailles, 1977), Lecture Notes in Math., vol. 704, Springer, Berlin, 1979.

23. ____, *Estimations fines de coefficients homogeneises* (P. Kree, ed.), Pitman Research Notes in Mathematics, 1986.

24. S. Venakides, *The zero dispersion limit of the KdV equation with periodic initial data*, preprint.

25. _____, *The zero dispersion limit of the KdV equation with nontrivial reflection coefficient*, Comm. Pure Appl. Math. **38** (1985), 125–155.

26. H. Flaschka, M. G. Forest, and D. W. McLaughlin, *Multiphase averaging and the inverse spectral solution of the Korteweg-de Vries equation*, Comm. Pure Appl. Math. **33** (1980), 739–784.

27. D. W. McLaughlin, *Modulation of KdV wavetrains*, Physica 3D **1** (1981), 343–355.

UNIVERSITY OF CALIFORNIA, BERKELEY, CALIFORNIA 94720, USA

Proceedings of the International Congress of Mathematicians
Berkeley, California, USA, 1986

Quasiconvexity and Partial Regularity in the Calculus of Variations

LAWRENCE C. EVANS

1. Introduction. I will describe here some recent research [13–15], much of it conducted jointly with R. F. Gariepy, concerning the stability and smoothness of minimizers for certain problems in the calculus of variations. The main point I want to emphasize is that "quasiconvexity," the necessary and sufficient hypothesis on the nonlinearity identified by Morrey in his investigation of the existence theory for such problems, is also, slightly strengthened, the basic hypothesis for a perturbation and regularity theory. Thus, very loosely speaking, "the existence of minimizers generally implies their stability and smoothness."

We will address problems in the calculus of variations of the following form. Let n, N denote positive integers, and suppose $\Omega \subset \mathbf{R}^n$ is open, bounded, and smooth. Then for sufficiently regular functions $v\colon \Omega \to \mathbf{R}^N$ consider the functional

$$I[v] \equiv \int_{\Omega} F(Dv)\, dx, \tag{1.1}$$

where

$$Dv \equiv \left(\left(\frac{\partial v^i}{\partial x_\alpha}\right)\right) \qquad (1 \le \alpha \le n,\ 1 \le i \le N)$$

is the gradient matrix of v and $F\colon M^{n \times N} \to \mathbf{R}$ is given, $M^{n \times N}$ denoting the space of real, $n \times N$ matrices. We are interested in ascertaining the existence, and then other properties, of minimizers of $I[\cdot]$ among all appropriate functions satisfying certain given, but here unspecified, boundary conditions.

2. Existence. The real breakthrough for the existence theory in the case $n, N > 1$ was Morrey's 1952 paper [27] which isolated a property of F both necessary and sufficient in many circumstances for the weak sequential lower semicontinuity of $I[\cdot]$ on appropriate Sobolev spaces. The condition is that F be *quasiconvex,* which means

$$F(A) \le \frac{1}{|O|} \int_O F(A + D\phi)\, dx \tag{2.1}$$

for all open sets $O \subset \mathbf{R}^n$, all matrices $A \in M^{n \times N}$, and all $\phi \in C^1(O; \mathbf{R}^N)$ which vanish on ∂O.

Using Jensen's inequality we readily see that any convex function F is quasiconvex; but what's interesting is that there are nonconvex examples, the most important of which are *polyconvex* functions F, for which $F(A)$ has the form of a convex function of various determinants of square submatrices of A. For instance, when $n = N = 3$

$$F(A) \equiv G(A, \operatorname{cof} A, \det A) \qquad (A \in M^{3 \times 3}) \tag{2.2}$$

is polyconvex, provided G is convex. (Here $\operatorname{cof} A =$ cofactor matrix of A, $\det A =$ determinant of A.)

The importance of quasiconvexity in the existence theory for the calculus of variations is documented in the following theorem, which is a special case of results of Acerbi-Fusco [1], who refined some of Morrey's original techniques. We henceforth assume

$$F: M^{n \times N} \to \mathbf{R} \text{ is continuous, and } 0 \le F(P) \le C(1 + |P|^q) \quad (P \in M^{n \times N}) \tag{2.3}$$

for given constants $C > 0$, $1 \le q < \infty$.

THEOREM 1. *The function $I[\cdot]$ is weakly sequentially lower semicontinuous on the Sobolev space $W^{1,q}(\Omega; \mathbf{R}^N)$ if and only if F is quasiconvex.*

The proof of the necessity of quasiconvexity is relatively simple. Indeed, given A and ϕ as in inequality (2.1), we suppose, for simplicity, O is the unit cube in $\mathbf{R}^n$. For each $m = 1, \ldots$ we subdivide O into 2^{mn} equal subcubes and define u^m to equal on each subcube the plane Ay plus a rescaled copy of ϕ to that cube. This done properly, we see that $u^m \rightharpoonup Ay$ weakly in $W^{1,q}$, and $\int_O F(Du^m)\, dx = \int_O F(A + D\phi)\, dx$ $(m = 1, \ldots)$. Thus, if $I[\cdot]$ is weakly sequentially lower semicontinuous, inequality (2.1) must hold.

The converse statement is harder. Assuming now $u^m \rightharpoonup u$ weakly in $W^{1,q}$ and F is quasiconvex, we follow Morrey [27] and subdivide O into many small cubes on each of which Du is approximately constant. We then modify the u^m to agree on the boundary of each cube with the average value of Du on that cube, apply inequality (2.1), and control all the error terms. This last task is a bit subtle since the gradients of the u^m are bounded only in L^q and not L^∞: see [12, 24], or [15] for technical tricks to handle this problem.

When $1 < q < \infty$, Theorem 1 and an additional coercivity assumption of the form $F(A) \ge b|A|^q$ $(b > 0, A \in M^{n \times N})$ imply for appropriate Dirichlet boundary

---

*Morrey changed this terminology somewhat when writing §4.4 of his book [28]. I know of at least two completely different meanings of the word "quasiconvexity," and am disappointed that the important property (2.1) does not have a better name. I propose a one hundred years' moratorium on the use of prefix "quasi-" in mathematical nomenclature.

conditions the existence of at least one minimizer of $I[\cdot]$ (cf. [**28**, Theorem 4.4.7]). This result and its various extensions provide a very satisfactory existence theory, forming, for example, the basis for J. Ball's [**4–7**] work on nonlinear elasticity with incompressibility constraints. Other relevant papers are Marcellini [**24, 25**], Acerbi-Fusco [**1**], etc.

3. Weak and strong convergence, stability. As we have just seen, quasiconvexity is the natural, indeed the necessary and sufficient, condition on the nonlinearity required for lower semicontinuity theorems. The goal of this section and the next is to show further that quasiconvexity, or, more precisely, a slightly strengthened variant thereof, is a basic hypothesis for some rather different assertions in the calculus of variations.

First we note and make precise a heuristic principle that quasiconvexity "improves weak to strong convergence" by "damping out oscillations in the gradients" of certain sequences of functions. More precisely, let us fix some $1 < q < \infty$, and suppose F satisfies the structure hypotheses (2.1), (2.3). We then ask: if $\{u_m\}_{m=1}^\infty$ is a sequence of functions such that

$$u_m \rightharpoonup u \text{ weakly in } W^{1,q}(\Omega; \mathbf{R}^N) \text{ and } I[u_m] \to I[u], \tag{3.1}$$

can we conclude that

$$u_m \to u \text{ strongly in } W^{1,q}(\Omega; \mathbf{R}^N)? \tag{3.2}$$

We will see later why this is an interesting question.

Our answer will require a stronger hypothesis than (2.1), to eliminate possible degeneracies in F. We say $F: M^{n \times N} \to \mathbf{R}$ is *uniformly strictly quasiconvex* if for some $\gamma > 0$,

$$\int_O F(A) + \gamma |D\phi|^q \, dx \le \int_O F(A + D\phi) \, dx \tag{3.3}$$

for all O, A, etc., as in (2.1). The idea is that uniform strict quasiconvexity is to quasiconvexity as uniform strict convexity is to convexity.

THEOREM 2 [**15**]. *Assume (3.1). Suppose in addition to the aforementioned hypothesis, F is uniformly strongly quasiconvex. Then $u_m \to u$ strongly in $W^{1,q}_{\text{loc}}(\Omega; \mathbf{R}^N)$.*

The proof is based essentially upon Morrey's methods described in §2, but the basic idea is easily seen in the following special case. Suppose F has the form $F(A) \equiv \lambda |A|^q + G(A)$ $(A \in M^{n \times N})$ where G is quasiconvex, $0 \le G(A) \le C(1 + |A|^q)$, $\lambda > 0$, and $q > 1$. Suppose (3.1) holds. Then

$$\limsup_{m \to \infty} \lambda \int_\Omega |Du_m|^q \, dx = \lim_{m \to \infty} I[u_m] - \liminf_{m \to \infty} \int_\Omega G(Du_m) \, dx$$

$$\le I[u] - \int_\Omega G(Du) \, dx = \lambda \int_\Omega |Du|^q \, dx,$$

where we used Theorem 1 applied to G. The above estimate and the weak convergence imply strong convergence in $W^{1,q}$.

As a first application of these ideas we show the stability of minimizers under weak convergence.

THEOREM 3. *Suppose F is uniformly strongly quasiconvex and that the $\{u_m\}_{m=1}^{\infty}$ are minimizers of $I[\cdot]$ (not necessarily with respect to the same boundary conditions). Then if*

$$u_m \rightharpoonup u \text{ weakly in } W_{\text{loc}}^{1,q}(\Omega; \mathbf{R}^N), \qquad (3.4)$$

we have

$$u_m \to u \text{ strongly in } W_{\text{loc}}^{1,q}(\Omega; \mathbf{R}^N), \qquad (3.5)$$

and

$$u \text{ is a minimizer of } I[\cdot]. \qquad (3.6)$$

To establish (3.5) we need only show $I'[u_m] \to I'[u]$ where $I'[v] \equiv \int_{\Omega'} F(Dv)\, dx$ for $\Omega' \Subset \Omega$. Now $I'[u] \le \liminf_{m \to \infty} I'[u_m]$ according to Theorem 1. On the other hand, since u_m is a minimizer,

$$I'[u_m] \le I'[u] + o(1) \quad \text{as } m \to \infty, \qquad (3.7)$$

at least assuming Ω' is properly situated. To see this note that if by some chance $u_m = u$ on $\partial\Omega'$, then (3.8) is immediate (with no "$o(1)$" term); the general case follows by cutting off u near $\partial\Omega'$ to agree with u_m and controlling the resultant error: see [15]. But then (3.7) gives $\limsup_{m \to \infty} I'[u_m] \le I'[u]$, and the proof of (3.5) follows from Theorem 2. Assertion (3.6) is an easy consequence.

4. Partial regularity. We now address the question of possible smoothness of minimizers of the functional $I[\cdot]$. Assume $2 \le q < \infty$. We slightly strengthen (3.3) by assuming

$$\int_O F(A) + \gamma(1 + |D\phi|^{q-2})|D\phi|^2\, dx \le \int_O F(A + D\phi)\, dx \qquad (4.1)$$

for all ϕ, A, etc. as above.

THEOREM 4 [**13, 14**]. *Suppose F is C^2, satisfies (4.1), and*

$$|D^2 F(P)| \le C(1 + |P|^{q-2}) \quad (P \in M^{n \times N})$$

for some constant C. Let $u \in W^{1,q}(\Omega; \mathbf{R}^N)$ be a minimizer of $I[\cdot]$.

Then there exists an open set at $\Omega_0 \subset \Omega$ with $|\Omega - \Omega_0| = 0$ and $Du \in C^{\gamma}(\Omega_0; M^{n \times N})$ for each $0 < \gamma < 1$. If, furthermore, F is C^{∞}, then $u \in C^{\infty}(\Omega_0; \mathbf{R}^N)$.

The full details of the proof are too lengthy to reproduce here, and instead I will informally explain why quasiconvexity, which as we have seen above is a natural hypothesis for lower semicontinuity theorems, in fact forces partial regularity for minimizers.

FIRST PROOF OF THEOREM 4. For simplicity, set $q = 2$. We first introduce the notation

$$(Du)_{x,r} \equiv \frac{1}{|B(x,r)|} \int_{B(x,r)} Du(y)\, dy,$$

$$U(x,r) \equiv \frac{1}{|B(x,r)|} \int_{B(x,r)} |Du(y) - (Du)_{x,r}|^2\, dy,$$

so that $U(x,r)$ measures the average mean-squared deviation of Du from its average over the ball $B(x,r) \subset \Omega$.

LEMMA 1. *For each $L > 0$, there exist $0 < \varepsilon_0, \tau < 1$ such that*

$$U(x,r) < \varepsilon_0, \qquad |(Du)_{x,r}| \le L \tag{4.2}$$

imply

$$U(x,\tau r) \le \tfrac{1}{2} U(x,r) \tag{4.3}$$

for all $x \in \Omega$, $0 < r < \operatorname{dist}(x,\partial\Omega)$.

Lemma 1 asserts, roughly, that "Du cannot have singularities of size less than ε_0 in Ω." Indeed if (4.1) is satisfied for such x and r as above, we inductively conclude

$$U(y,\tau^k r) \le \frac{1}{2^k} U(y,r) \qquad (k = 1, 2, \ldots)$$

for all y sufficiently near x; whence some standard real analysis lemmas (see, for example, [18, p. 70]) imply Du is C^γ for some $\gamma > 0$ near x. On the other hand, by Lebesgue's Differentiation Theorem (4.2) holds for a.e. $x \in \Omega$. A refinement of Lemma 1 leads to the assertion that Du is C^γ near x for all $0 < \gamma < 1$.

Thus the proof of Theorem 4 devolves upon the proof of Lemma 1. And this we carry out using the "blow-up" method of DeGiorgi, Almgren, etc. The plan is this: assuming Lemma 1 false, there would exist (for some $0 < \tau < 1$ fixed) balls $B(x_m, r_m) \subset \Omega$ such that

$$U(x_m, r_m) \to 0, \qquad |(Du)_{x_m, r_m}| \le L$$

but

$$U(x_m, \tau r_m) > \tfrac{1}{2} U(x_m, r_m) \qquad (m = 1, \ldots). \tag{4.4}$$

We appropriately rescale the function u on $B(x_m, r_m)$ to functions v^m on the unit ball $B = B(0,1)$ $(m = 1, \ldots)$. The scaling is such that the $\{v^m\}_{m=1}^\infty$ are bounded in $W^{1,2}(B; \mathbf{R}^N)$, and so there exists a *weakly* convergent subsequence in that space. If needs be, we reindex and assume the full sequence satisfies

$$v^m \rightharpoonup v \text{ weakly in } W^{1,2}(B; \mathbf{R}^N). \tag{4.5}$$

Now it turns out that since u and thus the v^m satisfy appropriate nonlinear Euler-Lagrange equations, v satisfies a *linear* uniformly elliptic system. Thus v is smooth and satisfies therefore various "good" estimates. On the other hand, the v^m satisfy rescaled versions of the "bad" estimates (4.4). We shall thereby obtain the desired contradiction, *except that the convergence in (4.5) is too weak.* We must improve the weak to strong convergence in $W^{1,2}_{\text{loc}}(B; \mathbf{R}^N)$.

But this is precisely the issue addressed in §3, where we saw in Theorem 3 that under the principal assumption of uniform strict quasiconvexity on the nonlinearity, weakly convergent sequence of minimizers in fact converges strongly. The current setting is more complicated in that we are continually rescaling as $m \to \infty$, but the conclusions (see [14]) are correct: we have $v^m \to v$ strongly in $W_{\text{loc}}^{1,2}(B; \mathbf{R}^N)$. This compactness assertion allows us to prove Lemma 1 as above and thereby complete the proof of Theorem 4.

SECOND PROOF OF THEOREM 4. Again we assume that, for simplicity, $q = 2$. We first ask why uniform strict quasiconvexity of the nonlinear term F in $I[\cdot]$ should imply partial regularity of minimizers, and now look at the definition (3.3) for a clue. We then see from (3.3) that if u is a minimizer and if u happens to agree with a plane $\pi(y) \equiv Ay + a$ on the boundry of some ball $B(x,r) \subset \Omega$, then in fact u equals that plane in all of $B(x,r)$ (Proof: let $\phi = u - \pi$ in (3.3)). This suggests that the deviation of u from any plane in $B(x,r)$ may somehow be controlled by the deviation of u from the same plane on $\partial B(x,r)$ or perhaps on the annulus $B(x,r) - B(x,r/2)$. This, if true, is interesting since several techniques in elliptic equations, most notably the "hole-filling" device of Widman, derive analytic estimates in just such a situation.

The above is all rather vague, but it does suggest that it may be useful to compare u on each ball with the plane π as above. We accomplish this by setting $v \equiv \varsigma \pi + (1 - \varsigma)u$ where ς is a cutoff function vanishing near $\partial B(x,r)$. Since u is a minimizer, $I[u] \leq I[v]$; and this inequality, suitably exploited, leads to

LEMMA 2 [13]. *There exists a constant C such that*

$$\int_{B(x,r/2)} |Du - A|^2 \, dy \leq \frac{C}{r^2} \int_{B(x,r)} |u - Ay - a|^2 \, dy \qquad (4.6)$$

for all $B(x,r) \subset \Omega$, $A \in M^{n \times N}$, $a \in \mathbf{R}^N$.

Estimate (4.6) is a "Caccioppoli inequality" in the terminology of [18]. It is useful since it allows for just enough control to provide another proof of Lemma 1 as above.

5. Concluding comments. A. In recent work Giaquinta-Modica [19] and Fusco-Hutchinson [17] have extended the partial regularity assertion to more general problems of the form $I[v] \equiv \int_\Omega F(Dv, v, x) \, dx$.

B. There has been much recent interest in computing the "quasiconvex relaxation" of various nonconvex and nonquasiconvex functionals arising, for example, in problems of optimal design [22], phase transitions [11], and inverse problems [23]. The forthcoming paper [23] includes an informal discussion concerning the advantages for numerical analysis of quasiconvex relaxation.

C. Knops and Stewart [21] have shown uniqueness of smooth critical points (not just minimizers) for strictly quasiconvex functionals subject to linear boundary conditions.

D. Very recently, Acerbi and Fusco have shown that the growth condition on $D^2 F$ in Theorem 4 is unnecessary.

E. Much of theory described above is analogous to known results in geometric measure theory concerning the existence and partial smoothness of minimizers of elliptic integrals in the sense of Almgren, Federer, etc. See [2, 3, 8, 16], etc. and also [13] for a brief discussion concerning the similarity of these ideas.

REFERENCES

1. E. Acerbi and N. Fusco, *Semicontinuity problems in the calculus of variations*, Arch. Rational Mech. Anal. **86** (1984), 125–145.

2. W. K. Allard and F. J. Almgren, Jr., *An introduction to regularity theory for parametric elliptic variational problems*, Proc. Sympos. Pure Math., vol. 23, Amer. Math. Soc. Providence, R. I., 1973, pp. 231–260.

3. F. J. Almgren, Jr., *Existence and regularity almost everywhere of solutions to elliptic variational problems among surfaces of varying topological type and singularity structure*, Ann. of Math. **87** (1968), 321–391.

4. J. M. Ball, *Convexity conditions and existence theorems in nonlinear elasticity*, Arch. Rational Mech. Anal. **63** (1977), 337–403.

5. _____, *Strict convexity, strong ellipticity, and regularity in the calculus of variations*, Math. Proc. Cambridge Philos. Soc. **87** (1980), 501–513.

6. J. M. Ball and J. E. Marsden, *Quasiconvexity at the boundary, positivity of the second variation and elastic stability*, Arch. Rational Mech. Anal. **86** (1984), 251–277.

7. J. M. Ball and F. Murat, $W^{1,p}$-*quasiconvexity and variational problems for multiple integrals*, J. Funct. Anal. **58** (1984), 225–253.

8. E. Bombieri, *Regularity theory for almost minimal currents*, Arch. Rational Mech. Anal. **78** (1982), 99–130.

9. B. Dacorogna, *Quasiconvexity and relaxation of non-convex problems in the calculus of variations*, J. Funct. Anal. **46** (1982), 102–118.

10. _____, *Weak continuity and weak lower semicontinuity of non-linear functionals*, Lecture Notes in Math., vol. 922, Springer-Verlag, Berlin-New York, 1982.

11. _____, *A relaxation theorem and its applications to the equilibrium of gases*, Arch. Rational Mech. Anal.

12. E. De Giorgi, *Sulla convergenza di alcune successioni di integrali del tipo dell'area*, Rend. Math. **8** (1975), 277–294.

13. L. C. Evans, *Quasiconvexity and partial regularity in the calculus of variations*, Arch Rational Mech. Anal. (to appear).

14. L. C. Evans and R. F. Gariepy, *Blow-up, compactness, and partial regularity in the calculus of variations*, Indiana Univ. Math. J. (to appear).

15. _____, *Some remarks concerning quasiconvexity and strong convergence*, Proc. Royal Soc. Edinburgh Sect. A (to appear).

16. H. Federer, *Geometric measure theory*, Springer, New York, 1969.

17. N. Fusco and J. Hutchinson, *Partial regularity of functions minimizing quasiconvex integrals* (to appear).

18. M. Giaquinta, *Multiple integrals in the calculus of variations and nonlinear elliptic systems*, Princeton Univ. Press, Princeton, N. J., 1983.

19. M. Giaquinta and G. Modica, *Partial regularity of minimizers of quasiconvex integrals* (to appear).

20. E. Giusti and M. Miranda, *Sulla regolarità delle soluzioni deboli di una classe di sistemi ellittici quasi-lineari*, Arch. Rational Mech. Anal. **31** (1968), 173–184.

21. R. J. Knops and C. A. Stuart, *Quasiconvexity and uniqueness of equilibrium solutions in nonlinear elasticity*, Arch. Rational Mech. Anal.

22. R. Kohn and G. Strang, *Optimal design and relaxation of variational problems*, Comm. Pure Appl. Math.

23. R. Kohn and M. Vogelius, *Relaxation of a variational method for impedance computed tomography* (to appear).

24. P. Marcellini, *Approximation of quasiconvex functions and lower semicontinuity of multiple integrals*, Manuscripta Math. (to appear).

25. _____, *On the definition and lower semicontinuity of certain quasiconvex integrals* (to appear).

26. N. Meyers, *Quasiconvexity and lower semicontinuity of multiple variational integrals of any order*, Trans. Amer. Math. Soc. **119** (1965), 125–149.

27. C. B. Morrey, Jr., *Quasiconvexity and the lower semicontinuity of multiple integrals*, Pacific J. Math. **2** (1952), 25–53.

28. _____, *Multiple integrals in the calculus of variations*, Springer, New York, 1966.

UNIVERSITY OF MARYLAND, COLLEGE PARK, MARYLAND 20742, USA

The Problem of the Regularity of Minimizers

MARIANO GIAQUINTA

1. Introduction. Since the origin of Calculus of Variations, and for a long time, all the information concerning the *minima* of variational functionals

$$\mathcal{F}[u; \Omega] = \int_\Omega F(x, u, Du)\, dx \qquad (1.1)$$

were obtained by means of its Euler-Lagrange equation

$$\frac{\partial}{\partial x}\frac{\partial F}{\partial p} - \frac{\partial F}{\partial u} = 0.$$

It was in this century, after the pioneering works of Riemann, Arzelà, Hilbert, Lebesgue, and the strong contributions of Leonida Tonelli and Charles B. Morrey, that the so-called direct methods of Calculus of Variations established themselves as the main tool to deal with the problem of the existence of minima.

As a result of the use of direct methods, a sufficiently general theorem that we may now state because of the contributions of many authors is

THEOREM 1.1. *Let Ω be a bounded domain in $\mathbf{R}^n$; let $F(x, u, p): \Omega \times \mathbf{R}^N \times \mathbf{R}^{nN} \to R$ be measurable in x, continuous in (u, p), and quasiconvex; i.e., for almost every $x_0 \in \Omega$ and for all $u_0 \in \mathbf{R}^N$ and $p_0 \in \mathbf{R}^{nN}$ and all $\phi(x) \in C_0^\infty(\Omega, \mathbf{R}^N)$ we have*

$$\int_\Omega F(x_0, u_0, p_0)\, dx \le \int_\Omega F(x_0, u_0, p_0 + D\phi)\, dx. \qquad (1.2)$$

Moreover suppose that $|p|^m \le F(x, u, p) \le \lambda(1 + |p|^2)^{m/2}$ where $m > 1$ and $\lambda > 0$. Then the functional $\mathcal{F}$ attains its minimum in the class of functions $u \in W^{1,m}(\Omega, \mathbf{R}^N)$ with prescribed value at the boundary $\partial\Omega$.

The notion of *quasiconvexity*, which seems to be a global condition, was introduced by C. B. Morrey in 1952 [45], and Theorem 1.1, under some stronger assumptions, goes back to him (for recent improvements see [1, 3, 41]). We recall that in *the scalar case $N = 1$*, the condition of quasiconvexity is equivalent to convexity with respect to p, that is, if F is smooth, to

$$\sum_{\alpha,\beta=1}^n F_{p_\alpha p_\beta}(x, u, p)\xi^\alpha \xi^\beta \ge 0 \qquad \forall \xi \in \mathbf{R}^n, \qquad (1.3)$$

while in the *vector-valued case* $N > 1$ it is weaker than convexity and implies, again if F is smooth, the so-called Legendre-Hadamard condition

$$\sum_{i,j=1}^{N} \sum_{\alpha,\beta=1}^{n} F_{p_\alpha^i p_\beta^j}(x,u,p)\xi^\alpha \xi^\beta \eta_i \eta_j \geq 0 \qquad \forall \xi \in \mathbf{R}^n,\ \forall \eta \in \mathbf{R}^N. \qquad (1.4)$$

We notice that quasiconvexity appears naturally as "the necessary and sufficient condition" for the sequentially lower semicontinuity of $\mathcal{F}$ in $W^{1,m}$-weak.

Moreover we notice that the use of Sobolev spaces $W^{1,m}$, $m > 1$, is by now natural for the application of direct methods and is related to the need of working on a class of functions with a sufficiently weak topology, so that *minimizing sequences* do converge.

So the *regularity problem* arises in a natural way; it is the problem of showing, if possible, continuity or differentiability of the *minimum points* (Hilbert's nineteenth problem). Since we would like to avoid, as far as possible, any complications due to boundary conditions, we shall consider most of the time *minimizers of $\mathcal{F}$. A minimizer is a function $u \in W^{1,m}(\Omega, \mathbf{R}^N)$ such that*

$$\mathcal{F}[u; \operatorname{spt}\phi] \leq \mathcal{F}[u + \phi; \operatorname{spt}\phi] \qquad \forall \phi \in W^{1,m}(\Omega, \mathbf{R}^N)\ \operatorname{spt}\phi \Subset \Omega.$$

In dealing with the regularity problem again, the Euler-Lagrange equation in its weak formulation

$$\int_\Omega [F_{p_\alpha^i}(x,u,Du)D_\alpha\phi^i + F_{u^i}(x,u,Du)\phi^i]\,dx = 0 \qquad \forall \phi \in C_0^\infty(\Omega, \mathbf{R}^N) \qquad (1.5)$$

has always been the starting point even in recent times—for instance, in the classical and celebrated works of E. De Giorgi, J. Nash, J. Moser, O. A. Ladyzhenskaya and N. N. Ural'tseva (see, e.g., [**39**, **46**, **29**, **17**]).

Actually this is natural and in a sense necessary if we want to study the continuity of the second order (or higher order) derivatives of a minimizer; a classical result, obtained after a research span of fifty years, is the following one: *if the integrand F is C^∞, or analytic, and*

$$F_{p_\alpha^i p_\beta^j}(x,u,p)\xi^\alpha \xi^\beta \eta_i \eta_j > c(M)|\xi|^2 |\eta|^2 \qquad \forall \xi \in \mathbf{R}^n,\ \forall \eta \in \mathbf{R}^N$$

for $|u| + |p| \leq M$, with $c(M) > 0$, then the points of minimum are C^∞, or analytic, as soon as it is known that they are of class C^1. But this approach has some disadvantages and it appears somehow unnatural when studying the first two steps in the regularity theory: the Hölder-continuity and the differentiability of the minimizers.

In fact, first of all, even if F is smooth, (1.5) does not hold without assuming additional conditions on the behavior of F_u at infinity, in such a way that for $u \in W^{1,m}$, $F_u(x,u,Du)$ lies in L^1_{loc}. But, even assuming the so-called "natural growth conditions," which ensure the local integrability of $F_u(x,u,Du)$, the weak Euler-Lagrange equation (1.5) would still be of no use in order to prove regularity, without extra conditions on the minimizers themselves such as boundedness or even smallness in modulus (see for a discussion [**17**, **33**, **39**]). Roughly the point is that (1.5), *and therefore this approach, does not distinguish between minimizers and stationary points.* But, in general, *minimizers have better properties than just stationary points.*

An interesting geometric example is given by the integral

$$\mathcal{E}[u; D] = \int_D |Du|^2/(1 + |u|^2)^2 \, dx, \qquad u = (u^1, \ldots, u^n),$$

which, apart from a constant, represents in local coordinates (choosing stereographic coordinates on the sphere S^n minus a point) the energy of a map from the disk $D_n = B_1(0) = \{x \in \mathbf{R}^n : |x| < 1\}$ into S^n-{point}. S. Hildebrandt, H. Kaul, K. O. Widman [34] have shown, as a consequence of a more general result, that the minimizers of $\mathcal{E}$, whose image lies strictly in a hemisphere (in our system of coordinates this corresponds to $|u| \le k < 1$) are regular, while the "equator map $u^*: x \to x/|x|$," which is obviously not strictly contained in a hemisphere, is a stationary point for $\mathcal{E}$. W. Jager, H. Kaul [35] then showed that u^* is actually a minimizer if $n \ge 7$ and it is not even stable, so in particular it is not a minimizer, for $n \le 6$. In general we have that *any energy minimizing map from the disk D_n into a sphere S^n whose image lies (not necessarily strictly) into a hemisphere is regular provided $n < 7$* (see [54, 28]).

Another interesting example is due to J. Frehse [12] who pointed out that the functional

$$\int_{B(0,e^{-1})} (1 + e^u |\log |x||^{12})^{-1} |Du|^2 \, dx,$$

which obviously has 0 as minimum point in $W_0^{1,2}(B(0, e^{-1}))$, has among its stationary points the unbounded function of $W_0^{1,2}(B(0, e^{-1}))$ given by $u(x) = 12 \log \log |x|^{-1}$.

In the last five years there has been a strong attempt to develop a kind of *direct approach to the regularity*, working directly with the functional $\mathcal{F}$ instead of working with its Euler-Lagrange equation. We mention especially the works of M. Giaquinta, E. Giusti, in the general situation, and the works of R. Schoen, K. Uhlenbeck, concerning the regularity theory of harmonic mapping between Riemannian manifolds. Moreover we mention the earlier works of L. Tonelli [56] and C. B. Morrey [44] where the idea of a direct approach to the regularity is present and regularity results for minimizers of nondifferentiable functionals in dimensions 1 and 2 are proved under extremely weak assumptions.

It is the aim of this lecture to illustrate some of the results obtained in this direction.

In the following it is convenient to distinguish two levels of regularity: 1. regularity of the minimizer u; 2. regularity of the derivatives of the minimizer u. Of course our results will take a different form in the scalar and in the vector-valued case, respectively $N = 1$ and $N > 1$. In fact, while in the scalar case it is natural to expect, under suitable hypotheses, *regularity everywhere*, in the vector-valued case this is quite rare; minimizers are in general noncontinuous or have noncontinuous derivatives, as shown by well-known counterexamples of

E. De Giorgi, E. Giusti and M. Miranda, V. G. Mazya, J. Nečas, S. Hildebrandt and K. O. Widman, M. Giaquinta (see, e.g., [17]), and we may only expect *partial regularity*, that is, regularity except possibly on a singular closed set.

2. Basic regularity: quasiminima. In studying the first level of regularity the notion of *quasiminima*, introduced by M. Giaquinta and E. Giusti [18, 22], plays an important role for its *unifying* and *clarifying* feature (besides, of course, the fact that we can prove interesting results for quasiminima).

Consider the functional $\mathcal{F}$ in (1.1) and suppose, for simplicity, that

$$|p|^m \le F(x, u, p) \le \lambda(1 + |p|^m), \qquad m > 1. \tag{2.1}$$

We say that *a function* $u \in W^{1,m}_{loc}(\Omega, \mathbf{R}^N)$ *is a Q-minimum,* $Q \ge 1,$ *for the functional* $\mathcal{F}$ *if for every open set* $A \Subset \Omega$ *and for every* $v \in W^{1,m}_{loc}(\Omega, \mathbf{R}^N)$ *with* $v = u$ *outside* A *we have*

$$\mathcal{F}[u; A] \le Q\mathcal{F}[v; A] \tag{2.2}$$

or, equivalently, for any $\phi \in W^{1,m}(\Omega, \mathbf{R}^N)$ *with* spt $\phi \Subset \Omega$ *we have*

$$\mathcal{F}[u; \{x \in \Omega : \phi(x) \ne 0\}] \le Q\mathcal{F}[u + \phi; \{x \in \Omega : \phi(x) \ne 0\}]. \tag{2.3}$$

We notice that the comparison in (2.2) is made for all open sets A; we may think of choosing special classes of A's, for example, balls B_R. Then we say that $u \in W^{1,m}(\Omega, \mathbf{R}^N)$ is a spherical Q-minimum if for any ball $B_R \Subset \Omega$ and for any $\phi \in W^{1,m}_0(B_R, \mathbf{R}^N)$ we have

$$\mathcal{F}[u; B_R] \le Q\mathcal{F}[u + \phi; B_R]. \tag{2.4}$$

Obviously any Q-minimum is a spherical Q-minimum, but the opposite is not true; moreover, in the scalar case $N = 1$, spherical Q-minima may be unbounded in dimension $n \ge 3$ (see [22]).

Of course any minimizer of $\mathcal{F}$ is a Q-minimum for $\mathcal{F}$ with $Q = 1$; but it is also a Q-minimum for the simpler functional $\int_\Omega (1 + |Du|^m)\, dx$. This shows once more the special relevance of the "Dirichlet integral." But the class of Q-minima is much wider [22]:

(a) Weak solutions of elliptic systems with L^∞ (and not even symmetric) coefficients

$$\int_\Omega A^{\alpha\beta}_{ij}(x) D_\alpha u^i D_\beta \psi^j\, dx = 0, \qquad \psi \in W^{1,2}_0(\Omega, \mathbf{R}^N),$$

$$|A^{\alpha\beta}_{ij}(x)| \le L, \qquad A^{\alpha\beta}_{ij} \xi^\alpha_i \xi^\beta_j \ge |\xi|^2 \quad \forall \xi \in \mathbf{R}^{nN}$$

are Q-minima for the Dirichlet integral: In order to see that, it is sufficient to choose as test function $\psi = u - (u + \phi)$, $\phi \in W^{1,2}_0(\Omega, \mathbf{R}^N)$.

(b) In general "solutions" (here the word "solution" has to be understood in the right sense; depending on the situations, they have to be bounded or even small (see [22, 17])) of nonlinear elliptic systems $D_\alpha A^\alpha_i(x, u, Du) = B(x, u, Du)$ under "natural hypotheses" are Q-minima for suitable functionals.

(c) Minimizers of functionals in constrained classes are Q-minima of free functionals.

(d) Quasiconformal maps and, more generally, quasiregular maps, i.e., $u \colon \Omega \subset \mathbf{R}^n \to \mathbf{R}^n$, $u \in W^{1,n}(\Omega, \mathbf{R}^n)$ such that $|Du(x)|^n \leq c \det Du$ a.e. in Ω, are Q-minima for $\int_\Omega |Du|^n \, dx$.

We have [18, 22]

THEOREM 2.2. (i) *In the scalar case $N = 1$: let $u \in W_{\mathrm{loc}}^{1,m}(\Omega, \mathbf{R})$ be a Q-minimum for $\mathcal{F}$; then u is locally Hölder-continuous in Ω.*

(ii) *In the vector-valued case $N \geq 1$: let $u \in W_{\mathrm{loc}}^{1,m}(\Omega, \mathbf{R}^N)$ be a spherical Q-minimum for $\mathcal{F}$; then there exists an exponent $r > m$ such that $u \in W_{\mathrm{loc}}^{1,r}(\Omega, \mathbf{R}^N)$; moreover the gradient of u satisfies the following reverse Hölder inequality with increasing supports,*

$$\left(\fint_{B_{R/2}} (1 + |Du|^r) \, dx \right)^{1/r} \leq c \left(\fint_{B_R} (1 + |Du|^q) \, dx \right)^{1/q},$$

for all $B_R \Subset \Omega$ and all q, $0 < q \leq m$. In particular if $m = n$, then u is locally Hölder-continuous.

The proof of (i) uses De Giorgi's results and their extensions due to Ladyzhenskaya and Ural'tseva on what we now call De Giorgi's classes [7, 39], while the proof of (ii) uses a result of M. Giaquinta and G. Modica [24] on reverse Hölder inequalities related to a previous result of W. Gehring [15]. Both results rely on a Caccioppoli type inequality (see, e.g., [17]).

If we read Theorem 2.2 for minimizers, it states the basic and optimal regularity properties of minimum points. We notice that no hypothesis of ellipticity has been made, so the basic regularity follows from the *minimality* and the *growth condition* 2.1. It is worth noticing that the results of Theorem 2.2 are not true for stationary points of $\mathcal{F}$, even under ellipticity conditions.

On the other hand, Theorem 2.2 permits to recover essentially all the results of Hölder-continuity and higher integrability of the gradient known for "solutions" (under inverted commas!) of "nonlinear elliptic systems" as consequences of the minimality condition 2.2.

As mentioned, well-known counterexamples show that in the vector-valued case minimizers (and therefore Q-minima) are in general noncontinuous. J. Souček [55] has shown that solutions of linear elliptic systems with L^∞ coefficients may be discontinuous on a dense set. Therefore, because of (b), Q-minima of the Dirichlet integral, for $N > 1$, may be discontinuous in a dense set; so it is quite surprising that the gradient which lies in L^m is in fact p-summable with some p larger than m.

Finally we would like to remark that the higher degree of integrability in (ii) clearly appears as a consequence of a comparison on B_R with "harmonic functions" with the same boundary value as the Q-minimum u, so it can be considered as a result of a linear perturbation. On the contrary, the result (i) of Theorem 2.2 does not hold for spherical Q-minima [22], so it is to be considered, in a sense, as a purely nonlinear result: this, in particular, is true for De Giorgi-Nash theorem.

Most of the known properties of solutions of elliptic equations have been shown to hold for scalar Q-minima, as the weak maximum principle, Liouville's type theorems, and so on. In particular we mention the very interesting paper by E. Di Benedetto and N. S. Trudinger [8] where they prove, using some ideas of De Giorgi and N. V. Krylov-M. V. Safanov, that the classical result of J. Moser [48] on Harnack's inequality holds for nonnegative quasi(-super-)minima and the paper of W. P. Ziemer [58] where Wiener type conditions for the regularity of Q-minima at boundary points are given. We also mention that the notion of Q-minimum has proved to be very useful in various contexts (see, e.g., [17, 42, 49]) and finally that Theorem 2.2 is an essential step in proving "regularity" of the derivatives of a minimizer.

3. Regularity of the first derivatives of a minimizer. When studying the regularity of the first derivatives of a minimizer, "ellipticity" will clearly play an important role; but "growth conditions," as explained later, are important too, at least for the methods of proof. In any case it is not necessary to assume that our functional be Gateaux differentiable, and this is not the case under our assumptions on the integrand $F(x, u, p): \Omega \times \mathbf{R}^N \times \mathbf{R}^{nN} \to \mathbf{R}$ in (1.1) which will be the following:

HYPOTHESIS 1. *Growth conditions on F: for $m \geq 2$ and a positive constant Λ we have*

$$|p|^m \leq F(x, u, p) \leq \Lambda(1 + |p|^m) \tag{3.1}$$

where $m \geq 2$. Actually it is sufficient that (3.1) holds in the integrated form on small balls.

HYPOTHESIS 2. *F is twice continuously differentiable with respect to p and*

$$|F_{pp}(x, u, p)| \leq c_1(\mu^2 + |p|^2)^{(m-2)/2}; \tag{3.2}$$

in particular,

$$|F_p(x, u, p)| \leq c_0(\mu^2 + |p|^2)^{(m-1)/2} \tag{3.3}$$

where $\mu \geq 0$.

HYPOTHESIS 3. *The function $(1 + |p|^2)^{-m/2} F(x, u, p)$ is Hölder-continuous in (x, u) uniformly with respect to p.*

HYPOTHESIS 4. *Strict uniform quasiconvexity: for all $x_0 \in \Omega$, $u_0 \in \mathbf{R}^N$, $p_0 \in \mathbf{R}^{nN}$ and for any $\phi \in C_0^1(\Omega, \mathbf{R}^N)$ we have*

$$\int_\Omega [F(x_0, u_0, p_0 + D\phi) - F(x_0, u_0, p_0)] \, dx \geq \int_\Omega |V(p_0 + D\phi) - V(p_0)|^2 \, dx \tag{3.4}$$

where $V(p)$ is the vector-valued function defined by $V(p) = (\mu^2 + |p|^2)^{(m-2)/4} p$.

We notice that in the scalar case $N = 1$, Hypothesis 4 is equivalent to

HYPOTHESIS 4'. *Strict uniform ellipticity: for some positive ν*

$$F_{p_\alpha p_\beta}(x_0, u_0, p_0)\xi_\alpha \xi_\beta \geq \nu(\mu^2 + |p_0|^2)^{(m-2)/2}|\xi|^2 \qquad \forall \xi \in \mathbf{R}^n, \qquad (3.5)$$

so if $\mu = 0$ we have degeneration at the points where the gradient of our minimizer is zero.

We now have the following *partial regularity* result,

THEOREM 3.1. *Suppose Hypotheses 1–4 hold with $\mu > 0$, $N \geq 1$. Let $u \in W^{1,m}_{\mathrm{loc}}(\Omega, \mathbf{R}^N)$ be a minimizer of (1.1). Then there exists an open set $\Omega_0 \subset \Omega$ where the first derivatives of u are locally Hölder-continuous. Moreover the Lebesgue measure of the singular set $\Omega - \Omega_0$ is zero.*

Theorem 3.1 was first proved by C. B. Morrey, E. Giusti and M. Miranda, and E. Guisti essentially in the case $F = F(p)$ and under the stronger assumption of uniform ellipticity

$$F_{p_\alpha^i p_\beta^j}(x, u, p)\xi_i^\alpha \xi_j^\beta \geq (1 + |p|^2)^{(m-2)/2}|\xi|^2 \qquad \forall \xi \in \mathbf{R}^{nN} \qquad (3.6)$$

by means of an indirect argument. Still under condition (3.6) it was then proved, by a direct argument, in the general case $F = F(x, u, p)$ by M. Giaquinta and E. Guisti for $m = 2$ and M. Giaquinta and P. A. Ivert for $m \geq 2$ in 1983–1984 (see, e.g., [17, 19, 23]). Under the strict quasiconvexity assumption, Theorem 3.1 was proved by L. C. Evans [10] (see also [11]) in the case $F = F(p)$, again by an indirect argument (see also [51]); in the form given above it was proved in 1986 independently by M. Giaquinta and G. Modica [25], and N. Fusco and J. Hutchinson [13].

I shall not insist on Theorem 3.1 and I refer to the talk of L. C. Evans at the 1986 International Congress of Mathematicians. I would only like to remark that no result seems to be available, in this generality, when $1 < m < 2$ and in the degenerate case $\mu = 0$, and that the result of Theorem 3.1 uses in a strong way the minimality property (see also [27, 51]). It is worth recalling that our functional $\mathcal{F}$ need not be differentiable.

As already stated, in the vector-valued case $N > 1$ the singular set is in general nonempty; on the contrary, we expect that it will be empty and the minimizers regular everywhere in the scalar case. This is actually true, even in the degenerate case $\mu = 0$, and we have

THEOREM 3.2. *In the scalar case $N = 1$, suppose that Hypotheses 1–3 and Hypothesis 4' hold with $\mu \geq 0$. Then any minimizer $u \in W^{1,m}_{\mathrm{loc}}(\Omega)$ of (1.1) has Hölder-continuous first derivatives in Ω.*

Theorem 3.2, in case $F = F(p)$ and $m = 2$, is a consequence of the celebrated De Giorgi-Nash Theorem [7] and, for $m > 2$, $\mu \geq 0$, $F = F(p)$, it was proved in a very interesting paper by K. Uhlenbeck [57]. For $F = F(x, u, p)$ it was proved for $m = 2$ by M. Giaquinta and E. Giusti [19] and, in the general case $m \geq 2$, $\mu \geq 0$, by M. Giaquinta and G. Modica [26].

It is worth noticing again that under the assumptions of Theorems 3.1 and 3.2, the functional $\mathcal{F}$ in (1.1) is not differentiable, and that the result of Theorem 3.2 is not true for "stationary points." Finally we notice that in the degenerate case $\mu = 0$, minimizers have not, in general, continuous second derivatives even in the simple situation $F = |p|^m$, $m > 2$ (see, e.g., [26, 40]); and in the general situation $F = F(x, u, p)$, it seems that there is an upper bound for the Hölder exponent which is strictly less than the Hölder exponent of the function $F(\cdot, \cdot, p)$ (see, e.g., [20, 50]).

4. Further contributions. Ever since the first results of partial regularity of minimizers were proved, many questions have been raised; and most of them still have no answer. In this final section I shall state a few of these questions and discuss some contributions.

1. Under which conditions are vector-valued minimizers regular everywhere? In this direction surely the most interesting result is due to K. Uhlenbeck [57] who showed that *under the assumptions of Theorem 3.1, if moreover we assume that $F(x, u, p) = G(|p|^2)$ with G smooth, then minimizers are regular everywhere.*

Everywhere regularity can also be proved if the functional $\mathcal{F}$ "is not far" from a quadratic functional or more generally from a functional whose minimizers are regular (see, e.g., [38]).

2. Nothing is known on the structure of the singular set nor even on its stability or instability with respect to perturbations of the data. More simply, we may ask whether we can improve the estimate of the dimension of the singular set. "Optimal" results have been proved for quadratic functionals.

Consider the variational integral

$$\mathcal{A}[u; \Omega] = \int_\Omega A_{ij}^{\alpha\beta}(x, u) D_\alpha u^i D_\beta u^j \, dx \tag{4.1}$$

where $A_{ij}^{\alpha\beta}$ are bounded and smooth functions satisfying the ellipticity condition

$$A_{ij}^{\alpha\beta} \xi_i^\alpha \xi_j^\beta \geq |\xi|^2 \qquad \forall \xi \in \mathbf{R}^{nN}. \tag{4.2}$$

We notice that $\mathcal{A}$ is not differentiable. M. Giaquinta and E. Giusti [18] showed that *if $u \in W_{\text{loc}}^{1,2}(\Omega, \mathbf{R}^N)$ is a minimizer of (4.1), and (4.2) holds, then the first derivatives of u are Hölder-continuous except possibly on a closed singular set whose Hausdorff dimension is strictly less than $n - 2$.*

Furthermore we have (see [26] and also [14]): *under the assumptions of Theorem 3.1, suppose moreover that*

$$F(x, u, p) = G(x, u, a^{\alpha\beta}(x) g_{ij}(u) p_\alpha^i p_\beta^j). \tag{4.3}$$

Then any minimizer has Hölder-continuous first derivatives in an open set Ω_0 and the singular set $\Omega - \Omega_0$ has Hausdorff dimension strictly less than $n - m$.

A "special" case of both (4.1) and (4.3) is given by the variational integral

$$\mathcal{E}[u; \Omega] = \int_\Omega a^{\alpha\beta}(x) g_{ij}(u) D_\alpha u^i D_\beta u^j \sqrt{a(x)} \, dx \tag{4.4}$$

where $(a^{\alpha\beta})$ and (g_{ij}) are smooth symmetric positive definite matrices and $a(x) = \det(a_{\alpha\beta})$, $(a_{\alpha\beta}) = (a^{\alpha\beta})^{-1}$, $\mathcal{E}$ in (4.4) represents in local coordinates the *energy* of a map between two Riemannian manifolds $M^n \to M^N$ with metric tensors respectively $a_{\alpha\beta}$ and g_{ij}. In this situation we have (see [21, 52]): *bounded minimizers of* (4.4) *can have at most isolated interior singularities in dimension $n = 3$ and, in general, the singular set has Hausdorff dimension not larger than $n - 3$*; while (see [36, 53]) *no singularity can occur at the boundary, provided, of course, the boundary datum is smooth*. We notice that it seems instead reasonable to expect singularities at the boundary for stationary points and even minimizers of (4.1) (see [16]).

The previous results apply to energy minimizing maps between two Riemannian manifolds $U: M^n \to M^N$ only if we know a priori that the image of U lies on a coordinate chart of M^N. This is of course a strong restriction. A general regularity theory for energy minimizing harmonic maps between Riemannian manifolds, which gives analogous results, has been developed by R. Schoen and K. Uhlenbeck [52, 53]. These results have been extended to the case of target manifolds M^N with boundary in [9]. In this context we also mention the recent work of H. Brezis, J. M. Coron, G. H. Lieb [5] where singular energy minimizing maps from a domain of R^n into S^{n-1} are studied, and the work of R. Hardt, D. Kinderlehrer, F. H. Lin [32] in connection with the theory of liquid crystals (we refer to the talk of R. Hardt at the 1986 International Congress of Mathematicians).

3. Are growth conditions really necessary? A theorem of L. Tonelli [56], in dimension $n = 1$, states: *if u is a minimizer of the functional* (1.1) *on an interval I where the integrand F is a C^∞ function satisfying $F_{pp}(x, u, p) > 0$, then there exists a closed set Σ with* meas $\Sigma = 0$ *such that $u \in C^\infty(I - \Sigma)$*. A more recent result of F. H. Clarke and R. B. Vinter [6], still in dimension $n = 1$, says that *in the autonomous case, i.e. $F = F(u, \dot{u})$, if F is convex in u and $F(u, \dot{u}) \geq \phi(|\dot{u}|)$ where $\phi(r)/r \to +\infty$ as $r \to +\infty$, then any minimizer is Lipschitz-continuous everywhere* (we refer to [4, 6] for more information). How far can we go in this direction in more than one variable? Not much is known even in terms of examples and counterexamples.

We shall therefore confine ourselves to a few remarks on Theorem 3.1. Hypotheses 1–3 are surely quite reasonable if we assume the uniform ellipticity in (3.6). But they are strong under the uniform strict quasiconvexity (3.4). Already L. C. Evans [10] pointed out that the estimate from below in (3.1) is not necessary if F does not depend explicitly on x and u. Recently Hong Min-Chin [43] has shown that, in the general situation, (3.1) can be substituted by

$$F^0(p) \leq F(x, u, p) \leq M|p|^m + 1$$

where $F^0(p)$ is a strictly quasiconvex function with $|F_{pp}^0(p)| \leq 1 + |p|^{m-2}$, so that functionals of the type

$$\int \{A(x, u) \, Du \, Du + L \det Du\} \, dx, \qquad n = N = 2,$$

are included. But still the control in (3.2) on the second derivatives of F is too strong as shown, for instance, by the functional

$$\int \{|Du|^2 + \sqrt{1 + (\det Du)^2}\}\, dx.$$

E. Acerbi and N. Fusco [2] have proved that (3.2) is not necessary for the partial regularity of the minimizers and that in fact it is sufficient to assume that (3.3) holds. So that we may state that *the conclusion of Theorem 3.1 holds under the weaker assumptions that* (i) F *is of class* C^2 *with respect to* p, $(1 + |p|^2)^{m-2} F(x, u, p)$ *be Hölder-continuous in* (x, u) *uniformly with respect to* p, (ii) $|F(x, u, p)| \leq c(1 + |p|^m)$, (iii) $F(x, u, p)$ *be strictly quasiconvex, and finally that there exists a strictly quasiconvex function* $F^0(p)$ *such that* $F(x, u, p) \geq F^0(p)$ (see [2]).

REFERENCES

1. E. Acerbi and N. Fusco, *Semicontinuity problems in the calculus of variations*, Arch. Rational Mech. Anal. **86** (1984), 125–145.

2. ——, *A regularity theorem for minimizers of quasiconvex integrals*, preprint, 1986.

3. J. M. Ball, *Convexity conditions and existence theorems in nonlinear elasticity*, Arch. Rational Mech. Anal. **63** (1977), 337–403.

4. J. M. Ball and V. J. Mizel, *One-dimensional variational problems whose minimizers do not satisfy the Euler-Lagrange equation*, preprint, 1985.

5. H. Brezis, J. M. Coron, and E. H. Lieb, *Harmonic maps with defects*, preprint, 1986.

6. F. H. Clarke and R. B. Vinter, *Regularity properties to the basic problem in the calculus of variations*, Trans. Amer. Math. Soc. **289** (1985), 73–98.

7. E. De Giorgi, *Sulla differenziabilità e l'analiticità delle estremali degli integrali multipli regolari*, Mem. Accad. Sci. Torino (3) **3** (1957), 25–43.

8. E. Di Benedetto and N. Trudinger, *Harnack inequalities for quasi-minima of variational integrals*, Ann. Inst. H. Poincaré Anal. Non Linéaire **1** (1984), 295–308.

9. F. Duzaar and M. Fuchs, *Optimal regularity theorems for variational problems with obstacles*, preprint, 1986.

10. L. C. Evans, *Quasi convexity and partial regularity in the Calculus of Variations*, preprint, 1984.

11. L. C. Evans and R. F. Gariepy, *Blow-up, compactness and partial regularity in the calculus of variations*, preprint, 1985.

12. J. Frehse, *A note on the Hölder continuity of solutions of variational problems*, Abh. Math. Sem. Univ. Hamburg **43** (1975), 59–63.

13. N. Fusco and J. Hutchinson, $C^{1,\alpha}$ *partial regularity of functions minimizing quasiconvex integrals*, Manuscripta Math. **54** (1985), 121–143.

14. ——, *Partial regularity for minimizers of certain functionals having nonquadratic growth*, preprint, 1986.

15. F. W. Gehring, *The L^p integrability of partial derivatives of a quasi conformal mapping*, Acta Math. **130** (1973), 265–277.

16. M. Giaquinta, *A counterexample to the boundary regularity of solutions to elliptic quasilinear systems*, Manuscripta Math. **14** (1978), 217–220.

17. ——, *Multiple integrals in the calculus of variations and nonlinear elliptic systems*, Ann. of Math. Studies, no. 105, Princeton Univ. Press, Princeton, N.J., 1983.

18. M. Giaquinta and E. Giusti, *On the regularity of minima of variational integrals*, Acta Math. **148** (1982), 31–46.

19. ——, *Differentiability of minima of nondifferentiable functionals*, Invent. Math. **72** (1983), 285–298.

20. ——, *Sharp estimates for the derivatives of local minima of variational integrals*, Boll. Un. Mat. Ital. A (6) **3** (1984), 239–248.

21. ——, *The singular set of the minima of certain quadratic functionals*, Ann. Scuola Norm. Sup. Pisa **11** (1984), 45–55.

22. ——, *Quasi-minima*, Ann. Inst. H. Poincaré Anal. Non Linéaire **1** (1984), 79–107.

23. M. Giaquinta and P. A. Ivert, *Partial regularity for minima of variational integrals*, preprint, FIM, ETH Zürich, 1984.

24. M. Giaquinta and G. Modica, *Regularity results for some classes of higher order nonlinear elliptic systems*, J. Reine Angew. Math. **311/312** (1979), 145–169.

25. ——, *Partial regularity of minimizers of quasiconvex integrals*, Ann. Inst. H. Poincaré Anal. Non Linéaire **3** (1986), 185–208.

26. ——, *Remarks on the regularity of the minimizers of certain degenerate functionals*, Manuscripta Math. (to appear).

27. M. Giaquinta and J. Souček, *Caccioppoli's inequality and Legendre-Hadamard condition*, Math. Ann. **270** (1985), 105–107.

28. ——, *Harmonic maps into a hemisphere*, Ann. Scuola Norm. Sup. Pisa **12** (1985), 81–90.

29. D. Gilbarg and N. S. Trudinger, *Elliptic partial differential equations of second order*, Springer-Verlag, Heidelberg, 1977.

30. E. Giusti, *Regolarità parziale delle soluzioni di sistemi ellittici quasi lineari di ordine arbitrario*, Ann. Scuola Norm. Sup. Pisa **23** (1969), 115–141.

31. E. Giusti and M. Miranda, *Sulla regolarità delle soluzioni deboli di una classe di sistemi ellittici quasilineari*, Arch. Rational Mech. Anal. **31** (1968), 173–184.

32. R. Hardt, D. Kinderlehrer, and F. H. Lin, *Existence and partial regularity of static liquid crystal configurations*, Comm. Math. Phys. (to appear).

33. S. Hildebrandt, *Nonlinear elliptic systems and harmonic mappings*, Proc. Beijing Sympos. Diff. Geo. and Diff. Eq., vol. 1, Science Press, Beijing, 1982, pp. 481–515.

34. S. Hildebrandt, H. Kaul, and K. O. Widman, *An existence theorem for harmonic mappings of Riemannian manifolds*, Acta Math. **138** (1977), 1–16.

35. W. Jäger and H. Kaul, *Rotationally symmetric harmonic maps from a ball into a sphere and the regularity problem for weak solutions of elliptic systems*, J. Reine Angew. Math. **343** (1983), 146–161.

36. J. Jost and M. Meier, *Boundary regularity for minima of certain quadratic functionals*, Math. Ann. **262** (1983), 549–561.

37. N. V. Krylov and M. V. Safanov, *Certain properties of solutions of parabolic equations with measurable coefficients*, Izv. Akad. Nauk SSSR **40** (1980), 161–175; English transl. in Math. USSR Izv. **16** (1981).

38. A. I. Koshelev and S. I. Chelkak, *Regularity of solutions of quasilinear elliptic systems*, Teubner-Texte zur Math. **77** (1985).

39. O. A. Ladyzhenskaya and N. N. Ural'tseva, *Linear and quasilinear elliptic equations*, Izdat. "Nauka", Moscow, 1964; English transl., Academic Press, New York, 1968.

40. J. L. Lewis, *Smoothness of certain degenerate elliptic equations*, Proc. Amer. Math. Soc. **80** (1980), 259–265.

41. P. Marcellini, *Approximation of quasiconvex functionals and lower semicontinuity of multiple integrals*, Manuscripta Math. **51** (1985), 1–28.

42. P. Marcellini and C. Sbordone, *On the existence of minima of multiple integrals of the calculus of variations*, J. Math. Pures Appl. **62** (1983), 1–9.

43. Min-Chun Hong, *Existence and partial regularity in the calculus of variations*, preprint, 1986.

44. C. B. Morrey, Jr., *Existence and differentiability theorems for the solutions of variational problems for multiple integrals*, Bull. Amer. Math. Soc. **46** (1940), 439–458.

45. ——, *Quasi-convexity and the lower semicontinuity of multiple integrals*, Pacific J. Math. **2** (1952), 25–53.

46. ——, *Multiple integrals in the calculus of variations*, Springer-Verlag, New York and Heidelberg, 1966.

47. ——, *Partial regularity results for nonlinear elliptic systems*, J. Math. Mech. **17** (1968), 649–670.

48. J. Moser, *On Harnack's theorem for elliptic differential equations*, Comm. Pure Appl. Math. **14** (1961), 577–591.

49. ____, *Minimal solutions of variational problems on a torus*, Ann. Inst. H. Poincaré Anal. Non Linéaire **3** (1986), 229–272.

50. K. Phillips, *A minimization problem in the presence of a free boundary*, Indiana Math. J. **32** (1983), 1–17.

51. V. Scheffer, *Regularity and irregularity of solutions to nonlinear second order elliptic systems of partial differential equations and inequalities*, Ph.D. Dissertation, Princeton Univ., Princeton, New Jersey, 1974.

52. R. Schoen and K. Uhlenbeck, *A regularity theory for harmonic maps*, J. Differential Geom. **17** (1982), 307–335.

53. ____, *Boundary regularity and the Dirichlet problem for harmonic maps*, J. Differential Geom. **18** (1983), 253–268.

54. ____, *Regularity of minimizing harmonic maps into the sphere*, Invent. Math. **78** (1984), 89–100.

55. J. Souček, *Singular solutions to linear elliptic systems*, Comment. Math. Univ. Carolin. **25** (1984), 273–281.

56. L. Tonelli, *Fondamenti di calcolo delle varizioni*, 2 vol., Zanichelli, Bologna, 1921–1923.

57. K. Uhlenbeck, *Regularity for a class of nonlinear elliptic systems*, Acta Math. **38** (1977), 219–240.

58. W. P. Ziemer, *Boundary regularity for quasiminima*, Arch. Rational Mech. Anal. **92** (1986), 371–382.

UNIVERSITÀ DI FIRENZE, ITALIA

Proceedings of the International Congress of Mathematicians
Berkeley, California, USA, 1986

Estimates for a Number of Negative Eigenvalues
of the Schrödinger Operator with Singular Potentials

VICTOR IVRII

There are presented estimates from above and from below for a maximal dimension $N \leq \infty$ of a linear subspace $\mathcal{L} \subset C_0^\infty(X)$ on which a quadratic form

$$Q(u) = \int [g^{jk}(D_j - V_j)u \cdot \overline{(D_k - V_k)u} + V|u|^2]\,dx$$

is negative definite. Here X is a domain in $\mathbf{R}^d$, $d \geq 3$, components of the metric tensor $g^{jk} = g^{kj}$ and potentials V_j, V are real-valued, and such that Q is correctly defined (we use Einstein's summation rule). If Q generates a selfadjoint operator $A = (D_j - V_j)g^{jk}(D_k - V_k) + V$ in $L^2(X)$ (with the Dirichlet boundary condition), then N is a dimension of its invariant negative subspace. The emphasis of this work is that these estimates are highly uniform and, in the case when Q depends on parameters, these estimates imply asymptotics of N. In §§1 and 2 we use Weyl's rule for calculation of the principal parts of the estimates and asymptotics, and these parts do not depend on the magnetic potentials V_j, the presence of which can only deteriorate the remainder estimates. But the situation changes in §§3 and 4 when we consider the case of the intensive magnetic field and find that Weyl's rule is no longer applicable and must be replaced by Tamura's formula.

1. Let us assume that

$$c^{-1} \leq |\xi|^{-2}g^{jk}\xi_j\xi_k \leq c \qquad \forall x \in X, \xi \in \mathbf{R}^d\backslash 0, \tag{H$_1$}$$

V_j, $g^{jk}V_jV_k + V \in L^1_{\text{loc}}(X)$, and there are given functions ς, γ on X such that

$$\varsigma \geq 0, \qquad \gamma \geq 0, \qquad |\gamma(x) - \gamma(y)| \leq |x - y|, \tag{H$_2$}$$

and for every $y \in X' = \{X, \varsigma\gamma > 1\}$ in $X \cap B(y, \gamma(y))$ the following conditions are fulfilled:

$$c^{-1} \leq \varsigma(x)/\varsigma(y) \leq c, \qquad |D_j\varsigma| \leq c\varsigma/\gamma, \tag{H$_3$}$$

$$|D^\alpha g^{jk}| \leq c\gamma^{-|\alpha|}, \tag{H$_4$1}$$

$$|D^\alpha(D_jV_k - D_kV_j)| \leq c\varsigma\gamma^{-1-|\alpha|}, \tag{H$_4$2}$$

$$|D^\alpha V| \leq c\varsigma^2\gamma^{-|\alpha|} \qquad \forall\alpha\colon |\alpha| \leq K, \tag{H$_4$3}$$

$$\partial X \cap B(y, \gamma(y)) = \{x_k = z(x_{\hat{k}})\} \cap B(y, \gamma(y)) \tag{H$_5$}$$

with

$$|D^\alpha z| \leq c\gamma^{1-|\alpha|} \qquad \forall \alpha : |\alpha| \leq K$$

where $B(y, \gamma)$ is an open ball with center y and radius γ, $x_{\hat{k}} = (x_1, \ldots, x_{k-1}, x_{k+1}, \ldots, x_d)$, $k = 1, \ldots, d$, and $K = K(d) < \infty$.

Moreover, let us assume that

$$Q(u) \geq c^{-1} \int \left(|\nabla u|^2 - W|u|^2 \right) dx \qquad \forall u \in C_0^\infty(X'') \tag{H$_6$}$$

where $W \in L^1_{\mathrm{loc}}(X'')$, $W \geq 0$, and $X'' = \{X, \varsigma\gamma < 2\} \cup \{X, V \geq c^{-1}\varsigma^2\}$.

Our first principal assertion is

THEOREM 1. *If conditions* (H$_1$)$-$(H$_6$) *are fulfilled, then the following estimates hold:*

$$M - CR_1 \leq N \leq M + C(R_1 + R_2) \tag{1}$$

where

$$M = (2\pi)^{-d}\omega_d \int_{X'} V_-^{d/2} \sqrt{g}\, dx, \tag{2}$$

$g = \det(g^{jk})^{-1}$, $\sqrt{g}\, dx$ *is the Riemannian density on* X, ω_d *is the volume of the unit ball in* $\mathbf{R}^d$, $V_\pm = \max(\pm V, 0)$, $C = C(d, c)$, *and*

$$R_1 = \int_{\{X', V \leq 2c^{-1}\varsigma^2\}} \varsigma^{d-1}\gamma^{-1}\, dx, \qquad R_2 = \int_{X''} W^{d/2}\, dx.$$

Moreover, if condition (H$_6$) *is valid only for* $X'' = \{X, \varsigma\gamma < 2\}$, *then*

$$M - CR_1 \leq N \leq M + C(R_1 + R_2 + R_3) \tag{1$'$}$$

where

$$R_3 = \int_{X'} \varsigma^{d-s}\gamma^{-s}\, dx,$$

s *is arbitrary and* $K = K(d, s)$, $C = C(d, c, s)$ *here.*

In order to derive more refined estimates, one has need of hypotheses of a global nature. Let us introduce on $T^*\overline{X}'$ the Hamiltonian

$$H(x, \xi) = \varsigma^{-1}\gamma \left[g^{jk}(\xi_j - V_j)(\xi_k - V_k) + V \right],$$

the set $\Sigma = \{T^*\overline{X}', H = 0\}$, and the natural density $dx\, d\xi : dH$ on Σ. One can observe that the $(2d-1)$-dimensional measure of the set $\{H = dH = 0\}$ equals zero.

A Hamiltonian curve is a curve $(x(t), \xi(t)) \subset \Sigma \cap T^*X'$ along which $dx/dt = \partial H/\partial\xi$, $d\xi/dt = -\partial H/\partial x$. A Hamiltonian billiard is a curve $(x(t), \xi(t)) \subset \Sigma$ composed of Hamiltonian segments (i.e., segments whose interiors are Hamiltonian curves) such that if (x, ξ) and (x', ξ') are an endpoint and an origin of two successive segments then $x = x' \in \partial X$ and $\xi - \xi' = kn(x)$ where $n(x)$ is a unit normal to ∂X and $k \neq 0$, and such that each finite interval contains only a finite number of segments.

Let Λ_0 be a subset of Σ, $T > 0$, and $\delta > 0$ such that

$$(x, \xi) \in \Lambda_0 \Rightarrow B(x, \delta\gamma(x)) \subset X', \tag{H$_7$}$$

and through every point $(\bar{x}, \bar{\xi}) \in \Lambda_0$ there passes a Hamiltonian billiard $(x(t), \xi(t))$ with either $t \in [0, T]$ or $t \in [-T, 0]$ and with $x(0) = \bar{x}$, $\xi(0) = \bar{\xi}$, along which the following estimates hold:

$$\gamma^{-1}|dx/dt| + \varsigma^{-1}|d\xi/dt| \geq \delta,$$

$$\operatorname{dist}(x, \partial X) \leq \delta\gamma(x) \Rightarrow |d\operatorname{dist}(x, \partial X)/dt| \geq \delta\gamma(x),$$

$$\gamma^{-1}|x - \bar{x}| + \varsigma^{-1}|\xi - \bar{\xi}| \geq \delta \min(|t|, 1).$$

That is, all these billiards are uniformly nonstationary, transversal to ∂X, and nonclosed.

Our second principal assertion is

THEOREM 2. *If conditions* (H_1)–(H_7) *are fulfilled, then*

$$M + M' - CR_4 - C'R_5 \leq N \leq M + M' + C(R_2 + R_4) + C'R_5 \qquad (3)$$

where

$$M' = -\frac{1}{4}(2\pi)^{1-d}\omega_{d-1} \int_{\partial X \cap \overline{X}'} V_-^{(d-1)/2} dS, \qquad (4)$$

dS *is a Riemannian density on* $\partial X \cap \overline{X}'$ *induced by the Riemannian metrics* g^{jk} *on* X,

$$R_4 = \int_{\Sigma \setminus \Lambda_0} dx\, d\xi : dH + \frac{1}{T}R_1, \qquad R_5 = \int_{\{X', V \leq 2c^{-1}\varsigma^2\}} \varsigma^{d-2}\gamma^{-2}\, dx,$$

$C' = C'(d, c, T, \delta)$. *Moreover, if condition* (H_6) *is valid only for* $X'' = \{X, \varsigma\gamma < 2\}$, *then one can recover these estimates by adding* CR_3 *to the right-hand expression.*

2. Let us apply Theorems 1 and 2 to derive the eigenvalue asymptotics. In this section we replace $(H_4)_2$ by the stronger condition

$$|D^\alpha V_j| \leq c_\varsigma\gamma^{-|\alpha|} \qquad \forall \alpha : |\alpha| \leq K \qquad (H_4)_2^*$$

and here $(H_4) = (H_4)_1 + (H_4)_2^* + (H_4)_3$. Let us replace V_j by $h^{-1}V_j$ and V by $h^{-2}(V - \lambda)$, where $h \in (0, 1]$ is a quasiclassical parameter and λ is a spectral parameter. Taking $\varsigma_{\lambda, h} = h^{-1}(\sqrt{|\lambda|} + \varsigma)$ we obtain

THEOREM 3. *If conditions* (H_1)–(H_5) *are fulfilled for every* $y \in \overline{X}^+ = \{\overline{X}, \gamma > 0\}$, *if for almost every* y $\gamma(y) = 0 \Rightarrow V(y) \geq 0$, $|V_j(y)| \leq c\sqrt{V(y)}$, *and if* ς^d, $\varsigma^{d-1}\gamma^{-1} \in L^1(\overline{X}^+)$, *then*

$$N(\lambda, h) = M(\lambda)h^{-d} + O(h^{1-d}) \quad as\ \lambda \leq 0,\ h \to +0 \qquad (5)$$

uniformly with respect to λ *where*

$$M(\lambda) = (2\pi)^{-d}\omega_d \int (\lambda - V)_+^{d/2} \sqrt{g}\, dx. \qquad (6)$$

Moreover, if there exists a set of measure zero $\Lambda \subset \Sigma = \{T^*\overline{X}^+, H = 0\}$ *such that through every point of* $\Sigma \setminus \Lambda$ *there passes a Hamiltonian infinitely long nonperiodic billiard, then*

$$N(\lambda, h) = M(\lambda)h^{-d} + (M' + o(1))h^{1-d} \quad as\ \lambda \to -0,\ h \to +0. \qquad (7)$$

Asymptotics (5), (7) with $\lambda = 0$ combined with the Birman-Schwinger principle imply the eigenvalue asymptotics for certain spectral problems [1, 5]

In what follows, either spectral parameter λ increases to $+\infty$ or to the lower bound of the essential spectrum, or λ decreases to $\inf V \geq -\infty$. We obtain the following assertions:

THEOREM 4. *If conditions* (H_1)–(H_5) *are fulfilled for every* $y \in \overline{X}^+$ *and if* ς^d, $\varsigma^{d-1}\gamma^{-1}$, γ^{-1}, $1 \in L^1(X)$, *then*

$$N(\lambda, h) = M(\lambda)h^{-d} + O(\lambda^{(d-1)/2}h^{1-d}) \quad as \ \lambda \geq 1, \ h \in (0, 1], \ \lambda/h \to +\infty, \quad (8)$$

where $M(\lambda) = \mathfrak{x}\lambda^{d/2} + O(\lambda^{(d-1)/2})$ *as* $\lambda \to +\infty$ *and* $\mathfrak{x} = (2\pi)^{-d}\omega_d \operatorname{vol} X$. *Moreover, if the set of all the points of* $\mathbf{S}^*\overline{X}^+ = \{T^*\overline{X}^+, H = g^{jk}\xi_j\xi_k - 1 = 0\}$ *periodic with respect to the broken geodesic flow has measure zero, then*

$$N(\lambda, h) = M(\lambda)h^{-d} + (\mathfrak{x}' + o(1))\lambda^{(d-1)/2}h^{1-d} \quad as \ \lambda \to +\infty, \ h \in (0, 1] \quad (9)$$

where $\mathfrak{x}' = -\frac{1}{4}(2\pi)^{1-d}\omega_{d-1}\operatorname{vol}'(\partial X \cap \overline{X}^+)$.

For example, (8) (and (9) under condition to geodesic flow) holds if $X = B(0, 1)$, $g^{jk} = g^{jk}(\theta)$, $V_j = v_j(\theta)r^m$, $V = v(\theta)r^{2m}$, where (r, θ) are spherical coordinates, g^{jk}, v_j, $v \in C^K$, and g^{jk} satisfy (H_1) and $m > -1$. Moreover, Theorems 1 and 2 show that $m = -1$ is also admissible provided $h = 1$ and $\min v(\theta) > -\beta$, $\beta = \beta(d, g^{jk}) > 0$. We refer to [6, 9] for certain examples of a geometric nature such as polyhedral domains, domains with conical singularities of the boundaries, domains with the cusps, etc.

Let us consider strongly singular coercive potentials.

THEOREM 5. *If conditions* (H_1)–(H_5) *and condition*

$$V \geq c^{-1}\varsigma^2 \tag{H_8}$$

are fulfilled for every $y \in \overline{X}^+$ *and if* $\gamma^{-1}, 1 \in L^1(X)$, *then asymptotics* (8) *holds and* $M(\lambda) = \mathfrak{x}\lambda^{d/2}(1 + o(1))$ *as* $\lambda \to +\infty$. *Moreover, if the set of all the points of* $\mathbf{S}^*\overline{X}^+$ *periodic with respect to the broken geodesic flow has measure zero, then asymptotics* (9) *holds.*

For example, (8) holds if $X = B(0, 1)$, $g^{jk} = g^{jk}(\theta)$, and either $V_j = v_j(\theta)r^m$, $V = v(\theta)r^{2m}$ or $V_j = v_j(\theta)\exp(r^{-p})$, $V = v(\theta)\exp(2r^{-p})$, where g^{jk}, v_j, $v \in C^K$, g^{jk} satisfy (H_1), $p < d - 1$, and $\min v > 0$. We refer to [2, 5] for more refined examples.

Let us consider the essentially nonbounded domains when the spectrum is discrete only because of the presence of coercive potential V.

THEOREM 6. *If* X *is a nonbounded connected domain and if conditions* (H_1)–(H_5), (H_8) *are fulfilled with* $\gamma = c^{-1}(|x| + 1)$, $\rho = (|x| + 1)^m$, $m > 0$, *then*

$$N(\lambda, h) = M(\lambda)h^{-d} + O(\lambda^{l(d-1)}h^{1-d}) \quad as \ \lambda \geq 1, \ h \in (0, 1],$$

$$and \ \lambda/h \to +\infty \quad (10)$$

where $M(\lambda) \sim \lambda^{ld}$ as $\lambda \to +\infty$, $l = (1 + 1/m)/2$. Moreover, if $X \cap \{|x| \geq c\} = X^0 \cap \{|x| \geq c\}$ where X^0 is a conical domain with $\partial X^0 \backslash 0 \in C^K$, if

$$D^\alpha(g^{jk} - g^{jk0}) = o(|x|^{-|\alpha|}), \qquad D^\alpha(V_j - V_j^0) = o(|x|^{m-|\alpha|}),$$
$$D^\alpha(V - V^0) = o(|x|^{2m-|\alpha|}) \quad \forall \alpha: \; |\alpha| \leq 1 \qquad (11)$$

as $|x| \to \infty$ where g^{jk0}, V_j^0, V^0 are C^K functions positively homogeneous of degrees $0, m, 2m$ respectively, and if the set of all the points of $\Sigma^0 = \{(\overline{X^0}\backslash 0) \times \mathbf{R}^d, \; H^0 = 0\}$ periodic with respect to the broken Hamiltonian flow generated by the Hamiltonian

$$H^0 = g^{jk0}(\xi_j - V_j^0)(\xi_k - V_k^0) + V^0 - 1 \qquad (12)$$

has measure zero, then

$$N(\lambda, h) = M(\lambda)h^{-d} + (\mathfrak{x}' + o(1))\lambda^{l(d-1)}h^{1-d} \quad \text{as } \lambda \to +\infty, \; h \in (0,1] \qquad (13)$$

and $M(\lambda) = \mathfrak{x}\lambda^{ld}(1 + o(1))$ with positive $\mathfrak{x}$.

We refer to [2, 5] for more refined examples of positively homogeneous potentials degenerating along certain directions, probably perturbed by indefinite potentials of a more moderate growth at infinity, and by examples of potentials of an exponential and logarithmic growth at infinity.

Let us now consider the quasiclassical asymptotics of the sets of the lowest eigenvalues.

THEOREM 7. If $0 \in \overline{X}$ and if conditions (H_1)–(H_5), (H_8) are fulfilled with $\gamma = c^{-1}|x|$, $\varsigma = |x|^m$, $m > 0$, then asymptotics (10) holds as $\lambda \to +0$ and $h = o(\lambda^l)$. Moreover, if $X \cap \{|x| \leq c^{-1}\} = X^0 \cap \{|x| \leq c^{-1}\}$ where X^0 is a conical domain with $\partial X^0 \backslash 0 \in C^K$, if stabilization condition (11) holds as $|x| \to 0$, and if the set of all the points of Σ^0 periodic with respect to the broken Hamiltonian flow generated by the Hamiltonian (12) has measure zero, then asymptotics (13) holds as $\lambda \to +0$, $h = o(\lambda^l)$, and $M(\lambda) = \mathfrak{x}\lambda^{ld}(1 + o(1))$, $\mathfrak{x} > 0$.

Let us consider now asymptotics of the negative spectrum.

THEOREM 8. (i) If X is a nonbounded connected domain and if conditions (H_1)–(H_5) are fulfilled with $\gamma = c^{-1}(|x| + 1)$, $\varsigma = (|x| + 1)^m$, $0 > m > -1$, then asymptotics (10) holds as $\lambda \in (-1, 0)$, $h \in (0, 1)$, $\lambda h \to -0$ with $M(\lambda) = O(|\lambda|^{ld})$ as $\lambda \to -0$; moreover, $M(\lambda) \sim |\lambda|^{ld}$ provided $V \leq -c^{-1}\varsigma^2$ along a certain ray in X.

(ii) Moreover, if $X \cap \{|x| \geq c\} = X^0 \cap \{|x| \geq c\}$ where X^0 is a conical domain with $\partial X^0 \backslash 0 \in C^K$, if stabilization condition (11) holds as $|x| \to \infty$, and if the set of all the points of Σ^0 periodic with respect to the broken Hamiltonian flow generated by the Hamiltonian

$$H^0 = |x|^{1-m}[g^{jk0}(\xi_j - V_j^0)(\xi_k - V_k^0) + V^0 + 1] \qquad (14)$$

has measure zero, then asymptotics (13) holds as $\lambda \to -0$, $h \in (0, 1]$, and $M(\lambda) = \mathfrak{x}|\lambda|^{ld}(1 + o(1))$.

(iii) *If conditions* (H_1)–(H_5) *are fulfilled with* $\gamma = c^{-1}(|x| + 1)$, $\varsigma = (|x| + 1)^{-1} \ln^p(|x| + 2)$, $p > 0$, *then*

$$N(\lambda, h) = M(\lambda)h^{-d} + O(|\ln|\lambda||^{p(d-1)+1}h^{1-d})$$

$$\text{as } \lambda \in (-1, 0), \ h \in (0, 1], \ \lambda h \to -0 \tag{15}$$

where $M(\lambda) = O(|\ln|\lambda||^{pd+1})$; *moreover*, $M(\lambda) \sim |\ln|\lambda||^{pd+1}$ *provided* $V \leq -c^{-1}\varsigma^2$ *along a certain ray in* X.

(iv) *Moreover, if* $X \cap \{|x| \geq c\} = X^0 \cap \{|x| \geq c\}$ *where* X^0 *is a conical domain with* $\partial X^0 \backslash 0 \in C^K$, *if* g^{jk}, $V_j \ln^{-p}|x|$, $V \ln^{-2p}|x|$ *satisfy stabilization condition* (11) *with* g^{jk0}, V_j^0, $V^0 \in C^K$ *positively homogeneous functions of degrees* 0, -1, -2 *respectively, and if the set of all the points of* Σ^0 *periodic with respect to the broken Hamiltonian flow generated by the Hamiltonian*

$$H^0 = |x|^2[g^{jk0}(\xi_j - V_j^0)(\xi_k - V_k^0) + V^0] \tag{16}$$

has measure zero, then

$$N(\lambda, h) = M(\lambda)h^{-d} + (\mathfrak{x}' + o(1))|\ln|\lambda||^{p(d-1)+1}h^{1-d} \tag{17}$$

as $\lambda \to -0$, $h \in (0, 1]$, *and either* $p > 0$ *or* $h \to 0$, *and*

$$M(\lambda) = \mathfrak{x}|\ln|\lambda||^{pd+1}(1 + o(1)).$$

We refer to [3, 5] for more refined examples of positively homogeneous potentials singular along certain directions and of potentials which decay as $\ln^{-1}|x|$ at infinity. Let us consider again the quasiclassical asymptotics of the sets of the lowest eigenvalues.

THEOREM 9. *If* $0 \in \overline{X}$ *and if conditions* (H_1)– (H_5) *are fulfilled with* $\gamma = c^{-1}|x|$, $\varsigma = |x|^m$, $0 > m > -1$, *then asymptotics* (10) *holds as* $\lambda \to -\infty$, $h = o(|\lambda|^l)$, *and* $M(\lambda) = O(|\lambda|^{ld})$; *moreover*, $M(\lambda) \sim |\lambda|^{ld}$ *provided* $V \leq -c^{-1}\varsigma^2$ *along a certain ray in* X. *Moreover, if* $X \cap \{|x| \leq c^{-1}\} = X^0 \cap \{|x| \leq c^{-1}\}$ *where* X^0 *is a conical domain with* $\partial X^0 \backslash 0 \in C^K$, *if stabilization condition* (11) *holds as* $|x| \to 0$, *and if the set of all the points of* Σ^0 *periodic with respect to the broken Hamiltonian flow generated by the Hamiltonian* (14) *has measure zero, then asymptotics* (13) *holds as* $\lambda \to -\infty$, $h = o(|\lambda|^l)$.

We refer to [3, 5] for applications of Theorem 1 to calculation of the density of states and its asymptotics.

3. Let us consider now the case of the intensive magnetic field. We do not assume in this and in the following sections that condition $(H_4)_2$ is fulfilled and now $(H_4) = (H_4)_1 + (H_4)_3$. To characterize the magnetic field, let us introduce its vector intensity $B^j = i\mathcal{E}^{jkl}(D_kV_l - D_lV_k)/2$ and its scalar intensity $B = (g_{jk}B^jB^k)^{1/2}$ where $d = 3$, $\mathcal{E}^{jkl}$ is the skew-symmetric pseudotensor with $\mathcal{E}^{123} = 1/\sqrt{g}$, and $(g_{jk}) = (g^{jk})^{-1}$; these intensities are invariant with respect to the gradient transform $V_j \mapsto V_j - iD_jf$ with a real-valued function f. Our third principal assertion is the following theorem.

THEOREM 10. *Let $d = 3$, let conditions (H_1)–(H_6) be fulfilled with $X'' = \{X, \varsigma\gamma < 2\}$, and for every $y \in X'$ in $X \cap B(y, \gamma(y))$, let*

$$|D^\alpha B^j| \leq c\beta\gamma^{-|\alpha|} \qquad \forall\alpha : |\alpha| \leq K \tag{H_4}'_2$$

where functions β, γ satisfy (H_2), (H_3) and $\beta \geq \varsigma/\gamma$. Moreover, let us assume that for every $y \in X' \cap \{\beta > c\varsigma/\gamma\}$

$$B(y, \gamma(y)) \subset X \tag{H_5}'$$

and in $B(y, \gamma(y))$

$$B \geq c^{-1}\beta, \qquad V \leq -c^{-1}\varsigma^2. \tag{H_8}'$$

(i) *Finally, let us assume that for every $y \in X' \cap \{\beta\gamma^2 > c(\varsigma\gamma)^{4/3-\sigma}\} \cap \{V + B \leq c^{-1}\varsigma^2\}$ at every point of $B(y, \gamma(y))$*

$$|\nabla(B/V)| \geq c^{-1}\beta\varsigma^{-2}\gamma^{-1} \quad or \quad \|(\text{Hess } (B/V))^{-1}\| \leq c\beta^{-1}\varsigma^2\gamma^2. \tag{H_9}$$

Then the following estimates hold:

$$M^* - CR_6 \leq N \leq M^* + C(R_2 + R_6 + R_7) \tag{18}$$

where

$$M^* = \frac{1}{2\pi^2} \sum_{n=0}^{\infty} \int_{X'} (V + (2n+1)B)_-^{1/2} B\sqrt{g}\, dx,$$

$$R_6 = \int_{\{X', V+B \leq 2c^{-1}\varsigma^2\}} \varsigma^2\gamma^{-1}\, dx,$$

$$R_7 = \int_{\{X', V+B \geq c^{-1}\varsigma^2\}} \beta^{3/2-s}\gamma^{-2s}\, dx. \tag{19}$$

Here and in what follows $K = K(s, \sigma)$, $C = C(s, \sigma, c)$, and s, σ are arbitrary positive numbers.

(ii) *On the other hand, let us assume instead of (H_9) that for every $y \in X' \cap \{\beta\gamma^2 > c(\varsigma\gamma)^{4/3-\sigma}\} \cap \{V+B \leq c^{-1}\varsigma^2\}$ in $B(y, \gamma(y))$ the following conditions are fulfilled :*

$$V + 3B \geq c^{-1}\varsigma^2, \tag{H_8}''$$

$$|D^\alpha(V + B)| \leq c\beta'^2\gamma^{-|\alpha|} \qquad \forall\alpha : |\alpha| \leq K, \tag{H_4}''_2$$

$$either \ |V + B| \geq c^{-1}\beta'^2 \ or \ |\nabla(B/V)| \geq c^{-1}\beta'^2\varsigma^{-2}\gamma^{-1}$$
$$or \ \|(\text{Hess}(B/V))^{-1}\| \leq c\beta'^{-2}\varsigma^2\gamma^2, \tag{H_9}'$$

where β', γ satisfy (H_2), (H_3) and $\varsigma\gamma \geq \beta'\gamma \geq 1$ in X'. Then

$$M^* - CR_8 \leq N \leq M^* + C(R_2 + R_7 + R_8 + R_9) \tag{20}$$

where

$$R_8 = \int_{\{X', V+B \leq 2c^{-1}\beta'^2\}} \varsigma^2\gamma^{-1}\, dx,$$

$$R_9 = \int_{\{X', V+B \geq c^{-1}\beta'^2\}} \varsigma^2\beta'^{-s}\gamma^{-1-s}\, dx.$$

COROLLARY. *If the conditions of Theorem 10 are fulfilled and if $X \cap \{\beta\gamma^2 > c(\varsigma\gamma)^{2-\sigma}\} = \varnothing$, then one can replace M^* by M in* (18); *i.e., in this case Weyl's rule is applicable.*

4. Let us apply Theorem 10 to derive the eigenvalue asymptotics in the case where there are no electric fields. We replace V by $-h^{-2}\lambda$ and V_j by $h^{-1}V_j$ and then Theorem 10(i) implies the following assertions.

THEOREM 11. (i) *Let $X = \mathbf{R}^3$ and let conditions* (H_1), $(H_4)_1$, $(H_4)'_2$, $(H_8)'$, (H_9) *be fulfilled with $\gamma = c^{-1}(|x|+1)$, $\beta = (|x|+1)^m$, $m > 0$, $v = -h^{-2}\lambda$, $\varsigma = h^{-1}\sqrt{\lambda}$. Then*

$$N(\lambda, h) = M^*(\lambda/h)h^{-3/2} + O(\lambda^{1+2/m}h^{-2-2/m}) \qquad (21)$$

as $\lambda/h \geq c^{-1}$, $h \in (0, 1]$, and either $\lambda \to +\infty$ or $h \to +0$, where

$$M^*(\mu) = \frac{1}{2\pi^2} \sum_{n=0}^{\infty} \int (\mu - (2n+1)B)_+^{1/2} B\sqrt{g}\, dx \qquad (22)$$

and $M^*(\mu) \sim \mu^{3/2+3/m}$ as $\mu \to +\infty$.

(ii) *Moreover, if (H_9) is fulfilled only at $\{|x| \geq c\}$ then asymptotics* (21) *holds provided $\lambda^{2-\sigma} \geq h^{1-2\sigma}$, and if $(H_8)'$, (H_9) are fulfilled only at $\{|x| \geq c\}$ then this asymptotics holds provided $\lambda \to +\infty$.*

THEOREM 12. *Let $0 \in X$ and let conditions* (H_1), $(H_4)_1$, (H_5), $(H_4)'_2$, $(H_8)'$, (H_9) *be fulfilled with $\gamma = c^{-1}|x|$, $\beta = |x|^m$, $m > 0$, $V = -h^{-2}\lambda$, $\varsigma = h^{-1}\sqrt{\lambda}$. Then asymptotics* (21) *holds as $h \to +0$, $\lambda/h \to +0$, and $h = o(\lambda^{(m+2)/(2m+2)})$ with $M^*(\mu) \sim \mu^{3/2+3/m}$ as $\mu \to +0$.*

One can easily prove similar assertions in the case of positively homogeneous g^{jk} and V_j with B degenerating along certain directions and in the case of an exponential growth of B at infinity.

On the other hand, Theorem 10(ii) implies

THEOREM 13. *Let $X = \mathbf{R}^3$ and let conditions* (H_1), $(H_4)_1$, $(H_4)'_2$ *be fulfilled with $\beta = 1$, $\gamma = c^{-1}(|x|+1)$. Moreover, let us assume that conditions $(H_4)''_2$, $(H_9)'$ are fulfilled with $\beta' = (|x|+1)^m$, $0 > m > -1$, and $V = -1$, $\varsigma = 1$ at $\{|x| \geq c\}$. Then*

$$N(1 - \eta, 1) = E(\eta) + O(\eta^{1/m}) \quad \text{as } \eta \to +0 \qquad (23)$$

where

$$E(\eta) = \frac{1}{2\pi^2} \int (1 - \eta - B)_+^{1/2} B\sqrt{g}\, dx \qquad (24)$$

and $E(\eta) = O(\eta^{1/2+3/2m})$. Moreover, $E(\eta) \sim \eta^{1/2+3/2m}$ provided $B \leq 1 - c^{-1}\beta'^2$ along a certain ray. On the other hand, if conditions $(H_4)''_2$, $(H_9)'$ are fulfilled with $\beta' = (|x|+1)^{-1} \ln^p(|x|+2)$, $p > 0$, and $V = -1$, $\varsigma = 1$ at $\{|x| \geq c\}$, then

$$N(1 - \eta, 1) = E(\eta) + O(\eta^{-1}|\ln \eta|^{2p}) \quad \text{as } \eta \to +0 \qquad (25)$$

and $E(\eta) = O(\eta^{-1}|\ln \eta|^{3p})$; moreover, $E(\eta) \sim \eta^{-1}|\ln \eta|^{3p}$ provided $B \leq 1 - c^{-1}\beta'^2$ along a certain ray.

5. The demonstrations of our principal theorems, 1, 2 [**4**] and 10, are based on the dilatation method applied to precise quasiclassical asymptotics of the spectral function for the Schrödinger operator with a small parameter h and with regular nondegenerated potentials; I replace h by $(\varsigma\gamma)^{-1}$ and satisfy nondegeneracy conditions by means of the appropriate choice of ς, γ. Moreover, the demonstration of Theorem 2 uses the fact that generic oscillations front sets of the solutions to nonstationary Schrödinger equation propagate along Hamiltonian billiards, and demonstration of Theorem 10 uses the microlocal canonical form of nonnegative symbols with noninvolutive double characteristic variety of codimension 3. Moreover, if $X'' \neq \varnothing$ then the proofs of the estimates from above use Rosenblyum's variational estimate [**10**]. Our arguments are applicable when deriving the eigenvalue asymptotics for higher-order operators [**7**] and even for nonsemibounded operators such as Dirac operator [**8**].

It should be noted that Theorems 1 and 2 remain true in the case $d = 2$ after a slight modification connected with Rosenblyum's estimate, but the case $d = 1$ is essentially specific. I cannot generalize Theorem 10 to other dimensions without certain additional assumptions; in the case $d > 3$, the expression for M^* is more sophisticated.

Our results remain true in the case of the quadratic form Q considered on $C_0^\infty(\mathbf{R}^d)|_X$ (this leads to natural boundary condition $g^{jk}n_j(D_k - V_k)u = 0$ on ∂X for operator A) provided $\varsigma\gamma > 1$ on ∂X. In this case one should replace M' by $-M'$. It should also be noted that Q is semibounded from below on L^2 and hence generates a selfadjoint operator provided one of the right-hand expressions in the estimates (1), (1)$'$, (18), (20) is finite.

REFERENCES

1. V. Ivrii, *Estimations pour le nombre de valeurs propres négatives de l'opérateur de Schrödinger avec potentiels singuliers*, C. R. Acad. Sci. Paris Sér. I Math. **302** (1986), 467–470.

2. _____, *Estimations pour le nombre de valeurs propres négatives de l'opérateur de Schrödinger avec potentiels singuliers et application au comportement asymptotique des grandes valeurs propres*, C. R. Acad. Sci. Paris Sér. I Math. **302** (1986), 491–494.

3. _____, *Estimations pour le nombre de valeurs propres négatives de l'operateur de Schrödinger avec potentiels singuliers et application au comportement asymptotique des valeurs propres s'accumulant vers −0, aux asymptotiques á 2 paramètres et à la densité des états*, C. R. Acad. Sci. Paris Sér. I Math. **302** (1986), 535–538.

4. _____, *Estimates for a number of the negative eigenvalues of the Schrödinger operator with singular potentials*. I, *Principal theorems* (to appear in Russian).

5. _____, *Estimates for a number of the negative eigenvalues of the Schrödinger operator with singular potentials*. II, *Applications to eigenvalue asymptotics* (to appear in Russian).

6. _____, *Asymptotic Weyl's formula for the Laplace-Beltrami operator in Riemannian polyhedra and in domains with conical singularities of boundary*, Dokl. Akad. Nauk SSSR, **288** (1986), 35–38. (Russian)

7. ____, *On precise eigenvalue asymptotics for two classes of differential operators in* $\mathbf{R}^d$, Dokl. Akad. Nauk SSSR **276** (1984), 268–270; English transl. in Soviet Math. Dokl. **29** (1984), 464–466.

8. ____, *On asymptotics of discrete spectrum for certain operators in* $\mathbf{R}^d$, Funktsional. Anal. i Prilozhen. **19** (1985), 73–74. (Russian)

9. V. Ivrii and S. Fedorova, *Dilatations and eigenvalue asymptotics for spectral problems with singularities*, Funktsional. Anal. i Prilozhen. **20** (1986). (Russian)

10. G. Rosenblyum, *Distribution of discrete spectrum for singular differential operators*, Izv. Vyssh. Uchebn. Zaved. Mat. (1976), 75–86. (Russian)

11. H. Tamura, *Asymptotic distribution of eigenvalues for Schrödinger operators with magnetic fields*, Preprint, 1985, 33pp.

MAGNITOGORSK MINING-METALLURGICAL INSTITUTE, MAGNITOGORSK 455000, USSR

Two-Dimensional Geometric Variational Problems

JÜRGEN JOST

It is the purpose of this paper to describe some recent results of the author and his colleagues on conformally invariant variational problems in two dimensions and to put these results in a systematic context that developed from our work.

Such conformally invariant variational problems arise in geometry, and solutions to such problems include (in increasing generality):

• conformal maps between surfaces,

• minimal surfaces in Euclidean space, or, more generally, in Riemannian manifolds,

• harmonic maps from a surface into a Riemannian manifold,

• surfaces of prescribed mean curvature, or, more generally, solutions of the mean curvature system in Riemannian manifolds.

Let us briefly describe the corresponding variational integrals and Euler-Lagrange equations.

N will be a Riemannian manifold of dimension n with metric tensor given in local coordinates by $(g_{ij})_{i,j=1,\dots,n}$ and corresponding Christoffel symbols Γ^i_{jk}.

Σ is a surface (always equipped with a conformal structure) with local isothermal coordinates $z = x + iy$ and $u : \Sigma \to N$ a map of class $H^{1,2}$. The energy of u then is defined as

$$E(u) := \frac{1}{2} \int_\Sigma g_{ij}(u(z))(u_x^i u_x^j + u_y^i u_y^j)\, dx\, dy \tag{1}$$

where, in local coordinates, $u = (u^1, \dots, u^n)$, $u_x^i := \frac{\partial u^i}{\partial x}$, etc., and the standard summation convention is employed. This expression is independent of the choice of local coordinates involved, and, also, we need not specify a metric on Σ as $E(u)$ depends only on the conformal structure of Σ.

The corresponding Euler-Lagrange equations are

$$\Delta u^i + \Gamma^i_{jk}(u(z))(u_x^j u_x^k + u_y^j u_y^k) = 0 \qquad (i = 1, \dots, u). \tag{2}$$

A solution u of (2) is called harmonic, and if, in addition, u is conformal, i.e.,

$$g_{jk}(u_x^j u_x^k - u_y^j u_y^k - 2i u_x^j u_y^k) = 0, \tag{3}$$

The author acknowledges support from SFB 72 (Bonn) and Stiftung Volkswagenwerk.

then u is a (parametric) minimal surface in N (possibly with branch points).

In general, for a solution u of (2), (3) need not be satisfied, but it turns out that the expression

$$g_{jk}(u_x^j u_x^k - u_y^j u_y^k - 2iu_x^j u_y^k)(dx^2 - dy^2 + 2i\,dx\,dy) \tag{4}$$

is a holomorphic quadratic differential. This already holds for a critical point u of $E(u)$, provided u is also stationary w.r.t. variations of the independent variables; i.e., if $\tau_\varepsilon : \Sigma \to \Sigma$ is a family of diffeomorphisms depending smoothly on the parameter ε with $\tau_0 = \mathrm{id}$, then we require that

$$\frac{d}{d\varepsilon}E(u \circ \tau_\varepsilon)_{|\,\varepsilon=0} = 0. \tag{5}$$

The fact that (5) implies that (4) is holomorphic depends on the invariance of (1) under changes of the independent coordinates, and hence the holomorphicity of (4) expresses the fact that our variational problem is conformally invariant.

For surfaces of prescribed mean curvature in a three-dimensional manifold, one looks at the H-surface functional

$$I(u) := \frac{1}{2}\int (g_{ij}(u)\nabla u^i \cdot \nabla u^j + Q(u)(u_x \wedge u_y)) \tag{6}$$

with $\operatorname{div} Q(u) = 4H(u)$, $\nabla u^i = (u_x^i, u_y^i)$, with Euler-Lagrange equations

$$\Delta u^i + \Gamma^i_{jk}(u)\nabla u^j \cdot \nabla u^k = 2H(u)\sqrt{g}g^{ij}(u_x \wedge u_y)^j \tag{7}$$

with $(g^{ij}) = (g_{ij})^{-1}$, $g = \det(g_{ij})$.

If u is conformal, i.e., if (3) holds, and u solves (7), then $H(u)$ is the mean curvature of the surface in N described by u. Again, for a general solution of (7), (4) is holomorphic.

Conversely, Grüter [G2] observed that the most general (positive definite) conformally invariant two-dimensional variational integral is already of the form

$$I(u) = \int (g_{ij}(u)\nabla u^i \cdot \nabla u^i + b_{ij}(u)\det(\nabla u^i, \nabla u^j)) \tag{8}$$

where $\nabla u^i = (u_x^i, u_y^i)$, (g_{ij}) is symmetric, and (b_{ij}) is skew-symmetric. If the dimension of N is three, this reduces to (6).

Σ will always be a compact surface. If $\partial\Sigma = \varnothing$, one looks for closed solutions of our problems, whereas in case $\partial\Sigma \neq \varnothing$, one can impose various boundary conditions, namely Dirichlet conditions, Plateau conditions (where $\partial\Sigma$ has to be mapped monotonically onto a given Jordan curve in N), or free boundary conditions where $\partial\Sigma$ has to be mapped into a (sufficiently regular) submanifold L of N, typically a hypersurface, and Σ has to meet L orthogonally along $\partial\Sigma$.

Regularity. Grüter [G1] showed that (under appropriate regularity assumptions for the coefficients of (8)) a weak critical point u of a conformally invariant integral is continuous if it is weakly conformal, i.e., if (3) holds almost everywhere. Here, u is called a weak critical point if it is a weak solution of the corresponding Euler-Lagrange equations. For example, for the H-system this means that $u \in H^{1,2}$ and that if $\phi \in \overset{\circ}{H}\,{}^{1,2} \cap L^\infty(\Sigma, \mathbf{R}^3)$ and $u(\operatorname{supp} \phi \backslash$ null set$)$ is contained in a coordinate neighborhood, then

$$\int_\Sigma (\nabla u^i \nabla \phi^i - \Gamma^i_{jk} \nabla u^i \nabla u^k \phi^i + 2H(u)\sqrt{g}g^{ij}(u_x \wedge u_y)^j \phi^i). \qquad (9)$$

The crucial observation in the proof of this regularity result is that the conformality relation (3) allows us to pull back the standard monotonicity formula of geometric measure theory under u.

Schoen [**Sch**] then reduced the case of a not necessarily conformal stationary point u of such an integral to the preceding case by using the holomorphic differential (4) to construct a weakly conformal map from u, to which Grüter's result applies. "Stationary" here means that the analogue of (5) has to hold because this is needed to show that (4) is holomorphic.

Likewise, by improving earlier results of Grüter-Hildebrandt-Nitsche [**GHN1**] and Dziuk [**D**], the author proved continuity of minimal surfaces at a free boundary, if the free boundary is a submanifold of N of class C^2 with bounded second fundamental form and a uniform neighborhood where the nearest point is smooth (cf. [**J5**]). The same result actually already holds if the free boundary is only piecewise C^2, provided the various pieces (of, possibly, various dimensions) satisfy the above assumptions and the angle between different pieces is uniformly bounded away from zero (cf. [**J6**]).

Boundary regularity for surfaces of prescribed mean curvature at a free boundary was obtained by Grüter-Hildebrandt-Nitsche [**GHN2**].

For interior regularity of solutions of the H-surface equation in Euclidean space ($\Delta u = 2H(u)u_x \wedge u_y$) one does not need (5), as shown by Heinz [**H**] under the condition

$$\sup_{\mathbf{R}^3}(|H(u)| + (1 + |u|)|\nabla H(u)|) < \infty.$$

Standard results from elliptic PDE then imply higher regularity of continuous solutions. The question of the dependence of these regularity results on the geometric quantities involved was then solved by Jost and Karcher [**JK**]. Their results yield $C^{2,a}$ bounds for a solution of (9) depending only on a C^α-bound for H and a bound for the absolute value of the sectional curvature of the image N, a lower bound for the injectivity radius of N, and the dimension of N. Likewise, for a higher dimensional domain M, where the geometry of M also enters into the harmonic map equation, $C^{2,\alpha}$ bounds for a harmonic map then also depend on the corresponding geometric quantities of M (sectional curvature, injectivity radius, and dimension). An essential point in the proof is the construction of harmonic coordinates on balls of controlled size, i.e., with a lower bound for the radius depending on the mentioned geometric quantities. Manifolds where these quantities are controlled play a prominent role in geometry—in particular, in Gromov's work—and the constructions of [**JK**] are useful for the investigation of such manifolds.

Existence of minimal surfaces. After Douglas and Radó had produced an area minimizing disk as a solution to Plateau's problem, minimal surfaces spanning a given Jordan curve in $\mathbf{R}^3$ that are of higher genus or unstable, were then studied by Douglas, Courant, Shiffman, and Morse-Tompkins. Today, however, one often finds that their arguments are not completely satisfactory, and therefore this "classical" theory was reworked by Struwe, who developed a Morse theory for disk and annulus type minimal surfaces [**S3**, **S4**], and by the author, who gave a solution to the so-called Plateau-Douglas problem in a Riemannian manifold [**J3**]. Here, one shows that a Jordan curve $\gamma \subset N$ spans a minimal surface of genus g if

$$\inf\{\text{Area of surfaces } \Sigma \text{ of genus } g \text{ with } \partial\Sigma = \gamma\} \tag{10}$$
$$< \inf\{\text{Area of surfaces } \Sigma' \text{ of genus } g - 1 \text{ with } \partial\Sigma' = \gamma\}.$$

Combining these results and going beyond the classical theory, a Morse theory for minimal surfaces of arbitrary genus spanning a given Jordan curve in $\mathbf{R}^3$ is being developed jointly by Struwe and the author. It also distinguishes between embedded and nonembedded surfaces.

A condition different from (10), formulated in terms of the existence of a suitable barrier of nonnegative mean curvature, that guarantees the existence of a minimal surface of given genus g, was found by Tomi-Tromba [**TT**]. The author [**J4**] showed that this condition actually implies the existence of an embedded minimal surface of genus g.

If $S \subset \mathbf{R}^3$ is diffeomorphic to S^2, then Struwe [**S1**] showed the existence of an (unstable) minimal disk with a free boundary on S. Here, however, stronger and physically and geometrically more satisfactory results could be obtained by using methods of geometric measure theory. Using the fundamental constructions of Pitts [**P**] and Simon-Smith [**SS**] and building upon joint work with Grüter on the regularity of varifolds with free boundaries, the author was able to show [**J4**] that if A is a subset of a three-dimensional Riemannian manifold, diffeomorphic to the closed ball, then there exists an unstable embedded minimal surface Σ in A of genus zero (but possibly higher connectivity) with a free boundary $\partial\Sigma$ on ∂A in case $\partial\Sigma \neq \varnothing$. If ∂A has nonnegative mean curvature w.r.t. the interior normal, then a simply connected such Σ can be obtained.

Also, if $A \subset \mathbf{R}^3$, ∂A strictly convex, and if the ratio between the outer and inner diameter of A does not exceed $\sqrt{2}$, then A contains at least three embedded minimal disks meeting ∂A orthogonally along their boundary. It is conjectured that the above restriction on the shape of ∂A is not necessary, however. If ∂A is a tetrahedron, this problem had been solved by Smyth [**Sm**].

Combining techniques from geometric measure theory, conformal representation of surfaces of higher genus with boundary, and the parametric methods used for solving the Plateau-Douglas problem, one can also obtain embedded minimal surfaces with a free boundary and of prescribed genus under appropriate assumptions [**J7**].

Existence of surfaces of constant mean curvature. It had been shown by Hildebrandt that if γ is a Jordan curve in $\mathbf{R}^3$ contained in a ball of radius R, and H is a constant with $|H|R \leq 1$, then there exists a surface Σ of constant mean curvature H with $\partial\Sigma = \gamma$. (Σ minimizes the H-surface functional.) Building upon fundamental work of Wente [**W**], Brézis-Coron [**BC**] then showed that if $|H|R < 1$, then there actually exist two such surfaces, thus solving the so-called Rellich conjecture, with an independent (but slightly weaker) solution by Struwe [**S2**] and Steffen [**Sf**]. (If γ is a plane circle, then Hildebrandt's solution corresponds to the small spherical cap, whereas the second one corresponds to the large spherical cap with mean curvature H and boundary γ.) Struwe [**S5**] then improved this result and showed that if one has a strict local minimum for the H-surface functional for given γ and constant H, then there also exists a second solution.

Existence of harmonic maps. Let Σ be a compact surface (for simplicity without boundary, although corresponding results hold for the Dirichlet problem, too), N a compact Riemannian manifold ($\partial N = \varnothing$), A compact, typically $A = [0,1]^d$. Let $h_0 \colon \Sigma \times A \to N$ be continuous, and let H be the class of all maps h homotopic to h_0, and

$$M := \inf_{h \in H} \sup_{t \in A} E(h(\cdot, t)),$$

and we fix $h_{|\Sigma \times \partial A}$ (in case $\partial A \neq \varnothing$) so that the supremum in the above expression cannot be obtained on ∂A. Then there exists a harmonic map $u_0 \colon \Sigma \to N$ and also, possibly, some (nontrivial) conformal harmonic maps $u_i \colon S^2 \to N$, $i = 1, \dots, m$, with

$$E(u_0) + \sum_{i=1}^{m} E(u_i) = M; \tag{11}$$

and $(u_0, u_1, \dots, u_m)$ represents a saddle point corresponding to H in the following sense: There exist sequences $(h_n) \subset H$ and $(t_n) \subset A$, and possibly points $x_1, \dots, x_k \in \Sigma$, $k \leq m$, with

$E(h_n(\cdot, t_n)) \to M,$

$h_n(\cdot, t_n) \to u_0$ weakly in $H^{1,2}$,

$h_n(\cdot, t_n) \rightrightarrows u_0$ uniformly on each compact subset of

$$\Sigma \backslash \{x_1, \dots, x_k\},$$

and

$$h_n(x/\lambda_n, t_n) \to u_i$$

where $\lambda_n \to 0$ and x/λ_n is interpreted as rescaling in polar coordinates centered at some appropriate x_j ($j \in \{1, \dots, k\}$).

This result (cf. [**J8**]) improves the theorem of Sacks-Uhlenbeck [**SU**] by obtaining equality in (11) (in [**SU**], one only has "$\leq$"). The method of proof refines the idea employed in [**J1**]. (11) should be interpreted as a version of strong convergence that allows for changes of the topological type, or accounts for energy jumps. This result will serve as the basis for a Morse theory for harmonic maps.

Properties and applications of harmonic maps. In [**JS**], Schoen and the author showed that given a diffeomorphism ϕ between compact surfaces Σ_1 and Σ_2 (without boundary), there exists a harmonic diffeomorphism u homotopic to ϕ. Here, on Σ_2, a Riemannian metric has to be given, but the result holds for any (smooth) metric. The proof depends on deep a priori estimates from below for the Jacobian of harmonic diffeomorphisms, due to Heinz. The result improves the earlier theorems of Schoen-Yau and Sampson which hold for image metrics of nonpositive curvature.

In a similar vein, the author also developed variational techniques for the conformal representation of surfaces (cf. [**J1, J3**]).

Now let S be a compact oriented topological surface without boundary of genus $g > 1$, Σ, Σ' surfaces equipped with conformal structures homeomorphic to S. On Σ' we also introduce the hyperbolic metric determined by the conformal structure. We then look at the harmonic map $u(\Sigma, \Sigma')\colon \Sigma \to \Sigma'$ homotopic to the identity of S. As noted in the beginning, u induces a holomorphic quadratic differential $\rho(\Sigma, \Sigma')$ on Σ. Wolf, in his thesis [**Wf**], managed to compute the effect of variations of the conformal structure of the image Σ' on u and ρ, whereas the author [**J8**] then studied the effect of variations of the domain Σ. With the help of these computations, one can recover all the basic structures of Teichmüller space (the space of conformal structures on a surface S with fixed topological marking, which in our situation corresponds to looking only at maps homotopic to the identity of S), namely, the topological ("Teichmüller's theorem"), differentiable, complex, Riemannian (Weil-Petersson metric), and Kählerian structure, and also to obtain a simple proof of Tromba's curvature result and Wolpert's curvature formula for the Weil-Petersson metric. Thus, the results of [**Wf**] and [**J8**] furnish a new approach to Teichmüller theory, replacing quasiconformal by harmonic maps.

REFERENCES

[**BC**] H. Brézis and J.-M. Coron, *Multiple solutions of H-systems and Rellich's conjecture*, Comm. Pure Appl. Math. **37** (1984), 149–187.

[**D**] G. Dziuk, C^2 *regularity for partially free minimal surfaces*, Math. Z. **189** (1985), 71–79.

[**G1**] M. Grüter, *Regularity of weak H-surfaces*, J. Reine Angew. Math. **329** (1981), 1–15.

[**G2**] ____, *Conformally invariant variational integrals and the removability of isolated singularities*, Manuscripta Math. **47** (1984), 85–104.

[**GHN1**] M. Grüter, S. Hildebrandt, and J. Nitsche, *On the boundary behavior of minimal surfaces with a free boundary which are not minima of the area*, Manuscripta Math. **35** (1981), 387–410.

[**GHN2**] ____, *Regularity for stationary surfaces of constant mean curvature with free boundaries*, Acta Math. **156** (1986), 119–152.

[**H**] E. Heinz, *Über die Regularität schwacher Lösungen nichtlinearer elliptischer Systeme*, Nachr. Akad. Wiss. Gött., Nr. 1, 1986.

[**J1**] J. Jost, *Harmonic maps between surfaces*, Lecture Notes in Math., Vol. 1062, Springer-Verlag, Berlin and New York, 1984.

[**J2**] ____, *Harmonic mappings between Riemannian manifolds*, Proc. Centre Math. Anal. Austral. Nat. Univ., Vol. 4, Austral, Nat. Univ., Canberra, 1984.

[J3] ____, *Conformal mappings and the Plateau Douglas problem in Riemannian manifolds*, J. Reine Angew. Math. **359** (1985), 37–54.

[J4] ____, *Existence results for embedded minimal surfaces of controlled topological type*. I, Ann. Scuola Norm. Sup. Pisa (Ser. IV), **13** (1986), 15–50; II, III, Ann. Scuola Norm. Sup. Pisa (to appear).

[J5] ____, *On the regularity of minimal surfaces with free boundaries in Riemannian manifolds*, Manuscripta Math. **56** (1986), 279–291.

[J6] ____, *Continuity of minimal surfaces with piecewise smooth free boundaries*, Preprint.

[J7] ____, *On the existence of embedded minimal surfaces of higher genus with free boundaries in Riemannian manifolds*, Variational Methods for Free Surface Interfaces, P. Concus and R. Finn (eds.), Springer, New York, 1987.

[J8] ____, *Two-dimensional geometric variational problems* (in preparation).

[JK] J. Jost and H. Karcher, *Geometrische Methoden zur Gewinnung von a-priori Schranken für harmonische Abbildungen*, Manuscripta Math. **40** (1982), 27–77.

[JS] J. Jost and R. Schoen, *On the existence of harmonic diffeomorphisms between surfaces*, Invent. Math. **66** (1982), 353–359.

[P] J. Pitts, *Existence and regularity of minimal surfaces on Riemannian manifolds*, Princeton Univ. Press, Princeton, N. J., 1981.

[SU] J. Sacks and K. Uhlenbeck, *The existence of minimal immersions of 2-spheres*, Ann. of Math. **113** (1981), 1–24.

[Sch] R. Schoen, *Analytic aspects of the harmonic map problem*, Seminar on Nonlinear Partial Differential Equations, S. S. Chern, Editor, Springer, 1984.

[SS] L. Simon and F. Smith, *On the existence of embedded minimal 2-spheres in the 3-sphere, endowed with an arbitrary metric* (to appear).

[Sm] B. Smyth, *Stationary minimal surfaces with boundary on a simplex*, Invent. Math. **76** (1984), 411–420.

[Sf] K. Steffen, *On the nonuniqueness of surfaces with prescribed constant mean curvature spanning a given contour*, Arch. Rational Mech. Anal. (to appear).

[S1] M. Struwe, *On a free boundary problem for minimal surfaces*, Invent. Math. **75** (1984), 547–560.

[S2] ____, *Nonuniqueness in the Plateau problem for surfaces of constant mean curvature*, Arch. Rational Mech. Anal. (to appear).

[S3] ____, *On a critical point theory for minimal surfaces spanning a wire in $\mathbf{R}^n$*, J. Reine Angew. Math. **349** (1984), 1–23.

[S4] ____, *A Morse theory for annulus type minimal surfaces*, J. Reine Angew. Math. (to appear).

[S5] ____, *Large H-surfaces via the mountain pass lemma*, Math. Ann. **270** (1985), 441–459.

[TT] F. Tomi and A. Tromba, *On Plateau's problem for minimal surfaces of higher genus in $\mathbf{R}^3$*, Bull. Amer. Math. Soc. (N.S.) **13** (1985), 169–171.

[W] H. Wente, *Large solutions to the volume constrained Plateau problem*, Arch. Rational Mech. Anal. **75** (1980), 59–77.

[Wf] M. Wolf, *The Teichmüller theory of harmonic maps*, Thesis, Stanford Univ., 1986.

RUHR-UNIVERSITÄT, D-4630 BOCHUM, FEDERAL REPUBLIC OF GERMANY

Some New Results in the Theory
of Nonlinear Elliptic and Parabolic Equations

N. V. KRYLOV

1. Introduction. Nonlinear elliptic and parabolic equations arise in the theory of partial differential equations as well as in numerous applications. Up to now the wide class of such equations known as quasilinear equations has been studied almost in the same detail as the class of linear equations. Many mathematicians contributed to the theory of quasilinear equations and their results are described in books [1–4].

The second-order quasilinear equations are linear with respect to second derivatives. As to the equations which are nonlinear with respect to second derivatives and which are the main object of the recent report, only fragmentary results were available during a long period. There did not exist a more or less general theory with the small exception of the case with two spatial variables for which the general theory of solvability was developed in $W_p^{I,2}$, W_p^2 for all p close to 2 (cf. [5]).

In the absence of a general theory some classes of nonlinear equations were studied by specialists in those fields of mathematics where these equations arose. The first nonlinear elliptic equation subjected to intensive investigations was the Monge-Ampère equation. This equation arises in the theory of convex surfaces, and it was studied with the help of this theory. Up to 1971 the smoothness of its generalized solutions introduced by Aleksandrov [6] was proved only for two variables [7]. Then Pogorelov [8] in 1971 proved the interior regularity in the multidimensional case (cf. also [9]). The smoothness up to the boundary for the multidimensional Monge-Ampère equation was proved only in 1982 after the general theory of nonlinear equations was developed [10–14].

One more source of nonlinear elliptic and parabolic equations gives the theory of controlled diffusion processes. The equations arising in this theory are the so-called Bellman equations. The corresponding theory started to develop much later than the theory of the Monge-Ampère equations. In the case of noncontrolled diffusion, this theory can be based on the theory of quasilinear equations; see [15] and the references there.

The author [16, 17] using probability methods in 1972 showed the solvability of general degenerating Bellman equations in the whole space in the class of functions with bounded derivatives. This result seems to be very important not only because the wide class of nonlinear equations is treated there but also because of the fact proved there that the Monge-Ampère equations are a partial case of the Bellman equations. Thus appeared the possibility of constructing a general theory of nonlinear equations including the Monge-Ampère equations. The approach to control theory suggested by the author was developed further by other mathematicians (cf., e.g., [18–24]). In particular, Pragarauskas [23, 24] developed a general theory of nonlinear integro-differential equations. The main results of the probabilistic treatment of the Bellman equations in the whole space were gathered in the author's book [25] in 1977 and in the author's report to the International Congress of Mathematicians in Helsinki in 1978.

Further evolution of probabilistic methods leads to consideration of the Bellman equations in bounded domains. Here I want to indicate not only the above-mentioned articles of Safonov and Lions but also my two papers [26, 27] of 1981, where a general method of how to reduce the Dirichlet problem to a problem on a compact smooth manifold without boundary is suggested. In particular in [27] the solvability is proved in the class of continuous functions with bounded derivatives u_x, u_{xx}, v_t, v_x, v_{xx} of the problems

$$\det(u_{xx}) = (f_+(x))^d, \quad (u_{xx}) \geq 0 \quad \text{(a.e. } D), \quad u|_{|x|=1} = 0,$$

$$v_t \det(v_{xx}) = (1 - |x|^2)(g_+(t,x))^{d+1}, \quad v_t \geq 0, \quad (v_{xx}) \geq 0 \quad \text{(a.e. } [0,T] \times D),$$

$v(T,x) = 0$ if $|x| \leq 1$, $v(t,x) = 0$ if $t \in [0,T]$, $|x| = 1$, where $D = \{x \in E_d : |x| < 1\}$, $f(x)$ is twice continuously differentiable in $\overline{D}$, $g(t,x)$ is twice continuously differentiable in $[0,T] \times \overline{D}$, $g(T,x) = 0$ for $|x| \leq 1$.

Approximately up to 1979 the probabilistic methods played the main role in the theory of the Bellman equations. In 1979 Brezis and Evans [28] considered the case of the Bellman equation with two elliptic operators and proved its solvability in $C^{2+\alpha}$. Probabilistic methods give results only in W_∞^2. Near that time the articles [29–31] appeared where PDE methods were also used, though the solutions were of class W_∞^2 and these results were not stronger than earlier-obtained probabilistic results.

In the general case the solvability of the Bellman equations only in W_∞^2 was known till 1982. Later I will discuss that very great progress in the theory of nonlinear equations which is recorded starting with the time of the last International Congress of Mathematicians in Warsaw in 1982 and which is mainly presented in the author's book [32]. In 1982 Evans [33] proved interior $C^{2+\alpha}$ regularity for elliptic Bellman equations with constant coefficients. Independently in the same year the author [34, 10] received the same result for elliptic and parabolic Bellman equations with variable coefficients and proved $C^{2+\alpha}$ regularity up to the boundary for the elliptic case. The basis of the works [33, 34, 10] are the results of Krylov and Safonov [35, 36] on the estimate of the Hölder constant for solutions of linear equations with measurable coefficients.

I want to stress that the investigations in [**33, 34, 10**], in the author's book [**32**], and in the articles of Safonov [**37–39**] are entirely based on the methods of the linear theory. One can say that a good linear theory breeds a good nonlinear theory in contradiction with the known claim that "linearity breeds contempt."

2. Uniformly nondegenerate equations. Fix some constants $K \geq \varepsilon > 0$, $T > 0$, a domain $D \subset E_d$, and define $Q = (0, T) \times D$, $Q(\kappa) = (0, T - \kappa) \times D(\kappa)$ with $D(\kappa) = \{x \in D \colon \operatorname{dist}(x, \partial D) > \kappa\}$.

DEFINITION 2.1. For all $(t, x) \in Q$ and real u_{ij}, u_i $(i, j = 1, \ldots, d)$, u, let a real function $F(u_{ij}, u_i, u, t, x)$ be defined. We write $F \in \mathcal{F}(\varepsilon, K, Q)$ if for every t the function F is twice continuously differentiable with respect to (u_{ij}, u_i, u, x); F is continuously differentiable with respect to all its arguments, and for all $(t, x) \in Q$, $\lambda, \tilde{x} \in E_d$, $r = 1, \ldots, d$, and real $u_{ij} = u_{ji}$, $\tilde{u}_{ij} = \tilde{u}_{ji}$, u_i, $\tilde{u}_i$, u, $\tilde{u}$, the following inequalities hold:

$$\varepsilon |\lambda|^2 \leq F_{u_{ij}} \lambda^i \lambda^j \leq K |\lambda|^2,$$

$$|F - F_{u_{ij}} u_{ij}| \leq M_1^F(u) \left(1 + \sum |u_i|^2\right),$$

$$|F_{u_r}| \left(1 + \sum |u_i|\right) + |F_u| + |F_{x^r}| \left(1 + \sum |u_i|\right)^{-1}$$
$$\leq M_1^F(u) \left(1 + \sum |u_i|^2 + \sum |u_{ij}|\right), \tag{2.1}$$

$$[M_2^F(u, u_k)]^{-1} F_{(\eta)(\eta)} \leq \sum |\tilde{u}_{ij}| \left[\sum |\tilde{u}_i| + \left(1 + \sum |u_{ij}|\right)(|\tilde{u}| + |\tilde{x}|)\right]$$
$$+ \sum |\tilde{u}_i|^2 \left(1 + \sum |u_{ij}|\right)$$
$$+ \left(1 + \sum |u_{ij}|^3\right)(|\tilde{u}|^2 + |\tilde{x}|^2),$$

$$|F_t| \leq M_3^F(u, u_k) \left(1 + \sum |u_{ij}|^2\right),$$

where M_i^F are some continuous functions increasing in $|u|$, $u_k u_k$; $F_{(\eta)(\eta)}$ is a quadratic form with respect to η which is equal to the second directional derivative of F along η, $\eta = (\tilde{u}_{ij}, \tilde{u}_i, \tilde{u}, \tilde{x})$.

This definition introduces the class of operators $\mathcal{F}(\varepsilon, K, Q)$. Naturally the functions M_i^F can differ for various elements of this class. The general property of the functions from $\mathcal{F}(\varepsilon, K, Q)$ is their concavity in (u_{ij}) on the set of all symmetric matrices (u_{ij}). This fact easily follows from (2.1) where on the right there are no terms of the second order with respect to $\tilde{u}_{ij}$.

DEFINITION 2.2. Let a function $F(u_{ij}, u_i, u, t, x)$ be defined for $(t, x) \in Q$ and real $u_{ij} = u_{ji}$, u_i $(i, j = 1, \ldots, d)$, u. We write $F \in \overline{\mathcal{F}}(\varepsilon, K, Q)$ if there exists a sequence $F_n \in \mathcal{F}(\varepsilon, K, Q(1/n))$ converging to F for all $(t, x) \in Q$, $u_{ij} = u_{ji}$, u_i, u and such that
(a) $M_i^{F_1} = M_i^{F_2} = \cdots =: \tilde{M}_i^F$, $i = 1, 2, 3$,
(b) there exists constants $\delta_0 =: \delta_0^F > 0$, $M_0 =: M_0^F > 0$ such that for all

$n \geq 1$, $(t, x) \in Q(1/n)$, and symmetric $(u_{ij}) \geq 0$,

$$F_n(u_{ij}, 0, -M_0, t, x) \geq \delta_0, \qquad F_n(-u_{ij}, 0, M_0, t, x) \leq -\delta_0. \qquad (2.2)$$

Note that we use (2.2) in a priori estimation of $|u|$. Let us consider the typical element of $\overline{\mathcal{F}}(\varepsilon, K, Q)$. Let Ω be a set and for every $\omega \in \Omega$ for $z = (t, x) \in Q$ and real $u_1, \ldots, u_d$, u let the functions $a^{ij}(\omega, u_k, u, z)$, $i, j = 1, \ldots, d$, $a(\omega, u_k, u, z)$ be defined. Suppose that for each $\omega \in \Omega$ the functions a^{ij}, a are continuously differentiable in (u_k, u, z) and for every t twice continuously differentiable with respect to (u_k, u, x). Suppose that above-mentioned derivatives are bounded on $\{(\omega, u_k, u, z): \omega \in \Omega, \sum |u_k| + |u| \leq M, z \in Q\}$ for every $M < \infty$. Finally suppose that for all $\omega \in \Omega$, $z \in Q$, $s, r, p = 1, \ldots, d$, $\lambda \in E_d$, u_i, $i = 1, \ldots, d$, u the following inequalities hold

$$\varepsilon |\lambda|^2 \leq a^{ij} \lambda^i \lambda^j \leq K|\lambda|^2, \qquad |a| \leq M_1(u)\left(1 + \sum |u_i|^2\right),$$

$$|a_{u_r}^{sp}| \left(1 + \sum |u_i|\right) + |a_u^{sp}| + |a_{x^r}^{sp}| \left(1 + \sum |u_i|\right)^{-1} \leq M_1(u),$$

$$|a_{u_r}| \left(1 + \sum |u_i|\right) + |a_u| + |a_{x^r}| \left(1 + \sum |u_i|\right)^{-1} \leq M_1(u) \left(1 + \sum |u_i|^2\right),$$

$$a(\omega, 0, -M_0, z) \geq \delta_0, \qquad a(\omega, 0, M_0, z) \leq -\delta_0,$$

where $M_1(u)$ is a continuous function, the constants $\delta_0 > 0$, $M_0 > 0$.

Then it appears that

$$F := \inf_{\omega \in \Omega} [a^{ij}(\omega, u_k, u, z)u_{ij} + a(\omega, u_k, u, z)] \in \overline{\mathcal{F}}(\varepsilon, K, Q).$$

Next we go to the main results about parabolic equations with $F \in \overline{\mathcal{F}}$.

THEOREM 2.1. *Let $\alpha \in (0, 1)$, $\psi \in C^{2+\alpha}(E_d)$, $D = \{x: \psi(x) > 0\}$ be a bounded nonempty domain, $|\psi_x| \geq 1$ on D, $F \in \overline{\mathcal{F}}(\varepsilon, K, Q)$, $\varphi \in C^{1+\alpha/2, 2+\alpha}(\overline{Q})$, $|\varphi| \leq M_0^F$ on Q. Then the problem*

$$u_t + F(u_{x^i x^j}, u_{x^i}, u, z) = 0 \quad in \ Q, \qquad (2.3)$$

$$u = \varphi \quad on \ \partial'(Q) \qquad (2.4)$$

($\partial' Q$ is the parabolic boundary of Q) admits the unique solution of the class $C(\overline{Q})$ with continuous bounded derivatives u_x, u_{xx}, u_t in Q.

Moreover, $|u| \leq M_0^F$ in Q and for every $n \geq 1$ we have

$$u \in C^{1+\beta/2, 2+\beta}([0, T] \times \overline{D}(n^{-1}) \cup [0, T - n^{-1}] \times \overline{D}),$$

where $\beta = \min(\alpha, \alpha_0)$, $\alpha_0 = \alpha_0(d, K, \varepsilon) \in (0, 1)$, and the norm of u in this space is bounded by a constant depending only on n, d, K, ε, α, M_0^F, $\tilde{M}_i^F$, and on the norms of ψ, φ in $C^{2+\alpha}(E_d)$, $C^{1+\alpha/2, 2+\alpha}(\overline{Q})$ respectively. Furthermore, if φ satisfies the first matching condition then $u \in C^{1+\beta/2, 2+\beta}(\overline{Q})$ and the norm of u in this space can be estimated only in terms of the same objects mentioned in the preceding sentence with the exception of n.

THEOREM 2.2. *Let $F \in \overline{\mathcal{F}}(\varepsilon, K, Q)$ and D satisfy an exterior sphere condition at each boundary point. Suppose that $\varphi \in C(\overline{Q})$, $|\varphi| \leq M_0^F$ on Q. Then*

the problem (2.3), (2.4) *has the unique solution* $u \in C(\overline{Q})$ *such that* $|u| \leq M_0^F$
on Q, $u \in C^{1+\alpha_0/2,2+\alpha_0}(\overline{Q}(\kappa))$ *for any* $\kappa > 0$, *and the norm of* u *in this space
is bounded by a constant depending only on* d, K, ε, M_0^F, κ, $\tilde{M}_i^F$.

Analogous results are true for elliptic equations.

THEOREM 2.3. *Let the conditions of Theorem 2.1 be satisfied and let* F, φ
be independent of t. *Then the problem*

$$F(u_{x^i x^j}, u_{x^i}, u, x) = 0 \quad in \; D, \tag{2.5}$$

$$u = \varphi \quad on \; \partial D \tag{2.6}$$

has a solution $u \in C^{2+\beta}(\overline{D})$ *for which* $|u| \leq M_0^F$ *on* $\overline{D}$, *and the norm of* u *in
$C^{2+\beta}(\overline{D})$ is bounded as is indicated in the last assertion of Theorem 2.1.*

THEOREM 2.4. *Let the conditions of Theorem 2.2 be satisfied, and let* F, φ
be independent of t. *Then the problem* (2.5), (2.6) *has a solution* $u \in C(\overline{D})$ *for
which* $|u| \leq M_0^F$ *on* D, *and the norm of* u *in* $C^{2+\alpha_0}(\overline{D}(\kappa))$ *for every* $\kappa > 0$ *is
estimated by the constant from Theorem 2.2.*

These results are taken from the book [32]. Various versions of Theorems
2.3 and 2.4 may be found also in [4, 33, 40]. Note that Theorems 2.1–2.4
do not contain the corresponding results from the linear theory because of the
superfluous smoothness assumptions on F. In the theory of quasilinear equations
[1–4] the smoothness assumptions on F are usually stronger than in the linear
theory. Nevertheless they are weaker than ours.

It appears that there exists a $C^{2+\alpha}$-theory of nonlinear equations which contains the corresponding results of the linear theory, and moreover in the case
of linear equations it is simpler than ordinarily used. Below in this section I
present the remarkable results due to Safonov (see [37–39]). As is mentioned
above, they are obtained by the methods universally applicable both for linear
and nonlinear equations and which differ, for instance, from the potential theory
methods. Let us consider only the oblique derivative problem. Analogous results
are proved by Safonov for the Dirichlet problem.

THEOREM 2.5. *Let* $\alpha \in (0,1)$, *let* D *be a bounded domain in* E_d, $\partial D \in$
$C^{2+\alpha}$, *and let a function* $F(u_{ij}, u_i, u, z)$ *be defined for all* $z = (t,x) \in Q$ *and real
u_{ij}, u_i, u. Let*
   (a) F *be convex in* (u_{ij}),
   (b) *for all* $\xi \in E_d$

$$\varepsilon|\xi|^2 \leq F(u_{ij} + \xi^i \xi^j, u_i, u, z) - F(u_{ij}, u_i, z) \leq K|\xi|^2,$$

   (c) F *satisfy the Lipschitz condition with respect to* (u_i, u) *with the constant
K, and*
   (d) *the norm of* $F(u_{ij}, u_i, u, \cdot)$ *in* $C^{\alpha/2,\alpha}(\overline{Q})$ *be finite and less than*

$$K\left(\sum |u_{ij}| + \sum |u_i| + |u| + 1\right).$$

Define $S = (0, T) \times \partial D$ and suppose that $\varphi \in C^{2+\alpha}(\overline{D})$, $|b^i n^i| \geq \varepsilon$ on S, $b^i, b, \Phi \in C^{(1+\alpha)/2, 1+\alpha}(\overline{S})$, where $n = (n^i)$ is the unit normal vector to ∂D. Finally suppose that

$$b^i \varphi_{x^i} + b\varphi = \Phi \quad \text{on } \{T\} \times \partial D.$$

Then equation (2.3) with the boundary condition

$$u = \varphi \quad \text{if } t = T, \qquad b^i u_{x^i} + bu = \Phi \quad \text{on } S$$

has the unique solution $u \in C^{1+\beta/2, 2+\beta}(\overline{Q})$.

3. Degenerate elliptic equations in a domain. Fix the constants $\delta \in (0, 1)$, $K > 0$, a function $\psi \in C^4(E_d)$, the integers $d_1, d_2 \geq 1$. Let $D := \{x \in E_d : \psi(x) > 0\}$ be a nonempty bounded domain and $|\psi_x| > 0$ on ∂D. Let Ω be a compact metric space and for all $\omega \in \Omega$, $p \in E_{d_2}$, $x \in E_d$, $i = 1, \ldots, d$, $k = 1, \ldots, d_1$, let real $\sigma^{ik}(\omega, p, x)$, $b^i(\omega, p, x)$, $c(\omega, p, x)$, and $f(\omega, p, x)$ be defined. Suppose that $c \leq 0$, the functions σ, b, c, f are continuous in (ω, p, x), twice continuously (in (p, x)) differentiable with respect to (p, x) for every ω, and their second-order derivatives are bounded on $\Omega \times E_{d_2} \times D$. Define

$$a^{ij} = \tfrac{1}{2}\sigma^{ik}\sigma^{jk}, \quad a = (a^{ij}), \quad \sigma^k = (\sigma^{ik}), \quad b = (b^i),$$

$$L = L(\omega, p, x) = a^{ij}(\omega, p, x)\frac{\partial^2}{\partial x^i \partial x^j} + b^i(\omega, p, x)\frac{\partial}{\partial x^i} + c(\omega, p, x),$$

$$F(u_{ij}, u_i, u, x) = \inf_{\omega \in \Omega}[a^{ij}(\omega, p, x)u_{ij} + b^i(\omega, p, x)u_i + c(\omega, p, x) + f(\omega, p, x)]$$

and suppose that the last infimum is independent of p.

We also need a matrix function $B(x) = (B^{ij}(x))$ of the dimension $d \times d$ such that $B = B^*$, a real function $\tilde{u}(x)$, and a function $\pi(\omega, x, \xi)$ on $\Omega \times \partial D \times E_d$ with values in E_{d_2}. Define

$$v_{(\pi,\xi)} = \sum_{i=1}^{d_2} v_{p^i} \pi^i + \sum_{i=1}^{d} v_{x^i} \xi^i, \qquad v_{(\xi)} = v_{x^i} \xi^i,$$

$$G = \{(\omega, p, x, \xi) : \omega \in \Omega, \ p = 0, \ x \in D, \ \xi \in E_d\},$$

$$\Gamma = \{(\omega, p, x, \pi, \xi) : \omega \in \Omega, \ p = 0, \ x \in \partial D, \ \xi \perp \psi_x(x),$$
$$(a(\omega, 0, x)\psi_x(x), \psi_x(x)) = 0, \ \pi = \pi(\omega, x, \xi)\}.$$

In the following theorem we also suppose that $B \in C^2(\overline{D})$, $\tilde{u} \in C^2(\overline{D})$,

$$\tilde{u} \geq \delta, \quad L\tilde{u} \leq -\delta, \quad (B\xi, \xi) \geq \delta|\xi|^2 \quad \text{on } G,$$

$$L(\omega, 0, x)\psi(x) \leq -\delta \quad \text{on } \Omega \times \partial D,$$

$$\tilde{u}\left[2(\xi, b_{(\xi)}) + \sum |\sigma^k_{(\xi)}|^2\right] + 2\tilde{u}_{(\sigma^k)}(\xi, \sigma^k_{(\xi)})$$
$$\leq -(1 - \delta)|\xi|^2 L\tilde{u} + K(a\xi, \xi) \quad \text{on } G,$$

$$\sum |(\sigma^k, \psi_x)_{(\pi,\xi)}|^2 + 4(B\xi, (a\psi_x)_{(\pi,\xi)})$$
$$\leq -(1 - \delta)(B\xi, \xi)L\psi \quad \text{on } \Gamma,$$

$$|\pi| \leq K|\xi| \quad \text{on } \Gamma.$$

Note that the two last inequalities are satisfied if $\Gamma = \varnothing$.

THEOREM 3.1. *Under above assumptions there exists a unique continuous in $\overline{D}$ function u with the following properties*:

(a) $u = 0$ *on* ∂D;

(b) *for every direction in E_d the right-hand side first-order directional derivative of u exists at every point in D, is bounded in D, and is almost everywhere differentiable*;

(c) $F(u^0_{x^i x^j}, u_{x^i}, u, x) = 0$ *(a.e.) in D, where u_{x^i} is the right-hand side derivative of u with respect to x^i, and $u^0_{x^i x^j} dx^j$ is its differential*;

(d) *for a constant N the function $u - N|x|^2$ is concave in every convex subdomain of D*;

(e) $L(\omega, 0, x) u(x) + f(\omega, 0, x) \geq 0$ *in D in the sense of distributions for every $\omega \in \Omega$.*

Moreover, if a domain $D' \subset D$, $l \in E_d$, $|l| = 1$, and

$$\inf_{\xi:\,(l,\xi)=1} \sup_{\omega} a^{ij}(\omega, 0, x) \xi^i \xi^j \geq \delta \quad \text{in } D',$$

then the Sobolev derivative $u_{(l)(l)}$ is bounded in D' (a.e.). If

$$\inf_{|\xi|=1} \sup_{\omega} a^{ij}(\omega, 0, x) \xi^i \xi^j \geq \delta \quad \text{in } D',$$

then all Sobolev derivatives $u_{x^i x^j}$ are bounded in D' (a.e.) and u satisfies (2.5) in D' (a.e.).

This result is a generalization of the *corresponding* results of [32, 21, and 41]. Let us consider the simplest case of Theorem 3.1.

THEOREM 3.2. *Let D be strictly convex, σ independent of p, x, $b = c = 0$, $\operatorname{tr} a = 1$ identically. Then all the assertions of Theorem 3.1 hold true.*

REFERENCES

1. O. A. Ladyzhenskaya and N. N. Ural'tseva, *Linear and quasilinear elliptic equations*, 2nd. ed., "Nauka", Moscow, 1973; English transl. of 1st ed., Academic Press, New York, 1968.

2. O. A. Ladyzhenskaya, V. A. Solonnikov, and N. N. Ural'tseva, *Linear and quasilinear equations of parabolic type*, "Nauka", Moscow, 1967; English transl., Amer. Math. Soc., Providence, R.I., 1968.

3. A. V. Ivanov, *Quasilinear degenerate and nonuniformly elliptic and parabolic equations of second order*, Trudy Mat. Inst. Steklov. **160** (1982)=Proc. Steklov Inst. Math. **1984**, no. 1(160).

4. D. Gilbarg and N. S. Trudinger, *Elliptic partial differential equations of second order*, Springer-Verlag, Berlin and New York, 1983.

5. N. V. Krylov, *Bounded homogeneous nonlinear elliptic and parabolic equations in the plane*, Mat. Sb. **82(124)** (1970), 99–110=Math. USSR Sb. **11** (1970), 89–99.

6. A. D. Aleksandrov, *The Dirichlet problem for the equation* $\operatorname{Det}\|z_{ij}\| = \varphi(z_1, \ldots, z_n, z, x_1, \ldots, x_n)$, Vestnik Leningrad. Univ. **1958**, no. 1 (Ser. Mat. Mekh. Astr. vyp. 1), 5–24. (Russian)

7. I. Ya. Bakel'man, *Geometric methods for solving elliptic equations*, "Nauka", Moscow, 1965. (Russian)

8. A. V. Pogorelov, *The Dirichlet problem for the n-dimensional analog of the Monge-Ampère equation*, Dokl. Akad. Nauk **201** (1971), 790–793=Soviet Math. Dokl. **12** (1971), 1727–1731.

9. Shiu-Yuen Cheng and Shing-Tung Yau, *On the regularity of the Monge-Ampère equation* $\det(\partial^2 u/\partial x_i \partial x_j) = F(x, u)$, Comm. Pure Appl. Math. **30** (1977), 41–68.

10. N. V. Krylov, *On nonlinear elliptic equations in a domain*, Uspekhi Mat. Nauk **37** (1982), no. 4(226), 91. (Russian)

11. _____, *On degenerate nonlinear elliptic equations*, Mat. Sb. **120(162)** (1983), 311–330=Math. USSR Sb. **48** (1984), 307–326.

12. I. M. Ivochkina, *Classical solvability of the Dirichlet problem for the Monge-Ampère equation*, Zap. Nauchn. Sem. Leningrad. Otdel. Mat. Inst. Steklov (LOMI) **131** (1983), 72–79-J. Soviet Math. **30** (1985), no. 4, 2287–2291.

13. L. Caffarelli, L. Nirenberg, and J. Spruck, *The Dirichlet problem for nonlinear second order elliptic equations*. I, Comm. Pure Appl. Math. **37** (1984), 369–402.

14. J. I. E. Urbas, *Elliptic equations of Monge-Ampère type*, Thesis, Australian Nat. Univ., Canberra, 1984.

15. W. H. Fleming and R. W. Rishel, *Deterministic and stochastic optimal control*, Springer-Verlag, 1975.

16. N. V. Krylov, *Control of the solution of a stochastic integral equation*, Teor. Veroyatnost. i Primenen. **17** (1972), 111–128=Theor. Probab. Appl. **17** (1972), 114–131.

17. _____, *On control of the solution of a stochastic integral equation with degeneration*, Izv. Akad. Nauk SSSR Ser. Mat. **36** (1972), 248–261=Math. USSR Izv. **6** (1972), 249–262.

18. M. Nisio, *Remarks on stochastic optimal controls*, Japan J. Math. **1** (1975), 159–183.

19. M. V. Safonov, *On the control of diffusion processes in a multidimensional cylindrical domain*, Abstr. Comm. Internat. Sympos. Stoch. Diff. Eq., Vilnius, 1978, pp. 168–172.

20. P. L. Lions, *Equations de Hamilton-Jacobi-Bellman dégénérées*, C. R. Acad. Sci. Paris Sér. A **289** (1979), 329–332.

21. _____, *Optimal control of diffusion processes and Hamilton-Jacobi-Bellman equations*. Part 3, Collége de France Seminar, Vol. 5, Pitman, London, 1983.

22. P. L. Lions and J. L. Menaldi, *Optimal control of stochastic integrals and Hamilton-Jacobi-Bellman equations*. I, II, SIAM J. Control Optim. **20** (1982), 58–95.

23. G. [H.] Pragarauskas, *On the Bellman equation in the structure of measures for general controlled random processes*. II, Litovsk. Mat. Sb. **22** (1982), no. 1, 138–145=Lithuanian Math. J. **22** (1982), 68–73.

24. _____, *Uniqueness of the solution of the Bellman equation in the case of general controlled random processes*, Litovsk. Mat. Sb. **22** (1982), no. 2, 137–149=Lithuanian Math. J. **22** (1982), 160–168.

25. N. V. Krylov, *Controlled diffusion processes*, "Nauka", Moscow, 1977; English transl., Springer-Verlag, Berlin and New York, 1980.

26. _____, *On controlled diffusion processes with unbounded coefficients*, Izv. Akad. Nauk SSSR Ser.Mat. **45** (1981), 734–759=Math. USSR Izv. **19** (1982), 41–64.

27. _____, *On the control of a diffusion process up to the time of first exit from a region*, Izv. Akad. Nauk SSSR Ser. Mat. **45** (1981), 1029–1048=Math. USSR Izv. **19** (1982), 297–313.

28. H. Brèzis and L. C. Evans, *A variational inequality approach to the Bellman-Dirichlet equation for two elliptic operators*, Arch. Rational Mech. Anal. **71** (1979), 1–13.

29. L. C. Evans and A. Friedman, *Optimal stochastic switching and the Dirichlet problem for the Bellman equation*, Trans. Amer. Math. Soc. **253** (1979), 365–389.

30. L. C. Evans and P. L. Lions, *Résolution des equations de Hamilton-Jacobi-Bellman pour des opérateurs uniformément elliptiques*, C. R. Acad. Sci. Paris Sér A-B **290** (1980), 1049–1052.

31. P. L. Lions, *Résolution analytiques des problèmes de Bellman-Dirichlet*, Acta Math. **146** (1981), 151–166.

32. N. V. Krylov, *Second-order nonlinear elliptic and parabolic equations*, "Nauka", Moscow, 1985. (Russian)

33. L. C. Evans, *Classical solutions of fully nonlinear convex second order elliptic equations*, Comm. Pure Appl. Math. **35** (1982), 333–363.

34. N. V. Krylov, *Boundedly nonhomogeneous elliptic and parabolic equations*, Izv. Akad. Nauk SSSR Ser. Mat. **46** (1982), 487–523=Math. USSR Izv. **20** (1983), 459–492.

35. N. V. Krylov and M. V. Safonov, *An estimate of the probability that a diffusion process hits a set of positive measure*, Dokl. Akad. Nauk SSSR **245** (1979), 18–20=Soviet Math. Dokl. **20** (1979), 253–255.

36. ____, *A certain property of solutions of parabolic equations with measurable coefficients*, Izv. Akad. Nauk SSSR Ser. Mat. **44** (1980), 161–175=Math. USSR Izv. **16** (1981), 151–164.

37. M. V. Safonov, *On the classical solution of Bellman's elliptic equation*, Dokl. Akad. Nauk SSSR **278** (1984), 810–813=Soviet Math. Dokl. **30** (1984), 482–485.

38. ____, *On model problems for Bellman's elliptic equation*, Uspekhi Mat. Nauk **39** (1984), no. 4(238), 111–112. (Russian)

39. ____, *On the classical solution of second-order nonlinear elliptic equations*, Zap. Nauchn. Sem. Leningrad. Otdel. Mat. Inst. Steklov. (LOMI) (to appear)=J. Soviet Math. (to appear).

40. L. Caffarelli, J. J. Kohn, L. Nirenberg, and J. Spruck, *The Dirichlet problem for nonlinear second order equations*. II, Comm. Pure Appl. Math. **38** (1985), 209–252.

41. L. Caffarelli, L. Nirenberg, and J. Spruck, *The Dirichlet problem for nonlinear second order elliptic equations*. III, Acta Math. **155** (1985), no. 3–4, 261–301.

MOSCOW STATE UNIVERSITY, 117234 MOSCOW, USSR

A Self-Focusing Solution
to the Navier-Stokes Equations
with a Speed-Reducing External Force

VLADIMIR SCHEFFER

Reference [2] contains a proof of the following theorem: *Let $v: R^3 \to R^3$ be an L^2 function with* $\text{div}(v) = 0$. *Then there exists a function* $u: R^3 \times [0, \infty) \to R^3$ *that satisfies (weakly) the Navier-Stokes equations of incompressible fluid flow with viscosity* $= 1$ *and initial condition* v, *and has the following regularity property: There exists a closed set* $S \subset R^3 \times (0, \infty)$ *such that* u *is continuous outside of* S *and the Hausdorff dimension* S *(called* $\dim(S)$*) does not exceed* 2.

The proof in [2] actually yields more information. It implies that S must satisfy $\dim(S \cap (R^3 \times \{t\})) \leq 1$ for all $t > 0$. This result was improved by L. Caffarelli, R. Kohn, and L. Nirenberg in [1]. They showed, among other things, that the dimension of S cannot exceed 1.

One may ask whether this is the best possible estimate for the dimension of S. The question is answered in the affirmative if, instead of the Navier-Stokes equations, we consider the Navier-Stokes *inequality*. Roughly speaking, the Navier-Stokes inequality consists of these four relations:

$$\frac{\partial u_i}{\partial t} = -\sum_{j=1}^{3} u_j \frac{\partial u_i}{\partial x_j} - \frac{\partial p}{\partial x_i} + \Delta u_i + f_i,$$

$$\sum_{i=1}^{3} \frac{\partial u_i}{\partial x_i} = 0, \qquad \sum_{i=1}^{3} \frac{\partial f_i}{\partial x_i} = 0, \qquad \sum_{i=1}^{3} f_i u_i \leq 0.$$

The above is formally equivalent to the relations

$$\int_0^\infty \int_{R^3} |\nabla u|^2 \phi \leq \int_0^\infty \int_{R^3} (2^{-1}|u|^2 + p)u \cdot \nabla \phi + \int_0^\infty \int_{R^3} 2^{-1}|u|^2 \left(\frac{\partial \phi}{\partial t} + \Delta \phi \right)$$

if the test function ϕ satisfies $\phi \geq 0$,

$$\sum_{i=1}^{3} \frac{\partial u_i}{\partial x_i} = 0, \qquad \Delta p = -\sum_{i=1}^{3}\sum_{j=1}^{3} \frac{\partial u_j}{\partial x_i} \frac{\partial u_i}{\partial x_j}.$$

This justifies the term "inequality." Of course, the Navier-Stokes equations are the case $f = 0$. One can think of f as a divergence-free force that pushes against the flow at every point of space-time.

The method of proof of [1] and [2] works just as well for the Navier-Stokes inequality. The only important change that has to be made is this: Regularity of u outside of S means that u is locally essentially bounded on the complement of S.

There is an example that shows that $\dim(S) \leq 1$ is the best possible estimate for the singular set S for the Navier-Stokes inequality. It is written up in [3, 4]. The theorem below is the precise statement of the example.

THEOREM. *If $\varsigma < 1$ then there exist*
(1) $S \subset R^3 \times \{1\}$, $u\colon R^3 \times [0,\infty) \to R^3$, $p\colon R^3 \times [0,\infty) \to R$ *satisfying the following properties*:
(2) *there is a compact set $K \subset R^3$ such that $u(x,t) = 0$ for all $x \notin K$,*
(3) *for fixed t, the function $u_t\colon R^3 \to R^3$ defined by $u_t(x) = u(x,t)$ is a C^∞ function,*
(4) $\sum_{i=1}^3 \partial u_i / \partial x_i = 0$,
(5)

$$ p(x,t) = \int_{R^3} \sum_{i=1}^{3} \sum_{j=1}^{3} \frac{\partial u_j}{\partial x_i}(y,t) \frac{\partial u_i}{\partial x_j}(y,t)(4\pi|x-y|)^{-1}\, dy, $$

(6) $\|u_t\|_2$ *is a bounded function of t (with u_t defined in (3)),*
(7) $|\nabla u|^2$, $|u|^3$, *and $|u|\,|p|$ are integrable.*
(8) *if $\phi\colon R^3 \times (0,\infty) \to R$ is C^∞ with compact support and $\phi \geq 0$, then*

$$ \int_0^\infty \int_{R^3} |\nabla u|^2 \phi \leq \int_0^\infty \int_{R^3} (2^{-1}|u|^2 + p) u \cdot \nabla \phi + \int_0^\infty \int_{R^3} 2^{-1}|u|^2 \left(\frac{\partial \phi}{\partial t} + \Delta \phi\right), $$

(9) *u is not essentially bounded on any neighborhood of any point of S,*
(10) *the Hausdorff dimension of S is greater than ς.*

We know that the inequality $\dim(S) \leq 1$ is a consequence of the conditions (1)–(9) (and of the weaker conditions which are detailed in [1, 2] and which do not assume (2)).

References [3, 4] contain an explicit construction of the example. In order to give an overview of this construction, it is convenient to look at the large-scale appearance first and then to work down to smaller details. Since [3, 4] reverse this order of exposition, the notation used below is different from the notation of those articles. The remainder of this presentation describes the appearance of u under increasing magnification.

With low magnification, we see certain compact sets $K^m \subset R^3$, where m belongs to an index set that can be described as follows: There is a fixed positive integer Y, we set $M(Z) = \{m = (m_1, m_2, \ldots, m_z)\colon m_i \in \{0, 1, \ldots, Y\}\}$, and we take $m \in \bigcup_{Z=1}^\infty M(Z)$. If $m \in M(Z)$, $n \in M(Z+1)$, and $m_i = n_i$ for $i \leq Z$ then we have $K^n \subset K^m$. If Z is fixed then $\{K^m\colon m \in M(Z)\}$ is a collection of disjoint sets. There is an infinite sequence $T_1 < T_2 < T_3 < \cdots$ of times with

$\lim_{Z \to \infty} T_Z = 1$. The restriction of u to $K^m \times [T_Z, T_{Z+1}]$, where $m \in M(Z)$, is called u^m. The functions u^m resemble each other qualitatively, but they differ in fine detail. We have $\|u^m\|_\infty \approx C\tau^{-Z}$ (with $m \in M(Z)$), where τ is a constant satisfying $0 < \tau < 1$. As time t approaches 1 from the left, a singularity develops on the set $\bigcap_{Z=1}^\infty \bigcup_{m \in M(Z)} K^m \times \{1\}$.

Now we increase the magnification and examine an individual u^m, where $m \in M(Z)$. Inside, we see vortices $V_1, V_2, W_j, W_j', W_j''$ for $j \in \{0, 1, \ldots, Y\}$ that share the same axis of symmetry. The vortices V_1 and V_2 are the largest of these, and they are widely separated. For each j, the vortices W_j, W_j', W_j'' are close together, they lie inside V_2, and they form a decreasing progression that points at a certain region D_j of V_2. As time passes from T_Z to T_{Z+1}, all of these vortices oscillate very rapidly. While V_1, W_j, W_j', W_j'' oscillate in unison, the vortex V_2 oscillates differently. Farther away, the analogous vortices belonging to other u^m (with the same Z) are also oscillating. However, the beat of the oscillation varies with m. There is a wide range of frequencies for these oscillations. The vortices affect each other through pressure forces. The geometry of the vortices and the nature of the oscillation have the following net effect: Vortex V_1 takes energy from a portion of V_2 and concentrates it into another portion, called A, of V_2. For every $j \in \{0, 1, \ldots, Y\}$, W_j takes advantage of the increase on A and concentrates it further into a smaller portion B_j of A. At the same time, W_j' pumps energy from B_j into a smaller C_j. Finally, W_j'' takes energy from C_j and puts it into D_j, which was mentioned earlier. This transfer of energy is completed by time T_{Z+1}. Our fixed K^m contains $K_0, K_1, \ldots, K_Y$, where K_j is the K^n with $n = (m_1, m_2, \ldots, m_Z, j) \in M(Z+1)$. Each K_j consists of a portion of A and a portion of D_j. The portion of A gives rise to the V_1 vortex of K_j in the next iteration (with Z replaced by $Z + 1$). The portion of D_j is the seed of the future vortices V_2, W_k, W_k', W_k'' of K_j.

We look more closely at the edge of u^m. Starting from the boundary of K^m and working our way in, we first encounter a region that does not oscillate. Here, u^m starts out being zero and increases as we go deeper into K^m. Then we meet a zone with gentle time decay of $|u^m|$ and still no oscillation. Later, we encounter a mild oscillation that increases steadily and becomes very intense as we travel farther from the boundary of K^m. This oscillation is superimposed on slow time decay.

REFERENCES

1. L. Caffarelli, R. Kohn and L. Nirenberg, *Partial regularity of suitable weak solutions of the Navier-Stokes equations*, Comm. Pure Appl. Math. **35** (1982), 771–831.

2. V. Scheffer, *Hausdorff measure and the Navier-Stokes equations*, Comm. Math. Phys. **55** (1977), 97–112.

3. ____, *A solution to the Navier-Stokes inequality with an internal singularity*, Comm. Math. Phys. **101** (1985), 47–85.

4. ____, *Nearly one dimensional singularities of solutions to the Navier-Stokes inequality* (to appear).

RUTGERS UNIVERSITY, NEW BRUNSWICK, NEW JERSEY 08903, USA

Proceedings of the International Congress of Mathematicians
Berkeley, California, USA, 1986

Free Boundary Problems
and Problems in Noncompact Domains
for the Navier-Stokes Equations

V. A. SOLONNIKOV

1. In this communication we are concerned mainly with free boundary problems for an incompressible viscous fluid. In these problems the domain $\Omega_t \subset \mathbb{R}^n$, $n = 2, 3$, occupied by the fluid at the moment $t > 0$ is to be determined together with the velocity vector field $\mathbf{v}(x, t) = (v_1, \ldots, v_n)$ and with the pressure $p(x, t)$ satisfying the Navier-Stokes equations

$$\mathbf{v}_t + (\mathbf{v} \cdot \nabla)\mathbf{v} - \nu \nabla^2 \mathbf{v} + \nabla p = \mathbf{f}(x, t), \quad \nabla \cdot \mathbf{v} = 0; \quad x \in \Omega_t,\ t > 0, \quad (1)$$

and the initial and boundary conditions

$$\begin{aligned}
\mathbf{v}|_{t=0} &= \mathbf{v}_0(x), \quad x \in \Omega_0 \equiv \Omega, \\
\mathbf{v}|_{x \in \Sigma} &= 0, \quad \mathsf{T}\mathbf{n} - \sigma H \mathbf{n}|_{x \in \Gamma_t} = 0.
\end{aligned} \quad (2)$$

Here ν and $\sigma > 0$ are constant coefficients of viscosity and of the surface tension, $\mathbf{f}(x, t)$ $(x \in \mathbb{R}^n, t > 0)$ is a vector field of external forces, Σ and Γ_t are two disjoint components of the boundary $\partial \Omega_t$ (Σ is fixed and Γ_t is free), $\mathbf{n}(x)$ is the unit outward normal vector to Γ_t at the point x, $\mathsf{T} = -p\mathbb{I} + \nu \mathsf{S}(\mathbf{v})$ is the stress tensor, $\mathsf{S}(\mathbf{v})$ is the strain tensor with the elements $S_{ij} = \partial v_i/\partial x_j + \partial v_j/\partial x_i$, $H/n - 1$ is the mean curvature of Γ_t. The sign of H is chosen in such a way that $H\mathbf{n} = \Delta(t)x$, where $\Delta(t)$ is the Laplacian on Γ_t.

According to a kinematic boundary condition at the free surface, $\Gamma_t = \{x = x(\xi, t),\ \xi \in \Gamma\}$, where $x(\xi, t)$ is a solution of the Cauchy problem

$$\partial \mathbf{x}/\partial t = \mathbf{v}(x, t), \quad \mathbf{x}(0) = \xi, \quad (3)$$

and it follows that $\Omega_t = \{x = x(\xi, t),\ \xi \in \Omega\}$. In particular, if Γ_t is given by $F(x, t) = 0$, then (3) is equivalent to $\partial F/\partial t + \mathbf{v} \cdot \nabla F|_{F=0} = 0$.

Sometimes additional conditions are necessary, for instance, conditions at infinity in the case of a noncompact Ω_t.

As a typical example, consider the free boundary problem governing the motion of a finite mass of a liquid bounded by a free surface [1–3]. In this problem $\overline{\Omega}_t$ is compact, $\partial \Omega_t = \Gamma_t$, $\Sigma = \varnothing$. To avoid the difficulties connected with the presence of a variable unknown surface $\Gamma_t \subset \partial\Omega_t$, it is convenient to choose $\xi \in \Omega$ as new independent variables. These so-called Lagrangian coordinates are connected with the Eulerian coordinates $x \in \Omega_t$ by relationship (3) or by an equivalent formula $\mathbf{x} = \boldsymbol{\xi} + \int_0^t \mathbf{u}(\xi, \tau)\, d\tau \equiv X_u(\xi, t)$ where $\mathbf{u}(\xi, t) = \mathbf{v}(X_u, t)$. In the Lagrangian coordinates, the problem (1)–(3) takes the form

$$\mathbf{u}_t - \nu\nabla_u^2\mathbf{u} + \nabla_u q = \mathbf{f}(X_u, t), \qquad \nabla_u \cdot \mathbf{u} = 0$$

$$(\xi \in \Omega,\ t > 0), \qquad (4)$$

$$\mathbf{u}|_{t=0} = \mathbf{v}_0(\xi), \qquad \mathsf{T}_u\mathbf{n} - \sigma\Delta(t)X_u|_{\xi\in\Gamma} = 0.$$

Here

$$q(\xi, t) = p(X_u(\xi, t), t), \qquad \nabla_u = \mathcal{A}\nabla = \left\{\sum_m A_{im}\frac{\partial}{\partial\xi_m}\right\}_{i=1,\dots,n},$$

$$\mathsf{T}_u = -q\mathsf{I} + \nu\mathsf{S}_u(\mathbf{u}), \qquad (S_u)_{ij} = \sum_m\left(A_{im}\frac{\partial u_j}{\partial\xi_m} + A_{jm}\frac{\partial u_i}{\partial\xi_m}\right)$$

and $\mathcal{A}$ is the matrix with the elements $A_{im} = \partial\xi_m/\partial x_i|_{x=X_u(\xi,t)}$, that may be computed as algebraic adjuncts of $a_{ij} = \partial x_i/\partial\xi_j = \delta_{ij} + \int_0^t \partial u_i/\partial\xi_j\, d\tau$. Finally, $\mathbf{n} = \mathbf{n}(X_u) = \mathcal{A}\mathbf{n}_0/|\mathcal{A}\mathbf{n}_0|$, where $\mathbf{n}_0(\xi)$ is the unit outward normal to $\partial\Omega = \Gamma$.

Let $r > 0$ and let $W_2^r(\Omega) = H_r(\Omega)$, $W_2^r(\Gamma)$, $W_2^r(0, T)$ be S. L. Sobolev–L. N. Slobodetskiĭ spaces of functions defined in Ω, on Γ, and in $(0, T)$, respectively. By $W_2^{r,r/2}(Q_T)$, $Q_T = \Omega \times (0, T)$, we mean the space $L_2((0, T); W_2^r(\Omega)) \cap L_2(\Omega; W_2^{r/2}(0, T))$ equipped with the norm

$$\|u\|_{W_2^{r,r/2}(Q_T)}^2 = \int_0^T \|u\|_{W_2^r(\Omega)}^2\, dt + \int_\Omega \|u\|_{W_2^{r/2}(0,T)}^2\, dx.$$

We also define the space $W_2^{r,r/2}(G_T)$ on the manifold $G_T = \Gamma \times (0, T)$ as $L_2((0, T); W_2^r(\Gamma)) \cap L_2(\Omega; W_2^{r/2}(0, T))$.

The spaces of vector fields with the components in $W_2^{r,r/2}(Q_T)$ or $W_2^{r,r/2}(G_T)$ are denoted by $\mathbf{W}_2^{r,r/2}(Q_T)$ or by $\mathbf{W}_2^{r,r/2}(G_T)$.

In the study of the free boundary problem (1)–(3) a considerable role is played by a corresponding linearized problem. We observe that the boundary conditions in (4) can be written as

$$\Pi_0\Pi S_u(\mathbf{u})\mathbf{n} = \Pi_0 S_u\mathbf{n} - \Pi_0\mathbf{n}(\mathbf{n}\cdot S_u\mathbf{n}) = 0, \qquad \mathbf{n}_0\cdot \mathsf{T}_u\mathbf{n} - \sigma\mathbf{n}_0\cdot\Delta(t)X_u = 0,$$

where $\Pi_0\mathbf{g} = \mathbf{g} - \mathbf{n}_0(\mathbf{n}_0\cdot\mathbf{g})$, $\Pi\mathbf{g} = \mathbf{g} - \mathbf{n}(\mathbf{n}\cdot\mathbf{g})$, and consider a linear problem

$$\mathbf{w}_t - \nu\nabla_u^2\mathbf{w} + \nabla_u s = \mathbf{f}(\xi, t), \qquad \nabla_u\cdot\mathbf{w} = \rho,$$

$$\mathbf{w}|_{t=0} = \mathbf{w}_0(\xi), \qquad \Pi_0\Pi S_u(\mathbf{w})\mathbf{n}|_{\xi\in\Gamma} = \Pi_0\mathbf{d}, \qquad (5)$$

$$\mathbf{n}_0\cdot \mathsf{T}_u(\mathbf{w}, s)\mathbf{n} - \sigma\mathbf{n}_0\cdot\Delta(t)\int_0^t \mathbf{w}\, d\tau|_{\xi\in\Gamma} = b + \sigma\int_0^t B\, d\tau.$$

THEOREM 1.   *Let* $l \in (\frac{1}{2}, 1)$, $\Gamma \in W_2^{3/2+l}$, $\mathbf{u} \in \mathbf{W}_2^{2+l,1+l/2}(Q_T)$, *and*

$$T^{1/2}\|\mathbf{u}\|_{\mathbf{W}_2^{2+l,1+l/2}(Q_T)} \le \delta$$

with a small $\delta > 0$, $\mathbf{f}, \nabla \rho \in \mathbf{W}_2^{l,l/2}(Q_T)$, $\mathbf{v}_0 \in \mathbf{W}_2^{1+l}(\Omega)$, $\mathbf{d} \in \mathbf{W}_2^{1/2+l,1/4+l/2}(G_T)$, $b \in W_2^{1/2+l,1/4+l/2}(G_T)$, $B \in W_2^{l-1/2,l/2-1/4}(G_T)$. *Assume that* $\rho = \nabla \cdot \mathbf{R}$, $\mathbf{R}_t \in L_2(\Omega; W_2^{l/2}(0,T))$, *and that the compatibility conditions* $\nabla \cdot \mathbf{w}_0 = \rho(x,0)$, $\Pi_0 S(\mathbf{w}_0)\mathbf{n}_0|_\Gamma = \Pi_0 \mathbf{d}(\xi,0)$ *hold. Then the problem* (5) *has the unique solution* $\mathbf{w} \in \mathbf{W}_2^{2+l,1+l/2}(Q_T)$, $\nabla s \in \mathbf{W}_2^{l,l/2}(Q_T)$, *and*

$$\|\mathbf{w}\|_{\mathbf{W}_2^{2+l,1+l/2}(Q_T)} + \|\nabla s\|_{\mathbf{W}_2^{l,l/2}(Q_T)} + \|s\|_{W_2^{1/2+l,1/4+l/2}(G_T)}$$

$$\leq C(T) \left(\|\mathbf{f}\|_{\mathbf{W}_2^{l,l/2}(Q_T)} + \|\nabla \rho\|_{\mathbf{W}_2^{l,l/2}(Q_T)} + \|\mathbf{w}_0\|_{\mathbf{W}_2^{1+l}(\Omega)} \right. \tag{6}$$

$$+ \|\mathbf{d}\|_{\mathbf{W}_2^{1/2+l,1/4+l/2}(G_T)} + \|b\|_{W_2^{1/2+l,1/4+l/2}(G_T)}$$

$$\left. + \|B\|_{W_2^{l-1/2,l/2-1/4}(G_T)} + \left(\int_\Omega \|\mathbf{R}_t\|^2_{W_2^{l/2}(0,T)} \, dx \right)^{1/2} \right).$$

We now turn to existence theorems for our free boundary problem in the three-dimensional case.

THEOREM 2.   *Let* $n = 3$, $l \in (\frac{1}{2},1)$, $\Gamma \in W_2^{5/2+l}$, $\mathbf{v}_0 \in W_2^{1+l}(\Omega)$, $\nabla \cdot \mathbf{v}_0 = 0$, $\Pi_0 S(\mathbf{v}_0)\mathbf{n}_0|_\Gamma = 0$, *and let* $\mathbf{f}, \mathbf{f}_{x_\kappa}$ *satisfy the Lipschitz condition with respect to* x *and the Hölder condition with the exponent* $\frac{1}{2}$ *with respect to* t. *Then the problem* (4) *has the unique solution* $\mathbf{u} \in \mathbf{W}_2^{2+l,1+l/2}(Q_{T_1})$, $\nabla q \in \mathbf{W}_2^{l,l/2}(Q_{T_1})$ *and* $q|_{G_{T_1}} \in W_2^{1/2+l,1/4+l/2}(G_{T_1})$. *The magnitude of* $T_1 < \infty$ *depends on* $\|\mathbf{v}_0\|_{\mathbf{W}_2^{1+l}(\Omega)}$,

$$\|\mathbf{f}\|_{T_1} = \sup_{x,t \leq T_1} |\mathbf{f}(x,t)| + \max_\kappa \sup |\mathbf{f}_{x_\kappa}| + \max_\kappa \sup_{x,t,\tau} \frac{|\mathbf{f}(x,t) - \mathbf{f}(x,\tau)|}{|t-\tau|^{1/2}},$$

and on $\sigma\|H(\xi,0)\|_{W_2^{1/2+l}(\Gamma)}$.

THEOREM 3.   *Assume that* $l \in (\frac{1}{2},1)$ *and that* Ω *is diffeomorphic to a ball:* $\Gamma = \{|x| = R(\omega), |\omega| = 1\}$. *Then the solution* $(\mathbf{u},q)$ *of the problem* (4) *satisfies the inequality*

$$\|\mathbf{u}\|_{\mathbf{W}_2^{2+l,1+l/2}(Q_{T_1})} + \|\nabla q\|_{\mathbf{W}_2^{l,l/2}(Q_{T_1})} + \|q - q_0\|_{W_2^{1/2+l,1/4+l/2}(G_{T_1})}$$

$$\leq C \left(\|\mathbf{f}\|_{T_1} + \|\mathbf{v}_0\|_{\mathbf{W}_2^{1+l}(\Omega)} + \left\| H(\xi,0) + \frac{2}{R_0} \right\|_{W_2^{1/2+l}(\Gamma)} \right), \tag{7}$$

where $q_0 = 2\sigma/R_0$ *and* $R_0 = (3|\Omega|/4\pi)^{1/3}$ *is the radius of a ball with the volume* $|\Omega|$.

The number T_1 *grows without limits as the norms in the right-hand side of* (7) *approach zero. If* $\mathbf{f} = 0$ *and* $\|\mathbf{v}_0\|_{\mathbf{W}_2^{1+l}(\Omega)} + \|R(\omega) - R_0\|_{W_2^{5/2+l}(S_1)} \leq \varepsilon$ *for sufficiently small* $\varepsilon > 0$, *then the solution of the problem* (1)–(3) *is defined for all* $t > 0$ *and the norms* $\|\mathbf{v}_t\|_{\mathbf{W}_2^{l}(\Omega_t)}$, $\|\mathbf{v}\|_{\mathbf{W}_2^{2+l}(\Omega_t)}$, $\|p - q_0\|_{W_2^{1+l}(\Omega_t)}$, *and* $\|R - R_0\|_{W_2^{5/2+l}(S_1)}$ ($S_1 = \{|\omega| = 1\}$) *are bounded uniformly with respect to* $t \geq t_0 > 0$.

The local existence theorem follows from the coercive estimate (6) for the linear problem (5). To prove the global existence theorem, we invoke conservation laws for the problem (1)–(3)

$$\int_{\Omega_t} \mathbf{v}(x,t) \cdot \boldsymbol{\eta}(x)\, dx = \int_{\Omega} \mathbf{v}_0 \cdot \boldsymbol{\eta}\, dx, \qquad \boldsymbol{\eta} = \mathbf{a} + \mathbf{b} \times \mathbf{x}, \tag{8}$$

$$\frac{d}{dt}\left(\int_{\Omega_t} |\mathbf{v}(x,t)|^2\, dx + \sigma |\Gamma_t| \right) + \nu \int_{\Omega_t} \sum_{i,j=1}^{3} \left(\frac{\partial v_i}{\partial x_j} + \frac{\partial v_j}{\partial x_i} \right)^2 dx = 0 \tag{9}$$

where $|\Gamma_t|$ is the area of Γ_t and $\mathbf{a}, \mathbf{b}$ are arbitrary constant vectors. If Ω_t is homeomorphic to a ball and Γ_t is given by $r = R(\omega,t)$ in the spherical coordinates (r,ω) with the origin in the center of gravity of Ω_t, then (9) implies

$$\int_{\Omega_t} |\mathbf{v}|^2\, dx + C\sigma \int_{S_1} (|R(\omega,t) - R_0|^2 + |\nabla_\omega R|^2)\, d\omega \le \int_{\Omega} |\mathbf{v}_0|^2\, dx + |\Gamma| - 4\pi R_0^2,$$

provided that $|R - R_0|^2 + |\nabla_\omega R|^2$ is small. The estimates of higher-order derivatives of the solution may be found from the local bounds for the problem (4). This makes it possible to extend the solution of the problem (1)–(3) defined for $t \in (0, T_1)$ into the intervals $t \in (jT_1, (j+1)T_1)$, $j = 1, 2, \ldots$, applying Theorem 2 to the problem (1)–(3) written for $t > jT_1$ in the Lagrangian coordinates $\xi^{(j)} \in \Omega_{jT_1}$.

Theorem 2 and the formulae (8), (9) hold in the case $\sigma = 0$ under the assumption $\Gamma \in W_2^{3/2+l}$. In this case $\int_{\Omega_t} |\mathbf{v}|^2\, dx = \int_{\Omega} |\mathbf{u}|^2\, d\xi$ decays exponentially for large $t > 0$. This is a consequence of (9) and of Korn's inequality

$$\int_{\Omega_t} \sum_{i,j=1}^{3} \left(\frac{\partial v_i}{\partial x_j} + \frac{\partial v_j}{\partial x_i} \right)^2 dx \ge C \int_{\Omega_t} (|D\mathbf{v}|^2 + |\mathbf{v}|^2)\, dx$$

which holds for arbitrary $\mathbf{v}$ orthogonal to $\boldsymbol{\eta} = \mathbf{a} + \mathbf{b} \times \mathbf{x}$. Therefore the solution of (1), (2) with $\sigma = 0$, $\mathbf{f} = 0$ can also be extended to all $t \ge 0$, provided $\int_{\Omega} \mathbf{v}_0 \cdot \boldsymbol{\eta}\, dx = 0$. The free boundary Γ_t tends to a limiting surface $\Gamma_\infty = \{\mathbf{x} = \xi + \int_0^\infty \mathbf{u}(\xi, \tau)\, d\tau, \ \xi \in \Gamma\}$.

Similar results can be obtained in the two-dimensional case.

V. Bytev [4] and O. Lavrent'eva [5] considered the problem (1)–(3) in the case when Ω_t is a ring with both boundaries free and $\mathbf{v}_0$ is axially symmetric. It was found [4] that in the case $\sigma = 0$ Ω_t expands to the infinity, if the condition $\int_{\Omega} \mathbf{v} \cdot \boldsymbol{\eta}\, dx = 0$ is violated. For $\sigma > 0$ it may also happen that Ω_t takes the shape of a circle at a certain moment $t = t_0 < \infty$ and it keeps this shape for $t > t_0$ [5].

The papers of T. Beale [6, 7] and G. Allain [8] are devoted to the flow of a heavy viscous fluid over an infinite bottom $\Sigma = \{x_3 = -b(x_1, x_2) < 0\}$ with a free boundary $\Gamma_t = \{x_3 = \varphi(x_1, x_2, t)\}$. The flow is described by the free boundary problem (1)–(3) with $\mathbf{f} = -g(0, 0, 1)$ and

$$H = \sum_{\alpha=1}^{2} \frac{\partial}{\partial x_\alpha} \frac{\varphi_{x_\alpha}}{\sqrt{1 + \varphi_{x_1}^2 + \varphi_{x_2}^2}},$$

$g = \text{const} > 0$. It is proved that this problem has a unique solution in a finite interval of time $(0, T_1)$ for $\sigma = 0$ [6] and in the infinite interval $t > 0$ for $\sigma > 0$, if the data $\mathbf{v}_0$ and $\varphi(x_1, x_2, 0)$ are small [7]. The local existence theorem for $\sigma > 0$, $n = 2$ is established in [8].

2. In the stationary case the flow does not depend on time and is described by the system

$$-\nu\nabla^2\mathbf{v} + (\mathbf{v}\cdot\nabla)\mathbf{v} + \nabla p = \mathbf{f}(x), \qquad \nabla\cdot\mathbf{v} = 0, \tag{10}$$

$$\mathbf{v}|_\Sigma = 0, \quad \mathbf{v}\cdot\mathbf{n}|_\Gamma = 0, \quad \mathbb{T}\mathbf{n} - \sigma H\mathbf{n}|_\Gamma = 0 \tag{11}$$

that often should be completed by additional relationships such as conditions at infinity or at $M = \overline{\Sigma}\cap\overline{\Gamma}$ (if $M \neq \varnothing$) where the so-called "wetting angle," i.e., the angle between Σ and Γ, is usually prescribed. Sometimes $|\Omega|$ is fixed, etc.

V. V. Puknachov [9] has studied a two-dimensional flow of a heavy viscous fluid over the bottom Σ: $x_2 = 0$ with sources and drains distributed on Σ periodically, so that the boundary condition at Σ is

$$\mathbf{v}|_{x_2=0} = \mathbf{a}(x_1), \quad \mathbf{a}(x_1 + T) = \mathbf{a}(x_1), \quad \int_0^T a_2\, dx_1 = 0.$$

The free boundary Γ is supposed to be given by $x_2 = \varphi(x_1)$ and

$$\int_0^T \varphi(x_1)\, dx_1 = h > 0.$$

It is proved in [9] that in the case of small $\mathbf{a}$ the stationary free boundary problem with these conditions has a solution $(\varphi, \mathbf{v}, p)$ in some Hölder classes of functions periodic with respect to x_1. The proof is based on coercive Schauder estimates for the linear problem

$$-\nu\nabla^2\mathbf{v} + \nabla p = \mathbf{f}, \quad \nabla\cdot\mathbf{v} = \rho \qquad (x\in\Omega,\ \varphi(x_1+T) = \varphi(x_1)),$$

$$\mathbf{v}|_\Sigma = \mathbf{a}, \quad \mathbf{v}\cdot\mathbf{n}|_\Gamma = b, \quad \Pi S(\mathbf{v})\mathbf{n}|_\Gamma = \Pi\mathbf{d},$$

$$\mathbf{v}(x_1 + T, x_2) = \mathbf{v}(x), \quad p(x_1 + T, x_2) = p(x)$$

and consists in the application of the contraction mapping principle to the equation

$$\mathbf{n}\cdot\mathbb{T}\mathbf{n}|_{x_2=\varphi(x_1)} = \sigma H = \sigma\frac{d}{dx_1}\frac{\varphi'}{\sqrt{1+\varphi'^2}}.$$

A number of other stationary free boundary problems were investigated in a similar way. They described such phenomena as the motion of a thin film of a heavy liquid on the surface of rotating cylinder [10], the motion of a drop in a symmetric force field [11, 12], the motion of a liquid on the surface of a sphere [13], the viscous flow past a liquid drop, and a steady fall of a drop [14, 15]. Two- and three-dimensional free boundary problems with nonempty $M = \overline{\Sigma}\cap\overline{\Gamma}$ are studied in [16–23]. They are slightly more complicated since M is a wedge or the union of angular points on $\partial\Omega$. For the investigation of problems of this type a special technique was developed in weighted Hölder and Sobolev spaces, but this technique is not necessary in the case of small wetting angles θ or $\theta = \pi/2$ [16, 17]. A different approach to free boundary problems in the two-dimensional

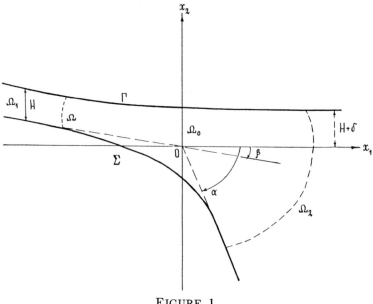

case based on the theory of functions of a complex variable was proposed by
L. K. Antanovskiĭ [**24**], who proved in particular the analyticity of the free
boundary. In the three-dimensional case this result is due to J. Bemelmans and
A. Friedman [**25**]. Some problems with a noncompact free boundary are studied
by R. Piletskas [**26**].

Consider the two-dimensional flow of a heavy viscous fluid down a wall Σ into
an infinite pool. Here Σ is an infinite line consisting of two straight lines $x_2 =
x_1 \tan \beta$, $x_1 < -R$, and $x_2 = x_1 \tan \alpha$, $x_2 < -R$, $0 > \beta > \alpha$, connected with a
smooth curve. The free boundary $\Gamma \colon x_2 = \varphi(x_1)$ is noncompact. The domain
Ω (see Figure 1) has two "exits at infinity," Ω_1 and Ω_2, i.e., $\Omega = \Omega_0 \cup \Omega_1 \cup \Omega_2$,
$\Omega_0 = \{x \in \Omega \colon |x| \le R_0\}$, $\overline{\Omega}_1 \cap \overline{\Omega}_2 = \varnothing$. The flow is described by (10), (11) with
$\mathbf{f} = -g(0,1)$ plus conditions at infinity

$$\varphi(x_1) - (x_1 \tan \beta + H) \to 0, \qquad x_1 \to -\infty,$$

$$\varphi(x_1) - (H + \delta) \to 0, \qquad x_1 \to +\infty,$$

$$\mathbf{v} - \mathbf{v}^{(-)} \to 0, \; p - p^{(-)} \to 0, \qquad |x| \to \infty, \; x \in \Omega_1,$$

$$\mathbf{v} - \mathbf{v}^{(+)} = O\left(\frac{1}{|x|^b}\right), \quad p - p^{(+)} = O\left(\frac{1}{|x|^{b+1}}\right), \qquad b > 1, \; |x| \to \infty, \; x \in \Omega_2,$$

where H is a given positive constant, δ is a certain unknown constant, $\mathbf{v}^{(-)}, p^{(-)}$
correspond to the Poiseuille flow in the strip $x_1 \tan \beta \le x_2 \le x_1 \tan \beta + H$, and

$\mathbf{v}^{(+)}, p^{(+)}$ represent the Jeffrey-Hamel flow in the angle $\alpha < \theta < 0, r > 0$:

$$\mathbf{v}^{(-)} = (-\cos\beta, \sin\beta)\frac{g\sin\beta}{2\nu}y(2H - y), \qquad p^{(-)} = -g\cos\beta(y - H\cos\beta),$$

$$y = x_2\cos\beta - x_1\sin\beta,$$

$$\mathbf{v}^{(+)} = \frac{6\nu U(\theta)}{r}\mathbf{i}_r, \qquad p^{(+)} = \frac{6\nu^2}{r^2}(2U(\theta) - C) + g(H + \delta - x_2).$$

The term $g(H + \delta - x_2)$ corresponds to a hydrostatic pressure, $U(\theta)$ is a smooth function, $U(\alpha) = 0$, and $U'(0) = 0$. The fluxes of $\mathbf{v}^{(+)}$ and $\mathbf{v}^{(-)}$ coincide:

$$6\nu\int_\alpha^0 U(\theta)\,d\theta = \int_{x_1\tan\beta}^{x_1\tan\beta+H} v_2^{(-)}\,dx_2 = -\frac{g}{3\nu}H^3\cos^3\beta\sin\beta.$$

This problem has a solution, if β is small enough and $\alpha > -\alpha_0/2$ where $\alpha_0 \in (\pi, 3\pi/2)$ is a root of the function $\tan t - t$, and $\varphi(x_1) - H - x_1\tan\beta$, $\mathbf{v} - \mathbf{v}^{(-)}$, $p - p^{(-)}$ decay exponentially as $x_1 \to -\infty$.

We present here the theorem on the solvability of a linear problem

$$-\nu\nabla^2\mathbf{v} + \nabla p = \mathbf{f}, \qquad \nabla\cdot\mathbf{v} = \rho, \qquad x \in \Omega,$$
$$\mathbf{v}|_\Sigma = 0, \qquad \Pi S(\mathbf{v})\mathbf{n}|_\Gamma = 0, \qquad \mathbf{v}\cdot\mathbf{n}|_\Gamma = 0.$$

(12)

Let $l = [l] + \lambda$, $\lambda \in (0, 1)$, $a, b > 0$, and let $C^l(\Omega, a, b)$ be the space of functions with the norm

$$|u|_{C^l(\Omega,a,b)} = |ue^{-ax_1/\cos\beta}|_{C^l(\Omega_1)} + |u|_{C^l(\Omega_0)}$$
$$+ \sum_{|\alpha|<l}\sup_{\Omega_2}|D^\alpha u(x)|x|^{|\alpha|+b}| + \sum_{|\alpha|=[l]}[D^\alpha u(x)|x|^{l+b}]_{\Omega_2}^{(\lambda)},$$

where $|u|_{C^l(\Omega_0)}$ is the usual Hölder norm and

$$[v]_{\Omega_2}^{(\lambda)} = \sup_{x,y\in\Omega_2}|x - y|^{-\lambda}|v(x) - v(y)|$$

is a Hölder constant in Ω_2. By $\mathbf{C}^l(\Omega, a, b)$ we mean the space of vector fields with components in $C^l(\Omega, a, b)$.

THEOREM 4. Let $0 > \alpha > -\alpha_0/2$, $\Sigma \in C^{l+2}$, $\Gamma \in C^{l+3}$, $a \in (0, \operatorname{Re} z)$, $b \in (1, \operatorname{Re}\varsigma)$, where z and ς are roots of $\operatorname{sh}(2zH\cos\beta) = 2zH\cos\beta$ and $\sin 2\varsigma\alpha = \varsigma\sin\alpha$, $\varsigma \neq 1$, with minimal real parts. For arbitrary $\mathbf{f} \in \mathbf{C}^l(\Omega, a, b+2)$, $\rho \in C^{l+1}(\Omega, a, b+1)$, $\int_\Omega \rho\,dx = 0$ the problem (12) has the unique solution $\mathbf{v} \in \mathbf{C}^{l+2}(\Omega, a, b)$, $\nabla p \in \mathbf{C}^l(\Omega, a, b+2)$ and $p \to 0$ as $|x| \to \infty$, $x \in \Omega_2$, $p \to p_0$ as $|x| \to \infty$, $x \in \Omega_1$. The solution satisfies the inequality

$$|\mathbf{v}|_{\mathbf{C}^{l+2}(\Omega,a,b)} + |\nabla p|_{\mathbf{C}^l(\Omega,a,b+2)} + |p_0|$$
$$\le C(|\mathbf{f}|_{\mathbf{C}^l(\Omega,a,b+2)} + |\rho|_{C^{l+1}(\Omega,a,b+1)}).$$

3. The problem (12) provides an example of a boundary value problem for the Stokes equations in a domain with a noncompact boundary. In a more general case it is supposed that $\Omega \subset \mathbf{R}^n$, $n = 2, 3$, has $m > 1$ "exits to infinity," i.e., $\Omega = \Omega_0 \cup \Omega_1 \cup \cdots \cup \Omega_m$ where $\Omega_0 = \{x \in \Omega: |x| \le R_0\}$ and $\Omega_1, \ldots, \Omega_m$ are

disjoint unbounded subdomains of Ω. Consider a boundary or initial-boundary value problem for the Navier-Stokes equations in Ω with adherence boundary conditions

$$\mathbf{v}|_{x\in\partial\Omega} = 0. \tag{13}$$

As shown in [27], the classical formulation of these problems is not quite complete, and it should be supplemented by some auxiliary conditions. In our case they can be taken in the form

$$\int_{\Sigma_i} \mathbf{v} \cdot \mathbf{n}\, dS = \alpha_i, \qquad i = 1, \ldots, m-1, \tag{14}$$

where $\int_{\Sigma_i} \mathbf{v} \cdot \mathbf{n}\, dS$ is a total flux of $\mathbf{v}$ in Ω_i; α_i may depend on time in a nonstationary case. Other variants of auxiliary conditions are discussed in [27–29].

It may happen [30], if Ω_i does not blow up too much at infinity, that all divergence-free vector fields with a finite Dirichlet integral $\int_{\Omega_i} |D\mathbf{v}|^2\, dx$ vanishing on $\partial\Omega_i \cap \partial\Omega$ have zero flux in Ω_i. It is natural therefore to consider boundary value problems with arbitrary α_i in the class of vector fields with an infinite Dirichlet integral $\int_\Omega |D\mathbf{v}|^2\, dx$. A weak solution of the problem (10), (13), (14) can be defined as a divergence-free vector field $\mathbf{v}$ with a finite Dirichlet integral in every finite subdomain $\Omega' \subset \Omega$ vanishing on $\partial\Omega$ and satisfying (13), (14), and the integral identity

$$\int_\Omega \left[\nu \sum_i \mathbf{v}_{x_i} \cdot \boldsymbol{\eta}_{x_i} - (\mathbf{v} \cdot \nabla)\boldsymbol{\eta} \cdot \mathbf{v} \right] dx = \int_\Omega \mathbf{f} \cdot \boldsymbol{\eta}\, dx$$

for any smooth divergence-free $\boldsymbol{\eta}$ with a compact support in Ω. The problem (10), (13), (14) is considered in [31–36] in different classes of vector fields and under various assumptions on the domain Ω. It is proved in particular [34] that it has at least one weak solution in the class of vector fields satisfying the conditions $\int_{\Omega(\lambda)} |D\mathbf{v}|^2\, dx \le q(\lambda)$, where $\Omega(\lambda)$ is a family of bounded domains exhausting Ω as $\lambda \to +\infty$ and $q(\lambda)$ is a function depending on $\mathbf{f}$ and on the geometry of Ω. This is a natural generalization of the well-known results of J. Leray and O. A. Ladyzhenskaya for bounded and exterior domains.

The initial boundary value problem for the Navier-Stokes equations in Ω is studied in [37–39]. It is proved both for $n = 3$ and for $n = 2$ that this problem has a unique solution in an interval $(0, T_1)$, $T_1 < \infty$, that belongs at least to $W_2^{1,1/2}(Q'_{T_1})$ in any $Q'_{T_1} = \Omega' \times (0, T_1)$, and $T_1 = \infty$ if the data $\mathbf{v}_0(x)$, $\mathbf{f}$, α_i are small. The problem of the existence of a global solution in the two-dimensional case is still open.

For more details see [37].

REFERENCES

1. V. K. Andreev and V. V. Pukhnachev, *The motion of a finite mass of fluid*, Zh. Prikl. Mekh. i Tekhn. Fiz. **1979**, no. 2, 25–43=J. Appl. Mech. Tech. Phys. **20** (1979), 144–157.

2. V. A. Solonnikov, *Solvability of a problem on the motion of a viscous incompressible fluid bounded by a free surface*, Izv. Akad. Nauk SSSR Ser. Mat. **41** (1977), 1388–1424= Math. USSR Izv. **11** (1977), 1323–1358.

3. ____, *Solvability of the problem of evolution of an isolated volume of a viscous incompressible capillary fluid*, Zap. Nauchn. Sem. Leningrad. Otdel. Mat. Inst. Steklov. (LOMI) 140 (1984), 179–186= J. Soviet Math. 32 (1986), 223–228.

4. V. O. Bytev, *Unsteady motion of a rotating ring of viscous incompressible fluid with a free boundary*, Zh. Prikl. Mekh. i Tekhn. Fiz. 1970, no. 3, 82–88= J. Appl. Mech. Tech. Phys. 11 (1970), 432–438.

5. O. M. Lavrent'eva, *Motion of a rotating ring of a viscous incompressible fluid*, Manuscript No. 7562–84, deposited at VINITI, 1984. (Russian)

6. J. Thomas Beale, *The initial value problem for the Navier-Stokes equations with a free boundary*, Comm. Pure Appl. Math. 34 (1981), 359–392.

7. ____, *Large-time regularity of viscous surface waves*, Arch. Rational Mech. Anal. 84 (1983/84), 307–352.

8. Geneviève Allain, *Small-time existence for the Navier-Stokes equations with a free surface*, Rapport Interne No. 135, École Polytechnique, Paris, 1985.

9. V. V. Pukhnachev, *The plane steady-state problem with a free boundary for the Navier-Stokes equations*, Zh. Prikl. Mekh. i Tekhn. Fiz. 1972, no. 3, 91–102= J. Appl. Mech. Tech. Phys. 13 (1972), 340–349.

10. ____, *The motion of a liquid film on the surface of a rotating cylinder in a gravitational field*, Zh. Prikl. Mekh. i Tekhn. Fiz. 1977, no. 3, 78–88= J. Appl. Mech. Tech. Phys. 18 (1977), 344–351.

11. V. G. Osmolovskiĭ, *The free surface of a drop in a symmetric force field*, Zap. Nauchn. Sem. Leningrad. Otdel. Mat. Inst. Steklov. (LOMI) 52 (1975), 160–174= J. Soviet Math. 9 (1978), 792–803.

12. J. Bemelmans, *Gleichgewichtsfiguren zäher Flüssigkeiten mit Oberflachengspannung*, Analysis 1 (1981), 241–282.

13. O. A. Ladyzhenskaya and V. G. Osmolovskiĭ, *On the free surface of a fluid layer over a solid sphere*, Vestnik Leningrad. Univ. 1976, no. 13 (Ser. Mat. Mekh. Astr. vyp. 3), 25–30= Vestnik Leningrad Univ. Math. 9 (1981), 197–203.

14. V. Ya. Rivkind, *A study of the problem of steady motion of a drop in the flow of a viscous imcompressible fluid*, Dokl. Akad. Nauk SSSR 227 (1976), 1071–1074= Soviet Phys. Dokl. 21 (1976), 185–187.

15. J. Bemelmans, *Liquid drops in a viscous fluid under the influence of gravity and surface tension*, Manuscripta Math. 36 (1981/82), 105–123.

16. D. H. Sattinger, *On the free surface of a viscous fluid motion*, Proc. Roy. Soc. London Ser. A 349 (1976), 183–204.

17. M. Jean, *Free surface of the steady flow of a Newtonian fluid in a finite channel*, Arch. Rational Mech. Anal. 74 (1980), 197–217.

18. V. A. Solonnikov, *Solvability of the problem of the plane motion of a heavy viscous incompressible capillary liquid partially filling a container*, Izv. Akad. Nauk SSSR Ser. Mat. 43 (1979), 203–236= Math. USSR Izv. 14 (1980), 193–221.

19. ____, *Solvability of three-dimensional problems with a free boundary for a system of steady-state Navier-Stokes equations*, Zap. Nauchn. Sem. Leningrad. Otdel. Mat. Inst. Steklov. (LOMI) 84 (1979), 252–285= J. Soviet Math. 21 (1983), 427–450.

20. V. G. Maz'ya, B. A. Plamenevskiĭ, and L. I. Stupyalis [Stupelis], *The three-dimensional problem of steady motion of a fluid with a free surface*, Differentsial'nye Uravneniya i Primenen.—Trudy Sem. Protsessy Optimal. Upravleniya I Sektsiya Vyp. 23 (1979); English transl., Amer. Math. Soc. Transl. (2) 123 (1984), 171–268.

21. V. A. Solonnikov, *Solvability of the three-dimensional boundary value problem with a free surface for the stationary Navier-Stokes system*, Partial Differential Equations (Warsaw, 1978; B. Bojarski, editor), Banach Center Publ., vol. 10, PWN, Warsaw, 1983, pp. 361–403.

22. ____, *On the Stokes equations in domains with nonsmooth boundaries, and on viscous incompressible flow with a free surface*, Nonlinear Partial Differential Equations and Their Applications (Collège de France Sém., Vol. III (1980/81); H. Brézis and J.-L. Lions, editors), Res. Notes in Math., vol. 70, Pitman, 1982, pp. 340–423.

23. I. B. Erunova, *Investigation of the problem of steady motion of two fluids in a vessel*, Dokl. Akad. Nauk SSSR 279 (1984), 55–58= Soviet Phys. Dokl. 29 (1984), 875–876.

24. L. K. Antanovskiĭ, *A complex representation of solutions of Navier-Stokes equations*, Dokl. Akad. Nauk SSSR **261** (1981), 829–832= Soviet Phys. Dokl. **26** (1981), 1120–1121.

25. Joseph [Josef] Bemelmans and Avner Friedman, *Analyticity for the Navier-Stokes equations governed by surface tension on the free boundary*, J. Differential Equations **55** (1984), 135–150.

26. K. I. Piletskas [Pileckas], *Solvability of a problem on the two-dimensional motion of a viscous incompressible fluid with a free noncompact boundary*, Differentsial'nye Uravneniya i Primenen.—Trudy Sem. Protsessy Optimal. Upravleniya I Sektsiya Vyp. 30 (1981), 57–95. (Russian)

27. John G. Heywood, *On uniqueness questions in the theory of viscous flow*, Acta Math. **136** (1976), 61–102.

28. V. A. Solonnikov, *On the solvability of boundary and initial-boundary value problems for the Navier-Stokes system in domains with noncompact boundaries*, Pacific J. Math. **93** (1981), 443–458.

29. L. V. Kapitanskiĭ, *Steady-state solutions of Navier-Stokes equations in periodic pipes*, Zap. Nauchn. Sem. Leningrad. Otdel. Mat. Inst. Steklov. (LOMI) **115** (1982), 104–113= J. Soviet Math. **28** (1985), 689–695.

30. V. A. Solonnikov and K. I. Piletskas [Pileckas], *On some spaces of solenoidal vectors and the solvability of a boundary value problem for systems of Navier-Stokes equations in domains with noncompact boundaries*, Zap. Nauchn. Sem. Leningrad. Otdel. Mat. Inst. Steklov. (LOMI) **73** (1977), 136–151= J. Soviet Math. **34** (1986), 2101–2111.

31. Charles J. Amick, *Properties of steady Navier-Stokes solutions for certain unbounded channels and pipes*, Nonlinear Anal. **2** (1978), 689–720.

32. C. J. Amick and L. E. Fraenkel, *Steady solutions of the Navier-Stokes equations representing plane flow in channels of various types*, Acta Math. **144** (1980), 83–151.

33. O. A. Ladyzhenskaya and V. A. Solonnikov, *On the solvability of boundary and initial-boundary value problems for the Navier-Stokes equations in regions with noncompact boundaries*, Vestnik Leningrad. Univ. **1977**, no. 13 (Ser. Mat. Mekh. Astr. vyp. 3), 39–47= Vestnik Leningrad Univ. Math. **10** (1982), 271–279.

34. _____, *Determination of solutions of boundary value problems for steady-state Stokes and Navier-Stokes equations having an unbounded Dirichlet integral*, Zap. Nauchn. Sem. Leningrad. Otdel. Mat. Inst. Steklov. (LOMI) **96** (1979), 117–160= J. Soviet Math. **21** (1983), 728–761.

35. V. A. Solonnikov, *Solutions of a steady-state system of Navier-Stokes equations that have an infinite Dirichlet integral*, Zap. Nauchn. Sem. Leningrad. Otdel. Mat. Inst. Steklov. (LOMI) **115** (1982), 251–263= J. Soviet Math. **28** (1985), 792–799.

36. L. V. Kapitanskiĭ and K. I. Piletskas [Pileckas], *On spaces of solenoidal vector fields and boundary value problems for the Navier-Stokes equations in domains with noncompact boundaries*, Trudy Mat. Inst. Steklov. **159** (1983), 5–36= Proc. Steklov Inst. Math. **1984**, no. 2 (159), 3–34.

37. V. A. Solonnikov, *Stokes and Navier-Stokes equations in domains with noncompact boundaries*, Nonlinear Partial Differential Equations and Their Applications (Collège de France Sém. Vol. IV (1981/82); H. Brézis and J.-L. Lions, editors), Res. Notes in Math., vol. 84, Pitman, 1983, pp. 240–349.

38. Olga A. Ladyzhenskaya, Vsevolod A. Solonnikov, and Hans True, *Résolution des équations de Stokes et Navier-Stokes dans des tuyaux infinis*, C. R. Acad. Sci. Paris Sér. I Math. **292** (1981), 251–254.

39. O. A. Ladyzhenskaya and V. A. Solonnikov, *On an initial-boundary value problem for linearized Navier-Stokes equations in domains with noncompact boundaries*, Trudy Mat. Inst. Steklov. **159** (1983), 37–40= Proc. Steklov Inst. Math. **1984**, no. 2 (159), 35–39.

LENINGRAD BRANCH, V. A. STEKLOV MATHEMATICAL INSTITUTE, 191011 LENINGRAD, USSR

Gauge Theories and
Nonlinear Partial Differential Equations

CLIFFORD HENRY TAUBES

I. Variational equations. There are partial differential equations which arise in geometry and physics and which are variational equations; this means that they arise as the Euler-Lagrange equations of some "energy" or "action" functional on the space of possible field configurations, for example, the Laplace equation,

$$\Delta\varphi \equiv -(\partial^2\backslash\partial x_1^2 + \partial^2\backslash\partial x_2^2 + \partial^2\backslash\partial x_3^2)\varphi = \rho. \tag{1.1}$$

Here ρ is some given, unknown function. This equation arises in the study of electrostatic phenomena; ρ represents the volume density of the electric charge, and φ is the electrostatic potential. The electric field is the vector $E = \nabla\varphi$.

The preceding equation is, formally, the variational equation of the following "function" (called a *functional*) on the space of all smooth functions on Euclidean space,

$$\varepsilon(\varphi) = \int_{\mathbf{R}^3} (|\nabla\varphi|(x)^2 + 2 \cdot \rho(x) \cdot \varphi(x)) \cdot d^3x. \tag{1.2}$$

Formally, φ obeys Laplace's equation if and only if the functional $\varepsilon(\cdot)$ has the property that for any smooth, compactly supported function, η,

$$(d/dt)(\varepsilon(\varphi + t\eta))|_{t=0} = 0. \tag{1.3}$$

A second example comes from magnetostatics; a divergence-free vector field $J(x)$ on $\mathbf{R}^3$ (which represents the volume density of electric current) induces a magnetic field which is computed from the vector potential, $A(x)$, by solving the vector equation on $\mathbf{R}^3$

$$\nabla \times \nabla \times A = J. \tag{1.4}$$

The magnetic field is $B = \nabla \times A$. The equation for A is, formally, the variational equation of the following functional on the space of all vector fields on $\mathbf{R}^3$:

$$\varepsilon(A) = \int_{\mathbf{R}^3} (|\nabla \times A|^2(x) + 2 \cdot A(x) \circ J(x)) \cdot d^3x. \tag{1.5}$$

Formally, A obeys (1.4) if and only if the functional in (1.5) has the property that

$$(d/dt)(\varepsilon(A + ta))|_{t=0} = 0 \qquad (1.6)$$

for all compactly supported vectors a on $\mathbf{R}^3$.

The equations above can be solved using the Green's function. This is because the equations are linear equations. Indeed, a typical linear equation requires solving for a matrix of functions, ψ, which obey

$$L\psi = \sigma, \qquad (1.7)$$

where σ is a given matrix of functions, and where

$$L = L(x_1, x_2, \ldots; \partial/\partial x_1, \partial/\partial x_2, \ldots)$$

is a matrix of functions of the derivatives and the coordinates, $(x_1, x_2, \ldots)$. To solve such a linear equation, the first strategy that one might attempt would be to find the appropriate Green's function for L.

Given that the function ρ in (1.1) is well behaved, a solution of (1.1) can be written using the Green's function for the Laplacian,

$$\varphi(x) = (4\pi)^{-1} \int_{\mathbf{R}^3} |x - y|^{-2} \cdot \rho(y) \cdot d^3 y. \qquad (1.8)$$

Equation (1.4) can also be solved using a Green's function if the current density J is well behaved,

$$A(x) = -\nabla \times \int_{\mathbf{R}^3} |x - y|^2 \cdot J(y) \cdot d^3 y. \qquad (1.9)$$

Unfortunately, a Green's function alone rarely resolves life's problems. The real world is nonlinear; particles interact with each other. Nonlinear equations are everywhere, and need to be understood.

When a nonlinear equation arises as the Euler-Lagrange variational equations of a functional on an appropriate configuration space, it is especially tempting to prove that solutions exist by using the methods from the calculus of variations.

Certain partial differential equations have been recently "understood" using the calculus of variations. The remainder of this article comprises one description of some of the basic ideas and strategies that lie behind this new understanding. The ideas and strategies that are related below were developed in parts by many workers in the field; the author only presumes to summarize. References to a sample of interesting research papers are provided at the end.

II. The calculus of variations. Consider a finite-dimensional calculus of variations problem: Let f be a smooth function on the circle, S^1. The circle is the analog of the configuration space of functions or vector fields or whatever; the set of objects in which one hopes to find a solution to the given partial differential equation. The function f is the analog of the energy functional; the functional whose variational equations on the configuration space yield the differential equation in question. The critical points of the function f are the points in S^1 where the gradient of f vanishes; these correspond to the configurations which actually solve the differential equation. Thus, the requirement of a point $p \in S^1$ that $\nabla f_p = 0$ is the finite-dimensional analog of the requirement of a configuration that it solve the differential equation.

In this finite-dimensional example, the analog of proving that there are solutions to the partial differential equation is the problem of proving that there are points on S^1 where the gradient of the function vanishes. Now, observe that S^1 is compact and every function on a compact manifold must have at least one maximum and one minimum, so there must be at least two points where the gradient of the function f vanishes. It is the global topology of the configuration manifold that forces the gradient of f to vanish somewhere on S^1.

This case is simple, but typical of a general principle, Morse theory [**Bo**]. (For an exercise which illustrates the power of Morse theory, try to establish the fact that every smooth function on the torus $(S^1 \times S^1)$ has at least 3 critical points. One can do this by minimizing the maximum of a given function over noncontractible loops in the torus.)

The observation that the global topology of the configuration space can force the existence of critical points is the crucial motivating idea behind the theory of the calculus of variations. For finite-dimensional problems, this is Morse theory and the starting point for a huge branch of mathematics, differential topology (cf. [**M1, M2**]).

The generalization of these finite-dimensional arguments to an energy functional for a differential equation is more complicated because the configuration space is going to be infinite-dimensional; typically, it is some space of functions, or vector fields or such. The relationships between the topology of the configuration space and the critical points of the given energy functional may be tenuous.

The crucial problem with infinite-dimensional spaces is that they tend to be noncompact in any sort of reasonable topology. In some sense, the pathologies which can arise in these infinite-dimensional variational problems are due to the noncompactness more than any other single quality of infinite-dimensional spaces.

Indeed, even on the simplest noncomapct spaces, the calculus of variations may not work as desired. For example, the simplest such space is the real line, $\mathbf{R}^1$. Consider the function $f(t) = e^{-t}/(1 + e^{-t})$. Here is a smooth, bounded function, but there are no points on the line where its gradient vanishes. Indeed even though $\min(f) = 0$, one has $f(t) > 0$ for all t. A sequence of points $\{t_i\}$ such that $f(t_i) > f(t_{i+1}) \to 0$ will not converge. This sequence of points moves off towards ∞.

Of course, there are "good" functions even on noncompact spaces. In the preceding example, one could perturb the function f slightly to obtain a function with the required critical point. Consider, for instance, the function $f_\delta(t) = (e^{-t} - \delta e^{-t/2})(1 + e^{-t})$. Note that for any positive δ, there is a point in $\mathbf{R}^1$ where the gradient of $f_\delta(t)$ vanishes.

The functions f and f_δ differ in their behavior near ∞ on $\mathbf{R}^1$, and this is an important lesson. It is a well-known fact among topologists that the topology of a noncompact manifold cannot be determined solely by restricting attention to a compact set.

III. Infinite dimensions. Here, one must suppose that one is given a linear, or nonlinear, partial differential equation which arises as the (formal) variational equations of a functional on some appropriate space of functions or vector fields or whatever. Let $\mathcal{B}$ denote the space in question, and let $\varepsilon \colon \mathcal{B} \to \mathbf{R}$ denote the functional. Let $\|\cdot\|$ denote a norm on the tangent space to $\mathcal{B}$. (Typically, $\|\cdot\|$ will come from some Hilbert space inner product.)

The "ends" of the space, the analog of ∞ on the real line, are determined by the functional. Quite formally, one can define for numbers $\delta > 0$ and E, the set

$$\mathrm{End}(E,\delta) \equiv \{c \in \mathcal{B} : \varepsilon(c) \in (E-\delta, E+\delta) \text{ and } \|\nabla \varepsilon_c\| < \delta\}. \qquad (3.1)$$

To have a calculus of variations for the functional ε on the space $\mathcal{B}$, one is forced to understand the topology of $\mathrm{End}(E,\delta)$ relative to the set

$$\mathrm{End}(E,\delta)^- \equiv \{c \in \mathrm{End}(E,\delta) : \varepsilon(c) < E\}. \qquad (3.2)$$

In fact, only an appropriate limit as $\delta \to 0$ of this relative topology is relevant. (Because $\mathrm{End}(E, \delta/2) \subset \mathrm{End}(E,\delta)$ such a limit can be considered.)

$\mathrm{End}(E,\delta)$ contributes to the Morse theory of the functional ε on $\mathcal{B}$ when, for all $\delta' < \delta$, the inclusion map,

$$i \colon (\mathrm{End}(E,\delta'), \mathrm{End}(E,\delta')^-) \to (\mathrm{End}(E,\delta), \mathrm{End}(E,\delta)^-), \qquad (3.3)$$

has nontrivial image in the relative homology of the pair $(\mathrm{End}(E,\delta), \mathrm{End}(E,\delta)^-)$.

For example, suppose that f is a smooth function on a compact manifold. In this case, classical Morse theory applies (see, for example, [**M1**]). If E is a regular value of f, then the relative homology of $(\mathrm{End}(E,\delta), \mathrm{End}(E,\delta)^-)$ vanishes for all δ sufficiently small. Suppose that f has a single nondegenerate critical point with critical value E. Then $\bigcap_\delta \mathrm{End}(E,\delta)$ is the critical point in question. Let $m \geq 0$ denote the number of negative eigenvalues of the matrix of 2nd derivatives of the function f at this critical point. Then, for all δ sufficiently small, the relative homology over $\mathbf{Z}$ of $(\mathrm{End}(E,\delta), \mathrm{End}(E,\delta)^-)$ is trivial in all dimensions but dimension m, where it is isomorphic to $\mathbf{Z}$. And, for all δ sufficiently small, the inclusion map i, above, induces an isomorphism in relative homology.

IV. Choosing a topology. Reconstruction of a topology on $\mathcal{B}$ from the topology of the spaces in (3.3) must be made. Here, a word of warning. The functional in question may not "see" the topology on the underlying point set which is implicit in the Banach manifold structure. Since the functional is assumed to be smooth, it will see the same local topology. But, unless some uniformity assumptions are made on the derivatives of the functional, there is no guarantee about the global topology. For the set where the gradient of the functional is small to have any meaning, it is necessary that the gradient of the functional (the first-order Taylor's expansion) approximate the variation in the functional in some uniform way through the Banach space.

The following example illustrates this sort of pathology. Consider the half line, $\{t \in (-\infty, 0)\}$, topologized by the metric

$$ds^2 = dt^2.$$

This is a smooth Banach manifold, though it is not complete as a metric space. Consider the function

$$f(t) = -t.$$

The infimum of f on $(-\infty, 0)$ is 0, but this infimum is not achieved. Since the norm of ∇f is identically 1, $\text{End}(E, \delta)$ is empty for all values of E and for all values of $\delta < 1$.

The pathology which is illustrated above has as root, the basic relationship between the topology of manifolds and the set of functions on them: Why should the derivative of a function tell anything at all about the manifold on which it sits?

There is a uniformity condition on the first derivatives of the functional which insures that the underlying Banach manifold topology is "seen" by the functional. Given this uniformity condition on the first derivatives, the topology of $\mathcal{B}$ can be reconstructed by excision from knowledge of the effects of the inclusion map in (3.3) on the relative homotopy or homology groups. This subject is discussed in the appendix of [Ta1].

V. Ellipticity. In a given variational problem, understanding the topology of $\text{End}(E, \delta)$ relative to $\text{End}(E, \delta)^-$ amounts to understanding the limiting behavior of sequences of configurations,

$$\{c_i \subset \text{End}(E, 2^{-i})\}. \tag{5.1}$$

What does the sequence of configurations look like as the index $i \to \infty$? Does it converge in some nice sense? If not, what are the singularities that can arise? What sort of space parametrizes the set of limits? And what does the functional ε look like as the limits are approached? The preceding questions require for answering a detailed knowledge of the functional ε.

There are two approaches towards studying the convergence of sequences in infinite-dimensional spaces. The first is the Arzela-Ascoli theorem which asserts that a bounded, equicontinuous sequence of functions has a convergent subsequence with a continuous function as a limit. The second is the Banach-Alaoglu theorem which asserts that the unit sphere in a reflexive Banach space is compact in the weak topology. (See, for example, [R].) In applying the first tool, one would consider convergence in various C^k-topologies on the configuration space, $\mathcal{B}$. In applying the second tool, one would consider the convergence in a reflexive Banach-space topology on $\mathcal{B}$. This of course means Sobolev spaces (see [Ad]). Indeed, one would invent, in hindsight, the Sobolev spaces precisely in order to have a reflexive Banach space topology for the space of smooth functions on a finite-dimensional manifold. For the uninitiated, the Sobolev L_k^p-norm of a function f on a Riemannian manifold M is the number

$$\left(\sum_{j=1,\ldots,k} \int_M |\nabla^{(j)} f|^p \cdot d\,\mathrm{vol} \right)^{1/p}.$$

To apply either the Arzela-Ascoli approach or the Banach-Alaoglu approach, information must be squeezed from the functional ε. For a configuration $c \in \mathrm{End}(E,\delta)$, the value of the energy, ε, is bounded; and the gradient of ε at c, $\nabla\varepsilon_c$, is small in an appropriate sense. From these facts, the analysis proceeds.

The condition that $\nabla\varepsilon_c$ is small means that c almost obeys the differential equation of interest. Write this fact in the schematic form

$$D^{(k)}c = (\text{nonlinearities}) + O(\delta), \tag{5.2}$$

where $D^{(k)}$ is the schematic for the term in the equation with the highest order (some integer, $k > 0$) derivatives. The term above which is denoted "nonlinearities" contains only terms with lower orders of differentiation.

Read (5.2) as saying that certain linear combinations of the kth derivatives of c are equal to functions of the derivatives of c of order less than k, modulo an error which is small. If it is assumed that c starts out to be $k-1$ times differentiable, then (5.2) asserts that certain linear combinations of the kth derivatives of c are continuous, modulo an error.

The equations are said to be *elliptic* when the fact that such a linear combination of kth derivatives are continuous means that *all* of the kth derivatives are continuous. (Whether or not a set of equations is elliptic can be reduced to a purely algebraic question about the term with the highest number of derivatives; see [**H**].)

When the variational equations are *elliptic*, the fact that c is in $\mathrm{End}(E,\delta)$, together with additional a priori bounds on certain norms of c will allow one to bound c, a priori, in some strictly stronger norm. A priori bounds on the norms of the mth derivatives of configurations in $\mathrm{End}(E,\delta)$ will imply via Arzela-Ascoli that the sequences in (5.1) have Cauchy subsequences in the C^{m-1} topology on $\mathcal{B}$. Alternately, a priori bounds on a reflexive Banach space norm for configurations in $\mathrm{End}(E,\delta)$ will imply via Banach-Alaoglu that the sequences in (5.1) have Cauchy sequences in the appropriate weak Banach space topology on $\mathcal{B}$.

Unfortunately, the additional a priori bounds that were alluded to in the preceding paragraph are not available unless they can be obtained solely from information about ε. Indeed, for c in $\mathrm{End}(E,\delta)$, only bounds on $\varepsilon(c)$ and $\nabla\varepsilon_c$ are provided. Typically, the functional ε is a functional of c and its derivatives up to some order. Thus, knowledge that $\varepsilon(c)$ is bounded can be interpreted as giving information about the size of certain combinations of the derivatives of c. The goal at this point is to parlay the information that was alluded to in the preceding paragraph into a bound on some Banach space norm of the configuration c. It may or may not be possible to achieve a suitable bound.

VI. The physicist's view. The control on the derivatives of c which is obtained by the bound on $\varepsilon(c)$ will imply that certain linear combinations of derivatives of c remain bounded almost everywhere. But these linear combinations may be large over sets of small measure.

A physicist might view a configuration, c, in $\mathrm{End}(E, \delta)$ as having "small field" regions, where the derivatives of c are small, and where the nonlinear equation is approximately linear; and then there are the large field regions, where the derivatives are large, and where the nonlinearities cannot be ignored.

To a physicist, configurations in $\mathrm{End}(E, \delta)$ would look like a gas of particles, or extended objects (one- or higher-dimensional structures) which interact according to a force law which is determined by the linearized equations. This view of $\mathrm{End}(E, \delta)$ is obtained by decomposing the energy, $\varepsilon(c)$, as a sum of terms; first, there is the contribution from the large field regions, each component contributes a part. To a physicist, this contribution can be interpreted as the "self" energy of the extended particles which are described by the large field regions. Then there is the contribution from the small field region. This last is interpreted as an interaction energy between the large field regions.

The rigorous justification for such an interpretation might be obtained from the following heuristic argument: A configuration in $\mathrm{End}(E, \delta)$ almost solves the differential equation. In the small field regions, the nonlinearities are, by definition, small. The constraint that the equations be solved, or almost solved, is approximately a linear constraint on the configuration over this region. When the equations are elliptic, the temptation is to view these linear fields as the force fields which are generated by "charge" distributions on the components of the large field regions. In this way, the energy contribution from the small field regions can be interpreted as an interaction energy due to charge densities on the extended particles which are described by the large field regions.

The region $\mathrm{End}(E, \delta)^-$ of (3.2) may be thought of as containing configurations whose large field regions have negative interaction energy.

After expanding the energy, ε, into large and small field contributions, the topology of the spaces in (3.3) might, in principle, be analyzed.

VII. Justification. One can see this interaction energy strategy in some recent research papers in differential geometry. Actually, this puts the cart before the horse, because it is in the articles below (among others) where the interaction energy strategy can be said to have evolved. To name the phenomenon is not to discover it.

The results below all involve functionals and Banach spaces for which the strong field regions, which were alluded to above, can be shown to be configurations of points.

Aubin's [**Au**] and Schoen's [**Sc**] solution to the Yamabe conjecture can be interpreted as interaction energy analysis. The classification of compact Kahler manifolds with positive bisectional curvature by Siu and Yau [**Si-Y**] employs a step which might be called energy analysis. The work of Siu and Yau depends

to a large extent on Sacks and Uhlenbeck's [Sa-U] work on harmonic maps from
2-spheres. This work by Sacks and Uhlenbeck gives the first detailed description
of the sets $\text{End}(E, \delta)$ for a geometric variational problem. In particular, one
sees here the indication that the sorts of singularities which are appearing in
configurations in $\text{End}(E, \delta)$ have important geometric content.

In Yang-Mills theory, K. Uhlenbeck [U1, U2] did the seminal work which
led to a description of the singularities which are approached by configurations
in $\text{End}(E, \delta)$. (See also the work of Sedlacek [Se].) The author's analysis of
the self-dual moduli spaces in Yang-Mills theory [Ta2, Ta3] makes crucial use
of Uhlenbeck's work, and it uses the sort of interaction energy analysis which
was described above. The use of this interaction energy analysis by the author
resulted in a complete solution of the variational problem for the Yang-Mills-
Higgs theory on $\mathbf{R}^3$ in [Ta4, Ta5].

Simon Donaldson's applications of gauge theories to study the differential
topology of 4-dimensional manifolds exploits the topology of the $\text{End}(E, \delta)$'s in
a spectacular fashion [D1–D4, F-U].

Work by Brezis and Nirenberg [Br-N] can be said, in hindsight, to use this
analysis.

Recently, Bahri and Coron [Ba-C] have analyzed the Yamabe problem on
topologically nontrivial domains in $\mathbf{R}^3$ using their own, roughly similar, strategy.
Bahri is applying his approach to other variational problems [Ba].

VIII. Example. The strong forces are the forces in nature which are responsible for atomic energy. These are the forces which hold atoms together. The
weak force is also a force which is felt by subatomic particles. It is responsible
to a certain extent for the glow of a radium-dialed wrist watch.

In a simplified model of these forces, the static fields are described by the
data (A, Φ); where $A \equiv (A_1, A_2, A_3)$ is a vector on $\mathbf{R}^3$ whose components are
2×2, trace zero, anti-hermitian matrices of functions. The field Φ is a 2×2,
trace zero, anti-hermitian matrix of functions also. (A trace zero, anti-hermitian
matrix has complex number σ_{ij} in the ith row and jth column, and the set $\{\sigma_{ij}\}$
obeys $\Sigma_i \sigma_{ii} = 0$, and $\sigma_{ji}^* = \sigma_{ij}$.) The equations generalize Maxwell's equations
from the introduction: The equation for A is

$$\nabla \times (\nabla \times A + A \times A) + A \times (\nabla \times A + A \times A) + (\nabla \times A + A \times A) \times A \quad (7.1a)$$
$$+ \Phi(\nabla \Phi + A\Phi - \Phi A) - (\nabla \Phi + A\Phi - \Phi A)\Phi = 0.$$

The equation for Φ is

$$\nabla \circ (\nabla \Phi + A\Phi - \Phi A) + A \circ (\nabla \Phi + A\Phi - \Phi A) - (\nabla \Phi + A\Phi - \Phi A) \circ A = 0. \quad (7.1b)$$

These are the variational equations of the Yang-Mills-Higgs energy functional
on the space of configurations (A, Φ):

$$\mathcal{E}(A, \Phi) = \int_{\mathbf{R}^3} |\nabla \times A + A \times A|^2 (x) \cdot d^3 x + \int_{\mathbf{R}^3} |\nabla \Phi + A\Phi - \Phi A|^2 (x) \cdot d^3 x. \quad (7.2)$$

Solutions of these equations with finite energy are expected to describe particles which can be found in nature. Indeed, if there are solutions and no particles, then these equations are not an accurate description of reality.

The configuration space for $\mathcal{E}$ is the set of smooth (A, Φ) such that $\mathcal{E}(A, \Phi) < \infty$ and such that the function $(1 - |\Phi|)^6$ is integrable on $\mathbf{R}^3$. This configuration space is so topologically convoluted [**Ta4**] that a "good" Morse function should have infinitely many critical points on each path component. (There are countably many path components [**G**].)

In this variational problem, a configuration in $\text{End}(E, \delta)$ "looks like" up to some $n(E)$ lumps of finite size but with large separation on $\mathbf{R}^3$. Each lump looks like a smooth solution to (7.1) (see [**Ta5**]). The noncompactness of the infinite-dimensional configuration space has been reduced to the finite-dimensional noncompactness of $\times_{n(E)} \mathbf{R}^3$.

Where $\varphi \equiv 1 - |\Phi|$ and A are small, the fields obey the linearized equations, which are precisely (1.2) and (1.3), the equations of electrostatics and magnetostatics. Thus, the large field regions will interact with each other as would bona fide particles with electric charges and magnetic charges. From the linearized equations (Maxwell's equations) and the corresponding Green's functions (in (1.8) and (1.9)), the interaction energy of a configuration in $\text{End}(E, \delta)$ is computed as an expansion (the usual multipole expansion) in the separation of the lumps:

$$\varepsilon(c) = \sum_i \varepsilon(i) - \sum_{i<j} \frac{(m_i m_j - q_i q_j)}{|x_i - x_j|} + O\left(\sum_{i<j} |x_i - x_j|^{-2}\right). \qquad (7.3)$$

Here, the sum is over the set of lumps, and m_i is a "mass" for the ith lump, while q_i is a charge for the ith lump.

The charges $\{m_i, q_i\}$ obey $m_i > 0$ and $|q_i| \le m_i$. This fact implies that the interaction energy is always attractive or it vanishes identically. In the latter case, the configuration c is within δ of an absolute minimum of $\mathcal{E}$. In the former case, this fact implies that for large lump separation, the relative homotopy group $\pi_k(\text{End}(E, \delta), \text{End}(E, \delta)^-)$ is trivial for all $k \ge 0$.

The facts of the preceding paragraph plus knowledge of the topology of the configuration space imply that there are solutions to (7.1) which are both minimal and nonminimal critical points of the energy functional in (7.2). Details are provided in [**Ta5**].

REFERENCES

[Ad] R. A. Adams, *Sobolev spaces*, Academic Press, New York, 1975.

[Au] T. Aubin, *Equations differentielles non lineaire et probleme de Yamabe concernant la courbure scailaire*, J. Math. Pures Appl. (9) **55** (1976), 269.

[Ba] A. Bahri (to appear).

[Ba-C] A. Bahri and J. C. Coron, *Une theorie des points critiques a l'infini pour l'equation de Yamabe et le probleme de Kazdan-Warner*, C. R. Acad. Sci. Paris Sér. I. Math. **3001** (1985), 513.

[Bo] R. Bott, *Lectures on Morse theory, old and new*, Bull. Amer. Math. Soc. (N.S.) **7** (1982), 331.

[Br-N] H. Brezis and L. Nirenberg, *Positive solutions of non-linear elliptic equations involving critical exponents*, Comm. Pure Appl. Math. **36** (1983), 437.

[D1] S. K. Donaldson, *An application of gauge theory to the topology of 4-manifolds*, J. Differential Geom. **18** (1983), 279.

[D2] ____, *Connections, cohomology and the intersection forms of 4-manifolds*, J. Differential Geom. (to appear).

[D3] ____, *Irrationality and the h-cobordism conjecture*, Preprint, 1986.

[D4] ____, *The orientation of Yang-Mills moduli spaces and 4-manifold topology*, Preprint, 1986.

[F-U] D. Freed and K. K. Uhlenbeck, *Instantons and four-manifolds*, Springer-Verlag, New York, 1984.

[G] D. Groisser, *Integrality of the monopole number in* SU(2) *Yang-Mills Higgs theories on* $\mathbf{R}^3$, Comm. Math. Phys. **93** (1984), 367.

[H] L. Hormander, *Linear partial differential operators*, Springer-Verlag, New York, 1963.

[M1] J. Milnor, *Morse theory*, Ann. of Math Studies, vol. 51, Princeton Univ. Press, Princeton, N.J., 1963.

[M2] ____, *Lectures on the h-cobordism theorem*, Princeton Univ. Press, Princeton, N. J., 1965.

[R] H. L. Royden, *Real analysis*, Macmillan, New York, 1968.

[Sa-U] J. Sacks and K. K. Uhlenbeck, *The existence of minimal 2-spheres*, Ann. of Math. (2) **113** (1981), 1.

[Sc] R. Schoen, *Conformal deformation of a Riemannian metric to constant scalar curvature*, J. Differential Geom. **20** (1984), 479.

[Se] S. Sedlacek, *A direct method for minimizing the Yang-Mills functional over 4-manifolds*, Comm. Math. Phys. **86** (1982), 515.

[Si-Y] Y. T. Siu and S. T. Yau, *Compact Kahler manifolds of positive bisectional curvature*, Invent. Math. **59** (1980), 189.

[Ta1] C. H. Taubes, *A framework for Morse theory for the Yang-Mills functional*, Preprint, 1986.

[Ta2] ____, *Self-dual connections on 4-manifolds with indefinite intersection matrix*, J. Differential Geom. **19** (1984), 517.

[Ta3] ____, *Path-connected Yang-Mills moduli spaces*, J. Differential Geom. **19** (1984), 337.

[Ta4] ____, *Monopoles and maps from* S^2 *to* S^2; *the topology of the configuration space*, Comm. Math. Phys. **95** (1984), 345.

[Ta5] ____, *Min-max theory for the Yang-Mills-Higgs equations*, Comm. Math. Phys. **97** (1985), 473.

[U1] K. K. Uhlenbeck, *Connections with* L^p-*bounds on curvatures*, Comm. Math. Phys. **83** (1982), 31.

[U2] ____, *Chern classes of Sobolev connections*, Comm. Math. Phys. **101** (1985), 449.

HARVARD UNIVERSITY, CAMBRIDGE, MASSACHUSETTS 02138, USA

Proceedings of the International Congress of Mathematicians
Berkeley, California, USA, 1986

Recent Progress on the Classical
Problem of Plateau

A. J. TROMBA

Introduction. In 1931 Jesse Douglas and, simultaneously, Tibor Rado solved the famous problem of Plateau, namely, that every Jordan wire in R^n bounds at least one disc-type surface of least area. For this work Douglas was one of the two first Fields medalists in 1936 (the other was Lars Ahlfors). By this time he had shown that his methods would allow one to prove that there exist minimal surfaces of genus zero and connectivity k spanning k Jordan curves $\Gamma_1, \ldots, \Gamma_k$ in R^n provided that one such surface exists having strictly less area than the infimum of the areas of all disconnected genus zero surfaces spanning $\Gamma_1, \ldots, \Gamma_k$ (see Figure 1). Somewhat later he announced and published proofs of theorems giving similar sufficient but essentially unverifiable conditions that guarantee the existence of a minimal surface of nonzero genus spanning one or more wires in Euclidean space. The ideas of Douglas, being of great historical significance, deserve some description and we shall begin with an analytic formulation of the problem.

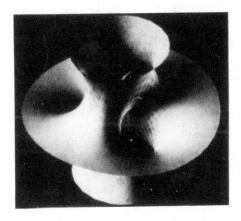

FIGURE 1

Let Γ be a Jordan curve in $\mathbf{R}^n$ and $D \subset \mathbf{R}^n$ the closed unit disc. The classical problem of Plateau asks that we minimize the area integral

$$A(u) = \int \sqrt{EG - F^2}\, dx\, dy$$

among all differentiable mappings $u \colon D \to \mathbf{R}^n$ such that

$$u \colon \partial D \to \Gamma \text{ is a homeomorphism.} \tag{1}$$

Here we have used the traditional abbreviations

$$E = \sum_{k=1}^{n} \left(\frac{\partial u^k}{\partial x}\right)^2, \quad G = \sum_{k=1}^{n} \left(\frac{\partial u^k}{\partial y}\right)^2, \quad F = \sum_{k=1}^{n} \frac{\partial u^k}{\partial x}\frac{\partial u^k}{\partial y}.$$

The Euler equations of this variational problem form a system of nonlinear partial differential equations expressing the condition that the surface u have mean curvature zero, i.e., it is a minimal surface. One may, however, try to take advantage of the fact that the area integral is invariant under the diffeomorphism group of the disc and to transform these equations into a particularly simple form by using special coordinate representations. Following Riemann, Weierstrass, H. A. Schwarz, and Darboux one introduces isothermal coordinates

$$E = G, \qquad F = 0, \tag{2}$$

which in fact linearize the Euler equations of least area; namely, they reduce to Laplace's equation

$$\Delta u = 0. \tag{3}$$

One is thus led to the definition of a classical disc-type minimal surface as a map $u \colon D \to \mathbf{R}^n$ that fulfills conditions (2) and (3).

For unknotted curves Garnier was able to prove the existence of solutions of (2) and (3) subject to the boundary condition (1) by function-theoretic methods. The general case evaded researchers until the work of Douglas and Rado. They both used the direct method of the calculus of variations and thus obtained and area-minimizing solution, while Garnier's solution might be unstable. In applying the direct method one now replaces the complicated area functional by the simpler Dirichlet functional $\mathcal{D}$ where

$$\mathcal{D}(u) = \frac{1}{2}\int (E + G)\, dx\, dy.$$

It is important to note that

$$A(u) \le \int \sqrt{EG}\, dx\, dy \le \frac{1}{2}\int (E + G)\, dx\, dy = \mathcal{D}(u)$$

with equality holding if and only if $E = G$, $F = 0$. This and the analogy with the length and energy functionals of geodesics make it plausible that minima of $\mathcal{D}$ should be minima of A. This is, in fact, the case. In his prize-winning paper however, Douglas did not explicitly attempt to find a minimum for Dirichlet's integral but another functional H, which is now called the Douglas functional.

Using Poisson's integral formula for harmonic functions, Douglas obtained the expression

$$H(u) = \frac{1}{4\pi} \int_0^{2\pi} \int_0^{2\pi} \frac{(u(\cos\alpha, \sin\alpha) - u(\cos\beta, \sin\beta))^2 \, d\alpha \, d\beta}{4\sin^2 \frac{1}{2}(\alpha - \beta)},$$

which equals $D(u)$ if u is harmonic. The replacement of $D(u)$ by $H(u)$ transforms a variational problem involving derivatives to one that does not, an important feature of Douglas's existence proof. In the case of two contours Γ_1 and Γ_2 in $\mathbf{R}^n$, where the domain of our mappings is an annulus, the functional $H(u)$ is similar. However, in the general case of surfaces of connectivity k and genus $p > 0$, one is forced to take as parameter domain a Riemann surface of genus p bounded by k circles, and the construction of $H(u)$ becomes not only less elementary, but from our point of view incredibly complicated. Douglas was able to accomplish this generalization by making essential use of the theory of Abelian functions on Riemann surfaces, the theory of theta functions defined on their Jacobi varieties, and their dependence on the moduli of the underlying Riemann surfaces. Namely, in order to obtain minimal surfaces through the minimization of D or H, it is necessary to minimize over all conformal classes of Riemann surfaces. This was carried out by Douglas at a point in mathematical history[1] when the structure of such conformal classes was not understood. That Douglas's work was a tour de force of classical function theory is an understatement.

According to Constance Reid's book *Courant*, Douglas gave a lecture at New York University in 1936 which stimulated Courant's interest in Plateau's problem and its generalizations to higher topological structure.

Roughly at about the same time that Courant became interested in Plateau's problem, Marston Morse also took up the problem. He had already successfully developed what we now call a *Morse theory* for geodesics and was attempting to extend these ideas to variational problems in more than one variable. This program was never truly successful. Plateau's problem for disc minimal surfaces provided an ideal and pleasing test case for such a theory. Unfortunately only very partial results were obtained and only in the disc case. Morse did not have the notion of differentiable critical point for Plateau's problem and consequently could never speak of the first nor the second variation of energy. Moreover he had no idea of how to extend his generic nondegeneracy results for geodesics to any multivariable situation and had no idea whether or not he could ever have a finite number of solutions in any situation other than where the solutions were known not to be unique.

Nevertheless, as he had done on other occasions he built in a definition in order to get some result out. He defined the notion of *homotopy critical point* in such a way that if you pass a critical level the topology would change as if you had a true critical point of some finite index. Assuming only finitely many such homotopy critical points, one could prove the existence of Morse inequalities.

---

[1] It is interesting to note that Teichmüller's pioneering work was appearing at about the same time.

We do not mean to imply that he had no success at all. To the contrary, he and Tompkins and independently Shiffman showed that the existence of two disc minimal surfaces u_1 and u_2 spanning a given contour, both of which provide a strict relative minimum for Dirichlet's functional D, imply the existence of a third disc minimal surface. This was also the first time that the "mountain pass lemma" so popular in nonlinear analysis appeared.

In §3 we shall see that the Morse theory for disc minimal surfaces has now been successfully completed. In addition, questions of finiteness, the existence of first and second variations, and the nondegeneracy of solutions, all left unresolved by early workers, have now been successfully answered. We consider these in §2.

These results are a consequence of a line of attack taken in the recent development of the classical variational approach to minimal surfaces by the author, Reinhold Böhme, and Friedrich Tomi. The honor of giving this talk belongs equally to them.

Finally we wish to make some remarks on the relation of the classical theory of Plateau's problem to geometric measure theory. This theory was mainly designed to attack the higher-dimensional form of Plateau's problem, a realm inaccessible to the classical theory. But, admittedly, also in the classical case of two-dimensional surfaces in $\mathbf{R}^3$, the geometric measure theory approach yields beautiful results which could not easily—if at all—be obtained within the classical theory, like the following one (due to Bob Hardt and Leon Simon): any sufficiently smooth Jordan curve in $\mathbf{R}^3$ spans an embedded (up to the boundary) minimal surface of some (unknown) topological type. Geometric measure theory in our opinion is, however, not well suited to questions where one is interested in surfaces of a prescribed topological type. We are therefore convinced that the classical theory continues to hold its place within the general theories of minimal surfaces.

1. Formulation of the problem. For the purposes of exposition we shall only consider the case of one boundary contour $\Gamma \subset \mathbf{R}^n$, bounding only oriented surfaces. Let $\alpha \colon S^1 \to \mathbf{R}^n$ be a smooth embedding with smoothness in the Sobolev class H^r, $r \geq 7$. Let $\Gamma = \alpha(S^1)$ and let M be a Riemann surface of genus p with ∂M diffeomorphic to S^1 via some map $\beta \colon S^1 \to \partial M$. Let $\mathcal{N}_\alpha(p)$ be the C^{r-s-1} manifold of H^{s+1} maps u of M into $\mathbf{R}^n$ taking ∂M to Γ and such that $u \circ \beta$ is homotopic to α. Let $\mathcal{M}$ be the space of C^∞ metrics on M having ∂M as a geodesic. We define Dirichlet's functional $\mathcal{D}_\alpha \colon \mathcal{M} \times \mathcal{N}_\alpha(p) \to R$ by

$$\mathcal{D}_\alpha(g, u) = \frac{1}{2} \sum_{i=1}^{n} \int_M g(x)(\nabla_g u^i, \nabla_g u^i) \, d\mu_g, \tag{1}$$

where $u = (u^1, \ldots, u^n)$, $\nabla_g u^i$ is the gradient of u^i w.r.t. the metric g, and $d\mu_g$ is the volume measure associated to g. Dirichlet's functional is conformally invariant, which means the following.

Let λ be any positive function on M. Then it is easy to see that

$$\mathcal{D}_\alpha(\lambda g, u) = \mathcal{D}_\alpha(g, u). \tag{2}$$

Moreover, we can take the Riemann surface M and form its double $2M$ by gluing an exact copy $\hat{M}$ along ∂M. Each point z of M has an associated conjugate point $\hat{z} \in \hat{M}$. The double $2M$ has a complex structure with the property that the map $S: z \to \hat{z}$ is antiholomorphic. The metric $g \in \mathcal{M}$ then extends to a metric g_s on $2M$ which is symmetric in the sense that S is an isometry for g_s. Let $f: 2M \hookleftarrow$ be a symmetric C^∞ diffeomorphism $(S^*f = f \circ S = f)$. Then we also have the invariance

$$\mathcal{D}_\alpha(f^*g, u \circ f) = \mathcal{D}_\alpha(g, u). \tag{3}$$

Let $\mathcal{P}$ be the space of C^∞ positive symmetric real-valued functions on $2M$, and let $\mathcal{D}_0$ be those C^∞ symmetric diffeomorphisms that are homotopic to the identity. Then (2) and (3) imply that Dirichlet's functional can be thought of as a map

$$\mathcal{D}_\alpha: \mathcal{T}(p) \times \mathcal{N}_\alpha(p) \to R,$$

where $\mathcal{T}(p) = \mathcal{M}/\mathcal{P}/\mathcal{D}_0$ is the quotient space of metrics on $\mathcal{M}$ factored out by the action of $\mathcal{P}$ and $\mathcal{D}_0$. This is precisely Teichmüller's moduli space, which carries naturally the structure of a C^∞ smooth finite-dimensional manifold.

Since Teichmüller's space for a disc D is a single point, it follows that if $M = D$ expression (1) reduces to the classically known expression for $\mathcal{D}_\alpha$, namely,

$$\mathcal{D}_\alpha(u) = \frac{1}{2}\sum_{i=1}^{n}\int_D \nabla u^i \cdot \nabla u^i = \frac{1}{2}\sum_{i=1}^{n}\int_D \left\{\left(\frac{\partial u^i}{\partial x}\right)^2 + \left(\frac{\partial u^i}{y}\right)^2\right\} dx\,dy. \tag{4}$$

The basic result (originally established in a totally different context by Douglas) is that the critical points of $\mathcal{D}_\alpha: \mathcal{T}(p) \times \mathcal{N}_\alpha(p) \to R$ consist of pairs (τ_0, u_0), where τ_0 represents a conformal equivalence class of metrics (or equivalently a complex structure on M) and a conformal map $u_0: M \to \mathbf{R}^n$; i.e., $u_0^*g_n = \lambda g_0$, where g_n is the $\mathbf{R}$ Euclidean metric and g_0 is any metric representative in the conformal class represented by τ_0.

For each α, at least for each α which is real analytic, we would like to be able to say that the number of critical points for $\mathcal{D}_\alpha$ is finite. This problem is, as yet, unsolved. However, in the case $M = D$, the answer is known in the generic case. This is the subject of our next section.

2. The index theorem for disc surfaces. If $M = D$ then, as remarked earlier, Dirichlet's functional takes the form $\mathcal{D}_\alpha: \mathcal{N}_\alpha \to R$, where

$$\mathcal{D}_\alpha(u) = \frac{1}{2}\sum_{i=1}^{n}\int_D \nabla u^i \cdot \nabla u^i$$

and for $p = 0$ we denote $\mathcal{N}_\alpha(p)$ simply by $\mathcal{N}_\alpha$.

Let us begin by asking the question of determining the structure of the set of *all* minimal surfaces (area-minimizing or not). In this direction let $\mathcal{A}$ be the open set (in H^r) of all embeddings α, and $\mathcal{N} = \bigcup_\alpha \mathcal{N}_\alpha$. Then $\mathcal{N}$ is a smooth fibre bundle over $\mathcal{A}$ with projection map $\pi: \mathcal{N} \to \mathcal{A}$ given by $\pi(u) = \alpha$ if $u \in \mathcal{N}_\alpha$. Let $\Sigma \subset \mathcal{N}$ denote the set of all minimal surfaces.

One might, at first glance, conjecture that Σ is a submanifold as would be the case for the geodesic problem with fixed endpoints. Surprisingly Σ is not a manifold but an "infinite-dimensional algebraic variety" composed of manifold strata determined by the singularity structure of the minimal surface. To be more precise let $u: D \to \mathbf{R}^n$ be a minimal surface. Then $F = \partial u/\partial x - i\partial u/\partial y$ is a holomorphic map of D into $\mathbf{C}^n$. A point z_0 where u fails to be an immersion is a zero for F. We say that $z_0 \in D$ is a *branch point* of u of order λ if $F(z) = (z - z_0)^\lambda G(z)$, where $G(z_0) \neq 0$. Clearly every interior branch point has some finite order. If α is sufficiently smooth, it can be shown that all boundary branch points have a finite order.

DEFINITION. Let $\lambda \in \mathbf{Z}^p$ and $\nu \in \mathbf{Z}^q$, $\lambda = (\lambda_1, \ldots, \lambda_p)$, $\nu = (\nu_1, \ldots, \nu_q)$, be tuples of integers. We say that a minimal surface u has branching type (λ, ν) if u has p (arbitrarily located) interior branch points $z_1, \ldots, z_p \in D^0$ of orders $\lambda_1, \ldots, \lambda_p$ and q boundary branch points $\xi_1, \ldots, \xi_q$ of orders $\nu_1, \ldots, \nu_q$.

Let Σ_ν^λ denote all the minimal surfaces in the bundle $\mathcal{N}$ of branching type (λ, ν). Let $\pi_\nu^\lambda = \pi|\Sigma_\nu^\lambda$ be the restriction of the bundle projection map to Σ_ν^λ. The following is then known as the *Index Theorem for Minimal Surfaces* [1].

THEOREM. *Each Σ_ν^λ is a manifold with Σ_0^λ a submanifold of $\mathcal{N}$. The map $\pi_\nu^\lambda: \Sigma_\nu^\lambda \to \mathcal{A}$ is Fredholm of index*

$$I(\lambda, \nu) = 2|\lambda|(2 - n) + |\nu|(2 - n) + 2p + q + 3, {}^2$$

where $|\lambda| = \Sigma_i^\lambda$, $|\nu| = \Sigma \nu_i$.

This index result is the basis of proving the generic (open-dense) finiteness and stability of minimal surfaces of disc type. The open dense set will be the set of regular values for the map $\pi_\Sigma = \pi|\Sigma$. In addition, this theorem leads us to a new definition of nondegeneracy in critical point theory. A minimal surface $u \in \Sigma_\nu^\lambda$ is nondegenerate if $I(\lambda, \nu) = 3$ (and hence $\nu = 0$) and the map π_0^λ restricted to a neighborhood of u in Σ_0^λ is a local diffeomorphism.

If $n > 3$ the nondegenerate minimal surfaces are immersed up to the boundary and if $n = 3$ they are either immersed or simply branched ($\lambda = (1, \ldots, 1)$). We should re-emphasize, at this point, that we are considering not only area-minimizing minimal surfaces bounded by $\alpha \in \mathcal{A}$ but all critical points. There are some additional, surprising consequences of this index formula. First (by Gulliver-Ossermann and Alt), area-minimizing minimal surfaces in $\mathbf{R}^3$ are free of interior branch points, whereas most minimal surfaces which are either immersed or have simple interior branch points are stable w.r.t. perturbations of the boundary. Second, for $n \geq 4$ minimal surfaces in $\mathbf{R}^n$ may have branch points even if they are area-minimizing but for such n no minimal surface in $\mathbf{R}^n$ is stable under perturbations of the boundary.

This index theorem has been generalized to k-connected regions by Karl Schüffler and Ursula Thiel [4, 6]. Recently Gulliver and Hildebrandt [2] have

---

[2]The number 3 arises from the conformal invariance of the problem under the action of the 3-dimensional conformal group of the disc. The number 3 would disappear if, for example, we imposed a three-point condition on our maps u.

given a beautiful example of three coaxial circles that bound a continuum of 2-connected minimal surfaces of genus zero.

3. Morse theory. As a consequence of the index theorem we know that the generic curve α in $\mathbf{R}^4$ bounds only a finite number $u_1, \ldots, u_m$ of disc minimal surfaces. For each u_i one can show that there is a finite Morse index θ_i for the Hessian of $\mathcal{D}_\alpha$ at u_i. The integer θ_i is the dimension of the largest subspace on which $D^2 \mathcal{D}_\alpha(u_i)$ is negative definite. If a complete Morse theory held for this problem (a handle body decomposition of $\mathcal{N}_\alpha$ in terms of the critical points of $\mathcal{D}_\alpha$) one would have, in particular, the Morse equality

$$\sum_i (-1)^{\theta_i} = 1. \tag{1}$$

This equality was established by the author [8, 9].

The difficulty with directly applying the Morse-theoretic ideas of Palais and Smale is that $\mathcal{D}_\alpha$ does not satisfy their connection C.

Nevertheless in a beautiful recent paper Michael Struwe [5] has shown that by restricting $\mathcal{D}_\alpha$ to the "convex" set $M_\alpha \subset \mathcal{N}_\alpha$ of these maps u which are monotonic on the boundary, one does have a full Morse theory for disc minimal surfaces spanning contours in $\mathbf{R}^4$.

In $\mathbf{R}^3$ the generic wire still bounds only finitely many minimal surfaces of disc type. Moreover, as it turns out, these minimal surfaces are zeros of a Fredholm vector field

$$\mathbf{X}_\alpha \colon \mathcal{N}_\alpha \to T\mathcal{N}_\alpha.$$

This permits one to define a rotation number of $\mathbf{X}_\alpha$ about each zero and hence the total Euler characteristic $\chi(\mathbf{X}_\alpha)$ of $\mathbf{X}_\alpha$ (which is the sum of the rotation numbers). The corresponding theorem in $\mathbf{R}^3$ is that

$$\chi(\mathbf{X}_\alpha) = 1. \tag{2}$$

Finally, there is another interpretation of (1) in $\mathbf{R}^4$. One can show that the main stratum Σ_0^0 of Σ is an (infinite-dimensional) oriented manifold. This allows one to define the degree of the map $\pi_\Sigma = \pi|\Sigma$. Then

$$\deg \pi_\Sigma = 1. \tag{3}$$

4. Existence of higher genus surfaces. Let us consider the pictures of physical soap films, as shown in Figure 2.

Although Douglas showed that any Jordan contour always bounds a disc minimal surface, no criterion was known, until recently, on a contour Γ which guaranteed the existence of a genus $p > 0$ minimal surface spanning Γ. All of the above physical examples of soap films can now be partially[3] explained by the following existence theorem [7] due to F. Tomi and the author.

---

[3] We use the word partially because we do not yet know whether our solutions are embedded as in the soap film examples.

FIGURE 2

THEOREM. *Let M be a surface of genus $p \geq 1$ with one boundary component, and let Γ be a rectifiable Jordan curve in $\mathbf{R}^3$. We assume that*

(i) *there exists a solid q-torus T of class C^3 in $\mathbf{R}^3$ with positive inward mean curvature such that $\Gamma \subset T$, where $q = 2p$ if M is orientable and $q = p$ if not, and*

(ii) *with respect to suitably chosen base points and generators the class of Γ in $\pi_1(T)$ is represented by the same word as the class of ∂M in $\pi_1(M)$, respectively. Then there exists a minimal surface $f\colon M \to \mathbf{R}^3$ mapping ∂M topologically onto Γ and $f(M)$ contained in T.*

The hypothesis on Γ is fulfilled in the following specific cases:

(i) M is oriented of genus p, and Γ is homotopic in T to the commutator product $\prod_{j=1}^{p}[\alpha_j, \beta_j]$ for some set $\alpha_1, \ldots, \beta_p$ of free generators of $\pi_1(T)$;

(ii) M is nonorientable of genus p and Γ is homotopic in T to $\prod_{j=1}^{p} \alpha_j^2$, where $\alpha_1, \ldots, \alpha_p$ are free generators of $\pi_1(T)$;

(iii) M is nonorientable of genus $p = 2k + 1$ and Γ is homotopic in T to $(\prod_{j=1}^{k}[\alpha_j, \beta_j])\gamma^2$, where $\alpha_1, \ldots, \alpha_k, \gamma$ are free generators of $\pi_1(T)$;

(iv) M is nonorientable of genus $p = 2k$, $k \geq 1$, and Γ is homotopic in T to $(\prod_{j=1}^{k-1}[\alpha_j, \beta_j])\alpha_k\beta_k\alpha_k^{-1}\beta_k$, for generators $\alpha_1, \ldots, \beta_k$ of $\pi_1(T)$.

5. Teichmüller theory and Plateau's problem.
The difficulty in obtaining the existence of minimal surfaces of genus $p > 0$ spanning a given contour Γ in Euclidian space by the direct method of the calculus of variations is that one cannot, in general, show that a minimizing sequence (τ_n, u_n), $\mathcal{D}_\alpha(\tau_n, u_n) \to \inf \mathcal{D}_\alpha$, has a convergent subsequence in some reasonable topology. The problem is that a minimizing sequence might be "degenerating" in some sense to a surface of lower genus. For the existence theorem of the last section a way, given the hypotheses, is found to prevent degeneration.

However if we would like to develop a general index theory or a Morse theory covering all genera, or at least be able to prove generic finiteness, one must allow, and in fact be able to completely understand, such degeneration. This amounts to passing through some hypothetical boundary of Teichmüller space $T(p)$ to another Teichmüller space $T(p')$, $p' < p$. From this point of view it would be nice if we had a universal Teichmüller space on which to define our problem. Such spaces have been investigated by Bers.

Since much of Plateau's problem is differential-geometric in flavor, one would like to approach this question along the lines of differential geometry. One could ask if there was a "natural" Riemannian structure on Teichmüller space whose curvature could be computed. If the sectional curvature were negative and the associated metric complete, one could then compactify along the lines developed by Eberline and O'Niell in their theory of visibility manifolds.

A Riemannian structure on Teichmüller space originally arising in number theory was introduced by Weil and then studied by Ahlfors, the so-called Weil-Petersson metric. This metric was shown by Ahlfors to have negative Ricci and holomorphic sectional curvature. Wolpert and Thu showed that the metric was incomplete and therefore Eberline-O'Niell compactness would not follow should the metric be negatively curved. However, from the point of view of minimal surface this incompleteness rather than being a defect makes the metric even more interesting.

Another motivation for studying the Weil-Petersson metric comes from General Relativity, where this metric has appeared in a totally different context.

We have proved the following result [10].

THEOREM. *The sectional curvature of the Weil-Petersson metric is negative.*

Since this result, Wolpert [11] and then later Jost [3] have computed the Riemann curvature tensor of this metric. It appears that the geodesics of $T(p)$ with respect to this metric may provide a mechanism to attain a compactification of Teichmüller space via differential geometry.

How this will impact, if at all, the development of Morse theory, index theory, and generic finiteness for higher genus awaits future developments.

REFERENCES

1. R. Böhme and A. J. Tromba, *The index theorem for classical minimal surfaces*, Ann. of Math. (2) **113** (1981).

2. R. Gulliver and S. Hildebrandt, *Boundary configurations spanning continua of minimal surfaces*, Manuscripta Math. **54** (1986), 323–347.

3. J. Jost, *Harmonic maps and Teichmüller theory*, Preprint.

4. K. Schüffler, *Eine globalanalytische Behandlung des Douglas'schen Problems*, Manuscripta Math. **48** (1984), 189–226.

5. M. Struwe, *A critical point theory for minimal surfaces spanning a curve in* $\mathbf{R}^n$, J. Reine Angew. Math. **349** (1984), 1–23.

6. U. Thiel, *The index theorem for k-fold connected minimal surfaces* (to appear).

7. F. Tomi and A. J. Tromba, *On Plateau's problem for minimal surfaces of higher genus in* $\mathbf{R}^3$, Bull. Amer. Math. Soc. (N.S.)**13** (1985), 169–171.

8. A. J. Tromba, *Degree theory on oriented infinite dimensional varieties and the Morse number of minimal surfaces spanning a curve in* $\mathbf{R}^n$. I: $n \geq 4$, Trans. Amer. Math. Soc. **290** (1985), 383–413.

9. ____, *Degree theory on oriented infinite dimensional varieties and the Morse number of minimal surfaces spanning a curve in* $\mathbf{R}^n$. II, Manuscripta Math. **48** (1984), 139–161.

10. ____, *On a natural algebraic affine connection on the space of almost complex structures and the curvature of Teichmüller space with respect to its Weil-Petersson metric*, Manuscripta Math. (to appear).

11. S. Wolpert, *The topology and geometry of the moduli space of Riemann surface*, Arbeitstagung, Bonn, 1984; Lecture Notes in Math., vol. 1111, Springer-Verlag, Berlin and New York, pp. 431–451.

UNIVERSITY OF CALIFORNIA, SANTA CRUZ, CALIFORNIA 95064, USA

Estimates of Derivatives of Solutions
of Elliptic and Parabolic Inequalities

N. N. URAL'TSEVA

The first part of this paper contains results by O. A. Ladyzhenskaya and the author obtained recently [1–3] for uniformly elliptic and parabolic equations of the second order of general (nondivergent) form. These results concern the solvability of boundary value problems in Sobolev spaces for equations with singularities with respect to independent variables. We have also substantially improved our previous results on the classical solvability of the first boundary value problem for parabolic equations. The results are based on our new exact a priori estimates of Hölder norms for the first derivatives of solutions on the boundary. In addition, we have weakened the conditions for interior estimates. All these estimates hold for an arbitrary function which satisfies elliptic or parabolic inequality. Let us consider some of these estimates.

LEMMA. *Let Ω be a convex domain in $\mathbf{R}^n$ and let $\Omega(\rho)$ be the ρ-neighborhood of the boundary in Ω, $\rho > 0$. If $u \in W^2_{n,\mathrm{loc}}(\overline{\Omega(\rho)})$, $u|_{\partial\Omega} = 0$, then*

$$\sup_{x \in \Omega(\rho)} \frac{u(x)}{\mathrm{dist}\{x; \partial\Omega\}} \le c_1 \left[\rho^{-1} \sup_{\Omega(\rho)} u + \rho^\delta \|(\mathcal{L}u)_-\|^*_{q,\Omega(\rho)} \right]. \tag{1}$$

Here $\delta = 1 - n/q > 0$;

$$\mathcal{L} = a_{ij}(x) \frac{\partial^2}{\partial x_i \partial x_j} + b_i(x) \frac{\partial}{\partial x_i}$$

is an elliptic operator with measurable coefficients a_{ij} satisfying in $\Omega(\rho)$ the inequalities

$$\nu \xi^2 \le a_{ij}\xi_i\xi_j \le \mu\xi^2, \qquad \nu,\mu = \mathrm{const} > 0, \ \forall \xi \in \mathbf{R}^n, \tag{2}$$

$$b_i \in L_{q,\mathrm{loc}}(\overline{\Omega(\rho)}); \qquad c_1 = c_1 \left(\nu^{-1}, \mu, \delta^{-1}, \rho^\delta \left\| \sqrt{\sum_{i=1}^n b_i^2} \right\|^*_{q,\Omega(\rho)} \right)$$

is a continuous increasing function of its arguments; $\| \cdot \|^*_{q,\Omega(\rho)}$ is defined as $\|v\|^*_{q,\Omega(\rho)} = \sup_{x^0 \in \partial^*\Omega} \|v\|_{q,\Omega \cap \pi(x^0,\rho)}$, where $\| \cdot \|_{q,\mathcal{D}}$ is the norm in $L_q(\mathcal{D})$; $\partial^*\Omega$

is the set of all points of $\partial\Omega$ for which the normal to $\partial\Omega$ exists. $\pi(x^0, \rho)$ is cylinder in $\mathbf{R}^n$ of the height ρ where the axis is parallel to the inner normal to $\partial\Omega$ at x^0. The base of the cylinder is the $(n-1)$-dimensional ball with the radius $c_0\rho$, $c_0 = (n\mu\nu^{-1})^{-1}$.

It is not assumed in the lemma that Ω is strictly convex or bounded; $\partial\Omega$ may be nonsmooth. The lemma is important for estimating the derivatives u_{x_i} on the boundary of an arbitrary domain. The proof of the lemma is based on a new "iterative-barrier" method [2]. One of the implications of the lemma combined with A. D. Aleksandrov's theorem [4] is the estimate

$$\frac{u(x)}{\mathrm{dist}\{x; \partial\Omega\}} \le c_2 \cdot (\mathrm{diam}\,\Omega)^\delta \|(\mathcal{L}u)_-\|_{q,\Omega}, \qquad \forall x \in \Omega, \tag{3}$$

for any function $u \in W_n^2(\Omega)$, $u|_{\partial\Omega} = 0$, in a convex domain Ω. Here

$$c_2 = c_2\left(\nu^{-1}, \mu, \delta^{-1}, (\mathrm{diam}\,\Omega)^\delta \cdot \left\|\sqrt{\sum_{1=i}^{n} b_i^2}\right\|_{q,\Omega}\right).$$

All the following statements deal with local estimates.

THEOREM 1. *Assume that Γ is a manifold of class W_q^2, Γ is an open subset of $\partial\Omega$. Let $u \in W_n^2(\Omega)$, $u|_\Gamma = 0$; u satisfies the inequalities (2) in Ω with $a_{ij} = a_{ij}(x, u(x), u_x(x))$ and the inequality*

$$-a_{ij}(x, u, u_x)u_{x_i x_j} \le \mu_1 |u_x|^2 + \phi(x)[1 + |u_x|] \tag{4}$$

with $\mu_1 = \mathrm{const} \ge 0$, $\phi \in L_q(\Omega)$. Then for any $x^0 \in \Gamma$, $x \in \Omega$, the estimate

$$\frac{u(x)}{|x - x^0|} \le c_3[1 + (\mathrm{dist}\{x^0; \partial\Omega \backslash \Gamma\})^{-1} + \|\phi\|_{q,\Omega}] \tag{5}$$

holds with a constant c_3 completely specified by $\nu^{-1}, \mu, \mu_1, \delta^{-1}, \sup_\Omega u$ and Γ. In particular, if $\Gamma = \partial\Omega$, then

$$\mathrm{ess}\sup_{\partial\Omega} \frac{\partial u}{\partial\gamma} \le c_3[1 + \|\phi\|_{q,\Omega}] \tag{6}$$

where $\partial/\partial\gamma$ is the inner normal derivative.

In the estimates of Hölder constants of u_{x_i} on $\partial\Omega$, unilateral condition (4) should be replaced by

$$|a_{ij}(x, u, u_x)u_{x_i x_j}| \le \mu_1 |u_x|^2 + \phi(x)[1 + |u_x|], \qquad x \in \Omega. \tag{7}$$

Taking into account the fact that the conditions (2), (4), and (7) are invariant with respect to the mappings of W_q^2-class, the following theorem is formulated for the special case $\Omega = B_1^+$, where $B_\rho^+ = \{x \in \mathbf{R}^n : |x| < \rho; \ x_n > 0\}$.

THEOREM 2. *Let the function u, $u \in W_n^2(\Omega)$, $\Omega = B_1^+$, $u|_{S_1} = 0$, satisfy inequalities (2), (7) with $\phi \in L_q(\Omega)$, $q > n$; $S_1 = \{x \in \partial B_1^+ : x_n = 0\}$. Then*

$$\mathrm{osc}\left\{\frac{u(x)}{x_n}; B_\rho^+\right\} \le c_4\rho^\alpha, \qquad \forall\rho \le \frac{1}{2}, \tag{8}$$

where the constant $\alpha \in (0,1)$ *depends only on* $\nu^{-1}, \mu, \delta^{-1}$ *and*

$$c_4 = c_4(\nu^{-1}, \mu, \delta^{-1}, \mu_1, \sup_\Omega |u|, \|\phi\|_{q,\Omega}).$$

Theorem 2 leads to an estimate of the Hölder norm of u_{x_i} on $\partial\Omega$ for the solutions $u \in W_n^2(\Omega)$ of inequalities (2), (7) in Ω if $u|_{\partial\Omega} = 0$, $\partial\Omega \in W_q^2$, $q > n$. For the particular case when inequality (7) takes the form $|a_{ij}u_{x_i x_j}| \le \text{const}$, Theorem 2 has been proved earlier by N. V. Krylov and M. V. Safonov [5, 6].

Estimates similar to those of Theorems 1 and 2 respectively are also established in [1, 2] for solutions of parabolic inequalities

$$u_t - a_{ij}(x, t, u, u_x)u_{x_i x_j} \le \mu_1 |u_x|^2 + \phi(x, t)[1 + |u_x|] \tag{9}$$

and

$$|u_t - a_{ij}(x, t, u, u_x)u_{x_i x_j}| \le \mu_1 |u_x|^2 + \phi(x, t)[1 + |u_x|] \tag{10}$$

in the cylinder $Q = \Omega \times (0, T)$.

Let us now present one of the results for solutions $u(x, t)$ of (10). It is assumed that $u, u_{x_i}, u_{x_i x_j}, u_t \in L_{n+1}(Q)$, i.e., $u \in W_{n+1}^{2,1}(Q)$, and u vanishes on the parabolic boundary $\partial'Q$ of the cylinder Q. To formulate the theorem we need some notations. We say $\phi \in L_{\bar{q}, \bar{s}}(Q)$, $\bar{q} = (q_1, \ldots, q_N)$, $\bar{s} = (s_1, \ldots, s_N)$, if $\phi(x, t) = \sum_{k=1}^N \phi_k(x, t)$ with

$$\|\phi_k\|_{q_k, s_k, Q} = \left[\int_0^T dt \left(\int_\Omega |\phi_k(x, t)|^{q_k} \, dx\right)^{s_k/q_k}\right]^{1/s_k} < \infty;$$

write $\|\phi\|_{\bar{q}, \bar{s}, Q} = \sum_{k=1}^N \|\phi_k\|_{q_k, s_k, Q}$.

THEOREM 3. *Let* $\Omega \subset \mathbf{R}^n$, $u \in W_{n+1}^{2,1}(Q)$, $u|_{\partial'Q} = 0$, $\partial\Omega \in W_{n+1}^2$, *and let the function* $u(x, t)$ *satisfy the inequalities* (2), (10) *in* Q *with* $\phi \in L_{\bar{q}, \bar{s}}(Q)$ *and where* $\bar{q} = (q_1, \ldots, q_N)$, $\bar{s} = (s_1, \ldots, s_N)$,

$$q_k \ge n+1, \quad s_k \ge n+1, \quad \frac{n}{q_k} + \frac{2}{s_k} \le 1 - \delta < 1, \quad k = 1, \ldots, N. \tag{11}$$

Then the derivatives u_{x_i} *are Hölder continuous on* $\partial''Q = \partial\Omega \times [0, T]$ *and for any points* $(x, t), (y, \tau) \in \partial''Q$,

$$|u_{x_i}(x, t) - u_{y_i}(y, \tau)| \le c_5[|x - y|^2 + |t - \tau|]^{\alpha/2}. \tag{12}$$

Here $\alpha \in (0,1)$ *depends only on* $\nu^{-1}, \mu, \delta^{-1}$ *and* c_5 *is specified by* ν^{-1}, μ, δ^{-1}, μ_1, $\operatorname{ess\,sup}_Q |u|$, $\|\phi\|_{\bar{q}, \bar{s}, Q}$, $\partial\Omega$.

Until now we have discussed the boundary estimates. To obtain the interior estimates of u_{x_i} we shall assume the differentiability of functions $a_{ij}(x, u, p)$, $(a_{ij}(x, t, u, p))$ with respect to x, u, p. In addition to (2) we impose the conditions

$$\left|\frac{\partial a_{ij}}{\partial p_l} - \frac{\partial a_{il}}{\partial p_j}\right| \le \mu_2(1 + |p|)^{-1};$$

$$\left|\frac{\partial a_{ij}}{\partial u}p^2 - \frac{\partial a_{kj}}{\partial u}p_k p_i + \frac{\partial a_{ij}}{\partial x_k}p_k - \frac{\partial a_{kj}}{\partial x_k}p_i\right| \le \mu_2 p^2 + \phi[1 + |p|]. \tag{13}$$

THEOREM 4. *Let* $u \in W_n^2(\Omega)$ *be a solution of* (7) *for* $x \in \Omega$, $u = u(x)$, *and* $p = u_x(x)$, *and let the conditions* (2) *and* (13) *hold with* $\phi \in L_q(\Omega)$, $q > n$. *Then for any* $\Omega' \Subset \Omega$, $\operatorname{ess\,sup}_{\Omega'} |u_x| \le c_6$,

$$c_6 = c_6(\nu^{-1}, \mu, \mu_1, \mu_2, \delta^{-1}, \|\phi\|_{q,\Omega}, \sup_\Omega |u|, (\operatorname{dist}\{\Omega'; \partial\Omega\})^{-1}).$$

Moreover, if $\partial\Omega \in W_q^2$ *and* $u|_{\partial\Omega} = 0$ *then* $\operatorname{ess\,sup}_\Omega |u_x| \le c_7$ *where* $c_7 = c_7(\nu^{-1}, \mu, \mu_1, \mu_2, \delta^{-1}, \|\phi\|_{q,\Omega}, \sup_\Omega |u|, \partial\Omega)$.

A similar result is also valid for solutions of (10).

The above estimates make it possible to prove the solvability of the Dirichlet problem for quasilinear elliptic and parabolic equations of nondivergent form under substantially weaker conditions than those used so far. First, it is not required that the functions $a(x, u, p)$ in an elliptic equation $a_{ij}(x, u, u_x)u_{x_i x_j} + a(x, u, u_x) = 0$ and $a(x, t, u, p)$ in a parabolic equation $-u_t + a_{ij}(x, t, u, u_x)u_{x_i x_j} + a(x, t, u, u_x) = 0$ are differentiable. Second, the functions forming the equations may be singular with respect to independent variables. Finally, the nonsmooth $\partial\Omega$ are permitted.

THEOREM 5. *Suppose that the following conditions are satisfied:*

(a) Ω *is a bounded domain in* $\mathbf{R}^n$, $\partial\Omega \in W_q^2$, $q > n$.

(b) *There exists a constant* M_0 *independent on* $u \in W_q^2(\Omega)$ *and on* $\tau \in [0, 1]$, *such that every* $\tau \in [0, 1]$ *solution of the Dirichlet problem*

$$[\tau a_{ij}(x, u, u_x) + (1 - \tau)\delta_i^j]u_{x_i x_j} + \tau a(x, u, u_x) = 0, \tag{14}$$
$$x \in \Omega, \ u = 0, \ x \in \partial\Omega,$$

satisfies $\operatorname{ess\,sup}_{x \in \Omega} |u(x, \tau)| \le M_0$.

(c) *For* $x \in \overline{\Omega}$, $|u| \le M_0$ *and for arbitrary* $p \in \mathbf{R}^n$ *the functions* $a_{ij}(x, u, p)$ *are differentiable and satisfy the inequalities* (2), (13), *and the measurable function* $a(x, u, p)$ *satisfies the inequality* $|a(x, u, p)| \le \mu_1 p^2 + \phi(x)[1 + |p|]$, $\phi \in L_q(\Omega)$. *This condition guarantees the estimate* $\operatorname{ess\,sup}_\Omega |u_x| \le M_1$.

(d) *On the set* $\mathcal{M}_1 = \{(x, u, p): x \in \overline{\Omega}, |u| \le M_0, |p| \le M_1\}$ *the conditions*

$$\left|\frac{\partial a_{ij}}{\partial p_k}\right| \le \mu_3; \qquad \left|\frac{\partial a_{ij}}{\partial u}\right|, \left|\frac{\partial a_{ij}}{\partial x_k}\right| \le \phi_1(x), \quad \phi_1 \in L_q(\Omega), \tag{15}$$

hold.

(e) *The function* $a(\cdot, u, p)$ *defined on the set* $\mathcal{M}_2 = \{(u, p): |u| \le M_0, |p| \le M_1\}$ *belongs to* $C(\mathcal{M}_2; L_q(\Omega))$.

Then for every $\tau \in [0, 1]$, *the Dirichlet problem* (14) *is solvable in* $W_q^2(\Omega)$. *If, additionally,* $\partial\Omega \in C^{2+\alpha}$, $a \in C^\alpha(\mathcal{M}_1)$, *then a solution* $u(\cdot, \tau)$ *of* (14) *belongs to* $C^{2+\beta}(\overline{\Omega})$, $\beta = \min\{1 - n/q; \alpha\}$.

THEOREM 6. *Let us suppose that the following conditions are satisfied:*

(a) Ω *is a bounded domain in* $\mathbf{R}^n$, $\partial\Omega \in W_{q+2}^2$, $q > n$, $Q = \Omega \times (0, T)$.

(b) *For all possible solutions* $u(\cdot, \tau)$ *in* $W_{q+2}^{2,1}(Q)$ *of the problem*

$$-u_t + [\tau a_{ij}(x, t, u, u_x) + (1 - \tau)\delta_i^j]u_{x_i x_j} + \tau a(x, t, u, u_x) = 0 \quad in \ Q, \tag{16}$$
$$u = 0 \quad on \ \partial' Q,$$

the estimate $\operatorname{ess\,sup}_Q |u(\cdot, \tau)| \le M_0$, $\forall \tau \in [0,1]$, *is valid.*

(c) *For* $(x,t) \in Q$, $|u| \le M_0$, $p \in \mathbf{R}^n$ *the functions* $a_{ij}(x,t,u,p)$ *are differentiable with respect to* x, u, p *and satisfy* (2), (13) *and a measurable function* $a(x,t,u,p)$ *satisfies the inequality* $|a(x,t,u,p)| \le \mu_1 p^2 + \phi(x,t)[1 + |p|]$ *with* $\phi \in L_{q+2}(Q)$. *This guarantees the estimate* $\operatorname{ess\,sup}_Q |u_x(\cdot, \tau)| \le M_1$, $\forall \tau \in [0,1]$.

(d) *On the set* $\mathcal{M}_3 = \{(x,t,u,p): x \in \overline{\Omega}, \ t \in [0,T], \ |u| \le M_0, \ |p| \le M_1\}$ *the inequalities* (15) *hold with* $\phi_2(x,t)$ *instead of* $\phi_1(x)$, $\phi_2 \in L_{q+2}(Q)$.

(e) $a_{ij} \in C(\mathcal{M}_3)$, $a \in C(\mathcal{M}_2; L_{q+2}(Q))$.

Then for every $\tau \in [0,1]$ *there exists at least one solution* $u(\cdot, \tau) \in W_{q+2}^{2,1}(Q)$ *of the problem* (16).

If, additionally, $\partial\Omega \in C^{2+\alpha}$, $a_{ij} \in C^{\alpha, \alpha/2, \alpha, \alpha}(\mathcal{M}_3)$, $a \in C^{\alpha, \alpha/2, \alpha, \alpha}(\mathcal{M}_3)$, *then* $u(\cdot, \tau) \in C^{2+\alpha, 1+\alpha/2}(\overline{Q})$.

Let us turn now to the second part of this paper. Here we want to describe the results on the smoothness of solutions of variational inequalities (VI) found by the author and by her students. We consider the VI

$$u \in K, \quad \int_\Omega [a_{ij} u_{x_j} (v-u)_{x_i} + (b_i u_{x_i} + b_0 u - f)(v - u)] \, dx \ge 0, \quad \forall v \in K, \quad (17)$$

and the corresponding evolution inequalities connected with elliptic and parabolic equations (and systems) under various convex constraints on solutions on the boundary. In the scalar case $u \colon \Omega \to \mathbf{R}^1$ (one equation) we assume that condition (2) holds.

In a problem with one obstacle on the boundary, i.e., if

$$K = K_1 = \{v \in W_2^1(\Omega): v(x) \ge 0 \text{ for } x \in \partial\Omega\},$$

the best possible smoothness up to the boundary of solutions of (17), i.e., the Hölder continuity of u_{x_i} in $\overline{\Omega}$, was proved by different methods by L. A. Caffarelli [7] and D. Kinderlehrer [8] under the assumptions $a_{ij} \in C^2(\overline{\Omega})$; $b_i, b_0, f \in C^1(\overline{\Omega})$, $\partial\Omega \in C^3$. In the author's papers [9, 10] these conditions were replaced by the minimal restrictions

$$a_{ij} \in W_q^1(\overline{\Omega}); \quad b_i, b_0, f \in L_q(\Omega); \quad \partial\Omega \in W_q^2, \quad q > n. \quad (18)$$

Such conditions are necessary for the Hölder continuity of u_{x_i} even in the case of the Dirichlet problem, that is, for $K = \overset{\circ}{W}_2^1(\Omega)$. In [10] the result was generalized to a quasilinear case, more precisely to inequalities (17) with $a_{ij} = a_{ij}(x, u)$, $f = f(x, u, u_x)$. In [10] it was assumed that $|\partial a_{ij}/\partial x_k|, |\partial a_{ij}/\partial u| \le \mu_2$; a_{ij} satisfies (2) and $|f(x,u,p)|, |\partial f/\partial p_k|, |\partial f/\partial x_k|, |\partial f/\partial u| \le \mu_2 p^2 + \phi(x)$, $\phi \in L_q(\Omega)$.

Evidently all these results are also valid for the problem with two obstacles on $\partial\Omega$ when

$$K = K_2 = \{v \in W_2^1(\Omega): \varphi(x) \le v(x) \le \psi(x), \ x \in \partial\Omega\}$$

if one supposes $\varphi, \psi \in W_q^2(\Omega)$, $\psi(x) > \varphi(x)$ on $\partial\Omega$. In [11] the two-obstacle problem is investigated for the case when $\varphi(x) \le \psi(x)$ on $\partial\Omega$ and a set $\{x \in \partial\Omega: \varphi(x) = \psi(x)\} \ne \varnothing$, that is, the two obstacles are partially stuck together.

It is shown that in such a case $u_{x_i} \in C^\alpha(\overline{\Omega})$ if $\varphi, \psi \in W_q^2(\Omega)$ and conditions (18) are also fulfilled.

The above results hold also for the case of obstacles in the whole domain Ω, more precisely if $K = K_3 = \{v \in W_2^1(\Omega) : \varphi(x) \le v(x) \le \psi(x),\ x \in \Omega\}$. Unlike the classical problem with an obstacle in the domain, K is here a subset of $W_2^1(\Omega)$ but not of $\overset{\circ}{W}{}_2^1(\Omega)$; i.e., the boundary values of u are not prescribed.

The best possible smoothness of solutions was proved also for other problems with constraints on surfaces: obstacle on the interface of two media, semipenetrable membrane [12], etc. The fact that in all of these cases the restrictions on smoothness of the data are minimal enables one to prove the Hölder continuity of u_{x_i} for the parabolic case as well [9, 11].

The possible convex constraints on $u: \Omega \to \mathbf{R}^\mathcal{N}$, $\mathcal{N} > 1$, i.e., on the solutions of elliptic systems, are more various. In this case K may be of the form

$$K = K_4 = \{v \in [W_2^1(\Omega)]^\mathcal{N} : v(x) \in \mathcal{K}(x) \text{ on } \partial\Omega\}$$

with $\mathcal{K}(x)$ being a convex closed set in $\mathbf{R}^\mathcal{N}$. For the so-called diagonal systems when $a_{ij}(x) \in \mathbf{R}^1$, $f(x) \in \mathbf{R}^\mathcal{N}$, b_i, b_0 are $(\mathcal{N} \times \mathcal{N})$-matrices for $x \in \Omega$, and $\mathcal{K}$ does not depend on x, the regularity results also established. They are of the same degree of generality as for the scalar case.

For the VI connected with general strongly elliptic systems of the second order, only the question of the existence of the second derivatives of solutions is discussed in the literature. Such problems are of special interest in unilateral problems of elasticity and fluid mechanics. For example, the equilibrium of the elastic plate under the pressure of a stamp is governed by inequalities (17) for strongly elliptic system and the set

$$K = K_5 = \{v \in [W_2^1(\Omega)]^\mathcal{N} : v(x) \cdot \gamma(x) + h(x) \ge 0, x \in \Omega\}$$

where $\gamma(x) \in \mathbf{R}^\mathcal{N}$, $h(x) \in \mathbf{R}^1$ are defined for $x \in \Omega$. It is proved in [13] that the solution of such a problem belongs to $[W_2^2(\Omega)]^\mathcal{N}$. We mention also the paper [14] where the existence of the second derivatives for the solutions of a wide class of VI, including contact problems of elasticity, is proved.

BIBLIOGRAPHY

1. O. A. Ladyzhenskaya and N. N. Ural'tseva, *Solvability of the first boundary value problem for quasilinear elliptic and parabolic equations in the presence of singularities,* Dokl. Akad. Nauk SSSR **281** (1985), no. 2, 275–279.

2. _____, LOMI Preprints, P-I-85, Leningrad, 1985, pp. 1–43.

3. _____, *Estimates of* max $|u_x|$ *for solutions of quasilinear elliptic and parabolic equations of general type, and some existence theorems,* Boundary Value Problems of Mathematical Physics and Related Problems in the Theory of Functions, 16, Zap. Nauchn. Sem. Leningrad. Otdel. Mat. Inst. Steklov. (LOMI) **138** (1984), 90–107.

4. A. D. Aleksandrov, *On majorants of solutions and uniqueness conditions for elliptic equations,* Vestnik Leningrad. Univ. **21** (1966), no. 7, 5–20; English transl. in Amer. Math. Soc. Transl. (2) **68** (1968), 144–161.

5. N. V. Krylov, *Boundedly inhomogeneous elliptic and parabolic equations in a domain,* Izv. Akad. Nauk SSSR Ser. Mat. **48** (1983), no. 1, 75–108.

6. M. V. Safonov, Uspekhi Mat. Nauk **38** (1983), 5 (233), 146–147.

7. L. A. Caffarelli, Comm. Partial Differential Equations **4** (1979), 65–97.

8. David Kinderlehrer, *The smoothness of the solution of the boundary obstacle problem*, J. Math. Pures Appl. (9) **60** (1981), 193–212.

9. N. N. Ural'tseva, Dokl. Akad. Nauk SSSR **280** (1985), no. 3, 563–565; English transl. in Soviet Math. Dokl. **31** (1985), no. 1, 135–138.

10. ____, Problemy Math. Analys., Izd. Leningrad. Univ. **10** (1986), 92–105.

11. A. A. Arhipova and N. N. Ural'tseva, Vestnik. Leningrad. Univ. **1** (1986), 3–9.

12. N. N. Ural'tseva and I. A. Denisova, Vestnik. Leningrad. Univ. **8** (1985), 36–42.

13. S. P. Vodjana, Depon. in Ukr. VIINTI 2 Sept. 1985, No. 2015 Uk-85.

14. G. V. Jakunina, Problemy Math. Analys., Izd. Leningrad. Univ. **8** (1981), 213–220.

LENINGRAD STATE UNIVERSITY, LENINGRAD, USSR

Families of One-Dimensional Maps
and Nearby Diffeomorphisms

M. V. JAKOBSON

1. Introduction. In the last decade, one-dimensional dynamical systems have been studied intensely from different points of view. One of the reasons for this interest is that the results about one-dimensional systems prove to be useful when studying multidimensional systems which act contractively in all but one direction. We shall call such systems near to one-dimensional. In this paper we present some results concerning one-dimensional maps and nearby diffeomorphisms. Consider some one-dimensional map $f \in C^k(\mathbf{R}, \mathbf{R})$, $1 \le k \le \infty$, and let a singular map $\tilde{f} \in C^k(\mathbf{R}^m, \mathbf{R}^m)$,

$$\tilde{f}\colon (y_0, y_1, \ldots, y_{m-1}) \mapsto (y_1, f(y_1), 0, \ldots, 0),$$

correspond to f. We shall also consider families of one-dimensional maps $f_t\colon \mathbf{R} \to \mathbf{R}$ continuously dependent on $t \in [0,1]$ in the C^k-topology and the corresponding families of singular maps $\tilde{f}_t \in C^k(\mathbf{R}^m, \mathbf{R}^m)$. We say that $F \in \mathrm{Diff}^k(\mathbf{R}^m)$ is near to a singular one-dimensional map $\tilde{f}$ if F is close to $\tilde{f}$ in $C^k(\mathbf{R}^m, \mathbf{R}^m)$. Here is an example:

$$F\colon (y_0, y_1, \ldots, y_{m-1}) \mapsto (y_1, f(y_1) + \varepsilon_1 y_0, \varepsilon_2 y_2, \ldots, \varepsilon_{m-1} y_{m-1}).$$

Similarly, a family of diffeomorphisms F_t is near to a singular family $\tilde{f}_t$ if F_t and $\tilde{f}_t$ are close in the topology $C([0,1], C^k(\mathbf{R}^m, \mathbf{R}^m))$.

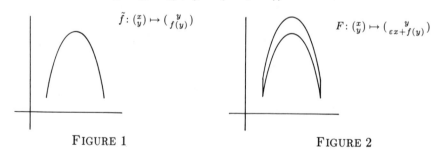

$$\tilde{f}\colon \begin{pmatrix} x \\ y \end{pmatrix} \mapsto \begin{pmatrix} y \\ f(y) \end{pmatrix}$$

$$F\colon \begin{pmatrix} x \\ y \end{pmatrix} \mapsto \begin{pmatrix} y \\ \varepsilon x + f(y) \end{pmatrix}$$

FIGURE 1   FIGURE 2

2. Structurally stable maps. (a) Let I be an interval of the real axis. We denote by $\mathfrak{A}_k \subset C^k(I, I)$ a set of mappings f with the following properties:

$$I = \Delta(f) \cup \Sigma(f),$$

where $\Delta(f)$ is an open f- and f^{-1}-invariant set which is the union of domains of attraction to a finite number of attracting cycles; $\Sigma(f)$ is a hyperbolic repelling set

$$|df^n/dx|_{\Sigma(f)} > ac^n, \qquad a > 0, \ c > 1;$$

and $f \mid \Sigma(f)$ is topologically conjugate to a subshift of finite type. For $f \in \mathfrak{A}_k$, the nonwandering set $\Omega(f)$ is the union of a finite number of attracting cycles and a repelling set $\Omega_1(f) \subset \Sigma(f)$. The mappings $f \in \mathfrak{A}_k$ are Ω-stable. If $k \geq 2$ and the critical points c_i, $i \in [1, m]$, are nondegenerate and $f^k(c_i) \neq f^l(c_j)$ for any $k \neq l$, $i \neq j$, then f is structurally stable. For $k \geq 2$ the Lebesgue measure of $\Sigma(f)$ equals zero and therefore the asymptotic behavior is periodic for almost all points $x \in I$.

For $k = 1$ the set $\mathfrak{A}_k$ is dense in $C^k(I, I)$ [6]. For $k > 1$ the question remains unsolved. The last results in this direction are due to Mañé [9]. In particular he proved that for any $k \in \mathbf{N}$, $\mathfrak{A}_k$ is dense in the space of C^k-immersions of the circle.

For a family $f_t \colon I \to I$, $t \in [0, 1]$, let us consider the set $S = \{t \colon f_t \in \mathfrak{A}_k\}$. It is not known if S is dense in $[0, 1]$ for "natural" one parameter families—in particular for quadratic family $q_t \colon x \mapsto 4tx(1 - x)$. For this family, the density of S in some neighborhood $\delta_\varepsilon = (1 - \varepsilon, 1)$ was announced in [1]. We point out here that the proof of such a statement is really equivalent to the proof of the density of S in $[0, 1]$. Indeed for any $\varepsilon > 0$ there exists an interval $\delta_{\varepsilon n} \subset \delta_\varepsilon$ such that for $t \in \delta_{\varepsilon n}$ there exist n intervals $l_0^{(n)}(t), l_1^{(n)}(t), \dots, l_{n-1}^{(n)}(t)$ which are cyclically permuted by f_t, and when t varies in $\delta_{\varepsilon n}$ the kneading-invariant of $f_t^n \mid l_0^{(n)}(t)$ varies from the maximal to the minimal one. When $n \to \infty$, the family $f_t^n \mid l_0^{(n)}(t)$ after the corresponding renormalization tends to the quadratic family. This universal low was discovered by Milnor and investigated by several authors. Thus the interval $t \in \delta_{\varepsilon n}$ plays the same role for $f_t^n \mid l_0^{(n)}(t)$ as $t \in [0, 1]$ plays for the whole quadratic family.

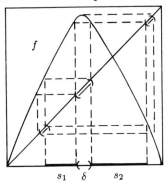

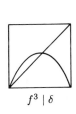

$$fs_1 = s_2$$
$$fs_2 \supset s_1 \cup s_2 \quad \Rightarrow \left(\begin{smallmatrix} 0 & 1 \\ 1 & 1 \end{smallmatrix}\right)$$

$f^3 \mid \delta$

FIGURE 3

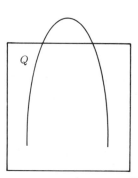

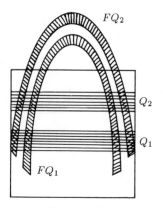

FIGURE 4. Singular map $\tilde{f}$ has an invariant Cantor set inside Q.

FIGURE 5. Nearby diffeomorphism F has a horseshoe.

(b) If $f: I \to I$ is hyperbolic, and if $\tilde{f}$ is the corresponding singular multidimensional map and F is a diffeomorphism near to $\tilde{f}$, then according to a theorem of S. Van Strien [13] F possesses an invariant locally maximal hyperbolic set $\Omega_1(F)$ in a small neighborhood of $\Omega_1(\tilde{f})$, and a continuous surjection $p: \Omega_1(F) \to \Omega_1(\tilde{f})$ is defined which is an injection on the set of periodic points. If $\tilde{f}_t$ is a family of singular maps and if F_t is a nearby family of diffeomorphisms, then a similar correspondence $p_t: \Omega_1(F_t) \to \Omega_1(\tilde{f}_t)$ exists for t belonging to a finite number of intervals of Ω-stability. Moreover, if the domain of parameter variation contains a finite number of t_i corresponding to the doubling bifurcations, then similar bifurcations occur for F_t (see [13]).

(c) Let $f \in \mathfrak{A}_k$, and let $\Omega_1(f)$ be the repelling part of $\Omega(f)$. It follows from the theorem of McCluskey and Manning [10] that $\dim_H \Omega_1(f) = t_0$ coincides with the unique root of the equation $P(t\varphi) = 0$ where $\varphi(x) = -\log|df/dx| \, | \, \Omega_1(f)$. P is the topological pressure, $P(\chi) = \sup(\mu(\chi) + h_\mu(f))$ where the sup is taken over all Borel f-invariant measures with the support inside $\Omega_1(f)$. The Hausdorff t_0-measure coincides with the unique measure μ_0 satisfying $\mu_0(t_0\varphi) + h_{\mu_0}(f) = P(\varphi \cdot t_0) = 0$. Thus $t_0 = h_{\mu_0} / \int \log|df/dx| \, d\mu_0$.

For the singular map $\tilde{f}$ we have $\dim_H \Omega_1(\tilde{f}) = \dim_H \Omega_1(f)$. One can deduce from the results of [10] that any diffeomorphism F ε-close to $\tilde{f}$ in $C^k(\mathbf{R}^2, \mathbf{R}^2)$ satisfies

$$\dim_H \Omega_1(F) \leq t_0 + \delta(\varepsilon)$$

where $\lim_{\varepsilon \to 0} \delta(\varepsilon) = 0$.

3. The consequences of universal behavior. If an infinite sequence of period doubling bifurcations occurs for a singular family $\tilde{f}_t$, then we can use Feigenbaum's theory to prove that the same holds for a nearby family of diffeomorphisms F_t.

Let ϕ_0 be the fixed point of doubling transformation $\mathcal{T}$ and let ϕ_t, $t \in [-1, 1]$, be the one-dimensional unstable manifold of ϕ_0 (see [4]). We choose t so that the

ϕ_t, with $t > 0$, have zero topological entropy. The mappings ϕ_t are of the form $\phi_t(x) = g_t(x^2)$ where g_t is analytic on a neighborhood of $[0,1]$. Now consider the singular family (see [4]) which is given by $\tilde{g}_t \colon (x,y) \mapsto (g_t(x^2 - y), 0)$. As far as I know, the following statement is not explicitly formulated in the literature, though it may be proved following [4, 14].

The doubling transformation $\mathcal{T}'$ in the space of one parameter families of analytic maps $F_t \colon \mathbf{R}^2 \to \mathbf{R}^2$ may be defined so that $\mathcal{T}'\tilde{g}_t = \tilde{g}_t$ and there exists an open neighborhood $\mathcal{U}$ of $\tilde{g}_t$ such that $\tilde{g}_t$ is an attractive fixed point of $\mathcal{T}'$ in $\mathcal{U}$.

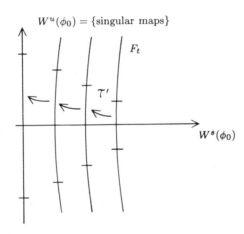

FIGURE 6

Thus if $F_t \subset \mathcal{U}$ is a family of diffeomorphisms, then $\mathcal{T}'^m F_t$ converges with an exponential rate toward the singular family $\tilde{g}_t$. This implies that a full sequence of period doubling bifurcations occur for F_t. The following concerns the behavior of complex eigenvalues of $DF_t^{2^n}$ at the points of 2^n-cycles (compare [16]).

Consider the family $G_t = \mathcal{T}'^m F_t$. For n sufficiently large, G_t has a cycle $\{A_1(t), A_2(t)\}$ of period 2 – the image of 2^{n+1}-cycle of F_t, which is close to the cycle $\{a_1(t), a_2(t)\}$ of the singular map $\tilde{g}_t$. We use $\lambda(t)$ to denote the universal function equal to the derivative of $D\tilde{g}_t^2$ at $a(t)$. $\lambda(t)$ is analytic and strictly monotone. Let $\lambda_1(t)$ be that eigenvalue of DG_t^2 at $A(t)$, which is close to $\lambda(t)$, and let $\lambda_2(t)$ be the second eigenvalue and Tr_t the trace of DG_t^2 at $A(t)$. Taking into account the fact that the family G_t is exponentially close to $\tilde{g}_t$, we find that parameter values $t_n^{(1)}$, $t_n^{(2)}$, given by $\lambda_1(t_n^{(1)}) = 1$, $\lambda_1(t_n^{(2)}) = -1$, are exponentially close to the parameter values $t^{(1)}$, $t^{(2)}$ of the first and second bifurcations for $\tilde{g}_t$. On the other hand, G_t are diffeomorphisms, and eigenvalues of DG_t cannot be zero. Thus they become complex numbers $\lambda_1(t) = \overline{\lambda_2(t)}$. If F_t is sufficiently close to the singular family $\tilde{g}_t$ we have $|\mathrm{Jac}\, F_t| < \varepsilon_0 < 1$. Thus $|\mathrm{Jac}\, F_t^{2^{n+1}}| < \varepsilon_0^{2^{n+1}}$ and the modulus of eigenvalues of $DF_t^{2^{n+1}}$ at the points of 2^{n+1}-cycle (these eigenvalues are just $\lambda_{1,2}(t)$) is less than $\varepsilon_0^{2^n}$.

FIGURE 7. Computer experiments show that for large n, $\operatorname{Re} \lambda_1^{(n)}(t)$ behaves, after rescaling, like universal function $\lambda(t)$.

Now $\lambda_{1,2}(t)$ are complex for t satisfying $\operatorname{Tr}_t^2 < 4 \operatorname{Jac} F_t^{2^{n+1}}$. For all t, Tr_t is exponentially close to $\lambda(t)$ with all derivatives. Thus in a neighborhood of a value t_0 defined by $\lambda(t_0) = 0$, Tr_t is almost linear. This implies

PROPOSITION 1. *Let F_t be a family of diffeomorphisms near to the singular family $\tilde{g}_t$. Then between two successive period doubling bifurcations, the eigenvalues of $DF_t^{2^n}$ at 2^n-cycle become complex for t belonging to an interval Δ_n such that*

$$|\Delta_n|/|t_n^{(2)} - t_n^{(1)}| < c \cdot \varepsilon^{2^{n-1}}.$$

In the complex plane, $\lambda_i(t)$ are moving along a curve which is close up to $O(r^2 \cdot 2^n)$ to some circle of radius $r < \operatorname{const} \varepsilon^{2^{n-1}}$. Besides, $d(\operatorname{Re} \lambda_i(t))/dt$ is exponentially close to the universal velocity $\frac{1}{2} d\lambda(t)/dt$.

4. Mappings with absolutely continuous invariant measure. (a) We consider $f \in C^k(I, I)$, $k \geq 2$, which admit an invariant measure absolutely continuous with respect to the Lebesgue measure (a.c.i.m.). The theorem of Ledrappier [8] states that a map f with an ergodic a.c.i.m. μ with positive entropy $h_\mu(f)$ has strong stochastic properties: μ is Bernoullian with respect to some power f^{k_0}.

The properties of smooth mappings with a.c.i.m. are similar to those of expanding maps, though the presence of critical points implies that the expansion is nonuniform. This nonuniformity may be compensated for by constructing some expanding piecewise monotone map T which has a countable number of laps, and on any lap coincides some interate of f. We specify it by the following construction. Suppose there are a finite number of intervals $L_1, L_2, \ldots, L_m$ such that $L_k \cap L_l = \varnothing$ and any L_k may be represented as a disjoint union

$$L_k = \left(\bigcup_{i=1}^{\infty} \Delta_{ki} \right) \cup \mathcal{O}_k$$

where $\mathcal{O}_k$ has Lebesgue measure zero, Δ_{ki} are intervals, and for any Δ_{ki} there exist $n = n(k, i) \in \mathbf{N}$ and $l = l(k, i) \in (1, m)$ such that f^n maps Δ_{ki} diffeomorphically onto L_l. Let us define the map T on $X = \bigcup_{k=1}^m L_k \backslash \mathcal{O}_k$ by $T \mid \Delta_{ki} = f^{n(k,i)}$.

The map T generates a subshift of finite type on the alphabet $1, 2, \ldots, m$ with the matrix $A = (a_{kl})$, $a_{kl} \in \{0, 1\}$, given by $a_{kl} = 1 \Leftrightarrow \exists \Delta_{ki} \colon T\Delta_{ki} = L_l$. Without loss of generality one can suggest that $A^{n_0} > 0$ for some $n_0 \in \mathbf{N}$. Otherwise there exists an A^{n_0}-invariant subset of $1, 2, \ldots, m$ such that the restriction of

A^{n_0} on this subset is positive. We call the union of L_k corresponding to this subset a component of primitivity.

Suppose that the following inequalities hold:

$$|dT^n/dx| > ac^n, \qquad a > 0, \ c > 1, \tag{1}$$

$$\sup_{k,i} \ \sup_{x \in \Delta_{ki}} \left| \frac{d^2T}{dx^2} \middle/ \frac{dT}{dx} \right| \cdot |\Delta_{ki}| < C_1. \tag{2}$$

Then using a theorem of Walters [15] we obtain that on any component of primitivity, T^{n_0} has a unique a.c.i.m. $\nu(dx) = h(x)\,dx$ where $h(x) \geq h_0 > 0$ is continuous and the natural extension of (T^{n_0}, ν) is isomorphic to a Bernoulli shift.

If $\Sigma_{k,i} n(k,i)|\Delta_{ki}| < \infty$ then an f-invariant ergodic a.c.i.m. $\mu(dx)$ corresponds to $\nu(dx)$.

DEFINITION. If f generates a map T with properties (1) and (2), then we call f expansion-inducing.

(b) For several one parameter families of one-dimensional maps, the set of parameter values corresponding to the maps with a.c.i.m. has positive measure [3, 7, 5, 1]. One possible approach to this problem is to prove that the maps under consideration are expansion-inducing.

Let $f_t\colon I \to I$ be a family of maps continuously dependent on t in C^2-topology. We consider t within an interval $[t_0 - \varepsilon, t_0]$ and formulate some conditions which imply that for t belonging to a set of positive measure, with t_0 being a density point of this set, f_t is expansion-inducing. The main conjecture is that some iterate of any critical point of f_{t_0} falls into an invariant hyperbolic repelling Cantor set. We shall specify it by formulating some conditions which are structurally stable under C^2-perturbations.

We assume that all critical points of f_t are nondegenerate for any t and denote them by c_k, $k \in [1, m]$.

The definition of an expansion-inducing map involves a piecewise expanding map T continuous on some $\Delta_{ki} \subset L_k$. Our first condition defines these L_k.

I. For any $t \in [t_0 - \varepsilon, t_0]$ there are m open intervals

$$L_i(t) = (x_i(t), y_i(t)), \qquad i \in [1, m],$$

such that $L_i(t) \cap L_j(t) = \varnothing$, $c_i(t) \subset L_i(t)$, $x_i(t)$, $y_i(t)$ are eventually periodic, and $f_t c_i(t) \notin \bigcup_{j=1}^{m} L_j(t)$. We assume that for any $i, j \in [1, m]$ the set $f_t^{-1} L_i(t) \cap L_j(t)$, unless empty, is a union of diffeomorphic preimages of $L_i(t)$, and $L_i(t)$ and these preimages vary continuously with t. We shall denote by $L(t)$ the union of $L_i(t)$.

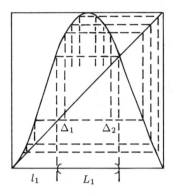

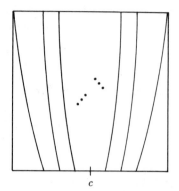

FIGURE 8. Map f_{t_0}, where $l_1 = \bigcup_1^\infty \overline{L}_1^{-n} \cup \{0\}$, $Y = \{0\}$, and $fL_1 \subset (l_1 \cup L_1)^{-1}$.

FIGURE 9. Induced map $T_{t_0} \mid L(t_0)$.

The following conditions define a hyperbolic Cantor set.

II. For any $t \in [t_0 - \varepsilon, t_0]$ there are p closed intervals $l_j = l_j(t) = [u_j(t), v_j(t)]$, $j \in [1, p]$, such that $L_i \cap l_j = \varnothing$, $l_j \cap l_k = \varnothing$, u_j, v_j are eventually periodic, and for any $j \in [1, p]$ there exists a $k_j \in \mathbf{N}$ such that f^{k_j} maps l_j diffeomorphically on its image

$$f^{k_j} l_j = (l_{s_1} \cup l_{s_2} \cup \cdots \cup l_{s_\nu}) \cup (\overline{L}_{r_1} \cup \overline{L}_{r_2} \cup \cdots \cup \overline{L}_{r_\mu})$$

where $\nu \geq 1$, $\mu \geq 1$, and the items vary continuously with t.

Let us denote the union of l_j by $l = l(t)$. Then we define $F_t(x)$ on l by $F_t(x) \mid l_j = f_t^{k_j}(x)$ and $F_{tj}^{-1}(y) = f_t^{-k_j}(y) \cap l_j$. We require the following hyperbolicity condition: there exist $a > 0$, $b > 1$ such that for all $t \in [t_0 - \varepsilon, t_0]$ and for any $j_1, j_2, \ldots, j_n$

$$\left| \frac{d}{dx} (F_{tj_1}^{-1} \circ F_{tj_2}^{-1} \circ \cdots \circ F_{tj_n}^{-1}) \right| < ab^{-n}. \tag{3}$$

The above conditions imply that for $t \in [t_0 - \varepsilon, t_0]$

$$l(t) = \bigcup_{i=1}^{m} \bigcup_{n=1}^{\infty} \overline{L}_i^{-n}(t) \cup Y_t \tag{4}$$

where we use $L_i^{-n}(t)$ to denote the preimages $F_{tj_1}^{-1} \circ F_{tj_2}^{-1} \circ \cdots \circ F_{tj_n}^{-1} L_i(t)$ and Y_t is an F_t-invariant locally maximal hyperbolic repelling Cantor set, meas $Y_t = 0$. In (4), $L_i^{-n}(t) \cap Y_t = \varnothing$ and the endpoints x_i^{-n}, y_i^{-n} of $L_i^{-n}(t)$ belong to Y_t if and only if they are nonwandering for $F_t \mid l$. The points of Y_t which are given by

$$y_{i_1 i_2 \cdots i_n \cdots}(t) = l_{i_1}^{-1}(t) \cap l_{i_2}^{-1}(t) \cap \cdots \cap l_{i_n}^{-(n-1)}(t) \cap \cdots$$

are differentiable functions of t.

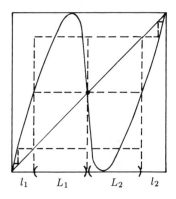

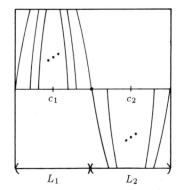

FIGURE 10. $l_1 = \bigcup_1^\infty \overline{L}_1^{-n} \cup \{0\}$,
$l_2 = \bigcup_1^\infty \overline{L}_2^{-n} \cup \{1\}$, $Y = \{0\} \cup \{1\}$.

FIGURE 11. $fL \subset l_1 \cup \overline{L}_1 \cup \overline{L}_2 \cup l_2$.

The next condition relates $L(t)$ and $l(t)$ and defines the structure of the induced map on $L(t)$.

III. There exists an $N \in \mathbf{N}$ such that for any $t \in [t_0 - \varepsilon, t_0]$,

$$f_t L(t) \subset \bigcup_{i=1}^m \bigcup_{k=0}^N \overline{L}_i^{-k}(t) \cup \bigcup_{j=1}^p \bigcup_{k=0}^N l_j^{-k}(t)$$

where $\overline{L}_i^{-k} \subset f^{-k}\overline{L}_i$, $l_j^{-k} \subset f^{-k}l_j$ are diffeomorphic preimages of $\overline{L}_i$ and l_j, and the items depend continuously on t.

Let us denote by $\mathcal{C}$ the map induced on L by f. It follows from I–III that the domain of $\mathcal{C}$ consists of a finite or a countable union of intervals Δ−preimages of L_i and that the restriction of $\mathcal{C}$ on any interval Δ belongs to one of the two possible types. If Δ contains a critical point c_k (the number of such Δ does not exceed m), then $\mathcal{C}: \Delta \to \mathcal{C}\Delta$ is a two-fold covering. Otherwise $\Delta = L_i^{-k}$ is a diffeomorphic preimage of L_i. In all cases $\mathcal{C}$ extends to a C^2-map on $\overline{\Delta}$. Since meas $Y = 0$, $\mathcal{C}$ is defined on a subset of full measure in L.

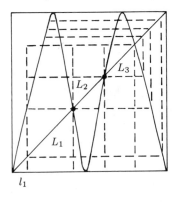

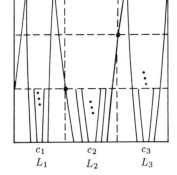

FIGURE 12. $l_1 = \bigcup_1^\infty \overline{L}_1^{-n} \cup \{0\}$,
$Y = \{0\}$.

FIGURE 13. $fL \subset (l_1 \cup \overline{L}_1 \cup \overline{L}_2 \cup \overline{L}_3)$
$\cup (l_1 \cup \overline{L}_1 \cup \overline{L}_2 \cup \overline{L}_3)^{-1}$.

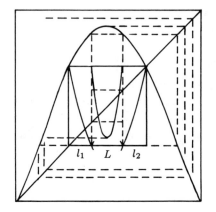

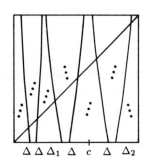

FIGURE 14. $l_1 \cup l_2 = \bigcup_1^\infty \overline{L_1}^{-n} \cup Y$,   FIGURE 15. $T \mid \Delta_1 = f^3$,
Y is a Cantor set, $T \mid \Delta_2 = f^3, T \mid \Delta = f^{2n+3}$.
$fL \subset (l_1 \cup L \cup l_2)^{-2}$.

The next conditions concern $t = t_0$.

IV. For any critical point $c_i(t_0) \in L_i(t_0)$ there exists some $k_i \leq N$ such that $f_{t_0}^{k_i}(c_i(t_0)) = y_i(t_0) \in Y_{t_0}$ and $y_i(t_0)$ is a limit point for $y \in Y_{t_0} \cap f_{t_0}^{k_i} L_i(t_0)$.

For $t = t_0$ the map T defined above coincides with C_{t_0}. Condition IV implies that the domain of C_{t_0} on $L(t_0)$ is an infinite union of intervals Δ such that C_{t_0} maps Δ diffeomorphically on some L_i. One can deduce from II that C_{t_0} is expanding on all but a finite number of Δ. Thus the following condition concerns the properties of C_{t_0} on a finite number of intervals.

V. There exist $A > 0$, $q > 1$ such that $\forall n \in \mathbf{N}$ $|dC_{t_0}^n/dx| > Aq^n$.

In various examples, condition V may follow from some special property of f such as analyticity or negative Schwarzian derivative, or may be checked numerically.[1]

The last condition says that for $t = t_0$ the critical values move with respect to Y with nonzero velocity.

VI.

$$\frac{d}{dt}\left(f_t^{k_i}c_i(t) - y_i(t)\right)\bigg|_{t=t_0} \neq 0, \qquad i \in [1, m].$$

THEOREM 1. *If a family f_t satisfies conditions I–VI then there exists a set of positive measure $\mathfrak{A} \subset [t_0 - \varepsilon, t_0]$ such that for $t \in \mathfrak{A}$ the map f_t is expansion-inducing and has an a.c.i.m. μ_t with positive entropy; t_0 is a density point of $\mathfrak{A}$, the Lyapunov exponent $\chi(\mu_t)$ is positive, and the entropy formula holds: $h(\mu_t) = \int \log |df_t/dx| \, d\mu_t$.*

(c) If f_t is a family of unimodal maps satisfying I–VI and g_t is C^2-close to f_t, then g_t also satisfies I–VI for $t \in [t_0' - \varepsilon', t_0']$ and there exists a similar set $\mathfrak{A}' \subset [t_0' - \varepsilon', t_0']$ with t_0' being a density point of $\mathfrak{A}'$. Besides, the following holds.

---

[1] Any example may be realized on a periodic interval.

PROPOSITION 2. *For any* $\varepsilon > 0$ *there exist* $\delta_1, \delta_2 > 0$ *such that for any family* g_t *lying in a* δ_1*-neighborhood of* f_t *and for any* $\delta \leq \delta_2$,

$$\delta^{-1} \cdot \text{meas}\{\mathfrak{A}' \cap [t_0' - \delta, t_0']\} > 1 - \varepsilon.$$

We use Proposition 2 in the following situation. Let $W^u(\phi_0)$ be the unstable one-dimensional manifold of Feigenbaum's fixed point ϕ_0, $W^s(\phi_0)$ the stable manifold of codimension 1. We consider some family f_t transversely intersecting $W^s(\phi_0)$ and suggest that the iterations of f_t under the action of the doubling transformation (acting on families) converges to $W^u(\phi_0)$. Let $f_0 = f_t \cap W^s(\phi_0)$ and suppose that f_t with $t > 0$ have zero topological entropy.

PROPOSITION 3. *For any* $t_0 < 0$, $\text{meas}\{t \in [t_0, 0]: f_t$ *has an a.c.i.m. of Theorem* 1$\}/t_0 > c$ *where* $c > 0$ *is independent of* t_0.

(d) Here we show that one parameter families of maps with m extrema are useful when studying m-parameter families. Consider

$$f_{q\omega} : x \mapsto x + \omega + (q/2\pi)\sin 2\pi x, \qquad x \in \mathbf{R}/\mathbf{Z}.$$

For $q \leq 1$ these mappings were studied in connection with the problem of disappearance of invariant tori in KAM theory. For $q > 1$, $f_{q\omega}$ is noninvertible. There is a conjecture that $f_{q\omega}$ with an a.c.i.m. exist for q infinitely close to 1. In [2] some points (q_n, ω_n) were constructed such that in any neighborhood of (q_n, ω_n) there exists a subset $X_n = \{(q, \omega): f_{q\omega}$ has an a.c.i.m.$\}$ with positive Lebesgue measure. Although (q_n, ω_n) constructed in [2] satisfy $q_n > q_* \approx 1.17$, it is possible that a similar method allows us to come nearer to 1. The points (q_n, ω_n) are characterized by the following property: there exists a periodic interval $[z_1(q_n, \omega_n), z_2(q_n, \omega_n)] = \tau_n$ such that $f_{q_n\omega_n}^n \mid \tau_n$ is a three-fold map of τ_n onto τ_n, with two critical points $c_1(q_n, \omega_n)$, $c_2(q_n, \omega_n)$ where

$$f_{q_n\omega_n}^n c_1(q_n, \omega_n) = z_2(q_n, \omega_n), \qquad f_{q_n\omega_n}^n c_2(q_n, \omega_n) = z_1(q_n, \omega_n).$$

Besides (q_n, ω_n) is the point of transversal intersection of two curves

$$\gamma_1 = \{(q, \omega): f_{q\omega}^n c_1(q, \omega) = z_2(q, \omega)\}, \qquad \gamma_2 = \{(q, \omega): f_{q\omega}^n c_2(q, \omega) = z_1(q, \omega)\}$$

which divide the neighborhood of (q_n, ω_n) into four parts. In one of these parts, characterized by $f_{q\omega}^n c_1(q, \omega) \in [z_1(q, \omega), z_2(q, \omega)] \ni f_{q\omega}^n c_2(q, \omega)$, any smooth curve passing though (q_n, ω_n) generates a one parameter family $f_{q(t)\omega(t)}$ satisfying the conditions of Theorem 1. Thus, using Fubini theorem we conclude that the set of (q, ω) corresponding to the stochastic behavior has positive Lebesgue measure on the plane of parameters.

(e) For a dissipative diffeomorphism, an analogue of a.c.i.m. is an invariant measure on some attractor Λ which generates on unstable manifolds W_x^u, $x \in \Lambda$, conditional measures absolutely continuous with respect to the length, i.e., Sinai-Ruelle-Bowen (S-R-B) measure.

There are no results about S-R-B measures analogous to the above results about a.c.i.m. for one-dimensional mapping. On the contrary, a theorem of

Newhouse [11] which has no one-dimensional analogue asserts the persistent tangency of stable and unstable manifolds and appearance of attractive cycles. However the set of parameter values complementary to the set of Newhouse may have positive measure (see [12]).

We formulate a hypothesis in this direction.

HYPOTHESIS. *Let f_t be a family of one-dimensional maps satisfying the conditions of Theorem 1, and let F_t be a family of diffeomorphisms near to f_t in C^2-topology. Then for a set of t of positive measure, F_t has an attractor carrying S-R-B measure.*

A possible approach to the proof of this hypothesis is related to the notion of diffeomorphisms inducing hyperbolicity—a modification of the notion of expansion-inducing one-dimensional maps.

ACKNOWLEDGMENT. I wish to thank J. Guckenheimer for his kind consent to deliver this address.

REFERENCES

1. M. Benedicks and L. Carleson, *On iterations of $1 - ax^2$ on $(-1, 1)$*, Ann. of Math. **122** (1985), 1–25.

2. P. M. Blecher and M. V. Jakobson, *Absolutely continuous invariant measures for some maps of the circle*, Progr. in Physics **10** (1985), 303–316.

3. P. Collet and J.-P. Eckmann, *On the abundance of aperiodic behavior for maps of the interval*, Comm. Math. Phys. **73** (1980), 115–160.

4. P. Collet, J.-P. Eckmann, and H. Koch, *Period doubling bifurcations for families of maps on $\mathbf{R}^n$*, J. Statist. Phys. **25** (1981), 1–14.

5. J. Guckenheimer, *Renormalization of one dimensional mappings and strange attractors*, Proceedings Lefschetz Symposium (Mexico City, 1984), Contemp. Math., Vol. 58, Amer. Math. Soc., Providence, R.I (to appear).

6. M. V. Jakobson, *On smooth mappings of the circle into itself*, Mat. Sb. **85** (1971), 163–188.

7. ____, *Absolutely continuous invariant measures for one parameter families of one dimensional maps*, Comm. Math. Phys. **81** (1981), 39–88.

8. F. Ledrappier, *Some properties of absolutely continuous invariant measures on an interval*, Ergodic Theory Dynamical Systems **1** (1981), 77–94.

9. R. Mañe, *Hyperbolicity, sinks and measure in one dimensional dynamics*, Comm. Math. Phys. **100** (1985), 495–524.

10. H. McCluskey and A. Manning, *Hausdorff dimension for horseshoes*, preprint, Univ. of Warwick, 1982.

11. S. Newhouse, *The abundance of wild hyperbolic sets and nonsmooth stable sets for diffeomorphisms*, Inst. Hautes Études Sci. Publ. Math. **50** (1979), 101–151.

12. J. Palis and F. Takens, *Cycles and measure of bifurcation sets for two-dimensional diffeomorphisms*, preprint, IMPA, 1984.

13. S. Strien, *On the bifurcations creating horseshoes*, Lecture Notes in Math. vol. 898, Springer, New York, 1981, pp. 316–351.

14. E. B. Vul, Ya. G. Sinai, and K. M. Khanin, *Feigenbaum universality and thermodynamic formalism*, Uspekhi Mat. Nauk **39** (1984), 3–37.

15. P. Walters, *Invariant measures and equilibrium states for some mappings which expand distances*, Trans. Amer. Math. Soc. **236** (1978), 121–153.

16. G. R. W. Quispel, *Scaling of the superstable fraction of the 2D period-doubling interval*, Phys. Lett. A **112** (1985), no. 8, 353–356.

P. P. SHIRSHOV INSTITUTE OF OCEANOLOGY, ACADEMY OF SCIENCES USSR, 117218 MOSCOW, USSR

Phenomena of Nonintegrability in Hamiltonian Systems

V. V. KOZLOV

In the last ten or fifteen years mathematicians have again become interested in problems related to the integrability of equations in classical dynamics, which, as a rule, are Hamiltonian equations. New, completely integrable systems have been found (multidimensional analogs of classical problems among them), and various algebro-geometric constructions have been suggested which elucidate the causes for the existence of "hidden" conservation laws. It is also useful to consider the peculiarities of the behavior of phase trajectories of nonintegrable Hamiltonian systems and present strict proofs of their nonintegrability. This paper is dedicated to the analysis of various phenomena of qualitative nature which hinder the integration of Hamiltonian equations.

1. Let us first recall the definition of a Hamiltonian dynamic system. Assume that M^{2n} is an even-dimensional manifold (phase space), ω is a closed nondegenerate 2-form on M (symplectic structure), H is a real function on M (Hamiltonian). Since ω is nondegenerate, the function H can be associated with a unique vector field v_H which is defined by the equation $\omega(v_H, \cdot) = dH$.

This field generates a Hamiltonian system on M, i.e.,

$$\dot{x}(t) = v_H(x(t)), \qquad x \colon \mathbf{R}_t \to M. \tag{1}$$

In suitable local coordinates $x_1, \ldots, x_n, y_1, \ldots, y_n$ (known as canonical coordinates) the form ω reduces to the form $\sum dy_s \wedge dx_s$ (Darboux's theorem). In the canonical coordinates x, y the form of Hamiltonian equations (1) is more customary:

$$\dot{x}_s = -\partial H/\partial y_s, \quad \dot{y}_s = \partial H/\partial x_s, \qquad 1 \le s \le n. \tag{2}$$

We have often to consider nonautonomous Hamiltonian systems in which the Hamiltonian H explicitly depends on time.

If differential equations are not of the form (2), this does not yet mean that they are not Hamiltonian. By virtue of this remark an interesting problem arises concerning the identification of Hamiltonian dynamic systems with an invariant measure and first integrals (in the autonomous case). Here is a simple example. Let

$$\dot{x} = Ax, \qquad x \in \mathbf{R}^n, \tag{3}$$

be a linear system with constant coefficients, which possesses a quadratic integral $f = (Bx, x)$, $B^{\mathrm{T}} = B$. If the operators A and B are nondegenerate, then system (3) is Hamiltonian with a Hamilton's function f. In particular, n is even. In $\mathbf{R}^n$ the symplectic structure is defined by the formula

$$\omega(x', x'') = (BA^{-1}x', x'').$$

2. Dynamic systems (Hamiltonian systems in particular) are customarily classified into integrable and nonintegrable, various definitions of integrability being possible, each with a certain intrinsic theoretic interest. A system which is integrable in the sense of one definition may prove to be nonintegrable in the sense of another definition. I will give examples later on. It is customary to associate the concept of an integrable system with a sufficiently large number of independent integrals ("conservation laws"). Thus, for a complete integrability of Hamiltonian equations with n degrees of freedom (on M^{2n}), it is sufficient to know n independent integrals $F_1, \ldots, F_n$, which are pairwise in involution: Poisson's brackets $\{F_i, F_j\} = \omega(v_{F_i}, v_{F_j})$ are zero. It is well known that compact energy surfaces $H = h$ of a completely integrable Hamiltonian system are stratified into multidimensional tori with a quasiperiodic motion.

If we have a nonautonomous Hamiltonian system with a Hamiltonian $H \colon M^{2n} \times \mathbf{R}_t \to \mathbf{R}$, then, for it to be completely integrable, it is sufficient to have n independent integrals $F_s \colon M^{2n} \times \mathbf{R}_t \to \mathbf{R}$ $(s = 1, \ldots, n)$, which are in involution for all values of t. The case when the Hamiltonian H and the integrals $F_1, \ldots, F_n$ are periodic with respect to t with the same period p is the most important for applications. Then it is natural to take $M^{2n} \times \mathbf{T}^1$ $\{\mathrm{mod}\, p\}$ rather than $M^{2n} \times \mathbf{R}$ as an extended phase space. If the integral surfaces $\{(z, t) \in M^{2n} \times \mathbf{T}^1 : F_s(z, t) = c_s, \ 1 \le s \le n\}$ are compact, then they are $(n + 1)$-dimensional tori with a quasiperiodic motion.

With due regard for the theorem on the straightening out of trajectories, it is reasonable to discuss the integrability of a dynamic system either in the neighborhood of the equilibrium position or in a sufficiently large region of a phase space where trajectories are recurrent.

Before investigating the integrability of specific systems, we must elaborate the concept of a set of independent integrals. We shall deal exclusively with analytic Hamiltonian systems. In that case it is natural to consider sets of analytic integrals which are independent at least at one point (then they are independent almost everywhere). We must bear in mind, however, that an analytic Hamiltonian system may possess integrals of the class C^r but at the same time not possess integrals of the class C^{r+1}. (We do not exclude the value $r = 0$: we consider a continuous function to be an integral when it is locally nonconstant and assumes constant values on each trajectory.) We shall consider the canonical Hamiltonian equations (2) with a Hamiltonian $H = \alpha y + f(x, t)$ as an example, where α is a real parameter, f is a 2π-periodic analytic function with respect to x and t [1]. Since the function H is periodic with respect to the variables x and

t, it is natural to take a direct product $\mathbf{R} \times \mathbf{T}^2 = \{y; x, t \bmod 2\pi\}$ as an extended phase space. We write equation (2) in the explicit form

$$\dot{x} = \alpha, \qquad \dot{y} = -\partial f/\partial x = -F(x,t). \tag{4}$$

We seek the integral of this system in the form $y + g(x,t)$, where $g \colon \mathbf{T}^2 \to \mathbf{R}$ is a smooth or analytic function which must satisfy the equation

$$\partial g/\partial t + \alpha \partial g/\partial x = F(x,t). \tag{5}$$

Equation (5) is well known in the theory of small denominators ([2], see also [3]). Assume that

$$F = {\sum}' F_{mn} e^{i(mx+nt)}, \qquad g = {\sum}' g_{mn} e^{i(mx+nt)}.$$

Then

$$g_{mn} = \frac{F_{mn}}{i(m\alpha + n)}.$$

For almost all α the numbers g_{mn} are Fourier coefficients of a certain analytic function. Now if the irrational numbers α can be sufficiently rapidly approximated by rational numbers, then equation (5) can have a solution of only finite smoothness or have no solutions. Generalizing these observations, we can show that for a certain $f \in C^\omega(\mathbf{T}^2)$ there are sets $M_\omega, M_\infty, \dots, M_k, \dots, M_0, M_\varnothing$, dense everywhere in $\mathbf{R}$, such that for $\alpha \in M_\omega$ equations (4) have an analytic integral, for $\alpha \in M_\infty$ there is a smooth integral but there is no analytic integral, $\dots$, for $\alpha \in M_k$ there is an integral of the class C^k but there are no integrals of the class $C^{k+1}, \dots$, for $\alpha \in M_0$ equations (4) possess only a continuous invariant function, for $\alpha \in M_\varnothing$ there are even no continuous integrals. We can derive the density of the set $M_\varnothing$ in $\mathbf{R}$ from the result obtained by A. B. Krygin concerning the ergodicity of cylindrical cascades [3]. Note that if we consider equations (4) in $\mathbf{R}^3 = \{x, y, t\}$ rather than in $\mathbf{R} \times \mathbf{T}^2$, then this system turns out to be completely integrable. It should be emphasized that system (4) can be explicitly integrated by simple quadratures for all values of α, but its behavior as a whole depends considerably on the Diophantine properties of the number α.

3. When mathematicians realized the impossibility of solving equations of classical dynamics in a closed form, strict results appeared concerning their nonintegrability. The first of those results was, evidently, Liouville's theorem (1841) stating that the equation $\ddot{x} + tx = 0$ cannot be solved by quadratures (see [4]). In 1887 Bruns stated that there are no algebraic integrals in the problem of three bodies independent of the classical ones (see [5]). This theorem was generalized by Painlevé to the case when integrals are algebraic with respect to the velocities of three gravitating bodies [6]. These classical results are of no importance for dynamics, however, since they do not take into account the peculiarities of the behavior of phase trajectories. Equations of motion may happen to be completely integrable but do not have, say, integrals which are polynomial with respect to velocities. Here is a simple example [7]. The motion of a point charge along a "plane" torus $\mathbf{T}^2 = \{x, y \bmod 2\pi\}$ in a constant magnetic field is described by the equations

$$\ddot{x} + \Omega\dot{y} = 0, \quad \ddot{y} - \Omega\dot{x} = 0; \qquad \Omega = \text{const.} \tag{6}$$

They have an energy integral $\dot{x}^2 + \dot{y}^2 = h$. It can be shown [7] that equations (6) do not have an additional integral, polynomial with respect to velocities, with smooth and single-valued coefficients on $\mathbf{T}^2$. System (6) is completely integrable, however: the function $\sin(\dot{x} + \Omega y)$ is an additional integral, for instance. The integral $\dot{x} + \Omega y$ is linear with respect to velocities, but it is a multivalued function in the phase space $\mathbf{R}^2 \times \mathbf{T}^2$.

Poincaré was the first to pose a problem on the nonintegrability of the Hamiltonian equations as a whole and to get some results in this respect [8]. He investigated Hamiltonian differential equations of the following kind:

$$\begin{aligned}
&\dot{x}_s = -\partial H/\partial y_s, \quad \dot{y}_s = \partial H/\partial x_s, \qquad 1 \le s \le n, \\
&H = H_0(x_1, \ldots, x_n) + \varepsilon H_1(x_1 \cdots x_n, y_1 \cdots y_n) + \cdots.
\end{aligned} \tag{7}$$

The Hamiltonian H is a power series with respect to ε, and its coefficients are analytic functions in $\mathbf{R}^n \times \mathbf{T}^n = \{x; y \bmod 2\pi\}$. For $\varepsilon = 0$ we have a completely integrable system. Differential equations (7) are often encountered in applications, and therefore Poincaré considered the problem of their investigation to be the "basic problem of dynamics." Poincaré tried to find out whether equations (7) have first integrals $F(x, y, \varepsilon)$, which are analytic in $D \times \mathbf{T}^n \times (-\varepsilon_0, \varepsilon_0)$, where D is a domain in $\mathbf{R}^n = \{x\}$. It is shown in [1] that it is more expedient to consider a problem on the existence of formal integrals in the form of power series $\sum F_s(x, y)\varepsilon^s$ with coefficients analytic in the domain $D \times \mathbf{T}^n$. This problem is closely connected with the possibility of realizing the classical scheme of perturbation theory.

The problem of the existence of analytic integrals of system (7) for fixed values of the parameter $\varepsilon \ne 0$ is more complicated. One of the most popular problems of this kind is the investigation of the complete integrability of the Hamiltonian system near a stable equilibrium. The formal analysis of this problem dates back to Birkhoff [9] and Siegel, who presented strict proofs [10] (for the discussion of these problems, see [11]).

In the majority of the integrated problems of classical mechanics the known first integrals are extended to a complex domain of variation of the phase variables as single-valued holomorphic (or meromorphic) functions. In connection with this remark an interesting problem arises concerning the complete "complex" integrability of a holomorphic Hamiltonian system. In this case we must bear in mind that the absence of holomorphic integrals in a complex domain does not yet mean that the Hamiltonian system is not integrable in a real sense. Here is a simple example. The linear Hamiltonian system $\ddot{z} + (\alpha^2 + \beta\varphi(t))z = 0$ possesses an analytic integral $f(\dot{z}, z, t)$, which is periodic with respect to t, in a real domain (Floquet-Lyapunov theorem). For almost all α and β, however, this system does not have a holomorphic integral in a complexified phase space (see [12]).

In "complex" completely integrable Hamiltonian systems the level surfaces of involute integrals often prove to be not simply real tori $\mathbf{T}^n$, but, being extended to a complex domain, to be Abelian manifolds $\mathbf{T}^{2n}$. In this case the general solution is expressed by the ϑ-functions of complex time. Systems possessing these properties are often said to be "algebraically integrable." When we seek the necessary conditions for algebraic integrability, we usually follow the method of Kovalevskaya, which she applied in 1888 to the dynamics of a rigid body. For the present-day state of these problems see [13, 14, 15] (see also the report made by P. Van Moerbeke at the International Congress of Mathematicians in Warsaw in 1982).

4. In recent years Poincaré's ideas have been further elaborated, and new phenomena in the behavior of Hamiltonian systems hindering their integrability have been discovered. This made it possible to present strict proofs of the nonintegrability of a number of significant problems of Hamiltonian mechanics (a heavy asymmetric top, a rigid body in ideal fluid, the problem of four-point vertices, etc.). What lies behind nonintegrability consists in the following. There are an infinite number of resonant tori filled with periodic trajectories in the phase space of an unperturbed completely integrable system. These tori become disintegrated when perturbation is added. The families of periodic solutions located on them yield pairs of nondegenerate periodic solutions. The first integrals are dependent on the trajectories of the nondegenerate periodic solutions. The disintegrated resonant tori accumulate, as a rule, by a pair of doubled separatrices (asymptotic surfaces) of the unperturbed problem. When a perturbation is added, the separatrices themselves split and, as a rule, intersect, forming a rather tangled network. The nondegenerate periodic solutions of the perturbed problem, being extended to the plane of complex time, are not single-valued functions, and their branching impedes the presence of holomorphic first integrals in the complexified phase space. For the necessary details see [1], which also contains the review of the achievements in this field covering the period up to 1983. In what follows we discuss new problems pertaining to the analysis of the phenomena of nonintegrability.

5. One of the problems of this kind consists in the investigation of perturbed integrable Hamiltonian systems, where at each stage of perturbation theory only a finite number of resonant invariant tori disintegrate. We consider as a model example Hamiltonian equations

$$\dot{x} = -H'_y, \quad \dot{y} = H'_x, \quad H = H_0(x) + \varepsilon H_1(y), \qquad x \in \mathbf{R}^n, \ y \in \mathbf{T}^n, \qquad (8)$$

where $H_0 = \frac{1}{2}(Ax, x)$ is a nondegenerate quadratic form with respect to the variables x, and H_1 is a trigonometric polynomial:

$$H_1 = \sum_{m \in \mathbf{Z}^n} h_m e^{i(m, y)}, \qquad h_m = \text{const.} \qquad (9)$$

Just as Poincaré did [**8**], we shall discuss the fact that system (8) has an additional integral as a formal series

$$\sum_{s \geq 0} F_s(x, y)\varepsilon^s$$

with single-valued analytic coefficients in $\mathbf{R}^n \times \mathbf{T}^n$. Since the Fourier series of the perturbation function (9) contains only a finite number of harmonics, Poincaré's results and their known generalizations cannot be applied to systems with a Hamiltonian (8).

We can treat Hamiltonian equations (8) as equations of motion of a mechanical system with a configurational space $\mathbf{T}^n$, kinetic energy H_0, and a small potential εH_1. It should be emphasized that a positive definiteness of the quadratic form H_0 is not presupposed.

Let us agree on some designations. Let $\xi, \eta \in \mathbf{R}^n$. We set $\langle \xi, \eta \rangle = (A\xi, \eta)$. We designate by $\mathfrak{M}$ a finite set of integer-valued vectors $m = (m_1, \ldots, m_n)$ for which $h_m \neq 0$. Since $H_1 \neq \text{const}$, it follows that $\mathfrak{M}$ contains at least two elements. Assume that $i_1, i_2, \ldots, i_n$ is any permutation of the indices $1, 2, \ldots, n$. We set

$$\alpha_{i_1} = \max_{\mathfrak{M}} m_{i_1}, \alpha_{i_2} = \max_{\substack{\mathfrak{M} \\ m_{i_1} = \alpha_{i_1}}} m_{i_2}, \ldots, \alpha_{i_n} = \max_{\substack{\mathfrak{M} \\ m_{i_1} = \alpha_{i_1} \\ \cdots \\ m_{i_{n-1}} = \alpha_{i_{n-1}}}} m_{i_n}. \qquad (10)$$

We term the vector $\alpha = (\alpha_1, \ldots, \alpha_n)$ the vertex $\mathfrak{M}$. Formulas (10) yield $n!$ vertices of the set $\mathfrak{M}$, but they are not all different. If we replace $\mathfrak{M}$ in formulas (10) by $\mathfrak{M} \backslash \{\alpha\}$, then we get an integer-valued vector β which is a vertex of the set $\mathfrak{M} \backslash \{\alpha\}$.

THEOREM 1. *We assume that the set $\mathfrak{M}$ has a vertex α such that:*
(i) *the vertices α and β are linearly independent,*
(ii) *$m\langle \alpha, \alpha \rangle + 2\langle \alpha, \beta \rangle \neq 0$ for all integers $m \geq 0$.*
Then the Hamiltonian system (8) does not have a complete set of independent integrals representable as a power series $\sum F_s(x, y)\varepsilon^s$ with coefficients analytic in $\mathbf{R}^n \times \mathbf{T}^n$.

This theorem was established by V. Kozlov and D. Treshchev. Note that when all the coefficients $h_m \neq 0$ in the Fourier expansion (9), the nonintegrability of equations (8) follows from the classical result of Poincaré [**8**]. For $n = 2$ we can assert still more: if in the expansion (9) there are an infinite number of vectors $m \in \mathbf{Z}^n$, which are pairwise independent, then equations (8) are nonintegrable either (see [**1**]). The proof of Theorem 1 is based on the analysis of the classical scheme of perturbation theory applied to the Hamiltonian equations (8).

There is the following corollary of Theorem 1.

COROLLARY. *If $n = 2$ and the functions H_0 and H_1 satisfy the hypothesis of Theorem 1, then the equations with the Hamiltonian $H_0 + H_1$ do not possess* *an additional integral in the form of a polynomial with respect to the momenta x with analytic and single-valued coefficients on* $\mathbf{T}^2$.

This statement is an addition to the classical results concerning the conditions of the existence of polynomial integrals with respect to momenta (see [6]). It should be emphasized that the potential H_1 need not be necessarily small here.

6. One more problem, which we shall discuss here, is connected with topological and geometric conditions for a complete integrability of Hamiltonian systems, which we come across in classical mechanics. Assume that M^n is a complete analytic Riemannian manifold and Σ^{2n-1} is a foliated space of unit tangent vectors. The Riemannian metric defines on Σ^{2n-1} a dynamic system which is a geodesic flow. From the viewpoint of mechanics a geodesic flow describes the motion of a particle along M^n by inertia with unit velocity. The famous principle of Mopertuis reduces the motion under the action of potential forces to a geodesic flow.

THEOREM 2 [16]. *If M is a compact two-dimensional surface of genus larger than one, then the geodesic flow on* Σ *does not possess a nonconstant analytic integral.*

The proof of this theorem is based on the analysis of the set of unstable periodic trajectories. Since the genus of M is greater than one, the Gaussian curvature is negative in the mean. If the curvature is negative everywhere, then the flow on Σ is Anosov's system [17]. In that case all periodic trajectories are unstable, they densely fill Σ everywhere, and the geodesic flow does not even possess a continuous integral. The hyperbolic behavior of phase trajectories lies at the basis of the proof of the nonintegrability of the restricted problem of three bodies advanced by Alekseev [18]. Note that a curvature negative in the mean is not always negative everywhere.

Theorem 2 has been generalized in different directions. Taimanov has proved the absence of a complete involute set of analytic integrals of a geodesic flow in a multidimensional case when one of the following additional conditions is satisfied [19]:

(1) $\dim M < \operatorname{rank} H_1(M, \mathbf{Z})$,

(2) the fundamental group $\pi_1(M)$ does not contain a commutative subgroup of a finite index.

The first condition was proved in [16] for $n = 2$, and was formulated as a hypothesis by the author in [20]. The second condition is a new one. It would be interesting to find other topological obstacles hindering complete integrability.

Another possible way of achieving generalization is to consider domains with a geodesically convex boundary. Assume that $\hat{M}$ is a compact submanifold with the boundary on the analytic surface M^2. Let $\hat{\Sigma}$ denote the set of all points of Σ which are taken by the projection $\pi \colon \mathbf{T}M^2 \to M^2$ into points of $\hat{M}$. We say that $\hat{M}$ is geodesically convex if for any two close points of the boundary $\partial\hat{M}$, the shortest geodesic of the Maupertuis metric joining the points lies entirely in $\hat{M}$.

THEOREM 3. *If* rank $H_1(\hat{M}, \mathbf{Z}) > 2$ *and* $\hat{M}$ *is geodesically convex, then the dynamic system on* Σ *does not possess a nonconstant analytic first integral. Moreover, there is no analytic integral even in the neighborhood of the set* $\hat{\Sigma} \subset \Sigma$.

S. Bolotin has found that the condition rank $H_1(M, \mathbf{Z}) > 2$ can be replaced by a weaker condition $\chi(M) < 0$, where χ is an Eulerian characteristic [21]. He has also found an interesting application of the generalized Theorem 3 to the problem of the motion of a point in the gravitational field of n fixed centers. Let $z_1, \ldots, z_n$ be different points of a complex plane $\mathbf{C}$. The Hamiltonian of a plane problem of n centers has the form

$$H = \tfrac{1}{2}|p|^2 + V(z), \qquad (z, p) \in U \times \mathbf{C},$$

where $U = \mathbf{C} \backslash \{z_1, \ldots, z_n\}$ is a configurational space, V is a gravitational potential of attraction of a moving particle z by the stationary points $z_1, \ldots, z_n$, i.e.,

$$V(z) = -\sum_{i=1}^{n} \mu_i |z - z_i|^{-1}, \qquad \mu_i > 0.$$

THEOREM 4 [21]. *Assume* $n > 2$. *Then the equations of the problem dealing with* n *centers do not have an analytic integral on the surface* $\{(z, p) \in U \times \mathbf{C} \colon H(z, p) = h \geq 0\}$.

Note that the conditions $n = 1, 2$ correspond to Keplerian and Eulerian integrable problems. When we prove Theorem 4, we make use of the Levi-Civita regularization. Assume that M is the Riemannian surface of the function $\sqrt{(z - z_1) \cdots (z - z_n)}$ and $\pi \colon M \to \mathbf{C}$ is a projection. It turns out that the Levi-Civita regularization reduces a phase flow on the surface $H = h$ to a geodesic flow on M with some complete metric. If D is a disc in a complex plane $\mathbf{C}$ of a sufficiently large radius, then the set $\hat{M} = \pi^{-1}(D)$ is compact, geodesically convex, and homotopically equivalent to M. By the Riemann-Hurwitz formula $\chi(M) = 2 - n < 0$ for $n > 2$.

We can generalize Theorems 2 and 3 to an irreversible case when additional hyroscopic forces act on a system [7, 22]. The origin of these forces differs; i.e., they appear, for instance, upon a transition to a rotating reference system and in the description of the motion of charged particles in a magnetic field. We consider as an example a plane-restricted problem of many bodies: n huge gravitating bodies which are in relative rest, rotating, as a rigid body, about their barycenter with a constant angular velocity, where an additional body of an infinitesimal mass moves under the action of gravitational forces in the plane of the circular orbits of huge bodies. One can show [22] that for $n > 2$ the equations of motion of this problem do not possess an analytic integral on the energy surface $H = h > 0$. This statement has not been proved for a restricted problem of three bodies (when $n = 2$). Weaker theorems can be found in [8, 23, 24]. Note that, in accordance with Chazy's hypothesis (see [25]), the problem of three bodies is completely integrable on the level surface of the energy integral for $H > 0$. This hypothesis is related to a more general idea: in the problem of scattering with a noncompact configurational space the data at infinity are candidates for integrals. However, the realization of this idea is hindered by some difficulties of principle connected with the domain of definition and smoothness of the "integral of scattering." One of these difficulties is the possibility of capture in the problems of many interacting particles.

REFERENCES

1. V. V. Kozlov, *Integrability and non-integrability in Hamiltonian mechanics*, Uspekhi Mat. Nauk **38** (1983), 3–67.

2. H. Poincaré, *Sur les courbes définies par les équations differentielles* (4), J. Math. Pures Appl. (4) **2** (1886), 151–217.

3. A. B. Krygin, *Examples of ergodic automorphisms of cylinder*, Mat. Zametki **16** (1974), 981–991.

4. I. Kaplansky, *An introduction to differential algebra*, Hermann, Paris, 1957.

5. E. T. Whittaker, *A treatise on the analytical dynamics*, Cambridge Univ. Press, 1927.

6. P. Painlevé, *Sur la méthode de Bruns*, Bull. Astr. **15** (1898), 81–113.

7. S. V. Bolotin, *On first integrals of Hamiltonian systems with hyroscopic forces*, Vestnik Moskov. Univ. Ser. I Mat. Mekh. **6** (1984), 75–82.

8. H. Poincaré, *Les méthodes nouvelles de la mécanique céleste*. Vol. 1, Gauthier-Villars, Paris, 1892.

9. G. D. Birkhoff, *Dynamical systems*, Amer. Math. Soc. Colloq. Publ., vol. 9, Amer. Math. Soc., Providence, R.I., 1927.

10. C. L. Siegel, *Über die Existenz einer Normalform analytischer Hamiltonscher Differentialgleichungen in der Nähe einer Gleichgewichtslösung*, Math. Ann. **128** (1954), 144–170.

11. J. Moser, *Lectures on Hamiltonian systems*, Mem. Amer. Math. Soc. No. 81 (1968), 1–60.

12. S. L. Ziglin, *Branching of solutions and nonexistence of first integrals in Hamiltonian mechanics*, Funktsional. Anal. i Prilozhen. I, **16** (1982), 30–41; II, **17** (1983), 7–23.

13. B. A. Dubrovin, V. B. Matveev, and S. P. Novikov, *Nonlinear equations of Korteweg-de Vries type, finite-zone operators and Abelian varieties*, Uspekhi Mat. Nauk **31** (1976), 55–136.

14. M. Adler and P. van Moerbeke, *Linearization of Hamiltonian systems, Jacobi varieties and representation theory*, Adv. in Math. **38** (1980), 267–317.

15. O. I. Bogoyavlensky, *New integrable problem of classical mechanics*, Comm. Math. Phys. **94** (1984), 255–269.

16. V. V. Kozlov, *Topological obstacles to the integrability of natural mechanical systems*, Dokl. Akad. Nauk SSSR **249** (1979), 1299–1302.

17. D. V. Anosov, *Geodesic flows on closed Riemannian manifolds with negative curvature*, Trudy Mat. Inst. Steklov **90** (1967), 3–210.

18. V. M. Alekseev, *Quasirandom dynamical systems*, Mat. Sb. (N.S.) I, **76** (1968), 72–134; II, **77** (1968), 545–601; III, **78** (1969), 3–50.

19. I. A. Taimanov, *Nonsimpleconnected manifolds with nonintegrable geodesic flow*, Izv. Akad. Nauk SSSR Ser. Mat. (to appear).

20. V. I. Arnold, V. V. Kozlov, and A. I. Neyshtadt, *Mathematical aspects of classical and celestial mechanics*, Contemporary Problems in Mathematics, Fundamental Directions 3, VINITI, Moscow, 1985.

21. S. V. Bolotin, *Nonintegrability of the n-center problem for n > 2*, Vestnik Moskov. Univ. Ser. I Mat. Mekh. **3** (1984), 65–68.

22. _____, *The effect of singularities of the potential energy on the integrability of mechanical systems*, Prikl. Mat. Mekh. **48** (1984), 356–362.

23. C. L. Siegel, *Über die algebraischen Integrale des restringierten Dreikörperproblems*, Trans. Amer. Math. Soc. **39** (1936), 225–233.

24. J. Llibre and C. Simo, *Oscillatory solutions in the planar restricted three-body problem*, Math. Ann. **248** (1980), 153–184.

25. V. M. Alekseev, *Final motions in the three-body problem and symbolic dynamics*, Uspekhi Mat. Nauk **36** (1981), 161–176.

MOSCOW STATE UNIVERSITY, MOSCOW 117234, USSR

Proceedings of the International Congress of Mathematicians
Berkeley, California, USA, 1986

Optimization of the Ensured Result
for the Dynamical Systems

A. V. KRYAZHIMSKII

This report deals with a special class of optimal control problems with uncertainty. Uncertainty appears, if the control system dynamics depends, besides
controls, on some uncontrolled parameters (disturbances). Disturbances may in
particular be controls of the controller's opponent whose aim is contrary to that
of the controller. This situation is studied in the theory of differential games. We
shall concentrate here upon the positional approach to differential games suggested by N. N. Krasovskii [1] and worked out in [2–6] and other publications;
the ultimate results are summed up in [4]. The positional formalization does not
restrict the laws of forming disturbances to some definite class of strategies of
the controller's opponent. That leads to considering, instead of game problems
with an opponent, more general problems of optimization of the ensured result,
i.e., the worst result for a given control law.

A brief outline of the positional approach, with formulations of the main results
concerning existence and construction of solutions, is given below. Central points
of a solution method discussed here are extremal shift and stochastic program
maximin. The former shows general structure of optimal closed-loop (positional)
controls; the latter establishes a connection between closed-loop game problems
and special open-loop (program) maximin problems of stochastic control.

From the point of view of applications the question of stability of the ensured
result with respect to the errors of current state measurements is of special interest.
In [2] a principle of stable control was suggested. It is based on including an
auxiliary control system—a "guide" or a "model"—into the control process. This
principle combined with a modification of the extremal shift method was applied
in [7] to constructing stable solutions for some ill-posed inverse problems for
control systems. Here we consider such problems as those of optimization of the
ensured result.

1. The problem of optimization of the ensured result. Let at each
time t a state of a control dynamical system be given by a vector $x[t] \in R^n$. The
evolution of $x[t]$ depends on a control parameter $u[t]$ as well as on an uncontrolled

parameter $v[t]$—we shall call it *a disturbance*. Namely, if an initial time t_* and an initial state x_* are fixed, then $x[t]$ ($t \geq t_*$) is a solution of the Cauchy problem

$$\dot{x} = f(t, x, u[t], v[t]), \qquad x(t_*) = x_*. \tag{1.1}$$

We assume $u[t]$ and $v[t]$ to be elements of compacta $P \subset R^p$ and $Q \subset R^q$, respectively. Measurable functions $u[\,\cdot\,] \colon [t_1, t_2] \mapsto P$ and $v[\,\cdot\,] \colon [t_1, t_2] \mapsto Q$ will be called *a control realization at $[t_1, t_2]$* and *a disturbance realization at $[t_1, t_2]$*, respectively. Denote the sets of all such $u[\,\cdot\,]$'s and $v[\,\cdot\,]$'s by $U[t_1, t_2]$ and $V[t_1, t_2]$, respectively. Let a final time instant ϑ be fixed. The aim of the controller is to form in real time a control realization at $[t_*, \vartheta]$ so as to optimize a control process, without knowing the disturbance realization. The natural way to solve such an uncertain problem is to optimize the ensured result, i.e., the result corresponding to the "worst" disturbance realization. That is the approach considered below.

From now on we assume that (i) $f(\,\cdot\,)$ is continuous, (ii) $|f(t, x, u, v)| \leq c_f(1+x)$ for a certain $c_f > 0$, and (iii) for each bounded set $G \subset R \times R^n$ there exists $c_{f,G} > 0$ such that

$$|f(t, x_1, u, v) - f(t, x_2, u, v)| \leq c_{f,G}|x_1 - x_2| \quad \text{for } (t, x_1), (t, x_2) \in G;$$

here and below $|\,\cdot\,|$ denotes the Euclidean norm. The assumptions (i)–(iii) guarantee existence and uniqueness of the solution $x[\,\cdot\,]$ of the Cauchy problem (1.1) at $[t_*, \vartheta]$ for every $t_* \leq \vartheta$, $x_* \in R^n$, $u[\,\cdot\,] \in U[t_*, \vartheta]$, and $v[\,\cdot\,] \in V[t_*, \vartheta]$; a triplet $(x[\,\cdot\,], u[\,\cdot\,], v[\,\cdot\,])$ will be called *a process* (at $[t_*, \vartheta]$); $x[\,\cdot\,]$ will be called *a trajectory* (at $[t_*, \vartheta]$); if we want to emphasize that $x[t_*] = x_*$, we will write $x[\,\cdot\, | \, x_*]$ instead of $x[\,\cdot\,]$ and say that the process corresponds to the initial state x_*. We fix a $t_0 < \vartheta$ and a bounded set $G_* \subset [t_0, \vartheta] \times R^n$ and assume further that $(t_*, x_*) \in G_*$.

Let γ be a functional on the set of all processes; it provides the optimality criterion to be minimized. Taking into account that $v[\,\cdot\,]$ is not known to the controller we admit the latter to form $u[t]$ on the basis of the past of a trajectory up to time t. Such a way of controlling may be defined formally as a mapping $\mathcal{U} \colon x[\,\cdot\,] \mapsto u[\,\cdot\,] = \mathcal{U}x[\,\cdot\,]$ satisfying the following *physical realizability condition* (*PR-condition*): if $x_1[\tau] = x_2[\tau]$ for all $\tau \in [t_*, t]$, then $u_1[\tau] = u_2[\tau]$ for all $\tau \in [t_*, t]$, where $u_i[\,\cdot\,] = \mathcal{U}x_i[\,\cdot\,]$, $i = 1, 2$. We will also require $\mathcal{U}$ to be compatible with any disturbance realization. Introduce the following *compatibility condition* (*C-condition*): for each x_* and each $v[\,\cdot\,] \in V[t_*, \vartheta]$ there exists $x[\,\cdot\,]$ such that $(x[\,\cdot\,], \mathcal{U}x[\,\cdot\,], v[\,\cdot\,])$ is a process at $[t_*, \vartheta]$ corresponding to x_*. A mapping $\mathcal{U} \colon C([t_*, \vartheta], R^n) \mapsto U[t_*, \vartheta]$ that satisfies PR- and C-conditions will be called *a control law* (at $[t_*, \vartheta]$). Control law is the controller's tool to minimize γ.

Suppose that t_* and x_* are fixed and a control law $\mathcal{U}$ is chosen. So far as each disturbance realization $v[\,\cdot\,]$ may be expected a priori, each process of the form $(x[\,\cdot\, | \, x_*], \mathcal{U}x[\,\cdot\, | \, x_*], v[\,\cdot\,])$ is admissible; further $S_*(\mathcal{U}; t_*, x_*)$ denotes the set of all such processes. Thus the value

$$\rho(\mathcal{U}; t_*, x_*) = \sup_{r[\cdot] \in S_*(\mathcal{U}; t_*, x_*)} \gamma(r[\,\cdot\,]) \tag{1.2}$$

describes the worst result expected a priori. We will call (1.2) *the ensured result for* $\mathcal{U}$ *at* (t_*, x_*). Now we formulate the optimization problem mentioned above as that of minimizing the ensured result by choosing a control law. The value $\inf_{\mathcal{U} \in \Gamma} \rho(\mathcal{U}; t_*, x_*)$, where Γ is a nonempty class of control laws at $[t_*, \vartheta]$ will be called *the optimal ensured result at* (t_*, x_*) *in* Γ.

2. Closed-loop control laws. Differential game. We will consider the problem for γ given by

$$\gamma(x[\ \cdot\ ], u[\ \cdot\ ], v[\ \cdot\ ]) = \sigma(x[\vartheta]) + \int_{t_*}^{\vartheta} \chi(t, x[t], u[t], v[t])\, dt,$$

where $\sigma(\cdot)$ is Lipschitz and $\chi(\cdot)$ is continuous and satisfies the condition similar to (iii) with constant $c_{\chi, G}$.

A function $u_*(\cdot) = u_*(t, x, \varepsilon)\colon R \times R^n \times [0, \infty) \mapsto P$ will be called *a* (*positional or closed-loop*) *strategy;* Δ will stand for an arbitrary partition $(\tau_i)_{i=0}^m$, $t_* = \tau_0 < \cdots < \tau_m = \vartheta$, of the interval $[t_*, \vartheta]$; the notations

$$d(\Delta) = \sup\{\tau_{i+1} - \tau_i : i = 0, \ldots, m-1\}, \qquad b_i = [\tau_i, \tau_{i+1})$$

will also be used. We will identify a triplet $(u_*(\ \cdot\ ), \Delta, \varepsilon)$ where $\varepsilon > 0$ with the control law $\mathcal{U}\colon x[\ \cdot\ ] \mapsto u[\ \cdot\ ]$ such that $u[t] = u_*(\tau_i, x[\tau_i], \varepsilon)$ for $t \in b_i$ ($\mathcal{U}$ is defined uniquely). We will call $\mathcal{U}$ a *closed-loop control law*. The class $\Gamma_c[t_*, \vartheta]$ of closed-loop control laws at $[t_*, \vartheta]$ is rather restricted. However, a considerable expansion of this class does not decrease the optimal ensured result. Namely, say that a control law $\mathcal{U}$ at $[t_*, \vartheta]$ is *admissible* if for any x_*, any function $v_*(\ \cdot\ ) = v_*(t, x, u)\colon R \times R^n \times P \mapsto Q$ Borel in u and any Δ there exists a process $(x[\ \cdot\ \mid x_*], u[\ \cdot\ ], v[\ \cdot\ ])$ at $[t_*, \vartheta]$ such that $u[\ \cdot\ ] = \mathcal{U}x[\ \cdot\ \mid x_*]$ and $v[t] = v_*(\tau_i, x[\tau_i \mid x_*], u[t])$ for $t \in b_i$. The class $\Gamma_a[t_*, \vartheta]$ of admissible control laws is considerably wider than $\Gamma_c[t_*, \vartheta]$. Hence

$$\rho_c^0(t_*, x_*) \geq \rho_a^0(t_*, x_*), \tag{2.1}$$

where $\rho_c^0(t_*, x_*)$ and $\rho_a^0(t_*, x_*)$ denote optimal ensured results at (t_*, x_*) in classes $\Gamma_c[t_*, \vartheta]$ and $\Gamma_a[t_*, \vartheta]$, respectively. In fact the equality is true:

PROPOSITION 2.1. $\rho_c^0(t_*, x_*) = \rho_a^0(t_*, x_*)$.

Now we introduce the optimization problem for strategies. If $u_*(\ \cdot\ )$ is a strategy, then the value

$$\rho_s(u_*(\ \cdot\ ); t_*, x_*) = \overline{\lim_{\varepsilon \to 0}} \lim_{\delta \to 0} \sup_{d(\Delta) \leq \delta} \rho((u_*(\ \cdot\ ), \Delta, \varepsilon); t_*, x_*)$$

will be called *the ensured result for* $u_*(\ \cdot\ )$ *at* (t_*, x_*); the value $\rho_s^0(t_*, x_*) = \inf\{\rho_s(u_*(\ \cdot\ ); t_*, x_*)\colon u_*(\ \cdot\ )$ is a strategy$\}$ will be called *the optimal ensured result at* (t_*, x_*); a strategy $u^*(\ \cdot\ )$ such that $\rho_s(u^0(\ \cdot\ ); t_*, x_*) = \rho_s^0(t_*, x_*)$ for all $(t_*, x_*) \in G_*$ will be called *optimal* (it is worthwhile to emphasize that $u^0(\ \cdot\ )$ does not depend on (t_*, x_*)). Obviously,

$$\rho_s^0(t_*, x_*) \geq \rho_c^0(t_*, x_*). \tag{2.2}$$

In fact the following proposition is true:

PROPOSITION 2.2. (1) $\rho_s^0(t_*, x_*) = \rho_c^0(t_*, x_*)$;
(2) *there exists an optimal strategy.*

Propositions 2.1 and 2.2 result from the positional game-theoretic approach [4]. According to this approach the minimization problem in the class of strategies is supplied by an analogous counter-problem of maximization. Namely, define *a counter-strategy* as a function $v_*(\,\cdot\,) = v_*(t, x, u, \varepsilon)\colon R \times R^n \times P \times [0, \infty) \mapsto Q$ Borel in u and *a closed-loop disturbance law* at $[t_*, \vartheta]$ as a triplet $\mathcal{V} = (v_*(\,\cdot\,), \Delta, \varepsilon)$; identify the latter with a mapping $\mathcal{V}\colon (x[\,\cdot\,], u[\,\cdot\,]) \mapsto v[\,\cdot\,] = \mathcal{V}(x[\,\cdot\,], u[\,\cdot\,])\colon C([t_*, \vartheta], R^n) \times U[t_*, \vartheta] \mapsto V[t_*, \vartheta]$, where $v[t] = v_*(\tau_i, x[\tau_i], u[t], \varepsilon)$ for $t \in b_i$; introduce *the ensured counter-result for* $\mathcal{V}$ at (t_*, x_*):

$$\pi(\mathcal{V}; t_*, x_*) = \inf_{r[\,\cdot\,] \in S^*(\mathcal{V}; t_*, x_*)} \gamma(r[\,\cdot\,]),$$

where $S^*(\mathcal{V}; t_*, x_*)$ is the set of all processes at $[t_*, \vartheta]$ of the form $(x[\,\cdot\,\mid x_*], u[\,\cdot\,], \mathcal{V}(x[\,\cdot\,\mid x_*], u[\,\cdot\,]))$; introduce the *ensured counter-result for* $v_*(\,\cdot\,)$ at (t_*, x_*):

$$\pi_s(v_*(\,\cdot\,); t_*, x_*) = \varlimsup_{\varepsilon \to 0} \lim_{\delta \to 0} \inf_{d(\Delta) \le \delta} \pi((v_*(\,\cdot\,), \Delta, \varepsilon); t_*, x_*);$$

define *the optimal ensured counter-result* at (t_*, x_*):

$$\pi_s^0(t_*, x_*) = \sup\{\pi_s(v_*(\,\cdot\,); t_*, x_*)\colon v_*(\,\cdot\,) \text{ is a counter-strategy}\};$$

call a counter-strategy $v^0(\,\cdot\,)$ *optimal*, if $\pi_s(v^0(\,\cdot\,); t_*, x_*) = \pi_s^0(t_*, x_*)$ for all $(t_*, x_*) \in G_*$.

The initial problem (find $u^0(\,\cdot\,)$) and the counter-problem (find $v^0(\,\cdot\,)$) form a differential game. It should be noted that if $v[\,\cdot\,]$'s are control realizations for a real controller's opponent, then the counter-problem has an actual sense. If

$$\rho_s^0(t_*, x_*) = \pi_s^0(t_*, x_*) = \nu(t_*, x_*) \quad \text{for all } (t_*, x_*) \in G_*, \tag{2.3}$$

then the function $\nu(\,\cdot\,)$ (on G_*) will be called *the value of the game*. If the value exists, then a pair $(u^0(\,\cdot\,), v^0(\,\cdot\,))$, where $u^0(\,\cdot\,)$ and $v^0(\,\cdot\,)$ are respectively an optimal strategy and an optimal counter-strategy, will be called *a saddle point*. The main result is

THEOREM 2.1. (1) *There exists the value* $\nu(\,\cdot\,)$ *of the game*;
(2) *there exists a saddle point.*

Note that Theorem 2.1 implies obviously statement (2) of Proposition 2.2. Further, it is easy to show that each admissible control law $\mathcal{U}$ and each closed-loop disturbance law $\mathcal{V}$ are compatible; i.e., for any x_* there exists a process $(x[\,\cdot\,\mid x_*], u[\,\cdot\,], v[\,\cdot\,])$ such that $u[\,\cdot\,] = \mathcal{U}x[\,\cdot\,\mid x_*]$ and $v[\,\cdot\,] = \mathcal{V}(x[\,\cdot\,\mid x_*], u[\,\cdot\,])$. That gives immediately that $\rho_a(\mathcal{U}; t_*, x_*) \ge \pi_s(\mathcal{V}; t_*, x_*)$. This inequality and (2.1)–(2.3) lead obviously to Proposition 2.1 and statement (1) of Proposition 2.2.

Now we describe the extremal shift method specifying the form of a saddle point. Taking into account condition (iii), suppose without loss of generality that $G_* = \{(t,x): t \in [t_0, \vartheta], |x| \leq \kappa(t)\}$, where $\kappa(\cdot)$ is continuous and for every $(t_*, x_*) \in G_*$, $u[\cdot] \in U[t_*, \vartheta]$, and $v[\cdot] \in V[t_*, \vartheta]$ the solution $x[\cdot]$ of (1.1) at $[t_*, \vartheta]$ is such that $|x[t]| \leq \kappa(t)$ for all $t \in [t_*, \vartheta]$. Let $c_* = \max\{c_{f,G_*}, c_{\chi, G_*}\}$, $\eta(t, \varepsilon) = (\varepsilon + \varepsilon(t - t_0))^{1/2} \exp c_*(t - t_0)$, and $l'g$ denotes the scalar product of vectors l and g. Define $u^0(t, x, \varepsilon)$ and $v^0(t, x, u, \varepsilon)$ to be, respectively, solutions of the extremal problems

$$\max_{v \in Q}(l'_1 f(t, x, u, v) + s_1 \chi(t, x, u, v)) \to \min, \qquad u \in P, \qquad (2.4)$$

$$l'_2 f(t, x, u, v) + s_2 \chi(t, x, u, v) \to \min, \qquad v \in Q, \qquad (2.5)$$

where (l_1, s_1) and (l_2, s_2) are, respectively, solutions of the extremal problems

$$\nu(t, x - l) - s \to \min, \qquad (t, x - l) \in G_*, \ |(l, s)| \leq \eta(t, \varepsilon),$$

$$\nu(t, x - l) - s \to \max, \qquad (t, x - l) \in G_*, \ |(l, s)| \leq \eta(t, \varepsilon).$$

Note that such strategy $u^0(\cdot)$ and counter-strategy $v^*(\cdot)$ actually exist.

PROPOSITION 2.3. $(u^0(\cdot), v^0(\cdot))$ is a saddle point.

REMARK. Suppose that for each $l \in R^n$ and $s \in R^n$

$$\min_{u \in P} \max_{v \in Q}(l'f(t, x, u, v) + s\chi(t, x, u, v))$$
$$= \max_{v \in Q} \min_{u \in P}(l'f(t, x, u, v) + s\chi(t, x, u, v)). \qquad (2.6)$$

Then Theorem 2.1 remains true provided counter-strategies do not depend on u. The problem (2.5) turns in this case into a problem similar to (2.4). Proposition 2.1 remains true if the $v_*(\cdot)$'s in the definition of an admissible control law do not depend on u.

3. Stochastic program maximin. Specify now the form of $\nu(\cdot)$. As it was shown in [2] $\nu(\cdot)$ is in some particular, *regular*, cases given by

$$\nu(t_*, x_*) = \max_{v[\cdot] \in V[t_*, \vartheta]} \min_{u[\cdot] \in U[t_*, \vartheta]} \gamma(x[\cdot \mid x_*], u[\cdot], v[\cdot]). \qquad (3.1)$$

The right-hand side of (3.1) is called *program maximin*. In the general case $\nu(t_*, x_*)$ is determined by an analogous expression with $v[\cdot]$'s and $u[\cdot]$'s depending (besides time) on some probability element. Namely, fix temporarily a partition Δ. Consider the probability space $(\Omega, \mathcal{B}_\Omega, \mu)$, where Ω is the product of m exemplars of the interval $[0, 1)$ (recall that $m + 1$ is the number of τ_i's in Δ) and μ is the standard Borel measure on $\mathcal{B}_\Omega$; here and below $\mathcal{B}_Z$ denotes the Borel σ-algebra on a metric space Z. Functions $\overline{u}[\cdot] = \overline{u}[t, \omega]: [t_*, \vartheta] \times \Omega \mapsto P$ and $\overline{v}[\cdot] = \overline{v}[t, u, \omega]: [t_*, \vartheta] \times P \times \Omega \mapsto Q$, $\mathcal{B}_{[t_*, \vartheta] \times \Omega}$- and $\mathcal{B}_{[t_*, \vartheta] \times P \times \Omega}$-measurable, respectively, will be called *programs* (for control and disturbance, respectively) if $\overline{u}[t, \omega]$ and $\overline{v}[t, u, \omega]$ considered as functions of $\omega = (\omega_1, \ldots, \omega_m)$ depend for $t \in b_i$ only on the first $i + 1$ coordinates of the vector ω. Denote the sets of all such $\overline{u}[\cdot]$'s and $\overline{v}[\cdot]$'s by $\overline{U}[t_*, \vartheta]$ and $\overline{V}[t_*, \vartheta]$, respectively. Each $x_*, \overline{u}[\cdot]$, and $\overline{v}[\cdot]$

determine the unique $\mathcal{B}_{[t_*,\vartheta]\times\Omega}$-measurable function $\overline{x}[\,\cdot\,\mid x_*; \overline{u}[\,\cdot\,], \overline{v}[\,\cdot\,]] = \overline{x}[\,\cdot\,\mid$ $x_*] = \overline{x}[t, \omega \mid x_*]\colon [t_*, \vartheta] \times \Omega \mapsto R^n$ such that $(\overline{x}[\,\cdot\,, \omega \mid x_*], \overline{u}[\,\cdot\,, \omega], \overline{v}[\,\cdot\,, \omega])$ is (a.e. on Ω) a process at $[t_*, \vartheta]$ corresponding to x_*. We set

$$\overline{\rho}_\Delta(t_*, x_* \mid \beta) \tag{3.2}$$
$$= \sup_{\overline{v}[\,\cdot\,]\in\overline{V}[t_*,\vartheta]} \inf_{\overline{u}[\,\cdot\,]\in\overline{U}[t_*,\vartheta]} \mu(\{\omega \in \Omega\colon \gamma(\overline{x}[\,\cdot\,, \omega \mid x_*], \overline{u}[\,\cdot\,, \omega], \overline{v}[\,\cdot\,, \omega]) > \beta\}),$$

$$\overline{\rho}_\Delta(t_*, x_*) = \sup\{\beta\colon \overline{\rho}_\Delta(t_*, x_* \mid \beta) > 0\}. \tag{3.3}$$

Further (Δ_k) is an arbitrary sequence of partitions of $[t_*, \vartheta]$ with $d(\Delta_k) \to 0$.

THEOREM 3.1. $\nu(t_*, x_*) = \lim_{k\to\infty} \overline{\rho}_{\Delta_k}(t_*, x_*)$.

If

$$f(t, x, u, v) = A(t)x + g(t, u, v), \tag{3.4}$$

where $A(t)$ is an $n \times n$-matrix-function continuous in t,

$$\sigma(\cdot) \text{ is convex} \quad \text{and} \quad \chi(t, x, u, v) = \chi(t, u, v), \tag{3.5}$$

then approximations of $\nu(t_*, x_*)$ may be expressed in terms of mathematical expectations. Namely, let

$$\hat{\rho}_\Delta(t_*, x_*) = \sup_{\overline{v}[\,\cdot\,]\in\overline{V}[t_*,\vartheta]} \inf_{\overline{u}[\,\cdot\,]\in\overline{U}[t_*,\vartheta]} \hat{\gamma}(x_*; \overline{u}[\,\cdot\,], \overline{v}[\,\cdot\,]), \tag{3.6}$$

where

$$\hat{\gamma}(x_*; \overline{u}[\,\cdot\,], \overline{v}[\,\cdot\,])$$
$$= E\left[\sigma(\overline{x}[\vartheta, \omega \mid x_*; \overline{u}[\,\cdot\,], \overline{v}[\,\cdot\,]]) + \int_{t_*}^{\vartheta} \chi(t, \overline{u}[t, \omega], \overline{v}[t, \omega])\,dt\right]$$

and E is mathematical expectation on the probability space $(\Omega, \mathcal{B}_\Omega, \mu)$ corresponding to Δ. Note that (3.6) is analogous to (3.1).

THEOREM 3.2. *If* (3.3) *and* (3.4) *are fulfilled, then*

$$\nu(t_*, x_*) = \lim_{k\to\infty} \hat{\rho}_{\Delta_k}(t_*, x_*).$$

REMARK. If (2.6) takes place, then the results remain true for $\overline{v}[\,\cdot\,]$'s not depending on u. For programs $\overline{u}[\,\cdot\,]$ and $\overline{v}[\,\cdot\,]$ other definitions may be given; in particular $\overline{u}[\,\cdot\,]$ and $\overline{v}[\,\cdot\,]$ may be defined as nonanticipatory functions with respect to a Brownian process at $[t_*, \vartheta]$; in this case the values (3.2), (3.3), and (3.6) do not depend on Δ, and the limits in Theorems 3.1 and 3.2 are not necessary.

In some particular cases Theorem 3.2 allows us to reduce the problem of computing $\nu(t_*, x_*)$ to finite-dimensional problems of convex programming. The following statement is true:

THEOREM 3.3. *Let* (a) $f(t, x, u, v) = A(t)x + B(t)u + C(t)v$ *where* $A(t)$, $B(t)$, *and* $C(t)$ *are, respectively,* $n \times n$-, $p \times n$-, *and* $q \times n$-*matrix-functions continuous in t,*

(b) $\sigma(x) = |x|$, $\chi(t, x, u, v) = u'\phi(t)u - v'\psi(t)v$, *where* $\phi(t)$ *and* $\psi(t)$ *are, respectively,* $p \times p$- *and* $q \times q$-*matrix-functions continuous in t and positively defined, and*

(c) $P = \{u\colon |u| \leq c\}$, $Q = \{v\colon |v| \leq c\}$, *where c is sufficiently large.*

Then

$$\nu(t_*, x_*) = \max_{|l| \leq 1}(l'X(\vartheta, t_*)x_* + l'F(t_*)l - \lambda^*|l|^2 + \lambda^*); \qquad (3.7)$$

here $X(t, \tau)$ *is the fundamental matrix for the system*

$$\dot{x} = A(t)x,$$

$$\lambda^* = \max\left\{\max_{|l|\leq 1} l'F(\tau)l\colon \tau \in [t_*, \vartheta]\right\},$$

$$F(\tau) = \int_\tau^\vartheta N(t)\,dt,$$

$$N(t) = \tfrac{1}{4}X(\vartheta, t)(C(t)\psi^{-1}(t)C'(t) - B(t)\phi^{-1}(t)B'(t))X'(\vartheta, t).$$

REMARK. The lower bound for c can be written down explicitly. Formula (3.7) allows us to simplify essentially the form of the optimal strategy $u^0(\,\cdot\,)$ (see Proposition 2.3).

4. Ill-posed inverse problems for dynamical systems.

Let the dynamical system be uncontrolled, i.e., $f(t, x, u, v) = f(t, x, v)$. Here we keep all assumptions imposed in §1 and fix the initial time t_*. Suppose that at each time $t \in [t_*, \vartheta]$ the state $x[t]$ is measured. The result $\xi[t]$ of the measurement satisfies the inequality $|\xi[t] - x[t]| \leq h$. The aim of the observer is to form in real time an approximation of $v[\,\cdot\,]$ close to $v[\,\cdot\,]$ for h sufficiently small. Here the L^2-distance ($L^2 = L^2([t_*, \vartheta], R^q)$) between the approximation (denoted further $u[\,\cdot\,]$) and $v[\,\cdot\,]$ will be considered. Since a trajectory $x[\,\cdot\,]$ may correspond to various $v[\,\cdot\,]$'s, we suppose that it is admissible to approximate any one of these $v[\,\cdot\,]$'s. Therefore the problem consists in constructing a stable (regularizing) algorithm for an ill-posed [8] inverse problem of dynamics [9]. Since $u[\,\cdot\,]$ is to be formed in real time, $u[t]$ may depend only on the past measurements (on $\xi[\tau]$ for $\tau \leq t$). We define a law of forming $u[\,\cdot\,]$ to be a mapping $\mathcal{U}\colon \xi[\,\cdot\,] \mapsto u[\,\cdot\,]$ satisfying the PR-condition (see §1): if $\xi_1[\tau] = \xi_2[\tau]$ for all $\tau \in [t_*, t]$, then $u_1[\tau] = u_2[\tau]$ for all $\tau \in [t_*, t]$, where $u_i[\,\cdot\,] = \mathcal{U}\xi_i[\,\cdot\,]$, $i = 1, 2$. Such a mapping $\mathcal{U}\colon \Xi \mapsto V[t_*, \vartheta]$, where Ξ is the set of all functions $\xi[\,\cdot\,]\colon [t_*, \vartheta] \mapsto R^n$ will be called *an approximation law.*

Introduce some notations. If we want to emphasize that a trajectory $x[\,\cdot\,]$ corresponds to a disturbance realization $v[\,\cdot\,]$, we will write $x[\,\cdot\, | v[\,\cdot\,]]$ instead of $x[\,\cdot\,]$. Further X is the set of all trajectories at $[t_*, \vartheta]$, $V = V[t_*, \vartheta]$, $V(x[\,\cdot\,])$ is the set of all $v[\,\cdot\,] \in V$ such that $x[\,\cdot\,] = x[\,\cdot\, | v[\,\cdot\,]]$, $\text{dist}(u[\,\cdot\,]; x[\,\cdot\,])$ is the L^2-distance between $u[\,\cdot\,]$ and $V(x[\,\cdot\,])$, and $\Xi_h(x[\,\cdot\,])$ is the set of all $\xi[\,\cdot\,] \in \Xi$ such that $|\xi[t] - x[t]| \leq h$ for all $t \in [t_*, \vartheta]$.

If $(\mathcal{U}_h)$, $h > 0$, is a family of approximation laws, then the value

$$\rho((\mathcal{U}_h)) = \sup_{x[\,\cdot\,] \in \mathcal{X}} \overline{\lim_{h \to 0}} \sup_{\xi[\,\cdot\,] \in \Xi_h(x[\,\cdot\,])} \operatorname{dist}(u[\,\cdot\,]; x[\,\cdot\,])$$

will be called *the ensured result for* $(\mathcal{U}_h)$. We shall say that a family $(\mathcal{U}_h)$ is *stable*, if $\rho((\mathcal{U}_h)) = 0$. Note that if each trajectory $x[\,\cdot\,]$ corresponds to a single $v[\,\cdot\,]$ (considered as an element of L^2) and $(\mathcal{U}_h)$ is stable, then for $u_h[\,\cdot\,] = \mathcal{U}_h \xi_h[\,\cdot\,]$, $\xi_h[\,\cdot\,] \in \Xi_h(x[\,\cdot\,])$, $|u_h[\,\cdot\,] - v[\,\cdot\,]|_{L^2} \to 0$ holds as $h \to 0$.

We shall choose a stable family within a special class of approximation laws identified with closed-loop control procedures with a model [2]. Fix a dynamical system (a model)

$$\dot{w} = g(t, \xi[t], u[t], z[t]), \qquad w(t_*) = \xi[t_*], \tag{4.1}$$

with $u[t] \in Q$ and $z[t] \in R^n$ standing for controls ($g(\,\cdot\,) \colon R \times R^n \times Q \times R^k \mapsto R^n$ is continuous). Let us modify the definition of a strategy and apply it to the model. A collection $(w[\,\cdot\,], \xi[\,\cdot\,], u[\,\cdot\,], z[\,\cdot\,])$, where $\xi[\,\cdot\,] \in \Xi$, $u[\,\cdot\,] \in V$, $z[\,\cdot\,] \colon [t_*, \vartheta] \mapsto R^k$, and $w[\,\cdot\,]$ is the solution of the Cauchy problem (4.1) at $[t_0, \vartheta]$, will be called *a modelling process*. Define *a strategy* as a pair $(u_*(\,\cdot\,), z_*(\,\cdot\,))$, $u_*(\,\cdot\,) = u(t, \xi, w) \colon R \times R^n \times R^n \mapsto Q$, $z_*(\,\cdot\,) = z_*(t, \xi, w) \colon R \times R^n \times R^n \mapsto R^k$. A triplet $(u_*(\,\cdot\,), z_*(\,\cdot\,), \Delta)$ will be identified with the approximation law $\mathcal{U} \colon \xi[\,\cdot\,] \mapsto u[\,\cdot\,]$ such that for the modelling process $(w[\,\cdot\,], \xi_\Delta[\,\cdot\,], u[\,\cdot\,], z[\,\cdot\,])$, where $\xi_\Delta[t] = \xi[\tau_i]$ for $t \in b_i$, $u[t] = u_*(\tau_i, \xi[\tau_i], w[\tau_i])$ and $z[t] = z_*(\tau_i, \xi[\tau_i], w[\tau_i])$ for $t \in b_i$ ($\mathcal{U}$ is determined uniquely); we will call $\mathcal{U}$ *a modelling law*. If $g(t, \xi, u, z)$ does not depend on z, then we identify modelling processes with triplets $(w[\,\cdot\,], \xi[\,\cdot\,], u[\,\cdot\,])$ and modelling laws with pairs $(u_*(\,\cdot\,), \Delta)$.

Suppose that

$$Q \text{ is convex} \quad \text{and} \quad f(t, x, v) = f_1(t, x) + f_2(t, x)v, \tag{4.2}$$

where a function $f_1(\,\cdot\,)$ and a matrix-function $f_2(\,\cdot\,)$ are continuous and satisfy the condition similar to (iii) (see §1). The construction of a stable family suggested below is based on a combination of the Tikhonov regularization method [8] and a modification of the extremal shift method.

PROPOSITION 4.1. *Let* (4.2) *be fulfilled,* $g(t, \xi, u, z) = f(t, \xi, u)$, *and a family* $((u_h(\,\cdot\,), \Delta_h))$ *of modelling laws be given by the conditions:*
(a) $d(\Delta_h) \leq c^0 h$ *for a certain* $c^0 > 0$,
(b) $u_h(t, \xi, w)$ *is a solution of the extremal problem*

$$(w - \xi)' f_2(t, \xi) u + \alpha(h)|u|^2 \to \min, \qquad u \in Q,$$

where

$$\alpha(h) > 0, \quad \alpha(h) \to 0, \quad h/\alpha(h) \to 0 \quad as \ h \to 0. \tag{4.3}$$

Then the family $((u_h(\,\cdot\,), \Delta_h))$ *is stable.*

REMARK. If each trajectory $x[\,\cdot\,]$ corresponds to a single $v[\,\cdot\,]$, the Proposition 4.1 remains true without assuming convexity of Q.

REMARK. For each $x[\,\cdot\,]$ there exists the single $v_*[\,\cdot\,] = v_*[\,\cdot\mid x[\,\cdot\,]]$ minimizing the norm $|\cdot|_{L^2}$ on $V(x[\,\cdot\,])$. If we put $\operatorname{dist}(u[\,\cdot\,]; x[\,\cdot\,]) = |u[\,\cdot\,] - v_*[\,\cdot\mid x[\,\cdot\,]]|_{L^2}$ (keeping all other notations), then Proposition 4.1 also remains true.

Let (4.2) be not true. Fix $K > 0$ such that $|f(t,x,v)| \le K$ for all $(t,x) \in G_*$ and $v \in Q$ and consider the auxiliary system

$$\dot{x} = \tilde{v}[t], \qquad x(t_*) = x_* . \tag{4.4}$$

with disturbance realizations $\tilde{v}[\,\cdot\,]$ taking values within the ball $\tilde{Q} = \{v \in R^n : |v| \le K\}$. Since (4.4) satisfies condition (4.2) (with Q and $f(t,x,v)$ replaced by $\tilde{Q}$ and $\tilde{f}(t,x,v) = v$, respectively), Proposition 4.1 gives an algorithm to form an L^2-approximation $z[\,\cdot\,]$ of $\tilde{v}[\,\cdot\,]$. If $\tilde{v}[t] = f(t, x[t], v[t])$ and $t \in b_i$, it is natural to take a minimum point of $\varsigma(u) = |z[\tau_i] - f(\tau_i, \xi[\tau_i], u)|$ as an approximation $u[t]$ of $v[t]$. This method indeed provides a solution.

PROPOSITION 4.2. *Let* $g(t, \xi, u, z) = z$ *and a family* $((u_h(\,\cdot\,), z_h(\,\cdot\,), \Delta_h)$ *of modelling laws be given by the conditions:*

(a) $d(\Delta_h) \le c^0 h$ *for a certain* $c^0 > 0$,

(b) $z_h(t, \xi, w)$ *and* $u_h(t, \xi, w)$ *are, respectively, solutions of the extremal problems*

$$(w - \xi)'z + \alpha(h)|z|^2 \to \min, \qquad z \in \tilde{Q},$$

where $\alpha(h)$ *satisfies* (4.3), *and*

$$|z_h(t, \xi, w) - f(t, \xi, u)| \to \min, \qquad u \in Q.$$

Then the family $(u_h(\,\cdot\,), z_h(\,\cdot\,), \Delta_h)$ *is stable.*

REFERENCES

1. N. N. Krasovskii, *Game problems of meeting of motions*, "Nauka", Moscow, 1970.

2. N. N. Krasovskii and A. I. Subbotin, *Positional differential games*, "Nauka", Moscow, 1974.

3. A. I. Subbotin and A. G. Chentsov, *Optimization of the guarantee in control problems*, "Nauka", Moscow, 1981.

4. N. N. Krasovskii, *Controlling of a dynamical system. Problem of the minimum of the ensured result*, "Nauka", Moscow, 1985.

5. ____, *Differential game for a positional functional*, Dokl. Akad. Nauk SSSR **253** (1980), 1303–1307.

6. N. N. Krasovskii and V. E. Tret'jakov, *Stochastic program synthesis for a positional differential game*, Dokl. Akad. Nauk SSSR **259** (1981), 24–27.

7. A. V. Kryazhimskii and Ju. S. Osipov, *On control modelling for a dynamical system*, Izv. Akad. Nauk SSSR, Tekhn. Kibernet. No. 2 (1983), 51–60.

8. A. N. Tikhonov and V. Ja. Arsenin, *Solution methods for ill-posed problems*, "Nauka", Moscow, 1979.

9. B. N. Petrov and P. D. Krutko, *Inverse problems of control system dynamics. Nonlinear models*, Izv. Akad. Nauk SSSR, Tekhn. Kibernet. No. 5 (1980), 149–155.

INSTITUTE OF MATHEMATICS AND MECHANICS, URALS SCIENCE CENTER, SVERDLOVSK 620219, USSR

Generalized Hamiltonians and Optimal Control: A Geometric Study of the Extremals

IVAN A. K. KUPKA

I. Introduction to geometric optimal control.

Introduction. Optimal control problems are generalizations of classical problems in the calculus of variations. A typical one can be stated as follows: given a smooth (C^∞ or real analytic C^ω) manifold M, a compact smooth manifold U (possibly with a boundary), a smooth vector-field $E\colon M \times U \to TM$ (tangent space of M) on M, parametrized by U, a smooth function $c\colon M \times U \to R$, and two points A, B in M, let $\mathrm{Tr}(A, B)$ be the set of all pairs $(x, u)\colon [a, b] \to M \times U$, such that: (1) x is absolutely continuous and u measurable; (2) $dx(t)/dt = E(x(t), u(t))$ a.e.; (3) $x(a) = A$, $x(b) = B$.

The problem is to find a pair $(\overline{x}, \overline{u})\colon [\overline{a}, \overline{b}] \to M \times U$ such that

$$\int_{\overline{a}}^{\overline{b}} c(\overline{x}(t), \overline{u}(t))\, dt = \inf \left[\int_a^b c(x(t), u(t))\, dt \mid (x, u)\colon \right.$$

$$\left. [a, b] \to M \times U, \text{ belonging to } \mathrm{Tr}(A, B)\right].$$

A pair in $\mathrm{Tr}(A, B)$ is called a trajectory of the system; the pair (x, u) is called an optimal trajectory.

It is well known that such a pair (x, u) is the projection on $M \times U$ of an "extremal." An extremal is the generalization of its namesake of classical calculus of variations. In the present situation, there are two families $\mathcal{E}_\lambda$, $\lambda = 0$ or 1, of extremals: a couple $(z, u)\colon [a, b] \to T^*M \times U$ belongs to $\mathcal{E}_\lambda$ if it satisfies the following conditions:

(1) $dz/dt = \overrightarrow{H^\lambda(z(t), u(t))}$ for almost all t in $[a, b]$, $H^\lambda(z, u) = \langle z, E(x, u)\rangle - \lambda c(x, u)$, x being the projection of z onto M; $\langle\ ,\ \rangle$ denotes the canonical pairing $TM \times_M T^*M \to R$, and $\overrightarrow{H^\lambda}$ is the hamiltonian field associated to H^λ considered as a function on T^*M parametrized by u.

(2) For almost all $t \in [a, b]$, $H^\lambda(z(t), u(t)) = K^\lambda(z(t))$, where $K^\lambda(z) = \sup\{H^\lambda(z, v)|v \in U\}$.

The family $\mathcal{E}_1$ is called ordinary, the family $\mathcal{E}_0$ exceptional.

As in the classical calculus of variations, one tries to solve the optimal control problems using the extremals. Two methods have been exploited up to now. The first one, which could be called the direct method, is being developed by H. Sussmann and his collaborators. It has yielded some important results in the case of dim M being two or three.

The second method, the singularity method, was introduced by I. Ekeland [**E**] in the special case when

$$M = U = R \quad \text{and} \quad E(x,u) = u.$$

More recently F. Klok pushed this analysis further in the same case [**Kl**]. Our approach belongs to this last line of thought.

1. *Preliminary considerations on extremals.* The main difference between the classical calculus of variations and our case is that in condition (2) of the definition of extremals the maximum K^λ can be attained for several distinct u's. This allows for the phenomenon of "switching," that is, the extremal changing its policy $u(t)$ abruptly at some time T. Mathematically this translates into the fact that z is not differentiable at t. Let us formalize this.

DEFINITION 0. A point $z(s)$ (resp. s) on an extremal $z\colon [a,b] \to T^*M$ is called a switching point (resp. a switching time) if s belongs to the closure of the set of all t's where z is not differentiable.

NOTATION. The set of all possible switching points is a subset of T^*M, called the switching surface.

The notion of switching points is crucial in the study of extremals. They determine the structure of these curves. What can we say about this structure? H. Sussmann has noticed that: (a) in the C^∞ case, any absolutely continuous curve in M is optimal for some appropriate system (E,c); (b) in the C^ω case, given a system (E,c), if there exists an optimal trajectory joining two given points A, B in M, then there exists another optimal trajectory joining A to B, which is analytic on an open dense subset of times. Since any optimal trajectory is the projection on M of an extremal, this shows that in order to get any reasonable theory, we have to put some restrictions on the system (E,c).

Now, even for a generic system (E,c), the extremals are not smooth in general. A consequence of our results is the fact that for an extremal to have an infinite number of switching points is a very stable property. Let us note here, that the structure of the general extremal in the generic case is not known.

Finally, the extremals would be the trajectories of the hamiltonian field associated to K^λ, if K^λ were smooth, which it is not, in general. Let us mention that generalizations of the concept of hamiltonian field to include this case have been put forward.

2. *Regular points of finite multiplicity.* From now on we drop the superscript λ in H^λ. Let us denote by S the subset in $T^*M \times U$ of all couples $(z, \overline{u})$ such that $\overline{u}$ is a local maximum point of the function $H\colon v \in U \to H(z,v)$. It is clear that if $(z,u)\colon [a,b] \to T^*M \times U$ is an extremal, then for almost every $t \in [a,b]$,

$(z(t), u(t))$ belongs to S. Let $p\colon S \to T^*M$ denote the restriction to S of the canonical projection: $T^*M \times U \to T^*M$.

Without making a formal statement, it is clear that, for a generic pair, there exist stratifications of T^*M and S such that: (1) p is stratified; (2) for any stratum A, $p\colon p^{-1}(A) \to A$ is a finite covering; (3) for any open stratum A of T^*M, for any z in A, all points in $p^{-1}(z)$ are either nondegenerate quadratic singular points of H_z, or are regular points of H_z belonging to the boundary of U, which are nondegenerate quadratic singular points for the restriction of H_z to the boundary of U.

On the lower-dimensional strata, a branching of singularities takes place. Since we are mainly interested in the switching phenomena, we shall not go into branching but concentrate on the open strata. This motivates the following definition. In it, we do not assume that S and p satisfy the conditions (1)–(3) above.

DEFINITION 1. A point q in T^*M is called a regular point of multiplicity m if there exists a neighborhood V of q in T^*M, such that:

(1) The restriction $p\colon p^{-1}(V) \to V$ is a trivial finite covering.

(2) Let J be the set of all sections $\varphi\colon V \to S$ of this covering such that q belongs to the closure $\Gamma(\varphi)$ of the set

$$\{z \in V, z \neq q, H(\varphi(z)) = K(z)\}.$$

Then for any two sections $\varphi, \psi \in J$ the germs at q of the restrictions of $\mathrm{Ho}\,\varphi$ and $\mathrm{Ho}\,\psi$ to $\Gamma(\varphi)$ and $\Gamma(\psi)$ respectively, are not equal. m is the cardinal of J.

The case $m = 1$ corresponds to the classical theory of the calculus of variations. Near q, the extremals are the trajectories of a hamiltonian vector-field. We have studied the cases when m is 2 or 3. The structure of the extremals near q depends essentially on the structure of the contacts of the hamiltonian vector-fields $\mathrm{Ho}\,\varphi$, $\varphi \in J$, with the switching surface and certain subsets of it defined by these contacts. This vague statement can be given a precise formulation using the Lie algebra, generated by the set of functions $[\mathrm{Ho}\,\varphi/\varphi \in J]$, under the Poisson bracket. The complexity of the contact structure at a point q is measured by the minimum of the length of the brackets not zero at q.

In the remainder of this paper, we shall discuss two of our results. Both deal with the case $m = 2$.

II. Statement of the results.

1. *Notations and auxiliary concepts.* Let q be a regular point of multiplicity 2. J contains two sections φ_+, φ_-. The associated functions $\mathrm{Ho}\,\varphi_+$ and $\mathrm{Ho}\,\varphi_-$ will be denoted by H_+ and H_- respectively. In a neighborhood of q, the switching surface Σ is defined by $H_+ - H_- = 0$. We shall make the following assumption for the remainder of this paper:

$$0 = dH_+(q) \wedge dH_-(q) \wedge d\{H_+, H_-\}(q), \text{ where } \{H_+, H_-\} \text{ denotes}$$
the Poisson bracket of H_+ and H_-, that is, the Lie derivate of $\qquad (*)$
H_+ in the direction of $\overrightarrow{H_-}$ (see [A–M]).

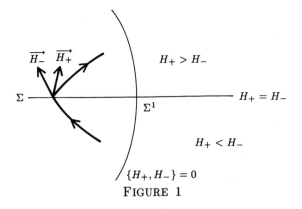

FIGURE 1

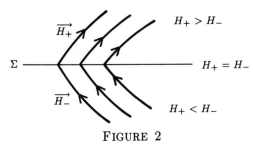

FIGURE 2

(∗) implies that the switching surface is a smooth manifold in a neighborhood of q and that the same is true for the set $\Sigma^1 = \Sigma \cap [\{H_+, H_-\} = 0]$, where $\overrightarrow{H_+}$ and $\overrightarrow{H_-}$ are tangent to Σ (see Figure 1).

If q does not belong to Σ^1 then, in a neighborhood of it, the extremals are the trajectories of a piecewise smooth flow, having a tangential discontinuity along Σ (see Figure 2). Hence the only interesting cases are when q belongs to Σ^1.

NOTATION. (1) For simplicity, let us denote by f, g the brackets

$$\{H_-\{H_+, H_-\}\} \quad \text{and} \quad \{H_+\{H_-, H_+\}\}$$

respectively.

(2) Given an open subset W in T^*M, a subset N of W will be called invariant in W if any extremal contained in W and meeting N is contained in N.

2. *First theorem—Fold points.*

DEFINITION 3. A regular point of multiplicity 2, q, satisfying the assumption (∗), is called a fold point if $f(q)$ and $g(q)$ are both nonzero. If they are both positive (resp. negative) q is called hyperbolic (resp. elliptic). If they have opposite signs, q is parabolic.

NOTATION. (3) In the elliptic case, the following vector-field R, defined on Σ^1, plays an important role:

$$R = \frac{f\overrightarrow{H_+} + g\overrightarrow{H_-}}{f + g}.$$

It is clear that R is the unique convex combination of $\overrightarrow{H_+}$ and $\overrightarrow{H_-}$ tangent to Σ^1 at the points of Σ^1.

THEOREM 1.   *Let q be a fold point. There exists a neighborhood W of q such that:*

(e) *If q is elliptic, we have a generalized "flow-box" result; there exist a ball B of codimension 1 in T^*M, an interval $I = [-a, a]$, and a continuous mapping $z \colon I \times B \to W$, with the properties:*

(i) *$z(0, b) = b$ if $b \in B$, and z is a homeomorphism onto W, piecewise smooth on any $I \times K$, K compact subset of $B - \Sigma^1$.*

(ii) *For $b \in \Sigma^1 \cap B$, the curve $z_b \colon I \to W$ is the trajectory of R, passing at time 0 through b.*

(iii) *For $b \in B - \Sigma^1$, z is an extremal.*

(iv) *For any subinterval d of I, any $b \in B - \Sigma^1$, let $N(d, b)$ be the number of switching times of z_b in d. When b tends to Σ^1, $N(d, b)$ tends to ∞ and $N(d, b) \cdot h(z(t, b))$ tends to the length of d, for any $t \in d$. h is the function $2 \cdot \{H_+, H_-\} \cdot [1/f + 1/g]$.*

(h) *If q is hyperbolic, it behaves somewhat like a hyperbolic singular point of a vector-field. W contains two smooth hypersurfaces $S(+)$ and $S(-)$, having a contact of first order with Σ along Σ^1, with the following properties:*

(i) *$W - S(+) \cup S(-)$ has four connected components W_+, W_-, W_r, W_l. W_+ and $S(+)$ (resp. W_- and $S(-)$) are located in $[H_+ \geq H_-]$ (resp. $[H_+ \leq H_-]$). W_k (resp. W_l) is located in $\{H_+, H_-\} > 0$ (resp. $\{H_+, H_-\} < 0$).*

(ii) *The sets $S(+) \cup S(-)$, W_+, W_-, W_r, W_l are invariant in W.*

(iii) *In W_+ (resp. W_-) the extremals are trajectories of $\overrightarrow{H_+}$ (resp. $\overrightarrow{H_-}$). They do not switch.*

(iv) *In W_r (resp. W_l), the extremals switch exactly once and they are the trajectories of a piecewise smooth flow.*

(v) *In $S(+) \cup S(-)$, the extremals either do not switch and then they are the trajectories of $\overrightarrow{H_+}$ in $S(+)$ or of $\overrightarrow{H_-}$ in $S(-)$, or they switch once and cross from $S(+)$ to $S(-)$ or vice versa.*

(p) *If q is parabolic, let us assume that $f(q) > 0$ and $g(q) < 0$. W contains a smooth hypersurface Sp with the following properties:*

(i) *$W - \mathrm{Sp}$ has two connected components, W_+ and W_-.*

(ii) *W_+ is contained in $[H_+ > H_-]$, $\mathrm{Sp} \cup W_-$ in $[H_+ \leq H_-]$.*

(iii) *Sp, W_+, W_-, are invariant in W.*

(iv) *In $\mathrm{Sp} \cup W_-$, the extremals do not switch and they are the trajectories of $\overrightarrow{H_-}$. In W_+ they switch twice and are the trajectories of a piecewise smooth flow. The case $f(q) < 0$, $g(q) > 0$ is similar.*

(For these results see Figure 3.)

This theorem calls for some remarks:

(1) The field R is called the residual field. It also shows up in some work of Arnold (see [**Ar**]).

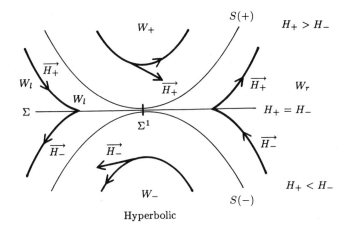

$H_+ > H_-$

$H_+ = H_-$

$H_+ < H_-$

Hyperbolic

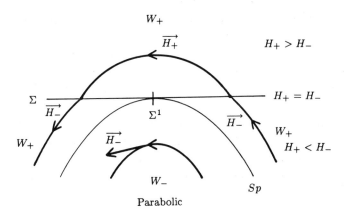

$H_+ > H_-$

$H_+ = H_-$

$H_+ < H_-$

Parabolic

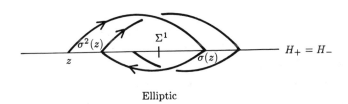

$H_+ = H_-$

Elliptic

Figure 3

(2) The only really interesting and nontrivial part of the preceding theorem is the elliptic case. Near an elliptic point the number of switching points on an extremal is not bounded, and this occurrence is stable. The extremals "spiral" around Σ^1, and as they tend to Σ^1, they pick up more and more switching points, so that, in the limit, they become smooth.

(3) The residual field R is also defined in the hyperbolic domain and it is important in the study of relaxed trajectories. The trajectories of R are relaxed trajectories of the system (E, c). In the elliptic case, these curves are more expensive than the nearby extremals. But in the hyperbolic case, they are cheaper and together with the extremals in $S(+) \cup S(-)$, they can be used to construct local optimal control synthesis of the "turnpike" type.

(4) The proof of the above result and some other ones will appear in the Transactions of the American Mathematical Society.

In agreement with our remarks at the end of paragraph 2 of §1 the preceding result dealt with the case where the bracket of length 2 of the Lie algebra L generated by (H_+, H_-) is zero at q, but those of length 3 are not. The situation when these latter are zero but those of length 4 are not, we shall not examine here and instead pass to the next stage where new phenomena appear. In our preceding considerations, a single extremal switched a finite number of times only. The question is: is it possible to have a general system (E, c) such that above each point from an open subset of M, there passes an extremal that switches an infinite number of times in a finite time-interval? We shall answer this question next.

3. *The Fuller curves.* Let q be a point in T^*M as in Definition 3.

DEFINITION 4. A pair of smooth arcs of curves, $C(+)$ and $C(-)$, contained in $\Sigma - \Sigma^1$, having both q as extremity and no other point in common, is called a Fuller pair if it has the following properties: (1) there is a continuous function

$$T\colon C(+) \cup C(-) \to \mathbf{R},$$

such that any extremal z starting at a point s in $C(+) \cup C(-)$, is defined on the interval $[0, T(s)]$ and $z(t)$ tends to q when t tends to $T(s)$.

(2) Let $s \in C(u)$, $u = +$ or $-$. The switching times of z form an increasing sequence $[0, t_1, \ldots, t_n, \ldots]$ such that $z(t_n)$ belongs to $C(u)$ (resp. $C(-u)$) if n is even (resp. odd).

(3) There exist a constant $k > 1$, depending only on the pair $(C(+), C(-))$ and a continuous function $D\colon C(+) \cup C(-) \to \mathbf{R}$, such that $k^n(t_{n+1} - t_n)$ tends to $D(s)$ as n tends to infinity (see Figure 4).

To state our second theorem we need one more notation.

NOTATION 4. Let Γ denote the vector space of all smooth functions on an unspecified open subset of T^*M. $\alpha\colon \Gamma \times \Gamma \to \Gamma \times \Gamma \times \cdots \times \Gamma$ (6 times) is the mapping defined as follows: $\alpha(f, g) = (\alpha_1(f, g), \ldots, \alpha_6(f, g))$ where $\alpha_1, \ldots, \alpha_6$ is the ordered set of all elements of length 5 from a Hall basis built on the set (f, g) ordered by $f \prec g$, ad $f(g) = \{f, g\}$.

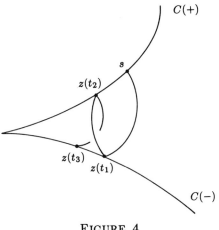

FIGURE 4

EXAMPLE.

$$\alpha_1(f,g) = \mathrm{ad}^4 f(g), \quad \alpha_2(f,g) = \mathrm{ad}\, g\, \mathrm{ad}^3 f(g), \quad \alpha_3(f,g) = \mathrm{ad}^2 g\, \mathrm{ad}^2 f(g),$$
$$\alpha_4(f,g) = -\mathrm{ad}^4 g(f), \quad \alpha_5(f,g) = \{\mathrm{ad}\, f(g), \mathrm{ad}^2 f(g)\},$$
$$\alpha_6(f,g) = \{\mathrm{ad}\, f(g), -\mathrm{ad}^2 g(f)\}.$$

THEOREM 2. *There exists a semialgebraic set $\mathcal{F}$ (explicit) in R^6, with non-empty interior having the property: let q be a regular point of multiplicity 2, satisfying the assumption (*). If the couple (H_+, H_-) has the properties (a)–(b) below, then there is a Fuller pair passing through q.*

(a) *All the Poisson brackets of length 2, 3, and 4, built on H_+, H_-, are zero at q.*

(b) *$\alpha(H_+, H_-)$ belongs to $\mathcal{F}$.*

COMMENT. This result shows that the presence of a Fuller pair is a remarkably stable phenomenon. Using it, one can show that on any smooth manifold M, there is an open (C^∞-topology) set of systems (E, c), such that, for any one of these systems, there exists an open subset O of M with the property that above any point x in O, there is a point q with a Fuller pair passing through it.

III. **Short review of the techniques used.** Essentially, three types of techniques are used: (i) discrete dynamical systems, (ii) partial normal forms, and (iii) blowing up procedures.

1. *Discrete dynamical systems.* To each point q regular of multiplicity 2, satisfying the assumption (*), we associate a discrete dynamical system (DDS for short), σ, as follows: the domain of σ, $\mathrm{dom}(\sigma)$, is the set of all z in Σ, such that:

(1) $\{H_+, H_-\}(z) \neq 0$. Let u be the sign of this number.

(2) The trajectory of $\overrightarrow{H_u}$, starting at z, meets Σ again, at z_u for the first time.

(3) $u\{H_+, H_-\}(z_u) < 0$.

(2) implies that: $u\{H_+, H_-\}(z) \leq 0)$. If z belongs to $\mathrm{dom}(\sigma)$, we set $\sigma(z) = z_u$.

The DDS σ is useful in keeping track of the switching points, and it determines the behavior of the extremals: if $z\colon [a,b] \to W - \Sigma^1$ is such a curve, its set of switching points is discrete. Let y be the first (timewise) of these points. Then the set of them, ordered by increasing time, is a partial orbit of $y\colon [y, \sigma(y), \sigma^2(y), \ldots, \sigma^N(y)]$.

As an example, the main ingredient in the proof of the elliptic case of Theorem 1 is a very convenient normal form for the associated DDS σ.

2. *Partial normal forms.* Using symplectic coordinate transformations, we determine partial normal forms for the pair (H_+, H_-). Such a form is the sum of a normal form for some set of (H_+, H_-) and a remainder term. More precisely: let

$$F = \tfrac{1}{2}(H_+ + H_-), \qquad G = \tfrac{1}{2}(H_+ - H_-).$$

We determine a symplectic system of coordinates, centered at q, $(x_1, \ldots, x_d, p_1, \ldots, p_d)$, such that $G = p_1$, $F = F_0 + F_1$, where F_0 is the normal form and F_1 the remainder. There is a gradation on the coordinates such that F_0 is a homogeneous polynomial and the order of F_1 at q (degree of the lowest degree terms in the Taylor series of F_1 at q) is greater than the degree of F_0. This gradation is intimately linked with the structure at q of the Lie algebra generated by H_+ and H_-. It defines a local group action of the multiplicative group of all positive reals on a neighborhood of q, for which F_0 and G are semi-invariants. In the next proposition, $w(P)$ will denote the degree of the homogeneous polynomial P, and $\mathrm{ord}(h)$, the order of h at q.

PROPOSITION 1.   (i) *If* $\{H_+, H_-\}(q) = 0$, *but not all brackets of length* 3 *are zero at* q, *then*

$$F_0 = p_2 + x_1(\tfrac{1}{2}ax_1 + bx_2),$$
$$w(x_n) = 1 \text{ and } w(p_n) = 2 \quad \text{if } n = 1 \text{ or } 2,$$
$$w(x_n) = w(p_n) = 3 \quad \text{if } n > 3,$$
$$w(F_0) = 2, \quad \mathrm{ord}(F_1) \geq 3,$$

$a = \{G\{G,F\}\}(q)$, $b = \{F\{G,F\}\}(q)$. *This is the fold case.*

(ii) *If all the Poisson brackets of length* 2 *and* 3 *of* H_+, H_- *are* 0 *at* q *but not all brackets of length* 4:

$$F_0 = p_2 + x_1(p_3 + \tfrac{1}{6}ax_1^2 + \tfrac{1}{2}bx_1x_2 + \tfrac{1}{2}cx_2^2),$$
$$w(x_n) = 1 \text{ and } w(p_n) = 2 \quad \text{if } n = 1 \text{ or } 2,$$
$$w(x_n) = w(p_n) = 3 \quad \text{if } n \geq 3,$$
$$W(F_0) = 3, \quad \mathrm{ord}(F_1) \geq 4.$$

$a = -\mathrm{ad}^3 G(F)(q)$, $c = \mathrm{ad}^3 F(G)(q)$, $b = -\mathrm{ad}\,F\,\mathrm{ad}^2 G(F)(q)$. *This is the cusp case.*

(iii) *If all the brackets of length* 2, 3, *and* 4 *of* H_+, H_- *at* q *are* 0 *but not all of length* 5:

$$F = p_2 + x_1(p_3 + \tfrac{1}{2}bx_1^2 x_3 + ax_2 x_3 - \tfrac{1}{6}c_4 x_2^3 + \tfrac{1}{4}c_3 x_1 x_2^2 + \tfrac{1}{6}c_2 x_1^2 x_2 + \tfrac{1}{4}c_1 x_1^3),$$

$$w(x_n) = 1 \text{ and } w(p_n) = 4 \quad \text{if } n = 1 \text{ or } 2,$$

$$w(x_3) = 2, \quad w(p_3) = 3, \quad w(x_n) = w(p_n) = 3 \quad \text{if } n \geq 4,$$

$$w(F_0) = 4, \quad \text{ord}(F_1) \geq 5, \quad c_n = \alpha_1(G, F)(q) \quad \text{if } i \leq n \leq 4,$$

$$b = \alpha_5(G, F)(q), \quad a = \alpha_6(G, F)(q).$$

3. *Blowing up technique.* Using the action of the multiplicative group of the positive reals, $\mathbf{R}_+^*$, we can blow up the point q on the manifold T^*M. This is not the classical blowing up procedure but a "weighted" version of it, q is replaced by the quotient, Q, of $V - q$ under the action of $\mathbf{R}_+^*$, V being a suitable neighborhood of q.

4. *Sketch of the proof of Theorem* 2. Let σ and σ_0 denote the associated DDS to the couples (H_+, H_-) and $(F_0 + G, F_0 - G)$. Using the blowing up technique, the stability theorem of hyperbolic manifolds we reduce the problem of finding a Fuller pair for the couple (H_+, H_-) to that same problem for the second couple. A suitable neighborhood of Q in the blown up space is fibered by the orbits of the $\mathbf{R}_+^*$-action. Since the couple $(F_0 + G, F_0 - G)$ is semi-invariant, the lifting $\tilde{\sigma}_0$ of σ_0 to the blown-up space preserves Q and this fibration. Assume we can find a fixed point h of the restriction of $\tilde{\sigma}_0^2$ to Q. The fiber $O(h)$ above h is then invariant under $\tilde{\sigma}_0^2$. If we can choose h in such a way that $O(h)$ is a contracting curve for $\tilde{\sigma}_0^2$, then the projection on T^*M of the pair $(O(h), \sigma_0(O(h)))$ is a Fuller pair for the couple $(F_0 + G, F_0 - G)$. This ends the "proof" of Theorem 2.

REFERENCES

[A–M] R. Abraham and J. Marsden, *Foundations of mechanics*, Benjamin, 1967.

[Ar] V. Arnold, *Lagrangian manifolds with singularities asymptotic rays and the open swallowtail*, Functional Anal. Appl. **15** (1981), 235–246.

[E] I. Ekeland, *Discontinuite des champs hamiltoniens et existence de solutions optimale en calcul des variations*, Inst. Hautes Études Sci. Publ. Math. **43** (1977), 5–32.

[Kl] F. Klok, *Broken solutions of homogeneous variational problems*, J. Differential Equations **55** (1984), 101–134.

[Ku] I. Kupka, *Geometric theory of extremals in optimal control problems* I: *The fold and Manwell case*, Trans. Amer. Math. Soc. **299** (1987), 225.

[Ku2] _____, *The ubiquity of the Fuller phenomenon*, Preprint, Inst. Fourier (1986).

INSTITUT FOURIER, 38402 SAINT MARTIN D'HÈRES, FRANCE

Proceedings of the International Congress of Mathematicians
Berkeley, California, USA, 1986

Dynamics of Area Preserving Maps

JOHN N. MATHER

Poincaré initiated the study of the dynamics of area preserving mappings, in his studies of celestial mechanics [20]. He showed that the study of the dynamics of the restricted three body problem (two positive masses, one zero mass) could be reduced to the study of the dynamics of an area preserving mapping. He showed, moreover, that even in this case, which is nearly the simplest nontrivial case of Hamiltonian mechanics, the dynamics is so complicated that there is no hope of "solving" the n-body problem (or even the restricted three body problem), in the sense of finding exact expressions of the trajectories as a function of time.

As a consequence of Poincaré's pioneering work, the focus of mathematical studies related to celestial mechanics has shifted to the more topological and analytical approach which Poincaré dubbed "dynamical systems." The books of Arnold and Avez [1] and Moser [15] and the articles of Kolmogorov [10] and Smale [21] present overviews of modern developments in the theory of dynamical systems.

One of the main questions of dynamical systems is the extent to which they display randomness or stability. Many studies in the past century have dealt with these questions. The KAM (Kolmogorov, Arnold, Moser) theory shows that small Hamiltonian perturbations of integrable Hamiltonian systems display a great deal of stability. Invariant tori on which the flow is conjugate to a linear flow exist and fill up most of phase space in the sense of Lebesgue measure. (See, e.g., Moser [15].) In contrast, hyperbolic systems exhibit a great deal of randomness, as is discussed, for example, in Hadamard [8], Anosov [2], Smale [21], Bowen [6], and Pesin [19]. But, even small Hamiltonian perturbations of integrable Hamiltonian systems have regions of instability or randomness alongside the regions of stability. This instability was discovered by Poincaré, further explored by Birkhoff, and given a very transparent form by Smale [21] in terms of "horseshoes."

All this work shows that, typically, one finds a pattern of stability and instability mixed together in a complicated way. But there are many unresolved questions. In the Newtonian n-body problem are the unbounded trajectories dense in phase space (where the center of mass is fixed at the origin)? Newton integrated the 2-body problem and thereby showed that for $n = 2$, the answer to this question is no. The restricted 3-body problem is a Hamiltonian system in 2 degrees of freedom and it is possible to deduce from KAM theory that the answer is no in that case, too (Moser [16]). But all other cases are unsolved.

Is it generically the case that Hamiltonian systems on a smooth compact symplectic manifold are topologically transitive? Here, one must specify what one means by "generically." A popular notion of genericity is that a property of Hamiltonian systems is C^r generic if the set of C^r Hamiltonian systems on the given manifold having that property is a residual set in the C^r topology, in the sense of Baire category. We recall that a dynamical system is said to be topologically transitive if it has a dense orbit. Here again, KAM theory shows that the answer is no for systems in 2 degrees of freedom (i.e., on a 4-manifold) and r sufficiently large, but other cases are unsolved. The KAM theory resolved the analogous problem for "topologically transitive" replaced by "ergodic" and r sufficiently large, the answer being no, contrary to what was expected.

These problems are very difficult and no solution is in sight. In this article, I will report on some recent progress on Hamiltonian systems in two degrees of freedom and the closely related subject of area preserving mappings. Even for such an apparently simple case, there are many difficult unresolved questions, and these questions have attracted engineers, who have recently done a great deal of numerical work on them (surveyed in [11]), as well as inspired mathematicians to obtain deep results (e.g., [9]).

In this article, I will report on one aspect of recent work on dynamics of area preserving mappings, based on variational methods. Although these methods do not apply to all area preserving homeomorphisms, they apply to a large class of such homeomorphisms, the monotone tilt homeomorphisms. This work is an extension of earlier work of Aubry [3] and myself [12]. Bangert [5], Chenciner [7], and Moser [17, 18] have given very complete expositions of this earlier work and related matters, so I will use this opportunity to announce extensions of this earlier work, which are not yet published.

For simplicity, I will confine the discussion to C^1 monotone twist (area preserving) diffeomorphisms, of an annulus. There is no loss of generality in considering only positive twist diffeomorphisms, since the inverse of such a diffeomorphism is a negative twist diffeomorphism. This is the class of mappings considered, for example, in [12].

A mapping in this class is a C^1 diffeomorphism $\bar{f}$ of the annulus

$$\bar{A} = S^1 \times [0, 1]$$

onto itself which maps each boundary component to itself, preserves area and orientation, and has the "positive twist" property, i.e., for each $\theta \in S^1$, the mapping $y \mapsto \mathrm{pr}_1 \bar{f}(\theta, y)$ has positive derivative at each point, where pr_1 denotes the projection of $S^1 \times [0, 1]$ on its first factor. We let f be a lift of $\bar{f}$ to the universal cover $A = \mathbf{R} \times [0, 1]$ of $\bar{A}$. Then the *rotation interval* $(\rho(f_0), \rho(f_1))$ of

f is defined, where $f_i = f|\mathbf{R} \times i$, $i = 0, 1$, and $\rho(f_i)$ is the Poincaré rotation number of f, i.e., $\rho(f_i) = \lim_{n \to \pm\infty} f_i^n(x)/n$ for any $x \in \mathbf{R}$.

We let $h: B \to \mathbf{R}$ be a "generating function" of f, i.e., $B = \{(x, x') \in \mathbf{R}^2:$ there exists $y \in [0, 1]$ with $\mathrm{pr}_1 f(x, y) = x'\}$, and h is the function defined (up to addition of a constant) by $f(x, y) = (x', y')$ if and only if $y = -\partial_1 h(x, x')$ and $y' = \partial_2 h(x, x')$.

We let $\mathcal{C}$ denote the subset of $\mathbf{R}^{\mathbf{Z}}$ consisting of bi-infinite sequences $x = (x_i)_{i \in \mathbf{Z}} \in \mathbf{R}^{\mathbf{Z}}$ with $(x_i, x_{i+1}) \in B$. We let $\mathcal{M}$ denote the set of $x \in \mathcal{C}$ such that for all $m, n \in \mathbf{Z}$ with $m < n$, we have that $x' \in \mathcal{C}$, $x'_m = x_m$, and $x'_n = x_n$ imply

$$\sum_{i=m}^{n} h(x_i, x_{i+1}) \le \sum h(x'_i, x'_{i+1}).$$

Elements of $\mathcal{C}$ are called *configurations* and elements of $\mathcal{M}$ are alled *minimal energy configurations*. Aubry and Le Daeron [3] have developed a more or less complete theory of minimal energy configurations. See Bangert [5] for a complete exposition of this theory. If x is a minimal energy configuration, the rotation number $\rho(x) = \lim_{i \to \pm\infty} x_i/i$ exists and lies in $[\rho(f_0), \rho(f_1)]$. We define the *rotation symbol* $\tilde{\rho}(x)$ of x to be $\rho(x)$ if $\rho(x)$ is irrational or if $\rho(x) = p/q$ and $x_{i+q} = x_i + p$ for all $i \in \mathbf{Z}$. If $\rho(x) = p/q$ and $x_{i+q} > x_i + p$ for all $i \in \mathbf{Z}$, we set $\tilde{\rho}(x) = p/q+$. If $\rho(x) = p/q$ and $x_{i+q} < x_i + p$ for all $i \in \mathbf{Z}$, we set $\tilde{\rho}(x) = p/q-$. According to the theory of Aubry and Le Daeron, one of these three possibilities always holds for $x \in \mathcal{M}$. If ω is a rotation symbol, we let $\mathcal{M}_\omega$ denote the set of minimal energy configurations of rotation symbol ω. We let $\Phi_\omega = p_0(\mathcal{M}_\omega)$, where $p_0(x) = x_0$. If $\omega \in \mathbf{R}$, then Φ_ω is a closed subset of $\mathbf{R}$. Also $\mathrm{cl}\,\Phi_{p/q+} = \Phi_{p/q+} \cup \Phi_{p/q}$, $\mathrm{cl}\,\Phi_{p/q-} = \Phi_{p/q-} \cup \Phi_{p/q}$, where cl means "closure." These results are due to Aubry and Le Daeron [3]. Bangert [5] has explained them clearly.

This machinery permits us to define "Peierls's energy barrier" $P_\omega(\xi)$ for a real number ξ and a rotation symbol ω, whose underlying number is in the rotation interval of f. If $\xi \in \mathrm{cl}\,\Phi_\omega$, we set $P_\omega(\xi) = 0$. Otherwise, we let (a, b) be the complementary interval of $\mathrm{cl}\,\Phi_\omega$ which contains ξ. By the theory of Aubry and Le Daeron, there exist $x, y \in \mathcal{M}_\omega$ such that $x_0 = a$ and $y_0 = b$. Moreover, $y_i > x_i$ for all $i \in \mathbf{Z}$, and $\sum_{i \in A} y_i - x_i \le 1$, where $A = \mathbf{Z}$ if ω is an irrational number or of the form $p/q+$ or $p/q-$ and $A = \{0, \ldots, q-1\}$ if $\omega = p/q$. We set

$$P_\omega(\xi) = \min\left\{\sum_{i \in A} h(z_i, z_{i+1}) - h(x_i, x_{i+1})\right\},$$

where z ranges over all configurations such that $x_i \le z_i \le y_i$ and $z_0 = \xi$. This was defined and called Peierls's energy barrier in Aubry, Le Daeron, and André [4]. See also Mather [13], where the basic properties of $P_\omega(\xi)$ are developed. I defined a closely related quantity ΔW_ω in [14], where I showed that as a function of the number ω, this quantity is continuous at irrationals, although it is discontinuous at rational ω, for generic f. The definition of ΔW_ω may be extended to rotation symbols ω, and then the functions $\omega \mapsto \Delta W_\omega$ and

$\omega \mapsto P_\omega(\xi)$ are continuous on the space of rotation symbols. We provide the set of rotation symbols with the topology associated to the obvious order. In this topology, rational numbers are isolated points. The intervals $[p/q+, p/q+\varepsilon)$ in the set of rotation symbols form a basis of neighborhoods of $p/q+$, where ε ranges over all positive numbers. The continuity of these functions follows from [14] or slight extensions of the results of [14]. Its importance derives from the fact I proved in [14] that there is an invariant circle for f of rotation number ω (where ω is irrational) if and only if $\Delta W_\omega = 0$, or equivalently, P_ω vanishes identically.

Recently, I have improved these results, to give moduli of continuity for ΔW_ω or $P_\omega(\xi)$, as functions of ω. It is easy to see that there exists $C > 0$ such that $|P_\omega(\xi) - P_\omega(\xi')| \leq C|\xi - \xi'|$, for all $\xi, \xi' \in \mathbf{R}$. The dependence on ω, however, is more complicated. For $P_\omega(\xi)$, we have

$$|P_{p/q}(\xi) - P_\omega(\xi)| \leq C(q^{-1} + |q\omega - p|),$$

where C depends only on f. Moreover,

$$|P_{p/q+}(\xi) - P_\omega(\xi)| \leq C|q\omega - p|, \quad \text{if } \omega > p/q,$$
$$|P_{p/q-}(\xi) - P_\omega(\xi)| \leq C|q\omega - p|, \quad \text{if } \omega < p/q.$$

There are similar estimates for ΔW_ω.

Using these estimates, I have been able to prove that if ω is a Liouville number, then there is a dense set $\mathcal{D}$ in the space of C^∞ monotone twist diffeomorphisms such that a homotopically nontrivial invariant circle of a diffeomorphism in $\mathcal{D}$ has rotation number ω. This is a converse of well-known results in KAM theory. The proof of this result is based on the theorem of Mather [14] that f has an invariant circle of rotation number ω if and only if $\Delta W_\omega = 0$, and this holds if and only if P_ω vanishes identically.

In another direction, I have shown that in a certain sense it is possible to "shadow" minimal energy orbits in a fixed Birkhoff region of instability by local minimal energy orbits. Recall that a minimal energy configuration x is *stationary*, in the sense that $\partial_1 h(x_{i-1}, x_i) + \partial_2 h(x_i, x_{i+1}) = 0$, and therefore if we set $y_i = -\partial_1 h(x_i, x_{i+1})$, we have $f(x_i, y_i) = (x_{i+1}, y_{i+1})$. Thus, to every minimal energy configuration, we may associate an orbit, and we call the resulting orbit a *minimal energy orbit*. Consider two homotopically nontrivial invariant circles which do not intersect, so they bound an annulus. If the annulus which they bound contains no invariant circle, then the region between the circles is called a *Birkhoff region of instability*. A *local minimal energy configuration* x minimizes in the same sense that a minimal energy configuration minimizes, but only for small perturbations of x. Local minimal energy orbits are the orbits corresponding to local minimal energy configurations. Then we have the following result: given a sequence $(\mathcal{O}_i)_{i\in\mathbf{z}}$ of minimal energy orbits, all in the same Birkhoff region of instability, and numbers $\varepsilon_i > 0$, there is a local minimal energy orbit $\mathcal{O} = (P_j)_{j\in\mathbf{z}}$ and an increasing sequence $(n_i)_{i\in\mathbf{z}}$ of integers, such that dist.$(P_{n(i)}, \mathcal{O}_i) < \varepsilon_i$, i.e., $\mathcal{O}$ comes as close as we please to each orbit $\mathcal{O}_i$ in turn. Proofs will appear elsewhere.

REFERENCES

1. V. I. Arnold and A. Avez, *Ergodic problems of classical mechanics*, Benjamin, New York, 1968.

2. D. V. Anosov, *Geodesic flows on closed Riemann manifolds with negative curvature*, Proc. Steklov Inst. Math., vol. 90, Amer. Math. Soc., Providence, R.I., 1969.

3. S. Aubry and P. Y. Le Daeron, *The discrete Frenkel-Kontorova model and its extensions*, Phys. D **8** (1983), 381–422.

4. S. Aubry, P. Y. Le Daeron, and G. André, *Classical ground-states of a one-dimensional model for incommensurate structures*, Preprint, 1982.

5. V. Bangert, *Mather sets for twist maps and geodesics on tori*, Preprint, 1968.

6. R. Bowen, *Equilibrium states and the ergodic theory of Anosov diffeomorphisms*, Lecture Notes in Math., vol. 470, Springer-Verlag, Berlin-New York, 1975.

7. A. Chenciner, *La dynamique au voisinage d'un point fixe elliptique conservatif: de Poincaré et Birkhoff á Aubry et Mather*, Sém. Bourbaki, Exp. 622, Astérisque, no. 121/122, Soc. Math. France, Paris, 1985, pp. 147–170.

8. J. Hadamard, *Les surfaces á courbures opposés et leurs lignes géodesiques*, J. Math. **5** (1898), 27–73.

9. M. R. Herman, *Introduction á l'étude des courbes invariantes par les diffeomorphismes de l'anneau*, Astérisque, no. 103/104, Soc. Math. France, Paris, 1983.

10. A. N. Kolmogorov, *Théorie générale des systémes dynamiques et mecanique classique*, Proc. Internat. Congr. Math. (Amsterdam, 1954), Vol. 1, North-Holland, Amsterdam, 1957, pp. 315–333.

11. A. J. Lichtenberg and M. A. Lieberman, *Regular and stochastic motion*, Springer-Verlag, New York, 1983.

12. J. N. Mather, *Existence of quasi-periodic orbits for twist homeomorphisms of the annulus*, Topology **21** (1982), 457–467.

13. ____, *More Denjoy minimal sets for area preserving diffeomorphisms*, Comment. Math. Helv. **60** (1985), 508–557.

14. ____, *A criterion for the non-existence of invariant circles*, Inst. Hautes Études Sci. Publ. Math. (to appear).

15. J. Moser, *Stable and random motions in dynamical systems*, Ann. of Math. Studies, no. 77, Princeton Univ. Press, Princeton, N.J., and Univ. of Tokyo Press, Tokyo, 1973.

16. ____, *Quasi-periodic solutions in the three-body problem*, Bull. Astr. **3** (1968), 53–59.

17. ____, *Break-down of stability*, Preprint, Forschungsinst. für Math. ETH Zürich, March, 1985.

18. ____, *Recent developments in the theory of Hamiltonian systems*, Preprint, Forschungsinst. für Math. ETH Zürich, September, 1985.

19. Ya. B. Pesin, *Characteristic Lyapunov exponents and smooth ergodic theory*, Russian Math. Surveys **32** (1977), 55–114.

20. H. Poincaré, *Les méthodes nouvelles de la mécanique céleste*, Paris, 1893.

21. S. Smale, *Differentiable dynamical systems*, Bull. Amer. Math. Soc. **73** (1967), 747–817.

PRINCETON UNIVERSITY, PRINCETON, NEW JERSEY 08540, USA

Ergodic Properties and Dimensionlike
Characteristics of Strange Attractors
That Are Close to Hyperbolic

YA. PESIN

1. Introduction. At present there is a rather widespread opinion that instability is one of the main reasons for stochasticity in completely deterministic dynamical systems. It is based on rigorous results in studies of stochasticity of hyperbolic and some quasihyperbolic attractors (such as the Lorenz attractor, the Lozi attractor, etc.), and also on an analysis of many of the physical origins where the stochasticity was found numerically. In this connection, for solving the problem whether an attractor Λ of a dynamical system f is stochastic, one can propose the following scheme:

(1) At first the Lyapunov exponents are calculated for a generic initial point (with respect to phase volume) in the neighborhood of Λ. It is worthwhile to mention that numerical procedures for the calculation of Lyapunov exponents which have been elaborated are often unreliable because Lyapunov exponents are only measurable and not continuous functions when we pass from one trajectory to another. It is well known that one can produce by a digital computer only ε-trajectories (for some ε) of the dynamical system and it is not clear in general whether there exists a "real" trajectory near the numerical one and which exponents it will have.

(2) Suppose that for a generic initial point the Lyapunov exponents are nonzero. In order to derive from this the instability of the trajectories on Λ it is necessary to prove a result of the following type: Let μ be a limit measure for a sequence of measures

$$\mu_n = \frac{1}{n} \sum_{k=0}^{n-1} f_*^k \nu,$$

where ν is a smooth initial distribution in the neighborhood of Λ (μ is obviously concentrated on Λ). Then Lyapunov exponents for μ-almost every point $x \in \Lambda$ are not equal to zero. However one can hope in general that the convergence takes place only for the restrictions of measures μ_n to subsets A in the neighborhood of Λ having positive ν-measure (compare with our Theorem 4).

(3) If a variant of the assertion formulated above holds then one can hope to show that the limit measure μ is Bowen-Ruelle-Sinai measure. This fact just implies the stochasticity of Λ.

(4) In systems of ordinary differential equations, describing the models of real physical systems, it is convenient to pass from the phase flow to a Poincaré map of a certain cross-section surface. However, this map is as a rule discontinuous, which creates additional complications for the investigations.

It turns out that the instability producing the stochasticity is in fact in many cases rather strong. This allows us to give some modifications of the scheme described above.

In the present work we introduce and study a new class of maps having the so-called generalized hyperbolic attractors. They are rather strongly unstable. In the linear approximation their instability is as strong as it is in classical hyperbolic attractors. However the maps considered here are discontinuous on some closed set (this set is usually the union of a finite number of submanifolds). There are trajectories which come very often "anomalously near" the discontinuity set. Although the set of such trajectories is small, their existence implies a weakening of hyperbolicity. In fact the hyperbolicity of our maps is as weak as one encounters in systems with nonzero Lyapunov exponents.

Our class of maps is described by some axioms. It is important to point out that in principle most of our conditions can be checked by a digital computer. For example, the hyperbolicity conditions are given by means of an invariant system of cones (i.e., by requirements on differentials of maps). Such an approach is due to V. M. Alekseev [1], Ya. G. Sinai [16], and D. V. Anosov [2] and was developed for the attractors by V. S. Afraimovich, V. V. Bikov, and L. P. Shilnikov in [3]. There are some conditions estimating the rate of growth of the differential of the system in the neighborhood of the discontinuity set. They have the same meaning as the analogous requirements in the definition of general systems with singularities (cf. [8]).

The aim of our work is to describe the ergodic properties of the dynamical systems having the generalized hyperbolic attractors. In particular we will prove the existence of the so-called Gibbs u-measures—the invariant measures for which induced conditional measures on unstable layers are absolutely continuous with respect to natural Lebesgue measure on these layers. Thus we will show that our systems are really stochastic.

We also consider a very interesting problem of the calculation of the dimension of the attractors. At present there are many dimensionlike characteristics which were introduced for the description of topological-geometric structure of the attractors. In [13] we gave a general construction, a generalization of the famous Carathéodory construction, which allows us to get a functional family of the dimensionlike characteristics. Among them are both well-known characteristics (for example, Hausdorff dimension, capacity) and new ones (for example, the so-called dimension with respect to the dynamical system). If μ is an invariant probability measure on the attractor then one can introduce the dimensionlike characteristics with respect to μ (measure Hausdorff dimension, measure capacity, cf. §5). We will formulate some results connecting them with Lyapunov exponents and measure theoretical entropy. These results are in agreement with a popular hypothesis which was suggested by H. Mori (cf. [11]) and discussed in [7]. It can be formulated as

$$\dim_H \mu = L(\mu), \tag{1}$$

where $\dim_H \mu$ is measure Hausdorff dimension and $L(\mu)$ is the so-called Lyapunov dimension which is uniquely defined by the spectrum of Lyapunov exponents of μ (we assume that μ is ergodic). The formula (1) was proven for the two-dimensional case by L.-S. Young in [17]. I do not know whether formula (1) is true for the multidimensional case but at any rate it is not true in general for the discontinuous maps, even for three-dimensional attractors. This can be explained by the following arguments. The topological structure of the hyperbolic attractor is that it is made of whole unstable layers of its points and its intersection with every stable layer is a Cantor set type.

However, because of the difference in the rates of the contractions in different directions of stable layers, the corresponding sets of n-rank are the "strong stretched ellipsoids." When we calculate the Huasdorff dimension we replace such ellipsoids by choosing balls in the appropriate way and making "best" packings of them. If ellipsoids are distributed in the stable layers "uniformly" enough, then this procedure of the calculation of the dimension allows us to get formula (1). For discontinuous maps this is not true in general. But we can use another method and introduce a new notion of the dimension taking as the "best" packings just ones made of the ellipsoids. Such an approach leads us to the dimension with respect to a dynamical system mentioned above. We will obtain a formula for it similar to (1). One should remark that this dimension is not pure geometric characteristic because it depends on the dynamical system.

2. Definition of generalized hyperbolic attractors; local properties.
Let M be a smooth compact Riemannian manifold, K be an open subset in M, and N^+ be a closed subset in K. Let also $f\colon K\backslash N^+ \to K$ be a map satisfying the following hypotheses:

(H1) f is a C^2-diffeomorphism from $K\backslash N^+$ onto its image $f(K\backslash N^+)$;

(H2) There exist $C_i > 0$, $\alpha_i > 0$, $i = 1, 2, 3, 4$, such that for any $x \in K\backslash N^+$

$$\|df_x\| \le C_1\rho(x, N^+)^{-\alpha_1}, \qquad \|d^2 f_x\| \le C_2\rho(x, N^+)^{-\alpha_2},$$

and for any $x \in f(K\backslash N^+)$

$$\|df_x^{-1}\| \le C_3\rho(x, N^-)^{-\alpha_3}, \qquad \|d^2 f_x^{-1}\| \le C_4\rho(x, N^-)^{-\alpha_4},$$

where ρ is a distance in M and $N^- = \{y \in K$: there exist $z \in N^+$ and $z_n \in K\backslash N^+$ such that $z_n \to z$, and $f(z_n) \to y\}$.

We set $N = N^+ \cup N^- \cup \partial K$ and define by induction the sets $K_0 = K$, $K_n = f(K_{n-1}\backslash N^+)$, $n = 1, 2, \dots$. Set

$$D = \bigcap_{n\ge 0} K_n, \qquad \Lambda = \overline{D}.$$

It is easy to see that $D \cap N = \varnothing$, the maps f, f^{-1} are defined on D, and $f(D) = D$, $f^{-1}(D) = D$.

Consider the continuous function $\varphi(z) = \rho(z, N)$, $z \in D$, and define

$$\chi_\varphi^\pm(z) = \lim_{n \to \pm\infty} \sup \frac{1}{n} \ln \varphi(f^n(z)).$$

It is obvious that functions $\chi_\varphi^\pm(z)$ are both f and f^{-1}-invariant and $\chi_\varphi^\pm(z) \leq 0$. We define the sets

$$D^\pm = \left\{ z \in D \colon \chi_\varphi^\pm(z) = 0 \right\}, \qquad D_0 = D^+ \cap D^-,$$

which are obviously both f and f^{-1}-invariant.

For $A \subset \Lambda$ denote $f^{-1}(A) = \{ z \in \Lambda \backslash N^+ \colon f(z) \in A \}$. A measure μ on Λ is called f-invariant if $\mu(A) = \mu(f^{-1}(A))$ for any $A \subset \Lambda$. Denote by M_f the set of all f-invariant Borel probability measures concentrated on Λ. We say that $\mu \in M_f$ is regular if

$$(1) \quad \mu(N) = 0, \qquad (2) \quad \left| \int_\Lambda \ln^+ \varphi(z) \, d\mu(z) \right| < \infty.$$

If $x \in D$ is a periodic point with a period p then the measure having the value $1/p$ at every point $f^i(x)$, $i = 1, \ldots, p$, is obviously regular.

PROPOSITION 1. *If $\mu \in M_f$ is a regular measure then $\mu(D_0) > 0$.*

We say that Λ is a regular attractor if there is a regular measure $\mu \in M_f$. Define the sets

$$\tilde{D}^\pm = \left\{ z \in D^\pm \colon \text{ there exists the limit } \chi_\varphi^\pm(z) \right.$$

$$\left. = \lim_{n \to \pm\infty} (1/n) \ln \varphi(f^n(z)) \right\},$$

$$\tilde{D}_0 = \tilde{D}^+ \cap \tilde{D}^-,$$

which are both f and f^{-1}-invariant. Moreover for any $\varepsilon > 0$ and $z \in \tilde{D}^+$ (respectively $z \in \tilde{D}^-$ or $z \in \tilde{D}_0$) there exists $C(\varepsilon, z) > 0$ such that for every $n > 0$ (respectively $n < 0$ or $n \in \mathbf{Z}$)

$$\rho(f^n(z), N) = \varphi(f^n(z)) \geq C(\varepsilon, z) \exp(-\varepsilon|n|).$$

PROPOSITION 2. *For any $\mu \in M_f$*
(1) $\mu(\tilde{D}^+) = \mu(\tilde{D}^-) = \mu(\tilde{D}_0) = \mu(D^+) = \mu(D^-) = \mu(D_0)$;
(2) *If Λ is a regular attractor then $\tilde{D}_0 \neq \varnothing$.*

By a cone $K(z, \alpha, P)$ ($z \in K$, $\alpha > 0$, P is a subspace in $T_z M$) we mean the set $\{ v \in T_z M \colon \measuredangle(v, P) = \min_{w \in P} \measuredangle(v, w) \leq \alpha \}$.

We call an attractor Λ *generalized hyperbolic* if there exist $C > 0$, $0 < \lambda < 1$, a continuous function $\alpha(z) > 0$ and two continuous families of subspaces $P^{(s)}(z)$, $P^{(u)}(z) \subset T_z M$, $z \in K$, such that
(1) $\dim P^{(s)}(z) = q$, $\dim P^{(u)}(z) = p$, $q + p = \dim M$;
(2) $K(z, \alpha(z), P^{(s)}(z)) \cap K(z, \alpha(z), P^{(u)}(z)) = \{0\}$;

(3) for any $z \in K \backslash N^+$

$$df(K(z, \alpha(z), P^{(u)}(z))) \subset K(f(z), \alpha(f(z)), P^{(u)}(f(z))),$$

and for any $z \in f(K \backslash N^+)$

$$df^{-1}(K(z, \alpha(z), P^{(s)}(z))) \subset K(f^{-1}(z), \alpha(f^{-1}(z)), P^{(s)}(f^{-1}(z)));$$

(4) for any $n > 0$, $z \in K \backslash N^+$, $v \in K(z, \alpha(z), P^{(u)}(z))$,

$$\|df^n v\| \geq C^{-1} \lambda^{-n} \|v\|,$$

and for any $n > 0$, $z \in K_n$, $v \in K(z, \alpha(z), P^{(s)}(z))$,

$$\|df^{-n} v\| \leq C \lambda^n \|v\|.$$

Now one can obtain in the usual way the uniformly hyperbolic structure on D given by two families of f-invariant and continuous stable and unstable subspaces $E^{(s)}(z)$, $E^{(u)}(z) \subset T_z M$, $z \in D$, where

$$E^{(s)}(z) = \bigcap_{n \geq 0} df^{-n} \left(K(f^n(z), \alpha(f^n(z)), P^{(s)}(f^n(z))) \right),$$

$$E^{(u)}(z) = \bigcap_{n \geq 0} df^n \left(K(f^{-n}(z), \alpha(f^{-n}(z)), P^{(u)}(f^{-n}(z))) \right).$$

Then using hypotheses (H1) and (H2) one can construct in the familiar way (cf. [10, 4]) for any $z \in \tilde{D}^+$ the local stable layer $V^{(s)}(z)$ and for any $z \in \tilde{D}^-$ the local unstable layer $V^{(u)}(z)$ (of course, for $z \in \tilde{D}_0$ both of these layers are defined). The stable layer $V^{(s)}(z)$ is characterized by the following condition: for any $y \in V^{(s)}(z)$, $n \geq 0$,

$$\rho(f^n(z), f^n(y)) \leq C(z) \mu^n \rho(y, z),$$

where $\mu \in (\lambda, 1)$ and $C(z) > 0$ are constants. The unstable layer $V^{(u)}(z)$ is characterized in the same way (taking $-n$ instead of n). We remark that the "sizes" of local layers are in general measurable and not continuous functions on D because f has the discontinuity set. Therefore, just as the hyperbolic structure on D for the linear approximation of f is uniform, the situation for the map f itself is similar to one in the systems with nonzero Lyapunov exponents and can be studied by the methods developed in the theory of such systems (cf. [14]). In this theory an important role belongs to the "sets with uniform estimations"; these are the sets $\tilde{D}_l^\pm \subset \tilde{D}^\pm$ such that for every point $x \in \tilde{D}_l^\pm$ the "size" of local layers are bigger than some $\delta_l > 0$ (and $\delta_l \to 0$ when $l \to \infty$ with a small exponential rate). One can show the following assertion.

PROPOSITION 3. $V^{(u)}(z) \subset D$ for every $z \in \tilde{D}^-$.

Thus an attractor Λ consists of local unstable layers (in this we suppose that $V^{(u)}(z) = \{z\}$ for $z \notin \tilde{D}^-$).

Let $A \subset \Lambda$. Define $\hat{f}(A) = f(A \backslash N^+)$, $\hat{f}^{-1}(A) = f^{-1}(A \backslash N^-)$. The sets $\hat{f}^n(A), \hat{f}^{-n}(A)$ for $n > 0$ are defined in the same way. For $z \in \tilde{D}^+$ we set

$$W^{(s)}(z) = \bigcup_{n \geq 0} \hat{f}^{-n} \left(V^{(s)}(f^n(z)) \right),$$

and for $z \in \tilde{D}^-$ we set

$$W^{(u)}(z) = \bigcup_{n \geq 0} \hat{f}^n \left(V^{(u)}(f^{-n}(z)) \right).$$

These sets are respectively called global stable and unstable layers at point z. They may be not connected in general.

3. Gibbs μ-measures: existence and ergodic properties. Let $z \in \tilde{D}^-$. Denote by $\nu^{(u)}(z)$ the measure in $W^{(u)}(z)$ induced by the restriction of the Riemannian metric to $W^{(u)}(z)$ (which is a smooth submanifold in M). We use the notation $J^{(u)}(z)$ for the Jacobian of the map $df \mid E^{(u)}(z)$. Fix $z \in \tilde{D}^-$, $y \in W^{(u)}(z)$, $n > 0$, and let

$$\kappa_n(z, y) = \prod_{k=0}^{n-1} \left[J^{(u)}(f^{-k}(z)) \right] \left[J^{(u)}(f^{-k}(y)) \right]^{-1}.$$

It is not difficult to show that the limit

$$\kappa(z, y) = \lim_{n \to \infty} \kappa_n(z, y) > 0$$

exists.

Let μ be a Borel probability measure on Λ. Following [15] we say that μ is Gibbs u-measure if the conditional measures induced by μ on $W^{(u)}(z)$ are absolutely continuous with respect to $\nu^{(u)}(z)$ and the corresponding density at the point $y \in W^{(u)}(z)$ is equal to $\kappa(z, y)$ (up to a normalized factor).

We shall be dealing with generalized hyperbolic attractors Λ that satisfy the following hypothesis:

(H3) There exist $C > 0$, $\varepsilon_0 > 0$, and $z \in \tilde{D}^-$ such that for any $\varepsilon \in (0, \varepsilon_0]$ and $n \geq 0$

$$\nu^{(u)}(V^{(u)}(z) \cap f^{-n}(U(\varepsilon, N^+))) \leq C\varepsilon,$$

where $U(\varepsilon, N^+)$ is an ε-neighborhood of N^+ in M.

In the following we will assume that the map f satisfies (H1) and (H2) and Λ is the regular generalized hyperbolic attractor for f, satisfying (H3).

THEOREM 1. *There exists Gibbs u-measure $\mu \in M_f$ for which $\mu(\tilde{D}^-) = 1$.*

In order to construct Gibbs u-measure we consider the point z for which hypothesis (H3) holds and measure ν on Λ given by

$$\nu(A) = \nu^{(u)} \left(A \cap V^{(u)}(z) \right) \quad \text{for } A \subset \Lambda.$$

Further, let

$$\mu_n(A) = \frac{1}{n} \sum_{k=0}^{n-1} \nu \left(f^{-k}(A) \right).$$

Then there is a sequence $n_i \to \infty$ such that the subsequence of measures μ_{n_i} converges (in the weak star topology) to f-invariant Gibbs u-measure.

Denote by M the class of all Gibbs u-measures $\mu \in M_f$ for which $\mu(\tilde{D}^-) = 1$. Ergodic properties of such measures follow from [9] (cf. also [14] where we considered the case when μ is a smooth measure; but using the easy modifications of arguments given there it is not difficult to get the corresponding results for Gibbs u-measures).

THEOREM 2. Let $\mu \in M_f^{(u)}$. Then the set Λ is decomposed on nonintersecting sets Λ_i, $i = 0, 1, 2, \ldots$, such that for $i > 0$

(1) $\Lambda_i \subset D$, $\mu(\Lambda_i) > 0$, $f(\Lambda_i) = \Lambda_i$, and $f \mid \Lambda_i$ is ergodic;

(2) the set Λ_i is decomposed on nonintersecting sets A_i^j, $j = 1, \ldots, n_i$, such that $f(A_i^j) = A_i^{j+1}$ for $j = 1, \ldots, n_i - 1$, $f(A_i^{n_i}) = A_i^1$, and $f^{n_i} \mid A_i^1$ is isomorphic to a Bernoulli automorphism.

THEOREM 3. Let $\mu \in M_f^{(u)}$. Then for measure theoretical entropy $h_\mu(f)$ the "formula for entropy" (cf. [9, 14])

$$h_\mu(f) = -\int_\Lambda \sum_{i=1}^{s(x)} q_i(x)\chi_i(x)\,d\mu(x)$$

holds, where $\chi_1(x) > \cdots > \chi_{s(x)}(x)$ are positive values of Lyapunov exponents at x and $q_i(x)$ is the multiplier of the value $\chi_i(x)$.

Following [12] we call the set

$$W^{(s)}(A) = \bigcup_{z \in \tilde{D}^+ \cap A} W^{(s)}(z)$$

the realm of attraction of the set $A \subset \Lambda$.

THEOREM 4. Let $\mu \in M_f^{(u)}$ be an ergodic measure. Then

(1) $\operatorname{meas}(W^{(s)}(\Lambda)) > 0$ (meas denotes Riemannian volume in M);

(2) there exists $A \subset \Lambda$ such that $\mu(A) = 1$ and for any $z \in W^{(s)}(A)$ and any continuous function φ on K

$$\lim_{n \to \infty} \frac{1}{n} \sum_{k=0}^{n-1} \varphi\left(f^k(z)\right) = \int_\Lambda \varphi\,d\mu.$$

This assertion means that μ is Bowen-Ruelle-Sinai measure.

THEOREM 5. The set Λ is decomposed on nonintersecting sets Λ_n, $n = 0, 1, 2, \ldots$, such that

(1) $\operatorname{meas}(W^{(s)}(\Lambda_n) \cap W^{(s)}(\Lambda_m)) = 0$ for $n \neq m$, $n, m > 0$;

(2) for $n > 0$ we have $\Lambda_n \subset D$, $f(\Lambda_n) = \Lambda_n$, and there exist $\mu_n \in M_f^{(u)}$ for which $\mu_n(\Lambda_n) = 1$ and $f \mid \Lambda_n$ is ergodic;

(3) if $\mu \in M_f^{(u)}$ then

$$\mu = \sum_{n \geq 1} \alpha_n \mu_n, \qquad \sum_{n \geq 1} \alpha_n = 1, \qquad \alpha_n \geq 0;$$

(4) if ν is a smooth initial distribution in a neighborhood of Λ and $\nu_n = \nu \mid W^{(s)}(\Lambda_n), n > 0$, then

$$\lim_{k \to \infty} \frac{1}{k} \sum_{i=1}^{k-1} f_*^i \nu_n = \mu_n$$

in the weak star topology.

4. Topological properties of generalized hyperbolic attractors. Our approach to the study of topological properties is unlike the usual one because it is based on the description of the metric properties given above and information on the structure of stable lamination.

THEOREM 6. *Suppose that lamination* $W^{(s)}$ *composed of layers* $W^{(s)}(z)$, $z \in \tilde{D}^+$, *is continuous. Then*

(1) *the set* Λ_n *(constructed in Theorem 3) for* $n > 0$ *is* $(\mu_n\text{-mod}\,0)$ *open (with respect to the topology in* Λ *induced by the topology in* M);

(2) *the set* $W^{(s)}(\Lambda_n)$, $n > 0$, *is* (meas-mod 0) *open*;

(3) *the map* $f \,|\, \Lambda_n$, $n > 0$, *is topologically transitive*;

(4) *periodic points of* f *are everywhere dense in* Λ_n *for* $n > 0$.

The sets $\overline{\Lambda}_n$, $n > 0$, are similar to basic sets for Axiom A diffeomorphisms. Theorem 6 is also true with the weaker assumption that $W^{(s)}$ is a $(\delta(x), k)$-continuous lamination (definition, cf. [**14**]).

It is worthwhile to point out that we describe topological behavior of f only on the "essential" part of Λ: we do not know the topological behavior of f on Λ_0.

It directly follows from Theorem 6 that if f is topologically transitive (on D) then there exists unique Gibbs u-measure on Λ. Nevertheless even in this case $W^{(s)}(\Lambda)$ being the open set (meas-mod 0) does not form in general a neighborhood of Λ. Therefore, if ν is a smooth initial distribution in the neighborhood of Λ then we can describe only the evolution of the restriction $\nu \,|\, W^{(s)}(\Lambda)$ (which converges to unique Gibss u-measure on Λ). However we know nothing about the evolution of ν elsewhere in the neighborhood.

5. Dimensionlike characteristics for generalized hyperbolic attractors. For $Z \subset \Lambda$ denote by $\dim_H Z$ Hausdorff dimension on Z and define upper and lower capacities of Z as follows:

$$\overline{C}(Z) = \limsup_{\varepsilon \to 0} \frac{\ln N(Z, \varepsilon)}{\ln(1/\varepsilon)}, \qquad \underline{C}(Z) = \liminf_{\varepsilon \to 0} \frac{\ln N(Z, \varepsilon)}{\ln(1/\varepsilon)},$$

where $N(Z, \varepsilon)$ is the smallest number of balls covering Z with radii $\leq \varepsilon$. It is easy to see that $\dim_H Z \leq \underline{C}(Z) \leq \overline{C}(Z)$.

Let μ be a Borel probability measure on Λ. Following [**17**] we define respectively measure Hausdorff dimension, measure upper and lower capacities by the formulae

$$\dim_H \mu = \inf \{\dim_H Z \colon Z \subset X, \, \mu(Z) = 1\},$$
$$\overline{C}(\mu) = \liminf_{\delta \to 0} \{\overline{C}(Z) \colon Z \subset X, \mu(Z) \geq 1 - \delta\},$$
$$\underline{C}(\mu) = \liminf_{\delta \to 0} \{\underline{C}(Z) \colon Z \subset X, \mu(Z) \geq 1 - \delta\}. \qquad (2)$$

Let ξ be a finite partition of Λ. We set

$$H_\mu(\xi) = -\sum \mu(C_\xi) \ln \mu(C_\xi),$$
$$H_\mu(\varepsilon) = \inf_\xi \{H_\mu(\xi) : \operatorname{diam} \xi \le \varepsilon\}$$

where $\operatorname{diam} \xi = \max \operatorname{diam} C_\xi$. We call

$$\overline{R}(\mu) = \limsup_{\varepsilon \to 0} \frac{H_\mu(\varepsilon)}{\ln(1/\varepsilon)}, \qquad \underline{R}(\mu) = \liminf_{\varepsilon \to 0} \frac{H_\mu(\varepsilon)}{\ln(1/\varepsilon)},$$

and call them respectively measure upper and lower information dimensions (Rényi dimensions).

For $x \in \Lambda$ we set

$$\bar{d}_\mu(x) = \limsup_{\delta \to 0} \frac{\ln \mu(B(x,\delta))}{\ln \delta},$$
$$\underline{d}_\mu(x) = \liminf_{\delta \to 0} \frac{\ln \mu(B(x,\delta))}{\ln \delta},$$

and call them respectively measure upper and lower pointwise dimensions. The following assertion is proved as in [17].

THEOREM 7. *Suppose that* $\dim M = 2$. *Then*

(1) *if* $\mu \in M_f$ *is an ergodic regular measure then for almost every* $x \in \Lambda$

$$\bar{d}_\mu(x) = \underline{d}_\mu(x) = h_\mu(f)/(1/\chi_\mu^1 - 1/\chi_\mu^2) \overset{\mathrm{def}}{=} d_\mu$$

where $\chi_\mu^1 > 0 > \chi_\mu^2$ *are the average values of Lyapunov exponents with respect to* μ; *moreover*

$$\dim_H \mu = \underline{C}(\mu) = \overline{C}(\mu) = \underline{R}(\mu) = \overline{R}(\mu) = d_\mu;$$

(2) *if* $\mu \in M_f^{(u)}$ *is an ergodic measure then*

$$d_\mu = 1 + \chi_\mu^1/|\chi_\mu^2| = L(\mu).$$

We consider the case when $t = \dim M > 2$. Now the above results are not true in general. Therefore we use dimensionlike characteristics, introduced in [13] (cf. also [4]).

Fix $\delta > 0$ and denote by

$$B_n(x,\delta) = \left\{ y \in \tilde{D}^- : \rho(f^{-k}(x), f^{-k}(y)) \le \delta, \ k = 0, \dots, n \right\}$$

the Bowen (n,δ)-ball at $x \in \tilde{D}^-$. For $l > 0$ we set

$$F_{\delta,l} = \left\{ B_n(x,\delta) : x \in \tilde{D}_l^-, \ n \in \mathbf{Z}^+ \right\}, \qquad \delta \le \delta_l,$$

and define for given $\alpha > 0$, $N > 0$, $Z \subset \tilde{D}_l^-$

$$M(\alpha, N, Z) = \inf_{G \subset F_{\delta,l}} \left\{ \sum_{B_n(x,\delta) \in G} \operatorname{meas}(B_n(x,\delta))^\alpha : \right.$$

$$\left. n \ge N, \ \bigcup_{B_n(x,\delta) \in G} B_n(x,\delta) \subset Z \right\}$$

where G is a finite or countable subset in $F_{\delta,l}$. It is obvious that the function $M(\alpha,\cdot,Z)$ does not decrease when $N \to \infty$. Therefore the limit

$$m(\alpha, Z) = \lim_{N \to \infty} M(\alpha, N, Z)$$

exists. One can show (cf. [4]) that there is α_0 such that $M(\alpha, Z) = \infty$ when $\alpha < \alpha_0$ and $M(\alpha, Z) = 0$ when $\alpha > \alpha_0$. We set

$$\dim_{\delta,l} Z = \alpha_0$$

and call

$$\dim_f Z = \sup_{l>0} \limsup_{\delta \to 0} \dim_{\delta,l} Z$$

the dimension of Z with respect to f.

We define for given $\alpha > 0$, $N > 0$, $Z \subset \tilde{D}_l^-$

$$R(\alpha, N, Z) = \inf_{G \subset F_{\delta,l}} \left\{ \sum_{B_N(x,\delta) \in G} \operatorname{meas}(B_N(x,\delta))^\alpha : \right.$$

$$\left. \bigcup_{B_N(x,\delta) \in G} B_N(x,\delta) \supset Z \right\}$$

where G is a finite or countable subset in $F_{\delta,l}$. Let

$$\bar{r}(\alpha, Z) = \limsup_{N \to \infty} R(\alpha, N, Z), \qquad \underline{r}(\alpha, Z) = \liminf_{N \to \infty} R(\alpha, N, Z).$$

Functions $\bar{r}(\cdot, Z)$, and $\underline{r}(\cdot, Z)$ have the following property: there are $\bar{\alpha}$ and $\underline{\alpha}$ such that

$$\bar{r}(\alpha, Z) = \begin{cases} \infty & \text{if } \alpha < \bar{\alpha}, \\ 0 & \text{if } \alpha > \bar{\alpha}; \end{cases} \qquad \underline{r}(\alpha, Z) = \begin{cases} \infty & \text{if } \alpha < \underline{\alpha}, \\ 0 & \text{if } \alpha > \underline{\alpha}. \end{cases}$$

We set

$$\overline{C}_{\delta,l}(Z) = \bar{\alpha}, \qquad \underline{C}_{\delta,l}(Z) = \underline{\alpha}$$

and call

$$\overline{C}_f(Z) = \sup_{l>0} \limsup_{\delta \to 0} \overline{C}_{\delta,l}(Z),$$

$$\underline{C}_f(Z) = \sup_{l>0} \limsup_{\delta \to 0} \underline{C}_{\delta,l}(Z)$$

upper and respectively lower capacities of Z with respect to f. It is easy to see that

$$\dim_f Z \leq \underline{C}_f(Z) \leq \overline{C}_f(Z). \tag{3}$$

Let μ be a Borel probability measure on Λ. We define now like (2)

$$\dim_f \mu = \inf \{ \dim_f Z : Z \subset \Lambda, \ \mu(Z) = 1 \},$$

$$\overline{C}_f(\mu) = \liminf_{\delta \to 0} \{ \overline{C}_f(Z) : Z \subset \Lambda, \ \mu(Z) \geq 1 - \delta \},$$

$$\underline{C}_f(\mu) = \liminf_{\delta \to 0} \{ \underline{C}_f(Z) : Z \subset \Lambda, \ \mu(Z) \geq 1 - \delta \},$$

which are called respectively measure dimension with respect to f, measure upper and lower capacities with respect to f. It follows from (3) that

$$\dim_f \mu \leq \underline{C}_f(\mu) \leq \overline{C}_f(\mu).$$

THEOREM 8. *Suppose that the Jacobian of f is uniformly bounded in K. Then*

(1) *if $\mu \in M_f$ is an ergodic regular measure then*

$$\dim_f \mu = \underline{C}_f(\mu) = \overline{C}_f(\mu) = -h_\mu(f) \bigg/ \sum_{i=s+1}^{t} \chi_\mu^i;$$

(2) *if $\mu \in M_f^{(u)}$ is an ergodic measure then*

$$\dim_f \mu = \underline{C}_f(\mu) = \overline{C}_f(\mu) = -\sum_{i=1}^{s} \chi_\mu^i \bigg/ \sum_{i=s+1}^{t} \chi_\mu^i,$$

where $\chi_\mu^1 \geq \cdots \geq \chi_\mu^s > 0 > \chi_\mu^{s+1} \geq \cdots \geq \chi_\mu^t$ are the average values of Lyapunov exponents with respect to μ.

6. Examples of generalized hyperbolic attractors.

1. *Lorenz type attractors.* Let $I = [-1, 1]$, B be a unit ball in $\mathbf{R}^p$, and $K = B \times I$. Let also $0 = a_0 < a_1 < \cdots < a_q < a_{q+1} = 1$. We set

$$P_i = B \times (a_i, a_{i+1}), \qquad i = 0, \ldots, q,$$

$$l = \{a_0, a_1, \ldots, a_q, a_{q+1}\} \times I.$$

Consider an injective map $T: K \backslash l \to K$ given in the form

$$T(x, y) = (f(x, y), g(x, y)), \qquad x \in B, \ y \in I,$$

where f, g satisfy the following conditions:

L1. f, g are continuous in $\overline{P}_i$ and

$$\lim_{y \uparrow a_i} f(x, y) = f_i^-, \qquad \lim_{y \uparrow a_i} g(x, y) = g_i^-,$$

$$\lim_{y \downarrow a_i} f(x, y) = f_i^+, \qquad \lim_{y \downarrow a_i} g(x, y) = g_i^+$$

where $f_i^\pm$ and $g_i^\pm$ do not depend on x, $i = 1, \ldots, q$.

L2. f, $g \in C^2$ in P_i and for $(x, y) \in K \backslash l$

$$C_2 \min |y - a_i|^{-\alpha} \geq |f_x(x, y)|, \ |f_y(x, y)|,$$

$$|g_x(x, y)|, \ |g_y(x, y)| \geq C_1 \min |y - a_i|^\alpha$$

where α, C_1, C_2 are some constants independent of x.

L3. The following inequalities are satisfied:

$$\|f_x\| < 1, \qquad \|g_y^{-1}\| < 1,$$

$$1 - \|g_y^{-1}\| \, \|f_x\| > 2\sqrt{\|g_y^{-1}\| \, \|g_x\| \, \|g_y^{-1} f_y\|},$$

$$\|g_y^{-1} f_y\| \, \|g_x\| < (1 - \|f_x\|)(1 - \|g_y^{-1}\|),$$

where $\| \cdot \| = \max_{i=0,\ldots,q} \sup_{(x,y) \in P_i} | \cdot |.$

L4. $\|f_{xx}\|$, $\|f_{xy}\|$, $\|g_{xy}\|$, $\|g_{xx}\| \leq$ const.

We give one example of a map satisfying these conditions.

THEOREM 9 (CF. [4]).   *Suppose that* $l = \{0\} \times I$ *and functions* $g(x, y)$ *and*

$$f(x, y) = (f_1(x, y), \dots, f_p(x, y)), \qquad x = (x_1, \dots, x_p),$$

are given by the equalities

$$\begin{cases} f_j = (-B_j |y|^{\nu_0} + B_j x_j |y|^{\nu_j} + 1)\mathrm{sgn}\ y, \\ g = ((1+A)|y|^{\nu_0} + A)\mathrm{sgn}\ y. \end{cases}$$

Assume that

$$0 < B_j < \tfrac{1}{2}, \quad \nu_j > 1, \quad j = 1, \dots, p,$$
$$(1+A)^{-1} < \nu_0 < 1, \quad 0 < A < 1.$$

Then T *satisfies conditions* L1–L4.

The following result is a consequence of [**3**, **4**].

THEOREM 10. T *satisfies* (H1) *and* (H2) *and the attractor* Λ *for* T *is the regular generalized hyperbolic attractor satisfying* (H3). *Moreover the lamination* $W^{(s)}$ *is extended to a* C^1-*continuous foliation in* K.

The existence of Gibbs u-measure in the case when the discontinuity set l consists of only one interval (and dim $K = 2$) was essentially proved in [**6**]. The arguments given there are based on the construction of a Markov partition for the Lorenz attractor (in the general case this construction is given in [**4**]). If the lamination $W^{(s)}$ is extended to a smooth foliation in K (in particular this happens when g is independent of x) then the existence of Gibbs u-measures follows from the results of the theory of one-dimensional piecewise monotonic maps (one can show that Λ is isomorphic to the inverse limit of some one-dimensional map which is monotone on the intervals (a_i, a_{i+1}), $i = 0, \dots, q$ (cf. [**4**]). For the arbitrary Lorenz type attractors one can show that any Gibbs u-measure has only a finite number of ergodic components of positive measure; there exists only a finite number of ergodic Gibbs u-measures and respectively a finite number of components of topological transitivity (cf. [**3**, **4**]).

One can prove the following assertion using the results obtained in [**4**].

THEOREM 11.   *For any smooth compact Riemannian manifold* M, dim $M \geq 3$, *there exists a vector field* X *having the following property: There is a smooth submanifold* S *such that the first return map* (*Poincaré map*) T *on* S *induced by the flow given by* X *satisfies the conditions* L1–L4 *and consquently has Lorenz type attractor.*

2. *Generalized Lozi attractors.* Let $c > 0$ and $I = [0, c]$, $K = I \times I$, $0 = a_0 < a_1 < \cdots < a_q < a_{q+1} = 1$. We set

$$l = \{a_0, a_1, \dots, a_q, a_{q+1}\} \times I.$$

Let $T\colon K\backslash l \to K$ be an injective continuous map given in the form

$$T(x,y) = (f(x,y), g(x,y)), \qquad x, y \in I,$$

and satisfying the following conditions:

Loz1. $T \,|\, (K\backslash l)$ is a C^2-diffeomorphism onto its image and both $T\,|\,(K\backslash l)$ and $T^{-1}\,|\,(K\backslash l)$ have bounded second derivative.

Loz2. $\mathrm{Jac}(T) < 1$.

Loz3.

$$\inf\left\{\left(\left|\frac{\partial f}{\partial x}\right| - \left|\frac{\partial f}{\partial y}\right|\right) - \left(\left|\frac{\partial g}{\partial x}\right| + \left|\frac{\partial g}{\partial y}\right|\right)\right\} \geq 0.$$

Loz4.

$$\inf\left\{\left|\frac{\partial f}{\partial x}\right| - \left|\frac{\partial f}{\partial y}\right|\right\} = u > 1.$$

Loz5.

$$\sup\left\{\frac{\left|\frac{\partial f}{\partial x}\right| + \left|\frac{\partial g}{\partial y}\right|}{(\left|\frac{\partial f}{\partial x}\right| - \left|\frac{\partial f}{\partial y}\right|)^2}\right\} < 1.$$

Loz6. There exists $N > 0$ such that $T^k(l) \cap l = \varnothing$ for $1 \leq k \leq N$ and $u^N > 2$.

This class of maps was introduced in [18] and includes the map

$$T(x,y) = (1 + by - a|x|, x) \tag{4}$$

which is obtained from the well-known Lozi map by a change of coordinates. It is easy to verify that there exist open intervals of a and b such that (4) takes some square $[0, c] \times [0, c]$ into itself and satisfies Loz1–Loz6. The ergodic properties of map (4) are described in [10]. The ergodic properties of maps satisfying Loz1–Loz6 follow from our Theorems 1–5 and the following result.

THEOREM 12. *The map T for which conditions* Loz1–Loz6 *hold satisfies* (H1) *and* (H2) *and the attractor Λ for T is the regular generalized hyperbolic attractor satisfying* (H3).

The proof of this theorem follows from [18] (where the existence of Sinai measure for T was also proved).

3. *Belykh attractor.* Let $I = [-1, 1]$, $K = I \times I$ and $l = \{(x,y)\colon y = kx\}$. Consider the map $T\colon K\backslash l \to K$ given in the following form:

$$T(x,y) = \begin{cases} (\lambda_1(x-1)+1, \lambda_2(y-1)+1) & \text{if } y > kx, \\ (\lambda_1(x+1)-1, \lambda_2(y+1)-1) & \text{if } y < kx. \end{cases}$$

This map was introduced in [5].

THEOREM 13. *Suppose that*

$$0 < \lambda_1 < \tfrac{1}{2}, \quad 1 < \lambda_2 < 2, \quad 1 - 2/\lambda_2 < |k| < 1, \quad k \neq 0.$$

Then T satisfies (H1) *and* (H2) *and the attractor Λ for T is the regular generalized hyperbolic attractor satisfying* (H3). *Moreover the lamination $W^{(s)}$ is extended to a continuous $(\delta(x,y), 1)$-foliation in K with some function $\delta(x,y), (x,y) \in K$.*

To prove this theorem it is sufficient to see that T is a piecewise linear map with the differential

$$dT = \begin{pmatrix} \lambda_1 & 0 \\ 0 & \lambda_2 \end{pmatrix}.$$

Moreover it is easy to verify that

$$W^{(s)}(x,y) = \begin{cases} \{(u,y)\colon\ -1 \le u < y/k\} & \text{if } y < kx, \\ \{(u,y)\colon\ y/k < u \le 1\} & \text{if } y > kx; \end{cases}$$
$$W^{(u)}(x,y) = \begin{cases} \{(x,v)\colon\ -1 \le v < kx\} & \text{if } y < kx, \\ \{(x,v)\colon\ kx < v \le 1\} & \text{if } y > kx. \end{cases}$$

V. Afraimovich has informed me that attractor Λ for T contains a countable number of periodic points and consequently Λ is the regular attractor. We also note that Λ is a Lorenz type attractor if $k = 0$. Using Theorems 1–7 and 13 we can obtain a description of the ergodic and topological properties and also dimensionlike characteristics for the Belykh attractor.

References

1. V. M. Alekseev, *Quasirandom dynamical systems*, Mat. USSR-Sbornik **7** (1969), 1–43.

2. D. V. Anosov, *On one class of invariant sets of smooth dynamical systems*, Trudy V Internat. Conf. on Nonlinear Oscillations, vol. 2, Inst. Mat. Akad. Nauk SSSR, Kiev, 1970, pp. 39–45.

3. V. S. Afraimovich, V. V. Bykov and L. P. Shilnikov, *On attracting nonstructurally stable limit sets of type of Lorenz attractor*, Trudy Moskov. Mat. Obshch. **44** (1983), 150–212.

4. V. S. Afraimovich and Ya. B. Pesin, *The dimension of Lorenz type attractors*, Soviet Math.-Phys. Reviews, vol. 6, Gordon and Breach, Harwood Acad. Publ., 1987 (in press).

5. V. M. Belykh, *Qualitative methods of the theory of nonlinear oscillations in point systems*, Gorki Univ. Press, 1980.

6. L. A. Bunimovich and Ya. G. Sinai, *Stochasticity of the attractor in the Lorenz model*, Nonlinear Waves (Proc. Winter School, Moscow), "Nauka", 1980, pp. 212–226.

7. J. D. Farmer, E. Ott and J. A. Yorke, *The dimension of chaotic attractors*, Phys. D **7D** (1983), 153–180.

8. A. Katok and J.-M. Strelcyn, *Smooth maps with singularities: invariant manifolds, entropy and billiards*, Lecture Notes in Math., Springer-Verlag (in press).

9. F. Ledrappier, *Some properties of absolutely continuous invariant measures on an interval*, Ergodic Theory Dynamical Systems **1** (1981), 77–94.

10. Y. Levy, *Ergodic properties of the Lozi map*, Lecture Notes in Math., vol. 1109, Springer-Verlag, 1985, pp. 103–116.

11. H. Mori, *Fractal dimension of chaotic flows autonomous dissipative systems*, Progr. Theoret. Phys. **63** (1980), 1044–1047.

12. J. Milnor, *On the concept of attractor*, Comm. Math. Phys. **99** (1985), 177–195.

13. Ya. B. Pesin, *A generalization of Carathéodory's construction for dimensional characteristic of dynamical systems*, Stat. Phys. and Dynam. Syst., Progr. Phys., Birkhäuser, 1985, pp. 191–202.

14. ____, *Lyapunov characteristic exponents and smooth ergodic theory*, Uspekhi Mat. Nauk **32** (1977), 55–114. (Russian)

15. Ya. B. Pesin and Ya. G. Sinai, *Gibbs measures for partially hyperbolic attractors*, Ergodic Theory Dynamical Systems **2** (1982), 417–438.

16. Ya. G. Sinai, *Several rigorous results concerning the decay of time-correlation functions*, Appendix in *Statistical irreversibility in nonlinear systems* by G. M. Zaslavski, "Nauka", Moscow, 1979.

17. L.-S. Young, *Dimension, entropy and Lyapunov exponents*, Ergodic Theory Dynamical Systems **2** (1982), 109–124.

18. ____, *Bowen-Ruelle measures for certain piecewise hyperbolic maps*, Trans. Amer. Math. Soc. **287** (1985), 41–48.

ALL-UNION EXTRAMURAL CIVIL ENGINEERING INSTITUTE, MOSCOW 109029, USSR

Symbolic Dynamics for Geodesic Flows

CAROLINE SERIES

The subject of symbolic dynamics is of central importance in the modern theory of dynamical systems. The use of symbolic sequences to study dynamical properties of geodesics originates in the work of Koebe [21, 22] and Morse [24, 25] and is already foreshadowed in Hadamard [15] and Jordan [19]. The method of Koebe and of Morse is to code a geodesic on a surface M of negative curvature by recording the order in which it traverses a given set of labelled curves on M. The treatment of Morse allows variable curvature but assumes at least two boundary components, whereas Koebe assumes constant curvature but allows infinite connectivity and nonorientability and treats the more difficult case of a closed surface. This last case is handled by recording crossings of a fixed pants decomposition, anticipating the Thurston parameterization of simple curves as described in [11]. Both Koebe and Morse used their codings to demonstrate the existence of countably many closed geodesics and of everywhere dense (transitive) geodesics. Morse further constructed the first nonsynthetic example of a recurrent nonperiodic discontinuous motion (in modern terminology, a minimal nowhere dense set). Later [26] Morse treated the special closed surfaces of genus g associated to tesselations of the disc $\mathbf{D}$ by regular $4g$-gons. Each edge of a region in such a tesselation may be labelled by the isometry which glues it to another side of the same region in forming the quotient surface M. The set of isometries appearing as labels generate $\pi_1(M)$. Thus any geodesic is coded as a doubly infinite sequence of generators of $\pi_1(M)$. The same method applies quite generally to tesselations associated to any Fuchsian group. We call the sequences thus obtained *cutting sequences* and refer to generating sets of this kind as *geometric*. The difficulty of course is to determine precisely the class of sequences which occur. For a surface with boundary, one obtains exactly reduced sequences in the generators. In general the problem is complicated, hence the difficulties encountered in [26] (see also Theorem 3 below).

There is another method of coding geodesics, using certain *boundary expansions* for points at infinity in the universal cover of M. For the modular surface $H/SL(2, \mathbb{Z})$ the appropriate expansions are continued fractions and for the symmetrical genus g surfaces above they are the *Nielsen boundary expansions* of [27].

Geodesics are coded by juxtaposing two semi-infinite expansions corresponding to the endpoints of suitable lifts to **H** or **D**. Such boundary expansions are connected to the geometry of the corresponding tesselations of hyperbolic space; for the regular $4g$-gons this is immediate from the construction, while the association of continued fractions to the geometry of the modular tesselation of **H** was already known and used in connection with the reduction theory of quadratic forms in the latter part of the nineteenth century by various authors including H. J. Smith [34] and Hurwitz [18].

Artin [2] used the continued fraction method to construct a topologically transitive geodesic on the modular surface and Hedlund [16, 17] established the more difficult result of ergodicity in this case and in the symmetrical genus g case using the same expansions. Martin [23] used Nielsen expansions to compute a bound for the Birkoff ergodic function for the geodesic flow. (The ergodic function is the minimum time taken for an orbit to cover the space within ε.)

Bowen was aware of these results and in 1977 proposed to me to construct analogous codings for general finitely generated Fuchsian groups. It was this work on which he was engaged at the time of his death. We proved [6]:

THEOREM 1. *Let G be a finitely generated Fuchsian group with no parabolic elements acting in* **D** *and let G_R be a geometric set of generators associated to a fundamental region R. Then there are disjoint open intervals $J(g) \subset \partial \mathbf{D}$, $g \in G_R$, and a map $f: \bigcup_{g \in G_R} \overline{J(g)} \to \partial \mathbf{D}$, $f|_{J(g)} = g^{-1}$, such that*

(i) $\bigcup_{g \in G_R} \overline{J(g)} \supset L_G$, *the limit set of G.*

(ii) f^n *is uniformly expanding for some n.*

(iii) *Each $J(g)$ is partitioned into a finite number of half open intervals $J_i(g)$ and $f(J_i(g)) \cap J_j(h) \neq \varphi \Rightarrow f(J_i(g)) \supset \overline{J_j(h)}$.*

(iv) $x \in \bigcup_{n \in \mathbf{Z}} f^n(y) \Leftrightarrow x \in G(y)$, $\forall x, y \in \bigcup \overline{J(g)}$.

Slight modifications to (i), (ii) *are needed in the case of parabolics in G.*

The boundary expansion of $x \in A$ is the sequence $(e_n)_{n=0}^{\infty}$, where $e_n = g$ if $f^n(x) \in J(g)$. By (iii), *these sequences can be recoded to sequences of finite type.*

(v) *Boundary expansions are shortest sequences in the word metric of G_R and the distinct finite blocks which occur run through the elements of G each exactly once.*

The question arises of the relationship between the two types of coding. Under a certain geometrical hypothesis on R, they are in fact almost identical. A region R has *even corners* if the net $G(\partial R)$ is a union of complete geodesics in **D**. The map f of Theorem 1 was first constructed under this hypothesis, and was essential for the Markov property (iii). The condition was introduced by Koebe [21].

THEOREM 2 [3]. *Let G, G_R, R be as above and assume R has even corners (with some extra restrictions if R has three or four sides). Then cutting sequences of geodesics are shortest in the word metric of G_R. Two shortest sequences representing the same element of G differ only by blocks which run along opposite* *sides of a line in $G(\partial R)$ passing through a number of consecutive vertex cycles in opposite directions.*

Theorem 2 is closely related to Dehn's solution of the word problem for Fuchsian groups. It is proved entirely from the geometry of the tesselation $G(\partial R)$. It gives algorithms for reducing words to their shortest form and for determining equality of two words. Further refinements characterize cyclically shortest words as shortest in their conjugacy class, subject to an extra condition on the side pairings of R.

THEOREM 3 [**31**]. *The boundary expansion of $x \in \partial\mathbf{D}$ coincides with the cutting sequence of any ray with initial point in R and endpoint at x, up to blocks of the type described in Theorem 2. There is an explicit geometrical conjugacy between the shift spaces of cutting sequences and of two-sided boundary expansions. The geodesic flow may be represented as a special flow over the shift on two-sided boundary expansions with height function corresponding to the time of crossing R.*

Theorem 3 solves the problem of characterizing the space of cutting sequences referred to above. In the special case where R has all vertices on $\partial\mathbf{D}$ the conjugacy is the identity map; in particular, by considering the subgroup $\Gamma(1)$ of $\mathrm{SL}(2,\mathbb{Z})$ one obtains a representation of the geodesic flow on the modular surface as a special flow over a two point extension of the continued fraction transformation [**30**]. Adler and Flatto [**1**] have described similar conjugacies in various special cases.

Applications.

1. *Structure of the group G.* According to Theorem 1(v) the elements of G are essentially the possible finite blocks occurring in a subshift of finite type. Application of the standard dynamical theory of zeta functions for subshifts shows that the growth function $P(z) = \sum_{n=0}^{\infty} N_n z^n$, where N_n is the number of elements of G of word length n, is rational, and can be explicitly computed in specific instances. Likewise conjugacy classes can be counted as the number of purely periodic sequences. This finite type structure appears in a related form in the work of Cannon [**8**] where the group graph (here the context is hyperbolic convex cocompact groups in any dimension) is shown to be built up recursively by a finite possible number of moves, giving another and more general proof of rationality of $P(z)$.

2. *Structure of L_G.* Theorem 1 gives a representation of the limit set L_G as a subshift of finite type. This specializes the result of Floyd [**12**] in which L_G is expressed as the completion of the space of finite words in G_R relative to a suitable metric. One can study the class of Gibbs measures on L_G; in particular, for $\varphi = -\delta \log |f'|$, δ the exponent of convergence of G, one obtains the Patterson-Sullivan measure [**28**]. This construction extends to quasi-Fuchsian groups and one obtains the Hausdorff measure on L_G in this way. Bowen used this representation to prove results on quasiconformal deformations using dynamical methods [**7**].

3. *Random walks on G.* The Martin boundary of the random walk on the free group F on two generators with one step transition probabilities was shown in [10] to be the space of reduced sequences in the generators. This is exactly the shift space corresponding to the boundary expansions of Theorem 1 when F is realized as the fundamental group of a one-holed torus. Theorem 2 may be viewed as the statement that the graph of a Fuchsian group G is "almost" a tree; based on this idea we showed in [29] that the Martin boundary of any finite-step random walk is L_G (with a double point corresponding to each parabolic cusp) and that the hitting distribution is Gibbs.

4. *Number theory.* (a) As mentioned above, the coding by continued fractions is related to the reduction theory of quadratic forms. S. Katok [20] has worked out an analogous reduction theory for general Fuchsian groups.

(b) In the diophantine approximation $|x - p/q| < c/q^2$ of $x \in \mathbb{R}$ by rationals p/q, the constant c can be interpreted as the distance of approach of a geodesic on the modular surface to the cusp. The best rational approximants are given by continued fractions, which code the path of the corresponding geodesics. This is exploited in [13, 14] to compute values of c when the rationals p/q are replaced by images of infinity under other zonal Fuchsian groups. In [33] the symbolic method is used to find an analogue of the Markoff spectrum in a certain Hecke group.

5. *Simple curves.* The cutting sequences of simple geodesics take a very special form and are related to the minimal nonperiodic motions of Morse mentioned above. An algorithm for detecting simple cutting sequences is given by Birman and the author in [4], which Cohen-Lustig have elaborated to count intersection numbers [9]. The rules characterizing simple sequences have a recursive nature and the number of admissible blocks of length n has only polynomial growth. A complete characterization has only been given in the case of a one-holed torus [32] and is already very interesting; the sequences which occur are then the Sturmian sequences of Hedlund and Morse and have risen in many contexts.

In general, simple curves can be parameterized by recording the number of occurrences of *adjacent pairs* of generators in the cutting sequence. This is analogous to the Thurston parameterization in [11]. The advantage of our version is its algebraic content, in that it can be read off immediately from any cyclically shortest representative of the curve in $\pi_1(M)$. There is a rather strange "algebraic linearity" theorem for the action of diffeomorphisms on these parameters [5].

REFERENCES

1. R. Adler and L. Flatto, *Cross section maps for the geodesic flow on the modular surface*, Contemp. Math., vol. 26, Amer. Math. Soc., Providence, R.I., 1984, pp. 9–24.

2. E. Artin, *Ein mechanisches System mit quasiergodischen Bahnen*, Abh. Math. Sem. Hamburg Bd. **3** (1924); Collected Papers, Addison-Wesley, Reading, Mass., 1965, pp. 499–501.

3. J. Birman and C. Series, *Dehn's algorithm revisited, with applications to simple curves on surfaces*, Proc. Conf. on Combinatorial Group Theory (Alta, Utah, 1984), Gersten and Stallings, eds., 1986.

4. ____, *An algorithm for simple curves on surfaces*, J. London Math. Soc. (2) **29** (1984), 331–342.

5. ____, *Algebraic linearity for an automorphism of a surface group*, J. Algebra (to appear).

6. R. Bowen and C. Series, *Markov maps associated to Fuchsian groups*, Inst. Hautes Études Sci. Publ. Math. **50** (1979), 153–170.

7. R. Bowen, *Hausdorff dimension of quasi-circles*, Inst. Hautes Études Sci. Publ. Math. **50** (1979), 259–273.

8. J. Cannon, *The combinatorial structure of cocompact discrete hyperbolic groups*, Geom. Dedicata **16** (1984), 123–148.

9. M. Cohen and M. Lustig, *Paths of geodesics and geometric intersection numbers.* I, Proc. Conf. on Combinatorial Group Theory (Alta, Utah, 1984), Gersten and Stallings, eds., 1986.

10. E. B. Dynkin and M. B. Malyutov, *Random walks on groups with a finite number of generators*, Soviet Math. Dokl. **2** (1961), 399–402.

11. A. Fathi, F. Laudenbach, V. Poénaru et al, *Trauvaux de Thurston sur les surfaces*, Astérisque **66–67** (1979).

12. W. Floyd, *Group completions and limit sets of Kleinian groups*, Invent. Math. **57** (1980), 205–218.

13. A. Haas, *Diophantine approximation on hyperbolic Riemann surfaces*, Acta Math. **156** (1986), 33–82.

14. A. Haas and C. Series, *The Hurwitz constant and Diophantine approximation on Hecke groups*, J. London Math. Soc. **33** (1986).

15. J. Hadamard, *Les surfaces a coubures opposées et leurs lignes géodesiques*, J. Math. Pures Appl. (5) **IV** (1898), 27–74.

16. G. Hedlund, *A metrically transitive group defined by the modular group*, Amer. J. Math. **57** (1935), 668–678.

17. ____, *On the metrical transitivity of geodesics on closed surfaces of constant negative curvature*, Ann. of Math. **35** (1934), 787–808.

18. A. Hurwitz, *Uber die Reduktion der binaren quadratischen Formen*, Math. Ann. **45** (1894), 85–117.

19. C. Jordan, *Des contours tracés sur les surfaces*, Journal de Mathématique **11** (1866), 110–130.

20. S. Katok, *Reduction theory for Fuchsian groups*, Math. Ann. **273** (1986), 461–470.

21. P. Koebe, *Riemannische Mannifaltigkeiten und nichteuklidische Raumformen*. IV, Sitzungberichte der Preussichen. Akad. der Wissenschaften (1929), 414–457.

22. ____, Urmanuskript (1917) der Preisschrift (Acta Math. **50** (1927), 157), deposited in Mittag-Leffler Institute.

23. M. Martin, *The ergodic function of Birkhoff*, Duke Math. J. **3** (1937), 248–278.

24. M. Morse, *A one-to-one representation of geodesics on a surface of negative curvature*, Amer. J. Math. **43** (1921), 33–51.

25. ____, *Recurrent geodesics on a surface of negative curvature*, Trans. Amer. Math. Soc. **22** (1921), 84–100.

26. ____, *Symbolic dynamics*, Institute for Advanced Study Notes, Princeton, 1966 (unpublished notes written in 1938).

27. J. Nielsen, *Untersuchungen zur Topologie der geschlossenen zweiseitige Flachen*, Acta Math. **50** (1927), 189–358.

28. S. J. Patterson, *The limit set of a Fuchsian group*, Acta Math. **136** (1976), 241–273.

29. C. Series, *Martin boundaries of random walks on Fuchsian groups*, Israel J. Math. **44** (1983), 221–242.

30. ____, *The modular surface and continued fractions*, J. London Math. Soc. (2) **31** (1985), 69–80.

31. ____, *Geometrical Markov coding of geodesics on surfaces of constant negative curvature*, Ergodic Theory Dynamical Systems **6** (1986).

32. ____, *The geometry of Markoff numbers*, Math. Intelligencer **7** (1985), no. 3, 20–29.

33. ____, *The Markoff spectrum in the Hecke group* $G(\lambda_5)$, University of Warwick, Preprint.

34. H. J. Smith, *Mémoire sur les equations modulaires*, Ac. dei Lincei Ser. III, Vol. I, 1877, pp. 136–149. (Coll. works, Vol. II, Chelsea, New York, 1965, pp. 224–241.)

WARWICK UNIVERSITY, COVENTRY CV4 7AL, UNITED KINGDOM

Proceedings of the International Congress of Mathematicians
Berkeley, California, USA, 1986

Quasiconformal Homeomorphisms in Dynamics, Topology, and Geometry

DENNIS SULLIVAN

Dedicated to R. H. Bing

This paper has four parts. Each part involves quasiconformal homeomorphisms. These can be defined between arbitrary metric spaces: $\varphi \colon X \to Y$ is K-quasiconformal (or K qc) if

$$H(x) = \limsup_{r \to 0} \frac{\sup |\varphi(x) - \varphi(y)| \quad \text{where } |x - y| = r \text{ and } x \text{ is fixed}}{\inf |\varphi(x) - \varphi(y)| \quad \text{where } |x - y| = r \text{ and } x \text{ is fixed}}$$

is at most K where $|\ |$ means distance. Between open sets in Euclidean space, $H(x) \le K$ implies φ has many interesting analytic properties. (See Gehring's lecture at this congress.)

In the *first part* we discuss Feigenbaum's numerical discoveries in one-dimensional iteration problems. Quasiconformal conjugacies can be used to define a useful coordinate independent *distance between real analytic dynamical systems* which is decreased by Feigenbaum's renormalization operator.

In the *second part* we discuss de Rham, Atiyah-Singer, and *Yang-Mills theory* in its foundational aspect *on quasiconformal manifolds*. The discussion (which is joint work with Simon Donaldson) connects with Donaldson's work and Freedman's work to complete a nice picture in the structure of manifolds—each topological manifold has an essentially unique quasiconformal structure in all dimensions—except four (Sullivan [21]). In dimension 4 both parts of the statement are false (Donaldson and Sullivan [3]).

In the *third part* we discuss the C-analytic classification of *expanding* analytic transformations near *fractal invariant sets*. The infinite dimensional Teichmüller space of such systems is embedded in the Hausdorff measure theories possible for the transformation on the fractal. These possible Hausdorff measure theories of fractals are nicely encoded in the theory of Gibbsian measure classes or Gibbs states.

In the *fourth part* we give a characterization of constant curvature among variable *negative curvature* in terms of a measure theoretical dynamical property equivalent to uniform quasiconformality for the geodesic flow. A dynamical equivalent of $-\frac{1}{4} < k \leq -1$ pinching is utilized.

I. Feigenbaum's renormalization operator and the quasiconformal Teichmüller metric.

Mitchell Feigenbaum [7] made some remarkable numerical discoveries concerning the iteration of families f_a of real quadratic-like functions, namely those which fold the line smoothly with a nondegenerate quadratic critical point, for example $f_a(x) = -x^2 + a$. These discoveries may be summarized as follows:

(i) If a parameter variation f_a creates basic $2, 4, 8, 16, \ldots$ period doubling, the periods actually double at parameter values a_n which converge at a definite geometric rate to a limit a_∞,

$$|a_\infty - a_n| \sim \text{constant}(4.6692\ldots)^{-n}.$$

(ii) The mapping for the limiting parameter value a_∞ has a Cantor set X to which almost all bounded orbits tend, and X has universal geometric properties like (a) Hausdorff dimension $X = .53\ldots$, and (b) X can be defined by an intersection of families of intervals $I_n = \{I_1^n, I_2^n, \ldots, I_{2^n}^n\}$ where ratios of sizes $|I_i^n|/|I_j^{n+1}|$ converge exponentially fast to universal ratios I_α, labeled by α which is any one-sided string of 0's and 1's.

Khanin, Sinai, and Vul [14] formulated the statement of Feigenbaum's convergence in this way—index the intervals of the nth level containing the critical point and critical value by 0 and 1, respectively. Index the remaining $2^n - 2$ intervals by time evolution ($f(I_k^n) = I_{k+1}^n$, $k + 1 < 2^n$), and think of the index in its 2-adic expansion, $k = \varepsilon_0 + \varepsilon_1 2 + \cdots + \varepsilon_{n-1} 2^{n-1}$.

Then the ratios $|I_{i_n}^n|/|I_{j_n}^{n+1}|$ converge where $j_n = i_n$ or $j_n = i_n + 2^n$ to universal ratios I_α where α is any 2-adic integer if the **final** coefficients of the expansions of $j_n = \varepsilon_0 + \varepsilon_1 2 + \cdots + \varepsilon_n 2^n$ agree on larger and larger **final** segments, and α is defined by $\alpha = \lim_{n\to\infty}(\varepsilon_n + \varepsilon_{n-1} 2 + \cdots + \varepsilon_0 2^n)$.

After making these first two numerical discoveries (and one more described below) Feigenbaum formulated a renormalization picture to describe these phenomena. The renormalization operator is obtained by iterating the transformation twice, restricting this iterate to an interval about the original critical point, and then renormalizing to obtain a real quadratic-like mapping on a fixed size interval. Studying this operator R numerically Feigenbaum found a third phenomenon:

(iii) $R^n f_\infty$ converges to a universal function g, where f_∞ denotes the mapping of the given family corresponding to the parameter value a_∞ mentioned above. The function g is a fixed point of the operator R,

$$g(x) = \lambda g \cdot g(x/\lambda), \qquad \lambda = -2.50290\ldots$$

(the Cvitanovic-Feigenbaum equation).

Since Feigenbaum's work, there has been more numerical work revealing similar phenomena in other dynamical situations, e.g., Cvitanovic—period tripling, etc, Shenker—circle mappings with critical point, Widom—boundaries of Siegel disks, Milnor—infinitely many points of the Mandelbrot set, and others. There has also been work trying to prove theoretical theorems modelling Feigenbaum's discovery (Campanino, Collett, Eckmann, Epstein, Khanin, Lanford, Ruelle, Sinai, Tresser, Vul, Wittwer, and many others). For example, see Epstein [5] for a function theory proof of the existence of the universal function g satisfying the Cvitanovic-Feigenbaum equation. Khanin, Sinai, and Vul [14] proved that for the universal function g the ratio of interval lengths converge exponentially fast to ratios I_α as indicated above. Lanford [15] proved (using rigorous numerical analysis) that the spectrum of the operator R linearized at g has one point outside the open unit circle $\{-4.6692\ldots\}$. Lanford's work yields Feigenbaum's picture of the renormalizaton operator R on *some neighborhood* $\mathcal{L}$ of the fixed function g. Also, it proves the point (i), $|a_\infty - a_n| \sim \text{constant}(4.6692\ldots)^{-n}$, for the $x \to -x^2 + a$, family

Several questions remain.

PROBLEM 1. Prove the second Feigenbaum discovery (ii) that the Feigenbaum attractor X has universal geometric structure for a general class of mappings. (This universal structure can be described by Gibbsian measure class as in Part III using $E\colon X \to X$ where E is an expanding map which is the union of $x \to \lambda x$ and $x \to \lambda g(x)$ on left and right pieces respectively of X.)

PROBLEM 2. Justify Feigenbaum's third numerical discovery (iii) by extending the local stable manifold (due to Lanford's work) of the Feigenbaum renormalization operator to a global stable manifold.

PROBLEM 3. Find a conceptual, more geometrical treatment of Feigenbaum's three points (i), (ii), (iii) yielding a new proof of Lanford's theorem on the spectrum and, hopefully, proofs for various generalized Feigenbaum phenomena heretofore only treated numerically.

We will study these problems using quasiconformal homeomorphisms to define a coordinate free distance between *complexifications* of real quadratic maps, a definition due to Douady and Hubbard [4]. A *complex quadratic-like mapping* is a pair (π, i) where $\pi\colon \overline{V} \to V$ is a two-sheeted covering with one branch point onto a simply-connected Reimannian surface V, and $i\colon \overline{V} \to V$ is a conformal embedding of $\overline{V}$ into V with compact closure. Given (π, i) we consider the quadratic-like mapping $f\colon V_1 \to V$ given by the composition $\pi \cdot i^{-1}$ where $V_1 = i\overline{V}$.

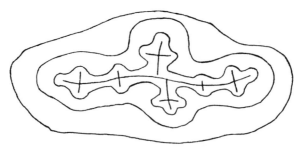

FIGURE 1

Note that $f^{-1}V = V_1$ and if $V_n = f^{-n}V$, then $K(f) = \bigcap_n V_n$ is a maximal and compact invariant set for the iteration of f on V_1.

We say two complex quadratic-like mappings f_1, f_2 are (i) *analytically equivalent* if they are complex analytically conjugate on neighborhoods of the compact invariant sets, and (ii) at *distance at most* $\log K$ if there is a K quasiconformal homeomorphism conjugating f_1 and f_2 between neighborhoods of the invariant sets. Define distance(f_1, f_2) as the infimum of such $\log K$.

THEOREM.   *If* distance$(f_1, f_2) = 0$, *then* f_1 *and* f_2 *are analytically equivalent on neighborhoods of their invariant sets.*

SKETCH OF PROOF. If the invariant set of f is connected one can associate (Douady and Hubbard [4]) a real analytic expanding map h of the circle (of exterior prime ends of the connected invariant set). The real analytic conjugacy class of h essentially determines the equivalence class of f and is in turn determined by the sizes of eigenvalues at its periodic points (see Part III of this paper or Shub and Sullivan [19]). However, the K qc conjugacy between f_1 and f_2 yields a Hölder continuous conjugacy φ between h_1 and h_2, $|\varphi(x) - \varphi(y)| \leq C|x - y|^{\alpha}$, where C depends on the size of the neighborhood and α only depends on K. If distance$(f_1, f_2) = 0$ this relationship implies the eigenvalues of h_1 equal to those of h_2. So h_1 and h_2 are real analytically conjugate and this (plus one more consideration which is automatic) implies f_1 and f_2 are $\mathbf{C}$ analytically conjugate on some neighborhood.

If the invariant set is not connected it is expanded by f and we may use Part III directly.   Q.E.D.

Let $\mathcal{F}$ be the metric space of equivalence classes of complex quadratic-like mappings at a finite distance from the Feigenbaum universal function (defined by $\lambda g \circ g(x/\lambda) = g(x)$.) (One knows the complexification of g is quadratic-like; see Epstein [6], for example.) The space $\mathcal{F}$ has rectifiable paths between any pair of points whose length is the distance (this uses the measurable Reimannian mapping as in the qc deformations section of Sullivan [20]).

Feigenbaum's renormalization operator may be defined on $\mathcal{F}$. Since g is quadratic-like, the equation $\lambda g \circ g(x/\lambda) = g$ shows $g \circ g$ is quadratic-like near the original critical point. Thus if f is qc conjugate to g, $f \circ f$ is qc conjugate to $g \circ g$ and so it is quadratic-like also on some disk around the critical point. It is easy to see that regarding $f \circ f$ as a quadratic-like mapping only depends (up to equivalence) on the critical point chosen and not on the disk. More generally, there are canonically defined renormalization operators defined on full path components of the space of quadratic-like mappings with the Teichmüller metric whenever one mapping of the component can be renormalized. Here is our result on Problem 2.

THEOREM.   *In the qc path component* $\mathcal{F}$ *of the universal Feigenbaum map, there is a canonical renormalization operator* $R\colon \mathcal{F} \to \mathcal{F}$ *defined on representatives by* $f \to f \circ f$ *(restricted to a neighborhood around the original critical point). The operator* R *is strictly distance decreasing for the above Teichmüller* *metric, $R[g] = [g]$, and for any $[f] \in \mathcal{F}$ the orbit of $[f]$ under iterates of R tends to $[g]$.*

NOTE. A real analytic mapping whose complexification is quadratic-like, and whose critical orbit has the kneading sequence of the Feigenbaum map, lies in the space $\mathcal{F}$.

COROLLARY. *The unique complex quadratic-like solution of the Cvitanovic-Feigenbaum functional equation is $[g]$.*

PROOF OF THEOREM. The nonincreasing distance property of R follows from the definitions. It follows from Lanford's results [15] that R is contracting on some neighborhood U of $[g]$. From the existence of distance paths connecting $[g]$ to any other point $[f]$ it follows $\mathcal{F} = \bigcup_n R^{-n}U$. Then the strictly decreasing property follows from Royden's interpretation of the Teichmüller metric as the Kobayaski metric.

DISCUSSION OF PROBLEM 1. Curt McMullen has observed (private communication) that the convergence of the theorem can be lifted to the level of representatives. To see this one looks at the size of $[f]$, the supremum of moduli of the annuli $V - V_1$ over representatives.

Feigenbaum and others have calculated the spectrum of R numerically. The part inside the unit circle satisfies $|\lambda| \leq (2.50290\ldots)^{-1}$. This inequality applies to R acting on representative functions up to linear rescaling.

The author and Feigenbaum [8] have proved that if on the level of representatives $R^n f \to g$ at an exponential rate $\leq (2.50290\ldots)^{-n}$ in the C^3 topology, then the Cantor set for f is $C^{1+\beta}$ diffeomorphic to that of g by a map which is a conjugacy on the Cantor sets.

SPECULATION. This summarizes our results about the third problem. Regarding the first two problems and others in iteration theory we conjecture that there is an infinite-dimensional *Teichmüller mapping* theorem in these dynamical contexts which may be used to show directly that all the renormalization operators strictly decrease distance and have fixed points (compare Milnor's conjectures [16]).

II. Analysis on quasiconformal manifolds and Yang-Mills fields. A quasiconformal manifold is a topological manifold provided with a maximal atlas of charts U_α where the overlap transformations $\varphi_{\alpha\beta}$ are quasiconformal homeomorphisms between open sets of Euclidean n-space. One knows that if $n \neq 4$ all topological n-manifolds have such qc charts. Also, if $n \neq 4$ the qc structure is unique up to homeomorphism arbitrarily close to the identity. (See Sullivan [21] for these and the same theorems for bi-Lipschitz homeomorphisms.)

In joint work with Simon Donaldson [3] we have tried to show enough global analysis exists on qc manifolds to replace the word smooth by the word quasiconformal in many of the latter's theorems. Then adding in Freedman's work one finds [3] many topological 4-manifolds do not have qc structures and many pairs of homeomorphic qc (even smooth or **C**-algebraic) 4-manifolds are not qc homeomorphic. We discuss that global analysis now.

If M is a smooth compact Riemannian n-manifold one has a $*$-operator from k-forms to $(n - k)$-forms, pointwise norms defined by

$$|w| = (w \wedge *w / \text{Riemannian volume})^{1/2}$$

and local and global L^p norms defined by $(\int |w|^p \, dm)^{1/p}$. The topological vector spaces so defined only depend on the underlying differentiable structure (even the Lipschitz structure). A set of these norms, one for each k, $\|\omega\| = (\int |\omega|^{n/k})^{k/n}$ where ω is a k-form, is unchanged if the metric is changed conformally. Also, $\|\omega \wedge \eta\| \leq \|\omega\| \cdot \|\eta\|$ by Hölder $(k/n + l/n = (k + l)/n)$ so we have a natural graded Banach algebra $\Omega(\|\cdot\|)$ of forms associated to the underlying conformal structure of the Riemannian manifold. It follows that the graded Banachable algebra Ω of forms is locally well defined under qc changes of coordinate using the fact that a K-quasiconformal homeomorphism φ (oriented) satisfies

(i) φ is differentiable a.e.;

(ii) Jacobian $\varphi > 0$ a.e.; and

(iii) $|d\varphi|^n \leq K$ Jacobian φ a.e.

The deeper fact (Gehring [9]) that Jacobian φ is locally p-summable for some $p = p(K) > 1$ implies the dense subalgebra $\Omega' \subset \Omega$ consisting of all k-forms with coefficients locally p-summable for some $p > n/k$ is also qc invariant.

Now we turn to the exterior differential.

PROPOSITION. *The unbounded operator defined in the distributional sense* $\Omega \xrightarrow{d} \Omega$ *in a local chart commutes with the action of qc homeomorphisms φ,*

$$d(\varphi^* \omega) = \varphi^* \, d\omega.$$

SKETCH OF PROOF. Using Chapter 3 of Morrey [17], Ziemer [28] shows the class of continuous functions f with $\|df\| < \infty$ (i.e., df is n-summable) is qc invariant, and $d(\varphi^* f) = \varphi^* \, df$ where df means the distributional total differential. Form the subalgebra of forms generated by such f, df. It has a d and may be used as a testing algebra to define a qc invariant distributional d. A smoothing argument (Vaisala [27, p. 80]) shows this qc differential d is the same as the smooth distributional d. Q.E.D.

Now define the *p-regular forms* on a qc manifold to be the set of forms in $\Omega' \subset \Omega$ whose exterior d is also in Ω', namely

$$p\text{-}regular \ forms \ on \ qc \ manifold = \Omega' \cap d^{-1}\Omega'.$$

Note that a p-regular function f is one such that df has coefficients in L^p for some $p > n$. Thus f is Hölder of exponent $(p/n - 1)$ (Morrey [17]). A similar result holds for h-regular functions defined below.

In a coordinate system consider a smoothing operator on forms $\omega \to \int t^* \omega \, d\mu$ $= R\omega$ where μ is a smooth measure on the translation group. Using [Lie derivative] $= [d(\text{contraction}) + (\text{contraction})d]$ one finds by integration a chain homotopy between R and the identity: $R - I = dS + Sd$, where S is (i) a derivation of degree -1, and (ii) a singular integral operator sending k-forms with p-summable coefficients into $(k-1)$-forms with coefficients having first partials p-summable, $p > 1$. Thus by the Sobolev embedding $L_1^p \subset L^q$ where $1/q = 1/p - 1/n$ we see that S carries Ω' into Ω', and Ω into Ω (except for Ω_n). This yields a Poincaré lemma and the following

THEOREM. *The de Rham cohomology of* $(\Omega' \xrightarrow{d} \Omega')$ *for a qc manifold agrees with the usual cohomology.*

Now consider a trivialized R^k-bundle over a chart U, gauge transformations $g: U \to O(k)$ which are regular (in the above sense and the sense below), $k \times k$ skew symmetric connection matrices of regular 1-forms θ, and the corresponding curvature forms $\Omega = d\theta + \theta \wedge \theta$. Changing the trivialization by a gauge transformation g induces the familiar changes $\theta \to g^{-1}dg + g^{-1}\theta g = \theta^g$, $\Omega \to g^{-1}\Omega g = \Omega^g$. If $g = \exp \xi t$, the infinitesimal change in θ is $d\xi - [\xi, \theta]$ where $[\ ,\ ]$ means commutator.

NONABELIAN POINCARÉ LEMMA. Take $\xi = S\theta$; then the infinitesimal change in θ is $dS\theta - [S\theta, \theta] = dS\theta + S\,d\theta - S\,d\theta - S(\theta \cdot \theta)$ (since S is a derivation)$= -\theta + R\theta - S\Omega$.

Using this we can show a regular connection form can be regauged to reduce $\|\theta\|$ so that it is dominated by a constant $\cdot \|\Omega\|$.

The second notion of regularity is h-regularity of ω which means the amount of $\|\ \|$ norm (for ω and $d\omega$) in a ball of radius r is at most Cr^α for some constants C, α. This notion of h-regularity is qc-invariant and the S operator preserves this h-regularity because Calderon Zygmund kernels do (Peetre [17.5]). Now consider locally h-regular connections on 4-manifolds whose curvatures satisfy, in addition, the *quasi-self duality condition*:

K-*quasi Yang-Mills condition* $|\Omega|^2 \le K \operatorname{tr} \Omega \wedge \Omega /$ volume element

relative to a measurable locally qc Euclidean metric.

If θ and $d\theta$ (equivalently θ, Ω) are h-regular in B^4-point, then in concentric annuli $\{2^{-n} \le r \le 2^{-n+1}\}$ one regauges the connection so that $\|\Omega\|$ controls $\|\theta\|$. If CS is the Chern Simons form $\operatorname{tr}(\Omega \wedge \theta - \frac{1}{3}\theta \cdot \theta \cdot \theta)$, one has $dCS = \operatorname{tr}\Omega \wedge \Omega$, and a Stokes theorem argument in the concentric annuli shows Ω, and thus the new regauged θ is h-regular over the point with C, α controlled by the norm $\|\ \|$ (assumed sufficiently small) of Ω on B^4-point.

These remarks allow one to have Karen Uhlenbeck's (compactness/noncompactness) picture [26] for any sequence of h-regular K-quasi Yang-Mills connections on a compact qc four manifold.

One may also develop an Atiyah-Singer Index theory for the signature operator with coefficients in a bundle over a qc M^{4l}. One uses the de Rham complex up to the middle

$$(\cdots \to \Omega^h \to \Omega^{h+1} \to \cdots \to \Omega^{2l-1} \to \Omega^{2l} \xrightarrow{*})$$

where the last arrow uses the measurable locally quasiconformally Euclidean metric to project $(d\Omega^{2l-1})$ onto $\frac{1}{2}$ the space Ω^{2l} (where $*\omega = \omega$). One may

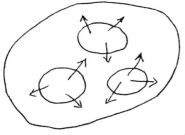 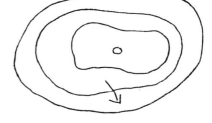

each little disk maps onto the larger disc f is a degree d covering

FIGURE 2

tensor with the bundle E and use S to restore the fact that one loses the $d^2 = 0$ property in the tensoring process.

In this way one obtains a Fredholm index provably independent of $*$ on M^{4k} and the connection in E (compare Teleman [24]).

With Uhlenbech's picture and the basic Atiyah-Singer theory in place one may develop at least one Donaldson type result (say a là Fintushel-Stern [8.5]) to show the qc theory of 4-manifolds is different from the theory of topological 4-manifolds.

ACKNOWLEDGEMENT. I am indebted to the participants in the quasiconformal gauge theory conference at the University of Texas, in particular Bill Beckner, John Gilbert, Bob Gompf, Gary Hamrick, and Bruce Palka.

REMARK. A relation between this section and dynamics is suggested by Alain Connes's theory of noncommunitative differential geometry where operators of Shatten class n/k are utilized.

III. A rigidity theorem for qc deformations of expanding systems and Gibbs states. A C-analytic expanding system is determined by a C-analytic map $f\colon U_1 \to U$ where U_1 is a domain properly embedded in the Riemann surface U and f is a $(d > 1)$-sheeted onto covering. Such systems are classified by analytic conjugacy near the compact invariant set $K_f = \bigcap_n f^{-n}U$. For example, see Figure 2.

To motivate the connections with quasiconformality we state the following

THEOREM. (1) *Any sufficiently small C-analytic perturbation of f near K_f defines another expanding system which is quasiconformally conjugate to f near the invariant set.*

(2) *Any topological conjugacy between the compact invariant sets of expanding systems agrees with a qc conjugacy between neighborhoods.*

(3) *All expanding systems qc conjugate near the invariant sets to a given one can be constructed by deforming the complex structures on the Riemann surface $U - U_1$ for some presentation (U, U_1) using the measurable Riemann mapping theorem. (See Sullivan [20] and [22].)*

The analytic classification of expanding systems of a given topological dynamics type on the invariant set is a kind of Teichmüller theory. The dimension is infinite because there are infinitely many complex moduli given by the complex eigenvalues of f-periodic points of K_f (which are dense in K_f and thus infinite; to avoid trivial cases and exceptions we suppose f/K_f is *leo* (locally eventually onto): for each neighborhood V in K_f, $f^n V \supset K_f$ for some n).

PROPOSITION. (i) *The δ-Hausdorff measure μ_δ of K_f is finite and positive;* (ii) *there is a unique real analytic conformal metric defined on K_f for which μ_δ is invariant by f, $f_* \mu_\delta = \mu_\delta$.*

SKETCH OF PROOF. One first shows the δ-Hausdorff measure is finite and positive where δ = dimension K_f, measured, say, in some real analytic metric (see Sullivan [22] or Ruelle [18]). If ν denotes this measure, one studies the density function ρ_n of $\nu_n = f_*^n \nu$ relative to ν. By calculation one sees that the ρ_n form a compact family of continuous functions and the limits are real analytic. (The value of ρ_n at x is a sum of d^n terms $\sum_{y \in f^{-n}x} \omega(y, fy, \dots)^\delta$ where ω is the product of reciprocal linear element derivatives of f along the orbit of y up to x. The Hausdorff dimension δ is the power that makes these sums finite and because f is expanding, the functions ω are absolute values of C-analytic functions with fixed domain and exponentially decreasing range.) Q.E.D.

We say an expanding system (f, K_f) is *linear* if

(i) the curvature of the natural metric (defined in the proposition) is identically zero near K_f;

(ii) the absolute value of the derivative of f in the metric is locally constant on K_f; otherwise we say the system is *nonlinear*.

EXAMPLE. $z \to z^n$ near $|z| = 1$ is linear.

REMARK. For linear systems there are flat C-analytic charts defined near K_f so that the complex derivative $f'(z)$ is locally constant near K_f.

Our rigidity theorems concern the *nonlinear* expanding systems.

THEOREM. *Let (f, K_f) and (g, K_g) be two C-analytic expanding systems not both linear. Then there is a C-analytic conjugacy between (f, K_f) and (g, K_g) which restricts to a given Borel map $h: K_f \to K_f$ satisfying $fh = hg$ if*

(i) *h is a homeomorphism and moduli of eigenvalues at a periodic points associated by h are equal; or if*

(ii) *h is a nonsingular transformation between Hausdorff measure classes on K_f and K_g, respectively (dim K_f = dim K_g is a consequence here not an assumption).*

REMARK. Both of these statements are false if both systems are linear.

COROLLARY. *The infinite-dimensional Teichmüller theory of complex analytic expanding systems is embedded in the Hausdorff measure theory of the fractal invariant sets.*

SKETCH OF PROOF. In the canonical metric consider the Jacobian of f, Jf relative to the invariant measure in the Hausdorff measure class. Consider the *Jacobian invariant*: domain of dynamics $\xrightarrow{J(f)}$ Hilbert cube defined by $x \to (Jf(x), Jf(fx), \dots)$. We show that J is locally injective somewhere unless we are in the linear case. We successively deduce (1) h is measure preserving by ergodicity so $Jf = Jg \circ h$; (2) h is somewhere locally Lipschitz; (3) h is continuous everywhere (and measure preserving); (4) h is real analytic; (5) h is complex analytic.

The idea of all these is to use the expanding dynamics (as in Part IV) to see the improved quality of h.

We reduce part (i) to part (ii) by showing h must preserve the Hausdorff measure class. Q.E.D.

The Hausdorff measure theory of such dynamical fractals can be understood in terms of Gibbsian measure classes and Gibbs states. For simplicity consider one topological model, the shift on the Cantor set X of one-sided strings on two symbols, $\{.\varepsilon_0 \varepsilon_1 \ldots\} \xrightarrow{T} \{.\varepsilon_1 \varepsilon_2 \ldots\}$. A (Hölder) Gibbsian measure class for $T: X \to X$ is a measure class determined by a probability measure ν on X so that (i) $\nu(A) > 0$ if and only if $\nu(TA) > 0$; and (ii) φ the log Jacobian of T rel ν (definable by (i)) satisfies $\sup |\varphi(x) - \varphi(y)| \leq c\alpha^{-n}$ whenever x and y agree for the first n-symbols. Let $C^{k,\alpha}$ denote $\{f: D^{k-1} \log Df$ is α-Hölder$\}$ where $k = 1, 2, 3, \ldots \alpha \in (0, 1]$.

THEOREM. *Let ν determine a (α-Hölder) Gibbsian measure class for $T: X \to X$. Then for each $\delta \in (0, 1)$ there is a Hölder continuous embedding of X in the real line $X \subset R$ and a $C^{1,\alpha}$ expanding map $f: R \to R$ defined on a neighborhood of X so that*

(i) X is the maximal invariant set of f in the neighborhood and f/X is the shift T;

(ii) ν is the δ-Hausdorff measure of X, and therefore δ is the Hausdorff dimension of X;

(iii) any such $C^{1,\alpha}$ geometric realization of $T, f: X \to X$ is determined up to bi-Lipschitz conjugacy by δ and the Gibbsian measure class determined by ν;

(iv) if $f: X \to X$ is $C^{k,\alpha}$ ($k \geq 2$, $k = \infty$, or $k = w$) and $f'' \neq 0$ (in the metric where Hausdorff measure is invariant) at some point of X, then $f: X \to X$ is determined up to $C^{k,\alpha}$ conjugacy near X by the Gibbsian measure class ν. (δ is determined also.)

REMARKS. (1) One knows (Bowen [1]) that a (Hölder) Gibbsian measure class is determined by the Jacobians at the periodic points. Also there is a canonical representative using the unique invariant measure. An important consequence of the first is that the set of Gibbsian measure classes is isomorphic to a locally closed subset in a Banach space. (2) All topological models based on one-sided subshifts of finite type can be similarly treated.

NOTE ADDED IN PROOF. The author has recently found that such $C^{1,\alpha}$ expanded Cantor sets C have a Hölder continuous *scale function* $\sigma: C^* \to R$ where $C^* = \{\ldots \varepsilon_3 \varepsilon_2 \varepsilon_1.\}$ if $C = \{\varepsilon_1 \varepsilon_2 \varepsilon_3 \ldots\}$, $\varepsilon_i = 0$ or 1. The scale function is independent of the smooth C^1 coordinate system being defined as asymptotic limits of ratios of lengths of intervals at stage n to lengths of containing intervals at stage $n - 1$. Two expanding systems which are $C^{k,\alpha}$ ($k = 1, 2, 3, \ldots \alpha$

in $(0,1]$) are $C^{k,\alpha}$ conjugate iff they have the same scale functions. Also any Hölder continuous function occurs as the scale function in some $C^{1,\alpha}$ expanding model. This caveat to the expanding theory is very useful for understanding the Feigenbaum discovery. (See Part I.)

IV. Quasiconformality in the geodesic flow of negatively curved manifolds. If M is a compact negatively curved $n+1$ manifold, then there is defined a topological n-sphere at infinity in the universal cover on which $\Gamma = \pi_1 M$ acts with every orbit dense. One knows that the sphere carries a qc manifold structure in which Γ *acts uniformly quasiconformally if and only if there is a constant negatively curved compact manifold of the same homotopy type as M.* Compare Gromov [11], Sullivan [23], and Tukia [25]. (More precisely, one shows the uniformly qc action is qc conjugate to a conformal action.) One hopes and conjectures that one may always find this quasiconformality when $n + 1 = 3$, and one knows for $n + 1 = 4, 5, 6, \ldots$ that it is generally impossible even for manifolds with sectional curvatures almost equal to -1 (Gromov and Thurston [12]).

Here we describe a necessary and sufficient condition for this quasiconformality in a more precise sense when the curvature is pinched $-\frac{1}{4} < k \leq -1$. In this case one knows the sphere at infinity has a natural C^1-structure; see Green [10] and Hirsch-Pugh [13] and the following text. (Gromov has asked if the sphere at infinity has a C^2 structure only in the locally symmetric case.)

The pinching condition implies that the horospheres in the universal cover have extrinsic curvatures satisfying $\frac{1}{2} < k \leq 1$. This implies that the geodesic flow has eigenvalues in its expanding manifolds satisfying $\frac{1}{2} < \log \lambda \leq 1$. An elementary calculation shows that composing expansions with these eigenvalue inequalities yields a composition F whose derivative only varies in Lipschitz manner (in $Gl(n)$) along an arc so that the length of its F image is ≤ 1. The point is that the Lipschitz constant is independent of the length of the composition.

We call this property of composed expansions the *quasilinearity principle*.

PROPOSITION.   *In a $-\frac{1}{4} < k \leq -1$ pinched Riemannian manifold the geodesic flow on its expanding horospherical foliation satisfies the quasilinearity principle.*

This proposition leads to the C^1-structure on the sphere at infinity. It also allows one to characterize uniform quasiconformality of the action of Γ on the C^1-sphere at infinity.

THEOREM.   *The following are equivalent in the $-\frac{1}{4} < k \leq -1$ pinched compact Riemannian manifold.*

(i) *The $\pi_1 M = \Gamma$ action on the C^1-sphere at infinity is uniformly quasiconformal.*

(ii) *The geodesic flow is uniformly quasiconformal on its expanding horospheres.*

(iii) *The Γ action on the tangent spaces of the sphere at ∞ is measurably irreducible.*

(iv) *The geodesic flow acting on the tangent spaces of the expanding horospheres is measurably irreducible.*

Measure irreducibility means there is no measurable field of proper subspaces of the tangent spaces which is a.e. invariant by the relevant action.

SKETCH OF PROOF. (1) The orbits of Γ on the sphere at infinity are in one-to-one correspondence with the leaves A_+ of forward asymptotic geodesics in the unit tangent bundle. Each leaf of the foliation A_+ is a family of horospheres swept out by the geodesic flow. Then the A_+ foliation is an R-extension of a foliation with polynomial growth leaves, the horospheres. For such foliations (yielding *amenable* equivalence relations) Zimmer [**29**] has shown any associated $Gl(n)$ cocyle has a *measurable reduction* or it is *measurably equivalent* to an *associated cocycle of similarities.*

(2) In the latter case one can show the measurable invariant similarity structure is continuous by expanding a small neighborhood of an almost continuity point (on the sphere at infinity). One uses (a) the quasilinear principle and (b) the existence of a natural metric on the similarity structures on one tangent space to enlarge a neighborhood with high percentage very small oscillation. (Because (b) is lacking for subspaces one cannot use this argument to show that the measurable reduction of (1) is continuous. In fact, this conclusion is certainly false for all the odd dimensional examples of Gromov-Thurston [**12**] because an even sphere has no continuous tangent subbundle.) If Γ preserves a continuous similarity structure the action is uniformly quasiconforml relative to the C^1-structure. This shows (iii)$\Rightarrow$(i).

(3) The rest of the implications do not use the $-\frac{1}{4} < k \le -1$ pinching: (i)$\Rightarrow$(ii) and (iv)$\Rightarrow$(iii) are formal, (ii)$\Rightarrow$(i) is a picture, and for hyperbolic manifolds (iv) is known, so (i)$\Rightarrow$(iv).

PROBLEM. Do the conclusions of the theorem imply the curvature is actually constant? (Part III suggests something of this sort.)

REFERENCES

1. R. Bowen, *Gibbs states...*, Lecture Notes in Math., Springer-Verlag.

2. D. Cooper, *Characterizing convex cocompact Kleinian groups*, Preprint.

3. S. Donaldson and D. Sullivan, manuscript in preparation.

4. A. Douady and J. Hubbard, *On the dynamics of polynomial like mappings*, Ann. Sci. École Norm. Sup. (4) **18** (1985), 287–343.

5. H. Epstein, *New proofs of the existence of the Feigenbaum function*, Comm. Math. Phys. (to appear); IHES/P/85/55.

6. ____, *Polynomial-like behavior of the Feigenbaum function*, IHES, 1986.

7. M. Feigenbaum, *The universal metric properties of nonlinear transformations*, J. Statist. Phys. **21** (1979), 669–706.

8. M. Feigenbaum and D. Sullivan, *Geometry of the period doubling attractor*, manuscript in preparation.

8.5. R. Fintushel and R. Stern, *Pseudofree orbifolds*, Ann. of Math. **122** (1985), 335–364.

9. F. W. Gehring, *The L^p integrability of the partial derivatives of a quasiconformal mapping*, Acta Math. **30** (1973), 265.

10. L. Green, *The generalized geodesic flow*, Duke Math. J. **41** (1974), 115–126.

11. M. Gromov, *Hyperbolic manifolds, groups and actions*, Riemann Surfaces and Related Topics: Proceedings of the 1978 Stony Brook Conference (State Univ. New York, Stony Brook, N.Y., 1978), Ann. of Math. Studies, No. 97, Princeton Univ. Press, Princeton, N.J., 1981.

12. M. Gromov and W. Thurston, *Variable negative curvature and constant negative curvature*, Invent. Math. (1987) (to appear).

13. M. Hirsch and C. Pugh, *Smoothness of horocycle foliations*, J. Differential Geom. **10** (1975), 225–238.

14. K. M. Khanin, Ya. G. Sinai, and E. B. Vul, *Feigenbaum universality and the thermodynamic formalisms*, Russian Math. Surveys **39** (1984), 1–40.

15. O. Lanford, *A computer assisted proof of Feigenbaum conjectures*, Bull. Amer. Math. Soc. **6** (1982), 427.

16. J. Milnor, *Self similarities of the Mandelbrot set*, IAS Preprint, 1986.

17. C. B. Morrey, *Multiple integrals in the calculus of variations*, Grundlehren Series, vol. 30, Springer-Verlag.

17.5. J. Peetre, *On convolution operators leaving $L^{p,\lambda}$ spaces invariant*, Ann. Math. Pura Appl. (4) **72** (1966), 295–304; J. Funct. Anal. **4** (1969), 71–87.

18. D. Ruelle, *Repellers for real analytic maps*, Ergodic Theory Dynamical Systems **2** (1982), 99–107.

19. M. Shub and D. Sullivan, *Expanding endomorphisms of the circle revisited*, Ergodic Theory Dynamical Systems (1985).

20. D. Sullivan, *Quasiconformal homeomorphisms and dynamics. I, Solution of the Fatou-Julia problem on wandering domains*, Ann. of Math. **122** (1985), 401–418.

21. ____, *Hyperbolic geometry and homemorphisms*, Geometric Topology (Proc. Georgia Topology Conf., Athens, Georgia, 1977), Academic Press, New York, 1979, pp. 543–555.

22. ____, *Conformal dynamical systems*, Geometric Dynamics (Proc. Internat. Conf. Dynamical Systems, Rio de Janeiro, 1981), Lecture Notes in Math., vol. 1007, Springer-Verlag, 1983, p. 725.

23. ____, *On the ergodic theory at infinity of an arbitrary discrete group of hyperbolic motions* (Stony Brook Conference on Riemann Surfaces, Stony Brook, N.Y., 1978), Princeton Univ. Press, Princeton, N.J., 1981.

24. N. Teleman, *Atiyah Singer theory for Lipschitz manifolds*, Inst. Hautes Études Sci. Publ. Math., 1984.

25. P. Tukia, *Groups of quasiconformal homeomorphisms...*, Acta Math.

26. K. Uhlenbeck, *Connections with L^p bounds on curvatures*, Comm. Math. Phys. **83** (1982), 11–29, 31–42.

27. J. Vaisala, *Lectures on n-dimensional quasiconformal mappings*, Lecture Notes in Math., vol. 229, Springer-Verlag.

28. W. P. Ziemer, *Absolute continuity and change of variables*, Trans. Amer. Math. Soc., 1968.

29. R. Zimmer, *Semisimple groups and ergodic theory*, Birkhauser, Basel, 1986.

30. C. Tresser and P. Coullet, *Iterations ... et groupe de renormalization*, J. Physique, Colloque C5, Supplement au no° 8, tome 39, 1978, pp. C5–25..

31. J. Guckenheimer, *Limit sets of s-unimodal maps with zero entropy*, Commun. Math. Phys. **110** (1987).

CITY UNIVERSITY OF NEW YORK, NEW YORK, NEW YORK 10036, USA

INSTITUT DES HAUTES ÉTUDES SCIENTIFIQUE, BURES-SUR-YVETTE, FRANCE

Homoclinic Bifurcations

FLORIS TAKENS

1. Introduction. We say that a one-parameter family of diffeomorphisms $\varphi_\mu\colon M \to M$, $\mu \in \mathbf{R}$, has a homoclinic bifurcation, or a homoclinic tangency, for $\mu = 0$ if φ_0 has an orbit of nontransverse intersection of a stable and an unstable manifold, both of the same hyperbolic fixed point (or periodic point), which splits, for $\mu > 0$, into two orbits of transverse intersection of these stable and unstable manifolds. Definitions will be recalled in §§2 and 3.

These orbits of intersection of stable and unstable manifolds of the *same* hyperbolic fixed points, or *homoclinic orbits*, often imply or are implied by complex dynamic behavior. So one may expect that at or near homoclinic bifurcations one will have transitions from simple to complex dynamic behavior and also (discontinuous) transitions between different kinds of complex dynamics. These transitions form the subject of this paper. This is a survey of recent work which was carried out mainly in collaboration with J. Palis and which is a continuation of the earlier work of S. Newhouse and J. Palis.

2. Homoclinic orbits, simple and complex dynamics, and hyperbolicity. We recall some definitions and basic results; see [21, 14, 19] for more details. Let $\varphi\colon M \to M$ be a diffeomorphism. A fixed point p of φ is called *hyperbolic* if $d\varphi(p)$ has no eigenvalue of norm 1. In that case, the *stable* and the *unstable manifolds* of p are injectively immersed submanifolds; these manifolds are defined by

$$W^s(p) = \left\{ x \in M \mid \lim_{n \to +\infty} \varphi^n(x) = p \right\}$$

and

$$W^u(p) = \left\{ x \in M \mid \lim_{n \to -\infty} \varphi^n(x) = p \right\},$$

respectively. A *homoclinic orbit* of such a hyperbolic fixed point p of φ is an orbit in $W^s(p) \cap W^u(p)\backslash\{p\}$.

The existence of a homoclinic orbit, especially when it is an orbit of transverse intersection of stable and unstable manifolds, implies complex dynamic behavior. We shall define what we understand here as complex dynamics in terms of the positive limit set. (This notion could also be defined in terms of the nonwandering set, the recurrent set, or other such sets, but the positive limit set seems to be more natural at least if the diffeomorphism decribes a time evolution: this positive limit set, together with its induced dynamics, determines the asymptotic dynamics of all orbits $\{\varphi^n(x)\}_{n=-\infty}^{+\infty}$ for $n \to +\infty$.) For a diffeomorphism $\varphi\colon M \to M$ and a point $x \in M$, one defines the ω-limit as $\omega(x) = \{x' | \exists n_i \to +\infty$ such that $\varphi^{n_i}(x) \to x'\}$; the α-limit is defined similarly with $n_i \to -\infty$ instead of $n_i \to +\infty$. Usually one assumes M to be compact in order to have $\omega(x)$ and $\alpha(x)$ nonempty. The *positive limit set* of φ is defined as $L^+ = \overline{\bigcup_{x \in M} \omega(x)}$. If we want to express the dependence of these sets on φ, we write $\omega(x, \varphi)$, $\alpha(x, \varphi)$, and $L^+(\varphi)$. We say that the *dynamics* of φ is *simple* if the positive limit set of φ is finite, and is *complex* if the set is infinite. From now on we restrict ourselves to diffeomorphisms on compact manifolds.

The existence of a transverse homoclinic orbit implies complex dynamics. One even conjectures that it is a generic property of diffeomorphisms (on a compact manifold) to have either simple dynamics or homoclinic orbits.

Independent of having simple or complex dynamics, a diffeomorphism may be *hyperbolic*. Let $K \subset M$ be a compact invariant set for a diffeomorphism $\varphi\colon M \to M$, and let $\|\ \|$ denote the norm of tangent vectors to M with respect to some Riemannian metric. We say that K is *hyperbolic* if there is a continuous splitting $T(M)|_K = E^u \oplus E^s$ of the tangent bundle of M restricted to K and if there are constants $C > 1$ and $\lambda > 1$ such that for all $v \in E^u$ and $n \geq 0$, $\|d\varphi^n(v)\| \geq C^{-1} \cdot \lambda^n$, and for all $v \in E^s$ and $n \geq 0$, $\|d\varphi^n(v)\| \leq C \cdot \lambda^{-n}$.

We say that K has a *cycle* if there is a finite sequence of points $x_0, x_1, \ldots, x_k = x_0$ in M, not all contained in K, such that for all $i = 0, 1, \ldots, k-1$, $\alpha(x_i) \subset K$, $\omega(x_i) \subset K$, and $\omega(x_i) \cap \alpha(x_{i+1}) \neq \varnothing$. Note that if K consists only of one hyperbolic fixed point, then a cycle is essentially the same as a homoclinic orbit. We say that a diffeomorphism $\varphi\colon M \to M$ is *hyperbolic* if $L^+(\varphi)$, as an invariant subset for φ is hyperbolic and has no cycles. Otherwise φ is called nonhyperbolic. Although the formulation of this definition differs from the corresponding definition in [16], it follows from [7, 22, 12] that the corresponding notions are equal.

A main reason for introducing the class of hyperbolic diffeomorphisms is that the topology of their dynamics, as far as the asymptotic behavior of orbits $\{\varphi^n(x)\}_{n=-\infty}^{+\infty}$ for $n \to +\infty$ is concerned, is persistent in the following sense. If $\varphi\colon M \to M$, M a compact manifold, is hyperbolic, then φ has a neighborhood $\mathfrak{U}$ in the space of C^1-diffeomorphisms such that for any $\Psi \in \mathfrak{U}$ there is a homeomorphism $h\colon L^+(\varphi) \to L^+(\Psi)$ such that

$$h \circ (\varphi | L^+(\varphi)) = (\Psi | L^+(\Psi)) \circ h.$$

This last statement is equivalent to the Ω-stability theorem [22]. If, for a diffeomorphism φ, there is a neighborhood $\mathfrak{U}$ as above, then we call φ *positive limit stable*.

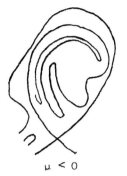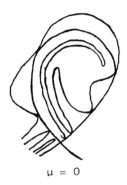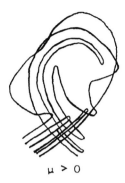

$\mu < 0$ $\mu = 0$ $\mu > 0$

FIGURE 1. Stable and unstable manifolds of a one-parameter family of 2-dimensional diffeomorphisms with a homoclinic tangency.

A homeomorphism h as above is called a *positive limit conjugacy*. In fact it may follow from recent work of Mañé [4, 5] that hyperbolicity is even equivalent with positive limit stability.

The topological structure and the ergodic properties of positive limit sets of hyperbolic diffeomorphisms have been studied extensively; e.g., see [2, 19]. On the other hand, many examples of so-called strange attractors probably only occur in nonhyperbolic diffeomorphisms with complex dynamics.

We observe that any diffeomorphism which has a homoclinic orbit of tangency is nonhyperbolic. This and Figure 1 indicate that near a homoclinic tangency there are very many nonhyperbolic diffeomorphisms with complex dynamics.

3. Homoclinic tangencies. As stated in the Introduction, we consider one-parameter families of diffeomorphisms $\varphi_\mu\colon M \to M$ which have a homoclinic tangency for $\mu = 0$. We have to impose, however, some further conditions for our considerations to be valid. In this section we discuss these extra conditions and formulate the main question which we want to investigate.

We assume the manifold M to be compact. Also, we assume that the diffeomorphisms φ_μ are C^2 for each μ and that their dependence on μ is at least C^1. For μ near zero, we assume that p_μ is a hyperbolic fixed point of φ_μ depending continuously on μ (for hyperbolic periodic points there is a similar theory), and that $W^u(p_\mu)$ and $W^s(p_\mu)$ have, for $\mu = 0$, one orbit of nontransverse intersection. We assume that this tangency is parabolic and unfolds generically; i.e., we assume that there are (μ-dependent) coordinates $x_1, \ldots, x_n$ so that locally on a neighborhood of a point of the orbit of tangency,

$$W^u(p_\mu) = \{(x_1, \ldots, x_n) | x_{u+1} = \cdots = x_n = 0\},$$
$$W^s(p_\mu) = \{(x_1, \ldots, x_n) | x_1 = \cdots = x_{u-1} = 0, \ x_{u+1} = x_u^2 - \mu\},$$

where $u = \dim(W^u(p_\mu)) = n - \dim(W^s(p_\mu))$; see Figure 2.

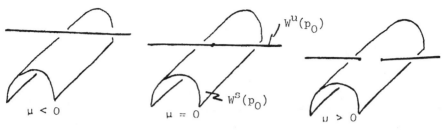

FIGURE 2

Other generic assumptions which we impose are:

• Except for the one orbit of tangency of φ_0, all periodic orbits of φ_0 are hyperbolic and all (other) intersections of stable and unstable manifolds are transverse (compare this with the Kupka-Smale theorem [14]).

• The eigenvalues of $d\varphi_0(p_0)$ have multiplicity one.

• If λ and μ are the norms of the weakest contracting and weakest expanding eigenvalues, respectively, of $d\varphi_0(p_0)$, then $\lambda \cdot \mu \neq 1$.

Next we have some nongeneric assumptions. First we assume that $\mu = 0$ is a "first bifurcation" in the sense that for some $\varepsilon > 0$, all φ_μ, with $-\varepsilon < \mu < 0$, are hyperbolic. This is as strong a restriction as can be seen from the analysis in [10, 11, 16]. The point is that in important classes of examples, homoclinic tangencies are preceded by infinite cascades of bifurcations; see [3] and [11]. These cascades may be related to strange attractors; see question 6 in [13].

Our second (nongeneric) assumption is that $\lim_{\mu \nearrow 0} L^+(\varphi_\mu)$ exists and is a hyperbolic set for φ_0; we denote this limit set by $\tilde{L}^+$. This second assumption is probably not completely independent of the first; at least, I do not know how to make a (generic) example which satisfies the first assumption but not the second.

Finally there is a last, somewhat technical assumption, but I do not know whether it is generic or not. Some further definitions are needed to state it, so we postpone it to §4. When $\dim(M) = 2$, this last condition can be omitted. The rest of this section does not depend on this last assumption.

Due to these assumptions, especially the second nongeneric assumption, the orbit of tangency will not be contained in $\tilde{L}^+$. This means that for small positive values of μ, $L^+(\varphi_\mu)$ consists of one part near $\tilde{L}^+$ (which is in fact topologically conjugated with $\tilde{L}^+$) and a part near the orbit of tangency. So the positive limit set $L^+(\varphi_\mu)$ explodes (i.e., becomes suddenly bigger) when μ passes through zero; we call a one-parameter family of diffeomorphisms φ_μ satisfying all the above conditions a *homoclinic limit explosion* (this is the same as what was called a homoclinic Ω explosion in [16]).

As observed before, for $\mu > 0$, φ_μ has complex dynamics. The main question we are interested in is whether or not φ_μ is hyperbolic for most small but positive values of μ. To make this more precise, let

$$B(\varphi_\mu) = \{\tilde{\mu} \geq 0 | \varphi_{\tilde{\mu}} \text{ is not hyperbolic}\}.$$

From §2 it follows that if $\overline{\mu} \notin B(\varphi_\mu)$ and μ' is near $\overline{\mu}$, then $\varphi_{\overline{\mu}}$ and $\varphi_{\mu'}$ are limit conjugated. Also, $B(\varphi_\mu)$ contains all the "Ω-bifurcations."

We are mainly interested in

$$\lim_{\delta \searrow 0} \frac{m(B(\varphi_\mu) \cap [0,\delta])}{\delta}$$

if it exists (m denotes the Lebesgue measure) and in the corresponding $\lim \sup$ and $\lim \inf$, otherwise.

The value of this limit depends on certain geometric properties of

$$\lim_{\mu \nearrow 0} L^+(\varphi_\mu) = \tilde{L}^+.$$

These properties are geometric in the sense that they are not preserved under topological conjugacies.

4. Limit capacity and thickness of hyperbolic invariant sets. Let $\varphi \colon M \to M$ be a C^2 diffeomorphism on a compact manifold M, let $L \subset M$ be a hyperbolic invariant set, and let $p \in L$ be a fixed point. We shall introduce a number of invariants associated to (φ, L, p). The main situation we have in mind here is the following: $\varphi = \varphi_0$, where φ_μ is a one-parameter family of diffeomorphisms with homoclinic limit explosion, $p = p_0$, where p_μ is the saddle point involved in the homoclinic tangency and $L = \tilde{L}^+ = \lim_{\mu \nearrow 0} L^+(\varphi_\mu)$ as defined in §3. We assume that p is not isolated in L, in the sense that both $W^s(p) \cap L$ and $W^u(p) \cap L$ have p as an accumulation point. Furthermore, we assume that the weakest contracting—and the weakest expanding—eigenvalues of $(d\varphi)(p)$ are real and positive, and have multiplicity one. This may seem like a strange condition, but with $\varphi = \varphi_0$, etc., as above, it can be shown that the weakest eigenvalues of $(d\varphi_0)(p_0)$ are real and positive whenever p_0 is not isolated in $\tilde{L}^+$ [11]. (This follows from the condition that $\mu = 0$ is a first bifurcation.) For the multiplicity one condition, see §3.

In this situation there is an invariant foliation, the strong stable foliation $\mathfrak{F}^{ss}$, in $W^s(p)$: its leaves have codimension one in $W^s(p)$, it is C^{k-1} if φ is C^k (so here it is at least C^1), and with the above properties it is unique. The leaf of $\mathfrak{F}^{ss}$ through p is denoted by $W^{ss}(p)$. With the present definitions we can formulate the *final condition* for homoclinic limit explosions in §3. This final condition is that $\tilde{L}^+ \cap W^{ss}(p_0) = \tilde{L}^+ \cap W^{uu}(p_0) = \{p_0\}$. Note that if p_0 is isolated in $\tilde{L}^+$ or if $\dim(M) = 2$, this condition is automatically satisfied. Also, from [11] and the other conditions imposed on homoclinic limit explosions it follows that $\tilde{L}^+ \cap W^s(p_0)$ must be contained in a closed half-space, bounded by $W^{ss}(p_0)$ (and the same for $W^u(p_0)$, $W^{uu}(p_0)$).

We return to the strong stable foliation $\mathfrak{F}^{ss}$ in $W^s(p)$. It defines a smooth projection π_s of $W^s(p)$ on a 1-dimensional manifold l_s, namely, on the space of leaves of $\mathfrak{F}^{ss}$. $L_s \subset l_s$ denotes the closure of $\pi_s(L \cap W^s(p))$. This subset $L_s \subset l_s$ admits a scaling: the map induced in l_s by $\varphi|W^s(p)$ is a hyperbolic contraction. Hence [23] there is a diffeomorphism $\kappa_s \colon l_s \to \mathbf{R}$ which linearizes this contraction. This linearized contraction is just multiplication with λ, the weakest contracting eigenvalue of $(d\varphi)(p)$. Since L is φ invariant, $\kappa_s(L_s) = \lambda \cdot \kappa_s(L_\lambda)$. (This implies that the global structure of L_s is determined by the intersection of L_s with a small neighborhood of $\pi_s(p)$.) Since the linearization κ_s is unique (up to a scalar multiplication) it defines a canonical metric in l_s (up to scalar multiplication).

If φ is hyperbolic, $L = L^+(\varphi)$, and $\dim(W^s(p)) = 1$ (so that $l_s = W^s(p)$ and $\pi_s = \mathrm{id}$), then L_s is a Cantor set. For $\dim(W^s(p)) > 1$, L_s may contain intervals; we do not know whether it may contain isolated points.

We define the *limit capacity* of L_s. For this we first define the limit capacity of a compact metric space (K, ρ). Let $n(K, \varepsilon)$ be the minimal number of ε-neighborhood necessary to cover K. The limit capacity of K is defined as

$$d(K) = \limsup_{\varepsilon \to 0} \frac{\ln(n(K, \varepsilon))}{-\ln \varepsilon}.$$

The limit capacity of L_s is now defined as the limit capacity, in the above sense, of a compact neighborhood of $\pi_s(p)$ in L_s with respect to the metric in l_s defined by the linearization. Due to the scaled structure of L_s and the fact that the limit capacity does not change when multiplying the metric ρ with a constant, the limit capacity $d(L_s)$ is independent of the choices of neighborhood and linearization.

We call $d(L_s)$ the *stable limit capacity of L at p*. The *unstable limit capacity of L at p* is similarly defined using the strong unstable foliation $\mathfrak{F}^{uu}$ in $W^u(p)$.

We define the notion of thickness for compact subsets $K \subset \mathbf{R}$.

A connected component of $\mathbf{R} - K$ is called a *gap*. A point $k \in K$ which is a boundary point of a finite gap of K is called a boundary point of K; the set of boundary points of K is denoted by ∂K. Let $k \in K$ be a boundary point of K. Let U be the gap such that $k \in \partial U$. Then there is a unique gap V such that:

(i) k is between U and V;

(ii) $l(V) \geq l(U)$ (V may be infinite);

(iii) if W is any gap between U and V, then $l(W) < l(U)$.

Let C be the closed interval between U and V. Then the thickness of K at k is

$$\tau(K, k) = l(C)/l(U).$$

The thickness of K is $\tau(K) = \inf \tau(K, k)$, where the infimum is taken over all boundary points $K \in \partial K$. Note that this definition is equivalent to the definition in [9]. The *stable thickness* of L at p is defined as the thickness of a neighborhood of $\pi_s(p)$ in L_s. Since the definition of thickness depends only on ratios of distances, the definition is independent of the linearization; since L_s has a scaling, this definition is independent of the size of the neighborhood of $\pi_s(p)$. The *unstable thickness* of L at p is similarly defined.

The main reason for these notions—limit capacity and thickness—are of importance to our present problem in the relation they have with the difference of two (Cantor) sets in $\mathbf{R}$, as expressed in the following propositions; the *difference* of two subsets $A, B \subset \mathbf{R}$ is defined as

$$A - B = \{t \mid A \cap (B + t) \neq \varnothing\}.$$

PROPOSITION [15]. *If $A, B \subset \mathbf{R}$ are closed sets and if the sum $d(A) + d(B)$ of their limit capacities is smaller than one, then the Lebesgue measure of $(A - B)$ is zero.*

PROPOSITION [6]. *If $A, B \subset \mathbf{R}$ are Cantor sets and if the product of their thicknesses $\tau(A) \cdot \tau(B)$ is bigger than one, then $A - B$ is a finite union of intervals.*

5. Results. Homoclinic orbits were studied by Poincaré in relation to celestial mechanics in [17] and [18]. He observed that they imply great dynamic complexity. This situation was further analyzed by Birkhoff [1]. A full description of the dynamic complexity due to a homoclinic orbit in a simple example (the horseshoe) was given by Smale [20], who also showed that near any transverse homoclinic orbit there is an invariant subset on which the dynamics is conjugated with that of the horseshoe.

Homoclinic tangencies were then investigated by Newhouse [6, 8, 9] in order to show that there are open sets of nonhyperbolic diffeomorphisms in the space of all C^2-diffeomorphisms of the 2-sphere. For our problem, this analysis leads to the following results.

THEOREM (NEWHOUSE). *Let $\varphi_\mu: M \to M$, $\dim(M) = 2$, be a one-parameter family of diffeomorphisms with a homoclinic limit explosion (see §3) involving the saddle point p_μ. Then there are arbitrarily-near-zero intervals $I \subset \mathbf{R}$ such that for $\mu \in I$, φ_μ is nonhyperbolic. If we also assume that the product of the stable and the unstable thicknesses of $\tilde{L}^+$ at p_0 are bigger than one, then there is some $\mu_0 > 0$ such that for any $0 < \mu < \mu_0$, φ_μ is nonhyperbolic.*

In higher dimensions there are corresponding but weaker results; it is not yet clear what the final result should be for general dimensions.

On the other hand we have

THEOREM (NEWHOUSE, PALIS, TAKENS). *Let $\varphi_\mu: M \to M$ be a one-parameter family of diffeomorphisms with a homoclinic limit explosion (see §3), involving the saddle point p_μ. If the sum of the stable and unstable limit capacity of $\tilde{L}^+$ at p_0 is smaller than one, then*

$$\lim_{\delta \searrow 0} \frac{m(B(\varphi_\mu) \cap [0, \delta])}{\delta} = 0;$$

see also §3.

The case that the dynamics of φ_μ, $\mu < 0$, is simple, was proved (in a somewhat weaker form) by Newhouse and Palis [10, 11]. The 2-dimensional case, without restricting the dynamics to be simple, was proved by Palis and Takens [16]. The proof for the higher dimensional case has not yet appeared. For corresponding results for cycles made by hetroclinic bifurcations, see [10, 11, 15].

Finally we mention that there are also generic examples of homoclinic limit explosions $\varphi_\mu: M \to M$, $\dim(M) \geq 3$, for which the expression

$$\frac{m(B(\varphi_\mu) \cap [0, \delta])}{\delta}$$

has both its lim sup and its lim inf for $\delta \to 0$ strictly contained in the open interval $(0, 1)$.

After this paper was completed I found a gap in the main proof of reference [9]; i.e., the thickness of $\Lambda_{22}(t_n')$ (page 138) is not proved in a convincing way.[1] This may affect the first statement in the theorem of Newhouse quoted in §5. The other statements are not affected by this. At this moment, C. Robinson claims to have solved this difficulty (added in proof, February 1987).

REFERENCES

1. G. D. Birkhoff, *Nouvelles recherches sur les systèmes dynamiques*, Mem. Pont. Acad. Sci. Novi Lyncaei **1** (1935), 85–216.

2. R. Bowen, *Equilibrium states and the ergodic theory of Anosov diffeomorphisms*, Lecture Notes in Math., Vol. 470, Springer-Verlag, Berlin-New York, 1975.

3. N. Gavrilov and L. Silnikov, *On three dimensional dynamical systems close to systems with a structurally unstable homoclinic curve*. I, Math. USSR-Sb. **17** (1972), no. 4, 467–485.

4. R. Mañé, *An ergodic closing lemma*, Ann. of Math. **116** (1982), 503–540.

5. _____, unpublished.

6. S. Newhouse, *Non-density of Axiom A(a) on S^r*, Proc. Sympos. Pure Math., vol. 14, Amer. Math. Soc., Providence, R.I., 1970, pp. 191–203.

7. _____, *Hyberbolic limit sets*, Trans. Amer. Math. Soc. **167** (1972), 125–150.

8. _____, *Diffeomorphisms with infinitely many sinks*, Topology **13** (1974), 9–18.

9. _____, *The abundance of wild hyperbolic sets and nonsmooth stable sets for diffeomorphisms*, Inst. Hautes Études Sci. Publ. Math. **50** (1979), 101–151.

10. S. Newhouse and J. Palis, *Bifurcations of Morse-Smale dynamical systems*, Dynamical Systems (Proc. Sympos., Univ. Bahia, Salvador, 1971), Academic Press, New York, 1973, pp. 303–366.

11. _____, *Cycles and bifurcations*, Astérisque, **31** (1976), 43–140.

12. J. Palis, *A note on Ω-stability*, Proc. Sympos. Pure Math., vol. 14, Amer. Math. Soc., Providence, R.I., 1970, pp. 221–222.

13. _____, *Homoclinic orbits, hyperbolic dynamics and dimension of Cantor sets*, Preprint I.M.P.A., Rio de Janeiro, 1986.

14. J. Palis and W. de Melo, *Geometric theory of dynamical systems, an introduction*, Springer-Verlag, 1982.

15. J. Palis and F. Takens, *Cycles and measure of bifurcation sets for two-dimensional diffeomorphisms*, Invent. Math. **82** (1985), 397–422.

16. _____, *Hyperbolicity and the creation of homoclinic orbits*, Preprint I.M.P.A., Rio de Janeiro, 1985.

17. H. Poincaré, *Sur le problème des trois corps et les equations de dynamique (Mem. couronné du Prix de S. M. le roi Oscar II de Suède)* Acta Math. **13** (1890), 1–270.

18. _____, *Les méthodes nouvelles de la méchanique céleste*. III, Gauthier-Villars, Paris, 1899.

19. M. Shub, *Stabilité global des systèmes dynamiques*, Astérisque **56** (1978).

20. S. Smale, *Diffeomorphisms with many periodic points*, Differential and Combinatorial Topology, Princeton Univ. Press, Princeton, N.J., 1965.

21. _____, *Differentiable dynamical systems*, Bull. Amer. Math. Soc. **73** (1967), 747–817.

22. _____, *The Ω-stability theorem*, Proc. Sympos. Pure Math., vol. 14, Amer. Math. Soc., Providence, R.I., 1970, pp. 289–298.

23. S. Sternberg, *Local contractions and a theorem of Poincaré*, Amer. J. Math. **79** (1957), 809–824.

I.M.P.A., RIO DE JANEIRO, BRASIL

MATHEMATISCH INSTITUUT, RIJKSUNIVERSITEIT, GRONINGEN, THE NETHERLANDS

---

[1]In June 1987, Newhouse showed me how to complete the proof; see Inst. Hautes Études Sci. Publ. Math., vol. 50.

Proceedings of the International Congress of Mathematicians
Berkeley, California, USA, 1986

The Arnold Conjecture for Fixed Points
of Symplectic Mappings and
Periodic Solutions of Hamiltonian Systems

E. ZEHNDER

It is the aim of this paper to describe some qualitative existence results for periodic solutions of Hamiltonian equations, which are related to the V. I. Arnold conjecture about fixed points of symplectic mappings. The conjecture originates in the circle of old questions of celestial mechanics related to the Poincaré-Birkhoff fixed point theorem.

1. V. I. Arnold conjecture; a history. We consider a compact symplectic manifold (M, ω), where ω is a distinguished closed and nondegenerate 2-form. To every smooth function $H \colon \mathbf{R} \times M \to \mathbf{R}$, one can associate a time-dependent vector field V_t on M defined by

$$\omega(\,\cdot\,, V_t) = dH_t(\,\cdot\,),$$

where $H_t(x) = H(t, x)$. This vector field is called the exact Hamiltonian vector field associated with the function H. The differential equation on M,

$$\frac{d}{dt}\phi^t = V_t \circ \phi^t, \qquad \phi^0 = \mathrm{id},$$

defines a family of diffeomorphisms on M which preserve the symplectic structure; i.e., for every $t \in \mathbf{R}$, $(\phi^t)^*\omega = \omega$, so that ϕ^t is a symplectic diffeomorphism.

DEFINITION. In the following we shall call a map ϕ on M Hamiltonian if it belongs to the flow ϕ^t of any time-dependent exact Hamiltonian vector field on M. We remark that one can show that the set of Hamiltonian maps is the subgroup $[G, G]$, where G is the one component of the group of all symplectic diffeomorphisms of M [**4**].

A Hamiltonian map is, in particular, homotopic to the identity and possesses, therefore, by Lefschetz theory, at least one fixed point if the Euler characteristic of M does not vanish, i.e., if $\chi(M) \neq 0$. The Lefschetz theory which applies to the class of all topological mappings is of no use for the Arnold conjecture. It is

Supported by the Stiftung Volkswagenwerk.

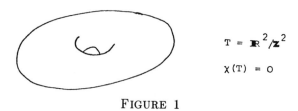

$$T = \mathbf{R}^2/\mathbf{Z}^2$$

$$\chi(T) = 0$$

FIGURE 1

the challenge of this conjecture to extend this theory to a Ljusternik-Schnirelman theory resp. to a Morse theory for fixed points for the more restricted class of Hamiltonian maps on symplectic manifolds.

ARNOLD CONJECTURE.. *A Hamiltonian map ϕ on a compact symplectic manifold (M,ω) possesses at least as many fixed points as a function on M has critical points.*

For example, in view of the Ljusternik-Schnirelman theory resp. the Morse theory for critical points of a function on M, we shall for the following restate the conjecture as

$$\# \ \{\text{fixed points}\} \geq \mathrm{CL}(M)$$

$$(\geq \mathrm{SB}(M) \ \text{if all the fixed points are nondegenerate}).$$

Here we abbreviate, by $\mathrm{CL}(M)$, the cuplength of a compact space X which is defined as the integer $1 + \sup\{k | \exists$ classes $\alpha_1,\ldots,\alpha_k \in H^*(X)\backslash\{1\}$ with $\alpha_1 \cup \alpha_2 \cup \cdots \cup \alpha_k \neq 0\}$. $\mathrm{SB}(X)$ stands for the sum of the Betti numbers of X. We also denote in the following the Poincaré polynomial of the space X by

$$p(t,x) = \sum \beta_k t^k, \qquad \beta_k = \dim H^k(X),$$

so that $\mathrm{SB}(X) = p(1,X)$, while the alternating sum $p(-1,X) = \chi(X)$ is the Euler characteristic of X.

Arnold was lead to the conjecture in his discussion [2] of the celebrated Poincaré-Birkhoff fixed point theorem for an area-preserving map on an annulus A in the plane [28, 5]. He showed that this theorem, which is not of topological nature, could be derived from a fixed point theorem for symplectic mappings $\phi \simeq \mathrm{id}$ on the 2-dimensional torus $T = \mathbf{R}^2/\mathbf{Z}^2$, at least in the differentiable case. (See Figure 1.)

The 2-torus is distinguished among the compact surfaces by the fact that the Lefschetz theory is not applicable, since $\chi(T) = 0$. As the translations show, the class of symplectic maps on T has to be restricted if it should possess fixed points. For ϕ belonging to the restricted class of Hamiltonian maps on T, Arnold formulated the following global fixed point theorem [1, 2], a special case of his conjecture, proved in 1983 [12].

$$\# \ \{\text{fixed points}\} \geq 3$$

$$(\geq 4 \ \text{if the fixed points are nondegenerate}).$$

The lower bounds are sharp. Clearly, this result is not an Euler-characteristic result; indeed $3 = \mathrm{CL}(T)$ and $4 = \mathrm{SB}(T)$. Under the additional condition that ϕ

is close to the identity map, i.e., $|\phi - \mathrm{id}|_{C^1}$ small, there is a direct and one-to-one relation between the fixed points of ϕ and the critical points of a function on T, a so-called generating function, as one verifies easily. In contrast, the above theorem is global and its proof is quite different.

In 2 dimensions a symplectic map simply preserves a volume form, and we illustrate the global consequences of this additional structure with two other examples. A homeomorphism of an open 2-disc in the plane does not necessarily have a fixed point; however, it always has at least one fixed point if it preserves a regular measure. This follows, e.g., by Brouwer's translation theorem. Similarly, a homeomorphism $\phi \simeq \mathrm{id}$ on S^2 possesses ≥ 1 fixed points by Lefschetz theory since $\chi(S^2) = 2 \neq 0$, but it may have only one. Under the additional assumption however that ϕ preserves a regular measure, it has ≥ 2 fixed points. In particular, ϕ satisfying $\phi^* \omega = \omega$ on S^2 has ≥ 2 fixed points. This was observed in 1974 by C. Simon [33] and N. Nikishin [26] using another strictly 2-dimensional argument.

It turns out that these global fixed point theorems are not a 2-dimensional phenomenon, as we shall illustrate next in the special case of the torus. The crucial observation is a dynamical interpretation of the Arnold conjecture. If the Hamiltonian function H depends periodically on time t, i.e., $H \colon S^1 \times M \to \mathbf{R}$, with period 1, then the 1-periodic solutions of the Hamiltonian vector field on M are obviously in one-to-one correspondence with the fixed points of the interpolated map ϕ which is the time 1 map of the flow. Instead of looking for fixed points of a map, we rather look for periodic solutions of a Hamiltonian equation. From this point of view the Arnold conjecture claims a lower bound for the number of periodic solutions of every periodic exact Hamiltonian vector field on (M, ω).

2. The torus; a variational principle.

(a) *Statement.* On the $2n$-dimensional torus $T = \mathbf{R}^{2n}/\mathbf{Z}^{2n}$, with its standard symplectic structure J, we consider any 1-periodic exact Hamiltonian vector field

$$\dot{x} = J\nabla H(t, x) \quad \text{on } T,$$

where $H \colon S^1 \times T \to \mathbf{R}$, and we look for 1-periodic solutions. The following result has been proved by C. Conley and the author in 1983 [12]:

$$\# \{\text{periodic solutions}\} \geq 2n + 1 = \mathrm{CL}(T)$$

$$\left(\geq 2^{2n} = \mathrm{SB}(T), \quad \begin{array}{l} \text{if all the periodic solutions} \\ \text{are nondegenerate} \end{array} \right).$$

The lower bounds are sharp. The statement is of qualitative nature. It requires no assumption on the Hamiltonian vector field. On the other hand, it gives no information about the position of the periodic solutions and about their linear and nonlinear Birkhoff invariants. This would be important in the investigation of the flow nearby using KAM theory. It turns out that all the periodic solutions claimed are contractible loops on T, and indeed other periodic solutions need not exist. This is in sharp contrast to the geometric problem of closed geodesics on T as a Riemannian manifold, where one easily finds a closed geodesic in every homotopy class of loops.

(b) *Idea of the proof.* The proof is based on a classical variational principle for which the critical points are precisely the periodic solutions we are looking for. Define on the loop space $\Omega = \Omega(T)$ of contractible loops of period 1, the function $f: \Omega \to \mathbf{R}$ by

$$f(x) = \int_0^1 \left\{ \frac{1}{2} \langle x, J\dot{x} \rangle + H(t, x) \right\} dt$$

which has the L_2-gradient

$$\nabla f(x) := J \frac{d}{dt} x + \nabla H(t, x), \qquad x \in \Omega,$$

so that the critical points $\nabla f(x) = 0$ are those loops in Ω which satisfy the Hamiltonian equation, since $J^2 = -1$. Searching for critical points of f one is confronted with the difficulty that this variational principle is degenerate; it is bounded neither from below nor from above so that standard variational techniques do not apply directly. For example, the Morse index of every possible critical point is infinite, hence topologically invisible at first sight. Only in the late seventies did P. Rabinowitz [29] demonstrate that such degenerate principles can be used effectively for existence proofs by means of subtle mini-max arguments. Our approach is different; the guiding principle is to study the set S of bounded solutions of the gradient equation

$$\frac{d}{ds} x = -\nabla f(x) \quad \text{on } \Omega.$$

Due to the gradient structure one expects the invariant set S to consist of the critical points of f as well as the orbits connecting them, and one hopes that S represents the topology of T. The Cauchy problem of this O.D.E. on Ω is not well-posed, since $J(d/dt)$ is a selfadjoint operator bounded neither from below nor from above. One could interpret a connecting orbit as a solution $u: \mathbf{R} \times S^1 \to T$ of the P.D.E.

$$\frac{\partial u}{\partial s} + J \frac{\partial u}{\partial t} = -\nabla H(t, u)$$

satisfying appropriate asymptotic conditions. In the case at hand, however, one can use the analytical device known as Liapunov-Schmidt reduction which allows us to reduce the problem of finding critical points for f on Ω to the equivalent problem of finding the critical points of a related function g, which is defined on a finite-dimensional submanifold $\hat{\Omega} \subset \Omega$, as $g: \hat{\Omega} := T \times \mathbf{R}^{2N} \to \mathbf{R}$, where $g(z) = f(u(z))$ with an embedding $u: \hat{\Omega} \to \Omega$. This reduction has a smoothing effect and makes the topology of S visible. The flow of the reduced equation $dz/ds = -\nabla g(z)$ on $\hat{\Omega}$ looks schematically as in Figure 2.

One verifies easily that the set S of bounded solutions, which consists of the critical points and the connecting orbits, has a Conley homotopy index [10] whose cohomology is equal to $H^*(T) \otimes H^*(\dot{S}^N)$, i.e., is equal to the cohomology

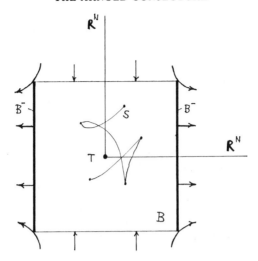

FIGURE 2. $B = T \times D_+ \times D_-$ has the exit set $B^- = T \times \partial D_+ \times D_-$.

of T shifted up in dimension by N. The Morse equation for S is, therefore, in the case of nondegenerate critical points,

$$\sum_{\nabla g(z)=0} t^{m(z)} = p(t, T) \cdot t^N + (1+t)Q(t)$$

with Q having nonnegative integer coefficients, $m(z)$ being the Morse index. Setting $t = 1$ one finds, in particular, # {critical points} $\geq \mathrm{SB}(T)$ as claimed. The Morse indices of the critical points depend on the reduction dimension; their differences, however, have an invariant interpretation in terms of the Maslov indices of the corresponding periodic solutions of the equation [11]; in fact, they depend only on the interpolated Hamiltonian map ϕ and the corresponding fixed points [34, 37].

The Conley index of S is not sufficient to determine a lower bound for the critical points in the degenerate case. But one can see that the topology of the invariant set S itself inherits the ring structure of T. There is a map $\alpha \colon \Omega \to T$ which induces an injective map $(\alpha|S)^* \colon H^*(T) \to H^*(S)$ in cohomology [18] so that, in particular, $\mathrm{CL}(S) \geq \mathrm{CL}(T)$. Since the flow on S is gradient-like, one therefore concludes that #{critical points} $\geq \mathrm{CL}(S)$ as claimed in the theorem. It is important in the proof that the Conley index is cohomologically not just a ring but a $H^*(T)$ module defined by the map α. This added structure is an additional and useful invariant in Conley's deformation theory [16].

The torus example illustrates the difference between the above degenerate variational principle and the variational problem for the geometric problem of closed geodesics on a Riemannian manifold. In the latter case the functional is bounded from below and every cohomology class of the loop space is represented in the index of some critical point. In the general Hamiltonian case this is clearly not true; only the cohomology of the underlying manifold itself has to be represented in the critical points.

3. Fixed point results. The Arnold conjecture is not yet proved in full generality. The variational approach described above has, however, been extended and lead to proofs of various special cases which we shall briefly summarize. We begin with a perturbation result. A. Weinstein [36] observed that the conjecture is true for every compact symplectic manifold (M, ω) provided the Hamiltonian map ϕ belongs to the flow of a sufficiently small Hamiltonian vector field (in the sup-norm).

The global cases for S^2, T^2, and (T^{2n}, ω^*) with the standard symplectic structure have already been mentioned. It should be said that for 2-dimensional surfaces it is sufficient to prove the conjecture for a convenient symplectic structure, since any two volume forms are equivalent by a diffeomorphism if they have the same total volume [25]. On a compact oriented surface of genus ≥ 1 there is a symplectic structure with an associated Riemann metric having nonpositive curvature, i.e.,

$$\omega(X, Y) = g(JX, Y)$$

with a Hermitian structure $J^2 = -\,\mathrm{id}$. This additional structure helps in proving the conjecture for all compact surfaces $F = F_g$ of genus $g \geq 1$: a Hamiltonian map ϕ on F has $\geq 3 = \mathrm{CL}(F)$ fixed points ($\geq 2 + 2g = \mathrm{SB}(F)$ if all the fixed points are nondegenerate). Proofs are due to A. Floer [17], J. C. Sikorav [32, 31], and also Ya. Eliashberg [13]. The first two authors proved the conjecture, in fact, for symplectic manifolds in arbitrary dimensions for which there is an associated Riemann metric with sectional curvature ≤ 0 and which satisfy additional technical conditions. The manifolds include, e.g., compact quotients of complex hyperbolic spaces $\mathbf{H}_\mathbf{C}^n$. The conjecture for the complex projective space $\mathbf{CP}^n$ with its standard symplectic structure has been verified by B. Fortune [19]; it extends $S^2 \cong \mathbf{CP}$ to higher dimensions.

By entirely different methods M. Gromov [20] established a fixed point for a Hamiltonian map ϕ on manifolds (M, ω) satisfying $\omega|\pi_2(M) = 0$. Most recently A. Floer [14] announced an extension of this existence result to a Morse-theory for fixed points of ϕ and proved [37–40] the Arnold conjecture for manifolds M with $\pi_2(M) = 0$ and the cohomology with respect to $\mathbf{Z}_2$-coefficients:

If the symplectic manifold (M, ω) satisfies $\pi_2(M) = 0$, then a Hamiltonian map φ has $\geq \mathrm{CL}_{\mathbf{Z}_2}(M)$ fixed points ($\geq \mathrm{SB}_{\mathbf{Z}_2}(M)$, if all the fixed points are nondegenerate).

Extending the underlying ideas of the torus case, Floer constructs an algebraic Conley index for the set S of critical points and connecting orbits of the corresponding gradient flow in the infinite-dimensional loop space. This index is shown to be isomorphic to $H^*(M, \mathbf{Z}_2)$ by a subtle continuation argument as ϕ is deformed to the identity map by means of the flow ϕ^t of the exact Hamiltonian vector field. It is very likely that the restriction $\pi_2(M) = 0$ can be dropped. The theorem follows from a general intersection theorem for Lagrangian manifolds described in the next section.

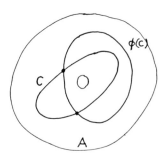

FIGURE 3

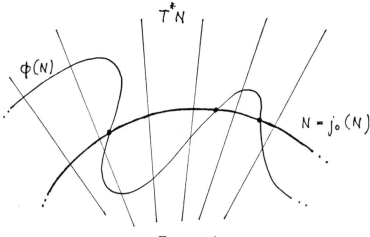

FIGURE 4

4. Intersections of Lagrangian manifolds.

As was first noticed by M. Chaperon [9], the variational approach is applicable also to other related global problems in symplectic geometry, in particular, to the special intersection problem of Lagrangian manifolds suggested by V. I. Arnold in [3] and originally prompted by the local Birkhoff-Lewis fixed point theorem for which we refer to [24]. On 2-dimensional manifolds the Lagrange-manifolds are simply curves. Recall that a closed and homotopically nontrivial curve C in the annulus A intersects its image curve $\phi(C)$ for a Hamiltonian map ϕ, which follows by Stoke's theorem. (See Figure 3.)

Similarly the zero-section of the cotangent bundle T^*S^1 cannot be disconnected from itself by a Hamiltonian map ϕ. In higher dimensions we consider the cotangent bundle T^*N of a compact manifold N, with the symplectic structure $\omega = d\lambda$, where $\lambda = $ "$p\,dq$" is the canonical 1-form. It turns out that the zero-section $j_0(N)$ of T^*N, which we identify with N, cannot be disconnected from itself by a Hamiltonian map ϕ on T^*N. (See Figure 4.)

More precisely, the following global intersection theorem holds:

$$\# \{\phi(N) \cap N\} \geq \mathrm{CL}(N)$$
$$(\geq \mathrm{SB}(N) \text{ if } \phi(N) \cap N \text{ is transversal}).$$

Proofs are due to M. Chaperon [9] for the torus and to H. Hofer [21] for a general compact manifold N. The intersection points are the critical points of the classical action-functional

$$f(x) = \int_0^1 \{x^*(\lambda) - H(t, x)\}\, dt,$$

with boundary conditions $x(0)$ and $x(1) \in j_0(N)$, where the function H defines an interpolating exact Hamiltonian vector field for ϕ. In their proofs F. Laudenbach and J. C. Sikorav [22] and J. C. Sikorav [31] use a more geometric approach similar to the generating function technique which is based on a different variational principle on finite-dimensional vector bundles over T^*N, prompted by [7, 8].

The submanifolds N and $\phi(N)$ of $(T^*N, d\lambda)$ are examples of exact Lagrangian manifolds. Recall that a submanifold $L \subset M$ of any symplectic manifold (M, ω) is called Lagrangian if $\omega|L = 0$ and if $\dim L = \frac{1}{2}\dim M$. In case $\omega = d\lambda$, a Lagrangian submanifold is called exact if $\lambda|L$ is an exact one-form. We point out that every exact Lagrangian embedding $j\colon N \to (T^*N, d\lambda)$ necessarily intersects the zero section $j_0(N)$; see Gromov [20].

Assume now (M, ω) to be a general, but compact symplectic manifold and consider a Lagrangian submanifold $L \subset M$. Then one could ask for intersections $\phi(L) \cap L$. Without further assumptions on (M, L), intersection points need not exist of course. For example, on the 2-dimensional torus a small loop bounding a 2-disc can easily be disconnected from itself by a Hamiltonian map, while for the homotopically nontrivial circle this is not possible. A general theorem has not been formulated up to now. There is, however, a partial result which has been most recently proved by A. Floer [37–40]:

Assume (M, ω) is a compact symplectic manifold and $L \subset M$ is a compact Lagrangian submanifold satisfying $\pi_2(M, L) = 0$. If ϕ is a Hamiltonian map on M, then the number of elements in $\phi(L) \cap L$ is $\geq \mathrm{CL}_{\mathbf{Z}_2}(L)$. If, moreover, $\phi(L) \cap L$ is transversal, then the number of intersection points is $\geq \mathrm{SB}_{\mathbf{Z}_2}(L)$.

REMARK. Repeated application of the Poincaré-Birkhoff fixed point theorem to higher iterates of the map gives, in one stroke, infinitely many periodic points. An analogous theorem for periodic points for a Hamiltonian map on a compact symplectic manifold has not yet been found. There is merely a partial result for the special case of the torus [41].

REFERENCES

1. V. I. Arnold, "Fixed points of symplectic diffeomorphisms", Problem XX in *Problems of Present Day Mathematics*, Mathematical Developments Arising from Hilbert Problems, Proc. Sympos. Pure Math., vol. 28, Amer. Math. Soc., Providence, R. I., 1976, p. 66.

2. ____, Appendix 9 in *Mathematical methods of classical mechanics*, Springer-Verlag, Berlin and New York, 1978.

3. ____, *Sur une propriété topologique des applications globalement canoniques de la mécanique classique*, C. R. Acad. Sci. Paris Sér. I Math. **261** (1965), 3719–3722.

4. A. Banyaga, *Sur la structure du groupe des difféomorphismes qui presérvent une forme symplectique*, Comment. Math. Helv. **53** (1978), 174–227.

5. G. D. Birkhoff, *Proof of Poincaré's geometric theorem*, Trans. Amer. Math. Soc. **14** (1913), 14–22.

6. ____, *The restricted problem of three bodies*, Rend. Circ. Mat. Palermo **39** (1915), 265–334.

7. M. Chaperon, *Une ideé du type "géodesiques brisées" pour les système hamiltoniens*, C.R. Acad. Sci. Paris Sér. I Math. **298** (1984), 293–296.

8. ____, *An elementary proof of the Conley-Zehnder theorem in symplectic geometry*, Dynamical Systems and Bifurcations, Lecture Notes in Math., vol. 1125, Springer-Verlag, Berlin and New York, pp. 1–8.

9. ____, *Quelques questions de géométrie symplectique*, Séminaire Bourbaki, 1982–83, Astérisque, 105–106, Soc. Math. France, Paris, 1983, pp. 231–249.

10. C. C. Conley, *Isolated invariant sets and the Morse index*, CBMS Regional Conf. Ser. in Math., no. 38, Amer. Math. Soc., Providence, R.I., 1978.

11. C. C. Conley and E. Zehnder, *Morse type index theory for flows and periodic solutions for Hamiltonian equations*, Comm. Pure Appl. Math. **37** (1984), 207–253.

12. ____, *The Birkhoff-Lewis fixed point theorem and a conjecture of V. I. Arnold*, Invent. Math. **73** (1983), 33–49.

13. Ya. M. Eliashberg, *Estimates on the number of fixed points of area preserving transformation*, Preprint, Syktyvkar, 1978. (Kissian)

14. A. Floer, *Morse theory for fixed points of symplectic diffeomorphisms*, Bull. Amer. Math. Soc. (N.S.) **16** (1987), 279–281.

15. ____, *Morse theory for Lagrangian intersections*, Preprint, SUNY Stony Brook, 1986.

16. ____, *A refinement of the Conley-Index and an application to the stability of hyperbolic invariant sets*, Ergodic Theory Dynamical Systems (to appear).

17. ____, *Proof of the Arnold conjecture for surfaces and generalizations for certain Kähler-manifolds*, Duke Math. J. **3** (1986), 1–32.

18. A. Floer and E. Zehnder, *Fixed point result for symplectic maps related to the Arnold-conjecture*, Dynamical Systems and Bifurcations, Lecture Notes in Math., vol. 1125, Springer-Verlag, Berlin and New York, 1985, pp. 47–64.

19. B. Fortune, *A symplectic fixed point theorem for CP^n*, Invent. Math. **81** (1985), 22–46.

20. M. Gromov, *Pseudo holomorphic curves in symplectic manifolds*, Invent. Math. **82** (1985), 307–347.

21. H. Hofer, *Lagrangian embeddings and critical point theory*, Ann. Inst. H. Poincaré. Anal. Non Linéaire **2** (1985), 407–462.

22. F. Laudenbach and J. C. Sikorav, *Persistence d'intersections avec la section nulle un cours d'une isotopie Hamiltonienne dans un fibre cotangent*, Invent. Math. **82** (1985), 349–357.

23. J. Milnor, *Lectures on the H-cobordism theorem*, Princeton Univ. Press, Princeton, N.J., 1965.

24. J. Moser, *Proof of a generalized form of a fixed point theorem due to G. D. Birkhoff*, Geometry and Topology, Lecture Notes in Math., vol. 597, Springer-Verlag, Berlin and New York, 1977, pp. 464–494.

25. ____, *On the volume elements on a manifold*, Trans. Amer. Math. Soc. **120** (1965), 286–294.

26. N. Nikishin, *Fixed points of diffeomorphisms on the twosphere that preserve area*, Funktsional. Anal. i. Prilozhen **8** (1974), 84–85.

27. H. Poincaré, *Methodes nouvelles de la mécanique céleste*, vol. 3, Chapter 28, Gauthier-Villars, Paris, 1899.

28. ____, *Sur un théorème de Géométrie*, Rend. Circ. Mat. Palermo **33** (1912), 375–407.

29. P. Rabinowitz, *Periodic solutions of Hamiltonian systems*, Comm. Pure Appl. Math. **31** (1978), 157–184.

30. J. C. Sikorav, *Sur les immersions Lagrangiennes dans un fibré cotangent admettant une phase gémératrice globale*, C.R. Acad. Sci. Paris Sér. I Math. **302** (1986), 119–122.

31. _____, *Problêmes d'intersections et de points fixes en géométrie hamiltonienne*, Comment. Math. Helv. **62** (1987), 62–73.

32. _____, *Points fixes d'une application symplectique homologue à l'identité*, J. Differential Geom. **22** (1985), 49–79.

33. C. P. Simon, *A bound for the fixed point index of an area preserving map with applications to mechanics*, Invent. Math. **26** (1974), 187–200.

34. C. Viterbo, *Intersection des sous variétés lagrangiennes et index des points critiques de la fonetionelle d'action*, manuscript, 1986.

35. A. Weinstein, *Lectures on symplectic manifolds*, CBMS Regional Conf. Ser. in Math., no. 29, Amer. Math. Soc., Providence, R.I., 1977.

36. _____, *On extending the Conley-Zehnder fixed point theorem to other manifolds*, Proc. Sympos. Pure Math., vol. 45, part 2, Amer. Math. Soc., Providence, R. I., 1986, pp. 541–544.

37. A. Floer, *A relative Morse index for the symplectic action*, Preprint, Courant Institute, N.Y.U., 1987.

38. _____, *Witten's complex for arbitrary coefficients and an application to Lagrangian intersections*, Preprint, Courant Institute, N.Y.U., 1987.

39. _____, *Cuplength estimates for Lagrangian intersections*, Preprint, Courant Institute, N.Y.U., 1987.

40. _____, *The unregularized gradient flow of the symplectic action*, Preprint, Courant Institute, N.Y.U., 1987.

41. C. Conley and E. Zehnder, *A global fixed point theorem for symplectic maps and subharmonic solutions of Hamiltonian equations on tori*, Proc. Sympos. Pure Math., vol. 45, part 1, Amer. Math. Soc., Providence, R. I., 1986, pp. 283–299.

INSTITUT FÜR REINE UND ANGEWANDTE MATHEMATIK, RWTH AACHEN, TEMPLERGRABEN 55, D-5100 AACHEN, FEDERAL REPUBLIC OF GERMANY

Actions of Semisimple Groups and Discrete Subgroups

ROBERT J. ZIMMER

1. Introduction. Let $V = \{\text{primes in } \mathbf{Z}\} \cup \{\infty\}$. As usual, $\mathbf{Q}_p$ will denote the p-adic numbers for p a finite prime, and we set $\mathbf{Q}_\infty = \mathbf{R}$. Let $S \subset V$ be a finite subset. For $p \in S$, let G_p be a connected semisimple algebraic group defined over $\mathbf{Q}_p$, $G_p(\mathbf{Q}_p)$ the group of $\mathbf{Q}_p$-points, and $G = \prod_{p \in S} G_p(\mathbf{Q}_p)$. Then G is a locally compact group. If $S = \{\infty\}$, we say that G is real, and in this case G is a semisimple Lie group. We say that G has *higher rank* if for all $p \in S$, the $\mathbf{Q}_p$-rank of every $\mathbf{Q}_p$-simple factor of G_p is at least 2. If G is real and simple, this is of course equivalent to the condition that the rank of the associated symmetric space be at least 2. Let $\Gamma \subset G$ be a lattice subgroup, i.e., Γ is discrete and G/Γ has a finite G-invariant measure. The basic examples of such groups are the S-arithmetic ones, and by results of Margulis [**M1**] (see also [**Z1**]), under the assumption of higher rank, every lattice is S-arithmetic. The finite-dimensional (continuous) representation theory (say over $\mathbf{C}$) of G is the same as that of $G_\infty(\mathbf{R})$, and hence is classical and well-understood. In higher rank, the finite-dimensional representation theory of Γ is now quite well understood as well, due to the work of Margulis [**M1**] on semisimple representations (Margulis's superrigidity theorem) (see also [**Z5**] and the work of Mostow [**Mo**] and Prasad [**P**]), Raghunathan [**R2, R3**] on general representations for the case in which Γ is not cocompact, and to the work of numerous authors (e.g., Weil [**W**], Matsushima-Murakami [**MM**], Raghunathan [**R1**], Borel [**B1, B2**], Borel-Wallach [**BW**], Garland [**G**], Kazhdan [**K**]) on cohomology vanishing theorems for semisimple representations, which (when combined with Margulis's results) yields essentially complete results in the cocompact case. (Much of this work is also valid under considerably weaker hypotheses than higher rank.) The fundamental conclusion of this theory is that all representations are essentially either orthogonal, or extend to G, or are built up from these two cases. The infinite-dimensional (say unitary) representation theory of G is of course now a highly developed and enormously rich subject. The corresponding theory for Γ is largely undeveloped, except for developments related to the discovery of Kazhdan [**K**] that in higher rank the trivial representation of Γ is isolated in the unitary dual.

Research partially supported by National Science Foundation Grant DMS-8301882.

This property is inherited from G, and is an extraordinarily powerful and flexible property of Γ.

In this report, we will discuss the much more recent program of understanding the realizations of G and Γ in another natural class of groups, namely as smooth transformation groups on compact manifolds. This has both finite-dimensional and infinite-dimensional features. We can view this as the nonlinear finite-dimensional theory, or as the study of homomorphisms into the infinite-dimensional group of diffeomorphisms of a compact manifold. The prevailing theme of the work to date on this program is that one sees very strong manifestations of the rigidity phenomena of the finite-dimensional linear theory in the present context. It does not appear out of the question that one could classify all volume preserving ergodic actions of G and Γ on compact manifolds.

(In many cases below, when considering actions of Γ or G, we will assume for simplicity of exposition that G has higher rank, even though in certain cases less restrictive hypotheses are sufficient.)

We now describe some basic examples:

(a) Let H be a connected Lie group and $\Lambda \subset H$ a cocompact subgroup such that H/Λ has a finite H invariant measure. Let $\rho\colon G \to H$ be a continuous homomorphism. Then G (and Γ) will act naturally on H/Λ. If H is semisimple, then under very mild assumptions, the fundamental theorem of Moore [**Mr1**] (see also [**Z1**]) implies that these actions are ergodic.

(b) Let K be a compact Lie group, $K_0 \subset K$ a closed subgroup, and suppose $\rho\colon \Gamma \to K$ is a homomorphism. Then Γ acts on K/K_0, preserving a Riemannian metric. The examples with $\rho(\Gamma)$ dense in K are exactly the isometric ergodic actions of Γ.

(c) Let N be a simply connected nilpotent Lie group on which H acts by automorphisms. Assume that $D \subset N$ is a lattice which is invariant under the action of Γ. Then Γ acts by automorphisms of the compact nilmanifold N/D. A basic example arises from arithmetic realizations of Γ. Namely, if $\rho\colon \Gamma \to \mathrm{GL}(n, \mathbf{Z})$ is a homomorphism, then Γ acts on the torus $\mathbf{R}^n/\mathbf{Z}^n$.

A basic question is to what extend these examples (and easy modifications of them) represent all the volume preserving examples in the case of higher rank, at least if we assume ergodicity. At present there are no known examples of volume preserving actions not derived from these fundamentally linear situations. While a complete classification does not appear within reach at present, the remainder of this report will describe the present understanding of the actions of these groups under various natural hypotheses: dimension restrictions on M, actions preserving geometric structures, deformations, growth conditions, etc.

2. Invariant geometric structures for semisimple groups. Let M be a compact, connected, n-manifold and $H \subset \mathrm{GL}(n, \mathbf{R})$ an algebraic subgroup. We recall that an H-structure on M is a reduction of the frame bundle of M to H, and hence is a principal H-bundle $P \to M$ contained in the frame bundle. We let $\mathrm{Aut}(P) \subset \mathrm{Diff}(M)$ be the subgroup of diffeomorphisms of M leaving P

invariant. Let G be as above, and assume G is real. In this section we describe results concerning actions of G preserving an H-structure, i.e., homomorphisms $G \to \mathrm{Aut}(P)$. All these results carry over to the case of higher order structures, i.e., to the case in which $H \subset \mathrm{GL}(n, \mathbf{R})^{(k)}$ is an algebraic subgroup, where $\mathrm{GL}(n, \mathbf{R})^{(k)}$ is the group of k-jets at 0 of diffeomorphisms of $\mathbf{R}^n$ fixing the origin.

THEOREM 2.1 [**Z4, Z7**]. *Suppose that G has no compact factors. If G acts effectively on a compact connected M preserving an H-structure where $H \subset \mathrm{SL}'(n, \mathbf{R})$ (the latter being the matrices with $|\det| = 1$), then there is an embedding of Lie algebras $\mathfrak{g} \to \mathfrak{h}$. In fact, this embedding is such that the representation $\mathfrak{g} \to \mathfrak{h} \to \mathfrak{sl}(n, \mathbf{R})$ contains $\mathrm{ad}_{\mathfrak{g}}$ as a direct summand.*

This result provides very strong obstructions for G to preserve such an H-structure. For example, if $H = \mathrm{O}(1, n - 1)$, so that we are considering actions preserving Lorentz metrics, we deduce

COROLLARY 2.2 [**Z4**]. *If a semisimple group G acts on a compact manifold (of any dimension) preserving a Lorentz metric, then G is locally isomorphic to $\mathrm{SL}(2, \mathbf{R}) \times K$ where K is a compact group.*

We remark that somewhat similar techniques can be used to analyze the solvable component of the automorphism group of a Lorentz manifold, although we shall not pursue this here. See [**Z4**].

The hypothesis of compactness and finite invariant measure on M are both necessary in Theorem 2.1. In case G is transitive on M, Theorem 2.1 can be deduced from the Borel density theorem (asserting that the stabilizer for a homogeneous space of G with finite invariant measure is Zariski dense in G). The proof in general actually makes use of Borel's theorem and the ideas surrounding it. Without the assumption of invariant measure (i.e., for H an arbitrary algebraic subgroup of $\mathrm{GL}(n, \mathbf{R})$), we have

THEOREM 2.3 [**Z7**]. *Suppose G is a semisimple group acting effectively on a compact manifold M preserving an H-structure where H is an algebraic group. Then $\mathbf{R}$-rank$(G) \leq \mathbf{R}$-rank(H).*

(We recall that $\mathbf{R}$-rank is the dimension of a maximal $\mathbf{R}$-split torus.)

3. Invariant geometric structures for discrete subgroups. Suppose now that G has higher rank. We can then state the following general conjecture for actions of lattices that preserve an H-structure.

CONJECTURE I. Suppose G has higher rank and that $\Gamma \subset G$ is a lattice. Suppose that M is a compact n-manifold and that $H \subset \mathrm{SL}'(n, \mathbf{R})$ is an algebraic subgroup. (For higher order structures, we assume $H \subset \mathrm{SL}'(n, \mathbf{R}) \cap \mathrm{GL}(n, \mathbf{R})^{(k)}$.) If Γ acts on M so as to preserve an H-structure, then either:
(i) there is a Γ-invariant Riemannian metric on M; or
(ii) there is a nontrivial Lie algebra homomorphism $\mathfrak{g}_\infty(\mathbf{R}) \to \mathfrak{h}$.

This conjecture can be viewed as a geometric, or nonlinear, version of Margulis's superrigidity theorem. Margulis's theorem implies (under the above hypotheses on G, Γ, and H) that for a homomorphism $\Gamma \to H$, either the image of Γ is precompact in H or there is a nontrivial Lie algebra homomorphism $\mathfrak{g}_\infty(\mathbf{R}) \to \mathfrak{h}$. We also remark with regard to conclusion (i) in the conjecture that the action is then given by a composition $\Gamma \to K \to \mathrm{Diff}(M)$, where K is a compact Lie group, and that Margulis has also described the compact Lie groups admitting a dense image homomorphism from Γ. In particular, for any such K we have that $\dim(K) \geq n(G)$, where $n(G) = \min\{\dim_{\mathbf{C}} G' | G'$ is a simple factor of $G_\infty\}$. (See [**Z5**].) If we further assume that the Γ action is ergodic, then in (i) we have that the action is on a homogeneous space of K. Before indicating what is known in the direction of the conjecture, we remark that the following conjecture would be an immediate consequence of Conjecture I, and the fact that the dimension of the isometry group of a compact Riemannian n-manifold is at most $n(n+1)/2$.

CONJECTURE II. Let G, Γ, M be as in Conjecture I. Let $d(G)$ be the smallest integer d for which there is a nontrivial Lie algebra representation $\mathfrak{g}_\infty(\mathbf{R}) \to \mathfrak{sl}(d, \mathbf{R})$. Suppose Γ acts on M so as to preserve a volume density. Then (with $\dim(M) = n$):

(a) If $n < d(G)$, then Γ preserves a Riemannian metric.

(b) If $n < d(G)$ and $\{n(n+1)/2\} < n(G)$, then the Γ action is finite, i.e., factors through a finite quotient of Γ.

For certain H-structures, Conjecture I is known to be true. We recall that E. Cartan has defined the notion of a H-structure of finite type. (See [**Ko**].) A connection is a (second order) structure of finite type, and any H-structure naturally defining a connection is of finite type. For example, any pseudo-Riemannian structure (i.e., $O(p, q)$-structure) is of finite type. The automorphism group of an H-structure of finite type is a Lie group; however the group of connected components may well be infinite.

THEOREM 3.1 [**Z5**]. *Conjecture I is true for H-structures of finite type (in the sense of E. Cartan). In particular, if $n < d(G)$, and $\{n(n+1)/2\} < n(G)$, then any action of Γ on M^n preserving a volume density and a H-structure of finite type is a finite action.*

As a concrete example we have

COROLLARY 3.2 [**Z5**]. *Let $G = \mathrm{SL}(n, \mathbf{R})$, $n \geq 3$, and M a compact manifold with $\dim(M) < n$. Then any action of Γ on M preserving a volume density and a connection is a finite action.*

This shows that the action of $\mathrm{SL}(n, \mathbf{Z})$ by automorphisms of the flat torus is a volume and connection preserving action of minimal dimension for lattices in $\mathrm{SL}(n, \mathbf{R})$. For $\dim(M) = n$, we have the following special result.

THEOREM 3.3 [**Z6**]. *If* Γ *is a lattice in* $\mathrm{SL}(n, \mathbf{R})$, $n \geq 3$, M *is a compact Riemannian n-manifold, and Γ acts on M preserving volume and the connection, then M is flat and Γ is commensurable with* $\mathrm{SL}(n, \mathbf{Z})$.

We now indicate two other situations in which Conjecture I is known, at least with some further hypotheses. We recall that a linear Lie algebra is called elliptic if it contains no matrices of rank 1. For any H-structure of finite type, $\mathfrak{h}$ will be elliptic.

THEOREM 3.4 [**Z5**]. *Conjecture I is true if $\mathfrak{h}$ is elliptic, provided we also assume that* $\mathrm{Aut}(P)$ *is transitive on M.*

A linear real algebraic group H is called distal if the reductive Levi component is compact.

THEOREM 3.5 [**Z3**]. *Conjecture I is true for H distal, provided we assume the Γ action on M is ergodic. Hence, any ergodic action of Γ preserving a distal structure is isometric.*

Theorem 3.5 is closely related to classical notions in dynamical systems. Namely, we recall that an action of a group Γ on a metric space M is called distal if $x, y \in M$, $x \neq y$, implies $\inf\{d(gx, gy)|g \in \Gamma\} > 0$. Furstenberg [**F**] has shown that the structure of every such minimal action can be explicitly described in terms of a tower of bundles, and in case M is a topological manifold, M. Rees [**Re**] showed that the tower consists of topological manifolds. If M is smooth, and the tower is smooth, then Γ will preserve a distal H-structure on M. (However an action may well preserve a smooth distal structure but not be distal in the above sense.) This then raises the purely topological question as to whether every distal action of Γ (where Γ is as in Conjecture I) will preserve a topological distance function. There are natural classes of groups for which this is known to be the case [**MZ**, **A**].

Finally, we remark that Conjecture I is true if assertion (i) is replaced by the weaker assertion that there is a measurable invariant Riemannian metric [**Z5**].

4. Perturbations, deformations, and cohomology. In this section we discuss the rigidity properties of some of the actions described in §2 under perturbations or deformations. Each group Γ under consideration is finitely generated. For each r, $1 \leq r \leq \infty$, the C^r-topology on Γ actions on M will be the topology of C^r-convergence in $\mathrm{Diff}(M)$ on a fixed finite generating set of Γ.

THEOREM 4.1 [**Z5**]. *Suppose that Γ acts isometrically on a compact Riemannian n-manifold M. For any nonnegaitve integer k, there is positive integer $r = r(n, k)$ (for $k = \infty$, let $r = \infty$), so that any action of Γ on M which*
  (i) *is sufficiently close to the original action in the C^r-topology,*
  (ii) *leaves a volume density invariant, and*
  (iii) *which is ergodic,*
must also leave a C^k-Riemannian metric invariant.

For $k = 0$, $r = n^2 + n + 1$. For $k \geq 1$, $r = n + k + 4 + \dim(\mathrm{GL}(n, \mathbf{R})^{(k+3)})$, where $\mathrm{GL}(n, \mathbf{R})^{(k)}$ is the group of k-jets at 0 of diffeomorphisms of $\mathbf{R}^n$ fixing the origin.

In other words, roughly speaking, a small perturbation of an isometric action will be isometric. This is of course in very sharp contrast to actions of $\mathbf{R}$ or free groups. It seems possible that the size of r stated above can be improved.

If M is a manifold, we let $\mathrm{Vect}(M)$ be the space of smooth vector fields on M. If Γ acts on M, then $\mathrm{Vect}(M)$ is naturally a Γ-module. The Γ action on M is called infinitesimally rigid if $H^1(\Gamma, \mathrm{Vect}(M)) = 0$, locally (r, k)-rigid if every action sufficiently close in the C^r-topology is conjugate to the original action by a C^k-homeomorphism, and r-structurally stable if it is locally $(r, 0)$-rigid.

Question. For G of higher rank, is every ergodic volume preserving Γ-action infinitesimally rigid and locally rigid? In particular, for $n \geq 3$, is the action of $\mathrm{SL}(n, \mathbf{Z})$ on $\mathbf{R}^n/\mathbf{Z}^n$ locally rigid?

For infinitesimal rigidity, we have the following result.

THEOREM 4.2 [**Z8**]. *Assume G is real, has higher rank, and that Γ is a cocompact lattice in G. Let H be a semisimple Lie group, $\Lambda \subset H$ a cocompact lattice, and suppose $\Gamma \to H$ is a homomorphism. Let $M = H/\Lambda$, so that Γ then acts on M. Then the Γ action on M is infinitesimally rigid in the following cases:*

(i) *the image of Γ is dense in H; or*

(ii) *$H = H_1 \times H_2$, Λ projects densely into both factors, and Γ maps densely into H_1 and trivially into H_2.*

We expect the techniques of proof to extend to eliminate the hypotheses that G is real, and to apply in at least some situations in which Γ is not cocompact. For Γ cocompact, most of the proof of Theorem 4.2 remains valid for all the ergodic examples of Γ actions considered in §1, and it is possible that the proof may extend to cover these cases as well. The arguments of the proof of Theorem 4.2 can be applied to compute the cohomology of Γ with coefficients in the space of smooth sections of other natural bundles. For example, under the hypotheses of Theorem 4.2 we can deduce that $H^1(\Gamma, C^\infty(M)) = 0$.

Problem. If G has higher rank, and Γ acts (perhaps ergodically) preserving a probability measure in the smooth measure class on a compact manifold M, is $H^1(\Gamma, C^\infty(M)) = 0$?

A positive answer to this question, combined with the techniques of proof of Theorems 3.5 and 4.1, would yield significant progress on a resolution of Conjectures I and II. From Kazhdan's property it follows that the map $H^1(\Gamma, C^\infty(M)) \to H^1(\Gamma, L^2(M))$ is 0.

For nonvolume preserving examples, we have the following theorems of Sullivan and Ghys.

THEOREM 4.3 [**S**]. *Let* $G = O(1, n + 1)$, $P \subset G$ *a minimal parabolic subgroup, so that* G/P *can be identified with* S^n *with the conformal action of* G. *If* $\Gamma \subset G$ *is a cocompact lattice, then the action of* Γ *on* S^n *is 1-structurally stable.*

THEOREM 4.4 [**Gh**]. *Let* $G = \mathrm{PSL}(2, \mathbf{R})$, *acting by conformal transformations of* S^1. *Let* $\Gamma \subset G$ *be a cocompact lattice. Then any smooth action of* Γ *on* S^1 *sufficiently* C^2-close *to the given conformal action on* S^1 *is conjugate (via a smooth diffeomorphism) to the action defined by a linear representation* $\Gamma \to G$.

It would of course be interesting to extend these results to a more general algebraic setting.

5. Actions with fixed points. For actions with a fixed point, or more generally with a finite orbit, we can prove a version of Conjecture II.

THEOREM 5.1 [**Z5**]. *Assume* G *has higher rank, and that* M *is a connected* n-manifold *(not necessarily compact) and that* Γ *acts on* M. *Assume* $n < d(G)$ *and* $\{n(n + 1)/2\} < n(G)$. *If there is at least one finite* Γ-orbit *in* M, *then the* Γ *action is finite.*

We remark, that in contrast to the conclusion of Theorem 5.1, the standard action of $\mathrm{SL}(n, \mathbf{Z})$ on the n-torus has a dense set of points with finite orbits.

D. Stowe [**St**] has shown for any group for which one has vanishing of the first cohomology group with coefficients in all finite-dimensional real representations that there is a persistence of fixed points under perturbations. Combined with the cohomological information alluded to in the introduction, we obtain the following instance of Stowe's theorem.

THEOREM 5.2. *Assume* G *has higher rank. Let* Γ *act on a manifold and assume that* p *is a fixed point for the action. Then any action sufficiently close in the* C^1-topology *to the original action has a fixed point near* p. *Without the assumption of higher rank, the same conclusion is true for actions of* G.

For compact group actions, a basic tool in the study of fixed points is the fact that one can linearize the action near the fixed point. The following result generalizes this to semisimple groups.

THEOREM 5.3 (GUILLEMIN-STERNBERG [**GS**]). *Suppose* G *acts on* M *with a fixed point* p. *If the action is real analytic, then in a neighborhood of* p, *the action is analytically equivalent to the representation of* G *on* TM_p.

6. Orbit structure. In the classical theory of flows, a significant role is played by the study of the phase portrait of the flow, or equivalently, studying the flow up to equivalence after time change. Similar ideas have played an important role in certain recent developments in ergodic theory [**Dy, CFW, Ru**]. If for $i = 1, 2$, G_i is a locally compact group acting on a Borel subset M_i of a complete separable metrizable space, preserving the null sets of a finite measure μ_i, we say that the actions are measurably orbit equivalent if there is a measure class preserving bijection (modulo null sets) $M_1 \to M_2$ taking G_1 orbits onto G_2 orbits. If G_i are discrete amenable groups acting ergodically (and with no orbit of full measure) and μ_i are invariant probability measures, then a result of Ornstein and Weiss (see also [**CFW**]), generalizing earlier work of Dye [**Dy**], asserts that the actions are measurably orbit equivalent. In contrast, we have the following result in the semisimple case.

THEOREM 6.1 [**Z1, Z2**]. *Assume G_i are as in the introduction and that G_1 has higher rank.*

(i) *If G_i acts ergodically and essentially freely on M_i, does not have a conull orbit, preserves a probability measure, and the G_1 action is measurably orbit equivalent to the G_2 action, then G_1 and G_2 are locally isomorphic. Further, the actions of the adjoint groups $G_i/Z(G_i)$ on $M/Z(G_i)$ are actually conjugate, after the identification of these adjoint groups via an isomorphism.*

(ii) *If $\Gamma_i \subset G_i$ are lattices and Γ_1 and Γ_2 have measurably orbit equivalent ergodic essentially free actions with finite invariant measure (and no conull orbit), then G_1 and G_2 are locally isomorphic.*

For certain isometric actions, Witte [**Wi**] has shown that the conclusion in (ii) can be strengthened to assert isomorphism of the lattices and conjugacy of the actions.

Roughly speaking, Theorem 6.1 asserts that the measurable orbit structure of the action determines both the semisimple group and (in case (i)) the action itself. This is diametrically opposed to the situation for amenable groups. It is not known whether or not this result is true for groups of split rank 1, although some information is available for groups with Kazhdan's property [**Z1**]. One can of course ask about orbit equivalence in the context of smooth actions, and ask that the orbit equivalence be smooth, or at least continuous. If G is real and acts on M, and $K \subset G$ is a maximal compact subgroup, then under suitable hypotheses the orbits of G will project to the leaves of a foliation of M/K, and these leaves will naturally carry the structure of an (infinite volume) locally symmetric space. The next result, which is joint work of P. Pansu and the author, is a result in the same spirit as that of Theorem 6.1 but formulated in the context of foliations.

THEOREM 6.2 [**PZ**]. *Let M be a compact manifold and $\mathcal{F}$ a foliation of M with a holonomy invariant transverse measure (which we assume to be positive on open sets of transversals, and finite on compact sets). For $i = 1, 2$, let ω_i be a Riemannian metric on the tangent bundle to $\mathcal{F}$, such that for $i = 1, 2$, each leaf is a locally symmetric space of negative curvature. Then there is a homeomorphism f of M, taking each leaf to itself diffeomorphically, such that $f^*(\omega_1) = \omega_2$.*

When the foliation has just one leaf, this reduces to a version of the Mostow rigidity theorem. It is natural to conjecture that the same result is true for symmetric spaces of higher rank, but this is not yet known. From Theorem 6.1, one can deduce the existence of a measurable bijection $f: M \to M$, taking each leaf to itself diffeomorphically, with $f^*(\omega_1) = \omega_2$. It is also natural to ask whether or not we can ensure that f in Theorem 6.2 can be taken to be smooth.

In the case of nonvolume preserving actions, we have the following result for some standard examples.

THEOREM 6.3 [**Z2**]. *For $i = 1, 2$, suppose G_i is semisimple and real, $H_i \subset G_i$ is an almost connected closed subgroup, and L_i is the maximal semisimple adjoint quotient group of H_i^0 with no compact factors. Assume L_1 has higher rank. If Γ_i is a lattice in G_i, and the action of Γ_1 on G_1/H_1 is measurably orbit equivalent to the action of Γ_2 on G_2/H_2, then L_1 and L_2 are locally isomorphic.*

A basic property that a measure class preserving action of a locally compact group might have is amenability. This is discussed at length in [**Z1**]. For essentially free actions (i.e., actions for which almost every stabilizer is trivial) amenability is an invariant of measurable orbit equivalence. One method of constructing amenable actions is to induce from an action of an amenable subgroup. I.e., if $A \subset G$ is a closed amenable subgroup, and X is a measurable A-space, then $M = (X \times G)/A$ will be an amenable G-space. For arbitrary G, this does not yield all amenable actions. However, we have

THEOREM 6.4 [**Z1**]. *If G is real and semisimple, any amenable ergodic action of G is measurably conjugate to an action induced from a maximal amenable algebraic subgroup.*

We remark that such subgroups have been classified by C. C. Moore [**Mr2**]. We also observe that for essentially free actions, while amenability depends only upon the action up to measurable orbit equivalence the property of being induced from an action of a subgroup from a given class of subgroups does not. Theorem 6.4 has immediate applications to the measurable cohomology theory of amenable actions of arbitrary groups and to amenable foliations. In this latter form this result has recently been applied by Hurder and Katok [**HK**] to prove a vanishing theorem for secondary characteristic classes of amenable foliations.

7. Restrictions to unipotent subgroups. In studying the linear representations of G, a basic role is played by the restriction of the representations to unipotent subgroups. Here we discuss some features of these restrictions in the case of actions. The following result was first proved by M. Ratner [**Ra1**] for $G = \mathrm{PSL}(2, \mathbf{R})$, and extended to more general semisimple groups by Witte [**Wi**].

THEOREM 7.1 [**Ra1, Wi**]. *For $i = 1, 2$, suppose G_i is a noncompact semisimple adjoint Lie group and that Γ_i is an irreducible lattice in G_i. Let $U_i \subset G_i$ be a one parameter unipotent subgroup. If the $\mathbf{R}$-actions defined by the U_i actions on G_i/Γ_i are measurably conjugate, then there is an isomorphism of G_1 with G_2 taking Γ_1 onto Γ_2.*

Witte also shows how the arguments of the proof lead to the following information about some of the examples considered in §2.

THEOREM 7.2 [**Wi**]. *Suppose that G is real and adjoint.*

(a) *For $i = 1, 2$, let G_i be a connected semisimple adjoint Lie group, and $\varphi_i \colon G \to G_i$ be an injective homomorphism. Let $\Gamma_i \subset G_i$ be an irreducible lattice, and suppose the G actions on G_1/Γ_1 and G_2/Γ_2 are measurably conjugate. Then there is an isomorphism $G_1 \cong G_2$ taking φ_1 to φ_2.*

(b) *Suppose further that G is of higher rank. Fix a nontrivial unipotent element $g \in G$. Then any measure preserving action of G on G_1/Γ_1 for which g acts by an element of G_1 is actually defined by a homomorphism $G \to G_1$.*

Ratner's theorem for PSL(2) was extended in another direction (concerning horocycle foliations in higher dimensional manifolds of negative curvature) by Flaminio [**Fl**]. For $\mathrm{PSL}(2, \mathbf{R})$, an analogous theorem to Theorem 7.1, assuming continuous orbit equivalence, rather than measurable conjugacy, was proved by Marcus [**Ms**]. This has been extended to a much broader context by Benardette [**Be**]. Further work on the restriction of transitive volume preserving actions to unipotent subgroups can be found in the work of Veech [**V**] and the extensive work of Dani [**D1**].

A basic result in the finite-dimensional linear theory is that for any linear representation of G, the image of a unipotent element is a unipotent matrix. The following problem asks for a generalization of this in the geometric context.

Problem. Assume G is real and of higher rank. Let M be a compact manifold, $E \to M$ a vector bundle of rank n on which G acts by vector bundle automorphisms. Suppose there is a probability measure on M, in the smooth measure class, which is invariant under G. Let U be a unipotent subgroup of G. Is there a distal algebraic subgroup $H \subset \mathrm{GL}(n, \mathbf{R})$ and a smooth U-invariant H-structure on E?

In all known examples, this question has an affirmative answer. In case G acts ergodically on M, it is known [**Z5**] that for some distal H there is a measurable U-invariant H-structure. A positive answer to the above question for some G, combined with the techniques of proof of Theorem 3.5, and results of Howe and Harish-Chandra on asymptotic behavior of matrix coefficients for unitary representations would yield a proof of Conjecture II in the ergodic case, although with weaker control on the dimension of M than that proposed in Conjecture II. See [**Z5**] for a complete discussion of this point and some partial results. We remark that in general one cannot find a U-invariant H-structure where H is a unipotent subgroup of $\mathrm{GL}(n, \mathbf{R})$.

8. Quotient actions. If a group Λ acts on a topological (resp., measure) space X, by a quotient action we mean a topological (resp. measure) space Y and a continuous (resp. measure class preserving) surjective Λ-map $X \to Y$. For some standard actions the quotient actions can all be identified. The following result of M. Ratner accomplishes this for horocycle flows of surfaces.

THEOREM 8.1 [**Ra2**]. *Let $G = \mathrm{PSL}(2, \mathbf{R})$ and let $\mathbf{R}$ act on G/Λ by the upper triangular unipotent subgroup, where Λ is a lattice in H. Then every* *measure theoretic quotient of this* **R**-*action is of the form* G/Λ', *where* $\Lambda' \supset \Lambda$ *is a larger lattice, and* **R** *acts via the same subgroup.*

For the case of actions on varieties, we have the following theorem, first proved in the measure theoretic case by Margulis [**M2**] (see also [**Z1**]), later in the topological case by Dani [**D2**].

THEOREM 8.2 [**M2, D2**]. *Assume G is adjoint and of higher rank and that $P \subset G$ is a parabolic subgroup. Then any quotient of the Γ action on G/P is of the form G/P', where P' is a parabolic subgroup containing P.*

9. Concluding remarks. We have said little about the proofs of the results stated above. In general they involve combinations of arguments of ergodic theory, algebraic groups, representation theory, differential geometry, and global analysis. A basic role is played by various ergodic theoretic generalizations of Margulis's superrigidity theorem [**Z1**], and Kazhdan's property. A number of the results above can be established assuming only Kazhdan's property (see [**Z9**], e.g.), and hence are applicable to certain split rank one groups. For some further topics we have not discussed, see [**Z1, Z5**], and the references therein.

REFERENCES

[**A**] H. Abels, *Which groups act distally?*, Ergodic Theory Dynamical Systems **3** (1983), 167–186.

[**Be**] D. Benardette, *Topological equivalence of flows on homogeneous spaces and divergence of one-parameter subgroups of Lie groups*, preprint.

[**B1**] A. Borel, *Stable real cohomology of arithmetic groups*, Ann. Sci. École Norm. Sup. **7** (1974), 235–272.

[**B2**] ——, *Stable real cohomology of arithmetic groups*. II, Progress in Math. vol. 14, Birkhauser, Boston, 1981, pp. 21–55.

[**BW**] A. Borel and N. Wallach, *Continuous cohomology, discrete subgroups, and representations of reductive groups*, Ann. of Math. Stud. **94** (1980).

[**CFW**] A. Connes, J. Feldman, and B. Weiss, *An amenable equivalence relation is generated by a single transformation*, Ergodic Theory Dynamical Systems **1** (1981), 431–450.

[**D1**] S. Dani, *Invariant measures and minimal sets of horocycle flows*, Invent. Math. **64** (1981), 357–385.

[**D2**] ——, *Continuous equivariant images of lattice actions on boundaries*, Ann. of Math. **119** (1984), 111–119.

[**Dy**] H. Dye, *On groups of measure preserving transformations*. I, Amer. J. Math. **81** (1959), 119–159.

[**Fl**] L. Flaminio, Thesis, Stanford University, California, 1985.

[**F**] H. Furstenberg, *The structure of distal flows*, Amer. J. Math. **85** (1963), 477–515.

[**G**] H. Garland, *p-adic curvature and the cohomology of discrete subgroups of p-adic groups*, Ann. of Math **97** (1973), 375–423.

[**Gh**] E. Ghys, *Actions localement libres du groupe affine* (to appear).

[**GS**] V. Guillemin and S. Sternberg, *Remarks on a paper of Hermann.*

[**HK**] S. Hurder and A. Katok, *Ergodic theory and Weil measures of foliations*, preprint.

[**K**] D. Kazhdan, *Connection of the dual space of a group with the structure of its closed subgroups*, Functional Anal. Appl. **1** (1967), 63–65.

[**Ko**] S. Kobayashi, *Transformation groups in differential geometry*, Springer, New York, 1972.

[**Ms**] B. Marcus, *Topological conjugacy of horocycle flows*, Amer. J. Math. **105** (1983), 623–632.

[M1] G. A. Margulis, *Discrete subgroups of motions of manifolds of non-positive curvature*, Amer. Math. Soc. Transl. **109** (1977), 33–45.

[M2] ____, *Quotient groups of discrete groups and measure theory*, Functional Anal. Appl. **12** (1978), 295–305.

[MM] Y. Matsushima and S. Murakami, *On vector bundle valued harmonic forms and automorphic forms on symmetric Riemannian manifolds*, Ann. of Math. **78** (1963), 365–416.

[Mr1] C. C. Moore, *Ergodicity of flows on homogeneous spaces*, Amer. J. Math. **88** (1966), 154–178.

[Mr2] ____, *Amenable subgroups of semisimple groups and proximal flows*, Israel J. Math. **34** (1979), 121–138.

[MZ] C. C. Moore and R. J. Zimmer, *Groups admitting ergodic actions with generalized discrete spectrum*, Invent. Math. **51** (1979), 171–188.

[Mo] G. D. Mostow, *Strong rigidity of locally symmetric spaces*, Ann. of Math. Stud. **78** (1973).

[PZ] P. Pansu and R. J. Zimmer (to appear).

[R1] M. Raghunathan, *On the first cohomology group of subgroups of semisimple Lie groups*, Amer. J. Math. **87** (1965), 103–139.

[R2] ____, *On the congruence subgroup problem*, Inst. Hautes Études Sci. Publ. Math. **46** (1976), 107–161.

[R3] ____, *On the congruence subgroup problem. II*, Invent. Math. **85** (1986), 73–118.

[Ra1] M. Ratner, *Rigidity of horocycle flows*, Ann. of Math. **115** (1982), 597–614.

[Ra2] ____, *Factors of horocycle flows*, Ergodic Theory Dynamical Systems **2** (1982), 465–489.

[Re] M. Rees, *On the structure of minimal distal transformation groups with topological manifolds as phase space*, preprint, 1977.

[Ru] D. Rudolph, *Restricted orbit equivalence*, Mem. Amer. Math. Soc. No. 323 (1985).

[St] D. Stowe, *The stationary set of a group action*, Proc. Amer. Math. Soc. **79** (1980), 139–146.

[S] D. Sullivan, *Quasi-conformal homeomorphisms and dynamics, II. Structural stability implies hyperbolicity for Kleinian groups*, Preprint, Inst. Hautes Études Sci. Publ. Math., 1982.

[V] W. A. Veech, *Unique ergodicity of horocycle flows*, Amer. J. Math. **99** (1977), 827–859.

[W] A. Weil, *On discrete subgroups of Lie groups*, Ann. of Math. **72** (1960), 369–384.

[Wi] D. Witte, *Rigidity of some translations on homogeneous spaces*, Invent. Math. **81** (1985), 1–27.

[Z1] R. J. Zimmer, *Ergodic theory and semisimple groups*, Birkhauser, Boston, 1984.

[Z2] ____, *Orbit equivalence and rigidity of ergodic actions of Lie groups*, Ergodic Theory Dynamical Systems **1** (1981), 237–253.

[Z3] ____, *Lattices in semisimple groups and distal geometric structures*, Invent. Math. **80** (1985), 123–137.

[Z4] ____, *On the automorphism group of a compact Lorentz manifold and other geometric manifolds*, Invent. Math. **83** (1986), 411–424.

[Z5] ____, *Lattices in semisimple groups and invariant geometric structures on compact manifolds*, Discrete Groups in Geometry and Analysis: Proceedings of a Conference in Honor of G. D. Mostow, Birkhauser (to appear).

[Z6] ____, *On connection preserving actions of discrete linear groups*, Ergodic Theory Dynamical Systems (to appear).

[Z7] ____, *Split rank and semisimple automorphism groups of G-structures*, J. Differential Geom. (to appear).

[Z8] ____, *Infinitesimal rigidity for smooth actions of discrete subgroups of Lie groups*, preprint.

[Z9] ____, *Kazhdan groups acting on compact manifolds*, Invent. Math. **75** (1984), 425–436.

UNIVERSITY OF CHICAGO, CHICAGO, ILLINOIS 60637, USA

Ultraviolet Stability Problems
in Quantum Field Theories

TADEUSZ BALABAN

We consider here only the case of pure gauge field theories. Later we will mention some other cases to which the method presented in this paper has been applied. The quantum gauge field theory, like any other quantum field theory in the so-called Euclidean formulation, is defined by expectation values given by functional integrals, i.e., by measures defined on spaces of functions. In the considered case, the space is the space of $\mathfrak{g}$-valued vector functions $A_\mu(x)$, $x \in R^d$, where $\mathfrak{g}$ is the Lie algebra of a compact semisimple Lie group G. It is identified with the space of $\mathfrak{g}$-valued 1-forms. The expectation values are given heuristically by the integrals

$$\langle F \rangle = Z^{-1} \int [dA] F(A) \exp[-S(A)] \tag{1}$$

where $[dA]$ denotes a fictitious Lebesgue measure on the space of 1-forms A, $F(A)$ is a function on the space, and $S(A)$ is the so-called gauge field action, which is the function

$$S(A) = \int dx \, \mathrm{tr} \, F^2(x), \qquad F(x) = dA(x) + gA(x) \wedge A(x), \tag{2}$$

or in components

$$F_{\mu\nu}(x) = \partial_\mu A_\nu(x) - \partial_\nu A_\mu(x) + g[A_\mu(x), A_\nu(x)]$$

($[X, Y]$ denotes the Lie algebra bracket).

There are several problems connected with the expressions of the type (1). An obvious one is that there is no measure $[dA]$, and this problem is common for all field theories. The other is that the action $S(A)$ is invariant with respect to the infinite-dimensional group of gauge transformations

$$A \to A^u, \qquad (A^u)_\mu(x) = u(x)A_\mu(x)u^{-1}(x) + i(\partial_\mu u(x))u^{-1}(x) \tag{3}$$

where $u(x)$ is a function on R^d with values in the group G. The integrand in (1) is constant on orbits of this group of transformations, therefore there is a problem with the definition of the integral. Such problems are studied by taking approximate expressions and integrals, which are mathematically well-defined, and then taking appropriate limits.

One such approximation was introduced by K. Wilson [1]. The continuous space R^d is replaced by the discrete lattice εZ^d with the lattice spacing $\varepsilon > 0$, or even by a finite subset Λ of this lattice. The continuous space variables $A_\mu(x)$ are replaced by the bond variables $U(\langle x, x + \varepsilon e_\mu \rangle) = U_\mu(x)$ with values in the group G (x is now a point of the lattice). The action $S(A)$ is replaced by $S^\varepsilon(U)$:

$$S_\Lambda^\varepsilon(U) = \sum_{p \subset \Lambda} (\varepsilon^{d-4}/g^2)[1 - \operatorname{Re} \operatorname{tr} U(\partial p)] \tag{4}$$

where the sum is over elementary squares (plaquettes) p of the lattice. If p is a plaquette with the boundary

$$\partial p = \langle x, y \rangle \cup \langle y, z \rangle \cup \langle z, w \rangle \cup \langle w, x \rangle,$$

then

$$\operatorname{tr} U(\partial p) = \operatorname{tr} U(\langle x, y \rangle) U(\langle y, z \rangle) U(\langle z, w \rangle) U(\langle w, x \rangle). \tag{5}$$

The action (4) is a good approximation of the action (2) in the sense that if we take a continuous space regular field $A_\mu(x)$, and we define the ε-lattice field $U_\mu(x)$ by the formula

$$U_\mu(x) = \exp i\varepsilon g A_\mu(x), \tag{6}$$

where exp is the exponential mapping of the Lie algebra $\mathfrak{g}$ into the Lie group G, then the lattice action (4) is convergent to the continuous space action (2) as ε goes to 0. The action (4) is also gauge invariant; i.e., it is invariant with respect to the group of the lattice gauge transformations

$$U \to U^u, \qquad U^u(\langle x, x' \rangle) = u(x) U(\langle x, x' \rangle) u^{-1}(x') \tag{7}$$

where $\langle x, x' \rangle$ is a bond of the lattice. The limit of such a transformation as $\varepsilon \to 0$, with the parametrization (6), yields the corresponding transformation (3).

Now we can define the expectation value for the lattice theory

$$\langle F \rangle_\Lambda^\varepsilon = (Z_\Lambda^\varepsilon)^{-1} \int dU F(U) \exp[-S_\Lambda^\varepsilon(U)], \tag{8}$$

where dU is the product of copies of the Haar measures on the group G, corresponding to bonds of the lattice contained in Λ. We integrate continuous functions on compact domains, hence the integrals (8) are well defined. We would like to define (1) as the limit of (8) as $\varepsilon \to 0$.

The integrals (8) are studied using the renormalization group approach in the so-called "block spin" form [2]. We will now describe very briefly some features of this approach. The original lattice is divided into blocks containing L^d points, where L is a fixed positive integer (e.g., $L = 3$ or $L = 15$). This determines a new lattice with the lattice spacing $L\varepsilon$, e.g., the lattice of centers of the blocks.

An averaging operation is defined, which transforms the gauge fields U on the bonds of the ε-lattice into gauge fields $\overline{U}$ defined on the bonds of the new

$L\varepsilon$-lattice. The definition is rather technical and we do not give it here. Having such an averaging operation we define the renormalization transformation, which transforms a density defined on gauge fields on the ε-lattice, into a density defined on gauge fields on the new $L\varepsilon$-lattice:

$$(T\rho)(V) = \rho_1(V) = \int dU \delta(V(\overline{U})^{-1})\rho(U). \tag{9}$$

This transformation satisfies the basic normalization property

$$\int dV \rho_1(V) = \int dV (T\rho)(V) = \int dU \rho(U). \tag{10}$$

It has also other important properties; for example, it transforms a gauge invariant density into a gauge invariant one. This transformation is iterated, and the sequence of densities is defined inductively by the formula

$$\rho_{n+1} = T\rho_n, \qquad n = 0, 1, \ldots, \qquad \rho_0 = \rho. \tag{11}$$

By the definition, the nth density is defined on gauge field configurations on the $L^n\varepsilon$-lattice obtained by applying n times the block construction. For the problems we want to investigate, it is enough to terminate the applications of the renormalization transformations in (11) when we reach the unit lattice, i.e., for the index N such that $L^N\varepsilon = 1$ (or, more generally, such that $L^N\varepsilon$ is closest to 1). We start with the density

$$\rho(U) = \exp[-S_\Lambda^\varepsilon(U) - E_\Lambda^\varepsilon]; \tag{12}$$

here E_Λ^ε is an appropriately defined constant, and we want to investigate this particular sequence of densities generated by applying (11). We would like to give as precise and effective description of the densities ρ_n as possible. Especially it should imply uniform bounds for these densities on the corresponding lattices. Such a description is a fundamental tool in investigation of the expectation values (8), and in proving the existence of the limit as $\varepsilon \to 0$. It was given in [3] for three-dimensional models, and in [4] a special case was considered for four-dimensional models. Unfortunately these descriptions, or inductive assumptions on the form of the densities, are very long and complicated and cannot be given here. Let us mention only that they involve decompositions of the space of gauge fields into subdomains according to some conditions (inequalities) satisfied by the fields in regions of the lattice. The lattice is divided into "small fields" and "large fields" regions, and such a division determines the subdomain in the gauge fields space. This is a common feature shared with other renormalization group methods [5, 6], and introduced by G. Gallavotti et al. [5]. Instead of the description of the densities, which is the real result of the presented approach, we formulate one of its simplest consequences in the form of the inequality.

THEOREM (ULTRAVIOLET STABILITY). *For $d = 3$ there exist constants E_-, E_+ independent of ε, Λ, such that*

$$X_n(U) \exp[-\sum_{p \subset \Lambda^{(n)}} \frac{1}{g^2 L^n\varepsilon}[1 - \operatorname{Re} \operatorname{tr} U(\partial p)] - E_- L^{-3n}|\Lambda|] \tag{13}$$

$$\leq \rho_n(U) \leq \exp[E_+ L^{-3n}|\Lambda|], \qquad n = 0, 1, \ldots, N,$$

where $X_n(U)$ is the characteristic function of the domain in the gauge fields space defined by the inequalities $|U(\partial p) - 1| < g(L^n \varepsilon)^{1/2}$ for $p \subset \Lambda^{(n)}$, $\Lambda^{(n)}$ denotes the domain obtained from the initial domain Λ in the ε-lattice by the n successive block constructions, and $|\Lambda|$ is the number of points in Λ.

The inequality (13) applied to $n = N$ (where $L^N \varepsilon = 1$) yields the bound uniform in the lattice spacing ε. This is the content of the ultraviolet stability concept. For $d = 4$ the obtained partial results imply the lower bound in (13). Let us stress once more that fundamental results of the described approach are contained in the precise inductive description of the densities. This can be applied to many problems. The above theorem is one of the possible applications, which has the simple enough formulation.

Similar results have been obtained for the three-dimensional Higgs model, i.e., the model of interacting Abelian gauge field with scalar fields, and for the class of two-dimensional nonlinear σ-models with G-valued fields. There is also another approach to gauge field theories in progress by P. Federbush [7].

REFERENCES

1. K. G. Wilson, *Confinement of quarks*, Phys. Rev. D **10** (1974), 2445–2459.

2. ____, *Quantum chromodynamics on a lattice*, New Developments in Quantum Field Theory and Statistical Mechanics (Proc. Cargèse Summer Inst., Cargèse, 1976), Nato Adv. Study Inst. Ser., Ser B: Physics, 26, Plenum, New York, 1977, pp. 143–172.

3. T. Balaban, *Ultraviolet stability of three-dimensional lattice pure gauge field theories*, Comm. Math. Phys. **102** (1985), 255–275.

4. ____, *Renormalisation group approach to lattice gauge field theories.* I, *Generation of effective actions in a small field approximation and a coupling constant renormalization in four dimensions*, Preprint HUTMP 85/B 189, Harvard Univ., Cambridge, Mass.

5. G. Benfatto, M. Cassandro, G. Gallavotti, F. Nicolo, E. Olivieri, E. Presutti, and E. Scaciatelli, *Ultraviolet stability in Euclidean scalar field theories*, Comm. Math. Phys. **71** (1980), 95–130.

6. K. Gawedzki and A. Kupiainen, *Rigorous renormalization group and asymptotic freedom*, Scaling and Self-similarity in Physics, Birkhäuser, Boston, 1983, pp. 227–262.

7. P. Federbush, *Phase-space cell approach to Yang-Mills field theories*, Preprint, University of Michigan.

NORTHEASTERN UNIVERSITY, BOSTON, MASSACHUSETTS 02115, USA

Proceedings of the International Congress of Mathematicians
Berkeley, California, USA, 1986

The Mechanism of Feigenbaum Universality

J.-P. ECKMANN

The story starts with the discovery by Feigenbaum [F] (see also Coullet, Tresser [CT] and Grossmann, Thomae [GT]) of the following striking phenomenon.

Consider the one-parameter family $\mu \to f_\mu(x) = 1 - \mu x^2$ of functions. For each $\mu \in [0, 2]$, f_μ maps $[-1, 1]$ into itself, and therefore we can consider $x \to f_\mu^n(x) = f_\mu \circ \cdots \circ f_\mu(x)$, n times. In this way, f_μ describes a dynamical system. There is a smallest value of μ_n for which the point 0 is periodic with period exactly 2^n. One says then that f_{μ_n} is *superstable* with period 2^n.

The discovery is that

(1) The sequence μ_n tends to a limit, $\mu_\infty \approx 1.40\ldots$.

(2) The numbers $\mu_n - \mu_\infty$ satisfy $\lim_{n \to \infty}(\mu_n - \mu_\infty)\delta_F^{-n} = A$, with $\delta_F \approx 4.66920\ldots$, $A \neq 0$.

(3) The points $y_n = f_{\mu_n}^{2^{n-1}}(0)$ (the point on the orbit of 0 under f_{μ_n} closest to, but different from, zero) satisfy

$$\lim_{n \to \infty} y_n \lambda_F^{-n} = B,$$

with $\lambda_F \approx -0.3995353\ldots$, $B \neq 0$.

Upon examining other one-parameter families of maps, such as $g_\mu(x) = \cos(\mu x)$, Feigenbaum discovered that (1), (2), (3) hold in many cases, with the *same* "universal" values of δ_F and λ_F (but, of course, not μ_∞). He conjectured that this holds "for all" one-parameter families, and he outlined an explanation of this conjecture.

Proofs of this conjecture have been worked out in part, or in full, in several references to be given later in this paper. I will outline here the idea of these proofs. It is most economical to describe the proof of this conjecture in *geometrical terms* in a *function space*. The theory [CEL] in fact centers on a renormalization, or doubling operator $\mathbf{N}$, acting on unimodal functions ϕ on $[-1, 1]$, satisfying $\phi(x) = \phi(-x)$, $\phi(0) = 1$, and defined by

$$(\mathbf{N}\phi)(x) = \lambda^{-1}\phi(\phi(\lambda x)), \qquad (*)$$

where $\lambda = \lambda(\phi) \equiv \phi(1)$. (The operator $\mathbf{N}$ preserves the normalization.)

We want to specify, in more detail, the function space $\mathbf{E}$ in which ϕ is supposed to lie. An adequate choice is the set of functions ϕ of the form $\phi(x) = h(x^r)$, $r = 2$, and h analytic and bounded on a disk, e.g., $D_{1,2.5} = \{z \in \mathbf{C} \mid |z-1| < 2.5\}$. The Feigenbaum conjectures rest on the following properties of $\mathbf{N}$ as an operator on $\mathbf{E}$.

P1. $\mathbf{N}$ has a fixed point g in $\mathbf{E}$.

P2. The operator $\mathbf{N}$ is differentiable and its derivative $D\mathbf{N}$ possesses at g an unstable invariant subspace of dimension 1, where $D\mathbf{N}$ reduces to multiplication by δ (with $|\delta| > 1$), and a stable invariant subspace of codimension one. (It follows that $\mathbf{N}$ has corresponding invariant manifolds, W_u and W_s.)

P3. The local unstable manifold at g can be extended to cut the surface $\Sigma = \{f \in \mathbf{E} \mid f(1) = 0, f(0) = 1\}$ transversally. (This surface is the set of functions in $\mathbf{E}$ having a superstable period of length 2.)

We shall call P1 and P2 *hyperbolicity* and P1, P2, and P3 *universality* for superstable periods 2^n.

The Feigenbaum conjectures follow easily from P1–P3, when $\delta = \delta_F$, and $\lambda(g) = \lambda_F$. The argument is about as follows. Take a one-parameter family $\mu \to g_\mu \in \mathbf{E}$ of maps in $\mathbf{E}$ cutting the stable manifold W_s of $\mathbf{N}$ transversally at $\mu = \mu_\infty$. The inverse images $\mathbf{N}^{-n}(\Sigma)$ of Σ under $\mathbf{N}$, viewed as sets, accumulate uniformly with the ratio $1/\delta^n$ at W_s. They intersect transversally with the curve g_μ at values μ_n, for large n. It is also easy to see from the definition of $\mathbf{N}$ that $\mathbf{N}^{-k}(\Sigma)$ is a set of functions having superstable period 2^k. Thus the intersection values μ_n, for which $g_{\mu_n} \in \mathbf{N}(\Sigma)$, satisfy (1) and (2). The assertion (3) follows by an explicit calculation, observing that $\lambda_F = g(1)$.

Other surfaces can be made to play a similar role as Σ; e.g., the surface $\Sigma^* = \{f \in \mathbf{E} \mid f^3(1) = -f(1)\}$ contains maps with absolutely continuous, invariant measures [**Mi**], and there is a sequence of $\hat{\mu}_n$ accumulating at μ_∞ (with rate δ, again) such that the critical point of g_μ falls on an (unstable) periodic point of period 2^n.

Thus, as we see, the Feigenbaum phenomenon is the consequence of the *hyperbolicity of the composition operator* $\mathbf{N}$. A proof of the Feigenbaum conjectures thus reduces to a proof of (1)–(3).

We can now *generalize* this idea to many other composition operators. In fact, there is a multitude of other situations where similar results hold.

(i) *Case of the interval.* Consider $r > 1$. The set $\mathbf{E}_r$ is defined by $\mathbf{E}_r = \{\phi, \phi$ is continuous on $[0,1]$ and $\phi(x) = h(x^r)$ for an *analytic* function h which is strictly decreasing with no critical points on $[0,1]\}$. For $r = 2$, there are numerous proofs of the existence of a fixed point for (∗) [**L1, CER, L3, EMO, EW1**]. Only [**L1**] contains proofs of P1 *and* P2 and only [**EW1**] proves P1–P3. For general $r > 1$, the map corresponding to (∗) has a fixed point [**E, CEL, EW**].

CONJECTURE 1. *It depends analytically on* r.

One can also ask for a fixed point of other, but related, operators $\mathbf{N}_p$:

$$(\mathbf{N}_p f)(x) = \lambda^{-1} f^p(\lambda x), \qquad \lambda = f^p(0).$$

The case discussed by Feigenbaum is $p = 2$. It satisfies hyperbolicity again, for all large p [**EEW**], with

$$\delta_p \approx A16^p, \qquad \lambda_p \approx B4^{-p}.$$

(ii) *Dissipative maps on* $\mathbf{R}^n$. By blowing up the situation of the original Feigenbaum operator, one can try to find a new operator $\mathbf{N}^{(n)}$ on maps on $\mathbf{R}^n \to \mathbf{R}^n$. It is

$$(N^{(n)}f) = \Lambda_f^{-1} \circ f \circ f \circ \Lambda_f,$$

where Λ_f is a diagonal, invertible coordinate transformation. One can show that this operator satisfies hyperbolicity again, with a $\delta_n \equiv \delta_F$. Also, the scaling in one of the directions is by a factor $\lambda_n \equiv \lambda_F$. Thus, the universal constant is *independent* of the dimension n of the ambient space in which we consider the dynamical system. It was pointed out [**CEK**], and later found to be true, that the constant δ_F should therefore be observed in many physical and numerical experiments, even if n is unknown.

(iii) *Area-preserving maps on* $\mathbf{R}^2$. For area-preserving maps of $\mathbf{R}^2$, a hyperbolic fixed point for a composition operator has been found [**EKW**] (but with $\delta \approx 8.721\ldots$), as well as two scalings in $\mathbf{R}^2$ (both different from λ).

(iv) *Circle maps*. The theory of circle maps with golden rotation number. This can be formulated in a related spirit [**ORSS, FKS**].

Hyperbolicity. The hyperbolicity picture P1–P3 has an extension, which is still conjectural, but which puts the dynamical aspects of Feigenbaum theory in a larger perspective. One possible form of the conjecture would be as follows. Consider the operators $\mathbf{N}_p$, as defined above, e.g., with $r = 2$.

Each of the operators $\mathbf{N}_p$ has (among others) as domain $\mathbf{D}_p$ those functions which "exchange" p disjoint intervals $J_0, J_1, \ldots, J_{p-1}$. That is, we have

$$f\colon J_i \to J_{i+1 \bmod p}.$$

Note that the domains $\mathbf{D}_p$ are disjoint for different p. We can now consider an operator $\mathbf{M}$ on "all" functions in $\mathbf{E}$ by setting

$$\mathbf{M}|_{\mathbf{D}_p} \equiv \mathbf{N}_p.$$

CONJECTURE 2. $\mathbf{N}$ *has a strange attractor that has a hyperbolic structure. So, e.g., if two functions f, q are in $\mathbf{D}_{p_0}$ and have the property that $\mathbf{M}^k f$ and $\mathbf{M}^k g$ are in the same $\mathbf{D}_{p_k}$ for all $k \geq 1$, then $\|\mathbf{N}^k f - \mathbf{N}^k g\| \to 0$ as $k \to \infty$. This conjecture has been forwarded by several authors, also in the context of circle maps* [**Rand, Bak, Lanford**, ...].

I have presented all of these examples to point out that there should be an underlying general theory, which is not well understood at present.

Some facts are known, however, about the fixed point function g, and they may be a starting point for further developments: see [**EL, E, EE**] and also [**DH, S**]. The idea is to consider the *inverse* function of g when expressed in the variable x^r. This function can be shown to be analytic in a cut plane and has the Herglotz property; i.e., it maps the upper halfplane to itself. Using these properties makes the existence proofs astonishingly straightforward and suggests that the class of fixed point functions encountered in the problems described in this talk may play a role similar to special function theory.

The list of references also contains some related work not mentioned in this paper.

REFERENCES

[CE] P. Collet and J.-P. Eckmann, *Iterated maps of the interval as dynamical systems*, Birkhäuser, Boston, 1980.

[CEL] P. Collet, J.-P. Eckmann, and O. E. Lanford, III, *Universal properties of maps on the interval*, Comm. Math. Phys. **76** (1980), 211–254.

[CEK] P. Collet, J.-P. Eckmann, and H. Koch, *Period doubling bifurcations for families of maps on $\mathbf{R}^n$*, J. Statist. Phys. **25** (1981), 1–14.

[CER] M. Campanino, H. Epstein, and D. Ruelle, *On Feigenbaum's functional equation*, Topology **21** (1982), 125–129; *On the existence of Feigenbaum's fixed point*, Comm. Math. Phys. **79** (1981), 261–302.

[CT] P. Coullet and C. Tresser, *Itération d'endomorphismes et groupe de renormalisation*, J. Phys. Colloq. C **539** (1978), C5-25; C. R. Acad. Sci. Paris Sér. A-B **287** (1978), no. 7, A577–A580.

[DH] A. Douady and J. H. Hubbard, *On the dynamics of polynomial-like mappings*, Ann. Sci. École Norm. Sup. (4) **18** (1985), 287–343.

[E] H. Epstein, *New proofs of the existence of the Feigenbaum functions*, Comm. Math. Phys. (to appear).

[EE] J-P. Eckmann and H. Epstein, *On the existence of fixed points of the composition operator for circle maps* (to appear).

[EEW] J.-P. Eckmann, H. Epstein, and P. Wittwer, *Fixed points of Feigenbaum's type for the equation $f^p(\lambda x) \equiv \lambda f(x)$*, Comm. Math. Phys. **93** (1984), 495–516.

[EKW] J.-P. Eckmann, H. Koch, and P. Wittwer, *A computer-assisted proof of universality for area-preserving maps*, Mem. Amer. Math. Soc. No. 47 (1984), 289.

[EL] H. Epstein and J. Lascoux, *Analyticity properties of the Feigenbaum function*, Comm. Math. Phys. **81** (1981), 437–453.

[EMO] J.-P. Eckmann, A. Malaspinas, and S. Oliffson-Kamphorst *A software tool for analysis in function spaces*, Preprint.

[EW] J.-P. Eckmann and P. Wittwer, *Computer methods and Borel summability applied to Feigenbaum's equation*, Lecture Notes in Phys., vol. 227, Springer-Verlag, Berlin, 1985.

[EW1] ____, *A complete proof of the Feigenbaum conjectures*, J. Stat. Phys. (in print).

[F] M. J. Feigenbaum, *Quantitative universality for a class of non-linear transformations*, J. Statist. Phys. **19** (1978), 25–52; *Universal metric properties of non-linear transformations*, J. Statist. Phys. **21** (1979), 669–706.

[FKS] M. J. Feigenbaum, L. P. Kadanoff, and S. J. Shenker, *Quasi-periodicity in dissipative systems: a renormalization group analysis*, Phys. D **5** (1982), 370–386.

[GT] S. Grossmann and S. Thomae, Z. Naturforsch. A **32** (1977), 1353.

[JR] L. Jonker and D. Rand, *Universal properties of maps of the circle with ε-singularities*, Comm. Math. Phys. **90** (1983), 273–292.

[L1] O. E. Lanford III, *Remarks on the accumulation of period-doubling bifurcations*, Mathematical Problems in Theoretical Physics, Lecture Notes in Phys., vol. 116, Springer-Verlag, Berlin, 1980, pp. 340–342; *A computer-assisted proof of the Feigenbaum conjectures*, Bull. Amer. Math. Soc. (N. S.) **6** (1984), 127.

[L2] ____, *Smooth transformations of intervals*, Séminaire N. Bourbaki 1980–1981, Lecture Notes in Math., vol. 563, Springer-Verlag, Berlin, 1981.

[L3] ____, *A shorter proof of the existence of the Feigenbaum fixed point*, Comm. Math. Phys. **96** (1984), 521–538.

[L4] ____, *Functional equations for circle homeomorphisms with golden ratio rotation number*, J. Statist. Phys. **34** (1984), 57–73.

[L5] ____, *Renormalization group methods for circle mappings*, Proceedings of the Conference on Statistical Mechanics and Field Theory: Mathematical Aspects (Groningen, 1985), Lecture Notes in Phys., Springer, Berlin-New York (to appear).

[LL] O. E. Lanford III and R. de la Llave (in preparation).

[M] B. Mestel, Ph.D. Dissertation, Department of Mathematics, Warwick University, 1985.

[MN] N. S. Manton and M. Nauenberg, *Universal scaling behavior for iterated maps in the complex plane*, Comm. Math. Phys. **89** (1983), 555.

[Mi] M. Misiurewicz, *Absolutely continuous measures for certain maps of the interval*, Inst. Hautes Études Sci. Publ. Math. **50** (1980).

[ORSS] S. Ostlund, D. Rand, J. Sethna, and E. Siggia, *Universal properties of the transition from quasi-periodicity to chaos in dissipative systems*, Phys. D **8** (1983), 303–342.

[S] D. Sullivan, *Quasi-conformal conjugacy classes and the stable manifold of the Feigenbaum operator*, Preprint, 1986.

[VSK] E. B. Vul, Ia. G. Sinai, and K. M. Khanin, *Feigenbaum universality and the thermodynamical formalism*, Uspekhi Mat. Nauk **39** (1984), 3–37.

[W] M. Widom, *Renormalization group analysis of quasi-periodicity in analytic maps*, Comm. Math. Phys. **92** (1983), 121–136.

UNIVERSITY OF GENEVA, GENEVA, SWITZERLAND

Renormalization Theory and Group
in Mathematical Physics

GIOVANNI GALLAVOTTI

1. Introduction. Starting in 1978 new techniques have been introduced in field theory in the attempt to transform the successful scale invariance ideas, developed in theoretical physics, into an algorithm useful in the mathematical problems of field theory ([1, 2]; for a more complete list of references see [26]).

In this way several problems received new solutions and new perspectives were opened. I mention here first a new derivation of the ultraviolet stability for superrenormalizable theories which led to the attack and solution of some new problems, including the first case of a three-dimensional gauge theory [3–6].

To be fair it should be stressed that the new approach was not born independently of the classical work on constructive field theory, where the ideas of scaling played a basic, although not very explicit and systematic, role [7–10].

Also the theory of renormalization received new impetus from new derivations of the basic results [11], of the recent $n!$-bounds, and of the convergence of the planar φ^4-theory [12–14].

In particular the theory of the convergence of the planar models led us to the understanding that the beta function could be defined in a mathematically rigorous way and thus used to construct a field theory which is not superrenormalizable (but renormalizable and asymptotically free).

The notion of beta function in [15, 16]* can in principle be extended to the planar nonrenormalizable theories to study their nontrivial realizations, or to the Gross-Neveu model in slightly more than 2 dimensions: two cases in which the above extensions of the beta function have a well-defined meaning, being expressed by convergent series, in the domain of interest [18–20].

The novelty of the approach even with respect to the classical problems of perturbation theory is that it made it possible to produce a rigorous proof of renormalizability for quantum electrodynamics (in 4 dimensions) together with natural bounds on the perturbation series coefficients: at least the possibility of

---

*See also §20 in the review paper [26]. Another model for which the beta function has the same properties of convergence as in the planar theory is the 2-dimensional Gross-Neveu model (discussed in [17]).

a renormalizability proof was well known, but a true proof was missing because the technical problems were hard to handle with classical tools [21].

Finally, the implications of the above techniques for the theory of the critical point and more generally for statistical mechanics seem to be under active investigation and far from being exhausted in their potentialities, even though statistical mechanics was the first field of mathematical physics where scaling ideas were applied [22–25].

2. The beta function. I cannot enter into too many details here but I wish to provide at least some of the ideas behind the above cited works recently dedicated to the theory of the beta function in the case of φ^4-theories in 4 dimensions (nonplanar or planar).

In Euclidean field theory the basic object is the free field: i.e., a gaussian random field on $\mathbf{R}^d$ with covariance operator

$$C = (1 - \Delta)^{-1}, \qquad \Delta = \text{Laplacian}, \tag{2.1}$$

which has a rather singular kernel (so that the sample fields of the corresponding process are distributions on $\mathbf{R}^d$ in $H^{\text{loc}}_{-d/2+1-\varepsilon}$, i.e., "far from ordinary functions").

Such singular fields φ admit a "scaling decomposition" into regular fields:

$$\varphi_x = \lim_{N \to \infty} \sum_{k=0}^{N} \varphi_x^{(k)} \equiv \lim_{N \to \infty} \varphi_x^{(\leq N)}, \qquad x \in \mathbf{R}^d, \tag{2.2}$$

where $\varphi_x^{(k)}$ has the same distribution as $\gamma^{(d-2)k/2} \varphi^{(0)}_{\gamma^k x}$, but is very smooth and essentially independently distributed on the scale γ^{-k}: here γ is an arbitrary prefixed parameter (usually one takes $\gamma = 2$), and furthermore the fields $\varphi^{(h)}, \varphi^{(k)}$ are independently distributed if $h \neq k$.

Intuitively one should think of $\varphi^{(k)}$ as a random field of large size $O(\gamma^{(d-2)k/2})$ but constant over cubes with side length γ^{-k}, i.e. "on scale" γ^{-k}, and furthermore with values independently distributed over different cubes.

The problem of field theory (for scalar fields) is to give a meaning to probability measures on the space of fields on $\mathbf{R}^d$ (i.e., on the space of distributions on $\mathbf{R}^d$) such as

$$\left(\exp \int_\Lambda V(\varphi_x) \, dx \right) P(d\varphi) \equiv \lim_{N \to \infty} \left(\exp \int_\Lambda V^{(N)}(\varphi_x^{(\leq N)}) \, dx \right) P(d\varphi^{(\leq N)}), \tag{2.3}$$

with Λ being a fixed volume, say a cube, and $V^{(N)}$ some suitable sequence of functions; P is the free field distribution.

We restrict our attention here to "φ^4-theories," i.e., to special V's which are fourth-order polynomials:

$$\begin{aligned}
V^{(N)}(\varphi_x^{(\leq N)}) &= \overline{\lambda}_N \varphi_x^{(\leq N)4} + \overline{\mu}_N \varphi_x^{(\leq N)2} + \overline{\nu}_N + \overline{\alpha}_N (\partial \varphi_x^{(\leq N)})^2 \\
&= \gamma^{4N}(\lambda_N H_4(X_x) + \mu_N H_2(X_x) + \nu_N + \alpha_N H_2(\partial X_x))
\end{aligned} \tag{2.4}$$

where $\bar{\lambda}_N, \ldots, \bar{\alpha}_N$ are arbitrary constants, $X_x = \varphi_x^{(\leq N)}/\sqrt{\gamma^{2N}}$ is a "normalized" field, and H_n are the Hermite polynomials ($H_0 = 1$, $H_1 = x$, $H_2 = x^2 - 1/2, \ldots$). The second way of writing the fourth-order polynomial, in terms of Hermite polynomials, is natural as is well known ("Wick ordering"). The factor γ^{4N} is inserted for convenience and is canceled when the integration over x is done if X_x is regarded as constant on the volume element, i.e., on the cubes of scale γ^{-N}.

The first basic idea is to find bounds on (2.3) by introducing the "effective potentials"

$$\exp A_k^{(N)}(\varphi^{(\leq k)}) \equiv \int \exp\left(\int_\Lambda V^{(N)}(\varphi_x^{(\leq N)})\,dx\right) P(d\varphi^{(k+1)}) \cdots P(d\varphi^{(N)})$$

(2.5)

with the purpose of proving the existence of the limit $\lim_{N\to\infty} A_k^{(N)}(\varphi^{(\leq k)})$, $k = 0, 1, \ldots$.

The second idea is that although $A_k^{(N)}$ is a considerably complicated functional of $\varphi \equiv \varphi^{(\leq k)}$, it consists in fact of a "simple" "relevant" part of the form (L stands for "local"):

$$A_k^{(N)L} \equiv \int_\Lambda (\bar{\lambda}_k \varphi_x^4 + \bar{\mu}_k \varphi_x^2 + \bar{\nu}_k + \bar{\alpha}_k (\partial \varphi_x)^2)\,dx,$$

(2.6)

which we always think of as written in Wick ordered form, i.e. in terms of the Hermite polynomials of $X_x = \varphi_x^{(\leq k)}/\sqrt{\gamma^{2k}}$ as in (2.4), plus a remainder $R_k^{(N)} = A_k^{(N)} - A_k^{(N)L}$.

The remainder is "irrelevant" in various senses; here we simply mean that it is expressible in terms of the "form factors" $\mathbf{c}_k = (\lambda_k, \alpha_k, \mu_k, \nu_k)$ which, in turn, are "self-sufficient" because they satisfy a recursion relation:

$$\mathbf{c}_k = L\mathbf{c}_{k+1} + B(\mathbf{c}_{k+1}) \equiv T(\mathbf{c}_{k+1})$$

(2.7)

where $B(\mathbf{c})$ is a formal power series in $\mathbf{c}$ and L is a diagonal matrix with diagonal $\mathrm{diag}(L) = (1, 1, \gamma^2, \gamma^4)$.

The main result of the renormalization group approach to renormalization theory is that the coefficients of B can be bounded; if B is

$$B(\mathbf{c}) = \sum_{p=2}^{\infty} \sum_{\substack{\mathbf{m}\in Z_+^4 \\ |\mathbf{m}|=p}} \beta(\mathbf{m};N)\mathbf{c}^{\mathbf{m}},$$

(2.8)

there is a constant $\beta > 0$ such that

$$|\beta(\mathbf{m};N)| \leq (p-1)!\beta^{p-1}, \qquad |\mathbf{m}| = p, \ \forall N,$$
$$\lim_{N\to\infty} \beta(\mathbf{m};N) = \beta(\mathbf{m}) \quad \text{exists } \forall \mathbf{m} \in Z_+^4.$$

(2.9)

The above bounds embody all the results of perturbation theory; once proved they imply the finiteness, as well as explicit bounds (the natural $n!$-bounds of [12]), of the coefficients of the formal power series expressing the "irrelevant" remainders in terms of the form factors $\mathbf{c}_0$ [15, 16, 26].

It seems reasonable (no proof, however, exists) that if one could overcome the problems of convergence of the series (2.8) for B, then "by the same argument," the corresponding problems for the series for R_k should disappear.

This indicates that the key questions seem to be:

(1) Give a summation rule for (2.7) which is meaningful for $\mathbf{c} \in \mathcal{D} \subset \mathbf{R}^4$ where $\mathcal{D}$ is some (a priori unknown) suitable domain.

(2) Show that $\mathcal{D}$ is invariant for the flow of the "renormalization map": $T^{-1}\mathcal{D} \subset \mathcal{D}$.

Without an answer to questions (1) and (2) the theory remains a purely formal perturbation theory, finite to every order but with open convergence problems (nevertheless even this order by order statement is rather nontrivial; it is completely solved by the above approach, which also yields explicit and best (to date) bounds).

(3) Check the compatibility of the resulting stochastic process, with the axioms that it should fulfill in order to be interpreted as a quantum field theory.

The last question is essential for the interest of the theory: it is in fact quite clear that the results in [27, 28] can be interpreted as solutions to (1) and (2) above which do not satisfy (3).

There are very few cases in which the above program can be carried through: basically they coincide with the cases where the series (2.8), or the corresponding one for models other than φ^4, admit bounds so much better than (2.9) that one is allowed to define unambiguously the beta function B because the series (2.8) converges.

In the planar φ^4-theory and in the 2-dimensional Gross-Neveu model the $(p-1)!$ can be replaced by 1; hence the series is convergent (see [15, 16, 26] for planar φ^4 and [17] for Gross-Neveu) for $|\mathbf{c}|$ small, thus answering unambiguously (1) with $\mathcal{D} = \{\mathbf{c} | \, |\mathbf{c}| < \delta\}$.

One can, in the latter cases, pass to the analysis of question (2) above (the third does not make sense for planar φ^4, while it is not hard in the case of the Gross-Neveu model because of the lack of ambiguity in answering (1)).

We then look for a set $\sigma \subset \mathcal{D}$ such that $T^{-k}\sigma \subset \mathcal{D}$, $\forall k \geq 0$. For instance a surface σ such that $T^{-k}\mathbf{c}_0 \underset{k \to \infty}{\longrightarrow} 0$ if $\mathbf{c}_0 \in \sigma$.

The great advantage of having a convergent beta function is that such a question can be easily answered by standard perturbation and bifurcation theory by truncating the series (2.8) to its second-order terms.

For instance, for planar φ^4-theory, $d = 4$, the series (2.8) truncated to second order becomes:

$$
\begin{aligned}
\lambda_k &= \lambda_{k+1} + \beta \lambda_{k+1}^2 + \delta \lambda_{k+1}\mu_{k+1}, \\
\alpha_k &= \alpha_{k+1} + \beta' \lambda_{k+1}^2 + \delta' \lambda_{k+1}\mu_{k+1}, \\
\mu_k &= \gamma^2 \mu_{k+1} + \beta'' \lambda_{k+1}^2 + \varsigma'' \mu_{k+1}^2 - \theta'' \alpha_{k+1}\mu_{k+1}, \\
\nu_k &= \gamma^4 \nu_{k+1} + \beta''' \lambda_{k+1}^2 + \varsigma''' \mu_{k+1}^2 + \theta''' \alpha_{k+1}\mu_{k+1} + \varepsilon''' \alpha_{k+1}^2,
\end{aligned}
\tag{2.10}
$$

where $\beta, \delta', \ldots, \varepsilon'''$ are positive, easily computable (see [26, (20.20), (20.21)]) constants.

Then an elementary analysis shows that for $\lambda_0 > 0$, $\alpha_0 = O(\lambda_0^2)$ suitably chosen, and λ_0 small, one can choose μ_0, ν_0 so that $\lambda_k = O(\lambda_0/(1 + \beta k \lambda_0))$, $\mu_k = O(\lambda_k^2)$, $\nu_k = O(\lambda_k^2)$, $\alpha_k = O(\lambda_k^2)$, solving (2) with $\mathcal{D}$ becoming now, for instance, restricted to the sequence $\mathbf{c}_k$.

In this case the check of the convergence of the remainders R_k is indeed, as expected, a very simple technical matter and one obtains in this way a complete construction of the planar φ^4-theory [15, 16]; unfortunately this is a little unsatisfactory because, a priori, the planar φ^4-theory is known to be unphysical and it is meaningless to ask question (3). The same ideas, however, can be applied to the physically meaningful Gross-Neveu model in 2 dimensions [17], and in this case question (3) is easily answered using the lack of ambiguity in problem (1).

The above technique can be extended to nonrenormalizable theories. However, it turns out that it is no longer possible to introduce even a formal recursion relation linking only $\mathbf{c}_k$ and $\mathbf{c}_{k+1}$: for instance, the simplest renormalizable theory is obtained by replacing the free field operator (2.1) by $C = (1 - \Delta)^{-1+\varepsilon/2}$, which also admits a scaling decomposition like (2.2) with $\varphi^{(k)}$ of order $O(\gamma^{(d-2+\varepsilon)k/2})$. Considering the φ^4-theory in 4 dimensions with this free field one finds that (2.7) is replaced by

$$\mathbf{c}_k = L\mathbf{c}_{k+1} + B(\mathbf{c}_{k+1}, \mathbf{c}_{k+2}, \dots, \mathbf{c}_N), \tag{2.11}$$

where B is a formal power series and L is a diagonal matrix,

$$\mathrm{diag}(L) = (\gamma^{2\varepsilon}, \gamma^\varepsilon, \gamma^{2+\varepsilon}, \gamma^4).$$

The relation (2.11) does not, of course, uniquely fix B: nevertheless B can be defined in a natural way and the coefficients of its formal power series can be bounded uniformly in N.

The main use of such series is again in the case of theories for which B is convergent when $\sup_k |\mathbf{c}_k|$ is small: the above φ^4-theory in the planar version is an interesting example. Felder [19] finds a nontrivial planar theory by simply proving that the equation

$$\mathbf{c} = L\mathbf{c} + B(\mathbf{c}, \mathbf{c}, \mathbf{c}, \dots) \tag{2.12}$$

has a solution $\mathbf{c}_0$ within the domain of convergence of the series for B (uniformly in N, of course). A similar situation is met in the Gross-Neveu model in $2 + \varepsilon$ dimensions [20].

3. The beta function and the tree expansion. The mathematical definition of the beta function is easily formulated in terms of the "tree expansion" [15, 16].

The recursive evaluation of the integrals defining the effective potential (2.5) can be represented graphically by suitably interpreting the formal expansion (Taylor expansion)

$$\int e^{A_N^{(N)}} P(d\varphi^{(N)}) = \exp \sum_{n=1}^{\infty} \frac{\mathcal{E}_N^T(A_N^{(N)}; n)}{n!} \equiv \exp A_{N-1}^{(N)},$$

$$\mathcal{E}_N^T(A_N^{(N)}; n) \equiv \frac{\partial^n}{\partial \theta^n} \log \int (\exp \theta A_N^{(N)}) P(d\varphi^{(N)})|_{\theta=0}. \tag{3.1}$$

One represents $A_N^{(N)}$ graphically as $_N$——$_L$ and $n!\mathcal{E}_N^T(A_N^{(N)};n)$ as

i.e., by a vertex with subscript N, "scale index," representing $\mathcal{E}_N^T$, and n lines, recalling the order n, emerging from it; the extra line ending with the scale index $N-1$ reminds us that the object represented graphically is a functional of $\varphi^{(\leq N-1)}$. The L reminds us that $A_N^{(N)}$ is "purely local", i.e., an integral of a function of $\varphi_x^{(\leq N)}$: see (2.6).

By means of a projection operator $\mathcal{L}_{N-1}$ we "extract" from each of the terms in the sum in (3.1) the "local part," i.e., a part which has the same form as (2.6), and collect all the results in a term which, of course, will have again the same form (2.6) with coefficients $(\lambda_{N-1}, \alpha_{N-1}, \mu_{N-1}, \nu_{N-1})$ and which we shall denote graphically $_{N-1}$——$_L$.

In this way we can write the sum in (3.1) in a graphical form:

$$\underset{N-1 \qquad L}{\text{———}} + \underset{N-1 \quad N}{\text{———}\!\!<}^{L}_{\;L} + \underset{N-1 \quad N}{\text{———}\!\!<}^{\;\;\;L}_{\;L}\; L + \cdots \quad (3.2)$$

where the term $_{N-1}$——^R_N——$_L$ is missing because $_{N-1}$——$_N$——$_L$ is already "local" by definition, and ^R_N symbolizes the operation $(1 - \mathcal{L}_{N-1})\mathcal{E}_N^T$.

There is some ambiguity in the choice of the projection operator $\mathcal{L}_{N-1}$: one can select it in such a way as to simplify the formalism. Basically it turns out that there is only one natural choice, and precisely one should select a projection $\mathcal{L}_{N-1}$ which "commutes" with the integrations over different scales; i.e., such that:

$$\mathcal{L}_h \mathcal{E}_{h+1} \cdots \mathcal{E}_q \equiv \mathcal{E}_{h+1} \cdots \mathcal{E}_q \mathcal{L}_q, \qquad \mathcal{E}_q(\cdot) \equiv \int \cdot P(d\varphi^{(q)}), \qquad (3.3)$$

as can be done (see [15, 16], and [26]).

Then, iterating the expansion (3.1), one reaches a representation of $A_k^{(N)}$ in terms of objects like the following:

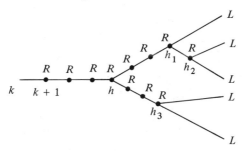

where each vertex bears a scale label representing an operation $\mathcal{E}_h^T$. The trivial vertices ——^R_h—— are obviously redundant and can be eliminated.

Thus one obtains a representation of $A_k^{(N)}$ in terms of "trees" τ with root scale k:

$$A_k^{(N)} = \underset{k\quad L}{\underline{\quad\quad}} + \sum_h \underset{k\quad h}{\overbrace{\quad}}^{R} \overset{L}{\underset{L}{<}} + \sum_h \underset{k\quad h}{\overbrace{\quad}}^{R}\overset{L}{\underset{L}{<}}\overset{L}{}$$

$$+ \cdots + \sum_{h,h_1} \underset{k\quad h}{\overbrace{\quad}}^{R}\underset{h_1}{<}\overset{R\;L}{\underset{L}{<}}\overset{L}{} + \cdots \tag{3.4}$$

$$\equiv \sum_{\substack{\text{trees } \tau \\ \text{root at } k}} A^{(N)}(\tau)$$

and the local part $A_k^{(N)L}$ is simply $\underset{k\quad L}{\underline{\quad\quad}}$ and can be defined graphically by:

$$\underset{k\quad L}{\underline{\quad\quad}} = \underset{k\quad k+1}{\underline{\quad\overset{L}{\bullet}\quad}}^{L} + \sum_{p=2}^{\infty} \sum_{\substack{\text{trees } \tau_1,\dots,\tau_p \\ \text{root } \tau_j \text{ at } k}} \underset{k\quad\quad k+1}{\underline{\quad\quad}}^{L}\overset{\tau_1}{\underset{\tau_2}{\underset{\tau_p}{<}}} \tag{3.5}$$

where $\underset{k+1}{\overset{L}{\big|}}$ represents the operation $\mathcal{L}_k \mathcal{E}_{k+1}^T$ and the terms with $p \geq 2$ are distinguished from those with $p = 1$ because the latter contribute the linear part in the recursion relation (2.6).

If one writes (3.5) explicitly, one finds a relation between the four coefficients $\mathbf{c}_k = (\lambda_k, \alpha_k, \mu_k, \nu_k)$ in $A_k^{(N)L}$ and the coefficients $\mathbf{c}_{k+1}, \mathbf{c}_{k+2}, \dots$ in the form of a formal power series:

$$\mathbf{c}_k = L\mathbf{c}_{k+1} + B(\mathbf{c}_{k+1}, \mathbf{c}_{k+2}, \dots, \mathbf{c}_N),$$

L is linear diagonal and $\mathrm{diag}(L) = (\gamma^\varepsilon, \gamma^{2\varepsilon}, \gamma^{2+\varepsilon}, \gamma^4)$,

$$B(\mathbf{c}_{k+1}, \dots) = \sum_{s=1}^{\infty} \sum_{\substack{h_1,\dots,h_s \\ h_i \geq k+1}} \sum_{\mathbf{m}_1,\dots,\mathbf{m}_s \in Z_+^4} \beta(h_1, \dots, h_s, \mathbf{m}_1, \dots, \mathbf{m}_s) \mathbf{c}_{h_1}^{\mathbf{m}_1} \cdots \mathbf{c}_{h_1}^{\mathbf{m}_s},$$

$$\tag{3.6}$$

with the β's independent on N and such that:

$$\sum_s \sum_{h_1,\dots,h_s} \sum_{\substack{\mathbf{m}_1,\dots,\mathbf{m}_s \\ \sum |\mathbf{m}_j| = p}} |\beta(h_1, \dots, \mathbf{m}_s)| \leq \beta^{p-1}(p-1)!. \tag{3.7}$$

Then the theory is renormalizable when the relation (3.6) can be solved recursively allowing us to express $\mathbf{c}_{k+1}, \mathbf{c}_{k+2}, \dots$ as a formal power series in $\mathbf{c}_k$ with coefficients uniformly bounded in N; nonrenormalizable otherwise. This happens respectively when the free field covariance is $(1 - \Delta)^{-1}$, i.e. $\varepsilon = 0$, or $(1 - \Delta)^{-1-\varepsilon/2}$, with $\varepsilon > 0$.

4. Applications to statistical mechanics. The tree expansion is useful in statistical mechanics too. In fact a beginning of the tree expansion can be found already in the proof of the Debye screening in the 3-dimensional Coulomb gas [29]. It has been applied to the identification of the phase transitions heralding the Kosterlitz-Thouless regime in the 2-dimensional Coulomb gas, in studying the smoothness properties of the pressure $p(\beta, \lambda)$ as a function of the inverse temperature β and of the activity λ, and in various problems on the "massive Yukawa gas."

If

$$V(x) \underset{x \to \infty}{\cong} \frac{1}{2\pi} \log \left(\frac{|x|}{r_0} \right)^{-1}$$

is the Coulomb potential and the interaction between the charges (assumed to be ± 1) is regularized at short distances, then one can show, by a tree expansion technique [30–32], that $p(\beta, \lambda)$ is $2n$ times continuously differentiable, at $\lambda = 0$, in λ if:

$$\beta > \beta_n = 8\pi(1 - 1/2n), \qquad n = 1, 2, \ldots. \tag{4.1}$$

Hence for $\beta > 8\pi$ it is C^∞ at the origin in λ; and in all cases the "Mayer expansion" (of $p(\beta, \lambda)$ in series of λ at $\lambda = 0$) is asymptotic up to the order for which it makes sense ($2n$ if $\beta > \beta_{2n}$).

The physical interpretation of the thresholds (4.1) is probably in terms of a sequence of phase transitions: for $\beta > \beta_n$ the gas contains a macroscopic fraction of charges bound into "stable" molecules with $2, 4, \ldots, 2n$ atoms. A long way, however, still remains towards a rigorous proof of the above picture; basically what is still missing is a macroscopic description of the equilibrium states above the thresholds (4.1).

Another open problem is whether the series for $p(\beta, \lambda)$ in powers of λ is in fact convergent (and not just asymptotic) for $\beta > 8\pi$. This has been sometimes suggested as possible. It is in fact true in a related model which is a hierarchical version of the Coulomb gas [33].

Some progress in the techniques that may be helpful in such a question, particularly if it turns out to have a positive answer, has been achieved in the recent work of Benfatto [35], where it is shown how to use the tree expansion to prove the analyticity of the Mayer expansion for $p(\beta, \lambda)$ in the Yukawa gas in 2 dimensions.

The latter gas is like the Coulomb gas except that now

$$V(x) \cong \frac{1}{2\pi} \log \left(\frac{|x|}{r_0} \right)^{-1} \qquad \text{for } x \to 0,$$

while $V(x)$ decays exponentially at ∞.

For this gas one can show [34] that $p(\beta, \lambda)$ is analytic at $\lambda = 0$ if $\beta < 4\pi$, by using classical cluster expansion techniques. In [35] the same result is derived by using the tree expansion; however, as remarked by T. Kennedy (private communication), more is proved in the paper [35], namely

$$p_2(\beta, \lambda) \equiv p(\beta, \lambda) - \frac{1}{2!} \frac{\partial^2 p}{\partial \lambda^2}(\beta, \lambda)|_{\lambda=0} \lambda^2 \tag{4.2}$$

is in fact analytic in β through the threshold $\beta_1 = 4\pi$ up to (but excluding) the next threshold at $\beta_2 = 6\pi$. This extension is obviously implicit in [35], but further extensions seem to involve problems of the same nature that one meets in trying to understand whether the Coulomb gas pressure is analytic at $\lambda = 0$ for $\beta > 8\pi$. The related question for the Yukawa gas seems to be: consider

$$p_{2n}(\beta, \lambda) = p(\beta, \lambda) - \sum_{j=0}^{n} \frac{1}{(2j)!} \frac{\partial^{2j} p(\beta, \lambda)}{\partial \lambda^{2j}}\Big|_{\lambda=0} \lambda^{2j}; \qquad (4.3)$$

is it analytically continuable in β up to $\beta < \beta_n$ for λ small?

It seems plausible that one could prove the analyticity in λ of p_4 at $\beta = \beta_2 = 6\pi$ included by using techniques similar to those in [35], combined perhaps with ideas based on the beta function; but the question for $\beta > 6\pi$ seems considerably harder.

The whole situation is quite open and slightly frustrating: for instance it is not clear whether the tree expansion technique is really suited for the above problems. There are cases where it does not seem to be capable of reproducing results known by other methods. I refer here to the case of the 2-dimensional dipole gas where [6] it is shown that $p(\beta, \lambda)$ is analytic in λ at small λ. The latter result does not seem to follow from the tree expansion in a straightforward way, as might be expected, which means that in some sense the tree expansion may be an "over-expansion" hiding some cancellations, even in models where things seem to be quite simple, like the hierarchical models.

It would be desirable to understand such cancellation mechanisms to incorporate them in the tree expansion techniques.

REFERENCES

1. Giovanni Gallavotti, *Some aspects of the renormalization problems in statistical mechanics and field theory*, Atti Accad. Naz. Lincei Mem. Cl. Sci. Fis. Mat. Natur. Sez. Ia (8) **15** (1978), 23–59.

2. G. Benfatto, *Some probabilistic techniques in field theory*, Comm. Math. Phys. **59** (1978), 143–166.

3. Tadeusz Bałaban, *(Higgs)$_{2,3}$ quantum fields in a finite volume. I. A lower bound*, Comm. Math. Phys. **85** (1982), 603–626.

4. _____, *(Higgs)$_{2,3}$ quantum fields in a finite volume. II. An upper bound*, Comm. Math. Phys. **86** (1982), 555–594.

5. _____, *(Higgs)$_{2,3}$ quantum fields in a finite volume. III. Renormalization*, Comm. Math. Phys. **88** (1983), 411–445.

6. Krysztof Gawędzki and Antti Kupiainen, *Block spin renormalization group for dipole gas and $(\nabla\phi)^4$*, Ann. Phys. **147** (1983), 198–243.

7. James G. Glimm, *Boson fields with the : Φ^1 : interaction in three dimensions*, Comm. Math. Phys. **10** (1968), 1–47.

8. James G. Glimm and Arthur Jaffe, *The $\lambda(\phi^4)_2$ quantum field theory without cutoffs. II. The field operators and the approximate vacuum*, Ann. of Math. (2) **91** (1970), 362–401.

9. Joel S. Feldman and Konrad Osterwalder, *The Wightman axioms and the mass gap for weakly coupled $(\phi^4)_3$ quantum field theories*, Ann. Phys. **97** (1976), 80–135.

10. J. Magnen and Roland Sénéor, *The infinite volume limit of the φ_3^4 model*, Ann. Inst. H. Poincaré Sect. A (N.S.) **24** (1976), 95–159.

11. Klaus Hepp, *Théorie de la renormalisation*, Lecture Notes in Physics, vol. 2, Springer-Verlag, Berlin and New York, 1969.

12. C. de Calan and V. Rivasseau, *Local existence of the Borel transform in Euclidean Φ_4^4*, Comm. Math. Phys. **82** (1981), 69–100.

13. Gerard 't Hooft, *Rigorous construction of planar diagram field theories in four-dimensional Euclidean space*, Comm. Math. Phys. **88** (1983), 1–25.

14. V. Rivasseau, *Construction and Borel summability of planar 4-dimensional Euclidean field theory*, Comm. Math. Phys. **95** (1984), 445–486.

15. Giovanni Gallavotti and Francesco Nicolò, *Renormalization theory in four-dimensional scalar fields. I*, Comm. Math. Phys. **100** (1985), 545–590.

16. ____, *Renormalization theory in four-dimensional scalar fields. II*, Comm. Math. Phys. **101** (1985), 247–282.

17. Krysztof Gawędzki and Antti Kupiainen, *Gross-Neveu model through convergent perturbation expansions*, Comm. Math. Phys. **102** (1985), 1–30.

18. Giovanni Felder and Giovanni Gallavotti, *Perturbation theory and nonrenormalizable fields*, Comm. Math. Phys. **102** (1985), 549–571.

19. Giovanni Felder, *Construction of a nontrivial planar field theory with ultraviolet stable fixed point*, Comm. Math. Phys. **102** (1985), 139–155.

20. Krysztof Gawędzki and Antti Kupiainen, *Renormalizing the nonrenormalizable*, Phys. Rev. Lett. **55** (1985), 363–365.

21. Joel S. Feldman et al., Univ. of British Columbia, Vancouver, Preprint, 1986.

22. Freeman J. Dyson, *An Ising ferromagnet with discontinuous long-range order*, Comm. Math. Phys. **21** (1971), 269–283.

23. P. M. Bleher and Ya. G. Sinaï, *Investigation of the critical point in models of the type of Dyson's hierarchical model*, Comm. Math. Phys. **33** (1973), 23–42.

24. ____, *Critical indices for Dyson's asymptotically hierarchical models*, Comm. Math. Phys. **45** (1975), 247–278.

25. Pierre Collet andd Jean-Pierre Eckmann, *A renormalization group analysis of the hierarchical model in statistical mechanics*, Lecture Notes in Physics, vol. 74, Springer-Verlag, Berlin and New York, 1978.

26. Giovanni Gallavotti, *Renormalization theory and ultraviolet stability for scalar fields via renormalization group methods*, Rev. Modern Phys. **57** (1985), 471–562.

27. Krysztof Gawędzki and Antti Kupiainen, Lecture notes at the Les Houches School, 1984 (K. Osterwalder and R. Stora, eds.).

28. Giovanni Gallavotti, *On the φ_4^4-problem*, Stochastic Problems in Classical and Quantum Systems, S. Albeverio and D. Merline, editors, Lecture Notes in Physics, vol. 262, Springer-Verlag, Heidelberg, 1986, pp. 278–295.

29. David C. Brydges, *A rigorous approach to Debye screening in dilute classical Coulomb systems*, Comm. Math. Phys. **58** (1978), 313–350.

30. Giovanni Gallavotti and Francesco Nicolò, *The "screening" phase transitions in the two-dimensional Coulomb gas*, J. Statist. Phys. **39** (1985), 133–156.

31. Francesco Nicolò, J. Renn and A. Steinmann, *The pressure of the two-dimensional Coulomb gas at low and intermediate temperatures*, Ann. Inst. H. Poincaré **44B** (1986), 211–261.

32. ____, *The sine-gordon equation in all regions of collapse*, Comm. Math. Phys. **105** (1986), 291–326.

33. G. Benfatto et al., *The dipole phase in the two-dimensional hierarchical Coulomb gas*, Comm. Math. Phys. **106** (1986), 277–288.

34. Jürg Frölich, *Classical and quantum statistical mechanics in one and two dimensions: two-component Yukawa- and Coulomb systems*, Comm. Math. Phys. **47** (1976), 233–268.

35. G. Benfatto, *An iterated Mayer expansion*, J. Statist. Phys. **41** (1986), 671–684.

II UNIVERSITÁ DEGLI STUDI DI ROMA, ROMA, ITALIA

Proceedings of the International Congress of Mathematicians
Berkeley, California, USA, 1986

Renormalization:
From Magic to Mathematics

K. GAWĘDZKI

Quantum Field Theory (QFT) is a physical scheme into which fit most of the theoretical attempts to understand the world of elementary particles and their interactions. From the mathematical point of view it may be perceived as a study of integrals over spaces of maps $\phi\colon \mathbf{R}^d \to \mathbf{R}^M$, formally given as

$$\int \prod_{j=1}^{J} \phi^{\alpha_j}(x_j) e^{-S(\phi)} \prod_{x,\alpha} d\phi^\alpha(x) \Big/ \int e^{-S(\phi)} \prod_{x,\alpha} d\phi^\alpha(x), \tag{1}$$

where x, $x_j \in \mathbf{R}^d$, $j = 1, \ldots, J$, α, α_j are integers between 1 and M and $S(\phi)$ is a local functional, e.g., in the simplest case of the so-called ϕ^4 theory, $M = 1$ and

$$S(\phi) \equiv S_{m,\lambda}(\phi) = \int_{\mathbf{R}^d} \left[\frac{1}{2} \sum_\mu (\partial_\mu \phi)^2 + \frac{1}{2} m^2 \phi^2 + \lambda \phi^4 \right]. \tag{2}$$

This formulation stresses the probabilistic aspect of QFT where the evaluation $\phi \mapsto \phi^\alpha(x)$ is treated as a random (distributional) field over a space of (generalized) functions.

The anticommuting version of (1) plays also an important role (for description of fermionic elementary particles). Then, for each $x \in \mathbf{R}^d$ and each α, $1 \leq \alpha \leq M$, $\phi^\alpha(x)$ is a Grassmann algebra generator and the anticommuting integration rule is defined, following [1], by setting $\int \phi^\alpha(x)\, d\phi^\alpha(x) = 1$, $\int d\phi^\alpha(x) = 0$. One of the simplest cases here is the Gross-Neveu (G-N) model [2], where $\phi(x) = (\overline{\psi}^r(x), \psi^s(x))$, $1 \leq r, s \leq N > 1$, $\overline{\psi}^r$ and ψ^s are Dirac spinors for each r, s, and

$$S_{m,\lambda}(\phi) = \int_{\mathbf{R}^d} \left[\sum_r \overline{\psi}^r i \partial\!\!\!/ \psi^r + m \sum_r \overline{\psi}^r \psi^r - \lambda \left(\sum_r \overline{\psi}^r \psi^r \right)^2 \right] \tag{3}$$

($\partial\!\!\!/$ denotes the Dirac operator). Expressions (1), called euclidean Green functions, encode physical properties of particles described by the field theory model.

If $S(\phi)$ is quadratic in ϕ, i.e., $\lambda = 0$ in the examples, then (1) reduces to a Gaussian integration over an infinite-dimensional space of functions (or a Grassmann algebra) whose mathematical structure is well understood [3, 1]. Such

Gaussian measures, as in the finite-dimensional case, are fully determined by the covariance $G_m(x_1, x_2)$, i.e., the second moment ((1) with $J = 2$). $G_m = (\Delta + m^2)^{-1}$ or $G_m = (i\partial + m)^{-1}$ for (2) and (3) respectively. Gaussian random fields describe however only noninteracting particles.

There are three sorts of difficulties when one attempts to include the interaction terms multiplied by the coupling constant λ into the integration:

(i) *Local regularity.* Since the Gaussian measure is supported by distributional ϕ's, even $\int_K \phi^4$ with K compact is not a well-defined random variable.

(ii) *Global regularity.* The interaction term involves the integration over whole $\mathbf{R}^d$ which probes the decay properties of typical ϕ's.

(iii) *Stability.* Large ϕ behavior of the interaction should clearly be crucial for the existence of integrals like (1) (in the commutative case). Compare the ϕ^4 theories with different signs of λ.

The traditional physicist's approach to the $\lambda \neq 0$ case consists of expanding (1) into powers of λ. Each power of this *perturbation expansion* (PE) involves only Gaussian integrals which are easily evaluated in terms of covariance G_m. For example, for the second moment ($J = 2$) of the ϕ^4 theory, we obtain

$$(1) \Big|_{J=2} = G_m(x_1, x_2) - 12\lambda \int_{\mathbf{R}^d} dx \, G_m(x_1, x)G_m(x, x)G_m(x, x_2)$$

$$+ 96\lambda^2 \int_{\mathbf{R}^{2d}} dx \, dy \, G_m(x_1, x)G_m(x, y)^3 G_m(y, x_2)$$

$$+ 144\lambda^2 \int_{\mathbf{R}^{2d}} dx \, dy \, G_m(x_1, x)G_m(x_2, x)G_m(x, y)^2 G_m(y, y)$$

$$+ 144\lambda^2 \int_{\mathbf{R}^{2d}} dx \, dy \, G_m(x_1, x)G_m(x, x)G_m(x, y)G_m(y, y)G_m(y, x_2)$$

$$+ O(\lambda^3). \tag{4}$$

(4) however diverges in 2 or more dimensions as

$$G_m(x, y) \underset{y \to x}{\sim} \begin{cases} O(\ln|x - y|), & d = 2, \\ O\left(\dfrac{1}{|x - y|^{d-2}}\right), & d > 2. \end{cases} \tag{5}$$

These short distance divergences which plague the PE in QFT are just a reflection of the lack of local regularity. The global regularity problem appears in PE only if $m = 0$, since $G_0(x, y)$ has a slow decay for $|x - y| \to \infty$ leading to long distance divergences in (4). The stability question is discarded altogether in the perturbative approach. In what follows, we shall not discuss the global regularity problem which in many cases can be treated by conventional methods of statistical mechanics.

There exists a celebrated procedure, called *renormalization*, which removes the short distance divergences from the field theoretic PE. Born as partly magic, partly suspicious manipulations on formal series with infinite coefficients, it led, when applied to Quantum Electrodynamics (a field theory describing interactions of electrons and photons), to finite results which were in spectacular agreement with experiment, see e.g. [4]. Subsequently the perturbative renormalization has obtained a precise formulation [5], whose main idea we shall sketch now.

Let us start by making the Gaussian integral underlying our theory more regular by replacing (2) by its cutoff version

$$S_{m,\lambda}^{\theta}(\phi) = \int_{\mathbf{R}^d} \left[\frac{1}{2} \sum_{\mu} \partial_{\mu}\phi e^{\Delta/\theta^2} \partial_{\mu}\phi + \frac{1}{2}m^2\phi^2 + \lambda\phi^4 \right] \tag{6}$$

(Δ stands for Laplacian) and similarly for (3) with $i\partial \!\!\!/ \mapsto i\partial \!\!\!/ e^{\Delta/\theta^2}$. This results in the change

$$G_m \mapsto G_m^{\theta} = \begin{cases} (\Delta e^{\Delta/\theta^2} + m^2)^{-1}, \\ (i\partial \!\!\!/ e^{\Delta/\theta^2} + m)^{-1}, \end{cases} \tag{7}$$

respectively with $G_m^{\theta}(x,x)$ finite. (Consequently the Gaussian measure becomes concentrated on functions.) Introduction of the cutoff allows replacing of the static point of view on the PE divergences by a dynamic one where the divergences may be observed in statu nascendi when $\theta \to \infty$.

Next step is to reparametrize the cutoff theory by means of another substitution.

$$S_{m,\lambda}^{\theta}(\phi) \mapsto S_{m_0,\lambda_0}^{\theta}(Z^{1/2}\phi), \tag{8}$$

where m_0, λ_0, and Z are given by formal power series

$$m_0 = m + \sum_{k=1}^{\infty} \alpha_k^{\theta}\lambda^k, \quad \lambda_0 = \lambda + \sum_{k=2}^{\infty} \beta_k^{\theta}\lambda^k, \quad Z = 1 + \sum_{k=2}^{\infty} \gamma_k^{\theta}\lambda^k. \tag{9}$$

This leads to a new, *renormalized PE* in powers of λ for the Green functions. The success of the perturbative renormalization may be summarized in

THEOREM 1. *For $d \le 4$ in the ϕ^4 theory or for $d \le 2$ in the G-N model there exist choices of α_k^{θ}, β_k^{θ}, γ_k^{θ} such that a (nontrivial) limit of the renormalized PE exists order by order.*

What happens is that the short distance divergences of the PE may be entirely absorbed into divergences of the coefficients α_k^{θ}, β_k^{θ}, γ_k^{θ} at $\theta \to \infty$. This result, although intuitive, is not easy to prove. It has a heavy analytic and combinatorial content. For a modern approach to the question of perturbative renormalizability, see [6] and the lecture of G. Gallavotti at this congress.

In general, one distinguishes three possibilities:

(i) The perturbative renormalization works with a finite number of diverging coefficients α_k^{θ}, β_k^{θ}, γ_k^{θ}—superrenormalizable models (like ϕ^4 in $d = 2,3$).

(ii) An infinite number of diverging coefficients is required—renormalizable models (like ϕ^4 in $d = 4$ and G-N in $d = 2$).

(iii) The perturbative renormalization does not work—nonrenormalizable models (like ϕ^4 in $d > 4$ and G-N in $d > 2$).

The important lesson to learn from the success of the perturbative renormalization is that one should consider θ- (and λ-) dependent m_0, λ_0 and Z in (8) if one wants to get finite results for the $\theta \to \infty$ limit of the Green functions.

Constructive QFT attempts a nonperturbative analysis of the choices of λ_0, m_0, Z which lead to finite nontrivial (i.e., describing interacting particles) results when $\theta \to \infty$ and of the properties of these limits. By a simple rescaling of the fields $\phi(x) \mapsto \phi'(x) \propto \phi(\theta x)$, the problem may be translated to the one where we probe the long distance asymptotics of the Green functions of the fixed cutoff theory and becomes, from the mathematical point of view, a search for limit theorems for certain strongly correlated random fields.

In the past twenty years, as a result of concentrated efforts [7], a good rigorous control of the superrenormalizable QFT models has been achieved. Here, we shall discuss the recent development in the field: the first mathematical constructions of renormalizable models.

THEOREM 2 [8, 9]. *For the G-N model in 2 dimensions, choose*

$$m_0 = m(\ln \theta)^{-(2N-1)/2(N-1)},$$

$$\lambda_0 = \left[\lambda^{-1} + \frac{2(N-1)}{\pi} \ln \theta - \frac{1}{\pi} \ln \left(1 + \frac{2(N-1)}{\pi} \lambda \ln \theta \right) \right]^{-1}, \qquad (10)$$

$$Z = 1$$

with $\lambda > 0$ *small and* $m > 0$ *large. Then the limit* $\theta \to \infty$ *of the Green functions exists and is nontrivial.*

Theorem 2 was proven independently by two groups. Both proofs are similar in spirit, although [8] exploits more heavily the conventional perturbative techniques. Our description of the proof will fit [9] better. The main idea here is a translation of the problem into the dynamical system language following the *renormalization group* philosophy of K. G. Wilson [10]. The Gaussian (Grassmann) cutoff field $\phi(x)$ corresponding to $m_0 = 0$, $\lambda_0 = 0$, i.e., with covariance $G_0 e^{-\Delta/\theta^2}$, is decomposed into $\sum_{n=0}^{n_0} \phi_n(x)$ ($\phi_n = (\overline{\psi}_n^r, \psi_n^s)$), where ϕ_n are independent fields with covariances

$$G_0 (e^{-L^{2n}\Delta/\theta^2} - e^{-L^{2(n+1)}\Delta/\theta^2}) \quad \text{for } n < n_0$$

and

$$G_0 e^{-L^{2n_0}\Delta/\theta^2} \quad \text{for } n = n_0.$$

(For independent fields the covariances sum up.) $L > 1$ is fixed and n_0 is chosen so that $L^{n_0}/\theta = O(1)$. The functional integral with the interaction term

$$S_I^0(\phi) = \int_{\mathbf{R}^2} \left[m_0 \sum_r \overline{\psi}^r \psi^r - \lambda_0 \left(\sum_r \overline{\psi}^r \psi^r \right)^2 \right] \qquad (11)$$

turned on may be analyzed inductively by integrating out ϕ_0 first, ϕ_1 next, and so on. In the commutative case this would correspond to subsequent conditioning with respect to the σ-algebra generated by $\phi^{\geq 1} \equiv \sum_{n \geq 1} \phi_n$, then $\phi^{\geq 2} \equiv \sum_{n \geq 2} \phi_n$, etc.

The effect of this operation may be summarized by the replacement of $S_I^0(\phi)$ by the effective interaction $S_I^1(\phi^{\geq 1})$, then $S_I^2(\phi^{\geq 2}),\ldots.$ This generates a (discrete) *flow* of the effective interactions

$$S_I^0 \mapsto S_I^1 \mapsto S_I^2 \mapsto \cdots \mapsto S_I^{n_0} \tag{12}$$

known in the physical literature under the name of renormalization group. The control of the flow in the infinite-dimensional space of effective interactions looks like a formidable problem. However, in the G-N model case, if the interactions are small (in the appropriate sense), then each S_I^{n+1} may be computed by a *convergent PE* in powers of S_I^n. This reflects the absence of the stability problem for anticommuting fields and allows an easy control of (12) in a neighborhood of zero. It can be shown inductively that each S_I^n has the form

$$S_I^n(\phi) = \int_{\mathbf{R}^2} \left[m_n \sum_r \overline{\psi}^r \psi^r - \lambda_n \left(\sum_r \overline{\psi}^r \psi^r \right)^2 \right] \tag{13}$$

$$+ \sum_{m=1}^{\infty} \int_{\mathbf{R}^{4m}} dx_1 \cdots dx_{2m} \Gamma_n^{2m}(x_1,\ldots,x_{2m}) \prod_{i=1}^{m} \overline{\psi}(x_i)\psi(x_{m+i}),$$

where $\int dx_2 \Gamma_n^2(x_1,x_2) = 0 = \int dx_2\, dx_3\, dx_4\, \Gamma_n^4(x_1,x_2,x_3,x_4)$ (the local quadratic and quartic terms have been separated out in (13)), and the series converges in appropriate sense.

The perturbative computation leads to the recursion

$$m_{n+1} = m_n \left(1 + \frac{2N-1}{\pi} \lambda_n \ln L \right) + o(m_n \lambda_n), \tag{14}$$

$$\frac{1}{\lambda_{n+1}} = \frac{1}{\lambda_n} - \frac{2(N-1)}{\pi} \ln L + \frac{2(N-1)}{\pi^2} \lambda_n \ln L + o(\lambda_n). \tag{15}$$

Notice that under (15), λ flows away from the *Gaussian fixed point* $\lambda = 0$. The flow of Γ_n's is driven by that of m_n and λ_n: Γ_n's correspond essentially to the stable directions of the flow. The ansätze (10) are just solutions for the initial values λ_0, m_0 which under n_0 iterations of (14), (15) produce $O(1)$ λ_{n_0} and m_{n_0} uniformly in θ. Upon this choice of the initial condition, (12) ends up when $\theta \to \infty$ on the *unstable manifold* of the Gaussian fixed point in the space of effective interactions. The fact that the coupling constant flows away from zero under the renormalization group transformations (called *asymptotic freedom*) is very important for the success of our rigorous analysis, because it guarantees smallness of effective interactions, in which we expand, for all inductive steps.

Although at each step the PE in powers of S_I^n converges, the renormalized PE (which may be identified as the expansion in powers of the "renormalized coupling constant" λ_{n_0} for the total functional integral) is still expected to diverge when $\theta \to \infty$. This can be illustrated best on an example of a function

$$f^\theta(\lambda) = \int_0^{\ln \theta} e^{a(s-\ln \theta)} u(\lambda(s))\, ds, \tag{16}$$

where $\lambda(s)$ is a solution of the differential equation

$$\frac{d\lambda}{ds} = \frac{2(N-1)}{\pi}\lambda^2 + \lambda^3 w(\lambda), \qquad (17)$$

$$\lambda(\ln\theta) = \lambda, \qquad (18)$$

with $u(\cdot)$ and $w(\cdot)$ analytic around zero. (17) gives a flow which underlies $f^\theta(\lambda)$ similarly as (12) does the Green functions which may be decomposed into a sum of contributions produced by subsequent effective interactions. For $a > 0$, f^θ clearly has a $\theta \to \infty$ limit for small positive $\lambda \geq 0$. Generically, this limit is not analytic around zero: its Taylor expansion diverges but is Borel summable, a typical singularity connected to the degeneracy of the fixed points of dynamical systems like $\lambda = 0$ for (17). The study of such and related singularities is the subject of the theory of resurgent functions [11].

Although a drastic truncation of the QFT problem, example (16) is still quite illustrative for the dependence of Green functions in a renormalizable model on the renormalized coupling constant. Specifically, we have

THEOREM 3 [8]. *The renormalized PE in the 2 dimensional G-N model is Borel summable.*

Actually, the Borel transform of the latter is expected to have similar analytic structure as that corresponding to (16) with branch-point singularities on the negative axis [12]. This appeals for a generalization of [11] to the case of infinite-dimensional flows.

The applicability of the inductive approach to QFT described above is limited neither to the anticommutative case nor to the renormalizable models, as the following results show:

THEOREM 4. 1.[13] *Consider the 4-dimensional cut-off ϕ^4 theory for small negative λ defined by analytic continuation of the positive λ case through the upper (lower) half-plane.*

2.[14] *Consider the 2-dimensional G-N model with the Gaussian covariance $(i\partial)^{-1}$ replaced by $(i\partial)^{-1}\Delta^\varepsilon$ for $\varepsilon > 0$ small.*

In both cases, for a suitable choice of m_0 and λ_0, (nontrivial) $\theta \to \infty$ limits of the Green functions exist.

The point of case 1 is that the negative λ ϕ^4 theory in $d = 4$ is not only perturbatively renormalizable but also asymptotically free, in contrast to the positive λ case which most probably does not have a nontrivial $\theta \to \infty$ limit [15]. Unlike for the G-N models, the control of case 1 required solving an involved stability problem. In case 2, the modification of the covariance renders the interacting model perturbatively nonrenormalizable. Nonperturbatively, the cut-off θ can be still removed, but it is a nonzero fixed point of the coupling constant recursion rather than the zero one which drives the asymptotic behavior of λ for large θ; see also [16] where a similar construction was done for another model. Unfortunately, the result of Theorem 4.1 is only of limited interest for QFT as the constructed Green functions are not expected to have the positivity properties required for the field-theoretic interpretation.

Although until now only the simplest case of a genuine renormalizable QFT has been mathematically controlled, it seems reasonable to expect that more realistic renormalizable models, containing 4-dimensional nonabelian gauge fields which share with the 2-dimensional G-N model the property of asymptotic freedom will be rigorously constructed before long; see T. Bałaban's lecture at this congress.

REFERENCES

1. F. A. Berezin, *The method of second quantization*, Academic Press, New York, 1966.

2a. P. Mitter and P. Weisz, *Asymptotic scale invariance in a massive Thirring model with $U(n)$ symmetry*, Phys. Rev. D **8** (1973), 4410, pp. 4410–4429.

2b. D. Gross and A. Neveu, *Dynamical symmetry breaking in asymptotically free field theories*, Phys. Rev. D **10** (1974), pp. 3235–3253.

3a. I. M. Gel'fand and N. Ya. Vilenkin, *Generalized functions. Vol. 4: Application of harmonic analysis*, Academic Press, New York, 1964.

3b. H.-H. Kuo, *Gaussian measures in Banach spaces*, Lecture Notes in Math., vol. 463, Springer-Verlag, Berlin, 1975.

4. C. Itzykson and J.-B. Zuber, *Quantum field theory*, McGraw-Hill, New York, 1980.

5a. N. N. Bogoliubow and O. S. Parasiuk, *Über die Multiplikation der Kausalfunktionen in der Quantentheorie der Felder*, Acta Math. **97** (1957), pp. 227–266.

5b. K. Hepp, *Théorie de la renormalisation*, Lecture Notes in Phys., vol. 2, Springer-Verlag, Berlin, 1969.

5c. W. Zimmermann, *Lectures on elementary particles and quantum field theory* (Summer Inst. Theoretical Phys., Brandeis Univ., 1970), M.I.T. Press, Cambridge, Mass., 1970, p. 395.

6a. J. Polchinski, *Renormalization and effective lagrangians*, Nuclear Phys. B **231** (1984), pp. 269–295.

6b. G. Gallavotti, *Renormalization theory and ultraviolet stability for scalar fields via renormalization group methods*, Rev. Modern Phys. **57** (1985), pp. 471–562.

7a. J. Glimm, *Analysis over infinite-dimensional spaces and applications to quantum field theory*, Proc. Internat. Congr. Math. (Vancouver, 1974), pp. 119–125.

7b. B. Simon, *The $P(\phi)_2$ euclidean (quantum) field theory*, Princeton Univ. Press, Princeton, N. J., 1974.

7c. J. Glimm and A. Jaffe, *Quantum physics. A functional integral point of view*, Springer-Verlag, New York, 1981.

8a. J. Feldman, J. Magnen, V. Rivasseau, and R. Sénéor, *Massive Gross-Neveu model: a rigorous perturbative construction*, Phys. Rev. Lett. **54** (1985), pp. 1479–1481.

8b. _____, *A renormalizable field theory: the massive Gross-Neveu model in two dimensions*, Comm. Math. Phys. **103** (1986), pp. 67–103.

9a. K. Gawędzki and A. Kupiainen, *Exact renormalization for the Gross-Neveu model of quantum fields*, Phys. Rev. Lett. **54** (1985), pp. 2191–2194.

9b. _____, *Gross-Neveu model through convergent perturbation expansions*, Comm. Math. Phys. **102** (1985), pp. 1–30.

10a. K. G. Wilson and J. Kogut, *The renormalization group and the ε expansion*, Phys. Rep **12C** (1974), pp. 75–200.

10b. K. G. Wilson, *The renormalization group and critical phenomena*, Rev. Modern Phys. **55** (1983), pp. 583–600.

11. J. Ecalle, *Fonctions resurgentes. I, II, III*, Publications Mathématiques d'Orsay: 81-05, 81-06, 85-06.

12. K. Gawędzki, A. Kupiainen, and B. Tirozzi, *Renormalons: a dynamical system approach*, Nuclear Phys. B **257** [**FS14**] (1985), pp. 610–628.

13. K. Gawędzki and A. Kupiainen, *Non-trivial continuum limit of a ϕ_4^4 model with negative coupling constant*, Nuclear Phys. B **257** [**FS14**] (1985), pp. 474–504.

14a. _____, *Renormalizing the nonrenormalizable*, Phys. Rev. Lett. **55** (1985), pp. 363–365.

14b. _____, *Renormalization of a non-renormalizable quantum field theory*, Nuclear Phys. B **262** (1985), pp. 33–48.

15a. M. Aizenman, *Proof of triviality of φ_d^4 field theory and some mean-field features of Ising models for $d > 4$*, Phys. Rev. Lett. **47** (1981), pp. 1–4.

15b. _____, *Geometric analysis of ϕ^4 fields and Ising models. I, II*, Comm. Math. Phys. **86** (1982), pp. 1–48.

15c. J. Fröhlich, *On triviality of $\lambda\varphi_d^4$ theories and the approach to the critical point in $d \geq 4$ dimensions*, Nuclear Phys. B **200** [**FS4**] (1982), pp. 281–296.

15d. C. Aragão de Carvalho, C. Caracciolo and J. Fröhlich, *Polymers and $g|\phi|^4$ theory in four dimensions*, Nuclear Phys. B **215** [**FS7**] (1983), pp. 209–248.

15e. M. A. Aizenman and R. Graham, *On the renormalized coupling constant and the susceptibility in ϕ_4^4 field theory and Ising model in four dimensions*, Nuclear Phys. B **225** [**FS9**] (1983), pp. 261–288.

16. G. Felder, *Construction of a non-trivial planar field theory with ultraviolet stable fixed point*, Comm. Math. Phys. **102** (1985), pp. 139–155.

CENTRE NATIONAL DE LA RECHERCHE SCIENTIFIQUE, INSTITUT DES HAUTES ÉTUDES SCIENTIFIQUE, 91440 BURES-SUR-YVETTE, FRANCE

Quantum Strings and Algebraic Curves

YU. I. MANIN

1. Quantum strings. Quantum theory of elementary constituents of matter
—quarks and leptons—starts from a classical image of point-particle. The first
quantization leads to the quantum mechanical picture of a complex wave function
(or amplitude) propagating in space. The second quantization takes into account
the nonconservation of the number of particles and, in the operator approach
to the quantum field theory, transforms wave functions into operator-valued
distributions.

Currently a different quantum field theory is being actively discussed, that of
one-dimensional elementary objects, called strings.

The theory was first suggested in the sixties as a device for describing strong
interactions—quantum strings modelling an image of a flux tube connecting
quarks in a hadron.

In the eighties the renewed interest in strings arose for several reasons. One
compelling idea was a radical change of physical interpretation of the theory.
Many people believe now that at the Planck energy scale the fundamental constituents of matter are quantum strings of a very particular type, whereas the
observable physics of low energy is a result of a hierarchy of symmetry break-
ings, including spontaneous compactification of six spatial dimensions of the
initial ten. The observed particles thus are described by phenomenological effective lagrangians taking into account only the lower excitation modes of the
quantum strings.

Since the ancient *high energy physics* has already led us to the verge of self-
destruction, one can only wonder where these new flights of fancy will lead our
descendants (if any).

Be that as it may, modern strings as mathematical objects can be endowed
with internal degrees of freedom and be spinning. In the latter case they are
also called superstrings. Quantum field theory of (super)strings exists in two versions, mathematically very different. The canonical, or operator, quantization,
pursued in many articles, looks like a very interesting new chapter of representation theory, in particular that of the Lie algebras of the Kac-Moody and

Virasoro type and their superanalogs. One flaw of this approach is the difficulty of encompassing interactions.

This difficulty does not arise in the Polyakov path integral approach [1], the interactions being accounted for by various topologies of string world sheets in the integration space.

After taking into account gauge invariances of the theory, the Polyakov integrals for amplitudes reduce to finite-dimensional integrals over moduli spaces of Riemann surfaces and their superextensions.

My talk is devoted to mathematical questions of this new theory, arising in the problem of the computation of the Polyakov partition function and its fermionic analogue. The results, presented here, were obtained partly in collaboration with A. A. Beilinson and A. S. Švarc. I am deeply grateful to them, and also to A. A. Belavin for very useful discussions.

2. Partition function of Polyakov's string. The partition function of the closed bosonic string is given by the perturbation theory series $Z = \sum_{g \geq 0} Z_g$:

$$
\begin{aligned}
Z_g &= e^{\beta(2-2g)} \int e^{-J(x,\gamma)} \, Dx \, D\gamma, \\
J(x,\gamma) &= \frac{1}{2} \int_N d^2 z \sqrt{|\gamma|} \gamma^{ab} \partial_a x^m \partial_b x^m.
\end{aligned}
\tag{1}
$$

Here N is a fixed compact oriented surface of genus g (*the number of vacuum loops*), z^a local coordinates on it, $x = (x^m)$ a map of N into a Euclidean space $\mathbf{R}^d$, $\gamma_{ab} \, dz^a \, dz^b$ a metric on N. The path integral (1) is taken over a space *of random Riemannian surfaces in* $\mathbf{R}^d$. The classical action $J(x,\gamma)$ is invariant with respect to the semidirect product of the diffeomorphism group of N and the group of conformal changes of a metric. The heuristic definition of the path integral (1) proceeds as follows (cf. Alvarez [2]). The Gaussian integral over x is defined by the standard formula

$$
(\det' \Delta_{0\gamma})^{-d/2},
$$

where $\Delta_{0\gamma}$ is the Laplacian with respect to γ, acting on functions, which is regularized, say, by the formula $\exp(-\varsigma'_\Delta(0))$. This regularization breaks the conformal invariance of the effective quantum action. The resulting conformal anomaly vanishes in the critical dimension 26, in which case the Faddeev-Popov trick allows us to reduce the integration over γ to a finite-dimensional integral over the moduli space M_g of Riemannian surfaces of genus g:

$$
Z_g = \text{const} \int_{M_g} d\nu \, \det' \Delta_{2\hat\gamma} (\det' \Delta_{0\hat\gamma})^{-13}, \qquad g \geq 2,
\tag{2}
$$

where $d\nu$ is the Petersson-Weil measure on M_g, $\Delta_{2\hat\gamma}$ the Laplacian on the quadratic differentials, $\hat\gamma$ is the metrics of the constant curvature -1 in a given conformal class (for $g = 1$ the formula is slightly different). We shall denote by $d\pi_g$ the Polyakov integration measure in (2).

3. Two methods for calculating Polyakov's measure. Using the Selberg trace formula, one can express the determinants in (2) as a product over the lengths of closed geodesics on a Riemannian surface (cf. [3, 4, 5] and other papers).

An essentially different approach is based upon a recent work of Belavin and Knizhnik [6], where it was established that $d\pi_g$ is the modulus squared of the Mumford holomorphic form. In the notation of [7] the Mumford form is a section of the sheaf $\lambda_2 \lambda_1^{-13} \cong O_{M_g}$, corresponding to 1 under this isomorphism. We shall denote it $d\mu_g$.

The Belavin-Knizhnik theorem follows from a more general result of Bismut and Freed on the determinants of the Dirac operators.

In my announcement [9], based upon these considerations, the Polyakov measure was expressed through the Riemann theta. To derive this formula, I have used in an essential way a technique of *arithmetical geometry* due to Arakelov [10, 11] and further developed by Faltings [11]. Specifically, Faltings's calculations of admissible metrics on λ_i, made for the proof of the arithmetic Noether theorem, were crucial for me. Below I shall state an improved version of this formula, this time for $d\mu_g$ rather than $d\pi_g$ only, which will be proved in the joint work of A. A. Beilinson and the author.

Note that the Bismut-Freed formula furnishes in the context of Hermitean complex geometry a precise formula of the Riemann-Roch-Grothendieck type for the Chern class c_1 of a direct image sheaf, on the level of forms and not only of cohomology classes. An extension of this result to higher c_i would be of great importance both for number theory and quantum anomalies in the spirit, advocated in Manin [13] and Atiyah [14].

4. Calculus and algebra on Riemannian surfaces. Let X be a smooth compact surface, or, which is the same thing, a projective algebraic curve over $\mathbf{C}$. Our immediate concern is to define certain almost canonical bases in the spaces of holomorphic differentials of weight j, which we denote $\Omega^{\otimes j}(X)$. Actually, it will be convenient to do this also for half-integral j. In order to do that we choose a typical odd theta characteristic $\Omega^{1/2}$, which is an invertible sheaf on X with $(\Omega^{1/2})^{\otimes 2} = \Omega^1$ and $h^0(\Omega^{1/2}) = h^1(\Omega^{1/2}) = 1$. We then put $\Omega^{\otimes j} = (\Omega^{1/2})^{\otimes 2j}$ for $j \in \frac{1}{2}\mathbf{Z}$.

(a) *Distinguished basis in $\Omega^{1/2}(X)$.* We choose a symplectic basis (a_i, b_j) in $H_1(X, \mathbf{Z})$. The Riemann basis $\omega_1, \ldots, \omega_g$ in $\Omega^1(X)$ is defined by the conditions $\int_{a_i} \omega_j = \delta_{ij}$. Put $\tau_{ij} = \int_{b_j} \omega_i$. Let α be a half-period, corresponding to the theta characteristics $\Omega^{1/2}$. (See, e.g., Fay [15].) Denote by $\theta(z, \tau)$ the Riemann theta function and set

$$\nu^2 = \varphi_0 = \sum_{j=1}^{g} \left(\frac{\partial}{\partial z_j} \theta[\alpha] \right) (0, \tau) \omega_j.$$

The half differential ν is the distinguished basis of $\Omega^{1/2}(X)$.

(b) *Distinguished basis in* $\Omega^1(X)$. Assume that $\operatorname{div} \varphi_0 = 2\sum_{j=1}^{g-1} P_j$ with pairwise different points P_j. Choose a local coordinate t_j at P_j in such a way that $\varphi_0 = t_j^2 \, dt_j$. The Lagrange type conditions

$$\varphi_j = (\delta_{jk} + a_{jk}t_k)dt_k + O(t_k^3 \, dt_k) \quad \text{near } P_k, \qquad j,k = 1,\ldots,g-1,$$

define the distinguished basis $(\varphi_0, \varphi_1, \ldots, \varphi_{g-1})$ in $\Omega^1(X)$.

(c) *Distinguished basis in* $\Omega^{\otimes 2}(X)$. For $g = 2$ or for a nonhyperelliptic curve of genus ≥ 3 we set

$$(w_1,\ldots,w_{3g-3}) = (\varphi_0^2, \varphi_0\varphi_1, \ldots, \varphi_0\varphi_{g-1}; \varphi_1^2, \ldots, \varphi_{g-1}^2;$$
$$a_{g-1,1}^{-1}\varphi_1\varphi_{g-1}, \ldots, a_{g-1,g-2}^{-1}\varphi_{g-2}\varphi_{g-1}).$$

It is essential that w_j are explicit bilinear combinations of φ_j and that the last group of 2-differentials fulfill the Lagrange conditions

$$w_{2g-1+j} = \delta_{jk}t_k \, dt_k^2 + O(t_k^2 \, dt_k^2) \quad \text{near } P_k, \qquad j,k-1 = 1,\ldots,g-2.$$

(d) *Distinguished bases in* $\Omega^{\otimes j}(X)$, $j > 2$. To define it, take the distinguished basis in the previous space (i.e., of weight $j - \frac{1}{2}$), multiply it by ν and add $g - 1$ differentials, satisfying the Lagrange conditions at $P_1, \ldots, P_{g-1}$.

Now we can describe the Polyakov measure.

5. THEOREM. (a) *The Mumford form equals*

$$d\mu_g = \operatorname{const}(\det B)^4 \frac{w_1 \wedge \cdots \wedge w_{3g-3}}{(\varphi_0 \wedge \cdots \wedge \varphi_{g-1})^{13}},$$

where $B = (B_i^j)$, $\varphi_i = B_i^j \omega_j$.

(b) *The Polyakov measure equals*

$$d\pi_g = \operatorname{const}|\det B|^{-18}(\det \operatorname{Im}\tau)^{-13} W_1 \wedge \overline{W}_1 \wedge \cdots \wedge W_{3g-3} \wedge \overline{W}_{3g-3}$$

where $W_i = k(w_i)$, k *the Kodaira-Spencer map, in this setting defined by* $k(\omega_k\omega_l)$ $= (1/2\pi i)\, d\tau_{kl}$.

6. Sketch of proof. Let $\pi: X \to S$ be a flat family of projective smooth algebraic curves. In order to calculate the Mumford form we utilize functorial properties of two algebro-geometric constructions.

(a) For each sheaf $\mathbf{L}$ on X, flat over S, one can construct a *multiplicative Euler characteristic*

$$d(L) = \det R\pi_* L = \bigotimes_i \det(R^i \pi_* L)^{(-1)^i},$$

where the last equality can be used, if all sheaves $R^i\pi_* L$ are locally free (see Knudsen-Mumford [**16**]).

(b) For each pair of invertible sheaves L, M on X, one can construct an invertible sheaf $\langle L, M \rangle$ on S, geometrizing the classical cross-ratio and Weil's pairing (see Deligne [**17**]).

The Mumford isomorphism

$$\lambda_{i+1} := d(\Omega^{\otimes i+1} X/S) \xrightarrow{\sim} \lambda_1^{6i^2+6i+1} = (d(\Omega^1 X/S))^{6i^2+6i+1}$$

becomes identical, if one trivializes the sheaves λ_i by means of exterior products
of the distinguished bases. This is proved by the detailed analysis of two auxiliary
isomorphisms

$$\langle \Omega^1 X/S, \Omega^1 X/S \rangle^{\otimes i} = \lambda_{i+1} \otimes \lambda_i^{-1}, \qquad \langle \Omega^1 X/S, \Omega^1 X/S \rangle = \lambda_1^8 \otimes \lambda_{1/2}^{-8}$$

as in [**12**].

7. More uses of distinguished bases. The distinguished bases, introduced
by A. A. Beilinson and the author, have other applications as well.

(a) They furnish a precise normalization of multicanonical embeddings of
algebraic curves and may be helpful in the understanding of Petri's construction
(see [**18**, pp. 123–135]).

(b) The determinants of Laplacians on the tensors of arbitrary weight on Riemann surfaces can be calculated by means of the Selberg trace formula. Combining this with the Belavin-Knizhnik method, one can express the values of the
Selberg zeta through the determinants of the Petersson-Weil metrics of the respective weight, calculated in distinguished bases. These formulae can be useful,
e.g., for p-adic interpolation.

8. Quantum superstrings in critical dimension 10. In order to define
the partition function of the fermionic Polyakov string, one considers a path
integral over the space of $2|2$-dimensional random supersurfaces, endowed with
a *super-Riemannian structure*. Due to lack of space, I shall omit the description
of differential geometric structures involved (see, e.g., [**19**]) and shall explain the
superanalytic picture which emerges after the reduction of path integral to the
sum of finite-dimensional ones. The rest of this talk is devoted therefore to some
background facts of the theory of $1|1$-dimensional superalgebraic curves with an
additional structure, or simply SUSY-curves (from supersymmetry).

Of particular importance for us are those constructions, which may be helpful
in defining and calculating the superanalogs of Mumford's and Polyakov's forms
on the appropriate moduli superspace, although many more results of the theory
of algebraic curves can be developed in this new context. Still, the theory is not
mature enough to allow a calculation of the partition function along the lines of
Theorem 5.

9. SUSY-families. Below we shall use some elementary notions of supergeometry as they are described in [**13**] and [**19**]. Let $\pi: X \to S$ be a submersion
of complex analytic supermanifolds. Its relative dimension is the rank of the
relative tangent sheaf $\mathcal{T} X/S$. A SUSY-structure on the family π is a locally free
locally direct subsheaf $\mathcal{T}^1 \subset \mathcal{T} X/S := \mathcal{T}$ of rank $0|1$; the rank of $\mathcal{T}$ itself being
assumed to be $1|1$. The sheaf $\mathcal{T}^1$ must be maximally nonintegrable, i.e., the
Frobenius form

$$\varphi: (\mathcal{T}^1)^{\otimes 2} \to \mathcal{T}^0 := \mathcal{T}/\mathcal{T}^1, \qquad \varphi(X,Y) = [X,Y] \bmod \mathcal{T}^1$$

must be an isomorphism. A family π together with a SUSY-structure on it will be called a SUSY-family of curves (not necessarily compact ones). A SUSY-structure is conveniently described by a family of odd vertical vector fields D on charts of an atlas of X, such that $\{D, D^2 = 1/2[D,D]\}$ form a local basis of $\mathcal{T}X/S$.

10. Examples. (a) Let $Z = (z, \varsigma)$ be a local relative coordinate system on X. Put $D_Z = \partial/\partial\varsigma + \varsigma\partial/\partial z$. Since $D_Z^2 = \partial/\partial z$, the subsheaf $O_X D_Z = \mathcal{T}^1$ defines a local SUSY-structure, associated with Z. If Z, Z' are associated with the same SUSY-structure, we set $D_Z = F_Z^{Z'} D_{Z'}$.

(b) Let $S = $ point, $X = P^{1|1}$ with homogeneous coordinates (z_1, z_2, ς_1). This projective superline is covered by two coordinate patches $U: Z = (z_1 z_2^{-1}, \varsigma_1 z_2^{-1}) = (z, \varsigma)$, $U': Z' = (-z_2 z_1^{-1}, \varsigma_1 z_1^{-1}) = (z', \varsigma')$. Coordinates Z, Z' define one and the same SUSY-structure on $U \cap U'$, since $D_{Z'} = z^{-1} D_Z$. We call it standard.

(c) Consider a $1|0$-dimensional complex curve, not necessarily compact, X_{red}, and an invertible sheaf I on it. We denote by X the $1|1$-dimensional supercurve with the structure sheaf $O_X = O_{X_{\mathrm{red}}} \oplus \Pi I$, where ΠI is an ideal with zero multiplication which is obtained from I by the parity change functor.

Assume now, that I is a theta characteristic; i.e., an isomorphism $\alpha: I^{\otimes 2} \to \Omega^1 X_{\mathrm{red}}$ is given. Then X is endowed with a natural SUSY-structure. It is associated with a family of local coordinates of the form (z, ς), where ς is a local section of ΠI such that $\alpha(\varsigma \otimes \varsigma) = dz$. Moreover, in this way we get a $(1,1)$-correspondence between SUSY-extensions of X_{red} and its theta characteristics. This fact generalizes to SUSY-families over pure even bases. In particular, for any SUSY-family its reduced family $X_{\mathrm{red}} \to S_{\mathrm{red}}$ is endowed with a theta characteristic which we shall call structural. Therefore, the notion of a SUSY-family is a natural extension of the classical notion of theta characteristics, which is richer than the latter since odd coordinates can emerge in the base spaces. Finally, we recall, that on Kähler manifolds square roots of canonical sheaf correspond to spinor structures. Hence the geometry of SUSY-curves actually describes fermionic degrees of freedom. In fact, it is an axiomatization of two-dimensional simple supergravity, modelled on the axiomatization of $D = 4$, $N = 1$ supergravity, as it was presented in [13] and in the last section of my book [20].

11. Geometry of SUSY-$P^{1|1}$ and superuniformization. The SUSY-curve $P^{1|1}$ plays the role of Riemannian supersphere, and its automorphism supergroup takes part of $\mathrm{PSL}(2, \mathbf{C})$ of the classical theory. To describe it, it is convenient to start with the group $C(2|1)$ of conformal automorphisms of symplectic $2|1$-form $(\tilde{Z}, \tilde{W}) = z_1 w_2 - z_2 w_1 - \varsigma_1 \nu_1$. It consists of matrices

$$T = \begin{pmatrix} a & b & \delta \\ c & d & \gamma \\ \beta & \alpha & e \end{pmatrix}, \qquad \begin{cases} e^2 = ad - bc + 3\alpha\beta, \\ e\gamma = d\beta - c\alpha, \\ e\delta = b\beta - a\alpha. \end{cases}$$

Such a matrix multiplies $(\tilde{Z}, \tilde{W})$ by $(\mathrm{Ber}\, T)^2 = (e - \alpha\beta e^{-1})^2$. Hence $\mathrm{SpO}(2|1) = \mathrm{SC}(2|1) \times \{\pm 1\}$. On the other hand, a direct computation shows, that $\mathrm{SC}(2|1)$

is the automorphism supergroup of $P^{1|1}$ with its standard SUSY-structure. We have

$$\dim SC(2|1) = 3|2, \qquad SC(2|1)_{\text{red}} = SL(2).$$

Consider two points $Z = (z, \varsigma)$, $W = (w, \nu)$, given by their associated affine coordinates. Set $(Z, W) = z - w - \varsigma \nu$. Then $(TZ, TW) = F_Z^{TZ} F_W^{TW}(Z, W)$. It follows, that the expression

$$((Z, W, Z', W')) = (Z, W)(Z', W')/(Z, Z')(W, W')$$

is $SC(2|1)$-invariant. It is a superextension of the classical cross ratio. We can now describe an analytical construction of SUSY-families, which generalize two classical uniformization methods: that of fuchsian groups and that of Schottky groups.

12. Superfuchsian uniformization. As is well known, the fundamental group Γ_g of a compact oriented surface of genus g can be given a presentation $(A_1, \ldots, A_g, B_1, \ldots, B_g)$, $[A_1, B_1] \cdots [A_g, B_g] = 1$. Consider a real analytic supermanifold S with the ring of real functions A. Assume that a representation $\rho \colon \Gamma_g \to SC(2|1, A)$ is given. It defines an action of Γ_g upon $P^{1|1} \times S$, compatible with the SUSY-structure and the evident real structure. Thus Γ_g acts on the relative upper half superplane $H^{1|1} \times S$. Assume that in the reduced picture, Γ_g acts discretely by hyperbolic fractional linear transformations. Then we can construct a factor family $\Gamma_g \backslash (H^{1|1} \times S)$ over a real analytic base.

We can extend the classical universal family of fuchsian groups, constructed by Fricke and Klein, to the similar family of superfuchsian groups. In this way we shall obtain a SUSY-family with local versality properties, at least at points, corresponding to typical structural theta characteristics.

Imitating the classical Selberg construction, we can define for such families the Selberg transform (with superparameters), Selberg's zeta function, and can compute with their help the determinants of certain operators, arising in the path integral for the partition function of the Polyakov superstring (see [21]).

However, in order to reproduce, in the fermionic case, the results of the first part of this talk, we need supermoduli spaces with complex, rather than real, analytic structure. To construct them explicitly, it is convenient to turn to the Schottky uniformization. Its usefulness for the calculation of the classical dual amplitudes was demonstrated already in the first period of the development of the string theory (cf. Alessandrini and coauthors [22, 23]). In the fermionic case it was recently investigated by Martinec.

Our point of view in this talk is based upon an experience of work with the p-adic Schottky uniformization, introduced by Mumford [25] and explored in Manin, Drinfeld [24].

13. Schottky's superuniformization. Consider now a Stein complex superspace S with the ring of analytic superfunctions A on it and denote by Γ a free group with g generators. Let $\rho \colon \Gamma \to SC(2|1, A)$ be such a representation, that at each point of S the image of Γ is a Schottky group, i.e., a discrete group, consisting of hyperbolic elements. Denote by Ω_S a superspace, which is an open supersubspace of $P^{1|1} \times S$, whose support is a fibrewise union of complements of sets of limit points of Γ. Then the factor space $\Gamma \backslash \Omega_S$ is a SUSY-family of compact curves of genus g.

By a universal construction of this kind, one can obtain a SUSY-family over a supermanifold M_g^f (f for fermionic) of dimension $3g - 3|2g - 2$ ($g \geq 2$), with even structural theta characteristics. It will be locally versal at least at points, where this theta characteristic is typical. At this point we must note that the general theory of local moduli deformations in complex analytic supergeometry was recently developed in the thesis by A. Yu. Vaintrob. For an announcement of a part of his results see [26].

The realization of the moduli superspace M_g^f by means of this construction has the same shortcomings as its classical counterpart, namely a complicated and badly understood structure of the boundary and a weird equivalence relation, induced by isomorphism of uniformized curves on the base space (cf. Hejhal [27]).

On the other hand, Schottky groups admit a beautiful theory of automorphic forms on the Riemannian sphere, which gives nice explicit expressions for holomorphic differentials, period matrix, and other important objects (see, e.g., Baker [28, Chapter XII]). String theory puts forward the problem of construction of the theory of modular forms on the Riemann moduli spaces and superspaces, which is quite underdeveloped as compared with the Siegel modular forms theory. One may hope that the Schottky uniformization will be of some help here.

I shall describe below several simple constructions on a SUSY-family, which is Schottky uniformized, taking [24] as a model.

14. Classical automorphic functions. Let Γ be a Schottky group, acting discretely on $\Omega \subset P^1$, $w_d(z)$ a rational function with divisor d in Ω. We choose a normalization point $z_0 \in \Omega \backslash \mathrm{supp}(d)$ and consider an infinite product

$$W_{d,z_0}(z) = \prod_{\gamma \in \Gamma} \frac{w_d(\gamma z)}{w_d(\gamma z_0)}.$$

It converges at least on an open subset of the Schottky domain, in a neighborhood of maximally degenerate curves. Using W_{d,z_0}, one can define the multiplier $\mu_d(\gamma)$ by the formula

$$W_{d,z_0}(\gamma z) = \mu_d(\gamma) W_{d,z_0}(z).$$

It does not depend on z_0 and is multiplicative in γ. It has especially interesting properties in the case $d = (\gamma' - 1)z_1$, i.e., $w_d(z) = (z - \gamma' z_1)(z - z_1)^{-1}$. From the symmetry properties of a cross ratio it follows that the expression

$$\langle \bar{\gamma}, \bar{\gamma}' \rangle := \mu_{(\gamma'-1)z_1}(\gamma), \qquad \bar{\gamma} = \gamma \bmod [\Gamma, \Gamma],$$

does not depend on z_1 and defines a symmetric scalar product on $\Gamma^{ab} = \Gamma/[\Gamma, \Gamma]$. Using it, one can construct the Jacobian of Γ as a factor of a torus by a multiplicative period lattice, generated by the columns of $\langle \bar{\gamma}_k, \bar{\gamma}_l \rangle = \exp(2\pi i \tau_{kl})$ for a basis of Γ^{ab}. The corresponding normalized abelian differentials are $\omega_{\bar{\gamma}} = d \log W_{(\gamma-1)z_1,z_0}$.

15. Superautomorphic functions. Now let a Schottky group act on a superspace Ω_S, where S is a parameter superspace. The principal construction of the previous section can be generalized if one takes instead of a cross ratio its superextension, defined in §11. Namely, put $w(Z) = (Z, T'Z_1)(Z, Z_1)^{-1}$ and consider an infinite product

$$W_{(T'-1)Z_1}(Z) = \prod_{T'' \in \Gamma} \frac{(T''Z, T'Z_1)}{(T''Z, Z_1)} \frac{(T''Z_0, Z_1)}{(T''Z_0, T'Z_1)}.$$

From SpO(2|1)-invariance of the standard SUSY-structure it follows that in the convergence domain we have

$$\mu_{(T'-1)Z_1}(T) = \mu_{(T^{-1}-1)Z_0}(T'^{-1}).$$

Therefore a scalar product is well defined

$$\exp(2\pi i \tau_{kl}) = \langle \overline{T}_k, \overline{T}_l \rangle, \qquad \overline{T}_k \in \Gamma^{ab},$$

with values in invertible even superfunctions on the parameter space S.

By the same token, one gets the lattice of even differentials of the first kind $\omega_T = d_{/S} \log W_T$ and its multiplicative period lattice.

16. Final remarks. To develop the algebro-geometric side of the fermionic string theory to the same level, as in the bosonic case, one needs several constructions in supergeometry. The formalism of $\det R\pi_*$ and $\langle L, M \rangle$ will hopefully be available in the near future. I am working on a proof of the analog of the Mumford isomorphism $\lambda_1^{13} = \lambda_2$ which in the fairly evident notation should take the form $\Lambda_1^5 = \Lambda_3$.

I would like also to mention the following natural problems.

(a) One should study an analog of superJacobians for SUSY-curves and superabelian varieties in general.

(b) An algebraic theory of moduli superspaces should be developed and the compactification by stably degenerate SUSY-curves should be constructed.

(c) A theory of vector superbundles over SUSY-curves, corresponding to Yang-Mills coupled with supergravity, should be formulated. In this way the heterotic string may be included in this context.

(d) What is a natural SUSY-structure on higher-dimensional supermanifolds, e.g., on various moduli superspaces?

(e) The same questions about the SUSY$_N$-curves corresponding to extended two-dimensional supergravities.

References

1. A. M. Polyakov, *Quantum geometry of bosonic strings*, Phys. Lett. B **103** (1981), 207–210.

2. O. Alvarez, *Theory of strings with boundaries: fluctuations, topology and quantum geometry*, Nuclear Phys. B **216** (1983), 125–184.

3. D. Ray and I. M. Singer, *Analytic torsion for complex manifolds*, Ann. of Math. **98** (1973), 154–180.

4. M. A. Baranov and A. S. Svarc, *On the multiloop contribution in the string theory*, Pisma ZETP **42** (1985), no. 8, 340–342.

5. E. D'Hoker and D. H. Phong, *On determinants of Laplacians on Riemann surfaces*, Preprint CUTP-334, Columbia Univ. Press, New York, 1986.

6. A. A. Belavin and V. A. Knizhnik, *Complex geometry and quantum string theory*, Preprint Landau Institute 6, 1986.

7. D. Mumford, *Stability of projective varieties*, Enseign. Math. **23** (1977), 39–100.

8. J.-M. Bismut and D. S. Freed, *The analysis of elliptic families*, Preprint 85 T 47, Univ. Paris-Sud, 1985.

9. Yu. I. Manin, *The partition function of the Polyakov string can be expressed through theta function*, Pisma ZETP **43** (1986), no. 4, 161–163.

10. S. J. Arakelov, *An intersection theory for divisors on an arithmetic surface*, Izv. Akad. Nauk SSSR Ser. Mat. **38** (1974), no. 6, 1179–1192.

11. ____, *Theory of intersections on arithmetic surface*, Proceedings of the International Congress of Mathematicians (Vancouver, B.C., 1974), Vol. 1, pp. 405–408.

12. G. Faltings, *Calculus on arithmetic surfaces*, Ann. of Math. **119** (1984), 387–424.

13. Yu. I. Manin, *New dimensions in geometry*, Lecture Notes in Math., Vol. 1111, Springer, 1985, pp. 59–101.

14. M. F. Atiyah, *Commentary on the article of Manin*, Lecture Notes in Math., Vol. 1111, Springer, 1985, pp. 103–109.

15. J. D. Fay, *Theta functions on Riemann surfaces*, Lecture Notes in Math., Vol. 352, Springer, 1973.

16. F. Knudsen and D. Mumford, *The projectivity of the moduli space of stable curves*, I: *Preliminaries on "det" and "Div"*, Math. Scand. **39** (1976), 19–55.

17. P. Deligne, *La formule de dualite globale*, Lecture Notes in Math., Vol. 305, Springer, 1970, pp. 481–587.

18. E. Arbarello, M. Cornalba, P. A. Griffiths, and J. Harris, *Geometry of algebraic curves*, Vol. 1, Springer, 1985.

19. M. A. Baranov, I. V. Frolov, and A. S. Svarc, *Geometry of two-dimensional superconformal field theories*, Teoret. Mat. Fiz. **16** (1985), 202–227.

20. Yu. I. Manin, *Gauge fields and complex geometry*, Nauka, Moscow, 1984 (English translation to appear in Springer).

21. M. A. Baranov, Yu. I. Manin, I. V. Frolov, and A. S. Svarc, *Multiloop contribution in fermionic string and Selberg trace formula*, Yadernaya Fiz. **43** (1986), no. 4, 1053–1056.

22. V. Alessandrini, *A general approach to dual multiloop diagrams*, Nuovo Cimento 2A **2** (1971), 321–352.

23. V. Alessandrini and D. Amati, *Properties of dual multiloop amplitudes*, Nuovo Cimento 4A **4** (1971), 793–844.

24. Yu. I. Manin and V. G. Drinfeld, *Periods of p-adic Schottky groups*, J. Reine Angew. Math. **262/263** (1973), 239–247.

25. D. Mumford, *An analytic construction of degenerating curves over complete local rings*, Compositio Math. **24** (1972), 129–174.

26. A. Yu. Vaintrob, *Deformations of complex structures on supermanifolds*, Funktsional. Anal. i Prilozhen. **18** (1984), no. 2, 59–60.

27. D. A. Hejhal, *On Schottky and Teichmüller spaces*, Adv. in Math. **15** (1975), 133–156.

28. H. F. Baker, *Abel's theorem and the allied theory (including the theory of theta functions)*, Cambridge Univ. Press, 1897.

STEKLOV MATHEMATICAL INSTITUTE, ACADEMY OF SCIENCES OF THE USSR, MOSCOW 117966, USSR

Spectral Properties of Metrically Transitive
Operators and Related Problems

L. A. PASTUR

I. Introduction. In the past decade the spectral properties of the Schrödinger equation and other differential and difference operators with random and almost periodic coefficients have attracted considerable and ever increasing interest. This interest arises not only from intrinsic logic of development of the spectral theory of operators, probability theory, and mathematical physics, but also from theoretical physics, primarily the theory of disordered condensed systems. And it was out of requirements of this theory that the study of such operators with random coefficients was commenced in the 50s and 60s by physicists Anderson, Lifshitz, and Mott, who formulated a variety of fundamental ideas and concepts. Theoretical physics even now remains a source of many problems and methods of the spectral theory of the present class of operators, and significantly influences the course of development of this branch of mathematics.

The problems, methods, and results of this branch generate mathematical interest, as I see it, for one more reason as well. Namely, they provide a quite natural and efficient formalization of intuitive concepts of the "typical operator," in particular the "typical differential or finite-difference operator with bounded coefficients." It is well known that, e.g., the spectral theory of the Schrödinger operator with a bounded potential, i.e., which neither increases nor decreases in any systematic way, amounts essentially to the case of the periodic potential. This is partly because of the difficulty of an adequately detailed and efficient description of an individual, more or less arbitrary bounded function from the point of view of spectral theory. It is natural therefore to use an approach dealing with a certain more or less broad ensemble of operators (equations) and finding facts and properties that are characteristic of all "typical" representations of such an ensemble. Mathematics knows at least two formalizations of the intuitive notion of "typical": categorical and probabilistic. The more productive and efficient approach in the area under discussion proved to be the probabilistic approach, in which the term "typical" is equivalent to "almost every with respect to a certain probability measure member of the ensemble" (though now there are results in the spectral theory which are formulated in terms of the categorical approach, where the "typical set" is equivalent to the "dense G_δ set," etc.). Within the probabilistic approach, differential and finite-difference operators, whose coefficients are homogeneous and ergodic, i.e., metrically transitive random fields in $\mathbf{R}^d$ or $\mathbf{Z}^d$, represent a natural ensemble for studying "typical" properties of the said operators with bounded coefficients. Such an approach has now been demonstrated to be rather productive and constructive as applied to a variety of problems, where traditional methods of spectral theory are not quite efficient.

This report will present some results obtained by this approach in recent years.

II. Abstract metrically transitive operators. A general abstract operator scheme convenient for the analysis of the spectral properties of the class of operators under discussion is as follows [41].

Let $(\Omega, \mathcal{F}, P)$ be a probability space, $\mathcal{T} = \{T\}$ a group of measure-preserving and metrically transitive (m.tr.) automorphisms of this space, $\mathcal{U}_T = \{U_T, T \in \mathcal{T}\}$ a unitary irreducible representation of this group in the Hilbert space $\mathcal{H}$. The function $A(\omega)$ on Ω with values in the set of closed operators in $\mathcal{H}$ will be called a metrically transitive operator (MTO), if there exists in $\mathcal{H}$ a dense linear manifold $\mathcal{D}$, entering into the domain of almost all $A(\omega)$ and such that

(a) $U_T \mathcal{D} \subset \mathcal{D}$;

(b) the quantities (u, Av) are measurable for all $u \in \mathcal{H}$, $v \in \mathcal{D}$;

(c) with probability 1 (p.1)

$$U_T A(\omega) v = A(T\omega) U_T v, \qquad \forall T \in \mathcal{T}, \forall v \in \mathcal{D}.$$

If, moreover, with p.1 $(u, Av) = (Au, v), \forall u, v \in \mathcal{D}$, then A will be said to be a symmetric MTO.

An interesting example, typical in many ways, of a symmetric MTO, is the Schrödinger operator

$$H = -\Delta + q(x, \omega), \qquad x \in \mathbf{R}^d,$$

acting in $\mathcal{L}^2(\mathbf{R}^d)$, with real-valued potential of the form $q(x, \omega) = Q(T_x \omega)$, where $Q(\omega)$ is a measurable function on Ω, $\{T_x, x \in \mathbf{R}^d\}$ is a m.tr. group, i.e., $q(x, \omega)$ is a m.tr. field. A class of such potentials proves to be quite rich, as is clear even from the following examples:

(a)

$$(\Omega, \mathcal{F}, P) = \left([0, 2\pi), \mathcal{B}_1, \frac{d\omega}{2\pi} \right)$$

where $\mathcal{B}_1$ is a Borel σ-algebra on a unit circle, $\mathcal{T} = \{T_x, x \in \mathbf{R}\}$ is the group of rotations of the circle: $T_x \omega = \omega + x \pmod{2\pi}$. In this case $q(x, \omega) = Q(x + \omega)$, where $Q(\omega)$ is a 2π-periodic function, i.e., $q(x, \omega)$ is a periodic potential with a random origin uniformly distributed over a period.[1]

---

[1] Recall that the origin shift used in the consideration of properties of the Schrödinger operator with a periodic potential proved to be a very useful procedure in the modern theory of nonlinear evolution equations with periodic initial data.

(b)

$$(\Omega, \mathcal{F}, P) = \left(\mathbf{T}^n, \mathcal{B}^n, \frac{d\omega_1 \cdots d\omega_n}{(2\pi)^n} \right), \qquad (T_x \omega)_i = \omega_i + \alpha_i x \pmod{2\pi},$$

where $\alpha_1, \ldots, \alpha_d$ are rationally independent real numbers. The corresponding potential is a quasiperiodic function with the frequency module generated by the numbers $\alpha_1, \ldots, \alpha_d$.

(c) $q(x, \omega)$ is an ergodic Markov process.

The above examples of m.tr. potentials correspond to three main classes of MTO: the periodic operators, whose spectral properties are sufficiently well understood, especially for the one-dimensional case; the almost-periodic operators; and the random operators, which are now being actively investigated.

Similar examples may be given also in the discrete case, i.e., in the space $l^2(\mathbf{Z}^d)$, where the analogue of the Schrödinger operator is the operator

$$(A\psi)(x) = - \sum_{|x-y|=1} \psi(y) + q(x, \omega)\psi(x), \qquad x \in \mathbf{Z}^d.$$

This operator, in the case where $q(x, \omega)$ is a family of independent identically distributed random variables, is known as the Anderson model.

Proceeding from the above abstract definition of MTO the following spectral properties may be proved:

(1) If $A(\omega)$ is a symmetric MTO, then its deficiency indices are both, with p.1, equal to zero or infinity [43].[2]

(2) The spectrum of the self-adjoint MTO (SMTO) $A(\omega)$, as all of its components (point, continuous, absolutely continuous ones, etc.), with p.1, are nonrandom sets [41, 31, 25].

(3) The spectrum of a SMTO is, with p.1, essential [41].

(4) The probability that a fixed point $\lambda \in \mathbf{R}$ is an eigenvalue of finite multiplicity of a SMTO is zero [41].

(5) The spectral multiplicity of any fixed interval is, with p.1, nonrandom [44].

III. Integrated density of states (IDS). For the important class of differential and matrix MTO defined by differential and finite-difference operations with the m.tr. coefficients in the spaces $\mathcal{L}^2(\mathbf{R}^d)$ and $l^2(\mathbf{Z}^d)$ respectively, an interesting spectral characteristic is the integrated density of states (IDS) $N(\lambda)$, which also plays an important part in physical applications. It is a nonnegative and nondecreasing function of the spectral parameter defined as follows. Let $\{\Lambda\}$ be a family of domains in $\mathbf{R}^d$ ($\mathbf{Z}^d$) which sufficiently regularly (e.g., homothetically) expand to the whole $\mathbf{R}^d$ ($\mathbf{Z}^d$),[3] and A_Λ an operator defined inside Λ by the differential or finite-difference operation, that defines A, and certain selfadjoint conditions on the boundary $\partial\Lambda$. Denote by $N_\Lambda(\lambda)$ the number of eigenvalues

---

[2]Hence, in particular, the one-dimensional Schrödinger operator with a m.tr. potential is always essentially selfadjoint.

[3]In the ergodic theory, these families of domains are called Föllner domains.

of A_Λ, not exceeding λ, divided by the measure (the number of points) of Λ. Then, under very mild conditions on the m.tr. coefficients of the operator A, there exists a nonrandom nondecreasing function $N(\lambda)$ such that $N(-\infty) = 0$ and

(a) with p.1, at every point of continuity,

$$\lim_{\Lambda \nearrow \mathbf{R}^d \, (\mathbf{Z}^d)} N_\Lambda(\lambda) = N(\lambda);$$

(b)

$$N(\lambda) = E\{\mathcal{E}_A(0, 0, \lambda)\}$$

where $\mathcal{E}_A(x, y, \lambda)$, $x, y \in \mathbf{R}^d \, (\mathbf{Z}^d)$, is the kernel (matrix) of the resolution of the identity $\mathcal{E}_A(\lambda)$ of the SMTO A.

The latter equality, which is sometimes taken as a definition of the $N(\lambda)$, relates this function to the spectrum of almost all realizations of SMTO. Here are some simple examples of this relationship:

(1) the spectrum of the SMTO A, with p.1, coincides with the set of growth points of $N(\lambda)$;

(2) $N(\lambda)$ is continuous at the point λ, if and only if the probability of λ being an eigenvalue is zero.

These facts were first formulated by Benderskiĭ and Pastur [5], Pastur [41], and Pastur and Figotin [43].

Property (1), combined with some arguments essentially equivalent to the Weyl criterion known in spectral theory, provides in many cases an answer to the important question of spectral analysis concerning the location of the spectrum [31, 27].

In particular, for the Anderson model,

$$\sigma = \sigma_0 + \operatorname{supp} F,$$

where σ and σ_0 are the spectra of this SMTO and the operator $-\Delta$ ($\sigma_0 = [-2d, 2d]$), and $F(dq)$ is the potential distribution. A similar relation is valid also for the Schrödinger operator.

A large number of results are associated with properties of the IDS $N(\lambda)$, its smoothness, asymptotic behavior near the spectrum boundaries, etc. Some of these are as follows:

(1) $N(\lambda)$ is a continuous function in the general one-dimensional [41] and the multidimensional discrete cases [9], while in the case of the discrete analogue of the multidimensional Schrödinger operator,

$$|N(\lambda_1) - N(\lambda_2)| \leq \operatorname{const}|\ln|\lambda_1 - \lambda_2||^{-1}$$

[8].

(2) In the Anderson model, when the distribution function $F(q)$ of the random variables $q(x)$, $x \in \mathbf{Z}^d$, is absolutely continuous: $F(dq) = p(q)dq$ with $p(q) \leq C$, $N(\lambda)$ is also absolutely continuous and the corresponding density is bounded by the same constant: $N(d\lambda) = \rho(\lambda)d\lambda$, $\rho(\lambda) \leq C$ [48].

(3) If the spectrum of a SMTO contains the point $+\infty$ and

$$N(\lambda) = \sum_{1}^{n} c_k \lambda^{a_k} + o(\lambda^{a_n}), \qquad \lambda \to +\infty, \ a_1 > a_2 > \cdots > a_n$$

(such asymptotic expansions may be obtained for elliptic MTO with smooth coefficients), then the length $\delta(\lambda)$ of spectral gaps located to the right of λ has the estimate $\delta(\lambda) = o(\lambda^{a_n - a_1 + 1})$ [46].

(4) For the Schrödinger operator, $H = -\Delta + q$, in the cases of:

(a) the Gaussian random potential with zero average and the correlation function $b(x)$ [40]

$$\ln N(\lambda) = -\frac{\lambda^2}{2b(0)}(1 + o(1)), \qquad \lambda \to -\infty;$$

(b) $q(x) = \sum_j u(x - x_j)$ where $u(x)$ is a smooth summable function and the points x_j are distributed in accordance with the Poisson law in $\mathbf{R}^d$ with the density n [42]

$$\ln N(\lambda) = (1 + o(1)) \begin{cases} u \le 0, \dfrac{\lambda}{u(0)} \ln \dfrac{\lambda}{u(0)}, \ \lambda \to -\infty, \\[2mm] u \ge 0, u = o(|x|^{-d-2}), |x| \to \infty, C(d)n\lambda^{-d/2}, \lambda \downarrow 0, \\[2mm] u \ge 0, u = \dfrac{u_0}{|x|^\alpha}(1 + o(1)), |x| \to \infty, \\[2mm] \quad d < \alpha < d + 2, C(d, \alpha)\lambda^{-d/\alpha - d}, \lambda \downarrow 0; \end{cases}$$

(c) $q(x) = \int_{\mathbf{R}^d} u(x - y)g^2(y)\,dg$, where $u(x) \ge 0$, $u(x) = o(|x|^{-d-2})$, $|x| \to \infty$ and $g(x)$ is the Gaussian field in $\mathbf{R}^d$ with a zero average and the correlation function $b(x) = b_0 \exp\{-\sum_1^d |x_i|\}$:

$$\ln N(\lambda) = -C(b_0, d)\lambda^{-d/2 - 1}(\ln \lambda^{-1})^{2(d-1)}(1 + o(1)), \qquad \lambda \downarrow 0$$

([11], where rather a general case is also considered).

An important point of the rigorous proof of the series of these asymptotic formulae are powerful results of Donsker and Varadhan on the asymptotic properties of the Wiener process for large times.

The above asymptotic formulae for $N(\lambda)$ are typical for the so-called fluctuation spectrum boundaries of random operators, where the spectrum is formed mainly by large deviations (fluctuations) of the random potential from its mathematical expectation. These fluctuations have the shape of wide potential wells (if the spectrum bottom is the point $\lambda = 0$, the size of the wells is of the order of $\lambda^{-1/2}$) separated by very large (of the order of $N^{-1/d}(\lambda)$) distances. Because of the random nature of the potential, all these wells differ in shape and thus, in terms of quantum mechanics, resonance tunneling between them is practically impossible. As a result, the asymptotic of IDS coincides, with logarithmic accuracy, with the probability of appearance in the potential realization of an "optimal" well, in which the given $\lambda \downarrow 0$ is the minimum eigenvalue. This "fluctuation" ideology was proposed by Lifshitz [33] and is the basis of the so-called optimal fluctuation method, widely used in disordered system physics and giving the asymptotic behavior of IDS and other physical and spectral quantities in the fluctuation region on the physical level of rigour (see, e.g., Gredeskul et al. [21]). It also provides quite convincing, though heuristic, reasons to believe the spectrum in this region to be pure point for any dimension $d \geq 1$ and the respective eigenfunctions to be concentrated essentially within the "optimal" well and exponentially decreasing outside of it. This heuristic picture of the spectral properties of the Schrödinger operator and its difference analogues in the fluctuation region of the spectrum has recently evolved into a rigorous mathematical theory in the remarkable works of Fröhlich, Spencer and their coworkers ([15, 16], and the references therein).

The fluctuation boundaries of a spectrum are a characteristic feature of random operators. Boundaries of another type, the so-called stable boundaries, may be found in any differential and finite-difference MTO. In the vicinity of such a boundary the spectrum is formed by the whole realization of m.tr. coefficients (not only its large deviation part) of the operator, and the IDS behaves asymptotically in the same way as the analogous function of a certain operator with constant coefficients, i.e., here occurs a kind of "homogenization."

Here are some examples of stable boundaries.

(1) The point $\lambda = +\infty$ for the Schrödinger operator with a m.tr. potential, where $N(\lambda) = N_0(\lambda)(1 + o(1))$, $\lambda \to +\infty$, and $N_0(\lambda) = C_d \lambda^{d/2}$ is the IDS of the Laplace operator $-\Delta$.

(2) The point $\lambda = 0$ for the Sturm-Liouville operator

$$-\frac{d}{dx}\left(p(x)\frac{d\psi}{dx}\right) = \lambda m(x)\psi,$$

where $p(x)$ and $m(x)$ are positive and bounded m.tr. random processes.

Here

$$N(\lambda) = N_0\left(\lambda\frac{\mu}{\kappa}\right)(1 + o(1)), \qquad \lambda \downarrow 0,$$

$$\mu = E\{m(x)\}, \qquad \kappa^{-1} = E\{p^{-1}(x)\}$$

(Lifshitz et al. [21]).

(3) The point $\lambda = 0$ for the multidimensional MTO

$$-\frac{\partial}{\partial x_i}\left(a_{ij}(x)\frac{\partial}{\partial x_j}\right) = \lambda\psi, \qquad x \in \mathbf{R}^d,$$

where the m.tr. matrix $a_{ij}(x)$ satisfies the strong ellipticity conditions. The asymptotic of the IDS analogous to those above was found for these operators and their finite-difference analogues by Kozlov (see [30] and the references therein).

Note that there are reasons to expect the spectrum in the vicinity of a stable boundary in the multidimensional case $d \geq 3$ to be continuous.

Finally, we may regard the spectrum of the SMTO A_Λ as a sequence of random point processes in $\mathbf{R}$. Then it is natural to consider, besides $N(\lambda)$ whose derivative is a limit density (as $\Lambda \nearrow \mathbf{R}^d$) of this sequence, its more detailed characteristics, such as, e.g., the statistics of distances between nearest neighbors.

This interesting question, related to a range of problems of solid state physics, has been studied for the one-dimensional Schrödinger operator with the Markov potential by Molchanov [37] and Molchanov et al. [22].

IV. Nature of the spectrum of random MTO.

The IDS $N(\lambda)$ characterized the geometry of a spectrum and its "thickness" in the case of differential and matrix MTOs of a very general form. However the nature of a spectrum (i.e., presence of a point, continuous component, etc.) cannot be found without additional information on the form of the operator and the properties of its coefficients (i.e., their smoothness, mixing, etc.). Most results here have been obtained for one-dimensional operators of the second order (discrete and continuous), where an important role belongs to the Lyapunov exponent

$$\gamma(\lambda, \omega, e) = \lim_{|x| \to \infty} |x|^{-1} \ln \|\Phi(x, \omega)e\|.$$

Here $\Phi(x, \omega)$ is a fundamental matrix (transfer matrix) of the respective equation of the second order and $e \in \mathbf{R}^2$ is the unit vector of initial data.

In the general case of a m.tr. potential, the following facts are true:

(1) The limit in the definition of $\gamma(\lambda, \omega, e)$ exists, with p.1, for every λ and all $e \in \mathbf{R}^2$, if $E\{|q(x)|\} < \infty$ in the continuous case and $E\{\ln(1 + |q(x)|)\} < \infty$ in the discrete case [38] (in stability theory, the equations for which this limit exists are called Lyapunov regular).

(2) The senior Lyapunov exponent $\gamma(\lambda) = \sup_{\|e\|=1} \gamma(\lambda, \omega, e)$ is nonnegative, nonrandom, and related to the IDS by the Thouless-Herbert-Jones formula:

$$\gamma(\lambda) = \gamma_0(\lambda) + \int_{\mathbf{R}} \ln|\lambda - \mu|[N(d\mu) - N_0(d\mu)],$$

where $\gamma_0(\lambda) = \max(0, -\lambda)^{1/2}$, $N_0(\lambda) = \pi^{-1}\max(0, \lambda)^{1/2}$ in the continuous Schrödinger case, and $\gamma_0 = N_0 = 0$ in the discrete case (Jacobi matrix) [2, 12, 8].

(3) The set $X \subset \mathbf{R}$ of the positive Lebesgue measure contains the absolutely continuous component of the spectrum of the respective MTO, if and only if for Lebesque almost all $\lambda \in X$, $\gamma(\lambda) = 0$ [24, 41] (the only if part); [25] (the if part).

But for random operators, much more is known. Namely, in the very general situation of independent or Markov coefficients, the Lyapunov exponent is positive on the spectrum [17, 24, 41, 36], and thus, such operators have no absolutely continuous spectrum (recall that in the case of a periodic potential the spectrum is pure absolutely continuous and coincides with the set $\{\lambda: \gamma(\lambda) = 0\}$).

If the potential in the one-dimensional Schrödinger operator is $q(x) = Q(\xi_x)$, where ξ_x is Brownian motion on a smooth compact Riemann manifold and $Q(\xi)$ is the smooth nonflat function on it, then the spectrum of such an operator, with p.1, is pure point, dense [19] and all the eigenfunctions are exponentially localized [36]. Similar facts in the discrete case have been established by Kunz and Souillard [31]. The hypothesis of the pure point nature of the spectrum was suggested in the early 60s by Mott and Twose.

It should be noted that the very possibility of existence of a dense point spectrum for the one-dimensional Schrödinger operator is provided by unique solvability of the inverse problem of spectral analysis for this operator. However, the traditional spectral theory, which dealt mainly with increasing or decreasing potentials, used to treat this type of spectrum rather as an exotic thing, an exception, than as a rule. Besides, the inverse problem technique is not powerful enough to give fairly constructive sufficient conditions on the class of potentials providing a dense point spectrum. In this context, it is noteworthy that the theory of spectral analysis inverse problems, which is of interest in quite a variety of respects, and in relation to integration of nonlinear evolution equations in particular, is rather poorly developed as regards MTOs. Besides the periodic case, thoroughly studied by Marčenko and Ostrovskii (see, e.g., the book by Marčenko [34]), more or less complete results are available only for the special class of limit periodic potentials [43]. For the general case of the m.tr. potential, there are only few, though very interesting, results [29], and little is known of random potentials.

The theorem of the pure point spectrum of the one-dimensional Schrödinger operator with a random potential proved by Goldsheĭd, Molchanov, and Pastur [19] has been generalized and proved more than once. The most complete and conceptually transparent form of this theorem is due to Kotani [28]. However, in all proofs the assumption of smoothness of probabilistic characteristics of the potential (existence of smooth densities, transition probabilities, etc.) was as essential as that of sufficient independence of its values for distant x. Let us illustrate this point by a fact following from the results of Kotani [28]: namely, if in the one-dimensional discrete Schrödinger operator the potential $q(x)$, $x \in \mathbf{Z}$, is a stationary Gaussian sequence, which is therefore very smooth in the present sense, with the spectral density $f(k)$, $k \in [0, 2\pi)$ (the Fourier transform of the correlation function $b(x) = E\{q(x)q(0)\}$), satisfying the condition

$$\int_0^{2\pi} f^{-1}(k)\, dk < \infty,$$

then the spectrum of this MTO is pure point and the eigenfunctions decrease exponentially. But this condition admits of very slow decrease of the correlation function (e.g., one can find $b(x) = \text{const} \cdot |x|^{-1}(1 + o(1))$, $|x| \to \infty$) and thus also very strong correlation between the potential values at distant points.

This is why, for example, the structure was unknown of the spectrum of the simple-looking operator of the one-dimensional Anderson model with the Bernoulli potential $q(x)$: $q(x) = 0$, q_0 with probabilities p_1 and p_2, $p_1 + p_2 = 1$ independently for different x. Recently, Carmona et al. [7], proceeding from the ideas of Fröhlich et al. [16] and the supersymmetric technique of combinatoric estimations, proved that the spectrum of this operator is pure point and $N(\lambda)$ has a singularly continuous component, at least at sufficiently large q_0.

These results, as well as many others obtained in recent years, lead to the conclusion that the spectral theory of one-dimensional random MTOs of the second order is at present a fairly well developed theory, which on one hand is in many cases comparable in completeness with the spectral theory of one-dimensional periodic operators and, on the other hand, provides many interesting and challenging unresolved problems.

As regards the spectral theory of multidimensional random MTOs, though it has made only its first steps, they represent substantial progress. These are the above-mentioned works by Fröhlich and Spencer proving the pure point character of the spectrum and the exponential decrease in eigenfunctions in the multidimensional Anderson model, whose potential has a distribution function with smooth density in the case of a sufficiently high potential amplitude or the spectral parameter sufficiently close to the spectrum edge (fluctuation region of the spectrum). These authors developed a new, very powerful technique, combining ideas of KAM theory and of the renormalization group method borrowed from quantum field theory and statistical physics.

There are, however, very many unresolved problems in spectral analysis of multidimensional random MTOs, since the structure of the spectrum of these operators is substantially richer and more diversified than in the one-dimensional case. In particular, it is generally accepted in physical literature, according to the Anderson hypothesis, that for the Schrödinger operator with $d \geq 3$ and the potential amplitude not too high, the spectrum, well away from fluctuation boundaries, is absolutely continuous. The problem of substantiating this hypothesis, the proof of absolute continuity of the spectrum of the Schrödinger operator for $d \geq 3$ and sufficiently large spectral parameter in particular, now seems to be one of the most important in the spectral analysis of random operators. Extremely interesting and poorly studied is the vicinity of the point separating a point spectrum from an absolutely continuous spectrum, which must be similar in many ways to the point of a phase transition (presumably of the second order). This point has not even been rigorously proved to exist at all, that is, it has not yet been proved that a point and absolutely continuous components cannot coexist in the same region of the spectrum of the random Schrödinger operator in $\mathcal{L}^2(\mathbf{R}^d)$ or $l^2(\mathbf{Z}^d)$ (one can construct examples of almost-periodic operators, quite similar to the Schrödinger operator, for which such coexistence is the case (see below)). There is practically nothing known of the structure of generalized eigenfunctions in the vicinity of the point, which must have little in common with Bloch functions of a periodic operator and may have possibly some kind of "fractal" form. It has not been proved either that the spectrum of random operators, with p.1, has no singular component.

Of much interest are problems of the spectral analysis of two-dimensional random and almost-periodic operators having various physical applications (weak localization effects, quantum coherent effects, in particular the integer Hall effect). Here, as in the one-dimensional case, the spectrum seems to be pure point, though more complicated in structure. A new and highly interesting set of problems is encountered in studying the two-dimensional Schrödinger operator with a magnetic field (uniform and random), where subtle effects may arise associated with infinitely multiple eigenvalues (Landau levels), including appearance of an absolutely continuous component in the vicinity of these levels, etc.

V. Almost-periodic operators. Random and periodic MTOs are, in some sense, extreme cases of the class of possible spectral types of MTOs (recall that in the one-dimensional case, periodic operators have always the absolutely continuous type, while random ones have, as a rule, the pure point type).[4] Various intermediate types of spectral behavior are demonstrated by almost-periodic MTOs. Such operators model the so-called incommensurable structures and a variety of other physical systems. In addition, the study of this class of MTO proved rather useful for the spectral theory of MTOs as a whole.

The main effect revealed here is, in my opinion, the essentially nonalgebraic nature of the spectral theory, in contrast to the periodic case, where commutativity of the operator with translation operators is enough to provide the basic facts of the spectral picture. This effect, and also some others, was discovered in the case of the equation

$$-\psi(x+1) - \psi(x-1) + 2g\cos(2\pi\alpha n + \omega)\psi(x) = \lambda\psi(x), \qquad x \in \mathbf{Z},$$

known as almost-Mathieu.

An important observation was made by Aubry and André [1] who proposed elegant, though not quite rigorous, arguments showing that for the Lyapunov exponent of the almost-Mathieu equation, with an irrational α, the inequality

$$\gamma(\lambda) \geq \ln|g|$$

is true. The arguments of Aubry and André were made rigorous in papers by Avron and Simon [2], and Pastur and Figotin [12, 13]. Hermann [23] gave an alternative proof of this and a variety of related inequalities. Thus, in virtue of what was stated in the preceding section, the spectrum of the corresponding operator H_{AM} for $|g| > 1$ has no absolutely continuous component σ_{ac} (α being rational, the spectrum of this operator is absolutely continuous for any g). Then, Bellissard et al. [3], generalizing the technique of Dinaburg and Sinai [10] and Belokolos [4], showed that $\sigma_{ac} \neq \varnothing$ for small $|g|$, while for large $|g|$, H_{AM} has a fairly massive point spectrum.[5] These results were obtained under the conditions

$$|\alpha - p/q| \geq \text{const} \cdot q^{-2-\varepsilon}, \qquad \varepsilon > 0, \ p, q \in \mathbf{Z}.$$

If, on the contrary, this condition is not true, in particular if for a certain sequence of rational numbers p_n/q_n

$$|\alpha - p_n/q_n| \leq \text{const} \cdot n^{-q_n}$$

---

[4]Though, at present, examples are known of one-dimensional random second-order operators, which in one region of parameters entering into a corresponding equation have a pure point dense spectrum and exponentially decreasing eigenfunctions and in another domain their spectrum is pure absolutely continuous (see, e.g., Bratus' et al. [6]).

[5]Sinai has recently shown that for large $|g|$ this operator (and even more general ones, when $q(x) = Q(2\pi\alpha x + \omega)$ where $Q(t)$ is the 2π-periodic C^2-function which has one nondegenerate maximum and one nondegenerate minimum) has a pure point spectrum and exponentially decreasing eigenfunctions.

then, as was shown by Avron and Simon [2], on the basis of results of Pastur
[41] and Gordon [20], the spectrum of H_{AM} for $|g| > 1$ is singularly continuous.

Thus, we see that the rather simple-looking almost-Mathieu equation demonstrates practically all possible types of spectral behavior, depending not only on
the arithmetic properties of the frequency α, but also on the amplitude g of the
potential.

Note that by Hermann's method [23] the analogue of the Aubry-André in-
equality $\gamma(\lambda) \geq \ln |g|$ may be proved for a broad class of quasiperiodic potentials.
Namely, let

$$q(x, \omega) = \sum_{\xi \in Q} b_\xi \exp(2\pi i(\alpha x + \omega) \cdot \xi), \qquad x \in \mathbf{Z}, \omega \in \mathbf{T}^l,$$

where Q is a finite set on the lattice $\mathbf{Z}^l$, $l \geq 1$, and $\alpha = (\alpha_1, \ldots, \alpha_l)$ are ra-
tionally independent numbers. Then $\gamma(\lambda) \geq \ln |b_{\xi^*}|$ where $\xi_i^* = \max_{\xi \in Q} |\xi_j|$,
$j = 1, \ldots, l$. A simple example of a potential of this form is

$$q(x, \omega) = g \prod_1^l 2 \cos 2\pi(\alpha_j x + \omega_j),^6$$

where $\gamma(\lambda) \geq \ln |g|$ again. A similar lower bound may be obtained for the
potential

$$q(x, \omega) = 2g \cos 2\pi[\alpha x^2 + (2\omega_1 - \alpha)x + \omega_2], \qquad x \in \mathbf{Z}, \omega \in \mathbf{T}^2,$$

which for irrational α is not a quasiperiodic function on $\mathbf{Z}$ (the corresponding
shift is the so-called complex skew shift on a torus, not its irrational winding).

In the remaining part of the report we shall discuss results of the spectral
analysis of two classes of almost-periodic operators for which a fairly detailed
picture of the spectrum can be given.

Let us consider an increasing sequence of numbers $\mathcal{A} = \{a_n, n \geq 1\}$ such that
$a_{n+1}/a_n \in \mathbf{N}\backslash\{1\}$ and let

$$\mathcal{R}_n = \left\{ r = \frac{\pi k}{a_n}, k \in \mathbf{Z} \right\}, \qquad \mathcal{R} = \bigcup_{n \geq 1} \mathcal{R}_n$$

and the sequence $\{q_r, r \in \mathcal{R}\}$ be such that

$$\lim_{n \to \infty} e^{Ca_{n+1}} \sum_{r \in \mathcal{R}\backslash \mathcal{R}_{n+1}} |q_r|^2 = 0, \qquad \forall C > 0.$$

Then the generalized Fourier series

$$q(x) = \sum_{r \in \mathcal{R}} q_r e^{2irx}, \qquad x \in \mathbf{R},$$

defines the limit periodic function which is superexponentially rapidly approximated by periodic functions in the Stepanov metric

$$\lim_{n \to \infty} e^{Ca_{n+1}} \|q - q_n\|_{S^2} = 0, \qquad \forall C > 0,$$

---

[6] But not $2g \sum_1^l \cos 2\pi(\alpha_j x + \omega_j)$.

where

$$q_n(x) = \sum_{r \in \mathcal{R}_n} e^{2irx} q_r,$$

$$\|q\|_{S^2}^2 = \sup_{x \in \mathcal{R}} \int_x^{x+1} |q(t)|^2 \, dt.$$

Denote the family of such functions by $Q_\infty(\mathcal{A})$. Also, denote by $\mathcal{K}_\infty(\mathcal{A})$ the family of complex-valued sequences $\{\kappa_r, r \in \mathcal{R}\}$, i.e., functions on $\mathcal{R}$ such that

$$\lim_{n \to \infty} e^{Ca_{n+1}} \sum_{r \in \mathcal{R} \setminus \mathcal{R}_n} r^2 |\kappa_r|^2 = 0, \qquad \forall C > 0.$$

Then the following facts hold [45]:

(i) Every potential $q \in Q_\infty(\mathcal{A})$ defines the unique element $\kappa \in \mathcal{K}_\infty(\mathcal{A})$ and vice versa, so that $\mathcal{K}_\infty(\mathcal{A})$ is the family of complete sets of independent spectral data for potentials from $Q_\infty(\mathcal{A})$.

(ii) The closed set $\sigma \subset \mathbf{R}$ is the spectrum of an operator $-d^2/dx^2 + q$, $Q_\infty(\mathcal{A})$ if and only if $\sigma = \{\lambda \colon \lambda \geq \lambda_0, \operatorname{Im} \theta(\sqrt{\lambda - \lambda_0}) = 0\}$, where the function $\theta(z)$ conformally maps the upper half-plane $\mathbf{C}^+$ onto $\Pi_+(\mathcal{A}) = \mathbf{C} \backslash \bigcup_{r \in \mathcal{R}} [r, r + i|\kappa_{|r|}|]$ under the conditions

$$\theta(0) = 0, \qquad \lim_{y \uparrow \infty} (iy)\theta(iy) = \pi, \qquad \lambda_0 = |\kappa_0|^2.$$

(iii)

$$\pi^{-1} \operatorname{Re} \theta(\sqrt{\lambda - \lambda_0}) = N(\lambda), \qquad \operatorname{Im} \theta(\sqrt{\lambda - \lambda_0}) = \gamma(\lambda),$$

where $N(\lambda)$ and $\gamma(\lambda)$ are the IDS and the Lyapunov exponent of the present MTO.

(iv) The spectrum of the Schrödinger operator with a potential from $Q_\infty(\mathcal{A})$ is absolutely continuous, has multiplicity two and is nowhere dense, if the support of κ_r is dense in $\mathcal{R}$.

We see that this class of MTOs is maximally close to periodic operators. The only difference is a nowhere dense (Cantor) spectrum though of an always positive Lebesgue measure.

The other class of almost-periodic operators we shall consider has spectral properties quite similar to random operators. These are the finite-difference operators in $l^2(\mathbf{Z}^d)$, $d \geq 1$, of the form

$$(A_d \psi)(x) = \sum_{y \in \mathbf{Z}^d} w_{x-y} \psi(y) + g \tan \alpha_x \cdot \psi(x), \qquad x \in \mathbf{Z}^d,$$

where

$$w_x^* = w_{-x}, \qquad |w_x| \leq C e^{-\rho|x|}, \qquad C, \rho > 0,$$

$$\alpha_x = \pi \alpha \cdot x + \omega, \qquad \alpha \in \mathbf{R}^d, \qquad \omega \in [0, \pi),$$

$$\omega \neq \pi/2 - \pi \alpha \cdot x.$$

For this family of operators, discovered by Fishman et al. [14], the following statements are valid (see [13]; similar results may be found in [47]).

(i) The IDS of the operators A_d is absolutely continuous and its density $\rho(\lambda)$ is

$$\rho(\lambda) = \int_{\mathbf{T}^d} \frac{g}{(\tilde{w}(k) - \lambda)^2 + g^2}\, dk$$

where $\tilde{w}(k)$ is the Fourier transform of the sequence w_x, and $\mathbf{T}^d$ is a d-dimensional torus.

(ii) If the vector of the frequencies $\alpha = (\alpha_1, \ldots, \alpha_d)$ is badly approximated by rationals, so that

$$|\alpha \cdot x - m| \geq \text{const} |x|^{-\beta}, \qquad \beta > 0, |x| \neq 0, m \in \mathbf{Z},$$

then the spectrum A_d is pure point, consists of solutions λ_x, $x \in \mathbf{Z}^d$, of the equation

$$N(\lambda) = \frac{1}{2} + \frac{\omega}{2} + \alpha \cdot x,$$

and by virtue of the monotonicity of $N(\lambda)$ is dense in $\mathbf{R}$. There is a unique eigenfunction, exponentially decreasing as $|x| \to \infty$, corresponding to every eigenvalue λ_x, $x \in \mathbf{Z}$.

(iii) For $d = 1$ and any irrational α, the spectrum of A_1 has no absolutely continuous component.

(iv) For $d = 1$ and irrational α such that

$$|\alpha - p_n/q_n| \leq \text{const} \cdot \exp(-q_n^{-1-\varepsilon}), \qquad \varepsilon > 0,$$

where p_n/q_n is a certain sequence of rational numbers, the spectrum A_1 is singularly continuous.

(v) Statement (ii) remains valid for the operator

$$\tilde{A}_d = A_d + \varepsilon B,$$

where

$$(B\psi)(x) = B(\exp(2\pi i\alpha \cdot x + 2i\omega)),$$

$\varepsilon \ll 1$, and $B(z)$, $z \in \mathbf{C}$, is a function analytic in the vicinity of the unit circle and mapping it onto the real axis.

Let us consider now in $\mathcal{L}^2(\mathbf{R}^d)$ the operator of the form

$$-\Delta + \sum_{n \in \mathbf{Z}^d} \tan(\pi\alpha \cdot x + \omega) P_n,$$

where P_n is an operator of orthogonal projection on the function $u(x+n)$, $x \in \mathbf{R}^d$, and assume that the support of the Fourier transform $\tilde{u}(k)$ of this function is the ball $|k|^2 \leq R < 1/2$. It may be shown that the spectrum of this operator consists of a point σ_p and an absolutely continuous σ_{ac} components, where $\sigma_p = \mathbf{R}$ and $\sigma_{ac} = [R, \infty)$. Thus, we have an example of an MTO for which a point and an absolutely continuous component coexist in the same region of the spectrum. Note also that the inclusion $\sigma_{ac} \supset [R, \infty)$ is valid in the general case too, where $\tan(\pi\alpha \cdot n + \omega)$ is replaced by an arbitrary m.tr. field on $\mathbf{Z}^d$.

In conclusion, I would like to note the following.

1. The variety of applications of the spectral theory of MTOs suggests the importance and desirability of studies not only of traditional problems of spectral theory, but of many other ones. I have in mind first of all a sufficiently constructive analysis of various quantities, constructed of eigenfunctions and eigenvalues of MTOs, $N(\lambda)$ being a very simple, yet very representative, example. Such quantities, not always with a direct spectral meaning, arise from consideration of problems in solid-state physics, optics, radiophysics, and mechanics of disordered media. These have as a rule quite a complicated structure, and thus this is a case primarily for an asymptotic or approximate study. A very important example of such a quantity is seen in the tensor of conductivity of the ideal Fermi gas in a random field, at zero temperature:

$$\sigma_{\alpha\beta}(\lambda, \nu) = E\left\{ \int_{\mathbf{R}^d} \frac{\partial}{\partial x^\alpha} e(x, y, \lambda + \nu) \frac{\partial}{\partial x_\beta} e(y, x, \lambda) \Big|_{x=0} dy \right\},$$

where $e(x, y, \lambda)$ is the kernel of the operator $d\mathcal{E}/d\lambda$ and $\mathcal{E}(\lambda)$ the resolution of the identity of the Schrödinger operator with a random potential, ν the a.c. electric field frequency, and λ the Fermi energy. It is very important to know the asymptotic behavior of $\sigma_{\alpha\beta}(\lambda, \nu)$ for $\nu \downarrow 0$ (the low frequency asymptotic) in various regions of λ and for various types of random potentials. Physicists have developed a variety of approximate methods to find such asymptotics (see, e.g., Gredeskul et al. [21]); still, rigorous asymptotic results and even rigorous estimations are here very few (see, e.g., [13], where it is shown that for the quasiperiodic operator Ad, $\sigma_{\alpha\alpha}(\lambda, \nu) \leq C_1(\lambda) \exp\{-C_2(\lambda)\nu^{-1/\beta}\}$).

2. There is also a different trend in the study of spectral properties of random symmetric matrices, initiated by Wigner [49], in connection with certain problems of nuclear physics (and also partly some problems of mathematical statistics). Here, as in many other problems of the spectral theory of random operators, we start with random symmetric matrices of other n, but such that all their entries are of the same order of magnitude (e.g., are identically distributed) and are independent. Therefore, the limiting operator for $n \to \infty$ does not exist here; but, there exists a limit IDS, i.e. a limit of the normalized distribution functions of eigenvalues. Such random matrices may be referred to as the matrix analogue of the scheme of series known in probability theory, whereas a MTO is the matrix analogue of an infinite sequence of independent, identically distributed random variables. An important feature of these studies is that they give, under very mild conditions, closed functional equations for the IDS, which provide an adequate description of the limit eigenvalue distribution, and even yield it explicitly in some cases [35]. In particular, for random matrices whose elements satisfy the Lindeberg condition, the corresponding IDS represents the so-called semicircle law [39]. A very general and detailed study of this subject has been carried out by Girko [18], and an interesting problem which has not yet received adequate attention is the statistics of spacings between eigenvalues.

References

1. Gilles André and Serge Aubry, *Analyticity breaking and Anderson localization in incommensurate lattices*, Group Theoretical Methods in Physics, Ann. Israel Phys. Soc. **3**, Hilger, Bristol, 1980, pp. 133–164.

2. Joseph E. Avron and Barry Simon, *Almost periodic Schrödinger operators*. II, *The integrated density of states*, Duke Math. J. **50** (1983), 369–391.

3. Jean Béllissard, Ricardo Lima and D. Testard, *A metal-insulator transition for the almost Mathieu model*, Comm. Math. Phys. **88** (1983), 207–234.

4. E. D. Belokolos, *A quantum particle in a one-dimensional deformed lattice. Estimates of lacunae dimensions in the spectrum*, Teoret. Mat. Fiz. **25** (1975), 344–357. (Russian)

5. M. M. Benderskiĭ and L. A. Pastur, *The spectrum of the one-dimensional Schrödinger equation with random potential*, Mat. Sb. (N.S.) **82(124)** (1970), 273–284. (Russian)

6. E. Bratus', S. A. Gredeskul, L. A. Pastur and V. S. Shumeĭko, *The absorption of stochastic acoustic signals in superconductors*, Soviet Low Temp. Phys. **12** (1986), 322–325.

7. René Carmona, Abel Klein and Fabio Martinelli, *Anderson localization for Bernoulli and other singular potential*, Univ. of California (Irvine) Preprint, 1986.

8. Walter Craig and Barry Simon, *Subharmonicity of the Lyapunov index*, Duke Math. J. **50** (1983), 551–560.

9. François Delyon and Bernard Souillard, *Remark on the continuity of the density of states of ergodic finite difference operators*, Comm. Math. Phys. **94** (1984), 289–291.

10. E. I. Dinaburg and Ya. G. Sinaĭ, *The one-dimensional Schrödinger equation with quasiperiodic potential*, Funktsional. Anal. i Prilozhen. **9** (1975), 8–21. (Russian)

11. A. L. Figotin, *Distribution of eigenvalues of the Schrödinger equation with random potential and asymptotic behavior for large times of a Wiener integral*, Dokl. Akad. Nauk Ukr. SSR **6** (1981), 27–29. (Russian)

12. A. L. Figotin and L. A. Pastur, *The positivity of the Lyapunov exponent and the absence of the absolutely continuous spectrum for the almost-Mathieu equation*, J. Math. Phys. **25** (1984), 774–777.

13. _____, *An exactly solvable model of a multidimensional incommensurate structure*, Comm. Math. Phys. **95** (1984), 401–425.

14. Shmuel Fishman, D. Grempel and R. Prange, *Localization in an incommensurate potential: an exactly solvable model*, Phys. Rev. Lett. **49** (1982), 833–836.

15. Jürg Fröhlich and Thomas Spencer, *Absence of diffusion in the Anderson tight binding model for large disorder or low energy*, Comm. Math. Phys. **88** (1983), 151–184.

16. Jürg Fröhlich, Fabio Martinelli, E. Scoppola and Thomas Spencer, *Constructive proof of localization in the Anderson tight binding model*, Comm. Math. Phys. **101** (1985), 21–46.

17. Harry Furstenberg, *Noncommuting random products*, Trans. Amer. Math. Soc. **108** (1963), 377–428.

18. V. L. Girko, *Theory of random determinants*, "Vyshcha Shkola" (Izdat. Kiev. Univ.), Kiev, 1980. (Russian)

19. I. Ya. Gol'dsheĭd, S. A. Molchanov and L. A. Pastur, *A random one-dimensional Schrödinger operator has a pure point spectrum*, Funktsional. Anal. i Prilozhen. **11** (1977), 1–10, 96. (Russian)

20. A. Ya. Gordon, *The point spectrum of the one-dimensional Schrödinger operator*, Uspekhi Mat. Nauk **31** (1976), 257–258.

21. S. A. Gredeskul, I. M. Lifshitz and L. A. Pastur, *Introduction to the theory of disordered systems*, "Nauka", Moscow, 1982; English transl., Wiley, New York.

22. L. N. Grenkova, S. A. Molchanov and Yu. N. Sudarev, *On the basic states of one-dimensional disordered structures*, Comm. Math. Phys. **90** (1983), 101–123.

23. Michael-Robert Herman, *Une méthode pour minorer les exposants de Lyapounov et quelques exemples montrant le caractère local d'un théorème de Arnol'd et de Moser sur le tore de dimension 2*, Comment. Math. Helv. **58** (1983), 453–502.

24. Kazushige Ishii, *Localization of eigenstates and transport phenomena in the one-dimensional disordered systems*, Progr. Theoret. Phys. Suppl. **53** (1973), 77–120.

25. Werner Kirsch and Fabio Martinelli, *On the ergodic properties of the spectrum of general random operators*, J. Reine Angew. Math. **334** (1982), 141–156.

26. Shinichi Kotani, *Ljapunov indices determine absolutely continuous spectra of stationary random one-dimensional Schrödinger operators*, Stochastic Analysis (Katata/Kyoto, 1982), North-Holland Math. Library, vol. 32, North-Holland, Amsterdam and New York, 1984, pp. 225–247.

27. ____, *Support theorems for random Schrödinger operators*, Comm. Math. Phys. **97** (1985), 443–452.

28. ____, *Lyapunov exponents and spectra for one-dimensional random Schrödinger operators*, Random Matrices and Their Applications, Contemp. Math., vol. 50, Amer. Math. Soc., Providence, R. I., 1985, pp. 277–286.

29. ____, *On the inverse problem for random Schrödinger operators*, Particle Systems, Random Media and Large Deviations, Contemp. Math., vol. 41, Amer. Math. Soc., Providence, R. I., 1985, pp. 267–280.

30. S. M. Kozlov, *The averaging method and walks in inhomogeneous environments*, Uspekhi Mat. Nauk **40** (1985), 61–120.

31. Hervé Kunz and Bernard Souillard, *Sur le spectre des opérateurs aux différences finies aléatoires*, Comm. Math. Phys. **78** (1980/81), 201–246.

32. ____, J. Phys. (Paris) Lett. **44** (1983), 411–.

33. I. M. Lifshitz, *On the structure of the energy spectrum and quantum states of disordered condensed systems*, Adv. in Phys. **13** (1964), 485–531.

34. V. A. Marchenko, *Sturm-Liouville operators and their applications*, "Naukova Dumka", Kiev, 1977; English transl., Reidel, Boston.

35. V. A. Marchenko and L. A. Pastur, Mat. Sb. **72** (1966), 430–.

36. S. A. Molchanov, Math. USSR-Izv. **42** (1978), 243–.

37. ____, *The local structure of the one-dimensional Schrödinger operator*, Comm. Math. Phys. **78** (1980/81), 429–446.

38. V. I. Oseledec, *A multiplicative ergodic theorem. Characteristic Ljapunov exponents of dynamical systems*, Trudy Moskov. Mat. Obshch. **19** (1968), 179–210. (Russian)

39. L. A. Pastur, *The spectrum of random matrices*, Teoret. Mat. Fiz. **10** (1972), 102–112.

40. ____, *Spectra of random selfadjoint operators*, Uspekhi Mat. Nauk **28** (1973), 3–64.

41. ____, *On the spectrum of the random Jacobi matrices and the Schrödinger equation with random potential on the whole axis*, Preprint FTINT, Kharkov, 1974 (see also Comm. Math. Phys. **75** (1980), 179–196).

42. ____, *The behavior of certain Wiener integrals as* $t \to \infty$ *and the density of states of Schrödinger equations with random potential*, Teoret. Mat. Fiz. **32** (1977), 88–95. (Russian)

43. L. A. Pastur and A. L. Figotin, *Ergodic properties of the distribution of the eigenvalues of certain classes of random self-adjoint operators*, Differentsialnye Uravneniai Metody Funktsionalnogo analiza, Naukova Dumka, Kiev, 1978, pp. 117–133. (Russian) (Translated in Selecta Math. Sov. **3** (1983/84), 69–86.)

44. ____, (1983) (unpublished).

45. L. A. Pastur and V. A. Tkachenko, *On the spectral theory of the one-dimensional Schrödinger operator with limit-periodic potential*, Dokl. Akad. Nauk USSR **274** (1984), 1050–1053. (Russian)

46. M. A. Shubin, *Almost periodic functions and partial differential operators*, Uspekhi Mat. Nauk **33** (1978), 3–47, 247. (Russian)

47. Barry Simon, *Almost periodic Schrödinger operators. IV. The Maryland model*, Ann. Phys. (USA) **159** (1985), 157–183.

48. Franz J. Wegner, *Bounds on the density of states in disordered systems*, Z. Phys. B **44** (1981), 9–15.

49. E. Wigner, *On the distribution of the roots of certain symmetric matrices*, Ann. of Math. (2) **67** (1958), 325–327.

INSTITUTE FOR LOW TEMPERATURE PHYSICS, UKR. SSR ACADEMY OF SCIENCES, KHARKOV 310164, USSR

Random and Quasiperiodic Schrödinger Operators

THOMAS SPENCER

1. Introduction. We shall discuss some recent results and open problems for a class of Schrödinger operators with random or quasiperiodic potentials v. For technical reasons, the Schrödinger operator $H = -\varepsilon^2 \Delta + v$ will frequently be considered on the lattice $\mathbf{Z}^d$ as well as on the continuum $\mathbf{R}^d$. H is defined as a selfadjoint operator on the Hilbert spaces $l^2(\mathbf{Z}^d)$ and $L^2(\mathbf{R}^d)$, respectively. The lattice Laplacian, defined by

$$(\Delta f)(j) = \sum_{i:|i-j|=1} [f(i) - f(j)], \qquad i,j \in \mathbf{Z}^d,$$

is a bounded operator whose spectrum is the interval $[0, 4d]$, and ε is a constant that will be specified later. At low energy, the spectral properties of H on the lattice and in the continuum are expected to be similar.

As we shall see below, our motivation for studying random and quasiperiodic potentials comes from both physics and dynamical systems. We shall primarily focus on the behavior of solutions to the equation $H\psi = E\psi$. The analysis of this equation for the class of potentials considered below generally involves overcoming small divisor problems. Roughly speaking, this is achieved by a multiscale analysis related to KAM (Kolmogorov, Arnold, Moser) methods and to renormalization group techniques.

Four classes of potentials that correspond to different physical situations will be considered.

(1) *Periodic potentials.* Here the hamiltonian describes, for example, the motion of electrons in a perfect crystal.

(2) *Random potentials on $\mathbf{Z}^d$.* We assume that $v(j)$ are independent random variables with a common bounded distribution density $g(v)\,dv$. For example, if g is uniformly distributed over the interval $[-w, w]$, then H is called the Anderson tight binding hamiltonian. Anderson introduced and analyzed this operator to study electrons in a random medium such as a crystal with impurities [1].

(3) *Quasiperiodic potentials on $\mathbf{Z}^1$ or $\mathbf{R}^1$.* We restrict ourselves to the case where there is just one incommensurate frequency α present and the potential

has the form:

$$\text{(a)} \quad v(x) = \cos(x) + \cos(\alpha x + \theta), \qquad x, \theta \in \mathbf{R},$$

$$\text{(b)} \quad v(j) = \cos 2\pi(\alpha j + \theta), \qquad j \in \mathbf{Z}.$$

The number α is assumed to be poorly approximated by rationals. More precisely, we assume

$$|\sin n\pi\alpha| \geq C_0/n^2. \tag{1}$$

The set of irrationals satisfying (1), for some C, has measure one. Although we shall state our results for the cosine, this function can be replaced by any smooth, even, periodic function with precisely two nondegenerate critical points.

Quasiperiodic potentials naturally arise if one linearizes about a quasiperiodic orbit. See (4) below. They also arise in the study of a quantum electron moving in a perfect crystal subjected to a periodic potential whose period is incommensurate with that of the lattice.

(4) *Nonlinear systems.* Let us consider, for example, the discrete time pendulum or standard map defined by the recursion relation

$$\varepsilon^2(\Delta x)_j \equiv \varepsilon^2(x_{j+1} + x_{j-1} - 2x_j) = \sin x_j. \tag{2}$$

This naturally defines an area-preserving transformation of the torus to itself. Note that x_j depends on the initial data (x_0, x_1), hence so does the potential defined by $v(j) = \cos(x_j)$, $j \in \mathbf{Z}$. If we wish to determine the sensitivity of the orbit x_j on x_0, we differentiate (2) with respect to x_0 and obtain the equation

$$H\psi = 0 \quad \text{where } \psi(j) = \partial x_j/\partial x_0.$$

In this case, we see that the Schrödinger equation arises from linearizing about a nonlinear orbit. Most of our comments on (2) will be conjectural.

We shall also present some recent results on an infinite chain of nonlinearly coupled oscillators whose hamiltonian has the form

$$\sum_i p_i^2 + w_i^2 q_i^2 + \varepsilon f(q_i - q_{i+1}), \tag{3}$$

where the w_i are assumed to be independent random variables and f is an analytic function which satisfies $f(q) = O(q^4)$ for small q.

2. Results and conjectures. (1) When v is periodic and $\varepsilon > 0$, it is well known that the spectrum of H is purely absolutely continuous. Moreover, all generalized eigenfunctions (i.e., polynomially bounded solutions to $H\psi = E\psi$) are Bloch waves

$$\psi(x) = p(x) \exp(ikx)$$

where p is a periodic function having the same period as v.

(2) If $v(j)$ are independent random variables, it is known that, with probability one, the spectrum of $H = [0, 4d\varepsilon^2] + \operatorname{supp} g$ where $\operatorname{supp} g$ is the support of g. Note that if $\varepsilon = 0$, the spectrum of H is pure point with eigenvalues $v(j)$ and eigenfunctions which are Kronecker δ functions. Goldsheid, Molehanov,

and Pastur [2] were the first to prove that a similar picture holds for all ε in one dimension. (Actually their results were for the continuum.) More precisely, they proved that, with probability one, the spectrum of H is pure point with eigenfunctions which decay exponentially fast. Physicists call this *localization* since the eigenfunctions are exponentially localized in space. Physically, localization implies that under time evolution the electron's wave function does not spread. Thus there is no conductivity or diffusion in the system. These results on localization in one dimension have been extended to potentials with singular distributions such as $v_j = \pm 1$ [3] and potentials which are "nondeterministic," i.e., the independence assumption has been weakened [4].

When $d \geq 2$ it is known that there is always an interval near the edge of the spectrum consisting only of point spectrum. Furthermore, if ε is small we have only point spectrum. These results were obtained independently by a number of authors: Simon and Wolff [5]; Delyon, Levy, and Souillard [6]; Fröhlich, Martinelli, Scippola, and Spencer [7]. When $d = 2$, all states are believed to be localized for all ε, but there are no rigorous results of this kind. When $d \geq 3$ there should be a band of absolutely continuous spectrum $[E_m, E'_m]$ corresponding to "extended" states provided ε is large. However, the only results of this kind are known when the lattice $\mathbf{Z}^d$ is replaced by the Bethe lattice [8]. E_m and E'_m are called mobility edges, since electrons should be mobile in this energy range and conduction should occur.

The basic estimate needed to establish the above results on localization is the exponential decay of the Green's function or, equivalently, when $d = 1$, the positivity of the Liapunov exponent. Let E be fixed. In one dimension we say that the Liapunov exponent $\gamma(E)$ is positive if, with probability one,

$$(H - E)\psi(j) = 0$$

only has solutions which grow at an exponential rate $\gamma(E)$ as j goes to $+$ or $-$ infinity. Equivalently, the Green's function satisfies

$$|G(E, 0, j)| \equiv |(E - H)^{-1}(0, j)| \leq C_v \exp -\gamma(E)|j| \tag{4}$$

where C_v is a v dependent constant which is finite with probability one. In one dimension, the positivity of γ is due to Furstenberg [9]. Recently Kotani showed that $\gamma(E) > 0$ for almost all E for a wide class of potentials which are "nondeterministic" [10].

For $d \geq 2$, the decay of G when either ε is small or E is near the edge of the spectrum is due to Fröhlich and Spencer [11].

The point of (4) is that it holds even when E belongs to the spectrum of H. With probability one, E is not an eigenvalue but there are eigenvalues E_i which get arbitrarily near E making the operator norm of $G(E)$ infinite. We refer to [12, 13] for mathematical reviews.

(3) *Quasiperiodic case.* Some time ago, Dinaburg and Sinai [14] proved that in the continuum there are always high energy Bloch type eigenfunctions

$$\psi(x) = q(x) \exp(-ik(E)x)$$

via KAM methods. Here q is a quasiperiodic function of x. On the lattice, in the special case $v(j) = \cos(\alpha j + \theta)$, Aubry [15] and Herman [16] have very elegant proofs of the positivity of $\gamma(E)$ for all E when $\varepsilon^2 < \frac{1}{2}$, and Bellissard et al. [17] showed that for small ε there are exponentially localized eigenstates. The following results are due to Fröhlich, Wittwer, and Spencer [18], and they hold for a set of θ of measure 1.

For case (3a) there are an infinite set of low energy eigenstates which decay exponentially fast provided ε is small.

For case (3b), if ε is small there are only localized states. In other words, exponentially decaying eigenstates form a basis. Furthermore, these eigenstates have 2^n peaks where $n = 0, 1, 2, \ldots$, which are *self-similar* under reflection. For example, if ψ is a wave function with two peaks at 0 and l, then there is a constant C such that $|j| \leq l/3$,

$$C\psi(j) = \psi(l - j) + O(\exp{-l}). \tag{5}$$

If the wave function ψ has four peaks at $0 < l_1 < l_2 - l' < l_2$, then $l' = l$ and $l_2 \geq \exp c l_1$. Moreover,

$$|v(0) - v(l_2)| \leq \exp{-\text{const.}\, l_2},$$

which by (1) and the evenness of cosine implies

$$|v(j) - v(l_2 - j)| \leq \exp{-cl_2}. \tag{6}$$

This relation implies (5) with l replaced by l_2.

REMARKS. Sinai has also established similar results for case (b).

We believe that the methods for the lattice can be extended to prove that for the continuum (case (a)):

(i) If ε is large, there are only Bloch type eigenstates.

(ii) If ε is small, there are only localized states at low energy. These states exhibit a self-similar structure as described for case (b).

(4) An outstanding problem for the standard map is to prove that the Liapunov exponent $\gamma = \gamma(E = 0) > 0$; i.e., $\psi(j) = \partial x_j / \partial x_0$ grows exponentially fast in j, for a set of (x_0, x_1) of positive measure. Equivalently, the Green's function $H^{-1}(0, j)$ decays exponentially fast. This means that there is sensitive dependence on initial data, and the theory of Pesin assures us that there is an ergodic component of positive measure on which the orbit moves "stochastically." Unfortunately, the only results of this kind are known for certain piecewise smooth nonlinearities, as in the case where the sine in (2) is replaced by a sawtooth function. We also expect that the spectrum of H consists of dense point spectrum—(localized states) for almost all (x_0, x_1) provided ε belongs to a Cantor set of positive measure and $|\varepsilon| << 1$. Hence for ε in the Cantor set, we expect $v(j) = \cos(x_j)$ to behave as in the random case. We remark that the Green's function G also has small divisors E_i^{-1}, since $E = 0$ is presumably in the spectrum of H. This reflects the nonuniform hyperbolicity of the dynamical system (2). However, these small divisors are of an entirely different kind from

those encountered in conventional KAM situations where one is searching for integrable or quasiperiodic motions.

Next we briefly describe some results for the infinite chain of oscillators (3) done in collaboration with Fröhlich, Wayne [19]. We show that if ε is small, there are infinite-dimensional, invariant tori of spacially localized, time almost periodic solutions to the equations of motion with high probability. In particular, for some fixed i, $|p_i^2(t) + q_i^2(t)| \geq$ Const. > 0 and the local energy does not go to zero. If the ω_j are all equal, then typically there are theorems which show that the local energy goes to zero since the wave packet can spread. The existence of these localized waves follows by using a variant of KAM methods. Much work remains to be done in understanding how general solutions to the equations of motion behave. Also, our methods cannot yet handle nonlinearities of the form $f(q) = O(q^2)$ for small q unless $f(q) = q^2$, in which case the problem is linear and equivalent to a random Schrödinger equation. See [20] for related results.

3. Remarks on the proofs. The estimates for the Green's function (4) and the eigenfunctions are obtained by a multiscale perturbation scheme. In the case where $v(j)$ is random, our approach might be described as a probabilistic Newton scheme. The 0th step of this scheme is obtained by expanding G in a formal series in ε.

$$G(E) = \sum_{n=0}^{\infty} \varepsilon^{2n} [(E - v)^{-1} \Delta]^n (E - v)^{-1}. \tag{7}$$

The small divisors occur at sites j where $|v(j) - E|$ is small. Since this difference can become arbitrarily small, one cannot expect (7) to converge even for small ε. The position of the small divisors at the 0th scale is defined by

$$S_0(E, v) = \{j : |v(j) - E| \leq \delta_0\}.$$

Now let G_Λ denote the Green's function for $H(\Lambda)$, i.e., H restricted to a box $\Lambda \subset \mathbf{Z}^d$ with Dirichlet boundary conditions on $\partial\Lambda$. If $\Lambda \cap S_0 = \varnothing$, then (7) converges if δ_0 is not too small, and in fact

$$|G_\Lambda(E; i, j)| \leq C r_0^{|i-j|}$$

where $r_0 = O(\varepsilon^2/\delta_0)$. When $v(j)$ are independent random variables it is easy to see that $\Lambda \cap S_0 = \varnothing$ with probability at least $1 - O(\delta_0)|\Lambda|$.

In [11], sets S_m were introduced to describe small divisors at the mth scale. The S_m are defined by induction and satisfy $S_0 \supset S_1 \supset S_2 \cdots$. These sets are defined in terms of a sequence of length scales l_i which grow very rapidly, and in terms of energy widths $\delta_i = \delta_0 \exp -\sqrt{l_i}$. A crude approximate definition of S_{m+1} is given as follows. Let $p \in S_m(E)$ and let $\Lambda_m(p)$ be a cube centered at p of side l_{m+1}. Then $p \in S_{m+1}(E)$ if, for some eigenvalue, E_m of $H(\Lambda_m(p))$ satisfies $|E_m - E| \leq \delta_{m+1}$. The difference $E_m - E$ is the small divisor at scale m. As m increases, the S_m typically describe increasingly rare events.

Now the estimate (4) on the Green's function can be shown to follow from the following two theorems.

THEOREM 3.1. *If* $\Lambda \cap S_m(E) = \varnothing$, *then*

$$|G_\Lambda(E'; i, j)| \le r^{|i-j|} \qquad where \ r = O(\varepsilon),$$

provided that $|i - j| \ge l_m^{2/3}$ *and* $|E' - E| \le \delta_m/2$.

This theorem is more precisely stated and proved in [11, 12] by induction on m. The main idea is that the exponential decay $r^{|i-j|}$, obtained in the previous induction step when $S_{m-1} \cap \Lambda = \varnothing$, dominates the divergence arising from the small divisor $\delta_m^{-1} = \exp \sqrt{l_m}$ provided $|i - j| \ge l_m^{2/3}$.

For the case where $v(j)$ are independent random variables, we have the following result for *fixed* E.

THEOREM 3.2. *For* ε *sufficiently small or* E *near the edge of the spectrum,*

$$S_m(E) \cap \Lambda_m = \varnothing \quad for \ all \ m \ge m(v)$$

where $m(v)$ *is finite with probability one and* Λ_m *is a cube centered at* 0 *of width* l_{m+1}.

If we allow E to vary for fixed v, $S_m(E)$ will not be empty, and, in fact, points in S_m represent possible peaks of the eigenfunctions whose eigenvalues are δ_m near E. In the quasiperiodic case, we prove that $S_m(E, \theta)$ has a very special structure: if $a, b \in S_m$ and $|a - b| \le l_m$, then

$$|v(a) - v(b')| \le \exp -\sqrt{|a - b|} \tag{8}$$

for some $b' \in S_m$ where $|b' - b| \le |a - b|/3$. Thus the small divisors can be identified just by looking at the potential. Notice that (1) and (8) imply that the potential near a and b' are nearly mirror images of each other. See (6). This is responsible for the self-similarity of the eigenfunctions (5) mentioned earlier. In the quasiperiodic case, our probability space is the circle and is parametrized by a single variable θ. Note that the θ which satisfy (8) is of very small measure when a and b are far apart. Hence, separated peaks of an eigenfunction are very unlikely. See [18] for details.

REFERENCES

1. P. W. Anderson, Phys. Rev. **109** (1958), 1492.
2. I. Gold'sheid, S. Molchanov, and L. Pastur, *A random homogeneous Schrödinger operator has a pure point spectrum*, Functional Anal. Appl. **11** (1977), no. 1.
3. R. Carmona, A. Klein, and F. Martinelli, Preprint.
4. S. Kotani and B. Simon, CMP (to appear).
5. B. Simon and T. Wolff, Comm. Pure Appl. Math. **39** (1986), 75.
6. F. Delyon, Y. Levy, and B. Souillard, *Anderson localization for multidimensional systems at large disorder or large energy*, Comm. Math. Phys. **100** (1985), 463–470.
7. J. Fröhlich, F. Martinelli, E. Scoppola, and T. Spencer, Comm. Math. Phys. **101** (1985), 21.
8. H. Kunz and B. Souillard, J. Phys. Lett. **44** (1983), L411.
9. H. Furstenberg, *Noncommuting random products*, Trans. Amer. Math. Soc. **108** (1963), 377–428.
10. S. Kotani, Proc. of Taniguchi Sympos., SA Katata, 1982.

11. J. Fröhlich and T. Spencer, *Absence of diffusion in the Anderson tight binding model for large disorder or low energy*, Comm. Math. Phys. **88** (1983), 151.

12. T. Spencer, Proceedings of the Les Houches Summer School 1984 (K. Osterwalder and R. Stora, eds.) North-Holland, Amsterdam, 1986.

13. R. Carmona, Ecole d'Ete de Probabilities XIV Saint Flour, 1984.

14. E. Dinaburg and Ya. Sinai, *The one-dimensional Schrödinger equation with quasi-periodic potential*, Functional Anal. Appl. **9** (1975), 279.

15. S. Aubry, Solid State Sci. **8** (1978), 264.

16. M. Herman, *A method for determining the lower bounds of Lyapunov exponents and some examples showing the local character of a Arnol'd-Moser theorem on the two-dimensional torus*, Comment. Math. Helv. **58** (1983), 453–502.

17. J. Bellissard, R. Lima, and D. Testard, *A metal-insulator transition for the almost Mathieu model*, Comm. Math. Phys. **88** (1983), 207–234.

18. J. Fröhlich, T. Spencer, and P. Wittwer (to appear).

19. J. Fröhlich, T. Spencer, and C. Wayne, J. Statist. Phys. **42** (1986), 247.

20. M. Vittot and J. Bellissard (to appear).

SCHOOL OF MATHEMATICS, INSTITUTE FOR ADVANCED STUDY, PRINCETON, NEW JERSEY 08540, USA

Multilevel Approaches to Large Scale Problems

ACHI BRANDT

1. Introduction. Most massive computational tasks facing us today have one feature in common: They are mainly governed by *local* relations in some low (e.g., 2 or 3) dimensional space or grid. Such are all differential problems, including flows, electromagnetism, magnetohydrodynamics, quantum mechanics, structural mechanics, tectonics, tribology, general relativity, etc., as well as statistical or partly differential, partly statistical problems (e.g., in statistical mechanics, field theory, turbulence), and many nondifferential problems like those in geodesy, multivariate interpolation, image reconstruction, pattern recognition, many design, optimization, and mathematical programming problems (e.g., traveling salesman, VLSI design, linear programming transportation), network problems, and so on. This common feature can be exploited very effectively by multilevel (multigrid) solvers, which combine local processing on different scales with various interscale interactions. Even when the governing relations are not strictly local (e.g., integral and integro-differential equations, x-ray crystallography, tomography, econometrics), any problem with a multitude of unknowns is likely to have some internal structure which can be used by multilevel solvers. In many cases, the computational cost of such solvers has been shown to be essentially as low as the cost can ever be; that is, the amount of processing is not much larger than the amount of real physical information.

This article is a brief survey of this field of study, emphasizing important recent developments and their implications. No attempt is made to scan the fast-growing multigrid literature. (A list of more than 600 papers will appear in [24]; see also the multigrid books [21, 25, 7, 28, 20, 26, 22].) A more detailed account will be given in [8].

Multigrid methods were first developed (see historical note, §16) as fast solvers for discretized linear elliptic PDEs (see §§3, 4, 5), then extended to nonelliptic (§6), nonlinear (§7) and time-dependent (§10) problems, and to more general algebraic systems (§§2, 11). The multigrid apparatus has also been used to

Research supported mainly by the Air-Force Office of Scientific Research, United States Air Force under Grants AFOSR-84-0070 and AFOSR-86-0127, and also by the United States Army Contract DAJA 45-84-C-0036.

obtain improved discretization schemes (§8), and is especially effective in treating compound problems and sequences of many similar problems (§9). Recently, mainly in response to current computational bottlenecks in theoretical physics, new types of multilevel methods have been developed for solving large lattice equations (e.g., Dirac equations in gauge fields (§11)); for calculating determinants (§12) and accelerating Monte-Carlo iterations (§14); and for discrete-state and highly nonquadratic minimization (§13), the latter being applicable to spin systems and also to image reconstruction, crystallography, and combinatorial minimization. Multilevel linear programming is reported in §15.

2. Slow components in matrix iterations. Consider the real linear system of equations

$$Ax = b, \qquad (2.1)$$

where A is a general $n \times m$ real matrix. For any approximate solution vector $\tilde{x}$, denote the error vector by $e = x - \tilde{x}$, and the vector of residuals by $r = Ae = b - A\tilde{x}$. Given $\tilde{x}$, it is usually easy to calculate r—especially when A is a sparse matrix, e.g., when A is based on local relations. One can then easily use these residuals to *correct* $\tilde{x}$, for instance, by taking one residual r_i at a time, and replacing $\tilde{x}$ by $\tilde{x} + (r_i/a_i a_i^T)a_i^T$, where a_i is the ith row of A (thus projecting $\tilde{x}$ onto the hyperplane of solutions to the ith equation). Doing this for $i = 1, \ldots, n$ is called a Kaczmarz *relaxation sweep*. It can be shown (Theorem 3.4 in [9]) that the convergence to a solution x (if one exists), of a sequence of such (or other) relaxation sweeps, should slow down only when

$$|\bar{r}| \ll |e|, \qquad (2.2)$$

where $\bar{r}$ is the *normalized* residual vector ($\bar{r}_i = a_i e/|a_i|$) and $|\cdot|$ is the Euclidean (l_2) norm. From the normalization of $\bar{r}$ it is clear that, for most error vectors, $|\bar{r}|$ is comparable to $|e|$; (2.2) can clearly hold only for special error vectors, dominated by special components (eigenvectors with small eigenvalues), whose number is small. Thus, when relaxation slows down, the error can be approximated by vectors in some much-lower dimensional space, called the *space of slow components*.

The concrete characterization of slowness depends on the nature of the problem, and is sometimes far from trivial (see, e.g., the "multiple representations" in §8). In many cases of interest, however, we will now see that slowness simply means smoothness (see §11 for a generalization).

3. Discretized differential equations. In case the system (2.1) represents a discretization of a stationary partial differential equation $Lu = F$ on some grid with meshsize h, we customarily rewrite it in the form

$$L^h u^h = F^h, \qquad (3.1)$$

where u^h is a grid function. Barring cases of alignment (see §6), such a system is numerically stable if and only if L^h has a good *measure of ellipticity on scale* h, inherited either from a similar h-ellipticity measure of L, or (e.g., in case L

is nonelliptic) from artificial ellipticity introduced either by "upstream" differencing or through explicit "artificial viscosity" terms. (Ellipticity measures on uniform grids, and their scale dependence, are discussed in [7, §2.1].)

For any h-elliptic operator L^h, relation (2.2) holds if and only if the error is smooth on the scale of the grid; i.e., iff its differences over neighboring grid points are small compared with itself. (This in fact is exactly the meaning of h-ellipticity.) The space of slow components can therefore be defined as the space of grid-h functions of the form $I_H^h v^H$, where v^H are functions on a *coarser grid*, with meshsize $H > h$, and I_H^h is an interpolation operator from grid H to grid h.

The coarse grid should not be too coarse; $H = 2h$ or so is about optimal: On one hand, it keeps H close enough to h, so that all errors which cannot be approximated on grid H are so highly oscillatory that their convergence by relaxation on grid h must be very fast (convergence factor .25 per sweep, typically). On the other hand, $H = 2h$ already yields a small enough number of coarse grid points, so that the work associated with the coarse grid (in the algorithms described below) is already just a fraction of the relaxation work on the fine grid.

Let $\tilde{u}^h$ be an approximation to the solution u^h, obtained for example after several relaxation sweeps. To define a coarse-grid approximation v^H to the smooth error $v^h = u^h - \tilde{u}^h$, one approximates the "*residual equation*"

$$L^h v^h = r^h \underset{\text{def}}{=} F^h - L^h \tilde{u}^h \tag{3.2}$$

by the coarse-grid equation

$$L^H v^H = I_h^H r^h, \tag{3.3}$$

where I_h^H is a fine-to-coarse interpolation (local averaging in fact, sometimes called "weighting" or "restriction") and L^H is a coarse grid approximation to L^h. One can either use the Galerkin-type approximation $L^H = I_h^H L^h I_H^h$, or derive L^H directly from L by differencing (replacing derivatives by finite differences) on grid H, which is less automatic but often also far less expensive in computer time and storage. A generally sensible approach is to use *compatible coarsening*, i.e., the Galerkin approach when L^h itself has been constructed by Galerkin (or variational) discretization, and direct differencing in case L^h itself is so derived, using the same discretization order and "double discretization" (see §10) as used by L^h, etc. (see discussion in [7, §11]).

A *coarse grid correction* is the replacement of $\tilde{u}^h$ by $\tilde{u}^h + I_H^h v^H$. Using alternately a couple of relaxation sweeps and a coarse grid correction is called a *two-grid cycle*.

4. Multigrid algorithms. There is no need of course to solve (3.3) exactly. Its approximate solution is most efficiently obtained by again alternately using relaxation sweeps (now on grid H) and corrections from a still coarser grid ($2H$). We thus construct a sequence of grids, each typically being twice as coarse as the former, with the coarsest grid containing so few equations that they can be solved (e.g., by Gaussian elimination) in negligible time.

A *multigrid cycle* for improving an approximate solution to (3.1) is recursively defined as follows: If h is the coarsest grid, solve (3.1) by whatever method. If not, denoting by H the next coarser grid, perform the following three steps: (A) ν_1 relaxation sweeps on grid h; (B) a coarse grid correction, in which (3.3) is approximately solved by starting with $\tilde{v}^H = 0$ and improving it by γ multigrid cycles; (C) ν_2 additional relaxation sweeps on grid h.

The *full multigrid algorithm* N-FMG for solving (3.1), when h is not the coarsest grid and H is the next coarser, is recursively defined as follows: (A) Solve $L^H u^H = F^H$ by a similar N-FMG algorithm, where $F^H = I_h^H F^h$. (F^H may also be derived directly from F.) (B) Start with the first approximation $\tilde{u}^h = \bar{I}_H^h u^H$, and improve it by N multigrid cycles. The solution interpolation $\bar{I}_H^h$ has usually a higher order than the correction interpolation I_H^h mentioned above.

For almost any discretized stationary PDE problem, a 1-FMG algorithm, employing cycles with $\nu_1 + \nu_2 = 2$ or 3 and $\gamma = 1$ or 2, is enough for solving (3.1) *to the level of truncation errors* (i.e., to the point where the approximate solution $\tilde{u}^h$ satisfies $\|\tilde{u}^h - u^h\| \leq \|u^h - u\|$, in any desired norm)—provided proper relaxation and interpolation procedures are used (see §5). Only when L^h has a high approximation order p, larger N-FMG may be required, with N growing linearly in p.

This means that the solution is obtained in just few L^h-work-units, where an L^h-work-unit is the number of computer operations involved in just *expressing* L^h at all grid-points. The only solvers with an almost comparable (but on large grids still inferior) speed are the direct solvers based on the Fast Fourier Transform (FFT), but they are essentially limited to equations with constant coefficients on rectangular domains and constant boundary operators. The FMG solver, by contrast, attains the same efficiency for general nonlinear, not necessarily elliptic, problems (see §§6, 7), for any boundary shape and boundary conditions, for compound problems (§9), for eigenproblems, and for problems including free surfaces, shocks, reentrant corners, discontinuous coefficients, and other singularities.

Moreover, the multigrid solvers can fully exploit very high degrees of parallel and/or vector processing. In case L^h is the standard 5-point approximation to the Laplacian, for example, (3.1) has been solved on the CDC CYBER 205 at the rate of 5 million equations per second [3]. Also, for little extra computer work these solvers can incorporate local grid adaptation (§7) or provide a sequence of extra solutions to a sequence of similar problems (§9).

5. Performance prediction, optimization, and rigorous analysis. The multigrid algorithms have many parameters, including their relaxation schemes, orders of interpolations, their treatment of boundaries and of the interior equations near boundaries, etc. To obtain their best performance, and to debug the programs, an analytical tool is needed which can predict, for example, the precise convergence factor per cycle. Such a tool is the following *local mode analysis*.

For equations with constant coefficients on infinite uniform grids, only a few (l, say) Fourier components of the error function $\tilde{u}^h - u^h$ are coupled at a time by the processes of the two-grid cycle, and it is thus easy to calculate (usually by a small computer program) the two-grid convergence factor (the largest among the spectral radii of the corresponding $l \times l$ transfer matrices). For general equations in a general domain, the *local two-grid convergence factor* is defined as the worst (largest) two-grid convergence factor for any "freezing" of the equation at any given point (extending that equation to the infinite domain).

For a general elliptic system of equations $Lu = F$ with continuous coefficients, discretized on a uniform (or continuously changing) grid in a general domain, it has been proved [10, 8] that for small meshsizes ($h \rightarrow 0$) the local two-grid convergence factor is actually obtained globally, provided the algorithm is allowed to be modified near boundaries, by adding there local relaxation sweeps that cost negligible extra work. Numerical tests clearly show that this local relaxation is indeed sometimes necessary, e.g., near re-entrant corners and other singularities [2, §4]. The performance of *multi*grid cycles can also be precisely predicted, either by perturbations to the two-grid analysis or by more complex (e.g., three-level) Fourier analyses (coupling more components at a time).

Moreover, it can also be proved that the two-grid convergence factor, λ, can itself be anticipated by the "smoothing factor" of the relaxation process, $\bar{\mu}$, which can be calculated by a much simpler local mode analysis. Namely, $\lambda = \bar{\mu}^s$ can always be obtained, provided s, the number of fine-grid relaxation sweeps per cycle, is not large, and provided suitable intergrid transfers (high enough interpolation orders) are used. Furthermore, in case of a complicated system of q differential equations, i.e., when L is a $q \times q$ matrix of differential operators, a relaxation scheme can always be constructed for which

$$\bar{\mu} = \max(\bar{\mu}_{L_1}, \ldots, \bar{\mu}_{L_k}),$$

where $L_1 \cdots L_k$ is a factorization, into first- and second-order scalar operators, of the h-principal part (the principal part on scale h) of the determinant of L, and $\bar{\mu}_{L_i}$ is the smoothing factor obtainable for a relaxation of L_i^h (see [7, §3.7]). Thus, the entire multigrid efficiency can be anticipated from the smoothing factors obtainable for simple scalar operators, and the practical task then is to construct the intergrid transfers so that λ indeed approaches $\bar{\mu}^s$, and then to adjust the boundary processes until the convergence factor per multigrid cycle indeed approaches λ.

In case of uniformly elliptic problems, for example, the factors of det L are usually Laplacians, for which the smoothing factor $\bar{\mu} = .25$ is obtainable, using the (fully parallelizable and extremely cheap) Gauss-Seidel relaxation in red-black ordering. Hence a multigrid cycle can be constructed with convergence factors .25 per fine-grid relaxation, or about .4 per work unit (taking coarse-grid overhead into account).

For highly discontinuous equations or discretizations, the theoretical treatment is far less precise, but practical approaches were developed [1], successful enough to yield fairly general black-box solvers [15].

Many situations are analyzed by *nonlocal* theories, developed over a vast literature; see, e.g., [20] and references therein. The trouble with the nonlocal approach is that its estimates are not realistically quantitative: the convergence factor per cycle *is* indeed shown to be bounded away from 1 independently of h, but its actual size is either not specified or is so close to 1 that it is useless for practical purposes (such as prediction and selecting, optimizing, and debugging the various processes), and no one believing it would use the algorithm. In fact, it led to several practical misconceptions [7, §14].

The theory in [9] gives rigorous realistic two-grid convergence estimates for very irregular cases, in fact for general symmetric algebraic systems without any grids or any other geometrical basis. This theory is nearly optimal for the crude (geometry-less) interpolations it considers. To extend it to the prediction of the *multi*grid rates obtainable with better (geometrically based) interpolations, it should be combined with some local analysis, not yet developed.

6. Nonellipticity and slight ellipticity. For nonelliptic differential equations (or equations with small ellipticity measures on scale h, which for numerical purposes is the same), it is a mistake to try to obtain uniformly fast convergence per cycle. Much simpler and more efficient algorithms are obtained by allowing components with larger truncation errors (such as the "characteristic components") to converge slower, insisting only that the 1-FMG algorithm still solves the problem well below truncation errors. That this can be obtained is shown by modified types of local mode analysis (infinite-space FMG analysis supplemented by half-space FMG analysis; see [6]).

Indeed, the usual FMG algorithm need only be modified in case of *consistent alignment*, i.e., in case the grid is *consistently* aligned with the characteristic directions. Such alignment is necessary when accuracy is desired in the "characteristic components," i.e., components which are smoother along than across characteristic lines. For obtaining that accuracy, L^h should be non-h-elliptic, and the usual point-by-point relaxation will then smooth the error only in the characteristic directions (in which semi-h-ellipticity is necessarily still maintained). One should therefore either modify relaxation, by *simultaneously* relaxing points along characteristic lines ("line relaxation"), or use "semicoarsening," i.e., a coarser grid whose meshsize is larger only in the characteristic directions. Semicoarsening, sometimes combined with line relaxation, is especially recommended in higher-dimensional situations where the alignment is not in lines but in planes.

Expensive procedures of *alternating-direction* line or plane relaxation are not needed in natural coordinates, since only *consistent* alignment matters in solving to the level of truncation errors. Such expensive procedures *will* however very *often* be needed if anisotropic coordinate transformations, and nonuniform gridline spacings in particular, are employed, thereby artificially creating excessively strong, grid-aligned discrete couplings. It is therefore generally not recommended to use global grid (or coordinate) transformations, but instead to create local refinements and local grid curvings in the multigrid manner (see §7).

For nonelliptic or slightly elliptic problems it is also recommended to use double discretization schemes (see §8), since some natural (e.g., only central) discretizations are good for smooth components but bad for non-smooth ones.

7. FAS: Nonlinear equations, local grid adaptation, τ extrapolation, small storage. In the *Full Approximation Scheme (FAS)* the coarse-grid unknown v^H is replaced by the unknown

$$u^H \underset{\text{def}}{=} \hat{I}_h^H \tilde{u}^h + v^H,$$

where $\hat{I}_h^H$ is another fine-to-coarse interpolation (or averaging). In terms of u^H, the coarse grid equation (3.3) becomes

$$L^H u^H = F^H + \tau_h^H, \tag{7.1}$$

where $F^H = I_h^H F^h$ and $\tau_h^H = L^H \hat{I}_h^H \tilde{u}^h - I_h^H L^h \tilde{u}^h$. This equation evidently has the form of a "defect correction" (correcting L^H by L^h, their difference being measured by $\tilde{u}^h$); hence it makes full sense even in the case that L is nonlinear.

Indeed, using FAS, nonlinear equations are solved as easily and fast as linear ones. No linearization is required (except for some local linearization, in relaxation, into h-principal terms, which in almost all cases means no linearization at all). The 1-FMG algorithm has solved, well below truncation errors, various flow problems, including compressible and incompressible Navier-Stokes and Euler equations, problems with shocks, constrained minimization problems (complementarity problems, with free surfaces), and many others. "Continuation" techniques, sometimes needed for reaching the solution "attraction basin," can be incorporated for little extra calculations (see §9).

In FAS, averages of the full solution are represented on all coarser grids (hence the name of the scheme). This allows for various advanced techniques which use finer grids very sparingly. For example, the fine grid may cover only part of the domain: outside that part (7.1) will simply be used without the τ_h^H term. One can use progressively finer grids at increasingly more specialized subdomains, effectively achieving a nonuniform discretization (needed near singularities) which still uses simple uniform grids, still has the very fast multigrid solver, and yet is very flexible. Grid adaptation can in fact in this way be incorporated into the FMG algorithm: On proceeding to finer levels the algorithm also defines their extent (see [5, 2]). Moreover, each of the local refinement grids may use its own local coordinate system, thus curving itself to fit boundaries, fronts, characteristic directions, or discontinuities (all whose locations are already approximately known from the coarser levels), with the additional possibility of using anisotropic meshsizes (e.g., much finer across than along the front). Since this curving is only local, it can be accomplished by a trivial transformation, which does not add substantial complexity to the basic equations (in contrast to global transformations).

The fine-to-coarse correction τ_h^H gives a rough estimate of the local discretization error. This can be used in *grid adaptation criteria*. It can also be used to *h-extrapolate* the equations, in order to obtain a higher-order discretization for little extra work. This extrapolation is more useful than the Richardson type, since it is local (extrapolating the equation, not the solution): it can, for example, be used together with any procedure of local refinements.

In view of (7.1), the role of grid h is really only to supply the defect correction τ_h^H to grid H. For that, only a local piece of the fine grid is needed at a time. Similarly, only a piece of grid $H = 2h$ is needed at a time, to supply τ_{2h}^{4h}, etc. This gives rise to algorithms that can do with *very small computer storage* (even without using external storage).

8. Multigrid discretization techniques. The above *local refinements, local coordinates, refinement criteria, local h extrapolations*, and *small-storage techniques* were examples of using the multilevel apparatus to obtain better *discretizations*, not just fast solvers. Other examples are:

Double discretization schemes. The discrete operator L^h used in calculating the residuals (3.2), for the global process of coarse grid corrections, does not need to coincide with the one used in the local process of relaxation. The latter should have good *local* properties, such as stability (possibly obtained by adding artificial viscosity) and admittance of sharp discontinuities (through suitable "limiters"), while the first should excel in *global* attributes, such as high accuracy (obtained by omitting artificial viscosities and possibly using higher-order differencing) and conservation (through conservative differencing). Such schemes do not converge to zero residuals, of course, but can approximate the differential equations much better than either of their constituent discretizations alone, especially in cases of conflicting requirements (cf. §6).

Multiple representation schemes. The coarse-grid solution representation does not need to coincide with that on the fine grid. For example, some nearly singular smooth components (typical in slightly indefinite problems) should on some coarser grids be singled out and represented by one parameter each (see [**14**]). Or, more importantly, highly oscillatory components showing small normalized residuals (typical in standing wave problems, as in acoustics, electromagnetism, Schrödinger equations, etc.) should be represented on coarser grids by their slowly varying amplitudes. The coarser the grid the more such "rays" should be separately represented. Grids fine enough to resolve the natural wavelength can be used only locally, near boundary singularities, where ray representations break down. This hybrid of wave equations and geometric optics can treat problems which neither of them can alone, in addition to supplying a fast solver for highly indefinite equations.

Global conditions and nonlocal boundary conditions (radiation conditions, flow exit boundaries, etc.) are easily incorporated, by transferring their residuals from fine grids and imposing them only at suitably coarser levels.

Treating large domains by placing increasingly coarser grids to cover increasingly wider regions.

Fast integrals. In case of integral equations discretized on n gridpoint, most of the $O(n^2)$ operations involved in just performing the integrations can be spared by performing them mainly on coarser grids, using suitable FAS versions [**7**, §8.6]. When their kernels are sufficiently smooth, the integral equations can this way be solved, to the level of truncation errors, in $O(n)$ operations. With the usual singular kernels, $O(n \log n)$ operations are required.

Finally, multigrid convergence factors always *detect bad discretizations*, especially when "compatible coarsening" is used (see §3). Several previously unnoticed flaws in widely accepted discretization schemes were so discovered. Furthermore, brief 1-FMG algorithms tend to *correct bad discretizations*, by being very slow in admitting ill-posed components (components showing small residuals compared with other components of comparable smoothness). For example, quasielliptic discretizations (resulting, e.g., from central differencing on nonstaggered grids of elliptic systems with first-order principal derivatives) are so solved with their highly oscillating bad components left out [**13**]. More generally, the FMG algorithm and the multilevel structure provide effective tools to deal with *ill-posed problems*, whether the ill-posedness is in the differential problem or only in its discretization: finer grids can be introduced (in the manner of §7) only where their scale does not admit ill-posed components; nonlinear controlling constraints, either global, local, or at any intermediate scale, are easily incorporated; etc.

9. Compound problems and problem sequences. A compound problem is one whose solution would normally involve solving several, or even many, systems of equations similar to each other. With multilevel techniques, the work of solving a compound problem can often be reduced to that of solving just one single system, or just a fraction more.

Take for example *continuation* (embedding) processes, in which a problem parameter is gradually changed in order to drive the approximate solutions into the attraction basin of the desired solution to some target nonlinear problem. Flow problems, for instance, are easily solved for the case of large viscosity, which can then gradually be lowered to the desired level, with the equations being solved at each step taking the previous-step solution to serve as a first approximation. This process is almost automatically performed by the FMG algorithm (§4) itself, since it starts on coarse levels, where a large artificial viscosity is introduced by the discretization, and then gradually works its way to finer grids with proportionately smaller viscosity. The process, by the way, can then be continued to still lower viscosity by using still finer levels only locally (see §7), at regions where the size of viscosity matters (i.e., where the flow is driven by viscosity), and eliminating viscosity elsewhere (e.g., by double discretization— see §8).

One 1-FMG algorithm, with no extra iterations, can even be directed to *locate limit points* (turning and bifurcation points) on solution diagrams; or to *optimize* some problem parameters, including optimization of boundary shapes, diffusion coefficients, control parameters, etc.; or to *trace* free boundaries, strong shocks, and other discontinuities; or to solve related *inverse problems* (e.g., system identification); and so on—all with accuracy below truncation errors.

In many cases, however, *repeated* applications of the FMG solver are still needed: cases of complicated bifurcation diagrams, interactive design situations, etc. Even then, the multigrid machinery generally provides for extremely cheap *re*-solving: one should only be careful to apply FMG to the *incremental* problem (calculating only the *change* from the old solution; using FAS this is easily done even in nonlinear problems) and to skip finer grids (or parts thereof) wherever they describe negligible high-frequency *changes*.

In designing a structure, for example, one often wants to *re*-solve the elasticity equations after modifying some part of the structure. The *changes* in the solution are then very smooth, except near the modified part. In *incremental* re-solving one therefore needs the fine grid h only near that part, while at other regions the coarser grid H can suffice—provided the τ_h^H correction (see (7.1)) is kept in those regions frozen at its previous (premodification) values (otherwise one ignores the high frequency components themselves, not just their changes). Similarly, at some larger distance from the modified part, grid $H = 2h$ itself can also be omitted, then grid $4h$, etc. In this way re-solving can be so inexpensive in computer time and storage as to allow on-line interactive design of complicated structures. Similar *frozen-τ techniques* can be used in continuation processes and in evolution problems.

10. Evolution problems. Some time-dependent problems may need no multileveling. These are hyperbolic schemes where all the characteristic velocities are comparable to each other, and their explicit discretization on one grid is therefore fully effective: the amount of processing is essentially equal to the amount of physical information. However, as soon as any stiffness enters, implicit discretization and multigrid techniques similar to those in §9 become desired.

Solving the sequence of implicit systems, the 1-FMG algorithm is all one needs per time step—provided it is consistently applied to the time *incremental* problem, since one needs to solve to the level of the incremental (not the cumulative) truncation errors. Moreover, in most cases, notably in parabolic problems, this work can vastly be reduced, because most of the time at most places the increment is very smooth, hence seldom requires fine-grid processing.

For example, it has been demonstrated for the heat equation

$$\partial u/\partial t = \Delta u + F$$

with steady boundary conditions and steady sources F that, given any initial conditions at $t = 0$, the solution at *any* target time T can be calculated, to the level of spatial truncation errors, in less than 10 work units, where the work unit here is the work invested in one *explicit* time step. To obtain the solution with that accuracy *throughout* the interval $0 \leq t \leq T$, the number of required work units is $O(\log(T/h^2))$.

By combining methods developed for such purely parabolic problems with the method of characteristics, it may be possible to obtain similar results for problems with *convection*, because the time increment can be described as a smooth change superposed on pure convection.

All *multigrid discretization techniques* (see §8) can be useful for time-dependent problems, too. One example: the popular Crank-Nicholson discretization, which offers superior accuracy for smooth components, has the disadvantage of badly treating high-frequency components at large time steps. This conflict is easily resolved by a double discretization scheme, which at some initial time steps, and only at the fine grids' relaxation process, replaces Crank-Nicholson by the Fully Implicit scheme. Other examples that were already used include local refinements, the τ refinement criteria, τ extrapolations, and a treatment of an ill-posed (the inverse heat) problem.

Time-periodic solutions, or more generally, solutions with the same solution growth w per time period, can inexpensively be computed, for any spatial grid h, by integrating basically on grid $2h$: once a steady growth ω^{2h} has been established on grid $2h$, a defect correction to w^{2h} can be found by integrating one period on grid h; then the calculations on grid $2h$ resume, with that defect added at each period, until a new steady growth is established. The calculations on grid $2h$ can similarly be done by integrating basically on grid $4h$, and so on. Each grid integration may of course also use the above frozen τ techniques.

11. Geometrically based problems. Integral equations. AMG. Most large systems, even those not derived from discretized continuous problems, still have a geometric basis; that is, each unknown has a location in some low- (usually at most 4-) dimensional underlying space—indeed, the unknowns are often still arranged in lattices—and the equations reflect this geometry, e.g., by more strongly coupling closer unknowns. Examples abound (see §1). Excluding for the moment probabilistic aspects (see §14), these systems can usually be cast as minimization problems: the solution vector u should minimize some functional $E(u)$, called "energy." This naturally leads to various Gauss-Seidel-type relaxation schemes, in which E is decreased as far as possible by changing one unknown (or one block of unknowns) at a time. (Kaczmarz relaxation in §2 can be viewed as Gauss-Seidel for $\bar{u}$, where $u = A^T \bar{u}$ and $E(u) = \frac{1}{2} u^T u - \bar{u}^T b$.)

Excluding now the case of discrete or partly discrete unknowns (see §13), in all such geometrically based systems the slow components (see §2) are either "smoothly representable" or ill-posed. A general smooth representation of components is for example by short sums of terms such as $a(x)\varphi(x)$, where $a(x)$ is smooth (at least in some directions) while $\varphi(x)$ may be highly nonsmooth but is fixed and known (or easily computable). A multilevel solver can then be constructed in which $a(x)$ is interpolated from coarser levels. The coarser level equations may be derived either variationally (i.e., from the requirement that $E(u)$ is lowered as far as possible by the interpolated $a(x)$), or by simulating direct differencing approximations.

A multigrid solver of the latter kind has been constructed for a simple case of lattice Dirac equations in a gauge field. In QED and QCD (quantum electrodynamics and chromodynamics) simulations, these types of equations should be solved at each Monte-Carlo iteration, consuming enormous computer resources (see, e.g., [18]). This solver, which employs itself also for updating $\varphi(x)$, exhibits the usual multigrid speed, and requires only a short cycle, costing far less than the rest of the calculations, per Monte-Carlo iteration. (See also §12.)

In many problems, including first-kind integral equations in fields like image reconstruction, tomography, and crystallography, there exist slow components which are not smoothly representable. Since they give large errors for small residuals without being smooth in any sense, they are by definition ill posed. Such error components are introduced only very slowly by the multigrid solvers. Hence they are harmful only in as far as their absence causes the solution to "look bad." Specifying what "looking bad" is can be done by augmenting $E(x)$ and/or by imposing nonlinear constraints. Such constraints, on any scale, can be incorporated in the multilevel solver (see [11]), even when they are discrete (see §13).

Integral equations of the *second* kind are easily solved by the usual multigrid algorithm, possibly using its structure also for fast integrations (see §8).

Multilevel solvers can be constructed even when the geometric basis is not explicit. In such *"algebraic multigrid"* (*AMG*) solvers the coarse-level variables are typically selected by the requirement that each fine-level variable is "strongly connected," by the fine-level equations, to at least some coarse-level variables. The coarse-to-fine and fine-to-coarse transfers may also be purely based on the algebraic equations, although geometrical information may be used too (see [9, 29]). AMG solvers are good as black boxes, even for discretized PDEs, since they require no special attention to boundaries, anisotropies and strong discontinuities, and no well-organized grids (allowing, e.g., general-partition finite elements).

12. Calculating determinants. At each Monte-Carlo iteration in QED and QCD simulations, what is really required is not to *solve* the lattice Dirac equations (see §11), but actually (if possible) to calculate $\delta \log \det Q$, where Q is the matrix of that system and δ denotes change per iteration. Since the steps are small, $\delta \log \det Q \approx$ trace of $Q^{-1}\delta Q$, for which calculations one needs to know $(Q^{-1})_{ij}$ for all pairs of neighboring (on the lattice) i and j. Now, it can be shown that by storing and updating similar information for coarse-grid approximations to Q (for which purpose one also needs to store and update the function $\varphi(x)$ mentioned in §11), all updates can immediately be done. The implied coarse-level work, including the coarsening of Q, is just a small overhead.

This approach leads to a general fast method for calculating determinants of lattice equations.

13. Discrete-state minimization: Multilevel annealing. In statistical physics, combinatorial optimization (e.g., traveling salesman, or integrated circuits design), pattern recognition, econometrics, and many other fields, the unknowns u_i, or part of them, may only assume discrete states. A typical example is Ising spins, where $u_i = \pm 1$. To minimize $E(u)$ in such problems is far more intricate than in continuous-state problems, since the relaxation process is not only slow, but is very likely to get trapped in a "local minimum"; i.e., in a configuration u which is not the true minimum but for which no allowable change of any one u_i, or even a small block of them, can lower E.

"*Simulated annealing*" is a general technique for trying to escape such local minima by assigning at each step a certain probability for the energy to grow. This is done by simulating thermal systems: to each configuration u the "Boltzmann probability"

$$P(u) = e^{-\beta E(u)}/Z(\beta) \qquad (13.1)$$

is assigned (physically $1/\beta$ is proportional to the absolute temperature and $Z(\beta)$ is a normalization factor), and the above strict-minimization relaxation sweeps are replaced by "Monte-Carlo iterations," in which each u_i change is governed by (13.1). Gradually and carefully β is increased (the system is "cooled") so that the Monte-Carlo process tends back to strict minimization. (See [23].)

In many cases, unfortunately, the global minimum is likely to be reached only if β is increased impractically slowly, requiring exponentially growing computer times, or else the process will be trapped in some local minimum with a large "attraction basin" (usually containing smaller-scale subbasins from which the process does escape). This difficulty is removed by *multilevel annealing*, based on the following principles:

(i) A hierarchy of changes is selected. In two-dimensional Ising spin lattices, for example, a change on level l is defined as the simultaneous flipping (sign reversal) of all the spins in a $2^l \times 2^l$ block. (ii) Each coarse-level change is decided only *after* recursively calculating its effects (i.e., minimizing around it) at all finer levels, starting from the finest. (iii) At each level a specific β, just large enough to escape local minima on that scale, is first employed; then, still at that level, strict minimization follows. (iv) A procedure (LCC) for keeping track of the so-far minimal configuration is added at each level. (See [12].)

These principles were applied to difficult two-dimensional lattice problems with N Ising spins. The global minimum has always been reached in $O(N^{3/2})$ to $O(N^2)$ computer operations. Similar algorithms are being developed for the traveling salesman problem. (The "statistical" TSP with N cities is solved in $O(N)$ operations.)

The above principles should also apply in many problems where the discrete-state nature is less obvious. Take, for example, XY spins or Heisenberg spins, where each u_i is a 2- or 3-dimensional vector *of length* 1. Although each u_i can change continuously, some large-scale topological features of the field of spins (such as the existence of closed curves along which the spins gradually rotate a full circle) can only change discretely. Similar situations arise in crystallography. Another example: in image reconstruction, each unknown u_i, representing the grey level in the ith pixel, can be considered continuous, but nonlinear constraints that should be added to the problem (cf. §11) may well include discrete elements, such as the appearance of an "edge." In each of these cases a certain combination of the multilevel annealing with classical multigrid should be used. More generally, coarse-level annealing should apply in any minimization problem with large-scale local minima, and *multilevel annealing is required whenever a hierarchy of attraction basins is involved.*

14. Statistical problems. Multilevel Monte-Carlo. The aim in statistical physics is to calculate various average properties of configurations governed by the probability distribution (13.1). This is usually done by measuring those averages over a sequence of "Monte-Carlo iterations," in which each u_i in its turn is randomly changed in a way that obeys (13.1) (using, e.g., Metropolis rule [**27**]). Unfortunately, in such processes statistical equilibrium is usually reached very slowly, and, more severely, even when it has been reached, some averages are still very slow to converge, especially those long-range correlations the physicist needs most.

These two troubles may be cured by multilevel Monte-Carlo techniques, in which coarse-level changes (changing the solution in preassigned blocks in preassigned patterns) are added and averaged over. In problems and at levels where the physical states may be considered continuous, this can be done quite straightforwardly and very efficiently: once per several coarse-level sweeps, the probabilities associated with coarse changes are defect-corrected by fine-level Monte-Carlo iterations (see [**12**]). In case of discrete states, principles similar to those in §13 should apply. Namely, the exact *pattern* of each coarse-level change, as well as the probabilities associated with it, are recursively decided by finer-level Monte-Carlo passes around it.

15. Linear programming (LP). A multilevel approach, called iterative aggregation, has been developed for LP problems (see [**16**, **31**]), especially for situations in which the planned system is naturally divided into a hierarchy of sectors and subsectors. This considerably speeds up the calculations, and also provides the manager with a very useful hierarchical view of the system.

For very large systems, to obtain the typical speed of *multigrid* solvers, more refined aggregations are needed. This can easily be done, for example, in problems with a geometrical basis (cf. §11), such as the LP transportation problem (see, e.g., [**19**]). Recent tests were made with a method that lumps together two (or so) neighboring destinations into a "block destination," two neighboring blocks into a super-block, etc. Shipping costs to a block are determined from the current intra-block marginal costs. It turns out that a 1-FMG-like algorithm gets very close (practically obtains) the solution. The required work is even smaller since many of the blocks that are supplied by one origin need no fine-level processing. Several orders of magnitude savings, compared to simplex solutions, were indicated.

16. Historical note. Various multilevel solution processes have independently occurred to many investigators (see partial list in [**5**]). The earliest we know is Southwell's acceleration of relaxation by "group relaxation" [**30**], a two-level algorithm. The first to describe a recursive procedure with more than two levels is Fedorenko [**17**]. Similar approaches were early introduced to economic planning (see §15). All these early works lacked full understanding of the real efficiency that can be obtained by multileveling, and how to obtain it, since they did not regard the fine-grid processes as strictly local, hence thought in terms of too-crude aggregations. Fedorenko's estimates of the work involved in solving simple Poisson equations are off by a factor 10^4, for example. Fully efficient multigrid algorithms, based on local analysis, were first developed at the Weizmann Institute in 1970–1972 (see [**4**]), leading then to most of the developments reported in the present article.

REFERENCES

1. R. E. Alcouffe, A. Brandt, J. E. Dendy, Jr., and J. W. Painter, *The multi-grid methods for the diffusion equation with strongly discontinuous coefficients*, SIAM J. Sci. Statist. Comput. **2** (1981), 430–454.

2. D. Bai and A. Brandt, *Local mesh refinement multilevel techniques*, SIAM J. Sci. Statist. Comput. (to appear).

3. D. Barkai and A. Brandt, *Vectorized multigrid Poisson solver*, Appl. Math. Comput. **13** (1983), 215–227.

4. A. Brandt, *Multi-level adaptive technique (MLAT) for fast numerical solutions to boundary value problems*, Proc. 3rd Internat. Conf. Numerical Methods in Fluid Mechanics (Paris, 1972), Lecture Notes in Physics, Vol. 18, Springer-Verlag, Berlin, pp. 82–89.

5. ___, *Multi-level adaptive solutions to boundary value problems*, Math. Comp. **31** (1977), 333–390.

6. ___, *Multi-grid solvers for non-elliptic and singular-perturbation steady-state problems*, Weizmann Institute of Science, December 1981.

7. ___, *Multigrid techniques: 1984 guide, with applications to fluid dynamics*, Available as GMD Studien Nr. 85, GMD-AIW, Postfach 1240, D-5205, St. Augustin 1, W. Germany, 1984. Main part appeared in [**21**].

8. ___, *Multigrid guide*. (An extension of [**7**], planned for 1987.)

9. ___, *Algebraic multigrid theory: The symmetric case*, Proc. 2nd Copper Mountain Multigrid Conf., S. McCormick, editor, Appl. Math. Comput. **19** (1986).

10. ___, *Rigorous local mode analysis*, Lecture at the 2nd European Conf. on Multigrid Methods, Cologne, October 1985.

11. A. Brandt and C. W. Cryer, *Multi-grid algorithms for the solution of linear complementarity problems arising from free boundary problems*, SIAM J. Sci. Statist. Comput. **4** (1983), 655–684.

12. A. Brandt, D. Ron, and D. J. Amit, *Multi-level approaches to discrete-state and stochastic problems*, Proc. 2nd European Conf. on Multigrid Methods, W. Hackbusch and U. Trottenberg, editors, Springer-Verlag (to appear).

13. A. Brandt and S. Ta'asan, *Multigrid solutions to quasi-elliptic schemes*, Progress and Supercomputing in Computational Fluid Dynamics, E. M. Murman and S. S. Abarbanel, editors, Birkhäuser, Boston, 1985, pp. 235–255.

14. _____, *Multigrid method for nearly singular and slightly indefinite problems*, Proc. 2nd European Conf. on Multigrid Methods, W. Hackbusch and U. Trottenberg, editors, Springer-Verlag (to appear).

15. J. E. Dendy, *Black box multigrid*, J. Comp. Phys. **48** (1982), 366–386.

16. L. M. Dudkin and E. B. Yershov, *Interindustries input-output models and the material balances of separate products*, Planned Economy **5** (1965), 54–63.

17. R. P. Fedorenko, *On the speed of convergence of an iteration process*, Zh. Vychisl. Mat. i Mat. Fiz. **4** (1964), 559–564.

18. F. Fucito and S. Solomon, *Concurrent pseudo-fermions algorithm*, Comput. Phys. Comm. **36** (1985), 141.

19. S. T. Gass, *Linear programming*, 3rd ed., McGraw-Hill, New York, 1969.

20. W. Hackbusch, *Multigrid methods and applications*, Springer-Verlag, 1985.

21. W. Hackbusch and U. Trottenberg (Editors), *Multigrid methods*, Lecture Notes in Math., Vol. 960, Springer-Verlag, 1982.

22. _____, *Proceedings of the Second European Conference on Multigrid Methods*, Springer-Verlag, 1986 (to appear).

23. S. Kirkpatrick, C. D. Gelatt, Jr., and M. P. Vecchi, *Optimization by simulated annealing*, Science **220** (1983), 671.

24. S. McCormick (Editor), *SIAM Frontiers in Applied Math.*, Vol. V (a book on multigrid methods) (to appear).

25. S. McCormick and U. Trottenberg (Editors), *Multigrid methods*, Appl. Math. Comput. **13** (1983), 213–474 (special issue).

26. S. McCormick (Editor), *Proceeding of the 2nd Copper Mountain Multigrid Conference*, Appl. Math. Comput. **19** (1986), 1–372 (special issue).

27. N. Metropolis, A. W. Rosenbluth, M. N. Rosenbluth, A. H. Teller, and E. Teller, *Equation of state calculations by fast computing machines*, J. Chem. Phys. **21** (1953), 1087.

28. D. J. Paddon and H. Holstein (Editors), *Multigrid methods for integral and differential equations*, Clarendon Press, Oxford, 1985.

29. J. Ruge and K. Stueben, *Algebraic multigrid* (*AMG*), SIAM Frontiers in Applied Math., Vol. V (to appear).

30. R. V. Southwell, *Stress calculation in frameworks by the method of systematic relaxation of constraints*. I, II, Proc. Roy. Soc. London Ser. A **151** (1935), 56–95.

31. I. Y. Vakhutinsky, L. M. Dudkin, and A. A. Ryvkin, *Iterative aggregation—a new approach to the solution of large-scale problems*, Econometrica **47** (1979), 821–841.

THE WEIZMANN INSTITUTE OF SCIENCE, REHOVOT 76100, ISRAEL

New Applications of
Mixed Finite Element Methods

F. BREZZI

1. Introduction. The aim of this paper is to present an introductory survey on mixed finite element methods. We shall deal first with the so-called mixed formulation of some problems arising in the applications. Then we shall analyze the difficulties connected with the choice of appropriate finite element discretizations for a mixed formulation. Finally we shall discuss some special techniques that are often helpful for solving the discretized problem.

By default, the notation will follow Ciarlet [**19**].

2. Mixed formulations. A precise and satisfactory definition of "mixed method" (or of "mixed formulation") does not exist. The term started in the engineering literature (Herrmann [**34**, **35**]; Hellan [**33**]) in connection with the elasticity theory, to denote methods, based on the Hellinger Reissner principle, in which both displacements and stresses were approximated simultaneously. Even among mathematicians, in the papers that can now be considered as pioneering, such as Glowinski [**30**], Babuška [**6**], Crouzeix-Raviart [**20**], Johnson [**37**], the term "mixed" was used only by Johnson, who dealt with plate bending problems. The term is now used in a much wider sense, mostly rather vague. Here we are going to live with such vagueness, and we shall not try a new unsatisfactory definition. Instead, we are going to present a few examples: the case of linear elliptic problems, the case of the Stokes equations for incompressible fluids, and the case of linear elasticity problems. We will deal with the first case in more detail because it is formally much simpler, while only a few essential points will be stressed for the other two cases.

Example 1. *Linear elliptic operators.* The use of mixed formulations for linear elliptic operators is rather recent and, as we shall see, is recommended only in some special case. However, its presentation is very simple and this makes it an ideal first example. Consider the model problem

$$\operatorname{div}(A(\mathbf{x}) \operatorname{\mathbf{grad}} u) = f \quad \text{in } D \subset \mathbf{R}^d, \tag{2.1}$$

$$(A(\mathbf{x}) \operatorname{\mathbf{grad}} u) \cdot \mathbf{n} = g_1 \quad \text{on } \Gamma_{\text{Neu}}, \tag{2.2}$$

$$u = g_0 \quad \text{on } \Gamma_{\text{Dir}}, \tag{2.3}$$

where (i) $\Gamma_{\text{Dir}} \cup \Gamma_{\text{Neu}} = \Gamma = \partial D$ is a splitting of ∂D; (ii) $A(\mathbf{x})$ is a smooth function on $\overline{D}$, with $A(\mathbf{x}) \geq \alpha > 0$ for every $\mathbf{x}$ in D; (iii) $\mathbf{n}$ is the unit outward normal to ∂D; and (iv) f, g_1, g_0 are given smooth functions in D and on Γ_{Neu}, Γ_{Dir}, respectively. Introducing the manifold

$$H^1_{,g}(D) = \{v \mid v \in H^1(D); v = g \text{ on } \Gamma_{\text{Dir}}\}$$

for $g = g_0$ or $g = 0$, we can write the variational formulation of (2.1)–(2.3) as follows: Find $u \in H^1_{,g}$ such that

$$\int_D A(\mathbf{x}) \operatorname{\mathbf{grad}} u \cdot \operatorname{\mathbf{grad}} v \, dx = -\int_D fv \, dx + \int_{\Gamma_{\text{Neu}}} g_1 v \, d\Gamma \quad \forall v \in H^1_{,0}(D).$$

In order to reach the mixed formulation of (2.1)–(2.3), we introduce the variable

$$\mathbf{p} = A(\mathbf{x}) \operatorname{\mathbf{grad}} u \quad \text{in } D, \tag{2.4}$$

so that (2.1) and (2.2) become, respectively,

$$\operatorname{div} \mathbf{p} = f \quad \text{in } D, \tag{2.5}$$

$$\mathbf{p} \cdot \mathbf{n} = g_1 \quad \text{on } \Gamma_{\text{Neu}}. \tag{2.6}$$

The formulation (2.3)–(2.6) is often called the mixed formulation of (2.1)–(2.3). There are now two possible reasonable variational formulations for (2.3)–(2.6). Let us look at both of them. The first one is: Find $u \in H^1_{,g_0}(D)$ and $\mathbf{p} \in (L^2(D))^d$ such that

$$\int_D (A(\mathbf{x}))^{-1}\mathbf{p} \cdot \mathbf{q} \, dx - \int_D \mathbf{q} \cdot \operatorname{\mathbf{grad}} u \, dx = 0 \quad \forall \mathbf{q} \in (L^2(D))^d, \tag{2.7}$$

$$-\int_D \mathbf{p} \cdot \operatorname{\mathbf{grad}} v \, dx = \int_D fv \, dx - \int_{\Gamma_{\text{Neu}}} g_1 v \, d\Gamma \quad \forall v \in H^1_{,0}(D). \tag{2.8}$$

In order to introduce the second variational formulation of (2.3)–(2.6), we define, for $g = g_1$ or $g = 0$, the manifold

$$H_{,g}(\operatorname{div}; D) = \{\mathbf{q} \mid \mathbf{q} \in (L^2(D))^d; \operatorname{div} \mathbf{q} \in L^2(D); \ \mathbf{q} \cdot \mathbf{n} = g \text{ on } \Gamma_{\text{Neu}}\}.$$

The second variational formulation of (2.3)–(2.6) is now: Find $u \in L^2(D)$ and $\mathbf{p} \in H_{,g_1}(\operatorname{div}; D)$ such that

$$\int_D (A(\mathbf{x}))^{-1}\mathbf{p} \cdot \mathbf{q} \, dx + \int_D u \operatorname{div} \mathbf{q} \, dx = \int_{\Gamma_{\text{Dir}}} g_0 \mathbf{q} \cdot \mathbf{n} \, d\Gamma \quad \forall \mathbf{q} \in H_{,0}(\operatorname{div}; D), \tag{2.9}$$

$$\int_D v \operatorname{div} \mathbf{p} \, dx = \int_D fv \, dx \quad \forall v \in L^2(D). \tag{2.10}$$

The difference between (2.7)–(2.8) and (2.9)–(2.10) is clearly a simple integration by parts (or, if you prefer, a Green's formula). However, it must be pointed out that the *regularity* a priori required for u and $\mathbf{p}$ is somehow interchanged. This implies that in discretizing (2.7)–(2.8) one has to use a continuous "u" and can use a discontinuous "$\mathbf{p}$," while in discretizing (2.9)–(2.10) one can use a discontinuous "u" but has to use a "$\mathbf{p}$" with divergence in $L^2(D)$ (and hence

$\mathbf{p} \cdot \mathbf{n}$ continuous at the interfaces). Note also the inversion in the way of dealing with the boundary conditions.

It is questionable whether (2.7)–(2.8) should be called a mixed formulation for problem (2.1)–(2.3). On the other hand, everybody seems to agree in calling the formulation (2.9)–(2.10) "mixed." In general, the original formulation (2.1)–(2.3) is preferred. It is more simple, it uses just one variable, and many extremely efficient methods are known for its approximation. However, in some applications the "auxiliary" unknown $\mathbf{p}$ defined in (2.4) is actually the more relevant physical variable and/or is the only information that has to be transferred into other equations that are coupled with (2.1)–(2.3). In such cases, the use of a mixed formulation might be preferred, as long as it provides (as it often does) a better accuracy for $\mathbf{p}$. In general, the formulation (2.9)–(2.10) is then used, since it deals with a smoother vector field $\mathbf{p}$. It is often said that the crucial feature in the mixed approach is that it averages $(A(\mathbf{x}))^{-1}$ instead of $A(\mathbf{x})$. This is surely a better thing to do, at least in one dimension, being connected with the homogenization theory. See, for instance, Babuška-Osborn [7]. However, dramatic improvements have been obtained by using (2.9)–(2.10) with a constant $A(\mathbf{x})$. See, for instance, Marini-Savini [41]. The *true* reason (if any) for the better behavior of the mixed formulations over the classical ones is still to be understood. Practical experiences suggest the use of a mixed formulation for "bad behaved" problems in which the variable $\mathbf{p}(\mathbf{x})$ is expected to be "smoother" than the variable $u(\mathbf{x})$. But clearly this is not the whole story.

Example 2. Incompressible fluids. The Stokes equations for incompressible fluids are of the type

$$- \Delta \mathbf{u} + \operatorname{grad} p = \mathbf{f} \quad \text{in } D \subset \mathbf{R}^d, \tag{2.11}$$

$$\operatorname{div} \mathbf{u} = 0 \quad \text{in } D. \tag{2.12}$$

Various kinds of boundary conditions can be used in connection with (2.11), (2.12). For the sake of simplicity we shall only consider the (physically uninteresting) Dirichlet boundary conditions

$$\mathbf{u} = 0 \quad \text{on } \Gamma = \partial D. \tag{2.13}$$

The natural variational formulation of (2.11)–(2.13) is: Find $\mathbf{u} \in (H_0^1(D))^d$ and $p \in L^2(D)$ such that

$$\int_D \operatorname{grad} \mathbf{u} : \operatorname{grad} \mathbf{v}\, dx - \int_D p \operatorname{div} \mathbf{v}\, dx = \int_D \mathbf{f} \cdot \mathbf{v}\, dx \quad \forall \mathbf{v} \in (H_0^1(D))^d, \tag{2.14}$$

$$\int_D q \operatorname{div} \mathbf{u}\, dx = 0 \quad \forall q \in L^2(D). \tag{2.15}$$

The formulation (2.14)–(2.15) has been used for years long before the term "mixed method" came into use. However, it is recognized now that (2.14)–(2.15) behaves like a mixed formulation as far as the difficulties in finding good approximations are concerned. We shall also see that (2.14)–(2.15) easily falls into the same abstract framework that is commonly used for mixed methods. Hence we are somehow allowed to consider (2.14)–(2.15) as a mixed formulation.

Example 3. Linear elasticity problems. For a vector-valued function $\mathbf{v}(\mathbf{x})$ we define $\boldsymbol{\varepsilon}(\mathbf{v})$ by

$$\varepsilon_{ij} = \frac{1}{2}\left(\frac{\partial v_i}{\partial x_j} + \frac{\partial v_j}{\partial x_i}\right) \qquad (i,j = 1,\ldots,d). \tag{2.16}$$

The linear elasticity equations are now

$$\boldsymbol{\sigma} = E\!:\!\boldsymbol{\varepsilon}(\mathbf{u}) \quad \left(\text{i.e., } \sigma_{ij} = \sum_{r=1}^{d}\sum_{s=1}^{d} E_{ijrs}\varepsilon_{rs}(\mathbf{u})\right), \tag{2.17}$$

$$\mathbf{div}\,\boldsymbol{\sigma} = \mathbf{f} \quad \text{in } D. \tag{2.18}$$

Substituting (2.16), (2.17) into (2.18) gives a second order linear elliptic system in the unknowns $\mathbf{u}$. Clearly E is the elasticity tensor and is assumed here to have constant coefficients (and nice "ellipticity" properties ...). Its inverse (compliance tensor) will be denoted by C. Hence

$$\boldsymbol{\tau} = E\!:\!\boldsymbol{\varepsilon}(\mathbf{v}) \Leftrightarrow \boldsymbol{\varepsilon}(\mathbf{v}) = C\!:\!\boldsymbol{\tau}. \tag{2.19}$$

We are again going to assume the simplified boundary conditions

$$\mathbf{u} = 0 \quad \text{on } \partial D. \tag{2.20}$$

This, of course, is strongly unrealistic: usually one has $\mathbf{u} = \tilde{\mathbf{u}}$ given on Γ_{Dir} and $\boldsymbol{\sigma}\cdot\mathbf{n} = \mathbf{t}$ given on Γ_{Neu}. However, the proper way of dealing with realistic boundary conditions coincides with the one used in Example 1; we chose then to give more details there (with simpler notations) and to simplify here.

One can notice that the splitting of the problem in more than one unknown is extremely natural here and has solid physical reasons. This is probably why the first mixed formulations were used in elasticity theory. We shall present here only one mixed formulation, which is similar to the formulation (2.9)–(2.10) for a single elliptic equation. We set

$$H(\mathbf{div}; D) = \{\boldsymbol{\tau} \mid \boldsymbol{\tau} \in (L^2(D))^{d^2};\ \tau_{ij} = \tau_{ji}\ \forall i,j;\ \mathbf{div}\,\boldsymbol{\tau} \in (L^2(D))^d\}$$

and we consider this problem: Find $\mathbf{u} \in (L^2(D))^d$ and $\boldsymbol{\sigma} \in H(\mathbf{div}; D)$ such that

$$\int_D (C\!:\!\boldsymbol{\sigma})\!:\!\boldsymbol{\tau}\,dx + \int_D \mathbf{u}\cdot\mathbf{div}\,\boldsymbol{\tau}\,dx = 0 \quad \forall \boldsymbol{\tau} \in H(\mathbf{div}; D), \tag{2.21}$$

$$\int_D \mathbf{v}\cdot\mathbf{div}\,\boldsymbol{\sigma}\,dx = \int_D \mathbf{f}\cdot\mathbf{v}\,dx \quad \forall \mathbf{v} \in (L^2(D))^d. \tag{2.22}$$

One can see that (2.21)–(2.22) practically coincide with the variational formulation of the Hellinger-Reissner principle. The use of this principle in the framework of finite elements can be traced back to the pioneering work of Herrmann [**34, 35**] and Hellan [**33**]. The interest in using the stress field $\boldsymbol{\sigma}$ as an independent variable is questionable in as simple a case as the present one, but it is clear in more general and more complicated problems involving nonlinearities, plasticity, and so on.

We are now going to state an abstract existence theorem that is a simplified version of a more general result, proved in [**12**].

THEOREM 1. *Let Ξ and Ψ be real Hilbert spaces, $a(\xi_1, \xi_2)$ a bilinear form on $\Xi \times \Xi$, and $b(\xi, \psi)$ a bilinear form an $\Xi \times \Psi$. Set*

$$K = \{\xi \mid \xi \in \Xi, b(\xi, \psi) = 0 \; \forall \psi \in \Psi\},$$

and assume that

$$\exists \alpha > 0 \quad s.t. \quad a(\xi, \xi) \geq \alpha \|\xi\|_\Xi^2 \quad \forall \xi \in K, \tag{2.23}$$

$$\exists \beta > 0 \quad s.t. \quad \sup_{\xi \in \Xi - \{0\}} \frac{b(\xi, \psi)}{\|\xi\|_\Xi} \geq \beta \|\psi\|_\Psi \quad \forall \psi \in \Psi. \tag{2.24}$$

Then for every $l_1 \in \Xi'$ and $l_2 \in \Psi'$ there exists a unique solution $(\overline{\xi}, \overline{\psi})$ of the problem

$$a(\overline{\xi}, \xi) + b(\xi, \overline{\psi}) = \langle l_1, \xi \rangle \quad \forall \xi \in \Xi, \tag{2.25}$$

$$b(\overline{\xi}, \psi) = \langle l_2, \psi \rangle \quad \forall \psi \in \Psi. \qquad \square \tag{2.26}$$

REMARK. Actually a stronger result is proved in [**12**]; namely, if problem (2.25), (2.26) has a unique solution for every $l_1 \in \Xi'$ and $l_2 \in \Psi'$, then (2.24) holds and the bilinear form $a(\xi_1, \xi_2)$, restricted to K, is nonsingular (in the sense that it induces an isomorphism from K onto K'). Clearly if one assumes that $a(\xi_1, \xi_2)$ is symmetric and positive semidefinite, then (2.23) and (2.24) are *necessary and sufficient* for the existence and uniqueness of the solution of (2.25)–(2.26).

REMARK. It is clear that if $a(\xi_1, \xi_2)$ is symmetric, the solution $(\overline{\xi}, \overline{\psi})$ of (2.25)–(2.26) minimizes the functional

$$J(\xi) = \tfrac{1}{2} a(\xi, \xi) - \langle l_1, \xi \rangle \tag{2.27}$$

on the subspace of Ξ,

$$K(l_2) = \{\xi \mid b(\xi, \psi) = \langle l_2, \psi \rangle \; \forall \psi \in \Psi\}, \tag{2.28}$$

and the formulation (2.25)–(2.26) corresponds to the introduction in (2.27)–(2.28) of the Lagrange multiplier $\overline{\psi}$.

3. Discretizing a mixed formulation. Let us deal first with the abstract framework (2.25)–(2.26). Assume that we are given two sequences $\{\Xi_h\}_{h>0}$ and $\{\Psi_h\}_{h>0}$ of subspaces of Ξ and Ψ, respectively. We set

$$K_h = \{\xi_h \mid \xi_h \in \Xi_h, \; b(\xi_h, \psi_h) = 0 \; \forall \psi_h \in \Psi_h\}. \tag{3.1}$$

We have the following approximation theorem [**12**].

THEOREM 2. *Assume that*

$$\exists \alpha_h > 0 \quad s.t. \quad a(\xi, \xi) \geq \alpha_h \|\xi\|_\Xi^2 \quad \forall \xi \in K_h, \tag{3.2}$$

$$\exists \beta_h > 0 \quad s.t. \quad \sup_{\xi \in \Xi_h - \{0\}} \frac{b(\xi, \psi)}{\|\xi\|_\Xi} \geq \beta_h \|\psi\|_\Psi \quad \forall \psi \in \Psi_h. \tag{3.3}$$

Then for every $l_1 \in \Xi'$ and $l_2 \in \Psi'$, and for every $h > 0$, the discrete problem

$$a(\overline{\xi}_h, \xi) + b(\xi, \overline{\psi}_h) = \langle l_1, \xi \rangle \quad \forall \xi \in \Xi_h, \tag{3.4}$$

$$b(\overline{\xi}_h, \psi) = \langle l_2, \psi \rangle \quad \forall \psi \in \Psi_h \tag{3.5}$$

has a unique solution. Moreover, there exists a constant $\gamma_h(\alpha_h, \beta_h) > 0$ such that

$$\|\overline{\xi} - \overline{\xi}_h\|_\Xi + \|\overline{\psi} - \overline{\psi}_h\|_\Psi \le \gamma_h \left(\inf_{\xi_h \in \Xi_h} \|\overline{\xi} - \xi_h\|_\Xi + \inf_{\psi_h \in \psi} \|\overline{\psi} - \psi_h\|_\Psi \right). \quad \square \quad (3.6)$$

The dependence of γ_h on α_h and β_h can be easily traced (see [12]). Clearly if (3.2) and (3.3) hold with constants $\overline{\alpha}$, $\overline{\beta}$ independent of h, then (3.6) holds with a constant $\overline{\gamma}$ independent of h. More general versions of Theorem 2 (and also of Theorem 1) can be found, for instance, in Falk-Osborn [23] or in Bernardi-Canuto-Maday [9].

We are now going to see the implications of Theorem 2 in the examples of the previous section.

Example 1h. *Discretizations of the mixed formulations for linear elliptic operators.* Many examples of successful discretizations of (2.9)–(2.10) are known. The first ones were introduced by Raviart and Thomas in [45] and then re-elaborated and extended to more general cases by Nedelec [42]. Other families of possible discretizations were introduced years later by Brezzi, Douglas, and Marini [16] and then re-elaborated and extended in several more recent papers (see, e.g., [14, 43, 15]). All of them share a very helpful property, the so-called "commuting diagram property," whose importance was first fully recognized in Douglas-Roberts [21]. Let us look at it in a particular case: the BDM element of degree 1 for two-dimensional problems ($D \subset \mathbf{R}^2$). Let $\mathcal{T}_h$ be a regular sequence of decompositions of D into triangles. We assume for the sake of simplicity that $\Gamma_{\text{Neu}} = \varnothing$ in (2.2) and $A(\mathbf{x}) = 1$. As a discretization of $H(\text{div}; D)$ and $L^2(D)$ respectively, we take

$$\Xi_h = \{\mathbf{q} \mid \mathbf{q} \in H(\text{div}; D); \ \mathbf{q}_{|T} \in (P_1)^2 \ \forall T \in \mathcal{T}_h\}, \quad (3.7)$$

$$\Psi_h = \{v \mid v \in L^2(D); \ v_{|T} \in P_0 \ \forall T \in \mathcal{T}_h\}. \quad (3.8)$$

Here and in the following, $P_k(S)$ (or simply P_k) will denote the set of polynomials of degree $\le k$ on the set S. We consider now the discretized problem: find $\mathbf{p}_h \in \Xi_h$ and $u_h \in \Psi_h$ such that

$$\int_D \mathbf{p}_h \cdot \mathbf{q} \, d\mathbf{x} + \int_D u_h \, \text{div} \, \mathbf{q} \, d\mathbf{x} = \int_{\partial D} g_0 \mathbf{q} \cdot \mathbf{n} \, d\Gamma \quad \forall \mathbf{q} \in \Xi_h, \quad (3.9)$$

$$\int_D v \, \text{div} \, \mathbf{p}_h \, d\mathbf{x} = \int_D f v \, d\mathbf{x} \quad \forall v \in \Psi_h. \quad (3.10)$$

We now define an operator M_h from $(H^1(D))^2$ into Ξ_h by

$$\int_e (\mathbf{q} - M_h \mathbf{q}) \cdot \mathbf{n} p_1 \, ds = 0 \quad \forall e \text{ edge in } \mathcal{T}_h, \ \forall p_1 \in P_1(e), \quad (3.11)$$

and an operator $\mathcal{P}_h$ from $L^2(D)$ into Ψ_h by

$$\int_T (v - \mathcal{P}_h v) \, d\mathbf{x} = 0 \quad \forall T \text{ triangle in } \mathcal{T}_h. \quad (3.12)$$

Let us check now that $\operatorname{div} M_h \mathbf{q} = \mathcal{P}_h \operatorname{div} \mathbf{q}$ for all $\mathbf{q} \in (H^1(D))^2$; actually, for all $v_h \in \Psi_h$ we have

$$\int_T v_h \operatorname{div} M_h \mathbf{q} \, d\mathbf{x} = \int_{\partial T} v_h (M_h \mathbf{q} \cdot \mathbf{n}) \, ds = \int_{\partial T} v_h \mathbf{q} \cdot \mathbf{n} \, ds \qquad (3.13)$$

$$= \int_T v_h \operatorname{div} \mathbf{q} \, d\mathbf{x} = \int_T v_h \mathcal{P}_h \operatorname{div} \mathbf{q} \, d\mathbf{x}.$$

It is also easy to check that the divergence operator is linear continuous and surjective from $(H^1(D))^2$ onto $L^2(D)$. This can be summarized in the following diagram:

$$\begin{array}{ccc}
(H^1(D))^2 & \stackrel{\operatorname{div}}{\rightarrow} & L^2(D) \quad \rightarrow 0 \\
M_h \downarrow & & \downarrow \mathcal{P}_h \\
\Xi_h & \stackrel{\operatorname{div}}{\rightarrow} & \Psi_h \quad \rightarrow 0
\end{array} \qquad (3.14)$$

It is easy to check that (3.14) implies, in particular, (3.2) and (3.3). But it is much more powerful than that. For instance, it implies

$$\|\mathbf{p} - \mathbf{p}_h\|_{(L^2(D))^2} \leq \gamma_1 \|\mathbf{p} - M_h \mathbf{p}\|_{(L^2(D))^2}, \qquad (3.15)$$

$$\|u - u_h\|_{L^2(D)} \leq \gamma_2 (\|\mathbf{p} - M_h \mathbf{p}\|_{(L^2(D))^2} + \|u - \mathcal{P}_h u\|_{L^2(D)}), \qquad (3.16)$$

with γ_1, γ_2 independent of h (whenever $\mathbf{p} \in (H^1(D))^2$). In particular, with the choice (3.7)–(3.8) this yields

$$\|\mathbf{p} - \mathbf{p}_h\|_{(L^2(D))^2} \leq \gamma_1 h^2 \|\mathbf{p}\|_{(H^2(D))^2}, \qquad (3.17)$$

$$\|u - u_h\|_{L^2(D)} \leq \gamma_2 h (\|u\|_{H^2(D)} + \|\mathbf{p}\|_{(H^1(D))^2}). \qquad (3.18)$$

Note that (3.17) does not follow from the abstract error estimate (3.6).

The commuting diagram property has other nice properties. For instance, it allows a simple proof of error estimates in dual norms, as in Douglas-Roberts [22] or inBrezzi-Douglas-Marini [16]. Error estimates in L^∞ norms are also available; see, for instance, Scholz [46, 47] and Gastaldi-Nochetto [27, 28].

The most popular scheme for (2.9)–(2.10), that is, the "lowest order Raviart-Thomas," can be obtained by using, instead of (3.7),

$$\Xi_h = \{\mathbf{q} \mid \mathbf{q} \in H(\operatorname{div}; D); \; \mathbf{q}_{|T} \in (P_1(T))^2 \; \forall T; \; \mathbf{q} \cdot \mathbf{n}_{|e} \in P_0(e) \; \forall e\}. \qquad (3.19)$$

Accordingly, one then uses $P_0(e)$ instead of $P_1(e)$ in (3.11). It is immediate that (3.13) still holds, and then (3.14) also holds. Clearly, only an $O(h)$ rate can now be achieved in (3.17).

Example 2h. *Discretizations of the Stokes equations.* Life is much harder when we go from (2.9)–(2.10) to (2.14)–(2.15). The only positive aspect is that now the bilinear form $a(\mathbf{u}, \mathbf{v})$ is such that (2.23) actually holds in the whole $(H_0^1(D))^2$ (our present Ξ) so that (3.2) also holds true regardless of the choice of the discretization. This might partially excuse all the Stokes-thinking people that consider (3.3) as *the* condition for mixed methods. If you try to discretize even the easy (2.9)–(2.10) with a scheme that does not satisfy (3.14) you will see that (3.2) can bite badly.

However, coming back to Stokes, it is true that the only condition to be satisfied by the discretization is (3.3), which now reads

$$\exists \beta_h > 0: \quad \sup_{v \in \Xi_h - \{0\}} \frac{\int q \operatorname{div} \mathbf{v} \, dx}{\|\mathbf{v}\|_1} \geq \beta_h \|q\|_{L^2(D)/\mathbf{R}} \quad \forall q \in \Psi_h \qquad (3.20)$$

with, if possible, β_h independent of h. A sufficient condition for it is the following so-called Fortin's trick [24]: we have to find a linear operator M_h from $(H^1(D))^2$ into Ξ_h such that

$$\|M_h \mathbf{v}\|_1 \leq c_1 \|\mathbf{v}\|_1 \quad \forall \mathbf{v} \in (H_0^1(D))^2, \qquad (3.21)$$

$$\int_D q_h \operatorname{div}(\mathbf{v} - M_h \mathbf{v}) \, dx = 0 \quad \forall \mathbf{v} \in (H_0^1(D))^2, \, \forall q_h \in \Psi_h. \qquad (3.22)$$

Let us see one example. Let $\mathcal{T}_h$ be a decomposition of D into rectangles R with sides parallel to the axes (the use of isoparametric elements is obviously also allowed, but more complicated to describe), and choose

$$\Xi_h = \{\mathbf{v} \mid \mathbf{v} \in (H_0^1(D))^2; \, \mathbf{v}_{|R} \in (Q_2(R))^2 \, \forall R \in \mathcal{T}_h\}, \qquad (3.23)$$

$$\Psi_h = \{q \mid q \in L^2(D); \, \int_D q \, dx = 0; \, q_{|R} \in P_1(R) \, \forall R \in \mathcal{T}_h\}. \qquad (3.24)$$

In (3.23), $Q_2(R)$ means the set of polynomials of degree ≤ 2 in each variable. Let us see how to construct the operator M_h at least for a smooth $\mathbf{v}$. To deal with a general $\mathbf{v}$ in $(H_0^1(D))^2$ is just technically more complicated but the philosophy is the same. In each R we set

$$M_h \mathbf{v} = \mathbf{v} \quad \text{at the vertices} \quad \text{(8 conditions)}, \qquad (3.25)$$

$$\int_e (M_h \mathbf{v} - \mathbf{v}) \, ds = 0 \quad \text{on each edge} \quad \text{(8 conditions)}, \qquad (3.26)$$

$$\int_R \operatorname{div}(M_h \mathbf{v} - \mathbf{v}) x_i \, dx = 0 \quad i = 1, 2 \quad \text{(2 conditions)}. \qquad (3.27)$$

We have a total of 18 conditions (note that the dimension of Q_2 is 9). It is easy to check that they are independent. Let us check (3.22); that is, let us check that

$$\int_R \operatorname{div}(M_h \mathbf{v} - \mathbf{v}) p_1 \, dx = 0 \quad \forall p_1 \in P_1(R). \qquad (3.28)$$

Clearly (3.27) implies that (3.28) holds for $p_1 = x_1$ and $p_1 = x_2$. We need only to check $p_1 \equiv 1$:

$$\int_R \operatorname{div}(M_h \mathbf{v} - \mathbf{v}) \, dx = \int_{\partial R} (M_h \mathbf{v} - \mathbf{v}) \cdot \mathbf{n} \, ds = 0, \qquad (3.29)$$

due to (3.26). We can now apply (3.6) and get

$$\|\mathbf{u} - \mathbf{u}_h\|_1 + \|p - p_h\|_{L^2(D)/\mathbf{R}} \leq ch^2(\|\mathbf{u}\|_3 + \|p\|_2). \qquad (3.30)$$

There are many other known choices available for getting a discretization of (2.14)–(2.15) that satisfies (3.20). An almost complete list of them can be found in Brezzi-Fortin [17] together with the references. In particular, Scott and Vogelius [48] proved that, under minor restrictions on the decomposition of D

into triangles, one can always use a continuous velocity field of local degree k and a discontinuous pressure field of local degree $k - 1$, provided $k \geq 4$. For the low degrees, a special headache is provided by the use of bilinear velocities and constant pressures; its convergence has been proved in a variety of cases (see Johnson-Pitkäranta [39], Stenberg [49], Pitkäranta-Stenberg [44]), but not yet in the general case. Anyway, a filtering of the pressure field is always required to eliminate the checkerboard modes. General strategies for constructing discretizations that fulfill (3.20) are given in Boland-Nicolaides [10] and Brezzi-Pitkäranta [18]. Modifications of the discrete equations that allow us to violate (3.3) were introduced in Brezzi-Pitkäranta [18], Hughes-Balestra-Franca [36], and Brezzi-Douglas [13].

For the use of more general boundary conditions, the basic reference is Verfürth [50]; see also the references contained therein.

Additional references for the Stokes and Navier-Stokes equations can be found in Glowinski-Pironneau [31], Glowinski [32], Girault-Raviart [29], and Brezzi-Fortin [17].

Example 3h. *Discretizations of linear elasticity problems.* It is difficult, in general, to find convenient finite element discretizations for equations (2.21)–(2.22). We shall briefly indicate here three possible ways for tackling the difficulty. The first possibility is to try to construct spaces that verify the commuting diagram property (as in (3.14)). This has been possible, up to now, only by means of *composite elements*; that is, each element is split into subelements and one uses trial functions that are polynomials in each subelement (plus suitable continuity requirements from one subelement to the other). Examples of this approach can be found in Johnson-Mercier [38] or Arnold-Douglas-Gupta [4]. A second possibility is to give up the symmetry condition that appears in the definition of $H(\mathbf{div}; D)$ and to enforce it a posteriori by means of a Lagrange multiplier. After discretization we deal then with stress fields having only a *weak symmetry*. This idea was first used by Fraeijs de Veubeke [25] and then modified and analyzed by Amara-Thomas [1] and Arnold-Brezzi-Douglas [3]. A third possibility is to change the "auxiliary function" and use a *different, nonsymmetric, tensor field* instead of $\boldsymbol{\sigma}$. This will, in general, produce some trouble at Γ_{Neu} (if $\Gamma_{\text{Neu}} \neq \varnothing$) that can be treated with the introduction of an additional Lagrange multiplier of Γ_{Neu}. We refer to Arnold-Falk [5] for more details on this approach.

It has to be pointed out that additional difficulties arise when dealing with nearly incompressible materials. In these cases (2.23) stops to hold (in the limit) in the whole space but still holds for free-divergence tensor fields. This implies that (3.2) must also be checked if the discretization is such that

$$K_h \not\subset K.$$

Additional references for the applications above (and many others) can be found in Brezzi-Fortin [17].

4. Numerical methods for solving the discretized problem. The major difficulty that arises in solving a linear problem such as (3.4), (3.5) is that the associated matrix

$$\begin{pmatrix} A & B \\ B^T & 0 \end{pmatrix} \tag{4.1}$$

is indefinite. There are many ways of getting around this difficulty, mostly using some particular feature of the problem under consideration in order to rewrite it in a different form. Here we shall briefly sketch two of them, one which is mostly used in Examples 1 and 3, and one which is used in Example 2.

The first technique, which is very old (see Fraeijs de Veubeke [26]), starts from the following simple observation. If the space Ξ_h is made of functions that are completely discontinuous from one element to the other, then the most natural choice of basis functions for Ξ_h will produce a matrix A in (4.1) which is block-diagonal. Then the inverse matrix A^{-1} can be easily computed explicitly. Solving (3.4) element by element for $\bar{\xi}_h$ and substituting into (3.5) (static condensation) leaves us with the final matrix $B^T A^{-1} B$ and the only unknown $\bar{\psi}_h$. Note that (3.2) and (3.3) will imply that $B^T A^{-1} B$ is symmetric and positive definite (if $a(\xi_1, \xi_2)$ is symmetric). Now if, instead, the functions in Ξ_h have some continuity properties from one element to the other (for instance, in Example 1h we had $\mathbf{p}_h \cdot \mathbf{n}$ continuous at the interfaces) this cannot be done. However, one can choose to work in a larger space, say $\tilde{\Xi}_h$, made of discontinuous functions, and then require the continuity by means of a Lagrange multiplier. Let us see the procedure in the particular case of Example 1h. We set

$$\tilde{\Xi}_h = \{\mathbf{q} \mid \mathbf{q} \in (L^2(D))^2; \ \mathbf{q}_{|T} \in (P_1)^2 \ \forall T \in \mathcal{T}_h\}, \tag{4.2}$$

$$\Lambda_h = \{\mu \mid \mu_{|e} \in P_1(e) \ \forall e, \text{ internal edge in } \mathcal{T}_h\}, \tag{4.3}$$

$$c(\mathbf{q}, \mu) = \sum_{T \in \mathcal{T}_h} \int_{\partial T} (\mathbf{q} \cdot \mathbf{n}) \mu \, ds. \tag{4.4}$$

Clearly, if $\mathbf{q} \in \tilde{\Xi}_h$ then

$$\mathbf{q} \in \Xi_h \Leftrightarrow c(\mathbf{q}, \mu) = 0 \quad \forall \mu \in \Lambda_h. \tag{4.5}$$

It is not difficult to check that the new problem, find $\tilde{\mathbf{p}}_h \in \tilde{\Xi}_h, \ \tilde{u}_h \in \Psi_h, \ \lambda_h \in \Lambda_h$ such that

$$\int_D \tilde{\mathbf{p}}_h \cdot \mathbf{q} \, dx + \sum_T \int_T \tilde{u}_h \, \mathrm{div} \, \mathbf{q} \, dx = \int_{\partial D} g_0 \mathbf{q} \cdot \mathbf{n} \, d\Gamma + c(\mathbf{q}, \lambda_h) \quad \forall \mathbf{q} \in \tilde{\Xi}_h, \tag{4.6}$$

$$\sum_T \int_T v \, \mathrm{div} \, \tilde{\mathbf{p}}_h \, dx = \int_D f v \, dx \quad \forall v \in \Psi_h, \tag{4.7}$$

$$c(\tilde{\mathbf{p}}_h, \mu) = 0 \quad \forall \mu \in \Lambda_h, \tag{4.8}$$

has a unique solution, and that $\tilde{\mathbf{p}}_h = \mathbf{p}_h, \ \tilde{u}_h = u_h$. Now both the unknowns $\tilde{\mathbf{p}}_h$ and $\tilde{u}_h$ are *a priori* discontinuous and they can be eliminated, at the element level, by static condensation. The final matrix, in the unknown λ_h, will be symmetric and positive definite. It is clear that λ_h itself should be an approximation of u at the interfaces, and as such it was used by the engineers. However, it was only rather recently that it was proved mathematically that λ_h converges to u and, in general, with a *better* order of convergence than u_h itself (see Arnold-Brezzi [2]). For instance, in the present case, once λ_h is known one can construct, element by element, a $u_h^* \in P_1(T)$ such that

$$\int_e (u_h^* - \lambda_h)\, ds = 0 \quad \forall e \text{ edge of } T \tag{4.9}$$

and show that

$$\|u - u_h^*\|_{L^2(D)} \leq O(h^2) \tag{4.10}$$

instead of (3.18) (for the proof of (4.10) see [2]). A similar result can also be achieved with the lowest order Raviart-Thomas element described at the end of Example 1h. However, the best way to compute the solution for this last element is to solve with the so-called P_1-nonconforming method and then use the postprocessing of Marini [40]. For additional results on the convergence of λ_h to u see Brezzi-Douglas-Marini [16], Brezzi-Douglas-Fortin-Marini [15], and Gastaldi-Nochetto [28].

This same idea (disconnect Ξ_h and use a Lagrange multiplier to force back the continuity) can be used also for elasticity problems and in many other cases. However it has not been possible, so far, to use it, for instance, for the Stokes equations (and more generally when continuity at the *vertices* is required in Ξ_h). Then one can use the following other trick that was first analyzed by Bercovier [8]. If the space Ψ_h is made of discontinuous functions (as was the case in our Example 2h), then one can perturb equation (3.5) into

$$b(\overline{\xi}_h, \psi) = \varepsilon(\overline{\psi}_h, \psi)_\Psi + \langle l_2, \psi \rangle \quad \forall \psi \in \Psi_h. \tag{4.11}$$

The corresponding matrix (for (3.4), (4.11)) becomes, roughly,

$$\begin{pmatrix} A & B \\ B^T & -\varepsilon I \end{pmatrix}. \tag{4.12}$$

Now the discontinuity in Ψ_h allows us to eliminate $\overline{\psi}_h$ at the element level. We obtain, in that way, a matrix $A + \varepsilon^{-1}BB^T$. If (3.2) and (3.3) are satisfied, this new matrix will be symmetric and positive definite (always if $a(\xi_1, \xi_2)$ is symmetric). Moreover, calling $(\xi_h^\varepsilon, \psi_h^\varepsilon)$ the solution of (3.4) and (4.11), one has

$$\|\overline{\xi}_h - \xi_h^\varepsilon\|_\Xi + \|\overline{\psi}_h - \psi_h^\varepsilon\|_\Psi = O(\varepsilon). \tag{4.13}$$

The method can also be applied when Ψ_h is made of continuous functions, provided that some lumping procedure is used in computing the inner product in (4.11). However, in such cases, one gets for $\varepsilon^{-1}BB^T$ a bandwidth that is generally larger than the one of A, and this is often a considerable drawback.

A different attempt to reduce (3.4), (3.5) to a single equation in the case of the Stokes equations can be found in Bramble [11].

REFERENCES

1. M. Amara and J. M. Thomas, *Equilibrium finite elements for the linear elastic problem*, Numer. Math. **33** (1979), 367–383.

2. D. N. Arnold and F. Brezzi, *Mixed and nonconforming finite element methods: implementation, postprocessing, and error estimates*, RAIRO Modél. Math. Anal. Numér. **19** (1985), 7–32.

3. D. N. Arnold, F. Brezzi, and J. Douglas, Jr., *PEERS: A new mixed finite element for plane elasticity*, Japan J. Appl. Math. **1** (1984), 347–367.

4. D. N. Arnold, J. Douglas, Jr., and C. P. Gupta, *A family of higher order finite element methods for plane elasticity*, Numer. Math. **45** (1984), 1–22.

5. D. N. Arnold and R. S. Falk, *A new mixed formulation for elasticity* (to appear).

6. I. Babuška, *The finite element method with Lagrangian multipliers*, Numer. Math. **20** (1973), 179–192.

7. I. Babuška and J. E. Osborn, *Numerical treatment of eigenvalue problems for differential equations with discontinuous coefficients*, Math. Comp. **32** (1978), 991–1023.

8. M. Bercovier, *Régularisation duale des problèmes variationnels mixtes*, Thèse de Doctorat d'Etat, Université de Rouen, 1976.

9. C. Bernardi, C. Canuto, and Y. Maday, *Generalized Inf-Sup condition for Chebyshev approximation of the Navier-Stokes equations*, SIAM J. Numer. Anal. (submitted).

10. J. Boland and R. Nicolaides, *Stability of finite elements under divergence constraints*, SIAM J. Numer. Anal. **20** (1983), no. 4, 722–731.

11. J. H. Bramble (to appear).

12. F. Brezzi, *On the existence uniqueness and approximation of saddle point problems arising from Lagrangian multipliers*, RAIRO **8-32** (1974), 129–151.

13. F. Brezzi and J. Douglas, Jr. *Stabilized mixed methods for the Stokes problem*, Numer. Math. (submitted).

14. F. Brezzi, J. Douglas, Jr., R. Duran, and M. Fortin, *Mixed finite elements for second order elliptic problems in three variables*, Numer. Math. (to appear).

15. F. Brezzi, J. Douglas, Jr., M. Fortin, and L. D. Marini, *Efficient rectangular mixed finite elements in two and three space variables*, RAIRO Modél. Math. Anal. Numér. (to appear).

16. F. Brezzi, J. Douglas, Jr., and L. D. Marini, *Two families of mixed finite elements for second order elliptic problems*, Numer. Math. **47** (1985), 217–235.

17. F. Brezzi and M. Fortin (book in preparation).

18. F. Brezzi and J. Pitkäranta, *On the stabilization of finite element approximations of the Stokes equations*, Report MAT-A 219, Helsinki University of Technology, 1984.

19. P. G. Ciarlet, *The finite element method for elliptic problems*, North-Holland, Amsterdam, 1978.

20. M. Crouzeix and P. A. Raviart, *Conforming and non-conforming finite element methods for solving the stationary Stokes equations*, RAIRO **R3** (1973), 33–76.

21. J. Douglas, Jr. and J. E. Roberts, *Mixed finite element methods for second order elliptic problems*, Mat. Apl. Comput. **1** (1982), 91–103.

22. ____, *Global estimates for mixed methods for second order elliptic equations*, Math. Comp. **44** (1985), 39–52.

23. R. S. Falk and J. E. Osborn, *Error estimates for mixed methods*, RAIRO Anal. Numér. **14** (1980), 309–324.

24. M. Fortin, *An analysis of the convergence of mixed finite element methods*, RAIRO Anal. Numér. **11** (1977), 341–354.

25. B. X. Fraeijs de Veubeke, *Stress function approach*, World Congress on the Finite Element Method in Structural Mechanics, Bornemouth, 1975.

26. ____, *Displacement and equilibrium models in the finite element method*, Stress Analysis (O. C. Zienkiewicz and G. Hollister, eds.), Wiley, New York, 1965.

27. L. Gastaldi and R. H. Nochetto, *Optimal L^∞-error estimates for nonconforming and mixed finite element methods of lowest order*, Numer. Math. (to appear).

28. ____, *On L^∞-accuracy of mixed finite element methods for second order elliptic problems*, Math. Comp. (submitted).

29. V. Girault and P. A. Raviart, *Finite element approximation of the Navier-Stokes equations*, Lecture Notes in Math., vol. 749, Springer-Verlag, Berlin-New York, 1979.

30. R. Glowinski, *Approximations externes par éléments finis d'ordre un et deux du problème de Dirichlet pour Δ^2*, Topics in Numerical Analysis. I (J. J. H. Miller, ed.), Academic Press, London, 1973, pp. 123–171.

31. R. Glowinski and O. Pironneau, *Numerical methods for the first biharmonic equation and for the two-dimensional Stokes problem*, SIAM Rev. **21** (1979), no. 2, 167–212.

32. R. Glowinski, *Numerical methods for nonlinear variational problems*, Springer Ser. Comput. Phys., Springer-Verlag, Berlin-New York, 1984.

33. K. Hellan, *Analysis of elastic plates in flexure by a simplified finite element method*, Acta Polytech. Scand. Math. Comput. Sci. Ser. **46** (1967).

34. L. R. Herrmann, *Finite element bending analysis for plates*, J. Eng. Mech. Div. ASCE EMS **93** (1967), 49–83.

35 ____, *A bending analysis for plates*, Proc. Conf. on Matrix Methods in Structural Mechanics, Air Force Inst. of Tech., Wright-Patterson, 1965, 577–604.

36. T. J. R. Hughes, L. P. Franca, and M. Balestra, *A new finite element formulation for computational fluid dynamics. V, Circumventing the Babuska-Brezzi condition: a stable Petrov-Galerkin formulation of the Stokes problem accomodating equal order interpolations*, Comput. Methods Appl. Mech. Engrg. **59** (1986), 85–101.

37. C. Johnson, *On the convergence of some mixed finite element methods for plate bending problems*, Numer. Math. **21** (1973), 43–62.

38. C. Johnson and B. Mercier, *Some equilibrium finite element methods for two-dimensional elasticity problems*, Numer. Math. **30** (1978), 103–116.

39. C. Johnson and J. Pitkäranta, *Analysis of some mixed finite element methods related to reduced integration*, Math. Comp. **38** (1982), 375–400.

40. L. D. Marini, *An inexpensive method for the evaluation of the solution of the lowest order Raviart-Thomas mixed method*, SIAM J. Numer. Anal. **22** (1985), 493–496.

41. L. D. Marini and A. Savini, *Accurate computation of electric field in reverse-biased semiconductor devices: a mixed finite element approach*, Compel **3** (1984), 123–135.

42. J. C. Nedelec, *Mixed finite elements in $\mathbf{R}^3$*, Numer. Math. **35** (1980), 315–341.

43. ____, *A new family of mixed finite elements in $\mathbf{R}^3$*, Numer. Math. **50** (1986), 57–82.

44. J. Pitkäranta and R. Stenberg, *Error bounds for the approximation of the Stokes problem using bilinear/constant elements on irregular quadrilateral meshes*, Preprint.

45. P. A. Raviart and J. M. Thomas, *A mixed finite element method for 2nd order elliptic problems*, Mathematical Aspects of Finite Element Methods (Proc. Conf., Consiglio Naz. delle Richerche, Rome, 1975), Lecture Notes in Math., vol. 606, Springer-Berlag, New York, 1977, pp. 292–315.

46. R. Scholz, *L^∞-convergence of saddle-point approximations for second order problems*, RAIRO Anal. Numér. **11** (1977), 209–216.

47. ____, *Optimal L^∞-estimates for a mixed finite element method for elliptic and parabolic problems*, Calcolo **20** (1983), 355–377.

48. L. R. Scott and M. Vogelius, *Norm estimates for a maximal right inverse of the divergence operator in spaces of piecewise polynomials*, RAIRO Modél. Math. Anal. Numér. **19** (1985), 111–143.

49. R. Stenberg, *Analysis of mixed finite element methods for the Stokes problem: a unified approach*, Math. Comp. **42** (1984), 9–23.

50. R. Verfürth, *Finite element approximation of stationary Navier-Stokes equations with slip boundary conditions*, Ruhr Universität Bochum, B.75, June 1986.

DIPARTIMENTO DI MECCANICA STRUTTURALE AND ISTITUTO DI ANALISI NUMERICA DEL C.N.R., PAVIA, ITALY

Proceedings of the International Congress of Mathematicians
Berkeley, California, USA, 1986

Vortex Methods and Turbulence Theory

ALEXANDRE J. CHORIN

Vortex methods have been developed over the last fifteen years, and have found numerous applications to incompressible and slightly compressible flow, combustion theory, reaction/diffusion equations, and boundary layer theory. Vorticity dominates the mechanics of most real flows and often concentrates in small subsets of the flow field, and thus vortex methods focus effort on the variables of most interest and the regions of most interest.

Vortex methods have no equal in several areas of application, for example in combustion theory, where small scale fluctuations have a major influence on the evolution of the chemical kinetics and have not been successfully resolved in any other way. The major interest in vortex methods comes however from their applications to turbulence theory, where they have revealed and explained important phenomena. Historically the study of turbulence has motivated a number of important mathematical developments, from the theory of stochastic processes to dynamical systems theory. Despite this long and otherwise very fruitful effort, turbulence in fluids has remained something of a mystery, and has only recently begun to yield some of its secrets. Vortex methods and related vortex models are among the leading causes of the advance.

Vortex methods. Vortex methods are most readily explained in the case of two-dimensional flow [8, 19, 24]. The Navier-Stokes equations that describe two-dimensional viscous incompressible flow can be written in the form

$$\partial_t \xi + (\mathbf{u} \cdot \nabla)\xi = R^{-1}\Delta\xi, \tag{1a}$$

$$\operatorname{div} \mathbf{u} = 0, \tag{1b}$$

$$\xi = \operatorname{curl} \mathbf{u}, \tag{1c}$$

where $\mathbf{u} = (u_1, u_2)$ is the velocity, ξ is the vorticity, R is the Reynolds number (i.e., the reciprocal of the viscosity measured in appropriate units), ∇ is the differentiation vector, $\Delta = \nabla \cdot \nabla$, and t is the time. There exists a stream function ψ such that $u_1 = -\partial_2\psi$, $u_2 = \partial_1\psi$, $\partial_i \equiv \partial/\partial x_i$, $\Delta\psi = -\xi$. If $-G$ is Green's function for Δ, then $\psi = G * \xi$, where $*$ denotes a convolution. If $\mathbf{K} = (\partial_2 G, -\partial_1 G)$, then $\mathbf{u} = \mathbf{K} * \xi$. In the special case of inviscid flow ($R^{-1} = 0$), a particle of fluid located at $\mathbf{x} = \mathbf{x}(t)$ moves according to the law

$$\frac{d\mathbf{x}}{dt} = \mathbf{u} = \mathbf{K} * \xi. \tag{2}$$

The vorticity moves with the fluid, and thus equations (2) form a closed system. The vortex method in this special case consists simply of approximating equation (2) for a collection of test particles whose support approximates the support of ξ. The convolution is approximated by a finite sum over the particles. Vorticity distributed over a finite number of point particles gives rise to a singular flow field, and thus $\mathbf{K}$ must be smoothed; the simplest smoothing results from the replacement of $\mathbf{K}$ by $\mathbf{K}_\delta = \mathbf{K}_\delta(\mathbf{x}) = \mathbf{K} * f$, where $f_\delta(\mathbf{x}) = \delta^{-2} f(|\mathbf{x}|/\delta)$, and f is a fixed function with integral 1. The accuracy of the vortex method depends on the choice of f and δ (see [5, 13, 19]). If the domain occupied by the fluid has a boundary, the evaluation of $\mathbf{u}$ requires in addition the solution of a Laplace equation, which can be found without introducing a grid. The vortex method is thus grid-free.

If $R^{-1} \neq 0$, equation (2) for the motion of a particle must be replaced by a stochastic differential equation

$$d\mathbf{x} = \mathbf{u}dt + R^{-1/2}d\mathbf{w} \tag{3}$$

where $d\mathbf{w}$ is normalized two-component Brownian motion. One obtains an equation of this type for each sample particle. Suitable approximations for (3) can readily be found. In addition, if $R^{-1} \neq 0$, an additional boundary condition must be satisfied at solid walls, and this is done through vorticity creation [8, 9], a process that has intriguing analogues in quantum and statistical mechanics [25].

In three-dimensional space the vorticity is a vector quantity; in addition, the vorticity associated with a given computational element is not a constant of the motion (because of vortex stretching, see below) and an updating strategy must be found. These facts complicate the method somewhat but do not change its basic structure. The method can also handle variations in density [2], combustion and energy release [15, 27], singular vorticity distributions [13, 20], and other phenomena [16]. In special problems, interesting variants exist, for example, for piecewise constant inviscid flow [26] and near boundaries [9]. An interesting recent development is the discovery that the amount of labor involved can be radically reduced through a divide/conquer strategy for counting vortex interactions [1, 18], and through an effective use of multiprocessors [4].

The convergence of vortex methods. The convergence of vortex methods has been the object of an elegant theory. In the inviscid case the early results are due to Hald [19] and a general theory that includes three-dimensional flows was given by Beale and Majda [5]. The theory of Beale and Majda provides guidance for the choice of the core function f. For a recent review of the inviscid theory, see [3].

The convergence of random vortex methods (i.e., $R^{-1} \neq 0$) has been proved by Goodman [17] and Long [22] among others. The basic vorticity creation algorithm has been analyzed by Benfatto and Pulvirenti [6].

A convergence proof is not always the best guide to the actual performance of an algorithm; a well chosen example is often at least as useful. The vortex method has been subjected to numerous and extensive computational tests on a variety of specific problems; for a recent and elaborate example, see [27]. This work demonstrates the ability of vortex methods to separate real and numerical diffusion and to approximate flows of great complexity.

Vortex stretching. Among the many problems examined with the help of vortex methods, we choose to describe briefly the analysis of vortex stretching in three dimensional inviscid flow. The problem is important because vortex stretching is one of the fundamental driving mechanisms of fluid turbulence (some say, *the* fundamental driving mechanism), and it is very poorly understood.

The basic phenomenon can be described as follows: the distance between points in a turbulent flow can increase. If two points belong to the same integral line of the vorticity field, that line will increase in length. The requirements of incompressibility and conservation of angular momentum will then cause an increase in the "enstrophy" $\int |\xi|^2 \, d\mathbf{x}$. Numerical experiment shows that increase to be dramatic for "most" initial data. The increase is accompanied by a complex pattern of bifurcation and randomization which is poorly understood.

A number of elaborate vortex calculations have been carried out to investigate this phenomenon (see, e.g., [10]). The calculations exhibit an interesting folding and binding phenomenon: vortex lines fold as they stretch and the ε-support of the vorticity (i.e., the support of all but an ε-fraction of the vorticity) shrinks, a phenomenon known as "intermittency." Furthermore, stretched vortex lines approach each other. The support of the vorticity must shrink as a result of conservation of angular momentum, but it must remain large enough so that the energy of the fluid remain bounded. In the analogous but not identical problem in electrostatics, the support of the charge must have positive capacity (and Hausdorff dimension > 1) if the corresponding field is to have a finite energy, and it is conjectured that a similar constraint must hold here. The process of folding is the process by which a support of finite capacity is produced. The calculations also suggest that the solution of the inviscid equations breaks down in finite time.

The calculations in [10] and in later work are imperfect; in particular they remain valid only for a finite time. This fact does not contradict the convergence theory in [5]: the smoothness assumptions made in [5] break down. The situation is somewhat analogous to what happens with difference approximations, where it was shown in [7] that in the presence of turbulence, the error remains tolerable until a certain threshold is reached, beyond which the error grows catastrophically.

In [12] a lattice vortex model was developed, with the double goal of examining the behavior of vortex methods and also of understanding the dynamics of vortex stretching beyond the limit of validity of "honest" vortex calculations. In this model, vorticity is assumed to be supported by cylindrical segments whose axes coincide with the bonds on a cubic lattice. These bonds are allowed to stretch and bend at random, subject to the constraints of conservation of volume, angular momentum, energy, connectivity, and an integral form of the kinematic relation that defines vorticity. Scaling transformations introduce successively smaller scales into the calculation. This model is a direct descendant of vortex calculations ([10] and the references therein), and it provides a substantial insight into vortex stretching and its effect in turbulence.

In particular, within the framework of the model it can be shown that energy conservation implies intermittency in a turbulent flow. The old Kolmogorov scaling law is verified, but the assumptions that led to it are not. The ε-support of the vorticity is seen to be a fractal set whose dimension is estimated, and the general structure of the energy cascade is exhibited. In particular, vortex folding is seen to shield the infinite vortex self-induction that results from vortex stretching in such a way that the effective self-induction remains finite. Turbulence in fluids is seen to be mathematically analyzable in terms of geometric measure theory and potential theory, as could really have been forecast from the general form of the equations of motion. The conservation of mass leads to an elliptic equation, and the conservation of circulation (= angular momentum), of which equation (1a) is the two-dimensional version, leads to a singular or nearly singular distribution of sources. There is an obvious analogy here to the mechanism of generation of unstable fronts in porous media [11].

Future prospects. The range of application of vortex methods is rapidly expanding, and they are likely to become one of the backbones of computational fluid dynamics and of the theory of fluid turbulence. The mathematical aspects of vortex stretching are only beginning to be explored; in particular, the question of blow-up for Euler's equation is still open (see, e.g., [22]) and the mathematical description of intermittency is in its infancy (but see [13]). As far as the practical modeling of turbulence is concerned, I would guess that the future belongs to a combination of vortex methods and small-scale modeling, possibly patterned on the lattice model just described.

REFERENCES

1. C. Anderson, *A method of local corrections for computing the velocity field due to a distribution of vortex blobs*, J. Comput. Phys. **62** (1986), 111–123.

2. ____, *A vortex method for flows with slight density variations*, J. Comput. Phys. **61** (1985), 417–428.

3. C. Anderson and C. Greengard, *On vortex methods*, SIAM J. Numer. Anal. **22** (1985), 413–444.

4. S. Baden, *Dynamic Load Balancing of a Vortex Calculation on Multiprocessors*, Ph.D. Thesis, Dept. of Computer Science, Univ. of California, Berkeley, 1986.

5. J. T. Beale and A. J. Majda, *Vortex methods. II; Higher accuracy in two and three dimensions*, Math. Comp. **39** (1982), 29–52.

6. G. Benfatto and M. Pulvirenti, *Convergence of Chorin's product formula in the half plane*, manuscript, University of Rome, 1986.

7. A. J. Chorin, *The convergence of discrete approximations to the Navier-Stokes equations*, Math. Comp. **23** (1969), 341–353.

8. ——, *Numerical study of slightly viscous flow*, J. Fluid Mech. **57** (1973), 785–796.

9. ——, *Vortex models and boundary layer instability*, SIAM J. Sci. Statist. Comput. **1** (1980), 1–24.

10. ——, *The evolution of a turbulent vortex*, Comm. Math. Phys. **83** (1982), 517–535.

11. ——, *The instability of fronts in a porous medium*, Comm. Math. Phys. **91** (1983), 103–116.

12. ——, *Turbulence and vortex stretching on a lattice*, Comm. Pure Appl. Math. (1986).

13. A. J. Chorin and P. Bernard, *Discretization of a vortex sheet with an example of roll-up*, J. Comput. Phys. **13** (1973), 423–429.

14. R. DiPerna and A. J. Majda, *Concentration and oscillations in solutions of the incompressible fluid equations*, manuscript, 1986.

15. A. F. Ghoniem, A. J. Chorin and A. K. Oppenheim, *Numerical modelling of turbulent flow in a combustion tunnel*, Philos. Trans. Roy. Soc. London Ser. A **304** (1982), 303–325.

16. A. F. Ghoniem and F. Sherman, *Grid-free simulation of diffusion using random walk methods*, J. Comput. Phys. **61** (1985), 1–38.

17. J. Goodman, *Convergence of the random vortex method*, Comm. Pure Appl. Math. (1986).

18. L. Greengard and V. Rokhlin, *A fast algorithm for particle simulations*, Research Report, Dept. of Computer Science, Yale University, 1986.

19. O. Hald, *The convergence of vortex methods*. II, SIAM J. Numer. Anal. **16** (1979), 726–755.

20. R. Krasny, *Computation of vortex sheet roll-up in the Trefftz plane*, manuscript, Courant Institute, 1986.

21. A. Leonard, *Vortex methods for flow simulations*, J. Comput. Phys. **37** (1980), 289–355.

22. D. G. Long, *Convergence of random vortex methods*, Ph.D. Thesis, Univ. of California, Berkeley, 1986.

23. A. J. Majda, *Mathematical foundations of incompressible fluid flow*, Lecture Notes, Princeton University, 1985.

24. C. Marchioro and M. Pulvirenti, *Vortex methods in two-dimensional fluid mechanics*, Lecture Notes in Physics, No. 206, Springer, New York, 1984.

25. J. Neu, *Fields and coherent structures*, manuscript, Dept. of Mathematics, Univ. of California, Berkeley, 1986.

26. E. Overman and N. Zabusky, *Evolution and merger of isolated vortex structures*, Phys. Fluids **25** (1982), 1297–1305.

27. J. Sethian and A. F. Ghoniem, *The validation of vortex methods*, J. Comput. Phys. (1986).

UNIVERSITY OF CALIFORNIA, BERKELEY, CALIFORNIA 94720, USA

Проблема гарантированной точности в численных методах линейной алгебры

С. К. ГОДУНОВ

Речь идет о задачах, в которых матрицы и векторы, входящие в условия, известны с некоторой относительной точностью в евклидовой метрике. Вместо A_0, f_0 считаются доступными лишь некоторые их приближения A, f такие, что

$$\|A - A_0\| < \varepsilon\|A_0\|, \qquad \|f - f_0\| < \varepsilon\|f_0\|,$$

где ε — характеризует степень точности. Типичные значения $\varepsilon = 10^{-5}, 10^{-10}, 10^{-30}, \dots$.

Сделанное предположение ограничивает круг задач, для которых можно разработать эффективные методы с гарантированной оценкой погрешности. Если погрешности порядка ε в условиях приводят к погрешностям порядка 1 в результате, то нельзя надеяться построить эффективный алгоритм. Такого рода ситуации характеризуются как «патологические». Иногда их можно избежать видоизменив постановку задачи.

Расчет спектра A часто используется для анализа устойчивости решений системы $\dot{x} = Ax$. Типичный пример патологической ситуации возникает при рассмотрении двухдиагональной $\mathcal{N} \times \mathcal{N}$ матрицы A_0, у которой на главной диагонали всюду стоят -1, а на верхней побочной — всюду 10. Все ее $\lambda(A_0)$ равны -1 и устойчивость имеет место. При $\mathcal{N} = 25$ у матрицы A_ω, отличающейся от A_0 заменой нуля в левом нижнем углу всего лишь на $\omega = -10 \cdot 8^{-25} \approx -2.6 \cdot 10^{-22}$, существует $\lambda(A_\omega) = +\frac{1}{4}$. На ЭВМ с $\varepsilon = 10^{-20}$ A_0 и A_ω неразличимы. Конечно, можно подвергнуть A_0 подобному преобразованию $W^{-1}A_0W = \tilde{A}_0$ с диагональной $W = \operatorname{diag}(1, 10^{-2}, 10^{-4}, \dots, 10^{-48})$ и вычислять $\lambda(\tilde{A}_0)$, совпадающие с $\lambda(A_0)$. При таком преобразовании, которое часто применяется, приходим к $\tilde{A}_0$ с чрезвычайно хорошо обусловленными собственными значениями, но законность преобразования сомнительна, так как W практически вырождена и $W^{-1}A_\omega W$ отличается от $W^1 A_0 W$ совсем не малым элементом $10^{24}\omega = -260$ в левом нижнем углу.

Мы предлагаем другой путь анализа устойчивости $\dot{x} = Ax$. Вместо $\lambda_i(A)$ рекомендуется [1] определять «качество устойчивости»

$$\kappa(A) = \sup_{x(0)} \left\{ \int_0^\infty \|x(t)\|^2 \, dt \Big/ \left[\int_0^\infty \exp(-2t\|A\|) \cdot \|x(0)\|^2 \, dt \right] \right\}$$

($\kappa(A) = \infty$, если $\dot{x} = Ax$ не асимптотически при $t \to +\infty$ устойчива). У матрицы A_0 $\kappa(A_0) > 10^{23}$, то есть $\dot{x} = A_0 x$ практически неустойчива, несмотря на то, что все $\lambda_i(A_0) = -1$. В самом деле, если $x_1(0) = x_2(0) = \cdots = x_{24}(0) = 0$, $x_{25}(0) = \xi = \sqrt{48\pi} \cdot 10^{-23} \approx 1.22 \cdot 10^{-22}$, то решение $x_j(t) = \xi(10^{25-j}/j!)t^{25-j}e^{-t}$ при $t = 24$ имеет компоненту $x_1(24) = (10 \cdot 24/e)^{24} \cdot (\sqrt{48\pi} \cdot 10^{-23}/24!) \approx 10$.

Интересно алгебраическое определение $\kappa(A) = \|H\|$ [2], где H положительно определенное решение уравнения $HA + A^*H = -2\|A\| \cdot I$. (Если оно неразрешимо, или если H не положительно определено, $\kappa(A) = \infty$.) Можно показать, что при $t > 0$:

$$\|e^{tA}\| \leq \sqrt{\kappa(A)} \exp[-t\|A\|/\kappa(A)] \tag{1}$$

и что при $\|B\|/\|A\| < 1/[10\kappa^2(A)]$:

$$|\kappa(A + B) - \kappa(A)| < 13\kappa^3(A)\|B\|/\|A\|. \tag{2}$$

Для вычисления H удобно пользоваться формулой Ляпунова $H = 2\|A\| \int_0^\infty e^{tA^*} e^{tA} \, dt$. Отметим, что если Π ортогональный проектор ($\Pi^* = \Pi$, $\Pi^2 = \Pi$, $\Pi A \Pi = A\Pi$) на некоторое инвариантное для A подпространство, в котором все собственные значения A отрицательны, то определены матричные интегралы ($C = C^*$):

$$y_k = \int_0^{2^k \tau} \Pi e^{t\Pi A^*} C e^{tA\Pi} \Pi \, dt \tag{3}$$

и их предел $y = \lim_{k\to\infty} y_k$, который для гурвицевой A при $\Pi = I$, $C = 2\|A\| \cdot I$, $\tau = 1/(2\|A\|)$) совпадает с H. Известная процедура Девисона и Мена [6] вычисления интегралов типа y_k начинается с определения по формуле Тейлора y_1, $B_1 = e^{\tau A\Pi}\Pi$ и состоит в использовании рекуррентных формул

$$B_k = B_{k-1}^2, \qquad y_k = y_{k-1} + B_{k-1}^* y_{k-1} B_{k-1}. \tag{4}$$

При исследовании устойчивости ($\Pi = I$, $C = 2\|A\| \cdot I$), пользуясь (1) и задавшись κ^* таким, что при $\kappa(A) > \kappa^*$ систему $\dot{x} = Ax$ надо считать практически неустойчивой, можно оценить через сколько шагов j процесса (4) ($\tau = 1/2\|A\|$) при некотором $1 > \rho > 0$ окажется выполненным неравенство

$$\|A^* y_j + y_j A + 2\|A\| \cdot I\| \leq 2\rho\|A\| \tag{5}$$

если только $\kappa(A) < \kappa^*$. Оно заведомо справедливо при $j > 1 + \{\log_2[\kappa^* \ln(2\kappa^*/\rho)]\}$. Более того, при таких j для любой X^j, для которой $\|X_j - y_j\| \leq \rho/2$,

$$\|A^* X_j + X_j A + 2\|A\| \cdot I\| \leq 2\rho\|A\|. \tag{6}$$

В качестве X_j можно использовать приближенное значение, полученное в результате расчета. Если (6) справедливо, то (см. [3]):

$$
\begin{aligned}
\|H - X_j\|/\|X_j\| &< \rho/(1 - \rho), \\
[(1 - 2\rho)/(1 - \rho)]\|X_j\| &< \kappa(A) < \|X_j\|/(1 - \rho).
\end{aligned}
\tag{7}
$$

В [3] использованы оценки [4] погрешности вычисления матричной экспоненты, полученные в предположении $\kappa(A) < \kappa^*$, основанные на (1). Сделав j шагов (4), если (6) не выполнено, устанавливается, что $\kappa(A) > \kappa^*$, либо вычисляется приближенное значение $\kappa(A)$. Итерационное уточнение позволяет найти $\kappa(A)$ с точностью, допускаемой ЭВМ. Попутно при $\kappa(A) < \kappa^*$ рассчитывается функция Ляпунова (Hx, x). Приведем $\kappa(A_0)$ для описанных выше A_0 при $3 \le \mathcal{N} \le 8$:

$$
\begin{array}{llll}
\mathcal{N} = 3, & \kappa = 4.05 \cdot 10^4; & \mathcal{N} = 6, & \kappa = 2.7 \cdot 10^{10}; \\
\mathcal{N} = 4, & \kappa = 3.4 \cdot 10^6; & \mathcal{N} = 7, & \kappa = 2.5 \cdot 10^{12}; \\
\mathcal{N} = 5, & \kappa = 3.0 \cdot 10^8; & \mathcal{N} = 8, & \kappa = 2.3 \cdot 10^{14}.
\end{array}
$$

Заметим еще, что число обусловленности μ линейных уравнений, составляющих $HA + A^*H = -C$, оценивается через $\kappa(A)$:

$$
\mu < \mathcal{N}\kappa^2(A).
$$

Обобщением проблемы Гурвица является задача о дихотомии матричного спектра. При заданном p требуется найти число корней уравнения $\det(A - \lambda I) = 0$ таких, что $\operatorname{Re} \lambda < p$ и таких, что $\operatorname{Re} \lambda > p$. В частности, если $p = 0$, надо выяснить сколько корней лежит в левой, а сколько в правой полуплоскости и есть ли чисто мнимые $\lambda = i\omega$. С задачей о дихотомии мы встречаемся, например, в задаче о построении матрицы Грина $G(t)$, ограниченной при $-\infty < t < +\infty$ и удовлетворяющей уравнению

$$
\frac{d}{dt}G(t) = AG(t) + \delta(t) \cdot I;
\tag{8}
$$

разрешимому лишь если на мнимой оси нет $\lambda_j(A)$. При этом

$$
\begin{aligned}
\Pi_+ &= G(-0)[G^*(+0)G(+0) + G^*(-0)G(-0)]^{-1}G^*(-0), \\
\Pi_- &= G(+0)[G^*(+0)G(+0) + G^*(-0)G(-0)]^{-1}G^*(+0)
\end{aligned}
\tag{9}
$$

— ортогональные проекторы инвариантных для A подпространств с $\lambda_j(A) > 0$ и с $\lambda_j(A) < 0$, соответственно, размерностей $\mathcal{N}_\pm = \operatorname{tr} \Pi_\pm$. Отметим еще равенства

$$
\begin{aligned}
e^{tA}G(-0) &= e^{tA\Pi_+}G(-0) = e^{tA\Pi_+}\Pi_+G(-0) & (t < 0), \\
e^{tA}G(+0) &= e^{tA\Pi_-}G(+0) = e^{tA\Pi_-}\Pi_-G(+0) & (t > 0).
\end{aligned}
\tag{10}
$$

Мы предложили [5] в качестве критерия дихотомии величину $\kappa(A) = 2\|A\| \cdot \|H\|$ $(H = H_+ + H_-)$, где

$$H_+ = \int_0^\infty G^*(t)G(t)\,dt = G^*(+0)y^{(+)}G(+0),$$

$$H_- = \int_{-\infty}^0 G^*(t)G(t)\,dt = G^*(-0)y^{(-)}G(-0),$$

(11)

$$y^{(+)} = \int_0^\infty \Pi_- e^{t\Pi_- A^*} e^{tA\Pi_-} \Pi_- \, dt,$$

$$y^{(-)} = \int_0^\infty \Pi_+ e^{-t\Pi_+ A^*} e^{-tA\Pi_+} \Pi_+ \, dt.$$

(12)

Если предварительно вычислить $G(\pm 0)$, $\Pi_\pm$, то для расчета интегралов (11), (12) можно использовать процесс (5).

Показано, что $\|G(t)\| \le \kappa(A)\exp[-|t|\,\|A\|/\kappa(A)]$. Вместо $G(t)$ удобно рассчитывать матрицы Грина $G_n(t)$ краевых задач на конечном интервале $|t| < n/(2\|A\|)$ такие, что $G_n'(t) = AG_n(t) + \delta(t) \cdot I$, $G_n[n/(2\|A\|)] = G_n[-n/(2\|A\|)]$. При этом

$$G_n(t) = \sum_{k=-\infty}^{k=+\infty} G(t + kn/\|A\|),$$

(13)

$$\|G_n(t) - G(t)\| \le 2\kappa(A)\exp\left[-\frac{n}{2\kappa(A)}\right] \cdot \left\{1 - \exp\left[-\frac{n}{\kappa(A)}\right]\right\}^{-1}$$

и если $\kappa(A) < \infty$, матрицы

$$\Pi_\pm^{(n)} = G_n(\mp 0)[G_n^*(+0)G_n(+0) + G^*(-0)G_n(-0)]^{-1}G_n^*(\mp 0)$$

(14)

стремятся при $n \to +\infty$ к проекторам $\Pi_\pm$.

Положив $\mathcal{P}_0 = \mathcal{Q}_0 = (1/\sqrt{2})I$, определим последовательно $\mathcal{P}_{i+1}$, $\mathcal{Q}_{i+1}$ и верхние треугольные $\mathcal{R}_{i+1}$ из следующих равенств ($\tau = 1/(2\|A\|)$):

$$\begin{bmatrix} e^{\tau A} & \mathcal{Q}_i \\ e^{-\tau A} & \mathcal{P}_i \end{bmatrix} = \begin{bmatrix} \mathcal{Q}_{i+1} \\ \mathcal{P}_{i+1} \end{bmatrix} \mathcal{R}_{i+1}, \qquad \mathcal{Q}_{i+1}^* \mathcal{Q}_{i+1} + \mathcal{P}_{i+1}^* \mathcal{P}_{i+1} = I, \qquad (15)$$

реализующих известный ортогонально степенной метод для составной клеточно диагональной матрицы удвоенного порядка с диагональными клетками $e^{\pm\tau A}$. Любое решение матричного уравнения $Z' = AZ$, определенного на отрезке $[-n\tau, n\tau]$ с выколотой точкой $t = 0$: $[-n\tau, 0) \cup (0, +n\tau]$, удовлетворяющее условию $Z(-n\tau) = Z(+n\tau)$ допускает представление

$$Z(t) = \begin{cases} e^{tA}\mathcal{Q}_n T_n & (-n\tau \le t < 0), \\ e^{tA}\mathcal{P}_n T_n & (0 < t < n\tau), \end{cases}$$

с некоторой T_n. В частности $Z(-0) = \mathcal{Q}_n T_n$, $Z(+0) = \mathcal{P}_n T_n$. Если положить $T_n = (\mathcal{P}_n - \mathcal{Q}_n)^{-1}$, то $Z(+0) - Z(-0) = I$, и, следовательно, $Z(t) = G_n(t)$. Поэтому

$$G_n(+0) = \mathcal{P}_n(\mathcal{P}_n - \mathcal{Q}_n)^{-1} = \mathcal{P}_n T_n,$$

$$G_n(-0) = \mathcal{Q}_n(\mathcal{P}_n - \mathcal{Q}_n)^{-1} = \mathcal{Q}_n T_n.$$

(16)

Оценки:

$$\|T_n^{-1}\| \le 2,$$

$$\|T_n\| \le 2\kappa(A) \left\{ 1 + 2\exp\left[-\frac{n}{2\kappa(A)}\right] \middle/ \left[1 - \exp\left[-\frac{n}{\kappa(A)}\right]\right] \right\}$$

гарантируют возможность вычисления $G_n(+0), G_n(-0)$, если только $\kappa(A)$ не чрезмерно велико.

Вычислительный процесс, основные этапы которого здесь были намечены, может быть организован также, как и описанный выше процесс исследования устойчивости. Он либо приведет к утверждению $\kappa(A) > \kappa^*$, либо определит $\kappa(A), H, \Pi_\pm$. Для расчета проекторов используются формулы:

$$\Pi_+^{(n)} = \mathcal{Q}_n \mathcal{Q}_n^*, \qquad \Pi_-^{(n)} = \mathcal{P}_n \mathcal{P}_n^*.$$

Приведем в качестве иллюстрации результаты расчета $\kappa(A - pI)$ для верхней двухдиагональной A порядка 21, на главной диагонали которой стоят собственные значения $-28, -27, \ldots, -16, -15, +21, +20, \ldots, +16, +15$, а на побочной $14, \ldots, 14$ (всего 13 раз), затем 0 и далее $7, \ldots, 7$ (всего 6 раз). Проекторы Π_+, Π_- здесь ортогональны ($\Pi_+\Pi_- = \Pi_-\Pi_+ = 0$, $\mathcal{N}_+ = 7$, $\mathcal{N}_- = 14$).

Оказалось, что $\kappa(A) = 8.08$ и что на интервалах $-31.7 < p < -10.5$, $14.8 < p < 21.3$, покрывающих, соответственно, отрицательную и положительную части спектра, $\kappa(A - pI)$ практически бесконечно ($\kappa[A - pI] > \kappa^* = 10^8$). Для p лежащих вне этих интервалов значения $\kappa(A - pI)$ следующие:

p	-60	-40	-32	-9.5	-8.5	-6.5
κ	3.99	27.9	$2.05 \cdot 10^5$	$5.13 \cdot 10^3$	$9.91 \cdot 10^2$	99.99
p	-4.5	-0.5	4.5	8.5	10.5	11.5
κ	28.2	8.8	7.1	20.4	62.9	$1.7 \cdot 10^2$
p	12.5	13.5	14.5	21.5	22.5	23.5
κ	$88 \cdot 10^2$	$1.3 \cdot 10^4$	$1.88 \cdot 10^5$	$2.4 \cdot 10^7$	$1.78 \cdot 10^4$	$1.56 \cdot 10^2$

Итак, дихотомия спектра A на части $\operatorname{Re}\lambda > 0$, $\operatorname{Re}\lambda < 0$ осуществима, а части спектра $\operatorname{Re}\lambda < 17.5$, $\operatorname{Re}\lambda > 17.5$ практически неразделимы, хотя при $\lambda = 17.5$ точек спектра нет.

Вопрос о том, как должны ставиться спектральные задачи возник из рассмотрения примеров спектров последовательностей разностных операторов [7]–[10]. Отметим также построенный недавно (см. [11]) пример матрицы, спектром которой с большой точностью могут считаться все точки двух плоских областей вычурной формы. Эти области не удается отделить друг от друга прямой или окружностью и, поэтому, нельзя построить отвечающие им проекционные операторы с помощью QR-алгоритма или ортогонально степенного метода.

Разобранный подход к устранению патологических ситуаций в задаче о расчете спектра произвольных матриц состоит в изменении формулировки задачи на другую, более определенную и, к тому же, специально ориентированную на вполне определенные приложения.

В ряде других задач линейной алгебры патологические ситуации возникают не потому, что решаемая задача плохо поставлена, а потому, что на первый взгляд естественный алгоритм ее решения обладает скрытым, не бросающимся в глаза дефектом. Его трудно обнаружить и поэтому трудно установить.

Эта ситуация будет проиллюстрирована на примере алгоритма исчерпывания — приведения симметричной трехдиагональной A к диагональному виду путем последовательного аннулирования элементов побочной диагонали. Предварительно нужно рассмотреть задачу о вычислении собственного вектора A при уже приближенно вычисленном $\lambda(A)$, которое, например, найдено методом бисекций. Эта последняя задача состоит в решении однородных уравнений

$$(d_1 - \lambda)u_1 - b_2 u_2 = 0,$$
$$-b_j u_{j-1} + (d_j - \lambda)u_j - b_{j+1}u_{j+1} = 0, \quad 2 \le j \le m-1,$$
$$-b_m u_{m-1} + (d_m - \lambda)u_m = 0.$$

У нас всегда все $b_i > 0$. Можно, например, положив $u_1 = 1$, найти, что $u_2 = (d_2 - \lambda)/b_2$, а все последующие u_k $(3 \le k \le m)$ найти из рекуррентных соотношений $u_k = [-b_{k-1}u_{k-2} + (d_k - \lambda)u_{k-1}]/b_k$. В этом расчете последняя строка $(-b_m, d_m)$ матрицы A не участвует и проверка того, равна ли нулю последняя компонента v_m вектора невязки $v = (A - \lambda I)u$ может использоваться для контроля вычислений.

Оказывается, что ничтожно малые погрешности определения собственного числа могут повлечь за собой катастрофическое невыполнение решаемых уравнений. Рассмотрим симметрическую якобиеву матрицу пятого порядка, на главной диагонали которой стоят 2, $1+\delta$, 2δ, $1+\delta$, 2, а на побочной -1, $-\delta$, $-\delta$, -1. У нее существует изолированное собственное значение в интервале $(0, 3\delta)$. Если δ очень мало, можно считать $\lambda \approx 0$. Положив $u_1 = 1$ и вычислив, как описано выше, остальные компоненты $u = (1, \ 2, \ 2 + 1/\delta, \ 2 + 2/\delta, \ 3 + 2/\delta)^T$, получим вектор невязки $v = (0, \ 0, \ 0, \ 0, \ 4 + 2/\delta)^T$. При этом, если $\delta \to 0$, то $\|v\|/\|u\| = (4\delta + 2)/\sqrt{22\delta^2 + 24\delta + 9} \to \frac{2}{3}$. Описанного парадокса можно избежать, если отношения $P_i = u_{i-1}/u_i$, образующие последовательность Штурма, сначала вычислить с некоторыми предосторожностями слева направо по формулам, в которых $g_i = d_i - \lambda$, $P_1' = b_1 = 0$, $P_i' = b_i/(g_{i-1} - b_{i-1}P_{i-1}')$, а затем еще раз справа налево: $b_{m+1} = 0$, $P_{m+1}'' = \infty$, $P_i'' = (g_i - b_{i+1}/P_{i+1}'')/b_i$ и, выбрав по правилу, которое мы приведем дальше, $i = i_0$, положить $P_i = P_i'$ если $i < i_0$, $P_i = P_i''$ если $i > i_0$. Прежде, чем описывать предосторожности, которые надо соблюдать при расчете последовательностей Штурма, и способ выбора i_0, надо ввести взаимнообратные специальные непрерывные монотонные функции $\omega = \omega(a, g, b, \gamma)$, $\gamma = \gamma(a, g, b, \omega)$, определенные при $-\infty < \gamma < +\infty$, $-\infty < \omega < +\infty$. С помощью этих функций $(a, b > 0)$ осуществляется униформизация соотношения $\operatorname{tg}\omega = b/(g - a\operatorname{tg}\gamma)$. Если положить $P_i' = \operatorname{tg}\varphi_i$, $P_i'' = \operatorname{tg}\psi_i$, то φ_i, ψ_i окажутся связанными равенствами $\varphi_1 = 0$, $\varphi_i = \omega(b_i, d_i - \lambda, b_{i-1}, \varphi_{i-1})$ $(i > 1)$, $\psi_m = (n - \frac{1}{2})\pi$, $\psi_i = \gamma(b_{i+1}, d_{i+1} - \lambda, b_i, \psi_{i+1})$ $(i < m)$.

Если λ — собственное значение $\lambda = \lambda_n$, то $\varphi_i = \psi_i$. Теорема Штурма, переформулированная в терминах φ_i, ψ_i, утверждает, что при увеличении λ каждое из φ_i возрастает, а каждое из ψ_i убывает. Можно эту теорему обобщить, указав направления изменения φ_i, ψ_i при направленном изменении любого параметра b_j, d_k, a_i. Пользуясь этим обобщением, нетрудно так организовать процесс вычисления P_i', P_i'', применяя направленные округления с избытком или недостатком, чтобы φ_i', ψ_i'', соответствующие вычисленным последовательностям Штурма, мажорировали $\varphi_i = \psi_i$ соответствующие точным P_i, отвечающим точному значению λ. Легко убедиться, что при этом найдется хотя бы один номер i_0 такой, что $\psi_{i_0-1}'' > \varphi_{i_0-1}'$, $\psi_{i_0}'' < \varphi_{i_0}'$, который и должен быть использован в качестве границы, начиная с которой P_i' в составной последовательности Штурма надо заменять на P_i''. При реальном расчете вычислять φ_i', ψ_i'' нет необходимости, так как критерий выбора i_0 может быть переформулирован в терминах P_i', P_i'', но при этом он менее нагляден. Подробно описанная идея развита в [12, 13] где показано, что в оценке точности $\|Au - \lambda u\| < \varepsilon\|u\|$ мера погрешности ε зависит лишь от разрядной сетки ЭВМ и не зависит от порядка m якобиевой матрицы A. Даже если собственное значение оказалось почти кратным, величина ε такая же, как и для изолированного λ. Она пропорциональна с небольшим коэффициентом разности между единицей и ближайшим к ней и от нее отличным машинным числом.

Если использовать $P_i = u_{i-1}/u_i$, рассчитанные с помощью описанного варианта метода Штурма, при определении параметров цепочки двумерных вращений осуществляющей шаг исчерпывания, положив $c_1 = 1$, $c_i = (c_{i-1}\operatorname{sgn}P_i)/\sqrt{c_i^2 + P_i^2}$, $s_i = |P_i|/\sqrt{c_i^2 + P_i^2}$, то после ортогонального преобразования, описываемого этой цепочкой, якобиев оператор A опять примет трехдиагональный вид, в котором новые элементы главной диагонали $\bar{d}_i$ и побочной $\bar{b}_i$ будут: $(1 \le i \le m - 1)$

$$\bar{d}_i = d_{i+1} - (c_{i+1}c_i b_{i+1}/s_{i+1}) + (c_{i+2}c_{i+1}b_{i+2}/s_{i+2}),$$
$$\bar{d}_m = \lambda_n, \quad \bar{b}_m = 0, \quad b_i = s_i b_{i+1}/s_{i+1} \qquad (2 \le i \le m - 1).$$

С использованием этих формул можно построить алгоритм, обеспечивающий точность вычисления преобразованной матрицы

$$\|\overline{A}_{\text{выч}} - \overline{A}\| \approx m\sqrt{m}\,\varepsilon\|A\|.$$

При этом не остается места для парадоксов, состоящих в том, что при обычном последовательном определении параметров двумерных вращений [14] из условий, обеспечивающих сохранение якобиевого вида у преобразованной матритсы, иногда возникают патологические ситуации, при которых $\bar{b}_m$ оказывается отличным от нуля. На экспериментах выяснилось, что эти парадоксы возникают как раз в тех случаях, когда последовательное определение компонент собственного вектора не позволяет удовлетворить последнему уравнению $-b_m u_{m-1} + (d_m - \lambda)u_m = 0$.

Аналогично тому, как описанный прием приводит к безотказному варианту исчерпывания трехдиагональных матриц [15, 16, 18], он может быть использован и в сингулярном исчерпывании двухдиагональных матриц. Здесь этот прием состоит в предварительном определении отношений компонент сингулярных векторов и в вычислении через эти отношения параметров цепочек двумерных вращений. В результате получается новый вариант SVD-алгоритма, похожий на алгоритм Голуба-Кахана [19], но допускающий строгое обоснование. Отметим, что как правило, при реализациях SVD-алгоритма применяется итерационное уточнение, с помощью которого в большинстве случаев удается избежать патологических ситуаций. В алгоритме из [15, 18] требуется затратить предварительную работу на возможно точный расчет сингулярного числа, но зато сам процесс исчерпывания осуществляется безитерационно.

ЛИТЕРАТУРА

1. S. K. Godounov [Godunov] and A. J. Boulgakov [A. Ya. Bulgakov], *Difficultés calculatives dans le problème de Hurwitz et méthodes à les surmonter* (*aspect calculatif du problème de Hurwitz*), Analysis and Optimization of Systems (Proc. Fifth Internat. Conf., Versailles, 1982), Lecture Notes in Control and Information Sci., vol. 44, Springer-Verlag, 1982, pp. 846–851. English abstract, ibid., p. 845.

2. А. Я. Булгаков, *Эффективно вычисляемый параметр качества устойчивости систем линейных дифференциальных уравнений с постоянными коэффициентами*, Сиб. мат. ж., 1980, т. 21, № 3, с. 32–41.

3. А. Я. Булгаков и С. К. Годунов, *Расчет положительно определенных решений уравнения Ляпунова*, Вычислительные методы линейной алгебры, Наука, Новосибирск, 1985, с. 17–38.

4. А. Я. Булгаков, *Вычисление экспоненты от асимптотически устойчивой матрицы*, Вычислительные методы линейной алгебры, Наука, Новосибирск, 1985, с. 4–17.

5. А. Я. Булгаков и С. К. Годунов, *Параметр дихотомии спектра матрицы и схема его расчета*, Б. и., Новосибирск, 1985, с. 20 (Препринт/ИМ СО АН СССР № 28).

6. E. J. Davison and F. T. Man, *The numerical solution of $A^*Q + QA = -C$*, IEEE Trans. Automatic Control **AC-13** (1968), 448–449.

7. С. К. Годунов и В. С. Рябенький, *Введение в теорию разностных схем*, Физматгиз, Москва, 1962, 340 стр.

8. ____, *Спектральные признаки устойчивости краевых задач для несамосопряженных разностных уравнений*, Успехи мат. наук., 1963, т. 18, № 3, с. 3–14.

9. S. K. Godounov and V. S. Ryabenki, *Theory of difference schemes: an introduction*, North-Holland, Amsterdam, and Interscience, New York, 1964.

10. R. D. Richtmyer and K. W. Morton, *Difference methods for initial-value problems*, 2nd ed., Interscience, 1967.

11. В. И. Костин и Ш. И. Раззаков, *О сходимости ортогонально степенного метода расчета спектра*, Вычислительные методы линейной алгебры, Наука, Новосибирск, 1985, с. 55–84.

12. С. К. Годунов, В. И. Костин и А. Д. Митченко, *Вычисление собственного вектора симметрической трехдиагональной матрицы*, Б. и., Новосибирск, 1983, 58 с.

13. ____, *Вычисление собственного вектора симметрической трехдиагональной матрицы*, Сиб. мат. ж., 1985, т. 26, № 5.

14. H. Rutishauser, *On Jacobi rotation patterns*, Experimental Arithmetic, High Speed Computing and Mathematics, Proc. Sympos. Appl. Math., vol. 15, Amer. Math. Soc., Providence, R. I., 1963, pp. 219–239.

15. А. Д. Митченко, *Алгоритмы исчерпывания трехдиагональных симметрических и двухдиагональных матриц*, Б. и., Новосибирск, 1984, с. 42 (Препринт/ИМ СО АН СССР; 59).

16. ____, *Учет вычислительных погрешностей в алгоритме исчерпывания симметрической трехдиагональной матриц*, Б. и., Новосибирск, 1984, 44 с. (Препринт/ИМ СО АН СССР; № 60).

17. ____, *Учет вычислительных погрешностей в алгоритме исчерпывания симметрической трехдиагональной матриц*, Б. и., Новосибирск, 1984, 46 с. (Препринт/ИМ СО АН СССР; № 61).

18. ____, *Алгоритмы исчерпывания трехдиагональных симметрических и двухдиагональных матриц с гарантированной оценкой точности*, Вычислительные методы линейной алгебры, Наука, Новосибирск, 1985, с. 110–161.

19. G. Golub and W. Kahan, *Calculating the singular values and pseudo-inverse of a matrix*, J. Soc. Indust. Appl. Math. Ser. B Numer. Anal. **2** (1965), 205–224.

Институт математики Сибирского отделения АН СССР, Новосибирск 630090, СССР

Zeros of Epstein Zeta Functions and Supercomputers

DENNIS A. HEJHAL

In Memory of G. Pólya

The aim of this paper is to report on an experiment which sheds new light on the possible application of supercomputers (such as the CRAY 1 and 2) in analytic number theory.* The work described here is partly a collaboration with Enrico Bombieri.

I. Introduction. History shows that the exploration of virgin territory often calls for improvisation and risk-taking. Mathematics is no exception. In the absence of theory, intelligent experimentation is often the surest road to progress.

> "If you cannot solve the proposed problem, try to solve first some related problem... Do not forget that human superiority consists in going around an obstacle that cannot be overcome directly, in devising some suitable auxiliary problem when the original one appears insoluble." [**37**, p. 114]

The search for truth can take many forms. Under the right conditions, theory and experiment can be symbiotic.

Number-theorists have occasionally used "Rechenmaschinen" in attempts to gain insight into various types of phenomena. This paper continues on in that tradition.

The advantage to using a *super*computer is that it will typically enable a researcher to "see" several orders of magnitude beyond the range of a smaller machine. In number theory, this capability can be very important. Essential features are sometimes masked by working in too small a regime.

Let there be no misunderstanding, however. Tools are tools. And risks are risks. Any tool has limitations. It often pays to have a healthy skepticism regarding the "miracles" of modern technology. The safest approach is to make certain that "pure thought" is able to pick up where the computer leaves off. Cf. [**51**].

---

*This paper corresponds rather closely to the author's ICM-86 address. As such, the style is a bit informal.

II. The Riemann zeta-function. A few words about $\varsigma(s)$ and the Riemann Nachlass will help set the stage.

Everyone is familiar with the Riemann Hypothesis [**41**, p. 148]. In stating his conjecture, Riemann forms $\Xi(t)$ (effectively $\varsigma(\frac{1}{2}+it)$) and notes that the number of zeros having $0 \le \operatorname{Re}(t) \le T$ is asymptotic to $(T/2\pi)\log(T/2\pi e)$. Cf. [**54**]. He goes on to say:

> "Man findet nun in der Tat etwa so viel reelle Wurzeln innerhalb dieser Grenzen, und es ist sehr wahrscheinlich, dass alle Wurzeln reelle sind."

> i.e., "One finds in fact approximately this number of real roots within these bounds, and it is very probable that all roots are real."

The basis for the "in fact" clause is not clear, even today. Cf. [**49**, p. 276, line 9] and [**7**, pp. 164–166]. The essential point for us, however, is that Riemann *appears* to have based the RH on some sort of explicit computation. This is strikingly confirmed by studying the relevant folder in the Riemann Nachlass. Cf. [**49**, pp. 275–276] and Figures 1–3.

The man noted for his general ideas is also seen to be a master at computation: [**49**, p. 276, paragraph 3].

[**49**] is based on Siegel's careful examination of the "zeta" folder. By *reworking* several of Riemann's (fragmented) formulae, Siegel came up with what is now known as the Riemann-Siegel formula for $\varsigma(s)$. He later extended this work to the case of Dirichlet L-series [**50**].

The Riemann-Siegel formula is ideally suited for modern computers. Recently it was used to verify that the first $1\frac{1}{2}$ billion zeros of $\varsigma(s)$ lie along the critical line. Cf. [**29**]. This feat required over 1000 cpu hours on a CYBER 205.

As far as the Nachlass itself goes, the largest numerical value of t actually considered by Riemann seems to be about 100.

In 1942, Selberg [**43**] showed that a positive proportion of the zeros of $\Xi(t)$ were real. Selberg's constant was later improved to 33% by N. Levinson [**26**].

III. The Epstein zeta function. By definition,

$$Z_Q(s) = \sum_{mn}{}' Q(m,n)^{-s} \qquad (3.1)$$

where $(m,n) \in \mathbf{Z}^2 - \{0\}$ and $Q = au^2 + buv + cv^2$ is a positive definite quadratic form. In order to have absolute convergence, one takes $\operatorname{Re}(s) > 1$. The function $Z_Q(s)$ can then be continued analytically to $\mathbf{C} - \{1\}$. The point $s = 1$ is a simple pole. One finds that

$$\left(\frac{\sqrt{|\Delta|}}{2\pi}\right)^s \Gamma(s)Z_Q(s) \equiv \left(\frac{\sqrt{|\Delta|}}{2\pi}\right)^{1-s} \Gamma(1-s)Z_Q(1-s), \quad \text{where } \Delta = b^2 - 4ac.$$

$$(3.2)$$

FIGURE 1

The Riemann-Siegel formula is visible on the lower 2/3 of this page.
(Figures 1–3 reproduced with the permission of the Niedersächsische Staats-
und Universitätsbibliothek, Handschriftenabteilung, Göttingen.)

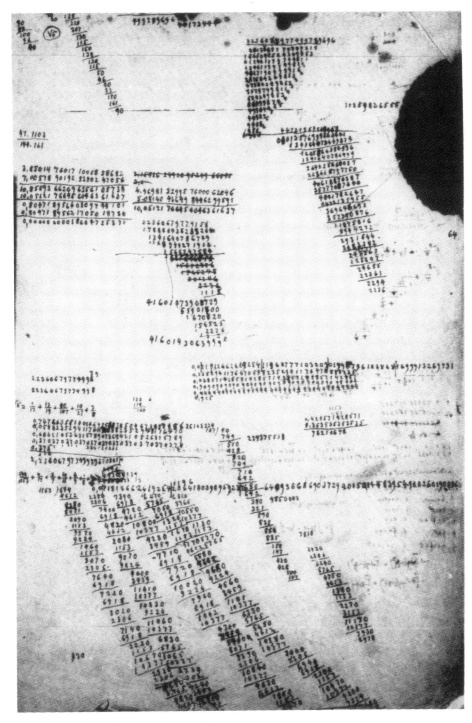

FIGURE 2
The Riemann Nachlass contains many pages like this.

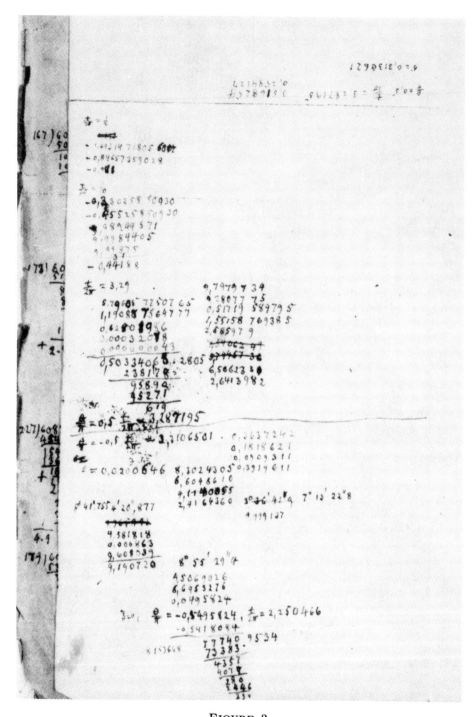

FIGURE 3

Riemann's computation of the first zero of the Riemann zeta function.
($\frac{t}{2\pi} = 2.250466$ corresponds to $t \cong 14.140$.)

Various classical zeta functions are "included" as special cases of $Z_Q(s)$; e.g.,

$$
\begin{aligned}
u^2 + v^2 &\Rightarrow Z_Q(s) = 4\varsigma(s)L(s, \chi_{-4}), \\
u^2 + 4v^2 &\Rightarrow Z_Q(s) = 2\varsigma(s)L(s, \chi_{-4})(1 - 2^{-s} + 2^{1-2s}), \\
u^2 + uv + 3v^2 &\Rightarrow Z_Q(s) = 2\varsigma(s)L(s, \chi_{-11}), \\
u^2 + 7v^2 &\Rightarrow Z_Q(s) = 2\varsigma(s)L(s, \chi_{-7})(1 - 2^{1-s} + 2^{1-2s}).
\end{aligned} \tag{3.3}
$$

χ_k denotes the Kronecker symbol $(k/*)$.

Investigating how the zeros of $Z_Q(s)$ move as the parameters a, b, c vary seems like a very natural problem. This is especially true if one hopes to understand the zeros of, say, $\varsigma(s)$ from a wider (more dynamic) perspective.

One is reminded here of identities like $\pi^2/6 = \sum_{n=1}^{\infty} n^{-2}$ which become completely transparent only when viewed in the correct "ambient" space.

In the case at hand, let $\mathcal{R}(T) = \{10 < \operatorname{Im}(s) < T\}$ and

$$
\begin{aligned}
N(T) &= \text{the number of zeros of } Z_Q(s) \text{ in } \mathcal{R}(T), \\
N_\delta(T) &= \text{the number of zeros of } Z_Q(s) \text{ in } \mathcal{R}(T) \cap \{|\operatorname{Re} s - \tfrac{1}{2}| > \delta\}, \\
N_{\mathrm{crit}}(T) &= \text{the number of zeros of } Z_Q(s) \text{ in } \mathcal{R}(T) \cap \{\operatorname{Re}(s) = \tfrac{1}{2}\}.
\end{aligned}
$$

Quite generally, one knows that

$$
N(T) = \frac{T}{\pi} \log(AT) + O(\log T) \tag{3.4}
$$

where $A = |\Delta|^{1/2}(2\pi e\lambda)^{-1}$, $\Delta = b^2 - 4ac$, $\lambda = \min\{Q(m, n): m^2 + n^2 \neq 0\}$. In addition:

$$
\lim_{T \to \infty} N_{\mathrm{crit}}(T) = \infty; \tag{3.5}
$$

cf. [22].

When Q has rational (equivalently: integral) coefficients, we can assert that

(i) $N_{\mathrm{crit}}(T) > (\text{constant})T$;

(ii) $N_\delta(T) = o[N(T)]$ for every $\delta > 0$; indeed $N_\delta(T) = O(T)$;

(iiii) unless $Z_Q(s)$ admits a natural Euler product, $N_{1/2}(T) > (\text{constant})T$.

Cf. [6, 25, 33, 39, 56]. Assertion (iii) shows that the "typical" $Z_Q(s)$ has many zeros *off* the critical line. Assertion (ii), on the other hand, shows that the overwhelming majority of the zeros lie either on or else very close to $\operatorname{Re}(s) = \tfrac{1}{2}$.

In cases where $Z_Q(s)$ admits an Euler product (cf. (3.3)) we certainly *expect* that $N_{\mathrm{crit}}(T) \sim N(T)$. The generalized Riemann Hypothesis would (effectively) say that $N_{\mathrm{crit}}(T) \equiv N(T)$.

Prior to studying trajectories of off-line zeros, it makes sense to ask what can be said about

$$
\lim_{T \to \infty} \frac{N_{\mathrm{crit}}(T)}{N(T)} \quad \text{for general } Q. \tag{3.6}
$$

The importance of improving estimate (i) was already noted by A. Selberg [45, p. 196 (bottom)] in 1946.

Modification of a recent work by S. Voronin [57] would seem to say that something like $N_{\mathrm{crit}}(T) > c_1 T \exp[c_2 (\log \log \log \log T)^{1/2}]$ holds for a wide class of rational Q. Voronin's techniques are an extension of [42]. [1]

This is where matters stood in December of 1983. In short, no one had any clear notion of what might ultimately happen even for the simplest non-Euler cases. [2]

Influenced by [16, pp. 105–106], Bombieri had suggested somewhat earlier that we might wish to try some computer experiments in this area. See also [23, 35].

IV. The Fourier series approach. For arbitrary values of a, b, c, $Z_Q(s)$ can be computed by relating $Z_Q(s)$ to an appropriate Eisenstein series $E_0(z; s)$ on $\mathrm{PSL}(2, \mathbf{Z})$; i.e.,

$$E_0(z; s) = \frac{1}{2} \sideset{}{'}\sum_{m,n} \frac{y^s}{|mz + n|^{2s}}, \qquad \{z = x + iy,\ y > 0\}. \tag{4.1}$$

The choice of z is determined by Q. The function $E_0(z; s)$ is an eigenfunction of the non-Euclidean Laplacian with respect to z. From the Selberg theory [15, 47] one knows how to expand $E_0(z; s)$ as a Fourier series. Specifically,

$$\sqrt{\frac{\pi}{y}}\pi^{-s}\Gamma(s)E_0(z; s) = \left(\frac{y}{\pi}\right)^{s-1/2}\Gamma(s)\varsigma(2s) + \left(\frac{y}{\pi}\right)^{1/2-s}\Gamma(1-s)\varsigma(2-2s)$$
$$+ 4\sqrt{\pi}\sum_{n=1}^{\infty} n^{-A-iR}\sigma_{2A+2iR}(n)K_{A+iR}(2\pi ny)\cos(2\pi nx). \tag{4.2}$$

Here $s = \frac{1}{2} + A + iR$ and

$$K_\nu(\xi) = \frac{1}{2}\int_{-\infty}^{\infty} e^{-\xi\cosh t}e^{\nu t}\,dt = \text{the familiar Bessel function.} \tag{4.3}$$

There is no loss of generality in assuming that $R > 0$. We denote the LHS of (4.2) by $\mathcal{E}(z; s)$. [3]

Bombieri and I began our collaboration by implementing (4.2) on a fast scalar machine, a CYBER 845, at the University of Minnesota. The terms $\Gamma(s)$ and $\varsigma(2s)$ were easily handled by means of Stirling's formula and Euler-Maclaurin summation. Cf. [8, pp. 3(2), 47(1)] and [7, 19].

The real difficulty is $K_{A+iR}(2\pi ny)$. Straightforward numerical integration of (4.3) does not work. This stems from the fact that $K_{A+iR}(2\pi ny)$ is ridiculously small. Indeed,

$$K_{A+iR}(X) = \text{roughly} \begin{cases} e^{-\pi R/2}, & 1 < X < R \\ e^{-R\arcsin(R/X)-\sqrt{X^2-R^2}}, & X > R \end{cases}. \tag{4.4}$$

---

[1] Selberg was aware of a result analogous to [57] when he wrote [42].

[2] For instance, $u^2 + 5v^2$.

[3] Note that $\mathcal{E}(z; s)$ is *real* for $\mathrm{Re}(s) = \frac{1}{2}$.

Here $A = O(1)$ and $R \gg 1$. Note that the brace is dominated by $e^{-\pi R/2}$.

The source of the difficulty is now apparent. Loosely put, think of A and X as fixed. The integrand in (4.3) effectively varies from 0 to 1 in absolute value. Because of (4.4), massive cancellation must somehow occur. Simpson's rule is additive, however. Each term consists of a certain number of bits plus garbage. When these terms are all added together, the true answer "cancels out" leaving only garbage behind (i.e., no significant digits). This effect can already be seen at $R = 40$.

To solve this problem, we deform the path of integration in a manner similar to stationary phase. This approach enables us to compute $\exp(\frac{\pi}{2}R)K_{A+iR}(X)$ quite accurately ($10 \sim 12$ places) for R out to around 100,000. Compare [24].[4]

Let $\mathcal{E}_1 = \exp(\frac{\pi}{2}R)\mathcal{E}$. We now take $A \equiv 0$ in (4.2). After 4 to 5 weeks of additional code optimization, the program for $\mathcal{E}_1(z; \frac{1}{2}+iR)$ was finally implemented[5] on the CRAY-1 at the University of Minnesota.

The basic idea was to "graph" $\mathcal{E}_1(z; \frac{1}{2}+iR)$ on randomly selected R-intervals for a variety of z. We proposed to study $N_{\text{crit}}(T)/N(T)$ by evaluating:

$$\frac{N_{\text{crit}}[R_1, R_2]}{L_z(R_2) - L_z(R_1)} \quad \text{with } L_z(t) = \frac{t}{\pi} \log\left(\frac{ty}{\pi e}\right). \tag{4.5}$$

Cf. (3.4). [6,7]

The sample intervals $[R_1, R_2]$ should obviously be taken sufficiently long, though how long is not immediately clear. Certain indications can be obtained by reviewing the "error term" in (3.4). Note that $O(\log T)$ is more properly written as $O[\log(Ty)]$. Cf. [5, pp. 82, 102].

We tentatively decided to use "chunks" of 100.

The computation of $\mathcal{E}_1(z; \frac{1}{2}+iR)$ requires essentially $R/2\pi y$ numerical integrations. Because of vectorization, the CRAY-1 is able to perform such integrations approximately $13 \sim 25$ times faster than the 845. The 845 proceeds at roughly $2 \sim 3$ million operations per second.

This program (for the CRAY-1) worked quite well out to around $R = 80000$. Its accuracy could be checked by taking $z = 2i$ (for instance) and then noting that the zeros of $1 - 2^{-s} + 2^{1-2s}$ are precisely

$$s = \frac{1}{2} \pm i\left(\frac{2n\pi + \alpha}{\log 2}\right), \qquad \alpha = \arctan(\sqrt{7}). \tag{4.6}$$

Cf. (3.3). At $R = 80000$ we were still achieving $6 \sim 7$ place accuracy.

The primary reason for stopping at 80000 was a combination of memory size + expense. The computation of $\mathcal{E}_1(z; \frac{1}{2}+iR)$ takes approximately 4.9 seconds

---

[4]Further work may be useful on iterative methods. Cf. [53] and [28]. The latter algorithm suffers from numerical instability when $R \gg 100$.

[5]In Fortran 77.

[6]To keep $\lambda \equiv 1$ for $|mz + n|^2$, we take z in the standard polygon for $PSL(2, \mathbf{Z})\backslash H$.

[7]The on-line zeros are counted by making a Lagrange interpolation of high degree and then having the computer look for changes-of-sign. The basic R-increment was typically chosen to be a small fraction of the average zero-spacing.

when $R = 80000$ and $y = 2$. In 1984, CRAY-1 time was running around \$0.25 per *second*. Expenses were cut by doing several x in parallel. Cf. (4.2).

Here then is a sample of our results. (In cases where several numbers are listed, this means that several contiguous intervals were done.)

R	$x=0$			$x=\frac{1}{2}$				
	y_1	y_2	y_3	y_1	y_2	y_3	y_4	y_5
14000	.649 .615	.619 .632	.661 .600	.611	.622	.600	.664 .617	.584
25000	.608 .595	.595 .598	.594 .625	.611 .595	.621	.632	.641 .617	.641
50000	.610	.625 .580	.623 .612	.616	.553 .586	.570 .617	.653 .641 .641	.624 .601
75000		.645	.621		.616	.595		.635
av.%	62	61	62	61	60	60	64	62

R	$x=x_1$					$x=x_2$					$x=\frac{1}{4}$				
	y_1	y_2	y_3	y_4	y_5	y_1	y_2	y_3	y_4	y_5	y_1	y_2	y_3	y_4	y_5
14000	.704	.693	.687	.713 .666	.795	.642	.727	.681	.713 .705	.700	.670	.683	.674	.788 .781	.713
25000	.708 .653	.675	.720	.694 .722	.679	.685 .672	.697	.660	.737 .725	.685	.698 .643	.694	.663	.740 .685	.710
5000	.688	.691 .646	.689 .659	.690 .728 .687	.664 .682	.667	.691 .684	.644 .659	.714 .699 .687	.702 .685	.727	.667 .672	.683 .671	.728 .722 .702	.743 .717
75000		.711	.689		.702		.662	.712		.708		.668	.674		.705
av.%	69	68	69	70	70	67	69	67	71	70	68	68	67	74	72

TABLE 1. $y_1 = \sqrt[4]{23}$, $y_2 = 2.3$, $y_3 = 2.6$, $y_4 = \sqrt{11}$, $y_5 = \sqrt{10}$, $x_1 = .12831481774213$, $x_2 = .21891437812894$

R	$x=0$ $y=\sqrt{5}$	$\frac{1}{2}$ $\frac{1}{2}\sqrt{5}$	0 $\sqrt{6}$	0 $\frac{1}{2}\sqrt{6}$	0 $\sqrt{10}$	0 $\frac{1}{2}\sqrt{10}$	0 $\sqrt{13}$	$\frac{1}{2}$ $\frac{1}{2}\sqrt{13}$	0 $\sqrt{11}$	0 $\sqrt{14}$
3000	.719	.737							.797	
14000	.733 .654	.759	.759	.733	.716	.744	.765	.670	.778 .790	.746
25000	.735 .719	.760 .767	.715	.725	.734	.708	.718	.688	.842 .790	.738
50000	.761 .686	.758	.722 .755	.760	.734 .705	.738	.736 .736	.682	.844 .797 .814	.729 .740
75000					.719		.727			
av.%	72	76	74	74	72	73	74	68	81	74

TABLE 2

It should be borne in mind here that these figures represent the first steps into virgin territory. In the absence of any apriori model, it is dangerous to draw any more than tentative conclusions. Having said this, we restrict ourselves to 5 remarks.

(i) $N_{\text{crit}}(T)$ seems to grow more like $T \log T$ than T.

(ii) Taking intervals of length 100 was *not* sufficient to remove local fluctuations from (4.5). Formation of averages is therefore necessary.

(iii) There seems to be evidence of a probabilistic structure (or "trend") operating in Table 1. E.g., compare the averages for $x = x_1, x_2, \frac{1}{4}$. Then for $x = 0, \frac{1}{2}$. Out of all possible integers from 1 to 100, why these?? Bear in mind that the averages themselves have fuzz on the order of ± 1.

(iv) The first 4 chunks of Table 2 correspond to fundamental discriminants $-20, -24, -40, -52$. Cf. [**18**, p. 456]. In each case, the class number is 2 and the associated $Z_Q(s)$ splits into $cA^s(L_1 L_2 \pm L_3 L_4)$ where the L_j are suitable $L(s, \chi_k)$. Examples of this type are the simplest kind of Epstein zeta function where GRH does not apply. As such, they are of obvious interest. There is again evidence of a probabilistic structure (cf. (iii)) though

$$\frac{1}{2} + \frac{1}{2} i \sqrt{5} \quad \text{and} \quad \frac{1}{2} + \frac{1}{2} i \sqrt{13}$$

are a bit disturbing.

(v) $i\sqrt{14}$ corresponds to fundamental discriminant -56 and $h = 4$. $i\sqrt{11}$ corresponds to discriminant -44 and $h = 3$. By comparing these columns with the others, it becomes apparent that h in itself is not the controlling feature.

Though these results are promising, further investigation would certainly seem necessary. A skeptic (in the back row) simply has to mutter $T \log T / \log \log T$.

V. The Riemann-Siegel approach. There is now an obvious need to proceed along two fronts; specifically,

(a) to extend Tables 1 and 2 to larger R-values;

(b) to develop some sort of heuristic explanation for what is being observed.

Direction (a) causes serious problems. In short, one needs to find a much faster way of computing $\mathcal{E}_1(z; \frac{1}{2} + iR)$.

The obvious thought is to apply some type of "Riemann-Siegel" formula. To obtain any significant savings, however, it is *essential* that the basic sum have length $c\sqrt{R}$ not cR. In general, this will only be possible in simple examples like §IV(iv) (where products of Dirichlet L-series appear). Cusp forms create havoc. Cf. [**2, 11**] and §IX (first paragraph).

After weighing the various possibilities, Bombieri and the author decided to restrict their attention to examples like IV(iv), at least temporarily. The simplicity of such examples would certainly seem advantageous for item (b) as well.

In September of 1985, the author proceeded to implement the Riemann-Siegel formula for $L(s, \chi_k)$ on the CRAY-1 in Sweden. It was decided to focus on 5

D. A. HEJHAL

pairs of forms [18, p. 456]:

$$
\begin{array}{c|c|c|c|c}
u^2 + 5v^2 & u^2 + 6v^2 & u^2 + 10v^2 & u^2 + 13v^2 & u^2 + 22v^2 \\
2u^2 + 2uv + 3v^2 & 2u^2 + 3v^2 & 2u^2 + 5v^2 & 2u^2 + 2uv + 7v^2 & 2u^2 + 11v^2 \\
\Delta = -20 & \Delta = -24 & \Delta = -40 & \Delta = -52 & \Delta = -88.
\end{array}
\tag{5.1}
$$

The CRAY-1 has an effective memory size of just over 900,000 words. This restriction came into play (on our $L_1 L_2 \pm L_3 L_4$ program) beginning around $R = 5 \times 10^8$. The results from Sweden are listed in the *first half* of Table 3.

Testing 5×10^8 (with length 1000) takes between 4000 and 7000 seconds depending on the size of Δ. [8,9,10]

This experiment was continued several months later on the recently installed CRAY-2 at the University of Minnesota.

On a good Fortran program, the CRAY-2 can run $2 \sim 3$ times as fast as the CRAY-1. (We obtained 2.) More importantly, however, the effective memory size is something like 256,000,000 words!!

The results from the CRAY-2 are listed in the *second half* of Table 3. Numbers in parentheses refer to interval lengths.

Testing 10^{11} with length 500 takes between 6 and 15 hours depending on the size of Δ. We decided to skip $\Delta = -88$.

interval length= (1000) unless otherwise specified

R	$x = 0$ $y = \sqrt{5}$	$\frac{1}{2}$ $\frac{1}{2}\sqrt{5}$	0 $\sqrt{6}$	0 $\frac{1}{2}\sqrt{6}$	0 $\sqrt{10}$	0 $\frac{1}{2}\sqrt{10}$	0 $\sqrt{13}$	$\frac{1}{2}$ $\frac{1}{2}\sqrt{13}$	0 $\sqrt{22}$	0 $\frac{1}{2}\sqrt{22}$
$5 \cdot 10^5$	.724	.759	.742	.746	.747	.723	.737	.708	.754	.696
$2 \cdot 10^6$	.745	.768	.745	.745	.750	.711	.745	.710	.769	.693
10^7	.739	.759	.739	.749	.735	.724	.737	.726	.754	.702
10^8	.751	.764	.737	.747	.744	.730	.751	.722	.752	.696
$5 \cdot 10^8$	.755	.763	.745	.738	.740	.730	.734	.734	.754	.714
10^9	.748	.761	.742	.753	.743	.740	.746	.732	.751	.708
10^{10}	.752	.760	.744	.755	.749	.728	.750	.736	.761	.713
10^{11}	.752	.772	.751	.743	.744 (500)	.738	.757 (500)	.738		

TABLE 3

---

[8] Our R-S program uses only the G_0, G_1, G_2 terms in [50, pp. 153, 158]. G_1 and G_2 are computed by means of numerical differentiation. This necessitates (temporarily) passing to double precision. Other approaches are possible [10], at least for $k = 1$.

[9] Substitution of judiciously chosen values of $R \approx 50000$, 75000 yielded $6 \sim 7$ place agreement between R-S and §IV.

[10] Caution is necessary because the correction term in [50, pp. 157, 158] can vary in size by close to 7 orders of magnitude as η moves from $-\frac{1}{2}$ to $|k| - \frac{1}{2}$. Here $3 \leq |k| \leq 88$. The intervals in Table 3 were therefore chosen so that $|\eta/k|$ would always remain small.

There is now even stronger evidence of a probabilistic structure. Indeed, it is beginning to appear that

$$\lim_{T \to \infty} \frac{N_{\mathrm{crit}}(T)}{N(T)} = c \qquad (*)$$

may hold with the same c in all 10 examples. Specifically, $.75 \le c \le .77$. (But note the "slow-poke" on the outside track.)

VI. Heuristic lower bound. The experiments in §V were actually not done in the dark. Influenced by §IV, the author had already made some progress on §V(b).

Although limitations of space prevent us from giving complete details, we can provide a sketch. The following argument is heuristic.

Consider any Epstein zeta function of type $cA^s(L_1L_2 \pm L_3L_4)$. Cf. (5.1). In accordance with (9.1), we have $L_j = L(s, \chi_{k_j})$ and

$$-N = k_1 k_2 = k_3 k_4.$$

Form $e^{i\varphi_j} L_j$ as in [**50**, p. 158]. It is easily verified that $\varphi_1 + \varphi_2 \equiv \varphi_3 + \varphi_4$. Let

$$Z_1 = e^{i(\varphi_1 + \varphi_2)} L_1 L_2, \qquad Z_2 = e^{i(\varphi_3 + \varphi_4)} L_3 L_4, \qquad (6.1)$$

$$x = \frac{t}{2\pi} \log\left(\frac{t\sqrt{N}}{2\pi e}\right). \qquad (6.2)$$

The functions Z_j are real along $\mathrm{Re}(s) = \frac{1}{2}$. By abusing the notation, we can write $Z_j(x)$ in place of $Z_j(\frac{1}{2} + it)$ for $t \gg 0$.

We want to study the zeros of $Z_1(x) - Z_2(x)$, say. To do so, *assume* GRH. Treat the zeros of Z_1 and Z_2 as points along the x-axis. Their average spacing (as $x \to \infty$) is $\frac{1}{2}$. Similarly for L_i though here the spacing is 1.

In both cases: it is useful to think of the associated sequence of x-values as defining a kind of point process [**3**].

Let a and b be successive zeros of $Z_1(x)$. For the sake of definiteness, assume that $Z_1(x) > 0$ for $x \in (a, b)$. Let $\xi_1, \ldots, \xi_l$ be the consecutive zeros of $Z_2(x)$ in (a, b). We need not concern ourselves with zeros at the endpoints.

If $l = 1$, a typical configuration is:

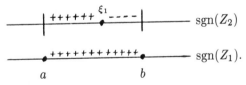

Apply the intermediate value theorem to $Z_2(x)/Z_1(x)$ on (a, ξ_1). We immediately obtain at least one zero of $Z_1 - Z_2$.

A similar analysis can be carried through for each l and each configuration of signs. In so doing, it is important to make a distinction between "accidental" and "forced" zeros. The zeros for $l = 1$ are "forced." They are inescapable. To illustrate the concept of "accidental" zero, look at $l = 0$ and the configuration

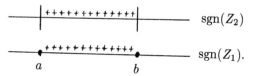

The equation $Z_2(x)/Z_1(x) = 1$ has a solution iff $\min_{a<x<b}(Z_2/Z_1) \le 1$. This may or may not be the case. Zeros arising from such cases are called "accidental."[11]

One can now do a careful bookkeeping of all the possibilities. The key point is how often each type of configuration takes place in the long run. In other words, what percentage of the $[a, b]$ have such-and-such property.

Let P_l be the percentage (i.e., fraction) of $[a, b]$ containing exactly l zeros of Z_2. The bookkeeping shows that, for $0 \le x \le X$, the number of "forced" zeros will asymptotically exceed $2\left(\sum_{k=1}^{\infty} P_k\right) X = 2(1 - P_0)X$. For $1 \le t \le T$, this corresponds to

$$(1 - P_0)\frac{T}{\pi} \log\left(\frac{T\sqrt{N}}{2\pi e}\right). \tag{6.3}$$

A strengthened form of Montgomery's pair correlation conjecture [32] now enters the picture. This conjecture essentially states that (statistically) the zeros of L_i will imitate eigenvalues of random Hermitian matrices of large dimension as $x \to \infty$. The latter statistical ensemble is known to mathematical physicists as GUE; it has been studied quite extensively [1, 30, 31, 38].

Let $\{x_n\}$ temporarily denote the consecutive zeros of L_i. Let $W_m(x)\Delta x$ be the fraction of n for which $x_{n+m} - x_n \in (x, x+\Delta x)$. Let $\mathring{W}_m(x)\Delta x$ be the corresponding fraction for GUE. By the strengthened PCC, we must have $W_m(x) \equiv \mathring{W}_m(x)$. The original PCC merely asserted that:

$$\sum_{m=1}^{\infty} W_m(x) = 1 - \left(\frac{\sin \pi x}{\pi x}\right)^2 = \sum_{m=1}^{\infty} \mathring{W}_m(x). \tag{6.4}$$

A. Odlyzko [34] has used supercomputers to tabulate large numbers of zeros of $\varsigma(s)$ and has found excellent agreement with the conjectured identities. See also §X(F).

In the present case we require Z_1 not L_i. This change presents little difficulty from the point-of-view of statistics. The essential point is that, heuristically, the functions L_1 and L_2 should be statistically independent; compare [12] and §VII. The same goes for their zeros. Hence the process associated with $L_1 L_2$ is simply the superposition of two independent processes, L_1 and L_2. The function analogous to $W_m(x)$ for $L_1 L_2$, call it $\mathcal{W}_m(x)$, can now be computed in terms of the distribution functions for the individual L_i. In particular:

$$\mathcal{W}_1(x) = W_1(x) \int_x^{\infty} (t - x)W_1(t)\, dt + \left(\int_x^{\infty} W_1(t)\, dt\right)^2. \tag{6.5}$$

---

[11]Accidental zeros (controlled by the size of some min or max) can occur for any $l \ne 1$.

Cf. [30, p. 217] or [38, p. 316].

But Z_1 and Z_2 are themselves independent as processes. We therefore have

$$P_l = \int_0^\infty W_1(x)\, \mathcal{V}_l(x)\, dx \tag{6.6}$$

where $\mathcal{V}_l(x)$ is the probability that a *random* interval of length x contains exactly l zeros of Z_2. The functions $\mathcal{V}_l(x)$ are known in statistics as Palm-Khinchin functions. Cf. [3, §2.4; 21, pp. 39–40; 31, p. 330]. There are simple identities relating $\mathcal{V}_l(x)$ to $W_m(x)$. E.g.,

$$W_1(x) = \frac{1}{2}\mathcal{V}_0''(x), \qquad \mathcal{V}_0(x) = \left[\int_x^\infty (\eta - x)W_1(\eta)\, d\eta\right]^2. \tag{6.7}$$

The physicists have very precise (but somewhat involved) ways of computing $\overset{\circ}{W}_m(x)$. A reasonable approximation can be obtained by taking:

$$\overset{\circ}{W}_1(x) = \frac{32}{\pi^2}x^2 \exp\left(-\frac{4}{\pi}x^2\right),$$

$$\tag{6.8}$$

$\overset{\circ}{W}_n(x) =$ the normal density with mean n,

$$\text{and standard deviation } \left[\mathrm{Var}(\overset{\circ}{W}_1) + \frac{1}{\pi^2}\log n\right]^{1/2}.$$

Cf. [1, 38]. It is now elementary to get approximate values for the first few P_l. One obtains

$$\begin{array}{lll} P_0 = .366, & P_1 = .345, & P_2 = .225, \\ P_3 = .054, & P_4 = .010, & P_5 = .001 \end{array} \tag{6.9}$$

We conclude that at least 63.4% of the zeros of $L_1L_2 - L_3L_4$ lie along the critical line. The same argument works for any $L_1L_2 + cL_3L_4$, $c \neq 0$. Note too that only forced zeros have been counted.

Needless to say, there is good agreement here with Tables 2 and 3.

VII. Some results of A. Selberg. Let $\mathcal{U}$ be a primitive character. Let $F(t) = L(\frac{1}{2} + it, \mathcal{U})$. By building on his earlier work [44, 46], Selberg was able to show (unpublished, 1949) that

$$\lim_{T\to\infty} \frac{1}{T} m\left[5 \leq t \leq T : \frac{\ln|F(t)|}{\sqrt{\pi\ln\ln t}} \in A, \ \frac{\arg F(t)}{\sqrt{\pi\ln\ln t}} \in B\right]$$
$$= \int_A \int_B e^{-\pi(u^2+v^2)}\, dv\, du, \tag{7.1}$$

$$\frac{1}{T}\int_T^{2T} |\log F(t+h) - \log F(t)|^{2k}\, dt = O[\log^k(2 + h\ln T)], \qquad 0 < h < 1. \tag{7.2}$$

These results are discussed (in part, at least) in [9, 55].

In connection with §VI, the author noted that similar results could be obtained for $(L(\frac{1}{2} + it, \mathcal{U}_1), \ldots, L(\frac{1}{2} + it, \mathcal{U}_n))$. Here $\mathcal{U}_j$ are primitive and distinct. The corresponding extension of (7.1) justifies our earlier remark about statistical independence.

The main identity in all of this is [44, p. 26(4.8), (4.9)]. Compare [54, p. 310 (top)].

VIII. A dilemma. The key question concerning §V(∗) is whether R has been taken large enough to eliminate any possibility of "masking." Cf. §I. The observed percentages might simply be a reflection of a common statistical "basis" which changes ever so slowly as $R \to \infty$. This is where having some sort of heuristic model is crucial. Right now all we have is a lower bound.

The problem therefore shifts to the accidental zeros. We need to estimate their contribution. This can be done, but imposition of certain working hypotheses soon becomes necessary. The simplest possibility is to represent $Z_1(x)$ on $[a, b]$ as a kind of "quasi-polynomial" $A(x)(x - a_N) \cdots (x - a_1)(x - b_1) \cdots (x - b_N)$. Here $a = a_1$, $b = b_1$. The natural hypothesis is that $A(x)$ is effectively constant and independent of $b - a$ (on average).[12]

Odlyzko performed some tests with his extensive data for $\varsigma(s)$ and found that the correlation between A and $b - a$ indeed disappeared as $N \uparrow$.

To the extent that N can be fixed (independently of x), something remarkable happens. Use of §VII allows one to *prove*, at least heuristically, that $c = 1$!

Skeptical, the author tried a 2nd approach based on Hadamard products[13] and the conjectured relation with GUE. Writing $A(x) = \exp[\Omega_N(x)]$, it was found that $\Omega_N(x)$ "should" have a Lipschitz constant like $\sqrt{\log N}/N$. A slight modification of the earlier bookkeeping showed that if c existed at all, it had to be 1.

These results tended to support a view expressed some years earlier by H. Montgomery—namely, that c should be 1.[14] The gist of Montgomery's argument was this: By Selberg's results, $|Z_1|$ dominates $|Z_2|$ approximately 50% of the time. Cf. §VII. *If* this dominance could be shown to occur over *long* intervals, a kind of real Rouché theorem[15] could then be used (at least, in principle) to force $c = 1$. [16]

With these developments, the possibility of masking had become very real.

We decided to "turn off" the CRAY. The machine had taken us as far as it could.

IX. The main theorem. Let $-N$ be a fundamental discriminant. Let $K = \mathbf{Q}(\sqrt{-N})$. The Z_Q associated with quadratic forms of discriminant $-N$ are essentially zeta functions $\varsigma(s, \mathcal{R})$ associated with ideal classes $\mathcal{R}$ (in K). Cf. [52,

---

[12]We continue to assume GRH.

[13]i.e., "polynomials of infinite degree."

[14]The author first learned of this conjecture from A. Odlyzko.

[15]Applied to the dominant Z_j.

[16]The word "dominate" causes problems here since both Z_1 and Z_2 are frequently 0. The idea is to work with envelopes instead. It is not immediately clear, however, that such envelopes exist (with probability 1). This fact requires proof. Montgomery used Hadamard products and PCC to convince himself that $\Omega_N(x)/\sqrt{\ln \ln x}$ would tend to change more and more gradually on $|x - a_1| \leq L$ as $N \uparrow$. Here $L = $ large constant and $x \to \infty$.

p. 70] and [**14**, p. 214]. One knows that

$$\varsigma(s, \mathcal{R}) = \frac{1}{h} \sum_{\psi} \overline{\psi(\mathcal{R})} L_K(s, \psi). \tag{9.1}$$

Here ψ ranges over all ideal class characters of K. Real-valued ψ correspond to genus characters. For such ψ,

$$L_K(s, \psi) = L(s, \chi_{d_1}) L(s, \chi_{d_2}) \tag{9.2}$$

with appropriate $d_1 d_2 = -N$. The remaining $L_K(s, \psi)$ correspond to cusp forms of dihedral type on $\Gamma_0(N)$. Cf. [**48**, pp. 208, 241] and [**13**, pp. 690, 831, 839 (Satz 24)].

Let $\psi_1, \ldots, \psi_a$ be distinct, nonconjugate, ideal class characters for K. Let $c_1, \ldots, c_a$ be nonzero reals. Assume GRH. Let

$$Z(s) = \sum_{j=1}^{a} c_j L_k(s, \psi_j). \tag{9.3}$$

We say that hypothesis $(\mathcal{H}_\alpha)$ holds for a given ψ_j if

$$\limsup_{T \to \infty} \frac{N[\gamma : 1 \leq \gamma, \gamma' \leq T, \ 0 < \gamma' - \gamma < c/\log T]}{T \log T} = O(c^\alpha) \tag{9.4}$$

for every $c \in (0, 1)$. Successive zeros of $L_K(s, \psi_j)$ are denoted here by $\frac{1}{2} + i\gamma$, $\frac{1}{2} + i\gamma'$. For notational simplicity we assume that all zeros have multiplicity one. α is understood to be positive.

Two possible ways of ensuring $(\mathcal{H}_1)$ are:

(a) to assume that the zeros of $L_K(s, \psi_j)$ define a stationary point process with continuous densities; or

(b) to impose the (appropriate) pair correlation conjecture for $L_K(s, \psi_j)$.

THEOREM. *Assume GRH and hypothesis $(\mathcal{H}_\alpha)$ for $\psi_1, \ldots, \psi_a$. Then:*

$$\lim_{T \to \infty} \frac{N_{\mathrm{crit}}(T)}{N(T)} = 1 \quad \text{holds for } Z. \tag{9.5}$$

In other words, almost all zeros of $Z(s)$ lie along the critical line.

The proof is an intricate application of §VII (making no direct use of computers). The idea for the proof arose out of a conversation in which Bombieri insisted that one should try to eliminate GUE [from method 2 in §VIII] by working directly with the primes ala [**44**]. Working together we were soon able to develop a heuristic proof. The key observation is that

$$\ln|F(t)| = \mathrm{Re} \sum_{p \leq z} \frac{\mathcal{U}(p)}{\sqrt{p}} p^{-it} + \sum_{|\gamma - t| \leq A/\log T} \ln|x(t) - x(\gamma)| \quad + \quad \mathcal{U}_A(t) \tag{9.6}$$

with $\int_T^{2T} |\mathcal{U}_A(t)|^{2k} dt = O_{kA}(T)$.[17] Cf. (6.2) regarding $x(t)$.

---

[17]Here $A =$ any positive constant and $z = T^{\varepsilon/k}$. There is no difficulty extending (9.6) to $L_K(\frac{1}{2} + it, \psi)$.

It took another month or so for the author to make things rigorous. This is where the "technicalities" occur. The trick is to define:

$$e^{i\varphi_j} L_K(\tfrac{1}{2} + iu, \psi_j) \equiv \exp[\Omega_{tj}(u)] \cdot \prod_{|\gamma - t| \le 10M/\log T}[x(u) - x(\gamma)] \qquad (9.7)$$

for $|u - t| \le 9M/\log T$. Here $t \in [T, 2T]$. Using (9.6), one can show that

$$\int_T^{2T} \left| \Omega_{tj}\left(t + \frac{n_1 M}{\log T}\right) - \Omega_{tj}\left(t + \frac{n_2 M}{\log T}\right) \right|^{2k} dt = O_{k,M}(T) \qquad (9.8)$$

for $|n_i| \le 5$, say. But $-\Omega_{tj}(u)$ is essentially convex by GRH. This implies that, apart from a t-set of small relative measure, the total variation of $\Omega_{tj}(u)$ on $|u - t| \le M/\log T$ is $O_{k,M}(1)$. This is enough to show the existence of a "probabilistic envelope." By the extension of (7.1), the differences $\operatorname{Re}\Omega_{tj}(t) - \operatorname{Re}\Omega_{tk}(t)$ are (in effect) normally distributed with standard deviation $c\sqrt{\ln\ln T}$. On most portions of $[T, 2T]$, one $\Omega_{tj}(u)$ will therefore dominate the rest. Hypothesis ($\mathcal{H}_\alpha$) and equation (7.2) are used to show that the corresponding $e^{i\varphi_j} L_K(\tfrac{1}{2} + it, \psi_j)$ has the right number of sufficiently high peaks. After paying careful attention to the exceptional sets, the "pieces" can be "sewn" together yielding $\underline{c} \ge 1 - M^{-\beta}$ for suitable $\beta > 0$. Here $\underline{c} \equiv \liminf_{T \to \infty}(N_{\mathrm{crit}}(T)/N(T))$. To finish the proof, one lets $M \to \infty$.

Incidentally, k can be fixed at $[\![(1 + \alpha)/2\alpha]\!] + 1$.

X. Concluding remarks.

(A) For us, Tables 1–3 were an important stimulus. There can be no question that the CRAY provided important insight.

(B) How long does it take to reach (say) 80% in Table 3??? Some indication can be obtained by replacing (4.5) by a better statistic. Specifically: using $2[x(R_2) - x(R_1)]$ in the denominator and then pair-averaging. This procedure seems less biased from the point-of-view of §VI. One obtains Table 4.

Revised Averages

R \ Δ	-20	-24	-40	-52	-88
$5 \cdot 10^5$	.721	.724	.716	.704	.707
$2 \cdot 10^6$	.738	.727	.714	.711	.715
10^7	.732	.728	.714	.716	.714
10^8	.743	.728	.724	.723	.711
$5 \cdot 10^8$	.746	.729	.723	.722	.722
10^9	.742	.735	.729	.727	.718
10^{10}	.745	.738	.728	.732	.727
10^{11}	.751	.737	.731	.738	

TABLE 4

Since $\sqrt{\ln \ln t}$ goes up so slowly (and $e^{-u^2/2}$ down so rapidly!) one informal estimate yielded something like $R \approx 10^{113}$. [18]

(C) It is reasonable to expect that $\Omega_{tj}(u)$ becomes less and less correlated with the individual zero spacings near t as $M \uparrow$. As mentioned in §VIII, this would give another proof for 100% (by bookkeeping).

(D) The 81% in Table 2 is a reflection of the fact that the corresponding $Z_Q(s)$ involves the L-series of a cusp form. Cf. (9.1) and [**36**, p. 69]. The associated $\mathcal{W}_m(x)$ must therefore be modified. Cf. §VI. $i\sqrt{14}$ is similar except that now there are 3 terms, only one of which is cuspidal. [§VI extends to sums of 3 or more $c_j Z_j$ but the resulting percentage gets smaller quite rapidly.]

(E) Work is continuing on many aspects of these questions. There is little doubt that §IX can be extended to nonfundamental discriminants and to more general Euler products. It should also be possible to formulate a kind of "$\min(\mathcal{C}_1, \ldots, \mathcal{C}_a)$" result when GRH is not assumed. Much more interesting, however, is what happens when "$h = \infty$." [19] Table 1 suggests that Euler product decompositions may *not* be the right approach to phenomena of this sort. At present: we are unable to establish even a positive lower bound (ala §VI). Ideas akin to rational approximation suggest that 100% may persist at least for certain kinds of "irrational" Q.

We are also:

(i) continuing our investigation of off-line zeros;

(ii) trying to get a better grip on the statistical model for $Z(s)$ in the "realm" of moderate $\log \log t$.

Cf. [**4**, **20**, **27**, **40**] for several ideas related to (ii).

(F) In connection with §§V and VI, we tabulated around 150000 zeros of various $L(s, \chi_k)$. The following plots (Figures 4 and 5) were made by A. Odlyzko using 100000 of these zeros. The continuous curve is $\overset{\circ}{W}_m(x)$, $m = 1, 2$.

Cf. (6.8) regarding $\overset{\circ}{W}_1$ and $\overset{\circ}{W}_2$.

---

[18] The CRAY-2 takes .42 seconds to compute one value of $Z_Q(\frac{1}{2} + it)$ with $t \approx 10^9$ and $\Delta = -20$. At this rate, 10^{113} would require something like 10^{44} years. Incidentally, $\ln \ln(10^9) \approx 3.03$, $\ln \ln(10^{113}) \approx 5.56$.

[19] i.e., for generic z.

D. A. HEJHAL

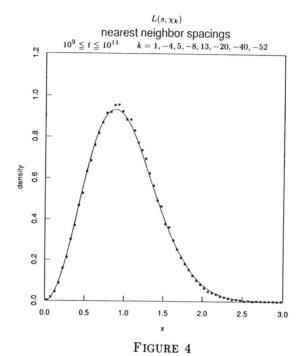

FIGURE 4

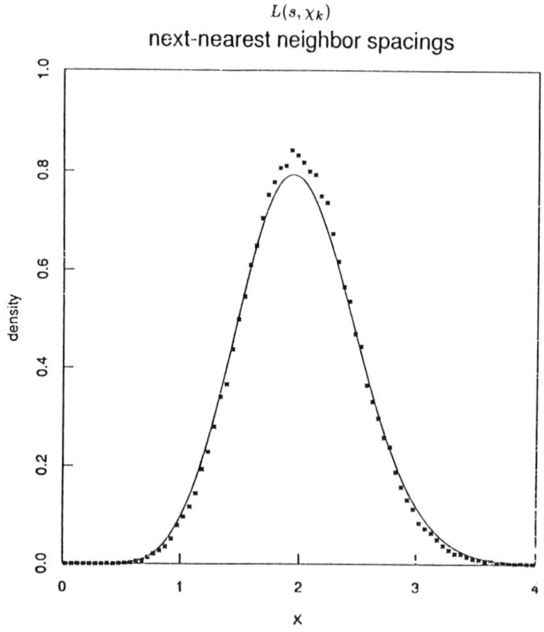

FIGURE 5

Odlyzko also made plots illustrating the spacing distribution of Z_1 and $Z_1 Z_2$. Cf. (6.1) and Figures 6 and 7. The continuous curves represent (6.5) [with $W_1 \longleftrightarrow \overset{\circ}{W}_1$] and its "2-fold" extension.

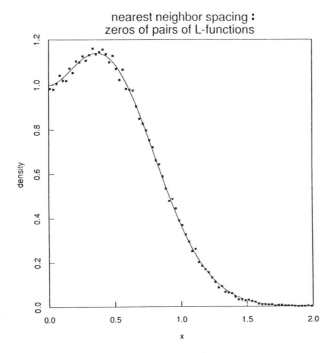

FIGURE 6

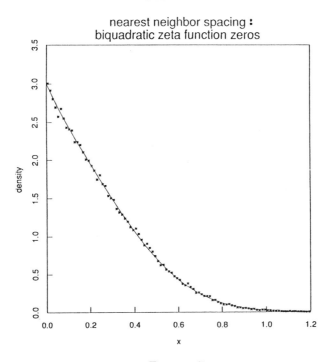

FIGURE 7

Finally, in a related direction, we computed P_ℓ empirically (using ≈ 53000 zeros of Z_1) and obtained

$$P_0 \cong .366, \qquad P_1 \cong .346, \qquad P_2 \cong .222,$$

$$P_3 \cong .055, \qquad P_4 \cong .010, \qquad P_5 \cong .001.$$

The agreement with (6.9) is striking.

(G) A more detailed account of all this material will appear elsewhere. Cf. [17] for further information about §IX.

(H) The author is grateful to the National Science Foundation, the Minnesota Supercomputer Institute, and Chalmers University of Technology (+ the Swedish NFR) for their generous support of this work.

REFERENCES

1. T. Brody, J. Flores, J. French, P. Mello, A. Pandey, and S. Wong, *Random matrix physics: spectrum and strength fluctuations*, Rev. Modern Phys. **53** (1981), 385–479.

2. K. Chandrasekharan and R. Narasimhan, *The approximate functional equation for a class of zeta functions*, Math. Ann. **152** (1963), 30–64.

3. D. Cox and V. Isham, *Point processes*, Chapman and Hall, London, 1980.

4. H. Cramer and M. R. Leadbetter, *Stationary and related stochastic processes*, Wiley, New York, 1967.

5. H. Davenport, *Multiplicative number theory*, 2nd ed., Springer, 1980.

6. H. Davenport and H. Heilbronn, *On the zeros of certain Dirichlet series*, J. London Math. Soc. **11** (1936), 181–185, 307–312.

7. H. M. Edwards, *Riemann's zeta function*, Academic Press, New York, 1974.

8. A. Erdélyi et al., *Higher transcendental functions*, vol. 1, McGraw-Hill, New York, 1953.

9. A. Fujii, *On the zeros of Dirichlet L-series* I, Trans. Amer. Math. Soc. **196** (1974), 225–235; Acta Arith. **28** (1976), 395–403.

10. W. Gabcke, *Neue Herleitung und explicite Restabschätzung der Riemann-Siegel-Formel*, Ph. D. dissertation, Göttingen, 1979, 191 pp.

11. A. Good, *Approximative Funktionalgleichungen und Mittelwertsätze für Dirichletreihen, die Spitzenformen assoziiert sind*, Comment. Math. Helv. **50** (1975), 327–361.

12. P. Hartman, E. van Kampen, and A. Wintner, *Asymptotic distributions and statistical independence*, Amer. J. Math. **61** (1939), 477–486.

13. E. Hecke, *Mathematische Werke*, Vandenhoeck & Ruprecht, Göttingen, 1959.

14. _____, *Vorlesungen über die Theorie der algebraischen Zahlen*, Akad. Verlag, Leipzig, 1923.

15. D. A. Hejhal, *The Selberg trace formula for* PSL(2, **R**), volume 2, Lecture Notes in Math., vol. 1001, Springer, 1983.

16. _____, *Some observations concerning eigenvalues of the Laplacian and Dirichlet L-series*, Recent Progress in Analytic Number Theory (Halberstam and Hooley, eds.), vol. 2, Academic Press, New York, 1981, pp. 95–110.

17. D. A. Hejhal and E. Bombieri, *Sur les zéros des fonctions zêta d'Epstein*, C. R. Acad. Sci. Paris Sér I Math. **304** (1987), 213–217.

18. L. K. Hua, *Introduction to number theory*, Springer, 1982.

19. J. I. Hutchinson, *On the roots of the Riemann zeta-function*, Trans. Amer. Math. Soc. **27** (1925), 49–60.

20. B. Jessen and H. Tornehave, *Mean motions and zeros of almost periodic functions*, Acta Math. **77** (1945), 137–279. See also: Acta Math. **80** (1948), 97–166.

21. A. Khintchine, *Mathematical methods in the theory of queueing*, Griffin, London, 1960.

22. H. Kober, *Nullstellen Epsteinscher Zetafunktionen*, Proc. London Math. Soc. **42** (1936), 1–8.

23. J. Lagarias and A. Odlyzko, *On computing Artin L-functions in the critical strip*, Math. Comp. **33** (1979), 1081–1095.

24. J. Lear and J. Sturm, *An integral representation for the modified Bessel function of the third kind (computable for large imaginary order)*, Math. Comp. **21** (1967), 496–498.

25. C. G. Lekkerkerker, *On the zeros of a class of Dirichlet series*, Van Gorcum & Comp. NV, Assen, 1955.

26. N. Levinson, *More than one third of the zeros of Riemann's zeta function are on* $\sigma = \frac{1}{2}$, Adv. in Math. **13** (1974), 383–436; **18** (1975), 239–242.

27. M. S. Longuet-Higgins, *Statistical properties of wave groups in a random sea state*, Philos. Trans. Roy. Soc. London **312** (1984), 219–250. See also: **249** (1957), 321–387; **254** (1962); 557–599.

28. Y. Luke, *Mathematical functions and their approximations*, Academic Press, New York, 1975, p. 366.

29. J. van de Lune, H. te Riele, and D. Winter, *On the zeros of the Riemann zeta function in the critical strip*, Math. Comp. **46** (1986), 667–681.

30. M. L. Mehta, *Random matrices*, Academic Press, New York, 1967.

31. M. L. Mehta and J. des Cloizeaux, *The probabilities for several consecutive eigenvalues of a random matrix*, Indian J. Pure Appl. Math. **3** (1972), 329–351.

32. H. Montgomery, *The pair correlation of zeros of the zeta function*, Proc. Sympos. Pure Math., vol. 24, Amer. Math. Soc., Providence, R.I., 1973, pp. 181–193.

33. _____, *Topics in multiplicative number theory*, Lecture Notes in Math., vol. 227, Springer, 1971, pp. 95–96.

34. A. Odlyzko, *On the distribution of spacings between zeros of the zeta function*, Math. Comp. **48** (1987), 273–308.

35. D. J. O'Leary, *Some zeros of the Epstein zeta function associated with* $x^2 + 5y^2$, Abstracts Amer. Math. Soc. **5** (1984), 204.

36. H. Petersson, *Modulfunktionen und quadratische Formen*, Springer, 1982.

37. G. Pólya, *How to solve it*, 2nd ed., Princeton Univ. Press, Princeton, N.J., 1957.

38. C. Porter (ed.), *Statistical theories of spectra: fluctuations*, Academic Press, New York, 1965.

39. H. S. Potter and E. C. Titchmarsh, *The zeros of Epstein's zeta functions*, Proc. London Math. Soc. **39** (1935), 372–384.

40. S. O. Rice, *Mathematical analysis of random noise*, Bell Sys. Tech. J. **23** (1944), 282–332; **24** (1945), 46–156.

41. B. Riemann, *Gesammelte Mathematische Abhandlungen*, Teubner, Leipzig, 1892.

42. A. Selberg, *On the zeros of Riemann's zeta-function on the critical line*, Arch. Math. og Naturv. **45** (1942), 101–114.

43. _____, *On the zeros of Riemann's zeta-function*, Skr. Norske Vid. Akad. Oslo I. (1942), no. 10, 59 pp.

44. _____, *Contributions to the theory of the Riemann zeta-function*, Arch. Math. og Naturv. **48** (1946), no. 5, 67 pp.

45. _____, *The zeta-function and the Riemann hypothesis*, C. R. Dixième Congrès Math. Scandinaves 1946, pp. 187–200.

46. _____, *Contributions to the theory of Dirichlet's L-functions*, Skr. Norske Vid. Akad. Oslo I. (1946), no. 3, 62 pp.

47. _____, *Harmonic analysis and discontinuous groups in weakly symmetric Riemannian spaces with applications to Dirichlet series*, J. Indian Math. Soc. **20** (1956), 47–87.

48. J.-P. Serre, *Modular forms of weight one and Galois representations*, Algebraic Number Fields (A. Fröhlich, ed.), Academic Press, London, 1977, pp. 193–268.

49. C. L. Siegel, *Über Riemanns Nachlass zur analytischen Zahlentheorie*, Quell. und Stud. zur Geschichte der Math. Astr. Phys. **2** (1932), 45–80; Gesammelte Abhandlungen **1**, 275–310.

50. _____, *Contributions to the theory of the Dirichlet L-series and the Epstein zeta functions*, Ann. Math. **44** (1943), 143–172.

51. _____, *Zu zwei Bemerkungen Kummers*, Göttingen Nachr. (1964), no. 6, 54 (bottom); Gesammelte Abhandlungen **3**, 439 (bottom).

52. _____, *Advanced analytic number theory*, 2nd ed., Tata Institute, Bombay, 1980.

53. H. Stark, *Fourier coefficients of Maass wave forms*, Modular Forms (R. A. Rankin, ed.), Ellis-Horwood, 1984, pp. 263–269.

54. E. C. Titchmarsh, *The theory of the Riemann zeta-function*, Clarendon Press, Oxford, 1951.

55. K. Tsang, *The distribution of the values of the Riemann zeta-function*, Ph. D. dissertation, Princeton Univ., 1984, 179 pp.

56. S. M. Voronin, *On the zeros of zeta-functions of quadratic forms*, Proc. Steklov Inst. Math. **142** (1979), 143–155.

57. _____, *On the zeros of some Dirichlet series lying on the critical line*, Math. USSR-Izv. **16** (1981), 55–82.

SCHOOL OF MATHEMATICS, UNIVERSITY OF MINNESOTA, MINNEAPOLIS, MINNESOTA 55455, USA

Computer-Assisted Proofs in Analysis

OSCAR E. LANFORD III

Computers are useful in many ways in mathematical research. They make it possible, for example, to perform experiments on a wide variety of mathematical objects and to carry out or check complicated algebraic manipulations. In this talk I will describe another service they can provide: The proof of *strict bounds* on the results of (possibly very complicated) computations on real numbers. I will try to illustrate, in a concrete example, how the availability of this kind of help in proving numerical bounds opens up a new way of approaching some qualitative questions which have proved hard to treat by more standard methods.

Interval arithmetic. The techniques to be described here rest on a standard and elementary method of numerical analysis known as *interval arithmetic*. To explain this method, we begin by reviewing the rudiments of how computers perform arithmetic.

Standard computing environments provide two ways of working with numbers; they may be treated either as *integers* or as *floating point numbers*. Since the computations we will discuss deal with general real numbers and not just with integers, integer arithmetic is not directly applicable to them, and we will accordingly concentrate on floating point arithmetic.[1] Floating point numbers are manipulated and stored in a sign-exponent-fraction representation with a fixed number of digits available for exponent and fraction. The details of the representation vary, but any given format is capable of representing only finitely many numbers; we shall refer to these as the *representable* numbers in that format. Elementary arithmetic operations with representable operands often produce results with too many digits to be representable.[2] When this happens, what is normally done is to "round off" the result, i.e., to return a representable number approximating the exact result. In clean computing environments, the returned

---

[1] A possible alternative approach, which we will not pursue here, is to work with rational numbers, represented as quotients of integers of "arbitrary" precision, on which arithmetic operations can be done exactly.

[2] It is also possible, of course, for a result to be too big or too small to be representable with the limited range of exponents available, but we will ignore these occurrences for the purposes of this schematic review.

result may indeed always be the exact result "correctly rounded," i.e., that representable number which is nearest to the exact result, but this is unfortunately far from being universally the case.

Each individual arithmetic operation, then, can—and usually will—introduce some amount of *round-off error.* The user is normally assured of some bound on the amount of round-off error produced in each operation relative to the size of the returned result. It is thus possible, at least in principle, to apply elementary methods of error propagation to derive strict bounds on the error in the result of any given sequence of arithmetic operations. In practice, except for simple computations or ones with particularly transparent structure, these theoretical error estimates are usually too complicated to be feasible.

Interval arithmetic, by contrast, is a method by which the computer is programmed to generate error bounds automatically and mechanically. The idea is to carry through the computation strict upper and lower bounds for all quantities encountered, i.e., intervals guaranteed to contain the corresponding exact quantities. The end points of these intervals are representable numbers. To propagate these error bounds a step at a time it is only necessary to have available procedures for "doing elementary operations $(+, -, \times, \div)$ on intervals." To see what this means, consider the case of multiplication. A procedure for multiplying intervals means one which, given two intervals $[a_1, b_1]$ and $[a_2, b_2]$ with representable endpoints, produces a third interval $[a_3, b_3]$, again with representable endpoints, such that

$$x \in [a_1, b_1] \text{ and } y \in [a_2, b_2] \Rightarrow x \times y \in [a_3, b_3].$$

Once a set of procedures for doing the elementary operations on intervals is available, a program for computing strict upper and lower bounds on the result of any given sequence of arithmetic operations can be constructed in a completely straightforward way by stringing together calls to these procedures.

Here is one way in which a procedure for multiplying intervals can be constructed. First form the four exact products $a_1 \times a_2, a_1 \times b_2, b_1 \times a_2, b_1 \times b_2$. If representable numbers have at most n digits, each of these products has at most $2n$ digits. To construct (the best possible) a_3, find the smallest of these exact products and round down to the next smaller representable number. Similarly, to construct b_3, find the largest and round up.

We have presented the above algorithm for multiplying intervals simply to show that there is nothing difficult or abstruse, in principle at least, about constructing such procedures. In practice, the method described would probably be replaced by another which is less transparent but more efficient. For example, unless both $a_1 < 0 < b_1$ and $a_2 < 0 < b_2$, it is only necessary to compute two of the four above products. Furthermore, instead of computing the full $2n$-digit products, it may be advantageous to compute n-digit approximations with controlled error—using, for example, floating point hardware—and to enlarge the interval found to compensate for the error. This latter approach is easy to program and produces fast-running code; although it need not give the smallest possible intervals, the slight loss in sharpness in the estimates obtained is not likely to be important. These practical considerations, very important a few years ago, are now fortunately becoming less so. The Institute of Electrical and Electronics Engineers (IEEE) has recently promulgated a standard for floating point arithmetic which provides, among many other things, facilities making it easy to construct interval arithmetic procedures which always produce the best possible results within the limitations imposed by the set of representable numbers. Floating point environments conforming to this standard are becoming more widely available (at least on small computers). This development promises to make good-quality and efficient interval arithmetic much more accessible than in the past.

There are a few warnings which need to be issued about interval arithmetic. The first is that, although it does give *correct* bounds, it sometimes gives *excessively pessimistic* ones. It is well known to numerical analysts, for example, that the error estimates given by interval arithmetic applied in a straightforward way to solving simultaneous linear equations can be unrealistically large and that a theoretical error analysis gives much more realistic bounds. We will not pursue this example, since it involves a long and sophisticated analysis, but will discuss briefly a simpler example (which I learned from S. de Gregorio) giving some idea of what can happen. The example is the computation of the sequence (x_n) defined by

$$x_{n+1} = x_n - h \times x_n \quad \text{where } x_0 \neq 0 \text{ and } 0 < h < 1,$$

i.e., the numerical solution of the differential equation $dx/dt = -x$ by Euler's method. It is easy to see that, if interval arithmetic is applied to this computation, the length of the interval obtained for x_{n+1} is at least $1 + h$ times the length of the interval for x_n, and that, even if x_0 and h are intervals of length zero, some x_n will have nonzero length. Thus, the interval obtained for x_n will have length which grows exponentially with n. On the other hand, a simple analysis shows that, under very weak assumptions about round-off error, the sequence obtained by computing the x_n's in the straightforward way using floating point arithmetic gets and stays very small (as does, of course, the exact sequence). Thus: A theoretical error analysis shows that the computed sequence is a reasonable approximation to the exact sequence, while interval arithmetic gives almost no useful information.

There is another lesson to be learned from this example. The defining formula can be rewritten as

$$x_{n+1} = (1 - h) \times x_n.$$

It is not difficult to see that applying interval arithmetic to this version of the formula *does* give good estimates. This illustrates a general phenomenon: A minor rearrangement in a formula can make a major difference in the sharpness of the estimates obtained by applying interval arithmetic to that formula. I don't know any nontrivial rules for distinguishing between good and bad ways of writing a formula, but the general principle is that it is advantageous to write them with as many cancellations as possible done algebraically rather than numerically.

Another caution: Not all programs doing floating point computation can be translated in a straightforward way into useful interval arithmetic programs producing strict bounds on the results. The translation is easy to do for any program *which contains no branches conditional on the result of comparing two floating point numbers.* Difficulty arises in transcribing a conditional branch because the two intervals being compared may overlap. In principle, this difficulty could be circumvented by following both branches, but repeated conditional branches might then make it necessary to trace very many different routes and thus result in an impractically slow and complex program. Since efficient and robust methods for such tasks as inverting matrices or finding roots of equations are full of conditional branches, this difficulty is a serious one. In the two cases mentioned, however, and in many others like them, there is an effective alternative approach: First do a computation with ordinary floating point arithmetic to find an approximate solution: then use interval arithmetic (together with some kind of perturbation theory) to find an interval around the computed solution which is guaranteed to contain an exact solution. In fact, even when the straightforward direct application of interval arithmetic is feasible, this latter approach often gives much better results.

The Feigenbaum fixed point. We turn now to a discussion, by example, of what is involved in using interval arithmetic to prove a qualitative mathematical result. The example to be discussed, which was the first application of the set of ideas being described here (see Lanford [6]), is the proof of the following

THEOREM. *The Feigenbaum-Cvitanović functional equation*

$$g(x) = -\frac{1}{\lambda} g \circ g(-\lambda x)$$

admits an even analytic solution

$$g(x) = 1 - 1.5276 \cdots x^2 + 0.1048 \cdots x^4 + 0.026 \cdots x^6 + \cdots.$$

We are not going to discuss here the application of this theorem to dynamical systems theory—see the contributions of Eckmann and Sullivan to these proceedings—but we will begin with a few orienting remarks about the functional equation.

First of all, it has a scale invariance: If $g(x)$ is a solution, so is $\gamma g(x/\gamma)$ for any nonzero constant γ. Hence, if we are interested only in solutions with $g(0) \neq 0$, there is no loss of generality in assuming that $g(0) = 1$. Once this normalization condition is imposed, putting $x = 0$ in the equation gives $\lambda = -g(1)$. Hence, we can rewrite the functional equation as

$$g(x) = \frac{1}{g(1)} g \circ g(g(1)x) \equiv \mathcal{T} g(x),$$

i.e., g is a fixed point for the operator $\mathcal{T}$.

It is a fact of experience that any reasonable approach to solving this equation numerically works. For example, take n distinct points in $(0, 1]$ and look for a polynomial of degree n in x^2 which is 1 at 0 and satisfies the equation at these n points. This is a system of n nonlinear equations in n unknowns to which Newton's method can be applied. In practice, Newton's method converges for any reasonable initial guess; what it converges to doesn't depend much on the choice of points; and the approximate solution obtained in this way shows every sign of converging as n goes to ∞.

This certainly suggests that the equation does have a solution. Nevertheless, it has turned out to be surprisingly difficult to prove the existence of this solution by standard qualitative methods. The first proof of existence used estimates proved by computer; it is this proof which will be described here. Subsequently, several successively less computational proofs have been given, and the most recent of these, due to H. Epstein [3], makes no use at all of a computer or even of a calculator. Thus, for the *existence* of a solution, computer assistance is no longer necessary. For applications to dynamical systems theory, however, it is not enough to know that the operator $\mathcal{T}$ has a fixed point g; one also needs to know that the derivative $D\mathcal{T}(g)$ of $\mathcal{T}$ at g is hyperbolic with one-dimensional expanding subspace. The original computer-assisted proof established this along with existence. Although a certain amount of progress has been made on this question by qualitative methods, no complete conceptual proof has so far been given.

The first step in proving the existence of g is to convert the functional equation to a fixed point problem for a *contraction*. (The operator $\mathcal{T}$ itself is not contractive near g; it has one expanding direction.) We use a simplified version of Newton's method. Newton's method for solving $\mathcal{T}g = g$ amounts to searching for a fixed point of

$$f \mapsto f - [D\mathcal{T}(f) - 1]^{-1}[\mathcal{T}(f) - f]$$

by iteration. We simplify this iteration by replacing $D\mathcal{T}(f)$ by a fixed, explicit linear operator Γ, and we observe that a fixed point of $\mathcal{T}$ is the same thing as a fixed point of

$$\Phi(f) = f - [\Gamma - 1]^{-1}[\mathcal{T}(f) - f].$$

A simple calculation shows, moreover, that

$$D\Phi(f) = [\Gamma - 1]^{-1}[\Gamma - D\mathcal{T}(f)],$$

and from this it is easy so see that

To show that Φ maps the ball of radius ρ about g_0 contractively into itself, it suffices to show that
    E1. $\|[\Gamma - 1]^{-1}[\Gamma - D\mathcal{T}(f)]\| \leq \kappa < 1$ *for* $\|f - g_0\| \leq \rho$.
    E2. $\|\Phi(g_0) - g_0\| \leq \rho(1 - \kappa)$.

Hence, to prove the existence of a fixed point g, all we have to do is to find g_0, Γ, ρ, and κ so that these two inequalities hold. Here, roughly, is how we choose them: We take for g_0 an accurate approximate fixed point; we then take Γ to be a good enough approximation to $DT(g_0)$ to make the estimate (E1) hold at least for $f = g_0$ and ρ small enough so that it continues to hold on the ball of radius ρ about g_0. Estimate (E2) can then be expected to hold provided that g_0 was a good enough approximate fixed point. Both inequalities are proved with the aid of computer estimates; for purposes of exposition, we will concentrate on the first of them.

The above analysis is abstract and general; it applies in any Banach space. The next step will be to choose the space in which we work. There is first of all a trivial reduction. The candidates f for the fixed point satisfy some conditions: They are even functions taking the value 1 at 0. We want to make a space with these conditions built in. Thus, we write

$$f(x) = 1 + x^2 h(x^2).$$

Replacing f by h as the "independent variable" is simply a linear change of variable, and we can reasonably allow h to vary over an open set in some Banach space. The operator T acting on functions f induces an operator acting on the corresponding h's, which we will also denote by T.

We are looking for an *analytic* fixed point, and so we will work in a space of analytic functions. In fact, we choose to work in a space of functions $h(t)$ analytic on a disk of radius 2.5 about 1. This choice cannot be justified on general grounds; it is made on the basis of a careful (but "nonrigorous") numerical study of the fixed point (together with considerations of convenience). Evidently, we would not have chosen this domain if we had not had good reason to believe that the fixed point whose existence we were trying to prove is analytic on $\Omega = \{x \colon |x^2 - 1| < 2.5\}$. This domain has another, less obvious, property which plays an essential role in making our argument work: The mapping $x \mapsto g(-\lambda x)$ sends a neighborhood of $\overline{\Omega}$ into Ω. One consequence is that the right-hand side of the functional equation

$$g(x) = -\frac{1}{\lambda} g \circ g(-\lambda x)$$

gives an analytic continuation of g from Ω to a strictly larger domain, and control over g on Ω automatically gives control on this larger domain. A second consequence is that T is differentiable on a neighborhood of g and the derivative $DT(f)$ is a compact operator if f is close enought to g on Ω.

Finally, we choose a norm; we put

$$\|h\| = \sum_{n=0}^{\infty} |h_n| \quad \text{where } h(t) = \sum_{n=0}^{\infty} h_n \left(\frac{t-1}{2.5}\right)^n.$$

The principal reason for using this l^1 norm, rather than the supremum norm, is that it makes it easy to estimate norms of operators: If T is a linear operator, then

$$\|T\| = \sup_n \|Te_n\| \quad \text{where } e_n(t) = \left(\frac{t-1}{2.5}\right)^n.$$

We now choose an approximation Γ to $DT(g)$ which, in the basis (e_n), is represented by a matrix with only finitely many nonzero entries. The choice of Γ, again, is guided by numerical computation. What we need to do, then, is to estimate

$$\sup_n \|[\Gamma - 1]^{-1}[DT(f) - \Gamma]e_n\|$$

for all f in some given (small) ball about an approximate fixed point g_0. We do this in two steps: We make a relatively crude estimate on

$$\|[\Gamma - 1]^{-1}[DT(f) - \Gamma]e_n\|$$

valid for all sufficiently large n; we then estimate this quantity by detailed computation for the finitely many n's not covered by the large-n estimate. Both estimates require the use of the computer. We will concentrate on the second step; the first is simpler, but depends on the specific form of $DT(f)$.

For any given n, and an explicit Γ, it is not difficult to write an explicit formula for $[\Gamma - 1]^{-1}[DT(f) - \Gamma]e_n$. This formula is complicated but is build up in a straightforward way from a sequence of elementary operations on functions—such operations as pointwise sums and products, composition, etc. A convenient way to organize the programming of bounds on $\|[\Gamma - 1]^{-1}[DT(f) - \Gamma]e_n\|$ is to devise a data structure, generalizing intervals for real numbers, giving a finite representation for a class of regions in function space, and to construct procedures for estimating the results of elementary operations on these regions. The formula for $[\Gamma - 1]^{-1}[DT(f) - \Gamma]e_n$ can then simply be transcribed into a sequence of calls to these procedures to get a program for estimating this quantity.

Roughly, we specify these regions in function space by giving upper and lower bounds on Taylor series coefficients up to some fixed order together with an upper bound on the sum of the absolute values of all coefficients of higher order. That is: We first pick an N, the degree up to which the coefficients will be bounded individually. Once N has been fixed, regions are specified by giving N pairs of representable numbers:

$$l_0 \le u_0, \; l_1 \le u_1, \; \ldots, \; l_{N-1} \le u_{N-1},$$

together with a nonnegative representable number ε. The corresponding region, which we denote by $\mathcal{U}(l_0, u_0, \ldots, l_{N-1}, u_{N-1}, \varepsilon)$ is the set of functions h with Taylor series

$$h(t) = \sum_{n=0}^{\infty} h_n \left(\frac{t-1}{2.5}\right)^n$$

such that

$$l_i \le h_i \le u_i \quad \text{for } i = 0, \ldots, N-1, \quad \text{and} \quad \sum_{n=N}^{\infty} |h_n| \le \varepsilon.$$

(In fact, although this representation would work, a more complicated one giving better estimates is usually used in practice.)

The sorts of procedures needed to estimate the results of elementary operations on these regions can be illustrated by the pointwise product. We need a procedure which, given two of these representable regions $\mathcal{U}_1$ and $\mathcal{U}_2$, produces a third $\mathcal{U}_3$ such that

$$u_1 \in \mathcal{U}_1 \text{ and } u_2 \in \mathcal{U}_2 \Rightarrow u_1 \times u_2 \in \mathcal{U}_3.$$

Constructing such a procedure, starting from ordinary interval arithmetic, is straightforward in principle and not very difficult in practice. Once these procedures are available for a set of about ten elementary operations on functions, it is easy to write a program estimating $\|[\Gamma - 1]^{-1}[D\mathcal{T}(f) - \Gamma]e_n\|$ for any given n, and this, as we have already noted, is just what is needed to complete the particular branch of the proof which we have been describing.

Other applications. Strict bounds proved by computer with the aid of interval arithmetic have been applied to a variety of problems besides the one described in the preceding section. Among these are:

1. Existence and properties of solutions of other functional equations arising from renormalization group analyses in dynamical systems theory: Eckmann, Koch, and Wittwer [1], Eckmann and Wittwer [2], Mestel [10], Lanford and de la Llave [7].

2. Existence of a non-Gaussian fixed point for a hierarchical lattice field theory: Koch and Wittwer [5].

3. Stability of a semirelativistic quantum mechanical model for matter (via estimates on the behavior of solutions of some explicit ordinary differential equations): Fefferman and de la Llave [4].

4. Nonexistence of invariant circles for area preserving maps: MacKay and Percival [9].

5. Realistic estimates on the sizes of Siegel linearization domains: de la Llave and Rana [8].

Concluding remarks. I would like to close with some general remarks about these techniques. Should an argument like the one described above be regarded as a valid proof? There are a number of reasons for reluctance to accept proofs using these methods. One of them arises from the widespread misconception that a "computer proof" is nothing more than a numerical experiment, i.e., that errors are controlled by empirical or heuristic methods rather than by strict mathematics. I hope that the preceding discussion has made it clear that this is not the case. It seems to me that the argument outlined above would certainly be accepted as a valid proof if all the verifications were carried out by hand (provided, of course, that all details in the analysis can be filled in and that no mistakes are made in carrying out the verifications). There remain, nevertheless, a number of genuine issues.

First: Using a computer to perform verifications raises practical problems of reliability. Computers are complicated devices, and many things can go wrong. It is certainly not possible to give absolute assurances that errors cannot happen. Nevertheless, with reasonable care, it is possible to reduce the *probability* of error to a very low level.

Second: There is a problem about how such a proof can be communicated and to what extent the result is reproducible, notably on computers different from the one on which it was first given. The communication problem can be solved by communicating the program, typically written in a high-level language. The reproducibility question has two answers, a formal one and an informal one. The strict formal answer is to specify completely the details of the floating point environment and the interval arithmetic operations. If this is done, it is possible in principle–albeit perhaps outrageously difficult in practice—to repeat the verification on a different computer.[3] The informal answer is that many results, such as the existence of the Feigenbaum fixed point, can be made to follow from estimates which are not very critical. In such a case, the high-level language program can be expected to verify these estimates successfully independent of the details of the floating point environment in which it is run. The program is thus a sketch of the proof; verifications on different computers are different ways of filling in the details. The expectation that any reasonable approach will succeed in proving the estimates is of course only heuristic; if it is not realized, it is always possible to fall back on the complete specification. It is, after all, only necesary to prove the desired estimates with *one* correct set of interval arithmetic operations.

Finally: It is possible to take the position that the only acceptable proofs are those in which all steps are carried out by humans, and thus to reject all computer-assisted proofs. This position seems to me to be perfectly defensible. Like other restrictive views of what techniques should be accepted in mathematics, however, it has to be weighted against the benefits derived from the use of these techniques. For example: Proving the existence and basic properties of the Feigenbaum fixed point was not an end in itself; it served as a starting point for other analyses, carried out by standard qualitative methods. In this case, the computer-assisted proof served to eliminate a bottleneck which would otherwise have prevented further progress in this area for many years.

Deciding to accept a computer-assisted proof as valid does not mean that one has to be *satisfied* with it. Proofs vary in the extent to which they are conceptual or computational. Computer-assisted proofs are certainly on the extreme computational end of this scale. It seems to me that it is always desirable to replace a computational proof by a more conceptual one, and especially so when the computational proof requires the assistance of a computer.

It is worth noting, nevertheless, a difference between a computer-assisted proof and a computational proof carred out by hand. In a computer-assisted proof, there is a clean separation between the conceptual part of the analysis and the part which is simply a mechanical verification. Furthermore, the steps to be performed in the mechanical part of the proof must be specified completely and unambiguously. It is thus possible to read and understand the theoretical analysis establishing that a certain computation suffices to prove some desired

---

[3] This kind of strict reproducibility becomes in fact quite practical if both computers provide a set of floating point operations conforming to the abovementioned IEEE standard.

result without actually going through that computation. While this can also be true of computational proofs done by hand, it need not be and in practice there is a strong tendancy to mix the analysis and the computation. Thus, a carefully constructed computer-assisted proof may be more transparent than a computational proof done by hand even if the amount of computation done in the former is much larger.

REFERENCES

1. J.-P. Eckmann, H. Koch, and P. Wittwer, *A computer-assisted proof of universality for area preserving maps*, Mem. Amer. Math. Soc. No. 47 (1984).

2. J.-P. Eckmann and P. Wittwer, *Computer methods and Borel summability applied to Feigenbaum's equation*, Lecture Notes in Physics, vol. 227, Springer-Verlag, Berlin, 1985.

3. H. Epstein, *New proofs of the existence of the Feigenbaum functions*, Comm. Math. Phys. **106** (1986), 393–426.

4. C. Fefferman and R. de la Llave, *Relativistic stability of matter*. I, Preprint, Princeton University, 1985.

5. H. Koch and P. Wittwer, *A non-Gaussian renormalization group fixed point for hierarchical scalar lattice field theories*, Comm. Math. Phys. **106** (1986), 495–532.

6. O. E. Lanford III, *A computer-assisted proof of the Feigenbaum conjectures*, Bull. Amer. Math. Soc. (N.S.) **6** (1982), 427–434.

7. O. E. Lanford III and R. de la Llave, *Solution of the functional equation for critical circle mappings with golden ratio rotation number*, in preparation.

8. R. de la Llave and D. Rana, *Proof of accurate bound in small denominator problems*, Preprint, Princeton University, 1986.

9. R. S. MacKay and I. C. Percival, *Converse KAM: theory and practice*, Comm. Math. Phys. **98** (1985), 469–512.

10. B. D. Mestel, *A computer assisted proof of universality for cubic maps of the circle with golden mean rotation number*, Ph.D. Thesis, University of Warwick, 1985.

INSTITUT DES HAUTES ÉTUDES SCIENTIFIQUE, 91440 BURES-SUR-YVETTE, FRANCE

Renormalization Group Analysis of Turbulence

STEVEN A. ORSZAG AND VICTOR YAKHOT

The direct interaction approximation (DIA), due to Kraichnan [1], was the first field-theoretical approach to the theory of turbulence. Formulated in terms of the Dyson equation, the DIA is characterized as the lowest-order approximation that includes nonlinear corrections to the propagator for the mode $\mathbf{v}(\mathbf{k}, \omega)$. It was shown [1] that, in the inertial range, the DIA gives the energy spectrum is $E(k) \propto k^{-3/2}$. This result contradicts both experimental data and the Kolmogorov theory of turbulence which gives $E(k) \propto k^{-5/3}$, perhaps with small corrections due to intermittency.

The source of this discrepancy between the DIA and the Kolmogorov theory has long been understood [2]. The DIA does not distinguish between dynamic and kinematic interactions between eddies of widely separated length-scales. Small eddies are convected by large eddies in a purely kinematic way, which should not lead to energy redistribution between scales. The spurious effect of large-scale convection on small scales has been removed from the DIA by use of a Lagrangian description of the flow. This Lagrangian History Direct Interaction Approximation (LHDIA) [3] leads to the Kolmogorov $\frac{5}{3}$-energy spectrum with the Kolmogorov constant $C_K = 1.77$ [see (11) below], which is in reasonable agreement with experiment [4]. However, application of the LHDIA to the problem of turbulent diffusion of a passive scalar does not lead to quantitive agreement with experimental data: The turbulent Prandtl number P_t calculated from the LHDIA [4] is roughly 0.14, much smaller than the experimentally observed $P_t \approx 0.7$–0.9.

In 1977 Forster, Nelson, and Stephen [5] used dynamic renormalization group (RNG) methods, originally developed for the description of the dynamics of critical phenomena [6] to derive velocity correlations generated by the Navier-Stokes equation with a random force term. The ideas expressed in [5] have been used by others in the context of hydrodynamic turbulence [7–10]. The problem is formulated as follows: Consider the d-dimensional space-time Fourier-transformed Navier-Stokes equation for incompressible flow

$$v_l(\hat{k}) = G^0 f_l(\hat{k}) - \frac{i\lambda_0}{2} G^0 P_{lmn}(\mathbf{k}) \int v_m(\hat{q})v_n(\hat{k} - \hat{q})\frac{d\hat{q}}{(2\pi)^{d+1}}, \qquad (1)$$

where the zero mean Gaussian random force $\mathbf{f}(k, \omega)$ is determined by its correlation function

$$\langle f_i(k, \omega) f_j(k', \omega') \rangle = (2\pi)^{d+1} 2 D_0 k^{-y} P_{ij}(\mathbf{k}) \delta(\hat{k} + \hat{k}'). \tag{2}$$

Here

$$G^0 = (-i\omega + \nu_0 k^2)^{-1}, \quad P_{ij}(\mathbf{k}) = \delta_{ij} - k_i k_j / k^2,$$
$$P_{ijk}(\mathbf{k}) = k_k P_{ij}(\mathbf{k}) + k_j P_{ik}(\mathbf{k}), \tag{3}$$

$\hat{k} = (\mathbf{k}, \omega)$, ν_0 is the kinematic viscosity, $\lambda_0 = 1$, and the constant $y > -2$. The problem (1)–(3) is formulated on the interval $0 < k \leq \Lambda_0$ and $-\infty < \omega < \infty$, where Λ_0 is a wavenumber beyond the dissipation wavenumber at which substantial modal excitations cease. The parameter D_0, which determines the intensity of the random force, is discussed below.

The RNG procedure consists of the elimination of modes $\mathbf{v}^>(\hat{k})$ with wavevectors satisfying $\Lambda_0 e^{-r} < k < \Lambda_0$ from the equations of motion for the modes $\mathbf{v}^<(\hat{k})$ with wavevectors from the interval $0 < k < \Lambda_0 e^{-r}$. At this stage, kinematic interactions are excluded by construction, and one can expect physically meaningful results in the limit $k \to 0$. Details of this RNG procedure are given elsewhere [5, 11].

The RNG scale-elimination procedure gives a correction to the bare viscosity ν_0 in terms of an effective viscosity which takes into account the effect of the eliminated modes. The result is

$$\nu_r = \nu_0 \left(1 + A_d \bar{\lambda}_0^{-2} \frac{e^{\varepsilon r} - 1}{\varepsilon} \right), \tag{4}$$

where $\varepsilon = 4 + y - d$, $A_d = \tilde{A}_d S_d / (2\pi)^d$, and

$$\tilde{A}_d = \frac{1}{2} \frac{d^2 - d - \varepsilon}{d(d+2)}, \quad S_d = \frac{(2\pi)^{d/2}}{\Gamma(\frac{1}{2}d)}. \tag{5}$$

The dimensionless expansion parameter $\bar{\lambda}_0$ (which is a Reynolds number) is defined as $\bar{\lambda}_0^2 = D_0 / \nu_0^3 \Lambda_0^\varepsilon$. As we shall see below, the choice of $y = d$ recovers the Kolmogorov scaling in the inertial range.

Varying the cut-off $\Lambda(r) = \Lambda_0 e^{-r}$ we derive differential recursion relations for $\bar{\lambda}(r) = (D_0 / \nu(r)^3 \Lambda(r)^\varepsilon)^{1/2}$ and $\nu(r)$ [11]:

$$d\nu/dr = A_d \nu(r) \bar{\lambda}^2(r), \quad d\bar{\lambda}^2/dr = \bar{\lambda}^2 (\varepsilon - 3 A_d \bar{\lambda}^2). \tag{6}$$

The solutions to (6) are

$$\bar{\lambda}(r) = \bar{\lambda}_0 e^{\varepsilon r/2} \left(1 + 3 A_d \bar{\lambda}_0^{-2} \frac{e^{\varepsilon r} - 1}{\varepsilon} \right)^{-1/2},$$

$$\nu(r) = \nu_0 \left(1 + 3 A_d \bar{\lambda}_0^{-2} \frac{e^{\varepsilon r} - 1}{\varepsilon} \right)^{1/3}.$$

In the limit $r \to \infty$ the coupling parameter $\bar{\lambda}$ (which is an effective Reynolds number) goes to the fixed point $\bar{\lambda}_* = (\varepsilon / 3 A_d)^{1/2}$ and $\nu(\Lambda) = (\frac{3}{4} A_d D_0)^{1/3} \Lambda^{-\varepsilon/3}$. Eliminating all modes with $q > k$, we set $\Lambda = k$ and obtain

$$\nu(k) = (\tfrac{3}{4} A_d D_0)^{1/3} k^{-\varepsilon/3} = 0.4217 (2 D_0 S_d / (2\pi)^d)^{1/3} k^{-4/3} \tag{7}$$

when $y = d = 3$. The coefficient $\tilde{A}_d$ is computed from (5) in the lowest order of ε-expansion ($\varepsilon \to 0$); thus $\tilde{A}_d = 0.2$ in the three-dimensional case $d = 3$.

The energy spectrum can be calculated to lowest order in ε from the equation: $\mathbf{v}(\hat{k}) = G(\hat{k})\mathbf{f}(\hat{k})$, where the propagator $G(\hat{k})$ is evaluated with the k-dependent viscosity (7). The result is

$$E(k) = 1.186(2D_0 S_d/(2\pi)^d)^{2/3} k^{-5/3}. \tag{8}$$

Thus the renormalization group procedure applied to randomly stirred fluid gives the Kolmogorov spectrum in the case $y = d$.

In order to complete the analysis, it is necessary to relate the parameter D_0 to observables. Consider a fluid described by the Navier-Stokes equation

$$\partial v/\partial t + (v \cdot \nabla)v = (1/\rho)\nabla p + \nu_0 \nabla^2 v \tag{9}$$

subject to initial and boundary conditions. We assume that strongly turbulent fluid is characterized in the inertial range of scales by statistically universal scaling laws (Kolmogorov spectrum, etc.) which are independent of initial and boundary conditions. Thus, the system in the universal regime can be described by equations of motion which does not involve any particular initial and boundary conditions, (1)–(2) for example, provided the random force in (1)–(2) is chosen in such a way that it generates velocity fluctuations which are statistically equivalent to the solutions of (9) subject to initial and boundary conditions. In other words, to describe the fluid in the inertial range we may replace (9) with the corresponding system (1)–(2) with a properly chosen force. In this case, it has been shown [12] that if we assume that solutions of equation (9) in the inertial range scale as

$$\nu(k) = N\mathcal{E}^{1/3} k^{-4/3} \tag{10}$$

and

$$E(k) = C_K \mathcal{E}^{2/3} k^{-5/3}, \tag{11}$$

then energy balance in analytical turbulence theory requires that the Kolmogorov constant C_K in (11) and the parameter N in (10) be related as $N/C_K^2 = 0.1904$. Here $\mathcal{E}$ is the rate of energy dissipation in the fluid. Demanding the equivalence of (7)–(8) with (10)–(11) in the inertial range gives

$$2D_0 S_d/(2\pi)^d = 1.594\mathcal{E}, \tag{12}$$

so that $C_K = 1.617$.

A similar RNG procedure [11] applied to the equation of a passive scalar gives the result that the turbulent Prandtl number P_t in the case $y = d = 3$ is

$$P_t^{-1} = \frac{1}{2}\left(-1 + \sqrt{1 + \frac{4(d-1)}{d}\hat{A}_3^{-1}}\right) = 1.3929,$$

so $P_t = 0.7179$. The Batchelor constant Ba is defined by the inertial range scalar fluctuation spectrum. Using energy balance in terms of the k-dependent viscosity at the fixed point, we find [11] $Ba = C_K P_t$, so $Ba = 1.161$. Another calculation [13] of Ba, based on an RNG modified version of the direct-interaction approximation, gives the same result. The results for the turbulent Prandtl number and the Batchelor constant are in close agreement with experimental data [14].

The renormalization group procedure can also be used for deriving averages of different nonlinear operators over the fluctuating velocity field [11]. For example, the skewness factor, which is a dimensionless measure of nonlinear transfer, is defined as

$$S = -\overline{\left(\frac{\partial v_1}{\partial x_1}\right)^3} \bigg/ \left(\overline{\left(\frac{\partial v_1}{\partial x_1}\right)^2}\right)^{3/2} \equiv -\frac{A}{B^{3/2}}, \qquad (13)$$

where

$$A = \overline{\left(\frac{\partial v_1}{\partial x_1}\right)^3} = -i \int q_1 p_1 (k - q - p)_1 v_1(\hat{q}) v_1(\hat{p}) v_1(\hat{k} - \hat{q} - \hat{p}) \frac{d\hat{q}\,d\hat{p}}{(2\pi)^{2d+2}}$$

in the limit $k \to 0$. Decomposing the velocity field into the components $v^>$ and $v^<$ and eliminating small scales using the forced Navier-Stokes equation (1)–(2) we find, in the lowest order in the ε-expansion, that [11]

$$A^< = -i \int q_1 p_1 (k - q - p_1) v_1^<(q) v_1^<(p) v_1^<(\hat{k} - \hat{q} - \hat{p}) \frac{d\hat{q}\,d\hat{p}}{(2\pi)^{2d+2}}$$

$$= -\frac{1}{420} \frac{2D_0 S_d}{(2\pi)^d} \frac{\mathcal{E}}{\nu^3 \Lambda^2}$$

in the limit $k \to 0$ $(r \to \infty)$. The same procedure applied to evaluation of B in (13) gives:

$$B^< = \frac{1}{20} \frac{2D_0 S_d/(2\pi)^d}{\nu}$$

in the limit $k \to 0$ $(r \to \infty)$. Thus

$$S^<(r) = -\frac{A^<}{(B^<)^{3/2}} = 0.1336 \left\{ \frac{2D_0 S_d/(2\pi)^d}{\nu^3 \Lambda^4} \right\}^{1/2} = 0.4878 \qquad (14)$$

when calculated at the fixed point of the RNG calculation. Since $S^<(r)$ does not depend on r in the limit $r \to \infty$, we assume that (14) holds everywhere in the inertial range, so $S = 0.4878$. It should also be noted that the same RNG procedure gives the exact result $S = 0$ in the two-dimensional case $d = 2$.

Another important relation can be derived from the Kolmogorov energy spectrum and formula (7) for the turbulent viscosity. It can be checked readily that the total kinetic energy K in the system is $K = 1.195\mathcal{E}/\nu\Lambda^2$, where Λ is the wave-vector corresponding to the integral scale of turbulence. Combining this relation with (7) and (12) we derive a relation between ν, kinetic energy K, and the mean dissipation rate, namely, $\nu = 0.0837K^2/\mathcal{E}$.

The RNG procedure can be used to evaluate each term of the equations of motion for kinetic energy and dissipation rate. This leads to a so-called K-$\mathcal{E}$ model of turbulence. It can be shown [11] that this RNG model implies that isotropic turbulence decays as $K \propto (t - t_0)^{-1.3807}$, which is close to the experimental data [14] and recent results of direct numerical simulations [15]. The same model, which does not involve any experimentally adjustable parameters, gives the von Karman constant [11] $\kappa = 0.372$ for the logarithmic velocity profile.

The good agreement of the RNG-predicted constants $(C_{\mathrm{K}}, \mathrm{Ba}, \mathrm{P_t}, S, \kappa)$ with experimental data is to some extent surprising since the RNG procedure does not take into account local interactions between eddies of similar size. However, it has been pointed out [8] that the ratio of time-constants corresponding to nonlocal and local interactions is $O(\varepsilon^{1/2})$. Thus, local interactions are weak if ε is assumed small. It remains to be explained why the lowest-order truncation of the RNG expansion in powers of $\varepsilon = 4$ works so well.

ACKNOWLEDGMENTS. We are greteful to Dr. A. Yakhot and Dr. W. Dannevik for numerous suggestions which influenced the course of this work. We would also like to acknowledge support by the Air Force Office of Scientific Research under Contract F49620-85-C-0026, the Office of Naval Research under Contracts N00014-82-C-0451 and N00014-85-K-0201, and the National Science Foundation under Grants MSM-8514128 and ATM-8414410.

REFERENCES

1. R. H. Kraichnan, J. Fluid Mech. **5** (1959), 497.

2. ———, Phys. Fluids **7** (1964), 1723.

3. ———, Phys. Fluids **8** (1965), 575.

4. ———, Phys. Fluids **9** (1966), 1728.

5. D. Forster, D. Nelson, and M. Stephen, Phys. Rev. A **16** (1977), 732.

6. S. K. Ma and G. Mazenko, Phys. Rev. **11** (1975), 4077.

7. C. De Dominicis and P. C. Martin, Phys. Rev. A **19** (1979), 419.

8. J. P. Fournier and U. Frisch, Phys. Rev. A **17** (1978), 747.

9. ———, Phys. Rev. A **28** (1983), 1000.

10. V. Yakhot, Phys. Rev. A **23** (1981), 1486.

11. V. Yakhot and S. A. Orszag, J. Sci. Comput. **1** (1986), 1.

12. R. H. Kraichnan, J. Fluid Mech. **47** (1971), 525.

13. W. Dannevik and V. Yakhot (in preparation).

14. A. S. Monin and A. M. Yaglom, *Statistical fluid mechanics*. Vol. 2, M.I.T. Press, Cambridge, Mass., 1975.

15. M. J. Lee and W. C. Reynolds, Report #TF-024, Department of Mechanical Engineering, Stanford Univ., 1985.

PRINCETON UNIVERSITY, PRINCETON, NEW JERSEY 08544, USA

Uniformity and Irregularity

J. BECK

Introduction. The purpose of this report is to draw attention to some non-trivial connections between the ("continuous") theory of irregularities of distribution and discrete mathematics. The object of the theory of irregularities of distribution is to measure the uniformity (or nonuniformity) of sequences and point sets. For instance: how uniformly can an arbitrary set of N points in the unit cube be distributed relative to a given family of "nice" sets (e.g., boxes with sides parallel to the coordinate axes, rotated boxes, balls, all convex sets). This theory lies on the border of many branches of mathematics (number theory, discrete geometry, combinatorics, etc.) and has very important applications, e.g., in numerical integration. Here we focus, of course, on the combinatorial aspects of the theory.

As an illustration, we shall discuss first two problems of a discrete nature which have fascinating connections with this "continuous" theory. The first question is concerned with balanced two-colorings of finite sets in a square (the problem is due to G. Tusnády). Let $\mathcal{P} = \{\mathbf{p}_1, \ldots, \mathbf{p}_N\}$ be a distribution of N points in the unit square $[0,1)^2$. Let $f: \mathcal{P} \to \{-1, +1\}$ be a "two-coloring" of $\mathcal{P}$. Let B denote any rectangle in $[0,1)^2$ with sides parallel to the coordinate axes (an *aligned* rectangle, in short). Consider the function

$$T(N) = \sup_{\mathcal{P}} \inf_{f} \sup_{B} \left| \sum_{\mathbf{p}_i \in \mathcal{P} \cap B} f(\mathbf{p}_i) \right|,$$

where the supremum is taken over all subsets $\mathcal{P} \subset [0,1)^2$, $\#\mathcal{P} = N$, and all aligned rectangles B in $[0,1)^2$, and the infimum is taken over all "two-colorings" f of $\mathcal{P}$. Tusnády conjectured that $N^{\varepsilon} > T(N) \to \infty$. A positive answer was obtained in [1]:

$$c_1 (\log N)^4 > T(N) > c_2 \log N.$$

The proof of the lower bound was nonconstructive. Recently Roth [19] investigated the following explicit construction: let

$$\mathcal{P}_\alpha = \{(\{n\alpha\}, n \cdot N^{-1}) \in \mathbf{R}^2 : 0 \le n \le N - 1\} \subset [0,1)^2,$$

where $\{\beta\}$ stands for the fractional part of the real number β and α is an irrational number whose continued fraction has bounded partial quotients. Roth's theorem states that given any two-coloring $f \colon P_\alpha \to \{-1, +1\}$ of the set P_α, one can find an aligned rectangle B with deviation

$$\left| \sum_{\mathbf{p} \in P_\alpha \cap B} f(\mathbf{p}) \right| > c(\alpha) \log N.$$

The set P_α is well studied in irregularities of distribution and belongs to the class of "most uniformly" distributed N-element sets relative to aligned rectangles. What is more, Roth's proof is based on the so-called "Roth-Halász orthogonal function method" in irregularities of distribution.

The second problem is as follows: For what set of N points on the unit sphere is the sum of all $\binom{N}{2}$ euclidean distances between points maximal, and what is the maximum?

Let S^k denote the surface of the unit sphere in $\mathbf{R}^{k+1}$. Let $P = \{\mathbf{p}_1, \dots, \mathbf{p}_N\}$ be a distribution of N points on S^k. Let $|\mathbf{p}_i - \mathbf{p}_j|$ denote the usual euclidean distance of $\mathbf{p}_i$ and $\mathbf{p}_j$. We define

$$L(N, k, P) = \sum_{1 \le i < j \le N} |\mathbf{p}_i - \mathbf{p}_j| \quad \text{and} \quad L(N, k) = \max_{P} L(N, k, P),$$

where the maximum is taken over all $P \subset S^k$, $\#P = N$.

The determination of $L(N, k)$ is a long-standing open problem in discrete geometry. For $k = 1$, the solution is given by the regular N-gon (see Fejes Tóth [13]). It is also known that for $N = k + 2$, the regular simplex is optimal. For $N > k + 2$ and $k \ge 2$, the exact value of $L(N, k)$ is unknown. The reason for this is that if N is sufficiently large compared to k, then there are no "regular" configurations on the sphere, so the extremal point system(s) is (are), as expected, quite complicated and "ad hoc."

Since the determination of $L(N, k)$ seems to be hopeless, it is natural to compare the discrete sum $L(N, k, P)$ with the integral (the solution of the "continuous relaxation" of the distance problem)

$$\frac{N^2}{2} \cdot \frac{1}{\sigma(S^k)} \int_{S^k} |\mathbf{x}_0 - \mathbf{x}| \, d\sigma(\mathbf{x}) = c_0(k) N^2, \tag{1}$$

where σ denotes the surface area, $d\sigma(\mathbf{x})$ represents an element of the surface area on S^k, $\mathbf{x}_0 = (1, 0, \dots, 0) \in \mathbf{R}^{k+1}$. Note that the constants $c_0(k)$ can be calculated explicitly (e.g., $c_0(1) = 2/\pi$, $c_0(2) = 2/3$), and on the left-hand side of (1) the correct coefficient is $N^2/2$ rather than $\binom{N}{2}$, since in the definition of $L(N, k, P)$ we can write $1 \le i \le j \le N$ in place of $1 \le i < j \le N$ without changing the value. In this way Stolarsky [25] has discovered a beautiful identity. It states, roughly speaking, that the discrete sum $L(N, k, P)$ plus a measure of how far the set P deviates from uniform distribution is constant. Thus the sum of distances is maximized by a well-distributed set of points. Combining

Stolarsky's identity with a result in irregularities of distribution, one can obtain further information on the order of magnitude of $L(N, k)$ (see [5]).

1. Measure-theoretic discrepancy. It is time to give a brief survey of irregularities of distribution. It was initiated by a conjecture of van der Corput and the work of van Aardenne–Ehrenfest, and owes its current prominence to the contribution of K. F. Roth and W. M. Schmidt. We refer the reader to Schmidt's book [22]; see also the forthcoming book by Chen and Beck [9].

Let $\mathcal{P} = \{\mathbf{p}_1, \mathbf{p}_2, \mathbf{p}_3, \ldots\}$ be a completely arbitrary infinite discrete set of points in euclidean k-space $\mathbf{R}^k$. (We can assume that $\mathcal{P}$ has density 1, otherwise the results below are trivial.) Let $B(\mathbf{c}, r) \subset \mathbf{R}^k$ be the ball with center $\mathbf{c}$ and radius r. In 1969 Schmidt [20] proved the following pioneering result: *Let $x > 1$. Then there exists a ball $B(\mathbf{c}, r) \subset \mathbf{R}^k$ with $r \leq x$ and*

$$\left| \sum_{\mathbf{p} \in \mathcal{P} \cap B(\mathbf{c}, r)} 1 - \mathrm{vol}(B(\mathbf{c}, r)) \right| > x^{(k-1)/2 - \varepsilon}.$$

Here the exponent $((k-1)/2 - \varepsilon)$ of x cannot be replaced by $((k-1)/2 + \varepsilon)$. Essentially improving on the earlier result of Schmidt, the following good localization of the ball was proved [6]: *Let $x > 1$. Then there exists a ball $B(\mathbf{c}, r) \subset [0, x)^k$ such that*

$$\left| \sum_{\mathbf{p} \in \mathcal{P} \cap B(\mathbf{c}, r)} 1 - \mathrm{vol}(B(\mathbf{c}, r)) \right| > x^{(k-1)/2 - \varepsilon}.$$

We mention next a far-reaching generalization of the case of balls. Given a compact and convex body $A \subset \mathbf{R}^k$, denote by $\sigma(\partial A)$ the surface area of the boundary ∂A of A. The following result shows that for convex bodies the "rotation discrepancy" is always large and behaves like the square-root of the surface area of the boundary (see [6]; see also Montgomery [16]): *Let $A \subset \mathbf{R}^k$ be a compact and convex body. Then there exists $A' = A(\tau', \mathbf{v}', \lambda')$ obtained from A by a similarity transformation of rotation $\tau' \in \mathrm{SO}(k)$, translation $\mathbf{v}' \in \mathbf{R}^k$, and contraction $\lambda' \in (0, 1]$ such that*

$$\left| \sum_{\mathbf{p} \in \mathcal{P} \cap A'} 1 - \mathrm{vol}(A') \right| > c(k)(\sigma(\partial A))^{1/2}. \tag{2}$$

Note that (2) is essentially the best possible (see [4]).

The next result answers an old question of Roth (see [3]): *Let $\mathcal{P}$ be an arbitrary finite set in the disc $B(\mathbf{0}, r) \subset \mathbf{R}^2$ of radius r. There exists a disc-segment A (i.e., an intersection of $B(\mathbf{0}, r)$ with a half-plane) such that*

$$\left| \sum_{\mathbf{p} \in \mathcal{P} \cap A} 1 - \mathrm{area}(A) \right| > c \cdot r^{1/2} \cdot (\log r)^{-7/2}. \tag{3}$$

Inequality (3) is also nearly sharp. If rotation is forbidden, the situation undergoes a complete change. To avoid the considerable technical difficulties caused by higher dimensions, we restrict ourselves to the two-dimensional case.

Again, let $\mathcal{P} = \{\mathbf{p}_1, \mathbf{p}_2, \mathbf{p}_3, \ldots\}$ be an infinite discrete set in $\mathbf{R}^2$. Given a compact and convex region $A \subset \mathbf{R}^2$, write

$$D[\mathcal{P}; A] = \sum_{\mathbf{p} \in \mathcal{P} \cap A} 1 - \text{area}(A).$$

For any real number $\lambda \in [-1, 1]$ and any vector $\mathbf{v} \in \mathbf{R}^2$, set $A(\mathbf{v}, \lambda) = \{\lambda\mathbf{x} + \mathbf{v}: \mathbf{x} \in A\}$. Clearly $A(\mathbf{v}, \lambda)$ is a homothetic image of A. (Note that reflection across the origin is allowed, as $-1 \leq \lambda \leq 1$.) Let

$$\Delta[\mathcal{P}; A] = \sup_{|\lambda| \leq 1, \mathbf{v}} |D[\mathcal{P}; A(\mathbf{v}, \lambda)]|$$

and define the (usual) *discrepancy* of A by

$$\Delta[A] = \inf_{\mathcal{P}} \Delta[\mathcal{P}; A],$$

where the infimum is taken over all infinite discrete sets $\mathcal{P} \subset \mathbf{R}^2$. In contrast to the "rotation discrepancy," the (usual) discrepancy $\Delta[A]$ of a convex region A depends mainly on the "smoothness" of its boundary arc (see [7]). If A is sufficiently smooth, then $\Delta[A]$ has essentially the same order of magnitude as for circular discs. If we have no assumption on the smoothness of A, we can guarantee only a much smaller discrepancy:

$$\Delta[A] > c(\log \text{area}(A))^{1/2}. \tag{4}$$

Inequality (4) probably remains true if we replace the exponent $1/2$ by the exponent 1. If true, this is best possible. The important particular case of squares was proved by Halász [14]. Note that Halász's theorem implies Schmidt's solution [21] of the classical van der Corput's conjecture. If A is a polygon of "few" sides, then (4) is not very far from the truth in the sense that we cannot expect larger discrepancy than a power of $(\log \text{area}(A))$.

To get further information of the intermediate cases, one can introduce the concept of an *approximability number* $\xi(A)$ (which describes how well a convex region A can be approximated by an inscribed polygon of few sides. The related results in [7] can be summarized as follows:

$$(\xi(A) + \log(\text{area}(A)))^{c_1} > \Delta[A] > (\xi(A) + \log(\text{area}(A)))^{c_2}.$$

Note that if we know the equation of the boundary arc of A, then the determination, or at least the estimation, of $\xi(A)$ is an easy elementary problem.

We now consider very briefly the case when both rotation and contraction are forbidden, i.e., we study the supremum of the discrepancy over the family $\{A + \mathbf{v}: \mathbf{v} \in \mathbf{R}^2\}$. In contrast to the previous cases, the discrepancy function does not necessarily tend to infinity as $\text{area}(A)$ tends to infinity. (Let, e.g., $\mathcal{P} = \mathbf{Z}^2$ and $A = [0, n)^2$, $n \geq 1$ integer.) However, in the case of circular discs, we can guarantee "large" discrepancy for any single value of the radius (see [8]—the problem is due to P. Erdős).

Finally, we mention the "Great Open Problem" of this field.

CONJECTURE. *Let P be an arbitrary finite set in the cube $[0, x)^k$, $x \geq 2$, $k \geq 3$. Does there exist an aligned box $B \subset [0, x)^k$ such that*

$$\left| \sum_{\mathbf{p} \in P \cap B} 1 - \mathrm{vol}(B) \right| > c(k)(\log x)^{k-1}?$$

Since 1954 the best lower bound is $(\log x)^{(k-1)/2}$ (see Roth [17]).

The proofs of all these lower bounds are based on tools in harmonic analysis (e.g., modified Rademacher functions, Riesz products, summability kernels). The common idea of all different approaches—to "blow up the trivial error." Note that the use of Fourier analysis in the opposite direction (i.e., to show the uniformity of sequences) is a classical idea and goes back to H. Weyl ("Weyl's criterion" and its quantitative versions, e.g., "Erdős-Turán inequality").

The proofs of the upper bounds are based on ideas from number theory, probability theory, and "combinatorial discrepancy theory."

2. Combinatorial discrepancy. The basic problem of the so-called "combinatorial discrepancy theory" is how to color with two colors a set as uniformly as possible with respect to a given family of subsets. What we want to achieve is that the coloring be nearly balanced in each of the subsets considered. As a beautiful example, we mention Roth's theorem on long arithmetic progressions [18]. Roth proved that coloring the integers from 1 to N red and blue in any fashion, there exists an arithmetic progression such that the difference of the numbers of red and blue terms in this progression has absolute value $> c \cdot N^{1/4}$. In this section we discuss some general upper bounds concerning hypergraphs. They are interesting on their own, but also they have applications to different structures.

Let X be an arbitrary finite set and $\mathcal{H} = \{Y_1, Y_2, Y_3, \ldots\}$ an arbitrary family of subsets of X. We would like to find a two-coloring $f : X \to \{-1, +1\}$ of the underlying set X such that $\max_{Y \in \mathcal{H}} |\sum_{x \in Y} f(x)|$ is as small as possible. In other words, let

$$\mathrm{dis}(\mathcal{H}) = \min_f \max_{Y \in \mathcal{H}} \left| \sum_{x \in Y} f(x) \right|,$$

where the minimum is taken over all $f : X \to \{-1, +1\}$. We call $\mathrm{dis}(\mathcal{H})$ the *combinatorial discrepancy* of the family $\mathcal{H}$.

Let $d(\mathcal{H})$ be the maximum degree of $\mathcal{H}$, i.e.,

$$d(\mathcal{H}) = \max_{x \in X} \#\{Y \in \mathcal{H} : x \in Y\}.$$

The following result gives an upper bound on $\mathrm{dis}(\mathcal{H})$ which depends only on

$d(\mathcal{H})$, i.e., the "local size" of $\mathcal{H}$ (see Fiala and Beck [**10**]): *For any finite family* $\mathcal{H}$,

$$\text{dis}(\mathcal{H}) < 2d(\mathcal{H}). \tag{5}$$

Note that the proof of (5) gives a good (polynomial time) algorithm which constructs the two-coloring. There is an important point to emphasize here: In many applications $d(\mathcal{H})$ is much less than $\#X$ and $\#\mathcal{H}$. As an illustration, we derive the upper bound $T(N) < c(\log N)^4$ (see Introduction), which is clearly equivalent to the following result: *Let* $A = (a_{ij})$, *where* $a_{ij} = 0$ *or* 1, *be a matrix of size* $N \times N$. *Then there exist "signs"* $\varepsilon_{ij} = \pm 1$ *such that*

$$\left| \sum_{i=1}^{s} \sum_{j=1}^{t} \varepsilon_{ij} a_{ij} \right| < c(\log N)^4 \quad \text{for all } s, t \in \{1, 2, \dots, N\}. \tag{6}$$

We can assume that $N = 2^l$, where l is an integer. For $0 \le p, q \le l$, we partition the matrix A into 2^{p+q} submatrices, splitting the horizontal side of the matrix into 2^p equal pieces and the vertical side of the matrix into 2^q equal pieces. There are $(l+1)^2 \sim (\log N)^2$ such partitions. Let us call a submatrix of A *special* if it occurs in one of these partitions, and let $\mathcal{H}$ be the collection of all these special submatrices. Then by (5), there exists an assignment of ± 1's so that the absolute value of the sum of signed entries in each of the special submatrices is less than $2d(\mathcal{H}) \le 2(l+1)^2$. Note, however, that any submatrix of A containing the lower left corner of A is the union of at most l^2 disjoint special submatrices, and (6) follows.

In higher dimensions, the same argument gives the following generalization of (6): *Let* $A = (a_{\mathbf{n}})$, *where* $a_{\mathbf{n}} = 0$ *or* 1, *be a k-dimensional matrix of size* $N \times \cdots \times N$. *Then there exist "signs"* $\varepsilon_{\mathbf{n}} = \pm 1$ *such that*

$$\left| \sum_{\mathbf{n}: \mathbf{n} \le \mathbf{m}} \varepsilon_{\mathbf{n}} a_{\mathbf{n}} \right| < c(k)(\log N)^{2k} \tag{7}$$

for all $\mathbf{m} = (m_1, \dots, m_k)$ *satisfying* $1 \le m_i \le N$ $(1 \le i \le k)$. *Here* $\mathbf{n} \le \mathbf{m}$ *if and only if* $n_i \le m_i$ *for all* $i \in [1, k]$.

We conjecture that inequality (5) can be improved to $\text{dis}(\mathcal{H}) < (d(\mathcal{H}))^{1/2+\varepsilon}$, where $d = d(\mathcal{H}) > d_0(\varepsilon)$. The following result justifies this conjecture when both $\#X$ and $\#\mathcal{H}$ are "subexponential" functions of $d = d(\mathcal{H})$ ([**7**]—see inequality (9) below).

Let X be a finite set and $\mathcal{H}$ a family of subsets of X. Suppose that there is a second family $\mathcal{G}$ of subsets of X such that

(i) *$d(\mathcal{G}) \le d$, and*

(ii) *every $Y \in \mathcal{H}$ can be represented as the union of at most t disjoint elements of $\mathcal{G}$.*

Then

$$\text{dis}(\mathcal{H}) < c(t \cdot d \cdot \log d \cdot \log \#\mathcal{H})^{1/2} \cdot (\log \#X). \tag{8}$$

In the particular case $\mathcal{G} = \mathcal{H}$, we obtain

$$\operatorname{dis}(\mathcal{H}) < c(d(\mathcal{H}))^{1/2} \cdot \log \#X \cdot \log \#\mathcal{H}. \tag{9}$$

We next apply (8) to improve on (7). Let $X = \{\mathbf{n}: a_{\mathbf{n}} = 1\}$, let $\mathcal{H}$ be the family of all submatrices $(a_{\mathbf{n}})$ $(\mathbf{n} \leq \mathbf{m})$, $\mathcal{G}$ be the family of all k-dimensional *special* submatrices, $N = 2^l$, $d = (l+1)^k$, and $t = l^k$. By (8) we have that

$$\left| \sum_{\mathbf{n}:\mathbf{n} \leq \mathbf{m}} \varepsilon_{\mathbf{n}} a_{\mathbf{n}} \right| \leq c(k, \varepsilon)(\log N)^{k+3/2+\varepsilon} \quad \text{for all } \mathbf{m}.$$

The proof of (8) is "nonconstructive." The new idea is a combination of probabilistic arguments with the pigeon-hole principle. As far as we know, the first application of this idea is in [2]. (It was shown that Roth's theorem on long arithmetic progressions is essentially the best possible.) Later the same method was utilized by Spencer and Beck (see [11, 12, 24]).

For further results in combinatorial discrepancy theory, see Vera Sós [23] and Lovász, Spencer, Vesztergombi [15].

Finally, we have to remark that there are no general lower bounds on the combinatorial discrepancy of hypergraphs. To illustrate the difficulties, we mention a more than fifty-year-old question of Erdős.

CONJECTURE (ERDŐS). *Let* $f(n) = \pm 1$ *be a function on the set of positive integers. Given arbitrary large constant* c *there is a* d *and an* m *so that* $|\sum_{i=1}^{m} f(i \cdot d)| > c$.

REFERENCES

1. J. Beck, *Balanced two-colorings of finite sets in the square*. I, Combinatorica **1** (1981), 327–335.

2. ____, *Roth's estimate of discrepancy of integer sequences is nearly sharp*, Combinatorica **1** (1981), 319–325.

3. ____, *On a problem of K. F. Roth concerning irregularities of point distribution*, Invent. Math. **74** (1983), 477–487.

4. ____, *Some upper bounds in the theory of irregularities of distribution*, Acta Arith. **43** (1984), 115–130.

5. ____, *Sums of distances between points on a sphere*, Mathematika **31** (1984), 33–41.

6. ____, *Irregularities of distribution*. I, Acta Math. (to appear).

7. ____, *Irregularities of distribution*. II, submitted to J. London Math. Soc.

8. ____, *On a problem of Erdős in the theory of irregularities of distribution*, submitted to Math. Ann.

9. J. Beck and W. Chen, *Irregularities of distribution*, Cambridge Univ. Press (to appear).

10. J. Beck and T. Fiala, *Integer-making theorems*, Discrete Appl. Math. **3** (1981), 1–8.

11. J. Beck and J. Spencer, *Integral approximation sequences*, Math. Programming **30** (1984), 88–98.

12. ____, *Well-distributed 2-colorings of integers relative to long arithmetic progressions*, Acta Arith. **43** (1984), 287–294.

13. L. Fejes Tóth, *On the sum of distances determined by a point set*, Acta Math. Acad. Sci. Hungar. **7** (1956), 397–401.

14. G. Halász, *Irregularities of distribution*, manuscript (1985).

15. L. Lovász, J. Spencer, and K. Vesztergombi, *Discrepancy of set-systems and matrices*, European J. Combin. **7** (1986), 151–160.

16. H. L. Montgomery *Irregularities of distribution by means of power sums*, submitted to the Proceedings of the Congress de Teoria de los Numeros, Bilbao, 1984.

17. K. F. Roth, *On irregularities of distribution*, Mathematika **1** (1954), 73–79.

18. ____, *Remark concerning integer sequences*, Acta Arith. **9** (1964), 257–260.

19. ____, *On a theorem of Beck*, Glasgow Math. J. **27** (1985), 195–201.

20. W. M. Schmidt, *Irregularities of distribution*. IV, Invent. Math. **7** (1969), 55–82.

21. ____, *Irregularities of distribution*. VII, Acta Arith. **21** (1972), 45–50.

22. ____, *Lectures on irregularities of distribution*, Tata Inst. Fund. Res., Bombay, 1977.

23. V. T. Sós, *Irregularities of partitions*, London Math. Soc. Lecture Note Series, vol. 82, Cambridge Univ. Press, 1983, pp. 201–246.

24. J. Spencer, *Six standard deviations suffice*, Trans. Amer. Math. Soc. **289** (1985), 679–706.

25. K. B. Stolarsky, *Sums of distances between points on a sphere*. II, Proc. Amer. Math. Soc. **41** (1973), 575–582.

EÖTVÖS LORÁND UNIVERSITY, BUDAPEST, HUNGARY

Proceedings of the International Congress of Mathematicians
Berkeley, California, USA, 1986

Face Numbers of Complexes and Polytopes

ANDERS BJÖRNER

Introduction. Let C be a finite *polyhedral complex*, i.e., a finite nonempty collection of convex polytopes in $\mathbf{R}^d$ such that (i) when $P, Q \in C$ then $P \cap Q$ is a face of both and (ii) if $P \in C$ and Q is a face of P then $Q \in C$. Let $f_i = f_i(C)$ be the number of i-dimensional members of C, $i \geq 0$, called the ith *face number*.

The discovery that the numbers f_i are governed by interesting relations is due to Euler, whose formula $f_0 - f_1 + f_2 = 2$ for the boundary complex of a 3-dimensional convex polytope, published in 1752, had great impact on the subsequent development of combinatorics and topology. By the second half of the nineteenth century it was "known" (without fully satisfactory proof, cf. [20]) that the boundary complex of a d-dimensional convex polytope satisfies the *generalized Euler relation*

$$f_0 - f_1 + f_2 - \cdots + (-1)^{d-1} f_{d-1} = 1 + (-1)^{d-1}. \tag{1}$$

An incompleteness in the 1852 proof of Schläfli was recently rectified by Brugges-ser and Mani [15]. The first complete proof of (1) is due to Poincaré [36, 37], as a special case of the following vast generalization, which is now usually called the *Euler-Poincaré formula*

$$f_0 - f_1 + f_2 - \cdots + (-1)^{d-1} f_{d-1} = 1 + \beta_0 - \beta_1 + \beta_2 - \cdots + (-1)^{d-1}\beta_{d-1}. \tag{2}$$

This relation holds for *any* finite polyhedral complex C with Betti numbers in reduced homology $\beta_i = \mathrm{rank}\, \tilde{H}_i(C, \mathbf{Z})$. As is well known, (2) is true also for more general cell complexes, but in this paper attention will be limited to finite polyhedral complexes, and in particular to the subclass of simplicial complexes, which are the two classes of greatest interest in combinatorics.

The sequence $f(C) = (f_0, f_1, \ldots, f_{d-1})$, now usually called the *f-vector*, has been intensively studied for boundary complexes of convex d-polytopes, and much information in addition to (1) is available (e.g., see [20, 21, 35, 45]). Part of the motivation for interest in polytopal f-vectors has come from linear programming in connection with efforts to better understand the combinatorial structure of solution sets to systems of linear inequalities (see [29, 51]). The

Partially supported by the National Science Foundation.

f-vectors $f(\mathcal{C})$ of more general classes of complexes have also received increasing attention in recent years (see [13, 41, 44]).

This paper will present a brief and selective account of some developments in the study of f-vectors over the last few years. Particular attention will be given to the fact that the linear relations (1) and (2) of Euler, Schläfli and Poincaré have been complemented by other linear and nonlinear relations to achieve a complete characterization of the f-vectors for *simplicial* convex polytopes (Billera, Lee, McMullen and Stanley; Theorem 4.1) and, similarly, a complete characterization of the compatible pairs of f-vectors and Betti sequences for finite *simplicial* complexes (Björner and Kalai; Theorem 2.1). In contrast, no such characterizations are presently known for nonsimplicial polytopes and for general polyhedral complexes.

1. M-sequences. In this section we establish some definitions and review two important combinatorial theorems. Without further mention, all complexes considered are assumed to be finite.

For any two integers $k, n \geq 1$, there is a unique way of writing

$$n = \binom{a_k}{k} + \binom{a_{k-1}}{k-1} + \cdots + \binom{a_i}{i},$$

so that $a_k > a_{k-1} > \cdots > a_i \geq i > 0$. Then define

$$\partial_{k-1}(n) = \binom{a_k}{k-1} + \binom{a_{k-1}}{k-2} + \cdots + \binom{a_i}{i-1},$$

and

$$\partial^{k-1}(n) = \binom{a_k - 1}{k-1} + \binom{a_{k-1} - 1}{k-2} + \cdots + \binom{a_i - 1}{i-1}.$$

Also, let $\partial_{k-1}(0) = \partial^{k-1}(0) = 0$. The number-theoretic functions ∂_k and ∂^k, $k \geq 0$, play the role of numerical minimal boundary operators for sets and multisets, respectively. $\mathbf{N}^{(\infty)}$ will denote the set of ultimately vanishing sequences of nonnegative integers.

The following result was found for simplicial complex by Kruskal [30] and Katona [25]. The extension to polyhedral complexes is due to Wegner [50].

THEOREM 1.1. *For $f = (f_0, f_1, \ldots) \in \mathbf{N}^{(\infty)}$ the following conditions are equivalent:*
 (i) *f is the f-vector of a simplicial complex;*
 (ii) *f is the f-vector of a polyhedral complex;*
 (iii) *$\partial_k(f_k) \leq f_{k-1}$, for all $k \geq 1$.*

By a *multicomplex* $\mathcal{M}$ we shall understand a nonempty collection of monomials in finitely many variables, such that if $m \in \mathcal{M}$ then every divisor of m is also in $\mathcal{M}$. Let $f_i(\mathcal{M})$ denote the number of members in $\mathcal{M}$ of degree i, $i \geq 0$, and call $f(\mathcal{M}) = (f_0, f_1, \ldots)$ the *f-vector* of $\mathcal{M}$. The next result is essentially due to Macaulay [31], cf. [42].

THEOREM 1.2. *For* $f = (f_0, f_1, \dots) \in \mathbf{N}^\infty$ *the following conditions are equivalent*:

(i) f *is the* f-*vector of a multicomplex*;

(ii) f *is the Hilbert function of a finitely generated graded $\underline{k}$-algebra $R = \bigoplus_{i \geq 0} R_i$, such that $R_0 \cong \underline{k}$ and R_1 generates R, $\underline{k}$ a field*;

(iii) $f_0 = 1$, *and* $\partial^k(f_{k+1}) \leq f_k$, *for all* $k \geq 1$.

A sequence $f = (f_0, f_1, \dots)$ will be called an *M-sequence* if it satisfies the conditions of Theorem 1.2. The formal similarity between parts of Theorems 1.1 and 1.2 is explained by a common generalization found by Clements and Lindström [**17**].

2. f-vectors and Betti numbers. It has been known at least since Mayer [**32**] that every linear relation between the face numbers and the Betti numbers of a simplicial complex is a multiple of the Euler-Poincaré formula (2). Also, it follows from a result of D. Sullivan (see [**1**, pp. 212, 223]) that if a function of simplicial complexes is topologically invariant and depends only on the f-vector, then it in fact depends only on the Euler characteristic. These results seem to cast doubt on the possible existence of general relations between face numbers and Betti numbers other than (2). However, it turns out that a complete set of relations can be formulated, as shown by the following result of Björner and Kalai [**13**, **14**].

THEOREM 2.1. *For* $f = (f_0, f_1, \dots) \in \mathbf{N}^{(\infty)}$ *and* $\beta = (\beta_0, \beta_1, \dots) \in \mathbf{N}^{(\infty)}$ *the following conditions are equivalent*:

(i) f *is the* f-*vector and* β *the sequence of Betti numbers (over an arbitrary coefficient field) of some simplicial complex*;

(ii) f *is the* f-*vector of some simplicial complex having the homotopy type of a wedge consisting for each $i \geq 0$ of β_i spheres of dimension i*;

(iii) *let* $\chi_{k-1} = \sum_{j \geq k}(-1)^{j-k}(f_j - \beta_j), k \geq 0$; *then*

$$\chi_{-1} = 1, \tag{3}$$

and

$$\partial_k(\chi_k + \beta_k) \leq \chi_{k-1}, \quad \textit{for all } k \geq 1. \tag{4}$$

Notice that condition (3) is the Euler-Poincaré formula. The characterization is obviously independent of field characteristic. For a d-dimensional simplicial complex C condition (4) gives d additional nonlinear relations satisfied by $f(C)$ and $\beta(C)$. The homological interpretation of these relations is that the space of k-cycles and the space of $(k-1)$-boundaries must satisfy $\partial_k(\dim Z_k) \leq B_{k-1}$, for all $k \geq 1$.

Theorem 2.1 provides means for the study of f-vectors which are realized by simplicial complexes having some specified sequence of Betti numbers. For instance, it can be shown that there exists a unique componentwise minimal such f-vector. Furthermore, this minimal f-vector is characterized by equality in relations (4) for all $k \geq 1$. Dually, if the f-vector is fixed, there exists a unique componentwise maximal compatible Betti sequence.

A collection $\mathcal{A}$ of nonempty subsets of a finite set E, $|E| = n$, having no proper inclusions among its members (i.e., $S, T \in \mathcal{A}$, $S \subseteq T$ implies $S = T$) is called a *Sperner family of rank n*. Letting f_i be the number of members of cardinality $i + 1$, we define the *f-vector* $f(\mathcal{A}) = (f_0, f_1, \ldots, f_{n-1})$.

THEOREM 2.2. *For $\beta = (\beta_0, \beta_1, \ldots, \beta_{n-1}) \in \mathbf{N}^n$ and a field $\underline{k}$, the following conditions are equivalent:*

(i) *β is the Betti sequence over $\underline{k}$ of some simplicial complex on at most $n + 1$ vertices;*

(ii) *β is the Betti sequence over $\underline{k}$ of some polyhedral complex on at most $n + 1$ vertices;*

(iii) *β is the f-vector of some Sperner family of rank n;*

(iv)

$$\beta = 0 \quad or \quad \partial_{j+1}(\cdots \partial_{n-2}(\partial_{n-1}(\beta_{n-1}) + \beta_{n-2}) + \beta_{n-3} \cdots) + \beta_j \leq \binom{n}{j+1}, \tag{5}$$

where j is minimal such that $\beta_j \neq 0$.

The equivalence of the first three conditions was shown by Björner and Kalai [13, 14]. The last two conditions were proved equivalent by Clements [16] and Daykin, Godfrey and Hilton [18]. A combinatorial theorem of Sperner [39] states that the maximum size of a Sperner family of rank n is $\binom{n}{[n/2]}$. Hence, we conclude that the same number gives the tight upper bound for the sum of Betti numbers of complexes on at most $n + 1$ vertices. (Recall that we are dealing with *reduced* Betti numbers, so $\beta_0 + 1$ rather than β_0 gives the number of connected components.)

As a small numerical example, consider the integer vectors $f = (15, 20, 19, 9)$, $\beta = (5, 2, 3, 2)$, and $\beta' = (8, 4, 2, 2)$. The pair (f, β) fails to satisfy relation (4) for $k = 2$. Hence, it is not realizable as the f-vector and Betti sequence of any simplicial complex. On the other hand, (f, β') satisfies (3) and (4) with equality throughout, so f is the minimal f-vector achieved by complexes having Betti sequence β'. Finally, β satisfies relation (5) for all $n \geq 11$, so the minimal number of vertices needed to build a complex which realizes β as a Betti sequence is 12.

3. h-vectors. For some classes of complexes the f-vectors lie in the positive integer span of a certain canonical basis, and more insight is obtained by studying the coefficients with respect to that basis than by working directly with the f-vectors themselves. Important examples will be given in this and the next section.

Fix $d > 0$, and let $(f_0, f_1, \ldots, f_{d-1})$ be the f-vector of a $(d-1)$-dimensional simplicial complex C. Also, let $f_{-1} = 1$. Define integers h_i by

$$\sum_{i=0}^{d} h_i x^{d-i} = \sum_{i=0}^{d} f_{i-1}(x-1)^{d-i}. \tag{6}$$

Then $h = (h_0, h_1, \ldots, h_d)$ is called the h-vector of C.

A simplicial complex C is said to be *Cohen-Macaulay* (*over the field $\underset{\tilde{}}{k}$*) if $\tilde{H}_i(lk\sigma, \underset{\tilde{}}{k}) = 0$ for all $i < \dim(lk\sigma)$ and all $\sigma \in C$ (including $\sigma = \phi$). Here, $lk\sigma = \{\tau \in C | \sigma \cup \tau \in C, \ \sigma \cap \tau = \phi\}$. Equivalently, by a theorem of Reisner [38], this condition holds if a certain commutative ring associated with C is a Cohen-Macaulay ring. See Stanley [41, 44] for details regarding this connection with commutative algebra, and for the following characterization result.

THEOREM 3.1. *For $h = (h_0, h_1, \ldots, h_d) \in \mathbf{Z}^{d+1}$ the following conditions are equivalent:*

(i) *h is the h-vector of a $(d-1)$-dimensional Cohen-Macaulay complex (over an arbitrary field);*

(ii) *h is an M-sequence.*

Equation (6) is equivalent to

$$f_{j-1} = \sum_{i=0}^{d} h_i \binom{d-i}{d-j}, \qquad 0 \le j \le d, \tag{7}$$

so upper bounds and lower bounds on the h-vector translate directly into corresponding bounds on the f-vector. By Theorems 1.2 and 3.1, the h-vector of a $(d-1)$-dimensional Cohen-Macaulay complex counts certain monomials in $h_1 = f_0 - d$ variables, so

$$(1, f_0 - d, 0, 0, \ldots, 0) \le h \le \left(1, f_0 - d, \binom{f_0 - d + 1}{2}, \ldots, \binom{f_0 - 1}{d}\right).$$

Part of this upper bound was used by Stanley [40] to prove the upper bound conjecture for spheres. The lower bound was used in a somewhat more general form by Björner [9, 10] to derive several inequalities for matroid complexes.

It is of interest to characterize the h-vectors for special subclasses of Cohen-Macaulay complexes, but beyond what is mentioned in the next section little in this direction is known. Some special properties of h-vectors of matroid complexes are given in [41] and [9]. Some results concerning h-vectors of polyhedral balls can be found in [8]. The lower bound $h_i \ge \binom{d}{i}$ for h-vectors of $(d-1)$-dimensional Cohen-Macaulay complexes admitting a free $\mathbf{Z}_2$-action is proven in [47].

4. g-vectors. We will consider in this section three classes of $(d-1)$-dimensional simplicial complexes:

(i) *polytopal spheres* (i.e., boundary complexes of d-dimensional simplicial convex polytopes), (ii) *spheres* (i.e., triangulations of the standard topological sphere of points of unit norm in $\mathbf{R}^d$), and (iii) *homology spheres* (i.e., complexes such that $\tilde{H}_i(lk\sigma, \underline{k}) \cong 0$ for $i < \dim(lk\sigma)$, and $\cong \underline{k}$ for $i = \dim(lk\sigma)$, for all faces σ, including $\sigma = \phi$; the field $\underline{k}$, once chosen, is irrelevant for the following discussion). Each class is known to be a proper subclass of its successor; e.g., the smallest sphere which is not polytopal is 3-dimensional and has 8 vertices (see [20]).

The h-vector $(h_0, h_1, \ldots, h_d)$ of a homology sphere satisfies the *Dehn-Sommerville equations*

$$h_i = h_{d-i}, \qquad 0 \le i \le d. \tag{8}$$

For polytopal spheres these relations were discovered by Dehn in 1905, for $d \le 5$, and in general by Sommerville in 1927. They were shown to extend to all Eulerian manifolds (a class of complexes somewhat larger than the homology spheres) by Klee [28].

Define the *g-vector* $(g_0, g_1, \ldots, g_{[d/2]})$ and the *g-polynomial* $g(x) = g_0 + g_1 x + \cdots + g_{[d/2]} x^{[d/2]}$ of a homology sphere C by

$$g_0 = h_0 \quad \text{and} \quad g_i = h_i - h_{i-1}, \quad \text{for } 1 \le i \le [d/2]. \tag{9}$$

Clearly, by (8) the g-vector determines the h-vector and hence also the f-vector of C. In fact, multiplication of formula (6) by $(x - 1)$ yields, in view of (8),

$$x^{d+1} g\left(\frac{1}{x}\right) - g(x) = \sum_{i=0}^{d} f_{i-1}(x - 1)^{d+1-i}, \tag{10}$$

or, equivalently,

$$f_{j-1} = \sum_{i=0}^{[d/2]} g_i \left(\binom{d+1-i}{d+1-j} - \binom{i}{d+1-j} \right), \qquad 0 \le j \le d. \tag{11}$$

Hence, the f-vectors of homology spheres are governed by their g-vectors much like the f-vectors of Cohen-Macaulay complexes are governed by their h-vectors.

The importance of g-vectors for polytopal spheres was first perceived by McMullen [34], who conjectured the following characterization result. Sufficiency was subsequently shown by Billera and Lee [6, 7] and necessity by Stanley [43].

THEOREM 4.1. *For* $g = (g_0, g_1, \ldots, g_{[d/2]}) \in \mathbf{Z}^{[d/2]+1}$ *the following conditions are equivalent:*

(i) g *is the g-vector of a $(d-1)$-dimensional polytopal sphere;*

(ii) g *is an M-sequence.*

We remark here that the generalized Euler formula (1) is equivalent to the Dehn-Sommerville relation $h_0 = h_d$, and so was coded into the preceding characterization at an early stage.

A reformulation of Theorem 4.1 incorporating (11) reveals the structure of f-vectors of polytopal spheres particularly well. Define a $([d/2] + 1) \times d$ matrix $M_d = (m_{ij})$ by $m_{ij} = \binom{d+1-i}{d-j} - \binom{i}{d-j}$, $0 \le i \le [d/2]$, $0 \le j \le d-1$. Then M_d has nonnegative entries, and the lower left corner $[d/2] \times [d/2]$ submatrix is upper triangular with ones on the diagonal. The reformulation is then: *The mapping*

$$\mu : g \mapsto g \cdot M_d \qquad\qquad (12)$$

gives a one-to-one correspondence from the set of f-vectors of multicomplexes of rank $\leq [d/2]$ on exactly t variables to the set of f-vectors of $(d-1)$-dimensional polytopal spheres on $d + 1 + t$ vertices, $t \geq 0$.

It was pointed out by McMullen [**34**] that Theorem 4.1 would imply many of the major results concerning f-vectors of simplicial polytopes which were at that time proven by separate means. The "*McMullen correspondence*" (12) gives quick access to some such results more or less by inspection. For instance: (i) the componentwise maximal f-vector of a polytopal $(d-1)$-sphere on $d+1+t$ vertices is

$$\mu\left(\left(1, t, \binom{t+1}{2}, \ldots, \binom{t + [d/2] - 1}{[d/2]}\right)\right),$$

which is the "upper bound theorem" of McMullen [**33**]; (ii) the componentwise minimal f-vector of a polytopal $(d-1)$-sphere on $d+1+t$ vertices is $\mu((1, t, 0, \ldots, 0))$, which is the "lower bound theorem" of Barnette [**3, 4**]. The validity of the upper bound theorem was subsequently extended to all homology spheres by Stanley [**40**], whereas Barnette's original method actually proves the lower bound theorem for all homology spheres (cf. [**23**]).

One of the outstanding open problems concerning f-vectors is to find a characterization of the f-vectors of spheres. It has been conjectured that the characterization in Theorem 4.1 extends to all spheres, and if so, most likely also to all homology spheres.

For certain subclasses of polytopal spheres there are geometric constraints which influence their f-vectors, so one is led to seek the underlying constraints on their g-vectors. The following such result was proven by Stanley [**47**] in response to a conjecture by the author.

THEOREM 4.2. *The g-vector of a centrally symmetric polytopal $(d-1)$-sphere satisfies $g_i \geq \binom{d}{i} - \binom{d}{i-1}$, for $0 \leq i \leq [d/2]$.*

The implied lower bounds for f-vectors (i.e., $f_{d-1} \geq 2^d + 2(n-d)(d-1)$, and $f_j \geq 2^{j+1}\binom{d}{j+1} + 2(n-d)\binom{d}{j}$ for $0 \leq j \leq d-2$) were conjectured by Bárány and Lovász [**2**]. It is an interesting open problem to find the sharp upper bounds to (or, even better, a characterization of) the g-vectors of centrally symmetric simplicial polytopes.

The rows of the M_d-matrix can be shown to increase along the first half and then to decrease along the last quarter of their length. A similar analysis can be carried out for general homology spheres leading to the following result of Björner [**11**].

THEOREM 4.3. *The f-vector of a $(d-1)$-dimensional homology sphere, $d \geq 3$, satisfies the following conditions.*

(a) $f_0 < f_1 < \cdots < f_{[d/2]-1} \le f_{[d/2]}$ and $f_{[3(d-1)/4]} > \cdots > f_{d-2} > f_{d-1}$, and the limits $[d/2]$ and $[3(d-1)/4]$ are best possible for such inequalities even among polytopal spheres.

(b) $f_i < f_{d-2-i}$ and $f_i \le f_{d-1-i}$, for $0 \le i \le [(d-3)/2]$.

5. Remarks.

(5.1) *Methods.* Within the limited space available there is unfortunately no room to discuss the variety of methods which were used to obtain the results that have been discussed. These include methods from combinatorics (extremal set theory, polyhedral techniques), from algebra (exterior algebra, commutative rings), from topology (simplicial homology), and from algebraic geometry (co-homology of toric varieties). The reader is referred to the references for more detailed information.

One noteworthy feature is that all the nonlinear relations which occur in the characterization results (as well as in several other results on f-vectors) are inequalities involving the two families of functions ∂_k and ∂^k. Their ubiquitous appearance in so many results of this kind gives a certain sense of unity to the theorems in the area, even though, at present, the methods of proof are quite diverse. A fundamental problem that seems to lie at the heart of the difficulty with a more unified approach is to understand how local homological information (i.e., Betti numbers of links) influences the global properties of the f-vector. For a more complete discussion, see [13].

(5.2) *Nonsimplicial complexes.* In principle, Theorem 2.1 characterizing compatible pairs (f, β) must imply Theorem 1.1 (projection on the first coordinate) and Theorem 2.2 (projection on the second coordinate) for *simplicial* complexes, but no direct proof of these implications is known. Based on the evidence that both these latter results hold for polyhedral complexes as well, Björner and Kalai [13] conjecture that also Theorem 2.1 extends to polyhedral complexes.

(5.3) *Nonsimplicial polytopes.* Very little is known about the f-vectors of boundary complexes of general d-dimensional convex polytopes. For $d = 3$ a complete characterization was found by E. Steinitz in 1906, and for general d it is known that the only linear relation satisfied by all such f-vectors is the generalized Euler formula (1) (see [20]).

Recently the Dehn-Sommerville equations, and hence the concept of g-vector, have been extended to general polytopes by Stanley [46, 48]. Unfortunately, the generalized g-vector does not alone uniquely determine the f-vector for nonsimplicial polytopes, but it raises some interesting theoretical questions which may in the end turn out to have substantial bearing on f-vectors. We end with a few comments about this.

Briefly, the generalized g-polynomials $g_P(x)$ are defined as follows via a recursive procedure over the face-lattices $L(P)$ of polytopes P, starting with the one-point lattice $L(\phi)$:

(i) $g_\phi(x) = 1$;

(ii) $\deg g_P(x) \le [\dim P/2]$;

(iii)

$$x^{\dim P+1} g_P\left(\frac{1}{x}\right) - g_P(x) = \sum_{\substack{Q \in L(P) \\ Q \neq P}} g_Q(x) \cdot (x-1)^{\dim P - \dim Q}. \quad (13)$$

Stanley [**46, 48**] shows that the polynomial $g_P(x)$ is well defined by this recursion. If Q is a simplex, then $g_Q(x) = 1$, so for simplicial polytopes this reduces to the usual g-polynomial (cf. equation (10)), and hence also reproves the Dehn-Sommerville equations.

What can be said about the coefficients g_i of $g_P(x)$ in general? Do they form an M-sequence? Stanley has asked whether at least $g_i \geq 0$. For *rational* polytopes (i.e., polytopes whose vertices have rational coordinates) this can be deduced from results of R. MacPherson and others on the intersection cohomology of toric varieties (see the references in [**46**]). However, as first shown by M. Perles (cf. [**20**]) there exist polytopes whose combinatorial type cannot be realized with rational coordinates. (Perles's example is 8-dimensional with 12 vertices; recently Sturmfels [**49**] constructed a 6-dimensional example with 13 vertices.) For nonrational, nonsimplicial polytopes very little is known about the generalized g-vector. It is easy to see that $g_1 = f_0 - d - 1 \geq 0$. The inequality $g_2 \geq 0$ was recently proved for general polytopes by Kalai [**23**]. Some additional results on generalized g-vectors and their relationship to f-vectors appear in Kalai [**24**].

There are some striking similarities between g-polynomials of polytopes and the Kazhdan-Lusztig polynomials of Coxeter groups, which should not be ignored. These latter polynomials are defined by a recursion similar to (13) over intervals in Bruhat order, and Kazhdan and Lusztig [**26**] conjecture that their coefficients are always nonnegative. For some results in this direction based on intersection cohomology, see [**27**]. Now, every interval in Bruhat order is known to be isomorphic to the poset of faces of a regular cell decomposition of a sphere (cf. [**12**]), and hence is structurally very similar to the face lattice of a polytope. It would be desirable to find a common approach to polynomials generated over spherical posets in this manner, leading to a common explanation for the nature of their coefficients (at best: a combinatorial interpretation). In terms of intrinsic interest and because of the far-reaching implications, ranging from the f-vectors of general polytopes to the representation theory of the symmetric groups, this is, in the author's opinion, one of the outstanding current problems in combinatorics.

REFERENCES

1. M. K. Agoston, *Algebraic topology*, Marcel Dekker, New York, 1976.
2. I. Bárány and L. Lovász, *Borsuk's theorem and the number of facets of centrally symmetric polytopes*, Acta Math. Hungar. **40** (1982), 323–329.
3. D. W. Barnette, *The minimum number of vertices of a simple polytope*, Israel J. Math. **10** (1971), 121–125.

4. ____, *A proof of the lower bound conjecture for convex polytopes*, Pacific J. Math. **46** (1973), 349–354.

5. L. J. Billera, *Polyhedral theory and commutative algebra*, Mathematical Programming: The State of the Art (A. Bachem, M. Grötschel, and B. Korte, editors), Springer-Verlag, Berlin, 1983, pp. 57–77.

6. L. J. Billera and C. W. Lee, *Sufficiency of McMullen's conditions for f-vectors of simplicial polytopes*, Bull. Amer. Math. Soc. (N.S.) **2** (1980), 181–185.

7. ____, *A proof of the sufficiency of McMullen's conditions for f-vectors of simplicial convex polytopes*, J. Combin. Theory Ser. A **31** (1981), 237–255.

8. ____, *The numbers of faces of polytope pairs and unbounded polyhedra*, European J. Combin. **2** (1981), 307–322.

9. A. Björner, *Homology of matroids*, Institut Mittag-Leffler, 1979; Combinatorial Geometries (The Theory of Matroids), Vol. 3 (N. White, editor), Cambridge Univ. Press (to appear).

10. ____, *Some matroid inequalities*, Discrete Math. **31** (1980), 101–103.

11. ____, *The unimodality conjecture for convex polytopes*, Bull. Amer. Math. Soc. (N.S.) **4** (1981), 187–188.

12. ____, *Posets, regular CW complexes and Bruhat order*, European J. Combin. **5** (1984), 7–16.

13. A. Björner and G. Kalai, *On f-vectors and homology* (Proceedings of the 3rd International Conference on Combinatorial Mathematics, New York, 1985), Ann. New York Acad. Sci. (to appear).

14. ____, *An extended Euler-Poincaré theorem*, Preprint.

15. M. Bruggesser and P. Mani, *Shellable decompositions of cells and spheres*, Math. Scand. **29** (1971), 197–205.

16. G. Clements, *A minimization problem concerning subsets*, Discrete Math. **4** (1973), 123–128.

17. G. Clements and B. Lindström, *A generalization of a combinatorial theorem of Macaulay*, J. Combin. Theory **7** (1969), 230–238.

18. D. E. Daykin, J. Godfrey, and A. J. W. Hilton, *Existence theorems for Sperner families*, J. Combin. Theory Ser. A **17** (1974), 245–251.

19. J. Eckhoff, *Über kombinatorisch-geometrische Eigenschaften von Komplexen und Familien konvexer Mengen*, J. Reine Angew. Math. **313** (1980), 171–188.

20. B. Grünbaum, *Convex polytopes*, Interscience, 1967.

21. ____, *Polytopes, graphs and complexes*, Bull. Amer. Math. Soc. **76** (1970), 1131–1201.

22. G. Kalai, *A characterization of f-vectors of families of convex sets in R^d*. Part I: *Necessity of Eckhoff's conditions*, Israel J. Math. **48** (1984), 175–195.

23. ____, *Rigidity and the lower bound theorem. I*, Invent. Math. (to appear).

24. ____, *A new basis of polytopes*, J. Combin. Theory Ser. A (to appear).

25. G. O. H. Katona, *A theorem of finite sets*, Theory of Graphs (Proc. Colloq., Tihany, 1966, P. Erdös and G. Katona, editors), Academic Press, New York, and Adadémia Kiadó, Budapest, 1968, pp. 187–207.

26. D. Kazhdan and G. Lusztig, *Representations of Coxeter groups and Hecke algebras*, Invent. Math. **53** (1979), 165–184.

27. ____, *Schubert varieties and Poincaré duality*, Proc. Sympos. Pure Math., vol. 36, Amer. Math. Soc., Providence, R.I., 1980, pp. 185–203.

28. V. Klee, *A combinatorial analogue of Poincaré's duality theorem*, Canad. J. Math. **16** (1964), 517–531.

29. ____, *Convex polyhedra and mathematical programming*, Proc. Internat. Congr. Math. (Vancouver, 1974), Vol. 1, Canad. Math. Congr., Montreal, 1974, pp. 485–490.

30. J. B. Kruskal, *The number of simplices in a complex*, Mathematical Optimization Techniques (R. Bellman, editor), Univ. of California Press, Berkeley–Los Angeles, 1963, pp. 251–278.

31. F. S. Macaulay, *Some properties of enumeration in the theory of modular systems*, Proc. London Math. Soc. **26** (1927), 531–555.

32. W. Mayer, *A new homology theory*. II, Ann. of Math. (2) **43** (1942), 594–605.

33. P. McMullen, *The maximum numbers of faces of a convex polytope*, Mathematika **17** (1970), 179–184.

34. _____, *The numbers of faces of simplicial polytopes*, Israel J. Math. **9** (1971), 559–570.

35. P. McMullen and G. C. Shephard, *Convex polytopes and the Upper Bound Conjecture*, London Math. Soc. Lecture Notes, vol. 3, Cambridge Univ. Press, London, 1971.

36. H. Poincaré, *Sur la généralisation d'un théorème d'Euler relatif aux polyèdres*, C. R. Acad. Sci. Paris **117** (1893), 144–145.

37. _____, *Complément à l'analysis situs*, Rend. Circ. Mat. Palermo **13** (1899), 285–343.

38. G. A. Reisner, *Cohen-Macaulay quotients of polynomial rings*, Advances in Math. **21** (1976), 30–49.

39. E. Sperner, *Ein Satz über Untermenge einer endlichen Menge*, Math. Z. **27** (1928), 544–548.

40. R. Stanley, *The upper bound conjecture and Cohen-Macaulay rings*, Studies in Appl. Math. **54** (1975), 135–142.

41. _____, *Cohen-Macaulay complexes*, Higher Combinatorics (M. Aigner, editor), Reidel, Dordrecht, 1977, pp. 51–62.

42. _____, *Hilbert functions of graded algebras*, Advances in Math. **28** (1978), 57–83.

43. _____, *The number of faces of simplicial convex polytopes*, Advances in Math. **35** (1980), 236–238.

44. _____, *Combinatorics and commutative algebra*, Birkhäuser, Boston, 1983.

45. _____, *The number of faces of simplicial polytopes and spheres*, Discrete Geometry and Convexity (J. E. Goodman et al., editors), Ann. New York Acad. Sci., vol. 440, New York Acad. Sci., New York, 1985, pp. 212–223.

46. _____, *Generalized h-vectors, intersection cohomology of toric varieties, and related results*, Proceedings of Symposium on Commutative Algebra and Combinatorics, Kyoto, 1985 (M. Nagata, editor), North-Holland (to appear).

47. _____, *On the number of faces of centrally symmetric simplicial polytopes*, Preprint.

48. _____, *Enumerative combinatorics*, Vol. 1, Wadsworth, Monterey, Calif., 1986.

49. B. Sturmfels, *Boundary complexes of convex polytopes cannot be characterized locally*, J. London Math. Soc. (to appear).

50. G. Wegner, *Kruskal-Katona's theorem in generalized complexes*, Finite and Infinite Sets, Vol. 2, Colloq. Math. Soc. János Bolyai, vol. 37, North-Holland, Amsterdam, 1984, pp. 821–827.

51. V. A. Yemelichev, M. M. Kovalev, and M. K. Kravtsov, *Polytopes, graphs and optimization*, "Nauka", Moscow, 1981; English transl., Cambridge Univ. Press, Cambridge, 1984.

MASSACHUSETTS INSTITUTE OF TECHNOLOGY, CAMBRIDGE, MASSACHUSETTS 02139, USA

UNIVERSITY OF STOCKHOLM, STOCKHOLM, SWEDEN

Intersection Theorems for Finite Sets
and Geometric Applications

PETER FRANKL

1. **Introduction.** Let X be an n-element set and $\mathbf{F} \subset 2^X$ a family of distinct subsets of X. Suppose that the members of $\mathbf{F}$ satisfy some conditions. What is the maximum (or minimum) value of $|\mathbf{F}|$—this is the generic problem in extremal set theory. There have been far too many papers and results in this area to be overviewed in such a short paper.

Therefore, we will only deal with some intersection theorems. The simplest is

THEOREM 0. *Suppose that* $\mathbf{F} \subset 2^X$ *satisfies*

$$F \cap F' \neq \varnothing \quad \text{for all } F, F' \in \mathbf{F}. \tag{1.1}$$

Then $|\mathbf{F}| \leq 2^{n-1}$ *holds.*

PROOF. If $F \in \mathbf{F}$ then $(X - F) \notin \mathbf{F}$ implying $|\mathbf{F}| \leq \frac{1}{2}2^n$. ∎

A family $\mathbf{F}$ satisfying (1.1) is called *intersecting*. More generally, for a positive integer t, $\mathbf{F}$ is called *t-intersecting* if any two of its members intersect in at least t elements.

The two central theorems concerning t-intersecting families are the Erdős-Ko-Rado theorem and the Katona theorem (cf. §§2 and 3).

The following general problem—by now rich in results and applications—was proposed by Vera T. Sós [S]:

Given a set $L = \{l_1, \ldots, l_s\}$ of nonnegative integers, a family $\mathbf{F}$ is called an *L-system* if $|F \cap F'| \in L$ holds for all distinct $F, F' \in L$.

Determine or estimate $m(n, L) = \max\{|\mathbf{F}|: \mathbf{F} \subset 2^X \text{ is an } L\text{-system}\}$.

After surveying some general results on L-systems in §4, in §5 we consider the case of one excluded intersection size. The interest in this special case was generated by two exciting conjectures of Erdős. The first of them said that if $\mathbf{F} \subset \binom{X}{k}$ avoids the intersection size l and $|\mathbf{F}|$ is maximal, then (for $n > n_0(k, l)$) $\mathbf{F}$ has all intersection sizes less than l (an l-packing) or all greater than l (($l+1$)-intersecting).

The second said that there exists a positive ε such that every $\mathbf{F} \subset 2^X$ with $|\mathbf{F}| > (2-\varepsilon)^n$ contains two sets with intersection of size $\lfloor n/4 \rfloor$. That is, excluding $\lfloor n/4 \rfloor$ brings down the maximum size of $\mathbf{F}$ exponentially.

Both conjectures were settled recently. Extensions of the second conjecture led to the following geometrical result. Let A be the vertex set of a nondegenerate simplex in R^d. There exists $\varepsilon = \varepsilon(A) > 0$ such that for every partition of R^n, $n \geq d$, into less than $(1+\varepsilon)^n$ classes, one of the classes contains a set A' congruent (isometric) to A. This and similar results are discussed in §6.

2. The Erdős-Ko-Rado theorem. Let us introduce the notation $\binom{X}{k} = \{A \subset X: |A| = k\}$.

For $A_0 \in \binom{X}{t}$ the family $\mathbf{A}_0 = \{A \in \binom{X}{k}: A_0 \subset A\}$ is t-intersecting and satisfies $|\mathbf{A}_0| = \binom{n-t}{k-t}$.

THEOREM 2.1 (ERDŐS-KO-RADO [**EKR**]). *Let $n > k > t \geq 1$ be integers and suppose that $\mathbf{F} \subset \binom{X}{k}$ is t-intersecting. Then for $n \geq n_0(k,t)$ one has*

$$|\mathbf{F}| \leq \binom{n-t}{k-t}. \tag{2.1}$$

REMARK 2.2. Erdős told me that they proved this theorem already before World War II, however, at that time there was very little interest in such results, that is why it was not published until 1961.

By now, the best possible bound for $n_0(k,t)$ is known:

$$n_0(k,t) = (k-t+1)(t+1).$$

This was proved by Erdős, Ko, and Rado for $t = 1$, by the author [**F1**] for $t \geq 15$, and recently by Wilson [**W1**] for all t.

To see that (2.1) is no longer true for $n < (k-t+1)(t+1)$, define families $\mathbf{A}_i$ for $0 \leq i \leq k-t$:

Let $A_i \in \binom{X}{t+2i}$ and $\mathbf{A}_i = \{A \subset \binom{X}{k}: |A \cap A_i| \geq t+i\}$.

Now $\mathbf{A}_i$ is t-intersecting and simple computation shows that $|\mathbf{A}_0| \gtreqless |\mathbf{A}_1|$ according as $n \gtreqless (k-t+1)(t+1)$ holds.

The following conjecture, if true, would determine the maximum size of t-intersecting families in general.

CONJECTURE 2.3 [**F1**]. *Suppose that $\mathbf{F} \subset \binom{X}{k}$ is t-intersecting. Then $|\mathbf{F}| \leq \max_{0 \leq i \leq k-t} |\mathbf{A}_i|$ holds.*

At present this conjecture appears hopelessly difficult in general. However, there are some partial results (cf. [**Hu, F1**, and **FF4**]).

It is natural to ask what families are achieving equality in (2.1). For $n > (k-t+1)(t+1)$, $\mathbf{A}_0$ is the unique family with this property (cf. [**HM**] for $t = 1$, [**W1**] for $t \geq 1$). In [**F1**] it is proved that for $t \geq 15$ even for $n = (k-t+1)(t+1)$ the only optimal families are $\mathbf{A}_0$ and $\mathbf{A}_1$. This is probably true for $2 \leq t \leq 14$ as well.

However, the case $t = 1$, $n = 2k$ is different. Then the statement of the Erdős-Ko-Rado theorem is very easy:

$$|\mathbf{F}| \leq \binom{2k-1}{k-1} = \frac{1}{2}\binom{2k}{k}.$$

Indeed, for every $A \in \binom{X}{k}$, at most one of the two sets $A, X - A$ can be present in an intersecting family. Since $n = 2k$, to every $A \in \binom{X}{k}$ there is a unique set $B \in \binom{X}{k}$ with $A \cap B = \varnothing$, namely $B = X - A$.

This observation makes possible the construction of very many intersecting families $\mathbf{F} \subset \binom{X}{k}$ with $|\mathbf{F}| = \frac{1}{2}\binom{2k}{k}$.

Let T and S be disjoint subsets of X, $|T| = t$, $|S| = k - t + 1$. Define

$$\mathbf{B}_1 = \left\{ B \in \binom{X}{k} : (T \subset B \text{ and } S \cap B \neq \varnothing) \text{ or } (B \subset T \cup S) \right\}.$$

Note that $\mathbf{B}_1$ is t-intersecting with

$$|\mathbf{B}_1| = \binom{n-t}{k-t} - \binom{n-k-1}{k-t} + t.$$

THEOREM 2.4. *Suppose that $\mathbf{F} \subset \binom{X}{k}$ is t-intersecting with $|\bigcap \mathbf{F}| < t$. Then for $n \geq n_1(k, t)$*

$$|\mathbf{F}| \leq \max\{|\mathbf{A}_1|, |\mathbf{B}_1|\} \quad holds. \tag{2.2}$$

REMARK 2.5. For $t = 1$, (2.2) was proved by Hilton and Milner [HM] along with $n_1(k, 1) = 2k$. A simple proof was given recently by Füredi and the author [FF3]. For other proofs cf. [A] and [M]. For $t \geq 2$, (2.2) was proved in [F2]. However, the value of $n_1(k, t)$ is still unknown.

An interesting class of problems was proposed by Erdős, Rothschild, and Szemerédi (cf. [E1]):

Let $0 < c < 1$ be a real number. Let $\mathbf{F} \subset \binom{X}{k}$ be an intersecting family.

Let $f(n, k, c)$ denote the maximum possible size of $\mathbf{F}$ if $\mathbf{F}$ satisfies the additional degree condition, namely that every element of X is contained in at most $c|\mathbf{F}|$ members of $\mathbf{F}$.

Füredi [Fü1] has many results on this problem. Let me cite one, which is particularly beautiful. Suppose that $l \geq 2$, $Y \subset X$, $|Y| = l^2 + l + 1$, and $\mathbf{L} \subset \binom{Y}{l+1}$ is the collection of all lines of a projective plane of order l on Y. Define the intersecting family $\mathbf{F}(\mathbf{L})$ by:

$$\mathbf{F}(\mathbf{L}) = \left\{ F \in \binom{X}{k} : F \cap Y \in \mathbf{L} \right\}.$$

Clearly, $|\mathbf{F}(\mathbf{L})| = (l^2 + l + 1)\binom{n-l^2-l-1}{k-l-1}$ and $\mathbf{F}$ satisfies the above condition with $c = (l+1)/(l^2 + l + 1)$.

THEOREM 2.6 (FÜREDI [Fü1]). *Suppose that $c = (l+1)/(l^2 + l + 1)$ where $l \geq 2$ is an integer, $l < k$. Then for $n > \tilde{n}(k, l)$ one of the following two possibilities occurs.*

(i) *There is no projective plane of order l and $f(n, k, c) = O(n^{k-l-2})$.*

(ii) *There exists a projective plane* **L** *of order l and $f(n, k, c) = |\mathbf{F}(\mathbf{L})|$.*

The determination of $\tilde{n}(k, l)$ appears difficult. In [**FF5**] $\tilde{n}(k, l) < c_l k$ is proved, where c_l is a constant depending only on l. In [**FF5**] many related problems are considered, e.g., for t-intersecting families.

3. The Katona theorem. What is the maximum size of a t-intersecting family? This problem was asked by Erdös, Ko, and Rado and it was completely solved by Katona [**K1**].

Let us define $\mathbf{K}_0 = \{K \subset X: |K| \geq (n + t)/2\}$. It is clear that for K, $K' \in \mathbf{K}_0$ one has $|K \cap K'| \geq |K| + |K'| - n \geq t$, i.e., $\mathbf{K}_0$ is t-intersecting. If $n + t$ is odd, then one can add further sets to $\mathbf{K}_0$ without destroying the t-intersecting property. For Y an $(n - 1)$-subset of X, define

$$\mathbf{K}_1 = \mathbf{K}_0 \cup \binom{Y}{(n - 1 + t)/2}.$$

One can describe $\mathbf{K}_1$ also by $\mathbf{K}_1 = \{K \subset X: |K \cap Y| \geq (|Y| + t)/2\}$.

THEOREM 3.1 (KATONA [**K1**]). *Suppose that* $\mathbf{F} \subset 2^X$ *is t-intersecting, $n \geq t \geq 1$. Then one of the following two possibilities occurs.*

(i) $n + t$ *is even and* $|\mathbf{F}| \leq |\mathbf{K}_0|$.

(ii) $n + t$ *is odd and* $|\mathbf{F}| \leq |\mathbf{K}_1|$.

REMARK 3.2. Note that the case $t = 1$ is simply Theorem 0 from the introduction. Katona proved also that $\mathbf{K}_0$ and $\mathbf{K}_1$ are the unique optimal families for $t \geq 2$.

For the proof of Theorem 3.1 Katona proved another important theorem, concerning the shadows of t-intersecting families. For a family $\mathbf{F}$ and an integer h let $\Delta_h(\mathbf{F})$ be the h-shadow of $\mathbf{F}$, i.e.,

$$\Delta_h(\mathbf{F}) = \left\{ H \in \binom{X}{h}: H \subset F, \text{ for some } F \in \mathbf{F} \right\}.$$

Let A be a $(2k - t)$-element set and $\mathbf{A} = \binom{A}{k}$. Then $\mathbf{A}$ is t-intersecting and clearly $|\Delta_h(\mathbf{A})| = \binom{2k-t}{h}$ holds for all $h \leq k$.

THEOREM 3.3 (KATONA [**K1**]). *Suppose that* $\mathbf{F} \subset \binom{X}{k}$ *is t-intersecting. Then for $k - t \leq h \leq k$ one has*

$$|\Delta_h(\mathbf{F})|/|\mathbf{F}| \geq \binom{2k - t}{h} \bigg/ \binom{2k - t}{k}. \tag{3.1}$$

Note that the RHS of (3.1) is 1 for $h = k - t$ or k but it is strictly greater otherwise. To illustrate the strength of this theorem, let us use it to prove the $t = 1$ case of the Erdös-Ko-Rado theorem. This proof is due to Katona [**K1**].

Suppose that $\mathbf{G} \subset \binom{X}{h}$, $\mathbf{G}$ is intersecting, $n \geq 2h$. We want to show that $|\mathbf{G}| \leq \binom{n-1}{h-1}$ holds. Consider $\mathbf{F} = \{X - G: G \in \mathbf{G}\}$. Since $\mathbf{G}$ is intersecting, $\Delta_h(\mathbf{F}) \cap \mathbf{G} = \varnothing$ holds. Also, for $G, G' \in \mathbf{G}$ we have $|(X - G) \cap (X - G')| = |X| - |G \cup G'| \geq n - 2h + 1$, i.e., $\mathbf{F}$ is $(n - 2h + 1)$-intersecting.

Applying Theorem 3.3 with $k = n - h$ and $t = n - 2h + 1$ gives

$$|\Delta_h(\mathbf{F})| \geq |\mathbf{F}| \binom{n-1}{h} \Big/ \binom{n-1}{n-h} = |\mathbf{G}|(n-h)/h.$$

Using $|\Delta_h(\mathbf{F})| + |\mathbf{G}| \leq \binom{n}{h}$, we infer $|\mathbf{G}| \leq (h/n)\binom{n}{h} = \binom{n-1}{h-1}$, as desired.

For a famiy $\mathbf{F} \subset \binom{X}{k}$ and $0 \leq h \leq k$ the hth containment matrix $M_h(\mathbf{F})$ is a $|\Delta_h(\mathbf{F})|$ by $|\mathbf{F}|$ matrix whose rows are indexed by the members G of $\Delta_h(\mathbf{F})$, the columns by $F \in \mathbf{F}$ and the (G, F)-entry is 1 or 0 according to whether $G \subset F$ or $G \not\subset F$ hold.

THEOREM 3.4 (FRANKL AND FÜREDI [FF1]). *Suppose that* $\mathbf{F} \subset \binom{X}{k}$, $0 \leq g < k$, *and* $\mathrm{rank}(M_g(\mathbf{F})) = |\mathbf{F}|$ *holds. Then*

$$|\Delta_h(\mathbf{F})|/|\mathbf{F}| \geq \binom{k+g}{h} \Big/ \binom{k+g}{k} \qquad \text{*holds for all* } g \leq h \leq k. \qquad (3.2)$$

REMARK 3.5. It can be shown that if $\mathbf{F}$ is t-intersecting then always

$$\mathrm{rank}(M_{k-t}(\mathbf{F})) = |\mathbf{F}|$$

holds and therefore Theorem 3.4 is a generalization of Theorem 3.3.

In this context we should mention the Kruskal-Katona theorem, one of the most important theorems in extremal set theory. Given m, k, and h, this theorem describes the minimum size of $\Delta_k(\mathbf{F})$ over all families $\mathbf{F}$, consisting of k-element sets and satisfying $|\mathbf{F}| = m$.

Let $\mathbf{N}$ be the set of positive integers. We define a total order (called the reverse lexicographic order) on $\binom{\mathbf{N}}{k}$ by setting $A < B$ if and only if $\max\{a\colon a \in (A - B)\} < \max\{b\colon b \in (B - A)\}$. E.g., $\{10, 11\} < \{1, 12\}$.

THEOREM 3.5 (KRUSKAL [Kr] AND KATONA [K2]). *The size of the h-shadow is minimized over all* $\mathbf{F} \subset \binom{\mathbf{N}}{k}$ *with* $|\mathbf{F}| = m$, *by taking the smallest m sets in the reverse lexicographic order.*

REMARK 3.6. The optimal families are not always unique in the Kruskal-Katona theorem. Füredi and Griggs [FG] characterized the values of (m, k, h) for which there is a unique optimal family. See Mörs [M] for some refinement of Theorem 3.5 and [F3] for a simple proof.

4. **Families with prescribed intersections.** An L-system $\mathbf{F} \subset \binom{X}{k}$ is called an (n, k, L)-*system.*

DEFINITION 4.1. $m(n, k, L) = \max\{|\mathbf{F}|\colon \mathbf{F}$ is an (n, k, L)-system$\}$.

Note that with this definition the Erdös-Ko-Rado theorem can be rephrased as

$$m(n, k, \{t, t+1, \ldots, k-1\}) = \binom{n-t}{k-t} \qquad \text{for } n \geq n_0(k, t). \qquad (4.1)$$

THEOREM 4.2 (RAY-CHAUDHURI AND WILSON [**RW**]).

$$m(n, k, L) \leq \binom{n}{|L|}. \tag{4.2}$$

For $n > n_0(, L)$ the inequality (4.2) was improved in the following way. Let $L = \{l_1, \ldots, l_s\}$ with $0 \leq l_1 < \cdots < l_s < k$.

THEOREM 4.3 (DEZA, ERDÖS, AND FRANKL [**DEF**]). *Suppose that* $n > n_0(k, L)$. *Then*

$$m(n, k, L) \leq \prod_{l \in L} \frac{n - l}{k - l}, \tag{4.3}$$

moreover, $m(n, k, L) = O(n^{s-1})$ *unless* $(l_2 - l_1)|(l_3 - l_2)| \cdots |(l_s - l_{s-1})|(k - l_s)$.

Note that (4.3) implies (4.1). Deza [**D**] proves that equality is possible in (4.3) for given $n, k, l_s, l_{s-1}, \ldots, l_1$ if and only if there exists a matroid on n vertices, of rank $s + 1$, in which all hyperflats have size k and all i-flats have size l_{i+1} for $i = 0, 1, \ldots, s - 1$. Such a matroid is called a perfect matroid design. Examples include affine, vector, and projective spaces, their truncations, Steiner-systems, etc.

It is a tantalizing open question to decide for which (k, L) does $m(n, k, L) > cn^s$ hold with some positive constant $c = c(k, L)$.

The first open questions are

$$k = 13, \ L = \{0, 1, 3\} \quad \text{and} \quad k = 11, \ L = \{0, 1, 2, 3, 5\}.$$

One cannot expect to determine $m(n, k, L)$ in general because, e.g.,

$$m(l^2 + l + 1, l + 1, \{0, 1\}) = l^2 + l + 1$$

if and only if a projective plane of order l exists.

In general, for $L = \{0, 1, \ldots, t - 1\}$ an easy consideration gives

$$m(n, k, \{0, 1, \ldots, t - 1\}) \leq \binom{n}{t} \Big/ \binom{k}{t} \quad \text{for } n \geq k \geq t \geq 1.$$

Moreover, equality holds if and only if there exists $\mathbf{S} \subset \binom{X}{k}$ with the property that every t-subset of X is contained in exactly one member of $\mathbf{S}$ (i.e., $\mathbf{S}$ is a (t, k, n)-Steiner-system). For $t = 1$ Steiner-systems exist iff $k|n$. For $t = 2$ a celebrated result of Wilson [**W2**] shows that the trivial necessary conditions $((k - 1)|(n - 1)$ and $\binom{k}{2}|\binom{n}{2})$ are sufficient for $n > n_0(k)$. However, very little is known for $t \geq 3$. In particular, no Steiner-system is known with $t \geq 6$.

The best lower bound on $m(n, k, \{0, 1, \ldots, t - 1\})$—obtained by an ingenious application of the probabilistic method—is due to Rödl.

THEOREM 4.4 (RÖDL [**R**]).

$$\lim_{n \to \infty} m(n, k, \{0, 1, \ldots, t - 1\}) \Big/ \binom{n}{t} = 1 \Big/ \binom{k}{t} \quad \text{for all } k \geq t \geq 1. \tag{4.4}$$

The following result is an improvement on (4.2).

THEOREM 4.5 (FRANKL AND WILSON [**FW**]). *Suppose that p is a prime and $q(x)$ is an integer-valued polynomial of degree d with $p \nmid q(k)$ but $p \mid q(l)$ for all $l \in L$. Then*

$$|\mathbf{F}| \leq |\Delta_d(\mathbf{F})| \leq \binom{n}{d} \quad holds. \tag{4.5}$$

To deduce (4.2) from (4.5) take $q(x) = \prod_{l \in L}(x - l)$. To prove (4.5) one shows that if $\mathbf{F}$ satisfies the assumptions of the theorem, then $M_d(\mathbf{F})$ has rank equal to $\mathbf{F}$. Since $M_d(\mathbf{F})$ has $|\Delta_d(\mathbf{F})|$ rows, this implies (4.5).

Let us recall that $f(n) = \Theta(n^\alpha)$ means that there exist positive constants c_1 and c_2 such that $\lim_{n \to \infty} \inf f(n)n^{-\alpha} > c_1$ and $\lim_{n \to \infty} \sup f(n)n^{-\alpha} < c_2$ holds.

It is trivial to see that $m(n, k, \{l\}) = \Theta(n)$. It follows from Theorem 4.3 that $m(n, k, \{l_1, l_2\}) = \Theta(n^2)$ or $\Theta(n)$ according to whether $(l_2 - l_1)|(k - l_2)$ or $(l_2 - l_1) \nmid (k - l_2)$ holds.

For $|L| = 3$ the situation is already much more complicated. In [**F4**] examples with $m(n, k, L) = \Theta(n^{3/2})$ and in [**Fü3**] with $m(n, k, L) = \Theta(n^{4/3})$ are exhibited (in both cases $|L| = 3$). On the other hand, in [**F4**] it is shown that for $|L| = 3$ either $m(n, k, L) = O(n^{3/2})$ or $m(n, k, L) > c(k, L)n^2$ holds.

THEOREM 4.6 [**F5**]. *For every rational number $r \geq 1$ there exists an infinity of choices of $k = k(r)$, $L = L(r)$ such that $m(n, k, L) = \Theta(n^r)$ holds.*

This theorem raises two problems.

Problem 4.7. Are there irrational numbers α such that $m(n, k, L) = \Theta(n^\alpha)$ holds for some k and L?

Problem 4.8. Given k, L does there always exist a real number $\alpha \geq 1$ such that $m(n, k, L) = \Theta(n^\alpha)$ holds?

5. Families with one forbidden intersection size. For $k > l \geq 0$ let us introduce the notation

$$m(n, k, \bar{l}) = m(n, k, \{0, 1, \ldots, k - 1\} - \{l\}).$$

That is, $m(n, k, \bar{l})$ is the maximum number of k-subsets of an n-set without two which intersect in exactly l elements.

There are two natural ways to avoid intersection size l. One is to take an $(l+1)$-intersecting family, e.g., the $\binom{n-l-1}{k-l-1}$ k-element sets through a fixed $(l+1)$-element set. The other is to take an l-packing, i.e., a collection of sets any two of which intersect in less than l elements. This could not produce more than $\binom{n}{l}/\binom{k}{l}$ k-subsets. One can do better.

Let $\mathbf{S} \subset \binom{X}{2k-l-1}$ be an $(n, 2k-l-1, \{0, 1, \ldots, l-1\})$-system of maximal size. Define $\mathbf{F} = \Delta_k(\mathbf{S})$.

By (4.4) we have

$$|\mathbf{F}| = (1 - o(1))\binom{2k - l - 1}{k}\binom{n}{l} \Big/ \binom{2k - l - 1}{l}, \tag{5.1}$$

and one can check that $\mathbf{F}$ contains no two sets with intersection of size l.

The next result, which was conjectured by Erdös [E2], shows that for $k > 2l+1$ the first construction is best possible.

THEOREM 5.1 (FRANKL AND FÜREDI [FF2]). *Suppose that $l \geq 0$, $k \geq 2l + 2$, and $\mathbf{F} \subset \binom{X}{k}$ satisfies $|F \cap F'| \neq l$ for all $F, F' \in \mathbf{F}$. Then*

$$|\mathbf{F}| \leq \binom{n-l-1}{k-l-1} \quad \text{holds for } n > n_2(k,l). \tag{5.2}$$

Moreover, equality holds in (5.2) if and only if

$$\mathbf{F} = \left\{ F \in \binom{X}{k} \colon T \subset F \right\} \quad \text{for some } T \in \binom{X}{l+1}.$$

Clearly, Theorem 5.1 is a strengthening of the Erdös-Ko-Rado theorem, except that the bounds we have for $n_2(k,l)$ are rather poor. For $k \leq 2l+1$ we have the following partial complement to (5.1).

THEOREM 5.2 [F6]. *Suppose that $\mathbf{F} \subset \binom{X}{k}$ contains no two sets whose intersection has size l, $k \leq 2l + 1$. Then*

$$|\mathbf{F}| \leq \binom{2k-l-1}{k}\binom{n}{l} \Big/ \binom{2k-l-1}{l} \quad \text{holds for } k - l \text{ a prime power.} \tag{5.3}$$

Moreover, if $k - l$ is a prime, then equality holds in (5.3) if and only if there exists an $(l, 2k - l - 1, n)$-Steiner-system, $\mathbf{S}$ with $\mathbf{F} = \Delta_k(\mathbf{S})$.

CONJECTURE 5.3. *The statement of Theorem 5.2 is true for all k and l satisfying $k \leq 2l + 1$.*

Let us mention that for $k = 2l + 1$ the value of the RHS of (5.3) is $\binom{n}{l}$ which is only slightly larger (for $n \to \infty$) than $\binom{n-l-1}{k-l-1}$.

Let us also mention, that one can weaken the assumption on $k - l$ in Theorem 5.2 to: $k - l$ has a prime power divisor greater than l (cf. [FS]).

Let us consider now the nonuniform case, i.e., there is no restriction on $|F|$ for $F \in \mathbf{F}$.

Let us define

$$m(n, \bar{l}) = \max\{|\mathbf{F}| \colon \mathbf{F} \subset 2^X, |F \cap F'| \neq l \text{ for all distinct } F, F' \in \mathbf{F}\}.$$

It is clear that $\mathbf{A}_0 = \{A \subset X \colon |A| < l \text{ or } |A| > (n+l)/2\}$ has no two members intersecting in l elements. As in the case of the Katona theorem, for $n+l$ even we can extend $\mathbf{A}_0$ to $\mathbf{A}_1 = \mathbf{A}_0 \cup \binom{Y}{(n+l)/2}$, $Y \in \binom{X}{n-1}$, without losing this property.

Using Theorem 3.4, Füredi and the author proved that for $n > n_0(l)$ these constructions are best possible.

THEOREM 5.4 [FF1]. *Suppose that $l \geq 1$ and $n \geq n_0(l)$. Then one of (i) and (ii) holds.*

(i) $n + l$ is odd, $m(n, \bar{l}) = |\mathbf{A}_0|$ and $\mathbf{A}_0$ is the only optimal family.

(ii) $n + l$ is even, $m(n, \bar{l}) = |\mathbf{A}_1|$ and $\mathbf{A}_1$ is the only optimal family.

Let us note that our proof only gives $n_0(l) \lesssim 3^l$. However, Conjecture 5.3 would imply $n_0(l) \lesssim 6l$.

For the case $l \neq o(n)$ we have no exact results. However, Rödl and the author proved, e.g., $m(n, \lfloor \bar{n}/4 \rfloor) < 1.99^n$, i.e., it is exponentially smaller than 2^n.

THEOREM 5.5 (FRANKL AND RÖDL [FR1]). *Suppose that α is a real number, $0 < \alpha < \frac{1}{4}$. Then there exists a positive real number $\varepsilon = \varepsilon(\alpha)$ such that*

$$m(n, \bar{l}) < (2 - \varepsilon)^n \quad \text{holds for all integers } l \text{ satisfying } \alpha n < l < (\tfrac{1}{2} - \alpha)n. \quad (5.4)$$

REMARK 5.6. This result was conjectured by Erdős [E2], and it appears in the list of his favorite problems, cf. [E3]. Let us note also, that Conjecture 5.3 would imply (5.4) with a better $\varepsilon = \varepsilon(\alpha)$.

For two families $\mathbf{A}$ and $\mathbf{B}$ let $i_l(\mathbf{A}, \mathbf{B})$ denote the number of pairs (A, B) satisfying $A \in \mathbf{A}$, $B \in \mathbf{B}$, and $|A \cap B| = l$.

THEOREM 5.7 (FRANKL AND RÖDL [FR1]). *For every positive δ there exists a positive $\varepsilon = \varepsilon(\delta)$ such that for all integers a, b, l and families $\mathbf{A} \subset \binom{X}{a}$, $\mathbf{B} \subset \binom{X}{b}$ satisfying $|\mathbf{A}| \, |\mathbf{B}| > \binom{n}{a}\binom{n}{b}(1 - \varepsilon)^n$ one has*

$$i_l(\mathbf{A}, \mathbf{B}) \geq (1 - \delta)^n i_l \left(\binom{X}{a}, \binom{X}{b} \right). \quad (5.5)$$

Clearly Theorem 5.7 is much stronger than Theorem 5.5. E.g., for $l = \lfloor n/4 \rfloor$ it implies the following.

COROLLARY 5.8. *For every $\delta > 0$ there exists $\varepsilon = \varepsilon(\delta) > 0$ such that $i_{\lfloor n/4 \rfloor}(\mathbf{F}, \mathbf{F}) > (4 - \delta)^n$ holds for every family $\mathbf{F} \subset 2^X$ and satisfying $|\mathbf{F}| > (2 - \varepsilon)^n$.*

6. Euclidean Ramsey theory.

A finite subset A of R^d is called *Ramsey* (cf. [EG]) if for every r there exists $n = n_0(A, r)$ with the property that for every partition $R^n = X_1 \cup \cdots \cup X_r$ there is some i and $A' \subset X_i$ with A' congruent to A. Trivially, if $|A| = 2$ then A is Ramsey with $n_0(A, r) \leq r$.

Solving an old problem of Hadwiger, with the help of Theorem 4.5, the following was proved.

THEOREM 6.1 [FW]. *Suppose that $r < 1.2^n$ and $R^n = X_1 \cup \cdots \cup X_r$. Then there exists an i such that for every positive real δ there are two points in X_i whose distance is exactly δ.*

Let us note that Larman and Rogers [LR] showed that one cannot replace 1.2 by 3.

Erdős, Graham, Montgomery, Rothschild, Spencer, and Straus [EG] showed that the vertex set of all d-dimensional rectangular parallelepipeds is Ramsey. On the other hand they proved that if A is Ramsey, then A is spherical; i.e., it has a circumcenter, a point at equal distance to all of them. This leaves open, e.g., the case of nondegenerate, obtuse triangles. Using Ramsey's theorem Rödl and the author [FR2] proved that all nondegenerate triangles were Ramsey.

Let $S_\rho^d = \{(x_0, \ldots, x_d) \in R^{d+1} : x_0^2 + \cdots + x_d^2 = \rho^2\}$ be the d-dimensional sphere of radius ρ.

Let us call a set $A \subset R^d$ *exponentially sphere-Ramsey* if there exist positive reals $\rho = \rho(A)$ and $\varepsilon = \varepsilon(A)$ such that for every partition $S_\rho^n = X_1 \cup \cdots \cup X_r$ with $n \geq d$, $r < (1 + \varepsilon)^n$, there exist an i and $A' \subset X_i$ with A' congruent to A.

THEOREM 6.2 [**FR1**]. *Suppose that A is the vertex set of a nondegenerate simplex or of a parallelepiped in R^d. Then A is exponentially sphere-Ramsey.*

REMARK 6.3. If A is a parallelepiped with circumradius γ then one can take $\rho(A) = \gamma + \delta$ for an arbitrary $\delta > 0$. This proves, in a stronger form, a conjecture of Graham [**G**].

The main tool in proving Theorem 6.2 is a rather complicated extension of Theorem 5.7 to families of partitions.

One can use Corollary 5.8 to prove the following strengthened version of a conjecture of Larman and Rogers [**LR**].

THEOREM 6.4 [**FR1**]. *To every $r \geq 2$ there exists $\varepsilon = \varepsilon(r) > 0$ such that in every family of more than $(2 - \varepsilon)^{4n}$ (± 1)-vectors of length $4n$ one can find r which are pairwise orthogonal.*

7. Multiple intersections. Let us call $\mathbf{F} \subset 2^X$ *r-wise t-intersecting* if $|F_1 \cap \cdots \cap F_r| \geq t$ holds for all $F_1, \ldots, F_r \in \mathbf{F}$. Thus 2-wise t-intersecting means simply t-intersecting. Let $g(n, r, t)$ denote the maximum size of an r-wise t-intersecting family $\mathbf{F} \subset 2^X$, $n \geq t$. By trivial considerations $g(n + 1, r, t) \geq 2g(n, r, t)$ holds. Therefore the limit $p(r, t) = \lim_{n \to \infty} g(n, r, t)2^{-n}$ exists.

Obviously, $p(r, t) \geq 2^{-t}$. The lower bound part of the Katona theorem shows that $p(2, t) = \frac{1}{2}$ for all $t \geq 1$. For $r \geq 3$ let β_r be the unique root of the polynomial $x^r - 2x + 1$ in the interval $(0, 1)$. It is easy to see that β_r is monotone decreasing in r and it is tending to $\frac{1}{2}$ exponentially fast.

THEOREM 7.1 [**F7**]. *There exists a constant $c = c(r)$ such that*

$$c\beta_r^t / \sqrt{t} \leq p(r, t) \leq \beta_r^t. \tag{7.1}$$

The lower bound is obtained via the following construction. Let $B_i \in \binom{X}{t+ri}$ and define $\mathbf{B}_i = \{F \subset X : |F \cap B_i| \geq t + (r - 1)i\}$. Clearly, $\mathbf{B}_i$ is r-wise t-intersecting.

CONJECTURE 7.2 [**F8**].

$$g(n, r, t) = \max_i |\mathbf{B}_i|. \tag{7.2}$$

Let us note that in [**F8**] (7.2) is proved for $t \leq 2^r r / 150$. In [**F7**] it is proved for $r = 3$, $t = 2$, giving $g(n, 3, 2) = 2^{n-2}$.

For the proof of these results a shifting technique, introduced by Erdös, Ko, and Rado, plays a crucial role. Let us mention that Kalai [**Ka**] defined a more powerful, algebraic shifting, which proved very useful in various situations.

REFERENCES

[A] N. Alon, Ph.D. Thesis, Hebrew University, 1983.

[D] M. Deza, *Pavage généralisé parfait comme généralisation de matroide configuration et de simple t-configuration*, Proc. Coll. Comb. (Orsay, 1976), CNRS, 1977, pp. 97–100.

[DEF] M. Deza, P. Erdös, and P. Frankl, *Intersection properties of systems of finite sets*, Proc. London Math. Soc. **36** (1978), 369–384.

[E1] P. Erdös, *Problems and results in combinatorial analysis*, Coll. Internazionale Sulle Teorie Combinatorie (Roma, 1973), vol. 2, Acad. Naz. dei Lincei, Rome, 1976, pp. 3–17.

[E2] ____, *Problems and results in graph theory and combinatorial analysis*, Proc. Fifth British Comb. Conf. (Aberdeen, 1975), Congress Numerantium, no. 15, Utilitas Math., Winnipeg, Man., 1976, pp. 169–192.

[E3] ____, *On the combinatorial problems which I would most like to be solved*, Combinatorica **1** (1981), 25–42.

[EG] P. Erdös, R. L. Graham, P. Montgomery, B. L. Rothschild, J. Spencer, and E. G. Straus, *Euclidean Ramsey theorems*. I, J. Combin. Theory Ser. A **14** (1973), 341–363.

[EKR] P. Erdös, C. Ko, and R. Rado, *Intersection theorems for systems of finite sets*, Quart. J. Math. Oxford **12** (1961), 313–320.

[F1] P. Frankl, *The Erdös-Ko-Rado theorem is true for n = ckt*, Colloq. Math. Soc. János Bolyai **18** (1978), 365–375.

[F2] ____, *On intersecting families of sets*, J. Combin. Theory Ser. A. **24** (1978), 146–161.

[F3] ____, *A simple proof of the Kruskal-Katona theorem*, Discrete Math. **48** (1984), 327–329.

[F4] ____, *Families of sets with three intersections*, Combinatorica **4** (1984), 141–148.

[F5] ____, *All rationals occur as exponents*, J. Combin. Theory Ser. A **42** (1986), 200–206.

[F6] ____, *An extremal set theoretic characterization of some Steiner-systems*, Combinatorica **3** (1983), 193–199.

[F7] ____, *Families of finite sets satisfying an intersection condition*, Bull. Austral. Math. Soc. **15** (1976), 73–79.

[F8] ____, *Families of finite sets satisfying a union condition*, Discrete Math. **26** (1979), 111–118.

[FF1] P. Frankl and Z. Füredi, *On hypergraphs without two edges intersecting in a given number of vertices*, J. Combin. Theory Ser. A **36** (1984), 230–236.

[FF2] ____, *Forbidding just one intersection*, J. Combin. Theory Ser. A **39** (1985), 160–176.

[FF3] ____, *Non-trivial intersecting families*, J. Combin. Theory Ser. A **41** (1986), 150–153.

[FF4] ____, *More on the Erdös-Ko-Rado problem: beyond stars* (in preparation).

[FF5] ____, *Finite projective spaces and intersecting hypergraphs*, Combinatorica **6** (1986), 373–392..

[FG] Z. Füredi and J. R. Griggs, *Families of finite sets with minimum shadows*, Combinatorica **6** (1986), 393–401.

[FR1] P. Frankl and V. Rödl, *Forbidden intersections*, Trans. Amer. Math. Soc. **299** (1987).

[FR2] ____, *All triangles are Ramsey*, Trans. Amer. Math. Soc. **297** (1986), 777–779.

[FS] P. Frankl and N. M. Singhi, *Linear dependencies among subsets of a finite set*, European J. Combin. **4** (1983), 313–318.

[FW] P. Frankl and R. M. Wilson, *Intersection theorems with geometric consequences*, Combinatorica **1** (1981), 357–368.

[Fü1] Z. Füredi, *Erdös-Ko-Rado type theorems with upper bounds on the maximum degree*, Colloq. Math. Soc. János Bolyai **25** (1981), 177–207.

[Fü2] ____, *On finite set-systems whose every intersection is the kernel of a star*, Discrete Math. **47** (1983), 129–132.

[Fü3] ____, *Set-systems with three intersections*, Combinatorica **5** (1985), 27–31.

[G] R. L. Graham, *Old and new euclidean Ramsey theorems*, Discrete Geometry and Convexity, Ann. New York Acad. Sci., vol. 440, 1985, pp. 20–30.

[H] H. Hadwiger, *Überdeckungssätze für den Euklidischen Raum*, Portugal. Math. **4** (1944), 140–144.

[Hu] M. Hujter, personal communication.

[HM] A. J. W. Hilton and E. C. Milner, *Some intersection theorems for systems of finite sets*, Quart. J. Math. Oxford **18** (1967), 369–384.

[K1] Gy. Katona, *Intersection theorems for systems of finite sets*, Acta Math. Hungar. **15** (1964), 329–337.

[K2] ____, *A theorem of finite sets*, Theory of Graphs (Proc. Colloq., Tihany, 1966), Akad. Kiadó, Budapest, 1968, pp. 187–207.

[Ka] G. Kalai, *Characterization of f-vectors of families of convex sets in R^d*, Israel J. Math. **48** (1984), 161–174.

[Kr] J. B. Kruskal, *The number of simplices in a complex*, Mathematical Optimization Techniques, Univ. of California Press, Berkeley, 1963, pp. 251–278.

[LR] D. G. Larman and C. A. Rogers, *The realization of distances within sets in Euclidean space*, Mathematika **19** (1972), 1–24.

[M] M. Mörs, *A generalization of a theorem of Kruskal*, Graphs Combin. **1** (1985), 167–183.

[R] V. Rödl, *On a packing and covering problem*, European J. Combin. **6** (1985), 69–78.

[RW] D. K. Ray-Chaudhuri and R. M. Wilson, *On t-designs*, Osaka J. Math. **12** (1975), 737–744.

[S] V. T. Sós, *Remarks on the connection of graph theory, finite geometry and block designs*, Colloq. Inter. Sulle Teorie Combinatorie (Rome, 1973), Tomo II, Accad. Naz. Lincei, Rome, 1976, pp. 223–233.

[W1] R. M. Wilson, *The exact bound in the Erdös-Ko-Rado theorem*, Combinatorica **4** (1984), 247–257.

[W2] ____, *An existence theory for pairwise balanced designs. I, II*, J. Combin. Theory Ser. A **13** (1972), 220–273; **18** (1975), 71–79.

CENTRE NATIONAL DE LA RECHERCHE SCIENTIFIQUE, PARIS, FRANCE

Polyhedral Combinatorics—
Some Recent Developments and Results

ALEXANDER SCHRIJVER

Polyhedral combinatorics deals with characterizing convex hulls of vectors obtained from combinatorial structures, and with deriving min-max relations and algorithms for corresponding combinatorial optimization problems. In this paper, after an introduction discussing the matching polytope (§1) and some algorithmic consequences (§2), we give some illustrations of recent developments (viz., applications of lattice and decomposition techniques (§§3 and 4)), we go into the relation to cutting planes (§5), and we describe some other recent results (§6).

1. A basic example: The matching polytope. We first describe a basic result in polyhedral combinatorics, due to Edmonds [**7**]. Let $G = (V, E)$ be an undirected graph (i.e., V is a finite set (of *vertices*) and E is a collection of pairs (*edges*) of vertices). A subset M of E is called a *matching* if $e' \cap e'' = \varnothing$ whenever $e', e'' \in M$, $e' \neq e''$. The *matching polytope* of G is the set conv.hull$\{\chi^M | M \text{ matching}\}$ in $\mathbf{R}^E$, where χ^M is the incidence vector of M (i.e., $\chi^M \in \mathbf{R}^E$ with $\chi^M(e) = 1$ if $e \in M$, and $= 0$ otherwise). Edmonds now showed:

THEOREM 1 (EDMONDS'S MATCHING POLYTOPE THEOREM). *The matching polytope of $G = (V, E)$ is equal to the set of vectors $x \in \mathbf{R}^E$ satisfying:*

$$
\begin{aligned}
&\text{(i)} \quad x_e \geq 0 && (e \in E), \\
&\text{(ii)} \quad \sum_{e \ni v} x_e \leq 1 && (v \in V), \\
&\text{(iii)} \quad \sum_{e \subseteq U} x_e \leq \left\lfloor \frac{1}{2}|U| \right\rfloor && (U \subseteq V, |U| \text{ odd}).
\end{aligned}
\tag{1}
$$

For proofs we refer to [**24, 30, 33**].

Edmonds's theorem has the following application. If we are given some "weight" function $c \in \mathbf{R}^E$, we can describe the problem of finding a matching M of maximum "weight" $\sum_{e \in M} c_e$ equivalently as the problem of maximizing $c^{\mathrm{T}} x$ over the matching polytope, that is, by Edmonds's theorem, over $x \in \mathbf{R}^E$ satisfying (1). This last is a linear programming problem, and we can apply

LP-techniques to solve this problem, and hence to solve the combinatorial optimization problem. Among others, with the help of the ellipsoid method, it can be shown that the maximum matching problem is solvable in polynomial time—see §2.

Another, theoretical, application of Edmonds's theorem is obtained with the duality theorem of linear programming. Let $Ax \le b$ denote the system (1). Then for any $c \in \mathbf{R}^E$

$$\max \left\{ \sum_{e \in M} c_e | M \text{ matching} \right\} = \max\{c^\mathrm{T} x | Ax \le b\}$$

$$= \min\{y^\mathrm{T} b | y \ge 0; y^\mathrm{T} A = c^\mathrm{T}\}. \qquad (2)$$

So we have a min-max relation for the maximum matching problem. It was shown by Cunningham and Marsh [6] that if c is integer-valued, then the minimum in (2) has an integer optimum solution y. The special case $c \equiv \mathbf{1}$ (the all-one function) is equivalent to the following *Tutte-Berge formula* [35, 1]: the maximum cardinality of a matching in a graph $G = (V, E)$ is equal to

$$\min_{U \subseteq V} \frac{|V| + |U| - \mathcal{O}(V \backslash U)}{2}, \qquad (3)$$

where $\mathcal{O}(V \backslash U)$ denotes the number of components of $\langle V \backslash U \rangle$ with an odd number of vertices ($\langle V \backslash U \rangle$ denotes the graph $(V \backslash U, \{e \in E | e \subseteq V \backslash U\})$).

Note that the constraint matrix A in (1) generally is not totally unimodular (a matrix is *totally unimodular* if all subdeterminants belong to $\{0, \pm 1\}$). If G is bipartite (i.e., V can be split into classes V' and V'' (the *color classes*) so that $E \subseteq \{\{v', v''\} | v' \in V', v'' \in V''\}$), then the inequalities (1)(iii) can be deleted as they are implied by the constraints (i) and (ii), as one easily checks. In that case, the theorem is due to Egerváry [9] and follows more simply from the fact that if M is totally unimodular and d is integer, then each vertex of the polyhedron determined by $Mx \le d$ is integer.

Similarly, for bipartite G, the Tutte-Berge formula above reduces to the well-known König-Egerváry theorem [21, 9].

2. Polyhedral combinatorics and polynomial solvability. Above we mentioned obtaining polynomial-time algorithms from polyhedral results with the ellipsoid method. In this section we describe this more precisely.

Suppose that for each graph $G = (V, E)$ we have a collection $\mathcal{F}_G$ of subsets of E. For example:

$$
\begin{aligned}
&\text{(i)} \quad \mathcal{F}_G = \{M \subseteq E | M \text{ is a matching}\};\\
&\text{(ii)} \quad \mathcal{F}_G = \{M \subseteq E | M \text{ is a spanning tree}\}; \qquad (4)\\
&\text{(iii)} \quad \mathcal{F}_G = \{M \subseteq E | M \text{ is a Hamiltonian circuit}\}.
\end{aligned}
$$

With any family $(\mathcal{F}_G | G \text{ graph})$ we can associate the following problem:

Optimization problem. Given a graph $G = (V, E)$ and $c \in \mathbf{Q}^E$, (5)
find $M \in \mathcal{F}_G$ maximizing $\sum_{e \in M} c_e$.

So if $(\mathcal{F}_G | G$ graph$)$ is as in (i), (ii), and (iii), respectively, problem (5) amounts to finding a maximum weighted matching, a maximum weighted spanning tree, and a maximum weighted Hamiltonian circuit, respectively. The last problem is the well-known traveling salesman problem (note that by replacing c by $-c$ (5) becomes a minimization problem).

Given a family $(\mathcal{F}_G | G$ graph$)$, we are interested in finding, for any graph $G = (V, E)$, a system $Ax \leq b$ of linear inequalities in $x \in \mathbf{R}^E$ so that

$$\text{conv.hull}\{\chi^M | M \in \mathcal{F}_G\} = \{x | Ax \leq b\}. \tag{6}$$

If (6) holds, then for any $c \in \mathbf{R}^E$:

$$\max\left\{\sum_{e \in M} c_e | M \in \mathcal{F}_G\right\} = \max\{c^{\mathrm{T}}x | Ax \leq b\}$$

$$= \min\{y^{\mathrm{T}}b | y \geq 0; y^{\mathrm{T}}A = c^{\mathrm{T}}\}, \tag{7}$$

thus formulating the combinatorial optimization problem as a linear programming problem.

The optimization problem (5) is said to be *solvable in polynomial time* or *polynomially solvable* if it is solvable by an algorithm whose running time is bounded above by a polynomial in the *input size* $|V| + |E| + \text{size}(c)$. Here $\text{size}(c) := \sum_{e \in E} \text{size}(c_e)$, where the size of a rational number p/q is equal to $\log_2(|p| + 1) + \log_2|q|$. So $\text{size}(c)$ is about the space needed to specify c in binary notation.

It has been shown by Karp and Papadimitriou [20] and Grötschel, Lovász, and Schrijver [16] that (5) is polynomially solvable if and only if the following problem is solvable in polynomial time:

> *Separation problem.* Given a graph $G = (V, E)$ and $x \in \mathbf{Q}^E$, (8) determine if x belongs to conv.hull$\{\chi^M | M \in \mathcal{F}_G\}$, and if not, find a separating hyperplane.

Again, "polynomial time" means: time bounded by a polynomial in $|V| + |E| + \sum_{e \in E} \text{size}(x_e)$.

THEOREM 2. *For any fixed family $(\mathcal{F}_G | G$ graph$)$, the optimization problem (5) is polynomially solvable if and only if the separation problem (8) is polynomially solvable.*

The theorem implies that with respect to the question of polynomial-time solvability, the approach described above (studying the convex hull) is more or less essential: a combinatorial optimization problem is polynomially solvable if and only if the corresponding convex hulls can be decently described—decently, in the sense of the separation problem.

As an application of Theorem 2, it can be shown that the system (1) of linear inequalities can be tested in polynomial time, although there exist exponentially many constraints (Padberg and Rao [28]). Hence, the maximum matching problem is polynomially solvable (in fact, this was shown directly by Edmonds [7]).

Theorem 2 can also be used in the negative: if a combinatorial optimization problem is not polynomially solvable (maybe the traveling salesman problem), then the corresponding polytopes have no decent description.

Theorem 2 is shown with the ellipsoid method, for which we refer to the books of Grötschel, Lovász, and Schrijver [17] and Schrijver [32]. The ellipsoid method does not give practical algorithms, but it may give insight in the complexity of a problem.

There are several variations of Theorem 2. For instance, a similar result holds if we consider collections $\mathcal{F}_G$ of subsets of the vertex set V, instead of subsets of the edge set E. Moreover, we may consider families $(\mathcal{F}_G | G \in \mathcal{G})$, where $\mathcal{G}$ is a subclass of the class of all graphs. Similarly, we can consider directed graphs.

3. Lattices and strongly polynomial algorithms. A first recent development in polyhedral combinatorics is the influence of lattice techniques, to a large extent due to the recently developed *basis reduction method* given by Lenstra, Lenstra, and Lovász [23]. In this section we give one illustration of this influence, due to Frank and Tardos [10].

The basis reduction method solves the following problem:

> Given a nonsingular rational $n \times n$-matrix A, find a basis $b_1, \ldots, b_n$ (9)
> for the lattice generated by the columns of A satisfying
> $$\|b_1\| \cdots \|b_n\| \leq 2^{n(n-1)/4} |\det A|,$$

in time bounded by a polynomial in size$(A) := \sum_{i,j} \text{size}(a_{ij})$. Here the *lattice generated by $a_1, \ldots, a_n$* is the set of vectors $\lambda_1 a_1 + \cdots + \lambda_n a_n$ with $\lambda_1, \ldots, \lambda_n \in \mathbf{Z}$. Any linearly independent set of vectors generating the lattice is called a *basis* for the lattice.

One of the many consequences is a polynomial-time algorithm for the following *simultaneous diophantine approximation problem*:

> Given $n \in \mathbf{N}$, $a \in \mathbf{Q}^n$, and ε with $0 < \varepsilon < 1$, find an integer   (10)
> vector p and an integer q satisfying $\|a - (1/q)p\| < \varepsilon/q$ and
> $1 \leq q \leq 2^{n(n+1)/4} \varepsilon^{-n}$.

This can be seen by applying the basis reduction method to the $(n+1) \times (n+1)$-matrix

$$A := \begin{pmatrix} I & a \\ 0 & 2^{-n(n+1)/4} \varepsilon^{n+1} \end{pmatrix}, \tag{11}$$

where I is the $n \times n$ identity matrix.

Frank and Tardos showed that this approximation algorithm yields so-called *strongly polynomial* algorithms. The algorithm for the optimization problem (5) derived from the ellipsoid method performs a number of arithmetic operations, which number is bounded by a polynomial in $|V| + |E| + \text{size}(c)$. (*Arithmetic operations* here are: addition, subtraction, multiplication, division, comparison.)

It would be preferable if the size of the weight function c only influences the sizes of the numbers occurring when executing the algorithm, but not the *number* of arithmetic operations. Therefore, one has defined an algorithm for (5) to be *strongly polynomial* if it consists of a number of arithmetic operations, bounded by a polynomial in $|V| + |E|$, on numbers of size bounded by a polynomial in $|V| + |E| + \text{size}(c)$.

Frank and Tardos however showed the equivalence of the two concepts when applied to (5):

THEOREM 3. *For any family* $(\mathcal{F}_G|G$ *graph), there exists a polynomial-time algorithm for the optimization problem* (5) *if and only if there exists a strongly polynomial algorithm for* (5).

Their result was obtained by constructing a strongly polynomial algorithm for the following problem:

$$\text{Given } n \in \mathbf{N} \text{ and } c \in \mathbf{Q}^n, \text{ find } \tilde{c} \in \mathbf{Z}^n \text{ such that } \|\tilde{c}\|_\infty \leq 2^{9n^3} \qquad (12)$$
$$\text{and such that: } c^T x > c^T y \Leftrightarrow \tilde{c}^T x > \tilde{c}^T y, \text{ for all } x, y \in \{0,1\}^n.$$

With this method the size of c in the optimization problem can be reduced to $O(|E|^3)$, without changing the optimum solution. Hence any polynomial-time algorithm for the optimization problem yields a strongly polynomial algorithm.

As another interesting recent lattice result we mention Lovász's [25] characterization of the *perfect matching lattice* (i.e., the lattice generated by the incidence vectors of perfect matchings in a graph), in the same vein as Edmonds's matching polytope theorem.

4. The coclique polytope and decomposition techniques. As another recent development in polyhedral combinatorics we mention the propagation of decomposition techniques. Fundamental decomposition methods are described in Seymour's paper *Decomposition of regular matroids* [34]. Also Burlet, Fonlupt, and Uhry [2, 3] obtained deep decomposition results.

We illustrate the decomposition methods of Seymour by applying them to characterizing the "coclique polytope" of certain graphs. For any undirected graph $G = (V, E)$, a set $C \subseteq V$ is called a *coclique* if it does not contain any edge of G as a subset. The *coclique polytope* of G is the convex hull of the incidence vectors of cocliques in G, i.e., conv.hull$\{\chi^C|C$ coclique$\} \subseteq \mathbf{R}^V$.

The problem

$$\text{Given } G = (V, E) \text{ and } c \in \mathbf{R}^V, \text{ find a coclique } C \text{ in } G \text{ maximiz-} \qquad (13)$$
$$\text{ing } \textstyle\sum_{v \in C} c_v$$

is NP-complete, and hence probably not polynomially solvable. Therefore, by Theorem 2 (now in the variant with subsets of V instead of E), there is probably no polynomial-time algorithm for the separation problem for coclique polytopes. So we should not expect a decent description for coclique polytopes similar to Edmonds's matching polytope theorem.

For some classes of graphs, however, the coclique polytope has a decent description, e.g., for perfect graphs (including bipartite graphs, line graphs of bipartite graphs, comparability graphs, triangulated graphs, and their complements). Another class of graphs is described in the following theorem of Gerards and Schrijver [14]. An undirected graph $G = (V, E)$ is called *odd-K_4-free* if G has no subgraph homeomorphic to

where wriggled lines stand for paths, so that each face in this graph is enclosed by a circuit of odd length.

THEOREM 4. *For any odd-K_4-free graph $G = (V, E)$, the coclique polytope is equal to the set of vectors x in $\mathbf{R}^V$ satisfying:*

$$
\begin{aligned}
&\text{(i)} \quad 0 \le x_v \le 1 &&(v \in V), \\
&\text{(ii)} \quad x_v + x_w \le 1 &&(\{v, w\} \in E), \\
&\text{(iii)} \quad \sum_{v \in C} x_v \le \left\lfloor \frac{1}{2}|C| \right\rfloor &&(C \ \text{circuit with } |C| \text{ odd}).
\end{aligned}
\tag{14}
$$

(Here C is a circuit if $C = \{v_1, \ldots, v_k\}$ with $\{v_{i-1}, v_i\} \in E$ ($i = 1, 2, \ldots, k$) and $\{v_k, v_1\} \in E$.)

Note that if G is bipartite, then G has no odd circuit, and hence there are no constraints (iii). In that case the theorem reduces to a theorem of Egerváry [9].

The theorem implies, with the help of Theorem 2, that problem (13) is polynomially solvable for odd-K_4-free graphs. Indeed, the constraints (14) can be tested for any given $x \in \mathbf{R}^V$ in time bounded by a polynomial in $|V| + |E| + \text{size}(x)$, although there are exponentially many constraints. (The condition (iii) can be tested using a shortest path algorithm.)

We sketch how Theorem 4 can be shown using decomposition techniques (which also yield a direct combinatorial polynomial-time algorithm for the maximum coclique problem for odd-K_4-free graphs). It was shown by Seymour [34] that "each regular matroid is obtained by taking 1-,2-, and 3-sums of graphic matroids, cographic matroids, and R_{10}." *Regular* matroids are matroids representable over each field. By a theorem of Tutte [36], regular matroids are exactly those binary matroids not containing the Fano-matroid or its dual as a minor.

Seymour's theorem can be equivalently stated as: "Each totally unimodular matrix can be decomposed into network matrices and their transposes and into certain 5×5-matrices." It implies a polynomial-time test for the total unimodularity of matrices, and a polynomial-time algorithm for linear programs over totally unimodular matrices. It also has implications in geometry and graph theory. One of them described by Gerards, Lovász, Schrijver, Seymour, and Truemper [13] is as follows.

Consider the following four compositions of graphs $G' = (V', E')$ and $G'' = (V'', E'')$ into a new graph H. *Composition 1.* If $|V' \cap V''| \leq 1$, then $H := (V' \cup V'', E' \cup E'')$. *Composition 2.* If $V' \cap V'' = \{v_1, v_2\} \in E' \cap E''$ and G'' is bipartite, then $H := (V' \cup V'', (E' \cup E'') \backslash \{\{v_1, v_2\}\})$. *Composition 3.* If $V' \cap V'' = \{v_0, v_1, v_2\}$, $E' \cap E'' = \{\{v_0, v_1\}, \{v_0, v_2\}, \{v_1, v_2\}\}$, and v_0 has degree 2 both in G' and in G'', then $H := ((V' \cup V'') \backslash \{v_0\}, (E' \cup E'') \backslash (E' \cap E''))$. *Composition 4.* If $V' \cap V'' = \{v_0, v_1, v_2, v_3\}$, $E' \cap E'' = \{\{v_0, v_1\}, \{v_0, v_2\}, \{v_0, v_3\}\}$, v_0 has degree 3 both in G' and in G'', and G'' is bipartite, then

$$H := ((V' \cup V'') \backslash \{v_0\}, (E' \cup E'') \backslash (E' \cap E'')).$$

Moreover, consider the following operations on a graph $G = (V, E)$. *Operation 1.* If $\{v_0, v_1\}, \{v_1, v_2\}, \{v_2, v_3\} \in E$, where both v_1 and v_2 have degree 2, then

$$H := (V \backslash \{v_1, v_2\}, (E \backslash \{\{v_0, v_1\}, \{v_1, v_2\}, \{v_2, v_3\}\}) \cup \{v_0, v_3\}).$$

Operation 2. If $v_0 \in V$, where $\{v_0, v_1\}, \ldots, \{v_0, v_k\}$ are the edges of G containing v_0, then let $w_1, \ldots, w_k$ be "new" vertices and

$$H := (V \cup \{w_1, \ldots, w_k\}, (E' \backslash \{\{v_0, v_1\}, \ldots, \{v_0, v_k\}\})$$
$$\cup \{\{v_0, w_1\}, \ldots, \{v_0, w_k\}, \{w_1, v_1\}, \ldots, \{w_k, v_k\}\}).$$

THEOREM 5. *An undirected graph is odd-K_4-free if and only if it can be constructed by a series of compositions and operations above starting with the following graphs:*

(i) *graphs $G = (V, E)$ having a vertex v_0 so that the graph $(V \backslash \{v_0\}, E \backslash \{e | e \ni v_0\})$ is bipartite;*

(ii) *planar graphs having exactly two odd facets (an odd facet*   (15)
is a facet enclosed by an odd number of edges);

(iii) *the following graph :*

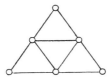

Sufficiency in this theorem is easy to see: each of the graphs in (i), (ii), and (iii) is odd-K_4-free. Moreover, each of the compositions and operations maintains the property of being odd-K_4-free. The content of the theorem is that in this way all odd-K_4-free graphs can be constructed.

In order to derive now Theorem 4, it suffices to prove that each of the graphs (15) has the property described in Theorem 4, and moreover that this property is maintained under each of the compositions and operations above. Showing this is not as hard as the original direct proof of Theorem 4.

If we let $Ax \leq b$ denote the system (14), then by Theorem 4 for odd-K_4-free graphs $G = (V, E)$ and $c \in \mathbf{R}^V$:

$$\max \left\{ \sum_{v \in C} c_v | C \text{ coclique} \right\} = \max\{c^T x | Ax \leq b\}$$

$$= \min\{y^T b | y \geq 0; y^T A = c^T\}. \qquad (16)$$

Using the above decomposition techniques, Gerards [12] showed that if c is integer-valued, the minimum has an integer optimum solution y. In particular, if $c \equiv \mathbf{1}$ (the all-one function) then the maximum size of a coclique is equal to

$$\min \left(|F| + \sum_{i=1}^{t} \left\lfloor \frac{1}{2}|C_i| \right\rfloor \right), \qquad (17)$$

where the minimum ranges over all subsets F of E and circuits $C_1, \ldots, C_t$ such that $V = \bigcup F \cup \bigcup_{i=1}^{t} C_i$. This forms an extension of a theorem of König [22] for bipartite graphs.

5. Cutting planes. Quite often the problem of characterizing the convex hull of certain $\{0, 1\}$-vectors amounts to characterizing, for some polyhedron P, the polyhedron

$$P_{\mathrm{I}} := \text{conv.hull}\{x \in P | x \text{ integral}\}. \qquad (18)$$

P_{I} is called the *integer hull* of P. E.g., if $G = (V, E)$ is a graph, and

$$P := \left\{ x \in \mathbf{R}^E | x_e \geq 0 \ (e \in E); \sum_{e \ni v} x_e \leq 1 \ (v \in V) \right\}, \qquad (19)$$

the integral vectors in P are exactly the incidence vectors of matchings, and hence P_{I} is equal to the matching polytope of G. Similarly, for

$$P := \left\{ x \in \mathbf{R}^V | x_v \geq 0 \ (v \in V); \sum_{v \in e} x_v \leq 1 \ (e \in E) \right\}, \qquad (20)$$

P_{I} is the coclique polytope of G.

For any rational polyhedron P, there is a procedure of deriving the inequalities determining P_{I} from those determining P—the *cutting plane method*, due to Gomory [15]. The following description is due to Chvátal [4] and Schrijver [29].

Clearly, if H is a *rational half-space*, i.e., H is of form

$$H = \{x \in \mathbf{R}^n | a^T x \leq \beta\}, \qquad (21)$$

where $a \in \mathbf{Q}^n$, $a \neq \mathbf{0}$, $\beta \in \mathbf{Q}$, we may assume without loss of generality that a is integral, and that the components of a are relatively prime integers. In that case:

$$H_{\mathrm{I}} = \{x \in \mathbf{R}^n | a^T x \leq \lfloor \beta \rfloor\}. \qquad (22)$$

H_{I} arises from H by shifting its bounding hyperplane until it contains integral vectors.

Now define for any set P in $\mathbf{R}^n$:

$$P' := \bigcup_{H \supseteq P} H_{\mathrm{I}}, \tag{23}$$

where H ranges over all rational half-spaces containing P. Since $H \supseteq P$ implies $H_{\mathrm{I}} \supseteq P_{\mathrm{I}}$, it follows that $P' \supseteq P_{\mathrm{I}}$. It can be shown that if P is a rational polyhedron (i.e., a polyhedron determined by linear inequalities with rational coefficients), then P' is a rational polyhedron again.

To P' we can apply this operation again, yielding P''. It is not difficult to find rational polyhedra with $P'' \neq P'$. Each rational polyhedron P thus gives a sequence of polyhedra containing P_{I}:

$$P \supseteq P' \supseteq P'' \supseteq P''' \supseteq \cdots \supseteq P_{\mathrm{I}}. \tag{24}$$

Denoting the $(t+1)$th set in this sequence by $P^{(t)}$, the following can be shown.

THEOREM 6. *For each rational polyhedron P there exists a number t such that $P^{(t)} = P_{\mathrm{I}}$.*

The theorem is the theoretical essence of the termination of the cutting plane method of Gomory. The equation $a^{\mathrm{T}}x = \lfloor \beta \rfloor$ defining H_{I}, or more strictly the hyperplane $\{x | a^{\mathrm{T}}x = \lfloor \beta \rfloor\}$, is called a *cutting plane*.

The smallest t for which $P^{(t)} = P_{\mathrm{I}}$ can be considered as a measure for the complexity of P_{I} relative to that of P. In a sense, P' is near to P, P'' to P', and so on.

Let us study some specific polyhedra related to graphs. Let $G = (V, E)$ be an undirected graph, and let P be the polytope (19), implying that P_{I} is the matching polytope of G. It is not hard to see that for each graph G, the polytope P' is the set of all vectors x in P satisfying

$$\sum_{e \subseteq U} x_e \leq \left\lfloor \frac{1}{2}|U| \right\rfloor \qquad (U \subseteq V, |U| \text{ odd}). \tag{25}$$

(Of course, there are infinitely many half-spaces H containing P, but the corresponding inequalities $a^{\mathrm{T}}x \leq \lfloor \beta \rfloor$ all are implied by the inequalities defining P and by (25).) So Edmonds's matching polytope theorem in fact tells us that $P' = P_{\mathrm{I}}$ for each graph G. ($P = P_{\mathrm{I}}$ for bipartite G, since in that case (25) is implied by the inequalities determining P.)

Next let, for any undirected graph $G = (V, E)$, P be the polytope (20), implying that P_{I} is the coclique polytope of G. It is not difficult to check that the polytope P' is the set of vectors x in P satisfying

$$\sum_{v \in C} x_v \leq \left\lfloor \frac{1}{2}|C| \right\rfloor \qquad (C \text{ odd circuit}). \tag{26}$$

So Theorem 4 states that $P' = P_{\mathrm{I}}$ if G is odd-K_4-free. By Egerváry's theorem $P = P_{\mathrm{I}}$ if and only if G is bipartite. Chvátal [5] has shown that there exists no fixed t so that $P^{(t)} = P_{\mathrm{I}}$ for each graph G.

An important computational application of cutting planes is to the traveling salesman problem, which we mention in the following section.

6. The traveling salesman problem and cuts. The well-known *traveling salesman problem* (in its directed, asymmetric form) can be formulated as an integer linear programming problem as follows, for given $n \in \mathbf{N}$ and $c = (c_{ij}) \in \mathbf{R}^{n \times n}$:

$$\text{minimize} \sum_{i,j=1}^{n} x_{ij},$$

such that

$$(*)\begin{cases} x_{ij} \geq 0 & (i, j = 1, \ldots, n); \\ \sum_{i \notin U, j \in U} x_{ij} \geq 1 & (\varnothing \neq U \subsetneq \{1, \ldots, n\}); \\ \sum_{j=1}^{n} x_{ij} = 1 & (i = 1, \ldots, n); \end{cases} \qquad (27)$$

$$x_{ij} \text{ integer} \qquad (i, j = 1, \ldots, n).$$

Let P be the polytope in $\mathbf{R}^{n \times n}$ determined by $(*)$. It is clear that P_{I} is the convex hull of the incidence vectors of traveling salesman routes. Since the traveling salesman problem is NP-complete, we may not expect a "decent" description of P_{I} in the sense of Theorem 2. In fact, if NP $\neq$ co-NP there is no fixed t such that $P^{(t)} = P_{\mathrm{I}}$ for each n.

On the other hand, cutting planes can be helpful in solving instances of the traveling salesman problem. The traveling salesman problem is equivalent to solving $\min\{c^{\mathrm{T}} x | x \in P_{\mathrm{I}}\}$, while solving $\min\{c^{\mathrm{T}} x | x \in P\}$ is not so difficult (it is polynomially solvable), and it yields a good lower bound for the traveling salesman optimum value (since $P \supseteq P_{\mathrm{I}}$). Good bounds are essential in branch-and-bound procedures for the traveling salesman problem.

Adding *all* cutting planes to $(*)$ to obtain P_{I} seems infeasible, but instead we could add *some* cutting planes in order to obtain a better lower bound. This is a basic ingredient in the recent successes of Crowder, Grötschel, and Padberg in solving large-scale traveling salesman problems (see [**18**, **27**]). Recently, Padberg was able to solve a symmetric 2392-"city" problem using cutting planes.

We shall not go into the details of solving the traveling salesman problem. We describe some theoretical results related to the above, which exhibit some of the connections of polyhedral results with combinatorial min-max relations.

Let C be a collection of subsets of $V := \{1, \ldots, n\}$ satisfying:

(i) $\varnothing \notin C$, $V \notin C$;

(ii) if $T, U \in C$, $T \cap U \neq \varnothing$, $T \cup U \neq V$, then $T \cap U \in C$ (28)
      and $T \cup U \in C$.

Such a collection is called a *crossing family*. Consider the polytope P consisting of all $x = (x_{ij}) \in \mathbf{R}^{n \times n}$ satisfying:

$$x_{ij} \geq 0 \quad (i, j = 1, \ldots, n), \qquad \sum_{i \notin U, j \in U} x_{ij} \geq 1 \quad (U \in C). \qquad (29)$$

Note that (∗) in (27) defines a facet of P, for $C = P(V) \setminus \{\varnothing, V\}$.

The following theorem was shown in [**31**].

THEOREM 7. *P has integral vertices if and only if*

$$there\ are\ no\ sets\ V_1, V_2, V_3, V_4, V_5\ in\ C\ such\ that\ V_1 \subseteq V_2 \cap V_3, \qquad (30)$$
$$V_2 \cup V_3 = V,\ V_3 \cup V_4 \subseteq V_5,\ V_3 \cap V_4 = \varnothing.$$

Note that if x is an integral vertex of P, then x is a $\{0, 1\}$-vector.

Theorem 7 can be put in a more combinatorial setting. Let $C \subseteq P(V)$ be a crossing family and let $D = (V, A)$ be a *directed* graph (i.e., V is a finite set and $A \subseteq V \times V$). Call a subset A' of A a *covering* (for C) if each $U \in C$ is entered by at least one arc in A' ($a = (v, w)$ enters U if $v \notin U$, $w \in U$). Call a subset A' of A a *cut* (induced by C) if $A' = \delta_A^-(U) := \{a \in A | a\ \text{enters}\ U\}$ for some $U \in C$. So each covering intersects each cut.

Consider the polyhedron in $\mathbf{R}^A$ determined by:

$$x_a \geq 0 \quad (a \in A), \qquad \sum_{a \in \delta^-(U)} x_a \geq 1 \quad (U \in C). \qquad (31)$$

Then Theorem 7 is equivalent to:

THEOREM 8. *Each vertex of the polyhedron determined by* (31) *is the incidence vector of a covering, for each directed graph $D = (V, A)$, if and only if* (30) *holds.*

Now we have the following: (30) holds $\Leftrightarrow$ the polyhedron determined by (31) has vertices coming from coverings and facets coming from cuts $\Leftrightarrow$ (by polarity) the polyhedron determined by

$$x_a \geq 0 \quad (a \in A), \qquad \sum_{a \in C} x_a \geq 1 \quad (C\ \text{covering}) \qquad (32)$$

has *vertices* coming from cuts and *facets* coming from coverings. So Theorem 8 is equivalent to:

THEOREM 9. *Each vertex of the polyhedron determined by* (32) *is the incidence vector of a cut, for each directed graph $D = (V, A)$, if and only if* (30) *holds.*

It follows that if (30) holds, and $c \in \mathbf{Z}_+^A$, then the linear programs of minimizing $c^T x$ over (31) and over (32), respectively, have integral optimum solutions, corresponding to a minimum-weighted covering and a minimum-weighted cut, respectively. In fact, it is shown in [**31**] that if (30) holds, then also the linear programs dual to these programs have integer optimum solutions. By LP-duality this means:

THEOREM 10. *Let C be a crossing family satisfying* (30), *let $c \in \mathbf{Z}^A$, and let $D = (V, A)$ be a directed graph. Then* (i) *the minimum weight of a covering is equal to the maximum number t of cuts $C_1, \ldots, C_t$ (repetition allowed) so that each arc a of D is in at most c_a of the cuts C_i;* (ii) *the minimum weight of a cut*

*is equal to the maximum number t of coverings $C_1, \ldots, C_t$ (repetition allowed)
so that each arc a of D is in at most c_a of the coverings C_i.*

We mention the following applications.

1. Let V be partitioned into classes V' and V'', let $\mathcal{C} := \{\{v\}|v \in V'\} \cup \{V\setminus\{v\}|v \in V''\}$, $A \subseteq V'' \times V'$, $c \equiv 1$. Then (i) in Theorem 10 is equivalent to a theorem of König [22]: the minimum number of edges covering all vertices in a bipartite graph is equal to the maximum size of a coclique. Similarly, (ii) is equivalent to a theorem of Gupta [19]: the minimum degree in a bipartite graph is equal to the maximum number of pairwise disjoint edge sets $E_1, \ldots, E_t$ each covering all vertices.

2. Let $r, s \in V$ be fixed, let $\mathcal{C} := \{U \subseteq V | r \notin U, s \in U\}$, $D = (V, A)$ arbitrary, and $c \equiv 1$. Then (i) in Theorem 10 is equivalent to the (easy) result that the minimum number of edges in a path from r to s in D is equal to the maximum number of pairwise disjoint cuts separating r from s. Assertion (ii) is Menger's theorem [26]: the minimum number of edges in a cut separating r from s is equal to the maximum number of pairwise edge-disjoint paths from r to s.

3. Let $r \in V$ be fixed, let $\mathcal{C} := \{U \subseteq V | r \notin U \neq \varnothing\}$, and let $D = (V, A)$ and c be arbitrary. Then (i) in Theorem 10 is equivalent to a theorem of Fulkerson [11]: the minimum weight of an r-branching (= a subset of A forming a rooted directed tree with root r) is equal to the maximum number t of r-cuts (= sets of form $\delta_A^-(U)$ with $U \in \mathcal{C}$) (repetition allowed) such that any arc a of D is in at most c_a of these r-cuts. If $c \equiv 1$, assertion (ii) is equivalent to a theorem of Edmonds [8]: the minimum size of any r-cut is equal to the maximum number of pairwise disjoint r-branchings.

References

1. C. Berge, *Sur le couplage maximum d'un graphe*, C. R. Acad. Sci. Paris Vie Académique **247** (1958), 258–259.

2. M. Burlet and J. Fonlupt, *Polynomial algorithm to recognize a Meyniel graph*, Progress in Combinatorial Optimization, W. R. Pulleyblank, editor, Academic Press, Toronto, 1984, pp. 69–99.

3. M. Burlet and J. P. Uhry, *Parity graphs*, Annals Discrete Math. **16** (1982), 1–26.

4. V. Chvátal, *Edmonds polytopes and a hierarchy of combinatorial problems*, Discrete Math. **4** (1973), 305–337.

5. ____, *Cutting-plane proofs and the stability number of a graph*, Report No. 84326, Institut für Operations Research, Universität Bonn, 1984.

6. W. H. Cunningham and A. B. Marsh III, *A primal algorithm for optimal matching*, Math. Programming Stud. **8** (1978), 50–72.

7. J. Edmonds, *Maximum matching and a polyhedron with 0, 1-vertices*, J. Res. Nat. Bur. Standards (B) **69** (1965), 125–130.

8. ____, *Edge-disjoint branchings*, Combinatorial Algorithms, B. Rustin, editor, Academic Press, New York, 1973, pp. 91–96.

9. E. Egerváry, *Matrizok kombinatorius tulajdonságairól*, Mat. és Fiz. Lapok **38** (1931), 16–28.

10. A. Frank and É. Tardos, *An application of simultaneous approximation in combinatorial optimization*, 26th Annual Symposium on Foundations of Computer Science, IEEE, New York, 1985, pp. 459–463.

11. D. R. Fulkerson, *Packing rooted directed cuts in a weighted directed graph*, Math. Programming **6** (1974), 1–13.

12. A. M. H. Gerards, *An extension of König's theorem to graphs with no odd-K_4*, Preprint.

13. A. M. H. Gerards, L. Lovász, A. Schrijver, P. D. Seymour, and K. Truemper (to appear).

14. A. M. H. Gerards and A. Schrijver, *Matrices with the Edmonds-Johnson property*, Combinatorica **6** (1986), 403–417.

15. R. E. Gomory, *Outline of an algorithm for integer solutions to linear programs*, Bull. Amer. Math. Soc. **64** (1958), 275–278.

16. M. Grötschel, L. Lovász, and A. Schrijver, *The ellipsoid method and its consequences in combinatorial optimization*, Combinatorica **1** (1981), 169–197.

17. ____, *Geometric algorithms and combinatorial optimization*, Springer-Verlag, Berlin (to appear).

18. M. Grötschel and M. W. Padberg, *Polyhedral theory*, The Traveling Salesman Problem, E. L. Lawler et al., editors, Wiley, Chichester, 1985, pp. 251–305.

19. R. P. Gupta, *A decomposition theorem for bipartite graphs*, Theory of Graphs, P. Rosenstiehl, editor, Gordon and Breach, New York, 1967, pp. 135–138.

20. R. M. Karp and C. H. Papadimitriou, *On linear characterizations of combinatorial optimization problems*, SIAM J. Comput. **11** (1982), 620–632.

21. D. König, *Graphok és matrixok*, Mat. és Fiz. Lapok **38** (1931), 116–119.

22. ____, *Über trennende Knotenpunkte in Graphen (nebst Anwendungen auf Determinanten und Matrizen)*, Acta Litt. Sci. Regiae Univ. Hungar. Francisco-Josephinae (Szeged), Sect. Sci. Math. **6** (1933), 155–179.

23. A. K. Lenstra, H. W. Lenstra Jr., and L. Lovász, *Factoring polynomials with rational coefficients*, Math. Ann. **261** (1982), 515–534.

24. L. Lovász, *Graph theory and integer programming*, Ann. Discrete Math. **4** (1979), 141–158.

25. ____, *The matching structure and the matching lattice*, Report MSRI 04118-86, Math. Sci. Res. Inst., Berkeley, Calif., 1986.

26. K. Menger, *Zur allgemeinen Kurventheorie*, Fund. Math. **10** (1927), 96–115.

27. M. W. Padberg and M. Grötschel, *Polyhedral computations*, The Traveling Salesman Problem, E. L. Lawler et al., editors, Wiley, Chichester, 1985, pp. 307–360.

28. M. W. Padberg and M. R. Rao, *The Russian method and integer programming*, Ann. Oper. Res. (to appear).

29. A. Schrijver, *On cutting planes*, Ann. Discrete Math. **9** (1980), 291–296.

30. ____, *Short proofs on the matching polyhedron*, J. Combin. Theory Ser. B **34** (1983), 104–108.

31. ____, *Packing and covering of crossing families of cuts*, J. Combin. Theory Ser. B **35** (1983), 104–128.

32. ____, *Theory of linear and integer programming*, Wiley, Chichester, 1986.

33. P. D. Seymour, *On multi-colourings of cubic graphs, and conjectures of Fulkerson and Tutte*, Proc. London Math. Soc. (3) **38** (1979), 423–460.

34. ____, *Decomposition of regular matroids*, J. Combin. Theory Ser. B **28** (1980), 305–359.

35. W. T. Tutte, *The factorization of linear graphs*, J. London Math. Soc. **22** (1947), 107–111.

36. ____, *A homotopy theorem for matroids. I, II*, Trans. Amer. Math. Soc. **88** (1958), 144–174.

TILBURG UNIVERSITY, 5000 LE TILBURG, THE NETHERLANDS

MATHEMATICAL CENTRE, 1098 SJ AMSTERDAM, THE NETHERLANDS

How to Prove a Theorem
So No One Else Can Claim It

MANUEL BLUM

Goldwasser, Micali, and Rackoff [**GMR**] define for us what it means for a theorem to have a "zero-knowledge proof." In brief, a *zero-knowledge proof* is an interactive probabilistic protocol that gives highly convincing (but not absolutely certain) evidence that a theorem is true and that the prover knows a proof (a "standard" proof in a given logical system), while providing *not a single additional bit of information* about the proof. GMR formalize this idea. We do not. Nevertheless, we hope that the reader who has not read their paper will still understand our proofs.

Goldreich, Micali, and Wigderson [**GMW**] take another leap forward. They show that if one makes a reasonable assumption (that one-way functions[1] exist), then it is possible to convert any standard constructive proof of any of the theorems in a large natural class of theorems[2] into a zero-knowledge proof that the theorem is true. GMW start by considering a particular NP-complete problem:

Graph 3-Colorability.

Instance. A graph G.

Question. Can G be "properly" 3-colored (each node colored by one of 3 given colors so that no two adjacent nodes receive the same color).

GMW show that a "prover" who knows how to 3-color a particular graph G can convince a verifier that (1) G is 3-colorable, and (2) the prover knows a 3-coloring, without giving away any additional information. In particular, the prover does not give away the slightest clue *how* to 3-color G.

Supported, in part, by National Science Foundation Grant DCR 85-13926.

[1]One-way functions are 1-1 functions from n-bit integers to n-bit integers that, informally, are easy to compute in the forward direction, but hard to invert on all but a small fraction of n-bit integers.

[2]These theorems, which arise frequently in mathematics and computer science, are the *yes*-instances of decision problems π in NP. A good reference to NP and the theory of NP-completeness is: Michael Garey and David Johnson, *Computers and intractability: a guide to the theory of NP-completeness*, Freeman and Company, 1979.

The essence of the GMW proof is to show the prover how to break up his proof into several pieces in such a way that

(1) the verifier can tell, by looking at any one piece of the proof, whether or not that piece has been properly constructed. Moreover, it should be clear to the verifier that if all the pieces are properly constructed, then the proof is valid, and

(2) the prover will not reveal any information about how the proof was constructed when he reveals any single piece of the proof.

To start, the prover hides each piece of the proof in its own locked safe, in reality a one-way function applied to the piece of proof. The verifier is permitted to point to any safe and ask the prover to open it. The fact that the piece of proof inside the safe is properly constructed is evidence to the verifier that *all* pieces are properly constructed, so the proof is valid. It will be evident to the verifier that the pieces can all be properly constructed by any prover who knows how to properly 3-color G, but at least one piece must be improperly constructed by a prover who does not.

Now, proofs can be broken up into pieces in many ways. The prover must select a sequence of breakups such that a piece from the first breakup plus another from the second, and so on, does not accumulate evidence to provide the slightest hint to the verifier about how to prove the theorem. It is even possible to continue the process indefinitely without ever providing a single additional bit of information about how to prove the theorem.

By repeatedly breaking up the proof and opening just one safe each time—whichever the verifier requests—the prover convinces the verifier that he is not cheating unless he is very *very* lucky.

GMW point out that because Graph 3-Colorability is an NP-complete problem, any problem in NP can be given a zero-knowledge proof, i.e., anyone who knows a polynomial length proof of a *yes*-instance of an NP problem can give a zero-knowledge proof of this fact. (For the reader familiar with the concepts of NP-completeness, this result is a consequence of Cook's theorem that *satisfiability* is NP-complete, the NP-completeness of 3-*colorability*, and the fact that the transformations used in these proofs are (many-one) Karp-reductions.)

Outline of the talk. In this talk we show the following:

(1) How a prover can give a zero-knowledge proof that he knows a Hamilton cycle in a graph. Since the proof is zero-knowledge, the prover does not give the verifier the slightest idea how to construct that cycle. The zero-knowledge proof is interactive (the prover breaks his proof up into pieces and the verifier requests to see a particular piece) along the lines of GMW's proof. It is, however, more efficient than GMW in terms of the number of requests the verifier must make to achieve any fixed level of confidence. To ensure that a cheater will pass the test with probability of cheating $\leq 1/2^k$, we require just k requests rather than the $k \cdot E$ requests required by GMW for graphs with E edges.

(2) How a prover can give a zero-knowledge proof that a graph G is 3-colorable. This serves to show that the Hamilton cycle problem is not special so far as zero-

knowledge is concerned. Again, the standard proof of 3-colorability is broken into just 2 pieces in every round, though at the price of polynomial growth in each piece. The reader who knows the GMW proof may find it instructive to compare their proof with ours.

(3) How the proof of any theorem whatsoever (e.g., Fermat's Last Theorem) whose proof has been formalized in a standard logical system (such as Whitehead and Russell's *Principia Mathematicae*), together with any integer upper bound on the length of the proof, can be translated into a zero-knowledge proof. The zero-knowledge proof shows that the theorem is very probably true and that the prover almost certainly knows a proof in the given logical system. It gives away no other information whatsoever.

Some zero-knowledge proofs. Let G be a graph. A *Hamilton cycle* in G is a cycle that passes through all the nodes of G exactly once. We show how a prover can convince a verifier that he knows a Hamilton cycle in graph G without giving the slightest additional clue about how to construct that cycle.

The theorem to be given a zero-knowledge proof is one of a class of theorems asserting the existence of a Hamilton cycle in a graph. Although we do not formalize it, the logical proof system in which the theorem is proved is one in which each proof is just a sequence of edges in the graph. If the edges form a cycle through all the nodes of G, then the proof (that G has a Hamilton cycle) is valid; otherwise, it is not.

In the following protocols, we assume that lockable boxes are available to the prover, and that only the prover has the key. Instead of locking information in a box, however, one can encrypt it. One-way functions serve this purpose, providing us the equivalent of digital lockable boxes. The one-way functions make it possible to pursue the following interactive protocol entirely on paper rather than by using the hardware of lockable boxes and keys.

A zero-knowledge protocol for proving that a graph G has a Hamilton cycle. The protocol is interactive and probabilistic. It is probabilistic because (1) both prover and verifier must have the capability of generating sequences of independent unbiased random bits, and (2) the successful outcome of the protocol ensures to the verifier that the prover is *probably* not cheating. On the other hand, the protocol absolutely—not just probably—guarantees the prover that no hint of the proof is divulged to the verifier.

Begin.

The n nodes of G are labeled $N_1, \ldots, N_n$.

Prover: Fix one Hamiltonian cycle.

The protocol has k rounds. Each round proceeds as follows:

Begin.

Prover: In secret (i.e., without letting the verifier know what you are doing), encrypt G with the boxes. Do this by randomly mapping n labeled nodes $N_1, \ldots, N_n$ 1-1 into n labeled boxes $B_1, \ldots, B_n$, in such a way that every one of the $n!$ permutations of the nodes into the boxes is equally probable. For every pair of boxes (B_i, B_j) prepare a box labeled B_{ij}. This box is to contain a 1 if the node placed in B_i is adjacent (linked by an edge) to the node in B_j; 0 otherwise. All $n + \binom{n}{2}$ boxes are then to be locked and presented to the verifier.

The verifier receives $n + \binom{n}{2}$ labeled boxes. He is now given a choice:

(1) If he wishes, the prover will unlock *all* the boxes. In this case, the verifier may check that the boxes contain a description of G. (For example, if N_1 is adjacent to both N_2 and N_5 but to no other nodes of G, and if N_1 is in B_i, N_2 in B_j, and N_5 in B_k, then there should be a 1 in both B_{ij} and B_{ik}, and a 0 in B_{ix} for every other value of x.)

(2) On the other hand, if the verifier so chooses, the prover will open exactly n boxes $B_{ij}, B_{jk}, B_{kl}, \ldots, B_{l'i}$ (note the cyclic subscripts), those containing the Hamilton cycle that the prover selected in G, and show that these boxes all contain a 1. This proves the existence of a Hamilton cycle (in whatever graph, if any, is represented by the boxes). Since the B_i are not opened, the sequence of node numbers defining the Hamilton cycle in G is *not* revealed.

Verifier: Select one of the 2 options (graph or Hamilton cycle) at random in such a way that both choices are equally likely.

Prover: Open the appropriate boxes.

End.

Verifier: *Accept* the proof if the prover complies and, in every case, correctly exhibits either the requested G or the requested Hamilton cycle. Otherwise, *reject* the proof.

End.

THEOREM 1 (PROVER PROBABLY CANNOT CHEAT VERIFIER). *If the prover does not know a proof of the theorem, his chances of convincing the verifier that he* does *know a proof are* $\leq 1/2^k$ *when there are k rounds.*

PROOF. If the prover does not know a proof, then to pass the test, he must quess in advance what the provee will request. He fails the test if he quesses wrong even once. Q.E.D.

THEOREM 2 (VERIFIER CANNOT CHEAT PROVER). *The verifier gets not the slightest hint of the proof (other than that "the theorem is true and the prover knows a proof in the given logical proof system"). In particular, the verifier cannot turn around and prove the theorem to anyone else without proving it from scratch himself.*

PROOF. (1) When the prover reveals G, what does the verifier get? Just one of the $n!$ random mappings of the n nodes of G into n labeled boxes, each instance of G having exactly the same probability as any other. The verifier could have constructed such instances for himself with the same probability distribution. So the prover is not giving the verifier any additional information.

(2) When the prover reveals the Hamilton cycle, what does the verifier get? Just a random n-cycle, every n-cycle being exactly as likely as any other. This is because (a) the prover is required to select a particular Hamilton cycle in G and to always reveal *this particular* cycle when so requested, and (b) every permutation of the n nodes into the n boxes is equally likely. Thus the verifier is being shown a random cycle. He could have created random cycles with this uniform distribution himself. Q.E.D.

The above theorem does *not* prove that the protocol is zero-knowledge. The formal definition of zero-knowledge requires one to show that a verifier can simulate the prover, that is, take the prover's part in the dialogue with the verifier. If so, then any efficient[3] probabilistic algorithm that enables the verifier to extract useful information from his conversation with the prover could just as well be used *without* the prover to obtain that information efficiently. Here is how a proof of zero-knowledge would go:

THEOREM 3. *The protocol above for proving that a graph G has a Hamilton cycle is zero-knowledge.*

PROOF. Suppose the verifier has an efficient probabilistic algorithm A to extract useful information from his conversation with the prover. Then the verifier can use his algorithm to extract the information even without the aid of the prover. In each round he does the following:

Begin.

Verifier simulates the *prover*: The verifier flips a fair coin and, according to the outcome of the coin, encrypts either the graph G or an arbitrary n-cycle. G is (randomly) encrypted the same way the prover would have done so. A cycle is (randomly) encrypted just the way the prover would have encrypted an n-cycle (in G). Then, acting as prover, he presents the encrypted information to the verifier. Now he takes the other side.

Verifier simulates the *verifier*: The verifier uses his algorithm A to compute (perhaps probabilistically) whether to request a graph or a cycle. Because the algorithm has no way to guess with any advantage whether the boxes contain a graph or a cycle, there is a 50% chance that A requests an option (graph or cycle) that the verifier, in the guise of prover, can supply. If not, the verifier backs up algorithm A to the state it was in at the start of this round and restarts the entire round (verifier simulating the *prover*).

End.

In an expected 2 passes through each round, the verifier will obtain the benefit of algorithm A *without* the help of the prover. Thus the algorithm does

---

[3]An *efficient* (probabilistic) algorithm is one that computes its output in (expected) time polynomial in the length of the input.

not help the verifier do something *with* the prover in expected polynomial time that he could not as well have done *without* the prover in expected polynomial time.   Q.E.D.

What is the difference between Theorems 2 and 3? Theorem 2 asserts that the verifier gets no hint of the proof of a theorem from the protocol (though he may get other information). Theorem 3 asserts that the verifier not only gets no hint of the proof but actually gets no information (that he couldn't equally well have generated efficiently for himself) whatsoever. It may be helpful to observe that a proof of zero-knowledge will be difficult if not impossible to obtain (i.e., I do not know how to obtain it) if the protocol for proving that G has a Hamilton cycle is modified so that its rounds are executed in parallel. In *parallel* means that the prover first presents all k graphs to the verifier, then the verifier makes his k requests all at once, and finally the prover opens the requested boxes. To prove that this parallel protocol is zero-knowledge is difficult because it is unclear how the verifier can simulate the prover's role in this interaction efficiently.

The Hamilton cycle problem is not the only one with zero-knowledge proofs. In fact, any logical proof of length n can be split into two pieces of length polynomial(n) along the lines shown above, so that k rounds will catch all but 1 in 2^k attempts to cheat. Moreover, as GMW have shown, and as we indicate in our own way in Theorem 4, the process of transforming logical proofs into zero-knowledge proofs can be entirely mechanized so that a computer program could do it efficiently. We now give another simple example of how to transform a standard proof, in this case a proof of 3-colorability, into a zero-knowledge proof. The reader who knows GMW's method for breaking up a proof of 3-colorability into E pieces, E being the number of edges in G, may find it interesting to compare that protocol to ours, which breaks a proof into just 2 pieces.

A zero-knowledge protocol for proving that a graph G is 3-colorable.
The n nodes of G are labeled $N_1, \ldots, N_n$. The colors of the nodes will be red, white, and blue. To start, the prover knows a proper 3-coloring of G, which we call the "standard" 3-coloring. If node N_i is colored red, we call it N_i^R; if white, N_i^W; if blue, N_i^B. A triangle might therefore have the proper coloring scheme N_1^R, N_2^B, N_3^W.

Begin.
The protocol has k rounds. Each round proceeds as follows:

Begin.
Prover: Prepare $3n$ *pairs* of boxes $\langle B_1^c, B_1 \rangle, \langle B_2^c, B_2 \rangle, \ldots, \langle B_{3n}^c, B_{3n} \rangle$. Without revealing to the verifier what you are doing, randomly map $3n$ nodes $N_1^R, \ldots, N_n^R, N_1^W, \ldots, N_n^W, N_1^B, \ldots, N_n^B$ 1-1 onto the $3n$ pairs of boxes. Do this in such a way that every one of the $(3n)!$ permutations mapping the $\{N_i^x\}$ onto the $\{\langle B_j^c, B_j \rangle\}$ is equally probable. Next, insert $N_{node-number}^{color}$ into the associated $\langle B_j^c, B_j \rangle$ by putting *color* into B_j^c and *node-number* into B_j.

For every pair of number-containing boxes (B_i, B_j), prepare a box labeled B_{ij}. This box is to contain a 1 if the prover's proper 3-coloring of G has colored the node of G in B_i with the color in B_i^c, the node of G in B_j with the color in B_j^c, and if the node in B_i is adjacent in G to the node in B_j; 0 otherwise.

All boxes are then to be locked and presented to the verifier.

The verifier is now given a choice:

(1) If the verifier so wishes, the prover will unlock *all* the boxes B_{ij} and *all* the number-containing boxes B_i, but *none* of the color-containing B_i^c. In this way, the prover reveals the graph G without revealing its coloring. The verifier checks that the boxes contain a correct description of G.

(2) On the other hand, if the verifier so chooses, the prover will open the $3n$ boxes $\{B_i^c\}$ to reveal the colors they contain, and then open just those boxes B_{ij} (joining B_i to B_j) such that B_i^c contains the same color as B_j^c. The opened boxes B_{ij} will all contain a 0 if and only if any 2 nodes that are colored the same are *not* adjacent in the graph represented by the boxes. This allows the verifier to check correct 3-coloring.

Verifier: The correct thing to do is select one of the 2 options at random in such a way that both choices are equally likely.

Prover: Open the requested boxes.

End.

Verifier: *Accept* if the prover correctly complies with all requests; *reject* otherwise.

End.

This protocol is zero-knowledge, and the probability that a fake prover can cheat a verifier is $1/2^k$.

A zero-knowledge protocol for proving any theorem. Impagliazzo [I] has given direct zero-knowledge protocols along the lines shown above for several problems including the *subset sum* problem, *satisfiability*, and the very general problem of proving that a given polynomial-time nondeterministic Turing machine accepts a given input.

THEOREM 4. *Given any logical proof system (such as Russell and Whitehead's very general system within which it is generally acknowledged that all mathematical theorems can be formulated and proved), given any theorem provable in that system, and given an upper bound, L, on the length of some proof of the theorem in the system, it is possible to efficiently transform that proof into a* zero-knowledge proof *of the theorem. This is an interactive probabilistic protocol* *whereby the prover persuades the verifier that with high probability,*

(1) *the theorem has a proof in the given proof system of length $\leq L$, and*

(2) *the prover knows such a proof.*

The probability that a cheater, i.e., a prover who does not know a proof, will pass this test $\leq 1/2^k$ for a protocol with k rounds.

IDEA OF PROOF. The *proof system* is defined by a nondeterministic TM (Turing machine) which, on input ⟨statement of *theorem*, 1^n⟩, guesses a proof of the *theorem* of length $\leq n$, checks if it is a valid proof within the system, and *accepts* if it is, *rejects* if not.

The prover gives the verifier a zero-knowledge proof that he, the prover, knows an accepting path for this TM for some n. The protocol for this is along the same lines as for *Hamilton cycle* in a graph [I]: one splits the computations into two pieces. The integer n must be chosen by the prover to be an upper bound on the length of his proof in the system.   Q.E.D.

REFERENCES

[GMR] Shafi Goldwasser, Silvio Micali, and Charles Rackoff, *The knowledge complexity of interactive proof-systems*, Proc. 17th ACM Sympos. on Theory of Computing,[4] 1985, pp. 291–304.

[GMW] Oded Goldreich, Silvio Micali, and Avi Wigderson, *Proofs that yield nothing but the validity of the assertion, and the methodology of cryptographic protocol design*, presented at a Workshop on Probabilistic Algorithms (Marseille, March 1986) organized by C. P. Schnorr; Proc. 27th IEEE Sympos. on the Found. of Computer Science,[5] 1986, pp. 174–187.

[I] Russell Impagliazzo, *A collection of direct zero-knowledge protocols for NP-complete problems*, Berkeley Computer Science Division Rept., Univ. of Calif., Berkeley, Calif., 1986.

[BC] Gilles Brassard and Claude Crepeau, *Non-transitive transfer of confidence: a perfect zero-knowledge interactive protocol for SAT and beyond*, Proc. 27th IEEE Sympos. on the Found. of Computer Science,[5] 1986, pp. 188–195.

[BCR] Gilles Brassard, Claude Crepeau, and Jean-Marc Robert, *Information theoretic reductions among disclosure problems*, Proc. 27th Sympos. on the Found. of Computer Science,[5] 1986, pp. 168-173.

[Y] Andrew Yao, *How to generate and exchange secrets*, Proc. 27th Sympos. on the Found. of Computer Science,[5] 1986, pp. 162–167.

UNIVERSITY OF CALIFORNIA, BERKELEY, CALIFORNIA 94720, USA

---

[4]The ACM Symposium on Theory of Computing (STOC) is held yearly in May. Its proceedings, which are refereed by a distinguished 10-member committee, contain the best of the previous six months computer science research in all areas of theoretical computer science. Proceedings may be ordered from the ACM Order Dept., P.O. Box 64145, Baltimore, MD 21264.

[5]The IEEE Symposium on Foundations of Computer Science (FOCS) is held yearly in October. Its proceedings, like STOC, are refereed by a distinguished 10-member committee. They contain the best of the previous six months computer science research in all areas of theoretical computer science. Proceedings may be ordered from IEEE Comp. Soc., P.O. Box 80452, Worldway Postal Center, Los Angeles, CA 90080.

Proceedings of the International Congress of Mathematicians
Berkeley, California, USA, 1986

Computational Complexity in Polynomial Algebra

D. YU. GRIGOR'EV

In recent years a number of algorithms have been designed for the "inverse" computational problems of polynomial algebra—factoring polynomials, solving systems of polynomial equations, or systems of polynomial inequalities, and related problems—with running time considerably less than that of the algorithms which were previously known. (For the computational complexity of the "direct" problems such as polynomial multiplication or determination of g.c.d.'s see [1, 16] and also [9].) It should be remarked that as a result a hierarchical relationship between the computational problems of polynomial algebra, from the point of view of computational complexity, has been elucidated. The successful design of these algorithms depended to a large degree on developing them in the correct order: first the algorithms for the problems which are easier in the sense of this hierarchy were designed, which were then applied as subroutines in the solutions of more difficult problems. So far problems of the type discussed here have been considered easier only when they are special cases of the more difficult ones; e.g., the solution of a system of polynomial equations is considered as a particular case of quantifier elimination.

A powerful impetus for this development came initially from the development of polynomial-time algorithms for factoring polynomials. On the other hand, a major role has been played by a new insight from the computational point of view: treating the solution of systems of polynomial equations in the framework of the determination of the irreducible components of an algebraic variety. This has made it possible to apply the polynomial factorization algorithm to this problem. In addition a successful reduction of the problem of solving systems of polynomial inequalities to the "nonspecial" case of this problem was achieved by means of an explicit use of infinitesimals in the calculations, and the "nonspecial" case was in turn reduced to the solution of a suitable system of polynomial equations. Finally, for the design of decision procedures for the first order theories of algebraically closed or real closed fields, appropriate solvability criteria for the corresponding systems with variable coefficients were produced which are "uniform" in the set of auxiliary parameters.

Since all the bounds on time complexity given in the present paper are only specified up to a polynomial, while on the other hand all reasonable models of computation (such as Turing machines or RAM's) are equivalent in the sense of polynomial time complexity, the choice of a particular model of computation is irrelevant to this paper. One may take the complexity measure below to be the number of bit operations executed. As usual, complexity is considered as a function of the size of the input data in the worst case. The terms "polynomial time" and "exponential time" will be used in this sense (see, e.g., [1]).

1. Factoring polynomials. Attempts to design procedures for factoring polynomials go back to Newton (for a historical survey see [16]). The Kronecker-Schubert algorithm for factoring polynomials from the ring $\mathbf{Q}[X_1, \ldots, X_n]$ is well known (see, e.g., [25]). This and similar algorithms have exponential running time, however. Thus the question arose as to whether a polynomial time algorithm for factoring polynomials exists.

In the case of polynomials $f \in F_p[X]$ in one variable over a finite field of characteristic p, a positive answer to this question was given by Berlekamp's algorithm (see, e.g., [16]), whose running time is polynomial in p, s and the degree $\deg_X(f)$. For a long time there was no significant progress in attempts to design fast algorithms for factoring polynomials, until finally in [18] an ingenious polynomial-time algorithm for factoring polynomials from the ring $\mathbf{Q}[X]$ was produced. In [18] the problem of factoring polynomials was reduced to one of finding a sufficiently short vector in a lattice, and in addition for the latter problem a polynomial-time algorithm was designed. The result of [18] was then generalized in [3] (see also [4, 5, 8]), where a polynomial-time algorithm for factoring polynomials $f \in F[X_1, \ldots, X_n]$ in many variables over a fairly large class of fields F was produced. We mention also that in [12, 13] an algorithm for factoring polynomials from the ring $\mathbf{Q}[X_1, \ldots, X_n]$ was designed, whose complexity is polynomial for a fixed number n of variables.

Before proceeding to an exact formulation of the result from [3], we need to describe how a ground field F and a polynomial $f \in F[X_1, \ldots, X_n]$ are presented. Thus, we consider a field of the form $F = H(T_1, \ldots, T_e)[\eta]$, where $H = \mathbf{Q}$ or $H = \mathbf{F}_p$ (in other words H is a prime field), the elements $T_1, \ldots, T_e$ are algebraically independent over H, the element η is separably algebraic over the field $H(T_1, \ldots, T_e)$. Let $\varphi(Z) = \sum_{0 \leq i < \deg_Z(\varphi)} (\varphi_i^{(1)}/\varphi^{(2)}) Z^i \in H(T_1, \ldots, T_e)[Z]$ be the minimal polynomial of η over the field $H(T_1, \ldots, T_e)$ with the leading coefficient $\mathrm{lc}_Z(\varphi) = 1$, where the polynomials $\varphi_i^{(1)}, \varphi^{(2)} \in H[T_1, \ldots, T_e]$ and the degree $\deg(\varphi^{(2)})$ is the least possible. Any polynomial $f \in F[X_1, \ldots, X_n]$ can be uniquely represented in the form

$$f = \sum_{0 \leq i < \deg_Z(\varphi); i_1, \ldots, i_n} (a_{i, i_1, \ldots, i_n}/b) \eta^i X_1^{i_1} \cdots X_n^{i_n}$$

where the polynomials $a_{i, i_1, \ldots, i_n}, b \in H[T_1, \ldots, T_e]$ and the degree $\deg(b)$ is the least possible. Define the degree

$$\deg_{T_j}(f) = \max_{i,i_1,\dots,i_n} \{\deg_{T_j}(a_{i,i_1,\dots,i_n}), \deg_{T_j}(b)\}.$$

Another measure of the size of a representation of a polynomial is the (bit) length of its coefficients (from the field H). Namely, if $H = \mathbf{Q}$ and $\alpha/\beta \in \mathbf{Q}$, where α, β are relatively prime integers, then the length $l(\alpha/\beta)$ is defined by $\log_2(|\alpha\beta| + 2)$; if $H = \mathbf{F}_p$ then the length $l(\alpha)$ for any element $\alpha \in \mathbf{F}_p$ is defined as $\log_2 p$. The length $l(f)$ of the coefficients of a polynomial f is defined as the maximal length of the coefficients from H of the monomials in the variables $T_1,\dots,T_e$ occurring in the polynomials $a_{i,i_1,\dots,i_n}, b$. Finally, as the size $L_1(f)$ of a polynomial f we take here the value

$$\left(\max_{1\le i\le n} \deg_{X_i}(f) + 1\right)^{n+e} \left(\max_{1\le j\le e} \deg_{T_j}(f) + 1\right)^e (\deg_Z(\varphi) + 1)l(f),$$

analogously

$$L_1(\varphi) = \left(\max_{1\le j\le e} \deg_{T_j}(\varphi) + 1\right)^e (\deg_Z(\varphi) + 1)l(\varphi).$$

The size of a polynomial provides an estimate for the sum of the bit lengths of all its coefficients.

We use the notation $g_1 \le g_2 P(g_3,\dots,g_s)$ for functions $g_1,\dots,g_s$ to mean that for a suitable polynomial P the following inequality holds:

$$|g_1| \le |g_2| P(|g_3|,\dots,|g_s|).$$

THEOREM 1. *One can factor a polynomial f over the field F within time polynomial in $L_1(f)$, $L_1(\varphi)$, p. Moreover for any normalized divisor $f_1 \in F[X_1,\dots,X_n]$ of the polynomial f the following bounds are valid:*

$$\deg_{T_j}(f_1) \le \deg_{T_j}(f) P\left(\max_{1\le i\le n} \deg_{X_i}(f), \max_{1\le j\le e} \deg_{T_j}(\varphi), \deg_Z(\varphi)\right),$$

$$l(f_1) \le (l(f) + l(\varphi) + e \max_{1\le j\le e} \deg_{T_j}(f) + n)$$

$$\cdot P\left(\max_{1\le i\le n} \deg_{X_i}(f), \max_{1\le j\le e} \deg_{T_j}(\varphi), \deg_Z(\varphi)\right).$$

First Theorem 1 was proved in [3] for finite fields F, where in order to reduce the multivariable case to the case of two variables an effective version of Hilbert's Irreducibility Theorem was given.

Theorem 1 has various applications (see, e.g., [4]) to absolute polynomial factorization, to constructing a primitive element in a field extension, and to finding the Galois group of a polynomial.

2. Solving a system of polynomial equations. Let the polynomials

$$f_1, \ldots, f_\kappa \in F[X_1, \ldots, X_n]$$

be given for a field of the same form as in §1. Assume for the present section that the following bounds are fulfilled:

$$\deg_{X_1,\ldots,X_n}(f_i) < d, \qquad \deg_{T_1,\ldots,T_e,Z}(\varphi) < d_1, \qquad \deg_{T_1,\ldots,T_e}(f_i) < d_2,$$

$$l(\varphi) \le M_1, \qquad l(f_i) \le M_2, \qquad 1 \le i \le \kappa.$$

A way to decide the solvability of a system of the form $f_1 = \cdots = f_\kappa = 0$ over the algebraic closure $\overline{F}$ of a field F was given in the nineteenth century relying on elimination theory (see, e.g., [25]). The time complexity of this procedure, however, is nonelementary (in particular, it grows faster than any tower of a fixed number of exponential functions). In [22] (see also [11]) a method was devised with the help of which one can solve systems within time $(M_2 \kappa d)^{2^{O(n)}}$ when either $F = \mathbf{Q}$ or F is finite. In [17] an algorithm was produced for solving a system of homogeneous equations in the case when the projective variety of all its roots (defined over the field $\overline{F}$) consists of a finite number of points, and the running time of this algorithm is polynomial in M_2, κ, d^n, p if the ground field F is finite of characteristic p. In [4] (see also [5, 8]) an algorithm for solving systems of polynomial equations was designed, whose running time can be bounded by a polynomial in M_2, κ, d^{n^2}, p in the case when either the field $F = \mathbf{Q}$ or F is finite.

Actually, the algorithm from [4] finds the irreducible components V_i of the variety $V = \bigcup_i V_i \subset \overline{F}^n$ of all the roots of the system $f_1 = \cdots = f_\kappa = 0$. Furthermore, the algorithm represents each component in two ways: by a generic point, and secondly by a certain system of polynomials, whose associated variety coincides with the component.

In this connection, a generic point of a variety $W \subset \overline{F}^n$ of dimension $\dim(W) = n - m$ which is both defined and irreducible over the perfect closure $F^{p^{-\infty}}$ of the field F [27] is an effective version of the usual notion of generic point in algebraic geometry (an embedding of the field of rational functions on the variety). Thus we now define a generic point to be a field isomorphism of the following form:

$$F(t_1, \ldots, t_{n-m})[\theta] \simeq F(X_{j_1}, \ldots, X_{j_{n-m}}, X_1^{p^\nu}, \ldots, X_n^{p^\nu}) \subset F^{p^{-\infty}}(W) \quad (1)$$

where $t_1, \ldots, t_{n-m}$ are algebraically independent over the field F, and in addition $F^{p^{-\infty}}(W)$ is a field of rational functions on the variety W over the field $F^{p^{-\infty}}$, and the exponent $\nu \ge 0$ (we adopt the convention that $p^\nu = 1$ when $\mathrm{char}(F) = 0$); furthermore the element θ is the image under the isomorphism (1) of a linear function $\sum_{1 \le j \le n} c_j X_j$ for certain natural numbers $c_1, \ldots, c_n$. Under the isomorphism (1) the coordinate function X_{j_i} is mapped into t_i, for $1 \le i \le n - m$. The algorithm represents a generic point by specifying the coefficients $c_1, \ldots, c_n$, the exponent p^ν, the minimal polynomial $\Phi(Z) \in F(t_1, \ldots, t_{n-m})[Z]$ of the element θ, and the images under the isomorphism (1) of the functions $X_j^{p^\nu}$ in the field $F(t_1, \ldots, t_{n-m})[\theta]$. In the formulations of the theorems below we use the notations introduced in (1), and we define the degrees and the lengths of the coefficients of the functions $X_j^{p^\nu}$ as the degrees and the lengths of the coefficients of their images.

THEOREM 2. *For given polynomials $f_1, \ldots, f_\kappa$ one can find all irreudcible components V_i of the variety $V \subset \overline{F}^n$ of all the roots of the system $f_1 = \cdots = f_\kappa = 0$ within time polynomial in $M_1, M_2, (d^n d_1 d_2)^{n+e}, \kappa, p$.*

Moreover, for each component V_i the algorithm yields a generic point for it (see (1)) and a family of polynomials $\Psi_1^{(i)}, \ldots, \Psi_N^{(i)} \in F[X_1, \ldots, X_n]$ such that V_i coincides with the variety of all roots of the system $\Psi_1^{(i)} = \cdots = \Psi_N^{(i)} = 0$. Denote $m = \operatorname{codim} V_i$, $\theta_i = \theta$, $\Phi_i = \Phi$. Then the following bounds hold:

$$p^\nu \leq d^{2m}, \qquad c_j \leq \deg_Z(\Phi_i) \leq \deg V_i \leq (d-1)^m, \qquad N \leq m^2 d^{4m};$$

$$\deg_{T_1, \ldots, T_e, t_1, \ldots, t_{n-m}}(\Phi_i), \deg_{T_1, \ldots, T_e, t_1, \ldots, t_{n-m}}(X_j^{p^\nu}) \leq d_2 P(d^m, d_1);$$

$$l(\Phi_i), l(X_j^{p^\nu}), l(\Psi_s^{(i)}) \leq (M_1 + M_2 + (e+n-m)d_2) P(d^m, d_1);$$

$$\deg_{X_1, \ldots, X_n}(\Psi_s^{(i)}) \leq d^{2m}; \qquad \deg_{T_1, \ldots, T_e}(\Psi_s^{(i)}) \leq d_2 P(d^m, d_1).$$

Theorem 2 allows us to answer the principal questions about the variety of roots of a system of polynomial equations, namely, whether the variety is empty, and what its dimension is. Provided that the variety consists of a finite number of points, the algorithm enumerates all of them; otherwise if the variety is not zero-dimensional then the algorithm allows us to pick out any desired number of roots of the system.

Evidently, the time-bound in Theorem 2 cannot be considerably improved in general, if one desires to find all the irreducible components of a variety, since the size of a presentation of a component with dimension near $n/2$ is of the order $M_2 d^{n^2}$ in the case when either $F = \mathbf{Q}$ or F is finite.

The algorithm from Theorem 1 is involved essentially in the proof of Theorem 2. On the other hand, polynomial factorization is a particular case (when $\kappa = 1$) of the problem of finding all the irreducible components of a variety.

As a corollary of Theorem 2 one can find all the absolutely irreducible components of a variety within the same time-bound as in Theorem 2 [4].

Note that the methods discussed do not allow us to recognize within the same time-bound, whether a polynomial f belongs to an ideal $(f_1, \ldots, f_\kappa) \subset F[X_1, \ldots, X_n]$ (by means of Theorem 2 one can test, however, whether a polynomial f belongs to the radical $\operatorname{rad}(f_1, \ldots, f_\kappa)$).

3. Quantifier elimination in the first-order theory of algebraically closed fields.

Quantifier elimination in the first-order theory of algebraically closed fields is a generalization of the problem of solving systems of polynomial equations. Thus, consider a formula of this theory of the form

$$\exists X_{1,1} \cdots \exists X_{1,s_1} \forall X_{2,1} \cdots \forall X_{2,s_2} \cdots \exists X_{a,1} \cdots \exists X_{a,s_a} (\Pi) \qquad (2)$$

where Π is a quantifier-free formula of the theory containing κ atomic subformulas of the sort $(f_i = 0)$, $1 \leq i \leq \kappa$, here the polynomials

$$f_i \in F[X_1, \ldots, X_{s_0}, X_{1,1}, \ldots, X_{a,s_a}]$$

(we assume the field F and the polynomials f_i satisfy the same bounds as in the beginning of the previous section). Denote by $n = s_0 + s_1 + \cdots + s_a$ the total number of variables (including free ones $X_1, \ldots, X_{s_0}$), and by a the number of quantifier alternations in the formula (in the presentation of the formula (2) a is odd, but this is not essential).

In [23] (see also [21]) a quantifier elimination procedure was described, which for a given formula of the form (2) yields an equivalent quantifier-free formula. The time-bounds of these procedures, however, were nonelementary. In [11] a quantifier elimination method is described, having time-bound $(M_2 \kappa d)^{2^{O(n)}}$ in the case when either the field $F = \mathbf{Q}$ or F is finite (when $F = \mathbf{Q}$ the same time-bound follows from the methods of [6, 26]). In [5] a quantifier elimination algorithm is produced with time-bound polynomial in $M_2, (\kappa d)^{(O(n))^{2a}}$ in the case when either $F = \mathbf{Q}$ or F is finite, more exactly the following is valid.

THEOREM 3. *For a given formula of the form* (2) *one can construct an equivalent quantifier-free formula of the first-order theory of algebraically closed fields*

$$\bigvee_{1 \leq i \leq \mathcal{N}} \left(\underset{1 \leq j \leq \mathcal{K}}{\&} (g_{ij} = 0) \& (g_{i,0} \neq 0) \right)$$

within time polynomial in $M_1, M_2, (\kappa d)^{(O(n))^{2a} e}, (d_1^a d_2)^{n+e}, p$. *Moreover the polynomials* $g_{ij} \in F[X_1, \ldots, X_{s_0}]$ *satisfy the following bounds*:

$$\deg_{X_1, \ldots, X_{s_0}} (g_{ij}) \leq (\kappa d^n)^{(82(n+3)(n+2a)/a)^a} = \mathcal{M};$$

$$\deg_{T_1, \ldots, T_e} (g_{ij}) \leq d_2 P(\mathcal{M}, d_1^a);$$

$$l(g_{ij}) \leq (M_1 + M_2 + ed_2) P(\mathcal{M}, d_1^a); \qquad \mathcal{N}, \mathcal{K} \leq \mathcal{M}.$$

The main auxiliary subroutine for proving the theorem is the projection (with respect to many variables) of a quasiprojective variety, based on Theorem 2. Furthermore, a bound on the degree of a projection of a constructible set is obtained. For a constructible set $\mathcal{W} \subset \overline{F}^n$ we say that its degree $\deg(\mathcal{W}) \leq \mathcal{D}$, provided that there is a representation $\mathcal{W} = \bigcup_i (\mathcal{V}_i \backslash \mathcal{U}_i)$, where $\mathcal{V}_i, \mathcal{U}_i$ are closed sets (in the Zariski topology [26]) such that $\sum_i (\deg(\mathcal{V}_i) + \deg(\mathcal{U}_i)) \leq \mathcal{D}$. The method from [5] entails the following bound. If $\pi \colon \overline{F}^n \to \overline{F}^{n-m}$ is a linear projection, then $\deg(\pi(\mathcal{W})) \leq (\deg(\mathcal{W}))^{O(nm+1)}$.

The time-bound in Theorem 3 is significantly lower than time-bounds from [6, 26, 20, 11] for small a. We remark, on the other hand, that an exponential lower bound for the complexity of a decision procedure for the first-order theory of algebraically closed fields was obtained in [7] (see also [2]) for a succession of formulas in which the number a of quantifier alternations grows linearly with the number n of variables. From this remark and from Theorem 3 one can conclude that the parameter a gives the most significant contribution to the complexity of quantifier elimination in a formula of the theory.

4. Solving system of polynomial inequalities. Let a system of polynomial inequalities

$$f_1 > 0, \ldots, f_m > 0, f_{m+1} \geq 0, \ldots, f_\kappa \geq 0 \tag{3}$$

be given, where the polynomials $f_i \in \mathbf{Q}[X_1, \ldots, X_n]$ satisfy the bounds

$$\deg_{X_1, \ldots, X_n}(f_i) < d, \quad l(f_i) \leq M, \quad 1 \leq i \leq \kappa.$$

Decidability (over the field $\mathbf{R}$) of systems of the form (3) was proved in [23] (see also [21]). The time-bounds of the procedures from [23, 21], however, were nonelementary. In [6, 26] the algorithms for solving systems of inequalities were designed with time-bound $(M\kappa d)^{2^{O(n)}}$ (also, an algorithm with a worse elementary time-bound was described in [20]). In [24] an algorithm for this problem was produced with time-bound polynomial in $M(\kappa d)^{n^2}$.

We mention that in the case when $\deg(f_i) = 1$ for $1 \leq i \leq \kappa$ (linear programming) a polynomial time algorithm was described for the first time in [15] (a more practical polynomial time method was described in [14]).

For the exact formulation of the result [24] we introduce the notion of a representative set for a semialgebraic set. The set consisting of all real points satisfying a system of inequalities of the form (3), is a semialgebraic set $S \subset \mathbf{R}^n$, which can be represented as a union $S = \bigcup_i S_i$ of its connected components (in the euclidean topology), each S_i being in its turn a semialgebraic set [23]. We say that a finite family of points $\mathcal{T} \subset S \subset \mathbf{R}^n$ is a representative set for the system of inequalities (3) (or for the semialgebraic set S) if $\mathcal{T} \cap S_i \neq \varnothing$ for every i.

Observe that unlike §2, where an algebraic point from $\overline{F}^n$ was given by the algorithm actually as an element of a class of points conjugate over the field F, to represent a real algebraic point $\alpha = (\alpha_1, \ldots, \alpha_n) \in \mathbf{R}^n$ one needs to specify an interval containing a unique root of the minimal polynomial of a primitive element of the field $\mathbf{Q}(\alpha_1, \ldots, \alpha_n)$. Namely, $\alpha_i = \sum_j \alpha_i^{(j)} \theta^j$ where $\alpha_i^{(j)} \in \mathbf{Q}$ and $\theta \in \mathbf{R}$ is a root of a polynomial $\Phi(Z) \in \mathbf{Q}[Z]$ which is irreducible over $\mathbf{Q}$, furthermore $\theta = \sum_{1 \leq i \leq n} c_i \alpha_i$ for some natural numbers $c_1, \ldots, c_n$; the algorithm gives $\Phi, \alpha_i^{(j)}, c_i$ and in addition an interval $(\beta_1, \beta_2) \subset \mathbf{R}$ with rational endpoints $\beta_1 < \beta_2$, containing only one root θ of the polynomial Φ. Below in the formulation of Theorem 4 we utilize the same notation.

THEOREM 4. *For a given system of inequalities of the kind* (3) *one can construct a representative set* $\mathcal{T}$ *containing* $(\kappa d)^{O(n^2)}$ *points within time polynomial in* $M(\kappa d)^{n^2}$. *Moreover, for any point* $\alpha = (\alpha_1, \ldots, \alpha_n) \in \mathcal{T}$ *the following bounds are valid:*

$$c_i \leq \deg(\Phi) \leq (\kappa d)^{O(n)}; \qquad l(\Phi), l(\alpha_j^{(i)}), l(\beta_1), l(\beta_2) \leq M(\kappa d)^{O(n)}.$$

We remark that the number of connected components S_i of a semialgebraic set S does not exceed $(\kappa d)^{O(n)}$ (see, e.g., [19]).

The proof of Theorem 4 involves essentially Theorem 2.

5. Deciding Tarski algebra. Similarly to the case of algebraically closed fields (§3) we now consider the first-order theory of real closed fields (or in other words, Tarski algebra). Namely, consider a formula of the form

$$\exists X_{1,1} \cdots \exists X_{1,s_1} \forall X_{2,1} \cdots \forall X_{2,s_2} \cdots \exists X_{a,1} \cdots \exists X_{a,s_a} (\Omega) \qquad (4)$$

where Ω is a quantifier-free formula of Tarski algebra, containing κ atomic subformulas of the kind $(f_i \geq 0)$, $1 \leq i \leq \kappa$; here the polynomials $f_i \in \mathbf{Q}[X_{1,1}, \ldots, X_{a,s_a}]$. As in §3 a is the number of quantifier alternations. Unlike §3 we consider only closed formulas (without free variables) in the present section; denote by $n = s_1 + \cdots + s_a$ the number of all variables. As in §4 assume that $\deg(f_i) < d$, $l(f_i) \leq M$, $1 \leq i \leq \kappa$.

In [23] (see also [21]) a quantifier elimination procedure for Tarski algebra was described, which implies its decidability. The time-bounds for these procedures, however, were nonelementary. In [6, 26] quantifier elimination methods for Tarski algebra were described with running time $(M\kappa d)^{2^{O(n)}}$. (Also in [20] a certain method was described having an elementary, but worse time-bound.) In [10] the following theorem is claimed.

THEOREM 5. *There is a decision algorithm for Tarski algebra with running time for formulas of the form* (4) *polynomial in* $M(\kappa d)^{(O(n))^{4a-2}}$.

In the proof of Theorem 5, Theorems 3, 4 are involved essentially. Observe that as in §3 one can draw the conclusion that the parameter a makes the most significant contribution to the complexity of the decision procedure.

As a corollary of Theorem 5 one can calculate the dimension of a semialgebraic set $S \subset \mathbf{R}^n$ consisting of the solutions of a system of the kind (3) within time polynomial in $M(\kappa d)^{(O(n))^{10}}$.

Note in conclusion that it would be possible to design a quantifier elimination procedure for Tarski algebra with the same time-bound as in Theorem 5, provided that one could solve within time e.g. $P(M(\kappa d)^{n^2})$ at least one of two following computational problems. First: elimination of a single quantifier in a formula of Tarski algebra. Second: for a given semialgebraic set $S \subset \mathbf{R}^n$ to find its connected components S_i, i.e., to find quantifier-free formulas Ω_i of Tarski algebra such that S_i coincides with the set of points in $\mathbf{R}^n$ satisfying Ω_i.

REFERENCES

1. A. Aho, J. Hopcroft, and J. Ullman, *The design and analysis of computer algorithms*, Addison-Wesley, Reading, Mass., 1976.

2. L. Berman, *The complexity of logical theories*, Theoret. Comput. Sci. 11 (1980), 71–77.

3. A. L. Chistov and D. Yu. Grigor'ev, *Polynomial-time factoring of polynomials over a global field*, Preprint LOMI E-5-82, Leningrad, 1982.

4. ____, *Subexponential-time solving systems of algebraic equations*. I, II, Preprints LOMI E-9-83, E-10-83, Leningrad, 1983.

5. ____, *Complexity of quantifier elimination in the theory of algebraically closed fields*, Lecture Notes in Comput. Sci. **176** (1984), 17–31.

6. G. E. Collins, *Quantifier elimination for real closed fields by cylindrical algebraic decomposition*, Lecture Notes in Comput. Sci. **33** (1975), 134–183.

7. M. J. Fischer and M. O. Rabin, *Super-exponential complexity of Presburger arithmetic*, Complexity of Computations (Proc. SIAM-AMS Sympos., New York, 1973), SIAM-AMS Proc., Vol. 7, Amer. Math. Soc., Providence, R.I., 1974, pp. 27–41.

8. D. Yu. Grigor'ev and A. L. Chistov, *Fast decomposition of polynomials into irreducible ones and the solution of systems of algebraic equations*, Soviet Math. Dokl. **29** (1984), 380–383.

9. D. Yu. Grigor'ev, *Multiplicative complexity of a pair of bilinear forms and of the polynomial multiplication*, Lecture Notes in Comput. Sci. **64** (1978), 250–256.

10. ____, *Complexity of deciding the first-order theory of real closed fields*, Proc. All-Union Conf. Applied Logic, Novosibirsk, 1985, pp. 64–66. (Russian)

11. J. Heintz, *Definability and fast quantifier elimination in algebraically closed fields*, Theoret. Comput. Sci. **24** (1983), 239–278.

12. E. Kaltofen, *A polynomial reduction from multivariate to bivariate integral polynomial factorization*, Proc. 14 ACM Sympos. Th. Comput. (May, 1982), pp. 261–266.

13. ____, *A polynomial-time reduction from bivariate to univariate integral polynomial factorization*, 23rd Annual Sympos. on Foundations of Comput. Sci. (Chicago, Ill., 1982), IEEE, 1982, pp. 57–64.

14. N. Karmarkar, *A new polynomial-time algorithm for linear programming*, Proc. 16 ACM Sympos. Th. Comput. (May, 1984), pp. 302–311.

15. L. G. Khachian, *A polynomial algorithm in linear programming*, Soviet Math. Dokl. **20** (1979), 191–194.

16. D. Knuth, *The art of computer programming*, vol. 2, Addison-Wesley, Reading, Mass., 1969.

17. D. Lazard, *Resolution des systemes d'équation algebriques*, Theoret. Comput. Sci. **15** (1981), 77–110.

18. A. K. Lenstra, H. W. Lenstra, and L. Lovasz, *Factoring polynomials with rational coefficients*, Math. Ann. **261** (1982), 515–534.

19. J. Milnor, *On the Betti numbers of real varieties*, Proc. Amer. Math. Soc. **15** (1964), 275–280.

20. L. Monk, *An elementary-recursive decision procedure for* $\text{Th}(R, +, \cdot)$, Ph.D. Thesis, Univ. of California, Berkeley, 1974.

21. A. Seidenberg, *A new decision method for elementary algebra*, Ann. of Math. **60** (1954), 365–374.

22. ____, *Constructions in algebra*, Trans. Amer. Math. Soc. **197** (1974), 273–314.

23. A. Tarski, *A decision method for elementary algebra and geometry*, Univ. of California Press, Berkeley, Calif., 1951.

24. N. N. Vorob'ev, Jr. and D. Yu. Grigor'ev, *Finding real solutions of systems of polynomial inequalities in subexponential time*, Soviet Math. Dokl. **32** (1985), 316–320.

25. B. L. van der Waerden, *Moderne Algebra*, B. I, II, Springer-Verlag, 1930, 1931.

26. H. Wüthrich, *Ein Entscheidungsverfahren für die Theorie der reell-abgeschlossenen Körper*, Lecture Notes in Comput. Sci. **43** (1976), 138–162.

27. O. Zariski and P. Samuel, *Commutative algebra*, vols. 1, 2, Van Nostrand, Princeton, N.J., 1958, 1960.

LENINGRAD BRANCH, STEKLOV MATHEMATICAL INSTITUTE, ACADEMY OF SCIENCES OF THE USSR, LENINGRAD 191011, USSR

Retrieval and Data Compression Complexity

R. E. KRICHEVSKY

Some recent results of the retrieval theory are discussed. Among them: the ABC word-order is not always the best for a dictionary; there is a threshold for redundancy. On one side of it, the retrieval complexity is a linear; on the other, an exponential function of word-length.

1. Introduction. It is rather hard for a participant of an International Congress of Mathematicians to remember his (or her) room number. So, there ought to be a computer supplied with a program. A name being entered, the computer produces the corresponding number. The problem is to accommodate the participants so as to make the shortest program.

There are variants of the problem. E.g., the computer either must or must not tell a participant from an outsider. A participant either must or must not be given a separate room. In the latter case beds make a line in a room, and the quality of accommodation is judged by the average time to reach one's bed.

A result: the name–room number transformation is rather easy if the number of rooms exceeds the number of participants squared. In the other case it is quite difficult for almost any set of participants.

More formally, let D be a subset of the set E^n of all n-length binary words, $n > 0$. Such a set can be called a dictionary or combinatorial source (in information theory). Its per letter entropy equals

$$H(D) = \frac{\log |D|}{n} \qquad (1)$$

($\log x$ stands for $\log_2 x$).

A map $f\colon D \to E^l$, $l > 0$, is called either a hash-function (in computer science) or a compression (in information theory). It is characterized, on the one hand, by its redundancy $\rho(f, D)$

$$\rho(f, D) = \frac{l/n - H(D)}{H(D)} \qquad (2)$$

and by its related load factor $\alpha(f, D)$

$$\alpha(f, D) = |D|/2^l, \qquad \alpha(f, D) = |D|^{-\rho(f, D)}. \qquad (3)$$

On the other hand, it is characterized by its collision index $I(f, D)$

$$I(f, D) = \frac{1}{|D|} \sum_{k \in \operatorname{Im} f} |f^{-1}(k)|^2 - 1, \tag{4}$$

where $\operatorname{Im} f$ is the range of f, $|f^{-1}(k)|$ is the number of words mapped to k, $k \in \operatorname{Im} f$. If D is the set of names, then $\operatorname{Im} f$ is the set of room numbers; the average time to find one's bed is $t(f, D)$,

$$t(f, D) = \frac{1}{|D|} \sum_{k \in \operatorname{Im} f} (1 + \cdots + |f^{-1}(k)|).$$

Actually, just $t(f, D)$ is used in computer programming [10] to measure the performance of hashing with separate chaining. We use for this end the index $I(f, D)$ which is linearly related to $t(f, D)$:

$$I(f, D) = 2(t(f, D) - 1).$$

If the index $I(f, D)$ of a map f equals zero, then the map f is called a perfect hash function, or enumeration, or injection.

An enumeration f is called strong for D if $f(x) = \varnothing$, $x \notin D$. Such an enumeration can distinguish between members and nonmembers of D.

A word x is colliding by f if there is a word x_1, $x_1 \in D$, such that $f(x_1) = f(x)$. If e stands for the number of colliding words, then

$$I(f, D) \geq e/|D|. \tag{5}$$

Hence, the inequality $I(f, D) < 1/|D|$ implies the injectivity of f on D.

A binary word P is said to be a computing program of a map f if, being fed with P and a word x, an initially empty computer produces $f(x)$. The shortest bitlength of computing programs of a map f is called the program or Kolmogorov complexity of f and denoted by $L(f)$. Program complexity is asymptotically computer independent unlike running time which does depend on the choice of a computer.

Given numbers ρ and a, $L(D, \rho, a)$ stands for the minimal program complexity of maps f whose index $I(f, D)$ and redundancy $\rho(f, D)$ on D do not exceed a and ρ, correspondingly. (If $a > 0$, we restrict ourselves for simplicity's sake to uniform maps which assume each value the same number of times.) Given ρ, a, n, and τ, $0 < \tau < 1$, $L(\tau, \rho, a)$ stands for the maximal value of $L(D, \rho, a)$ over all dictionaries $D \subseteq E^n$, $H(D) = \tau$. The similar quantity for strong enumerations ($\rho = 0$, $a = 0$) is denoted by $L^s(\tau)$.

Both injective and hash retrieval methods are widely used, see [10]. However, their complexity have not yet been compared with the theoretically optimal value $L(\tau, \rho, a)$ and that value has not even been known. We are going to make the comparison in §4, Table 1. Some new optimal or nearly optimal retrieval programs will be exhibited. Although their performance is good, they are rather hard to tailor for a given dictionary. On the contrary, conventional programs are easy to construct but lengthy. Programs combining good features of both are looked for.

Optimal programs make use of universal hash-sets which for any dictionary of a given size contain a good hash-function. Such sets are discussed in §§2, 3. Correcting codes are employed in §3. In its turn, the retrieval theory yields a statement for the theory of error-correcting codes.

There is a saying that theorems of interest are proved at least twice. It is just the case with the retrieval theory.

2. Universal hash-sets. Given a word length n, a load factor α, an entropy value τ, and an index level a, a set M of uniform maps $E^n \to [0, m-1]$, $m = (1/\alpha)2^{n\tau}$, is called a universal a-hash-set, if for any D, $D \subseteq E^n$, $H(D) = \tau$, there is a map f, $f \in M$, $I(f, D) \le a$. Let $N(\tau, \alpha, a)$ be the cardinality of a minimal a-hash-set. Then there is a threshold value of the load factor α, equal to the index level a. If $a < \alpha$, then $\log N(\tau, \alpha, a)$ is $O(T)$, $T = 2^{n\tau}$, but if $a \ge \alpha$, then $\log N(\tau, \alpha, a)$ is $O(\log T)$, see [16]. It is of interest to develop nearly minimal a-hash-sets. For a dictionary D one can choose elements of such a set at random until a map f with $I(f, D) \le a$ is found.

A 0-hash set is called a universal set of enumerators. In other words, a set $U(X, T, \alpha) = \{f \colon X \to [0, T/\alpha - 1]\}$ of maps from a set X to the segment $[0, T/\alpha - 1]$ of natural numbers is called a universal set of enumerators if for any D, $D \subseteq X$, $|D| = T$, there is an injection $f \in U(X, T, \alpha)$. Obviously, $|U(X, T, \alpha)| = N(\log T/n, \alpha, 0)$. That definition was introduced in [13] and detailed in [14]. It is proved there that

$$\frac{C_{|X|}^T}{C_{T/\alpha}^T} \left(\frac{T}{\alpha|X|}\right)^T \ln C_{|X|}^T \ge |U(X, T, \alpha)| \ge \frac{C_{|X|}^T}{C_{T/\alpha}^T} \left(\frac{T}{\alpha|X|}\right)^T \tag{6}$$

(T/α supposed to be an integer). If $T/\alpha|X| \to 0$, then

$$\log |U(X, T, \alpha)| \sim T \log_2 e(1 + (1/\alpha - 1)\ln(1 - \alpha)). \tag{7}$$

If, moreover, $\alpha \to 0$, then

$$\log |U(X, T, \alpha)| \sim T^{1-\rho+o(1)}, \tag{8}$$

where ρ is found by (3), $|D| = T$, $\rho(f, D) = \rho$. Upper and lower bounds (6)–(8) were proved in [13, 14] by random coding and volume methods, correspondingly. Just the same bounds are proved the same way in [4]. The set $U(X, T, \alpha)$ is called the $(T, T/\alpha)$-separating system there. There are also some improvements of bounds (6)–(7) there. A relation between universal sets of enumerators (separating systems) and the entropy of graphs is found in [11].

Let $X = \{x_1, \ldots, x_{|X|}\}$ be a set of boolean variables,

$$U(X, T, 1) = \{f_1, \ldots, f_{|U(X,T,1)|}\},$$

$f_i \colon X \to [0, T-1]$. The formula

$$\bigvee_{i=1}^{U(X,T,1)} \bigwedge_{j=0}^{T-1} \bigvee_{\alpha, x_\alpha \in f_i^{-1}(j)} x_\alpha \tag{9}$$

is a basis $(\vee, \wedge)$ realization of the monotone symmetric threshold T boolean function of $|X|$ variables. Its size is $|X| \cdot |U(X, T, 1)|$. Connection (9) between universal sets of enumerators and those functions was wittily exploited in [9] to yield nearly minimal formulae. More constructive (without exhaustive search) methods to make good formulae for those functions were later developed in [6, 12]. Relation (9) may yet be of value for the retrieval theory.

3. Code-based universal hash-sets. A source of retrieval programs is the following statement from [16]:

THEOREM 1. *Let $\Phi = \{\varphi_1, \ldots, \varphi_{|\Phi|}\}$ be a set of maps from E^n to $[0, m-1]$, $n > 0$, $m > 0$, $\bigotimes \Phi$ be a map which takes a word $x \in E^n$ to the concatenation $\varphi_1(x) \cdots \varphi_{|\Phi|}(x)$, $r = \min_{x,y \in E^n} \rho(\bigotimes \Phi x, \bigotimes \Phi y)$, and let ρ be the Hamming distance. Then for any $D \subseteq E^n$ there is a map φ_0 with a not very big colliding index:*

$$I(\varphi_0, D) \leq (|D| - 1)(1 - r/|\Phi|).$$

Any map $\varphi_i \in \Phi$ for which the sum of distances $\sum_{x,y \in D} \rho(\varphi_i(x), \varphi_i(y))$ is not less than the average distance $(1/|\Phi|) \sum_{x,y \in D} \rho(\bigotimes \Phi x, \bigotimes \Phi y)$ may be chosen as φ_0.

So the set Φ is an a-hash-set for T-size dictionaries, $a \leq (T-1)(1 - r/|\Phi|)$. A polynomial map φ_b is specified by a vector b, $|b| = \mu$. Subdivide a vector x, $|x| = n$, into μ-length subvectors: $x = x_1, \ldots, x_{\lceil n/\mu \rceil}$ (the last subvector is supplemented by zeros to the length μ, if necessary). The vectors b, $x_1, \ldots, x_{\lceil n/\mu \rceil}$ may be considered members of $GF(2^\mu)$. Let

$$\varphi_b(x) = \sum_{i=1}^{\lceil n/\mu \rceil} x_i b^{i-1}. \tag{10}$$

It is supposed to take our computer a unit of time to find $\varphi_b(x)$. A computing program for $\varphi_b(x)$ consists of the vector b and several computing instructions. Hence, $\varphi_b(x) = \mu(1 + o(1))$, $\mu \to \infty$.

COROLLARY 1. *For any $D \subseteq E^n$ there is a vector b, $|b| = \mu$, such that*

$$I(\varphi_b, D) \leq \frac{|D| \cdot n}{\mu \cdot 2^\mu}(1 + o(1)).$$

COROLLARY 2. *Let*

$$\mu = \lceil 2 \log |D| \rceil + \left\lceil \log \frac{n}{2 \log |D|} \right\rceil.$$

For any $D \subseteq E^n$ there is an injective on D polynomial map $\varphi_b \colon E^n \to E^\mu$.

Corollary 2 follows from Corollary 1 and (5).

A linear map ψ_b is specified by an n-length vector b and a number μ, which is a divisor of n (for simplicity).

We let

$$\psi_b(x) = \sum_{i=1}^{n/\mu} x_i b_i, \tag{11}$$

where x_i and b_i are μ-length subvectors of x and b, and calculations are made in $GF(2^\mu)$.

COROLLARY 3. *For any $D \subseteq E^n$ there is a linear map ψ_b: $E^n \to 2^\mu$,*
$L(\psi_b) = n(1 + o(1))$, $T(\psi_b, D) \leq |D|/2^\mu = \alpha$.

At last, let the map φ_i take a word x to its ith letter, $i = 1, \ldots, n = |x|$, $D(1, \sigma)$ be the subset of words of a dictionary D, whose lth letter is σ, $l = 1, \ldots, n$, $\sigma = 0, 1$.

COROLLARY 4. *For any dictionary D there is l, $0 < l \leq n$, such that $D(l, 0)$ and $D(l, 1)$ are of equal size to within $O(n)$ factors:*

$$\max_{1 \leq i \leq n} \min_{\sigma = 0,1} |D(i, \sigma)| \geq \frac{|D|}{4n}.$$

4. Retrieval methods. Characteristics of retrieval methods are displayed in Table 1 (running time depends on the type of a computer). We will comment on it.

First, the bounds of $L(\tau, \rho, a)$ and $L^s(\tau)$ from [13–17]. The redundancy threshold equals minus the logarithm of the index. For perfect hashing, it equals 1, see (5). More precisely, $L(\tau, \rho, a) = O(T)$ if $a > \alpha$, $L(\tau, \rho, a) = O(\log T)$, if $a \leq \alpha$. Here $n \to \infty$, $T = 2^{n\tau}$, $\alpha = T^{-\rho}$, $0 < \tau < \frac{1}{2}$ (conditions and details are in [16]). For perfect hashing:

$$L(\tau, \rho, 0) = \begin{cases} T(\log_2(1 + (1/\alpha - 1)\ln(1 - \alpha))), & \alpha = \text{const}, \\ T^{1-\rho+o(1)}, & 0 < \rho < 1, \\ O(n), & \rho > 1. \end{cases}$$

Hence, the size of the best name–room number transforming program equals $T \log_2 e \simeq T \cdot 1.4$ bits if both numbers of participants and rooms available equal T.

There is an entropy doubling effect: the complexity of injective maps $D \to E^\mu$, $D \subseteq E^n$, is exponential if $\mu < n(2H(D) - \varepsilon)$, and it is linear if $\mu > n(2H(D) + \varepsilon)$, for the majority of dictionaries.

For the strong retrieval $L^s(\tau) \sim T(n - \log T)$, $0 < \tau < 1$, $n \to \infty$.

The lower bounds of $L(\tau, \rho, a)$ follow the lower bounds of universal hash sets. Proceed to Table 1.

An unordered search program consists of a dictionary and some computer instructions. The search time $O(T)$ can be decreased to $O(\log T)$ if the words are ordered lexicographically.

When making a digital search one moves from one node of a retrieval tree to another according to letters of a word. There are $\log T$-bit pointers (links). Those pointers make the bulk of the program size [10].

Linear and polynomial hashings are defined by (10) and (11). One sees comparing their program size with the lower bounds that they are nearly optimal. However, it may be difficult to find a suitable vector b, $|b| = \log m$. Weak enumeration methods depend on the redundancy desired. If $\rho > 1$, such a method

R. E. KRICHEVSKY

TABLE 1. The complexity of retrieval algorithms

Retrieval algorithm	Program size (bits)	Running time	Colliding index I	Strong(s) or weak(w)
Retrieval in an unordered table	Tn	CT	0	s
Logarithmic retrieval in lexicographically ordered table [10]	Tn	$C\log T$	0	s
Digital search [10]	$CT\log T$	Cn	0	w
Enumerative encoding [2]	$T(n - \log T)$	$C2^n$	0	s
Quasidichotomous digital search [15]	$CT\log n$	Cn	0	w
As.opt.strong digital search [15]	$T(n - \log T)$	Cn	0	s
Weak optimal enumeration, $\rho > 1$ [17]	$O(n)$	$O(1)$	0	w
Weak optimal enumeration, $0 < \rho < 1$ [17]	$T^{1-\rho+o(1)}$	Cn	0	w
Weak optimal enumeration, $\alpha = $ const [13, 14]	$T \cdot \log_2 e(1 + (1/\alpha - 1)\ln(1 - \alpha))$	Cn	0	w
K. Mehlhorn's method A [18]	$O(\alpha T)$	$O(e^T)$	0	w
K. Mehlhorn's method C [18]	$O(T\log T)$	$\ln^* 2^n$	0	w
Double FKS-hashing [5]	$O(T)$	$O(1)$	0	w
Linear hashing [16]	n	$O(1)$	α	w
Polynomial hashing [16]	$\log(T/\alpha)$	$O(1)$	$\alpha n/\log m$	w

A T-size dictionary, containing n-length binary words, $0 < (\log T)/n < 1$, $n \to \infty$, is to be stored in an m-size table, $\alpha = T/m$, $\rho = -\log \alpha/\log T$. Words with the same address make a list. Average search time equals $\frac{1}{2}I + 1$. If $I = 0$, we obtain an injection. Strong retrieval must distinguish the words of D from the ones outside D.

is yielded by Corollary 2 (results of [8, 19] may also be used). If $0 < \rho < 1$, a polynomial map f_1, $I(f_1, D) < T^{-\rho+o(1)}$, is chosen according to Corollary 1. Colliding words are resolved lexicographically. Their number is not very great, thanks to (5). The case of a constant load factor is the most complicated. It was settled in [13, 14]. A simplified version of the algorithm is to appear in *Information and Computation*. A sketch: numbers r and R, $0 < r < R < \log T$, are fixed and r steps of lexicographical search are made. For any subdictionary of still nonresolved words an injective map $E^n \to E^\mu$ (Corollary 2) is selected and $R - r$ additional steps of lexicographical search of μ-length words are made. At last, a map from $U(E, T/2^R, \alpha)$ is used.

The method from [5] may be called double FKS-hashing, by the initials of its authors. First, a hashing is performed. Second, the still colliding words are resolved. It is possible to use, first, polynomial hashing and, second, Corollary 2, although the authors employ division by primes. There is a report on a successful application of FKS-hashing in [1].

The program of the quasidichotomous digital weak search is not optimal, but is very easy to obtain. Select a letter, dividing a dictionary D into two subdictionaries of nearly equal size (Corollary 4). Go on with each subdictionary, etc., until subdictionaries of $O(n^2)$ size are reached.

Keep making digital search in those dictionaries, but use shorter pointers within them. Details are in [15]. A small modification yields an optimal strong retrieval program. Neither of the optimal algorithms use the ABC word-order.

The papers [20, 21] are closely related to the subject.

5. Channels with arbitrary additive noise.

Such a channel is specified by a set D of binary n-length words, $n > 0$. When transmitting an n-length word x, any word $x + y$, $y \in D$, may be received. A set $K \subseteq E^n$ is called a D-correcting code if for any $x_1, x_2 \in K$ and any $y_1, y_2 \in D$, $x_1 + y_1 \neq x_2 + y_2$. For any group D-correcting code generalized Hamming inequality

$$H(K) \leq 1 - H(D)$$

holds [3].

For any $\varepsilon > 0$ and any D, $H(D) < \frac{1}{2}$, there is the Varshamov-Gilbert code $(n \to \infty)$:

$$H(K) \geq 1 - 2H(D) - \varepsilon,$$

see [3, 7].

It is deduced in [17] from the bounds for the cardinality of universal sets of enumerators that for nearly any set D, the Varshamov-Gilbert codes are the best group ones. In other words, not the Hamming inequality, but the Varshamov-Gilbert one is tight, as a rule.

ACKNOWLEDGMENT. The author wishes to thank Mrs. Korneev for editing and Mrs. Zuzkov for typing.

References

1. G. V. Cormac, R. Horspool, and M. Kaiserwerth, *Practical perfect hashing*, Comput. J. **28** (1985), p. 54–58.

2. T. M. Cover, *Enumerative source encoding*, IEEE Trans. **JT-18** (1973), 73–77.

3. M. E. Deza, *Effectiveness of detecting or correcting the noise*, Problemy Peredachi Informatsii **1** (1965), no. 3, 29–39. (Russian)

4. M. L. Fredman and J. Komlos, *On the size of separating systems and families of perfect hash functions*, SIAM J. Algebraic Discrete Methods **5** (1984), 61–68.

5. M. L. Fredman, J. Komlos, and E. Szemeredi, *Storing a sparse table with $O(1)$ worst case access time*, J. Assoc. Comput. Mach. **31**(1984), 538–544.

6. J. Friedman, *Constructing $O(n \log n)$ size monotone formulae for the kth elementary symmetric polynomial of n boolean variables*, 25th Annual Symposium on Foundations of Computer Science, IEEE, 1984, pp. 506–515.

7. V. D. Goppa, *Correcting arbitrary noise by irreducible codes*, Problemy Peredachi Informatsii **10** (1974), no. 3, 118–119. (Russian)

8. _____, *Algebraic-geometric codes*, Izv. Akad. Nauk SSSR Ser. Mat. **46** (1982), 762–780. (Russian)

9. L. S. Khasin, *Complexity bounds for the realization of monotonic symmetrical functions by means of formulas in the basis V, &, $\rceil$*. Soviet Phys. Dokl. **14** (1970), 1149–1151.

10. D. Knuth, *The art of computer programming*. Vol. 3: *Searching and sorting*, Addison-Wesley, Reading, Mass., 1973.

11. J. Körner, *Fredman-Komlos bounds and information theory*, Preprint 17/1985, Math. Inst. Hungar. Acad. Sci.

12. M. Kleiman and N. Pippenger, *An explicit construction of short monotone formulae for the monotone symmetric functions*, Theoret. Comput. Sci. **7** (1978), 325–332.

13. R. E. Krichevsky, *The complexity of enumerating a finite set of words*, Soviet Math. Dokl. **17** (1976), 700–703.

14. _____, *Optimal information retrieval*, Problemy Kibernet., No. 36 (1979), 159–180. (Russian)

15. _____, *Digital enumeration of binary dictionaries*, Soviet Math. Dokl. **19** (1978), 469–473.

16. _____, *Optimal hashing*, Inform. and Control **62** (1984), 64–92.

17. _____, *Entropy doubling effect*, Dokl. Akad. Nauk SSSR Ser. Mat. **284** (1985), 795–798.

18. K. Mehlhorn, *On the program size of perfect and universal hash functions*, 23rd Annual Symposium on Foundations of Computer Science (Chicago, Ill., 1982), IEEE, New York, 1982, pp. 170–175.

19. L. A. Sholomov, *Functional gates implementation of partial boolean functions*, Problemy Kibernet, No. 21 (1969), 215–227. (Russian)

20. R. E. Tarjan and A. C. C. Yao, *Storing a sparse table*, Comm. ACM **21** (1979), 606–611.

21. A. C. C. Yao, *Should tables be sorted?*, J. Assoc. Comput. Mach. **28** (1981), 615–628.

INSTITUTE OF MATHEMATICS, SIBERIAN DIVISION OF THE USSR ACADEMY OF SCIENCES, NOVOSIBIRSK 630090, USSR

Reliable Computation in the Presence of Noise

NICHOLAS PIPPENGER

1. Introduction. This talk concerns computation by systems whose components exhibit noise (that is, errors committed at random according to certain probabilistic laws). If we aspire to construct a theory of computation in the presence of noise, we must possess at the outset a satisfactory theory of computation in the absence of noise. A theory that has received considerable attention in this context is that of the computation of Boolean functions by networks (with perhaps the strongest competition coming from the theory of cellular automata; see [**G**] and [**GR**]).

The theory of computation by networks associates with any two sets Q and R of Boolean functions a number $L_Q(R)$ (the "size" of R with respect to Q), defined as the minimum number of "gates," each computing a function from the basis Q, that can be interconnected to form a "network" that computes all of the functions in R. This theory has many pleasant properties, among which is the fact that if Q and Q' are finite and "complete," then

$$L_Q(R) \leq C_{Q,Q'} L_{Q'}(R), \tag{1.1}$$

for some constant $C_{Q,Q'}$ independent of R (see [**M**]). Thus, if one is unconcerned with constant factors, one may drop the subscript Q and consider $L(R)$ as a measure of the complexity of computing the functions in R. Another pleasant property, however, is the existence of an exquisitely precise theory of the complexity of "generic" functions. Thus for "almost all" functions f of degree n (that is, depending on n arguments), one has

$$L_Q(f) \sim C_Q 2^n/n \tag{1.2}$$

as $n \to \infty$, where C_Q is a constant independent of n (see [**L**]).

The theory of computation by networks in the presence of noise was founded by von Neumann [**N**]. Firstly, von Neumann showed that reliable computation in the presence of noise is possible. If a network N contains L gates, each of which fails with probability at most ε, then N fails with probability at most $L\varepsilon$. This crude bound becomes uninformative, however, if L grows while $\varepsilon > 0$ remains fixed. It was proved by von Neumann that N can be replaced by a network N', with a larger number L' of gates, so that N' fails with probability at most δ, where $\delta < 1/2$ is fixed (independent of L and L') when ε is sufficiently small and the gates of N' fail independently with probability ε.

Let $L'_{Q,\varepsilon,\delta}(R)$ denote the counterpart to $L_Q(R)$ when the network must fail with probability at most δ, given that each gate fails independently with probability ε. A heuristic argument to the effect that

$$L'_{Q,\varepsilon,\delta}(R) = O(L_Q(R) \log L_Q(R)) \tag{1.3}$$

was given by von Neumann; this was proved rigorously by Dobrushin and Ortyukov [**DO1**]. They also gave, in [**DO2**], a sequence f_n of functions such that

$$L_Q(f_n) = O(n),$$

but

$$L'_{Q,\varepsilon,\delta}(f_n) = \Omega(n \log n),$$

so that the estimate (1.3) is, in general, the best possible. On the other hand, I have shown in [**P**] that the estimate

$$L'_{Q,\varepsilon,\delta}(f) = O(L_Q(f)) \tag{1.4}$$

holds not only for many specific functions, but also for "almost all" functions in the sense of (1.2). Results such as (1.3) and (1.4), and others not mentioned here, form the core of a theory with many of the properties typified by (1.1) and characterized by a lack of concern for constant factors. A theory with results like (1.2), however, seems far beyond our grasp at this time.

My goal in this talk is to sketch a theory in which results like (1.2) may be within reach, though they have not yet been obtained. My proposal is to consider formulae, which behave rather more simply than networks, and to consider depth, which behaves rather more simply than size.

Let $\mathbf{B}$ denote the Boolean algebra with 2 elements. These elements will be denoted 0 ("false") and 1 ("true"); the operations will be denoted $(x, y) \mapsto x \wedge y$ ("and," or conjunction), $(x, y) \mapsto x \vee y$ ("or," or disjunction) and $x \mapsto \bar{x}$ ("not," or negation).

By a *Boolean function* we shall mean a map $f \colon \mathbf{B}^n \to \mathbf{B}$, for some n which is called the *degree* of f. Let $x_1, \ldots, x_n$ be indeterminates, and let $\mathbf{B}(x_1, \ldots, x_n)$ denote the extension of $\mathbf{B}$ by $x_1, \ldots, x_n$. The Boolean functions of degree n are in an obvious one-to-one correspondence with the elements of $\mathbf{B}(x_1, \ldots, x_n)$, which will therefore also be called Boolean functions. Boolean functions of various degrees are thereby identified in accordance with the filtration $\mathbf{B} \subseteq \mathbf{B}(x_1) \subseteq \mathbf{B}(x_1, x_2) \subseteq \cdots \subseteq \mathbf{B}(x_1, \ldots, x_n) \subseteq \cdots$.

By a *formula* on $x_1, \ldots, x_n$ over Q we shall mean an expression of one of three kinds. The first kind, a *source*, is an expression c, where $c \in \mathbf{B}$; it has *depth* 0 and computes the constant function $c \in \mathbf{B}(x_1, \ldots, x_n)$. The second kind, an *input*, is an expression x_m, where $1 \leq m \leq n$; it has depth 0 and computes the projection function $x_m \in \mathbf{B}(x_1, \ldots, x_n)$. The third kind, a *gate*, is an expression $g(N_1, \ldots, N_k)$, where $g \in Q$ and $N_1, \ldots, N_k$ are formulae on

$x_1, \ldots, x_n$ over Q; if $N_1, \ldots, N_k$ have depths $d_1, \ldots, d_k$, respectively, and compute the functions $f_1(x_1, \ldots, x_n), \ldots, f_k(x_1, \ldots, x_n)$, respectively, then it has depth $1 + \max\{d_1, \ldots, d_k\}$ and computes the function

$$g(f_1(x_1, \ldots, x_n), \ldots, f_k(x_1, \ldots, x_n)) \in \mathbf{B}(x_1, \ldots, x_n).$$

A set Q is *complete* if every Boolean function is computed by some formula over Q. If Q is complete, define $D_Q(f)$ to be the minimum possible depth of a formula over Q that computes f. If R is finite, define $D_Q(R)$ to be the maximum of $D_Q(f)$ over $f \in R$.

It is easy to see that

$$D_Q(RS) \leq D_Q(R) + D_Q(S), \tag{1.5}$$

where RS denotes the set of functions obtained by substituting functions from S for the arguments of functions from R. We also have

$$D_Q(S) \leq D_Q(R)D_R(S), \tag{1.6}$$

which is the counterpart to (1.1) for depth.

To discuss computation by formulae in the presence of noise, we must adopt probabilistic assumptions about the errors, then reconsider what it means for a formula to "compute" a function. For technical reasons it is convenient to work not with probabilities of incorrect behavior, ε and δ, but with probabilities of correct behavior, $\rho = 1 - \varepsilon$ and $\sigma = 1 - \delta$. The assumptions we shall make are not the simplest ones, but they have the merit that they yield counterparts to (1.5) and (1.6).

Consider the evaluation of a function $f(x_1, \ldots, x_n)$ at a point $c_1, \ldots, c_n \in \mathbf{B}^n$ by a formula N. We shall say that $f(c_1, \ldots, c_n)$ is the *correct* value for N. Let M be a subformula of N. If M is a source c, it produces the correct value, c. If M is an input x_m, it produces the correct value, c_m, if M is *proper*; otherwise it produces $\overline{c_m}$. If $M = g(M_1, \ldots, M_k)$ is a gate, and if the subformulae $M_1, \ldots, M_k$ produce the values $m_1, \ldots, m_k$ (correct or not), then it produces $g(m_1, \ldots, m_k)$ (correct or not) if M is proper; otherwise it produces $\overline{g(m_1, \ldots, m_k)}$. We shall assume that each input is proper with probability at least α and each gate is proper with probability at least ρ, even when these probabilities are conditioned on other inputs or gates being proper or improper; these probabilities may also depend on $c_1, \ldots, c_n$. If in this situation N produces the correct value with probability at least β for all $c_1, \ldots, c_n$, we shall say that N (ρ, α, β)-*computes* f.

Let $D^*_{Q, \rho, \alpha, \beta}(f)$ denote the minimum possible depth of a formula over Q that (ρ, α, β)-computes f, and let $D^*_{Q, \rho, \alpha, \beta}(R)$ denote the maximum of $D^*_{Q, \rho, \alpha, \beta}(f)$ over $f \in R$.

It is clear that $D^*_{Q, \rho, \alpha, \beta}(R)$ is decreasing in Q, ρ, and α, and increasing in R and β, and that $D^*_{Q, \rho, \alpha, \beta}(R) \geq D_Q(R)$. We have

$$D^*_{Q, \rho, \alpha, \gamma}(RS) \leq D^*_{Q, \rho, \alpha, \beta}(R) + D^*_{Q, \rho, \beta, \gamma}(S), \tag{1.7}$$

which is the counterpart to (1.5). This inequality suggests that $D^*_{Q,\rho,\sigma,\sigma}(R)$ behaves particularly simply. Indeed,

$$D^*_{Q,\rho,\tau,\tau}(S) \leq D^*_{Q,\rho,\sigma,\sigma}(R) D^*_{R,\sigma,\tau,\tau}(S),$$

which is the counterpart to (1.6).

Let $D^*_{Q,\rho,\sigma}(d)$ denote the maximum of $D^*_{Q,\rho,\sigma,\sigma}(f)$ over all functions f such that $D_Q(f) \leq d$. A subadditivity argument based on (1.7) shows that

$$\lim_{d \to \infty} D^*_{Q,\rho,\sigma}(d)/d$$

exists; the limit represents the factor by which computations take longer in the presence of noise than in its absence.

2. An upper bound. We shall start with an exemplary theorem, due in essence to von Neumann [**N**]. All formulae will be over the complete basis {minor} (where

$$\text{minor}\{x, y, z\} = \overline{(x \wedge y) \vee (x \wedge z) \vee (y \wedge z)}$$

denotes the minority of its three arguments), so we shall drop subscripts indicating the basis.

LEMMA 2.1. *Let N be a formula that (ρ, α, ξ)-computes f. Then the formula* minor$\{N, N, N\}$ $(\rho, \alpha, F(\xi))$-*computes $\bar{f}$, where $F(\xi) = \rho(3\xi^2 - 2\xi^3)$.*

PROOF. The formula minor$\{N, N, N\}$ produces the correct value if at least two of its immediate subformulae produce the correct value and if the gate is proper. $\square$

LEMMA 2.2. *Let N_1, N_2, and N_3 be formulae that (ρ, α, ξ)-compute f_1, f_2, and f_3, respectively. Then* minor$\{N_1, N_2, N_3\}$ *is a formula that $(\rho, \alpha, G(\xi))$-computes* minor$\{f_1, f_2, f_3\}$, *where $G(\xi) = \rho\xi^2$.*

PROOF. When f_1, f_2, and f_3 all assume the same value, we are in the situation of Lemma 2.1, and $F(\xi) \geq G(\xi)$. Otherwise, two of these functions assume a common value and the third assumes the complementary value. The formula minor$\{N_1, N_2, N_3\}$ produces the correct value provided that the corresponding two immediate subformulae produce the correct value and the gate is proper. $\square$

If $\rho = (10/9)(5/6)^{2/3} = 0.9839\ldots$, $\sigma = (9/10)(6/5)^{1/3} = 0.9563\ldots$, and $\tau = 9/10$, then $\sigma = F(\tau)$ and $\tau = G(\sigma)$. (This value of ρ is the smallest for which such values of σ and τ can be found; it is a root of the discriminant of $F(G(\xi)) = \xi$.)

THEOREM 2.3. *Let $\rho = (10/9)(5/6)^{2/3}$ and $\sigma = (9/10)(6/5)^{1/3}$. For any Boolean function f,*

$$D^*_{\rho,\sigma,\sigma}(f, \bar{f}) \leq 2D(f) + 1.$$

PROOF. We proceed by induction on $D(f)$. If $D(f) = 0$, then f is a constant or a projection, and the claim follows from Lemma 2.1. Otherwise, f and $\bar{f}$ are each of the form minor$\{f_1, f_2, f_3\}$ where $D(f_1, f_2, f_3) \leq D(f) - 1$. The claim follows by applying the inductive hypothesis to f_1, f_2, and f_3, then applying Lemma 2.2, and finally applying Lemma 2.1. □

The foregoing theorem shows that reliable computation in the presence of noise is possible, at least if $\rho \geq \rho_2 = (10/9)(5/6)^{2/3}$ and if we are willing to spend about twice the time. This is done by alternating "correcting steps" (Lemma 2.1) with "computing steps" (Lemma 2.2). If $\rho > \rho_2$, we might hope to perform more than one computing step per correcting step, and thus to obtain

$$D^*_{\rho,\sigma,\sigma}(f) \leq C_\rho D(f) + o(D(f)), \qquad (*)$$

with $C_\rho \to 1$ as $\rho \to 1$. If $\rho < \rho_2$, we still might hope to compute reliably by performing more than one correcting step per computing step, and thus to obtain (*) for some $C_\rho > 2$, at least if ρ is not too small. When $\rho \leq 1/2$, the value produced by a gate can be statistically independent of the values computed by its immediate subformulae, and reliable computation will certainly not be possible. Thus we must expect $C_\rho \to \infty$ as $\rho \to \rho_1$ for some $\rho_1 \leq 1/2$. In the remainder of this section we shall indicate how these hopes may be fulfilled.

Let us consider the action of the maps F and G on the interval $(0, 1]$. If $\rho < 1$, G is deflationary: $G(\xi) < \xi$. Thus if the damage done by a computation step is to be ameliorated by a correction step, there must be values $\xi \in (0, 1]$ for which F is inflationary: $F(\xi) > \xi$. This happens precisely when $\rho > \rho_0 = 8/9$ (this value is a root of the discriminant of $F(\xi) = \xi$). When $\rho > \rho_0$, the equation $F(\xi) = \xi$ has two roots:

$$\xi^{\pm} = \frac{3 \pm \sqrt{9 - 8/\rho}}{4}.$$

Under iteration of F, ξ^- is a repulsive fixed point and ξ^+ is an attractive one. Thus if the damage done by a computation step is to be undone by a finite number of correction steps, we must in fact have $G(\xi^+) > \xi^-$. This happens precisely when $\rho > \rho_1 = (10 + 4\sqrt{13})/27 = 0.904\ldots$. This is the lower limit to the reliability for which the scheme we are describing works.

Suppose then that $\rho_1 < \rho < 1$. Suppose further that $\xi^- < \sigma < \xi^+$.

Let $\{F, G\}^*$ be the free monoid generated by the symbols F and G, and let $\langle F(\xi), G(\xi) \rangle$ be the monoid of polynomials under composition generated by $F(\xi)$ and $G(\xi)$. For every $W \in \{F, G\}^*$, let $P_W(\xi) \in \langle F(\xi), G(\xi) \rangle$ be the image of W under the homomorphism that sends $F \mapsto F(\xi)$ and $G \mapsto G(\xi)$. Given d, let $M(d)$ denote the minimum possible number of symbols in a word $W \in \{F, G\}^*$ that contains d occurrences of the symbol G and satisfies $P_W(\sigma) \geq \sigma$. A subadditivity argument shows that $\lim_{d \to \infty} M(d)/d$ exists. This is the ratio C_ρ by which computations are slowed down by the scheme we are describing.

To determine the behavior of $M(d)$, it is helpful to transform the problem. Given c, d, and ξ, let $T(c, d, \xi)$ denote the maximum of $P_W(\xi)$ over all words $W \in \{F, G\}^*$ that contain c occurrences of the symbol F and d occurrences of the symbol G. Then $M(d) = \min\{m \geq d \colon T(m - d, d, \sigma) \geq \sigma\}$. It is clear that $T(0, 0, \xi) = \xi$, $T(c, 0, \xi) = T(c - 1, 0, F(\xi))$ for $c \geq 1$, and $T(0, d, \xi) = T(0, d - 1, G(\xi))$ for $d \geq 1$. The only problem arises when $c, d \geq 1$ and one must decide whether to apply $F(\xi)$ or $G(\xi)$ first. This problem can be resolved by considering the Poisson bracket: $[F, G](\xi) = F(G(\xi)) - G(F(\xi))$. If $[F, G](\xi) > 0$, it is more advantageous to apply $G(\xi)$ before $F(\xi)$. This happens precisely when $\xi > \xi_0$, where

$$\xi_0 = \frac{1}{1 + \sqrt{(1 - \rho)/3}}$$

(this value is a root of $F(G(\xi)) = G(F(\xi))$). A monotonicity argument shows that if $c, d \geq 1$, then $T(c, d, \xi) = T(c - 1, d, F(\xi))$ if $\xi \leq \xi_0$ and $T(c, d, \xi) = T(c, d - 1, G(\xi))$ if $\xi > \xi_0$. This recurrence, together with the boundary conditions given above, determines $T(c, d, \xi)$ and therefore $M(d)$.

For $\xi^- < \xi < \xi^+$, define $H(\xi)$ to be $F(\xi)$ if $\xi \leq \xi_0$ and to be $G(\xi)$ if $\xi > \xi_0$. The iteration of the map H generates the sequence of values of ξ that governs the recurrence for $T(c, d, \xi)$. Let H^* be the restriction of H to the interval $[G(\xi_0), F(\xi_0)]$ with the identification $F(\xi_0) = G(\xi_0)$. Then H^* is an orientation-preserving homeomorphism of the circle. Let θ be the rotation number of H^*; θ is the average number of cycles per step in the iteration of H^*. Since H^* has no fixed point, $\theta > 0$. Since a cycle must contain at least one application of $F(\xi)$ and one of $G(\xi)$, $\theta \leq 1/2$. If $\rho \leq \rho_2$, there is exactly one computation step per cycle; thus $C_\rho = 1/\theta$. If $\rho \geq \rho_2$, there is exactly one correction step per cycle; thus $C_\rho = 1/(1 - \theta)$.

The foregoing analysis describes the factor C_ρ in terms of the rotation number θ of a certain homeomorphism of the circle. Some further analysis yields the following asymptotic formulae:

$$C_\rho - 1 \sim \frac{2 \log 2}{\log \frac{1}{(1 - \rho)}},$$

as $\rho \to 1$, and

$$C_\rho \sim \frac{\left(\log \frac{7 + \sqrt{13}}{6}\right)\left(\log \frac{1}{(\rho - \rho_1)}\right)}{\left(\log \frac{11 - \sqrt{13}}{6}\right)\left(\log \frac{5 + \sqrt{13}}{6}\right)},$$

as $\rho \to \rho_1$.

3. A lower bound. I conjecture that the method described above is essentially optimal, in the sense that reliable computation is impossible if $\rho \leq \rho_1$ and takes C_ρ times as long if $\rho > \rho_1$. I have only succeeded in proving, however, that it is impossible if $\rho \leq 2/3$ and takes $1/(1 + \log_3(2\rho - 1))$ times as long if $\rho > 2/3$. We shall continue to confine our attention to formulae over the basis {minor}. An advantage of the argument we shall present is that it applies to formulae over any basis, with $2/3$ replaced by $(k + 1)/2k$ and $\log_3(2\rho - 1)$ replaced by $\log_k(2\rho - 1)$, where k is the largest of the degrees of the functions in the basis. The corresponding disadvantage is that it is unable to predict the threshold ρ_1 and the factor C_ρ, which undoubtedly depend on the particular functions present in the basis, and not merely on their degrees.

Let us say that f is a *subfunction* of g if f can be obtained from g by evaluation (substituting constants for indeterminates). Let $d \geq 2$ be even, let $n = 3^d$, and let f_d denote a function of degree n such that $D(f_d) = d$ and all n projections are subfunctions of f_d.

THEOREM 3.1. *Suppose that $\sigma > 1/2$ and N is a formula on $x_1, \ldots, x_n$ that (ρ, σ, σ)-computes f_d. Then $\rho > 2/3$ and $D(N) \geq d/(1 + \log_3(2\rho - 1))$.*

PROOF. For each input M of N, let $\Delta(M)$ denote the number of gates on the unique path from M to the root of N. Let $\Phi(\xi)$ denote the sum of $\xi^{\Delta(M)}$ over all inputs M of N.

We shall prove below that

$$\Phi(1/3) \leq 1/3 \qquad (3.1)$$

and

$$\Phi(2\rho - 1) \geq n. \qquad (3.2)$$

For now, let us see how these inequalities imply the theorem. Suppose first that $\rho > 2/3$. Let $r = 1 + \log_3(2\rho - 1)$, so that $0 < r \leq 1$, and recall Hölder's inequality:

$$\sum_M a_M^{1-r} b_M^r \leq \left(\sum_M a_M \right)^{1-r} \left(\sum_M b_M \right)^r.$$

By (3.2), Hölder's inequality (with $a_M = 1/3^{\Delta(M)}$ and $b_M = 1$) and (3.1) we have

$$n \leq \Phi(2\rho - 1) \leq \Phi(1/3)^{1-r} \Phi(1)^r \leq \Phi(1)^r.$$

Since $\Phi(1)$ is the number of inputs of N, and is thus at most $3^{D(N)}$, and since $n = 3^d$, taking logarithms yields $D(N) \geq d/r$, as claimed. Since this lower bound diverges as $\rho \to 2/3$, we conclude that $\rho > 2/3$ is necessary as well. $\square$

It remains to prove (3.1) and (3.2). To do this we shall write $\Phi_N(\xi)$ rather than $\Phi(\xi)$, to indicate the dependence on the formula N. If M is a source, then $\Phi_M(\xi) = 0$. If M is an input, then $\Phi_M(\xi) = 1$. If $M = \text{minor}\{M_1, M_2, M_3\}$, then $\Phi_M(\xi) = \xi(\Phi_{M_1}(\xi) + \Phi_{M_2}(\xi) + \Phi_{M_3}(\xi))$. Inequality (3.1) now follows immediately by induction on the structure of N.

To prove (3.2), let $\Phi^{(m)}(\xi)$ denote the sum of $\xi^{\Delta(M)}$ over all inputs M in N that compute the projection x_m. Since $\Phi(\xi) = \sum_{1 \leq m \leq n} \Phi^{(m)}(\xi)$, it will suffice to prove that

$$\Phi^{(m)}(2\rho - 1) \geq 1$$

for all $1 \leq m \leq n$. Since f_d contains all n projections as subfunctions, we can substitute sources for inputs in N to obtain a formula $N^{(m)}$ that (ρ, σ, σ)-computes the projection x_m and such that $\Phi^{(m)}(\xi) = \Phi_{N^{(m)}}(\xi)$. Thus it will suffice to show that if N is a formula on x that (ρ, σ, σ)-computes the projection x, then

$$\Phi_N(2\rho - 1) \geq 1. \qquad (3.3)$$

Let $K = 1 + \sigma \log_2 \sigma + (1 - \sigma) \log_2(1 - \sigma)$. Since $\sigma < 1/2$, $K > 0$. With each subformula M of N we shall associate a number Ψ_M with the following properties. If M is a source, then

$$\Psi_M = 0. \qquad (3.4)$$

If M is an input computing the projection x, then

$$\Psi_M = K. \tag{3.5}$$

If $M = \text{minor}\{M_1, M_2, M_3\}$, then

$$\Psi_M \leq (2\rho - 1)(\Psi_{M_1} + \Psi_{M_2} + \Psi_{M_3}). \tag{3.6}$$

These properties imply

$$\Psi_N \leq K\Phi_N(2\rho - 1),$$

by induction on the structure of N. We shall also prove that if N is a formula on x that (ρ, σ, σ)-computes the projection x, then

$$\Psi_N \geq K. \tag{3.7}$$

This will complete the proof of (3.3).

To define Ψ_M with the desired properties, we shall use Shannon's information theory. If X is a random variable assuming t distinct values with probabilities $p_1, \ldots, p_t$, define the *entropy* $H(X)$ by

$$H(X) = - \sum_{1 \leq s \leq t} p_s \log_2 p_s.$$

If X and Y are jointly distributed random variables, we shall write $H(X, Y)$ for $H((X, Y))$. The entropy satisfies the following properties: (A) $H(X) \geq 0$, and $H(X) = 0$ if and only if X is constant with probability one; (B) $H(X, Y) \geq H(X)$; and (C) $H(X, Y, Z) + H(X) \leq H(X, Y) + H(X, Z)$. These properties are immediate consequences of the fact that the logarithm is increasing, concave, and vanishes at unity. Define the *mutual information* $I(X; Y)$ by $I(X; Y) = H(X) + H(Y) - H(X, Y)$.

Let X be a random variable assuming values 0 and 1 with equal probability. Let N be a formula on x that (ρ, σ, σ)-computes the projection x, and let the random variable Y_M assume the value produced by the subformula M of N when x is assigned the value X, inputs have reliability σ (independently of X and of each other), and gates have reliability ρ (independently of X, of the inputs and of each other). Set $\Psi_M = I(X; Y_M)$.

With this definition, it is straightforward to verify properties (3.4)—(3.7); the proof of (3.6) is best broken into three parts: if $M = \text{minor}\{M_1, M_2, M_3\}$, then

$$I(X; (Y_{M_1}, Y_{M_2}, Y_{M_3})) \leq I(X; Y_{M_1}) + I(X; Y_{M_2}) + I(X; Y_{M_3});$$

$$I(X; \text{minor}\{Y_{M_1}, Y_{M_2}, Y_{M_3}\}) \leq I(X; (Y_{M_1}, Y_{M_2}, Y_{M_3}));$$

and

$$I(X; Y_M) \leq (2\rho - 1)I(X; \text{minor}\{Y_{M_1}, Y_{M_2}, Y_{M_3}\}).$$

These inequalities, and the other properties of Ψ_M, are easy consequences of (A), (B), and (C).

REFERENCES

[**DO1**] R. L. Dobrushin and S. I. Ortyukov, *Upper bound for the redundancy of self-correcting arrangements of unreliable functional elements*, Problems Inform. Transmission **13** (1977), 203–218.

[**DO2**] _____, *Lower bound for the redundancy of self-correcting arrangements of unreliable functional elements*, Problems Inform. Transmission **13** (1977), 59–65.

[**G**] P. Gács, *Reliable computation with cellular automata*, J. Comput. System. Sci. **32** (1986), 15–78.

[**GR**] P. Gács and J. H. Reif, *A simple three-dimensional real-time reliable cellular array*, ACM Sympos. Theory of Comp. **17** (1985), 288–395.

[**L**] O. B. Lupanov, *Ob Odnom Metode Sinteza Skhem*, Izv. Vyssh. Uchebn. Zaved. Radiofiz. **1** (1958), 120–140.

[**M**] D. E. Muller, *Complexity in electronic switching circuits*, IRE Trans. Electr. Comp. **5** (1956), 15–19.

[**N**] J. von Neumann, *Probabilistic logics and the synthesis of reliable organisms from unreliable components*, Automata Studies, C. E. Shannon and J. McCarthy, editors, Princeton Univ. Press, Princeton, N.J., 1956, pp. 43–98.

[**P**] N. Pippenger, *On networks of noisy gates*, IEEE Sympos. Found. of Comp. Sci. **26** (1985), 30–38.

IBM ALMADEN RESEARCH LABORATORY, SAN JOSE, CALIFORNIA 95120, USA

Нижние оценки монотонной сложности булевых функций

А. А. РАЗБОРОВ

Главная цель настоящего сообщения — осветить новые результаты по сверхполиномиальным нижним оценкам монотонной сложности естественных булевых функций, полученные за последние два года. Кроме того, мы вкратце упомянем некоторые старые результаты в этом направлении. В заключение мы коснемся другого ограничения на функциональные схемы — ограничения на глубину, так как для возникающих здесь задач за последние несколько лет также достигнут определенный прогресс.

Мы начинаем с основного определения. *Функциональной схемой* (circuit, network) *с n входами* называется последовательность булевых функций от n переменных

$$\{f_i(x_1, \ldots, x_n)\}_{i=1}^t, \tag{1}$$

в которой для любого $1 \le i \le t$ реализуется одна из трех следующих возможностей:

а) $\exists j \ (1 \le j \le n) \ f_i = x_j,$

б) $\exists i_1, i_2 < i \ \exists * \in \{\&, \vee\}(f_i = f_{i_1} * f_{i_2}),$ \hfill (2)

в) $\exists i_1 < i(f_i = \neg f_{i_1}).$

Схема (1) *вычисляет функцию* f, если $\exists i \ (1 \le i \le t) \ f_i = f$. *Размером схемы* (1) называется число t. *Схемная сложность* (combinational complexity) $L(f)$ функции f — минимальный возможный размер вычисляющей её функциональной схемы.

Интерес к задаче получения нижних оценок схемной сложности $L(f)$ естественных функций f во-многом стимулируется тем, что такие оценки повлекли бы нижние оценки тьюринговой сложности соответствующих языков. Более точно, пусть $\{f_n(x_1, x_2, \ldots, x_n)\}_{n=1}^\infty$ — последовательность булевых функций и $\mathcal{L}$ — язык в алфавите $\{0, 1\}$, определенный формулой

$$\varepsilon_1 \varepsilon_2 \cdots \varepsilon_n \in \mathcal{L} \Leftrightarrow f_n(\varepsilon_1, \varepsilon_2, \ldots, \varepsilon_n) = 1.$$

Тогда имеет место следующий простой факт, точное авторство которого, по-видимому, установить достаточно трудно:

ТЕОРЕМА 1. *Для любой машины Тьюринга* M, *распознающей язык* $\mathfrak{L}$, *справедливо*

$$T_M(n)S_M(n) \succcurlyeq L(f_n),$$

где T_M *и* S_M — *сигнализирующие времени и памяти соответственно.*

Однако, несмотря на значительные усилия, наилучшие нижние оценки величины $L(f_n)$ для естественных функций f_n, известные на сегодняшний день, всего лишь линейны по n. Отметим, что простые мощностные рассуждения показывают, что $L(f_n) \approx 2^n$ почти для всех f_n.

Пытаясь нащупать подходы к получению нижних оценок схемной сложности естественных булевых функций, исследователи рассматривают её упрощённые варианты. Один из возможных вариантов — оценки *совместной сложности* $L(f_1, \ldots, f_m)$ набора $\bar{f} = f_1, \ldots, f_m$ булевых функций, т.е. минимального возможного размера t схемы (1), вычисляющей все функции из рассматриваемого набора. Очевидно, что $L(f_1, \ldots, f_m) \geq m$, если все f_i различны, поэтому, как правило, ограничиваются случаем $m \preccurlyeq n$ (n — число переменных). К сожалению, на сегодняшний день нет нелинейных по n нижних оценок даже для совместной сложности естественных булевых функций.

Другое естественное упрощение состоит в рассмотрении лишь монотонных схем, т.е. схем, в построении которых не используется пункт в) определения (2). Разумеется, если схема (1) монотонна, то монотонны и все составляющие её функции. Для монотонной функции f её *монотонная сложность* $L^+(f)$ определяется аналогично схемной— единственное отличие состоит в том, что мы рассматриваем не все схемы, а только монотонные.

Нелинейные нижние оценки совместной монотонной сложности естественных функций известны довольно давно. Впервые они были получены Нечипоруком [3]. В этой работе рассмотрены наборы монотонных функций вида

$$f_i = \bigvee_{j \in F_i} x_j \quad (1 \leq i \leq m;\ F_i \subseteq \{1, 2, \ldots, n\}), \tag{3}$$

и доказано, что

$$\text{если } i_1 \neq i_2 \Rightarrow |F_{i_1} \cap F_{i_2}| \leq 1, \quad \text{то } L^+(f) \geq \sum_{i=1}^{m} |F_i| - m, \tag{4}$$

и на основании этого предъявлен явный пример системы $\bar{f} = f_1, f_2, \ldots, f_n$ вида (3) с $L^+(\bar{f}) \succcurlyeq n^{3/2}$. Впоследствии наборы вида (3) получили название *булевых сумм*. Развивая методы работы [3], Mehlhorn [18] и Pippenger [21] построили явные примеры булевых сумм с оценкой $L^+(\bar{f}) \succcurlyeq n^{5/3}$.

Оценка Нечипорука (4) была обобщена в работе [25], где Wegener установил, что если $i_1 \neq i_2 \Rightarrow |F_{i_1} \cap F_{i_2}| \leq k$, то

$$L^+(\bar{f}) \geq k^{-1} \cdot \sum_{i=1}^{m} |F_i| - m.$$

Григорьев [2] рассмотрел булевы суммы в случае, когда $n = 2^p$; $\{1, 2, \ldots, n\}$ — множество всех векторов p-мерного пространства над полем $\mathbf{F}_2$, а $\{F_1, F_2, \ldots, F_{n-1}\}$ — множество всех гиперплоскостей этого пространства, и доказал оценку $L^+(\bar{f}) \succcurlyeq n \log n$. Отметим, что здесь неприменимы методы только что упомянутых работ.

Монотонные симметрические функции, или *функции большинства* (threshold) определяются следующим образом:

$$T_k^n(x_1, \ldots, x_n) = 1 \Leftrightarrow \sum_{i=1}^{n} x_i \geq k. \tag{5}$$

Для них в работе [17] доказана нижняя оценка $L^+(T_1^n, T_2^n, \ldots, T_n^n) \succcurlyeq n \log n$, асимптотически совпадающая с верхней [8].

Обратимся теперь к рассмотрению *монотонных билинейных форм*, т.е. наборов вида

$$f_i = \bigvee_{p,q \in T_i} (x_p \,\&\, y_q) \quad (1 \leq i \leq m; \; T_i \subseteq [n_1] \times [n_2]). \tag{6}$$

Естественный пример такого набора — булево умножение матриц $m \times m$ ($f_{ij} = \bigvee_{k=1}^{m} (x_{ik} \,\&\, y_{kj})$; $1 \leq i, j \leq m$), для которого в работах [22, 20, 19] была доказана оценка $L^+(\bar{f}) \succcurlyeq m^3 = n^{3/2}$.

В [16] набор (6) назван полуразделенным (semidisjoint), если выполнены следующие условия:

а) $i_i \neq i_2 \Rightarrow T_{i_1} \cap T_{i_2} = \varnothing$,

б) $\forall 1 \leq i \leq m \; \forall 1 \leq j \leq n_1 \; (\langle j, k_1 \rangle, \langle j, k_2 \rangle \in T_i \Rightarrow k_1 = k_2)$,

в) $\forall 1 \leq i \leq m \; \forall 1 \leq k \leq n_2 \; (\langle j_1, k \rangle, \langle j_2, k \rangle \in T_i \Rightarrow j_1 = j_2)$.

В этой работе для *полуразделенных* наборов (6) доказана нижняя оценка

$$L^+(\bar{f}) \succcurlyeq \sum_{i=1}^{m} |T_i| \log |T_i| / \min(n_1, n_2).$$

В частности, эта оценка для так называемой *свертки* (convolution) $f_k = \bigvee \{(x_p \,\&\, y_q)| \; 1 \leq p, q \leq n$ и $p + q = k\}$ ($2 \leq k \leq 2n$) равна $\Omega(n \log n)$. Weiss в [27] доказал, что из полуразделенности набора (6) вытекает более сильная в ряде случаев оценка $L^+(\bar{f}) \geq \sum_{j=1}^{n_1} \sqrt{r_j}$, где r_j — число вхождений x_j в набор $\bar{f}$. В частности, для свертки он получил нижнюю оценку $\Omega(n^{3/2})$. Blum [11] доказал другими методами более слабую оценку $\Omega(n^{4/3})$.

В работе [**24**] Wegener рассмотрел наборы функций $f_{(\varepsilon_1,\dots,\varepsilon_{\log_2 n})} = \bigvee_{1 \le j \le n/(2\log_2 n)} \underset{1 \le i \le \log_2 n}{\&} x_{\varepsilon,ij}$, индексированные последовательностями $(\varepsilon_1, \varepsilon_2, \dots, \varepsilon_{\log_2 n})$ из 0 и 1 от множества переменных

$$\{x_{\varepsilon ij} \mid 0 \le \varepsilon \le 1,\ 1 \le i \le \log_2 n,\ 1 \le j \le n/(2\log_2 n)\}$$

и доказал для них оценку $L^+(\bar f) \succcurlyeq n^2/(\log^2 n)$, усиленную им же в [**26**] до $L^+(\bar f) \succcurlyeq n^2/\log n$.

Перейдём теперь к монотонной сложности отдельных функций. Среди функций большинства (5) выделяется функция *голосования* (majority) $\mathrm{MAJ}(n) = T^n_{n/2}$. Для неё была получена оценка $L^+(\mathrm{MAJ}(n)) \ge 3n$ [**10**], усиленная в [**13**] до $3,5n$. В [**12**] доказано, что $L^+(T^n_3) \ge 2,5n - 5,5$. Tiekenheinrich [**23**] определил функцию $F = T^n_{n-1} \vee (x_{n+1}\ \&\ T^n_2)$ и доказал, что $L^+(F) \ge 4n - 8$.

Сверхполиномиальные нижние оценки монотонной сложности были впервые получены докладчиком в работах [**4, 5**]. Функция $\mathrm{CLIQUE}(m,s)$ (*клика*) от переменных $\{x_{ij} \mid 1 \le i < j \le m\}$ определяется следующим образом:

$$\mathrm{CLIQUE}(m,s) = \bigvee_{\substack{I \subseteq [m] \\ |I| = s}} \underset{i,j \in I}{\&}\ x_{ij},$$

а функция $\mathrm{PERM}(m)$ (*перманент*) от переменных $\{x_{ij} \mid 1 \le i, j \le m\}$ —

$$\mathrm{PERM}(m) = \bigvee_{\sigma \in S_m} \overset{m}{\underset{i=1}{\&}}\ x_{i\sigma(i)}.$$

Обе эти функции имеют прозрачную интерпретацию в терминах теории графов, на которой мы здесь не останавливаемся. В [**4, 5**] доказаны следующие оценки:

ТЕОРЕМА 2 [**4**].
а) $L^+(\mathrm{CLIQUE}(m,s)) = \exp(\Omega(\log^2 m))$, *если* $s \asymp \log m$;
б) $L^+(\mathrm{CLIQUE}(m,s)) = \Omega(m^s/\log^{2s} m)$, *если* $s \asymp 1$.

ТЕОРЕМА 3 [**4, 5**]. $L^+(\mathrm{PERM}(m)) = \exp(\Omega(\log^2 m))$.

Андреев в [**1**] ввёл специальную монотонную функцию $\mathrm{POLY}(q,s)$ от переменных $\{x_{ij} \mid 1 \le i, j \le q\}$ (q — степень простого числа):

$$\mathrm{POLY}(q,s) = \bigvee \left\{ \overset{q}{\underset{i=1}{\&}}\ x_{if(i)} \mid f(z) \in \mathrm{GF}(q)[z]\ \text{и}\ \deg(f) \le s - 1 \right\}$$

и показал, что для неё методы работ [**4, 5**] дают экспоненциальную оценку:

ТЕОРЕМА 4 [**1**].
$$L^+(\mathrm{POLY}(q,s)) = \exp(\Omega(n^{1/8}/\sqrt{\log n})),$$

если $s \asymp q^{1/4}/\sqrt{\log q}$ ($n = q^2$ — *число переменных*).

Вскоре выяснилось, что для усиления оценок до экспоненциальных нет необходимости вводить экзотические функции: Alon и Boppana [**9**] усилили оценку теоремы 2 следующим образом:

ТЕОРЕМА 5 [9]. а) *Пусть* $3 \leq s_1 \leq s_2$; $\sqrt{s_1 s_2} \leq m/8 \log_2 m$ *и булева функция* f *такова, что* $\mathrm{CLIQUE}(m, s_1) \leq f \leq \mathrm{CLIQUE}(m, s_2)$. *Тогда* $L^+(f) = \Omega(\sqrt{s_1})$. *В частности, при* $s_1 = s_2 \asymp (m/8 \log m)^{2/3}$ *имеем*

$$L^+(\mathrm{CLIQUE}(m, s)) = \exp(\Omega(n^{1/6}/\log^{1/3} n)).$$

б) $L^+(\mathrm{CLIQUE}(m, s)) = \Omega(m^s/\log^s m)$, *если* $s \asymp 1$.

Кроме того, в этой работе усилена теорема 4:

ТЕОРЕМА 6 [9].

$$L^+(\mathrm{POLY}(q, s)) = \exp(\Omega(n^{1/4}\sqrt{\log n})), \quad \text{если } s \asymp \sqrt{q/\log q}.$$

Скажем несколько слов о методе, которым получены теоремы 2–6. Во-первых, операции $\&, \vee$ в определении схемы (2) заменяются на специально сконструированные операции $\sqcap$, $\sqcup$ обладающие тем свойством, что из переменных $x_1, x_2, \ldots, x_n$ применением операций $\sqcap, \sqcup$ можно получить не все монотонные булевы функции, а лишь функции из некоторого специального класса $\mathfrak{M}$. При этом $f \sqcup g \geq f \vee g$, $f \sqcap g \leq f \& g$ $(f, g \in \mathfrak{M})$.

Во-вторых, следующим образом вводятся численные меры отличия операций $\sqcap$, $\sqcup$, от $\&$, $\vee$:

$$\delta_+ = \max_{f,g \in \mathfrak{M}} |\{\varepsilon \in E_+ | (f \& g)(\varepsilon) = 1 \text{ и } (f \sqcap g)(\varepsilon) = 0\}|,$$

$$\delta_- = \max_{f,g \in \mathfrak{M}} \mathsf{P}[(f \sqcup g)(\varepsilon^*) = 1 \text{ и } (f \vee g)(\varepsilon^*) = 0],$$

где

$$E_+ = \{\varepsilon \in \{0,1\}^n | F(\varepsilon) = 1 \text{ и } \forall \delta < \varepsilon(F(\delta) = 0)\}$$

(F — та монотонная функция, сложность которой мы хотим оценить), а ε^* — некоторая случайная величина, распределенная на $\{0,1\}^n$.

Легко доказывается следующее утверждение: для любой монотонной функции f существует $\hat{f} \in \mathfrak{M}$ такая, что

$$\begin{aligned} |\{\varepsilon \in E_+ | \hat{f}(\varepsilon) = 0 \text{ и } f(\varepsilon) = 1\}| &\leq \delta_+ \cdot L^+(f), \\ \mathsf{P}[\hat{f}(\varepsilon^*) = 1 \text{ и } f(\varepsilon^*) = 0] &\leq \delta_- \cdot L^+(f). \end{aligned} \tag{7}$$

Чтобы получить $\hat{f}$, достаточно взять монотонную схему минимального размера, вычисляющую f, и заменить в ней операции $\&$, $\vee$ на $\sqcap$, $\sqcup$.

За счет выбора $\mathfrak{M}, \sqcup, \sqcap, \varepsilon^*$ можно получить все теоремы 2–6. Однако эти построения обладают рядом общих черт, которые мы сейчас отметим.

Построение $\mathfrak{M}$ основывается прежде всего на множестве $\mathfrak{A}$, состоящем из некоторых конъюнкций переменных. Фиксируется число r и на множестве $\mathfrak{A}$ следующим образом определяется правило вывода:

$$\mathfrak{K}_1, \mathfrak{K}_2, \ldots, \mathfrak{K}_r \vdash \mathfrak{K} \Leftrightarrow \forall i, j (1 \le i < j \le r \Rightarrow \mathfrak{K}_i \cap \mathfrak{K}_j \subseteq \mathfrak{K}).$$

(здесь под конъюнкцией понимается просто множество входящих в неё переменных). Подмножество $\mathfrak{A}_0 \subseteq \mathfrak{A}$ называется замкнутым, если оно замкнуто относительно правила $\vdash$. $\mathfrak{M}$ состоит из всех функций вида $\ulcorner \mathfrak{A}_0 \urcorner = \{ \bigvee \mathfrak{K} \mid \mathfrak{K} \in \mathfrak{A}_0 \}$, где $\mathfrak{A}_0$ пробегает все замкнутые подмножества множества $\mathfrak{A}$. Чтобы найти $f \sqcup g$ для $f = \ulcorner \mathfrak{A}_1 \urcorner$; $g = \ulcorner \mathfrak{A}_2 \urcorner$, надо взять

$$f \vee g = \ulcorner \mathfrak{A}_1 \cup \mathfrak{A}_2 \urcorner, \tag{8}$$

и добавить в качестве дизъюнктивных членов все новые конъюнкции $\mathfrak{K} \in \mathfrak{A}$, которые, возможно, выводимы из $\mathfrak{A}_1 \cup \mathfrak{A}_2$. Чтобы вычислить $f \sqcap g$, надо записать

$$f \mathbin{\&} g = \bigvee_{\mathfrak{K} \in \mathfrak{A}_1} \bigvee_{\mathfrak{L} \in \mathfrak{A}_2} (\mathfrak{K} \mathbin{\&} \mathfrak{L}), \tag{9}$$

и удалить те конъюнктивные члены $\mathfrak{K} \mathbin{\&} \mathfrak{L}$, которые не лежат в $\mathfrak{A}$ (замкнутость множества оставшихся конъюнкций легко проверяется). Строение случайной величины ε^* зависит от конкретного случая.

При использовании (7) главную роль играют следующие два комбинаторных утверждения:

если $s = \max_{\mathfrak{K} \in \mathfrak{A}} |\mathfrak{K}|$, то любое замкнутое $\mathfrak{A}_0 \subseteq \mathfrak{A}$ содержит не более $(r-1)^s$ минимальных конъюнкций; $\tag{10}$

если $\mathfrak{K}_1, \mathfrak{K}_2, \ldots, \mathfrak{K}_r \in \mathfrak{A}$ — непересекающиеся конъюнкции, то $\mathsf{P}[\forall i (1 \le i \le r) \mathfrak{K}_i(\varepsilon^*) = 0]$ мала. $\tag{11}$

В [9] доказано, что оценка $(r-1)^s$ в (10) точна.

ПРИМЕР. При доказательстве теоремы 3 следует положить $\mathfrak{A} = \{x_{i_1 j_1} \mathbin{\&} x_{i_2 j_2} \mathbin{\&} \cdots \mathbin{\&} x_{i_l j_l} \mid l \le s;\ i_1, i_2, \ldots, i_l$ попарно различны; $j_1, j_2, \ldots, j_l$ попарно различны$\}$, где $s = \frac{1}{8} \log_2 m$;

$$E_+ = \{\varepsilon(\sigma) \mid \sigma \in S_m\}, \quad \text{где } \varepsilon_{ij}(\sigma) = 1 \Leftrightarrow j = \sigma(i);$$

$$\varepsilon_{ij}^* = 1 \Leftrightarrow h^*(v_i) = h^*(w_j),$$

где h^* — равномерно распределенная на $\{0,1\}^{\{v_1, \ldots, v_m, w_1, \ldots, w_m\}}$ функция;

$$\delta_+ \le (s! r^s)^2 \cdot (m - s - 1)! \quad (\text{лемма 3 в [5]}),$$

$$\delta_- \le (1 - 2^{-s})^{\sqrt{r}/s} \cdot \sum_{i=0}^{s} \left(i! \binom{m}{i}^2 \right) \quad (\text{лемма 6 в [5]}),$$

где $r = [m^{1/4} (\log_2 m)^8]$.

В работе Андреева [1] используются другие обозначения, поэтому мы для удобства читателя приводим их возможный перевод на язык работ [4, 5, 9] (см. также доказательство теоремы 6 в [9]). Функция, которая операциями $\sqcap$, $\sqcup$ ставится в соответствие любой паре функций h_1, h_2 — это функция h_5 в доказательстве леммы 2 [1]. Класс $\mathfrak{M}$ в наших обозначениях — это класс $\mathfrak{R}^n_{r,u}$ (определение на стр. 1033 [1]). Распределение ε^* задается формулой [1, стр. 1034] $\mathsf{P}[\varepsilon^* = \tilde{\alpha}] = p^{n-\|\tilde{\alpha}\|}(1-p)^{\|\tilde{\alpha}\|}$. Оценка $\delta_+ \leq (u!r^u)^2$ у Андреева производится формулой (6) из [1] а оценка $\delta_- \leq (up)^r(r^u u!)^2$ — формулой (15) из [1]. Функция $\hat{f}$ — это функция g_s в доказательстве теоремы 1 из [1], а промежуточные функции $g_1, g_2, \ldots, g_s$ в том же доказательстве — в точности функции из схемы (1), в которой произведена замена операций $\&, \vee \to \sqcap, \sqcup$. Множество $\mathfrak{A}$ у Андреева состоит из всех конъюнкций длины $\leq u$, а применение правила вывода соответствует замене (s,r)-регулярной функции g на $x_1 \cdot x_2 \cdot \ldots \cdot x_s$ в доказательстве леммы 1 из [1]. Там же доказано свойство (10). Формулы (8) и (9) приобретают в доказательстве леммы 2 [1] вид $h_1 * h_2 = h_3 \vee h_4$, где h_3 — семейство всех конъюнкций, не лежащих в $\mathfrak{A}$; h_4 — лежащих в $\mathfrak{A}$ ($h_3 = \varnothing$, если $* = \vee$).

Взятие замыкания в случае (8) производится переходом от h_4 к h_5 в формуле (4) из [1]. Наконец, свойству (11) соответствует в доказательстве леммы 1 [1] формула $\rho_p(1, g_1) \leq (up)^r$.

В каком направлении могут развиваться дальше полученные в [4, 5, 1, 9] результаты? Наиболее интересный вопрос, разумеется, состоит в том, насколько полезными они могут быть для получения нижних оценок в полном базисе. Ещё в 1973 году Schnorr отметил, что $P \neq NP$ следовало бы из следующих двух утверждений:

а) для любой монотонной функции f имеем $L^+(f) \preccurlyeq p(L(f))$, где $p(t)$ — фиксированный полином;

б) $L^+(\mathrm{CLIQUE}(m, m/2))$ растёт сверхполиномиально от m.

Но предположение а) опровергается теоремой 3; хотя б) — подтверждается теоремами 2 и 5. Введём следующую функцию:

$$I(t) = \max\{L^+(f) \mid L(f) \leq t\},$$

где максимум берется по всем монотонным булевым функциям. Тогда из теоремы 3 следует $I(t) = \exp(\Omega(\log^2 t))$. Отметим, что если бы удалось доказать неравенство

$$\forall \varepsilon > 0 \ I(t) = O(\exp(t^\varepsilon)), \tag{12}$$

то любая из теорем 4–6 повлекла бы $P \neq NP$. Впрочем, неравенство (12) выглядит крайне маловероятным, и мы ставим следующую проблему.

Задача 1. Опровергнуть (12).[1]

Естественный подход к этой задаче — усилить теорему 3 или доказать экспоненциальную нижнюю оценку монотонной сложности для какой-либо другой последовательности монотонных булевых функций из класса P.

---

[1] *Примечание автора:* Неравенство (12) недавно было опровергнуто É. Tardos (Combinatorica, 1987, v. 7).

Что же касается рассмотренных методов, то автор выражает надежду, что идея замены основных операций на «мало отличающиеся» от них вспомогательные может оказаться полезной при рассмотрении схем в полном базисе; косвенное подтверждение этому — формулируемый ниже свежий результат из [6]. Развитая же комбинаторная техника, по-видимому, не переносится на немонотонный случай.

Отметим теперь, что теоремы 2б) и 5б) дают примеры последовательностей монотонных функций, монотонная сложность которых нелинейна, но ограничена сверху полиномом. Было бы интересно построить более естественные примеры таких функций. В частности, мы ставим следующие две задачи:

Задача 2. Верно ли, что

$$L^+(\mathrm{MAJ}(n)) \asymp n \log n?$$

Задача 3. Построить явные примеры монотонных билинейных форм состоящих из одной функции с нелинейной нижней оценкой монотонной сложности.

Завершающим аккордом в исследовании монотонной сложности мог бы стать удовлетворительный ответ на следующий вопрос:

Задача 4. Построить обозримый критерий применимости описанного выше метода, под который должны подпадать функции из теорем 2–6.

В целом же, по-видимому, имеет смысл переключить внимание на другие возможные ограничения на функциональные схемы.

В заключение настоящего сообщения мы очень кратко коснёмся результатов по ограниченной глубине. Определение схемы глубины k даётся индукцией по k. Схема глубины 0 есть элемент множества $\{x_1, \neg x_1, x_2, \neg x_2, \ldots, x_n, \neg x_n\}$. Схема глубины k есть непустое множество схем глубины $(k-1)$. Размер схемы C есть мощность транзитивного рефлексивного замыкания множества C, т.е. количество использованных при построении C схем. Для схемы C глубины k булева функция f_C определяется индукцией по k. Если $k = 0$, то $f_C = C$. Функция, которая реализуется схемой C глубины $k > 0$ есть

$$f_C = \underset{B \in C}{*} f_B, \tag{13}$$

где $* = \bigvee$, если k чётно, и $* = \&$, если k нечётно. Минимальный возможный размер схемы глубины $k \geq 2$, реализующей функцию f, обозначим через $L_k(f)$. Отметим, что по любой схеме глубины k размера t очевидным образом строится функциональная схема (1) размера $O(kt^2)$, вычисляющая ту же функцию, так что $L(f) \preccurlyeq k L_k^2(f)$.

В работах [14, 7, 28, 15] была рассмотрена функция сложения по mod 2 (parity) $x_1 \oplus x_2 \oplus \cdots \oplus x_n$ и для любого фиксированного k была доказана вначале сверхполиномиальная нижняя оценка для $L_k(x_1 \oplus x_2 \oplus \cdots \oplus x_n)$

[14, 7], а затем — экспоненциальная:

ТЕОРЕМА 7 [28].

$$L_k(x_1 \oplus \cdots \oplus x_n) = \exp(\Omega(n^{\lambda_k})), \qquad k = \text{const}.$$

В работе [6] рассматриваемый класс схем ограниченной глубины был расширен следующим образом: при определении схемы в (13) допускается ещё одна новая операция $* = \oplus$; при этом порядок, в котором применяются операции $\{\&, \vee, \oplus\}$, может быть любым. Обозначим через $L_k^{\oplus}$ модифицированную таким образом сложность. Верна следующая

ТЕОРЕМА 8 [6]. $L_k^{\oplus}(\text{MAJ}(n)) = \exp(\Omega(n^{\lambda_k}))$, $k = \text{const}$.

Отметим, что первый шаг в доказательстве теоремы 8, как и для монотонной сложности, состоит в замене основных операций на «мало отличающиеся» от них вспомогательные операции аналогично тому, как в [4, 5, 1, 9] операции $\&$ и $\vee$ заменялись на $\sqcap$ и $\sqcup$.

ЛИТЕРАТУРА

1. А. Е. Андреев, *Об одном методе получения нижних оценок сложности индивидуальных монотонных функций*, ДАН СССР, 1985, т. 282, № 5, с. 1033–1037 (Engl. transl. in: Sov. Math. Dokl. **31** (1985), 530–534).

2. Д. Ю. Григорьев, Зап. научн. семинаров ЛОМИ АН, 1977, т. 68, с. 19–25.

3. Э. И. Нечипорук, Проблемы кибернетики, 1970, вып. 23, с. 291–294 (Engl. transl. in Systems Res. Theory **21** (1971), 236–239).

4. А. А. Разборов, *Нижние оценки монотонной сложности некоторых булевых функций*, ДАН СССР, 1985, т. 281, № 4, с. 798–801 (Engl. transl. in: Sov. Math. Dokl. **31** (1985), 354–357).

5. ——, Матем. зам., 1985, т. 37, № 6, с. 887–900 (Engl. transl. in: Math. Notes Acad. Sci. USSR **37** (1985), 485–493).

6. ——, *Нижние оценки размера схем ограниченной глубины в базисе* $\{\&, \oplus\}$, 1986, препринт МИАН (см. также Матем. зам., 1987, т. 41, № 4).

7. M. Ajtai, Σ_1^1-*formulae on finite structures*, Ann. Pure Appl. Logic **24** (1983), 1–48.

8. M. Ajtai, J. Komlós and E. Szemerédi, Proc. 15th STOC (1983), 1–9.

9. N. Alon and R. B. Boppana, *The monotone circuit complexity of Boolean functions*, Preprint, 1985 (see also Combinatorica **7** (1), 1987).

10. P. Bloniarz, Techn. Rep. No. 238, Lab. Comput. Sci. MIT, 1979.

11. N. Blum, Theoret. Comput. Sci. **36** (1985), 59–69.

12. P. E. Dunne, Acta Inform. **22** (1985), 229–240.

13. ——, Proc. 22nd Ann. Allerton Conf. on Communication Control and Computing (1984), 911–920.

14. M. Furst, J. Saxe and M. Sipser, *Parity, circuits, and the polynomial-time hierarchy*, Math. Systems Theory **17** (1984), 13–27.

15. J. Hastad, Proc. 18th STOC (1986).

16. E. A. Lamagna, *The complexity of monotone networks for certain bilinear forms, routing problems, sorting, and merging*, IEEE Trans. Comput. **28** (1979), 773–782.

17. E. A. Lamagna and J. E. Savage, *Combinational complexity of some monotone functions*, Proc. 15th Ann. IEEE Symp. on Switching and Automata Theory (1974), 140–144.

18. K. Mehlhorn, *Some remarks on Boolean sums*, Acta Inform. **12** (1979), 371–375.

19. K. Mehlhorn and Z. Galil, *Monotone switching circuits and Boolean matrix product*, Computing **16** (1976), 99–111.

20. M. S. Paterson, *Complexity of monotone networks for Boolean matrix product*, Theoret. Comput. Sci. **1** (1975), 13–20. (рус. пер. в Киб. сб., нов. сер., вып. 15).

21. N. Pippenger, *On another Boolean matrix*, Theoret. Comput. Sci. **11** (1980), 45-56.

22. V. R. Pratt, *The power of negative thinking in multiplying Boolean matrices*, SIAM J. Comput. **4** (1975), 326–330.

23. J. Tiekenheinrich, *A 4n-lower bound on the monotone network complexity of a one-output Boolean function*, Inform. Process. Lett. **18** (1984), 201–202.

24. I. Wegener, *Switching functions whose monotone complexity is nearly quadratic*, Theoret. Comput. Sci. **9** (1979), 83–97. (рус. пер. в Киб. сб., нов. сер., вып. 18).

25. ____, *A new lower bound on the monotone network complexity of Boolean sums*, Acta Inform. **13** (1980), 109–114.

26. ____, *Boolean functions whose monotone complexity is of size $n^2 \log n$*, Theoret. Comput. Sci. **21** (1982), 213–224.

27. J. Weiss, *An $n^{3/2}$ lower bound on the monotone network complexity of the Boolean convolution*, Inform. and Control **59** (1983), 184–188.

28. A. C. Yao, Proc. 26th FOCS (1985), 1–10.

Математический институт им. В. А. Стеклова, Москва 117966, СССР

The Search for Provably Secure
Identification Schemes

ADI SHAMIR

1. Introduction. An identification scheme is a protocol which enables party A to prove his identity to party B in the presence of imposters C. This is one of the fundamental problems in cryptography, and it has numerous practical applications. In fact, whenever we present a driver's license, use a passport, pay with a credit card, enter a computer password, or punch a secret code into an automatic teller machine, we execute an identification protocol. The basic problem with these practical protocols is that A proves his identity by revealing to B a constant (in the form of a printed card or a memorized value). A sophisticated adversary C who cooperates with a dishonest B can use a xerox copy of the card or a recording of the secret value to misrepresent himself successfully as A at a later stage. Our goal in this paper is to survey some of the mathematical techniques developed to solve this problem, and to propose a new identification scheme which is provably secure if factoring is difficult, and orders of magnitude faster than previous schemes of this type.

The mathematical version of the identification problem assumes that A is distinguished by knowing some secret information s which no one else knows. A's goal is to prove to B that he knows s, and B's goal is to verify the correctness of A's proof. B is assisted by the public information v revealed in advance by A. Since v is also available to C, there should be no efficient algorithm for computing s from v, even though the two values are obviously related. The authenticity of v is guaranteed by publishing it in a public key directory or by attaching to it the digital signature of a trusted center, and thus we do not consider attacks in which C replaces A's real v by a modified v'.

An identification protocol specifies the sequence of steps executed by A and B during their interaction. The opponent C is allowed to misrepresent himself either as a prover or as a verifier, and then he can deviate from the specified protocol in an arbitrary manner. The original protocol is assumed to be public knowledge, but the modified protocol is usually known only to C. Typical examples of modified protocols include nonrandom choices of presumably random values, giving incorrect answers to the other party's questions, or reusing recorded portions from previous executions of the same protocol.

In this paper we denote the prover's real protocol by $\overline{A}$, the verifier's real protocol by $\overline{B}$, C's version of the prover's protocol by $\tilde{A}$, and C's version of the verifier's protocol by $\tilde{B}$ (the use of *straight* and *crooked* lines is mnemonic). We denote the execution of the two party protocol in which the prover's part is $Z' \in \{\overline{A}, \tilde{A}\}$ and the verifier's part is $Z'' \in \{\overline{B}, \tilde{B}\}$ by (Z', Z''). All these protocols are assumed to be polynomial time probabilistic algorithms with access to the public information v, but only $\overline{A}$ has access to the secret information s.

An identification protocol $(\overline{A}, \overline{B})$ is called secure if there are no $\tilde{A}$ and $\tilde{B}$ with the property that after polynomially many executions of $(\overline{A}, \tilde{B})$, C can execute $(\tilde{A}, \overline{B})$ with a nonnegligible (i.e., 1/polynomial) probability of success. In other words, C should not be able to impersonate A even after he witnesses, verifies, and influences (as $\tilde{B}$) polynomially many proofs of identity generated by the real $\overline{A}$.

Identification protocols are closely related to the notion of digital signatures (as introduced by Diffie and Hellman [2]), with the exception that there is no judge and no need to settle disputes at a later stage. The proof of identity should be either accepted or rejected in real time, and as a result the requested access or service should be granted or withheld. The absence of the judge is one of the main reasons the new identification scheme is faster than conventional signature schemes.

2. Previous identification schemes.

Almost all the identification and signature schemes proposed so far are based on the difficulty of solving polynomial equations when the factorization of the modulus is unknown. The public information is usually a modulus n which is the product of two large primes p and q, and the secret information is the factorization of n. A proves that he knows this factorization by solving a polynomial equation mod n which includes a random test value m chosen by B as one of its coefficients. B verifies the solution by substituting it into the equation, and this can be done without knowing the secret information.

The first scheme based on this principle was the RSA scheme (Rivest, Shamir, and Adleman [9]), which used the polynomial equation

$$x^e = m \ (\mathrm{mod} \, n)$$

with an exponent e which is relatively prime to $\varphi(n) = (p-1)(q-1)$. For any m this equation has a unique solution x which can be computed as $x = m^d$ $(\mathrm{mod} \, n)$ where d is the multiplicative inverse of $e \bmod \varphi(n)$. The difficulty of computing x from m when $\varphi(n)$ is not known is conjectured (but not proven) to be equivalent to the difficulty of factoring n. The main practical drawback of the RSA scheme is its relative inefficiency: When n is a 200 digit number, both the prover and the verifier have to perform almost 1000 modular multiplications of 200 digit numbers. Hardware implementations of the RSA scheme are too expensive and software implementations of this scheme are too slow for many commercial applications, and in particular the scheme cannot be used in the new generation of ID cards which are based on the emerging technology of smart cards.

These security and efficiency issues were addressed by Rabin [8], who published a scheme which looks superficially as a special case of an RSA scheme with the polynomial equation $x^2 = m \pmod{n}$. Since $e = 2$ is not relatively prime to $\varphi(n)$, this equation is solvable only for one quarter of the possible m's, and for any such m it has four distinct solutions (with some exceptional cases). Knowledge of any pair of solutions x_1, x_2 such that $x_1 \neq \pm x_2 \pmod{n}$ can be used to factor n in polynomial time. Rabin used this fact to show that square root extraction cannot be easier than factoring. In addition, the low exponent implies that B has to perform only one modular multiplication to verify the solution, but A has to perform essentially the same number of operations as in the RSA scheme.

Unfortunately, Rabin's proof of security does not apply to identification schemes in which C can actively play B's role. Since the equivalence proof between square root extraction and factoring is constructive, C can exploit A's willingness to extract such roots in order to factor n!

To solve this problem, Rabin suggested a modified protocol in which A perturbs the m sent by B before he extracts its square root. As far as we know, this solves the practical security problem but the modified scheme is no longer provably equivalent to factoring.

By using long sequences of square root extractions, Goldwasser, Micali, and Rivest [5] were able to obtain a more complicated scheme with a formal proof of security even when C was allowed to be active. However, the large number of arithmetic operations required to implement this scheme made it totally impractical.

A very efficient scheme was suggested by Ong, Schnorr, and Shamir [6]. It was based on the two variable equation

$$x^2 - vy^2 = m \pmod{n}.$$

When the square root s of $v \bmod n$ is known, it is possible to generate all the x, y solutions of this equation with just 3 modular multiplication/division operations. Without knowing s (or the factorization of n) it is not easy to see how to solve this equation, since all the obvious methods require square root extraction. It was thus a major surprise when Pollard [7] announced a polynomial time heuristic for solving modular binary quadratic equations even when the factorization of n is unknown.

Another approach was taken by Okamoto and Shiraishi [10], who based their scheme on the quadratic approximation $x^2 \approx m \pmod{n}$. When n is chosen as p^2q, its factorization is known, and the allowable difference between $x^2 \pmod{n}$ and m is bounded by $n^{2/3}$, x can be computed from m in just 4 modular multiplication/division operations. However, several months later Brickell and

DeLaurentis [1] showed how to solve quadratic approximation problems efficiently without using the factorization of n. The security of higher degree approximation problems remains unknown, and identification schemes based on them have neither been broken nor proven equivalent to factoring.

In the next section we describe a new solution to this problem which is provably secure against any passive or active attack if factoring is difficult. In addition, typical implementations of the scheme require less than 8 modular multiplications to generate or verify probabilistic proofs of security. As a result, the scheme is an ideal solution to a wide variety of applications ranging from smart ID cards and remote control devices to secure operating systems and telecommunication networks.

3. The Fiat-Shamir identification scheme. The new scheme (developed jointly with Amos Fiat from the Weizmann Institute of Science) is motivated by the notion of zero-knowledge interactive proofs introduced by Goldwasser, Micali, and Rackoff [5]. Its simplest incarnation uses a single modulus $n = p \cdot q$ chosen by a trusted center and made available to everyone. Unlike the RSA scheme, there is no need to know the factorization of n in order to generate a proof of identity, and thus the center can destroy the factorization of n as soon as it is chosen.

Each user chooses a sequence of k secret numbers $s_1, \ldots, s_k$, and publishes the k values $v_j = 1/s_j^2 \pmod{n}$. When B challenges A to prove his identity, they execute the following protocol:

1. A picks t random numbers $r_i \in [0, n)$ and sends their squares $x_i = r_i^2$ $\pmod{n}$ to B.
2. B sends to A a random boolean $t \times k$ matrix $[e_{ij}]$.
3. A sends to B the t numbers

$$y_i = r_i \prod_{e_{ij}=1} s_j \pmod{n}.$$

4. B accepts the proof if for all $1 \leq i \leq t$

$$x_i = y_i^2 \prod_{e_{ij}=1} v_j \pmod{n}.$$

LEMMA 1. *When the real two party protocol $(\overline{A}, \overline{B})$ is executed, $\overline{B}$ always accepts the proof as valid.*

PROOF. By the definitions of y_i and v_j,

$$y_i^2 \prod_{e_{ij}=1} v_j = \left(r_i \prod_{e_{ij}=1} s_j \right)^2 \left(\prod_{e_{ij}=1} v_j \right)$$

$$= r_i^2 \prod_{e_{ij}=1} \left(s_j^2 v_j \right) = r_i^2 = x_i \pmod{n}. \quad \square$$

LEMMA 2. *Consider the probability distribution of x_i and y_i values for any fixed choice of $[e_{ij}]$ matrix and random choice of r_i values. This distribution does not depend on the choice of square root s_j of v_j (mod n).*

PROOF. Let $x_1, \ldots, x_t, y_1, \ldots, y_t$ be some sequence of numbers sent by $\overline{A}$ for a fixed choice of s_j and e_{ij} values. Each y_i defines a unique r_i which generates it. When s_j is replaced by another square root s_j' of the same v_j, the corresponding r_i is multiplied by some square root of 1 (mod n), which leaves $x_i = r_i^2$ (mod n) unchanged. Since the same tuples of x_i, y_i numbers are generated with the same probabilities, even an adversary $\tilde{B}$ (who can choose the $[e_{ij}]$ matrices nonrandomly) cannnot distinguish between $\overline{A}$'s behavior before and after the change. $\square$

LEMMA 3. *There exists a polynomial time protocol $\tilde{A}$ for which the execution of $(\tilde{A}, \overline{B})$ succeeds with probability 2^{-kt}.*

PROOF. If $\tilde{A}$ can guess the correct $[e_{ij}]$ matrix, he can send to $\overline{B}$ the following values:

$$x_i = r_i^2 \cdot \prod_{e_{ij}=1} v_j \ (\text{mod } n), \qquad y_i = r_i.$$

These computations do not require knowledge of the s_j values, and it is easy to verify that $\overline{B}$ will accept them. However, the x_i values should be sent to $\overline{B}$ before $\tilde{A}$ receives the $[e_{ij}]$ matrix, and thus the success rate of this technique is only 2^{-kt}. $\square$

Our main result shows that this attack is essentially optimal, and the probability of cheating cannot be increased to $(1+\varepsilon)2^{-kt}$ for any $\varepsilon > 0$ unless factoring is easy. To prove this result, we need the following technical lemma.

LEMMA 4. *Let D be a boolean $h \times w$ matrix with at least $(1+\varepsilon)h$ 1's arranged in an arbitrary pattern. Then it is possible to find two 1's on the same row with probability larger than $\varepsilon(1 - e^{-1})/(1 + \varepsilon)$ by performing $((2 + \varepsilon)/(1 + \varepsilon))w$ local probes of D entries.*

PROOF. The obvious strategy of choosing random rows and probing all their entries is too expensive for certain patterns of ones. A better strategy for finding two 1's on the same row is

1. Probe $w/(1 + \varepsilon)$ random entries in D.
2. Pick the first 1 found in step 1, and probe all the entries in its rows.

The complexity of this strategy is exactly $((2 + \varepsilon)/(1 + \varepsilon))w$ probes. It can fail if step 1 produces no 1's or step 2 finds no other 1 along the row chosen in step 1. The probability of failure of the first type is bounded by e^{-1} since we probe $w/(1 + \varepsilon)$ random locations with a $(1 + \varepsilon)h/wh$ probability of success in each probe. The probability of failure of the second type is bounded by $1/(1+\varepsilon)$ since at most h out of the $(1+\varepsilon)h$ 1's can be unique in their rows. Since the two probabilities are independent, the probability of success of this probing strategy is at least $\varepsilon(1 - e^{-1})/(1 + \varepsilon)$. $\square$

We can now prove

THEOREM 1. *Assume that C can execute $(\overline{A}, \tilde{B})$ g times, and then execute $(\tilde{A}, \overline{B})$ once more and succeed with probability larger than $(1 + \varepsilon)2^{-kt}$. Then n can be factored in*

$$g \cdot |(\overline{A}, \tilde{B})| + ((2 + \varepsilon)/(1 + \varepsilon))2^{kt}|(\tilde{A}, \overline{B})|$$

operations with probability larger than $\varepsilon(1 - e^{-1})/2(1 + \varepsilon)$, where $|Z|$ denotes the complexity of protocol Z.

PROOF (SKETCH). Assume that such $\tilde{A}$ and $\tilde{B}$ exist. We show how the real user A (who knows the s_j values) can use them to factor the n chosen by the center. Since this factorization was never revealed by the center (either explicitly by publication or implicitly by its interaction with users), this contradicts the assumption that factoring is difficult.

To factor n, A simulates the $(\overline{A}, \tilde{B})$ protocol g times, and gives the complete transcript of the communication (but not the secret s_j values!) to $\tilde{A}$. The possible executions of $(\tilde{A}, \overline{B})$ at this stage can be summarized by a large boolean matrix whose rows correspond to all the possible random tapes of $\tilde{A}$ and whose columns correspond to all the possible random tapes of $\overline{B}$. The outcome of the execution of $(\tilde{A}, \overline{B})$ is uniquely determined once we fix their randomizations, and thus we can fill each entry with a 1 if $\overline{B}$ accepts $\tilde{A}$'s proof of identity and with a 0 if $\overline{B}$ rejects $\tilde{A}$'s proof.

Let h and w be the height and width of this matrix. Since $\overline{B}$ needs exactly kt random bits to choose the $[e_{ij}]$ matrix, we can assume that $w = 2^{kt}$. By assumption, the probability of success of $(\tilde{A}, \overline{B})$ is at least $(1 + \varepsilon)2^{-kt}$, and thus the $h \times 2^{kt}$ matrix contains at least $(1 + \varepsilon)h$ 1's in it. By Lemma 4, A can probe this matrix $((2 + \varepsilon)/(1 + \varepsilon))2^{kt}$ times (by repeatedly executing the $(\tilde{A}, \overline{B})$ protocol) and find two 1's along the same row with probability $\varepsilon(1 - e^{-1})/(1 + \varepsilon)$.

Let x_i', y_i' and x_i'', y_i'' be the values sent by $\tilde{A}$ to $\overline{B}$ in these two executions. They occur along the same row, and thus $\tilde{A}$ uses the same random tape in both cases. Since the x_i are chosen before $\overline{B}$'s modified choice of $[e_{ij}]$ matrix can influence $\tilde{A}$'s behavior, $x_i' = x_i''$. Since

$$y_i'^2/y_i''^2 = \prod_{j=1}^{k} v_j^{c_j} \pmod{n}$$

where $c_j \in \{-1, 0, 1\}$, y_i'/y_i'' is one of the four possible roots of $\prod_{j=1}^{k} v_j^{c_j} \pmod{n}$. However, $\overline{A}$ already knows one of these roots: $\prod_{j=1}^{k} s_j^{c_j} \pmod{n}$. By Lemma 2, $\tilde{A}$ cannot know which square roots $\overline{A}$ was using, and thus A can factor n with probability $\frac{1}{2}$ by combining the square root that $\overline{A}$ knows with the square root that $\tilde{A}$ computes. $\square$

Theorem 1 clearly demonstrates that when the ratio between the complexity of factoring and 2^{kt} grows nonpolynomially, the adversary cannot increase his probability of success (compared to the trivial attack outlined in Lemma 3) by using a polynomial time strategy. However, the nonasymptotic nature of Theorem 1 enables us to get concrete lower bounds on the complexities of $\tilde{A}$ and $\tilde{B}$ for a 200 digit n if we assume a reasonable lower bound on the complexity of factoring such moduli.

4. Practical considerations. The 2^{-kt} probability of cheating is an absolute constant, and thus there is no need to keep wide margins of safety against unforseen technological and algorithmic improvements. In most applications, a security level of 2^{-20} suffices to deter cheaters: No one will pay with a forged credit card at a department store or try to enter a restricted area with a forged ID badge if he knows that his probability of success is less than one in a million. Even if the only penalty for a failed attempt is the confiscation of the card by the verification device and smart cards cost only one dollar to manufacture, each success is expected to cost more than a million dollars.

To achieve a 2^{-20} level of security, use $k = 5$ and $t = 4$. The five s_j values can be stored in about 400 bytes, and each proof requires about 670 communicated bytes. The average number of multiplications each party has to perform is 14, which requires less than a second on today's smart cards. Even better performance with the same level of security can be obtained if A generates the s_j values pseudorandomly from a secret seed, sends to B hashed versions of the x_i, and receives from B sparse 2×18 $[e_{ij}]$ matrices with at most three 1's per row: the seed can be stored in less than 20 bytes, the number of communicated bytes is reduced to 200, and the average complexity drops to 7.6 modular multiplications.

An interesting modification can eliminate the public key directory and lead to a *keyless* identification scheme. It assumes the existence of a trusted center (a government, a credit card company, or a computer center) which issues the smart cards to users after properly checking their physical identity. No further interaction with the center is required either to generate or to verify proofs of identity. In this version of the scheme the center creates a string I which contains the user's name, address, ID number, physical description, digitized fingerprint, and any other information provers or verifiers may want to establish. The v_j numbers are defined pseudorandomly from the seed I, and the center (which knows the secret factorization of n but does not reveal it to anyone else) computes their corresponding s_j values and stores them in the card. When A wants to prove his identity, he sends I to B who can derive the v_j directly from I rather than from a public key directory. The actual proof of identity remains the same, and it convinces B and A knows the corresponding s_j. These values could only be computed by the trusted center when the real A requested a card.

In the modified scheme, the verification devices are simple standalone microcomputers which do not contain any secret information, are not connected to any database, and yet can verify the identity of anyone in the world. An unlimited number of users can join such a system without degrading its performance, and it offers excellent security for everyone involved: provers cannot cheat verifiers, verifiers cannot later misrepresent themselves as provers, and even coalitions of provers and verifiers cannot create new identities, modifying existing identities, or find out the secret factorization of n.

More information about the theoretical and practical aspects of this scheme as well as a (slightly slower) signature scheme based on similar principles can be found in Fiat and Shamir [3].

Acknowledgement. I would like to thank Mike Fischer, Oded Goldreich, Shafi Goldwasser, Silvio Micali, Charlie Rackoff, Calus Schnorr, and Avi Wigderson for inspiring many of the ideas presented in this paper.

BIBLIOGRAPHY

1. E. Brickell and J. DeLaurentis, *An attack on a signature scheme proposed by Okamoto and Shiraishi*, Proceedings of CRYPTO 85, Lecture Notes in Comput. Sci., no. 218, Springer-Verlag, 1985.

2. W. Diffie and M. Hellman, *New directions in cryptography*, IEEE Trans. Inform. Theory **IT-22** (1976), 644–654.

3. A. Fiat and A. Shamir, *How to prove yourself: practical solutions to identification and signature problems*, Technical Report, The Weizmann Institute of Science, Rehovot, Israel, 1986.

4. S. Goldwasser, S. Micali, and C. Rackoff, *The knowledge complexity of interactive proof systems*, 17th ACM Sympos. on Theory of Comput. May 1985, pp. 291–304.

5. S. Goldwasser, S. Micali, and R. Rivest, *A paradoxical solution to the signature problem*, 25th Sympos. on Found. of Comput. Sci., October 1984, pp. 441–448.

6. H. Ong, C. Schnorr, and A. Shamir, *An efficient signature scheme based on quadratic equations*, 16th ACM Sympos. on Theory of Comput., April 1984, pp. 208–216.

7. J. Pollard, private communication.

8. M. Rabin, *Digitized signatures as interactable as factorization*, MIT Technical Report MIT/LCS/TR-212, January 1979.

9. R. Rivest, A. Shamir, and L. Adleman, *A method for obtaining digital signatures and public key cryptosystems*, Comm. ACM **21** (1978), no. 2, 120–126.

10. T. Okamoto and A. Shiraishi, *A fast signature scheme based on quadratic inequalities*, Proc. of the 1985 Sympos. on Security and Privacy (Oakland, Calif., April 1985).

THE WEIZMANN INSTITUTE OF SCIENCE, REHOVOT 76100, ISRAEL

Markov Random Field Image Models
and Their Applications to Computer Vision

STUART GEMAN AND CHRISTINE GRAFFIGNE

1. Introduction. Computer vision refers to a variety of applications involving a sensing device, a computer, and software for restoring and possibly interpreting the sensed data. Most commonly, visible light is sensed by a video camera and converted to an array of measured light intensities, each element corresponding to a small patch in the scene (a picture element, or "pixel"). The image is thereby "digitized," and this format is suitable for computer analysis. In some applications, the sensing mechanism responds to other forms of light, such as in infrared imaging where the camera is tuned to the invisible part of the spectrum neighboring the color red. Infrared light is emitted in proportion to temperature, and thus infrared imaging is suitable for detecting and analyzing the temperature profile of a scene. Applications include automated inspection in industrial settings, medical diagnosis, and targeting and tracking of military objects. In single photon emission tomography, as a diagnostic tool, individual photons, emitted from a "radiopharmaceutical" (isotope combined with a suitable pharmaceutical) are detected. The object is to reconstruct the distribution of isotope density inside the body from the externally-collected counts. Depending on the pharmaceutical, the isotope density may correspond to local blood flow ("perfusion") or local metabolic activity. Other applications of computer vision include satellite imaging for weather and crop yield prediction, radar imaging in military applications, ultrasonic imaging for industrial inspection and a host of medical applications, and there is a growing role for video imaging in robotics.

The variety of applications has yielded an equal variety of algorithms for restoration and interpretation. Unfortunately, few general principals have emerged and no common foundation has been layed. Algorithms are by and large ad hoc; they are typically dedicated to a single application, and often critically tuned to the particulars of the environment (lighting, weather conditions, magnification, and so on) in which they are implemented. It is likely that a

---

Research partially supported by Army Research Office Contract DAAG29-83-K-0116, National Science Foundation Grant DMS-8352087, and the General Motors Corporation.

coherent theoretical framework would support more robust and more powerful algorithms. We have been exploring an approach based upon probabilistic image models, well-defined principals of inference, and a Monte Carlo computation theory. Exploiting this framework, we have recently obtained encouraging results in several areas of application, including tomography, texture analysis, and scene segmentation.

As an illustration of our approach, we shall discuss here the application to texture analysis. Other applications, and more complete discussions of the foundations, can be found in [**1**, **3**, **4**, **10**, **12**, **13**, **14**, **17**, **18**, **23**, **25**, and **27**]. In the section that follows, §2, we lay out, briefly, our paradigm in its general formulation. Then, in §3, the application to texture analysis is developed and illustrated by computer experiments. This application requires that we treat a somewhat unusual problem in parameter estimation, namely the estimation of parameters of a Markov random field from a single, large, sample. §4 details the estimation method used, and provides a proof of its consistency in the "large picture" limit, which is more appropriate than the usual "large sample size" limit.

2. Bayesian paradigm. In real scenes, neighboring pixels typically have similar intensities, boundaries are usually smooth and often straight, textures, although sometimes random locally, define spatially homogeneous regions, and objects, such as grass, tree trunks, branches and leaves, have preferred relations and orientations. Our approach to picture processing is to articulate such regularities mathematically, and then to exploit them in a statistical framework to make inferences. The regularities are rarely deterministic; instead, they describe correlations and likelihoods. This leads us to the Bayesian formulation, in which prior expectations are formally represented by a probability distribution. Thus we design a distribution (a "prior") on relevant scene attributes to capture the tendencies and constraints that characterize the scenes of interest. Picture processing is then guided by this prior distribution, which, if properly conceived, enormously limits the plausible restorations and interpretations.

The approach involves five steps, which we shall briefly review here (see [**13** and **18**] for more details). This will define the general framework, and then, in the following sections, we will concentrate on the analysis of texture, as an illustrative application.

Image models. These are probability distributions on relevant image attributes. Both for reasons of mathematical and computational convenience, we use *Markov random fields* (MRF) as prior probability distributions. Let us suppose that we index all of the relevant attributes by the index set S. S is application specific. It typically includes indices for each of the pixels (about 512×512 in the usual video digitization) and may have other indices for such attributes as boundary elements, texture labels, object labels on so on. Associated with each "site" $s \in S$ is a real-valued random variable X_s, representing the state of the corresponding attribute. Thus X_s may be the measured intensity at pixel s

(typically, $X_s \in \{0, \ldots, 255\}$) or simply 1 or 0 as a boundary element at location s is present or absent.

The kind of knowledge we represent by the prior distribution is usually "local," which is to say that we articulate regularities in terms of small local collections of variables. In the end, this leads to a distribution on $X = \{X_s\}_{s \in S}$ with a more or less "local neighborhood structure" (again, we refer to [**13** and **18**] for details). Specifically, our priors are Markov random fields: there exists a (symmetric) *neighborhood relation* $G = \{G_s\}_{s \in S}$, wherein $G_s \subseteq S$ is the set of neighbors of s, such that

$$\Pi(X_s = x_s | X_r = x_r, r \in S, r \neq s) = \Pi(X_s = x_s | X_r = x_r, r \in G_s).$$

$\Pi(a|b)$ is conditional probability, and, by convention, $s \notin G_s$. G symmetric means $s \in G_r \Leftrightarrow r \in G_s$. (Here, we assume that the range of the random vector X is discrete; there are obvious modifications for the continuous or mixed case.)

It is well known, and very convenient, that a distribution Π defines a MRF on S with neighborhood relation G if and only if it is Gibbs with respect to the same graph, (S, G). The latter means that Π has the representation

$$\Pi(x) = \tfrac{1}{z} e^{-U(x)} \tag{2.1}$$

where

$$U(x) = \sum_{c \in C} V_c(x). \tag{2.2}$$

C is the collection of all cliques in (S, G) (collections of sites such that every two sites are neighbors), and $V_c(x)$ is a function depending only on $\{x_s\}_{s \in c}$. U is known as the "energy," and has the intuitive property that the low energy states are the more likely states under Π. The normalizing constant, z, is known as the "partition function." The Gibbs distribution arises in statistical mechanics as the equilibrium distribution of a system with energy function U.

As a simple example (too simple to be of much use for real pictures) suppose the pixel intensities are known, a priori, to be one of two levels, minus one ("black") or plus one ("white"). Let S be the $N \times N$ square lattice, and let G be the neighborhood system that corresponds to nearest horizontal and vertical neighbors:

$$
\begin{array}{ccccccc}
\circ & - & \circ & - & \circ & \cdots \\
| & & | & & | \\
\circ & - & \circ & - & \circ & \cdots \\
| & & | & & | \\
\circ & - & \circ & - & \circ & \cdots \\
\vdots & & \vdots & & \vdots
\end{array}
$$

For picture processing, think of N as typically 512. Suppose that the only relevant regularity is that neighboring pixels tend to have the same intensities. An "energy" consistent with this regularity is the "Ising" potential:

$$U(x) = -\beta \sum_{\langle s,t \rangle} x_s x_t, \qquad \beta > 0,$$

where $\sum_{\langle s,t \rangle}$ means summation over all neighboring pairs $s, t \in S$. The minimum of U is achieved when $x_s = x_t$, $\forall s, t \in S$. Under (2.1), the likely pictures are therefore the ones that respect our prior expectations; they segment into regions of constant intensities. The larger β, the larger the typical region. Later we will discuss the issue of *estimating* model parameters such as β. (With energy (2.2), Π in (2.1) is called the Ising model. It models the equilibrium distribution of the spin states of the atoms in a ferromagnet. These spins tend to "line up," and hence the favored configurations contain connected regions of constant spins.)

One very good reason for using MRF priors is their Gibbs representations. Gibbs distributions are characterized by their energy functions, and these are more convenient and intuitive for modelling than working directly with probabilities. See, for example, [**12**, **13**, **14**, **18**, and **23**] for many more examples, and §3 below for a more complex and useful MRF model.

Degradation model. The image model is a distribution $\Pi(\cdot)$ on the vector of image attributes $X = \{X_s\}_{s \in S}$. *By design*, the components of this vector contain all of the relevant information for the image processing task at hand. Hence, the goal is to estimate X. This estimation will be based upon partial or corrupted observations, and based upon the prior information. In emission tomography, X represents the spacial distribution of isotope in a target region of the body. What is actually observed is a collection of photon counts whose probability law is Poisson, with a mean function that is an attenuated radon transform of X. In the texture labelling problem, X is the pixel intensity array and a corresponding array of texture labels. Each label gives the texture type of the associated pixel. The observation is only partial: we observe the pixels, which are just the digitized picture, but not the labels. The purpose is then to estimate the labels from the picture.

The observations are related to the image process (X) by a *degradation model*. This models the relation between X and the *observation process*, say $Y = \{Y_s\}_{s \in T}$. For texture analysis, we will define $X = (X^P, X^L)$, where X^P is the usual grey-level pixel intensity process, and X^L is an associated array of texture labels. The observed picture is just X^P, and hence $Y = X^P$: the degradation is a projection. More typically, the degradation involves a random component, as in the tomography setting where the observations are Poisson variables whose means are related to the image process X. A more simple, and widely studied (if unrealistic) example is additive "white" noise. Let $X = \{X_s\}_{s \in S}$ be just the basic pixel process. In this case, $T = S$, and for each $s \in S$ we observe $Y_s = X_s + \eta_s$ where, for example, $\{\eta_s\}_{s \in S}$ is Gaussian with independent components, having means 0 and variances σ^2.

Formally, the degradation model is a conditional probability distribution, or density, for Y given X: $\Pi(y|x)$. If the degradation is just added "white noise," as in the above example, then

$$\Pi(y|x) = \left(\frac{1}{2\pi\sigma^2} \right)^{|s|/2} \exp \left\{ -\frac{1}{2\sigma^2} \sum_{s \in S} (y_s - x_s)^2 \right\}.$$

For labelling textures, the degradation is deterministic: $\Pi(y|x)$ is concentrated on $y = x^P$, where $x = (x^P, x^L)$ has both pixel and label components.

Posterior distribution. This is the conditional distribution on the image process X given the observation process Y. This "posterior" or "a posteriori" distribution contains the information relevant to the image restoration or image analysis task. Given an observation $Y = y$, and assuming the image model ($\Pi(x)$) and degradation model ($\Pi(y|x)$), the posterior distribution reveals the likely and unlikely states of the "true" (unobserved) image X. Having constructed X to contain all relevant image attributes, such as locations of boundaries, labels of objects or textures, and so on, the posterior distribution comes to play the fundamental role in our approach to image processing.

The posterior distribution is easily derived from "Bayes's rule":

$$\Pi(x|y) = \frac{\Pi(y|x)\Pi(x)}{\Pi(y)}.$$

The denominator, $\Pi(y)$, is difficult to evaluate. It derives from the prior and degradation models by integration: $\Pi(y) = \int_x \Pi(y|x)\Pi(dx)$, but the formula is computationally intractable. Happily, our analysis of the posterior distribution will require only *ratios*, not absolute probabilities. Since y is fixed by observation, $1/\Pi(y)$ is a constant that can be ignored (see paragraph below on "computing").

As an example we consider the simple "Ising model" prior, with observations corrupted by additive white noise. Then

$$\Pi(x) = \frac{1}{z} \exp\left\{ -\beta \sum_{\langle s,t \rangle} x_s x_t \right\}$$

and

$$\Pi(y|x) = \left(\frac{1}{2\pi\sigma^2} \right)^{|S|/2} \exp\left\{ -\frac{1}{2\sigma^2} \sum_{s \in S} (y_s - x_s)^2 \right\}.$$

The posterior distribution is then

$$\Pi(x|y) = \frac{1}{z_p} \exp\left\{ -\beta \sum_{\langle s,t \rangle} x_s x_t - \frac{1}{2\sigma^2} \sum_{s \in S} (y_s - x_s)^2 \right\}.$$

We denote by z_p the normalizing constant for the posterior distribution. Of course, it depends upon y, but the latter is fixed. Notice that the posterior distribution is again a MRF. In the case of additive white noise, the neighborhood system of the posterior distribution is that of the prior, and hence local. For a wide class of useful degradation models, including combinations of blur, added or multiplicative "colored noise," and a variety of nonlinear transformations, the posterior distribution is a MRF with a more or less local graph structure. This is convenient for our computational schemes, as we shall see shortly. We should note, however, that exceptions occur. In tomography, for example, the posterior distribution is associated with a highly nonlocal graph. This situation incurs a high computational cost (see [**14**] for more details).

MAP estimate. In our framework, image processing amounts to choosing a particular image x, given an observation $Y = y$. A sensible, and suitably-defined optimal, choice is the "maximum a posteriori," or "MAP" estimate: choose x to maximize $\Pi(x|y)$. The MAP estimate chooses the most likely x, given the observation. In most applications, our goal is to identify the MAP estimate, or a suitable approximation. However, in some settings other estimators are more appropriate. We have found, for example, that the posterior mean ($\int x\Pi(dx|y)$) is more effective for tomography, at least in our experiments. Here, we concentrate on MAP estimation.

In most applications we cannot hope to identify the true maximum a posteriori image vector x. To appreciate the computational difficulty, consider again the Ising model with added white noise:

$$\Pi(x|y) = \frac{1}{z_p} \exp\left\{ -\beta \sum_{\langle s,t \rangle} x_s x_t - \frac{1}{2\sigma^2} \sum_{s \in S} (y_s - x_s)^2 \right\}. \tag{2.3}$$

This is to be maximized over all possible vectors $x = \{x_s\}_{s \in S} \in \{-1,1\}^{|S|}$. with $|S| \sim 10^5$, brute force approaches are intractable; instead, we will employ a Monte Carlo algorithm which gives adequate approximations.

Maximizing (2.3) amounts to minimizing

$$U_p(x) = -\beta \sum_{\langle s,t \rangle} x_s x_t - \frac{1}{2\sigma^2} \sum_{s \in S} (y_s - x_s)^2$$

which might be thought of as the "posterior energy." (As with z_p, the fixed observation y is suppressed in the notation $U_p(x)$.) More generally, we write the posterior distribution as

$$\frac{1}{z_p} \exp\{-U_p(x)\} \tag{2.4}$$

and characterize the MAP estimator as the solution to the problem "choose x to minimize $U_p(x)$." The utility of this point of view is that it suggests a further analogy to statistical mechanics, and a computation scheme for approximating the MAP estimate, which we shall now describe.

Computing. Pretend that (2.4) is the equilibrium Gibbs distribution of a real system. Recall that MAP estimation amounts to finding a minimal energy state. For many physical systems the low energy states are the most ordered, and these often have desirable properties. The state of silicon suitable for wafer manufacturing, for example, is a low energy state. Physical chemists achieve low energy states by heating and then slowly cooling a substance. This procedure is called *annealing*. Cerný [5] and Kirkpatrick [21] suggest searching for good minimizers of $U(\cdot)$ by *simulating* the dynamics of annealing, with U playing the role of energy for an (imagined) physical system. In our image processing experiments, we often use "simulated annealing" to find an approximation to the MAP estimator.

Dynamics are simulated by producing a Markov chain, $X(1), X(2), \ldots$ with transition probabilities chosen so that the equilibrium distribution is the posterior (Gibbs) distribution (2.4). One way to do this is with the "Metropolis algorithm" [24]. More convenient for image processing is a variation we call *stochastic relaxation*. The full story can be found in [13 and 18]. Briefly, in stochastic relaxation we choose a sequence of sites $s(1), s(2), \ldots \in S$ such that each site in S is "visited" infinitely often. If $X(t) = x$, say, then $X_r(t+1) = x_r$, $\forall r \neq s(t)$, $r \in S$, and $X_{s(t)}(t+1)$ is a sample from

$$\Pi(X_{s(t)} = \cdot | X_r = x_r, r \neq s(t)),$$

the conditional distribution on $X_{s(t)}$ given $X_r = x_r \ \forall r \neq s(t)$. By the Markov property,

$$\Pi(X_{s(t)} = \cdot | X_r = x_r, r \neq s(t)) = \Pi(X_{s(t)} = \cdot | X_r = x_r, r \in G^p_{s(t)})$$

where $\{G^p_s\}_{s \in S}$ is the *posterior* neighborhood system, determined by the posterior energy $U_p(\cdot)$. The prior distributions that we have experimented with have mostly had local neighborhood systems, and usually the posterior neighborhood system is also more or less local as well. This means that $|G^p_{s(t)}|$ is small, and this makes it relatively easy to generate, Monte Carlo, $X(t+1)$ from $X(t)$. In fact, if Ω is the range of $X_{s(t)}$, then

$$\Pi(X_{s(t)} = \alpha | X_r = x_r, r \in G^p_{s(t)}) = \frac{\Pi(\alpha, s(t) x)}{\sum_{\hat{\alpha} \in \Omega} \Pi(\hat{\alpha}, s(t) x)} \qquad (2.5)$$

where

$$(\alpha, s(t) x)_r = \begin{cases} \alpha, & r = s(t), \\ x_r, & r \neq s(t). \end{cases}$$

Notice that (fortunately!) there is no need to compute the posterior partition function z_p. Also, the expression on the right-hand side of (2.5) involves only those potential terms associated with cliques containing $s(t)$, since all other terms are the same in the numerator and the denominator.

To simulate annealing, we introduce an artificial "temperature" into the posterior distribution:

$$\Pi_T(x) = \frac{\exp\{-U_p(x)/T\}}{Z_p(T)}.$$

As $T \to 0$, $\Pi_T(\cdot)$ concentrates on low energy states of U_p. To actually find these states, we run the stochastic relaxation algorithm while slowly lowering the temperature. Thus $T = T(t)$, and $T(t) \downarrow 0$. $\Pi_{T(t)}(\cdot)$ replaces $\Pi(\cdot)$ in computing the transition $X(t) \to X(t+1)$. In [13] we showed that, under suitable hypotheses on the sequence of site visits, $s(1), s(2), \ldots$:

> If $T(t) > c/(1 + \log(1+t))$, $T(t) \downarrow 0$, then for all c sufficiently large $X(t)$ converges weakly to the distribution concentrating uniformly on $\{x : U(x) = \min_y U(y)\}$.

More recently, our theorem has been improved upon by many authors. In particular, the smallest constant c which guarantees convergence of the annealing algorithm to a global minimum can be specified in terms of the energy function U_p (see [**15** and **19**]). Also, see Gidas [**16**] for some ideas about faster annealing via "renormalization group" methods.

In the experiments with texture to be described here, MAP estimates are approximated by using the annealing algorithm. This involves Monte Carlo computer-generation of the sequence $X(1), X(2), \ldots$, terminating when the state ceases to change substantially.

3. Texture segmentation. Texture *synthesis* refers to computer generation of homogeneous patterns, usually intended to match a natural texture such as wood, grass, or sand. In many instances, Markov random fields provide good models, and Metropolis-like Monte Carlo methods yield respectable facsimiles of the real textures [**8, 9**]. Here we combine MRF texture models, for the pixel process, with an Ising-like "texture label process," in order to segment and label a scene consisting of patches of natural textures. The image model thereby involves both a pixel process, of grey level intensities, and a label process, whose components identify the texture type of each picture element in the scene. Our approach is similar to those of Derin and Elliott [**9**] and Cohen and Cooper [**7**], especially in our use of the two-tiered image model.

Image model. The image process comprises a pixel process and a label process, $X = \{X^P, X^L\}$. As usual, the pixes sites form an $N \times N$ square lattice, say S^P. For each pixel site there is a corresponding label site, and thus the graph associated with the image model has sites $S = S^P \cup S^L$, where S^L is just a copy of S^P. The elements of S^P and S^L index the components of X^P and X^L, respectively, so that $X^P = \{X_s^P\}_{s \in S^P}$ and $X^L = \{X_s^L\}_{s \in S^L}$. In the experiments reported here, the pixels were allowed sixteen possible grey levels $X_s^P \in \{0, 1, \ldots, 15\}$, $\forall s \in S^P$, whereas the range of the labels depended upon the actual number of textures in the scene, thus assuming this number to be known a priori. Let M be the number of textures that are to be modelled. Then $X_s^L \in \{1, 2, \ldots, M\}$, $\forall s \in S^L$.

We shall develop the image model by first assuming that the texture type is fixed, say "l" and constant over the scene. *Conditioned* on $X_s^L = l \in \{1, 2, \ldots, M\}$, $\forall s \in S^L$, the process X^P is a Markov random field:

$$\Pi(x^P | X_s^L = l, s \in S^L) = \frac{1}{z^{(l)}} \exp\{-U^{(l)}(x^P)\}$$

where $z^{(l)}$ is the usual normalizing constant $z^{(l)} = \sum_{x^P} \exp\{-U^{(l)}(x^P)\}$. Only pair-cliques appear in the energy $U^{(l)}$. There are six types of pair-cliques, as shown in Figure 1. These we index by $i \in \{1, 2, 3, 4, 5, 6\}$. We denote by $\langle s, t \rangle_i$ a pair of sites s, t which form a type i clique, and by $\sum_{\langle s, t \rangle_i}$ the summation over all such pairs. With these conventions, the (conditional) energy is

$$U^{(l)}(x^P) = -\sum_{i=1}^{6} \sum_{\langle s, t \rangle_i} \theta_i^{(l)} \Phi(x_s^P - x_t^P), \qquad \Phi(\Delta) \doteq (1 + (|\Delta|/\delta)^2)^{-1} \quad (3.1)$$

for some fixed $\delta > 0$. Notice that $\Phi(x_s^P - x_t^P)$ is larger when $x_s^P = x_t^P$, and is monotonic in $|x_s^P - x_t^P|$. Because of this, the texture-dependent parameters $\theta_1^{(l)}, \ldots, \theta_6^{(l)}$ determine the degree to which neighboring pixels, of a particular type of pair-clique, will tend to have similar grey-levels. In face, if $\theta_i^{(l)} > 0$, then for texture "l" we expect pixel pairs x_s and x_t, of clique type i, to typically have similar intensities. If $\theta_1^{(l)} < 0$ then the tendency is to be different. Of course, these simple rules are complicated by the actions of the other five types of pair-cliques.

FIGURE 1. Pair-cliques for texture model.

The parameters $\theta_i^{(l)}, i = 1, 2, \ldots, 6, l = 1, 2, \ldots, M$, are estimated from pictures of the M textures, as explained in the following section (§4). On the other hand, Φ, and indeed the neighborhood structure, is ad hoc. We have used Φ extensively in other applications in which our main concern is with the difference of intensities between neighboring pixels. Of course the quadratic, $\Phi(\Delta) = \Delta^2$, is simpler, but it unduly penalizes large differences. Having modeled the M textures, we now construct a composite Markov random field which accounts for both texture labels, $X^L = \{X_s^L, s \in S^L\}$, and grey-levels, $X^P = \{X_s^P, s \in S^P\}$. The joint distribution is

$$\Pi(X^P = x^P, X^L = x^L) = \frac{\exp\{-U_1(x^P, x^L) - U_2(x^L)\}}{z} \tag{3.2}$$

in which U_2 promotes label bonding (we expect the textures to appear in patches rather than interspersed) and U_1 specifies the interaction between labels and intensities. Specifically, we employ a simple Ising-type potential for the labels:

$$U_2(x^L) = -\beta \sum_{[s,t]} 1_{x_s^L = x_t^L} + \sum_{s \in S} w(x_s^L), \qquad \beta > 0. \tag{3.3}$$

Here β determines the degree of clustering, $[s, t]$ indicates a pair of nearest horizontal or vertical neighbors, and $w(\cdot)$ is adjusted to eliminate bias in the label probabilities (more on the choice of $w(\cdot)$ later).

To describe the interaction between labels and pixels we introduce the symbols $\tau_1, \tau_2, \ldots, \tau_6$ to represent the lattice vectors associated with the 6 pair-cliques (Figure 1). Thus s and $s + \tau_i$ are neighbors, constituting a pair with clique type i. The interaction is then given in terms of pixel-based contributions,

$$H(x^P, l, s) \doteq -\sum_{i=1}^{6} \theta_i^{(l)} \{\Phi(x_s^P - x_{s+\tau_i}^P) + \Phi(x_s^P - x_{s-\tau_i}^P)\} \tag{3.4}$$

and local sums of these called block-based contributions,

$$Z(x^P, l, s) \doteq \frac{1}{a} \sum_{t \in N_s} H(x^P, l, t). \tag{3.5}$$

Here, N_s is a block of sites centered at s (5 by 5 in all of our experiments), and the constant a is adjusted so that the sum of all block-based contributions reduces to $U^{(l)}$ (see (3.1)):

$$U^{(l)}(x^P) = \sum_{s \in S} Z(x^P, l, s). \tag{3.6}$$

This amounts to ensuring that each pair-clique appears exactly once ($a = 50$, for example, when N_s is 5 by 5). In terms of (3.4) and (3.5), the "interaction energy," $U_1(x^P, x^L)$, is written

$$U_1(x^P, x^L) = \sum_{s \in S} Z(x^P, x_s^L, s). \tag{3.7}$$

Because of (3.6), the model is consistent with (3.1) for homogeneous textures, $X_s^L = l$, $\forall s \in S$. The idea is that each local texture label, X_s^L, is influenced by the pixel grey levels in a neighborhood of s.

Finally, to clarify the bias correction term $w(\cdot)$, we briefly examine the local characteristics of the field, specifically the conditional distributions for the labels given all the intensity data and the values of the neighboring labels. (The actual neighborhoods of the Markov random field corresponding to (3.2) can be easily inferred from (3.3) and (3.7).) The log odds of texture type k to type j is

$$\log \left\{ \frac{\Pi(X_r^L = k | X_s^L = x_s^L, \ s \neq r; \ X_s^P = x_s^P, \ s \in S)}{\Pi(X_r^L = j | X_s^L = x_s^L, \ s \neq r; \ X_s^P = x_s^P, \ s \in S)} \right\}$$

$$= Z(x^P, j, r) - Z(x^P, k, r) + \beta \sum_{t:\,[t,r]} (1_{x_t^L = k} - 1_{x_t^L = j}) + w(j) - w(k)$$

$$= \frac{1}{z} \sum_{i=1}^{6} \sum_{s \in N_r} (\theta_i^{(k)} - \theta_i^{(j)}) \{\Phi(x_s^P - x_{s+\tau_i}^P) + \Phi(x_s^P - x_{s-\tau_i}^P)\}$$

$$+ \beta \sum_{t:\,[t,r]} (1_{x_t^L = k} - 1_{x_t^L = j}) + w(j) - w(k).$$

The first term imposes fidelity to the "data" x^P, and the second bonds the labels. The efficacy of the model depends on the extent to which the first term separates the two types k and j, which can be assessed by plotting histograms for the values of this quantity both for pure k and pure j data. A clean separation of the histograms signifies a good discriminator. However, since we are looking at log odds, we insist that the histograms straddle the origin, with positive (resp. negative) values associated with texture type k (resp. j). The function $w(\cdot)$ makes this adjustment.

Degradation model. The degradation is deterministic. The observation process is the pixel process $Y = X^P$, and hence the degradation is just the *projection* $(X^P, X^L) \to X^P$.

Posterior distribution. In this special case, the posterior energy is the same as the prior energy, but some of the components are fixed. In particular,

$$\Pi((x^P, x^L)|y) = \frac{1}{z_p} \exp\{-U_1(x^P, x^L) - U_2(x^L)\} 1_{x^P = y}.$$

FIGURE 2. Wood on plastic background.

Equivalently, we simply use $\Pi(x^L|x^P)$ as the posterior distribution:

$$\Pi(x^L|x^P) = \frac{1}{z_p} \exp\{-U_1(x^P, x^L) - U_2(x^L)\}.$$

MAP estimate. Given an observation, $X^P = x^P$, we shall seek x^L to minimize $U_1(x^P, x^L) + U_2(x^L)$.

Computing. We use stochastic relaxation, with simulated annealing, as described in §2. A convenient starting point is arrived at by "turning off" the Ising term in the label model (3.3): we set $\beta = 0$. Since this is the only label/label interaction term in the model, the MAP estimate of x^L, with $\beta = 0$, is determined by (locally) optimizing x_s^L at each $s \in S^L$. The computation time is negligible. Thereafter, we set β to the model value (see §4) and begin stochastic relaxation. In the experiments, each site was visited about 150 times.

Experimental results. Three experiments were done on texture discrimination, based on two images with two textures each and one with four. There are four textures involved: wood, plastic, carpet, and cloth. As mentioned above, the parameters were estimated from the pure types (see §4). There was no pre- or post-processing. In particular, no effort was made to "clean-up" the boundaries, expecting smooth transitions. The results are shown in Figures 2, 3, and 4; these correspond to (i) wood on plastic, (ii) carpet on plastic, and (iii) wood, carpet, and cloth on plastic background. In each figure, the left panel is the textured scene, and the right panel shows the segmentation, with texture labels coded by grey level. It is interesting to note that the grey-level histograms of the four textures are very similar (Figure 5); in particular, discrimination based on shading alone is virtually impossible.

FIGURE 3. Carpet on plastic background.

FIGURE 4. Wood, carpet, and cloth on plastic background.

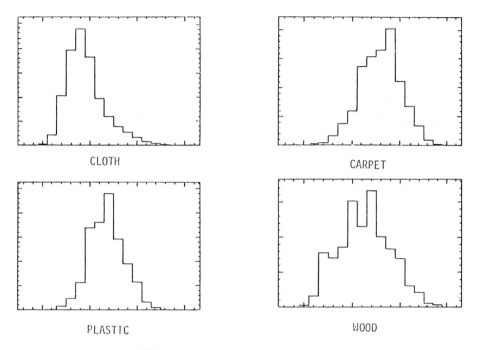

CLOTH   CARPET

PLASTIC   WOOD

FIGURE 5. Grey-level histograms.

The model is not really adequate for texture *synthesis*; samples generated from the model do not resemble the texture very well. Evidently, the utility of Markov random field models does not depend on their capacity for simulating real-world imagery. A more serious drawback of our model is that it is dedicated to a fixed repertoire of textures, viewed at a particular orientation and at a particular magnification, or range. The problem is easier if the goal is merely *segmentation*, without *recognition*. We are experimenting with segmentation algorithms that are scale and orientation independent. Indeed, there are no texture-specific parameters. These are built upon the same modelling/computing framework.

4. Parameter estimation.

Maximum pseudolikelihood. The performance of the model is not unduly sensitive to the choice of δ (see (3.1)) or β (see (3.3)), which were determined by trial and error. On the other hand, the pair-clique parameters $\theta_i^{(l)}$, $i = 1, 2, \ldots, 6$, $l = 1, 2, \ldots, M$, characterize the M textures, and critically determine the ability of the model to segment and label. Needless to say, these must be systematically estimated. Trial and error is not feasible.

We have estimated the parameters from samples of the M textures. These "training samples" contain only one texture each, and we used just one sample for each texture. For a fixed texture, say wood, and from a single sample, say $\tilde{x}^p$, the problem then is to estimate $\theta_1, \theta_2, \ldots, \theta_6$ in the model

$$\Pi(X^P = x^P; \theta) = \frac{\exp\{-U(x^P; \theta)\}}{z(\theta)}$$

where

$$U(x^P; \theta) = -\sum_{i=1}^{6} \sum_{\langle s,t \rangle_i} \theta_i \Phi(x_s^P - x_t^P)$$

and

$$z(\theta) = \sum_{x^P} \exp\{-U(x^P; \theta)\}.$$

(We include $\theta \doteq (\theta_1, \ldots, \theta_6)$ in Π, U, and Z to emphasize the dependencies on the unknown parameters.) The standard approach is to maximize the "likelihood": choose θ to maximize $\Pi(\tilde{x}^P; \theta)$. Of course, maximizing Π is equivalent to maximizing $\log \Pi$. It is easily demonstrated that the latter is *concave* in θ with gradient

$$\nabla \log \Pi(\tilde{x}^P; \theta) = \left\{ \sum_{\langle s,t \rangle_i} \Phi(\tilde{x}_s^P - \tilde{x}_t^P) - E_\theta \left[\sum_{\langle s,t \rangle_i} \Phi(X_s^P - X_t^P) \right] \right\}_{i=1,\ldots,6} \qquad (4.1)$$

where $E_\theta[\cdot]$ is expectation with respect to $\Pi(\cdot; \theta)$. This suggests a gradient ascent procedure, but the expectation $E_\theta[\cdot]$ is intractable, involving summation over the entire range of X^P. In our experiments, we used a 16 grey-level scale for the pixels, and 204×204 lattices: the expectation in (4.1) has 16^{204^2} terms. An alternative to brute force evaluation is to use stochastic relaxation (see §2), which produces an (asymptotically) *ergodic* sequence $X^P(1), X^P(2), \ldots$ for any given θ, and from which expectations can be approximated by appropriate time-averages. This, too, is computationally intensive, but feasible. In some settings we have found no alternative, and this Monte Carlo procedure has worked well, albeit slowly (see [22]). See also Hinton and Sejnowski [20] for a closely related algorithm, used to model learning in a theory of neuron dynamics.

For homogeneous random fields, such as our image models, Besag [2, 3] has proposed an ingenious alternative to maximum likelihood, known as "maximum pseudolikelihood." The pseudolikelihood function is

$$PL(\tilde{x}^P; \theta) \doteq \prod_{s \in S^P \backslash \partial S^P} \Pi(X_s^P = \tilde{x}_s^P | X_r^P = \tilde{x}_r^P, r \neq s; \theta)$$

where ∂S^P is the boundary of S^P under the neighborhood system determined by the energy U, and $S^P \backslash \partial S^P$ is the complement of ∂S^P relative to S^P. The "pseudolikelihood estimator" is the θ that maximizes $PL(\tilde{x}^P; \theta)$. In the next few pages we shall lend some analytic support, by establishing consistency of pseudolikelihood in the "large graph" limit. But first, we emphasize the overwhelming computational advantage. As with the log likelihood function, the log pseudolikelihood function, $\log PL(\tilde{x}^P; \theta)$, is concave, but this time the gradient is directly computable:

$$\nabla \log PL(\tilde{x}^P; \theta)$$

$$= \nabla \sum_{s \in S^P \setminus \partial S^P} \left\{ \sum_{i=1}^{6} \theta_i \{ \Phi(\tilde{x}_s^P - \tilde{x}_{s+\tau_i}^P) + \Phi(\tilde{x}_s^P - \tilde{x}_{s-\tau_i}^P) \} \right.$$

$$\left. - \log \sum_{\alpha} \exp \left\{ \sum_{i=1}^{6} \theta_i \{ \Phi(\alpha - \tilde{x}_{s+\tau_i}^P) + \Phi(\alpha - \tilde{x}_{x-\tau_i}^P) \} \right\} \right\}$$

(where $\sum_{\alpha}$ is summation over pixel grey levels, zero through fifteen in our experiments)

$$= \left\{ \sum_{s \in S^P \setminus \partial S^P} \{ \Phi(\tilde{x}_s^P - \tilde{x}_{s+\tau_i}^P) + \Phi(\tilde{x}_s^P - \tilde{x}_{s-\tau_i}^P) \right.$$

$$\left. - E_\theta [\Phi(X_s^P - X_{s+\tau_i}^P) + \Phi(X_s^P - X_{s-\tau_i}^P) | X_r^P = \tilde{x}_r^P, r \neq s] \} \right\}_{i=1,\ldots,6}.$$

This time, the expectation is tractable. The conditional distribution on X_s^P, given $X_r^P = \tilde{x}_r^P$, $r \neq s$, involves only those variables $\tilde{x}_r^P$ in the neighborhood of s. Furthermore, this time summation is over the range of X_s^P only, which has only sixteen values. In short, the gradient of the log pseudolikelihood is directly computable, and therefore gradient ascent is feasible without resorting to time-consuming Monte Carlo methods. For the experiments discussed in the previous section, the pair-clique parameters were estimated, for each texture type, by gradient ascent of the pseudolikelihood function.

Some modifications of maximum and pseudolikelihood have been recently introduced by Chalmond [6]. A third alternative was suggested by Derin and Elliott [9, 11], and has been studied and analyzed extensively by Possolo [26]. This involves a regression fit of the log of the local conditional probabilities, and works best when there are a small number of values in the range of the random variables. For example, the method is very effective for Ising-like models.

Consistency of pseudolikelihood. We will study parameter estimation from a single realization of a finite-graph Markov random field. The typical framework for establishing consistency is in the limit as the number of samples increases. But we have in mind estimation from a single sample of the random field, with the *number of sites* large (e.g., 512×512). To study estimation in this "large graph" setting, we will imagine a sequence of samples, $X(1), X(2), \ldots$, from a sequence of Markov random fields, $\Pi_1, \Pi_2, \ldots$, in which the latter are associated with an expanding sequence of regular graphs. We will assume that the sequence of distributions of these random fields has a common unknown parameter vector $\theta_0 \in R^m$. We will define the pseudolikelihood estimate, $\hat{\theta}_n = \hat{\theta}_n(X(n))$, for each sample, $X(n)$, and show that $\hat{\theta}_n \to \theta_0$ with probability one.

The samples $X(1), X(2), \ldots$ need not be independent. For example, we may wish to model the observations as subsamples from a single *infinite volume*

Gibbs state. Then, there is one infinite-volume process X, e.g., $X = \{X_s\}_{s \in S}$, $S \doteq \{(i,j): -\infty < i,j < \infty\}$, and the observations are associated with increasing subsets: $X(k) = \{X_s\}_{s \in S_k}$ with, e.g. $S_1 \subseteq S_2 \subseteq \cdots, \bigcup_{k=1}^{\infty} S_k = S$. The sequence of distributions, $\Pi_1, \Pi_2, \ldots$, is the sequence of *conditional distributions*, on $\{X_s\}_{s \in S_k}$, conditioned on $\{X_s\}_{s \in S \backslash S_k}, k = 1, 2, \ldots$. Under a suitable "homogeneity" (translation invariance) assumption for the Gibbs potential, the theorem applies, guaranteeing consistency of the pseudolikelihood estimate. This is regardless of *critical phenomena*, or *lack of spatial stationarity*, both of which can occur with infinite volume Gibbs states having translation-invariant potentials [28].

Henceforth, we specialize to regular square lattices: S will represent the d-dimensional infinite square lattice. (Generalizations are straightforward.) For each n, $S_n \subset S$ is a d-dimensional cube with sides length n. On S is a translation-invariant neighborhood system $G = \{G_s\}_{s \in S}$ ($s \notin G_s; s \in G_r \Leftrightarrow r \in G_s; s \in G_r \Leftrightarrow s + \tau \in G_{r+\tau} \, \forall s, r, \tau \in S$). We will assume "finite" interactions: $\exists R \ni s \in G_r \Rightarrow |s - r| \leq R$. We will denote the subgraph of (S, G) with sites S_n by (S_n, G). Associated with each n is a Markov random field, Π_n, on (S_n, G). The site variables, $\{X_s\}_{s \in S_n}$, are assumed to have common range Ω, with $|\Omega| < \infty$.

The distributions $\Pi_1, \Pi_2, \ldots$ are related by their dependencies on a common unknown parameter $\theta_0 \in R^m$. Pseudolikelihood exploits the dependencies of local conditional probabilities on this parameter. In particular, fix n and let $x \in \Omega^{S_n}$, the range of the random field with distribution Π_n. For each s, let $_sx = \{x_r : r \in G_s \cap S_n\}$. Actually, $_sx$ will be treated as a vector, in which the components are placed in some arbitrary order. "Local characteristics" of Π_n refers to the conditional probabilities $\Pi_n(X_s = x_s|_sX = {}_sx; \theta_0)$ for each $s \in S_n$, $x \in \Omega^{S_n}$. The distributions $\Pi_1, \Pi_2, \ldots$ are tied together by the assumption that these local characteristics, which depend upon θ_0, are independent of s and n, for all s in the interior of S_n. More precisely, letting $S_n^0 = S_n \backslash \partial S_n$ under G, we assume that there exists $\Psi(\cdot) = (\Psi_1(\cdot), \ldots, \Psi_m(\cdot))$ such that

$$\Pi_n(X_s = x_s|_sX = {}_sx; \theta_0) = \frac{\exp\{\theta_0 \cdot \Psi(x_s, {}_sx)\}}{\sum_{\alpha \in \Omega} \exp\{\theta_0 \cdot \Psi(\alpha, {}_sx)\}} \qquad (4.2)$$

for all n, $s \in S_n^0$, x_s, and $_sx$. Any homogeneous field with finite interactions is suitable, regardless of boundary conditions. Examples include the Ising model, and the texture model (for a single, homogeneous texture) developed in §3.

Whenever $s \in S_n^0, \Pi_n(X_s = x_s|_sX = {}_sx; \theta)$ does not depend on n. Since we will only be interested in local characteristics at interior sites, we henceforth drop the subscript n when writing conditional probabilities. Given $X = x$, a sample from Π_n, the pseudolikelihood function of $\theta \in R^m$ is

$$PL_n(x; \theta) \doteq \prod_{s \in S_n^0} \Pi(x_s|_sx; \theta)$$

$$= \prod_{s \in S_n^0} \frac{\exp\{\theta \cdot \Psi(x_s, {}_sx)\}}{\sum_{\alpha \in \Omega} \exp\{\theta \cdot \Psi(\alpha, {}_sx)\}}.$$

The pseudolikelihood estimate is the *set* $M_n(x)$, of θ that maximize $PL_n(x;\theta)$:

$$M_n(x) = \left\{ \theta \in R^m : PL_n(x,\theta) = \sup_{\phi \in R^m} PL_n(x,\phi) \right\}.$$

In establishing consistency for pseudolikelihood estimation we will assume *identifiability*, in the following sense:

DEFINITION. We will say that $\theta_0 \in R^m$, is *identifiable* if $\theta \neq \theta_0 \Rightarrow \exists x_s, {}_s x$, such that $\Pi(x_s|_s x;\theta) \neq \Pi(x_s|_s x;\theta_0)$.

THEOREM (CONSISTENCY OF PSEUDOLIKELIHOOD). *For each* $n = 1, 2,$ $\ldots$, *let* $X(n)$ *be a sample from the Markov random field* Π_n, *with local characteristics* (4.2). *If* θ_0 *is identifiable, then*

(a) $P(\log PL_n(X(n);\theta)$ *is strictly concave for all* n *sufficiently large* $) = 1$;

(b) $P(M_n(X(n))$ *is a singleton for all* n *sufficiently large* $) = 1$;

(c) $P(\sup_{\theta \in M_n(X(n))} |\theta - \theta_0| \to 0) = 1$.

REMARKS. (1) Extensions to more general graph structures and interaction potentials are possible, and mostly routine.

(2) More relevant to the problem of estimating θ_0 from a sample $X(n)$, with n large, is the following immediate corollary:

$$\lim_{n \to \infty} P\left(\sup_{\theta \in M_n(X(n))} |\theta - \theta_0| > \varepsilon \right) = 0 \quad \forall \varepsilon > 0.$$

PROOF OF THEOREM. Let $N_n = |S_n^0|$,

$$N_n(\beta) = \#\{s \in S_n^0 : {}_s X(n) = \beta\},$$

and

$$N_n(\alpha, \beta) = \#\{s \in S_n^0 : X_s(n) = \alpha, {}_s X(n) = \beta\},$$

using α and β as generic elements of Ω and $\Omega^{|G_s|}$, respectively. The proof can be divided into five steps, which we now state as lemmas.

LEMMA 1. $\liminf_{n \to \infty} N_n(\beta)/N_n > 0$ *a.s.*, $\forall \beta$.

LEMMA 2. $\lim_{n \to \infty} N_n(\alpha, \beta)/N_n(\beta) = \Pi(\alpha|\beta;\theta_0)$ *a.s.*, $\forall \alpha, \beta$.

LEMMA 3. *Let*

$$F_n(\theta) = \frac{1}{N_n}\{\log PL_n(X(n);\theta) - \log PL_n(X(n);\theta_0)\}$$

$$= \sum_\beta \frac{N_n(\beta)}{N_n} \sum_\alpha \frac{N_n(\alpha, \beta)}{N_n(\beta)} \log \frac{\Pi(\alpha|\beta;\theta)}{\Pi(\alpha|\beta;\theta_0)}.$$

$P(F_n(\cdot)$ *is strictly concave for all* n *sufficiently large* $) = 1$.

LEMMA 4. *Let*

$$G_n(\theta) = \sum_\beta \frac{N_n(\beta)}{N_n} \sum_\alpha \Pi(\alpha|\beta;\theta_0) \log \frac{\Pi(\alpha|\beta;\theta)}{\Pi(\alpha|\beta;\theta_0)}.$$

(a) *With probability one,* $\forall \varepsilon > 0 \; \exists \delta > 0 \ni$

$$\limsup_{n \to \infty} \sup_{|\theta - \theta_0| \le \varepsilon} \sup_{\phi \in R^m, \, |\phi|=1} \phi^t H(G_n(\theta))\phi < -\delta$$

where $H(G_n(\theta))$ *is the matrix of second derivatives (Hessian) of* $G_n(\theta)$ *with respect to* θ.

(b) $G_n(\theta) \le 0 \; \forall \theta, n$.

(c) $G_n(\theta_0) = 0 \; \forall n$.

LEMMA 5. $\forall \varepsilon > 0$,

$$\lim_{n \to \infty} \sup_{|\theta - \theta_0| \le \varepsilon} |F_n(\theta) - G_n(\theta)| = 0 \quad a.s.$$

With these pieces in place, we complete the proof as follows.

Fix $\varepsilon > 0$. From Lemma 4, conclude that

$$\liminf_{n \to \infty} \inf_{|\theta - \theta_0|=\varepsilon} (G_n(\theta_0) - G_n(\theta)) > 0 \quad a.s. \tag{4.3}$$

Since F_n is uniformly approximated by G_n (in the sense of Lemma 5), (4.3) also holds for F_n:

$$\liminf_{n \to \infty} \inf_{|\theta - \theta_0|=\varepsilon} (F_n(\theta_0) - F_n(\theta)) > 0 \quad a.s.$$

Since F_n is eventually strictly concave (Lemma 3), it eventually achieves its maximum, uniquely, in $\{\theta \colon |\theta - \theta_0| < \varepsilon\}$. Finally, since $\log PL_n(X(n); \theta) = N_n F_n(\theta) + \log PL_n(X(n); \theta_0)$, these same statements apply to $\log PL_n(X(n); \theta)$.

We now proceed to prove Lemmas 1–5.

PROOF OF LEMMA 1. The first two lemmas are based on the following version of the "strong law of large numbers":

PROPOSITION. *For each* $n = 1, 2, \ldots$, *let* $Z_1(n), Z_2(n), \ldots, Z_{m_n}(n)$ *be random variables and* $Y(n)$ *be a random vector. Assume*

(1) $\liminf_{n \to \infty} m_n/n > 0$.

(2) $Z_1(n), \ldots, Z_{m_n}(n)$ *are conditionally independent, given* $Y(n)$.

(3) $|Z_i(n)| \le B < \infty \; \forall i, n$.

Then

$$\left| \frac{1}{m_n} \sum_{i=1}^{m_n} (Z_i(n) - E[Z_i(n)|Y(n)]) \right| \to 0 \quad a.s.$$

PROOF. The methods here are standard. We will provide an outline only. Fix $\varepsilon > 0$ and let A_n be the event

$$A_n = \left\{ \left| \frac{1}{m_n} \sum_{i=1}^{m_n} (Z_i(n) - E[Z_i(n)|Y(n)]) \right| > \varepsilon \right\}.$$

Then the usual exponential bounds (but derived by first conditioning on $Y(n)$) give $P(A_n) = o(1/C^{m_n})$ for some $C > 1$. The rest follows from the Borel-Cantelli lemma: $P(A_n$ infinitely often $) = 0$.

Now back to the proof of Lemma 1: For any $s \in S$, let

$$B_s = \partial\{(s \cup G_s)^c\} = \{r \colon \exists t \in (s \cup G_s) r \in G_t, \; r \notin (s \cup G_s)\},$$

i.e., the neighborhood of $s \cup G_s$. For each n, choose $s_1, s_2, \ldots, s_{m_n} \in S_n$ such that

(1) $\liminf_{n \to \infty} m_n/N_n > 0$,
(2) $B_{s_i} \subseteq S_n$, $i = 1, \ldots, m_n$,
(3) $i \neq j \to (s_i \cup G_{s_i}) \cap B_{s_j} = \varnothing$,

(e.g., regularly partition S_n into large cubes, with sizes independent of n, and big enough to accommodate $s \cup G_s \cup B_s$, for some s).

Fix β and let $Y(n) = \{X_s(n): s \in \bigcup_{i=1}^{m_n} B_{s_i}\}$, and $Z_i(n) = 1_{s_i X(n)=\beta}$. By the Markov property, $Z_1(n), \ldots, Z_{m_n}(n)$ are conditionally independent, given $Y(n)$. Hence, by the proposition,

$$\left| \frac{1}{m_n} \sum_{i=1}^{m_n} (Z_i(n) - E[Z_i(n)|Y(n)]) \right| \to 0 \quad \text{a.s.}$$

Using again the Markov property,

$$E[Z_i(n)|Y(n)] = \Pi(_{s_i}X(n) = \beta | X_s(n), s \in B_{s_i}; \theta_0) \ *$$

which can have only a finite number of possible values (corresponding to the $|\Omega|^{|B_{s_i}|}$ configurations of $\{X_s(n)\}_{s \in B_{s_i}}$), all of which are positive. Hence, for some $\varepsilon > 0$,

$$\frac{1}{m_n} \sum_{i=1}^{m_n} E[Z_i(n)|Y(n)] > \varepsilon, \quad \forall n,$$

and

$$\liminf \frac{1}{m_n} \sum_{1}^{m_n} Z_i(n) \geq \varepsilon \quad \text{a.s.}$$

Since $N_n(\beta) \geq \sum_{1}^{m_n} Z_i(n)$, it also follows that $\liminf N_n(\beta)/m_n \geq \varepsilon$ a.s. Finally, since $\liminf m_n/N_n > 0$, $\liminf N_n(\beta)/N_n \geq \liminf N_n(\beta)/m_n \cdot \liminf m_n/N_n > 0$.

PROOF OF LEMMA 2. Let $C = \{c_i: i = 1, \ldots, n_c\}$ be a coloring of (S, G). In other words, $c_1, c_2, \ldots, c_{n_c}$ partition S, and $r, s \in c_i \to r \notin G_s$. Because (S, G) is regular, we can assume that C is chosen so that $\liminf |S_n^0 \cap c_i|/N_n > 0$, $i = 1, \ldots, n_c$.

For each $i \in \{1, \ldots, n_c\}$ define

$$N_n(\beta; c_i) = \#\{s \in S_n^0 \cap c_i: {}_sX(n) = \beta\},$$
$$N_n(\alpha, \beta; c_i) = \#\{s \in S_n^0 \cap c_i: X_s(n) = \alpha, {}_sX(n) = \beta\}.$$

Fix $i \in \{1, \ldots, n_c\}$, α and β, and let

$$Z_s(n) = 1_{X_s(n)=\alpha; {}_sX(n)=\beta} \quad \text{for each } s \in S_n^0 \cap c_i.$$

Let $B_n = \partial\{(S_n^0 \cap c_i)^c\}$ (the neighborhood of $S_n^0 \cap c_i$) and let $Y(n) = \{X_s(n): s \in B_n\}$. Given $Y(n)$, the random variables $Z_s(n), s \in S_n^0 \cap c_i$, are independent

---

*It is well known that the local characteristics (4.2) determine these conditional probabilities as well. Hence, this conditional distribution is independent of n.

(Markov property). By the proposition

$$\left| \frac{1}{|S_n^0 \cap c_i|} \sum_{s \in S_n^0 \cap c_i} (Z_s(n) - E[Z_s(n)|Y(n)]) \right| \to 0 \quad \text{a.s.}$$

Using again the Markov property: $E[Z_s(n)|Y(n)] = \Pi(\alpha|\beta; \theta_0) 1_{sX(n)=\beta}$. Since

$$\sum_{s \in S_n^0 \cap c_i} Z_s(n) = N_n(\alpha, \beta; c_i) \quad \text{and} \quad \sum_{s \in S_n^0 \cap c_i} 1_{sX(n)=\beta} = N_n(\beta; c_i),$$

$$\frac{1}{|S_n^0 \cap c_i|} |N_n(\alpha, \beta; c_i) - \Pi(\alpha|\beta; \theta_0) \cdot N_n(\beta; c_i)| \to 0 \quad \text{a.s.}$$

Finally, recalling that $\liminf N_n(\beta)/N_n > 0$, a.s.:

$$\left| \frac{N_n(\alpha, \beta)}{N_n(\beta)} - \Pi(\alpha|\beta; \theta_0) \right|$$

$$= \left| \sum_{i=1}^{n_c} \frac{N_n(\alpha, \beta; c_i)}{N_n(\beta)} - \Pi(\alpha|\beta; \theta_0) \sum_{i=1}^{n_c} \frac{N_n(\beta; c_i)}{N_n(\beta)} \right|$$

$$\leq \sum_{i=1}^{n_c} \frac{1}{N_n(\beta)} |N_n(\alpha, \beta; c_i) - \Pi(\alpha|\beta; \theta_0) N_n(\beta; c_i)|$$

$$= \sum_{i=1}^{n_c} \frac{N_n}{N_n(\beta)} \frac{|S_n^0 \cap c_i|}{N_n} \frac{1}{|S_n^0 \cap c_i|} |N_n(\alpha, \beta; c_i) - \Pi(\alpha|\beta; \theta_0) N_n(\beta; c_i)|$$

$$\to 0 \quad \text{a.s.}$$

PROOF OF LEMMA 3. Let $H(F_n(\theta))$ be the Hessian (matrix) of $F_n(\theta)$, and let $\phi \in R^m$. By routine calculation, we derive

$$\phi^t H(F_n(\theta))\phi$$

$$= -\sum_\beta \frac{N_n(\beta)}{N_n} \frac{\sum_{\tilde{\alpha} \in \Omega} (\phi \cdot (\psi(\tilde{\alpha}, \beta) - E_\theta[\psi(\alpha, \beta)|\beta]))^2 \exp\{\theta \cdot \psi(\tilde{\alpha}, \beta)\}}{\sum_{\tilde{\alpha} \in \Omega} \exp\{\theta \cdot \psi(\tilde{\alpha}, \beta)\}}$$

where $E_\theta[\cdot|\beta]$ is expectation on Ω with respect to $\Pi(\cdot|\beta; \theta)$. Obviously,

$$\phi^t H(F_n(\theta))\phi \leq 0, \quad \forall \phi,$$

and hence $F_n(\theta)$ is concave. By Lemma 1, with probability one, $\inf_\beta N_n(\beta)/N_n > 0$ for all n sufficiently large. Suppose $\inf_\beta N_n(\beta)/N_n > 0$ and $\phi^t H(F_n(\theta))\phi = 0$ for some θ and $\phi \neq 0$. Then, for all $\tilde{\alpha}$ and $\beta, \phi \cdot \psi(\tilde{\alpha}, \beta) = E_\theta[\psi(\alpha, \beta)|\beta]$. In particular, for every β, $\phi \cdot \psi(\alpha, \beta)$ is independent of α. This implies that $\Pi(\alpha|\beta; \theta + \phi) = \Pi(\alpha|\beta; \theta_0)$ for all α and β, which contradicts the identifiability assumption. Hence $F_n(\theta)$ is strictly concave whenever $\inf_\beta N_n(\beta)/N_n > 0$.

PROOF OF LEMMA 4. By the same argument used for Lemma 3, $G_n(\theta)$ is strictly concave, whenever $\inf_\beta N_n(\beta)/N_n > 0$. By Lemma 1, with probability one, there is a $\varsigma > 0$ such that $\inf_\beta N_n(\beta)/N_n \geq \varsigma$ for all n sufficiently large. Since $\phi^t H(G_n(\theta))\phi$ is jointly continuous in ϕ, θ, and the finite collection of variables $N_n(\beta)/N_n$, it must achieve its maximum on the compact set $|\phi| = 1$, $|\theta - \theta_0| \leq \varepsilon$, and $N_n(\beta)/N_n \in [\varsigma, 1]$ for all β. Part (a) of Lemma 4 now follows from the strict concavity of $G_n(\theta)$.

For part (b), apply Jensen's inequality:

$$\sum_\alpha \Pi(\alpha|\beta; \theta_0) \log \frac{\Pi(\alpha|\beta; \theta)}{\Pi(\alpha|\beta; \theta_0)} \le \log \sum_\alpha \Pi(\alpha|\beta; \theta_0) \frac{\Pi(\alpha|\beta; \theta)}{\Pi(\alpha|\beta; \theta_0)}$$

$$= \log \sum_\alpha \Pi(\alpha|\beta; \theta) = \log 1 = 0.$$

Part (c) follows immediately from the expression for $G_n(\theta)$.

PROOF OF LEMMA 5.

$$\limsup_{n \to \infty} \sup_{|\theta - \theta_0| \le \varepsilon} |F_n(\theta) - G_n(\theta)|$$

$$= \limsup_{n \to \infty} \sup_{|\theta - \theta_0| \le \varepsilon} \left| \sum_\beta \frac{N_n(\beta)}{N_n} \sum_\alpha \left(\frac{N_n(\alpha, \beta)}{N_n(\beta)} - \Pi(\alpha|\beta; \theta_0) \right) \log \frac{\Pi(\alpha|\beta; \theta)}{\Pi(\alpha|\beta; \theta_0)} \right|$$

$$\le |\Omega| \sup_{\alpha, \beta, |\theta - \theta_0| < \varepsilon} \left| \log \frac{\Pi(\alpha|\beta; \theta)}{\Pi(\alpha|\beta; \theta_0)} \right| \limsup_{n \to \infty} \sup_{\alpha, \beta} \left| \frac{N_n(\alpha, \beta)}{N_n(\beta)} - \Pi(\alpha|\beta; \theta_0) \right|.$$

By Lemma 2,

$$\limsup_{n \to \infty} \sup_{\alpha, \beta} \left| \frac{N_n(\alpha, \beta)}{N_n(\beta)} - \Pi(\alpha|\beta; \theta_0) \right| = 0 \quad \text{a.s.}$$

Since $\Pi(\alpha|\beta; \theta) \ne 0$ for any $\alpha, \beta, \theta \in R^m$, and is continuous in θ for each of the finite numbers of $\alpha \in \Omega$, $\beta \in \Omega^{|G_s|}$,

$$\sup_{\alpha, \beta, |\theta - \theta_0| < \varepsilon} \left| \log \frac{\Pi(\alpha|\beta; \theta)}{\Pi(\alpha|\beta; \theta_0)} \right|$$

is finite.

ACKNOWLEDGEMENTS. This work has benefited from a host of suggestions by D. Geman and B. Gidas. D. Geman has experimented extensively with MRF texture models, and with pseudolikelihood parameter estimation, and has shared his conclusions. B. Gidas has shared his insight and knowledge of Markov random fields, especially concerning the vagaries of infinite volume Gibbs states.

REFERENCES

1. E. Aarts and P. van Laarhoven, *Simulated annealing: a pedestrian review of the theory and some applications*, NATO Advanced Study Institute on Pattern Recognition: Theory and Applications, Spa, Belgium, June 1986.

2. J. Besag, *Spatial interaction and the statistical analysis of lattice systems (with discussion)*, J. Roy. Statist. Soc. Ser. B **36** (1974), 192–236.

3. ____, *Statistical analysis of non-lattice data*, The Statistician **24** (1975), 179–195.

4. ____, *On the statistical analysis of dirty pictures (with discussion)*, J. Roy. Statist. Soc. Ser. B **48** (1986).

5. V. Cerný, *A thermodynamical approach to the travelling salesman problem: an efficient simulation algorithm*, Preprint, Inst. Phys. and Biophys., Comenius Univ., Bratislava, 1982.

6. B. Chalmond, *Image restoration using an estimated Markov model*, Preprint, Mathematics Dept., University of Paris-Sud, Orsay, 1986.

7. F. S. Cohen and D. B. Cooper, *Simple parallel hierarchical and relaxation algorithms for segmenting noncausal Markovian random fields*, IEEE Trans. Pattern Anal. Machine Intell., PAMI-9 (1987), pp. 195–219.

8. G. R. Cross and A. K. Jain, *Markov random field texture models*, IEEE Trans. Pattern Anal. Machine Intell., PAMI-5 (1983), pp. 25–40.

9. H. Derin and H. Elliott, *Modelling and segmentation of noisy and textured images using Gibbs random fields*, IEEE Trans. Pattern Anal. Machine Intell., PAMI-9 (1987), pp. 39–55.

10. P. A. Devijver, *Hidden Markov models for speech and images*, Nato Advanced Study Institute on Pattern Recognition: Theory and Applications, Spa, Belgium, June 1986.

11. H. Elliott and H. Derin, *Modeling and segmentation of noisy and textured images using Gibbs random fields*, Tech. Report ECE-UMASS-SE84-15, Dept. Elec. Comput. Eng., Univ. of Massachusetts, Amherst, Mass.

12. D. Geman, S. Geman, and C. Graffigne, *Locating texture and object boundaries*, Pattern Recognition Theory and Application (P. Devijver, ed.), NATA ASI Series, Springer-Verlag, Heidelberg, 1986.

13. S. Geman and D. Geman, *Stochastic relaxation, Gibbs distributions, and the Bayesian restoration of images*, IEEE Trans. Pattern Anal. Machine Intell. **6** (1984), 721–741.

14. S. Geman and D. E. McClure, *Bayesian image analysis: an application to single photon emission tomography*, 1985, Statistical Computing Section, Proceedings of the American Statistical Association, 1985, pp. 12–18.

15. B. Gidas, *Non-stationary Markov chains and convergence of the annealing algorithm*, J. Statist. Phys. **39** (1985), 73–131.

16. _____, *A renormalization group approach to image processing problems*, Preprint, Division of Applied Mathematics, Brown University, 1986.

17. U. Grenander, *Lectures in pattern theory*, vols. I,II,III, Springer-Verlag, New York, 1976.

18. _____, *Tutorial in pattern theory*, Division of Applied Mathematics, Brown University, 1983.

19. B. Hajek, *Cooling schedules for optimal annealing*, Math. Oper. Res. (to appear).

20. G. E. Hinton and T. J. Sejnowski, *Optimal perceptual inference*, Proc. IEEE Conf. Comput. Vision Pattern Recognition, 1983.

21. S. Kirkpatrick, C. D. Gellatt, and M. P. Vecchi, *Optimization by simulated annealing*, Science **220** (1983), 671–680.

22. A. Lippman, *A maximum entropy method for expert system construction*, Ph.D. Thesis, Division of Applied Mathematics, Brown University, 1986.

23. J. Marroquin, S. Mitter, and T. Poggio, *Probabilistic solution of ill-posed problems in computational vision*, Artif. Intell. Lab. Tech. Report, M.I.T., 1985.

24. N. Metropolis, A. W. Rosenbluth, M. N. Rosenbluth, A. H. Teller, and E. Teller, *Equations of state calculations by fast computing machines*, J. Chem. Phys. **21** (1953), 1087–1091.

25. D. W. Murray, A. Kashko, and H. Buxton, *A parallel approach to the picture restoration algorithm of Geman and Geman on an SIMD machine*, Preprint, 1986.

26. A. Possolo, *Estimation of binary Markov random fields*, Preprint, Department of Statistics, Univ. of Washington, Seattle, 1986.

27. B. D. Ripley, *Statistics, images, and pattern recognition*, Canad. J. Statist. **14** (1986), 83–111.

28. D. Ruelle, *Thermodynamic formalism*, Addison-Wesley, Reading, Mass., 1978.

BROWN UNIVERSITY, PROVIDENCE, RHODE ISLAND 02912, USA

BROWN UNIVERSITY, PROVIDENCE, RHODE ISLAND 02912, USA

Proceedings of the International Congress of Mathematicians
Berkeley, California, USA, 1986

Equilibrium Analysis of Large Economies

WERNER HILDENBRAND

1. Introduction. One of the most fundamental problems in economics is to explain why in a competitive economy with private ownership where a large number of economic agents make decisions that are taken independently from each other and are motivated by self-interest, why in such a situation we do not observe chaos, but a state which definitely looks more like an equilibrium than total disorder.

This fundamental observation was formulated as early as 1776 by Adam Smith:

> The natural price, therefore, is as it were, the central price, to which the prices of all commodities are continually gravitating. Different accidents may sometimes keep them suspended a good deal above it, and sometimes force them down even somewhat below it. But whatever may be the obstacles which hinder them from settling in this centre of repose and continuance, they are constantly tending towards it. (*The Wealth of Nations*, Book I, Chapter VII)

This vision of Adam Smith was made more precise by nineteenth-century economists. I would like to quote Léon Walras (1874), the founder of general equilibrium theory, and Alfred Marshall (1890):

> Such an equilibrium is exactly similar to that of a suspended body of which the centre of gravity lies directly beneath the point of suspension, so that if this centre of gravity were displaced from the vertical line beneath the point of suspension, it would automatically return to its original position through the force of gravitation. This equilibrium is, therefore, stable. (*Elements of Pure Economics*, Lesson 7, §66)

> When demand and supply are in stable equilibrium, if any ac-
> cident should move the scale of production from its equilibrium
> position, there will be instantly brought into play forces tend-
> ing to push it back to that position; just as, if a stone hanging
> by a string is displaced from its equilibrium position, the force
> of gravity will at once tend to bring it back to its equilibrium
> position. (*Principles of Economics*, Book V, Chapter III)

The classical economists of the nineteenth century viewed the determination of prices in a market economy in analogy to classical mechanics as a kind of "mechanics of market prices";[1] the price system—like a body—"moves" in a force-field that is determined by excess demand, that is to say, the difference between demand and supply. The force-field was thought to be such that there is always a tendency towards equilibrium. Therefore economists concentrate their attention on situations where demand equals supply, that is to say, on equilibria.

Does economic theory today offer a sound explanation of this "stable equilibrium paradigm?" And more specifically, is mathematics useful in such an explanation? These questions I would like to discuss in this lecture.

Clearly, the first and crucial step is to define the basic concepts of demand and supply. "... it is actually the first step, on which everything else depends, which is the most dubious." (Hicks, *Value and Capital*, 1946, p. 11)

From a purely formal point of view our problem would be quite simple if one could express demand and supply, and hence excess demand, as a function of prices. Indeed, if there were given an *excess demand function* $E: D \to \mathbf{R}^l$, $D \subset \mathbf{R}^l_+$, then either one defines equilibrium as a solution of the system of equations $E(p) = 0$ or one studies the asymptotic behavior of a price-adjustment process, for example $\Delta p \sim E(p)$ or in differential form, $\dot{p} = E(p)$. For both cases there are well-developed mathematical theories.

Of course, demand and supply depend on other determinants than prices alone, yet in traditional microeconomics these other determinants are either assumed to be fixed or they are considered, in turn, as functions of prices. For example, individual preferences (tastes) are typically assumed to be fixed while individual wealth or income is defined as a function of prices.

If economic equilibrium theory were built on an ad hoc chosen excess demand function $E: D \to \mathbf{R}^l$, then the often-raised question, why such an analysis qualifies as economics, would, indeed, be well taken.

2. Excess demand of an economy with private ownership. From the introduction it should be clear that a careful definition of excess demand is required. However, here is not the place to give the necessary details. First one should clarify the type of commodities which are demanded or supplied; are these perishable commodities like food or labor, or durables like cars or houses?

---

[1]The last publication of Walras in 1909 had the title "Economie et Mécanique."

Also one has to specify whether one speaks of current (temporary) demand and supply or of plans which extend into the future. For details see [2, 7, 1].

In general equilibrium analysis one considers a finite number l of different commodities (which might be dated or conditional on certain events). All commodities are considered as infinitely divisible. A consumption—or production—plan can then be defined as a vector in $\mathbf{R}^l$.

The economy consists of a finite set A of consumers and of n production units. Every consumer $a \in A$ is described by a *consumption set* $X_a \subset \mathbf{R}^l$, a *preference relation* $\precsim_a$, and a *wealth-function* $b_a(\cdot)$: the wealth b_a depends on the price system p and the status of private ownership.

The decision of a consumer, called demand, and the decision of a production unit, called supply, are defined as a result of a maximization problem. Given the price system $p \in \mathbf{R}^l$ one defines the individual demand $\varphi^a(p)$ of consumer $a \in A$ by:

$\hat{x} \in \varphi^a(p)$ if and only if $\hat{x}$ belongs to the budget set, i.e.,

$\hat{x} \in \{x \in X_a | p \cdot x \le b_a(p)\}$ and $\hat{x}$ maximizes preferences, i.e.,

if $x \succ_a \hat{x}$ then $p \cdot x > b_a(p)$.

A production unit is described by its technology $Y_j \subset \mathbf{R}^l$; the technology describes the set of possible input-output combinations. Given a price system $p \in \mathbf{R}^l$ one defines the supply $\eta^j(p)$ of the production unit j by:

$\hat{y} \in \eta^j(p)$ if and only if $p \cdot \hat{y} \ge p \cdot y, \ y \in Y_j$.

An economy $\mathcal{E}$ with private ownership is defined by the commodity space $\mathbf{R}^l$, the characteristics of the consumers, $X_a, \precsim_a, b_a(\cdot)$, the characteristics of the production units, Y_j, and the vector of total initial endowments $\bar{e}$;

$$\mathcal{E} = \{\mathbf{R}^l, X_a, \precsim_a, b_a(\cdot), Y_j, \bar{e}\}.$$

The *excess demand* $E_{\mathcal{E}}$ of the economy $\mathcal{E}$ is defined by

$$E_{\mathcal{E}}(p) = \sum_{a \in A} \varphi^a(p) - \sum_{j=1}^{n} \eta^j(p) - \bar{e}.$$

In general the excess demand $E_{\mathcal{E}}(\cdot)$ is not a function but a correspondence (set-valued). With strict convexity (resp. smoothness) assumptions on preferences $\precsim_a$ and technology Y_j, the excess demand can be shown to be a continuous (resp. differentiable) function on a suitably defined domain of price-systems. For details see [2, 15].

The problem of existence of a competitive equilibrium for an economy $\mathcal{E}$, i.e., a solution of $0 \in E_{\mathcal{E}}(p)$, is well settled. The standard reference is [2]; see also [5] and the literature given there.

3. Uniqueness and stability. Let $E_{\mathcal{E}} : \mathbf{R}^l_{++} \to \mathbf{R}^l$ denote the *excess demand function* of an economy $\mathcal{E}$ with private ownership. It follows that the function $E_{\mathcal{E}}$ is *homogeneous*, i.e., $E_{\mathcal{E}}(\lambda p) = E_{\mathcal{E}}(p)$, $\lambda > 0$, and satisfies the

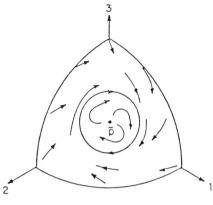

FIGURE 1

Walras identity $p \cdot E_{\mathcal{E}}(p) = 0$. Consequently an excess demand function $E_{\mathcal{E}}(\cdot)$ defines a vector field on

$$S = \left\{ x \in \mathbf{R}_{++}^l \mid \sum x_i^2 = 1 \right\}.$$

Which type of vector field on S is generated by economies? For example, are they gradient fields? Scarf (1960) has shown that the phase portrait in Figure 1 comes from a "nice" exchange economy.

The following result shows that there is nothing special about vector fields on S which are defined by excess demand functions $E_{\mathcal{E}}$ of exchange economies $\mathcal{E}$.

THEOREM. *For every continuous vector field f on S and every compact subset $K \subset S$ there exists an exchange economy*

$$\mathcal{E} = \{\precsim_a, e_a\}_{a \in A}$$

such that $E_{\mathcal{E}}(p) = f(p)$ on K. Furthermore the economy $\mathcal{E}$ has no pathological features.

That is, the individual characteristics $(\precsim_a, e_a)$ satisfy the assumptions that are traditionally made in microeconomics (one can even assume that all preferences $\precsim$ are identical and that endowments e_a are collinear). For details see, e.g., the survey paper by Schafer and Sonnenschein [16] and the references given there, in particular, [17, 4].

If one believes (not all economists share this belief!) that a good mathematical model of an economy should have at least one stable equilibrium or even stronger, should have a unique and stable equilibrium, then the above theorem shows that the model $\mathcal{E}$ considered up to now is quite unsatisfactory or, at least, is incomplete. On the other hand, one knows very well the *mathematical conditions* on the excess demand function $E_{\mathcal{E}}$ that lead to a unique or stable solution of $E(p) = 0$. But which economically meaningful and empirically supported hypothesis leads to these mathematical conditions? It is not the lack of mathematical theory but the unsatisfactory or incomplete economic modelling that creates the problem.

Here are two relevant mathematical results.

(I) An economy $\mathcal{E}$ is called *regular* if the excess demand function $\mathcal{E}$ is continuously differentiable and if the rank of the Jacobian matrix $\partial E_{\mathcal{E}}(p)$ is equal to $l-1$ for every equilibrium price $p \in \Pi_{\mathcal{E}}$.

INDEX THEOREM. *For a regular economy $\mathcal{E}$ the set $\Pi_{\mathcal{E}}$ of equilibrium prices is nonempty, and finite, furthermore*

$$\sum_{p \in \Pi_{\mathcal{E}}} \text{index}(p) = 1.$$

The index of p is $+1$ (resp. -1) if

$$\text{Det} \begin{vmatrix} -\partial E(p) & p \\ -p^T & 0 \end{vmatrix} \quad \text{is positive (resp. negative).}$$

The concept of a regular exchange economy is due to Debreu [3], who proved the finiteness of $\Pi_{\mathcal{E}}$. The index result is due to E. Dierker [6]. Extensions to economies with production have been given by Mas-Colell [14] and Kehoe [11]. For definitions and a detailed treatment of the index theorem see [15].

(II) WALRAS-STABILITY THEOREM. *The equilibrium $p^* \in \Pi_{\mathcal{E}}$ of an economy $\mathcal{E}$ with a continuously differentiable excess demand function is Walras stable (i.e., nearby solutions of $\dot{p} = E(p)$ approach p^* exponentially fast) if all eigenvalues of $\partial E(p^*)\colon H_{p^*} \to H_{p^*}$, where $H_{p^*} := \{x \in \mathbf{R}^l\colon p^* \cdot x = 0\}$ have negative real parts. (See, e.g., [10, Chapter 9].)*

As a consequence of these two theorems one can easily show that a *regular economy $\mathcal{E}$ has a unique and stable equilibrium if for every $p \in \Pi_{\mathcal{E}}$ the Jacobian matrix $\partial E(p)$ is negative definite on a hyperplane*, i.e., there is a vector $z \in \mathbf{R}_+^l$ such that the quadratic form $v \cdot \partial E(p)v < 0$ for every $v \neq 0$ and $v \cdot z = 0$.

For example, this condition is satisfied if the excess demand function $E(p)$ satisfies the assumption of "gross-substitution" (i.e., all off-diagonal elements of $\partial E(p)$ are positive). Yet we shall see later that in an economy with production "gross-substitution" is not a satisfactory assumption.

There are of course alternative assumptions on the excess demand function that would allow one of the above theorems to be applied. For example, if the Jacobian matrix (after deleting the jth row and column) has a negative dominant diagonal then all eigenvalues have negative real parts.

Thus we know the *mathematical conditions* for an excess demand function that imply that the economy has a unique and stable equilibrium, but we know also that these conditions do not follow from general assumptions on the individual primitive concepts which define the economy. Even if we were willing to make strong ad hoc assumptions on individual preferences (e.g., Cobb-Douglas utilities) in an economy with production we still cannot derive the desired properties of $\partial E(p)$ [12].

In the remainder of this paper we want to discuss the circumstances under which one might expect that the Jacobian matrix $\partial E(p)$ is negative definite on a hyperplane (or has this property at least "approximately").

Excess demand is defined as the difference between demand and supply. If supply $\eta(p)$ is derived by profit maximization, then one can easily verify the following inequality:

$$(p - q) \cdot (\eta(p) - \eta(q)) \geq 0 \quad \text{for every } p, q \in \mathbf{R}^l.$$

Thus if we are in a situation where supply is given by a differentiable function, then it follows that the Jacobian matrix $\partial \eta(p)$ is positive semidefinite. One can easily show that there is, in general, no uniform sign pattern for the off-diagonal elements. Thus if, in general, the Jacobian matrix $\partial E(p)$ is not negative definite on a hyperplane, then all difficulties must come from the demand sector of the economy.

Let us define the demand sector of an economy by a mean demand function F,

$$F(p) = \int f(p, b) \rho(p, b) \, db,$$

where $f(p, b)$ denotes the mean demand at the price vector p of all households with income b; $f(p, \cdot)$ is called an *Engel function*, and $\rho(p, \cdot)$ denotes the density of the income distribution (thus, we assume that we have a "large" consumption sector). Then we obtain

$$\partial E(p) = S(p) - A(p) + D(p),$$

where

$$S - A = \left(\int \rho(p, \cdot) \partial_{p_j} f_i(p, \cdot) \right)_{i,j},$$

$$D = \left(\int f_i(p, \cdot) \partial_{p_j} \rho(p, \cdot) \right)_{i,j},$$

$$A = \left(\int f_j(p, b) \partial_b f_i(p, b) \rho(p, b) \, db \right)_{i,j}.$$

We shall now discuss the properties of the matrices A and D.[2] The matrix S, the Slutzky substitution matrix of the Engel function, is known to be negative semidefinite if (and only if) the Engel function satisfies the weak axiom of revealed preference [13].

The matrix D. If the Engel functions $f_h(p, \cdot)$ are increasing, then one can show that the matrix D is positive (thus has a dominant positive eigenvalue). Otherwise it can have any structure; this depends on the distribution of private ownership in the economy. Thus the best we can hope is that the matrix D has a very low rank. Consequently, we have to make an assumption that specifies how prices influence the income distribution. This influence must be small. For example, let $B(p) = \int b \rho(p, b) \, db$ denote the mean income at price p. If the influence of the prices comes only through the mean income, i.e., $\rho(p, \cdot) = \mu(B(p), \cdot)$ for some density μ, then we would obtain that the rank of D is equal

---

[2]For more details see [9].

to 1. In general, the matrix D can be of full rank or, as in the above example, of rank 1. Whether prices influence the income distribution very strongly or not is an empirical question. Figures 2 and 3 show the evolution (from 1969 to 1981) of the normalized income distributions for the United Kingdom. Figure 2 shows a lognormal and Figure 3 a nonparametric (kernel) estimation. The densities are surprisingly stable over time.

The matrix A. The following lemma shows that the form of the income distribution $\rho(p, \cdot)$ plays an important role.

LEMMA (HILDENBRAND [8]). *If the density $\rho(p, \cdot)$ is decreasing on $(0, \beta)$ and if $f(p, 0) = 0$, then the matrix A is positive (semi)definite.*

This mathematical result is empirically not very relevant. We have seen in Figures 2 and 3 that income densities are not decreasing; in this case the form of the Engel functions come into play. These can be estimated from household income-expenditure data (see Figures 4 and 5). Thus the matrix A can be estimated and one can compute the eigenvalues of the symmetrized matrix. Calculations have shown that the matrix A has typically a few positive eigenvalues and all other eigenvalues are very small (compared with the largest positive one), but some are negative.

If one assumes that all Engel curves are of a certain functional form, then the question of positive definiteness of A reduces to a condition on the income density alone. For example, if all Engel functions are polynomial of degree n,

$$f_h(p, b) = \alpha_1 b + \cdots + \alpha_n b^n,$$

then the matrix A is positive definite if the matrix $((i + j)m_{i+j-1})_{i,j}$ is positive definite, where $m_k = \int b^k \rho(p, b)\, db$ denotes the kth moment of ρ. Since the moments of the income distribution can easily be estimated, one can check the positive definiteness of the matrix A provided Engel curves can be well approximated by polynomials of relatively low degree. For example, for $n \leq 5$, the matrix of moments of the densities in Figures 2 and 3 turned out to be positive definite. If one assumes in addition that the income distribution is of a lognormal type, then the matrix $((i + j)m_{i+j-1})_{i,j}$ is positive definite provided the variance of the normalized income distribution is large enough. For example if $n = 4$ then the variance of the normalized income distribution must be larger than 0.35; the estimated variance during the years 1969 to 1981 is always larger than 0.4.

Thus it turns out that the two matrices A and D, which, in principle, can be quite arbitrary, tend to be positive definite or of rank 1 provided one is willing to accept certain assumptions on the income distribution; after normalization the income distribution should not be too sensitive to price changes and it must have sufficient variance. Both requirements seem to be well supported by empirical evidence.

NETINCOME DENSITY IN SPACE VIEW
LOGNORMAL FIT. YEAR - 69 TO 81 BY 2

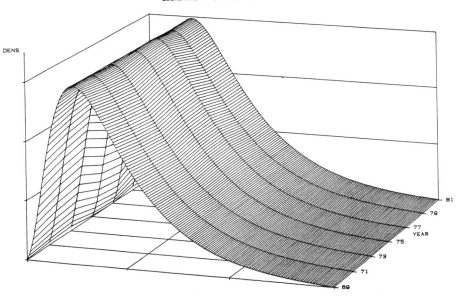

FIGURE 2

NETINCOME DENSITY IN SPACE VIEW
BANDWIDTH - 0. 2. YEAR - 1969 TO 1981 BY 2

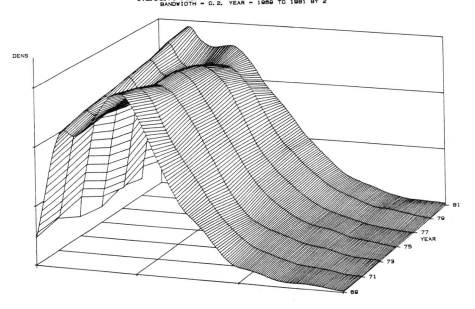

FIGURE 3

FOOD ENGEL CURVE
KERNEL ESTIMATOR+UNIFORM CONFIDENCE BAND

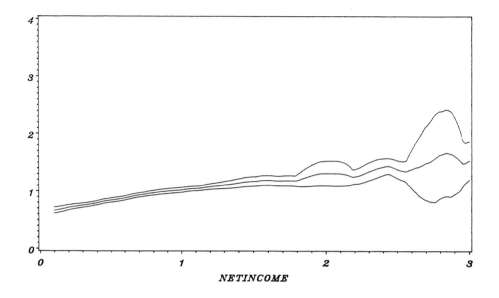

FIGURE 4

ALCOHOL+TOBACCO ENGEL CURVE
KERNEL ESTIMATOR+UNIFORM CONFIDENCE BAND

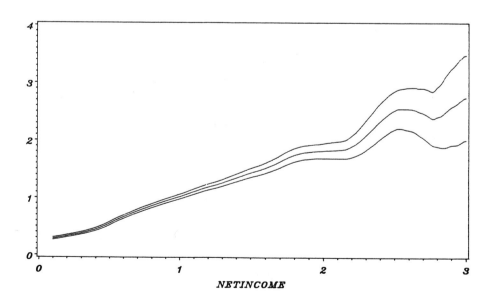

FIGURE 5

REFERENCES

1. A. Deaton and J. Muellbauer, *Economics and consumer behaviour*, Cambridge Univ. Press, 1980.

2. G. Debreu, *Theory of value*, Wiley, New York, 1959.

3. ____, *Economies with a finite set of equilibria*, Econometrica **38** (1970), 387–392.

4. ____, *Excess demand functions*, J. Math. Econom. **1** (1974), 15–21.

5. ____, *Existence of competitive equilibrium*, Chapter 15 in *Handbook of Mathematical Economics*, vol. II, edited by K. Arrow and M. Intriligator, North-Holland, 1982, pp. 71–95.

6. E. Dierker, *Two remarks on the number of equilibria of an economy*, Econometrica **40** (1972), 951–953.

7. J. M. Grandmont, *Temporary general equilibrium theory*, Econometrica **45** (1977), 535–572.

8. W. Hildenbrand, *On the "law of demand"*, Econometrica **51** (1983), 997–1020.

9. K. Hildenbrand and W. Hildenbrand, *On the mean income effect*, Chapter 4 in *Contributions to Mathematical Economics*, edited by W. Hildenbrand and A. Mas-Colell, North-Holland, 1986, pp. 247–268.

10. M. Hirsch and S. Smale, *Differential equations, dynamical systems, and linear algebra*, Academic Press, 1974.

11. T. Kehoe, *An index theorem for general equilibrium models with production*, Econometrica **48** (1980), 1211–1232.

12. ____, *Multiplicity of equilibria and comparative statics*, Quart. J. Econom. **100** (1985), 119–148.

13. R. Kihlstrom, A. Mas-Colell, and H. Sonnenschein, *The demand theory of the weak axiom of revealed preference*, Econometrica **44** (1976), 971–978.

14. A. Mas-Colell, *Regular, nonconvex economies*, Econometrica **45** (1977), 1387–1407.

15. ____, *The theory of general economic equilibrium*, Cambridge Univ. Press, 1985.

16. W. Shafer and H. Sonnenschein, *Market demand and excess demand functions*, Chapter 14 in *Handbook of Mathematical Economics*, vol. II, edited by K. Arrow and M. Intriligator, North-Holland, 1982, pp. 71–95.

17. H. Sonnenschein, *Do Walras' identity and continuity characterize the class of community excess demand functions?*, J. Econom. Theory **6** (1973), 345–354.

WIRTSCHAFTSTHEORIE II, ADENAUERALLEE 24–26, D 5300 BONN 1, FEDERAL REPUBLIC OF GERMANY

Repeated Games

JEAN-FRANÇOIS MERTENS

A. Origins and motivation of the problem

Interest by economists in repeated games was originally motivated by the theme that repetition enables cooperation (e.g., [18]). This theme is of fundamental importance in economics, far exceeding its obvious implications, e.g. to industrial organization and antitrust policy. For example, in the prisoner's dilemma

$$\begin{pmatrix} 2,2 & 0,3 \\ 3,0 & 1,1 \end{pmatrix}$$

players 1 and 2 simultaneously choose resp. a row and a column; the first (resp. second) number represents the resulting payoff to 1 (resp. 2).

The only rational outcome in noncooperative play ("Nash equilibrium point": given the other players' recommended strategies, no player has an advantage to deviate from the strategy recommended to himself) of the one-shot game is $(1,1)$. But in repeated play, they can achieve the cooperative outcome $(2,2)$ in equilibrium—for instance, by playing any mutually agreed sequence of moves leading to that payoff in the long-run average, while threatening each other to revert forever to the second strategy as soon as a deviation of the opponent is observed.

The above argument is perfectly general and constitutes the proof of the so-called "folk-theorem": every feasible and individually rational payoff of the one-shot game is achievable as an equilibrium of the repeated game (and vice versa; "individually rational" means that each player gets at least his minmax value—minimizing over the strategies of his opponents the maximum payoff he can achieve against them; the feasible vectors are those in the convex hull of the payoff vectors appearing in the normal form).

This interest had been preceded by the study by game theorists of a number of early examples—often with no explicit economic motivation, but sometimes a military paradigm. Those examples were all two-player zero-sum games (only two players, and for every pair of strategy choices the sum of their payoffs is zero; typically, in such a case one writes only the payoff of player 1, which is then paid by player 2). Such contributions include, for instance, [**28**, **29**, **44**, **45**, **48**, **85**, **108**, **110**]. Besides the development of a number of beautiful and most interesting techniques two important discoveries were made in this period. The first was that, for most realistic applications, current actions influence not only the current payoff, but also state variables of the problem—hence the definition of stochastic games: current state and actions jointly determine both a current payoff vector and a probability distribution to select the next state; players are fully informed of current state and past actions. The other was that the slightest alteration in this information pattern—even just some information lag—causes tremendous difficulties for the analysis.

Another area of concern was that some types of information may never become available to the players. The theory postulates that all players know the full description of the game—and this description includes the payoff function of all his opponents, i.e., in fact their utility function, their preferences. In a great many cases, this and other knowledge about the individual players' characteristics and strategic possibilities are not available to their opponents, who must therefore act according to some personal beliefs, prior distributions over those data. These beliefs themselves may be only imperfectly known to their opponents, and so on .... J. Harsanyi [**46**] showed how to cut through this knot of difficulties, which until then could not even be modeled, and how to describe such situations nevertheless as a proper game. In this model, each player can be of several possible types, known to him but unknown to his opponents. For each vector of types—one for each player—a payoff matrix (more generally, a game) is given. The game starts by a lottery selecting a state of nature (vector of types), after which the players play the resulting game, each one being informed only of his own type. Here the full description of the game, including the probabilities of the lottery, is common knowledge to all players. This is the model for games with incomplete information. This model had a tremendous impact, and now pervades almost all areas of economic theory, from finance to industrial organization and to labor economics. For instance, in the insurance literature the fact that the insurer is typically less informed than his client about the degree of risk involved in a specific contract is known as the adverse selection phenomenon: it has for effect that the insurer has to charge a premium based on the average degree of risk in his pool. Hence the better risks in the pool will find it worthwhile to forego insurance, so the quality of the pool will decrease and the premium increase—until in the limit only the worst risks will remain: good prospects cannot be insured. Similarly in markets where the seller has more information than the buyer on the quality of the product—say the used-car market—the above phenomenon will cause a collapse of the market, and that only the worst qualities are traded (Akerlof). Similar analysis can be made in the credit markets.

Assume prospective employees can be of several different quality types, undistinguishable by the employer. Assume even that university education was perfectly useless, but was much more painful to get through for low-quality students than for high-quality students. Then in labor market equilibria, people with a university degree would get higher wages—the duration of college would be just sufficiently long for the wage differential to make it worthwhile for high-quality students to suffer the pain of getting through college and not for low-quality students: a huge investment in this perfectly useless education is made by all high-quality youngsters, just in order to signal their quality to prospective employers (Spence). Such signalling equilibria of games with incomplete information have generated a vast literature on their own in many areas of economic theory.

When the game which is selected is itself repeated (without the initial lottery being repeated), one speaks of repeated games with incomplete information. Such models arise quite naturally, not only from the desire to extend the theme that repetition enables cooperation to the more realistic incomplete information setup, but also from the situations themselves that incomplete information models are meant to analyze. For example, insurance contracts are often long-time affairs, where the performance can be monitored year after year, and the contract adjusted accordingly (cf. also [107]); the same is often true for credit relationships and certainly for employer-employee (or more generally principal-agent: client–portfolio manager, client-lawyer, firm-subcontractor, etc.) relationships: in all those cases, the repetitive aspect is in fact essential. E.g., Milgrom and Roberts have shown that advertising by the subcontractor as a signal of his quality could be explained by the larger number of stages a profitable relationship would last for a high-quality subcontractor, thus making the investment more worthwhile for him.

Yet the original work on repeated games with incomplete information was in fact motivated by still another application field, and was done under contract with the U.S. Arms Control and Disarmament Agency [12, 13, 14, 15, 123].

More generally, repeated games with incomplete information may in some sense be considered as the proper approach to statistics for social science: when playing against an opponent instead of against nature, and trying to make inferences about his type from his actions, one cannot assume that, given the state of nature (his type), his actions are generated mechanically by some given stochastic process: the role of his preferences has to be recognized, and that he may be trying to mislead you—to bluff, or to signal something, or to outguess you, reasoning about your own rationality.... Clearly, an explicit game-theoretic treatment is required; and those are exactly the questions addressed in repeated games with incomplete information. This applies even in very down-to-earth problems like quality control.

We did not specify, when defining repeated games with incomplete information, what additional information the players receive after each stage of the game—which is necessary for a full description of the game. This can be done formally by specifying, for each state of nature and all strategy choices, in addition to the payoff vector a lottery (probability distribution) selecting a vector of signals—one to each player. More generally, the same joint lottery would select for each player both a payoff and a signal. (The lottery represents the compound effect of all moves of nature in the extensive form [57] from which the normal form is derived, and the signals arise from a new datum, the information partition on the terminal nodes of this extensive form.) Care has to be taken that only the signal, and not the payoff, is told to each player. An argument is sometimes made that, on first principles, a player knows at least his own current utility: "everybody knows how happy he feels." The utility meant in this argument certainly has to be the random payoff generated by the above lottery, i.e., the actual payoff written at the terminal node of the tree, and not the normal form payoff, which represents an average of such terminal payoffs over all possible moves of nature, given the strategy choice of all players: indeed, for all but the one move of nature that in fact materialized, the player won't have observed the actions of his opponents, and so will have no clue as to what their strategy implied in those cases. But even with this interpretation the argument does not hold water: the payoff that has to be written at a terminal node of a game tree is the Von Neumann–Morgenstern utility of that event occurring, and therefore does not represent at all "how happy the player feels" at that stage, but rather how happy he would feel if he was now further informed that exactly this terminal node has occurred. The way the player feels in fact depends also on his guesses about what other players have done, i.e., on his analysis of the game finally. To give a somewhat grim example, in the day-after-day bargaining by a father over the release and ransom of a kidnapped child, the father's utility for a given day (stage) most certainly depends on how his child is treated that day, even if he does not know it immediately—and in fact he may never know how badly the child was treated if it is later found murdered: it is not merely a matter of lag of information, but of lack of information.

The same is true in purely economic situations. Consider, for example, a principal-agent relationship (employer-employee, or firm-subcontractor), where the agent's role is to clean the premises every day. Definitely the principal's objective function—his payoff—is a function of the actual degree of cleanness of the premises that day. Yet he will not inspect them completely every day— at best, he may occasionally decide to do some quality-control sampling, which would give him statistical information about his utility, but even this is part of his decision problem, of his strategic behavior in the principal-agent game. Claiming utilities (i.e., essentially the actual outcomes) to be known is essentially negating the whole quality-control problem, which is an essential part of most principal-agent relationships—if not the single most important, judging from the size of the literature and of the budgets devoted to this, as compared with those related to the actual contracting part of the relationship.

It is however probably true that, in several situations of economic interest, the assumption that players know their own random utility is well justified—and sometimes extremely useful as shown for instance in the recent work of E. Lehrer [59].

This brings us to our last main theme: imperfect monitoring. In most early work, like in our above proof of the "folk-theorem," it is assumed that the pure strategy choices of all players—or at least their extensive form actions—become known to everybody after each stage. When this is not assumed, and instead the more general model we just described is used, one speaks of "imperfect monitoring." Typically, even the actions are unobservable: they are the effort-level of an employee, the number of man-hours and other resources devoted by a portfolio manager to your portfolio, etc. A number of important applications can be found, e.g., in the work of R. Radner and of Rubinstein and Yaari.

Let us finally observe that, in many of the above-described situations, there are in fact also state variables which are essential to the problem and may change during the course of the game according to the players' actions—i.e., there is a stochastic game aspect. For instance, in the client–portfolio manager relationship, the current value of the portfolio is the essential variable; it is clearly strongly affected by the players' past actions, and completely determines the future possibilities: it is best considered as a state variable. One might observe that, at least in this case, a minimal simplification is obtained in that regular statements make the state variable monitorable—but this leads us again to the problem of information lags, which are of crucial importance in this business. In the hostage story, life or death of the hostage is an obvious state variable. In tactical duels, like the bomber-battleship problem [28], the current position of the antagonists is one, while in the disarmament negotiations story, current military capabilities of the parties, what secret research programs they may have under way in their military labs, and what successes were booked by such programs clearly also constitute state variables. In credit markets, current financial position of the debtor is one, etc.

To integrate such effects into our previous model, it is sufficient to let the same state- and action-dependent lottery that selects signals and payoffs after every stage also select the next state of nature. If one wants in addition to incorporate the effect of information lags, the lottery should be allowed to depend also on past events (states, actions, signals, payoffs). This becomes apparently a pretty monstruous model; we will see later however (in §C.1) that it admits a very simple "normal form."

Note however that, at this stage, we have a perfectly general model for "stationary games" (i.e., with a description invariant under time shifts). Imagine, for example, that the games in the different states of nature were in fact extensive games, whose duration was a different number of physical periods according to the path—and where players maybe have to wait after completion a variable number of periods before entering the next game. It would be sufficient, in order to model this, that the payoffs and new state selected by the lottery contain qualifications as to how many periods the payoff has to be delayed and how many periods afterwards the new game is delayed: then introduce payoff vectors and states with such qualifications as a new set of states of nature, where the counters are decreased by one for every period until they reach zero. At periods with no payoff a zero payoff is given. Alternatively, if the extensive game has perfect recall [57], all positions of the extensive form (including the waiting positions) could be introduced as new states of nature. Thus our only restriction on our games is that players have some concept of physical time, of day and night—formally, we consider "multistage games" [37] (and our overall game is a general model for stationary multistage games). This is not a harmful restriction, and is inherent in the very concept of stationarity, and the fact that we are going to average utilities over time, or discount them.

To be completely explicit about the information structure of our games, we still have to specify that players remember all information they have received in the past, and to recall that it is immaterial whether one assumes or not that players remember in addition their own past actions or pure strategy choices [25]—so no assumption about this is necessary.

Let us also state explicitly here that we will systematically stick to the basic finiteness assumption, that all sets considered (set of states, set of players, action sets, sets of signals) are finite. All aforementioned problems, which are problems of substance, arise already in this case; when occasionally a continuous variable may seem natural, a discretization is conceptually harmless, and may in fact add to realism. It is therefore useless to complicate them further by mere technical generalizations. When occasionally we consider a continuous model, it will be stated explicitly, and will always be as a tool for the study of the above finite models.

As payoff function for the infinite game we will consider the Cesaro limit of the payoffs u_t in each stage t. Obviously this payoff need not be defined for every possible history (thus the overall game is technically speaking not a well-defined game), so care will have to be taken in the precise definitions of solutions, and one will have to show that they exist, are well defined, and depend on no further specification of the payoff function. Care will also have to be taken in those definitions to reflect the fact that we want this model also to serve for the study of long finite games (i.e., with payoff function $\overline{u}_T = (1/T)\sum_{t=1}^{T} u_t$ for fixed, large T) and for the study of discounted games with small discount factor λ [i.e., with payoff function $u_\lambda = \lambda \sum_{t=0}^{\infty}(1-\lambda)^t u_t$]. Note that the infinite game model is sufficiently flexible to include the finite games and the discounted games: for the discounted games, introduce for every payoff vector appearing in the description of the game a new, absorbing state where this payoff is paid forever, and replace every lottery by a new lottery that, in a first stage does the same as the old lottery, but in a second stage moves with probability λ instead to the absorbing state that corresponds to the payoff selected. (Different discount rates for each player could be similarly accommodated). For the finite games, the same thing can be done after having first replaced the old state by its product with $\{1, 2, \ldots, T\}$, using a t-dependent $\lambda = \lambda_t = 1/(T - t + 1)$.

Besides this flexibility, the reason for sticking with the Cesaro limits is that this payoff function preserves the stationary (shift-invariant) character of the game—one could think of other stationary functions of history, but additively separable utility is by far best understood, and seems the only interesting one in the applications.

As for solution concepts, in the two-person zero-sum case, an unambiguous solution is provided by the Von Neumann minimax value, together with a corresponding pair of optimal or of ε-optimal strategies. (A pure strategy of a player is a function telling him what action to take in every possible occurrence, as a function of all information available to him at that stage; the pure strategy set of a player in our games is a product of finite sets, and a (mixed) strategy is a probability distribution over this set—the player chooses secretly one of his pure strategies at random; it can equivalently be viewed [57] as a function specifying, for each possible occurrence, a probability distribution over the actions of that player at that occurrence, as a function of all information then available to him. In general, in the two-person zero-sum case, we will denote strategies of player 1 and player 2 respectively by letters σ and τ.)

One says then that $\underline{v}$ is the sup inf of the game (and that σ_ε is an ε-optimal strategy of player 1) iff

$$\forall \varepsilon > 0, \ \exists \sigma_\varepsilon, \ \exists T_0 \ : \ \forall \tau, \ \forall T \geq T_0 \quad E_{\sigma_\varepsilon, \tau}(\overline{u}_T) > \underline{v} - \varepsilon$$

and

$$\forall \varepsilon > 0, \ \forall \sigma, \ \exists \tau, \ \exists T_0 \ : \ \forall T \geq T_0 \quad E_{\sigma, \tau}(\overline{u}_T) < \underline{v} + \varepsilon.$$

($E_{\sigma, \tau}$ denotes the expectation under the probability distribution induced by σ and τ.) One defines similarly the inf sup $\overline{v}$ and ε-optimal strategies τ_ε of player 2. Whenever possible, as for instance for stochastic games, one also requires the above inequalities for $T = \infty$:

$$E_{\sigma_\varepsilon, \tau}\left(\liminf_{T \to \infty} \overline{u}_T\right) \geq \underline{v} - \varepsilon \quad \text{and resp.} \quad E_{\sigma, \tau}\left(\limsup_{T \to \infty} \overline{u}_T\right) \leq \underline{v} + \varepsilon.$$

A strategy is called optimal if it is ε-optimal for all ε. When $\underline{v} = \overline{v}$, this is called the value. Observe that such definitions are indeed independent of any further specification of the payoff function; on the other hand, with such definitions, even the existence of $\underline{v}$ and $\overline{v}$ has to be established. The definitions are set up so as to provide insight in the solution of long finite games and of the discounted games with small discount factor: an easy abelian argument shows that, for $\lambda \leq \lambda_0$ with some λ_0 depending only on T_0 and the maximum payoff appearing in the games, the u_λ will satisfy the same two inequalities as the $\overline{u}_T$; hence σ_ε guarantees a payoff $\underline{v} - \varepsilon$ in all sufficiently long finite games and in all games with sufficiently small discount factor, and in each case $\underline{v}$ is the best that can be achieved in this way with strategies that do not depend on the exact value of the discount factor or of the duration of the game.

Another reason for the specific interest in $\underline{v}$ and $\overline{v}$, even when they are different, is the following. We certainly study the two-person zero-sum case for its own sake, not only because many interesting applications already can be modeled appropriately in this framework, but also because the minmax theorem provides us here with clear, uncontestable solution concepts, which are further much more tractable than in the nonzero-sum case: it is a laboratory case to study the pure phenomena of information usage, unencumbered by considerations of threats, cooperation, etc. But another major reason stems from interest in nonzero-sum games: not only is the zero-sum case a particular case to be well understood before being able to tackle the general case, it is also a basic tool in the analysis of the general case. This can already be seen from our above proof of the "folk-theorem," where the set of equilibrium points of the repeated game is described by means of the players' individually rational levels, i.e., the value of the zero-sum game where the player plays against his set of opponents considered as a single player that tries to minimize his payoff. Given the theme that repetition enables cooperation, one may be interested in other cooperative solution concepts too ([7], cf. also [69]). But they too are described in terms of the characteristic function, i.e., finally of the values of the zero-sum games where one set of players plays against the complementary set, with payoff the sum of the payoffs over the set minus the sum of the payoffs of the other players (as a function of appropriate weight factors assigned to each player's utility).

Now, for such application, it is $\bar{v}$ which is important—even when not equal to $\underline{v}$. This can already be seen in our above proof of the "folk-theorem," with at least 3 players—although there the zero-sum games considered are not standard, because one of the players is not allowed to mix his strategies. Examples with two persons, where the corresponding zero-sum games are standard and fall into our model, can be found in F. Forges's characterization [36] of a class of correlated equilibria (see below) for repeated games with incomplete information on one side. (It cannot be seen in S. Hart's characterization [47] of Nash equilibria for the same model, because there the corresponding zero-sum games had $\underline{v} = \bar{v}$.)

Remark that this use of zero-sum games, where one player's payoff is in fact minus the payoff of his opponent in the true, nonzero-sum game, is an additional reason for not assuming that each player knows his own payoff—certainly in the zero-sum case.

When the game has a value v ($\underline{v} = \bar{v}$), we have seen that the definitions imply that the values v_T of the finite games and v_λ of the discounted games—which always exist by the standard minmax theorem—converge to v when $T \to \infty$ and $\lambda \to 0$; and that therefore ε-optimal strategies are also ε-optimal in any sufficiently long finite game and in any discounted game with sufficiently small discount factor.

But when $\underline{v} < \bar{v}$, another object of interest is the limit v_∞ of v_T when $T \to \infty$ and of v_λ when $\lambda \to 0$ (those limits exist and are the same in all cases analyzed up to now). And one would like to know relatively simple sequences of strategies σ_T and σ_λ which are asymptotically optimal (or asymptotically ε-optimal). Until now, in no case of significant generality are such sequences known.

In the nonzero-sum case, attention focuses basically on Nash equilibria—previously defined—and on correlated equilibria—introduced in [8]. Although correlated equilibria are an even larger set than the Nash equilibria, they are in some sense conceptually more appealing [10] and are mathematically better behaved and more tractable (defined by finitely many linear inequalities instead of via Brouwer's fixed point theorem). Further, they and their extensive form relatives [37] seem ideally suited to an incomplete information environment, because of the interplay made possible between the players' initial information from the game and the correlated information they obtain during preplay communication on each other from the correlated equilibrium.

But those concepts are subject to some criticism, and several refinements have been proposed [e.g., for Nash equilibria, Selten [109], with a later variant by Kreps and Wilson [56], which were sharply criticized in [52], where still another "improvement" is proposed].

Many further complications and distinctions arise because we consider infinite games, even more than in the zero-sum case. For example, one could consider the Nash equilibrium payoffs, or the payoffs which are for any $\varepsilon > 0$, ε-close to ε-equilibrium payoffs. In the zero-sum case, both coincide with the value whenever they exist—although in many cases (stochastic games, e.g.) the former may be empty while the latter still define the value. In general both could be nonempty and still different, and it might not be clear which one—if any—is the proper generalization of the value.

In the meantime, waiting for the conceptual dust to settle concerning finite nonzero-sum games, it seems best to limit efforts in the nonzero-sum case, as has been done until now, to those situations where, in the zero-sum case, there exists both a value and optimal strategies—i.e., essentially repeated games with incomplete information on one side, but including imperfect monitoring (and maybe information lags). Concentrating there on the infinite game extension of a couple of clear concepts for finite games, like correlated equilibria and their cousins and Nash equilibria, we may hope at the same time to clarify the above-mentioned methodological difficulties.

Finally, even in such cases, care has to be taken, in the definition of the concepts, since the payoff function is not everywhere defined. The approach used in the work of S. Hart [47] and of F. Forges [37] is the following: Fix a Branch limit $\mathcal{L}$, and define the payoff vector for any pair of mixed strategies σ and τ as $\mathcal{L}(E_{\sigma,\tau}(\overline{u}_T))$: this yields an everywhere well-defined game. Show that the set of Nash-equilibrium payoffs of this game does not depend on $\mathcal{L}$, and that each of those payoffs can also be obtained by a fixed strategy pair (σ, τ), which is in equilibrium for every $\mathcal{L}$.

This yields an unambiguous definition of the set of Nash equilibria of the infinite game. One proves in addition, like in the zero-sum case, some uniformity conditions on the convergence for this pair (σ, τ), to get ε-equilibria in the finite and the discounted games (cf. [47]) for details) allowing thus an interpretation of the "folk theorem" as advocated, e.g., in [98].

The above explains why we will largely concentrate in the sequel, like also the literature up to now, on the zero-sum case.

This survey is strongly oriented towards a mathematical audience in its way of presenting things; for a complementary, differently colored survey, the reader is urged to consult [9]—despite its being some five years old. Also, the author's bias as an economist has completely influenced this exposition, and caused a total neglect of other domains of application, like biology and ecology, where players become genes or species, and where apparently those models find more and more use. For such questions, I am incompetent, and the reader should consult Maynard Smith or R. Selten. Finally, since I have been asked to give this lecture under the framework of "Applications of mathematics in nonphysical sciences," I have deliberately omitted even touching the fast-growing literature concerning the applications of repeated games to mathematics itself—to logic, to topology, etc.—despite indications (hinted at in §C.3) of an increasing interplay between those methods and the problems here described.

B. A quick (and incomplete) guide to the literature

1. Single state of nature (and no initial information). Here the original name for the repeated game is "supergame," which we will retain because of its more specific meaning than "repeated game," which is currently used for almost any model falling under the general format of §C, and will accordingly be reserved for that general model. (I would have liked to use something like "Markov Games" if the name was not preempted.)

(a) In the N-person case, the theme that "repetition enables cooperation" prompted the question whether other solution concepts of the game, more cooperative than the full individually rational set, could be "justified" as the set of "equilibria" of some other type in the supergame. This was answered in the affirmative by Aumann [3, 5], who showed that the β-core of the game equals the set of strong equilibrium payoffs of the supergame. A nice survey and summary of this literature is contained in [7]. This core is the set of feasible payoff vectors such that every coalition can prevent the opposing coalition from improving the situation of all its members. A strong equilibrium point is a strategy vector, such that no coalition could improve the situation of all its members by deviating. The usual definition of the (β-) characteristic function, as required by cooperative theory, involves however correlated strategies by coalitions. With such a definition, the above theorem would no longer be valid in general for the corresponding core. It remains valid however, when the single stage game is a cooperative game [Aumann, loc. cit.], and when it is a game with pefect information [4]—since in this case, the required correlated strategies become pure, being minmax strategies (Zermelo).

This β-characteristic function coincides in fact both with the α- and the β-characteristic function of the supergame [69]; this implies much more than the inclusion of strong equilibrium points in the β-core—and has similar reinterpretations in the case of cooperative games or of games with perfect information.

E.g., it implies also that with the usual definition, the β-core equals the set of strong correlated equilibria of the supergame (if the payoff does not belong, the deviating coalition can ignore everything learned from the correlation device and play a pure strategy, using at every stage of the approachability strategy a best reply instead of an optimal strategy; the argument highlights an unpleasant feature of strong equilibria (Bernheim, Peleg, and Whinston), i.e., that members of a deviating coalition in the limit may have to deviate just in those cases where, given current information, deviation causes them a loss): the relationship with the "right" cooperative concept is obtained with correlated equilibria (just as in the case of the "folk theorem")—in this respect too, correlated equilibria seem better than Nash equilibria.

All the above is under assumptions of perfect monitoring, and essential use is made of Blackwell's minmax theorem for vector payoffs [23], like in several other applications below.

(b) Still under perfect monitoring, improvements in the folk theorem were sought in the following directions:

• The proof involves punishing a deviator forever—which may no longer be optimal for the punishers once the deviation has occurred. A concept of "perfect equilibrium point" would not have this defect. Aumann and Shapley [16] and Rubinstein [102] have shown that the folk theorem still holds with perfect equilibrium points, and Rubinstein [104] similarly perfected Aumann's β-core theorem (see above) (essentially the assumption of observability of mixed strategies can apparently be fixed).

• To interpret the supergame as a model for long finite games or for discounted games with small discount factors, one would like the set of equilibrium payoffs of those games to converge (Hausdorff) to that of the supergame. Those results are however false. For discounted games, convergence can be proved for a generic class of games [119]; an exception was constructed by Forges, Mertens, and Neyman [40]. For the finite games, convergence appears not to hold typically, e.g., in the finitely repeated prisoners' dilemma (see above), (1,1) is still the only equilibrium payoff (some care is needed: backwards induction does not hold and dominance is not sufficient—but a proof follows from [119]). This last paper contains much additional information related to this topic. In particular, it shows that what is essential for the prisoners' dilemma analysis to hold is that the single stage game has a single equilibrium payoff which is also the minmax point. Benoît and Krishna [19] show that, on the contrary, if for each player there exists a (perfect) equilibrium in the 13-stage game giving him more than his individually rational payoff, then the Hausdorff convergence of the (perfect) equilibrium payoff set in finite games to that of the infinite game does hold (under generic conditions). The idea is essentially to alternate sufficiently many times at the end of the finite game between the assumed equilibria for each player: in this way, each player accumulates at the end of the game a sufficiently high payoff in excess of his minmax payoff such that the threat of losing it is sufficient to scare him away from deviating in earlier stages.

• Even in the prisoners' dilemma case, Kreps, Milgrom, Roberts, and Wilson [54] showed that if one player assigned some ε-prior probability to his opponent being an automaton that plays the " 'tit-for-tat' " strategy (play cooperative at the first stage, next always the same as your opponent at the previous stage), then perfect equilibrium payoffs of the N-times repeated game would converge to the cooperative payoff (2,2). This is a beautiful result, implying in some sense that not only does repetition enable cooperation, it even forces it in equilibrium. (In fact, this ε-probability would seem largely justified in view of Axelrod's experiments [18].)

The high expectations generated by this result were however soon quelled to some extent by a result of Fudenberg and Maskin [43], which showed that the above was due to the very specific nature of the automaton, rather than to the idea of "ε-incomplete information": they showed under generic conditions that if for each player some ε-probability is assigned to him by his opponents having a different payoff function, then perfect equilibrium payoffs of the finitely repeated games and of the discounted games would converge to a set, which, by varying those different payoff functions, would cover the whole feasible and individually rational set—thus restoring "the folk-theorem."

• But the K-M-R-W result provided another important insight—the reputation effect. By acting only a couple of times like the automaton, instead of as prescribed by different hypothesized equilibrium strategies, the player very quickly would "build up a reputation for being the " 'tit-for-tat' player"—and increase the prior probability of ε assigned by his opponent to this event to a very substantial posterior. This idea, of the importance of building a reputation even when the prior probability is very small (as related to the previously mentioned signalling idea of Spence) was shown to be very significant in a wide variety of situations (e.g., [55, 84]).

• Finally, Aumann and Sorin [17] recently succeeded in getting some results indicating that the conclusion that repetition leads to cooperation might be saved from the Fudenberg-Maskin objection: in games with a Pareto-dominant outcome, if one puts a small positive probability on all possible automata, pure strategy equilibria of the repeated game will lead to the Pareto-dominant outcome.

(c) *Imperfect monitoring.* R. Radner [95, 96, 97], Radner, Myerson, and Maskin [100], and Rubinstein and Yaari [107] studied a number of important problems of this nature. Abreu, Pearce, and Stachetti [2] study the discounted game where all players receive the same signals, and each player's payoff depends only on the signal and his own action—under additional assumptions. Major progress was achieved when a characterization of the set of equilibrium payoffs of the supergame was obtained by E. Lehrer [59] in the two-person case, assuming in addition that each player knows his own payoffs: it is the individually rational part of the convex hull of the payoffs to those strategy pairs in the one-shot game, where no player has (statistically) undetectable profitable deviations which increase (weakly) his information. The set includes in particular all efficient points in the individually rational set, thus extending significantly the folk theorem. Substantial progress has been made in [60] towards removing the assumption that players know their own payoffs.

2. Stochastic games. Milnor and Shapley [85] and Everett [29] present most interesting particular models of stochastic games, whose analysis is not yet subsumed by later more general results. The initial definition of stochastic games was given in [110], which assumed a positive stopping probability at each stage (i.e., like discounting) and proved the existence of a value. Gilette [44] introduced the general model with zero-stop probabilities, as currently accepted, and posed the question of existence of the value. Blackwell and Ferguson [24] proved this existence for a particular example—the "Big Match"—which would be basic for further developments. Kohlberg [49] extended this result to a whole class of games, and Bewley and Kohlberg [20] proved the convergence of the values of the discounted games and of the finitely repeated games. The existence of a value—in a very strong sense in addition—was proved by Mertens and Neyman [72], allowing further for general state and action sets, under conditions which are, by Bewley and Kohlberg [20], always satisfied for finite state and action sets. We comment later, in §C, a bit more precisely on those last two results.

The method of [20] extends directly to the n-person case, yielding the existence of stationary equilibria as an algebraic function of the discount factor—hence in particular convergent [70]. One might have hoped to use those limits, as in the zero-sum case, as candidates for an existence proof of equilibria in the infinite game. Such a hope was however destroyed in [120], which showed that, in a nonzero-sum variant of Blackwell and Ferguson's "Big Match" [24], the discounted and the finite games had only one equilibrium point (the threat point), far away in payoff space from the (limits of) ε-equilibrium payoffs of the infinite game (the Pareto surface). (This of course does not facilitate a "folk theorem" interpretation for stochastic games.) Still, the infinite game solutions are also ε-equilibria in all sufficiently long finite games and all games with sufficiently low discount factor, and it is not clear at all in the example which of the two solutions is more appropriate for such games.

In the discounted case (hence also in the finitely repeated case), existence (and a characterization) of equilibria was shown by Mertens and Parthasarathy [74] with general state and action sets.

3. Repeated games with incomplete information. Those are repeated games where the initially chosen state of nature remains fixed forever, and in addition there are no information lags.

In this domain, the historical papers are in the Mathematica series [63].

(a) *Incomplete information on one side* (*zero sum*). In particular, Aumann and Maschler [14] considered the zero-sum case "with incomplete information on one side," where player 1 is fully informed of the initial state of nature and player 2 not at all. They proved the existence of a value and of optimal strategies, roughly with the following approach.

For each initial probability p on the states of nature, define NR(p)—the set of nonrevealing strategies of player 1—as those behavioral strategies in the one-shot game which, if used, would cause the posterior probability of player 2 to remain at p, whatever action he takes. Then define $u(p)$ as the value of the one shot game $\Delta(p)$, where player 1 is restricted to strategies in NR(p) $(u(p) = -\infty$ if NR$(p) = \varnothing)$. Denote by v the smallest concave function on the simplex which majorizes u—the "concavification" of u, Cav u.

Player 1 can easily guarantee himself $u(p)$ in the repeated game $\Gamma(p)$, by just repeating at every stage his optimal strategy in NR(p). A particular case of Lemma 4 in §C would show that whatever he can guarantee himself is concave in p; hence player 1 can get $v(p)$. The following argument is however more constructive: let $v(p) = \sum_i \alpha_i u(p_i)$ with $\alpha_i > 0$, $\sum \alpha_i = 1$, $\sum \alpha_i p_i = 1$. Then there exists a type-dependent lottery for player 1 on the set of indices i such that, if used, the total probability of outcome i would be α_i and the posterior probability of player 2, if informed of the outcome i, would be p_i. Let player 1 use this lottery; then play, if i, such as to get $u(p_i)$ in $\Gamma(p_i)$. Even if player 2 was informed of the outcome, player 1 would still guarantee himself by this "splitting strategy" $\sum_i \alpha_i u(p_i) = v(p)$—so a fortiori in $\Gamma(p)$.

Next, convergence of the values $v_T(p)$ of the T-stage games $\Gamma_T(p)$ to $v(p)$ is established, bascially by a martingale argument on the sequence of posteriors p_t of player 2: the martingale property implies that $E(\sum(p_{t+1} - p_t)^2) \leq 1$—or any vector-variant. Hence, with high probability, p_{t+1} is very close to p_t for all but a bounded number of t.

Let player 2 compute the p_t's, and play roughly in the following way: whenever p_t is far from the region R where $u(p) > -\infty$, let him choose an action uniformly. Otherwise let him play with probability $(1 - \varepsilon)$ an optimal strategy in $\Delta(\bar{p})$, for some $\bar{p} \in R$ close to p_t, and with probability ε a uniform strategy.

Since whenver p_t is far from R, player 2 playing a uniform strategy will cause on the average much information to be revealed, i.e., p_{t+1} far from p_t, this will happen only a few times. The rest of the time, p_{t+1} will most frequently be very close on the average to p_t—say ε^2-close—which means, since player 2 plays with probability ε a uniform strategy, that player 1 is playing close to NR$(\bar{p})$ for some $\bar{p} \in R$ close to p_t. Hence, since player 2 is using with probability $(1 - \varepsilon)$ an optimal strategy in $\Delta(\bar{p})$, the expected payoff will approximately be smaller than $u(\bar{p})$, hence than $v(\bar{p})$, hence approximately smaller than $v(p_t)$. Since $E(p_t) = p$, concavity implies that $E(v(p_t)) \leq v(p)$: thus player 1's expected payoff will be (approximately) smaller than $v(p)$. Hence $v(p)$ is approximately the maxmin of $\Gamma_T(p)$—our strategy for player 2 depends on player 1's through the p_t's— but since finite games have a value this implies (together with the previous result) the convergence of $v_n(p)$ to $v(p)$. Now let player 2 play during 1 stage his optimal strategy in $\Gamma_1(p)$, then forget everything and play for 2 stages his optimal strategy in $\Gamma_2(p)$, then again for $\Gamma_3(p)$, and so on: he guarantees himself a payoff of $v_n(p)$ in the nth block; the convergence of $v_n(p)$ to $v(p)$ implies that he guarantees himself $v(p)$ in $\Gamma(p)$: this is his optimal strategy, and $v(p)$ is the value of $\Gamma(p)$—hence the previously described strategy of 1 is optimal too.

(Obviously, all difficulties in establishing the bounds and approximations have been glossed over here.)

Assuming that in addition player 2's signals are player 1's actions, Aumann and Maschler also exhibited a much simpler and easier optimal strategy for player 2, based on Blackwell's approachability strategy [23]. This was then extended in [49] to the full generality of the present model.

N. Megiddo [65] showed that if we assume that player 2 is informed of his payoffs, then he has a strategy (independent of p) that guarantees him the value of the true game. He used a direct proof, but the result can also be deduced from the above Aumann-Maschler result: indeed, under that assumption one checks immediately that $\operatorname{Cav} u$ is linear. Since the amount guaranteed by a strategy of player 2 is a linear function of p, everywhere above $\operatorname{Cav} u$, this implies that an optimal strategy at some interior p is optimal at all p—and guarantees the value of the true game. To get finally a strategy which works against all games, and not just a given finite set, enumerate the games with rational payoffs and play successively a sufficiently long time against each initial segment. Being optimal against all rational games, this strategy is optimal against any game.

In the discounted case, Mayberry [64] exhibited an example where the value, as a function of p—which is necessarily concave by the splitting argument—was nondifferentiable on a dense set.

(b) *Incomplete information on two sides (zero sum)*. Stearns [123] showed that, as soon as there was incomplete information on both sides, the value ceased to exist in general. This is true even under the assumptions that players have each two types, which are chosen independently with probabilities p and q for players 1 and 2 respectively, and that they are just informed after each stage of each other's actions.

The basic idea is as follows. If a player knows his opponent's strategy, it is better for him to wait, before revealing any information himself, until the other has revealed as much as possible—the finite amount of time lost in waiting has no influence on long term averages, and doing so will enable him to adjust what he reveals to what he learned about his opponent's type. But this turns the game into a game like "picking the largest integer," which has no value.

In fact, this "waiting" strategy uses a somewhat similar martingale argument as above, and goes roughly as follows: fix a strategy of player 2. Player 1 maximizes the expected quadratic variation of his martingale of posteriors on player 2's types (the "quantity of information" he gets) over all his own nonrevealing (i.e., here: independent of his own type) strategies. He uses this strategy until all information is almost exhausted—he has then a posterior q_t over player 2's types, and knows that, whatever he does later, q_t will no longer move very much. Hence at that stage, he is essentially in a situation where player 2's type has been selected with q_t, but player 2 will play nonrevealingly—i.e., independently of his own type; but then the selection of the type of player 2 could as well be done at the end of the game: we are in a situation with incomplete information on one side. Hence player 1 can get $(\text{Cav}_p\, u)(p, q_t)$ from that stage on, where $u(p, q)$ is the value of the one-shot game where both players play nonrevealingly (i.e., independently of their type). ($\text{Cav}_p\, u$ denotes the concavification of u in p, similarly for Vex_q.) This quantity is more than $(\text{Vex}_q\, \text{Cav}_p\, u)(p, q_t)$, and since $E(q_t) = q$ the convexity implies that the expected payoff is $\geq (\text{Vex}_q\, \text{Cav}_p\, u)(p, q)$.

Since player 2 can clearly guarantee himself $(\text{Vex}_q\, \text{Cav}_p\, u)(p, q)$ (as seen above, the result on incomplete information on one side implies that, ignoring his own type, he can get $(\text{Cav}_p\, u)$; hence by the previously described "splitting strategy" he can also guarantee himself $\text{Vex}_q\, \text{Cav}_p\, u$); this implies that the minmax is $(\text{Vex}_q\, \text{Cav}_p\, u)(p, q)$. Similarly, the maxmin is $(\text{Cav}_p\, \text{Vex}_q\, u)(p, q)$, and the two are distinct in general—even when one player knows the true payoff matrix, as long as he has some uncertainty about the other's prior [122].

Due to space and time limitations, I will have to be much more sketchy about later developments, limiting myself to a couple of current state-of-the-art results, and without being able to even sketch the ideas of the proofs. An introduction to the subject can however be found in [113]. The above-described ideas are basic for almost all later work; nevertheless, for much of it, it would probably not be possible to give a fair idea of the proofs in a reasonable amount of space as above.

(b.1) When the signals received by the players after each stage are independent of the state of nature, $v_n(p)$ and $v_\lambda(p)$ nevertheless converge, in those games of "picking the largest integer," say to $v(p)$.

p denotes the initial distribution on the states of nature. Convex subsets $\Pi_{\mathrm{I}}(p)$ of the simplex are defined as those points in the simplex where the conditional distributions of player 1 on the state of nature given his type are the same as at p. For a function f on the simplex, $\text{Cav}_{\mathrm{I}}\, f$ denotes the smallest function larger or equal to f which is concave on each set $\Pi_{\mathrm{I}}(p)$. Permuting players, one defines similarly $\Pi_{\mathrm{II}}(p)$ and $\text{Vex}_{\mathrm{II}}\, f$.

Nonrevealing strategies of player 1 are those strategies where, for every action of 2, the distribution of player 2's signals is the same in every state of nature. Similarly for player 2, and $u(p)$ is defined as the value of the one-shot game where both players are restricted to nonrevealing strategies. Then v is characterized as the smallest solution of

$$v \geq \underset{\mathrm{I}}{\text{Cav}}\, \underset{\mathrm{II}}{\text{Vex}}\, \max(u, v)$$

or as the largest solution of

$$v \leq \underset{\mathrm{II}}{\text{Vex}}\, \underset{\mathrm{I}}{\text{Cav}}\, \min(u, v).$$

It satisfies in fact $v = \text{Cav}_I \min(u, v) = \text{Vex}_{\mathrm{II}} \max(u, v)$.

Those results were established in [67] (the last part was truncated in the printing process and appeared in [68]) after a particular case was solved in [75].

The set of "u-functions" is dense in the space of continuous functions—hence the above implies the existence of a continuous solution v to those functional equations and inequalities for any continuous u. A direct, analytic proof of this was given in [116] (after the particular case of independent types (thus $u(p,q)$, Cav_p, and Vex_q) in [79]).

A careful reading of the above-mentioned proof yields an error term

$$|v_n(p) - v(p)| \leq Kn^{-1/3},$$

and this is best possible by an example of Zamir [128] (when signals are the pair of actions, the error term was $O(n^{-1/2})$).

Even for fully rational games $v(p)$ can be transcendental [81], so no algebraic approach like for stochastic games [20] is possible.

Building on the techniques developed in this proof, the argument of Stearns [123] was also extended [80] to this model: maxmin exists and is equal to $(\mathrm{Cav}_{\mathrm{I}}\,\mathrm{Vex}_{\mathrm{II}}\,u)(p)$—and dually for minmax.

In the papers, the results are presented for deterministic signals, but this is not crucial to the idea of the proofs. (Formally, the same construction as in §C.1(e) could be used to reduce the case of random, correlated signals to the deterministic case.)

(b.2) In the symmetric case, where both players always have the same information (thus the same initial information about state of nature—this being common knowledge, one may assume no initial information—and the same signals after every stage, including each other's actions), the existence of a value was proved in [33] by reduction to a stochastic game with continuous state space (the principles of such a solution are developed, for our general model, in §C.3). The existence of a value for the stochastic game was proved by a direct, ad hoc discretization method. This work provided the basic insight that games could be analyzed ("recursively") by defining a corresponding stochastic game on the space of all initial distributions (cf. §C.3).

Here the case of random signals cannot be transformed trivially as in the preceding model to the case of deterministic signals; this was solved earlier in [53], where the method of reduction to stochastic games originates. In that case however, it led to *finite* stochastic games (this shows the basic difference) of a specific structure previously solved in [49] with this application in mind. This is because the above-mentioned transformation method of §C.1(e) would destroy the hypothesis that players know each other's actions.

(b.3) Whenever signals are deterministic and include the pair of actions, the "recursive structure"—as explained in §C.3—involves no general entrance laws, but can be simply expressed by probabilities on the states of nature. Thus such models might be amenable to a similar, explicit analysis as were the previous cases, while at the same time they would contain most of the basic difficulties both of cases (b.1) and (b.2) simultaneously. This unifying case presents a great challenge—among others probably to construct one strategy doing at the same time the job of the approachability strategy and that of the "Big Match" strategy. It was a major success when S. Sorin [118] managed to analyze a—I hope typical—class of examples of this type. To do this, he had to build on a previous "tour de force" of his—which was motivated by the example—where he studied a class of "stochastic games with incomplete information": he considered there "Big Match"–type games where either player 1 [117] or player 2 [115] is uncertain about the payoff matrix; and in both cases he obtained the minmax, the maxmin, and $\lim_{n\to\infty} v_n$—in particular, the better informed player could guarantee himself $\lim_{n\to\infty} v_n$ (cf. §C.3).

(b.4) In more general cases, no recursive structure seemed available—hence the titles of the papers: essentially, in §C.3, I am going to show how wrong one of my own titles was. To avoid unmanageable pure strategy sets, attention was restricted in what is currently still exploratory work to games where, at each stage, players either got a "blank" signal, or were told the true game—the same, deterministic signal to both players. In addition, there was no initial information. Analysis of those models is done by normal form methods. The idea is to construct an auxiliary one-shot game Γ: introduce a class of strategies that correspond closely to some strategies of the repeated game, and define the payoff in Γ as the corresponding asymptotic payoff in the repeated game, neglecting second-order terms. Different classes of strategies have to be introduced for minmax, for maxmin, and for $\lim v_n$. Show that Γ has a value, and use appropriate optimal strategies in Γ to reconstruct either optimal strategies or best replies in the original game: the class of strategies guessed was sufficient, and each of the three quantities is given by the value of an appropriate auxiliary one-shot game ("on the square"). The strategies in Γ involve certain exceptional moves; to assign a (random) date to those in the original game, a separate—and more artificial—application of the minmax theorem is done (with as strategy spaces compact subsets of L_∞ on the space of histories), even just for constructing appropriate best replies in the original game.

After an example in [76], C. Waternaux [125, 126] succeeded in obtaining minmax and maxmin for the above-described games. (Her first paper, which considered particular cases, is not entirely subsumed by the second: the algebraicity of the solution, for example, followed from the more precise results obtained for all particular cases, and no proof is yet available in the general case.) S. Sorin [121] recently proved that $\lim_{n\to\infty} v_n$ also exists in this model—but no "formula" is (yet) available like in the case of maxmin and minmax.

(c) The "error term" $e_n(p) = v_n(p) - \lim_{n\to\infty} v_n(p)$ was analyzed mainly in the case of incomplete information on one side, when the signals are the pair of actions—and, say, with two states of nature (the upper bound provided by the convergence theorem (b.1) is $K\sqrt{p(1-p)/n}$). The order of magnitude was studied in [127, 128] for many examples and showed a wide variety of behavior. The exact expression of the second-order term $\lim_{n\to\infty} \sqrt{n}e_n(p)$ was obtained for one example in [77]: it was $\phi(p)$, the standard normal density function evaluated at its own p-quantile. This was intimately related to the fact that $\phi(p)$ is also the limit of the maximum of $(1/\sqrt{n})\sum_1^n E|X_i - X_{i-1}|$ when (X_i) varies over all martingales with expectation p and with values in $[0,1]$ [78]. More recent results of those authors show that this example is not accidental, and that, when each player has only two actions, the second-order term is always expressed by means of the same function ϕ. Still, the appearance of the normal distribution here remains a complete mystery to me.

(d) Cases of specific, finite extensive forms (often with perfect information) with incomplete information about their payoffs do not really fall within our subject. Fascinating insights and interesting operational methods can however be found in the work of Ponssard and of Ponssard et al., and Sorin [112]. Not only does this work provide insights similar to some above-mentioned results more clearly and at a much lower technical cost, it also implies extremely efficient methods for analyzing finitely repeated games with incomplete information, when the game tree to be repeated has perfect information.

(e) *Nonzero-sum games.* Those were only analyzed in the case of 2-person games with incomplete information on one side, assuming in addition that the signals were the last pair of actions.

(e.1) Existence of Nash equilibria is one of the main open problems since [15]. It was finally proved in the case of 2 states of nature by S. Sorin [114].

A characterization of the set of Nash equilibrium payoffs (as defined in §A) was obtained by S. Hart [47]. Denote by $(A^s, B^s)_{s \in S}$ the payoff matrices to players 1 and 2, and by Δ^S the simplex of probabilities on S. $p \in \Delta^S$ is the initial probability. Define as individually rational a payoff vector $(a_s)_{s \in S}$ of player 1 if

$$\langle q, a \rangle = \sum_s q_s a_s \geq \text{value}\left(\sum q_s A^s\right) \quad \text{for all } q \in \Delta^S,$$

and a payoff β of player 2 if

$$\beta \geq \left[\underset{q}{\text{Vex}}\left[\text{Value}_2\left(\sum q_s B^s\right)\right]\right](p),$$

where the value to 2 of a payoff matrix B of player 2, $\text{Value}_2(B)$, equals $-\text{value}(-B)$.

Those conditions are obtained from the zero-sum theory (B.3(a)), and express respectively that player 2 has a strategy preventing any type s of player 1 from exceeding a_s, and that player 1 has a strategy that prevents player 2 from exceeding β.

Now define the feasible set F as the convex hull in $R^S \times R^S$ of all payoff vectors $[(a^s, b^s)_{s \in S}]_{i,j}$ appearing in the matrices. G is then defined as those triplets (a, β, p), where a is individually rational for 1 and β is for 2 at p, $p \in \Delta^S$, and such that, for some $(c, d) \in F$, one has $a \geq c$, $\langle p, a \rangle = \langle p, c \rangle$, and $\beta = \langle p, d \rangle$.

In other words, G consists of the individually rational payoffs which are feasible in nonrevealing strategies (those are clearly equilibria).

Define G^* as the set of initial values g_1 of bounded martingales $(g_t) = (a, \beta, p)_{i=1}^\infty$ with limit in G that satisfy for each t a.s. either $a_{t+1} = a_t$ or

$p_{t+1} = p_t$, and g_t takes only finitely many values. Then G^* is the set of equilibrium payoffs.

The interpretation of such equilibria is the following. Whenever $p_{t+1} \neq p_t$, player 1 uses the appropriate "splitting strategy" (see above). When $a_{t+1} \neq a_t$, players uses a "jointly controlled lottery" to select the value of g_{t+1}: they have to choose one of finitely many values, say with probabilities $(\alpha_1, \ldots, \alpha_n)$. Let

$$\beta_i = \sum_{j \leq i} \alpha_j \qquad (i = 0, \ldots, n).$$

Let players in successive stages select one of their two first strategies at random (so the theorem assumes each player has at least two actions), to select the successive bits of a random number in $[0, 1]$: the bits are set to one if the choices match, to zero otherwise. Players stop this process as soon as it is clear in which interval (β_{i-1}, β_i) the random number will fall, and then select the ith value of g_{t+1}. Once g_{t+1} is known, say in k stages, players play for kt periods a nonrevealing strategy leading to their current expectations of the limiting $(c, d)_\infty \in F$ (associated with $g_\infty \in G$). If a deviation is ever detected, the deviator is punished forever to his individually rational level.

Neglecting the β-coordinate, g_t becomes a bounded vector-valued martingale $(X, Y)_t$, such that at each stage only one of either X_t or Y_t changes. Such processes were called "bimartingales" by Aumann and Hart [11], who studied their properties and their relationship with the concepts of biconvex sets and of biconvex functions. In particular, this study leads to a characterization of the set G^* in terms of G by separation-like properties.

Few other properties of the set of equilibria are known—it is known neither whether it is closed, nor whether $G^* = G^{**}$, nor whether it is also the set of ε-equilibrium payoffs ($\varepsilon \to 0$) (those questions may be related). Such answers are provided in [11] for arbitrary sets G; it is when those sets arise from some game that answers are unknown.

Similarly, no example is known where a bimartingale is needed which does not stay constant from some (random) time on.

It is known however that infinite martingales are needed; they could not stay constant from some deterministic time on. In particular the set of equilibria is larger than the biconvex hull of G. A generic example to this effect was constructed in [35]; in the example, the payoffs are even independent of player 1's actions, which serve thus a pure signalling purpose. The model can thus be interpreted as a pure "signalling game": player 1 gets some private information, next players 1 and 2 can freely send messages to each other, as long as they wish, next player 2 takes an action. So even in this very restricted model, an unbounded number of signalling stages ("splittings") of player 1, separated by jointly controlled lotteries which require messages of 2, are required to achieve some equilibrium payoffs, which are in fact far superior to anything that could be achieved with only a bounded number of signalling stages—and a fortiori with no messages of the uninformed player.

(e.2) *Correlated equilibria.* A correlated equilibrium (R. J. Aumann, [8]) of a game is a Nash equilibrium of an extension of the game where players may observe, before the start of the game (e.g., during preplay communication) correlated random variables, say generated by a correlation device.

Similar concepts for multistage games were introduced by F. Forges [37]:

A "communication equilibrium" of a multistage game is a Nash equilibrium of an extension of the game by a "communication device" that, before every stage of the game, receives inputs from the players and then transmits to the players a vector of outputs—random variables depending on all past inputs and outputs.

An "r-device" is a device that receives only the first r inputs—hence an r-communication equilibrium. For $r = \infty$, one obtains the above definition; $r = 0$ yields the "autonomous devices" and the "extensive form correlated equilibria."

Denote by C and D_r the set of payoffs to correlated equilibria and to r-communication equilibria.

It was shown (loc. cit.) that there is no loss in restricting the outputs from stage 1 on to be a sequence of independent random variables, uniformly distributed on a finite set and public to all players. It was shown also that in addition, for finite multistage games, a single finite device of the above type (i.e., all inputs and outputs are in finite sets) was sufficient to generate the whole of C or of D_r—which were shown to be compact convex polyhedra—provided a single, continuous random variable, independent of all the rest, can be observed by the players before the start of the game. She also proved that even this could be dispensed with and replaced by a geometric random variable with sufficiently high expectation. This can be generated by a finite automaton containing a fixed lottery mechanism—outputting a "1" every time a success is scored until the first failure. Joining this to the device, one obtains a single finite automaton, containing a fixed, finite lottery mechanism, and that will generate the whole of C or of D_r as pure strategy equilibria.

She characterized those equilibrium sets in [36] (with as payoff to player 1 his vector payoff, as in Hart), were characterized in [36], for the model of repeated games with incomplete information we consider here. Proofs are exceedingly subtle and complex, and there is no way to convey the slightest idea here about what is going on. Just note that only the addition of initial, private information to both players turns the game in effect into a game with incomplete information on both sides (with a continuum of types), and that after a couple of stages, player 2's posterior on the state of nature will be unknown to player 1. And already in the zero-sum case, the analysis of incomplete information on both sides was very difficult, with minmax $\neq$ maxmin $\neq \lim_{n\to\infty} v_n$. A most interesting interplay is occurring here between the initial information of the players in the game and the information supplied by (and to) the device.

The results imply that points in D_∞ can be generated by a finite device to which player 1 initially tells his type (truly or not), after which the device has only an "alarm" input (and formally a "no alarm" input) for player 2. The device always gives the same output to both players: initially, a feasible payoff vector (i.e., a point in F) and as soon as the alarm button is hit, a punishment strategy of player 2, in the form of an individually rational payoff vector x of player 1—at all other times the device remains silent.

Finiteness of the device means that only finitely many payoff vectors and vectors x will ever be chosen. The players' strategies are to follow a fixed sequence of action pairs leading to the recommended payoff vector, to punish the opponent (with x for player 2, and to player 2's individually rational level for player 1) as soon as x is heard, for 2 to push on the alarm button iff a deviation by 1 occurs, and, for 1 to report truthfully his type. The equilibrium conditions on those strategies translate into a nice analytic characterization of D_∞, which is compact and convex.

Similarly, the sets D_r $(r = 1, 2, 3, \ldots)$ are all equal to a compact, convex set N generated by a device to which player 1 first transmits his type, after which the device transmits to both players a feasible payoff vector (i.e., in F) to be achieved, and to each one separately a punishment against his opponent in case of deviation. Player 2's punishment is a vector x as before, while player 1's is a convex function (in a bounded set in Lipschitz norm) that majorizes $b(p) = \text{Value}_2(\sum p_s B_s)$ on Δ^S. (Corresponding punishments are constructed with Blackwell's approachability strategy [23].) An analytic characterization of N is similarly obtained. It would be interesting to know a game with $N \neq D_\infty$. In the simpler case where payoffs do not depend on player 1's actions, which she analyzed previously in [38], all sets from C to D_∞ were equal.

Denote by N_f the subset of N corresponding to finite devices—as defined above. It is not known whether $N_f = N$. But it is shown (loc. cit. [36]) that any point in N_f where player 1's payoff is strictly individually rational lies in D_0—it is not known whether those conditions are necessary. Finally, if b is convex or if $\text{Vex}\, b$ is linear, then it is shown that $N_f = N = C$—implying equality of all sets but D_∞. (This includes the case where player 1's actions do not influence payoffs.)

A similar "boundary" difficulty as for D_0 occurred in [39], where she considered the case of "verifiable types," i.e., the Selten model rather than the Harsanyi model for games with incomplete information.

4. Information lags; stochastic games with incomplete information. This concerns some examples.

(a) In Gleason's game [45], three payoffs 1, 2, and (-3) are arranged in cyclic order, and alternately each player tells the referee whether he wants to move 1 step clockwise or counterclockwise, after which the referee tells him his position. It is clear that, for large n, $v_n(p)$ is almost independent of the initial probability p, since each player can garble this information almost completely by randomizing sufficiently during a couple of moves. [Formally, the least favorable initial information for 1 is when only 2 knows the initial state and p is common knowledge. Let $v_n(p)$ denote the value of this game. Let

$$\varphi_A(q) = v_n(0, q, 1-q), \qquad \varphi_B(q) = v_n(1-q, 0, q), \qquad \varphi_C(q) = v_n(q, 1-q, 0),$$

denote by $\varphi(q)$ the median of those 3 numbers, and let

$$w_n = \min_q \varphi(q),$$

the minimum being achieved at q_0. Let S $(= A, B,$ or $C)$ denote player 1's last position. For a given strategy of 2 in $v_n(p)$, player 1 can after each move compute the probability q that player 2 is going to use next; hence player 1 will in effect start the next move at $\varphi_S(q)$. Let him thus, as soon as $\varphi_S(q) \geq \varphi(q)$, start playing optimally in that game, and at all other earlier times play $(\frac{1}{2}, \frac{1}{2})$: he will start playing optimally in bounded expected time, and thus guarantees himself $w_n - k/n$. Similarly, in the most favorable position for 1, when only he knows the initial state, assume for instance $\varphi_A(q_0) > \varphi(q_0)$: player 2 can play alternatively $(\frac{1}{2}, \frac{1}{2})$ and $(q_0, 1 - q_0)$ until informed that this last move $(q_0, 1 - q_0)$ was done at position B or C, and then start playing optimally in remaining game: in bounded expected time he will start playing optimally; hence this strategy guarantees him, for any initial information, a payoff $\leq w_n + k/n$. Finally, the payoff guaranteed by an optimal strategy of 1 for $v_n(\frac{1}{3}, \frac{1}{3}, \frac{1}{3})$ is a linear function of p, everywhere $\leq v_n(p)$. Since $w_n - k/n \leq v_n(p) \leq w_n + k/n$, we obtain that this payoff must be for every $p \geq w_n - 5k/n$. Hence, whatever be the initial information, both players can guarantee themselves w_n up to k/n by strategies independent of this initial information. Those are used in the next argument.]

Let

$$\bar{v} = \limsup_{n \to \infty} v_n(p),$$

and let n_i be a corresponding subsequence. By playing first n_2 times optimally in Γ_{n_1}, then n_3 times in Γ_{n_2}, then n_4 in Γ_{n_3} and so on, player 1 guarantees himself a payoff at least equal to $\bar{v}$. Similarly for player 2: the game has a value v and optimal strategies. [This type of argument can be made in a wide variety of situations—e.g., some of the following examples too. For example, it yields immediately for games with incomplete information on one side (case (a) above) the convergence of v_n, and that player 2 can guarantee the limit. This observation reduces the complete solution of this case to a proof that if v_n converges, then player 1 can guarantee the limit. An important generalization of this is suggested below in §C.3.] One wants to know v and a pair of stationary optimal strategies. [Stationary means, at each stage, the probability used by a player is always the same function of the infinite sequence of all past positions (except the current one)—those are known to him. This should be optimal whatever infinite past he imagines before the starting position. I.e., something like $\lim_{\lambda \to 0} \lim_{t \to \infty} \sigma_t^\lambda$.]

This game seems to have resisted all attempts since ~ 1949. Recently, however, Ferguson and Shapley [31] have developed an efficient approach to the discounted case, using the methods of Scarf and Shapley [108] (in fact, the recursive structure as developed in C.3), and might be close to a solution of this game.

(b) Dubins's [28] and Karlin's [48] "Bomber-Battleship duel" are also examples of games with information lags.

(c) The previously mentioned "Big Matches with incomplete information" of S. Sorin [115, 117] fall naturally under the heading of stochastic games with incomplete information.

(d) Other examples under this heading include the work of Melolidakis [66] and a game analyzed by Ferguson, Shapley, and Weber [32]—which was already quite difficult, and the slightest variant of which seems to become utterly intractable.

Clearly, in this area we are still only scratching the surface.

C. Some remarks about the general model

We just want to give here some easy results and proofs to enable the reader to get a better feel for the problem.

1. A normal form for repeated games.

(a) *Getting rid of information lags.* In accordance with our basic finiteness assumption, information lags are described by a finite-state machine. Its initial state is selected together with the initial state of nature by same joint lottery, about whose outcome the players receive some private information. After every stage, the machine selects a new internal state at random according to some probability distribution depending on the previous internal state, the last actions of the players, and the new state of nature, payoffs, etc. selected in the game. The lottery determining the signals received by the players (and, if desired, also their payoffs and the new state of nature) is now allowed to depend in addition also on the new internal state of the machine.

One sees immediately that, by redefining a new set of states as the product of the old set of states and the set of internal states of the machine, we obtain a model as previously described, i.e., without information lags.

(b) To get rid of the initial lottery and the initial information of the players, just add a new initial state, where the players have only one action to take and receive a zero payoff, and where a new state and signals to the players are selected as in the initial lottery. (At worst this pushes all payoffs one step later in time, but it has definitely no effect on long term averages.)

(c) Now for each player add a constant to his payoff function so as to make it nonnegative, and divide all payoffs by a positive constant so as to make the sum of all players' payoffs everywhere smaller than or equal to one. This change has no strategic effect, and every payoff vector selected by the lotteries is now the barycenter of a probability distribution on the unit vectors $((1, 0, 0, \ldots),$ $(0, 1, 0, \ldots)$, etc.) and the zero vector. Replacing those 2 lotteries by the compound lottery gives us a game where all nonzero payoff vectors are unit vectors. This change has no information or other effects on the game; the players' expected payoffs at each stage will remain the same, and further, an appropriate form of the strong law of large numbers can be invoked to show that also, for any strategies of the players, the difference between u_T in the original formulation and in the reformulation will converge a.s. to zero: none of the quantities in which we are interested will be affected.

(d) Define now as a new set of states the initial state (with its zero payoff) together with all vectors (consisting of a new state, a vector of signals to all players, and a payoff vector (zero or unit)), which are chosen with positive probability by some lottery. We now have a model where the game starts in a given initial state, and after every stage a new state is selected at random as a function of the last state and the actions taken by the players in that state. Players then receive signals and a payoff which are a function of the new state.

Shifting payoffs one period further in time will not change long-term averages, and yields us now a model where the payoff (zero or unit) of each stage is only a function of the current state. (We have kept a zero payoff at the initial state).

Similarly, one player's signals can be viewed as a partition of the set of states minus the initial state. Add to this partition the singleton consisting of the initial state (this just amounts to adding before the start of the game the redundant information "you will now start in the initial state"); each player has now a partition of the set of states, such that *before* each stage, each player is told in which element of his partition the true state falls.

Finally, for the game to be well defined, it is necessary that each player know his action set whenever he has to play. It could be however that this information was given earlier to him—for instance in the beginning of the game if the state does not change. In that case, repeat this redundant information together with the signal at every stage, before computing his partition. Now his action set is the same in all states that belong to the same element of his partition. If necessary, duplicate some actions for some of those partition elements such that all those action sets have the same cardinality. Use now arbitrary one-to-one correspondences between those sets to identify them: each player has now a single action set.

To recapitulate, our game is now described by the following elements:

- a set of states S,
- an initial state s_0 in S,
- an action set A_n for each n in the player set $N = \{1, 2, \ldots, N\}$,
- a transition probability P from $S \times \prod_{n \in N} A_n$ to S,
- $\forall n \in N$, a partition P_n of S,
- a partition $W = \{W_0, W_1, \ldots, W_n\}$ of S. (W_0 is the set of states with zero-payoff vector, W_n that with the nth unit vector).

Before each stage, every player n is told in which element of P_n the current state s is. He then chooses an action in A_n, receives a payoff of 1 if $s \in W_n$ and zero otherwise; then P selects a new state and the game proceeds to the next stage. (There is no loss of generality, but it is not necessary to assume that $\{s_0\} \in P_n$ for all n.)

Informally, W_n is the set where player n wins the stage, and W_0 corresponds to a draw. Every player wants to maximize his expected winning frequency.

In the zero-sum case, to keep the zero-sum property, one scales only player 1's utility to lie between zero and one, thinking of it as a probability on the two points zero and one. One obtains in this way a subset W of S, where the payoff is one: W is the winning set of player 1, and its complement the winning set of player 2—there is no draw. (In fact, as will be seen later, a better—and equivalent—model would be where the information about the new current state occurs after each stage rather than before. The information before the initial state is indeed redundant, and recursive computations become easier. This will be corrected below in §C.1(f).

(e) *Getting rid of probabilities.* Here we assume that all elements of P are rational, and we want to show how to reduce the problem to P deterministic, i.e., a function. The assumption entails no real loss of generality and is in line with our basic finiteness assumption; it is further extremely likely that any proof under this assumption generalizes immediately to the original case. It is anyway a very small price to pay to get rid of any trace of randomness and of numbers in the model, and to obtain a completely combinatorial model. Any such simplification that enables us to focus better on the essential problem is worthwhile.

We will have to treat separately the two-person zero-sum case and the general case. Let m denote the smallest common denominator of all rationals appearing in P. If we had at every stage a new independent random variable uniformly distributed on Z_m, then P could be replaced by a function of actions and this random variable (and the current state).

In the zero-sum case, just let each player choose at each stage, besides his action some element in Z_m, and take as random variable the sum (in Z_m) of those two choices (no player is ever informed of any of those past choices of his opponent): we have replaced the action sets by their product with Z_m, and P is now a function from states and action-pairs to states. There remains to show that none of the quantities in which we are interested changes in this transformation.

Let us show that the question of existence and the value of $\underline{v}$ are the same in both games, and that ε-optimal strategies of player 1 (and the optimal strategies if existing) correspond in both games. The same results for $\overline{v}$, for v_λ, and for v_T will be a consequence, or follow from the same argument.

For any ω in the compact group

$$(Z_m)^\infty = \prod_{t=1}^\infty (Z_m)_t,$$

and for any strategy σ of player 1 in the new game, denote by σ^ω the same strategy, where ω_t is added to his choice in Z_m at time t. Define similarly τ^ω, and note that $(\sigma^\omega, \tau^{-\omega})$ induces the same distribution on histories as (σ, τ). σ^ω can be viewed as a transition probability from our compact group to pure strategies. Denote Haar measure on Z_m^∞ by μ, and let $\overline{\overline{\sigma}} = \int \sigma^\omega \mu(d\omega)$ (a mixed strategy in the new game), while $\bar{\sigma}$ denotes the corresponding mixed strategy (marginal distribution) in the old game (i.e., use $\bar{\bar{\sigma}}$ and forget the choices in Z_m). Finally, for any strategy $\bar{\sigma}$ in the old game, denote by $\tilde{\sigma}$ the corresponding strategy $\mu \otimes \bar{\sigma}$ in the new game. $\bar{\bar{\sigma}}$ and $\tilde{\sigma}$ obviously induce the same behavioral strategies, hence the same distribution on histories for every strategy τ of player 2 [**57**]. Note finally that, for any ω, the map $\sigma \to \sigma^\omega$ corresponds to a permutation of the strategy set, and that the map $\sigma \to \bar{\sigma}$ is onto.

Therefore, for any bounded random variable f defined on the space of histories —like $\bar{u}_T$ or $\liminf_{T\to\infty} \bar{u}_T$—we have for any σ:

$$\inf_\tau E_{\sigma,\tau}(f) = \inf_\tau E_{\sigma^\omega,\tau^{-\omega}}(f) = \inf_\tau E_{\sigma^\omega,\tau}(f) = f_\sigma$$

say, and hence

$$E_{\bar{\bar{\sigma}},\tau}(f) = \int \mu(d\omega) E_{\sigma^\omega,\tau}(f) \geq f_\sigma;$$

thus also

$$E_{\tilde{\sigma},\tau}(f) \geq f_\sigma, \quad \text{i.e.} \quad f_{\tilde{\sigma}} \geq f_\sigma.$$

But $f_{\tilde{\sigma}} = f_{\bar{\sigma}}$ (i.e., computed with strategies $\bar{\tau}$ in the old game), because $\tilde{\sigma}^\omega = \tilde{\sigma}$ (definition of Haar measure) implies that

$$E_{\tilde{\sigma},\tau}(f) = E_{\tilde{\sigma}^{(-\omega)},\tau^\omega}(f) = E_{\tilde{\sigma},\tau^\omega}(f) = \int \mu(d\omega) E_{\tilde{\sigma},\tau^\omega}(f)$$

$$= E_{\tilde{\sigma},\bar{\tau}}(f) = E_{\tilde{\sigma},\tilde{\tau}}(f) = E_{\bar{\sigma},\bar{\tau}}(f)$$

(and similarly $E_{\tilde{\sigma},\tau}(f) = E_{\bar{\sigma},\bar{\tau}}(f)$) (the last equality because, under $(\tilde{\sigma}, \tilde{\tau})$, the random variables daily generated in Z_m are really uniform and independent among themselves and from anything else in the game).

The relations $f_{\tilde{\sigma}} \geq f_\sigma$ and $f_{\tilde{\sigma}} = f_{\bar{\sigma}}$ imply immediately that the highest value of $\underline{v}$ for which the first condition in the definition of $\underline{v}$ is satisfied is the same in both games, and that ε-optimal and optimal strategies correspond under $\bar{\sigma} \leftrightarrow \tilde{\sigma}$.

The equality $E_{\tilde{\sigma},\tau} = E_{\bar{\sigma},\bar{\tau}}$ implies now also that the lowest value (of $\underline{v}$) for which the second condition in the definition is satisfied is in the new game lower than or equal to that in the old game; and the equality

$$E_{\tilde{\sigma},\tau} = E_{\bar{\sigma},\bar{\tau}}$$

shows similarly the reverse inequality (the $\tilde{\sigma}$ form a subset of strategies of the new game). In particular $\underline{v}$ will exist in one game if and only if it exists in the other, and will have the same value.

In the N-person case, we describe the construction only informally: we number the periods of the original game by even integers instead of all integers and introduce some action at odd periods. At those times, the original players' actions have no effect whatsoever (not even informational) and their payoffs are zero. Those are the times at which two additional players A and B will play a game of their own and receive their payoffs. They receive zero payoffs at even periods. A and B receive no information whatsoever about what happens in the original game; conversely the original players receive no information whatsoever about what happens at odd periods. The only link is that the actions of players A and B will determine a sequence of points in Z_m that will be used in the true game to make the random choices. The whole thing will be set up such that, in any Nash equilibrium of the full game, this sequence is a sequence of uniform random variables, independent among themselves and from anything else in the game. Because of the absence of linkage, any equilibrium of the full game induces an equilibrium of the game between A and B. We will set this game up as a constant sum game, so it will be a pair of optimal strategies. The action set of player B will be Z_m, while the action set of player A will be Z_m plus a fixed "waiting" strategy; and it is player B's actions which are going to be used to make the randomizations of the original game. Player A is informed of all past choices of player B, while player B is not informed of anything. Whenever player A uses his waiting strategy, the payoff is zero to him and one to player B. As soon as player A ever chooses a point in Z_m, the game moves to an absorbing state (without player B being informed), where the payoff is forever one to A and zero to B if player A could match player B's choice at that stage, and otherwise zero to A and one to B. It is easy to see that this game between A and B has optimal strategies—e.g., A playing at the first stage a random point in Z_m and B playing a sequence of independent random points in Z_m, and that this is the only optimal strategy of B.

Remark that our construction is less satisfactory in the general case than in the zero-sum case: we had to increase the number of players, which is annoying when one wants to focus on 2 person nonzero-sum games, since those often have specific features; also our argument shows only that Nash equilibria correspond, and not, as in the zero-sum case, any other quantity in which one might be interested. This might be improvable; however, as is it shows already that even in the general case, restricting one's attention to deterministic P's entails no substantial loss of generality, and should lead to results that extend rather directly to general P. Since anyway our main focus in the sequel is on zero-sum cases, we leave it here.

(f) *Final cleanup.* Here we want to give a somewhat easier and more transparent combinatorial form to our description—more "extensive form"–like.

First, rank players in cyclic order, player 1 following player N, and subdivide each stage into N substages, player n choosing his action at the nth substage. The new set of states is now partitioned into subsets $S_1, \ldots, S_N$, where only player n moves in S_n, and a point in S_n consists of a state of the original game (the same as the first substage) together with the actions lastly taken by players $1, \ldots, n-1$. Player n's partition becomes now a partition of S_n—the one determined by his original partition and the state coordinate in S_n. After substage N, the deterministic P is used to compute the new point in S_1 given the old point in S_1 and the N actions—i.e., given the point in S_N and the action of player N. All players' payoffs are a function of the point in S_1. Since this point is recorded in each S_n, it is convenient to delay player n's payoff until S_n

is reached—and to give zero payoff to other players in S_n. This determines the new winning set W_n of player n as a subset of S_n.

Each player's action set A_n can now be identified with a set of functions from S_n to S_{n+1}. There is some redundancy here, since before choosing the function he gets the information from his partition P_n. To remove this redundancy, replace A_n by $A_n^{P_n}$—allowing him to choose an action as a function of the current element of P_n, and give him the information about the current element of P_n only after he has chosen his action.

For symmetry, we can now allow the initial position to be an arbitrary point in S—not only in S_1.

In the zero-sum case, we could insist that payoffs in S_2 are zero and that both players receive their payoffs in S_1, but for similar symmetry reasons it is more convenient to allow for payoffs in both substages: if the state falls in W ($\subseteq S$), player 1 receives 1 from player 2.

This leads us to the following combinatorial form:

• the player set is Z_N;

• the state space S is partitioned into N subsets $(S_n)_{n \in Z_N}$;

• for each n, a partition P_n of S_n is given, as well as a set F_n of functions from S_n to S_{n+1};

• an initial state s_0 in S and a subset W of S are fixed.

The game proceeds as follows. It starts at s_0. When the current state s is in S_n, player n selects a function f in F_n and, if $s \in W$, he receives 1 from the referee (player 1 from player 2 in the zero-sum case). He is then told to what element of P_n s belongs and the game moves to the new state $f(s)$. (Recall that players are not informed of payoffs during the course of the game.) Players want to maximize their expected winning frequencies (in the zero-sum case, player 2 wants to minimize player 1's expected winning frequency). This description of the game is known to all players.

An even more economical model is now finally obtained by letting F denote a set of functions from S to S, whose restriction to each set S_n with F_n coincides (up to duplications), and by denoting by P the partition $\bigcup_n P_n$ of S. The partition $(S_n)_{n \in Z_N}$ is no longer needed; players just play in cyclical order, choosing each time an element of F before being informed of an element of P.

REMARK. The above combinatorial form has several natural extensions that preserve the property of having only finitely many combinatorial forms for a given cardinality of S. For instance, one could allow the payoff and the partition of each player (i.e., his signal finally) to depend also on his action. Or one could reintroduce numbers, and allow for transition probabilities and real-valued payoffs. I believe however that the above model is useful to focus on essentials; also going back and forth with existing proofs through the above-described normalization steps and the natural extensions of those normalizations, trying at each stage to find the natural assumptions, may help in improving substantially existing results, and in understanding them in a common framework. Finally, the model may be quite helpful in the (systematic?) search for interesting, small examples—note for instance that Gleason's game (see above) fits exactly in this framework (except for payoffs being small integers instead of just zeros and ones—one could consider as well the case of zero-one payoffs); S has 3 elements there and the partition is the maximal partition.

2. Entrance laws. One may also think that a more natural formulation of the above model would be to allow the initial state to be selected by an arbitrary lottery, known to all players. This would however not yet provide the adequate family of entrance laws, as shown below, and therefore it seems more appropriate to stick with the above fully combinatorial model.

In [**82**], it was shown that, for a given finite (or compact) set S and a given number of players N, there exists a canonical compact metric space T, canonically homeomorphic to the space of probabilities (weak*-topology) on $S \times T^{N-1}$. Hence a point t in T induces a probability t_1 on S; T_1 is the set of all those; then t induces a probability t_2 on $S \times T_1^{N-1}$; T_2 is the set of all those, and so on. T is in fact constructed as the projective limit of the T_k. Hence, whenever we have measurable spaces $\Omega_1, \ldots, \Omega_N$, and for each point ω in Ω_n a probability $t^n(\omega)$ on $S \times \Pi_{k \neq n}\Omega_k$, we can compute $t_1^n(\omega)$ for all n, then $t_2^n(\omega)$ etc.: in the projective limit we get a canonical image $\bar{t}^n(\omega) \in T$: Ω_n is canonically mapped to T (i.e., $\bar{t}^n(\omega)$, viewed as a probability, is the image probability of $t^n(\omega)$ under the maps $t^k \to \bar{t}^k$ for all $k \neq n$), in such a way that, if we interpret points in Ω_n as types of player n and $t^n(\omega)$ as his corresponding beliefs on S and the types of the other players, then the relevant beliefs (those on S) are preserved under the mapping, hence also the beliefs on relevant beliefs of other players, etc.

If in particular, there exists a probability distribution on $S \times \Pi_n\Omega_n$, of which the $t^n(\omega)$ are the conditional probabilities (which is true in all our models), then this probability is mapped canonically to a probability P on $S \times T^N$, of which the points in each factor T are the conditional probabilities—i.e., for every continuous function $f(s; t^1, t^2, \ldots, t^N)$ and every n we have

$$\int f\, dP = \int dP^n(t^n) \int f\, dt^n(s; t^1, t^2, \ldots, t^{n-1}, t^{n+1}, \ldots, t^N).$$

Such probabilities on $S \times T^N$ were called "consistent probabilities." I had always vaguely hoped that that study would lead to appropriate "entrance laws" or state variables for the problem considered in this paper. But I was repelled by the high level of abstractness and the nontransparency of the resulting concepts. Now I think however it is worth giving it a try.

Note first from the above formula that a consistent probability is fully determined by its marginal on any factor T. Denote by $\mathcal{P}$ the set of consistent probabilities. Convexity of $\mathcal{P}$ is immediately apparent from the above formula, and its (weak*-) compactness too—recall that the map from t^n to the corresponding probability is weak*-continuous. It should also be provable that any probability P in $\mathcal{P}$ is a weak*-limit of a sequence of probabilities with finite support in $\mathcal{P}$, whose supports converge (Hausdorff) to the support of P—considering for example an increasing sequence of borel partitions B_i of T, such that the maximum diameter of the elements of B_i converges to zero, and such that on each element of B_{i+1}, $t(A)$ varies by less than i^{-1}, for any subset A of $S \times (B_i)^{N-1}$: the restriction P_i of P to $S \times (B_i)^N$ is then a probability on a product $S \times \prod_n \Omega_n$ as we considered previously (conditionals are immediately computed), hence has a canonical image $\tilde{P}_i$ on $S \times T^N$. Such a sequence should do (a similar property was proved loc. cit.).

Let us now examine the relationship with our games. Consider for example a Nash equilibrium—the same analysis could be done for ε-equilibria, or for correlated equilibria, or any other solution concept: in each one, the strategy vector is given. The given strategy vector determines (together with the description of the game, including initial conditions) a probability distribution over the space H of all possible histories, which the players can compute. Provided we have performed at least normalization step (a), at any given time t, the future behavior of the game depends only on the current state s_t in S. s_t is a random variable defined on H, and players have in addition accumulated some private information on H—their pure strategy choices and their observations in the game during the first t periods. Let Ω_n be the set of values of such information for player n: we also have random variables from H to each Ω_n. From what we have seen previously, we get therefore a $P_t \in \mathcal{P}$ as canonical image of our probability on H, which contains all "relevant" information about the past: the future in the game should be as if it started from scratch at time t, with initial state and initial information chosen by P_t.

Strictly speaking, such an argument would be valid only in the zero-sum case. In general, one cannot forget the "irrelevant" information, because players might use it in order to coordinate, to correlate future behavior—say, to choose among different equilibria—and may have generated it for that purpose. It is then necessary to consider another random variable on H besides s_t to represent those—like the payoff expected by the players in the future in the work of S. Hart [47] and in that of F. Forges [38, 36] (or also Mertens and Parthasarathy [74]).

For an easy example, consider the battle of sexes:

$$\begin{pmatrix} 3,1 & 0,0 \\ 0,0 & 1,3 \end{pmatrix}.$$

An equilibrium (with payoff $(2,2)$) of the repeated game consists in both players choosing in the first stage by tossing a fair coin, then always playing their first or their second strategy according to whether their first day choices matched or not: while the information generated in the first day is irrelevant w.r.t. the unique state of nature, it is not so for analyzing the later stages.

Anyway, such considerations may lead in the nonzero-sum case to expand somewhat the set of state variables [at worst to replace S by $S \times \Sigma_1 \times \Sigma_2 \times \cdots \times \Sigma_N$ (and look in a second stage for a more compact representation: typically, a strategy n-tuple for the future matters only to the extent of the feasible payoff vector it generates—i.e., expected payoff to each player, indexed by initial state of nature), where Σ_n is the pure strategy space of player n (in the game starting from scratch at time t, i.e., his pure strategies for the future): this is still a fixed compact metric space, independent of t. Such an approach would be somewhat reminiscent of F. Forges's study of correlated equilibria (loc. cit)]. But it does not change the basic thrust—that the current state is described by an element of an appropriate P. Since in addition our chief focus is here the zero-sum case, we leave it there.

It is true that typically the P_t's obtained will have special features, for instance, a finite support (except in the study of correlated equilibria)—because only finitely many observations could have happened in the past. It is easy however to convince oneself that typically the size of this support grows to infinity with t, and that in general no such state variable can be obtained in a fixed finite-dimensional space. Since, for studying the long-run average payoff, we are chiefly interested in the asymptotic behavior of P_t, it may be appropriate to go immediately to the closure, and to consider the whole of P as our space of state variables. (Obviously, other special features may be usefully taken into account—e.g., that in the combinatorial model, with starting position in S_1, P_t assigns probability 1 to $S_{t \bmod(N)}$.)

Before looking in the next paragraph to see whether the above considerations really hold water and seem to lead somewhere, let us still remark

(a) that P_{t+1} is "easily" computed knowing just P_t and the players' behavioral strategies (as a function of their type) at time t, and

(b) that the expected payoff $E_{\sigma,\tau}(u_t^n)$ at time t is easily computed from P_t if the normalization step (d) in the previous paragraph has been performed: it is then $P_t(W_n)$—and similarly even if the payoff function is an arbitrary function on S.

3. A recursive formula. In this section, we concentrate exclusively on the zero-sum case, although such a recursive approach also can be used in the general case (in simple cases at least), recursing on the set of equilibria, as mentioned in §C.2 (cf. the 3 papers mentioned there).

For readers familiar with the literature, the basic change in approach proposed here, besides the general formulation which entails working on the canonical, infinite-dimensional space of state variables P, is akin to working with the distribution of posteriors rather than with the posteriors themselves.

Our first aim is to establish a recursive formula for the values v_λ of the discounted games—as a function of their entrance law $P \in P$, thus rather $v_\lambda(P)$. Recall that we assume the normalization sub (a) in §1 has been done. We write shortly R for $S \times T^2$.

(3.a) Since P may have infinite support, we first establish for the sake of completeness that $v_\lambda(P)$ and $v_T(P)$ are well defined, that the games have optimal strategies, and that v_λ and v_T are continuous linear functionals on P.

In this framework, we use the term "strategy" in the sense of [6]; however, to ensure that the values are well defined, independently of any measure-theoretical conventions, we show that players also have finite mixtures of pure strategies as ε-optimal strategies [**71**].

For each $s \in S$ let A^s denote an $I \times J$ matrix.

This defines a 2-person zero-sum game (A, P), where in a first stage a triplet (s, t_1, t_2) in R is chosen according to P, next t_n is told to player n, and finally players simultaneously choose resp. a row $i \in I$ and a column $j \in J$, with resulting payoff a_{ij}^s. We first show the above-mentioned properties for the value $v_A(P)$ of the game (A, P).

LEMMA 1. (A, P) has a value $v_A(P)$; more precisely, for every $\varepsilon > 0$ both players have finite convex combinations of pure strategies that are ε-optimal.

Both players also have optimal strategies, as transition probabilities from T to I and to J.

PROOF. Let P^n denote the marginal of P on the space of types T of player n. The mixed strategy space is identified with a (weak*-) compact, convex subset of $[L_\infty(T, P^1)]^I$ and $[L_\infty(T, P^2)]^J$.

The payoff becomes then a random variable in the unit ball of $L_\infty(R, P)$, and its expectation is a bilinear, separately continuous function on the mixed strategy spaces. By a standard minmax theorem [cf., e.g., [**71**] for references] it has a value $v_A(P)$ and optimal strategies. This establishes the second point. For the first point, note that the expected payoff is also a continuous function from player 1's mixed strategy space endowed with convergence in probability to the space of functions on player 2's strategy space, endowed with uniform convergence. Hence any good approximation in probability to an optimal strategy will be ε-optimal. Choose a step function: this can be rewritten as a finite convex combination of pure strategies. Point 1 is also established.

LEMMA 2. $v_A(P)$ is a continuous affine functional on (the compact, convex set) $\mathcal{P}$.

PROOF. For the continuity, let $P_k \in \mathcal{P}$ converge to P, and denote by $\sigma = (\sigma_t(i))_{i \in I, \ t \in T}$ an ε-optimal strategy of player 1 under P. By the approximation argument in Lemma 1, we can assume σ is a continuous function of t. If player 1 uses σ, and player 2 is of type t and uses his strategy $j \in J$, he expects a payoff of

$$\int_{S \times T} \left(\sum_i \sigma_{\tilde{t}}(i) a_{ij}^s \right) dt(s, \tilde{t}),$$

which is a continuous function of t since the integrand is continuous and also the identification of t with the corresponding probability on $S \times T$.

A best reply of player 2 is therefore to choose, for each of his types t, that $j \in J$ which minimizes the above expression.

It follows that σ guarantees player 1 in (A, P_k) an expected payoff of

$$\int_{T^2} \left[\min_{j \in J} \int_{S \times T^1} \left(\sum_{i \in I} \sigma_{\tilde{t}}(i) a_{ij}^s \right) dt(s, \tilde{t}) \right] dP_k^2(t).$$

(Recall that P_k^2 denotes the marginal of P on the types of player 2—any consistent P is fully determined by its marginals.)

The integrand is continuous, as a minimum of continuous functions; hence the integral is a continuous function of P_k.

Since the integral is a lower bound for $v_A(P_k)$, and since σ is ε-optimal at P, this establishes the lower semicontinuity of v_A at P. Upper semicontinuity follows dually, by permuting the roles of the players. Thus continuity.

As for the linearity, consider 2 points P and P' in $\mathcal{P}$, and $0 < \alpha < 1$. Consider the game where (A, P) and (A, P') are first chosen with probabilities α and $1-\alpha$, next both players are informed of the result, and then they play the chosen game. If (A, P) has been chosen, they will have the same posteriors on $S \times T$ as in (A, P), and similarly if (A, P'). It follows that the canonical consistent probability representing this information scheme is precisely $\alpha P + (1 - \alpha)P'$. But the value of the above-described game is clearly $\alpha v_A(P) + (1 - \alpha)v_A(P')$. Using the following Lemma 3, this establishes the linearity of v_A, and hence the lemma.

In Lemma 1, we did not use the fact that (R, P) was the canonical space—it could just as well have been an arbitrary probability space (H, P), equipped with random variables s, ω_1, and ω_2 with values in S, Ω_1, and Ω_2 resp.—Ω_n still denoting the space of observations of player n.

We have therefore also values $v_A(H, P, s, \omega_1, \omega_2)$.

LEMMA 3. *If $\tilde{P}$ denotes the canonical consistent probability corresponding to $(H, P, s, \omega_1, \omega_2)$, then $v_A(H, P, s, \omega_1, \omega_2) = v_A(\tilde{P})$, and optimal or ε-optimal strategies in $(A, \tilde{P})$ induce by the canonical projection mapping optimal or ε-optimal strategies in $\Gamma = (A, (H, P, s, \omega_1, \omega_2))$.*

PROOF. Consider a strategy σ (ε-optimal to fix ideas) of player 1 in $(A, \tilde{P})$. Denote by $\tilde{\sigma}$ the induced (by the projection mapping) strategy in Γ. It will be sufficient to show that $\tilde{\sigma}$ guarantees as much in Γ to player 1 as σ in (A, P).

We obtain the same expression as in Lemma 2 for player 2's expected payoff in Γ from strategy j against $\tilde{\sigma}$, when of type ω_2—except that the measure $dt(s, \tilde{t})$ has to be replaced by his conditional probability given ω_2. But the integrand depends only on s and $\tilde{t}$, since a strategy $\tilde{\sigma}$ is used; and the marginal distribution on $(s, \tilde{t})$ of player 2's conditional is exactly $dt_{\omega_2}(s, \tilde{t})$, if t_{ω_2} denotes the canonical image in T of his type ω_2—as we have seen in §C.2. Thus his best replies in Γ at ω_2 against $\tilde{\sigma}$ are the same as his best replies in $(A, \tilde{P})$ at t_{ω_2} against σ, and yield the same payoff. This payoff depends only on t_{ω_2}; hence the integrals under P and under $\tilde{P}$ are also the same.

This proves Lemma 3.

In classical cases, Lemma 2 just says that $v_A(P)$ is the integral for P of a continuous function of the prior on S. The next lemma generalizes the classical concavity and convexity properties of this function w.r.t. the sets $\Pi_{\mathrm{I}}(p)$ and $\Pi_{\mathrm{II}}(p)$ respectively (cf. §B.3(b)).

LEMMA 4 ("COMPARISON OF EXPERIMENTS"). *Given P and $\tilde{P}$ in $\mathcal{P}$, define "$\tilde{P}$ is more informative to 1 than P" whenever P and $\tilde{P}$ are the canonical measures of two information schemes*

$$(H, P, s, \omega_1, \omega_2) \quad and \quad (H, P, s, (\omega_1, \tilde{\omega}_1), \omega_2)$$

(i.e., player 1 receives more information in the second scheme).

Then $v_A(\tilde{P}) \geq v_A(P)$ whenever $\tilde{P}$ is more informative to 1 (and less informative to 2) than P.

PROOF. By Lemma 3, it is sufficient to prove the property for the $v_A(H, P, \dots)$ —there it is obvious, since increasing or decreasing information of a player increases or decreases his strategy set without changing payoffs.

It would be interesting to know the relevant orderings on $\mathcal{P}$.

LEMMA 5. *Lemmas $1, 2, 3, 4$ apply word for word to the discounted games and to the finite games—just replace v_A by v_λ or v_T, etc. The optimal strategies are now behavioral strategies.*

PROOF. The T stages of the finite game can be put into one big normal form, as a function of the starting state s. To this we can apply the previous lemmas.

For the discounted games Γ_λ, define approximations $\Gamma_{\lambda,T}$ to Γ_λ by setting all payoffs to zero after stage T. The payoff function of $\Gamma_{\lambda,T}$ converges uniformly to the payoff function of Γ_λ when $T \to \infty$, and our results apply to $\Gamma_{\lambda,T}$ since, after elimination of duplicate strategies, it becomes a finite game. If T is chosen such that the payoff function of $\Gamma_{\lambda,T}$ is ε-close to the payoff function of Γ_λ, then ε-optimal strategies in $\Gamma_{\lambda,T}$ become 2ε-optimal in Γ_λ. It follows that Γ_λ has a value v_λ, and has ε-optimal strategies which are finite, convex combinations of pure strategies (the pure strategies themselves having finitely many values as functions on T).

Let Σ_n denote the pure strategy space of player n, ignoring the initial information from the entrance law. Σ_n is compact metric as a product of finite sets. Let $\tilde{\Sigma}_n$ denote the space of probabilities on Σ_n, with the weak*-topology. Finally let $\mathcal{F}_n$ denote the space of measurable functions on (T, P^n) with values in $\tilde{\Sigma}_n$, endowed with the "weak*-weak *" topology, i.e., the coarsest topology for which, for any $g \in L_1(T, P^n)$ and $h \in C(\Sigma_n)$, $\int_T [\int_{\Sigma_n} h(\sigma) \, df_t(\sigma)] g(t) \, dP^n(t)$ is a continuous function of $f \in \mathcal{F}_n$. $\mathcal{F}_n$ is compact convex (certainly well known—easily proved anyway, e.g., using the weak*-compactness of the unit ball of L_∞, the lifting theorem (dispensible) and the Riesz representation theorem).

The expected payoff of $\Gamma_{\lambda,T}$ is a continuous linear function on $\mathcal{F}_n$ for fixed strategy of the opponent: indeed the projection from $\mathcal{F}_n$ to strategies in the T-stage game is obviously continuous, when the latter strategy space is endowed with the topology of Lemma 1—and our expected payoff depends only on this projection. So our proof of Lemma 1 also yields this. By uniform approximation, the expected payoff of Γ_λ is therefore also linear and continuous on $\mathcal{F}_n$ for fixed strategy of the opponent. The same use of the minmax theorem as in Lemma 1 yields therefore the existence of optimal strategies. Now rewrite those optimal strategies (both for Γ_T and for Γ_λ) as behavioral strategies (using, e.g., [6]).

All other properties go through by uniform approximation.

(3.b) For the recursive formula, it is convenient to adopt a normalization as in §C.1(d), where the payoff at any given stage depends only on the current state. It is also somewhat easier (for defining first-stage behavioral strategies) to assume that information in the game is given after each stage, and not before—like for instance in our final combinatorial form, and as suggested at the end of (d).

Assume an entrance law $P \in \mathcal{P}$, and given behavioral strategies σ and τ of players 1 and 2 for the first stage of the game—i.e., measurable functions from T to probabilities over the action set. Denote by $P^{\sigma,\tau}$ the corresponding information scheme: first a triplet (s, t_1, t_2) is selected according to P, and player n is informed of t_n; they independently select an action a_n respectively according to $\sigma(t_1)$ and $\tau(t_2)$; next (random) signals (ω_1, ω_2) are selected for them in the game as a function of s (and the pair of the actions) together with a new state $\tilde{s}$.

Our information scheme $P^{\sigma,\tau}$ is the above-described probability space, say $(H, P_{\sigma,\tau})$, together with the random variables $\tilde{s}$, (t_1, a_1, ω_1) and (t_2, a_2, ω_2). Denote finally by $P(\sigma, \tau)$ the corresponding canonical consistent probability.

$E_P(u)$ denotes as usual $\sum_{s \in S} P(s) u(s)$, with u the payoff function.

PROPOSITION 1 (RECURSIVE FORMULA). (a) *Let $v(P)$ denote either $v_\lambda(P)$ or $v_T(P)$. Then both $\max_\sigma \min_\tau v(P(\sigma, \tau))$ and $\min_\tau \max_\sigma v(P(\sigma, \tau))$ exist and they are equal. Denote for short this saddle point value by* $\mathrm{Val}[v(P(\sigma, \tau))]$.

(b)

$$v_\lambda(P) = \lambda E_P(u) + (1 - \lambda)\mathrm{Val}[v_\lambda(P(\sigma, \tau))]$$

and

$$v_T(P) = \frac{1}{T} E_P(u) + \frac{T - 1}{T}\mathrm{Val}[v_{T-1}(P(\sigma, \tau))].$$

(c) *v_λ and v_T are uniquely determined (in the space of all bounded functions on $\mathcal{P}$) by the above formulas.*

PROOF. Let $\sigma = (\sigma^1, \sigma^2, \sigma^3, \dots)$ be an optimal behavioral strategy of player 1 in Γ_λ, and let $\tau = (\tau^1, \tau^2, \tau^3, \dots)$ be an arbitrary behavioral strategy of player 2 in Γ_λ. Let $\bar{\sigma}$ and $\bar{\tau}$ denote the sequences $(\sigma^2, \sigma^3, \dots)$ and $(\tau^2, \tau^3, \dots)$: $\bar{\sigma}$ and $\bar{\tau}$ are behavioral strategies in the game $\Gamma_\lambda(P^{\sigma_1, \tau_1})$ which starts with the information scheme P^{σ_1, τ_1}. Denote the corresponding expected payoff by $u_\lambda(P^{\sigma_1, \tau_1}, \bar{\sigma}, \bar{\tau})$. Then

$$v_\lambda(P) \leq u_\lambda(P, \sigma, \tau) = \lambda E_P(u) + (1 - \lambda)u_\lambda(P^{\sigma_1, \tau_1}, \bar{\sigma}, \bar{\tau})$$

(by writing the payoff as the sum of the first-stage payoff and the payoff for the rest of the game). Hence, $\bar{\tau}$ being arbitrary,

$$v_\lambda(P) \leq \lambda E_P(u) + (1 - \lambda) \inf_{\bar{\tau}} u_\lambda(P^{\sigma_1, \tau_1}, \bar{\sigma}, \bar{\tau}),$$

and a fortiori

$$v_\lambda(P) \leq \lambda E_P(u) + (1 - \lambda) \sup_{\bar{\sigma}} \inf_{\bar{\tau}} u_\lambda(P^{\sigma_1, \tau_1}, \bar{\sigma}, \bar{\tau}).$$

Now the $\sup_{\overline{\sigma}} \inf_{\overline{\tau}}$ is, by Lemmas 1 and 5, equal to $v_\lambda(P^{\sigma_1,\tau_1})$, and hence, by Lemmas 3 and 5, equal to $v_\lambda(P(\sigma_1,\tau_1))$. Thus

$$v_\lambda(P) \leq \lambda E_P(u) + (1-\lambda)v_\lambda(P(\sigma_1,\tau_1));$$

hence, τ_1 being arbitrary,

$$v_\lambda(P) \leq \lambda E_P(u) + (1-\lambda)\inf_{\tau_1} v_\lambda(P(\sigma_1,\tau_1)).$$

Consider now the λ-discounted game where player 1 is constrained to use σ_1 in the first period. This is easily described as a game in our family; hence it has a value v and optimal strategies (Lemma 5). Even knowing the optimal strategy of player 1—or just its first stage component—player 2 would have no advantage to deviate; so v and the optimal strategies apply also to the variant when player 1 would be informed of the $1°$ stage mixed strategy τ_1 of 2. The value of this variant is, by the above argument, equal to our right-hand member. Since player 2 has an optimal strategy, it follows that the inf over τ_1 is achieved for any σ_1:

$$v_\lambda(P) \leq \lambda E_P(u) + (1-\lambda)\min_{\tau_1} v_\lambda(P(\sigma_1,\tau_1)).$$

Let $\overline{\sigma}_1$ and $\overline{\tau}_1$ denote the first-stage components of optimal strategies of players 1 and 2 in $\Gamma_\lambda(P)$.

We have shown that $\forall \sigma_1$, $\mathrm{Min}_{\tau_1} v_\lambda(P(\sigma_1,\tau_1))$ exists and

$$v_\lambda(P) \leq \lambda E_P(u) + (1-\lambda)\min_{\tau_1} v_\lambda(P(\overline{\sigma}_1,\tau_1)).$$

Permuting the roles of the players, we obtain dually $\forall \tau_1$, $\mathrm{Max}_{\sigma_1} v_\lambda(P(\sigma_1,\tau_1))$ exists and

$$v_\lambda(P) \geq \lambda E_P(u) + (1-\lambda)\max_{\sigma_1} v_\lambda(P(\sigma_1,\overline{\tau}_1)).$$

Hence $\max_{\sigma_1} v_\lambda(P(\sigma_1,\overline{\tau}_1)) \leq \min_{\tau_1} v_\lambda(P(\overline{\sigma}_1,\tau_1))$.

This shows that $v_\lambda(P(\sigma,\tau))$ has a saddle point $(\overline{\sigma}_1,\overline{\tau}_1)$, and hence the two inequalities yield

$$v_\lambda(P) = \lambda E_P(u) + (1-\lambda)\mathrm{Val}[v_\lambda(P(\sigma,\tau))].$$

The results for v_T are established in exactly the same way. This proves (a) and (b).

As for (c), note first that v_T is uniquely determined given v_{T-1}, and no assumption on v_0 is required for v_1. For the v_λ, replace "value" by "$\sup_\sigma \inf_\tau$" and use Picard's contraction principle in the space of all bounded functions: there exists a unique solution to the corresponding equation. This proves the proposition.

In the major known cases, this recursive formula (or rather its particular incarnations) allows us to prove convergence of v_λ and of v_T —to the same limit. In the case of repeated games with incomplete information, it even allows us to establish precise error bounds (cf. [77, 78, 127, 128], with more extensive and systematic results in a forthcoming paper of Mertens and Zamir). In the case of stochastic games, Bewley and Kohlberg [20, 21] used the real-algebraic nature of the resulting incarnation, together with the existence of Puiseux expansions for real algebraic functions, to establish the convergence and the second order terms.

Could one work directly with the above equations? This leads to the first problem suggested here—which we want to state independently of the above development on entrance laws:

Problem 1. Do v_λ and v_T converge and have the same limit for all games normalized in §1?

This would immediately imply convergence for all entrance laws with finite support. To go to the closure—recall that this should be $\mathcal{P}$—and for other, later, purposes, we suggest

Problem 2. Are the $v_A(P)$ equicontinuous when A varies over all finite (R^S-valued) matrices with entries in $[-1, 1]$?

Using the above uniform approximation of discounted games by finite games, this would immediately establish the equicontinuity of the $v_\lambda(P)$ and the $v_T(P)$ —and hence, given Problem 1, their uniform convergence.

This equicontinuity or Lipschitz character is crucial in many papers, among others in the only paper that establishes the value for a class of repeated games with incomplete information by reducing the problem to a stochastic game with continuous state space [33]—which we will shortly show is exactly what the above proposition generalizes (cf. Proposition 2).

Before that, let us establish some plausibility for the above problem by considering a particular case. Assume that player 1 always gets at least as much information as player 2—thus in particular is told player 2's actions. Then a current state of information is completely described by player 2's distribution on player 1's current posterior on S—since this distribution is also known to 1. Assume for notational simplicity that S has only two points: we thus get distributions on the unit interval as extreme points of the relevant subset of $\mathcal{P}$, and for $\mathcal{P}$ all probability measures μ on those distributions. The continuity and linearity of $v_A(\mu)$ imply that $v_A(\mu) = \int r_A(P)\mu(dP)$, where r_A is a continuous function of the distribution P. Thus, establishing the equicontinuity of the v_A's amounts to establishing the equicontinuity of the r_A's, where $r_A(P)$ is the value of the game, where first some $q \in [0, 1]$ is selected according to P and told to player 1, and next a state s is selected with probability q after which the players play A^s ($s = 1, 2$). The equicontinuity is w.r.t. the weak*-topology on P's. By the minmax theorem, we have

$$r_A(P) = \min_y \int_0^1 \max_i \sum_j y_j \left(qa_{ij}^1 + (1-q)a_{ij}^2\right) dP(q).$$

Since $|a_{ij}^s| \leq 1$, each $qa_{ij}^1 + (1-q)a_{ij}^2$ has Lipschitz constant ≤ 2, hence also the $\sum_j y_j(qa_{ij}^1 + (1-q)a_{ij}^2)$, hence also the maximum over i of those functions. Endow the set of P's with the metric generated by the integrals of all functions with Lipschitz constant ≤ 1—this metric generates the weak*-topology. The integrals then become Lipschitz functions of P with constant 2, and hence also their minimum over y. Thus the equicontinuity.

Let us now return to our recursive formula. Denote by Σ^n the set of all borel functions from T to the probabilities over the action set A^n of player n.

Denote by $\tilde{\Gamma}$ the (stochastic) game with continuous state and action sets P and Σ^1, Σ^2, where, if P denotes the current state, and σ^1, σ^2 the actions selected by the players at P, the current payoff is $E_P(u)$ and the next state is $P(\sigma^1, \sigma^2)$— i.e., a "deterministic stochastic" game where the current payoff depends only on the current state. $\tilde{\Gamma}_\lambda(P)$ and $\tilde{\Gamma}_T(P)$ will denote the corresponding discounted and finite games, with initial state P.

PROPOSITION 2. $\tilde{\Gamma}_\lambda(P)$ and $\tilde{\Gamma}_T(P)$ have values $\tilde{v}_\lambda(P)$ and $\tilde{v}_T(P)$ and optimal pure strategies. $\tilde{v}_\lambda = v_\lambda$ and $\tilde{v}_T = v_T$.

PROOF. We prove also that the optimal pure strategies are to use at state P a borel version of the first stage of $\Gamma_\lambda(P)$ or $\Gamma_T(P)$ (with T the number of remaining stages).

For the finite games, the proof goes immediately by induction on T. For the discounted games, assume player 1 uses the above-described optimal strategy, and player 2 also: the recursive formula implies now immediately that the payoff is $v_\lambda(P)$. Assume now instead that player 2 is free to choose his strategy in the first T stages, but uses the optimal strategy after: at stage T, his optimal choice is clearly the optimal strategy, by the recursive formula. Hence by recursion his best choice is to use always the optimal strategy: he cannot get below $v_\lambda(P)$ by such strategies.

Since the payoff of the discounted game depends only up to ε on what happens after T for T sufficiently large, it follows that, when player 1 uses his optimal strategy, the payoff is always at least $v(P)$. The dual statement for player 2 now implies the result.

Thus we have reduced Problem 1 to the same problem for a specific class of stochastic games.

It was shown by Mertens and Neyman [73] that stochastic games (even with continuous state and action spaces) have a value (as defined in §A) as soon as (conditions are always satisfied for finite space and action spaces)

(1) payoffs are uniformly bounded;

(2) the v_λ exist; and

(3) for some decreasing sequence λ_i such that λ_i converges to zero and λ_{i+1}/λ_i to one we have $\sum_i \|v_{\lambda_{i+1}} - v_{\lambda_i}\| < \infty$ (the norm is the supnorm over the state space).

A careful reading of the proof shows even more: the T_0 in the definition of the value (§A) can be chosen independently of the initial state, hence uniform convergence of v_T and v_λ to the value.

The existence theorem for v implies the (uniform as just seen) convergence of v_T and of v_λ to v. Conditions (1) and (2) are satisfied in our case (Proposition 2). Condition (3) is "slightly" stronger than uniform convergence.

Problem 3. Can condition (3) above be replaced by uniform convergence?

Uniform convergence is not just suggested out of the blue sky, or because it would make the theorem much more usable: it seems intimately tied to the existence of a value, and is clearly necessary, as the following example shows. The state space consists of the integers plus an origin. At the origin, player 1 can choose the next state. From any integer the game proceeds to the next one. Player 2 just has to watch the game. Payoff is one on the negative integers, zero elsewhere. v_λ equals one at the origin, and converges to zero elsewhere—but not uniformly. Clearly the game starting at the origin has no value—at least with our definition, which is needed in every application.

The nature of the problem may be better understood by looking at its implications in some particular cases:

- For 1 player, it is the general model of dynamic programming.

- For 0 players, it is the general model of Markov chains. It becomes then a pointwise ergodic theorem that uses no recurrence assumption, but only assumptions on the potentials of the function of which one wants to take arithmetic averages.

- If in addition there is no randomness, one obtains a (relatively weak) Tauberian theorem implying the Cesaro convergence of a bounded sequence under assumptions of Abel convergence.

I found in old notes of mine: "One can see immediately that, in the last two cases, the proof does not use the third assumption, but only the uniform convergence of v_λ." I did not check this again.

A positive solution to Problem 3 would have strong implications for our model too. For instance, by what we said before, given also a positive solution to Problem 2, Problem 1 would be reduced to the following basic problem: Does v_λ converge for our combinatorial model?

Note that, if we had started with the final normalization of §1, players would still move alternately in the stochastic game: the stochastic game would in addition be a game with perfect information, hence has a value $\underline{v}$ if we define the payoff, e.g., as $\liminf_{T\to\infty}(1/T)\sum_1^T P_t(W)$ (and similarly $\overline{v}$)—using Martin's [62] result. [This is stated for borel winning sets, but extends immediately to borel payoff functions by considering the level sets: there exists a value, and if one player has no optimal strategy, the other has a "super-optimal" strategy guaranteeing an outcome better than the value. An observation of A. Neyman and the author shows that the result even implies its own extension to the nonzero-sum case: note first that, in the zero-sum case, when the payoff takes only finitely many values, optimal strategies exist, and they can be "improved" so as to be still optimal after any finite history. In the general case, keeping the assumption of finitely many values, select for each player, considering all others as one single opponent trying to minimize his payoff, one such pure optimal strategy, and also one for the opponent. Now instruct all players each to use his own optimal strategy, until one of them deviates: from that stage on, all his opponents use their selected optimal pure strategy for minimizing his payoff:

this is a (pure strategy) equilibrium. For a general, bounded payoff function, choose a borel approximation with finitely many values, uniformly ε-close, and apply the above argument to this approximation: we get a (pure strategy) ε-equilibrium. (Subgame perfectness is a completely different issue....)] In fact, in view of our definition of value, we would be more interested in $\lim_{T\to\infty} \underline{v}_T$, with $\underline{v}_{T_0}$ arising from the payoff function $\inf_{T\geq T_0}(1/T)\sum_1^T P_t(W)$: this even yields a Gale-Stewart game.... This perfect information aspect also "explains" the saddle point property. Thus, not only have all information aspects disappeared from the problem, most strategic aspects have also: this points to the basic "summability" aspect of the problem, as already noted after Problem 3.

Note in particular that this Gale-Stewart aspect immediately implies that, if the stochastic game has a value, it has pure ε-optimal strategies, depending only on current state and date: consider the values $v_t(a, P_t)$ of the games with starting state and date P_t and t and with payoff function

$$\varphi_t(a, P_t)(\sigma, \tau) = \inf_{T\geq T_0 \vee t} \frac{1}{T}\left(a + \sum_{s=t}^{T} P_s(W)\right).$$

Also let

$$a_t = \sum_{1\leq s<t} P_s(W), \quad \alpha_t = \underline{v}_{T_0} - \varepsilon \sum_{1\leq s<t} 2^{-s},$$

$$\underline{a}_t(P) = \min\{a \mid 0 \leq a \leq +\infty, \ v_t(a, P) \geq \alpha_t\},$$

and let $\sigma_t(P)$ denote the first stage of a pure strategy of player 1 in $\varphi_t(\underline{a}_t(P), P)$ guaranteeing him $\geq \alpha_{t+1}$ in this game. [The set of action sequences for which $\varphi_t(a, P) \geq x$ is closed in the product of the discrete topologies—hence the game with this winning set has a value $v_t(a, P, x)$ and pure optimal strategies (Gale-Stewart). Then

$$v_t(a, P) = \sup\{x \mid v_t(a, P, x) = 1\} = \inf\{x \mid v_t(a, P, x) = 0\} :$$

both players can defend $v_t(a, P)$ with pure δ-optimal strategies for any δ. Since the payoff function is uniformly continuous in a, the value is also; hence

$$v_t(\underline{a}_t(P), P) \geq \alpha_t > \alpha_{t+1} :$$

player 1 has such a strategy $\sigma_t(P)$.] Let σ denote the strategy of player 1 consisting of playing $\sigma_t(P_t)$ at every stage t.

If player 1 uses $\sigma_t(P_t)$ at stage t, then whatever action the opponent takes after him, at the next stage player 1 will still be able to get α_{t+1}:

$$v_{t+1}(\underline{a}_t(P_t) + P_t(W), P_{t+1}) \geq \alpha_{t+1};$$

hence $\underline{a}_t(P_t)+P_t(W) \geq \underline{a}_{t+1}(P_{t+1})$. Under σ, this holds for all t—hence summing over $t < T$, we get $a_T \geq \underline{a}_T(P_T)$ for all T (since $\underline{a}_1(P_1) = 0$), and thus for $t \geq T_0$,

$$a_{t+1}/t \geq v_t(a_t, P_t) \geq v_t(\underline{a}_t(P_t), P_t) \geq \alpha_t > \underline{v}_{T_0} - \varepsilon :$$

σ guarantees to player 1 that, $\forall T \geq T_0$, $(1/T)\sum_1^T P_t(W) > \underline{v}_{T_0} - \varepsilon$: it is ε-optimal.

(3.c) Let us finally give an example of another, stronger application of those problems. Consider our most general model of §1, including information lags etc., but assume that player 1 always has at least as much information as player 2. There is no loss in adding redundant information to the signals; thus we can assume that each player's signals include his last action and his action set for the next stage, and that player 1's signals include player 2's. Our assumption is preserved in normalization steps (a) to (d)—only take care when duplicating player 2's actions in step (d) to inform player 1 of which duplicate is used; this is harmless. Thus we get a two-person zero-sum game as described sub(d), with $\mathcal{P}^1$ finer than $\mathcal{P}^2$ (and each player's last action determined by his partition). Let us only take the variant, suggested at the end of (d), that it is *after* each stage that players are told in which partition element the next state falls—i.e., no information is given before the first stage.

We have already shown in this subsection that Problem 2 is solved affirmatively for those games. Thus if v_λ converges for those games, then $v_\lambda(P)$ will converge uniformly over $\mathcal{P}$. We have also seen how the particular structure implied a particular structure for the relevant $\mathcal{P}$—the same goes through for the description of the stochastic game:

Let Δ^S denote the simplex of probabilities on S, and M^S the space of probabilities on Δ^S: $\mathcal{P}$ should be the space of probabilities μ on M^S, but we have seen that we could express the values as integrals of (Lipschitz) functions on M^S. It will therefore also be more convenient to use M^S as the state space for our stochastic game—even if it is then no longer deterministic.

Note that, given P in M^S, the current payoff is given by $\int_{\Delta^S} q(W)\,dP(q)$, and that the distribution of the next P (recall this is the posterior of 2 on 1's posterior q) can be computed from the mixed strategy of player 1—say $x_q(i)$—and the pure stategy j of player 2, since this is common knowledge. Thus we can use player 2's action set of the original game as his pure strategy set in the stochastic game.

Thus our stochastic game $\tilde{\Gamma}$ has M^S as a state space, and as strategy sets the original action set for 2 and the behavioral strategies x_q for 1.

Now, if we know the convergence of v_λ, and Problem 3 is solved (at least for those stochastic games), we will know that the stochastic game has a value $v_\infty(P) = \lim v_\lambda(P) = \lim v_T(P)$ for arbitrary initial distributions P. Consider then an ε-optimal strategy σ of player 1 in the stochastic game. It tells at each period which function x_q to take, as a function of past and present states P and past actions x_q and j.

The dependence on the first x_q can be eliminated, since it is known from σ. Hence also the second x_q, since it now depends only on earlier P's and j's, and so on: σ can give each x_q just as a function of earlier j's and P's and current P. (This is the proof that knowledge about own past actions can always be dispensed with [25].) But past and present P's and q's can be computed by player 1 in the original game: since he knows—by induction—previous P's and j's, he knows from σ the last x_q he used, hence the next P since it depends only on current P and x_q and the signal of 2, which is known to him. Similarly the next q depends only on the current q and his own signal. Thus we can rewrite σ as a strategy $\tilde{\sigma}$ in the original game, giving at each stage a probability over player 1's actions as a function only of his own past signals and of the initial q.

Consider now, in Γ_T, a best reply $\tilde{\tau}$ of player 2 against $\tilde{\sigma}$. At every stage, as a function just of his own past signals—not of his past mixed strategy—he can compute the current P since he knows he plays against $\tilde{\sigma}$. He can similarly compute the distribution of next P's for each one of his actions. Since his expected payoff at each stage depends only on the current P, a little dynamic programming (backwards induction from the endstage T on) shows that we can assume his best reply will at each stage be a function only of the current P: it arises from a strategy τ in the stochastic game. The correspondence is such that the expected payoff from $(\tilde{\sigma}, \tilde{\tau})$ in the original T-stage game is the same as the payoff from (σ, τ) in the stochastic game. Hence $\tilde{\sigma}$ guarantees in each Γ_T $(T \geq T_0)$ an expected payoff at least $v_\infty(P) - \varepsilon$; this proves that the first condition for $v_\infty(P)$ to be the maxmin of our game is satisfied. (Ideas for this type of argument can be traced back to conversations with A. Neyman and S. Sorin.) The ε-optimal strategy of player 2 in the stochastic game can be used similarly:

Consider first an arbitrary strategy σ of player 1 in Γ: σ determines at each stage a mixed action of 1 as a function of past signals of 1 and initial $q = q_0$. Note that the successive q's are computed knowing just the initial q and the signals of 1. Since those contain the pair of actions, knowledge of the mixed strategies is not necessary. Conditionally to q_1 and the first signal of 2, q_0 and the first signal of 1 are independent of the future strategy of 2, of the current state of nature s_1, and of future choices of nature; and this conditional distribution can be computed knowing just σ and P—since the action of 2 is included in the given signal of 2. Thus, conditionally to q_1 and this first signal of 2, player 1 has a lottery over his own first signals, and hence over his future strategy inasmuch as it depends on those signals. The conditional independence properties stated imply that he can as well average his future strategy with this lottery—whatever player 2's strategy, this will not affect the expected payoff in any future period— and clearly also not in the first period. This averaged future strategy can now be recast as a behavioral strategy without affecting anything; we have obtained an equivalent strategy σ_1 of 1, which does not depend (after the first stage) on the initial q and the first signal of 1, but just on q_1 and the first signal of 2 (which in turn are functions of those data). The same can now be done with σ_1 at the next stage, to obtain a strategy σ_2, and so on: finally we obtain an equivalent strategy of 1, say $\tilde{\sigma}$, where at each stage the current mixed action x depends only on the current q and the past signals of 2—i.e., it gives at each date a function x_q depending only on past signals of 2. Any such $\tilde{\sigma}$ is a genuine strategy of 1, since current q's are functions only of initial q and past signals of 1, independent of the players' mixed strategies. Thus, in any question where we are interested only in expected payoffs, we can assume player 1 uses only strategies $\tilde{\sigma}$.

Now let τ be an ε-optimal strategy of player 2 in the stochastic game: at every date we have a mixed action as a function of past P's and past x_q's (past actions of 2 can be dispensed with as usual).

Consider an arbitrary strategy $\tilde{\sigma}$ of player 1 in the original game, and define as follows a strategy τ_σ of 2 in the original game: substitute in τ for the past x_q's their expression by $\tilde{\sigma}$ as a function of player 2's past signals; substitute also for the past posteriors P of 2 their expression (computable from $\tilde{\sigma}$) as a function of the past signals of 2: τ_σ is a genuine strategy of 2, giving at every stage a mixed action only as a function of player 2's past signals.

Let us now construct, from τ_σ and $\tilde{\sigma}$, a *mixed* strategy $\overline{\sigma}$ of 1, in the stochastic game, such that expected payoffs are the same in the stochastic game from τ and $\overline{\sigma}$ as in the original game from τ_σ and $\tilde{\sigma}$. Introduce (in the original game) a referee, who would perform truthfully the randomizations required by the game, and give the players their information (and secretly their payoffs) after having been told their chosen actions. But he can also, since q is a function only of the initial q and player 1's signal, compute the current q—so, if given x_q by player 1, he can perform the required randomization over player 1's actions himself. But then player 1 no longer needs his own signals—he just needs to be told of player 2's signals to select x_q. Similarly, to facilitate things, the referee could just inform player 2 of player 1's last x_q and of the next P, instead of giving him his detailed signals—player 2 would then use τ to obtain the same effect as τ_σ (this is how τ_σ was constructed). He could just give the players instead of their actual payoffs at each stage their expected payoffs given the current P—i.e., $P(W)$—since they are only interested in the latter. Now most randomizations he makes are solely for player 1's purpose—otherwise he could just, knowing the starting P, listen at each stage to player 2's action and to player 1's x_q, make a small randomization to select the next P, and inform both players of the new P, the last x_q, and the last action of 2. Instead, he must now make additional randomizations, conditionally to the above-mentioned, to select signals of 2 for player 1, which he will never need himself in the future for his own randomizations. Thus player 1 could as well take care of those himself, since he is being told at each stage of the new conditions of the conditional distribution (last action of 2, last x_q, and the new P). But this is the description of the stochastic game, where player 2 uses τ and player 1 randomizes at each stage, as a function of current and past information and of the results of past randomizations, to select a new "signal of 2," and hence a new x_q (by $\tilde{\sigma}$): player 1 uses some mixed strategy $\overline{\sigma}$ in the stochastic game.

Therefore, for T_0 sufficiently large, any $T \geq T_0$, and any σ, the expected payoff from τ and $\overline{\sigma}$ in the T-stage stochastic game is $\leq v_\infty + \varepsilon$ (ε-optimality of τ), hence also the payoff in the original game resulting from $\tilde{\sigma}$ and τ_σ—and thus from σ and τ_σ.

This proves the second condition for $v_\infty(P)$ to be the maxmin of our game—and in fact slightly more: the statement itself implies $\lim v_n = $ maxmin, since T_0 is independent of σ.

Here too we could obtain a stochastic game with perfect information, by performing the analog of normalization step (f) after (d) [(e) cannot be performed]. Just take care to let player 1 move before player 2, to keep his advantage in information.

To recapitulate: the solution of Problem 3 (at least for the relevant class of stochastic games) would imply the following: just convergence of v_λ for games (as described in §1), where player 1 is always better informed than player 2, would already imply that $v_\lambda(P)$ and $v_T(P)$ converge uniformly (over all initial distributions P), say to $v_\infty(P)$, and that $v_\infty(P)$ is the maxmin of the game: in particular, the better informed player always has good strategies which are independent of the (sufficient) length of the game or of the (small) discount factor.

This would be a very significant extension of many important result: the symmetric case (with random signals) in incomplete information; incomplete information on one side; S. Sorin's "Big Matches" with incomplete information, stochastic games, etc. Further, the assumption of one player being better informed than the other is typically met in many applications. Finally, even this assumption is not really necessary—it would be sufficient that after a fixed lag L, player 1 is always informed of whatever player 2 knew. Indeed, we can then write an equivalent game, also in our framework, and where the lag disappears: Consider a player 1bis, who will receive at date T the information of both players 1 and 2 at time T, and will have to choose on this basis, at time T, an action of player 1 for time $T + L$, as a function of the stream of signals (including own actions) of player 1 between T and $T + L$. Player 1bis also chooses at time 0 a normal form pure strategy of player 1 for the first L stages of the game. This equivalent game between 1bis and 2 can be normalized to fit our framework using the same methods as previously.

(3.d) I hope to have been able to convey to some extent

(1) in §A, the importance of the general model, many major applications requiring most of its features at the same time, and its conceptual interest;

(2) in §B, a vague idea about what is available in this field;

(3) in the first part of §C, that the general model has nevertheless a conceptually very simple and purely combinatorial equivalent; and

(4) in the last two parts, a few new tools and ideas that might be useful for further progress, illustrating at the same time some old ideas in a renewed framework as well as the high degree of interplay between various apparently unrelated topics within this general model.

REFERENCES

1. D. Abreu, *Infinitely repeated games with discounting: a general theory*, Discussion Paper 8103, Harvard University, Cambridge, Mass., 1984.

2. D. Abreu, D. Pearce, and E. Stacchetti, *Towards a theory of discounted repeated games with imperfect monitoring*, mimeograph, 1986.

3. R. J. Aumann, *Acceptable points in general cooperative n-person games*, Contributions to the Theory of Games. Vol. IV, Ann. of Math. Stud., vol. 40, Princeton Univ. Press, Princeton, N.J., 1959, pp. 287–324.

4. ____, *Acceptable points in games of perfect information*, Pacific J. Math. **10** (1960), 381–387.

5. ____, *The core of a cooperative game without side payments*, Trans. Amer. Math. Soc. **98** (1961), 539–552.

6. ____, *Mixed and behaviour strategies in infinite extensive games*, Advances in Game Theory, M. Dresher et al., editors, Ann. of Math. Stud., vol. 52, Princeton Univ. Press, Princeton, N.J., 1964, pp. 627–650.

7. ____, *A survey of cooperative games without side payments*, Essays in Mathematical Economics in Honor of Oskar Morgenstern, M. Shubik, editor, Princeton Univ. Press, Princeton, N.J., 1967, pp. 3–27.

8. ____, *Subjectivity and correlation in randomized strategies*, J. Math. Econom. **1** (1974), 67–95.

9. ____, *Repeated games*, Issues in Contemporary Microeconomics and Welfare, G. R. Feiwel, editor, Macmillan, New York, 1985, pp. 209–242.

10. ____, *Correlated equilibria as an expression of Bayesian rationality*, Econometrica (to appear).

11. R. J. Aumann and S. Hart, *Bi-convexity and bi-martingales*, Israel J. Math. **54** (1986), 159–180.

12. R. J. Aumann and M. Maschler, *Game-theoretic aspects of gradual disarmament*, Reports to the U.S. Arms Control and Disarmament Agency, ST-80 (1966), Chapter V, pp. 1–55.

13. ____, *Repeated games with incomplete information: a survey of recent results*, Reports to the U.S. Arms Control and Disarmament Agency, ST-116 (1967), Chapter III, pp. 287–403.

14. ____, *Repeated games of incomplete information: the zero-sum extensive case*, Reports to the U.S. Arms Control and Disarmament Agency, ST-143 (1968), Chapter III, pp. 37–116.

15. R. J. Aumann, M. Maschler, and R. Stearns, *Repeated games of incomplete information: an approach to the non-zero-sum case*, Reports to the U.S. Arms Control and Disarmament Agency, ST-143 (1968), Chapter IV, pp. 117–216.

16. R. J. Aumann and L. S. Shapley, *Long term competition—a game theoretic analysis*, mimeograph, 1976.

17. R. J. Aumann and S. Sorin, private communication, 1986.

18. R. Axelrod, *The evolution of cooperation*, Basic Books, New York, 1984.

19. J.-P. Benoît and V. Krishna, *Finitely repeated games*, Econometrica **53** (1985), 905–922.

20. T. Bewley and E. Kohlberg, *The asymptotic theory of stochastic games*, Math. Oper. Res. **1** (1976), 197–208.

21. ____, *The asymptotic solution of a recursion equation occurring in stochastic games*, Math. Oper. Res. **1** (1976), 321–336.

22. ____, *On stochastic games with stationary optimal strategies*, Math. Oper. Res. **3** (1978), 104–125.

23. D. Blackwell, *An analog of the minmax theorem for vector payoffs*, Pacific J. Math. **6** (1956), 1–8.

24. D. Blackwell and T. S. Ferguson, *The big match*, Ann. Math. Stat. **39** (1968), 159–163.

25. N. Dalkey, *Equivalence of information patterns and essentially determinate games*, Contributions to the Theory of Games. Vol. II, H. W. Kuhn and A. W. Tucker, editors, Ann. of Math. Stud., vol. 28, Princeton Univ. Press, Princeton, N.J., 1953, pp. 217–243.

26. M. Dresher, A. W. Tucker, and P. Wolfe (Editors), *Contributions to the theory of games*. Vol. III, Ann. of Math. Stud., vol. 39, Princeton Univ. Press, Princeton., N.J., 1957.

27. M. Dresher et al. (Editors), *Advances in game theory*, Ann. of Math. Stud., vol. 52, Princeton Univ. Press, Princeton, N.J., 1964.

28. L. E. Dubins, *A discrete evasion game*, Contributions to the Theory of Games. Vol. III, M. Dresher, A. W. Tucker, and P. Wolfe, editors, Ann. of Math. Stud., vol. 39, Princeton Univ. Press, Princeton, N.J., 1957, pp. 231–255.

29. H. Everett, *Recursive games*, Contributions to the Theory of Games. Vol. III, Ann. of Math. Stud., vol. 39, Princeton Univ. Press, Princeton, N.J., 1957, pp. 19–1025.

30. A. Federgruen, *On N-person stochastic games with denumerable state space*, Adv. in Appl. Probab. **10** (1978), 452–471.

31. T. S. Ferguson and L. S. Shapley, *On a game of Gleason*, private communication, 1986.

32. T. S. Ferguson, L. S. Shapley, and B. Weber, *A stochastic game with incomplete information*, mimeograph, 1970.

33. F. Forges, *Infinitely repeated games of incomplete information: symmetric case with random signals*, Internat. J. Game Theory **11** (1982), 203–213.

34. ____, *A first study of correlated equilibria in repeated games with incomplete information*, CORE Discussion Paper 8212, Louvain-la-Neuve, Belgium, 1982.

35. ____, *A note on Nash equilibria in repeated games with incomplete information*, Internat. J. Game Theory **13** (1984), 179–187.

36. ____, *Communication equilibria in repeated games with incomplete information*. Parts I, II, III, CORE Discussion Papers 8406, 8411, 8412, Louvain-la-Neuve, Belgium, 1984; Math. Oper. Res. (to appear).

37. ____, *An approach to communication equilibria*, Econometrica **54** (1986), no. 6, 1375–1385.

38. ____, *Correlated equilibria in a class of repeated games with incomplete information*, Internat. J. Game Theory **14** (1985), 129–150.

39. ____, *Correlated equilibria in repeated games with lack of information on one side: a model with verifiable types*, Internat. J. Game Theory **15** (1986), 65–82.

40. F. Forges, J. F. Mertens, and A. Neyman, *A counterexample to the folk-theorem with discounting*, Econom. Lett. **20** (1986), 7.

41. J. Friedman, *Oligopoly and the theory of games*, North-Holland, Amsterdam, 1977.

42. D. Fudenberg and D. Levine, *Subgame-perfect equilibria of finite and infinite horizon games*, J. Econom. Theory **31** (1983), 251–268.

43. D. Fudenberg and E. Maskin, *The folk theorem in repeated games with discounting and with incomplete information*, Econometrica **54** (1986), 533–554.

44. D. Gillette, *Stochastic games with zero stop probabilities*, Contributions to the Theory of Games. Vol. III, M. Dresher, A. W. Tucker, and P. Wolfe, editors, Ann. of Math. Stud., vol. 39, Princeton Univ. Press, Princeton, N.J., 1957, pp. 179–187.

45. A. Gleason, *Unpublished example*, oral tradition, 1949?

46. J. C. Harsanyi, *Games of incomplete information played by Bayesian players*. Parts I, II, III, Management Sci. **14** (1967–68), 159–182, 320–334, 486–502.

47. S. Hart, *Nonzero-sum two-person repeated games with incomplete information*, Math. Oper. Res. **10** (1985), 117–153.

48. S. Karlin, *An infinite move game with a lag*, Contributions to the Theory of Games. Vol. III, M. Dresher, A. W. Tucker, and P. Wolfe, editors, Ann. of Math. Stud., vol. 39, Princeton Univ. Press, Princeton, N.J., 1957, pp. 255–272.

49. E. Kohlberg, *Repeated games with absorbing states*, Ann. Statist. **2** (1974), 724–738.

50. ____, *Optimal strategies in repeated games with incomplete information*, Internat. J. Game Theory **4** (1975), 7–24.

51. ____, *The information revealed in infinitely-repeated games of incomplete information*, Internat. J. Game Theory **4** (1975), 57–59.

52. E. Kohlberg and J.-F. Mertens, *On the strategic stability of equilibria*, Econometrica **54** (1986), no. 5, 1003–1037.

53. E. Kohlberg and S. Zamir, *Repeated games of incomplete information: the symmetric case*, Ann. Statist. **2** (1974), 1010–1041.

54. D. Kreps, F. Milgrom, J. Roberts, and R. Wilson, *Rational cooperation in the finitely-repeated prisoner's dilemma*, J. Econom. Theory **27** (1982), 245–252.

55. D. Kreps and R. Wilson, *Reputation and imperfect information*, J. Econom. Theory **27** (1982), 253–279.

56. ____, *Sequential equilibria*, Econometrica **50** (1982), 863–894.

57. H. W. Kuhn, *Extensive games and the problem of information*, Contributions to the Theory of Games, H. W. Kuhn and A. W. Tucker, editors, Ann. of Math. Stud., vol. 28, Princeton Univ. Press, Princeton, N.J., 1953, pp. 193–216.

58. H. W. Kuhn and A. W. Tucker (Editors), *Contributions to the theory of games*. Vol. II, Ann. of Math. Stud., vol. 28, Princeton Univ. Press, Princeton, N.J., 1953.

59. E. Lehrer, *Two players repeated games with non observable actions*, mimeograph, 1985.

60. ____, *Lower equilibrium payoffs in two-players repeated games with non-observable actions*, The Hebrew University, Jerusalem, CRIME and GT, R.M. 72, 1986.

61. A. Maitra and T. Parthasarathy, *On stochastic games*, J. Optim. Theory Appl. **5** (1970), 289–300.

62. D. Martin, *Borel determinacy*, Ann. of Math. **102** (1975), 363–371.

63. Mathematica, Reports to the U.S. Arms Control and Disarmament Agency, prepared by Mathematica, Inc., Princeton, N.J., ST-80 (1966), ST-116 (1967), ST-140 (1968).

64. J.-P. Mayberry, *Discounted repeated games with incomplete information*, Reports to the U.S. Arms Control and Disarmament Agency, ST-116 (1967), Chapter V, pp. 435–461.

65. N. Meggido, *On repeated games with incomplete information played by non-Bayesian players*, Internat. J. Game Theory **9** (1980), 157–167.

66. C. A. Melolidakis, *On stochastic games with lack of information on one side*, mimeograph, 1984.

67. J.-F. Mertens, *The value of two-person zero-sum repeated games: the extensive case*, Internat. J. Game Theory **1** (1972), 217–225.

68. ____, *A note on "The value of two-person zero-sum repeated games: the extensive case"*, Internat. J. Game Theory **2** (1973), 231–234.

69. ____, *A note on the characteristic function of supergames*, Internat. J. Game Theory **9** (1980), 189–190.

70. ____, *Repeated games: an overview of the zero-sum case*, Advances in Economic Theory, W. Hildenbrand, editor, Cambridge Univ. Press, London and New York, 1982.

71. ____, *The minmax theorem for u.s.c.-l.s.c. payoff functions*, Internat. J. Game Theory **15** (1986), no. 4, 237–250.

72. J.-F. Mertens and A. Neyman, *Stochastic games*, Internat. J. Game Theory **10** (1981), 53–56.

73. ____, *Stochastic games have a value*, Proc. Nat. Acad. Sci. U.S.A. **79** (1982), 2145–2146.

74. J.-F. Mertens and T. Parthasarathy, *Existence and characterisation of Nash equilibria for discounted stochastic games*, CORE Discussion Paper, Louvain-la-Neuve, Belgium (to appear).

75. J.-F. Mertens and S. Zamir, *The value of two-person zero-sum repeated games with lack of information on both sides*, Internat. J. Game Theory **1** (1971–72), 39–64.

76. ____, *On a repeated game without a recursive structure*, Internat. J. Game Theory **5** (1976), 173–182.

77. ____, *The normal distribution and repeated games*, Internat. J. Game Theory **5** (1976), 187–197.

78. ____, *The maximal variation of a bounded martingale*, Israel J. Math. **27** (1977), 252–276.

79. ____, *A duality theorem on a pair of simultaneous functional equations*, J. Math. Anal. Appl. **60** (1977), 550–558.

80. ____, *Minmax and Maxmin of repeated games with incomplete information*, Internat. J. Game Theory **9** (1980), 201–215.

81. ____, *Incomplete information games with transcendental values*, Math. Oper. Res. **6** (1981), 313–318.

82. ____, *Formulation of Bayesian analysis for games with incomplete information*, Internat. J. Game Theory **14** (1985), 1–29.

83. P. Milgrom and J. Roberts, *Limit pricing and entry under incomplete information*, Econometrica (1982), 443–460.

84. ____, *Predation, reputation and entry deterrence*, J. Econom. Theory **27** (1982), 280–312.

85. J. Milnor and L. S. Shapley, *On games of survival*, Contributions to the Theory of Games. Vol. III, M. Dresher, A. W. Tucker and P. Wolfe, editors, Ann. of Math. Stud., vol. 39, Princeton Univ. Press, Princeton, N.J., 1957, pp. 15–45.

86. T. Parthasarathy, *Discounted, positive, and noncooperative stochastic games*, Internat. J. Game Theory **2** (1973), 25–37.

87. J.-P. Ponssard, *Zero-sum games with 'almost' perfect information*, Management Sci. **21** (1975), 794–805.

88. ____, *A note on the L-P formulation of zero-sum sequential games with incomplete information*, Internat. J. Game Theory **4** (1975), 1–5.

89. ____, *On the subject of nonoptimal play in zero-sum extensive games: the trap phenomenon*, Internat. J. Game Theory **5** (1976), 107–115.

90. J.-P. Ponssard and S. Sorin, *The LP formulation of finite zero-sum games with incomplete information*, Internat. J. Game Theory **9** (1980), 99–105.

91. ____, *Some results on zero-sum games with incomplete information: the dependent case*, Internat. J. Game Theory **9** (1980), 233–245.

92. ____, *Optimal behavioral strategies in zero-sum games with almost perfect information*, Math. Oper. Res. **7** (1982), 14–31.

93. J.-F. Ponssard and S. Zamir, *Zero-sum sequential games with incomplete information*, Internat. J. Game Theory **2** (1973), 99–107.

94. R. Radner, *Collusive behavior in non-cooperative epsilon-equilibria in oligopolies with long but finite lives*, J. Econom. Theory **22** (1980), 136–154.

95. ____, *Monitoring cooperative agreements in a repeated principal-agent relationship*, Econometrica **49** (1981), 1127–1147.

96. ____, *Repeated principal-agent games with discounting*, Econometrica **53** (1985), 1173–1198.

97. ____, *Optimal equilibria in a class of repeated partnership games with imperfect monitoring*, Rev. Econom. Stud. **53** (1986), 43–58.

98. ____, *Can bounded rationality resolve the prisoners' dilemma*, Essays in Honor of Gerard Debreu, A. Mas-Colell and W. Hildenbrand, editors, North-Holland, Amsterdam, 1986.

99. ____, *Repeated moral hazard with low discount rates*, Essays in honor of K. Arrow, W. Heller, R. Starr, and D. Starrett, editors, Cambridge Univ. Press (to appear).

100. R. Radner, R. B. Myerson, and E. Maskin, *An example of a repeated partnership game with discounting and with uniformly inefficient equilibria*, Rev. Econom. Stud. **53** (1986), 59–70.

101. T. E. S. Raghavan, *Algorithms for stochastic games: a survey*, mimeograph, 1985.

102. A. Rubinstein, *Equilibrium in supergames*, The Hebrew University, Jerusalem, CRIME and G. T., R.M. 25, 1977.

103. ____, *Equilibrium in supergames with the overtaking criterion*, J. Econom. Theory **21** (1979), 1–9.

104. ____, *Strong perfect equilibrium in supergames*, Internat. J. Game Theory **9** (1980), 1–12.

105. ____, *A bargaining model with incomplete information about time preferences*, Econometrica **53** (1985), 1151–1172.

106. A. Rubinstein and A. Wolinski, *Equilibrium in a market with sequential bargaining*, Econometrica **53** (1985), 1133–1150.

107. A. Rubinstein and M. Yaari, *Repeated insurance contracts and moral hazard*, J. Econom. Theory **30** (1983), 74–97.

108. H. Scarf and L. S. Shapley, *Games with partial information*, Contributions to the Theory of Games. Vol. III, Ann. of Math. Stud., vol. 39, Princeton Univ. Press, Princeton, N.J., 1957, pp. 213–229.

109. R. Selten, *Reexamination of the perfectness concept for equilibrium points in extensive games*, Internat. J. Game Theory **4** (1975), 25–55.

110. L. S. Shapley, *Stochastic games*, Proc. Nat. Acad. Sci. U.S.A. **39** (1953), 1095–1100.

111. M. Sobel, *Non cooperative stochastic games*, Ann. of Math. Stat. **42** (1971), 1930–1935.

112. S. Sorin, *A note on the value of zero-sum sequential repeated games with incomplete information*, Internat. J. Game Theory **8** (1979), 217–223.

113. ____, *An introduction to two-person zero-sum repeated games with incomplete information*, IMSSS-Economics, Stanford University, TR312, 1980; French transl. in Cahiers du Groupe de Mathématiques Economiques. 1, Paris.

114. ____, *Some results on the existence of Nash equilibria for non-zero-sum games with incomplete information*, Internat. J. Game Theory **12** (1983), 193–205.

115. ____, *'Big match' with lack of information on one side*. Part I, Internat. J. Game Theory **13** (1984), 201–255.

116. ____, *On a pair of simultaneous functional equations*, J. Math. Anal. Appl. **98** (1984), 296–303.

117. ____, *'Big match' with lack of information on one side*. Part II, Internat. J. Game Theory **14** (1985), 173–204.

118. ____, *On a repeated game with state dependent signalling matrices*, Internat. J. Game Theory **14** (1985), 249–272.

119. ____, *On repeated games with complete information*, Math. Oper. Res. **11** (1986), 147–160.

120. ____, *Asymptotic properties of a non-zero sum stochastic game*, Internat. J. Game Theory **15** (1986), 101–107.

121. ____, *On repeated games without a recursive structure: existence of* $\lim v_n$, Math. Sci. Res. Inst. Publ. 08817-86, 1986.

122. S. Sorin and S. Zamir, *A 2-person game with lack of information on* $1\frac{1}{2}$ *sides*, Math. Oper. Res. **10** (1985), 17–23.

123. R. E. Stearns, *A formal information concept for games with incomplete information*, Reports to the U.S. Arms Control and Disarmament Agency, ST-116 (1967), Ch. IV, pp. 405–433.

124. A. W. Tucker and R. D. Luce (Editors), *Contributions to the theory of games*. Vol. IV, Ann. of Math. Stud., vol. 40, Princeton Univ. Press, Princeton, N.J., 1959.

125. C. Waternaux, *Solution for a class of games without recursive structure*, Internat. J. Game Theory **12** (1983), 129–160.

126. ____, *Minmax and maxmin of repeated games without a recursive structure*, CORE Discussion Paper 8313, Louvain-la-Neuve, Belgium; Internat. J. Game Theory (to appear).

127. S. Zamir, *On the relation between finitely and infinitely-repeated games with incomplete information*, Internat. J. Game Theory **1** (1971–72), 179–198.

128. ____, *On repeated games with general information function*, Internat. J. Game Theory **2** (1973), 215–229.

129. ____, *On the notion of value for games with infinitely many stages*, Ann. Statist. **1** (1973), 791–796.

UNIVERSITÉ CATHOLIQUE DE LOUVAIN, B-1348, LOUVAIN-LA-NEUVE, BELGIUM

Proceedings of the International Congress of Mathematicians
Berkeley, California, USA, 1986

A Formal Classification of Bursting Mechanisms in Excitable Systems

JOHN RINZEL

1. Introduction. Burst activity is characterized by slowly alternating phases of near steady state behavior and trains of rapid spike-like oscillations; examples of bursting patterns are shown in Figure 2. These two phases have been called the silent and active phases respectively [2]. In the case of electrical activity of biological membrane systems the slow time scale of bursting is on the order of tens of seconds while the spikes have millisecond time scales. In our study of several specific models for burst activity, we have identified a number of different mathematical mechanisms for burst generation (which are characteristic of classes of models). We will describe qualitatively some of these mechanisms by way of the schematic diagrams in Figure 1.

The basic idea is that there are slow processes which modulate the faster spike generating dynamics. For the models we describe here, the slow subsystem, however, does not act independently as a forcing function to the fast subsystem. There are two-way interactions between the fast and slow subsystems; the fast variables play a crucial role in the slow dynamics. For membrane electrical activity, the membrane potential is an important shared variable between the fast and slow processes.

Our understanding of these systems has come from extensive numerical calculations and investigation of broad parameter ranges in a number of explicit models [22–27]. For the most part, these models were formulated to mimic experimental data in a semiquantitative way. For example, rates of slow processes were chosen to match the time scales of burst patterns. The fast variables account for macroscopic features of spike shape and frequency while some aspects of spike shapes were disregarded to prevent introducing nonessential details which frequently do not alter qualitative properties. Also, each such detail places a further burden on numerical calculations.

This paper, with slight differences, also appears in *Proceedings of International Symposium on Mathematical Biology* (E. Teramoto, editor), Springer-Verlag (in press).

In §2 we will present schematically our qualitative view of a number of different mechanisms for bursting. To understand these mechanisms we exploit the different time scales. We first identify the fast and slow subsystems. Then the fast dynamics are considered with the slow variables treated as parameters. A full description of the steady state and periodic solution sets to the fast subsystem yields the slow manifold; i.e., this step is essentially a global bifurcation analysis of the fast subsystem with the slow variables treated as parameters. We find several different bifurcation structures with which we identify and correlate features of different observed burst patterns. For example, the various types of transition behavior between steady state and oscillation branches: subcritical Hopf bifurcation, large amplitude homoclinic orbits, and degenerate homoclinic orbits (which contact saddle-node singularities) lead to different spike frequency characteristics at the beginning or end of the active phase. To complete this lowest order approximation we then consider the flow determined by the slow dynamics on the branches of the slow manifold. By varying parameters of the slow dynamics one obtains a variety of burst patterns and other activity which correspond to various experimental findings. Our formal analysis is essentially the first step in a systematic singular perturbation treatment of these complex oscillators. Perhaps our presentation will motivate analysts to formulate and consider rigorously questions suggested by the phenomena described here.

In §3, we will discuss some of the explicit models for excitable membrane behavior and offer biophysical interpretations of the theoretical results.

2. A qualitative catalog of burst generating mechanisms. For this discussion we will consider a model of the form

$$\dot{x} = F(x,y), \quad x \in R^n \quad \text{(FAST)},$$
$$\dot{y} = \varepsilon G(x,y), \quad y \in R^m \quad \text{(SLOW)},$$

in which there are n fast variables x and m slow variables y; here, $0 < \varepsilon << 1$. We suppose that the components of G and F are $O(1)$ except in small neighborhoods around zeros of these vector functions. Our qualitative catalog with graphical representations is based on the limit of ε near zero.

In this limit, we focus on the steady state, $x_{ss}(y)$, and periodic, $x_{osc}(t;y)$, solutions to FAST (with y as a vector of parameters) which satisfy:

$$0 = F(x_{ss},y),$$

and

$$\dot{x}_{osc} = F(x_{osc},y), \qquad x_{osc}(t+T;y) = x_{osc}(t;y),$$

respectively, where $T = T(y)$ is the period of x_{osc}.

Specific model equations, with identification of physiological slow and fast variables, are given in §3. For example, the Chay-Keizer equations (which exhibit a burst pattern like that of Figure 2B) are a model for the electrical activity of insulin-secreting β-cells of the pancreas [6, 7].

In some types of bursters the fast dynamics exhibit bistability in which, for certain ranges of the slow variables, there are two different attracting branches of the slow manifold, for example, a (pseudo) steady state and an oscillation which would correspond to the silent and active phases respectively. In such models, the slow dynamics cause y to sweep back and forth through this regime, and the fast variables sample alternately the steady state branches of the slow manifold. The burst trajectory essentially traces a hysteresis loop.

The simplest example of such hysteresis behavior is illustrated in Figures 1A and 2A. In this case there is one slow variable, $y \in R^1$, and the fast subsystem exhibits three steady states over a certain range of y: two, upper and lower, which are stable, and one unstable saddle. Without being explicit, let us suppose that the slow dynamics are such that when x is in the upper state then $\dot{y} > 0$ (y is produced slowly) and when x is in the lower state then $\dot{y} < 0$ (y is depleted slowly). The full system then generates a relaxation oscillation (Figure 1A (right) and Figure 2A). We might identify this waveform as a burst without spikes or slow wave. It is important to realize that we have not hypothesized a mechanism which requires the slow system to oscillate independently through the hysteresis zone. The sign of $\dot{y}$ depends crucially on the values of x, and the dependence is autonomous. Without the hysteresis or bistability of FAST this slow wave would not exist; if instead of the Z-curve relation, x_{ss} vs. y were monotonically decreasing, then we would not have bursting. In this sense the fast subsystem, and its hysteresis, drives the slow variable y so as to produce the slow oscillation.

A minimal model for the slow wave of Figure 1A might involve only a single fast variable, $x \in R^1$. In this case the full system would be a classical second-order relaxation oscillator. Such a model could not account for an active phase with spiking. For this, the fast subsystem must have at least two variables. For example, suppose again that $y \in R^1$ and that FAST still had a steady state Z-shaped bifurcation diagram; for comparison purposes, let it be identical to that in Figure 1A. But suppose the upper state, instead of being stable, is unstable and is "surrounded" by an oscillation over some y-range (Figure 1B). An illustrative example of a phase portrait (in the case $x \in R^2$) is shown in Figure 3. In this model, the branch of oscillations emerges via Hopf bifurcation at HB and terminates at HC. At termination, the periodic orbit makes contact with the saddle point of the middle branch of the Z-curve; the period becomes infinite at HC, and the orbit is called homoclinic. Beyond HC, the lower steady state is globally attracting. Let us again hypothesize slow dynamics. Suppose, as above, that y is decreasing when x is in the lower steady state, and, analogous to the above, when x is in the upper attractor, i.e., the oscillatory mode, we suppose that y experiences a net increase for each cycle of the fast x-oscillation x_{osc}. This means that $\overline{G}(x_{osc}, y)$ is $O(1)$ and positive, where

$$\overline{G}(x_{osc}, y) = T^{-1} \int_0^T G(x_{osc}(t; y), y)\, dt, \qquad T = T(y). \qquad (2.1)$$

Under these hypotheses, one can intuitively predict that the full system generates a burst pattern (schematized in the right panel of Figure 1B) with the time course of Figure 2B. The upper oscillation branch of FAST generates the repetitive spike pattern of the burst's active phase. The fork (maximum and minimum of x_{osc} over a period) of the bifurcation diagram becomes the envelope for this train of spikes. In comparing Figures 2A and 2B (and recalling that the Z-curves are identical), we see that the spikes of the active phase (left panels) in Figure 2B oscillate around the plateau level in Figure 2A; also, the y excursion is larger in Figure 2A than in Figure 2B since the slow wave trajectory progresses to the right knee of the Z-curve (cf. Figure 1A) but the burst trajectory (cf., Figure 1B) does not. We may view the burst of Figures 1B and 2B as a generalized relaxation oscillation, i.e., repetitive visitation to overlapping coexistent branches of the slow manifold. But, in contrast to the classical relaxation oscillator in which both branches are steady state branches, here one branch is oscillatory. From Figure 1B, we reach a qualitative conclusion about the spike pattern during the burst: the instantaneous spike frequency drops dramatically at the end of the active phase (see Figure 2B) as the trajectory passes near the homoclinic orbit of FAST.

Variations on this general hysteresis-based mechanism will lead to different burst patterns. For example, if the supercritical Hopf bifurcation point were rightward of the Z-curve's left knee, then the active phase would exhibit an initial portion of near (upper) steady state behavior followed by growing oscillations (around the upper steady state) whose frequency drops suddenly as the active phase ends. One could also predict qualitatively the burst pattern in case the Hopf bifurcation is subcritical.

We mentioned, for the case of Figure 1A, that if hysteresis in FAST is destroyed, say, by stretching the "Z" out of the slow manifold, then the slow (relaxation) wave is lost. Similarly here, if hysteresis is precluded between the upper oscillatory and low steady state attractors of FAST, then bursting (in a robust way) is lost. For example, by adjusting parameters of FAST the homoclinic termination of the oscillation branch can be moved leftward until it meets the knee of the Z-curve (Figure 1C); in §3, we describe how this can be accomplished in the Chay-Keizer model. In this situation, FAST has a unique global attractor for each value of y; hysteresis is lost. The response will be either time independent, if $G(x_{ss}, y) = 0$, and $\partial G/\partial y < 0$ for some y rightward of HC with x_{ss} on the lower steady state branch, or continuous periodic spiking, if $\overline{G}(x_{osc}, y) = 0$, and $\partial \overline{G}/\partial y < 0$ for some y leftward of HC (with x_{osc} on the oscillatory branch). The time courses of two such different response patterns are shown in Figure 2C, dashed and solid, respectively.

For the fast dynamics represented by Figure 2C, the full model would exhibit no robust parameter range for bursting as parameters of SLOW are tuned through a range which includes steady state behavior at one extreme to continuous spiking at the other. On the other hand, if such tuning were done (dynamically) in a smooth repetitive manner, then a burst pattern would be generated nonautonomously. This burst trajectory would cross the HC boundary both at the beginning and at the end of the active phase, and consequently, a plot of instantaneous spike period versus spike number would appear as a

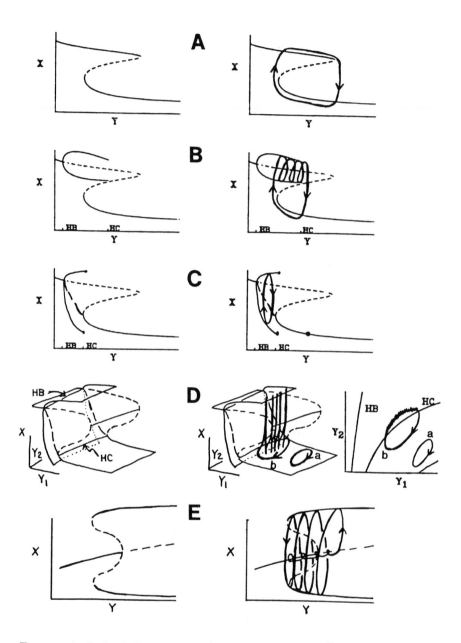

FIGURE 1. Left: Bifurcation diagram (compact description) of periodic and steady state solutions to fast subsystem (FAST) with slow variables as parameters. Maximum and minimum values of some solution component, or its time average over a period (long dashes), is plotted. Unstable solutions indicated by short dashes. Right: Schematic representation (heavy curves) of slow wave, burst, or continuous spiking trajectory as projected on bifurcation diagram (and its projection, case D, far right) of corresponding left panel.

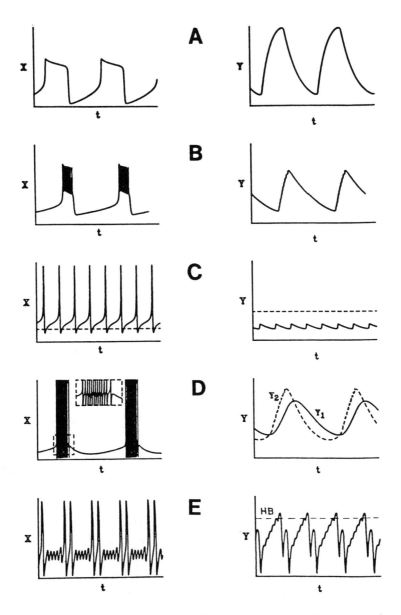

FIGURE 2. Time course of a fast (left) and slow (right) variable for corresponding schematics of Figure 1. These are for computed solutions of specific excitable membrane models. Cases A, B, C: V and Ca from (3.1)–(3.3) with parameter changes: $\lambda_n = 2.0, f = 0.08$ (A); $\lambda_n = 0.1$, and $k_{Ca} = 0.006$ (solid) or $k_{Ca} = 0.0005$ (dashed) (C). Case D: V, Ca (right, solid), and slow calcium conductance, x (right, dashed) for Plant's model (as given in [**25**] but with $\tau_x = 1.88 \times 10^4, K_C = 5.525 \times 10^{-3}, \rho = 0.9 \times 10^{-4}$). Case E: v and y for (3.4)–(3.6) with $\varepsilon = 0.0008$ (note, ε not too small here, so only 2 pulses per burst and y-increments not small).

concave-upward parabola. In the neurobiology literature such a waveform has been called a parabolic burst pattern [1]. To generate this behavior in an autonomous manner requires at least two slow variables. In the case $y \in R^2$, the slow manifold may be represented schematically as the surface in Figure 1D; it includes both steady state and oscillation branches. With the additional degree of freedom in y, one can formulate slow dynamics which lead to autonomous oscillatory behavior of y even when x is restricted to the lower steady state branch of the slow manifold. In this case, we would find a slow wave (Figure 1D, right panels, case (a)) but it would not be of relaxation type as in Figure 1A; both x and y would vary smoothly on the slow time scale.

To understand how a burst pattern may be generated, we note first that variations in the parameters of SLOW do not alter the slow manifold (the surfaces in Figure 1D) in any way. Thus, the parameters of G could be varied appropriately to move the slow wave toward the HC boundary. With sufficient parameter variation, a portion of the slow wave would just cross into the oscillatory regime of FAST and a parabolic burst trajectory would be obtained (Figure 1D, right, case (b), and Figure 2D). This intuitive description supposes that the trajectory will return to the steady state branch and then recycle through the silent phase. This is reasonable to expect. Since, for y near the HC boundary, periodic solutions to fast spend most of their time near the HC steady state, the (fast time) "averaged" flow of $\overline{G}(x_{\text{osc}}, y)$ will be continuously extended off of the steady state branch and therefore will inherit the (slow) oscillatory properties which G exhibits on the lower steady state branch. In this type of burst mechanism one can identify the underlying slow wave; it appears to drive the spikes. We note, again, that this mechanism for bursting does not require the fast subsystem to exhibit multiple *stable* states for any y in the bursting region.

A common feature of the above examples is that, over (at least part of) the y-range for bursting, the fast subsystem exhibits multiple steady states. In the preceding example of Figure 1D, at most one steady state is stable. A different burst generating mechanism, illustrated in Figure 1E, does not require multiple steady states and still can be realized with only a single slow variable. It is based on subcritical Hopf bifurcation in FAST, so it relies on hysteresis. To obtain bursting, we hypothesize for SLOW that y exhibits a net decrease over a cycle of the large amplitude fast oscillation, i.e., $\overline{G}(x_{\text{osc}}, y) < 0$ for $x_{\text{osc}}(t; y)$ on the "outer" branch. Thus, during the active phase, y decreases toward the turning point or knee of the fast oscillation branch where the stable and unstable fast periodic solutions coalesce and disappear. When y reaches the knee, x falls into the domain of attraction of the stable steady state.

In a neighborhood of this steady state we suppose that y, on the average, increases. Thus we have slow rightward movement along the steady state branch during the silent phase. If y passes inside HB, this pseudo steady state becomes unstable and x cannot continue to track it. Therefore x returns to the oscillatory mode of FAST to initiate the active phase. We have found that the silent phase does not necessarily end immediately when y passes inside HB but that there may be some delay as small oscillations in x grow slowly; we have estimated analytically this escape time [**27**]. This behavior is seen in the schematic phase space projection of Figure 1E (right) and in the time courses in Figure 2E. In this type of burst pattern one does not see the underlying smooth slow wave as in Figures 1D and 2D nor the relaxation pattern of Figures 1B and 2B. Here, the fast oscillations of the active phase surround the steady state of the silent phase. Also, since the pseudo steady state may behave as a damped oscillation in the entire overlap range of the subcritical Hopf bifurcation, one may expect to find small decaying and then growing oscillations in the silent phase.

3. Bursting in excitable membrane systems. Models for excitable membrane electrical activity usually involve modifications to the classical Hodgkin-Huxley description [**15**] of nerve impulse generation in the squid axon membrane. In models of bursting, the fast subsystem is based on such modified Hodgkin-Huxley spike dynamics. The associated fast variables include the membrane potential V, and activation/inactivation variables for ionic channel currents. These variables describe membrane properties. The fast subsystem may involve two or more variables. The rate-limiting dynamics of the slow processes typically involve at least one nonmembrane quantity, e.g., intracellular free calcium concentration Ca.

A biophysical description, e.g., in the case of insulin-secreting β-cells [**2**], would be the following. The membrane has ion-selective channels which activate, and possibly inactivate, to V-dependent levels and with V-dependent rates. For example, calcium channels allow inward current flow which increases $\dot{V}$ while potassium currents are outward and their activation decreases $\dot{V}$. When inward current kinetics are fast and outward currents are slower, then these two opposing currents can lead to oscillatory spike activity. In the β-cell system, these dynamics are such that, depending on Ca, the membrane can remain with V at the lower silent phase potential ($\doteq -55$ mV) or oscillate with V in a range (-40 to -25 mV) well above the silent phase potential. During the active phase of a burst, each spike causes a small net increase in Ca; this is because only a small fraction f of the entering calcium is free—most of it is bound rapidly to high affinity binding sites inside the cell. Here, f determines the slow time scale of the burst pattern. To modulate the fast membrane variables, there must be a feedback site for calcium at the membrane. Such feedback to the fast dynamics gives the bifurcation structure of Figure 1B (with V on the ordinate and Ca along the abscissa). In a number of models, this site is hypothesized to be a calcium-activated potassium channel, and it is usually considered distinct from the more classical Hodgkin-Huxley K^+-channel. Thus, as Ca slowly accumulates, it activates the conductance g_{K-Ca} (which is often treated as V-independent). As g_{K-Ca} rises during the active phase, so does the threshold for spike activity. This threshold corresponds to the saddle point on the middle branch of the Z-curve in Figure 1B. This rising threshold meets the trajectory for repetitive spike activity, and the active phase terminates at this point; the trajectory falls below threshold and V drops to the lower (pseudo steady state) silent phase potential. At this low V, the calcium channels are not active, and so there is no influx of calcium. The removal of calcium dominates during the silent phase; g_{K-Ca} slowly decreases, while the pseudo steady state V increases and the threshold falls. When V meets the threshold, then the active phase is reentered and the cycle repeats.

Explicit equations for the β-cell system were formulated originally by Chay and Keizer [6] (based upon the biophysical model of [2]) and subsequently modified [7]. A FAST/SLOW analysis of the original model was presented in [23] and a simplified model was treated in [24]. The equations of the simplified model are given by:

$$C_m \dot{V} = -\bar{g}_{Ca} m_\infty^3(V) h_\infty(V)(V - V_{Ca}) - \left[\bar{g}_K n^4 + \bar{g}_{K-Ca}\frac{Ca}{1 + Ca}\right](V - V_K)$$
$$- \bar{g}_L(V - V_L) + I_{app}, \tag{3.1}$$

$$\dot{n} = \lambda_n[n_\infty(V) - n]/\tau_n(V), \tag{3.2}$$

$$\dot{Ca} = f[\alpha\bar{g}_{Ca} m_\infty^3(V) h_\infty(V)(V_{Ca} - V) - k_{Ca}Ca], \tag{3.3}$$

where,

$$j_\infty(V) = \alpha_j(V)/[\alpha_j(V) + \beta_j(V)], \qquad j = m, h, \text{ or } n,$$
$$\tau_n(V) = 1/[\alpha_n(V) + \beta_n(V)],$$
$$\alpha_m(V) = 0.1(-V - 25)/[\exp\{0.1(-V - 25)\} - 1],$$
$$\alpha_h(V) = 0.07 \exp\{(-V - 50)/20\},$$
$$\alpha_n(V) = 0.01(-V - 20)/[\exp\{0.1(-V - 20)\} - 1],$$
$$\beta_m(V) = 4 \exp\{(-V - 50)/18\},$$
$$\beta_h(V) = 1/[\exp\{0.1(-V - 20)\} + 1],$$
$$\beta_n(V) = 0.125 \exp\{(-V - 30)/80\}.$$

The functions $m_\infty(V)$ and $n_\infty(V)$ are monotone increasing with V and they saturate with $m_\infty, n_\infty \to 1$ as $V \to \infty$, and $m_\infty, n_\infty \to 0$ as $V \to -\infty$; they represent the (steady state) activation of the calcium and potassium currents, respectively. The function $h_\infty(V)$ corresponds to calcium current inactivation; it is monotone decreasing, and $h_\infty \to 0$ as $V \to \infty$, and $h_\infty \to 1$ as $V \to -\infty$.

The variables n and Ca are dimensionless; t, V's, and $\bar{g}$'s have units of msec, mV, mmhos/cm^2, respectively. Parameter values used here are:

$$C_m = 1[\mu F/cm^2], \quad V_{Ca} = 100, \quad V_K = -75, \quad V_L = -40,$$
$$\bar{g}_{Ca} = 1.79934, \quad \bar{g}_K = 1.69765, \quad \bar{g}_{K-Ca} = 0.0104998,$$
$$\bar{g}_L = 0.00698514, \quad \lambda_n = 0.3, \quad k_{Ca} = 0.00513,$$
$$f = 0.0058, \quad \text{and} \quad \alpha = 0.0259102$$

(which involves Faraday's constant and the cell radius).

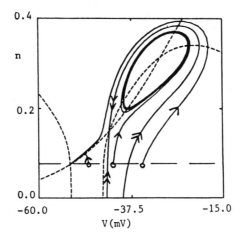

FIGURE 3. Phase plane portrait for fast subsystem, (3.1)–(3.2), of Chay-Keizer model for a fixed value of Ca, 0.7 μM. V and n-nullclines are shown dashed. Three singular points. Double arrowheads denote stable manifolds of saddle point. Responses for three different initial conditions (open circles) show bistability with stable lower steady state and stable oscillation surrounding unstable upper steady state.

In this simplification, the calcium current depends instantaneously upon V while the potassium current activation, n, has the time constant $\tau_n(v)$. Here, the $V - n$ equations form the fast subsystem in which the slow variable Ca appears. The phase portrait of FAST with Ca fixed at a value in the bursting range is shown in Figure 3. With the parameters given above, this model has FAST/SLOW dynamics and burst behavior as described by Figures 1B and 2B. Several features of experimentally observed burst patterns are consistent with the model. The spike frequency is seen to drop sharply near the end of the active phase. Changes in glucose do not alter spike envelopes or the silent phase potentials [5]. This is found theoretically when the effect of increasing glucose is modeled by increasing k_{Ca}; such changes affect only SLOW but not FAST. On the other hand, glucose alters the relative duration of the active and silent phases [5]; very low glucose leads to a stable rest state at low V, and very high glucose yields continuous spiking. This behavior can be understood theoretically by superposing the Ca-nullcline onto Figure 1B (see [23, 24]) and noting that changing k_{Ca} repositions the nullcline relative to the oscillatory and steady state branches. Other biophysical insights from the model's burst behavior are described in [4, 6, 7, 23]. An idealized model (without identifiable biophysical variables) for neuronal burst activity of this general class has also been formulated and studied by Hindmarsh and Rose [14]. Also, Martiel and Goldbeter [19] have applied our formalism to understand complex oscillations in a model for cyclic AMP signaling in *Dictyostelium* amoebae.

If, in (3.1)–(3.3), n acts too rapidly (i.e., λ_n large enough), then the upper steady state, for fixed Ca, is stable and the model behaves as in Figures 1A and 2A. This parameter range has not been found experimentally; relaxation slow waves have not been observed. Finally, if n acts too slowly, then the FAST dynamics of the Chay-Keizer model correspond to Figures 1C and 2C. This would predict that, as glucose increases, the response would evolve from steady state (Ca to right of HC) to continuous spiking (average Ca to left of HC) without passing through a bursting regime. Such conditions have been induced experimentally with TEA (see Figure 4 of [3]).

A well-studied neuronal pacemaker (see, e.g., [1]) exhibits parabolic bursting. Plant [21] formulated a model for such a system; our FAST/SLOW analysis [24, 25] reveals that its structure is that of Figures 1D and 2D. In this model, the inward current for fast spike generation is carried by sodium ions rather than calcium ions. There is also experimental evidence for an additional, and very slow, inward current with a substantial calcium component. In the model, this means there are two slow variables: the above-mentioned slow membrane conductance for calcium (an equation like $\dot{n}$ but with a much slower rate) and the intracellular free calcium concentration (with dynamics as in (3.3)). Parameter adjustments in the model (in particular, merely in the slow subsystem, as described qualitatively in §2) can lead to bursting, to smooth slow wave activity without spiking, or to continuous spiking. In Plant's formulation, as well as some others (see [1]), the biophysical mechanism for calcium feedback was modeled by a calcium-activated potassium (outward) channel. This view has been reconsidered, and an alternative hypothesis (see [1], and its references) involves calcium inactivation of the slow inward current. We have also shown how this second hypothesis can be formulated and explored, without altering the fast subsystem [25].

Plant's original model was also studied by Honerkamp et al. [16]. Kopell and Ermentrout [18] also consider a mechanism for parabolic bursting in which a degenerate homoclinic connection plays a key role. Their hypotheses however are not satisfied by Plant's model, and they did not require feedback from FAST to produce an oscillation in SLOW (in this sense, the slow subsystem would generate an oscillation independently, and thus it acts more like a driving force for FAST).

Models of bursting of the type represented by Figures 1E and 2E have been studied in the context of the Belousov-Zhabotinskii oscillating chemical reaction (see [24, 25] and references therein). Such bursting has not been exposed for a quantitative biophysical model or widely studied experimental preparation for excitable membranes. On the other hand, the classical Hodgkin-Huxley model exhibits a bifurcation structure like Figure 1E (left), with y representing the applied external current [22]. Experiments show that the squid axon behaves this way in appropriate regimes [11]. This bifurcation structure can be seen also with variations in other parameters of the Hodgkin-Huxley model, and presumably if such parameters can be appropriately treated as slow autonomous, dynamic variables then such bursting could be generated. The "skip runs" observed experimentally by Guttman and Barnhill [12] are similar in appearance to Figure 2E, and are suggestive of this type of mechanism.

An idealized nerve membrane model which exhibits bursting as represented by Figures 1E and 2E was formulated and studied numerically (FitzHugh and Rinzel, 1976, unpublished). The fast subsystem for this model is the classical FitzHugh-Nagumo equation [10, 20]. For appropriate parameter values, it has the subcritical Hopf bifurcation structure of Figure 1E (left) with applied current I as the parameter [17, 30]. By formulating a suitable slow dynamics for I, one generates the desired burst behavior. The model takes the following form:

$$\dot{v} = v - v^3/3 - w + y + I, \tag{3.4}$$

$$\dot{w} = \phi(v + a - bw), \tag{3.5}$$

$$\dot{y} = \varepsilon(-v + c - dy), \tag{3.6}$$

in which I is fixed and y describes slow modulation of the current. Figures 4A and 4B illustrate the time course of a burst pattern for the parameter values: $I = 0.3125$, $a = 0.7$, $b = 0.8$, $c = -0.775$, $d = 1.0$, $\phi = 0.08$, and $\varepsilon = 0.0001$. These plots illustrate the characteristic decay and growth of small oscillations during the silent phase. Observe also that, when the silent phase ends, y typically has progressed considerably above the value corresponding to the Hopf bifurcation (represented by the horizontal dashed line in Figure 4B) in the fast subsystem. Projection of an active and silent phase of this solution onto the y-v plane (Figures 4C and 4D) yields a comparison with the FAST/SLOW prediction. Notice also, over the time interval for Figures 4A and 4B, that the response does not appear to be periodic; its period could be long or the solution may be chaotic. We did not explore this in detail, but we found that such behavior is not uncommon for this model. This contrasts with our experience [26, 27] with specific models having this general structure for Belousov-Zhabotinskii oscillations where periodicity occurred more typically. We are uncertain about the primary factors which contribute to the solution behavior for (3.4)–(3.6). One possibility is that the attraction of the pseudo steady state, to the left of HB, is weak; its exponential rate is approximately 0.008, only $1/10$ of ϕ, the rate of w. This weak attraction likely contributes to irregularity and to premature reentry of the active phase before y reaches HB. Since small oscillations of the silent phase are not damped rapidly, the v-w trajectory (as y moves rightward toward HB) may cross the unstable periodic solution (dashed in Figures 4C and 4D) of FAST and then proceed to the "outer" stable oscillatory attractor before y reaches HB; this is premature reentry. (Note, corresponding phenomena were observed in the Chay-Keizer model, an example of the class represented by Figures 1B and 2B). The model, (3.4)–(3.6), can also exhibit bistability in the burst response, i.e., two different stable burst patterns for the same parameter values. One can hope that this model is simple enough so that further insight might be obtained analytically. Honerkamp et al. [16] have also studied a model of this sort.

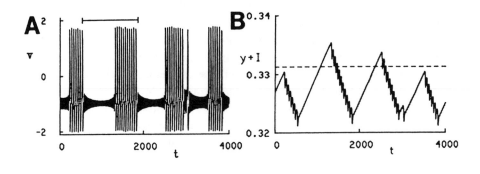

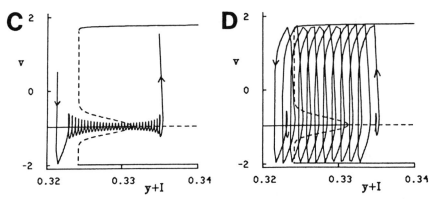

FIGURE 4. Solution to the bursting model, (3.4)–(3.6); parameters given in text. (A), (B): time course of v (fast variable) and y (slow variable), respectively. Projection of silent phase (C) and active phase (D) of a burst (time interval indicated by bar in (A)) onto bifurcation diagram of fast subsystem (cf. Figure 1E).

4. Discussion. We have outlined our formal approach to understanding qualitatively the different mechanisms for bursting in a number of models. Our FAST/SLOW dissection is essentially the first step in a systematic singular perturbation treatment of these complex oscillations. We have computed numerically, for some explicit models, the bifurcation structures (slow manifolds) corresponding to each of the cases schematized in Figure 1. In such efforts the automatic capabilities of AUTO [**9**] for branch-tracking, bifurcation, and stability analysis of stable and unstable, steady state, and/or periodic solutions have been extremely useful. Solutions to the full models were usually obtained by Gear's method of numerical integration for stiff equations [**13**]. We hope that these explicit examples will motivate analysts to provide a more rigorous basis and insight for our formal approach. The early results of Tikonov [**29**] and others (see [**31**] for an introduction and references) provide some groundwork for relaxation oscillators like that of Figure 1A. Such work must be extended to cases in which the slow manifold has more complex solution sets. (For recent analytic work in this direction, see [**28**].) In Figures 1B, 1C, and 1D, the branch of oscillatory solutions terminates in a homoclinic orbit. One could imagine even more complex attractors for FAST. The examples we have discussed are minimal in this regard. Also, we have not addressed the details of solution structure to the full model as parameters are varied, for example, of how the transition is made of a burst pattern of n-spikes to one of $(n+1)$-spikes, of how bursting goes into continuous spiking, or of how chaos arises (see, for example, [**8**]).

Questions of mathematical rigor also arise in regard to applying the method of "averaging" to obtain further reduced descriptions of burst activity. As we have seen, for small ε, each rapid oscillation during the active phase of a burst produces a small net increment in the slow variables. For given values of the slow variables, this increment is approximately equal to (ε times) the expression given in (2.1). With this information, one may obtain an approximation to the time course of the slow variables. The "averaged" dynamics for the active phase are

$$\dot{y} = \varepsilon \overline{G}(x_{\mathrm{osc}}, y), \tag{5.1}$$

and, for the silent phase, we have

$$\dot{y} = \varepsilon G(x_{\mathrm{ss}}, y). \tag{5.2}$$

Note, the quantity $\overline{G}(x_{\mathrm{osc}}, y)$ may be computed during the numerical determination of the oscillatory solution branch of the global bifurcation structure (e.g., as in Figure 1B, right). We [**24**] have applied successfully this averaged description, in certain parameter ranges, to the simplified Chay-Keizer model, (3.1)–(3.3). The method, however, did not predict accurately the behavior in the transition from bursting to continuous spiking. More rigorous analysis is needed here.

We do not view our catalog of bursting mechanisms as complete. Also, we appreciate that the classifications are not rigorous but that they convey the qualitative essence. Furthermore, our experience shows, for a model with given nonlinearities, by varying the time scale of merely one of the fast variables that different response classes can be obtained. For example, different values of λ_n in (3.1)–(3.3) can yield the behavior of Figures 1A, 1B, or 1C.

The qualitative categorizations of Figures 1 and 2 allow us to realize better the limitations and richness of specific models. By identifying the fast and slow time scales, and the underlying structure, we can identify parameters which affect certain aspects, but not others, of solution behavior. The work does not stop here, however. For explicit models one can ask about absolute quantities and values. How small must ε be to match the time scale of experimentally observed patterns and to guarantee that our FAST/SLOW analysis is valid? How robust is the bursting mechanism to parameter variations in the physiologic range?

ACKNOWLEDGMENT. I thank Steven M. Baer for computational help with Figure 4 and for reading the manuscript carefully.

I appreciate the hospitality of the Institute for Nonlinear Science, University of California, San Diego, where part of this contribution was prepared.

1592   JOHN RINZEL

REFERENCES

1. W. B. Adams and J. A. Benson, *The generation and modulation of endogenous rhythmicity in the Aplysia bursting pacemaker neurone* R15, Progr. Biophys. Molec. Biol. **46** (1985), 1–49.

2. I. Atwater, C. M. Dawson, A. Scott, G. Eddlestone, and E. Rojas, *The nature of the oscillatory behavior in electrical activity for pancreatic β-cell*, J. Hormone and Metabolic Res. Suppl. **10** (1980), 100–107.

3. I. Atwater, B. Ribalet, and E. Rojas, *Mouse pancreatic β-cells: tetraethylammonium blockage of the potassium permeability increase induced by depolarization*, J. Physiol. **288** (1979), 561–574.

4. I. Atwater and J. Rinzel, *The β-cell bursting pattern and intracellular calcium*, Ionic Channels in Cells and Model Systems, R. Latorre, editor, Plenum, New York, 1986.

5. P. M. Beigelman, B. Ribalet, and I. Atwater, *Electrical activity of mouse pancreatic beta-cells. II: Effects of glucose and arginine*, J. Physiol. (Paris) **73** (1977), 201–217.

6. T. R. Chay and J. Keizer, *Minimal model for membrane oscillations in the pancreatic β-cell*, Biophys. J. **42** (1983), 181–190.

7. T. R. Chay, *Chaos in a three-variable model of an excitable cell*, Phys. D **16** (1985), 233–242.

8. T. R. Chay and J. Rinzel, *Bursting, beating, and chaos in an excitable membrane model*, Biophys. J. **47** (1985), 357–366.

9. E. J. Doedel, *Software for continuation problems in ordinary differential equations*, Tech. Report. Applied Math. Dept., Calif. Inst. Tech., Pasedena, Calif., 1986.

10. R. FitzHugh, *Impulses and physiological states in models of nerve membrane*, Biophys. J. **1** (1961), 445–466.

11. R. Guttman, S. Lewis, and J. Rinzel, *Control of repetitive firing in squid axon membrane as a model for a neuroneoscillator*, J. Physiol. (London) **305** (1980), 377–395.

12. R. Guttman and R. Barnhill, *Oscillation and repetitive firing in squid axons: Comparison of experiments with computations*, J. Gen Physiol. **55** (1970), 104–118.

13. A. C. Hindmarsh, *Ordinary differential equations systems solver*, Report UCID-30001, Lawrence Livermore Lab, Livermore, Calif., 1974.

14. J. L. Hindmarsh and R. M. Rose, *A model of neuronal bursting using three coupled first order differential equations*, Proc. Roy. Soc. London Ser. B **221** (1984), 87–102.

15. A. L. Hodgkin and A. F. Huxley, *A quantitative description of membrane current and its application to conduction and excitation in nerve*, J. Physiol. (London) **117** (1952), 500–544.

16. J. Honorkamp, G. Mutschler, and R. Seitz, *Coupling of a slow and a fast oscillator can generate bursting*, Bull. Math. Biol. **47** (1985), 1–21.

17. I. Hsu and N. K. Kazarinoff, *An applicable Hopf bifurcation formula and instability of small periodic solutions of the Field-Noyes model*, J. Math. Anal. Appl. **55** (1976), 61–89.

18. N. Kopell and G. B. Ermentrout, *Subcellular oscillations and bursting*, Math. Biosci. **78** (1986), 265–291.

19. J. L. Martiel and A. Goldbeter, *Origin of bursting and biorhythmicity in a model for cyclic AMP oscillations in Dictyostelium cells*, Proc. Kyoto Internat. Sympos. on Math. Biol., E. Teramoto, editor, Springer-Verlag (to appear).

20. J. S. Nagumo, S. Arimoto, and S. Yoshizawa, *An active pulse transmission line simulating nerve axon*, Proc. IRE **50** (1962), 2061–2070.

21. R. E. Plant, *Bifurcation and resonance in a model for bursing nerve cells*, J. Math. Biol. **11** (1981), 15–32.

22. J. Rinzel, *On repetitive activity in nerve*, Federation Proc. **37** (1978), 2793–2802.

23. ____, *Bursting oscillations in an excitable membrane model*, Ordinary and Partial Differential Equations, B. D. Sleeman and R. J. Jarvis, editors, Lecture Notes in Math., vol. 1151, Springer-Verlag, 1985.

24. J. Rinzel and Y. S. Lee, *On different mechanisms for membrane potential bursting*, Nonlinear Oscillations in Biology and Chemistry (H. G. Othmer, editor), Lecture Notes in Biomath., vol. 66, Springer-Verlag, New York, 1986, pp. 19–33.

25. ____, *Dissection of a model for neuronal parabolic bursting*, Preprint.

26. J. Rinzel and I. B. Schwartz, *One variable map prediction of Belousov-Zhabotinskii mixed mode oscillations*, J. Chem. Phys. **80** (1984), 5610–5615.

27. J. Rinzel and W. C. Troy, *Bursting phenomena in a simplified Oregonator flow system model*, J. Chem. Phys. **76** (1982), 1775–1789.

28. J. Siska, L. Kubinova, and I. Dvorak, *Time hierarchy in systems with general attractors*, Proc. Forth Internat. Conf. on Math. Modeling (Zurich, 1983), Pergamon (to appear).

29. A. N. Tikhonov, *Systems of differential equations containing a small parameter multiplying the highest derivatives*, Mat. Sb. (N.S.) **31** (**73**) (1952), 575–585. (Russian)

30. W. C. Troy, *Oscillation phenomena in nerve conduction equations*, Doctoral thesis, State Univ. of New York at Buffalo, 1974.

31. W. Wasow, *Asymptotic expansions for ordinary differential equations*, Interscience, New York, 1965, pp. 297–303.

MATHEMATICAL RESEARCH BRANCH, NIDDK, NATIONAL INSTITUTES OF HEALTH, BETHESDA, MARYLAND 20892, USA

Motion Planning and Related
Geometric Algorithms in Robotics

J. T. SCHWARTZ AND M. SHARIR

1. Introduction. Research on theoretical problems in robotics looks ahead to a future generation of robots substantially more autonomous than present robotic systems, whose algorithmic and software capabilities remain rather primitive. The capabilities which this research aims to create can be grouped into three broad categories: sensing, planning, and control. Of these three, planning involves the use of an environment model to carry out significant parts of a robot's activities automatically. The aim is to allow the robot's user to specify a desired activity in very high level, general terms, and then have the system fill in the missing low-level details. For example, the user might specify the end product of some assembly process and ask the system to construct a sequence of assembly substeps; or, at a less demanding level, to plan collision-free motions which pick up individual subparts of an object to be assembled, transport them to their assembly position, and insert them into their proper places.

Studies in this area have shown it to have significant mathematical content; tools drawn from classical geometry, topology, algebraic geometry, algebra, and combinatorics have all proved relevant. This work relates closely to work in computational geometry, an area which has also progressed very rapidly during the last few years.

2. Statement of the problem. In its simplest form, the motion planning problem can be defined as follows. Let B be a robot system consisting of a collection of rigid subparts (some of which may be attached to each other at certain joints, while others may move independently) having a total of k degrees of freedom, and suppose that B is free to move in a two- or three-dimensional space V amidst a collection of obstacles whose geometry is known to the robot system. The *motion planning problem* for B is: Given an initial position Z_1 and a desired final position Z_2 of B, determine whether there exists a continuous

Work on this paper has been supported by Office of Naval Research Grant N00014-82-K-0381, National Science Foundation Grant NSF-DCR-83-20085, and by grants from the Digital Equipment Corporation and the IBM Corporation.

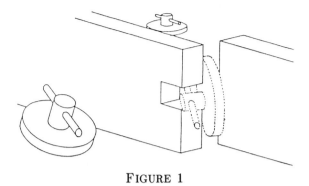

FIGURE 1

obstacle-avoiding motion of B from Z_1 to Z_2, and if so plan such a motion. See Figure 1 for an illustration of this problem.

This problem has been studied in many recent papers (cf. [**LPW, Mo1, Ud, Re, SS1, SS2, SS3, SA, SS4, Ya1, OY, OSY1, OSY2, LS1, LS2, LS3, KS1, KS2, KLPS, SiS, HJW1, HJW2, HW, HSS2, JP, Ya3, Ya4**]). Interesting heuristic and approximating approaches to the problem have also been developed by Lozano-Perez, Brooks, Mason, Taylor, and their collaborators; see [**Br1, Br2, BLP, LP, LP2**]. It is equivalent to the problem of calculating the path-connected components of the (k-dimensional) space FP of all *free positions* of B (i.e., the set of positions of B in which B does not contact any obstacle) and is therefore a problem in "computational topology." In general FP is a high-dimensional space with irregular boundaries and is thus hard to calculate efficiently.

Studies of the motion planning problem tend to make heavy use of many algorithmic techniques in computational geometry. Various motion-planning related problems in computational geometry will be reviewed in §5.

3. Motion planning in static and known environments. As above, let B be a moving robot system, k be its number of degrees of freedom, V denote the two- or three-dimensional space in which B is free to move, and FP denote the space of free positions of B, as defined above. The space FP is determined by the collection of algebraic inequalities which express the fact that at position Z the system B avoids collision with any of the obstacles present in its workplace. We will denote by n the number of inequalities needed to define FP and call it the "geometric (or combinatorial) complexity" of the given instance of the motion planning problem. As noted, we make the reasonable assumption that the parameters describing the degrees of freedom of B can be chosen in such a way that each of these inequalities is algebraic. Indeed, the group of motions (involving various combinations of translations and rotations) available to a given robot can ordinarily be given algebraic representation, and the system B and its environment V can typically be modeled as objects bounded by a collection of algebraic surfaces (e.g., polyhedral, quadratic, or spline-based).

(a) *The general motion planning problem.* Assuming then that FP is an algebraic or semialgebraic set in E^k, it was shown in [**SS2**] that the motion planning problem can be solved in time polynomial in the number n of algebraic constraints defining FP and in their maximal degree, but doubly exponential in k. The general procedure described uses a decomposition technique due to Collins [**Co**] and originally applied to Tarski's theory of real closed fields.

Collins's key definitions and theorems are as follows:

DEFINITION 1. For any subset X of Euclidean space, a *decomposition* of X is a finite collection K of disjoint connected subsets Y of X whose union is X. Such a decomposition is a *Tarski decomposition* if each subset Y is a Tarski set, i.e., a ("semialgebraic") set described by finitely many polynomial equalities and inequalities.

DEFINITION 2. A *cylindrical algebraic decomposition* of E^r is defined as follows. For $r = 1$ such a decomposition is just a partitioning of E^1 into a finite set of algebraic numbers and into the finite and infinite open intervals delimited by these numbers. For $r > 1$, a cylindrical algebraic decomposition of E^r is a decomposition K obtained recursively from some cylindrical algebraic decomposition K' of E^{r-1} as follows. Regard E^r as the Cartesian product of E^{r-1} and E^1 and accordingly represent each point p of E^r as a pair $[x, y]$ with $x \in E^{r-1}$ and $y \in E^1$. Then K must be defined in terms of K' and an auxiliary polynomial $P = P(x, y)$ with rational coefficients, in the following way:

(i) For each $c \in K'$, let $c \times E^1$ designate the *cylinder over c*, i.e., the set of all $[x, y]$ such that $x \in c$.

(ii) For each $c \in K'$ there must exist an integer n, such that for each $x \in c$ there are exactly n distinct real roots $f_1(x), \ldots, f_n(x)$ of $P(x, y)$ (regarded as a polynomial in y), and these roots must vary continuously with x. We suppose in what follows that these roots have been enumerated in ascending order. Then each one of the cells of K which intersects $c \times E^1$ must have one of the following forms:

(ii.a) $\{[x, y]: x \in c,\ y < f_1(x)\}$ (lower semi-infinite "segment" of $c \times E^1$).

(ii.b) $\{[x, f_i(x)]: x \in c\}$ ("section" of $c \times E^1$).

(ii.c) $\{[x, y]: x \in c,\ f_i(x) < y < f_{i+1}(x)\}$ ("segment" of $c \times E^1$).

(ii.d) $\{[x, y]: x \in c,\ f_n(x) < y\}$ (upper semi-infinite "segment" of $c \times E^1$).

All these cells are said to have c as their *base cell* in K'; K' is said to be the *base decomposition*, and P the *base polynomial*, of K.

It follows easily by induction that each of the sets constituting a cylindrical algebraic decomposition K of E^r is topologically equivalent to an open cell of some dimension $k \leq r$. We can therefore refer to the elements $c \in K$ as the (open) *Collins cells of the decomposition* K.

DEFINITION 3. Let S be a set of functions of r variables, and K a cylindrical algebraic decomposition of E^r. Then K is said to be *S-invariant* if, for each c

in K and each f in S, one of the following conditions holds uniformly for $x \in c$: either

(a) $f(x) = 0$ for all $x \in c$; or

(b) $f(x) < 0$ for all $x \in c$; or

(c) $f(x) > 0$ for all $x \in c$.

DEFINITION 4. A point $p \in E^r$ is *algebraic* if each of its coordinates is a real algebraic number. A *defining polynomial* for p is a polynomial with rational coefficients whose set of roots includes all the coordinates of p.

THEOREM 1 (COLLINS). *Given any finite set S of polynomials with rational coefficients in r variables, one can effectively construct an S-invariant cylindrical algebraic decomposition K of E^r into Tarski sets such that each $c \in K$ contains an algebraic point. Moreover, defining polynomials for all these algebraic points, and quantifier-free defining formulae for each of the sets $c \in K$, can also be constructed effectively.*

The theorem is proved by inductive consideration of polynomial resultants and is hence constructive (though by no means suggestive of an efficient algorithm). [SS2] supplements this result with the following technical definition and addendum:

DEFINITION 5. A Collins decomposition K is said to be *well-based* if the following condition holds. Let K' be the base decomposition and $P(b, x)$ the base polynomial of K. Then we require that $P(b, x)$ should not be identically zero for any $b \in E^{r-1}$. Moreover, we require that this same condition apply recursively to the base decomposition K'.

THEOREM 2. *The collection of compact cells of a (well-based) Collins decomposition of E^r forms a regular cell complex.*

COROLLARY 3. *For each j, the (singular) homology group $H_j(V)$ of the real algebraic variety V defined by any set Π of polynomial equations $P(x_1, \ldots, x_n) = 0$ with rational coefficients can be computed in a purely rational manner from the coefficients of the polynomials P.*

The techniques used to prove this theorem relate closely to the method used by Hironaka [Hi] to prove triangulability of real algebraic varieties. Constructive determination of the cell incidence relationships needed to prove Corollary 3 is achieved by considering the Laurent series for solutions of polynomial equations derived from the cell decomposition.

Corollary 3 yields an effective procedure for calculating the 0th homology group, i.e., the connected components, of any semialgebraic configuration space FP. The complexity of this procedure is shown in [SS2] to be of the same order of magnitude of Collins's original procedure, and as stated above, it is polynomial in the number of constraints defining FP and in their maximal degree, but is doubly exponential in k. Though hopelessly inefficient in practical terms, this result nevertheless serves to calibrate the computational complexity of the motion planning problem.

(b) *Lower bounds.* The result just cited suggests that motion planning becomes harder rapidly as the number k of degrees of freedom increases; this conjecture has in fact been proved for various model "robot" systems. Specifically, Reif [**Re**] proved that motion planning is PSPACE-hard for a certain 3-D system involving arbitrarily many links and moving through a complex system of narrow tunnels. Since then PSPACE-hardness has been established for simpler moving systems, including 2-D systems of mechanical linkages (Hopcroft, Joseph, and Whitesides [**HJW2**]), a system of 2-D independent rectangular blocks sliding inside a rectangular box (Hopcroft, Schwartz, and Sharir [**HSS2**]), and a single 2-D arm with many links moving through a 2-D polygonal space (Joseph and Plantinga [**JP**]). Several weaker results establishing NP-hardness for still simpler systems have also been obtained.

The Hopcroft-Joseph-Whitesides result is established by showing that, given an arbitrary Turing machine T with a fixed bounded tape memory, one can construct a planar linkage L whose motions simulate the actions of T, so that L can only move from a specified initial to a specified final configuration if the Turing machine T eventually halts. The size of the linkage L constructed is polynomially bounded by the size of T's state table and memory tape. One proceeds by noting that the actions of an arbitrary T can easily be characterized by a set of polynomial constraints, and then by using the (classical) Kempe, 1876 result [**Ke**] which shows how to construct a mechanical linkage capable of representing any specified multivariate polynomial $P(x_1, \ldots, x_n)$.

In more detail, a *planar linkage* is a mechanism consisting of finitely many rigid rods, of prespecified lengths, joined together at some of their endpoints by hinge-pins about which they are free to rotate. Any number of rod-ends are allowed to share a common hinge-pin; and particular pins can be held at specified points by being "fastened to the plane." Aside from this, the hinge-pins and rods are free to move in the plane, and it is assumed that the motion of one rod does not impede the motion of any other (i.e., the rods are allowed to "pass over" each other). Such a linkage is said to represent the multivariate polynomial $P(x_1, \ldots, x_n)$ if there exist n hinge-pins $p_1, \ldots, p_n$ which the linkage constrains to move along the real axis, and an $(n+1)$st hinge-pin p_0 which the linkage contrains to lie at the real point $P(x_1, \ldots, x_n)$ whenever $p_1, \ldots, p_n$ are placed at the real points $x_1, \ldots, x_n$. (It is assumed that the linkage leaves $p_1, \ldots, p_n$ free to move independently over some large interval of the real axis.)

The existence of a linkage representing an arbitrary polynomial P in the case just explained is established by exhibiting linkages which realize the basic operations of addition, multiplication, etc. and then showing how to represent arbitrary combinations of these operations by fastening sublinkages together appropriately.

The Hopcroft-Schwartz-Sharir result on PSPACE-hardness of the coordinated motion planning problem for an arbitrary set of rectangular blocks moving inside a rectangular frame is proved similarly. It is relatively easy to show that the actions of an arbitrary tape-bounded Turing machine can be imitated by the motions of a collection of similarly sized nearly rectangular "keys" whose edges bear protrusions and indentations which constrain the manner in which these "keys" can be juxtaposed, and hence the manner in which they can move within a confined space. A somewhat more technical discussion then shows that these keys can be cut appropriately into rectangles without introducing any significant possibilities for motion of their independent parts that do not correspond to motions of entire keys.

(c) *The "projection method."* In spite of these negative worst-case results, several acceptably efficient algorithms for planning the motions of various simple robot systems have been developed. These involve various general approaches to the design of motion planning algorithms. The so-called *projection method* uses ideas similar to those appearing in the Collins decomposition procedure described above. One fixes some of the problem's degrees of freedom (for the sake of exposition, suppose just one parameter y is fixed, and let $\bar{x}$ be the remaining parameters); then one solves the resulting restricted $(k-1)$-dimensional motion planning problem. This subproblem solution must be such as to yield a discrete combinatorial representation of the restricted free configuration space (essentially, a cross-section of the entire space FP) that changes only at a finite collection of "critical" values of the final parameter y. These critical values of y are then calculated; they partition the entire space FP into connected cells, and by calculating relationships of adjacency between these cells one can describe the connectivity of FP by a discrete *connectivity graph CG*. This graph has the aforesaid cells as vertices and has edges which represent relationships of cell adjacency in FP. The connected components of FP correspond in a one-to-one manner to the connected components of CG, reducing the problem to a discrete path searching problem in CG.

This relatively straightforward technique was applied in a series of papers by Schwartz and Sharir on the "piano movers" problem, to yield polynomial time motion planning algorithms for various specific systems. These initial solutions were coarse and not very efficient; subsequent refinements have improved their efficiency substantially.

A typical example that has been studied extensively is the case of a line segment B (a "rod") moving in two-dimensional polygonal space whose boundary consists of n segments ("walls"). Here the configuration space FP is three-dimensional, and it can be decomposed into cells efficiently using a modified projection technique developed by Leven and Sharir [**LS1**].

In this approach one starts by restricting the motion of B to a single degree of freedom of translation along its length. For this trivial subproblem the restricted FP simply consists of an interval which can be represented by a discrete label $[w_1, w_2]$ consisting of the two walls against which B stops when moving backwards or forwards from its given placement.

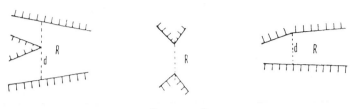

FIGURE 2

Next one admits a second degree of freedom by allowing arbitrary translational motion of B. The restricted FP can now be decomposed into "noncritical" trapezoidal regions, over each of which the label $[w_1, w_2]$ for the restricted 1-D subspace remains constant.

Finally one introduces the final rotational degree of freedom θ. Again one can show that topologically significant changes only occur when θ crosses certain *critical orientations*, at which the combinatorial characterization of the left or right boundary of a noncritical 2-D region R changes discontinuously (some of these critical orientations are illustrated in Figure 2).

Leven and Sharir show that the number of critical orientations is at most $O(n^2)$ and that, assuming B and the walls are in "general position," each critical orientation delimits only a small constant number of cells. Thus the total number of cells in FP is also $O(n^2)$. [**LS1**] presents a fairly straightforward algorithm for constructing these cells and for establishing their adjacency in FP, which runs in time $O(n^2 \log n)$, a very substantial improvement of the $O(n^5)$ algorithm originally presented in [**SS1**].

O'Rourke [**OR**] has recently shown that for certain configurations there exist two placements of B reachable from one another, but they are such that any motion between them must consist of $\Omega(n^2)$ different simple submotions, proving that the Leven-Sharir algorithm is nearly optimal in the worst case.

(d) *The "retraction method" and other approaches to the motion planning problem.* Several other important algorithmic motion planning techniques were developed subsequent to the simple projection technique originally considered. The so-called *retraction method* proceeds by retracting the configuration space FP onto a lower-dimensional (usually a 1-dimensional) subspace N, so that two system positions in FP lie in the same connected component of FP if and only if their retractions to N lie in the same connected component of N. This reduces the dimension of the problem, and if N is 1-dimensional the problem becomes one of searching a graph.

O'Dunlaing and Yap [**OY**] introduced this retraction technique in the simple case of a disc moving in 2-D polygonal space. Here the subspace N can be taken to be the Voronoi diagram associated with the set of given polygonal obstacles. Their technique yields an $O(n \log n)$ motion planning algorithm. After this first paper O'Dunlaing, Sharir, and Yap [**OSY1, OSY2**] generalized the retraction approach to the case of a rod moving in 2-D polygonal space by defining a variant Voronoi diagram in the 3-D configuration space FP of the rod and by retracting

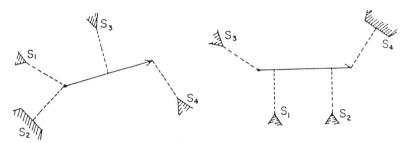

FIGURE 3

onto this diagram. This diagram consists of all placements of the rod at which it is simultaneously nearest to at least two obstacles. The Voronoi diagram defined by a set of obstacles in general position can readily be divided into 2-D Voronoi sheets (placements in which the rod is simultaneously nearest to two obstacles), which are bounded by 1-D Voronoi edges (placements in which the rod is nearest to three obstacles), which in turn are delimited by Voronoi vertices (placements in which the rod is nearest to four obstacles; cf. Figure 3). The algorithm described in [**OSY1, OSY2**] actually constructs a 1-D subcomplex within the Voronoi diagram; this complex consists of the Voronoi edges and vertices plus some additional connecting arcs. It is shown in [**OSY1**] that this Voronoi "skeleton" characterizes the connectivity of FP, in the sense that each connected component of FP contains exactly one connected component of the skeleton. A fairly involved geometric analysis given in [**OSY2**] shows that the total number of Voronoi vertices is $O(n^2 \log^* n)$ and that the entire "skeleton" can be calculated in time $O(n^2 \log n \log^* n)$ (a substantial improvement of the original projection technique, but nevertheless a result shortly afterward superceded by [**LS1**]).

A similar retraction approach was used by Leven and Sharir [**LS2**] to obtain an $O(n \log n)$ algorithm for planning the purely translational motion of a simple convex object B amidst polygonal barriers. This last result uses another generalization of the Voronoi diagram, known as the B-Voronoi diagram, which is defined as follows. Let O be a reference point within B, which we assume to lie initially at the origin, and define a generalized distance function d_B by

$$d_B(p, q) = \min\{\lambda : q \in p + \lambda B\}.$$

(The generalized distance function d_B satisfies the triangle inequality but is not symmetric in general.) Define the B-Voronoi diagram of the given set S of obstacles to consist of all points p for which there exist at least two obstacles s_1, s_2 such that

$$d_B(p, s_1) = d_B(p, s_2) \le d_B(p, s)$$

for all $s \in S$; see Figure 4 for an illustration of such a diagram.

Though of a more complex structure than standard Voronoi diagrams, the B-diagram retains most of the useful properties of standard Voronoi diagrams.

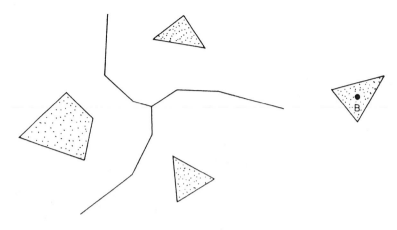

FIGURE 4

In particular, its size is linear in the number of obstacles in S, and, if B has sufficiently simple shape, can be calculated in time $O(n \log n)$, using a variant of the technique described in [**Ya2**].

Next, let N be the portion of the B-diagram consisting of points whose B-distance from the nearest obstacle is greater than 1. Then any translate of B in which the reference point O on B is placed at a point in N is a free placement of B. It is proved in [**LS2**] that N characterizes the connectivity of the free configuration space of B, in the sense defined above, so that, for purpose of planning, motion of B can be restricted to have the reference point O move only along N. This yields an $O(n \log n)$ motion planning algorithm for this case. (A somewhat simpler $O(n \log^2 n)$ algorithm, based on a general technique introduced by Lozano-Perez and Wesley [**LPW**], was described somewhat earlier by Kedem and Sharir [**KS1**] (cf. also [**KLPS**]); this last result makes use of an interesting topological property of intersecting planar Jordan curves.)

Recently Sifrony and Sharir [**SiS**] have devised another retraction-based algorithm for the motion of a rod in 2-D polygonal space. Their approach is to construct the 1-D network of arcs within FP consisting of all the 1-D edges on the boundary of FP (each such edge consists of semifree placements in which the rod simultaneously makes two specific contacts with the obstacles), plus several additional arcs which connect different components of the boundary of FP that bound the same connected component of FP. Again, this network characterizes the connectivity of FP, so that a motion planning algorithm need only consider motions proceeding within this network. The Sifrony-Sharir approach generates motions in which the rod is in contact with the obstacles and is thus conceptually somewhat inferior to the Voronoi-diagram based techniques, which aim to keep the moving system between obstacles, not letting it get too close to any single obstacle. However the network in [**SiS**] is much simpler to analyze and to construct. Specifically, it is shown in [**SiS**] that this network has $O(n^2)$ vertices and edges and can be constructed in time $O(n^2 \log n)$. (Actually, the network

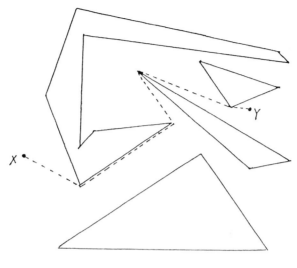

FIGURE 5

size is bounded by the number K of pairs of obstacles lying at distance less than
or equal to the length of the moving rod, and the complexity of the algorithm
is bounded by $O(K \log n)$. Thus if the obstacles are not too badly cluttered
together, the Sifrony-Sharir algorithm will run quite efficiently; this makes the
approach in [**SiS**] more attractive than the previous solutions in [**OSY2, LS1**].)

4. Variants of the motion planning problem.

(a) *Optimal motion planning.* The only optimal motion planning problem
which has been studied extensively thus far is that in which the moving system
is represented as a single point, in which case one aims to calculate the shortest
Euclidean path connecting initial and final system positions, given that specified
obstacles must be avoided. Most existing work on this problem assumes that the
obstacles are either polygonal (in 2-space) or polyhedral (in 3-space).

The 2-D case is considerably simpler than the 3-D case; see Figure 5. When
the free space V in 2-D is bounded by n straight edges, it is easy to calculate
the desired shortest path in time $O(n^2 \log n)$. This is done by constructing a
visibility graph VG whose edges connect all pairs of boundary corners of V which
are visible from each other through V, and then by searching for a shortest path
through VG (see [**ShS**] for a sketch of this idea). This procedure was improved
to $O(n^2)$ by Asano et al. [**AAGHI**], by Welzl [**We**], and by Reif and Storer [**RS**],
using a cleverer method for constructing VG. Their quadratic time bound has
been improved in certain special cases. However, it is not known whether shortest
paths for a general polygonal space V can be calculated in subquadratic time.
Among the special cases allowing more efficient treatment, the most important is
that of calculating shortest paths inside a simple polygon P. Lee and Preparata
[**LeP**] gave a linear time algorithm for this case, assuming that a triangulation
of P is given in advance. (As a matter of fact, a recent algorithm of Tarjan

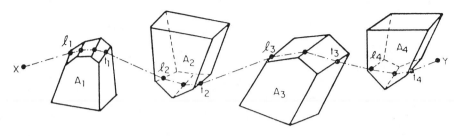

FIGURE 6

and Van Wyk shows that triangulation in $O(n \log \log n)$ time is always possible.) The Preparata-Lee result was recently extended by Guibas et al. [**GHLST**], who gave a linear time algorithm which calculates the shortest paths from a fixed source point to all vertices of P.

The 3-D polyhedral shortest path problem is substantially more difficult. To calculate shortest paths in 3-space amidst polyhedral obstacles bounded by n edges, we can begin by noting that any such path must be piecewise linear with corners lying on the obstacle edges, and that it must subtend equal incoming and outgoing angles at each such corner; see Figure 6. These remarks allow shortest path calculation to be split into two subproblems: (i) Find the sequence of edges through which the desired path must pass; (ii) Find the contact points of the path with these edges. A recent result of Canny and Reif [**CR**] indicates that the 3-D problem is NP-hard.

One of the reasons the problem is difficult in the general polyhedral case is that consecutive edges crossed by the shortest path can be skewed to one another. There are, however, some special cases in which this difficulty does not arise, and they admit efficient solutions. One such case is that in which we aim to calculate shortest paths lying along the surface of a single convex polyhedron having n edges. In this case subproblem (ii) can easily be solved by "unfolding" the polyhedron surface at the edges crossed by the path, thereby transforming the path to a straight segment connecting the unfolded source and destination points (cf. [**ShS**]). Extending this observation, Mount [**Mo2**] has devised an $O(n^2 \log n)$ algorithm, which proceeds by sophisticating an algorithmic technique originally introduced by Dijkstra to find the shortest path in graphs, specifically by maintaining and updating a combinatorial structure characterizing shortest paths from a fixed initial point to each of the edges of the polyhedron (cf. also [**ShS**] for an initial version of this approach).

This result has recently been extended in several ways. A similar $O(n^2 \log n)$ algorithm for shortest paths along a (not necessarily convex) polyhedral surface is given in [**MMP**]. [**BS**] considers the problem of finding the shortest path connecting two points lying on two disjoint convex polyhedral obstacles, and the authors report a nearly cubic algorithm, which makes use of the Davenport-Schinzel sequences described below. The case of shortest paths which avoid a fixed number k of disjoint convex polyhedral obstacles is analyzed in [**Sh3**], which describes an algorithm that is polynomial in the total number of obstacle edges, but is exponential in k. Finally, an approximating pseudopolynomial scheme for the general polyhedral case is reported in [**Pa**]; this involves splitting each obstacle edge by adding sufficiently many new vertices and by searching for the shortest piecewise linear path bending only at those vertices.

5. Results in computational geometry relevant to motion planning. The various studies of motion planning described above make extensive use of efficient algorithms for the geometric subproblems which they involve, for which reason motion planning has encouraged research in computational geometry. Problems in computational geometry whose solutions apply to robotic motion planning include the following:

(a) *Intersection detection.* The problem here is to detect intersections and to compute shortest distances, e.g., between moving subparts of a robot system and stationary or moving obstacles. Simplifications which have been studied include that in which all objects involved are circular discs (in the 2-D case) or spheres (in the 3-D case). In a study of the 2-D case of this problem, Sharir [**Sh1**] developed a generalization of Voronoi diagrams for a set of (possibly intersecting) circles and used this diagram to detect intersections and to compute shortest distances between discs in time $O(n \log^2 n)$. (An alternative approach to this appears in [**IIM**]). Hopcroft, Schwartz, and Sharir [**HSS1**] present an algorithm for detecting intersections among n 3-D spheres which also runs in time $O(n \log^2 n)$. However, this algorithm does not adapt in any obvious way to allow proximity calculation or other significant problem variants.

Other intersection detection algorithms appearing in the computational geometry literature involve rectilinear objects and use multidimensional searching techniques for achieving high efficiency (see [**Me**] for a survey of these techniques).

(b) *Generalized Voronoi diagrams.* The notion of the Voronoi diagram has proven to be a useful tool in the solution of many motion planning problems. The discussion given previously has mentioned the use of various variants of Voronoi diagram in the retraction-based algorithms for planning the motion of a disc [**OY**], or of a rod [**OSY1**, **OSY2**], or the translational motion of a convex object [**LS2**], and in the intersection detection algorithm for discs mentioned above [**Sh1**]. The papers just cited and some related works [**Ya2**, **LS4**] describe the analysis of these diagrams and the design of efficient algorithms for their calculation.

(c) *Davenport-Schinzel sequences.* Davenport-Schinzel sequences are combinatorial sequences of n symbols which do not contain certain forbidden subsequences of alternating symbols. Sequences of this sort appear in studies of efficient techniques for calculating the lower envelope of a set n continuous functions, if it is assumed that the graphs of any two functions in the set can intersect in some fixed number of points at most. These sequences, whose study

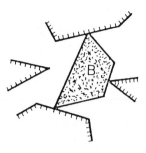

FIGURE 7

was initiated in [**DS**, **Da**], have proved to be powerful tools for analysis (and design) of a variety of geometric algorithms, many of which are useful for motion planning.

More specifically, an (n, s) Davenport-Schinzel sequence is defined to be a sequence U composed of n symbols, such that (i) no two adjacent elements of U are equal, and (ii) there do not exist $s + 2$ indices $i_1 < i_2 < \cdots < i_{s+2}$ such that $u_{i_1} = u_{i_3} = u_{i_5} = \cdots = a$, $u_{i_2} = u_{i_4} = u_{i_6} = \cdots = b$, with $a \neq b$. Let $\lambda_s(n)$ denote the maximal length of an (n, s) Davenport-Schinzel sequence. An early study by Szemeredi [**Sz**] of the maximum possible length of such sequences shows that $\lambda_s(n) \leq C_s n \log^* n$, where C_s is a constant depending on s. Improving on this result, Hart and Sharir [**HS**] proved that $\lambda_3(n) = \Theta(n\alpha(n))$, where $\alpha(n)$ is the very slowly growing inverse of the Ackermann function. In [**Sh2**, **Sh5**] Sharir established the bounds

$$\lambda_s(n) = O(n\alpha(n)^{O(\alpha(n)^{s-3})})$$

and

$$\lambda_s(n) = \Omega(n\alpha^{\lfloor (s-1)/2 \rfloor}(n))$$

for $s > 3$. These results show that, in practical terms, $\lambda_s(n)$ is an almost linear function of n (for any fixed s).

Recently, numerous applications of these sequences to motion planning have been found. These include the following:

(i) Let B be a convex k-gon translating and rotating in a closed 2-D polygonal space V bounded by n edges. The *polygon containment* problem calls for determining whether there exists any free placement of B, i.e., a placement in which B lies completely within V. Some variants of this problem have been studied by Chazelle [**Cha**], who showed that if such a free placement of B exists, then there also exists a *stable* free placement of B, namely, a placement in which B lies completely within V and makes three simultaneous contacts with the boundary of V (see Figure 7). Using Davenport-Schinzel sequences, Leven and Sharir have shown in [**LS3**] that the number of such free stable placements is at most $O(kn\lambda_6(kn))$, and that they can all be calculated in time $O(kn\lambda_6(kn) \log kn)$. Thus, within the same time bound, one can determine whether P can be placed inside Q.

Based on this result, Kedem and Sharir [**KS2**] have produced an $O(kn\lambda_6(kn)\log kn)$ algorithm for planning the motion of a convex k-gon B in a 2-D polygonal space bounded by n edges.

(ii) an $O(mn\alpha(mn)\log m\log n)$ algorithm for separating two interlocking simple polygons by a sequence of translations [**PSS**], where it is assumed that the polygons have m and n sides respectively.

(iii) an $O(n^2\lambda_{10}(n)\log n)$ algorithm for finding the shortest Euclidean path between two points in 3-space avoiding the interior of two disjoint convex polyhedra having n faces altogether [**BS**].

Other applications are found in [**At, HS, CS, SL, OSY2**].

(d) *Topological results related to motion planning.* Hopcroft and Wilfong [**HW**] have derived interesting qualitative results concerning the motion planning problem using ideas drawn from homology theory. Their basic idea is roughly as follows. Consider a connected rigid planar body B moving in the complement of a connected obstacle O. Both the body and the obstacle are assumed to be closed bounded regions. The space S of all positions of the body is defined by two translational and one rotational parameter and can be viewed as a topological equivalent of 3-dimensional Euclidean space provided that we distinguish between otherwise identical body positions differing from each other by 360° rotations (i.e., pass to the simply connected covering space of the ordinary space of positions of B). Let X be the (closed) set of all positions in which B contacts O (possibly overlapping O, i.e., with contact between the interior of B and the interior of O), and let Y be the (also closed) set of positions in which B does not overlap O (but contact between the boundaries of B and of O is allowed). Then plainly $S = X \cup Y$, while $X \cap Y$ is the space of all positions in which B contacts, but does not penetrate, O.

The set X is easily seen to be connected. Indeed, take some point $x_0 \in B$, and consider a body position $p_1 \in X$ for which some other point $x_1 \in B$ also belongs to O. Then since both B and O are connected, a translational motion of B connects the position p_1 to a body position p_0 in which x_0 lies at some standard point within O, and then by rotating B about x_0 we can bring B to a canonical position.

The relative homology groups of $H_i(S,X)$ and $H_i(Y,X\cap Y)$ are isomorphic; and since X is connected, the natural injection of $H_0(X)$ into $H_0(S)$ is an isomorphism. Hence $H_1(S,X)$ is zero, from which it follows that $H_1(Y,X\cap Y)$ is zero, as is $H_0(Y,X\cap Y) = H_0(S,X)$. Thus the natural injection $H_0(X\cap Y) \to H_0(Y)$ is an isomorphism also; i.e., two body positions p_1, p_2 in which B contacts O (without overlap) can be connected by a path along which B never overlaps O if and only if there exists such a path along which the boundaries of B and O remain continuously in contact.

Much the same argument applies in 3 dimensions, since the covering space of the space of all positions of a 3-dimensional body is topologically the product of Euclidean 3-space by a 3-sphere, whose only nontrivial homology is in dimension 3. Hopcroft and Wilfong show that these very general topological arguments extend also to robots capable of deforming as they move, explore the additional difficulties which arise if the obstacle O is not simply connected, etc.

REFERENCES

[**AAGHI**] T. Asano, T. Asano, L. Guibas, J. Hershberger, and H. Imai, *Visibility polygon search and Euclidean shortest paths*, Proc. 26th IEEE Sympos. on Foundations of Comp. Sci., IEEE, New York, 1985, pp. 155–164.

[**At**] M. Atallah, *Dynamic computational geometry*, Proc. 24th IEEE Sympos. on Foundations of Comp. Sci., IEEE, New York, 1983, pp. 92–99.

[**BS**] A. Baltsan and M. Sharir, *On Shortest paths between two convex polyhedra*, Tech. Rept. 180, Comp. Science Dept., Courant Institute, New York University, Sept. 1985.

[**Br1**] R. A. Brooks, *Solving the find-path problem by good representation of free space*, IEEE Trans. Systems Man Cybernet. **SMC-13** (1983), pp. 190–197.

[**Br2**] ____, *Planning collision-free motions for pick-and-place operations*, Internat. J. Robotics Res. **2** (1983), no. 4, 19–40.

[**BLP**] R. A. Brooks and T. Lozano Perez, *A subdivision algorithm in configuration space for findpath with rotation*, AI Memo 684, M.I.T., December 1982.

[**Ca**] J. Canny, *Constructing roadmaps of semi-algebraic sets*, Preprint, 1986.

[**CR**] J. Canny and J. Reif, private communication.

[**Cha**] B. Chazelle, *The polygon containment problem*, Advances in Computing Research. I, Computational Geometry (F. P. Preparate, ed.), JAI Press, 1983, pp. 1–33.

[**Ch**] L. P. Chew, *Planning the shortest path for a disc in $O(n^2 \log n)$ time*, Proc. ACM Sympos. on Computational Geom., ACM, New York, 1985, pp. 214–223.

[**CS**] R. Cole and M. Sharir, *Visibility of a polyhedral surface from a point*, in preparation.

[**Co**] G. E. Collins, *Quantifier elimination for real closed fields by cylindrical algebraic decomposition*, Second GI Conference on Automata Theory and Formal Languages, Lecture Notes Comp. Sci. vol. 33, Springer-Verlag, Berlin, pp. 134–183.

[**Da**] H. Davenport, *A combinatorial problem connected with differential equations, II*, Acta Arith. **17** (1971), 363–372.

[**DS**] H. Davenport and A. Schinzel, *A combinatorial problem connected with differential equations*, Amer. J. Math. **87** (1965), 684–694.

[**FWY**] S. Fortune, G. Wilfong, and C. K. Yap, *Coordinated motion of two robot arms*, Proc. 1986 IEEE Internat. Sympos. on Robotics and Automation, IEEE, New York, 1986, pp. 1216–1223.

[**GHLST**] L. Guibas, J. Hershberger, D. Leven, M. Sharir, and R. E. Tarjan, *Linear time algorithms for shortest path and visibility problems inside simple polygons*, Proc. 2nd ACM Sympos. on Computational Geom., ACM, New York, 1986, pp. 1–13.

[**HS**] S. Hart and M. Sharir, *Nonlinearity of Davenport-Schinzel sequences and of generalized path compression schemes*, Combinatorica **6** (1986), 175–201.

[**HJW1**] J. E. Hopcroft, D. A. Joseph, and S. H. Whitesides, *On the movement of robot arms in 2-dimensional bounded regions*, SIAM J. Comput. **14** (1985), 315–333.

[**HJW2**] ____, *Movement problems for 2-dimensional linkages*, SIAM J. Comput. **13** (1984), 610–629.

[**HSS1**] J. E. Hopcroft, J. T. Schwartz, and M. Sharir, *On the complexity of motion planning for multiple independent objects; PSPACE hardness of the 'Warehouseman's Problem'*, Internat. J. Robotics Res. **3** (1984), no. 4, 76–88.

[**HSS2**] ____, *Efficient detection of intersections among spheres*, Internat. J. Robotics Res. **2**(1983), no. 4, 77–80.

[**HSS3**] J. E. Hopcroft, J. T. Schwartz, and M. Sharir (Editors), *Planning, geometry and complexity of robot motion*, Ablex Pub. Co. (in press).

[**HW**] J. E. Hopcroft and G. Wilfong, *On the motion of objects in contact*, Tech. Rept. 84-602, Comp. Science Dept., Cornell University, May 1984.

[**IIM**] H. Imai, M. Iri, and K. Murota, *Voronoi diagram in the Laguerre geometry and its applications*, SIAM J. Comput. **14** (1985), 93–105.

[**JP**] D. A. Joseph and W. H. Plantinga, *On the complexity of reachability and motion planning questions*, Proc. ACM Sympos. on Computational Geom., 1985, pp. 62–66.

[**KLPS**] K. Kedem, R. Livne, J. Pach, and M. Sharir, *On the union of Jordan regions and collision-free translational motion amidst polygonal obstacles*, Discrete and Computational Geom. **1** (1986), 59–71.

[**KS1**] K. Kedem and M. Sharir, *An efficient algorithm for planning collision-free translational motion of a convex polygonal object in 2-dimensional space amidst polygonal obstacles*, Proc. ACM Sympos. on Computational Geom., 1985, pp. 75–80.

[**KS2**] ____, *An efficient motion-planning algorithm for a convex polygonal object in two-dimensional polygonal space*, in preparation.

[**Ke**] A. B. Kempe, *On a general method of describing plane curves of the n-th degree by linkwork*, Proc. London Math. Soc. **7** (1876), 213–216.

[**LeP**] D. T. Lee and F. P. Preparata, *Euclidean shortest paths in the presence of rectilinear barriers*, Networks **14** (1984), 393–410.

[**LS1**] D. Leven and M. Sharir, *An efficient and simple motion planning algorithm for a ladder moving in two-dimensional space amidst polygonal barriers*, Proc. ACM Sympos. on Computational Geom. 1985, pp. 221–227.

[**LS2**] ____, *Planning a purely translational motion for a convex object in two-dimensional space using generalized Voronoi diagrams*, Tech. Rept. 34/85, The Eskenasy Institute of Computer Science, Tel Aviv University, June 1985.

[**LS3**] ____, *On the number of critical free contacts of a convex polygonal object moving in 2-D polygonal space*, Tech. Rept. 187, Computer Science Dept., Courant Institute, New York University, October 1985.

[**LS4**] ____, *Intersection and proximity problems and Voronoi diagrams*, Advances in Robotics. Vol. I (J. Schwartz and C. Yap, eds.), Lawrence Erlbaum Associates, 1986 (to appear).

[**LP**] T. Lozano-Perez, *Spatial planning: a configuration space approach*, IEEE Trans. Comput. **C-32** (1983), 108–119.

[**LP2**] ____, *A simple motion planning algorithm for general robot manipulators*, Tech. Report, AI Lab, M.I.T., 1986.

[**LPW**] T. Lozano-Perez and M. Wesley, *An algorithm for planning collision-free paths among polyhedral obstacles*, Comm. ACM **22** (1979), 560–570.

[**LuS**] V. J. Lumesky and A. Stepanov, *Path planning strategies for a traveling automaton in an environment with uncertainty*, Tech. Rept. 8504, Center for Systems Science, Yale University, April 1985.

[**MY**] S. Maddila and C. K. Yap, *Moving a polygon around the corner in a corridor*, Proc. 2nd ACM Sympos. on Computational Geom., 1986, pp. 187–192.

[**Me**] K. Mehlhorn, *Data structures and algorithms. III, Multidimensional searching and computational geometry*, Springer-Verlag, New York, 1984.

[**MMP**] J. Mitchell, D. Mount, and C. Papadimitriou, *The discrete geodesic problem*, Tech. Rept., Dept. of Operations Research, Stanford University, 1985.

[**MP**] J. Mitchell and C. Papadimitriou, *The weighted region problem*, Tech. Rept., Dept. of Operations Research, Stanford University, 1985.

[**Mo1**] H. P. Moravec, *Robot rover visual navigation*, Ph. D. Dissertation, Stanford University (UMI Research Press, 1981).

[**Mo2**] D. M. Mount, *On finding shortest paths on convex polyhedra*, Tech. Rept., Computer Science Department, University of Maryland, October 1984.

[**OY**] C. O'Dunlaing and C. Yap, *A 'retraction' method for planning the motion of a disc*, J. Algorithms **6** (1985), 104–111.

[**OSY1**] C. O'Dunlaing, M. Sharir, and C. Yap, *Generalized Voronoi diagrams for a ladder. I, Topological analysis*, Tech. Rept. 139, Computer Science Dept., Courant Institute, New York University, November 1984; Comm. Pure Appl. Math. **39** (1986), 423–483.

[**OSY2**] ____, *Generalized Voronoi diagrams for a Ladder. II, Efficient construction of the diagram*, Tech. Rept. 140, Computer Science Dept., Courant Institute, New York University, November 1984 (to appear in Algorithmica).

[OR] J. O'Rourke, *Lower bounds on moving a ladder*, Tech. Rept. 85/20, Dept. of EECS, The Johns Hopkins University, 1985.

[Pa] C. Papadimitriou, *An algorithm for shortest path motion in three dimensions*, Inform. Process. Lett. **20** (1985), 259–263.

[PSS] R. Pollack, M. Sharir, and S. Sifrony, *Separating two simple polygons by a sequence of translations*, Tech. Rept. 215, Computer Science Dept., Courant Institute, New York University, April 1986.

[Re] J. H. Reif, *Complexity of the mover's problem and generalizations*, Proc. 20th IEEE Sympos. on Foundations of Comp. Sci., 1979, pp. 421–427.

[RSh] J. H. Reif and M. Sharir, *Motion planning in the presence of moving obstacles*, Tech. Rept. 39/85, The Eskenasy Institute of Computer Science, Tel Aviv University, August 1985.

[RS] J. H. Reif and J. A. Storer, *Shortest paths in Euclidean space with polyhedral obstacles*, Tech. Rept. CS-85-121, Computer Science Dept., Brandeis University, Waltham, Mass., April 1985.

[dRLW] P. J. deRezende, D. T. Lee, and Y. F. Wu, *Rectilinear shortest paths with rectangular barriers*, Proc. ACM Sympos. on Computational Geom., 1985, pp. 204–213.

[SS1] J. T. Schwartz and M. Sharir, *On the piano movers' problem. I, The case of a two-dimensional rigid polygonal body moving amidst polygonal barriers*, Comm. Pure Appl. Math. **36** (1983), 345–398.

[SS2] ___, *On the "piano movers" problem. II, General techniques for computing topological properties of real algebraic manifolds*, Advances in Appl. Math. **4** (1983), 298–351.

[SS3] ___, *On the piano movers' problem. III, Coordinating the motion of several independent bodies: the special case of circular bodies moving amidst polygonal barriers*, Internat. J. Robotics Res. **2**(1983), no. 3, 46–75.

[SS4] ___, *On the piano movers' problem. V, The case of a rod moving in three-dimensional space amidst polyhedral obstacles*, Comm. Pure Appl. Math. **37** (1984), 815–848.

[SS5] ___, *Efficient motion planning algorithms in environments of bounded local complexity*, Tech. Rept. 164, Comp. Science Dept., Courant Institute, New York University, June 1985.

[SS6] ___, *Mathematical problems and training in robotics*, Notices Amer. Math. Soc. **30** (1983), 478–481.

[Sh1] M. Sharir, *Intersection and closest-pair problems for a set of planar discs*, SIAM J. Comput. **14** (1985), 448–468.

[Sh2] ___, *Almost linear upper bounds on the length of general Davenport-Schinzel sequences*, Tech. Report 29/85, The Eskenasy Institute of Computer Sciences, Tel Aviv University, February 1985 (to appear in Combinatorica).

[Sh3] ___, *On shortest paths amidst convex polyhedra*, Tech. Rept. 181, Comp. Science Dept., Courant Institute, New York University, September 1985 (to appear in SIAM J. Comput.).

[Sh4] ___, *On the two-dimensional Davenport Schinzel problem*, Tech. Rept. 193, Comp. Science Dept., Courant Institute, New York University, December 1985.

[Sh5] ___, *Improved lower bounds on the length of Davenport Schinzel sequences*, Tech. Rept. 204, Comp. Sci. Dept., Courant Institute, New York University, February 1986.

[SA] M. Sharir and E. Ariel-Sheffi, *On the piano movers' problems. IV, Various decomposable two-dimensional motion planning problems*, Comm. Pure Appl. Math. **37** (1984), 479–493.

[SL] M. Sharir and R. Livne, *On minima of functions, intersection patterns of curves, and Davenport-Schinzel sequences*, Proc. 26th IEEE Sympos. on Foundations of Computer Science, 1985, IEEE, New York, pp. 312–320.

[ShS] M. Sharir and A. Schorr, *On shortest paths in polyhedral spaces*, SIAM J. Comput. **15** (1986), 193–215.

[SiS] S. Sifrony and M. Sharir, *An efficient motion planning algorithm for a rod moving in two-dimensional polygonal space*, Tech. Rept. 85–40, The Eskenasy Institute of Computer Sciences, Tel Aviv University, August 1985 (to appear in Algorithmica).

[**SY**] P. Spirakis and C. K. Yap, *Strong NP-hardness of moving many discs*, Inform. Process. Lett. **19** (1984), 55–59.

[**SM**] K. Sutner and W. Maass, *Motion planning among time dependent obstacles*, Preprint, 1985.

[**Sz**] E. Szemeredi, *On a problem by Davenport and Schinzel*, Acta Arith. **25** (1974), 213–224.

[**TV**] R. E. Tarjan and C. Van Wyk, *An $O(n \log \log n)$ time algorithm for triangulating simple polygons*, Preprint, August 1986.

[**Ud**] S. Udupa, *Collision detection and avoidance in computer-controlled manipulators*, Ph. D. Dissertation, Calif. Inst. of Technology Pasadena, Calif., 1977.

[**We**] E. Welzl, *Constructing the visibility graph for n line segments in $O(n^2)$ time*, Inform. Process. Lett. **20** (1985), 167–172.

[**Ya1**] C. K. Yap, *Coordinating the motion of several discs*, Tech. Rept. 105, Computer Science Dept., Courant Institute, New York University, February 1984.

[**Ya2**] ____, *An $O(n \log n)$ algorithm for the Voronoi diagram of a set of simple curve segments*, Tech. Rept. 161, Computer Science Dept., Courant Institute, New York University, May 1985.

[**Ya3**] ____, *Algorithm motion planning*, Advances in Robotics. Vol. I (J. Schwartz and C. Yap, eds.), 1986 (to appear).

[**Ya4**] ____, *How to move a chair through a door*, Preprint, 1984.

COURANT INSTITUTE, NEW YORK UNIVERSITY, NEW YORK, NEW YORK 10012, USA

SCHOOL OF MATHEMATICAL SCIENCES, TEL AVIV UNIVERSITY, RAMAT AVIV, TEL AVIV, ISRAEL

Диофантовы уравнения и эволюция алгебры

И. Г. БАШМАКОВА

По традиции диофантовы уравнения принято связывать в истории математики с развитием теории чисел, а эволюцию алгебры — по крайней мере до середины XIX века — с исследованием определенных алгебраических уравнений, главным образом, с проблемой решения их в радикалах. При такой точке зрения остается, однако, неясным, чем же обуславливалось развитие алгебры с III в. до н.э. и до XVI в. н.э. — ведь к III веку до н.э Евклид подвел итоги исследованиям квадратных уравнений, а уравнения степеней 3 и 4 были решены только в XVI веке. Между тем, за этот период числовая область была расширена до поля **Q** рациональных чисел и была введена буквенная символика, сначала для обозначения неизвестного и его степеней, а затем и для параметров задачи. Алгебра за это время освободилась от геометрического облачения и стала строиться на базе арифметики. Чем же можно это объяснить? Каковы были побудительные причины всех этих изменений?

Мы постараемся показать, что эволюция алгебры, начиная с первых шагов ее, обуславливалась не только исследованием определенных, но и *неопределенных* или *диофантовых* уравнений, а именно диофантовых уравнений с рациональными коэффициентами, решение которых искали в поле рациональных чисел **Q** (или только положительных рациональных чисел **Q**$^+$). Задача нахождения всех рациональных решений диофантова уравнения, а, тем облее, изучения структуры множества решений еще не ставилась. Достаточно было найти одно решение или, если возможно, бесконечно много решений.

Элементы алгебры мы находим впервые в древнем Вавилоне за 2 тысячи лет до н.э. Многочисленные глиняные таблицы, расшифрованные в 20-х годах нашего века, свидетельствуют о высокой математической культуре вавилонян. Крупнейшим достижением этого времени было решение квадратных уравнений в радикалах; иначе говоря, было найдено выражение для корней квадратного уравнения через его коэффициенты при помощи четырех действий арифметики и операции извлечения квадратного корня.

Сделаем несколько замечаний. Во-первых, у вавилонян не было символики для обозначения неизвестных и параметров, поэтому «формула решения» не записывалась в общем виде (как это делается у нас), а демонстрировалась на многочисленных однотипных числовых примерах. Во-вторых, более точно говорить не о «формуле решения», а об эквивалентном ей *алгоритме*, состоящем в перечислении последовательных операций, которые надо произвести над коэффициентами, чтобы получить значение корня уравнения.

Естественно предположить, что для получения этого алгоритма вавилоняне должны были преобразовывать уравнения или часто встречающиеся в таблицах системы

$$ax^2 \pm bx = c, \qquad x \pm y = a, \quad xy = b$$

к виду $u^2 = B$. Для этого они должны были знать некоторые общие свойства операций сложения и умножения, а также уметь делать подстановки. Другие примеры, имеющиеся в таблицах, подтверждают эту гипотезу. Можно считать установленным, что вавилоняне знали свойства и правила, которые мы выражаем сейчас с помощью формул:

$$(a \pm b)c = ac \pm bc \tag{1}$$

$$(a \pm b)^2 = a^2 + b^2 + 2ab \tag{2}$$

$$(a + b)(a - b) = a^2 - b^2 \tag{3}$$

Итак, за 2 тысячи лет до н.э вавилоняне знали уже свойства некоторых законов композиции (формула (1) есть не что иное, как закон дистрибутивности умножения по отношению к сложению), осуществляли подстановки и решали с помощью алгебраических методов квадратные уравнения и эквивалентные им системы. Все это позволяет говорить о наличии у них элементов алгебры. Этот первый этап в развитии алгебры будем условно называть *числовой алгеброй*.

Однако и в этот наиболее древний период возникновение и развитие алгебраических методов были связаны не только с решением квадратных уравнений. В самых древних текстах рассматривается решение неопределенных уравнений. Одним из первых было:

$$x^2 + y^2 = z^2. \tag{4}$$

Решение его искали в рациональных числах. Вавилоняне умели находить его решения (x, y, z), которые впоследствии получили название «пифагоровых троек». Правда, не совсем ясно, знали ли они общие формулы решения этого уравнения, но достоверно, что вавилоняне связывали «пифагоровы тройки» с решением другого неопределенного уравнения:

$$u^2 + v^2 = 2w^2. \tag{5}$$

А именно, было установлено, что если (x, y, z) — решение уравнения (4), то решение уравнения (5) получается так:

$$u = x + y, \quad v = x - y, \quad w = z.$$

Но самое удивительное, что, зная одно решение уравнения (5) (u_0, v_0, w_0), они умели находить бесконечно много других его решений вида (u_n, v_n, w_0). Для этого вавилоняне пользовались формулой (см. [7]):

$$(p^2 + q^2)(\alpha^2 + \beta^2) = (\alpha p - \beta q)^2 + (\alpha q + \beta p)^2 = (\alpha p + \beta q)^2 + (\alpha q - \beta p)^2, \quad (6)$$

которую теперь обычно называют формулой композиции форм вида $x^2 + y^2$. Эта формула, как мы увидим, сыграла выдающуюся роль в математике народов средневекового Востока и Европы.

Формула (6), как и другие, устанавливалась вавилонянами для конкретных чисел. Однако, древние вычислители понимали, что вместо этих чисел можно подставить любые другие рациональные значения.

Пусть теперь u_0, v_0, w_0 — решение уравнения (5); чтобы получить новое решение, вавилоняне применяли формулы (6), взяв такие α, β, что $\alpha^2 + \beta^2 = 1$ (например, $\alpha = \frac{3}{5}$, $\beta = \frac{4}{5}$), $p = u_0$, $q = v_0$. Тогда

$$2w_0^2 = 2z_0^2(\alpha^2 + \beta^2) = (u_0^2 + v_0^2)(\alpha^2 + \beta^2) = (\alpha u_0 - \beta v_0)^2 + (\alpha v_0 + \beta u_0)^2$$
$$= (\alpha u_0 + \beta v_0)^2 + (\alpha v_0 - \beta u_0)^2,$$

т.е. получим еще два решения:

$$u_1 = \alpha u_0 - \beta v_0, \quad v_1 = \alpha v_0 + \beta u_0,$$
$$u_2 = \alpha u_0 + \beta v_0, \quad v_2 = \alpha v_0 - \beta u_0.$$

Беря теперь эти новые значения в качестве p и q, получим новые решения и т.д. Можно также брать зиачения для α и β. Их можно получить из любой пифагоровой тройки $(\overline{x}, \overline{y}, \overline{z})$, разделив все члены на $\overline{z}$.

Итак, уже на первом этапе развития алгебры на нее оказывали влияние как проблемы решения определенных уравнений, так и исследование и решение диофантовых уравнений.

Второй этап развития алгебры совпадает с расцветом греческой математики (V–II вв. до н.э.). В это время математические знания, накапливаемые веками, были преобразованы в абстрактную науку, основанную на системе доказательств — математику. Если для вавилонян (которые, несомненно, тоже применяли отдельные выводы) центр тяжести лежал *не в методе получения результата*, а в *самом результате*, то теперь центр тяжести переносится на *метод и доказательство*. При этом доказательство служит не только для установления истины того или иного предложения, но и для выяснения его *сущности*, которая раскрывается путем установления его связи с другими предложениями, теми, на которые оно опирается, и теми, которые из него следуют. Эту сторону доказательств подчеркивал уже Аристотель во «Второй аналитике».

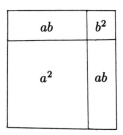

Рисунок 1

Такое глубокое преобразование нашей науки сказалось на всех частях ее, в том числе на арифметике и алгебре. К концу V в. до н.э. были построены основы элементарной теории чисел и теория положительных рациональных чисел, которые рассматривались как отношения целых (мы бы теперь сказали — как пары целых), однако в классическую эпоху эти отношения числами не считались. Первоначально основой математики служила арифметика; с ее помощь строилась теория музыки (гармония), астрономия и частично геометрия (учение о подобии). Это была первая попытка арифметизации математики и математического естествознания, которая нашла свое выражение в крылатом афоризме ранних пифагорейцев «все есть число». Открытие несоизмеримых отрезков (отношение которых не выражается отношением целых чисел) привело к кризису арифметической концепции. Вскоре более общей наукой стала считаться геометрия, которая сделалась основой и языком античной математики. Не позднее конца V в. до н.э. в геометрические доспехи облеклась и алгебра. Греки перевели все арифметические операции на геометрический язык и начали непосредственно оперировать с геометрическими объектами: отрезками, площадями и объемами, не прибегая к числам. Этот этап развития алгебры обычно называют, следуя Г. Г. Цейтену, *геометрической алгеброй*. Отрезки были первой областью объектов этой алгебры, их можно было складывать («приставлять один к другому») и вычитать из большего отрезка меньший. Произведением двух отрезков назывался прямоугольник, построенный на них: произведение трех отрезков представлялось построенной на них прямоугольной призмой. Произведение более трех отрезков было непредставимым, говорить о нем имело не больше смысла, чем о пространстве четырех или пяти измерений.

В геометрической алгебре можно было уже доказывать те свойства и тождества, которые были известны еще вавилонянам. И действительно, в «Началах» Евклида мы находим геометрическое доказательство формул (1)–(3). Так, формула $(a + b)^2 + b^2 + 2ab$ доказывалась геометрическим путем рассмотрения квадрата, сторона которого равна сумме величин a и b (рисунок 1). Задачи на квадратные уравнения также

формулировались геометрически, например задача, эквивалентная уравнению $x^2 = ab$, звучала так: «преобразовать данный прямоугольник
(т.е. ab) в квадрат». Евклид в «Началах» решает квадратные уравнения самого общего вида $\alpha x(a - x) = S$ и $\alpha x(a + x) = S$ и для первого
из них формулирует ограничение, которое надо наложить на α, a, S,
чтобы корень был действительным положительным. Все доказательства
проводились общим образом и годились как для соизмеримых, так и
для несоизмеримых величин.

В «Началах» рассматриваются также задачи, которые сводятся
к последовательному решению нескольких квадратных уравнений.
Так, для нахождения ребра правильного додекаэдра через диаметр D
описанной сферы надо решать биквадратное уравнение, которое можно
было свести в двум квадратным. И вообще, средствами геометрической
алгебры (иначе говоря, циркулем и линейкой) можно решить любое
алгебраическое уравнение, корни которого выражаются через вещ-
ественные квадратные радикалы (или еще: которое может быть сведено
к последовательному решению цепочки квадратных уравнений таких,
что коэффициенты каждого последующего являются рациональными
функциями от корней предыдущего). Таким образом, проблема решения
определенных уравнений в радикалах выступает на втором этапе развития алгебры как проблема решения уравнений в квадратных радикалах.
Например, уравнения $x^4 + x^3 + \cdots + x + 1 = 0$ и $x^{16} + x^{15} + \cdots + x + 1 = 0$
будут разрешимы средствами геометрической алгебры, а уравнение
$x^6 + x^5 + \cdots + x + 1 = 0$ — неразрешимо.[1]

В V в. появились первые «неразрешимые» задачи, т.е. неразрешимые
циркулем и линейкой; это знаменитая задача удвоения куба, которая
сводится к уравнению $x^3 = 2a^3$, а также задача трисекции угла. Отдельные кубические уравнения решал и исследовал Архимед. Их корни,
как и корни уравнения удвоения куба, были найдены с помощью
пересечения гипербол и парабол. Таким образом, проблема решения
уравнений в кубических радикалах в античности не ставилась, тем более
в радикалах более высокой степени.

В теоретических сочинениях V–III вв. до н.э. встречается два вида
неопределенных уравнений. Это пифагорово уравнение $x^2 + y^2 = z^2$ и
уравнение

$$y^2 = ax^2 + 1, \tag{7}$$

где a — целое неквадратное число, получившее впоследствии название
уравнения Пелля-Ферма. В обоих случаях речь шла об отыскании целых
положительных решений.

Изучение пифагорова уравнения проходит красной нитью через всю
античную математику. Формулы для его решения предлагали ранние
пифагорейцы (VI–V вв. до н.э.), Платон (IV в. до н.э.), наконец, самое

---

[1]Это было доказано К. Ф. Гауссом в его *Disquisitiones arithmeticae* (1801).

общее его решение содержится в «Началах» Евклида, где доказано, что все решения можно представить в виде:

$$x = p^2 - q^2, \qquad y = 2pq, \qquad z = p^2 + q^2; \qquad p, q \in \quad .$$

Уравнение (7) для $a = 2$ также рассмотрено в «Началах», где доказывается, что если x_{n-1}, y_{n-1} — решение уравнения (7) при $a = 2$, то новое решение x_n, y_n можно получить по формулам

$$\begin{cases} x_n = x_{n+1} + y_{n-1}, \\ y_n = 2x_{n-1} + y_{n-1}. \end{cases}$$

Последующая история неопределенных уравнений свидетельствует о том, что в классический период решались и многочисленные другие виды неопределенных уравнений, но их исследование оставалось за рамками теоретической науки и на развитие геометрической алгебры влияния не оказывало.

Третий, очень важный этап развития алгебры начинается в первые века нашей эры и заканчивается в конце XVI–начале XVII в. Это было время, когда алгебра обрела собственный, присущий ей язык: буквенное исчисление.

Геометрический язык стеснял развитие алгебры, он был для нее неестествен: во-первых, он не давал возможности рассматривать произведения более трех величин, а значит, и алгебраические уравнения выше третьей степени, во-вторых, геометрическое облачение делало алгебру громоздкой и малооперативной. Поэтому, когда в I в. до н.э. наступил упадок классической греческой математики и — что очень важно — традиция была прервана, ученые отказались от геометрической алгебры. Правда, еще долгое время сохранялись геометрические доказательства алгебраических правил и формул (такие доказательства можно найти еще в XVI в.), но сама алгебра уже не была скована геометрическим панцирем. Она приобрела новую форму.

Возврат к числовой алгебре мы находим уже у Герона Александрийского (I в.н.э.), в сочинениях которого решаются численные квадратные уравнения, а также новые интересные задачи на неопределенные уравнения. Заметим, что в «формуле Герона» для вычисления площади треугольника по трем его сторонам подкоренное выражение содержит произведение четырех величин, что в классическую эпоху было бы невозможно. Но особенно ярко новые тенденции нашли свое выражение в творчестве Диофанта Александрийского (III в.н.э.), который и положил начало буквенной алгебре.

До нас дошло два произведения Диофанта (оба не полностью) — «Арифметика» (шесть книг из тринадцати) и отрывки из трактата «О многоугольных числах». Для нас наибольший интерес представляет первое из них.

«Арифметика» начинается с введения, которое является, по существу, первым изложением основ алгебры. В нем строится поле рациональных чисел и вводится буквенная символика. Здесь же формулируются правила действий с многочленами и уравнениями.

Уже у Герона положительные рациональные числа получили права гражданства (в классической античной математике числами назывались множества единиц, т.е. натуральные числа). Диофант делает следующий решительный шаг — он вводит отрицательные числа. Только после этого он получает область, замкнутую относительно четырех действий арифметики, т.е. поле. Без этого развитие буквенной алгебры было бы невозможно. Но как же Диофант вводит эти новые объекты? Каким путем вообще это можно сделать? Для этого Диофант выбирает метод, который мы бы теперь назвали аксиоматическим: он определяет новый объект, который он называет «недостатком» ($\lambda\varepsilon\tilde{\iota}\psi\iota\varsigma$), формулируя правила действий с ним, а именно: если охарактеризовать новые объекты символом $(-)$, а произвольное рациональное положительное число знаком $(+)$, то Диофант постулирует:

$$(-)\cdot(-) = (+), \qquad (-)\cdot(+) = (-).$$

Правил сложения и вычитания для новых чисел он не формулирует, но в своих книгах свободно их применяет. Возможно, эти операции применялись и до него.

Далее, Диофант вводит неизвестное число, которое он обозначает специальным символом ς, а также шесть его первых положительных и шесть отрицательных степеней. Этим он окончательно порывает с геометрической алгеброй, в которой все степени выше третьей непредставимы. Замечательно, что он вводит также обозначение для нулевой степени неизвестного и специально выделяет два правила, которые отвечают двум из аксиом, определяющих группу:

$$x^m \cdot 1 = x^m, \qquad x^m \cdot x^{-m} = 1.$$

Для произведений остальных степеней неизвестного Диофант составляет «таблицу умножения», которую мы теперь можем коротко записать в виде

$$x^m x^n = x^{m+n}, \qquad -6 \le m+n \le 6.$$

Наконец, он вводит обозначения для знака минус и для знака равенства.

Все это дает возможность Диофанту записывать условия задачи в виде уравнения или системы уравнений. Собственно говоря, никаких уравнений, ни определенных, ни неопределенных, до Диофанта не было. Были только задачи, которые можно свести к уравнениям. Не более того.

Во введении Диофант формулирует два основных правила оперирования с уравнениями: 1) правило переноса члена уравнения из одной части в другую (с обратным знаком) и 2) правило приведения подобных членов. Это те самые правила, которые впоследствии получили широкую известность под арабскими названиями «аль джабр» и «аль-му-кабала».

Кроме того Диофант весьма искусно применяет правило подстановки, не формулируя его.

Итак, в книгах Диофанта мы находим уже четко записанные уравнения и чисто алгебраические преобразования. Но чему же посвящены эти книги? Для исследования и решения каких проблем Диофант ввел символику и расширил числовую область? Иначе говоря, с какими проблемами было связано рождение буквенной символики? Основное содержание «Арифметики» — это *решение неопределенных уравнений и систем* в рациональных числах, причем в ней действительно встречаются уравнения шестой степени. Что касается определенных уравнений, то Диофант решает линейные и квадратные уравнения, и притом такие, корни которых рациональны. Для их записи и решения вовсе не нужны были обозначения для степеней неизвестного выше второй. Очевидно, вовсе не проблема решения определенных уравнений побудила Диофанта к нововведениям в алгебре!

Рассмотрим вопрос о границах и возможностях символики Диофанта. Диофантовы уравнения обычно содержат несколько неизвестных, во всяком случае не меньше двух, между тем Диофант ввел символы только для одного неизвестного и его степеней. Как же он обходился при решении задач?

Задача всегда формулировалась общим образом, например: «Заданный квадрат разложить на сумму квадратов» (задача 8 книги II). После этого всем параметрам придавались конкретные числовые значения, в данном примере берется квадрат, равный 16. Затем выбиралось основное неизвестное, которое обозначалось введенным символом, другие же неизвестные Диофант выражал в виде линейных, квадратичных или более сложных рациональных функций через это основное и параметры. В приведенном выше примере второе неизвестное выражается через первое (t) в виде $kt - a$. При этом a принимается равным 4 (корню из 16), а в качестве коэффициента при t берется произвольное рациональное число; Диофант выбирает его равным 2, но оговаривает, что можно было бы взять и любое другое. Описанная нами схема является наиболее простой. Иногда оказывается, что на выбор параметров необходимо наложить некоторые ограничения, например надо выбрать такое целое число, которое представимо в виде суммы двух или трех квадратов. Тогда Диофант проводит анализ задачи и определяет, из какого множества чисел M может быть выбран тот или иной параметр.

Итак, конкретные числа несут в «Арифметике» двойную нагрузку: они выступают, во-первых, в роли обычных чисел, а во-вторых, в роли знаков для произвольных параметров. Эту вторую функцию числа будут выполнять в алгебре вплоть до конца XVI в.

Наконец, Диофант в процессе решения задачи может обозначать одним и тем же символом последовательно различные неизвестные (иногда такое «переименование» он применяет три-четыре раза). Таким образом, в рассматриваемый период развития алгебры само обозначение неизвестных требовало большого искусства.

Подведем итог. Диофант первый систематически сводил неопределенные и определенные задачи к уравнениям. Можно сказать, что для обширного круга задач арифметики и алгебры он сделал то же, что впоследствии сделал Декарт для обширного класса задач геометрии, а именно свел их к составлению и решению алгебраических уравнений. В самом деле, и Диофант, и Декарт для решения задачи — арифметической или геометрической — составляли алгебраическое уравнение, которое затем преобразовывали и исследовали по правилам алгебры. При этом производимые преобразования (исключение неизвестных, приведение подобных членов, различные подстановки) не имели непосредственного арифметического или геометрического смысла. Только окончательный результат этих формальных выкладок получал соответствующую интерпретацию и давал решение поставленной задачи. Этот важный шаг мы привыкли относить к созданию аналитической геометрии Декартом, однако задолго до того он был сделан Диофантом в его «Арифметике». Замечательно, что никому из европейских ученых XIII–XVI вв. до знакомства с «Арифметикой» не приходило в голову применить алгебру к решению теоретико-числовых задач.

Но основной проблемой для Диофанта было исследование и решение неопределенных уравнений. Именно с ними оказалось связанным рождение буквенной алгебры.

Заметим, что Диофант нашел основные методы решения в рациональных числах уравнения второй степени от двух неизвестных

$$F_2(x, y) = 0,$$

если для него известно одно рациональное решение. Еще более тонкие и интересные методы он применил для нахождения рациональных решений неопределенных уравнений 3-й и 4-й степени от двух неизвестных. Историю этих методов можно проследить до работ Анри Пуанкаре, относящихся к началу нынешнего века, в которых на основе этих методов была построена арифметика алгебраических кривых.

Дальнейшие успехи алгебры вплоть до X–XI вв. были также связаны с неопределенными уравнениями. К началу X в. относится перевод на арабский язык четырех книг арифметики, приписываемых Диофанту. Эти книги посвящены в основном решению диофантовых уравнений и, по-видимому, были составлены в Александрии в IV–V вв. В них содержатся как оригинальные задачи, которые можно приписать самому Диофанту, так и комментарии к ним. В этих книгах уже были введены 8-я и 9-я степени неизвестной.

Неопределенные уравнения занимали почетное место в алгебраических трактатах Абу Камила (начало X в.) и аль-Караджи (X–XI вв.). Методы исследования и решения они применяли те же, что и Диофант (за исключением решения неопределенных уравнений 3-й степени, метод решения которых остался им неизвестен). В книге «Аль-Фахри» ал-Караджи преобразует небольшое алгебраическое введение Диофанта в обширный трактат по алгебре. В нем он вводит бесконечно много положительных и бесконечно много отрицательных степеней неизвестного и определяет правила действия с ними. Для отрицательных чисел здесь формулируется не только правило умножения, как это было у Диофанта, но и правила их сложения и вычитания. В этом трактате аль-Караджи определяет по существу алгебру как науку о решении уравнений, *не делая различий между определенными и неопределенными уравнениями*. При решении задач он также чередует неопределенные задачи с определенными, причем последних у него больше, чем в «Арифметике» Диофанта.

Впоследствии Омар Хайям (XI в.) исследовал уравнения 3-й степени. Он находил корни таких уравнений как абсциссы точек пересечения конических сечений. Отметим также, что арабские математики по традиции наряду с «числовым» решением квадратного уравнения приводили и геометрическое обоснование этого решения. Однако у аль-Караджи имеется и чисто алгебраическое обоснование, основанное на выделении полного квадрата, которое он называет «решением по Диофанту».

Единственным, но весьма существенным отступлением арабских математиков от принципов Диофанта был отказ от алгебраической символики. Неизвестное и его степени они записывали с помощью специальных терминов.

Успехи алгебры в Европе относятся к XIII–XVI вв. Алгебраическая традиция передавалась двумя путями: из стран арабского Востока и из Византии. Первым крупным математиком был Леонардо Пизанский, или Фибоначчи (XIII в.), который оставил знаменитую *Либер абаци* (1202 г.), в которой вводил десятичную позиционную систему счисления, решал линейные системы и квадратные уравнения. Ему же принадлежит очень глубокое исследование неопределенных уравнений, которое он изложил в *Либер чуадраторум* (1225 г.). Чтобы дать представление об этих исследованиях, достаточно упомянуть, что он первый и за четыре века до Ферма утверждал, что площадь прямоугольного треугольника с рациональными сторонами не может быть квадратом. Это предложение эквивалентно Великой теореме Ферма для случая $n = 4$. Однако для трактовки неопределенных задач Леонардо не применял алгебраических методов, он решал их чисто арифметически. То же самое можно сказать и о книге Луки Пачоли *Сумма де аритхметица, геометриа, пропортиони ет пропорионалита* (опубликована в 1494 г. в Венеции), которая явилась энциклопедией математических знаний своего времени. В этой книге мы впервые встречаем постановку проблемы решения алгебраических уравнений 3-й степени и высших степеней в радикалах. Правда, Лука говорит о них как о задачах неразрешимых, таких же, как квадратура круга.

К моменту выхода *Сумма* европейские математики уже широко применяли символы для обозначения неизвестного и его положительных степеней, а также для операций сложения, вычитания, умножения и извлечения корня. Правда, они восприняли византийский метод обозначения степеней, который был гораздо менее удобен, чем способ Диофанта. В основе его лежал мультипликативный принцип: квадрато-куб обозначал не 5-ю степень, как у Диофанта, а 6-ю, 5-я степень обозначалась как $p^{o}r^{o}$ — *примо релато* (т.е. «первая невыразимая»), 7-я — как $2^{o}r^{o}$ — *сецундо релато* («вторая невыразимая»), 9-я степень называлась кубо-кубом и т.д. Таблицу такого обозначения степеней неизвестного мы находим в «Сумме». Однако вплоть до конца XVI в. не были введены ни обозначения для второго неизвестного, ни тем более для параметров.

Особые успехи выпали на долю алгебры в XVI в. Важнейшими из них были: 1) решение в радикалах уравнений 3-й и 4-й степени, 2) введение комплексных чисел и 3) создание первого буквенного исчисления.

Итак, проблема решения уравнений в радикалах появляется в XVI в. после перерыва более чем в 30 веков! За все это время побудительные импульсы шли в основном от неопределенных уравнений. Посмотрим, какое влияние они оказали на алгебру XVI в.

Мы не будем здесь касаться драматической истории, связанной с решением кубического уравнения. Об этом достаточно написано во всех руководствах по истории математики. Приведем лишь оценку этого решения, данную Г. Г. Цейтеном: «Дело здесь идет, таким образом, вовсе не об изобретении метода, но об открытии — именно открытии — формы иррациональности, которую должны иметь корни упомянутых уравнений» [**3**]. Какое же непосредственное влияние на развитие алгебры имело это открытие? Прежде всего следует отметить его большое психологическое значение. Оно показало, что «древние не все знали», укрепило веру европейских ученых в свои силы, придало им смелость, необходимую для дальнейших изысканий. Вскоре исследование уравнений

$$x^3 = px + q; \qquad p > 0, \; q > 0, \tag{8}$$

решение которых дается формулой

$$x = \sqrt[3]{\frac{q}{2} + \sqrt{\left(\frac{q}{2}\right)^2 - \left(\frac{p}{3}\right)^3}} + \sqrt[3]{\frac{q}{2} - \sqrt{\left(\frac{q}{2}\right)^2 - \left(\frac{p}{3}\right)^3}} \tag{9}$$

доставило случай применить эту смелость. Действительно, если $\left(\frac{q}{2}\right)^2 < \left(\frac{p}{3}\right)^3$, то в формуле (9) под знаком квадратных корней будет стоять отрицательное выражение $\left(\frac{q}{2}\right)^2 - \left(\frac{p}{3}\right)^3$. С другой стороны, нельзя было наложить ограничение $\left(\frac{q}{2}\right)^2 \geq \left(\frac{p}{3}\right)^3$, так как при $\left(\frac{q}{2}\right)^2 < \left(\frac{p}{3}\right)^3$ уравнение

(8) имеет действительные корни. Математики XVI в. обнаружили этот факт на примере уравнения $x^3 = 15x + 4$, которое имеет корень $x = 4$, причем и два других его корня действительны. Этот случай получил название «неприводимого». Ни Тарталья, ни Кардано, которым мы обязаны решением кубического уравнения, не смогли разрешить загадку «неприводимого» случая. Разгадку его смог дать Рафаэль Бомбелли — один из крупнейших алгебраистов Нового времени. Однако он смог это сделать лишь после того, как основательно изучил «Арифметику» Диофанта.

Мы подошли к величайшему событию в истории алгебры Нового времени, которое обычно недооценивается — к знакомству европейских математиков с «Арифметикой» Диофанта. Решительный поворот в развитии алгебры произошел именно в результате изучения этого замечательного произведения и освоения методов Диофанта. Впервые 143 задачи из «Арифметики» (всего она содержала 189 задач) появились в «Алгебре» Р. Бомбелли (1572 г.). Но дело было не только в этом «включении» задач. Итальянский историк науки Э. Бортолотти, изучавший рукописное наследие Бомбелли, заметил, что после знакомства с «Арифметикой» весь стиль книги Бомбелли изменился: задачи с «условием» в виде красочного рассказа псевдопрактического характера были заменены на задачи с сухими математическими формулировками, изменены обозначения степеней неизвестного в духе Диофанта; наконец, новые объекты — отрицательные и мнимые числа — были введены с помощью метода Диофанта. А именно: для введения отрицательных чисел Бомбелли определил для них правила сложения, вычитания, умножения и деления. После этого он сделал следующий шаг, замечательный по своей смелости — он ввел с помощью такого же приема и мнимые числа! Если мы, чтобы не утруждать читателя непривычной символикой, обозначим, как это теперь принято, мнимую единицу через i, то правила умножения, сформулированные Бомбелли, можно записать так:

$$1 \times i = i, \qquad i \times i = -1,$$
$$-1 \times i = -i, \qquad -i \times i = 1.$$

Таким образом: Р. Бомбелли понял сущность метода введения новых математических объектов и блестяще применил этот метод.

Введя мнимую единицу i, Бомбелли начал рассматривать символы вида $a + bi$, которые можно складывать и вычитать «покоординатно», т.е.

$$(a + bi) \pm (c + di) = a \pm c + (b \pm d)i$$

и умножать по правилу перемножения многочленов. По словам Н. Бурбаки, «это было первым появлением понятия линейной независимости». С помощью мнимых чисел Бомбелли сумел объяснить «неприводимый случай» кубического уравнения, который до этого ставил всех математиков в тупик. Книга Бомбелли, несомненно, является блестящим трактатом по алгебре.

Третья часть «алгебры» посвящена неопределенным уравнениям. Здесь Бомбелли воспроизводит задачи Диофанта, иногда меняя числовые параметры: он довел до конца многие задачи, решение которых было только намечено Диофантом или вовсе отсутствовало. Из решений видно, что Бомбелли хорошо понял методы Диофанта, относящиеся к неопределенным уравнениям, однако он нигде не делает попытки обобщить эти методы. Так, он первый решил задачу Диофанта:

$$x^3 + y^3 = a^3 - b^3,$$

при $a = 4, b = 3$, но не исследовал возможность ее решения при других значениях a и b.

Резюмируя, можно сказать, что хотя методы и даже обозначения Бомбелли сложились под влиянием «Арифметики» Диофанта, но интересы самого Бомбелли были связаны с проблемой решения уравнений в радикалах.

Рассматриваемый период в развитии алгебры нашел завершение в творчестве Франсуа Виета (1540–1603) — величайшего алгебраиста Нового времени. Со времен Диофанта математики совершенствовали обозначения для неизвестного и его степеней. Бомбелли ввел символы для любых целых положительных степеней в виде однородной последовательности ^{1}, ^{2}, ^{3}, ... Обозначения степеней ставились вверху справа при соответствующем коэффициенте, например $x^3 + 5x$ записывалось как $1^3 \; \tilde{p} \; 5^1$. Симон Стевин ввел аналогичные обозначения для степеней второго, третьего неизвестного. Однако никому не приходило в голову ввести обозначения для произвольных параметров. Задумав построить «аналитическое искусство» (так Виет именовал алгебру, чтобы не вводить чуждых латыни слов), которое обладало бы строгостью и глубиной геометрии древних и одновременно эффективностью и оперативностью алгебры, с тем, чтобы не осталось нерешенных проблем (nullum non problema solvere), Виет построил первое буквенное исчисление. А именно: он предложил обозначать гласными буквами неизвестные, а согласными — известные произвольные величины (параметры). Приняв, кроме того, классические знаки + и − для операций сложения и вычитания, а главное, приняв правило открытия скобок и правило подстановки вместо любой буквы выражения, полученного из букв при помощи правил операций, он получил возможность *записывать и выводить формулы*. Нам сейчас трудно представить, что до Виета *математика существовала без формул*, что все выкладки, которые мы сейчас проделываем *чисто механически*, нужно было делать в уме. Но это так. Только благодаря Виету мы можем теперь заменять некоторые рассуждения выкладками.

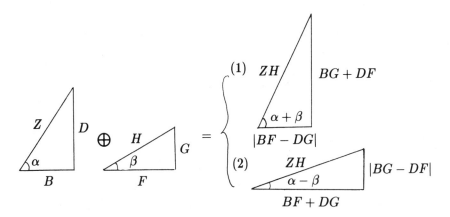

Рисунок 2

Исчисление Виета еще отличалось от принятого теперь в алгебре. Дело в том, что Виет, следуя принципам геометрической алгебры, учитывал размерность величин: складывать и вычитать можно было только величины одной размерности, при умножении размерности складывались, а при делении — вычитались. Современный вид буквенному исчислению был придан Рене Декартом (1596–1650).

Если мы теперь обратимся к сочинениям Виета, то увидим, что прежде всего он применил свое исчисление к трактовке диофантовых уравнений. В *Зететица* (особенно книги IV и V) он систематически алгебраизировал методы Диофанта и обобщил многие его задачи. Замечательно, что Виет, так же как Р. Бомбелли, ввел исчисление некоторых новых объектов, точнее, он построил исчисление треугольников, эквивалентное умножению комплексных чисел. Оно основывалось на формуле композиции (6) и носило название "genesis triangulorum".

Пусть даны два прямоугольных треугольника со сторонами B, D, Z и F, G, H, причем первой буквой обозначено основание, а последней — гипотенуза (рисунок 2). Результирующий треугольник должен был иметь гипотенузу, равную произведению гипотенуз ZH, а катеты его должны были рационально выражаться через катеты составляемых треугольников. Но по формуле (6)

$$Z^2 H^2 = (B^2 + D^2)(F^2 + G^2) = (BF - DG)^2 + (BG + DF)^2$$
$$= (BF + DG)^2 + (BG - DF)^2$$

Т.е. в результате композиции могут получиться два прямоугольных треугольника: первый — со сторонами $|BF - DG|, BG + DF, ZH$ — и второй — со сторонами $BF + DG, |BG - DF|, ZH$. Виет замечает, что острый угол при основании первого треугольника равен сумме острых углов при основаниях у компонируемых треугольников, а у второго — их разности. Если мы сопоставим каждому прямоугольному треугольнику комплексное число, первому — число $B + Di$, а второму — число $F + Gi$, тогда первый треугольник, полученный путем композиции, будет отвечать произведению чисел $(B+Di)(F+Gi)$, а второй — произведение первого числа на сопряженное ко второму — $(B + Di)(F - Gi)$. Но эти же треугольники можно характеризовать заданием гипотенузы и острого угла при основании. Первый треугольник можно условно записать как (Z, α), где α — острый угол при основании, а второй — соответственно (H, β). Тогда первый треугольник, полученный при композиции, будет $(ZH, \alpha + \beta)$, а второй — $(ZH, \alpha - \beta)$. Таким образом, построенное Виетом исчисление сразу же позволило обнаружить важнейшее свойство комплексных чисел: при умножении их модули перемножаются, а аргументы — складываются. Далее, Виет компонировал первый треугольник с собой, а затем последовательно с треугольниками, полученными путем композиции, и получал, таким образом, ряд треугольников: $(Z^2, 2\alpha), (Z^3, 3\alpha)$ и т.д. Это дало ему возможность вывести формулу, которая впоследствии получила имя формулы Муавра

$$[z(\cos \phi + i \sin \phi)]^m = z^m(\cos m\phi + i \sin m\phi).$$

С помощью этой формулы Виет получал выражения синусов и косинусов кратных дуг в виде многочленов от синусов и косинусов.

Сравним комплексные числа-символы Бомбелли с исчислением треугольников Виета. Обе эти системы имели свои достоинства и свои недостатки: числа-символы Бомбелли были удобны для производства четырех действий арифметики; по современной терминологии, они составляли поле, т.е. определенные для них два закона композиции обладали теми же «хорошими» свойствами, как и сложение и умножение рациональных чисел. Однако они не имели «тригонометрической формы», т.е. с ними не связывались понятия аргумента и модуля. Поэтому они были неудобны для выполнения операции извлечения корня, а также для приложений к тригонометрии. Исчисление треугольников Виета допускало и алгебраическую и тригонометрическую интерпретацию, поэтому оно сразу же было применено для получения ключевых формул тригонометрии. Виет пользовался им и при решении неопределенных уравнений. Однако это исчисление было малооперативным, над треугольниками был определен только один закон композиции (отвечающий умножению), короче, оно было построено еще в духе античной математики. Поэтому при дальнейшем развитии математики Нового времени предпочтение было отдано числам-символам Бомбелли. В XVIII в. они получили тригонометрическую интерпретацию, а в прошлом веке, особенно после того как Гаусс построил арифметику комплексных чисел, они приобрели полные права гражданства.

Итак, на протяжении всего третьего этапа развития алгебры основную роль играли задачи диофантова анализа, и только в конце его заметное влияние приобрели исследования определенных уравнений.

Четвертый этап истории алгебры начинается с 30-х годов XVII в. и продолжается до 70-х годов XVIII в. В это время математики прилагают усилия для решения уравнения 5-й степени и высших степеней в радикалах. Такие попытки мы находим у Эйлера, Безу, Варинга и многих других. Одновременно внимание привлекает основная теорема алгебры, утверждающая, что всякий многочлен с действительными коэффициентами может быть разложен на множители 1-й и 2-й степени. Различные ее доказательства были предложены Даламбером, Эйлером, Лагранжем, Лапласом и, наконец, Гауссом. В связи с этим развивается учение о группах подстановок, о симметрических функциях и рациональных функциях корней уравнения, остающихся неизменными при тех или иных подстановках корней. Все это подготавливало почву для будущей теории Галуа. Диофантовы уравнения хотя и продолжают по традиции рассматриваться в алгебре (весь второй том «Введения в алгебру» Эйлера посвящен неопределенному анализу), но, по существу, благодаря работам Ферма и Эйлера они сливаются с теорией чисел. С этого времени трудно различить импульсы, идущие от неопределенных уравнений как таковых, от импульсов, порождаемых теорией чисел. Исключение составляет, пожалуй, только Великая теорема Ферма. Уже доказательство ее для случая $n = 3$ потребовало расширения понятия целого числа, а именно перенесения понятия целого на выражения вида

$$a + b\sqrt{-3},$$

где a, b — целые рациональные.

Пятый этап в развитии алгебры — от 70-х годов XVIII до 70-х годов XIX в., казалось, весь прошел под знаком проблемы решения уравнений в радикалах. Достаточно напомнить о фундаментальных исследованиях Ж. Л. Лагранжа, К. Ф. Гаусса, Н. Г. Абеля и Э. Галуа, которые привели к введению таких важнейших понятий, как поле, группа, нормальный делитель, разрешимая группа и т.п. Но и здесь имеется другая линия развития, связанная с Великой теоремой Ферма, теорией квадратичных форм и законами взаимности. Для исследования этих проблем оказалось необходимым расширить понятие целого рационального числа до числа целого алгебраического, построить арифметику полей алгебраических чисел, что привело в свою очередь к введению понятий кольца, модуля и идеала. Все эти понятия являются фундаментальными в современной алгебре и применяются за ее пределами.

После 70-х годов прошлого века проблема решения уравнений в радикалах потеряла свое былое значение. Алгебра постепенно преобразовалась в науку о законах композиции, определенных для произвольных множеств объектов произвольной природы. И с современной точки зрения сама проблема решения уравнений в радикалах имеет несравненно меньшее значение, чем понятия группы и поля, которым она дала жизнь. Иначе обстоит дело с учением о диофантовых уравнениях. Это учение слилось в нынешнем веке с алгебраической геометрией, с одной стороны, и с теорией чисел — с другой. Роль их в истории алгебры далеко еще не исчерпана.

ЛИТЕРАТУРА

1. Diophanti Alexandrini, *Opera omnia* (P. Tannery, ed.), Lipsiae, 1893–95, 1–2 vol.

2. Диофант, *Арифметика* (пер. И. Н. Веселовского; предисловие и комментарии И. Г. Башмаковой), Москва, 1974.

3. Г. Г. Цейтен, *История математики*, XVII–XVIII вв., Москва-Ленинград, 1933.

4. И. Г. Башмакова и Е. И. Славутин, *История диофантова анализа (от Диофанта до Ферма)*, Москва, 1985.

5. F. Viete, *Opera mathematica*, Ludguni Botavarum, 1646.

6. R. Bombelli, *L'Algebra*, Bologna, 1572.

7. E. M. Bruins, *Reciprocal and pythagorean triads*, Physis, 1967.

Институт истории естествознания и техники, Москва 103012, СССР

Proceedings of the International Congress of Mathematicians
Berkeley, California, USA, 1986

The Concept of Construction
and the Representation of Curves
in Seventeenth-Century Mathematics

HENK J. M. BOS

Example 1. An exponential curve. My topic is best introduced by an example. I take it from the correspondence between Leibniz and Huygens in 1690–1691. Leibniz wrote about his new differential and integral calculus. Huygens was very skeptical and proposed problems for Leibniz to solve. In the course of this exchange Leibniz came to use an exponential equation to represent a curve. This was entirely new; the only curve equations used until then were algebraic ones. Huygens was even more skeptical about this novelty: he thought that Leibniz boasted, using fancy but empty symbolism. So Leibniz explained further. He took as an example the curve representing the relation between the time t and the velocity v of a body falling in a medium with resistance proportional to v^2. That curve, he said, was given by the following exponential equation:

$$b^t = \frac{1+v}{1-v}.\tag{1}$$

Huygens was still puzzled. He wrote:

> I must confess that the nature of that sort of supertranscendental lines, in which the unknowns enter the exponent, seems to me so obscure that I would not think about introducing them into geometry unless you could indicate some notable usefulness of them [11, Vol. 9, p. 537].

And somewhat later he wrote:

> I beg you to tell me whether you can represent the form of that curve by marking points on it or by whatever method [11, Vol. 9, p. 570].

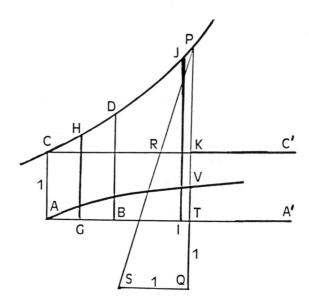

FIGURE 1

Leibniz's answer was affirmative. The equation, he wrote, implies the construction of points on the curve, and he gave the following

Construction [11, Vol. 10, pp. 14–15]. Draw (see Figure 1) parallel lines AA' and CC' with distance $AC = 1$. Take B on AA' with $AB = 1$. Take BD of arbitrary length b, perpendicular to AA'. Draw the *Logarithmica* through C and D with axis AA'. The *Logarithmica* is the curve with equation $y = b^x$. It was known to Huygens; not, of course, by its equation, but as the curve with the property that for every sequence of equidistant points on the axis, the corresponding ordinates are in geometrical progression. Hence if G is the middle of AB, then $GH = \sqrt{AC \cdot BD}$, which is constructible by ruler and compass. Again if $AB = BI$, then $IJ = (BD)^2/AC$, which is constructible as well. Thus this property implies a method to construct arbitrarily many points on the curve (by successive halving and doubling of segments on the axis and constructing the corresponding ordinates). It is to this pointwise construction of the curve that Leibniz refers in his explanation to Huygens. With the *Logarithmica* thus constructed, take P arbitrary on that curve, draw the ordinate PT, intersecting CC' in K, prolong to Q with $TQ = 1$. Take $QS = 1$ horizontally to the left, and connect S and P. SP intersects CC' in R. Take V on TP such that $TV = KR$. Then V is on the required curve. To find more points repeat this construction from other points P on the *Logarithmica*. □

Clearly, this is a rather complicated procedure to represent a curve. More surprising is the fact that for Huygens this method of marking points on the curve was much more enlightening than Leibniz's exponential equation. Indeed he wrote back:

> I have looked at your construction of the exponential curve
> which is very good. Still I do not see that this expression
> $b^t = \frac{1+v}{1-v}$ is a great help for that: I knew the curve already
> for a long time [11, Vol. 10, pp. 20–21].

Huygens's reaction shows that for him the exponential equation was not a sufficient representation of the curve; he only could understand, and indeed recognize, the curve when a construction of it was given. For him the canonical way of *giving* (and *understanding*) a curve was by a *construction procedure* for making points on it. The example, then, is about different views on the proper way of representing curves.

The representation of curves. I use the term "representation of a curve" as a technical term to denote a description of a curve that is sufficiently informative to consider the curve *known.*

In the seventeenth century, mathematicians were often confronted with the problem of how to represent curves, because they came upon many problems in which it was required to find hitherto unknown curves. Many of these problems were so-called "inverse tangent problems," equivalent to first-order differential equations and often arising from mechanical problems. Solving such problems required a convincing representation of the curve sought. As the analytical methods (analytic geometry, the calculus) were still very new, representation of a curve by its equation was often not considered sufficient (especially in the case of transcendental curves), and more geometrical ways of representation were required.

The representation of curves was an *informal practice*, without fixed criteria of adequacy. There was, at that time, no universally accepted definition of the concept of curve on which a formally determined way of representing curves could be based (nor, apparently, was a need for such a definition felt). Because it was an informal practice, it was subject to much debate; opinions about the proper representation of curves differed among mathematicians; and they changed over the period. These differences of opinion and the ensuing debates are interesting because they reveal much about the changing conceptions and aims within the mathematics of that period. In particular they reveal the complex process of the replacement of geometrical ways of thinking by analytical ones.

Geometrical construction. The example of the exponential curve illustrates that in the seventeenth century the representation of curves often relied on procedures of geometrical construction. At the beginning of the century this concept of construction had been central in a debate occurring within what may be called the *early modern tradition of geometrical problem solving*. The century between 1550 and 1650 was the time in which the classical Greek mathematics was taken up, understood, and elaborated. In particular, the early modern

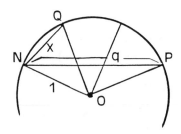

FIGURE 2

mathematicians took over the Greek interest in geometrical problems and their solution by construction.

In this practice they were confronted with two questions of method. The first was: *What means of construction should be used if problems cannot be constructed by ruler and compass?* Many problems (the classical ones as foremost cases) could not be constructed by ruler and compass. Obviously, they had to be solved, but by what means? More sophisticated instruments than ruler and compass? More complicated curves than straight lines and circles? Or should one adopt new postulates in addition to the Euclidean ones that are the basis of ruler and compass constructions? All these possibilities were considered and debated by early modern geometers.

The second methodological issue was *the search for analytic methods.* From the classical Greek geometrical works as they were known, about 1600 mathematicians inferred that the ancients had had a special method, called *analysis*, for finding proofs of theorems and constructions of problems, but that they had kept that method secret, or at least that works about the method had been lost. So the early modern geometers set themselves the task of recreating or creating such analytic methods.

Example 2. Trisection. Rather than discussing geometrical construction and the related methodological questions abstractly, I shall illustrate them by an example. It is taken from Descartes's *Géométrie* (1637) [**9**], and it concerns a classical problem, the trisection of the angle.

Let $\angle NOP$ (see Figure 2) be given, so that the chord $NP = q$ within the circle (radius 1) is known. It is required to construct $\angle NOQ = \frac{1}{3}\angle NOP$.

Descartes proceeded in two steps. He called x the chord NQ of the required angle, and he derived an equation for x. He found, by applying elementary Euclidean geometry:

$$x^3 - 3x + q = 0. \tag{2}$$

The second step was to geometrically construct a root x of the equation. Descartes gave the following

Construction [**9**, pp. 396–397]. With respect to perpendicular axes (see Figure 3) through O, draw a parabola with vertical axis, vertex in O, and passing through the point U, with coordinates of length 1. Take D on the vertical axis

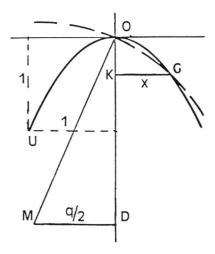

FIGURE 3

below O such that $OD = 2$. Take $DM = q/2$ horizontal to the left. Draw a circle with center M and radius MO; the circle intersects the parabola in O and in three other points, of which G is the one nearest to the vertex. Draw GK horizontally with K on the vertical axis. Then GK is the required root x; taking $NQ = GK$ in Figure 2 gives the required trisection. [The remaining roots occur as the ordinates of the other points of intersection of the circle and the parabola.] □

According to Descartes, this kind of construction was the canonical solution of an equation if it arose in a geometrical context. An algebraical solution (by a Cardano-type formula) would not be sufficient; the problem was geometrical and hence the solution had to be geometrical too. The example illustrates Descartes's particular answer to the methodological questions outlined above: Construction beyond ruler and compass was to be effectuated by the intersection of higher curves (here the parabola and the circle); the analytical method was algebra.

When is a problem solved? At this point the two examples enable me to state, somewhat slogan-like, the central theme of my research. It concerns the questions: When was a problem considered solved? When was an object considered known? In other words, what were the *criteria* for adequate solution and representation in seventeenth-century mathematics?

Such criteria evidently played a role in the mathematical practice of the period (as in fact in any period). They were not formalized, and they were controversial. Studying these criteria, the debates about them, and the changes they underwent often brings to light ways of mathematical thinking that were common and self-evident at the time but are very unfamiliar to us.

The criteria of adequacy have been little studied before by historians of mathematics. The reason for that neglect of an important part of seventeenth-century mathematics is that these criteria concern contemporary practice, whereas his-

torical research has often concentrated on the origin of modern ideas. Also the
criteria concern the mathematical *material*, the *objects* (like curves), and the
problems (construction problems or inverse tangent problems), whereas historical research has tended to concentrate on the *theories* and the *methods* (analytic
geometry, calculus) that were developed to deal with those objects and problems.

I have found a study of these criteria of adequacy very revealing and reward-
ing. In the remainder of this lecture I would like to mention some results of the
investigations around the theme outlined above, and give some examples.

Descartes's *Géométrie*. Let me begin with Descartes's *Géométrie* of 1637
(cf. [**7**]). This was without doubt the most influential book in seventeenth-
century mathematics; for one thing, it marked the beginning of analytic geometry. Through it, Descartes's particular choices (mentioned above) with respect to
the methodological issues in geometry, his criteria of adequacy, became paradigmatic for mathematicians after him. These choices largely determined the structure of the book and the conception of geometry behind it, as for instance the
restriction of geometry to algebraic relationships which Descartes advocated very
strongly. His methodological choices explain in particular what may be called
Descartes's program for geometry:

Given a geometrical problem, one calls x one of the line segments that have
to be constructed. One then derives an equation

$$H(x) = 0 \qquad\qquad (3)$$

for x, where H is a polynomial. Then, to determine x, the geometer's task is
to find *acceptable, simple* curves $\mathcal{F}$ and $\mathcal{G}$, such that the roots of $H(x) = 0$ are
equal to the ordinates of intersection points of $\mathcal{F}$ and $\mathcal{G}$. These curves are then
the *constructing curves* by which the problem is solved.

In Descartes's view of geometry, these curves should be algebraic. So, if
we write $F(x,y) = 0$ and $G(x,y) = 0$ for the equations of these curves, the
requirements are that $H(x)$ is a factor of the resultant of F and G:

$$\mathrm{Res}(F, G) = A(x) \cdot H(x), \qquad\qquad (4)$$

and that $\mathcal{F}$ and $\mathcal{G}$ are in some sense *acceptable* and *simple*. The procedure to
find such F and G for given H was called the "construction of the equation."

Descartes treated the construction of equations in general for equations $H(x)$
$= 0$ of degree 2–6. He showed that equations of degree 1 and 2 can be constructed
by circles and straight lines, equations of degree 3 and 4 by the intersection of
a conic and a circle (in fact, he showed that one fixed parabola is enough),
and equations of degree 5 and 6 by the intersection of a circle and a special
third-degree curve, the later so-called "Cartesian Parabola." Descartes did not
proceed to higher degrees; he simply stated at the end of his book that it would
be easy to go on. So he left a program for his successors: to work out a theory
of constructing equations.

A forgotten theory. Around 1650, the *Construction of Equations* (cf. [**8**]) was generally considered a sensible subject, a natural and legitimate interpretation of the program of finding exact constructions for geometrical problems of any degree of complexity. The theory attracted considerable attention; many books and articles about it appeared and mathematicians of first rank contributed to it, such as Descartes, Fermat, Newton, l'Hôpital, Riccati, Cramer, Euler, Lagrange. Descartes's opinion that the constructing curves should be algebraic was generally (though not universally) accepted, but there was much debate on the requirement that the curves be "simplest possible." Should the equation be simple? Or the shape of the curve? Or the movement by which it can be traced? Descartes had given little guidance here; he had only stated, without further argument, that a curve is simpler in as much as its degree is lower.[1]

The debates about these questions show how mathematicians struggled to formulate and fix the motivation and the aims of the theory. They often felt strongly about it; witness the legislative, almost moralistic overtones in the debate. Some quotations may illustrate this. Here, for instance, is Fermat:

> Certainly it is an offense against the more pure geometry if one assumes too complicated curves of higher degrees for the solution of some problem, not taking the simpler and more proper ones; for it has often been declared already, both by Pappus and by more recent mathematicians, that it is a considerable error in geometry to solve a problem by means that are not proper to it [**10**, Vol. 1, p. 121].

And Newton:

> Yet it is not its equation but its description which produces a geometrical curve.... It is not the simplicity of its equation but the ease of its description which primarily indicates that a line is to be admitted into the construction of problems.... Either, then, we are, with the ancients, to exclude from geometry all lines except the straight line and circle and may be the conics, or we are to admit them all according to the simplicity of their description [**14**, Vol. 5, pp. 425–427].

Many similar statements occur in the literature on the construction of equations. They use remarkable metaphors: geometry is seen as a lawful territory

---

[1] Around 1700, mathematicians had come to the following consensus about the degrees of the "best possible" constructing curves for an equation: If the degree of H is n, then constructing curves $F(x, y) = 0$ and $G(x, y) = 0$ can be found with degrees that are integer approximations of $\sqrt{n}$. The consensus was based on experience. Newton and l'Hôpital gave proofs, but these were incorrect. Euler and others accepted the result without questioning the proof. In modern terms the question is this: Given $H[X]$, a polynomial of degree $n = k \cdot l$; are there polynomials $F[X, Y]$ and $G[X, Y]$ with degrees k and l, such that $H = \text{Res}(F, G)$? It seems that this question is still open. I would be very thankful to any colleague who can give me more definite information about it.

that has to be protected and from which certain practices have to be excluded, or it is seen as a person, who can be offended and whose purity, one would almost say whose chastity, has to be defended. The issue was: to shape the proper rules of the subject and thereby to secure its status as a meaningful and sensible subject. The metaphors indicate that mathematicians felt strongly about it. Still, despite the strong words, the debates remained inconclusive; the questions about the aims of the field, and its proper procedures, could not be answered. After some time the debate died and so did the theory itself; after 1750 it quickly fell into oblivion.

The phenomenon of a theory that starts off as an evidently sensible enterprise and later dies amidst inconclusive discussions on its aims and motivations is a most interesting one. Why did the subject die? The answer turns out to be the following: The construction of equations originated as a sensible procedure within geometry. Purely algebraically, however, it does not make much sense. If a problem consists of a polynomial in one unknown, why should two polynomials in two unknowns constitute a solution? As the theory progressed, the techniques to find constructing curves became more and more algebraic. But the geometrical meaning of the subject—exact construction—and the geometrical criteria of adequacy—simplicity of the curves—refused translation into algebra in a natural way. The subject had a tendency to become algebraic, but its aims, criteria, and meaning proved untranslatable into algebra—it succumbed to this internal contradiction.

In the case of the construction of equations, we can follow in detail a process of development and decline of a mathematical field, a process whose causes were in the sphere of motivation, sense, and meaning. Such processes are little studied, although they are of evident interest for understanding the development of mathematics. The case also provides an informative example (or counterexample) with respect to theories about the historical development of scientific "research programs" as proposed by I. Lakatos and other recent philosophers of science.

Example 3. The *Elastica* and the *Paracentric Isochrone*. I now return to the representation of curves, about which methodological questions were raised remarkably similar to the ones discussed in connection with the construction of equations. Again, I can best illustrate these questions by an example. The example concerns two curves, the *Elastica* and the *Paracentric Isochrone*. In 1694 Jakob Bernoulli published in the *Acta Eruditorum* an article [2] about the form of elastic beams under tension. The beams (cf. Figure 4) are fixed vertically at the one end; a weight is attached such that the other end is bent horizontally. Bernoulli considered arbitrary relations between extension and force, but he devoted special attention to the case in which Hooke's law—extension proportional to force—applies. The *Elastica* is the form of the beam in that case.

Bernoulli derived the differential equation of the *Elastica*:

$$dy = \frac{x^2\,dx}{\sqrt{a^4 - x^4}},$$

(5)

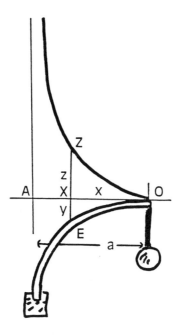

FIGURE 4

where a is the horizontal distance between the two ends of the beam. He represented the solution curve by means of the following

Construction. Take $OA = a$ along a horizontal X-axis (see Figure 4, positive values are taken to the left). Construct above the axis the curve with ordinates z satisfying the equation

$$z = \frac{ax^2}{\sqrt{a^4 - x^4}}. \tag{6}$$

[Bernoulli assumes his readers to be familiar with the construction of algebraic curves.] For any abscissa $OX = x$, determine y such that ay is equal to the area OXZ ($XZ = z(x)$). Take $XE = y$ vertically downwards. Then E is on the *Elastica.* More points on the curve are found by repeating the construction for other values of x. □

The construction is the geometrical equivalent of the analytical formula

$$ay = \int_0^x \frac{ax^2 \, dx}{\sqrt{a^4 - x^4}}. \tag{7}$$

Bernoulli could have written the solution of (5) in such an analytical form; it is important to note that he did not do so, but chose to represent the curve by this geometrical construction. It is a so-called "construction by quadrature," assuming (without explanation) that it is possible to determine a rectangle (ay) equal to an area under a given curve. Construction by quadrature was a common way to represent transcendental curves in the seventeenth century, but it was not considered the most desirable kind of representation.

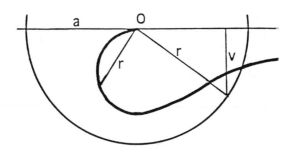

FIGURE 5

Bernoulli further calculated the differential of the arclength $s = OE$ of the *Elastica*:

$$ds = \frac{a^2\,dx}{\sqrt{a^4 - x^4}}. \tag{8}$$

This formula provided the link between the *Elastica* and the *Paracentric Iso-chrone*. The *Paracentric Isochrone* (see Figure 5) is the curve through a point O with the property that, in a vertical plane, a body moving under influence of gravity along the curve, recedes uniformly from O. That is, if $r(t)$ is the distance of the body to O and t the time, then $r(t)::t$.

Leibniz had challenged mathematicians to determine this curve. In an article [3] published together with the one on elastic beams, Bernoulli gave his solution. He derived the differential equation:

$$\frac{d(ar)}{2\sqrt{ar}} = \frac{a^2\,du}{\sqrt{a^4 - u^4}} \tag{9}$$

(with a depending on the initial velocity, r and v as in Figure 5, and $u^2 = av$). Bernoulli could now give a construction "by quadratures," the geometric equivalent of writing

$$\sqrt{ar} = \int_0^u \frac{a^2\,du}{\sqrt{a^4 - u^4}}. \tag{10}$$

Significantly, he did not do so. He recognized the right-hand differential in (9) as the arclength differential of the *Elastica*, and he concluded that this enabled him to give a construction "by rectification." It is as follows:

Construction. Assume an *Elastica* RQO given (see Figure 6). Draw a circle around 0 with radius $OB = a$. Take E arbitrary on OB and draw EQ vertically with Q on the *Elastica*. Take U on the circle such that $UV = OE^2/a$. Take W on OU such that $OW = (\text{arc } OQ)^2/a$. (Here it is assumed that the rectification of the *Elastica* can be performed.) Then W is on the *Paracentric Isochrone*. Repeat the construction for other points E to get arbitrarily many points on the required curve. □

The *best* representations. The remarkable thing about Bernoulli's construction is that according to him this was the *best* way of representing the solution of Leibniz's problem, better than the construction by quadratures which

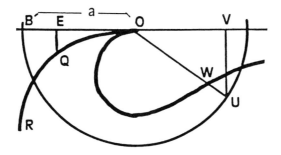

FIGURE 6

is implied in formulas (9)–(10). And this was not merely a curious idiosyncrasy of one mathematician. Shortly afterwards three further articles appeared, by Leibniz [13], Johann Bernoulli [6] (Jakob's brother), and Jakob himself [4], each containing reductions of the integral in (10) to an arclength of a curve. Indeed, while searching for a comparatively simple algebraic curve to reduce the integral, Jakob and Johann independently found the same curve. It was the *Lemniscate*, whose origin, therefore, lies in a preference for rectifications over quadratures in the representation of transcendental curves.

In the course of this exchange of solutions Jakob Bernoulli came to formulate explicitly [4, p. 608] his view on the proper representation of transcendental curves. He wrote that one should *at least* give a construction by quadrature of an algebraic curve. It was *better* to give a construction by rectification of an algebraic curve, or a "pointwise construction" (such as Leibniz's construction of the *Logarithmica*, see above). The *best* way to represent a curve, however, was a construction by curves "given in nature" (as the *Elastica* or, e.g., the *Catenary*). Bernoulli preferred rectifications over quadratures because, as he said, measuring length is easier than measuring area. He gave top preference to curves "given in nature" because if these can be found, all laborious construction of algebraic curves and their quadratures or rectifications could be avoided. These views of Jakob solicited several reactions, which I shall not further discuss; I use his statements here primarily to show that there was a debate and to illustrate its nature.

The debate shows striking similarities with the discussions about the construction of equations. In both cases analytical representation was seen as insufficient: a problem was considered solved only when a geometrical construction was given. The crucial point was the interpretation of "simplicity"; the constructing curves were considered better inasmuch as they are "simpler"; rectifications were preferred over quadratures because they were considered "simpler" to effectuate; construction by curves "given in nature" was advocated by Bernoulli because it provided "simpler," easier constructions. There were legislative overtones in both debates; Johann Bernoulli, for instance, uses terminology like "to sin against the laws of geometry" [6, p. 121]. And finally in both cases the debate remained inconclusive. With hindsight we can understand this; the relevant theories (equations, differential equations) became more and more analytical, but the concepts of geometrical simplicity could not be convincingly translated and formalized into analytical terms. The discussions were resolved by forgetting the problems.

Although these issues of construction and representation of curves were later forgotten, at the time they had a decisive influence on the development of mathematics. Analytic geometry originated in the context of geometrical construction by the intersection of curves. The first techniques for solving differential equations were elaborated with the aim of finding appropriate geometrical representations of the solution curves of inverse tangent problems. And, for instance, the early studies on elliptic integrals by Jakob Bernoulli, Fagnani, and others were a result of the effort to interpret integrals as arclengths.

Conclusion. I hope I have shown that the question of the *criteria of adequacy of representation and solution* provides an intriguing and fruitful way of looking at the mathematics of the seventeenth century. It provides new insights on three different levels.

On the *technical* level, an awareness of these issues leads to a better understanding of the terminology and the mental images of seventeenth-century mathematical practice. Curves were studied intensively in that period, but most of them (in particular the transcendental ones) could not be represented by equations. An understanding of the alternative ways of representation, of the reasons behind them, and of the mental images of mathematical objects which they presuppose is essential for understanding the texts of the period.

On the level of the *development* of mathematics, the approach helps in understanding certain directions and tendencies in seventeenth-century mathematical research, which would otherwise merely seem peculiar or superfluous, such as the interest in the geometrical construction of roots of equations or in representing integrals as arclengths.

Finally, on a more *general* level, a study of the criteria of adequacy is useful in understanding the processes of change in mathematics caused by the introduction of radically new methods (such as analysis in the seventeenth century) and the process of *habituation* to new ways of mathematical thinking. These processes operate both on the level of technique and on the level of motivation, meaning, and sense of the mathematical enterprise. They are not special to the seventeenth century, they belong to the mathematics of all times. They have received little attention until now; the research on which I am reporting may be of interest as an experiment in how these processes can be studied.

BIBLIOGRAPHY

1. Jakob Bernoulli, *Opera*, Geneva, 1744; reprint, Culture et Civilisation, Brussels, 1967.
2. ____, *Curvatura laminae elasticae*, Acta Eruditorum, 1694 (June); in Opera, pp. 576–600.

3. ____, *Solutio problematis Leibnitiani*, Acta Eruditorum, 1694 (June); in Opera, pp. 601–607.

4. ____, *Constructio curvae accessus et recessus aequabilis*, Acta Eruditorum, 1694 (Sept.); in Opera, pp. 608–612.

5. Johann Bernoulli, *Opera omnia*, Lausanne/Geneva, 1742; reprint, Olms, Hildesheim, 1968.

6. ____, *Constructio facilis curvae accessus et recessus aequabilis*, Acta Eruditorum, 1694 (Oct.); in Opera Omnia, Vol. 1, pp. 119–122.

7. H. J. M. Bos, *On the representation of curves in Descartes' Géométrie*, Arch. Hist. Exact Sci. **4** (1981), 295–338.

8. ____, *Arguments on motivation in the rise and decline of a mathematical theory; the "construction of equations", 1637–ca. 1750*, Arch. Hist. Exact Sci. **30** (1984), 331–380.

9. R. Descartes, *La géométrie*, Discours de la méthode, Leiden, 1637, pp. 297–413; English transl. in The Geometry of Rene Descartes, D. E. Smith and M. L. Latham, editors, Dover, New York, 1954.

10. P. De Fermat, *Oeuvres*, P. Tannery and C. Henry, editors, Gauthier-Villars, Paris, 1891–1912.

11. C. Huygens, *Oeuvres complètes*, M. Nijhoff, The Hague, 1888–1950.

12. G. W. Leibniz, *Mathematische Schriften*, C. I. Gerhardt, editor, Berlin and Halle, 1849–1863; reprint, Olms, Hildesheim, 1961–1962.

13. ____, *Constructio propria problematis de curva isochrona paracentrica*, Acta Eruditorum, 1694 (Aug.); in Mathematische Schriften., Vol. 5, pp. 309–318.

14. I. Newton, *The mathematical papers of Isaac Newton*, D. T. Whiteside, editor, Cambridge Univ. Press, London and New York, 1967–1981.

MATHEMATICAL INSTITUTE, STATE UNIVERSITY OF UTRECHT, UTRECHT, THE NETHERLANDS

Cayley's Counting Problem and
the Representation of Lie Algebras

THOMAS HAWKINS

On a simplistic level, it is easy to explain how the theory of the structure and representation of semisimple Lie algebras originated. In 1874 Sophus Lie introduced the notion of a Lie group and the associated infinitesimal group or Lie algebra. In 1888 Wilhelm Killing determined all simple and semisimple Lie algebras. His results were more rigorously established and somewhat extended by Élie Cartan in his doctoral dissertation of 1894. Then in 1913 Cartan introduced his theory of weights and determined all irreducible representations of simple and semisimple Lie algebras. Finally, in 1924 Hermann Weyl proved the complete reducibility theorem for semisimple Lie algebras and introduced the notion of a character to determine the dimension of an irreducible representation, which Cartan had been unable to do.

There is nothing factually incorrect with this account, but it is almost completely devoid of historical content. One would never know from this account that the works of Lie, Killing, Cartan, and Weyl are inextricably intertwined with a multitude of strands of nineteenth and early twentieth century mathematics involving many of the most prominent mathematical schools of that period. The challenge to the historian is to depict the origins of a mathematical theory so as to capture the diverse ways in which the creation of that theory was a vital part of the mathematics and mathematical perceptions of the era which produced it. The challenge is particularly formidable in the case of the structure and representation theory of Lie algebras because its historical roots reach out to an unusually broad and diverse spectrum of nineteenth and early twentieth century mathematical thought. But the challenge is for the same reason tantalizing because it affords the historian an extraordinary opportunity to bring to life the mathematics and mathematicians of the past.

In my work on the history of Lie algebras I have attempted to undertake the task of "revitalization" in various ways. One approach is to relate, when appropriate, a mathematician and his mathematics to the research programs and disciplinary ideals of the mathematical school in which he was trained or within which he subsequently worked. I have applied this approach to the groundbreaking work of Killing, who was trained in the Berlin School of Weierstrass [1, 2]. Another way in which I have attempted to bring to life the mathematics and mathematicians of the past is to identify a mathematical event, such as the articulation of a mathematical idea or problem which, when followed through time, provides an interesting historical mirror in which to view the developments that produced the representation theory of Lie algebras. In this manner one obtains, so to speak, a one-dimensional representation of a more complex historical reality which has the advantage that it focuses upon one type of link between the history of Lie algebras and the mathematics of bygone eras.

One example of such an event is provided by the geometrical Principle of Transfer which Otto Hesse introduced in 1866. Inspired by the duality between points and lines in the projective geometry of the plane, Hesse showed how to establish a correspondence between the projective geometry of n-dimensional space (with special attention to the case $n = 3$) and the projective geometry of the line. As time went on and Hesse's Principle was viewed in the light of Klein's *Erlanger Programm*, Lie's theory of transformation groups, the Clebsch-Gordan "series expansions" in the theory of invariants and the varieties introduced into algebraic geometry by Veronese and Segre, the Principle changed considerably. So much so that by 1913 Cartan viewed himself as applying Hesse's Principle when he showed how to construct all irreducible representations of a simple Lie algebra of rank l from l "fundamental" representations. That is, what is nowadays sometimes called the Cartan product of two irreducible representations was for Cartan himself a consequence of Hesse's Principle. The Principle had changed considerably, of course, and in the process played a role in the history of the theory of Lie algebra representations.

I have recently discussed Hesse's Principle in greater detail elsewhere [3]. Here I shall go into more detail regarding another "historical mirror" and the view of Lie algebra history it affords. This one is provided by a seminal paper on the new theory of invariants which Arthur Cayley published in 1854 and which introduced a problem I shall refer to as Cayley's Counting Problem. (I am grateful to Jacob Towber for calling my attention to Cayley's work on this problem.) Like Hesse's Principle, Cayley's Problem changed with time. As we shall see, Cayley's Problem is an interesting historical mirror in which to view some of the developments in the representation theory of Lie algebras not reflected in Hesse's Principle. An example is the discovery of the first proof of the complete reducibility theorem for semisimple Lie algebras. Hesse's Principle had encouraged Lie's colleague at Leipzig, Eduard Study, to interpret the series expansions of Clebsch and Gordan as special cases of this theorem and to conjecture the theorem for any semisimple Lie algebra. Cartan made the same conjecture, but, as he later explained, he lacked the tools to prove it. A proof was first discovered by Weyl in 1924. He made critical use of the tools provided by Issai Schur in a seminal paper published that same year. The historical background to Weyl's work is far too complex and diverse to consider fully here. Here the discussion will be restricted to the one-dimensional but nonetheless enlightening view of the background to Weyl's paper afforded by tracing the vicissitudes of Cayley's Problem.

The theory of invariants grew up around the general problem of describing, in one way or another, the in- and covariants associated with an n-ary form, i.e., a homogeneous polynomial in n variables. This general problem was first posed by Cayley, who was inspired to do so through some papers written by Boole in the 1840s [4]. Consider, for example, the case of a single binary form $F(a, x)$ of degree d: $F(a, x) = \sum_{j=0}^{d} \binom{d}{j} a_j x_1^{d-j} x_2^j$. A linear change of variable $x = Sx'$, $S \in$ GL(2), transforms F into a polynomial in x': $F(a, x) = F(a, Sx') = F(a', x')$, where the coefficients a_j' of the transformed F are linear homogeneous expressions in the a_j which thus define a linear transformation

$$a' = P_d(S)a, \qquad P_d(S) \in \text{GL}(d+1). \tag{1}$$

In the above definitions, GL(n) can be regarded as defined over the real or complex field, an ambiguity present in the nineteenth century literature. A polynomial $C(a, x)$ in the a_j and the x_i which is homogeneous of degree p in a and of degree q in x is a covariant with respect to $F(a, x)$ if

$$C(a', x') = (\det S)^w C(a, x) \quad \forall S \in \text{GL}(2).$$

An invariant is a covariant in the "limiting case" in which $q = 0$ so $C(a, x) = C(a)$.

The notation we have used in sketching the above definitions is historically misleading in the sense that it has a group theoretic flavor, whereas group theoretic considerations played no part in the theory of invariants until the 1890s. Nonetheless for our purposes it will be helpful to keep the group theoretic aspect in mind. For example, the correspondence $S \to P_d(S^{-1})$ defines a representation of GL(2).

As Cayley realized, one way to describe, say, the covariants associated with $F(a, x)$ would be to determine, if possible, a finite number of covariants such that every covariant can be represented as a polynomial in these. We shall refer to this as the Finiteness Problem. A related problem is *Cayley's Counting Problem: For fixed values of p and q, determine the number of linearly independent covariants $C(a, x)$ of degrees p and q in a and x respectively.* Cayley posed his Counting Problem in a paper written in 1852 for Crelle's journal [5]. By the time it appeared in print in 1854 he had solved it. The solution was presented in the second of Cayley's famous memoirs on "quantics" [6].

Cayley's solution was based upon his discovery that the in- and covariants of the binary form $F(a, x)$ can be characterized by means of differential equations. That is, if differential operators $\mathcal{X}$ and $\mathcal{Y}$ are defined by

$$\mathcal{X} = \sum_{j=1}^{d} j a_{j-1} \partial/\partial a_j, \qquad \mathcal{Y} = \sum_{j=0}^{d-1} (d-j) a_{j+1} \partial/\partial a_j,$$

then we have

THEOREM 1. $C(a, x)$ *is a covariant iff*

$$\mathcal{X}C = x_2 \partial C/\partial x_1 \quad and \quad \mathcal{Y}C = x_1 \partial C/\partial x_2.$$

$C(a)$ *is an invariant iff* $\mathcal{X}C = \mathcal{Y}C = 0$.

An immediate consequence of Theorem 1 is that if $C(a, x)$ is a covariant then $\mathcal{Z}(C) = 0$ where $\mathcal{Z}$ denotes the commutator

$$\mathcal{Z} = (\mathcal{X} - x_2 \partial/\partial x_1)(\mathcal{Y} - x_1 \partial/\partial x_2) - (\mathcal{Y} - x_1 \partial/\partial x_2)(\mathcal{X} - x_2 \partial/\partial x_1).$$

By writing $\mathcal{Z}$ in the simplified form

$$\mathcal{Z} = \mathcal{X}\mathcal{Y} - \mathcal{Y}\mathcal{X} + (-x_1 \partial/\partial x_2 + x_2 \partial/\partial x_1)$$

where $\mathcal{X}\mathcal{Y} - \mathcal{Y}\mathcal{X} = \sum_{j=0}^{d}(d - 2j)a_j \partial/\partial a_j$, Cayley discovered that $\mathcal{Z}C = 0$ imposes a condition on the terms of $C(a, x)$. That is, if

$$C(a, x) = \sum_{j=0}^{q} \binom{q}{j} B_j(a) x_1^{q-j} x_2^{j}, \tag{2}$$

where $B_j(a)$ is a homogeneous polynomial of degree p, viz.

$$B_j(a) = \sum_{m_0 + \cdots + m_d = p} \alpha_{m_0 \cdots m_d} a_0^{m_0} a_1^{m_1} \cdots a_d^{m_d},$$

then each term of B_j has the same "weight" $w_j = (pd - q)/2 + j$, where the weight of a term of $B_j(a)$ is defined to be $w_j = \sum_{i=0}^{d} im_i$. Later a homogeneous polynomial such that each term has the same weight was termed *isobaric*, a term we shall use. If x_1 and x_2 are assigned weights 1 and 0, respectively, so $x_1^{q-j} x_2^{j}$ is assigned weight $q - j$, then each term of $C(a, x)$ as given in (2) has the same weight so that $C(a, x)$ is also isobaric. That is, we have

THEOREM 2. *If* $C(a, x)$ *is a covariant then it must be isobaric of total weight* $(pd + q)/2$.

A mathematician reading Cayley's papers today will see immediately that their contents can be interpreted readily in terms of Lie algebras, although of course that concept was not introduced until twenty years later. Viewed in modern terms, the operators $\mathcal{X}, \mathcal{Y}$ and $[\mathcal{X}, \mathcal{Y}] = \mathcal{X}\mathcal{Y} - \mathcal{Y}\mathcal{X}$ act on the vector space of all homogeneous polynomials $B(a)$ of degree p, thereby making it into a module for the Lie algebra $sl(2, C)$. Indeed, as Cayley realized, if $B(a)$ has degree p and weight w, $(\mathcal{X}\mathcal{Y} - \mathcal{Y}\mathcal{X})B = (pd - 2w)B$. Thus $B(a)$ is a weight vector of Cartan weight $pd - 2w$. Furthermore, Cayley's solution to his Counting Problem for covariants $C(a, x)$ of degrees p and q in a and x respectively, involved, in effect, the irreducible representation of $sl(2, C)$ of highest weight q.

This follows from the fact that, if one writes out the two differential equations of Theorem 1 characterizing a covariant $C(a, x)$ they are equivalent, respectively, to

$$\mathcal{X}B_0 = 0, \ldots, \mathcal{X}B_j = jB_{j-1}, \ldots, \mathcal{X}B_q = qB_{q-1} \tag{3}$$

and

$$y B_q = 0, \ldots, y B_{q-j} = j B_{q-j+1}, \ldots, y B_0 = q B_1. \tag{4}$$

Since, by Theorem 2, $B_0(a)$ has Cayley weight $w = (pd - q)/2$, equations (3) and (4) imply to us that B_0 is a highest weight vector ($X B_0 = 0$) of Cartan weight $q = pd - 2w$ and that the B_j generated by B_0 under application of y as in (4) define a basis for the corresponding irreducible sl$(2, C)$ module. For Cayley the significance of equations (3) and (4) was that they showed that a covariant is determined by its leading coefficient B_0. Eventually he obtained the converse result contained in

THEOREM 3. *If $B_0(a)$ is any homogeneous polynomial of degree p and weight $w = (pd - q)/2$ which satisfies $X B_0 = 0$, then if $B_1, \ldots, B_q$ are defined by (4), the resulting polynomial $C(a, x)$ defined by (2) is a covariant.*

To prove Theorem 3, Cayley showed (3) and (4) were satisfied. This meant showing that $y B_q = 0$ (or equivalently that $y^{q+1} B_0 = 0$) and that $X B_j = j B_{j-1}$ for $j > 0$.

A homogeneous isobaric polynomial $B(a)$ such that $X B = 0$ was later called a *semi-invariant* (or *seminvariant*) since it satisfies half the conditions for an invariant given in Theorem 1. Adopting this terminology, we may say that Theorem 3, by setting up a correspondence between covariants and their leading coefficients, provided Cayley with the following

COROLLARY. *If $M(p, d, w)$ denotes the number of linearly independent seminvariants of degree p and weight $w = (pd - q)/2$, then $M(p, d, w)$ is the solution to the Counting Problem.*

Cayley discovered a simple characterization for the number $M(p, d, w)$:

THEOREM 4 (CAYLEY'S LAW). *If $N(p, d, w)$ denotes the number of nonnegative integral solutions $m_0, \ldots, m_d$ to the two equations $\sum_{j=0}^{d} m_j = p$, $\sum_{j=0}^{d} j m_j = w$, then $M(p, d, w) = N(p, d, w) - N(p, d, w - 1)$.*

The idea behind Cayley's proof is quite simple. Each simultaneous solution to the above Diophantine equations determines a monomial $a_0^{m_0} a_1^{m_1} \cdots a_d^{m_d}$ of degree p and weight w. Thus $N(p, d, w) = \dim V(p, d, w)$, where $V(p, d, w)$ denotes the vector space of all homogeneous isobaric polynomials in a of degree p and weight w. Since application of X lowers weight by 1, X is a linear transformation from $V(p, d, w)$ into $V(p, d, w - 1)$. The kernel of X consists precisely of the seminvariants of degree p and weight w. Thus if X is *surjective*, $\dim[\ker(X)] = \dim V(p, d, w) - \dim V(p, d, w - 1)$, and Cayley's Law follows. Cayley's actual proof was along these lines, albeit expressed in nineteenth century terminology. He felt it was evident that X was surjective. As he expressed it: if $B(a)$ is the "general" polynomial of degree p and weight w, then $X B$ will represent the general polynomial of degree p and weight $w - 1$ [**6**, p. 256]. Cayley was correct, but the proof turned out to be rather nontrivial.

Cayley's Law afforded him with a solution to the Counting Problem congenial to computations, which he relished. Although there is no simple formula for the number $N(p, d, w)$, it can be represented, as Cayley showed, as the coefficient of $x^w z^p$ in the series expansion of the generating function $\phi(x, z) = \left[\prod_{j=0}^{d}(1 - x^j z)\right]^{-1}$. From this representation Cayley obtained a generating function representation of the numbers $M(p, d, w)$ as well and worked out their values in extensive tables.

Were history a rational process, Cayley's work would have played a seminal role in the events that culminated in the publication of Cartan's theory of weights in 1913. Indeed, shortly after Cayley published his characterization of in- and covariants by differential equations (Theorem 1), his friend J. J. Sylvester rederived the equations using the notion of infinitesimal transformations [7], which is the way the elements in a Lie algebra were conceived in the nineteenth century. Sylvester also briefly considered the n-ary analog of Theorem 1, which involves $sl(n, C)$ and even suggested studying the invariants of an n-ary form defined by considering only orthogonal transformations. In fact, in a series of papers in the period 1887–1892, the Belgian mathematician Jacques Deruyts (1862–1945) did extend Cayley's theory of seminvariants to the context of the in- and covariants associated with an underlying n-ary form $F(a, x)$. (On Deruyts, see [8].) He even posed and solved Cayley's Problem within this context, although the solution is complicated. Lie algebras had been introduced by the time Deruyts did his work, but he made no reference to them, despite Lie's attempts in the 1880s to call attention to their relevance to invariant theory. This is unfortunate because Deruyts's papers involve, in effect, the irreducible modules for $sl(n, C)$ just as Cayley's had for $n = 2$.

History is not quite rational, however, and the work of Cayley, Sylvester, and Deruyts played no role in the developments that produced the Killing-Cartan theory of the structure and representation of semisimple Lie algebras. The irrational element in history actually makes it more fascinating than its "rational reconstructions." What happened is that Cayley's Problem and his solution did play a role in the origins of the representation theory of Lie algebras but not by influencing the developments that culminated in Cartan's theory of weights. Instead they played a role in the developments leading up to the proof of the complete reducibility theorem for semisimple Lie algebras—developments that came from outside the Killing-Cartan mathematical milieu. Since Cayley's Law can be, and nowadays is, proved using the complete reducibility theorem [9], it is perhaps fitting that, historically, its proof owes something to Cayley's Law. With this in mind, we consider the fate of Cayley's Counting Problem.

Although Cayley's Problem and his solution were omitted from the treatises on invariant theory by Clebsch (1872) and Gordan (1887)—the two leading continental practitioners of invariant theory during the 19th century—they were readily accessible to continental mathematicians through other publications. For example, Francesco Faà di Bruno, Professor of Mathematics at the University of Turin, included them in his *Théorie des formes binaires* (1876) which was translated into German in 1881. Faà di Bruno's discussion of Cayley's Law is especially interesting because he questioned its general validity in a footnote while giving essentially Cayley's proof in the text itself! Although his footnote lacked clarity, his remarks (especially as presented in the German edition of 1881) seem to indicate that he had put his finger on the lacuna in Cayley's proof—the matter of the surjectivity of X. To justify his caveat regarding the general validity of the theorem itself, he referred to Gordan's affirmative solution, in 1868, of the Finiteness Problem for binary forms. Gordan's solution was relevant to Cayley's Law because Cayley had deduced incorrectly from his generating function representations that the Finiteness Problem has a *negative* solution for the covariants associated with a binary form of degree five and the invariants of a binary form of degree seven [**6**, pp. 252–253, 268]. As Sylvester was to suggest in 1878 [**10**], Faà di Bruno had evidently concluded that Cayley's Law might be the source of Cayley's error, especially given the questionable proof. "Thus error breeds error," Sylvester lamented, "unless and until the pernicious brood is stamped out for good and all under the iron heel of rigid demonstration" (p. 117n). Sylvester then proceeded to provide such a demonstration.

Sylvester's proof was unfortunately not known to the editor of the German edition of Faà di Bruno's book, which consequently still contains the same criticisms of Cayley's solution to the Counting Problem. But in 1887 a young mathematician named David Hilbert removed all doubt about the veracity of Cayley's Law by providing another, simpler proof that X is surjective [**11**]. The proof made crucial use of the differential operator $\mathcal{H}$ defined on homogeneous isobaric polynomials $B(a)$ by

$$\mathcal{H}[B] = B - y\,X[B]/1!2! + y^2\,X^2[B]/2!3! - y^3\,X^3[B]/3!4! + \cdots. \qquad (5)$$

Hilbert had introduced the operator $\mathcal{H}$ in his doctoral thesis of 1885 on binary forms. Hilbert, who was to become famous for his affirmative solution, in 1888, to the Finiteness Problem for n-ary forms, thus put his blessing on Cayley's Problem and solution.

Hilbert's work on the Finiteness Problem was also instrumental in bringing about generalizations of Cayley's Problem that proved to be consequential for the representation theory of Lie algebras. His solution to this problem involved two steps, as he himself emphasized [**12**]. First of all, by virtue of his groundbreaking "Basis Theorem," it followed that if $\mathfrak{I}$ denotes the invariants $I(a)$ relative to the general n-ary form of degree d, then

THEOREM 5. *There exist* $I_1, \ldots, I_m$ *in* $\mathfrak{I}$ *such that for any* $I \in \mathfrak{I}$ *homogeneous polynomials* B_k *may be chosen so that*

$$I = B_1 I_1 + \cdots + B_m I_m. \qquad (6)$$

Theorem 5 constituted the first step. The second step consisted in deducing from (6) by means of various differential operators that

$$I = J_1 I_1 + \cdots + J_m I_m, \qquad (7)$$

where each J_k is an invariant or a constant. Thus each B_k is replaced by an invariant J_k. Then Theorem 5 can be applied to each nonconstant J_k and then the analog of (7) deduced from the analog of (6), and so on. Since, depending on the differential operator used, it can be arranged so that either the degrees or the weights of the J_k strictly decrease in this process, it must come to a halt after a finite number of steps: all the J_k at some stage will be constants and I will be expressed as a polynomial in the I_k. Thus the Finiteness Problem is solved affirmatively.

For n-ary forms with $n > 2$, Hilbert used what later became known as the omega process to go from (6) to (7). This was somewhat complicated, and for binary forms he showed that the operator $\mathcal{H}$ in (5) can be used. Indeed, his observations had the following implication. Let $\mathfrak{I}$ be any set of homogeneous polynomials which are invariants in some generalized sense, and suppose an operator $\mathcal{D}$ can be defined on homogeneous polynomials B with the following properties:

$$\mathcal{D}[B] \in \mathfrak{I}, \qquad \mathcal{D}[B_1 + B_2] = \mathcal{D}[B_1] + \mathcal{D}[B_2],$$
$$\mathcal{D}[I] = I \quad \text{and} \quad \mathcal{D}[BI] = \mathcal{D}[B]I \quad \text{for } I \in \mathfrak{I}. \tag{8}$$

Then Theorem 2 still applies, and if $\mathcal{D}$ is applied to (6), then, by virtue of the above properties, the result is (7). Thus the Finiteness Problem would be solved affirmatively for the "invariants" compromising $\mathfrak{I}$.

These implications of Hilbert's work stood out because he himself endorsed the idea of cultivating a theory of invariants based upon a broader conception of an invariant suggested, he said, by Klein's *Erlanger Programm* and by Lie's theory of transformation groups. Traditionally invariants of an n-ary form were defined by considering *all* transformations of GL(n) applied to the variables of the form. Hilbert proposed considering the invariants that arise when only the transformations of some subgroup G of GL(n) are applied to the variables of the n-ary form. Hilbert naturally focused on the Finiteness Problem for invariants in this generalized sense. He pointed out that his own method, using an analog of the omega operator, could be extended to deal with certain subgroups G connected with linear associative algebras; but these subgroups are very special. For example, Hilbert's method extends, as he showed, to the subgroup G of GL(3) consisting of all real orthogonal transformations with determinant $+1$, i.e., the rotation group in 3-space, which may be defined using quaternions. But his method did not extend to the rotation group in n-space for $n > 3$. As we shall see, Hilbert's former teacher and colleague, Adolf Hurwitz, took up the challenge of the rotation group inspired by the idea that there were other ways to create an appropriate $\mathcal{D}$. This idea had its origins in the work of Klein.

As noted, Hilbert had referred to Klein's *Erlanger Programm* to support the view of a group theoretic generalization of the notion of an invariant. The groups Klein had in mind in the *Erlanger Programm* were continuous groups, and it was probably such groups that Hilbert had in mind as well. But Klein never developed the ideas of the *Erlanger Programm* in his subsequent research. Instead he became involved with problems that involved "discontinuous groups" of various sorts. In particular, his interest in developing a generalization of Galois's theory of equations which would treat the solution of equations not solvable by radicals led him to stress the notion of an invariant of a finite subgroup G of GL(n), an invariant being defined in the modern manner as a homogeneous polynomial $I(x) = I(x_1, \ldots, x_n)$ such that $I(Sx) = I(x) \; \forall S \in G$. (For further details on this aspect of Klein's work and its influence on the representation theory of finite groups, see §§4–5 of [13].)

Thus among Klein's students and others familiar with this aspect of Klein's work, Hilbert's call for a group theoretic theory of invariants was taken to include finite groups as well. In this connection the following observations were made in the 1890s. Let J be defined on functions $F(x_1, \ldots, x_n)$ by

$$J[F] = \sum_{S \in G} F(Sx), \tag{9}$$

where G denotes a finite subgroup of GL(n). E. H. Moore of the University of Chicago was one of several mathematicians who observed that when F is a positive definite Hermitean form, then $J[F]$ is a G-invariant positive definite Hermitean form. This implied that the group G is equivalent to a group of unitary transformations. In 1898 Moore's colleague at Chicago, Heinrich Maschke, used this fact to prove in effect the complete reducibility theorem for finite groups. Although he did not relate this result to Frobenius's theory of group characters and representations which was just being created itself (starting in 1896), others did so later.

Hurwitz, another former student of Klein's, observed that for any F, $J[F]$, as given in (9), defines an invariant of G and that, in fact all G-invariants are so expressible. Hurwitz also realized that J has the properties described in (8), which implies that the Finiteness Problem has an affirmative solution for finite subgroups of GL(n). To these observations Hurwitz added the following fertile idea: J is defined by summation over the group G; for continuous groups such as the rotation group in n-space, the analog of J may be defined by replacing summation over the group by integration. In this way Hurwitz solved the Finiteness Problem for the n-space rotation group in 1897 [14].

Neither Maschke nor Hurwitz related their work to the theory of characters that Frobenius was just then creating. Another of Klein's former students, however, did just that. His name was Theodor Molien. Molien had the misfortune to independently introduce the basics of the theory of group characters and representations shortly after Frobenius. How this occurred has an interesting but rather involved history [13]. Here we simply point out that Molien, who had written something akin to a master's thesis under Klein's direction in 1885, wrote his doctoral thesis in 1892 at the University of Dorpat, Estonia, now part of the Soviet Union. His thesis advisor was Friedrich Schur, who had spent time at Leipzig in association with Lie during the years in which Killing was working on the structure of semisimple Lie algebras. In his thesis Molien used Killing's work as a paradigm for the study of linear associative algebras. Among other things, he obtained a criterion for when such an algebra is semisimple in the sense that it decomposes into a direct sum of complete matrix algebras. Motivated by the problem of, in effect, determining the representation of minimal degree for a finite group, a problem suggested by Klein's attempt to generalize Galois's theory, he later applied his theory to the group algebra, which always satisfies his criterion for semisimplicity. In this way he discovered independently the basics of Frobenius's theory, including the two orthogonality relations for characters of irreducible representations. These results were published by Molien in 1897 and quickly came to the attention of Frobenius, who informed Molien that he had obtained similar results. Soon after learning his discoveries had been anticipated by a prominent Berlin mathematician, Molien responded by submitting a paper through Frobenius to the Berlin Academy of Sciences in which he showed how Cayley's Counting Problem could be solved for finite groups by means of the theory of characters [15].

Expressed briefly and in more modern terms, Molien's idea was as follows. Let G be a finite subgroup of $\mathrm{GL}(n)$. The problem is to determine the number of linearly independent invariants of fixed degree p. To this end, consider the G-module of homogeneous polynomials $F(x)$ of fixed degree p. The complete reducibility theorem states that this G-module decomposes into irreducible G-modules; and the number of linearly independent invariants of degree p is precisely the number of times the trivial or 1-representation is contained in this decomposition. Let us denote this number by m_p. Using character theory Molien showed that

$$m_p = \frac{1}{M} \sum_{S \in G} a(S)_p, \tag{10}$$

where M is the order of G and $a(S)_p$ denotes the coefficient of the pth power of λ in the series expansion of the function

$$f(\lambda) = [\det(1 - \lambda S)]^{-1}. \tag{11}$$

For specific groups, including the icosahedral group, Molien showed how to compute m_p, the solution to Cayley's Counting Problem, using (10) and (11).

In summary: Molien showed how to solve Cayley's Problem for finite groups by using the complete reducibility theorem and character theory for such groups. Hurwitz's work suggested that results obtained for finite groups by summing over all elements of the group could be extended to continuous groups such as the rotation group with finite sums replaced by invariant integration. Suppose now that a mathematician was familiar with the above work of Hurwitz and Molien. Suppose in addition he realized that summation over the elements of a finite group can be made a basic operation from which the orthogonality relations and complete reducibility theorem can be derived for finite groups. Then he would be in a position to realize that they can be extended to the rotation group by replacing summation by integration so that the prospect of a solution to Cayley's Counting Problem presents itself for this group as well. Issai Schur was such a mathematician.

Schur had been a student at the University of Berlin during the years when Frobenius was working out his theory of group characters and representations. His doctoral dissertation of 1901 was a brilliant application of Frobenius's theory to the study of the representations of a continuous group, namely $GL(n)$. There Schur studied finite-dimensional polynomial representations of $GL(n)$, the representations implicit in invariant theory being examples, e.g., $S \to P_d(S^{-1})$ in the notation of (1). Schur showed that each homogeneous polynomial representation of degree d corresponds biuniquely to a representation of the symmetric group S_d so that polynomial representations can be studied by invoking the results of Frobenius's theory as applied to S_d. This indirect approach involved using the representation theory of S_d rather than directly attempting to build an analogous theory for $GL(n)$. But in 1905 Schur gave a simple derivation—using his now familiar Lemma—of the fundamental propositions of Frobenius's theory of group characters which made fundamental the process of summation over the group [**16**]. Shur's derivation left no doubt that some parts of Frobenius's theory, such as the orthogonality relations for characters and the complete reducibility theorem, could be established, via the Hurwitz integral, for the rotation group. These are precisely the elements of Frobenius's theory that Molien had used to solve Cayley's Problem for finite groups. Thus as early as 1905 it would seem that Schur was in a position to envision the possibility of an analogous solution for the rotation group. But it was not until 1924 that he published a solution. Why? Schur himself never provided an answer, but the following considerations make the apparent twenty year hiatus less of a surprise than it otherwise might seem.

First of all, it was only the *prospect* of an analogous solution to Cayley's Problem that presents itself by combining the ideas of Hurwitz and Molien. One obtains thereby the analog of Molien's formula (10) for the number m_p of G-invariants of degree p, but it involves an integral over the rotation group rather than a finite sum. The problem of expressing this integral in a more explicit form congenial to computation still remained. Thus *if* circa 1905 Schur realized the integral analog of (10), he may have lacked the motivation to try and make something useful out of it.

With regard to motivation, it should be noted that although in 1905 Schur certainly knew Molien's work and probably had some acquaintance with Hurwitz's solution to the Finiteness Problem, it is far from certain that he was familiar with Cayley's beautiful and explicit solution to the Counting Problem for $GL(2)$. If he knew only Molien's solution for finite groups, he would not have realized that this Problem had a tradition and elegant solution within classical invariant theory. That Schur's knowledge of the classical theory of invariants might have been limited in 1905 is quite likely because at Berlin there were no lectures on this subject. Frobenius regarded it to be of minor significance and most of the research done on it as "hack work." (See [**17**, p. 209].) Frobenius specifically applied that description to Hilbert's early work, which included his proof of Cayley's Law. Frobenius had great respect for Hilbert's solution to the Finiteness Problem, but he felt that Hilbert had thereby put an end to the subject. In fact, Frobenius felt that Hilbert had simultaneously founded and finished off the theory: By bringing the concepts of "higher algebra" to bear on the theory, he had laid the foundations for a legitimate mathematical theory; but by solving its central problem he had also finished it off!

So it is quite possible that Schur, although probably more open minded about invariant theory than his mentor, either did not know the classical literature on Cayley's Problem or was not inclined to devote much time to research in an area Frobenius deemed unimportant. In any case, other promising and important research projects were at hand. Many of these had to do with Frobenius's theory of group representations, such as the study of representations over algebraic number fields, which Frobenius also investigated, and the extension of Frobenius's theory to projective representations. The extension was motivated in part by the interest within Klein's school in the problem of determining finite subgroups of $\mathrm{PGL}(n)$, a problem treated computationally with little or no guiding theory. Frobenius was very pleased with the way Schur demonstrated what could be done using representation theory (see [**17**, p. 224]). Schur's fondness for linear algebra, a fondness he shared with Frobenius, also prompted him, starting in 1911, to contribute to the new theory of linear transformations on infinite dimensional vector spaces that Hilbert and his school had initiated. Thus in the decades following the publication of his doctoral thesis, Schur found many interesting and challenging problems to investigate. Furthermore, during these years, there was relatively little activity in the type of invariant theory that Hilbert had endorsed. Undoubtedly this was due in part to the fact that soon after solving the Finiteness Problem for $\mathrm{Gl}(n)$ Hilbert himself stopped doing research on the theory of invariants.

The situation was quite different in 1922, when Schur was elected to the Berlin Academy, having been appointed Professor at the University in 1919. Frobenius was gone, having died in 1917; and although Schur was still engaged in research on the arithmetical aspects of Frobenius's theory, his inaugural speech before the Academy [**18**] indicates that he was also looking for research projects that would combine algebra and analysis in a fruitful way. Furthermore a renewed interest in Hilbert-style invariant theory was now evident among mathematicians associated with Hilbert's school, such as Emil Fischer, Emmy Noether, and Alexander Ostrowski. Ostrowski in particular emphasized the group-theoretic aspects of future research in the theory of invariants, an emphasis which evidently appealed to Schur, who in fact collaborated with Ostrowski on a paper [**19**] which Ostrowski considered paradigmatic for his vision of a research program for invariant theory [**20**]. Schur's increasing interest in the theory of invariants is also reflected in the fact that during the Winter Semester of 1923–24 he took the unprecedented step of introducing a course of lectures on this subject into the curriculum at Berlin. A subsequent version of these excellent lectures was eventually published by Grunsky in 1968 [21], but we are also fortunate to have notes of the lectures as Schur first gave them. The notes were taken by the late Richard Brauer, one of Schur's most distinguished doctoral students. (I am grateful to the late Mrs. Richard Brauer and to Walter Feit and Jonathan Alperin for kindly making these notes available to me.)

Given the extensive literature on the theory of invariants what did Schur choose to present in his lectures? Aside from some necessary preliminaries, he focused primarily upon two topics: (i) Cayley's theory of seminvariants and its application to the Counting Problem, with Hilbert's proof of Cayley's Law; (ii) the Finiteness Problem for $GL(n)$, for finite subgroups and for the rotation group in n-space, with solutions via Hilbert's Basis Theorem and the use of both differential and integral operators to go from (6) to (7). The choice of the second topic is not surprising because Schur shared the view of Frobenius and the Hilbert school that Hilbert's solution of the Finiteness Problem was the most significant result in the theory of invariants. The choice of the first topic was not so obvious, however, and reflects Schur's appreciation for this product of the classical or pre-Hilbert stage of the theory.

Having immersed himself in these two topics, Schur saw clearly what he perhaps had overlooked or dismissed before: Hurwitz's integral method for solving the Finiteness Problem for the rotation group could be used to extend Molien's method of solving Cayley's Problem to this group, as well as the entire real orthogonal group. And now he took up the challenge of making Molien's solution workable for these groups. He succeeded, and on 10 January 1924 he presented his solution in a paper to the Berlin Academy [22]. To solve Cayley's Problem Schur of course first showed that an analog of Frobenius's theory of characters and representations, including orthogonality relations and a complete reducibility theorem (with Maschke's proof), could be developed for these groups. At the conclusion of his paper he pointed out that, since Hurwitz had showed how to define an invariant integral for more general Lie groups, the theory he had developed could be extended to other groups. He then added, as justification for the more limited scope of his paper, that the rotation and orthogonal groups "stand out, not only by virtue of the important role they play in applications but also by virtue of the fact that here the integral calculus provides a solution of the counting problem that is actually practically useful" (p. 208). It would only be worthwhile to extend the notions of representation theory to a more general class of groups if one could describe the irreducible representations. But when he wrote his paper Schur did not see how to do this even for the rotation and orthogonal groups (p. 197). It was the "elegant" solution to the Counting Problem for these groups that justified discussing the elements of the associated representation theory, and this is why Schur included it in his paper (p. 190). His paper was motivated by the Counting Problem rather than by the conviction that his paper would lead to the creation of a theory of representations applicable to a broad class of continuous groups.

Inspired by reading Schur's paper and by their ensuing correspondence, Weyl went on to show, with the aid of Cartan's results which were unfamiliar to

Schur, that a representation theory for semisimple groups could be worked out in considerable detail [**23, 24**]. In this manner the complete reducibility theorem for semisimple Lie algebras was first proved and the notion of a character added to what Killing and Cartan had created. But that is another story altogether and one that transcends the purview of the history of Lie algebra representations as reflected in Cayley's Counting Problem.

REFERENCES

1. T. Hawkins, *Non-Euclidean geometry and Weierstrassian mathematics: the background to Killing's work on Lie algebras*, Historia Math. **7** (1980), 289–324.

2. ____, *Wilhelm Killing and the structure of Lie algebras*, Arch. Hist. Exact Sci. **26** (1982), 127–192.

3. ____, *Hesse's principle of transfer and the representation of Lie algebras*, Preprint.

4. A. J. Crilly, *The rise of Cayley's invariant theory (1841-1862)*, Historia Math. (to appear).

5. A. Cayley, *Nouvelles recherches sur les covariantes*, J. Reine Angew. Math. **47** (1854), 75–143; Mathematical Papers **2** (1889), 164–178.

6. ____, *A second memoir upon quantics*, Philos. Trans. Roy. Soc. London **146** (1856), 101–126; Mathematical Papers **2**, 250–275.

7. J. J. Sylvester, *On the principles of the calculus of forms*, Cambridge and Dublin Math. J. **7** (1852), 52–97, 179–217; Mathematical Papers **1**, 284–363.

8. L. Godeaux, *Notice sur Jacques Deruyts*, Annuaire de l'Académie Royale de Belgique **115** (1949), 21–43.

9. T. A. Springer, *Invariant theory*, Lecture Notes in Math., vol. 585, Springer-Verlag, Berlin-New York, 1977.

10. J. J. Sylvester, *Proof of the hitherto undemonstrated fundamental theorem of invariants*, Philosophical Magazine **5** (1878), 178–188; Mathematical Papers **3** (1909), 117–126.

11. D. Hilbert, *Über eine Darstellungsweise der invarianten Gebilde in binären Formengebiete*, Math. Ann. **30** (1887), 15–29; Ges. Abhandlungen **2** (1933), 102–116.

12. ____, *Über die Theorie der algebraischen Formen*, Math. Ann. **36** (1890), 473–534; Ges. Abhandlungen **2**, 199–257.

13. T. Hawkins, *Hypercomplex numbers, Lie groups and the creation of group representation theory*, Arch. Hist. Exact Sci. **8** (1972), 243–287.

14. A. Hurwitz, *Über die Erzeugung der Invarianten durch Integration*, Göttingen Nachrichten (1897), 71–90; Math. Werke **2** (1963), 546–564.

15. T. Molien, *Über die Invarianten der linearen Substitutionsgruppen*, Sitzungsberichte Akademie der Wissenschaften Berlin (1898), 1152–1156.

16. I. Schur, *Neue Begründung der Theorie der Gruppencharaktere*, Sitzungsberichte Akademie der Wissenschaften Berlin (1905), 406–432; Ges. Abhandlungen **1**, 143–169.

17. K.-R. Biermann, *Die Mathematik und ihre Dozenten an der Berliner Universität 1810-1920*, Akademie-Verlag, Berlin, 1973.

18. *Antrittsrede des Hrn. Schur*, Sitzungsberichte Akademie der Wissenschaften Berlin (1922), lxxx-lxxxi; Ges. Abhandlungen **2**, 413–414.

19. A. Ostrowski and I. Schur, *Über eine fundamentale Eigenschaft der Invarianten einer allgemeinen binären Form*, Math. Z. **15** (1922), 81–105; I. Schur, Ges. Abhandlungen **2** (1973), 334–358; A. Ostrowski, Coll. Math. Papers **2**, 127–151.

20. A. Ostrowski, *Über eine neue Fragestellung in der algebraischen Invariantentheorie*, Jahresber. Deutsch. Math.-Verein. **33** (1924), 174–184; Coll. Math. Papers **2**, 152–162.

21. I. Schur, *Vorlesungen über Invariantentheorie* (H. Grunsky, ed.), Springer-Verlag, Berlin, 1968.

22. ____, *Neue Anwendungen der Integralrechnung auf Probleme der Invariantentheorie*, Sitzungsberichte Akademie der Wissenschaften (1924), 189–208; Ges. Abhandlungen **2** (1973), 440–459.

23. H. Weyl, *Zur Theorie der Darstellung der einfachen kontinuierlichen Gruppen*, Sitzungsberichte Akademie der Wissenschaften Berlin (1924), 338–345; Ges. Abhandlungen **2** (1968), 453–460.

24. ____, *Theorie der Darstellung kontinuierlicher halbeinfacher Gruppen durch lineare Transformationen*, Math. Z. **23** (1925), 271–309; **24** (1926), 328–395, 789–791; Ges. Abhandlungen **2** (1968), 543–647.

BOSTON UNIVERSITY, BOSTON, MASSACHUSETTS 02215, USA

Recent Studies of the
History of Chinese Mathematics

WU WEN-TSUN

1. Introduction. We shall restrict ourselves to the study of Chinese mathematics in ancient times, viz., from remote ancient times up to the fourteenth century. In recent years such studies were vigorously pursued both in China and in foreign countries. Much deeper understandings have since been gained about what Chinese ancient mathematics really was. The author will freely use their results but will be solely responsible for all points of view expressed in what follows.

Two basic principles of such studies will be strictly observed, viz.:

P1. All conclusions drawn should be based on original texts fortunately preserved up to the present time.

P2. All conclusions drawn should be based on reasonings in the manner of our ancestors in making use of knowledge and in utilizing auxiliary tools and methods available only at that ancient time.

For P1 we shall mention only [**AR, AN, SI, MA**], which will be referred to repeatedly in what follows.

For P2 we shall emphasize that the use of algebraic symbolic manipulations or parallel-line drawings should be strictly forbidden in any deductions of algebra or geometry since they were seemingly nonexistent in ancient Chinese classics. In fact, Chinese ancient mathematics had its own line of development, its own method of thinking, and even its own style of presentation. It is not only independent of, but even quite different from the western mathematics as descendents of Greeks. Before going into more details of concrete achievements, we shall first point out some peculiarities of Chinese ancient mathematics.

First, instead of calculations of pencil-paper type, the ancient Chinese made all computations in manipulating rods on counting boards. This was possible because the Chinese already possessed, in very remote times, the most perfect place-valued decimal system; it allowed them to represent the integers by properly arranged rods placed in due positions on the board. In particular, the number 0 in, or as, a decimal integer was just represented by leaving some empty place in the right position. In fact the word "arithmetic," the usual terminology for "mathematics," was just a literal translation of Chinese characters "Suan Shui" meaning "counting methods."

Secondly, results were usually presented in the form of separate problems, each of which was divided into several items, as follows. 1. Statement of the problem with numerical data. 2. Numerical answer to the problem. 3. "Shui," or the method of arriving at the result. It was most often just what we call today the "algorithm," sometimes also just a formula or a theorem. Note that the numerical values in Item 1 play no role at all in the method, which was so general that any other numerical values could be substituted equally well. Item 1 thus served just as an illustrative example. 4. Sometimes "Zhu," or demonstrations which explained the reason underlying the method in Item 3. In Song Dynasty and later, there was often added a further item: 5. "Cao," or drafts which contained details of the calculations for arriving at the final result.

2. Theoretical studies involving integers. In this section, by an integer we shall always mean a positive one.

In ancient Chinese mathematics there were no notions of prime number and factorization or its likeness. However, there was a *Mutual-Subtraction Algorithm*, for finding the GCD of two integers; its name literally meant *equal*. The algorithm ran as follows:

"Subract the less from the more, mutually subtract to diminue, in order to get the *equal*."

As a trivial example, the *equal* (:= GCD) of 24 and 15 is found to be 3 in the following manner:

$$(24, 15) \dashrightarrow (9, 15) \dashrightarrow (9, 6) \dashrightarrow (3, 6) \dashrightarrow (3, 3). \tag{2.1}$$

The underlying principle is, as pointed out by Liu Hui in [**AN**], that during the procedure the integers are steadily diminished in magnitudes while the *equal* duplicates remain the same.

In spite of the fact that the prime number concept was never introduced in our ancient times, there were some theoretical studies involving integers which were not at all trivial. We shall cite two of these mainly based on works of S. K. Mo at Nanking University and J. M. Li at Northwestern University, China.

The GouGu form (:= right-angled triangle) was a favorite object of study throughout the lengthy period of development of mathematics in ancient China. In particular, the triples of integers which can be attributed to 3 sides Gou, Gu, and Xuan (:= shorter arm, longer arm, and hypothenuse) of a GouGu form had been completely determined early in the classic [**AR**]. Thus, in the GouGu Chapter 9 of [**AR**] there appeared eight such triples, viz.,

$$(3, 4, 5), \qquad (5, 12, 13), \qquad (7, 24, 25), \qquad (8, 15, 17),$$
$$(20, 21, 29), \quad (20, 99, 101), \quad (48, 55, 73), \quad (60, 91, 109).$$

The occurence of such triples was not merely an accidental one. In fact, in Problem 14 of that chapter a method of general formation of such integer triples was implied. We record this problem.

"Two persons start from same position. A has a speed-rate 7 while B rate 3. B goes eastward while A goes first southward 10 units and then meets B in going northeasternwise. Find the units traversed by A and B."

The Shui (:= method or algorithm) for the solution was:

"Squaring 7, also 3, taking half of the sum, this will be the slantwise unit-ratio of A. Subtract this unit-ratio from square of 7, rest is the southern unit-ratio. Multiply 7 by 3 is eastern unit-ratio of B."

As already mentioned in §1, the particular numbers 7 and 3 in the problem serve merely as illustrations and we may equally well substitute these numbers by any pair of integers say m, n with $m > n > 0$. The Shui then says that the 3 sides are in the ratio

$$\text{Gou: Gu: Xuan} = [m^2 - (m^2 + n^2)/2]: m * n: (m^2 + n^2)/2.$$

The eight triples given above may then be determined by the pairs

$$(m, n) = (2, 1),\ (3, 2),\ (4, 3),\ (4, 1),\ (5, 2),\ (10, 1),\ (8, 3),\ (10, 3).$$

In Liu Hui's [**AN**] a demonstration or a proof of geometrical character was given which was based on some general *Out-In Complementary Principle*, and it will be explained in more detail in §3. We note here that Liu's proof showed also that $m : n$ is in reality the ratio of Gou + Xuan to Gu which will be a ratio in integers if and only if the three magnitudes Gou, Gu, Xuan are in ratio of integers. The Shui had thus given an exhaustive list of integer triples for the three sides of the GouGu form.

As a second example let us cite the *Seeking*-1 *Algorithm* which is now well known as the Chinese Remainder Theorem. Recent studies have shown that the algorithm originated in calendar-making since Hans Dynasty, and there was a sufficiently clear line of development until the appearance of the classic [**MA**] of Qin in 1247 A.D. In Qin's preface to his work he stated that the method was not contained in [**AR**] and no one knows how it was deduced, but it was widely applied by calendarists. The method was well-explained for the first time in the first part of [**MA**] and contained nine problems, ranging from calendar-making, dyke-erection, treasure-computing, tax-distribution, rice-selling, military-expedition, brick-architecture, up to even a case of stealing. All the problems were reduced to one which, in modern writings, would be of the form (:=: stands for "congruent to")

$$U :=: Uj \mod Mj, \qquad 1 \le j \le r, \tag{2.2}$$

with integers Uj, Mj known and U to be found. The integers Mj were called by Qin Ting-Mu (:= moduli), literally meaning fixed-denominators which were not necessarily prime to each other. Qin first gave an algorithm for reducing the problem to one with the moduli prime to each other two by two in applying successively the *Mutual-Subtraction Algorithm*. We shall therefore restrict ourselves, in what follows, to the case of Uj pairwise prime.

To a modern mathematician a solution to (2.2) would be found in the following manner (cf., e.g., [**AP**, p. 250]).

Let $\phi(N)$ be the Euler function of the integer N which can be determined from a factorization of N into prime numbers. Set

$$M = M1 * \cdots * Mr,$$
$$Nj = (M/Mj)^{\phi(Mj)}, \qquad 1 \le j \le r. \tag{2.3}$$

Then the solution of (2.2) will be given by

$$U :=: \sum_j Uj * Nj \quad \mod M.$$

Both the method and the result are really simple and elegant. However, in view of the difficulty of factorization and the amount of computation involved in (2.3), it would be rather difficult to get final answers to the nine problems in Qin's classic, even with the aid of modern computers.

On the other hand, the method of Qin ran as follows.

As a preliminary step let us take the remainder Rj of M/Mj mod Mj which was called Qi-Shu, literally meaning odd-number, but just some technical term. Now determine numbers Kj such that

$$Kj * Rj :=: 1 \quad \mod Mj. \tag{2.4}$$

The final answer to be found is then given by

$$U :=: \sum_j Uj * Kj * (M/Mj) \quad \mod M. \tag{2.5}$$

The integers Kj were called, by Qin Cheng-Lui, also a technical term literally meaning multiplication-rate (multiplier below). The algorithm for determining Kj to satisfy (2.4) was called, by Qin, *da-yan qiu-yi shui*, for which *qiu-yi* literally means *seeking*-1, while *da-yan* is some philosophical term of little interest to us. The first step of the Seeking-1 Algorithm consisted then in placing four known numbers 1, 0 (i.e., *empty*), Rj, Mj in the left-upper (LU), left-lower (LL), right-upper (RU), and right-lower (RL)

$$\text{corners of a square:} \quad \begin{array}{|cc|}\hline LU & RU \\ LL & RL \\\hline\end{array} = \begin{array}{|cc|}\hline 1 & Rj \\ & Mj \\\hline\end{array}$$

We remark that these four numbers verify the trivial congruences

$$LU * Rj :=: RU \quad \mod Mj, \qquad LL * Rj :=: -RL \quad \mod Mj. \tag{2.6}$$

The next steps of the algorithm consisted then of manipulating the four numbers in the square by steadily reducing their magnitudes while keeping the validity of congruences (2.6). After a finite number of steps the number, say RU, will be reduced to 1, and according to (2.6) the number LU is then the multiplier Kj to be found. The underlying principle of this Seeking-1 Algorithm, as listed below in details, is thus essentially the same as the Mutual-Subtraction Algorithm in finding the *equal* (:= GCD) of two integers, only much more complicated. The algorithm was:

"Put Qi at RU, and Ting at RL, and put Tian-Yuan-1 at LU. First divide RL by RU. Multiply quotient with 1 of LU and put it to LL. Next take the numbers RU and RL, mutually divide the more by the less. Then mutually multiply quotients to numbers in LU and LL. Stop until odd-1 in RU. Verify then the number in LU and take it as multiplier."

As a concrete example let us consider Problem 9 of Chin's classic which dealt with a stealing case. The judge in charge of the case was able to determine the amount of rice stolen by each of the three thiefs by means of the algorithm. For one of the thiefs the determination of the corresponding multiplier ran as follows:

$$\begin{bmatrix} 1 & 14 \\ & 19 \end{bmatrix} \rightarrow \begin{bmatrix} 1 & 14 \\ & 5 \end{bmatrix} 1 \rightarrow \begin{bmatrix} 1 & 14 \\ 1 & 5 \end{bmatrix} \rightarrow \begin{bmatrix} 1 & 4 \\ 1 & 5 \end{bmatrix} 2 \rightarrow \begin{bmatrix} 3 & 4 \\ 1 & 5 \end{bmatrix} \rightarrow$$

$$\begin{bmatrix} 3 & 4 \\ 1 & 1 \end{bmatrix} 1 \rightarrow \begin{bmatrix} 3 & 4 \\ 4 & 1 \end{bmatrix} \rightarrow \begin{bmatrix} 3 & 1 \\ 4 & 1 \end{bmatrix} 3 \rightarrow \begin{bmatrix} 15 & 1 \\ 4 & 1 \end{bmatrix} \rightarrow \text{stop: } k = 15.$$

One may compare this sequence of computations with the trivial one (2.1).

The numerical data in the above example is the simplest one among the nine problems of Chin's classic, but already not an easy one in using the mentioned method with Euler functions. The other eight problems will eventually involve astronomically large numbers which may be eventually out of reach of the Euler-function method, but were still done with ease by Qin in using the Seeking-1 Algorithm.

3. Geometry. In contrast to what one usually believes, geometry was intensely studied, in addition to being well-developed, in ancient China. The misunderstanding is likely due to the fact that Chinese ancient geometry was of a type quite different from that of Euclid, both in content and presentation. Thus, there were no deductive systems of euclidean fashion in the form of definition-axiom-theorem-proof. On the contrary, the ancient Chinese formulated, instead of a lot of axioms, a few general plausible principles on which various geometrical results were then discovered and proved in a deductive manner, as shown by Liu Hui [**AN**].

The points of emphasis in Chinese ancient geometry and in the geometry of Euclid were also quite different. Thus, the Chinese ancestors paid no attention at all to the parallelism but, on the contrary, showed great interest in orthogonality of lines. In fact, the GouGu form, or the right-angled triangle, had incessantly occupied a central position among the geometrical objects to be studied throughout thousands of years of development. Secondly, the Chinese ancestors showed little interest in angles but heavily emphasized distances. Thirdly, geometrical studies were always closely connected with applications so that measurements, determination of areas and volumes were among the central themes of study. Finally, geometry was always developed in step with algebra, which culminated in the algebrization of geometry in Song-Yuan Dynasties. This later discovery was rightly pointed out, e.g., by Needham to be the first important step (and indeed, the decisive step) toward the creation of analytical geometry.

We shall illustrate these points with a few examples.

EXAMPLE 1. The Sun-Height Formula. On the earth-level plane erect two gnomons G1, G2 of equal height with a certain distance apart. The sun-shadows of the gnomons are then measured and the sun's height over the level plane is given by

Sun-hgt = Gnomon-hgt * Gnomon-dist/Shadow-difference + Gnomon-hgt.

This formula, already depicted in some classic of early Hans Dynasty and cited very often in later calendarical works, was clearly too rough an estimate to rely on. Liu Hui had, however, translated the formula into earth measurements by replacing the sun by some sea-mountain, thus turning the Sun-Height Formula into a realistic Sea-Island Formula. His classic [SI] contained all nine such formulae beginning with the above one as the simplest. There were proofs as well as diagrams accompanying this classic; they are still mentioned in some classics of Song Dynasty but have since been lost. Based on fragments and incomplete colored diagrams of some classic by Zhao Shuang in 3c A.D., the author has reconstructed a proof of the above Sun-Height or Sea-Island Formula by rearranging the arguments in that classic as follows (Y =yellow, B =blue):

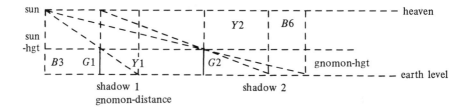

"$Y1$ and $Y2$ are equal in areas. $Y1$ connected with $B3$ and $Y2$ with $B6$ are also equal in areas. $B3$ and $B6$ are also equal in area. Multiply gnomon-distance by gnomon-height to be the area of $Y1$. Take shadow-difference as breadth of $Y2$ and divide, one gets height of $Y2$. The height rises up to same level as sun. From diagram gnomon-height is to be added."

With the accompanying diagram the proof of the formula is evident.

EXAMPLE 2. The Out-In Complementary Principle (OICP). In Example 1, various area-equalities were all consequences of a certain general Out-In Complementary Principle which was clearly formulated in the classic [AN] in very concise terms. It means simply that whenever a figure, planar or solid, is cut into pieces and moved to other places, then the sum of areas or volumes will remain unchanged. This seemingly most common-place principle had been applied successfully to problems of extreme diversity, sometimes unexpected, besides that of Example 1. As further examples consider the GouGu form with three sides: Gou, Gu, and Xuan. One may form various sums and differences from them

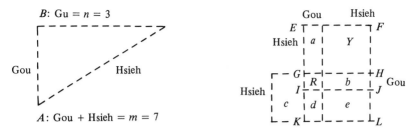

FIGURE 1

$c + d = \text{Hsieh}^2 - \text{Gou}^2 = d + e = \text{Gu}^2 = n^2,$

$2 * EFGH = EFKL = m^2 + n^2 = (\text{Gou}+\text{Hsieh})^2 + \text{Gu}^2,$

$a + Y = \text{Hsieh} * m = \text{Hsieh} * (\text{Gou}+\text{Hsieh}),$

$b + R = EFIJ - EFGH = \text{Gou} * m = \text{Gou} * (\text{Gou}+\text{Hsieh}) = m^2 - (m^2 + n^2)/2.$

like Gou-Gu sum, Gou-Xuan difference, etc. In the GouGu Chapter 9 of [**AR**], there were a number of problems for determining Gou, Gu, and Xuan from two of these nine entities, and all were solved by means of this principle. In particular, the general formula of Gou-Gu integers as described in §2 was obtained by applying the principle to Problem 14 by considering as known the ratio of Gou-Xuan sum to Gu. Liu Hui then demonstrated the result by OICP as shown in Figure 1 (R =red, Y =yellow).

In [**MA**] there was formula for determining the AREA of a triangle with three sides: the GReatest one, the SMallest one, and the MIDdle one in the form

$$4 * \text{AREA}^2 = \text{SM}^2 * \text{GR}^2 - [(\text{GR}^2 + \text{SM}^2 - \text{MID}^2)/2]^2.$$

This formula is clearly equivalent to the Heron one. It cannot be deduced from the latter since it is so ugly, in form, in comparison to the elegant latter formula. By applying some formula given in [**AN**] about Problem 14, based on OICP, the author has reconstructed a proof which is in accordance with Chinese tradition and leads naturally to Qin's formula.

We note that the Chinese ancient methods of (square and cubic) root-extraction and quadratic-equation solving were in fact all based on OICP geometrical in character. We also note that all the formulae in [**SI**], in quite intricate form, will be arrived at in a natural manner by applying OICP. On the other hand it seems difficult, or at least a roundabout, unnatural manner, to get these formulae if the euclidean method is to be used.

EXAMPLE 3. Volume of solids. With the OICP alone the areas of any polygonal form can be determined. This will not be the case for volumes of polyhedral solids, and Liu Hui was well aware of it. Liu Hui had, however, completely solved the problem in reasoning as follows. Let us cut a rectangular parallelopiped slantwise into two equal parts called Qiandu, and then cut the Qiandu slantwise into two parts called Yangma (a pyramid) and Bienao (a tetrahedron on special type). Using an ingenious reasoning corresponding to a certain limiting process,

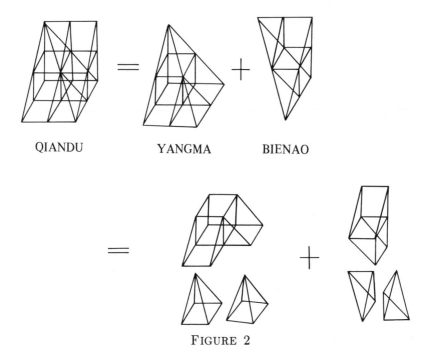

QIANDU   YANGMA   BIENAO

FIGURE 2

he made some assertion which the author has baptized as the *Liu Hui Principle*, viz.,

"Yangma occupies two and Bienao one, that's an invariable ratio."

Together with the OICP the volume of any polyhedral solid can then be determined, and a lot of beautiful formulae for various kinds of solids were determined in this way in the Sang-Gong Chapter 5 of [**AR**]. Liu Hui's demonstration of his principle, which was both elegant and rigorous, consisted of cutting a big QIANDU into smaller yangma's, etc., as in Figure 2.

From Figure 2 it is now clear that

$$1 \text{ YANGMA} - 2 \text{ BIENAO} = 2(1 \text{ yangma} - 2 \text{ bienao}).$$

Continuing, the right-hand side will become smaller and smaller and can be ultimately neglected, as argued by Liu Hui:

"The more they are cut into smaller halves, the smaller will be the remains. The ultimate smallness is infinitesimal, and infinitesimal is formless. Accordingly it is no need to take into account the remain."

For more details see [**WA**], a remarkable paper by Wagner.

Liu Hui had also considered the determination of curvilinear solids, notably that of a sphere. He showed that the solution will depend on the determination of the volume of a curious solid defined as the intersection of two inscribed cylinders in a cube. Liu Hui himself cannot solve this problem and left it, being rigorous in thinking and strict in attitude, to later generations, saying that

"Fearing loss of rightness, I dare to leave the doubts to gifted ones."

The keen observation of Liu Hui had been closely followed and ripened finally to a complete solution of the problem in 5c A.D. by Zu Geng, son of great mathematician, astronomer, and engineer Zu Congtze. In fact, Zu Geng had formulated a general principle which was equivalent to the later rediscovered Cavalieri Principle, viz.,

"Since areas in equal height are equal the volumes cannot be unequal."

We shall leave Zu Geng's beautiful proof about the formula of volume of sphere to other known works. On the other hand, this principle was, in reality, already used by Liu Hui himself in deriving formulae of volumes of various simple curvilinear solids treated in [AR], though without an explicit statement. For this reason the author has proposed to use the name *Liu-Zu Principle* instead of the name *Zu-Geng Principle* which is usually used by our Chinese colleagues.

In a word, the OICP, the Liu Hui Principle, and the Liu-Zu Principle were sufficient to edify the whole theory of solids, curvilinear or not, in a satisfactory manner as done by the Chinese ancestors.

4. Algebra. Algebra was no doubt the most developed part of mathematics in ancient China. It should be pointed out that algebra at that time was actually a synonym for method of equation-solving. The problems of equation-solving seem to come from two different sources. One of the sources was rudimentary commerce or goods-exchange which led to the *Excess-Deficiency Shui* in very remote times up to *Fang-Cheng Shui* as depicted in Chapter 8 of [AR]. This Chapter 8 dealt with methods of solving simultaneous linear equations along with the introduction of negative numbers. The title "Fang Cheng," the same terminology for "equations" used in Chinese texts nowadays, could be better interpreted as "square matrices." In fact, "Fang" literally means square or rectangle while "Cheng," as explained in Liu Hui's [AN], was just data arranged on the counting board in the form of a matrix, viz.

"Arranged as arrays in rows, so it is called Fang Cheng."

Furthermore, the method of solution was just manipulations of rows and columns as in elimination nowadays. Details of such stepwise reduction of arrays to normal forms in some examples can also be found in [AN].

A second source of equations was from measurements or geometrical problems. Thus, in the study of sun-heights there were formulae for both sun-height and sun's level distance from the observer. The sun-observer distance was then determined by means of the Gou-Gu Theorem, well known in quite remote times, which then required extraction of square roots. Both the proof of the Gou-Gu Theorem and the method of square root extraction were seemingly based on the OICP—so, also, for the cubic root extraction. Now in Gou-Gu Chapter 9 of [AR] there was also a problem which led naturally, by OICP, to a quadratic equation. There was some technical terminology for solving such an equation literally meaning "square-root extraction with an extra term Cong," which clearly implied the origin as well as the method of solving such equations.

The second line was developed further to solving cubic equations in early Tang Dynasty, at the latest, and culminated in the method of numerical solution of higher degree equations in Song Dynasty, identical, actually, to the later rediscovered Horner's method in 1819.

A discovery of utmost importance during Song-Yuan Dynasties (10–14c) was the introduction of the notion "Tian-Yuan," literally meaning "Heaven-Element," which was nothing but what we call an *unknown* nowadays. Though equation-solving occupied a central position in the development of mathematics for thousands of years already, this was perhaps the first time that precise notion and systematic use of *unknowns* were thereby introduced. The Chinese mathematicians at that time recognized very well the power of this method of Tian-Yuan as expressed in some classic of Zhu Szejze:

"To solve by Tian-Yuan not only is clear the underlying reasons and is versatile the method but also saved large amounts of efforts."

The method of Tian-Yuan was further developed in Song-Yuan Dynasties up to the solving of simultaneous high-degree equations involving four unknowns. Along with it, algebrization of geometry, manipulations of polynomials, and the method of elimination were also developed. The two lines of development of equation-solving thus merged into one which was closer to algebra in the modern sense. The limitation to four unknowns was largely due to the fact that all manipulations had to be carried out on counting boards with coefficients of different-type terms of a polynomial to be arranged in definite positions on the board. If one was to get rid of the counting board in adapting another system, as was fairly probable since communications with the outside arabic world were more influential than ever, then mathematics would face an exceedingly fertile era of flourishment. However, all further developments stopped and mathematics actually came to death since the end of Yuan Dynasty. When Matteo Ricci came to China at the end of Ming Dynasty, almost no Chinese high intellectuals knew about "Nine Chapters"!

5. Conclusion. We shall leave other achievements about limit concept, high-difference formulae, series summation, etc. owing to space limitation. In short, Chinese ancient mathematics was mainly constructive, algorithmic, and mechanical in character so that most of the *Shuis* can be readily turned into computerized programs. Moreover, it used to draw intrinsic conclusions from objective facts, then sum up the conclusions into succinct principles. These principles, plain in reasoning and extensive in application, form a unique character of ancient Chinese mathematics. The emphasis has always been on the tackling of concrete problems and on simple, seemingly plausible principles and general methods. The same spirit permeates even such outstanding achievements as the algebrization and the place-value decimal system of numbers. In a word, Chinese ancient mathematics had its own merits, and, of course, also its inherited deficiencies. It is surely inadmissible to neglect the brilliant achievements of our ancestors, as was the case in the Ming Dynasty. It would also be absurd not to absorb the superior techniques of the foreign world, as was the case of early Tang Dynasty. At that time the writing system of Indian numerals was imported, but its use as an alternative for the counting board system was rejected. In fully recognizing the powerfulness of our traditional method of thinking, and in absorbing at the same time the highly developed foreign techniques, we foresee a novel new era of achievements in Chinese mathematics.

BIBLIOGRAPHY

[**AN**] Liu Hui, *Annotations to "Nine chapters of arithmetic"*, 263 A.D.

[**AP**] Donald E. Knuth, *The art of computer programming*, Vol 2, Addison-Wesley, Reading, Mass., 1969.

[**AR**] *Nine chapters of arithmetic*, completed in definite form in 1c A.D.

[**MA**] Qin Jiushao, *Mathematical treatise in nine chapters*, 1247 A.D.

[**SI**] Liu Hui, *Sea island mathematical manual*, 263 A.D.

[**WA**] Donald B. Wagner, *An early Chinese derivation of the volume of a pyramid: Liu Hui, third century A.D.*, Historia Math. **6** (1979), 169–188.

INSTITUTE OF SYSTEMS SCIENCE, ACADEMIA SINICA, CHINA

Proceedings of the International Congress of Mathematicians
Berkeley, California, USA, 1986

The Centrality of Mathematics
in the History of Western Thought

JUDITH V. GRABINER

1. Introduction. Since this is a paper in the education section, let me start with a classroom experience. It happened in a course in which my students had read some of Euclid's *Elements of Geometry*. A student, a social-science major, said to me, "I never realized mathematics was like this. Why, it's like philosophy!" That is no accident, for philosophy is like mathematics. When I speak of the centrality of mathematics in western thought, it is this student's experience I want to recapture—to reclaim the context of mathematics from the hardware store with the rest of the tools and bring it back to the university. To do this, I will discuss some major developments in the history of ideas in which mathematics has played a central role.

I do not mean that mathematics has by itself caused all these developments; what I do mean is that mathematics, whether causing, suggesting, or reinforcing, has played a key role—it has been there, at center stage. We all know that mathematics has been the language of science for centuries. But what I wish to emphasize is the crucial role of mathematics in shaping views of man and the world held not just by scientists, but by everyone educated in the western tradition.

Given the vastness of that tradition, I will give many examples only briefly, and be able to treat only a few key illustrative examples at any length. Sources for the others may be found in the bibliography. (See also [**26**].)

Since I am arguing for the centrality of mathematics, I will organize the paper around the key features of mathematics which have produced the effects I will discuss. These features are the certainty of mathematics and the applicability of mathematics to the world.

2. Certainty. For over two thousand years, the certainty of mathematics— particularly of Euclidean geometry—has had to be addressed in some way by any theory of knowledge. Why was geometry certain? Was it because of the subject-matter of geometry, or because of its method? And what were the implications of that certainty?

Even before Euclid's monumental textbook, the philosopher Plato saw the certainty of Greek geometry—a subject which Plato called "knowledge of that which always is" [41, 527b]—as arising from the eternal, unchanging perfection of the objects of mathematics. By contrast, the objects of the physical world were always coming into being or passing away. The physical world changes, and is thus only an approximation to the higher ideal reality. The philosopher, then, to have his soul drawn from the changing to the real, had to study mathematics. Greek geometry fed Plato's idealistic philosophy; he emphasized the study of Forms or Ideas transcending experience—the idea of justice, the ideal state, the idea of the Good. Plato's views were used by philosophers within the Jewish, Christian, and Islamic traditions to deal with how a divine being, or souls, could interact with the material world [46, pp. 382–383; 51, pp. 17–40; 34, p. 305ff; 23, pp. 46–67]. For example, Plato's account of the creation of the world in his *Timaeus*, where a god makes the physical universe by copying an ideal mathematical model, became assimilated in early Christian thought to the Biblical account of creation [29, pp. 21–22]. One finds highly mathematicized cosmologies, influenced by Plato, in the mystical traditions of Islam and Judaism as well. The tradition of Platonic Forms or Ideas crops up also in such unexpected places as the debates in eighteenth- and nineteenth-century biology over the fixity of species. Linnaeus in the eighteenth, and Louis Agassiz in the nineteenth century seem to have thought of species as ideas in the mind of God [16, p. 34; 13, pp. 36–37]. When we use the common terms "certain" and "true" outside of mathematics, we use them in their historical context, which includes the long-held belief in an unchanging reality—a belief stemming historically from Plato, who consistently argued for it using examples from mathematics.

An equally notable philosopher, who lived just before Euclid, namely Aristotle, saw the success of geometry as stemming, not from perfect eternal objects, but instead from its method (Aristotle, *Posterior Analytics*, I 10–11; I 1–2 (77a5, 71b ff)) [19, vol. I, Chapter IX]. The certainty of mathematics for Aristotle rested on the validity of its logical deductions from self-evident assumptions and clearly stated definitions. Other subjects might come to share that certainty if they could be understood within the same logical form; Aristotle, in his *Posterior Analytics*, advocated reducing all scientific discourse to syllogisms, that is, to logically deduced explanations from first principles. In this tradition, Archimedes proved the law of the lever, not by experiments with weights, but from deductions à la Euclid from postulates like "equal weights balance at equal distances" [18, pp. 189–194]. Medieval theologians tried to prove the existence of God in the same way. This tradition culminates in the 1675 work of Spinoza, *Ethics Demonstrated in Geometrical Order*, with such axioms as "That which cannot be conceived through another must be conceived through itself," definitions like "By substance I understand that which is in itself and conceived through itself" (compare Euclid's "A point is that which has no parts"), and such propositions as "God or substance consisting of infinite attributes... necessarily exists," whose proof ends with a QED [48, pp. 41–50]. Isaac Newton called his famous three laws "Axioms, or Laws of Motion." His *Principia* has a Euclidean structure, and the law of gravity appears as Book III, Theorems VII and VIII [37, pp. 13–14, pp. 414–417]. The Declaration of Independence of the United States is one more example of an argument whose authors tried to inspire faith in its certainty by using the Euclidean form. "We hold these truths to be self-evident..." not that all right angles are equal, but "that all men are created equal." These self-evident truths include that if any government does not obey these postulates, "it is the right of the people to alter or abolish it." The central section begins by saying that they will "prove" King George's government does not obey them. The conclusion is "We, *therefore*,... declare, that these United Colonies are, and of right ought to be, free and independent states." (My italics) (Jefferson's mathematical education, by the way, was quite impressive by the standards of his time.)

Thus a good part of the historical context of the common term "proof" lies in Euclidean geometry—which was, I remind you, a central part of western education.

However, the certainty of mathematics is not limited to Euclidean geometry. Between the rise of Islamic culture and the eighteenth century, the paradigm governing mathematical research changed from a geometric one to an algebraic, symbolic one. In algebra even more than in the Euclidean model of reasoning, the method can be considered independently of the subject-matter involved. This view looks at the method of mathematics as finding truths by manipulating symbols. The approach first enters the western world with the introduction of the Hindu-Arabic number system in the twelfth-century translations into Latin of Arabic mathematical works, notably al-Khowarizmi's algebra. The simplified calculations using the Hindu-Arabic numbers were called the "method of al-Khowarizmi" or as Latinized "the method of algorism" or algorithm.

In an even more powerful triumph of the heuristic power of notation, François Viète in 1591 introduced literal symbols into algebra: first, using letters in general to stand for any number in the theory of equations; second, using letters for any number of unknowns to solve word problems [4, pp. 59–63, 65]. In the seventeenth century, Leibniz, struck by the heuristic power of arithmetical and algebraic notation, invented such a notation for his new science of finding differentials—an algorithm for manipulating the d and integral symbols, that is, a calculus (a term which meant to him the same thing as "algorithm" to us). Leibniz generalized the idea of heuristic notation in his philosophy [30, pp. 12–25]. He envisioned a symbolic language which would embody logical thought just as these earlier symbolic languages enable us to perform algebraic operations correctly and mechanically. He called this language a "universal characteristic," and later commentators, such as Bertrand Russell, see Leibniz as the pioneer of symbolic logic [45, p. 170]. Any time a disagreement occurred, said Leibniz, the opponents could sit down and say "Let us calculate," and—mechanically—

settle the question [**30**, p. 15]. Leibniz's appreciation of the mechanical element in mathematics when viewed as symbolic manipulation is further evidenced by his invention of a calculating machine. Other seventeenth-century thinkers also stressed the mechanical nature of thought in general: for instance, Thomas Hobbes wrote, "Words are wise men's counters, they do but reckon by them" [**21**, Chapter 4, p. 143]. Others tried to introduce heuristically powerful notation in different fields: consider Lavoisier's new chemical notation which he called a "chemical algebra" [**14**, p. 245].

These successes led the great prophet of progress, the Marquis de Condorcet, to write in 1793 that algebra gives "the only really exact and analytical language yet in existence.... Though this method is by itself only an instrument pertaining to the science of quantities, it contains within it the principles of a universal instrument, applicable to all combinations of ideas" [**9**, p. 238]. This could make the progress of "every subject embraced by human intelligence... as sure as that of mathematics" [**9**, 278–279]. The certainty of symbolic reasoning has led us to the idea of the certainty of progress. Though one might argue that some fields had not progressed one iota beyond antiquity, it was unquestionably true by 1793 that mathematics and the sciences had progressed. To quote Condorcet once more: "the progress of the mathematical and physical sciences reveals an immense horizon... a revolution in the destinies of the human race" [**9**, p. 237]. Progress was possible; why not apply the same method to the social and moral spheres as well?

No account of attempts to extend the method of mathematics to other fields would be complete without discussing René Descartes, who in the 1630s combined the two methods we have just discussed—that of geometry and that of algebra—into analytic geometry. Let us look at his own description of how to make such discoveries. Descartes depicted the building-up of the deductive structure of a science—proof—as a later task than analysis or discovery. One first needed to analyze the whole into the correct "elements" from which truths could later be deduced. "The first rule," he wrote in his *Discourse on Method*, "was never to accept anything as true unless I recognized it to be evidently such.... The second was to divide each of the difficulties which I encountered into as many parts as possible, and as might be required for an easier solution...." Then, "the third [rule] was to [start]... with the things which were simplest and easiest to understand, gradually and by degrees reaching toward more complex knowledge" [**10**, Part II, p. 12]. Descartes presented his method as the key to his own mathematical and scientific discoveries. Consider, for instance, the opening lines of his *Geometry*: "All problems in geometry can easily be reduced to such terms that a knowledge of the lengths of certain straight lines suffices for their construction. Just as arithmetic is composed of only four or five operations..., so in geometry..." [**11**, Book I, p. 3]. Descartes's influence on subsequent philosophy, from Locke's empiricism to Sartre's existentialism, is well known and will not be reviewed here. But for our purposes it is important to note that the thrust of Descartes's argument is that emulating the method successful in mathematical discovery will lead to successful discoveries in other fields [**10**, Part Five].

Descartes's method of analysis fits nicely with the Greek atomic theory, which had been newly revived in the seventeenth century: all matter is the sum of atoms; analyze the properties of the whole as the sum of these parts [**17**, Chapter VIII, esp. p. 217]. Thus the idea of studying something by "analysis" was doubly popular in seventeenth- and eighteenth-century thought. I would like to trace just one line of influence of this analytic method. Adam Smith in his 1776 *Wealth of Nations* analyzed [**47**, p. 12] the competitive success of economic systems by means of the concept of division of labor. The separate elements, each acting as efficiently as possible, provided for the overall success of the manufacturing process; similarly, each individual in the whole economy, while striving to increase his individual advantages, is "led as if by an Invisible Hand to promote ends which were not part of his original intention" [**47**, p. 27]—that is, the welfare of the whole of society. This Cartesian method of studying a whole system by analyzing it into its elements, then synthesizing the elements to produce the whole, was especially popular in France. For instance, Gaspard François de Prony had the job of calculating, for the French Revolutionary government, a set of logarithmic and trigonometric tables. He, himself, said he did it by applying Adam Smith's ideas about the division of labor. Prony organized a group of people into a hierarchical system to compute these tables. A few mathematicians decided which functions to use; competent technicians then reduced the job of calculating the functions to a set of simple additions and subtractions of pre-assigned numbers; and, finally, a large number of low-level human "calculators" carried out the additions and subtractions. Charles Babbage, the early nineteenth-century pioneer of the digital computer, applied the Smith-Prony analysis and embodied it in a machine [**1**, Chapter XIX]. The way Babbage's ideas developed can be found in a chapter in his *Economy of Machinery* entitled "On the Division of Mental Labour" [**1**]. Babbage was ready to convert Prony's organization into a computing machine because Babbage had long been impressed by the arguments of Leibniz and his followers on the power of notation to make much mathematical calculation mechanical, and Babbage, like Leibniz, accounted for the success of mathematics by "the accurate simplicity of its language" [**22**, p. 26]. Since Babbage's computer was designed to be "programmed" by punched cards, Hollerith's later invention of punched-card census data processing, twentieth-century computing, and other applications of the Cartesian "divide-and-solve" approach, including top-down programming, are also among the offspring of Descartes's mathematically-inspired method.

Whatever view of the *cause* of the certainty of mathematics one adopts, the *fact* of certainty in itself has had consequences. The "fact of mathematical certainty" has been taken to show that there exists *some* sort of knowledge, and thus to refute skepticism. Immanuel Kant in 1783 used such an argument to show that metaphysics is possible [**25**, Preamble, Section IV]. If metaphysics exists, it is independent of experience. Nevertheless, it is not a complex of tautologies. Metaphysics, for Kant, had to be what he called "synthetic," giving knowledge based on premises which is not obtainable simply by analyzing the premises logically. Is there such knowledge? Yes, said Kant, look at geometry. Consider the truth that the sum of the angles of a triangle is two right angles. We do not get this truth by analyzing the concept of triangle—all that gives us, Kant says, is that there are three angles. To gain the knowledge, one must make a construction: draw a line through one vertex parallel to the opposite side. (I now leave the proof as an exercise.) The construction is essential; it takes place in space, which Kant sees as a unique intuition of the intellect. (This example [**24**, II "Method of Transcendentalism," Chapter I, §I, p. 423] seems to require the space to be Euclidean; I will return to this point later on.) Thus synthetic knowledge independent of experience *is* possible, so metaphysics—skeptics like David Hume to the contrary—is also possible.

This same point—that mathematics is knowledge, so there *is* objective truth—has been made throughout history, from Plato's going beyond Socrates' agnostic critical method, through George Orwell's hero, Winston Smith, attempting to assert, in the face of the totalitarian state's overwhelming power over the human intellect, that two and two are four.

Moreover, since mathematics is certain, perhaps we can, by examining mathematics, find which properties *all* certain knowledge must have. One such application of the "fact of mathematical certainty" was its use to solve what in the sixteenth century was called the problem of the criterion [**43**, Chapter I]. If there is only one system of thought around, people might well accept that one as true—as many Catholics did about the teachings of the Church in the Middle Ages. But then the Reformation developed alternative religious systems, and the Renaissance rediscovered the thought of pagan antiquity. Now the problem of finding the criterion that identified the true system became acute. In the seventeenth and eighteenth centuries, many thinkers looked to mathematics to help find an answer. What was the sign of the certainty of the conclusions of mathematics? The fact that nobody disputed them [**43**, Chapter VII]. Distinguishing mathematics from religion and philosophy, Voltaire wrote, "There are no sects in geometry. One does not speak of a Euclidean, an Archimedean" [**49**, Article "Sect."]. What every reasonable person agrees about—that is the truth. How can this be applied to religion? Some religions forbid eating beef, some forbid eating pork—therefore, since they disagree, they both are wrong. But, continues Voltaire, all religions agree that one should worship God and be just; that must therefore be true. "There is but one morality," says Voltaire, "as there is but one geometry" [**49**, Miscellany, p. 225].

3. Applicability. Let us turn now from the certainty of mathematics to its applicability. Since applying mathematics to describe the world works so well, thinkers who reflect on the applicability of mathematics find that it affects their views not only about thought, but also about the world. For Plato, the applicability of mathematics occurs because this world is merely an approximation to the higher mathematical reality; even the motions of the planets were inferior to pure mathematical motions [**41**, 529d]. For Aristotle, on the other hand, mathematical objects are just abstracted from the physical world by the intellect. A typical mathematically-based science is optics, in which we study physical objects—rays of light—as though they were mathematical straight lines [*Physics* II, Chapter 2; 194a]. We can thus use all the tools of geometry in that science of optics, but it is the light that is real.

One might think that Plato is a dreamer and Aristotle a hard-headed practical man. But today's engineer steeped in differential equations is the descendant of the dreamer. From Plato—and his predecessors the Pythagoreans who taught that "all is number"—into the Renaissance, many thinkers looked for the mathematical reality beyond the appearances. So did Copernicus, Kepler, and Galileo [**7**, Chapters 3, 5, 6]. The Newtonian world-system that completed the Copernican revolution was embodied in a mathematical model, based on the laws of motion and inverse-square gravitation, and set in Platonically absolute space and time ([**6**]; cf. [**7**, Chapter 7]). The success of Newtonian physics not only strongly reinforced the view that mathematics was the appropriate language of science, but also strongly reinforced the emerging ideas of progress and of truth based on universal agreement.

Another consequence of the Newtonian revolution was Newton's explicit help to theology, strongly buttressing what was called the argument for God's existence from design. The mathematical perfection of the solar system—elliptical orbits nearly circular, planets moving all in the same plane and direction—could not have come about by chance, said Newton, but "from the counsel and dominion of an intelligent and powerful Being" [**37**, General Scholium, p. 544]. "Natural theology," as this doctrine was called, focussed on examples of design and adaptation in nature, inspiring considerable research in natural history, especially on adaptation, which was to play a role in Darwin's discovery of evolution by natural selection [**14**, pp. 263–266].

Just as the "fact of mathematical certainty" made certainty elsewhere seem achievable, so the "fact of mathematical applicability" in physical science inspired the pioneers of the idea of social science, Auguste Comte and Adolphe Quetelet. Both Comte and Quetelet were students of mathematical physics and astronomy in the early nineteenth century; Comte, while a student at the École polytechnique in Paris, was particularly inspired by Lagrange, and Quetelet, by Laplace. Lagrange's great *Analytical Mechanics* was an attempt to reduce all of mechanics to mathematics. Comte went further: if physics was built on mathematics, so was chemistry built on physics, biology on chemistry, psychology on biology, and finally his own new creation, sociology (the term is his) would be built on psychology [**8**, Chapter II]. The natural sciences were no longer (as they had once been) theological or metaphysical; they were what Comte called "positive"—based only on observed connections between things. Social science could now also become positive. Comte was a reformer, hoping for a better society through understanding what he called "social physics." His philosophy of positivism influenced twentieth-century logical positivism, and his ideas on history—"social dynamics"—influenced Feuerbach and Marx [**32**, Chapter 4]. Still, Comte only prophesied but did not create quantitative social science; this was done by Quetelet.

For Quetelet's conception of quantitative social science, the fact of applicability of mathematics was crucial. "We can judge of the perfection to which a science has come," he wrote in 1828, "by the ease with which it can be approached by calculation" (quoted by [**27**, p. 250]). Quetelet noted that Laplace had used probability and statistics in determining planetary orbits; Quetelet was especially impressed by what we call the normal curve of errors. Quetelet found empirically that many human traits—height, for instance—gave rise to a normal curve. From this, he defined the statistical concept, and the term, "average man" (*homme moyen*). Quetelet's work demonstrates that, just as the Platonic view that geometry underlies reality made mathematical physics possible, so having a statistical view of data is what makes social science possible.

Quetelet found also that many social statistics—the number of suicides in Belgium, for instance, or the number of murders—produced roughly the same figures every year. The constancy of these rates over time, he argued, indicated that murder or suicide had constant social causes. Quetelet's discovery of the constancy of crime rates raises an urgent question: whether the individuals are people or particles, do statistical laws say anything about individuals? or are the individuals free?

Laplace, recognizing that one needed probability to do physics, said that this fact did not mean that the laws governing the universe were ultimately statistical. In ignorance of the true causes, Laplace said, people thought that events in the universe depended on chance, but in fact all is determined. To an infinite intelligence which could comprehend all the forces in nature and the "respective situation of the beings who composed it," said Laplace, "nothing would be uncertain" [**28**, Chapter II]. Similarly, Quetelet held that "the social state prepares these crimes, and the criminal is merely the instrument to execute them" [**27**].

Another view was held by James Clerk Maxwell. In his work on the statistical mechanics of gases, Maxwell argued that statistical regularities in the large told you nothing about the behavior of individuals in the small [**33**, Chapter 22, pp. 315–316]. Maxwell seems to have been interested in this point because it allowed for free will. And this argument did not arise from Maxwell's physics; he had read and pondered the work of Quetelet on the application of statistical thinking to society [**44**]. The same sort of dispute about the meaning of probabilistically-stated laws has of course recurred in the twentieth-century philosophical debates over the foundations of quantum mechanics.

Thus discussions of basic philosophical questions—is the universe an accident or a divine design? is there free will or are we all programmed?—owe surprisingly much to the applicability of mathematics.

4. More than one geometry? Given the centrality of mathematics to western thought, what happens when prevailing views of the nature of mathematics change? Other things must change too. Since geometry had been for so long the canonical example both of the certainty and of the applicability of mathematics, the rise of non-Euclidean geometry was to have profound effects.

As is well known, in attempts to prove Euclid's parallel postulate and thus, as Saccheri put it in 1733, remove the single blemish from Euclid, mathematicians deduced a variety of surprising consequences from denying that postulate. Gauss, Bolyai, and Lobachevsky in the early nineteenth century each separately recognized that these consequences were not absurd, but rather were valid results in a consistent, non-Euclidean (Gauss's term) geometry.

Recall that Kant had said that space (by which he meant Euclidean 3-space) was the form of all our perceptions of objects. Hermann von Helmholtz, led in mid-century to geometry by his interest in the psychology of perception, asked whether Kant might be wrong: could we imagine ordering our perceptions in a non-Euclidean space? Yes, Helmholtz said. Consider the world as reflected in a convex mirror. Thus, the question of which geometry describes the world is no longer a matter for intuition—or for self-evident assumptions—but for *experience* [20].

What did this view—expressed as well by Bernhard Riemann and W. K. Clifford, among others—do to the received accounts of the relation between mathematics and the world? It detached mathematics from the world. Euclidean and non-Euclidean geometry give the first clear-cut historical example of two mutually contradictory mathematical structures, of which at most one can actually represent the world. This seems to indicate that the choice of mathematical axioms is one of intellectual freedom, not empirical constraint; this view, reinforced by Hamilton's discovery of a noncommutative algebra, suggested that mathematics is a purely formal structure, or as Benjamin Peirce put it, "Mathematics is the science which draws necessary conclusions" [40]—*not* the science of number (even symbolic algebra had been just a generalized science of number) or the science of space. Now that the axioms were no longer seen as necessarily deriving from the world, the applicability of mathematics to the world became turned upside down. The world is no longer, as it was for Plato, an imperfect model of the true mathematical reality; instead, mathematics provides a set of different models for one empirical reality. In 1902 the physicist Ludwig Boltzmann expressed a view which had become widely held: that models, whether physical or mathematical, whether geometric or statistical, had become the means by which the sciences "comprehend objects in thought and represent them in language" [3]. This view, which implies that the sciences are no longer claiming to speak directly about reality, is now widespread in the social sciences as well as the natural sciences, and has transformed the philosophy of science. As applied to mathematics itself—the formal model of mathematical reasoning—it has resulted in Gödel's demonstration that one can never prove the consistency of mathematics, and the resulting conclusion among some philosophers that there is no certainty anywhere, not even in mathematics [**2**, p. 206].

5. Opposition. The best proof of the centrality of mathematics is that every example we have given so far has provoked strong and significant opposition. Attacks on the influence of mathematics have been of three main types. Some people have simply favored one view of mathematics over other views; other people have granted the importance of mathematics but have opposed what they consider its overuse or extension into inappropriate domains; still others have attacked mathematics, and often all of science and reason, as cold, inhuman, or oppressive.

Aristotle's reaction against Platonism is perhaps the first example of opposition to one view of mathematics (eternal objects) while championing another (deductive method). Another example is Newton's attack on Descartes's attempt to use nothing but "self-evident" assumptions to figure out how the universe worked. There are many mathematical systems God could have used to set up the world, said Newton. One could not decide a priori which occurs; one must, he says, observe in nature which law actually holds. Though mathematics is the tool one uses to discover the laws, Newton concludes that God set up the world by free choice, not mathematical necessity [**35**, pp. 7–8; **36**, p. 47]. This point is crucial to Newton's natural theology: that the presence of order in nature proves that God exists.

Another example of one view based on mathematics attacking another can be found in Malthus's *Essay on Population* of 1798. He accepts the Euclidean deductive model—in fact he begins with two "postulata": man requires food, and the level of human sexuality remains constant [**31**, Chapter I]. His consequent analysis of the growth of population and of food supply rests on mathematical models. Nonetheless, one of Malthus's chief targets is the predictions by Condorcet and others of continued human progress modelled on that of mathematics and science. As in Newton's attack on Descartes, Malthus applied one view of mathematics to attack the conclusions others claimed to have drawn from mathematics.

Our second category of attacks—drawing a line that mathematics should not cross—is exemplified by the seventeenth-century philosopher and mathematician Blaise Pascal. Reacting against Cartesian rationalism, Pascal contrasted the "esprit géometrique" (abstract and precise thought) with what he called the "esprit de finesse" (intuition) [**39**, Pensée 1] holding that each had its proper sphere, but that mathematics had no business outside its own realm. "The heart has its reasons," wrote Pascal, "which reason does not know" [**39**, Pensée 277]. Nor is this contradicted by the fact that Pascal was willing to employ mathematical thinking for theological purposes—recall his "wager" argument to convince a gambling friend to try acting like a good Catholic [**39**, Pensée 233];

the point here was to use his friend's own probabilistic reasoning style in order to convince him to go on to a higher level.

Similarly, the mathematical reductionism of men like Lagrange and Comte was opposed by men like Cauchy. Cauchy, whom we know as the man who brought Euclidean rigor to the calculus, opposed both Lagrange's attempt to reduce mechanics to calculus and calculus to formalistic algebra [15, pp. 51–54], and opposed the positivists' attempt to reduce the human sciences to an ultimately mathematical form. "Let us assiduously cultivate the mathematical sciences," Cauchy wrote in 1821, but "let us not imagine that one can attack history with formulas, nor give for sanction to morality theorems of algebra or integral calculus" [5, p. vii]. Analogously, in our own day, computer scientist Joseph Weizenbaum attacks the modern, computer-influenced view that human beings are nothing but processors of symbolic information, arguing that the computer scientist should "teach the limitations of his tools as well as their power" [50, p. 277].

Finally, we have those who are completely opposed to the method of analysis, the mathematization of nature, and the application of mathematical thought to human affairs. Witness the Romantic reaction against the Enlightenment: Goethe's opposition to the Newtonian analysis of white light, or, even more extreme, William Wordsworth in the *The Tables Turned*:

> Sweet is the lore which Nature brings;
> Our meddling intellect
> Mis-shapes the beauteous forms of things:—
> We murder to dissect.

Again, Walt Whitman, in his poem "When I heard the learn'd astronomer," describes walking out on a lecture on celestial distances, having become "tired and sick," going outside instead to look "up in perfect silence at the stars."

Reacting against statistical thinking on behalf of the dignity of the individual, Charles Dickens in his 1854 novel *Hard Times* satirizes a "modern school" in which a pupil is addressed as "Girl number twenty" [12, Book I, Chapter II]; the schoolmaster's son betrays his father, justifying himself by pointing out that in any given population a certain percentage will become traitors, so there is no occasion for surprise or blame [12, Book III, Chapter VII]. In a more political point, Dickens through his hero denounces the analytically based efficiency of industrial division of labor, saying it regards workers as though they were nothing but "figures in a sum" [12, Book II, Chapter V].

The Russian novelist Evgeny Zamyatin, in his early-twentieth-century antiutopian novel *We* (a source for Orwell's *1984*), envisions individuals reduced to being numbers, and mathematical tables of organization used as instruments of social control. Though the certainty of mathematics, and thus its authority, has sometimes been an ally of liberalism, as we have seen in the cases of Voltaire and Condorcet, Zamyatin saw how it could also be used as a way of establishing an unchallengeable authority, as philosophers like Plato and Hobbes had tried to use it, and he wanted no part of it.

6. Conclusion. As the battles have raged in the history of western thought, mathematics has been on the front lines. What does it all—to choose a phrase—add up to?

My point is not that what these thinkers have said about mathematics is right, or is wrong. But this history shows that the nature of mathematics has been—and must be—taken into account by anyone who wants to say anything important about philosophy or about the world. I want, then, to conclude by advocating that we teach mathematics *not* just to teach quantitative reasoning, *not* just as the language of science—though these are very important—but that we teach mathematics to let people know that one cannot fully understand the humanities, the sciences, the world of work, and the world of man without understanding mathematics in its central role in the history of western thought.

ACKNOWLEDGMENTS. I thank the students in my Mathematics 1 classes at Pomona and Pitzer Colleges for stimulating discussions and suggestions on some of the topics covered in this paper, especially David Bricker, Maria Camareña, Marcelo D'Asero, Rachel Lawson, and Jason Gottlieb. I also thank Sandy Grabiner for both his helpful comments and his constant support and encouragement.

BIBLIOGRAPHY

1. Charles Babbage, *On the economy of machinery*, Charles Knight, London, 1832.

2. William Barrett, *Irrational man: A study in existential philosophy*, Doubleday, New York, 1958; excerpted in William L. Schaaf, editor, *Our mathematical heritage*, Collier, New York, 1963.

3. Ludwig Boltzmann, *Model*, Encyclopedia Britannica, 1902.

4. Carl Boyer, *History of analytic geometry*, Scripta Mathematica, New York, 1956.

5. A.-L. Cauchy, *Cours d'analyse*, reprinted in A.-L. Cauchy, *Oeuvres*, series 2, vol. 3, Gauthier-Villars, Paris, 1821.

6. I. Bernard Cohen, *The Newtonian revolution*, Cambridge Univ. Press, Cambridge, 1980.

7. _____, *The birth of a new physics*, 2nd edition, W. W. Norton, New York and London, 1985.

8. Auguste Comte, *Cours de philosophie positive*. vol. I, Bachelier, Paris, 1832.

9. Marquis de Condorcet, *Sketch for a historical picture of the progress of the human mind*, 1793; Transl. June Barraclough, In Keith Michael Baker (Editor), Condorcet: *Selected writings*, Bobbs-Merrill, Indianapolis, Ind., 1976.

10. René Descartes, *Discourse on method*, 1637; translated by L. J. Lafleur, Liberal Arts Press, New York, 1956.

11. _____, *La geometrie*, 1637; translated by D. E. Smith and Marcia L. Latham, *The geometry of René Descartes*, Dover, New York, 1954.

12. Charles Dickens, *Hard times*, 1854; Norton Critical Edition, George Ford and Sylvere Monod (Editors), W. W. Norton, New York and London, 1966.

13. Neal C. Gillespie, *Charles Darwin and the problem of creation*, Univ. of Chicago Press, Chicago and London, 1979.

14. Charles C. Gillispie, *The edge of objectivity*, Princeton Univ. Press, Princeton, N. J., 1960.

15. Judith V. Grabiner, *The origins of Cauchy's rigorous calculus*, M. I. T. Press., Cambridge, Mass., 1981.

16. John C. Greene, *Science, ideology, and world view: Essays in the history of evolutionary ideas*, Univ. of Calif. Press, Berkeley, Calif., 1981.

17. A. Rupert Hall, *From Galileo to Newton*, 1630–1720, Harper and Row, New York and Evanston, Ill., 1963.

18. T. L. Heath (Editor), *The works of Archimedes with the method of Archimedes*, Dover, New York, 1912.

19. ____, *The thirteen books of Euclid's Elements*. Vol. I, Cambridge Univ. Press, Cambridge, 1925; reprint, Dover, New York, 1956.

20. Hermann von Helmholtz, *On the origin and significance of geometrical axioms*, 1870; Reprint, Hermann von Helmholtz, Popular scientific lectures, Morris Kline, editor, Dover, New York, 1962.

21. Thomas Hobbes, *Leviathan, or the matter, form, and power of a commonwealth, ecclesiastical and civil*, 1651; reprint, *The English philosophers from Bacon to Mill*, E. Burtt, editor, Modern Library, New York, 1939.

22. Anthony Hyman, *Charles Babbage: Pioneer of the computer*, Princeton Univ. Press, Princeton, N. J., 1982.

23. Werner Jaeger, *Early Christianity and Greek Paideia*, Oxford, Univ. Press, London, Oxford and New York, 1961.

24. Immanuel Kant, *Critique of pure reason*, 1781; translated by F. Max Müller, Macmillan, New York, 1961.

25. ____, *Prolegomena to any future metaphysics*, 1783; reprint, Lewis W. Beck, editor, Liberal Arts Press, New York, 1951.

26. Morris Kline, *Mathematics in Western culture*, Oxford Univ. Press, New York, 1964.

27. D. Landau and P. Lazarsfeld, *Quetelet*, Internat. Encycl. Social Sciences, Vol. 13, Macmillan, New York, 1968, pp. 247–257.

28. Pierre Simon Laplace, *A philosophical essay on probabilities*, 1819; translated by F. W. Truscott and F. L. Emory, Dover, New York, 1951.

29. Desmond Lee (Editor), *Plato: Timaeus and Critias*, Penguin Books, London, 1965.

30. Leibniz, *Preface to the general science* and *Towards a universal characteristic*, 1677; reprint, *Selections*, Philip P. Wiener, editor, 1951, pp. 12–17, 17–25.

31. Thomas R. Malthus, *An essay on the principle of population, as it affects the future improvement of society: with remarks on the speculations of Mr. Godwin, M. Condorcet, and other writers*, 1798; often reprinted or excepted, e.g. in *Population, evolution and birth control*, Garrett Hardin, editor, Freeman, San Francisco, 1964, pp. 4–16.

32. Maurice Mandelbaum, "The search for a science of society: From Saint-Simon to Marx and Engels," Chapter 4 in *History, man, and reason*, Johns Hopkins Univ. Press, Baltimore, Md. and London, 1971, pp. 63–76.

33. James Clark Maxwell, *Theory of heat*, Longmans, Green, & Co., London, 1871.

34. Seyyed Hossein Nasr, *Science and civilization in Islam*, New American Library, New York, Toronto, and London, 1968.

35. Isaac Newton, Letter to Henry Oldenburg, July, 1672. In [**38**, pp. 7–8].

36. ____, Letter to Richard Bentley, December 10, 1692. In [**38**, pp. 46–50].

37. ____, *Sir Isaac Newton's mathematical principles of natural philosophy and his system of the world*, translated by Andrew Motte, revised and edited by Florian Cajori, Univ. of Calif. Press, Berkeley, Calif., 1934.

38. ____, *Newton's philosophy of nature: Selections from his writings*, H. S. Thayer, editor, Hafner, New York, 1951.

39. Blaise Pascal, *Pensées*, E. P. Dutton, New York, 1958.

40. Benjamin Peirce, *Linear associative algebra*, Amer. J. Math. **4** (1881).

41. Plato, *Republic*, many editions, e.g. *The Republic of Plato*, translated by A. D. Lindsay, E. P. Dutton, New York, 1950.

42. ____, *Timaeus*. In [**29**].

43. Richard H. Popkin, *The history of scepticism from Erasmus to Spinoza*, Univ. of Calif. Press, Berkeley and Los Angeles, Calif., 1979.

44. Theodore M. Porter, *A statistical survey of gases: Maxwell's social physics*, Hist. Stud. Phys. Sci., vol. 12, 1981, pp. 77–116.

45. Bertrand Russell, *A critical exposition of the philosophy of Leibniz*, 2nd ed., Allen and Unwin, London, 1937.

46. William G. Sinnigen and Arthur E. R. Boak, *A history of Rome to A.D. 565*, 6th ed., Macmillan, New York, 1977.

47. Andrew Skinner, *Adam Smith, the wealth of nations*, Penguin Books, London, 1974, pp. 11–97.

48. Benedict de Spinoza, *Ethics*, Preceded by *On the improvement of the understanding*, James Gutmann, editor, Hafner, New York, 1953.

49. François-Marie Aroute de Voltaire, *The portable Voltaire: Selections from Dictionnaire Philosophique*, Viking, New York, 1949, pp. 53–228.

50. Joseph Weizenbaum, *Computer power and human reason: From judgement to calculation*, Freeman, San Francisco, Calif., 1976.

51. Harry Austryn Wolfson, *What is new in Philo?* in *From Philo to Spinoza: Two studies in religious philosophy*, Behrman Hours, New York, 1977.

PITZER COLLEGE, THE CLAREMONT COLLEGES, CLAREMONT, CALIFORNIA 91711, USA

Enseignement Mathematique, Ordinateurs et Calculettes

JEAN-PIERRE KAHANE

Cet exposé s'inspire de l'étude de la C.I.E.M.[1] sur l'influence des ordinateurs et de l'informatique sur les mathématiques et leur enseignement. Il contient également des appréciations personnelles de ma part sur les calculettes et l'enseignement.

L'étude de la C.I.E.M. En 1983 la C.I.E.M. a décidé de mettre à l'étude des questions d'intérêt mondial, sur lesquelles une approche internationale pouvait apporter d'utiles mises au point. Le but, dans chaque cas, n'est pas de fournir des solutions garanties—C.I.E.M.; c'est de faire l'état de la question, en vue de permettre la poursuite de la réflexion et, lorsque c'est possible, des initiatives au plan régional, national ou institutionnel.

C'est dans ce cadre que s'est déroulée la première étude, en 1984 et 1985. Le comité de programme[2] a établi entre janvier et mars 1984 un document de discussion—rédigé d'abord en français puis en anglais, puis traduit en plusieurs langues—publié dans sa version anglaise par la revue *L'Enseignement Mathématique* [I1], distribué aux représentants nationaux, et appelant à des contributions à la discussion. Le document indiquait trois grands thèmes: l'influence des ordinateurs et de l'informatique sur les mathématiques en tant que science (leur développement, leurs concepts, leurs valeurs); les changements que les ordinateurs et l'informatique peuvent induire dans le contenu des programmes d'enseignement; l'aide qu'ils peuvent apporter dans l'enseignement lui-même.

---

[1] Il est bon de rappeler ce qu'est la C.I.E.M.—commission internationale de l'enseignement mathématique, alias I.C.M.I, International Commission on Mathematical Instruction. Créée par Félix Klein en 1907, c'est une commission de l'U.M.I. (Union Mathématique Internationale) dont le statut actuel a été établi sous la présidence d'Hassler Whitney, en 1980. Elle est constituée par un comité exécutif, élu par l'assemblée générale de l'U.M.I., et par des représentants nationaux—un par pays, désigné selon des formules différentes selon que le pays est membre ou non de l'U.M.I.

[2] Constitué de R. F. Churchhouse (Cardiff), B. Cornu (Grenoble), A. E. Eršov (Novosibirsk), A. G. Howson (Southampton), J. P. Kahane (Orsay), J. H. Van Lint (Eindhoven), F. Pluvinage (Strasbourg), A. Ralston (Buffalo), M. Yamaguti (Kyoto).

Pour mieux centrer la discussion, on se bornait à considérer le niveau universitaire et préuniversitaire (c'est-à-dire les élèves de plus de 16 ans).

Les contributions écrites ont été nombreuses (une cinquantaine), variées, et intéressantes. Le comité de programme a alors organisé une rencontre d'une semaine, à Strasbourg, fin mars 1985, pour discuter à la fois du document de base et des contributions. Dès la fin de la rencontre, le travail était bien préparé pour l'édition des documents finaux: les *Proceedings* (environ 160 pages) édités par Cambridge University Press [I2], et les *Supporting Papers* (équivalent à plus de 600 pages dactylographiées) éditées par l'I.R.E.M. de Strasbourg [I3].

Sur le déroulement du colloque et sur les résultats de l'étude, un excellent article est paru dans *Zentralblatt für Didaktik der Mathematik* [I4]. A côté d'appréciations flatteuses, je retiens une critique: malgré la publication dans *l'Enseignement Mathématique* et d'autres journaux, le document de discussion n'a atteint qu'une faible partie des collègues intéressés. Le présent congrès est l'occasion d'assurer une meilleur publicité aux travaux de la C.I.E.M.

Après le colloque de Strasbourg, la reflexion s'est poursuivie: un colloque de l'I.C.O.M.I.D.C. (International Committee on Mathematics in Developing Countries) à Monastir (Tunisie) en février 1986, sur l'informatique et l'enseignement des mathématiques avec 40 contributions très variées [I-I], et une rencontre internationale à Luminy (France) en janvier 1986, à l'invitation de la sous-commission française de la C.I.E.M., qui a abouti en particulier à la constitution d'une banque de logiciels d'enseignement mathématique au niveau universitaire [I-F].

A partir de maintenant, mon exposé s'inspirera librement de cet ensemble de travaux, sans chercher à en rendre compte.

L'informatique et la mathématique. L'informatique est partout. L'informatique influe sur toutes les sciences. Les ordinateurs sont utilisés dans toutes les disciplines. Directement, au plan du laboratoire et du travail scientifique. Indirectement, quand il s'agit de communiquer, de produire ou de consommer l'information. Les effets indirects sont déjà de grande portée: l'informatique a permis aux bases de données bibliographiques d'absorber la croissance exponentielle de la production scientifique (qui double tous les dix ans, si on la mesure en nombre d'articles publiés); elle permettra, sans doute, de faire face aux nouveaux besoins de publication et de communication rapide. Ces possibilités techniques produisent de nouvelles exigences intellectuelles. Pour prendre un exemple, les *Mathematical Reviews* sont devenues un outil de travail indispensable, mais elles ne jouent plus le rôle de guide et de critique qu'elles avaient il y a 30 ans. Pour se retrouver dans la littérature scientifique contemporaine, il faut des synthèses, des mises au points, des exposés historiques et critiques: on voit cette sorte d'articles scientifiques, qualifiés autrefois de "secondaires" prendre une place de premier plan. Le fait est général: *ni la puissance de calcul, ni la capacité de mémoire, ni les logiciels les plus élaborés, ni les systèmes experts n'éliminent l'activité intellectuelle; l'informatique déplace cette activité vers des champs nouveaux, et la stimule.*

L'informatique est entrée dans l'enseignement. L'usage des microordinateurs s'est largement répandu en Europe, aux Etats-Unis, au Japon [**I3**, pp. 14–16, 22–23, 39, 43–45]. L'Open University du Royaume Uni a introduit des graphiques animés produits par ordinateurs dans un cours de mathématiques de base dès 1971 [**I3**, p. 24]. En Union Soviétique, des cours de programmation pour étudiants en mathématiques existent depuis 1959, et en 1986 un cours sur les bases de l'informatique ("science de l'informatique et techniques de calcul") est introduit dans les écoles secondaires [**I3**, p. 8]. Avec des réticences diverses (notamment au Japon) les calculettes font une entrée en force dans les enseignements élémentaire et secondaire. Depuis 1978, elles sont autorisées pour tous les examens du "General Certificate of Education" en Ecosse [**ICR**]. A partir de 1986, elles figurent explicitement dans les programmes français de mathématiques au début des études secondaires (11-12 ans).

L'informatique est partout, mais inégalement distribuée. Les investissements et les frais de maintenance interdisent aux pays pauvres la diffusion massive des microordinateurs [**I-I**]. Par contre, une distribution massive de calculettes comme fournitures scolaires est envisageable—au même titre qu'une distribution massive de livres d'enseignement. C'est une raison, parmi d'autres, pour s'intéresser particulièrement au renouvellement possible de l'enseignement mathématique par l'usage des calculettes. En retour, les besoins de l'enseignement peuvent amener à de nouvelles spécifications pour les calculettes destinées aux fournitures scolaires.

Enfin l'informatique est doublement liée aux mathématiques. *Comme moyen nouveau de calcul et d'écriture, elle a, et elle aura de plus en plus, un impact sur les pratiques, les valeurs, et les concepts même des mathématiques. Comme outil à base mathématique, son histoire est liée à celle de la logique, et on doit s'attendre á un va et vient constant entre l'informatique (c'est-à-dire les ordinateurs et leurs usages), la logique, l'algèbre, et d'autres branches des mathématiques.*

Dans l'étude de la C.I.E.M., il était bon de commencer par là: l'influence de l'informatique sur la mathématique comme science. En particulier, sur une série d'exemples, on voit que les ordinateurs et l'informatique ont suscité de nouvelles recherches, remis à l'ordre du jour des questions étudiées il y a longtemps, et rendu possible l'étude de questions nouvelles. Ils ont multiplié brusquement nos possibilités d'observation et d'expérimentation en mathématiques. Ils ont valorisé tout ce qui peut se traduire en algorithmes. Au delà du calcul numérique, ils ont développé des possibilités de visualisation, et maintenant de calcul symbolique, qui sont de grande conséquence pour la recherche mathématique.

Cette influence est incontestable. Elle est déjà beaucoup plus profonde au niveau de la recherche que de l'enseignement. Pour certains, elle apparaît comme une menace. D'abord, une menace sur l'esprit même de la mathématique—comme science de l'ordre et des concepts unificateurs; la preuve du théorème des quatre couleurs au moyen de l'ordinateur peut être correcte, elle n'est pas "belle." Ensuite, une menace sur l'avenir du métier de mathématicien, concurrencé par l'appel des métiers de l'informatique, et par conséquent une menace sur l'héritage mathématique.

Au niveau de l'enseignement, on peut aussi énumérer les vues pessimistes:
- les élèves vont devenir paresseux
- ils ne vont plus savoir calculer à la main
- ils ne s'intéresseront plus qu'à l'informatique
- les enseignants ne pourront jamais s'adapter aux nouveaux outils
- ceux qui s'adapteront deviendront informaticiens
- si en plus on touche au contenu de l'enseignement, on court au même désastre qu'avec les "mathématiques modernes."

Ces dangers existent. Mais il faut également apprécier les chances nouvelles. Il y a de belles mathématiques à faire pour dominer l'usage des ordinateurs, et on peut attendre, dans l'avenir, un stimulant venant de l'informatique aussi important que le stimulant—classique—venant de la physique; aujourd'hui déjà, les "mathématiques discrètes" se trouvent ainsi stimulées et valorisées. D'autre part—et c'est là une raison essentielle d'être optimiste—les ordinateurs et même les calculettes ressuscitent de très belles mathématiques qui étaient oubliées ou négligées. J'illustrerai cela par quelques exemples tout à l'heure. Cette possibilité de réanimer des sujets dormants—parfois pendant des siècles—est un trait particulier des mathématiques dans l'ensemble des sciences, et c'est ce qui en fait un héritage extrêmement précieux. C'est une justification, pour le présent et pour l'avenir, d'une formation en grand nombre de jeunes mathématiciens.

On peut déjà dire qu'en face des ordinateurs, les élèves sont souvent actifs, intéressés et agiles—ils acquièrent l'usage des outils plus vite que leurs professeurs. Ils adoptent facilement l'attitude expérimentale. Mails ils ne peuvent pas découvrir seuls les bonnes voies où s'engager. L'expérience mathématique des enseignants est irremplaçable. Face à l'ordinateur, l'enseignant devient un conseiller. Plus encore que par le passé, l'essentiel est sa qualification comme mathématicien.

Après ces vues très générales, je vais évoquer des choses très anciennes sur lesquelles l'informatique, les ordinateurs et les calculettes font porter un regard neuf: les nombres, les figures, les symboles, les algorithmes.

Les nombres. Avec les cailloux (l'origine du "calcul") on a une conception claire de nombres entiers petits. Pour les nombres entiers plus grands, la vision qu'on en a dépend du mode de notation. Chez les Grecs de l'Antiquité, le système usuel permettait d'écrire 3 ou 700 avec une seule lettre, mais ne permettait pas d'écrire 2.100 (d'où la question de Socrate au savant Hippias: "toi qui es si savant, si on te demandait combien fait 3 fois 700, tu saurais répondre avec célérité et exactitude?"). Aujourd'hui, ne fût-ce que par la radio et la télévision, tous les enfants sont familiarisés avec des nombres entiers très grands; à chaque élection par exemple, on voit défiler des grands nombres, et des rapports exprimés en pourcentages. On change constamment d'échelle—le budget de la famille, le budget de la cité, les dépenses militaires dans le monde; l'âge de l'humanité, de la terre, de l'univers; les dimensions de l'atome, du système solaire, etc. Ce qui permet d'appréhender ces nombres et ces changements d'échelle, c'est l'écriture décimale et les puissances de 10. Ce qu'affiche une calculette, c'est justement une écriture décimale, et éventuellement (en virgule flottante) une puissance de 10.

L'enfant qui dispose d'une calculette se trouve immédiatement devant de grands nombres entiers, et aussi devant des développements décimaux assez longs. Des développements décimaux assez longs permettent d'imaginer naturellement des développements décimaux illimités. Ainsi la définition d'un nombre réel par un développement décimal illimité—esquissée par Simon Stevin il y a tout juste quatre siècles [**B**]—mérite d'être bien connue des enseignants. En France, un livre vient de paraître, sur les fondements de la géométrie, dont le premier chapitre est la théorie des réels à partir des développements décimaux illimités; c'est une présentation simple et complète, à l'intention des enseignants du secondaire, qui me paraît venir à son heure [**F**].

Cela ne supprime pas, mais au contraire valorise, la représentation du nombre réel positif comme rapport de deux longueurs. L'algorithme d'Euclide pour trouver une partie aliquote commune à deux longueurs, nous allons le retrouver comme algorithme des fractions continues.

La calculette ne se contente pas d'écrire. Elle permet de *traiter les données numériques*, elle calcule. La facilité du calcul doit permettre aux élèves, en arithmétique élémentaire comme en géographie, de traiter des données réelles, et beaucoup de suggestions intéressantes sont faites à ce sujet dans des ouvrages d'enseignement récents. Je ne m'y étend pas, quoiqu'il s'agisse d'une chose essentielle dans les écoles primaires. Ce traitement des données réelles oblige à réfléchir pour savoir quelles opérations faire, et par contre rend superflus les algorithmes d'opérations posées sur papier.

Ainsi l'addition, la multiplication, la soustraction, la division cessent d'être des opérations qu'on pose, pour devenir des opérations qu'on ordonne à la machine de faire. Il n'y a sans doute plus lieu, au début de l'enseignement élémentaire, d'insister sur les modes opératoires et les refrains traditionnels ("je pose, je retiens,..."). Par contre le *calcul mental*, et particulièrement le calcul des ordres de grandeurs, doivent permettre de deviner et de contrôler les résultats donnés par la machine. *Les modes opératoires traditionnels* ne disparaissent pas pour autant; mais ils prennent leur importance bien plus tard, quand on peut faire découvrir aux enfants les opérations en "multiprécision," c'est-à-dire quand on opère en base 10^n en utilisant la calculette comme table de multiplication.

La calculette fait instantanément les divisions. Plus précisément, elle donne immédiatement une écriture décimale à n chiffres (disons $n = 10$) en guise de quotient. Elle fait donc passer d'une fraction, disons 14/31, à un développement décimal, $0,451612903$. Ce développement décimal, illimité, serait périodique, mais on ne voit apparaître le fait qu'en multiprécision. Si l'on donne l'écriture décimale $0,451612903$, comment remonter à la fraction? On inverse, on prend la partie décimale et on recommence, et on trouve $1/(2 + 1/(4 + 1/(1 + 1/2)))$.[3] C'est donc l'algorithme des fractions continues qui permet de passer de l'écriture décimale à la fraction (ou à une fraction égale). Si un de vos collègues affirme que, dans sa classe, $45,16\%$ des élèves ont été capables de répondre à une question, vous pouvez, si vous avez une calculette sous la main, lui dire que sa classe a 31 élèves (ou un multiple de 31).

Les fractions continues expliquent pourquoi $22/7$ et $113/355$ sont de bonnes approximations de π. Elles se prêtent à des jeux (découvrir deux nombres entiers quand on donne leur quotient), à des expériences, à des découvertes. Personnellement, je n'ai pas découvert la formule

$$e^{1/x} = (1, x - 1, 1, 1, 3x - 1, 1, 1, 5x - 1, \ldots)$$

dans la littérature, mais en jouant avec une calculette. Les racines carrées des nombres entiers sont un vaste champ d'expériences. D'abord, on constate que la décomposition de

$$\sqrt{2} = 1,414\ldots = 1 + 1/2, 414\ldots$$

semble ne pas s'arrêter. L'ayant constaté, il est facile de le démontrer. C'est la preuve la plus naturelle que $\sqrt{2}$ est irrationnelle (sa version géométrique était bien antérieure, chez les Grecs, à la démonstration par l'absurde qui se trouve chez Euclide, et qui est sans doute due à Théétète). Les propriétés des décompositions de $\sqrt{n}$ se découvrent expérimentalement, et fournissent de jolis problèmes. C'est d'ailleurs un sujet d'actualité. On sait que la factorisation de nombres très grands est essentielle pour le décodage. Or, depuis la factorisation du septième nombre de Fermat F_7 ($= 2^{128} + 1$) par Morrison et Brillhart en 1970, il apparaît que les développements en fractions continues des $\sqrt{kN}$ peuvent fournir la clé de la factorisation de N, surtout si on les utilise en calcul parallèle [**W**].

J'insiste encore sur les fractions continues. Les fractions continues, familières à Euler et à Lagrange [**L**], importantes dans les applications et dans différentes branches des mathématiques, n'ont jamais fait partie de l'enseignement secondaire, et occupent encore aujourd'hui une place marginale dans l'enseignement supérieur. Elles me paraissent, grâce aux calculettes programmables, pouvoir être enseignées au lycée et devenir une connaissance commune.

De façon générale, ce qui est facile à programmer peut devenir facile à enseigner. Ainsi la sommation des séries (et ses pièges: pour éviter des additions du type $A+0+0+\cdots$, une série doit être sommée "à l'envers," ou par blocs). Ainsi la méthode de Newton pour le calcul des racines d'une équation (intéressante même pour $\sqrt{n}$). Ainsi la transformation en moyennes arithmétique et géométrique de

---

[3]En vérité, il faut identifier à un entier les nombres qui en sont assez voisins.

Gauss $T(a, b) = ((a+b)/2, \sqrt{ab})$, dont nous allons voir une application. Ces deux dernières méthodes donnent des approximations "quadratiques," c'est-à-dire, en gros, doublent le nombre de décimales exactes à chaque opération.

La calculette permet de se familiariser avec les nombres en général, et aussi avec des nombres particuliers: les nombres premiers, les carrés, les solutions d'équations diophantiennes; les racines carrées, le nombre e, le nombre π. Je m'attarderai un instant sur π.

La tradition, pour le calcul de π, est le calcul de séries, par l'intermédiaire de formules du type

$$\frac{\pi}{4} = 4 \operatorname{arctg} \frac{1}{5} - \operatorname{arctg} \frac{1}{239},$$

qui avait permis à J. Machin vers 1700 de calculer π avec 100 décimales; des formules analogues ont été introduites et exploitées par Euler [E], et elles ont servi dans les années 1960 pour le calcul de π sur ordinateur, avec 100.000 décimales. Il y a pourtant beaucoup mieux. Voici le meilleur procédé actuel— découvert par E. Salamin en 1976, repris par Borwein et Borwein en 1984 [BB], et perfectionné par D. J. Newman récemment [N]. La transformation de Gauss $(a, b) \rightarrow T(a, b) = ((a + b)/2, \sqrt{ab})$ laisse invariante l'intégrale

$$I(a, b) = \int_{-\infty}^{\infty} \frac{dx}{\sqrt{(x^2 + a^2)(x^2 + b^2)}}$$

et les $T^n(a, b)$ convergent très vite vers leur limite $m(a, b)$. Ainsi

$$I(a, b) = \int_{-\infty}^{\infty} \frac{dx}{x^2 + (m(a, b))^2} = \pi m(a, b).$$

Or

$$I(1, N) = 2 \int_0^{\sqrt{N}} \frac{dx}{\sqrt{(x^2 + 1)(x^2 + N^2)}}$$
$$= \frac{2}{N} \log 4N + O\left(\frac{1}{N^3}\right)$$

et le calcul de $m(1, N)$ avec $n \sim \log N$ décimales exige environ $\log n$ itérations de T. Si on veut se débarrasser du logarithme, on écrit

$$N((N + 1)m(1, N + 1) - Nm(1, N)) = \frac{2}{\pi} + O\left(\frac{1}{N}\right).$$

C'est un procédé remarquablement efficace, même avec une calculette, et qui peut intéresser des étudiants.

On peut donc faire beaucoup de choses intéressantes, à plusieurs niveaux, avec de simples calculettes. Tout l'aspect numérique—aussi bien de l'enseignement élémentaire que de l'analyse mathématique—se trouve ainsi valorisé.[4]

Les figures. Les possibilités de tracés par ordinateurs et de visualisation sur écran ont déjà été exploitées avec succès par des mathématiciens. J'évoquerai deux exemples.

Autour de 1920, Fatou et Julia, indépendamment, avaient mené très loin l'étude des itérations d'applications rationnelles de $\overline{C}$ dans $\overline{C}$ (plan complexe complété); en particulier, ils avaient étudié les diverses formes des orbites et des ensembles invariants, et ils avaient reconnu les deux situations typiques— ensembles parfaits totalement discontinus, ou ensembles connexes dont la frontière est généralement très irrégulière. Dans le cas des applications $z \rightarrow z^2 + c$ (c complexe), on a appelé ensembles F, ou ensembles de Julia, l'ensemble invariant constitué par les points dont l'orbite reste bornée; c'est B. Mandelbrot qui a eu l'idée de reprendre les recherches de Fatou et de Julia en utilisant les possibilités graphiques des ordinateurs. A. Douady et J. Hubbard, D. Sullivan et d'autres ont étudié de près les ensembles F, et aussi l'ensemble M (pour Mandelbrot) des points c pour lesquels F est connexe. On connait maintenant assez bien la dynamique de la transformation sur les ensembles F, et on a des formules pour la dimension de Hausdorff de F (dans le cas totalement discontinu) ou de sa frontière (dans le cas connexe), on sait que M est connexe et on a considéré sa frontière à la loupe, en y découvrant des invariants universels des applications rationnelles. Du coup, les mathématiciens retrouvent et comprennent mieux les phénomènes de bifurcation découverts par les physiciens en particulier, les invariants de Feigenbaum. Tout ce domaine—itérations et systèmes dynamiques—est le siège d'une compétition fructueuse entre physiciens et mathématiciens. Sans les ordinateurs, on n'aurait sans doute jamais tenté de reprendre le sujet si profondément exploré par Fatou et Julia (**13**, p. 48).

L'autre exemple concerne les surfaces dans $\mathbf{R}^3$. C'est un domaine où toutes les possibilités de visualisation (contour apparent, sections, agrandissements, rotations) se révèlent précieuses. En 1901, W. Boy, étudiant le plan projectif réel, avait montré qu'on pouvait l'immerger dans $\mathbf{R}^3$ sous la forme d'une surface dont la courbe d'autointersection est une hélice tripale. Dès cette époque, la question s'était posé de représenter une telle surface par une équation polynomiale $P(x, y, z) = 0$. En 1984, au terme d'une suite d'essais où la visualisation a joué un rôle majeur, F. Apéry est parvenu à résoudre la question, avec un polynôme P de degré 6. En fait, F. Apéry a utilisé non seulement les possibilités graphiques, mais un langage de programmation symbolique, pour traiter des calculs qui auraient été autrement impraticables (**13**, p. 46).

---

[4] J'ai désigné par calculette aussi bien l'ordinateur de poche que la calculette "4-opérations." Il faut porter grande attention à la qualité de ces outils et à leur maitrise par les élèves. L'éducation est un marché assez important pour qu'elle impose ses normes aux fabricants. Encore faut-il définir ces normes. La "transparence" semble recommandable. "Simplicity is a virtue" (W. Kahan, UCB).

Ainsi, au niveau de la recherche, la visualisation par ordinateur a au moins deux effets: (1) rendre accessibles des formes considérées classiquement comme étranges, introduire à ce que B. Mandelbrot appelle la géométrie fractale (2) développer l'intuition en matière de géométrie des surfaces, en particulier des surfaces algébriques.

En matière d'enseignement, on peut prévoir des conséquences du même ordre.

D'abord, intégrer à la culture mathématique—des mathématiciens, mais aussi des physiciens et des ingénieurs—des objets autrefois réputés étranges, tels que les ensembles parfaits totalement discontinus, les courbes simples sans tangentes, les courbes de Peano, les mesures singulières. Une manipulation simple permet de refaire l'expérience de Michelson découvrant avec surprise que les graphes des sommes partielles de la série de Fourier d'une fonction discontinue débordent sans cesse le graphe de la fonction (le "phénomène de Gibbs"). La non-dérivabilité des fonctions de Weierstrass $\sum a^n \cos b^n x$ ($0 < a < 1$, $ab > 1$) est évidente quand on zoome leurs graphes. La fonction partout non dérivable de Takagi $\sum \text{dis}(x, 2^{-n}\mathbf{Z})$ mérite d'être vue, d'autant plus qu'elle peut apparaître dans des problèmes physiques très naturels [**13**, p. 50]. La courbe de von Koch peut être dessinée à partir d'un programme très simple: remplacer le segment

par les quatre segments

et répéter. Le simple examen de cette courbe (et des courbes analogues où 1/3 est remplacé par une autre valeur) est une excellente introduction à la notion de dimension fractionnaire, ou fractale. La courbe de Peano la plus simple—et la plus utile—s'obtient en remplaçant 1/3 par 1/2. Au delà de ces exemples, c'est la théorie géométrique de la mesure à laquelle il conviendrait de faire une place dans l'enseignement, aux côtés de la mesure abstraite, de l'intégration et des probabilités. Les probabilités fournissent d'ailleurs les exemples les plus naturels de comportements non classiques: une trajectoire brownienne n'est pas difficile à programmer, et elle explique bien l'intuition de Jean Perrin, que le mouvement brownien fournit des fonctions nulle part dérivables. Toute cette géométrie fractale a aussi ses régularités, ses invariants (à commencer par la dimension de Hausdorff), et ses ressources esthétiques lui donnent un charme particulier.

Ensuite, réintégrer à la culture mathématique la géométrie dans l'espace, et en particulier la géométrie des surfaces. Autrefois, chaque département de mathématiques avait une "salle des modèles," avec des moulages de quadriques, de tores, de cônes, de rubans, de bouteilles et d'autres surfaces. L'informatique graphique a tous les avantages des moulages en platre, et beaucoup d'autres—comme d'aller inspecter de près les singularités. La géométrie dans l'espace sur ordinateurs se développe dans l'industrie, dans l'architecture, dans la médecine. En chimie, la géométrie des grosses molécules joue un rôle essentiel, et d'excellents logiciels de visualisation ont été développés. L'enseignement mathématique ne peut pas l'ignorer.

La visualisation sur écrans ou sur tables traçantes a été déjà largement utilisée dans l'enseignement mathématique. On connait les succès de Logo en géométrie élémentaire. De nombreux logiciels de géométrie plane sont maintenant commercialisés. Il faudrait y consacrer la même attention qu'aux manuels d'enseignement.

Dans l'enseignement supérieur, l'étude des fonctions et surtout l'étude des équations différentielles avec des moyens graphiques est un succès incontestable. Plusieurs équipes dans le monde ont des approches voisines: de l'exploration des figures naissent des conjectures et des problèmes, qui amènent à un travail mathématique beaucoup plus intéressant que la simple recherche (souvent illusoire!) d'une formule permettant de représenter les courbes intégrales. Etudier un système différentiel, c'est, dans cette approche, chercher quelle est l'allure globale et locale des solutions. Les notions relatives aux points critiques (cols, noeuds, foyers, centres) ou aux branches infinies (asymptotes, entonnoirs) apparaissent naturellement par l'inspection d'exemples. On peut obtenir des dessins surprenants à l'aide de systèmes très simples, tels que

$$\frac{dx}{dt} = \cos y, \quad \frac{dy}{dt} = \sin xy \qquad [\mathbf{I2}, \text{p. } 107; \mathbf{I3}, \text{p. } 49].$$

Equations différentielles et itérations—appelées parfois: équations aux différences—constituent les deux approches mathématiques les plus simples des systèmes dynamiques. Un excellent livre récent est consacré à leur expérimentation sur ordinateur [**Ko**]. L'itération des homéomorphismes du cercle serait d'ailleurs une bonne occasion de retrouver les fractions continues, et les rotations qu'elles définissent.

Les symboles. Dans l'aspect numérique et dans l'aspect graphique, on se préoccupe de valeurs approchées, de procédés d'approximation, d'étude qualitative. Pour la résolution des équations numériques ou des équations fonctionnelles, peu importent les formules exactes (résolution à l'aide de radicaux, ou résolution en termes de fonctions élémentaires) pourvu qu'on ait un moyen pratique de calculer ou de représenter la solution.

Il est remarquable que l'informatique réhabilite aussi les aspects algébriques, les manipulations de formules littérales, les solutions en forme finie. C'est ce qu'apportent les systèmes symboliques formels.

Remontons à un siècle et demi. En 1835, J. Liouville établissait un théorème général sur les intégrales indéfinies exprimables "sous forme finie." En voici l'énoncé: si $\int P \, dx$ est exprimable, sous forme finie, en fonction de $x, y, z, \ldots$ ($y', z', \ldots$ étant des fonctions algébriques de $x, y, z, \ldots$), comme superposition de fonctions algébriques, d'exponentielle et de logarithme, il est permis de poser

$$\int P \, dx = t + A \log u + B \log v + \cdots + C \log w,$$

$A, B, \ldots, C$ étant des constantes, et $t, u, v, \ldots, w$ des fonctions algébriques de $x, y, z, \ldots$

Le même énoncé vaut en remplaçant "algébrique" par "rationnelle," d'où, par exemple, l'impossibilité d'exprimer $\int e^{-x^2}\, dx$ sous forme finie. Ce théorème fut très apprécié en son temps, puis oublié.

A. Ostrowski dégagea de là en 1946 l'aspect algébrique, en définissant les "corps liouvilliens." Le théorème devient purement algébrique avec M. Rosenlicht (1968), algorithmique avec R. H. Risch (1969), et concrètement applicable au calcul formel avec J. H. Davenport (1982). Grâce à ces travaux, la machine peut décider si une intégrale compliquée se calcule ou non sous forme finie, et, dans l'affirmative, donne la forme en question [**I2**, p. 76].

La machine fait donc, beaucoup mieux, ce que peut faire un étudiant expert en calcul des intégrales. Cela signifie que l'agilité à calculer des intégrales pourrait être complètement supprimée des buts de l'enseignement. Si le temps laissé libre permet d'accéder au théorème de Liouville ou à ses versions modernes, tant mieux. Il faut préférer aux techniques ad hoc les énoncés les plus généraux et les plus puissants. Comme on le voit, l'ordinateur nous pousse vers l'abstraction!

Le calcul symbolique, ou calcul formel, n'était naguère accessible que sur de gros ordinateurs. Actuellement, un système tel que muMATH est accessible sur micro. Pour les élèves, le calcul littéral deviendra bientôt aussi facile à effectuer que le calcul numérique. Ce sont les principes les plus généraux du calcul littéral—les aspects algébriques—qui méritent donc d'être explicités [**I2**, p. 35].

Les algorithmes. Le concept majeur en informatique est celui d'algorithme. En mathématiques, les algorithmes ont toujours joué un rôle important. Cependant, dans les définitions et dans les preuves, les procédés non constructifs ont naguère été préférés aux procédés constructifs, à cause de leur élégance et de leur puissance (ainsi toutes les utilisations de l'axiome du choix). Les procédés constructifs reviennent à la mode, et l'élégance que nous leur attribuons dépend évidemment des outils dont nous disposons.

Par exemple, considérons la théorie et la définition du pgcd. En voici trois approches.

(1) L'algorithme des divisions successives. Si r est le reste de la division de a par b, on pose $(b, r) = T(a, b)$, et on répète T jusqu'à obtenir $(\rho, 0)$. Tous les diviseurs de a et b divisent ρ, et ρ divise a et b. C'est le pgcd.

(2) Le procédé des idéaux. L'ensemble $a\mathbf{Z} + b\mathbf{Z}$ a un plus petit élément > 0, soit ρ. On démontre $\rho\mathbf{Z} = a\mathbf{Z} + b\mathbf{Z}$, et ρ est le pgcd.

(3) La récurrence descendante:

$$\text{pgcd}(a, b) := \text{si } b = 0 \text{ alors } a$$
$$\text{sinon si } a > b \text{ alors pgcd}(a - b, b)$$
$$\text{sinon pgcd}(b, a) \quad [\mathbf{I3}, \text{p.32}].$$

Cette dernière définition est incontestablement la plus simple, pour le programme comme pour la théorie. On peut juger la rédaction rébarbative, et la changer pour la rendre conforme au bon usage. Mais on peut également se demander si le bon usage ne sera pas influencé par le langage de la programmation. Naturellement, même si la définition 3 est la plus simple (c'est l'algorithme des "soustractions successives"), les autres gardent leur valeur.

Voici un autre exemple, de définition par algorithme, emprunté à la vie politique française. Les dernières élections législatives ont eu lieu en appliquant dans chaque département "le système de représentation proportionnelle à la plus forte moyenne." Il s'agit de désigner D députés. Il y a L listes de candidats, chacune portant D noms. Chaque électeur vote pour une liste, sans rayure ni mélange. Le dépouillement du vote fait apparaître le nombre total de votants, N, et le nombre de voix pour chaque liste, $n_1, n_2, \ldots, n_L$ $(n_1 + \cdots + n_L = N)$. Voici l'algorithme qui définit le système. On calcule le "quotient électoral" $Q = [N/L]$. On attribue à la k-ième liste $[n_k/Q]$ députés. Si tous les sièges sont attribués, on s'arrête. Sinon, d_k étant le nombre de sièges attribués à la k-ième liste, on calcule $n_k/(d_k + 1)$ et on attribue un nouveau siège à la liste pour laquelle ce rapport (cette "moyenne") est le plus grand. On recommence jusqu'à ce que tous les sièges soient attribués.

Cet algorithme, très commode pour le calcul à la main (sauf au début, on ne divise que par de nombres petits, et le classement se fait d'un coup d'oeil), n'est pas agréable à programmer sur un ordinateur de poche. Il a pour autre inconvénient qu'on ne voit pas immédiatement à quoi il vise. C'est un autre algorithme, la résolution d'équation par dichotomie, qui convient à la machine: la fonction $\sum [n_k/q]$ est une fonction décroissante de q, prenant (on le suppose!) toutes les valeurs entières positives quand q varie sur $\mathbf{R}^+$. On choisit q de façon que cette somme soit D. Le nombre de sièges de la k-ième liste est alors $[n_k/q]$. Ce qu'on explique à la machine permet de comprendre ce qu'on cherche.

Les algorithmes disponibles ont un effet sur les définitions et sur les preuves, sur le style de rédaction, et sur la pensée mathématique elle-même. Je me suis déjà étendu sur les itérations, et en particulier sur les fractions continues. De manière générale, les algorithmes faciles à programmer peuvent avoir un contenu mathématique, et ce sont alors des candidats naturels pour entrer dans un programme d'enseignement. La complexité des algorithmes est un beau sujet mathématique dont on peut avoir rapidement une idée intuitive.

Le langage des algorithmes fait appel aux mathématiques discrètes. Depuis quelques années, des programmes de mathématiques discrètes sont proposés dans quelques universités des Etats-Unis sous la forme de cours d'un semestre ou de cours annuels, et il faut être attentif à leur évolution. Lorsqu'on en énumère les matières, on peut avoir l'impression d'un pot-pourri: ensembles et relations, logique élémentaire, induction et définitions récursives, combinatoire, équations aux différences, graphes et arbres, probabilités discrètes, matrices et programmation linéaire, systèmes de numération, groupes et anneaux, machines à états finis et leurs relations avec les langages et les algorithmes [**I2**, p. 16]. Mais ce pot-pourri constitue la culture commune aux étudiants en informatique et aux étudiants en mathématiques, comme les éléments d'analyse (fonctions d'une variable réelle, dérivées, intégrales, équations différentielles, fonctions de plusieurs variables) constitue la culture commune aux étudiants en physique et aux étudiants en mathématiques. Qu'on en fasse un cours séparé ou non, il est clair que des mathématiques discrètes doivent être enseignées dès la tranche d'âge 16–18 ans, même si c'est au détriment du "calculus." On peut économiser sur la virtuosité dans le calcul des intégrales ou la résolution des équations différentielles au moyen de fonctions élémentaires: la compréhension des méthodes générales, et de ce que peuvent apporter les systèmes de calcul formel, est maintenant plus importante que le calcul à la main.

★ ★ ★

La machine de Turing a tout juste 50 ans. C'est l'origine des machines à calculer modernes, et c'est aussi l'origine de la théorie moderne des algorithmes. L'idée de base de Turing est de réduire une activité mentale à une action mécanique: "according to my definition, a number is computable if its decimal can be written down by a machine." Naturellement, il ne s'agit pas là de l'activité d'invention, mais de l'exécution d'un programme de calcul.

La théorie des langages et celle des automates est sortie de là. Les automates à état fini sont devenus l'une des bases de l'informatique théorique. Les suites qu'ils engendrent—suites "automatiques"—interviennent en analyse et en théorie des nombres, soit comme outils, soit comme objets. Par exemple, la suite découverte indépendamment par H. S. Shapiro (1951) et W. Rudin (1959)

$$+ + + - + + - + + + + - - - + - \cdots$$

obtenue à partir du mot $abcd$ par l'application des règles $a \to ac$, $b \to dc$, $c \to ab$, $d \to db$ puis en remplaçant a et c par $+$, b et d par $-$ est très importante en analyse de Fourier, et ses généralisations donnent de bonnes approximations automatiques du mouvement brownien. Les constructions automatiques (traductions géométriques de suites automatiques, type pavage de Penrose) donnent aussi un large champ d'étude. Les automates devraient faire partie de la culture d'un mathématicien [**I3**, p. 33; **I2**, p. 69; **S**]. Peut-être, sur des exemples, conviendraient-ils aussi à de non-mathématiciens, comme modèles simples de ce que sait faire une machine.

L'idée de Turing s'applique également à la théorie de la preuve: une preuve est complètement formalisée si elle peut être vérifiée par une machine. C'est plus qu'une idée, c'est déjà une réalisation, avec le système Automath de N. de Bruijn. L'idée d'Automath, c'est d'expliquer les choses à une machine. Naturellement, ce n'est pas ainsi qu'il faut les expliquer à des étudiants. Mais, remarque de Bruijn, si on ne sait pas expliquer les choses à une machine, on peut avoir des difficultés à les expliquer à des étudiants [**I2**, p. 61].

Une remarque finale. La plupart des participants au colloque de Strasbourg avaient beaucoup plus d'expérience que moi en informatique et dans l'usage des ordinateurs. Si j'ai été invité à parler sur ce thème, ce n'est donc pas comme expert, mais comme amateur. Et c'est comme amateur que je crois pouvoir m'adresser aux collègues non-experts. On n'a pas besoin d'être informaticien pour apprécier le rôle de l'informatique sur la pensée mathématique—par contre, une telle appréciation est impossible à un non-mathématicien.

On doit réfléchir à ce rôle quand on pense à la place des mathématiques dans l'enseignement obligatoire. Quelles mathématiques enseigner, pourquoi et comment? Il y a maintenant de bons arguments pour les mathématiques discrètes, mais aussi pour la statistique, les probabilités, l'analyse numérique. On peut utiliser les outils de bien des façons—enseignement assisté par ordinateur ou expérimentation libre; calcul numérique, visualisation, calcul formel. Le champ des contenus possibles s'élargit autant que le champ des méthodes. Le champ des contenus souhaitables—fût-ce du seul point de vue de l'informatique—risque de croître encore plus vite.

L'enseignement des mathématiques—comme métier—n'est donc pas menacé par l'intrusion de l'informatique. Au contraire, dans presque tous les pays du monde, on trouvera qu'il est insuffisant—qu'il y faudrait consacrer plus d'heures, plus de moyens, plus d'enseignants.

Reste la question principale. Comment les enseignants peuvent-ils faire face à ce monde changeant, où il s'agit de renouveler et d'élargir les contenus et les méthodes? C'est une question très difficile, qui n'admet certainement pas de réponse universelle. Le signe le plus encourageant à cet égard, c'est l'ensemble des initiatives prises par le milieu lui-même pour élargir simultanément sa qualification mathématique et sa compétence en matière d'outils. Il y a maintenant d'excellents livres, où l'on trouve à la fois les règles d'utilisation d'un outil, d'un langage ou d'un système, X, et une série d'applications mathématiques de grand intérêt. Ainsi, le mathématicien désireux de s'initier à X a, du même coup, l'occasion de faire de bonnes mathématiques—et c'est actuellement dans des ouvrages mixtes de ce type (dont le plus avancé est sans doute [**G**]) qu'on trouve les meilleurs exposés des langages ou systèmes en usage [**Ki, G, CR, JL, Se, Ko**].

Nous devons être capables—et c'est un grand enjeu de société—de répondre aux nouvelles possibilités et aux nouveaux besoins qu'amènent les nouvelles technologies. L'apprentissage des mathématiques est un des moyens de développer l'intelligence humaine au même rythme que l'intelligence artificielle. L'enseignement des mathématiques n'a jamais été plus important et plus passionnant.

REFERENCES

[**I1**] *The influence of computers and informatics on mathematics and its teaching*, CIEM-ICMI Enseign. Math. **30** (1984), 159–172.

[**I2**] *The influence of computers and informatics on mathematics and its teaching*, ICMI Study Series, Proc. Strasbourg Symposium (1985), CUP, 1986.

[**I3**] *The influence of computers and informatics on mathematics and its teaching*, Supporting Papers (n° 1 à 50) from the Strasbourg Symposium (Strasbourg 25–30 March 1985), CIEM-ICMI, IREM 67084 Strasbourg Cedex.

[**I4**] *Report on the ICMI Symposium on "The Influence of computers and informatics on mathematics and its teaching"*, R. Biehler, R. Strässer, B. Winkelmann (Bielefeld), Zentralblatt für Didaktik der Mathematik (1986).

[I-F] *Logiciels et réalisations informatiques pour l'enseignement des mathématiques dans l'enseignement supérieur*, B. Cornu, Gazette des Mathématiciens **30** (1986), 147–164.

[I-I] *International Symposium on Informatics and the Teaching of Mathematics in Developing Countries*, Monastir (Tunisie), February 3–7, 1986, ICOMIDC-IFIP.

[ICR] *Working paper on hand-held calculators in schools*, Marilyn N. Suydam, SMEAC Information Reference Center, Ohio State Univ., Columbus (mars 1980), International Calculator Review.

[B] Nicolas Bourbaki, *Eléments de mathématiques*, livre III (topologie), chapitre 4 (nombres réels), Hermann, 1950.

[F] Jacqueline Ferrand, *Les fondements de la géométrie*, PUF, 1986.

[E] L. Euler, *Opera omnia*, I–14, p. 245, etc.

[L] J. Lagrange, *Additions à l'analyse indéterminée in Oeuvres d'Euler*, I, 1.

[W] M. C. Wunderlich, *Implementing the continued fraction factoring algorithm on parallel machines*, Math. Comp. **44** (1985), 251–260.

[BB] J. M. Borwein et P. B. Borwein, *The arithmetic and geometric mean and fast computation of elementary functions*, SIAM Rev. **26** (1984), 351–366.

[N] D. J. Newman, *A simplified version of the fast algorithms of Brent and Salamin*, Math. Comp. **44** (1985), 207–210.

[Ko] H. Koçak, *Differential and difference equations through computer experiments* (with diskettes containing Phaser: an animator/simulator for dynamical systems for IBM personal computers), Springer, 1986. (X = Phaser)

[S] A. Salomaa, *Computation and Automata*, Encyclopedia Math. Appl., vol. 25, CUP, 1985.

[G] U. Grenander, *Mathematical experiments on the computer*, Academic Press, New York, 1982. (X = APL)

[Ki] A. Kirch, *Elementary number theory: A computer approach*, Intext Educ. Publ., New York, 1974. (X = Fortran)

[CR] B. Cornu et C. Robert, (1) *Du calcul à la programmation*, Magnard, 1981 (X = TI 57); (2) *Mathématiques et calculatrice programmable au lycée et au baccalauréat*, Magnard, 1983. (X = TI 57 LCD)

[JL] D. Jakubowicz et H. Lehning, *Mathématiques par l'informatique individuelle* (I et II), Masson, Paris, 1982. (X = Basic)

[Se] R. Sedgewick, *Algorithms*, Addison-Wesley, Reading, Mass., 1983. (X = Pascal)

UNIVERSITÉ DE PARIS-SUD, MATHÉMATIQUE, 91405 ORSAY CEDEX, FRANCE

Verbal Problems in Arithmetic Teaching

ZBIGNIEW SEMADENI

When mathematics educators think of the role of *verbal problems* (or *word problems*, or *story problems*), most of them take it as problem-solving, that is, solving verbal problems prepared beforehand by the teacher or by a textbook author. There is an enormous literature on the subject (see, e.g., NCTM 1980 Year Book [6] or research surveys [5] and [11]). My talk, however, will concentrate on the following topics, which have received little attention:

I. Formulation of problems by a child.

II. Transformation of given problems by a child.

III. Dealing with intentionally ill-posed verbal problems.

They are prospective topics for inclusion in the child's mathematical learning.

I would like to stress that my talk will be entirely concerned with primary education, with special attention paid to the beginning of schooling. This is a crucial period for the child: the formal instruction starts; the child is in danger of getting lost in school math which seems to make no sense; the rote learning is likely to begin, with numerous cases and subcases, each with its own meaningless rules to be remembered. Children *can* learn mathematics successfully, if Polya's principle is (at least partially) implemented: "Let the students *discover by themselves as much as feasible* under the given circumstances" (cf. [11, p. 12]). However, most commonly teachers just follow the textbook and students are trained to give responses in just the form expected by the teacher; this is the root of students' failures. The issue is not easy to deal with, but if we want to improve the tragic situation of mathematics in schools, we must begin with the early grades. My own experience in the subject is based on my work during the last eighteen years in Poland and elsewhere.

For a decade the official Polish curriculum for grades 1–3 has encouraged teachers to include activities of the above three types in mathematics lessons, with special emphasis on problem-formulating. However, little is known about how these ideas are orchestrated in practice. I myself had the opportunity to organize and observe some experimental lessons in 1985 and 1986, and this taught me a great deal. In the literature, I could find scattered contributions from various countries but hardly any body of systematized knowledge.

The purpose of my talk is to call mathematicians' attention to the significance of these three topics for problem-solving. I believe that their present neglect is unwarranted. They are potentially useful for the child's learning, but are also likely to result in series of artificial, boring, meaningless exercises if children are not given opportunity to think and to have good enough communication with the teacher.

In the sequel, the term "problem" will mean a *verbal arithmetical problem.*

I. Formulation of problems by a child. Problem-solving is extremely important for mathematics learning and should begin very early. We do not have to wait until the children master addition to give them a story that will be of interest to them and can be solved by simulating the situation (or action) with, e.g, marbles, and simply counting them. There is abundant evidence that children can do this prior to formal instruction on problem-solving.

At first, problems should not be purely verbal: the teacher may use pictures or just enact the situation. And then, gradually, the child conceives an idea of what a problem is and what a solution is. To help to clarify the idea, the teacher should suggest that the students try to formulate problems themselves. "One way of involving students in problem solving is to have them formulate their own problems" (Jeremy Kilpatrick [**11**, p. 12]).

The idea of *problem-formulation* (or *problem-posing*) is an old one, though apparently every generation of educators has to discover it anew. Recent years have seen renewed interest in the topic (see, e.g., [**6**, p. 94]). Yet it may take several years for educators to appreciate fully the role of problem-formulation for the child's conception of what a verbal problem is all about.[1]

Problem-posing is hard for children. That is known (see, e.g., [**2**]). But it is so significant when it arises in their experience that it should accompany problem-solving almost from the beginning, when it need not mean much more than verbalizing stories, e.g., enacted by the teacher.

Later children may be invited to formulate problems to: (1) a given picture, situation, etc. (children tell the story); (2) a given theme (e.g., "Can you formulate a problem about boys and girls walking together?"); (3) a graphic scheme (e.g., consisting of three dots and four other dots, representing sweets and cookies, say); (4) a formula, e.g., $7 - 3 = 4$; (5) an equation, written as, e.g., $3 + x = 7$ or expressed in some other form; (6) a given question (e.g., "How old is Bob?"). The teacher may formulate a problem with blanks instead of numbers and ask children to provide numbers and then to solve the problem; or the students get a story with numbers and are asked to make up meaningful questions.

Activities in problem-posing should form a significant supplement to problem-solving. As we may expect, problems formulated by pupils usually are imitations

---

[1]There are several publications on problem-posing at the secondary level (e.g., those by M. I. Walter and S. I. Brown). Much less is known about problem-posing at the primary level, though some hints for teachers can be found in *Arithmetic Teacher.*

of problems given to them previously. Still, the most important is that children have the opportunity to speak, to express their thoughts, to exchange ideas.

II. Transformation of a given problem into another by a child. Problem-posing can be enriched with *problem-transforming*. This can be done in various ways.

1. *Having a child transform a given problem into an isomorphic one.* Two problems are called *isomorphic* if they differ only in their nonmathematical *context*, but have the same intrinsic mathematical *structure*. For instance, children first solve a problem of the type "3 apples + 5 apples = 8 apples" and are invited to formulate another problem which fits the same formula $3+5 = 8$. At first, the two problems may be very similar, almost identical, e.g., if "apples" are replaced by "oranges"; gradually, the children should learn to formulate problems which are less similar. (To make it easier, the teacher may suggest the context.)

> "It would be especially helpful for students to see that the same numbers in the same relationship can be provided with different 'cover stories,' and I would place more emphasis on the students actually working up to cover stories themselves" (J. Kilpatrick [**5**, p. 470]).

The object of such activities is to help the child learn what is mathematically essential in a problem.

2. *Transforming a simple problem* (with one operation) *into an inverse one.* Thus, the result of solving the given problem becomes a given number in the inverse problem, while a previously given number is now deemed to be unknown; the context is otherwise unchanged. For instance, a multiplication problem is converted into a division problem.

Such a change should be motivated. After having solved a problem, the child may see no point in posing an inverse problem; after all, the answer is known, so why pretend that we do not know it? The teacher may first tell a story of, say, a mother who wanted to know how many candies she should buy if each of her 4 daughters was to get 3 candies. Of course, she had to multiply the two numbers. When she came home, the girls noted that mummy brought 12 candies and wondered how many candies would be given to each of them. "Can you help the girls to answer the question?" the students may be asked. They divide 12 by 4 and find just the number that mummy had thought of.

Such activities are to contribute to the child's understanding of inverse operations and, more generally, inverse problems.

3. *Asking a new question* (to the same story, in case of a complex problem). This may help to develop the attitude: "What else can we ask?"

4. *Varying the given data* while keeping the story and the question otherwise unchanged. Rather than giving students a single problem (e.g., "*Mummy bought 10 rolls. Three of them were eaten. How many rolls were left?*") and jumping to another problem with unrelated story, once in a while the teacher should

extend the problem by just changing some data. E.g., keeping the information that mummy bought 10 rolls we may ask: "How many rolls were left when 4 rolls had been eaten?", "How many were left when 5 rolls had been eaten" and so on (including 2, 1 and zero rolls eaten). This gives a new dimension to the problem and prepares the child to think in terms of variables and functions (e.g., by noting that the more rolls are eaten, the fewer are left).

Similarly, in the problem "*A tailor must sew on* 3 *buttons to each of* 8 *shirts. How many buttons does he need?*" the number of shirts may vary. To record the successive answers, the pupils may fill out certain tables. They label rows (or columns) by words "shirts" and "buttons" which are—at a later stage— abbreviated to s and b, say. This leads to natural questions like "what number is s" and opens the way to the use of letters as variables.

Varying numbers in a verbal problem may help children to grasp some conceptually difficult extensions of arithmetical operations, where the ordinary intuition fails. This includes products of the form $0 \cdot n$, $n \cdot 0$, powers n^0, operations on fractions and on negative integers. A general approach of this sort, formulated as the *concretization permanence principle*, can be found in [**10**].

As an example suppose that students already know powers of the type $2^5 = 2 \cdot 2 \cdot 2 \cdot 2 \cdot 2$ and are to learn what are the powers 2^1 and 2^0. We need a concretization and choose a family tree, say, with one (unnamed) child, two parents, four grandparents, eight great grandparents, and so on. The students draw such a family tree, count the ancestors in successive generations and write the number of branches of the tree at each level: $1, 2, 4, 8, \ldots$. When they reach 32, say, they express the previously written numbers as powers: $32 = 2^5$, $16 = 2^4$ and so on. When they come down to the numbers 2 and 1, the pattern suggests that $2 = 2^1$ and $1 = 2^0$. If we asked "What number is 2^0?", we would invite the wrong answer that 2^0 is zero. The approach presented here enables us to ask "What power of 2 is 1?". What is crucial here is the fixed context with varying exponent.

5. *Transforming a given problem into a more complicated one* is an operation inverse to decomposing a multistep problem into simpler ones. For instance, the children first solve the problem: "*A hamburger costs* $1.50. *How much are* 2 *hamburgers?*" and then they are asked to formulate and solve more complicated problems (e.g., about the price of 2 hamburgers and a coke). In this way the children may learn to build up a multistep problem themselves and to recognize its structure. This should help them solve such problems and think in terms: "I've solved this problem and am going to formulate another, more interesting one!"

6. *Transforming a given series of isomorphic problems into another according to a single principle.* This can best be explained with an example. During an experimental lesson near Warsaw, eight- and nine-year-olds got the following problem: "*Daddy brought* 8 *kilograms of apples from his garden. He gave* 3 *kilograms of apples to uncle Peter. How much was left? Write a formula.*" The children solved the problem, wrote $8 - 3 = 5$, and then they were given the first

sentence of the next problem: "*In a pail there were 8 liters of milk.*" They were supposed to finish the story and to pose a question as to get a problem fitting the same formula $8 - 3 = 5$. When this had been done ("3 liters of milk was poured out to a can...") the students got first sentences of successive problems: "8 *cabbage-coles grew on a garden-bed,*" "*Henry had 8 meters of string,*" "*Stan had 8 zlotys*" (zloty is the unit of currency in Poland).

After this series of problems had been formulated, the teacher gave the children a new problem. Its context was the same as in the first problem of the series and there were the same numbers 8 and 3, but the operations were different: "*Daddy brought 3 boxes of apples from the garden. There were 8 kilograms of apples in each box. How many kilograms of apples did daddy bring?*" The children solved the problem and wrote the formula $3 \cdot 8 = 24$. Then the teacher asked them to formulate another problem so that the story would be about the same as in the second problem (that is, about milk in a pail) but the formula would be $3 \cdot 8 = 24$. A child proposed a problem about 3 pails with 8 liters of milk in each. And then the students were asked to tell how to reformulate the remaining problems: they were speaking of 3 garden-beds, 3 pieces of string, etc.

Such activities may hopefully contribute to the child's understanding of what is mathematically essential in a problem and to the flexibility of their thinking.

For more examples, see Janet H. Caldwell [**5**, pp. 393–404].

III. Dealing with intentionally ill-posed problems.

III. Dealing with intentionally ill-posed problems. In the sequel, a problem will be called *well-posed* if it is uniquely solvable and no numerical information given in the story is superfluous. Accordingly, a problem is *ill-posed*[2] if either (1) there is not enough information to answer the question uniquely, or (2) there is too much numerical data, or (3) the data is impossible (logically or physically). Thus, we distinguish three main categories of ill-posed problems: with *missing* (incomplete) *information*, with *surplus data*, with *impossible data*.[3]

There may be various types of problems within such a category. E.g., among problems without enough information we may distinguish: lack of some data in an otherwise standard problem; or a standard story with an irrelevant question; or a story with unrelated data from which no conclusion follows; or an ambiguous story, open to more than one interpretation; or a story with no reasonable arithmetical mathematization, e.g., in case of nonlinear growth. Surplus data may

---

[2]I hesitated whether to use such terms as "well-" or "ill-posed" verbal arithmetical problems. In fact, I believe that intentionally ill-posed problems may be didactically significant. But this talk is addressed to mathematicians who know what a well-posed or incorrectly-posed Cauchy problem is and may even know that incorrectly-posed problems are of increasing importance, e.g., in mathematical physics. As I do not have any better names, for the moment I am using these terms.

[3]The above categories are not sharply determined. One can even formulate (see [**9**]) a one-parameter family of problems $(P_s)_{s>0}$ with the same story, where s is a given variable datum, such that for certain values s_1, s_2, s_3 problem P_{s_1} is well-posed, P_{s_2} has incomplete information, whereas P_{s_3} has impossible data; therefore by changing the parameter s we may "continuously" pass from one category to another.

be *extraneous* (irrelevant) or *redundant* (duplication of information). Impossible data may be *contradictory* (inconsistent) or simply impossible in real life.

To avoid confusion, we should clearly distinguish three possible setups for an arithmetical problem:

(a) *Logico-mathematical setup.* The answer is to be deduced from what is said in the story (without using any additional information, except of general knowledge, e.g., that the value of a nickel is 5 cents). To solve the problem one may perform arithmetical operations, solve equations, simulate the situation with concrete materials (marbles, building blocks, drawings), and use some systematic trial-and-error procedure. (Unfortunately, many teachers interpret mathematical reasoning too strictly as performing some narrowly prescribed operations, not accepting simulation or systematic search.)

(b) *Applied-mathematics setup.* In order to solve a genuine problem in science or real-life one can use any valid outside source of pertinent data. Mathematics is one of the tools.

(c) *Real-life setup.* In order to answer a question involving numbers the solver can argue as in (a) or (b), but other options are also open: guessing, choosing a solution because of, e.g., the preference for the color blue over green, or ignoring the problem.

Both (a) and (b) are very important. In this paper, however, our attention is restricted to setup (a), and only to problems formulated so that a single quantity should be an answer. Thus, in case of missing data, the solver is supposed to say that the question cannot be answered because we do not know certain facts. A more sophisticated approach would be to think in terms of an open problem: The solver would have to note that several objects satisfy the given condition(s) and try to find all of them. The former approach is closer to the concrete thinking of the child; the latter is more formal and is less likely in early grades. (An intermediate approach would be: We do not know the number in question, it could be this or that.)

One of the goals of education should be that the students master suitable rudiments of (a) and (b). However, the school beginner thinks in terms of (c) and does not understand the tacit conventions of setup (a). The transition takes years of solving problems, reflecting upon them and developing the child's natural powers. Prematurely forcing the child to think in terms of (a) usually leads to rote learning and frustration. But if children have the opportunity to think meaningfully, freely, at their own pace, then they can make progress.

Within the above framework we shall discuss *intentionally ill-posed problems*, that is, problems with missing, surplus, or contradictory data given to students with the intent that they think whether the problem can be solved, explain its deficiency, and suggest how it could be emended so as to get a solvable problem with just enough data.

It is widely recognized that problem-solving is hard for children. Dealing with intentionally ill-posed problems is still harder. Research confirms this. Reports of various tests (formal and informal) of how children deal with intentionally ill-posed problems are disconcerting. Actually, most tests have concerned problems having superfluous data; it was found that—even in easy problems—these data confused many students. V. A. Kruteckiĭ [7], who made intensive psychological tests of how children reason when they are given (not necessarily arithmetical) problems with missing or superfluous data, found in particular that in a problem of a new type, the average pupil at first perceives only disconnected facts. Still worse is that during some tests a significant percentage (sometimes a majority) of students gave absurd answers. In the test organized in Grenoble [3], reported in Berkeley at ICME 1980, children aged 8 to 12 were given variations of the celebrated problem: "There are 26 sheep and 10 goats on a boat. How old is the captain?" Almost 75% of those aged 7 to 9 and 20% of those aged 9 to 11 used the given numbers to compute the age of the captain without expressing any doubt.

These appalling results (which could probably be confirmed in most countries) are usually explained by referring to rigid teaching routines which make mathematical operations meaningless for many children, and to the artificiality of numerous textbook problems. Children giving wrong answers to school problems may correctly solve mathematically equivalent problems when these actually arise in their own lives (see, e.g., [1] and [12]).

However, we should also take into account the natural epistemological obstacles that pupils may have to overcome while learning to solve problems. At the beginning of their schooling, most children do not understand what a verbal problem is; they consider it just as a puzzle and try to guess the answer. Later, as a result of inadequate teaching, they may develop distorted concepts instead of the clear ones that one might expect.

H. Freudenthal [4], in his profound analysis of various aspects of the Grenoble test, pointed out two more factors: 1° children may not have a clear concept of the age of a person; 2° something magical may be involved (note that for many children, and even for certain adults, performing mathematical operations is a ritual to be strictly observed though its meaning may not be understood).

Anyway, there seems to be enough evidence to support the stereotypic view that the use of intentionally ill-posed problems in school is a poor idea. Nevertheless, I think otherwise. Life is full of such problems and the children should learn to deal with them. But more important is the question of whether such problems may help children learn problem-solving.

One of the taboos of the traditional instruction is that nothing erroneous be ever shown to students (in fear that it might be imprinted in their memory). In case of mathematics learning, the opposite argument seems to be more convincing: One cannot protect children from making errors by always showing them the correct way. Mathematics requires understanding rather than memory. To overcome the conceptual difficulties the children should be given the opportunity to discuss suitable ill-posed problems and to find out themselves what is wrong with the problems.

I believe that—aside from the natural limitations of the child's thinking—the astonishing answers to tests with ill-posed problems resulted from the total absence of such problems in the instruction. The students had met only well-formulated, standard problems. They were surprised by the test. A task of a completely new type was given to them without any preparation, perhaps even without warning (a short formal warning would not help much).

School beginners first think that adults talk seriously. Then they get varying ideas about a game being played or think that all math problems are solvable. They learn that if the teacher gives them a problem to solve then they *must* perform some arithmetical operation. They look for clues. Sometimes they choose the right operation, sometimes not. Often they do not read the problem carefully for its meaning and just guess what operation to use. Now, suppose that they are exposed to an ill-posed problem. It is not the first problem that they find strange. Though the previous problems given them by the teacher were well-posed (from an educator's point of view), many of them appeared meaningless or inconsistent to children. The teacher kept telling them that they were supposed to try hard to solve a problem (even if they could see no sense in it). Consequently, when they get a problem with missing or contradictory data they follow the previous routine.

I have outlined the philosophy that underlies Polish attempts to incorporate ill-posed problems in regular classroom lessons in primary grades. The idea goes back to L. Jeleńska; half a century ago she advocated that very easy problems with missing or redundant data should be given to children to help them grasp the idea of a verbal problem. A year ago, during experimental lessons, children aged 7–9 were given a series of ill-posed problems of various types. In each class, the teacher read problems, explained them in the usual way, and asked children whether somebody wanted to say something. When I watched these lessons and made notes of what the pupils were saying, it struck me that they were more active than during lessons with other material taught previously. They frequently laughed after a problem had been read or upon hearing the naive answers of their colleagues. Many of those answers were as nonsensical as those in Grenoble, but they were followed by the laughter and protests of peers; this created a situation quite different from that of a single child answering test questions. Samples of problems and of childrens' reactions follow (for a more complete report with a survey of the literature, see [8]).

"*Johnnie bought* 2 *kilograms of apples and* 1 *kilogram of pears. How much did he pay?*" (Reaction: "We do not know the prices.") "*Mary invited* 5 *girls and* 3 *boys to her birthday party. How old is she?*" (Some pupils answered: "8 years"; others objected this.) "*A farmer had* 12 *pigs. He went to the market and sold* 4 *hens. How many pigs did he have left?*" (There were several answers "eight" remonstrated by peers: "Why, he sold hens, not pigs!"; some students suggested: "Maybe it's a slip.") "*At the market yesterday an egg cost* 15 *zlotys. Today an egg costs* 14 *zlotys. What will be the price of an egg tomorrow?*" ("We can't know," "We can. Because yesterday it cost 15, today 14, so tomorrow it'll cost 13. Because the price goes down!," "It may go up as well.") *"Jim and Mike are sitting at their desk in a classroom. There are girls standing at the blackboard. Jim sees 3 girls and Mike sees 3 girls. How many girls are standing at the blackboard?* (Several pupils said "six," but there were also answers "three," with explanation, e.g., "Because Jim and Mike saw the same girls.") *"It takes 5 minutes for a student to walk from a school to the nearest bus stop. How long does it take for 3 students to walk from the school to the bus stop?"* ("15 minutes," "5 minutes, because they go the same way," "The same, unless a student is slow.") *"Each day Olga puts money in her piggy-bank and keeps a record of how much she has there. On Monday she had 3 zlotys in her piggy-bank. On Tuesday she had 4 zlotys there. On Wednesday she had 8 zlotys in the piggy-bank. How much money has she put to her piggy-bank altogether?"* ("Fifteen," "Eight. One does not have to make computations, it is just said!") *"Anna is 7 years old and Bob is 10. How much older is Anna?"* ("Older??," "Three years older," "Zero years, for she is younger," "Why, how can a smaller number be greater?")

In some classes, the problems were presented in the following story-setting, involving a clever boy, nicknamed Gapcio, playing with his sister Dolly. One of them formulates a problem and the other is to solve it. "Today I'll give you a cunning problem", says Gapcio to Dolly. "Something will be wrong with it. You try to solve it and tell me what is wrong. We'll see whether you can be fooled!" The pupils were to help Dolly recognize what was wrong, and then to propose how to amend the problem. Sample: *"Gapcio says: 'I had 8 zlotys, and I bought 2 copy-books at 3 zlotys each, and candy costing 4 zlotys. How much change did I get?' What should Dolly say?"* ("He did not get any change at all," "Zero," "He even did not have enough for candies," "Or he had more money with him.")

Teachers who had volunteered to give students those problems were generally in favor of them. But the most important is that children liked the problems and in some classes they later urged the teacher to give them more of these "funny problems."

In conclusion, may I stress what is the main message of my talk: children in primary grades *can* learn mathematics successfully if they are encouraged to explore mathematics on their own (with the use of concrete material if they need it) rather than to memorize rules imposed upon them by the teacher. In particular, children should not only solve problems prepared by somebody else (and often meaningless to them), but should have the opportunity to formulate problems themselves, or to transform problems into new ones, and should learn to distinguish whether a problem (of their own or given to them) is well-posed. A lot of research is needed to learn more about how these ideas can be implemented, but I believe that it is worthwhile.

REFERENCES

1. Terezinha N. Carraher, David W. Carraher, and Analucia D. Schliemann, *Mathematics in the streets and in schools*, British J. Developmental Psych. **3** (1985), 21–29.
2. Erik De Corte and Lieven Verschaffel, *Beginning first graders' initial representation of arithmetic word problems*, J. Math. Behavior **4** (1985), 3–21.

3. Équipe Élementaire de l'IREM de Grenoble, *Quel est l'âge du capitaine?*, Bull. Assoc. Professeurs Math. Enseign. Publique **335** (1980).

4. Hans Freudenthal, *Fiabilité, validité et pertinence—critères de la recherche sur l'enseignment de la mathématique*, Ed. Stud. Math. **13** (1982), 395–408.

5. Gerald A. Goldin and C. Edwin McClintock (editors), *Task variables in mathematical problem solving*, Franklin Inst. Press, Philadelphia, Penn., 1984.

6. Stephen Krulik and Robert E. Reys (editors), *Problem solving in school mathematics*, Yearbook of National Council of Teachers of Math., Reston, Va., 1980.

7. Vadim Andreevič Kruteckiĭ, *The psychology of mathematical abilities in schoolchildren*, Prosveščenie Moscow, 1968; English transl., Univ. of Chicago Press, Chicago, Ill., 1976.

8. Ewa Puchalska and Zbigniew Semadeni, *Verbal problems with missing, surplus or contradictory data*, submitted to For the Learning of Mathematics (Montreal).

9. ____, *Structural categorization of verbal arithmetical problems with missing, surplus or contradictory data*, submitted to Journal für Mathematik-Didaktik.

10. Z. Semadeni, *A principle of concretization permanence for the formation of arithmetical concepts*, Ed. Stud. Math. **15** (1984), 379–395.

11. Edward A. Silver (editor), *Teaching and learning mathematical problem solving: multiple research perspectives*, L. Earlbaum, Hillsdale, N.J., 1985.

12. Hassler Whitney, *Taking responsibility in school mathematics education*, J. Math. Behavior **4** (1985), 219–235.

INSTITUTE OF MATHEMATICS OF THE POLISH ACADEMY OF SCIENCES, WARSAW, POLAND

UNIVERSITY OF WARSAW, WARSAW, POLAND

Index

* I indicates Volume I; II, Volume II.

ABCDEFGHIJ — 8987